Global Contamination Trends

of

Persistent Organic Chemicals

Global Contamination Trends

of

Persistent Organic Chemicals

Edited by

Bommanna G. Loganathan
Murray State University

Paul K. S. Lam
City University of Hong Kong

CRC Press
Taylor & Francis Group
Boca Raton   London   New York

CRC Press is an imprint of the
Taylor & Francis Group, an **informa** business

First published in paperback 2024

First published 2012
by CRC Press
2385 NW Executive Center Drive, Suite 320, Boca Raton FL 33431

and by CRC Press
4 Park Square, Milton Park, Abingdon, Oxon, OX14 4RN

CRC Press is an imprint of Taylor & Francis Group, LLC

© 2012, 2024 Taylor & Francis Group, LLC

Library of Congress Cataloging-in-Publication Data

Global contamination trends of persistent organic chemicals / editors, Bommanna G. Loganathan, Paul Kwan-Sing Lam.
p. cm.
Summary: "Composed by a diverse group of experts, this reference covers the history, present status, and projected future trends of environmental contamination from highly toxic synthetic chemical pollutants. The text informs practitioners on the effects of man-made chemical pollutants in a comprehensive manner not seen in existing textbooks and references. The book is relevant to professionals and students alike, providing timely coverage on the rising concern towards environmental issues including organic chemicals. Supplementary figures and equations are included for enhanced comprehension of the subject"-- Provided by publisher.
Includes bibliographical references and index.
ISBN 978-1-4398-3830-3 (hardback)
1. Organic compounds--Environmental aspects. 2. Persistent pollutants. I. Loganathan, Bommanna G. II. Lam, Paul K. S.

TD196.O73G58 2012
363.738'4--dc23   2011011802

ISBN: 978-1-4398-3830-3 (hbk)
ISBN: 978-1-032-91792-4 (pbk)
ISBN: 978-0-429-06377-0 (ebk)

DOI: 10.1201/b11098

Visit the Taylor & Francis Web site at
http://www.taylorandfrancis.com

and the CRC Press Web site at
http://www.crcpress.com

Contents

Preface...ix
Acknowledgments...xi
Editors.. xiii
Contributors xv

PART I   Classical and Emerging Persistent Organic Chemicals: Perspectives

Chapter 1   Global Contamination Trends of Persistent Organic Chemicals: An Overview..........3

Bommanna G. Loganathan

Chapter 2   Spatial and Temporal Trends of Polybrominated Diphenyl Ethers...........................33

Ying Guo, Susan D. Shaw, and Kurunthachalam Kannan

Chapter 3   Temporal Trends of Perfluorinated Compounds from Various Environmental Matrices..73

Yasuyuki Zushi and Shigeki Masunaga

Chapter 4   Synthetic Musks and Benzotriazole UV Stabilizers in Marine Ecosystems.............87

Haruhiko Nakata, Ryu-ichi Shinohara, Sayaka Murata, and Hiroshi Sasaki

Chapter 5   Global Environmental Distribution and Human Health Effects of Polycyclic Aromatic Hydrocarbons 97

Aramandla Ramesh, Anthony E. Archibong, Darryl B. Hood, Zhongmao Guo, and Bommanna G. Loganathan

PART II   Persistent Organic Chemicals in the Asia-Pacific Region

Chapter 6   Organohalogen Contamination Profiles and Trends in Australia 129

Munro R. Mortimer, Leisa-Maree Leontjew Toms, and Jochen F. Mueller

Chapter 7   Environmental Contamination Status of Polychlorinated Biphenyls in China........163

Ying Xing, Yonglong Lu, and Wenbin Liu

Chapter 8   Organochlorine Pesticides, Polychlorinated Biphenyls, and Polybrominated
Diphenyl Ethers in the Indian Atmosphere.. 179

Paromita Chakraborty and Gan Zhang

Chapter 9   Historical Trends of Dioxin Sources and Contamination in Japan.......................... 203

Shigeki Masunaga

Chapter 10 Polychlorinated Naphthalenes: Use and Contamination Trends in Japan and
China 215

*Sachi Taniyasu, Yuichi Horii, Nobuyasu Hanari, Nobuyoshi Yamashita,
Jing Pan, Yongliang Yang, and Bommanna G. Loganathan*

Chapter 11 Sources, Distributions, and Temporal Trends of Nonylphenols in South Korea...... 259

Seongjin Hong and Kyung-Hoon Shin

Chapter 12 Spatial and Temporal Trends of Persistent Organic Chemicals in Vietnam
Soils 279

Masahide Kawano and Vu Duc Thao

PART III   Persistent Organic Chemicals in Europe and Africa

Chapter 13 Contamination Profiles and Possible Trends of Organohalogen Compounds in
the Estonian Environment and Biota.. 307

*Ott Roots, Vladimir Zitko, Kurunthachalam Senthil Kumar, Kenneth Sajwan,
and Bommanna G. Loganathan*

Chapter 14 Chlorinated Hydrocarbons in Animal Tissues and Products of Animal Origin
from Poland 337

Alicja Niewiadowska and Jan Zmudzki

Chapter 15 Temporal Trends of Organohalogen Compounds in Mother's
Milk from Sweden.. 355

*Anders Glynn, Sanna Lignell, Marie Aune, Per Ola Darnerud,
and Anna Törnkvist*

Chapter 16 Temporal Trends of POPs in Human Milk from Italy 381

Annalisa Abballe, Anna Maria Ingelido, and Elena De Felip

Chapter 17 Contamination Status of Organochlorine Pesticides in Ghana............................... 393

William J. Ntow and Benjamin O. Botwe

PART IV Persistent Organic Chemicals in the Americas

Chapter 18 Contamination Profiles and Possible Trends of Persistent Organic Compounds
in the Brazilian Aquatic Environment 415

Gilberto Fillmann and Juliana Leonel

Chapter 19 Contamination Profiles and Temporal Trends of Persistent Organic Pollutants
in Oysters from the Gulf of Mexico 431

Jose L. Sericano and Terry L. Wade

Chapter 20 Using Sediment Cores to Assess Inputs of Organochlorines and Polycyclic
Aromatic Hydrocarbons in Coastal Georgia Estuaries 469

Clark R. Alexander, Allen D. Uhler, and Richard F. Lee

Chapter 21 Temporal Trends of Selected Chlorinated Hydrocarbon Pollutants
in Sediments of Kentucky Lake 481

Baki B. Sadi and Bommanna G. Loganathan

PART V Persistent Organic Chemicals in the Coastal, Oceanic, Arctic, and Antarctic Regions

Chapter 22 Temporal Trends of Polybrominated Diphenyl Ethers and
Hexabromocyclododecanes in Marine Mammals with Special Reference
to Hong Kong, South China 497

Ling Jin, James C.W. Lam, Margaret B. Murphy, and Paul K.S. Lam

Chapter 23 Status and Trends of POPs in Harbor Seals from the Northwest Atlantic 515

Susan D. Shaw, Michelle L. Berger, and Kurunthachalam Kannan

Chapter 24 Organohalogen Pollutants in Seabird Eggs from Northern Norway and Svalbard ... 547

Lisa Bjørnsdatter Helgason, Kjetil Sagerup, and Geir Wing Gabrielsen

Chapter 25 Contamination Profile and Temporal Trend of POPs in Antarctic Biota 571

Simonetta Corsolini

Chapter 26 Global Distribution of PFOS and Related Chemicals 593

*Nobuyoshi Yamashita, Leo W.Y. Yeung, Sachi Taniyasu, Karen Y. Kwok,
Gert Petrick, Toshitaka Gamo, Keerthi S. Guruge, Paul K.S. Lam, and
Bommanna G. Loganathan*

Index 629

Preface

The quality of life depends on the quality of our environment. Unexpected environmental disasters during the past few decades, such as the Gulf of Mexico oil spill in 2010, radiation leakage at the Daiichi nuclear power plant in Fukushima, Japan following the earthquake and tsunami in 2011, and uncertainties associated with global warming have prompted people to become more concerned about the future of the global environment than ever before. Environmental pollution by man-made persistent organic chemicals (POCs) has been a serious global issue for over half a century. This book addresses the past, present, and possible future trends of environmental contamination by POCs on the global scale. POCs in this book refer to any chemical species that are persistent in the environment and cause unintended effects on the environment and to the health of wildlife and humans. These chemicals are also referred to as PBTs (persistent, bioaccumulative, and toxic), POPs (persistent organic pollutants), and PLCs (persistent lipophilic contaminants), as their unique properties such as hydrophobicity and lipophilicity lead to their accumulation in biological tissues. Further, their low biodegradability and volatility (high vapor pressure) lead to their long-range transportation via atmosphere and water, resulting in widespread environmental contamination of biota and humans at sites remote from where these chemicals are produced and used. Exposure to certain POCs may result in serious health complications, including reproductive and immune dysfunction, birth defects, diminished intelligence, and certain types of cancers.

Given the long-range movement and dispersal of these chemicals, no one government acting alone can protect its citizens or the environment from POCs. Therefore, these chemicals have been the subject of an intensive regional, national, and international effort to limit their production and use, as well as the control and disposal of old or obsolete chemical stocks. In May 1995, the United Nations Environment Program (UNEP) determined that there were sufficient data on persistence and toxicity to eliminate or reduce emissions, or even in some cases halt production and use, of 12 POPs. Further, in 1998, 36 member states of the United Nations Economic Commission for Europe (UNECE), consisting of European countries, Canada, Russia, and the United States, agreed on a protocol that banned the production and use of some POPs. UNEP's Stockholm Convention on Persistent Organic Pollutants, a global treaty to protect human health and the environment, was adopted in 2001 and entered into force in 2004, requiring countries to take measures to eliminate or reduce the release of persistent chemicals into the environment. The countries that signed the treaty are expected to monitor and report trends in the occurrence and distribution of these chemicals in their national environments. New chemicals were added to the POPs list during the Stockholm Convention in 2009, emphasizing the importance of continual monitoring and assessment of environmental contaminants.

Studying the spatial and temporal trends of POCs has thus become an area of interest, since trend-monitoring data are useful in understanding past, present, and future trends of environmental contamination by these compounds. This book, *Global Contamination Trends of Persistent Organic Chemicals*, was prepared in order to provide the most comprehensive coverage of spatial and temporal trends of these "classical" pollutants, as well as emerging contaminants that have raised concerns around the world in recent years. These emerging chemicals constitute a diverse group, including halogenated compounds used in manufacturing, such as brominated flame retardants and perfluorinated surfactants, as well as personal care products, such as synthetic musks. Some of these chemicals have been studied for more than 10 years, while the environmental behavior of others is only beginning to be investigated. It is therefore important to have a greater understanding of how both "classical" and emerging chemicals are distributed in the global environment.

The content of this book is divided into five parts. Part I deals with perspectives on classical and emerging persistent chemicals, Part II presents POC trends in countries in the Asia-Pacific Region, Part III describes POC trends in countries from Europe and Africa, Part IV provides a detailed account of POC trends in the Americas, and Part V elucidates POC trends in remote regions such as coastal and open ocean environments as well as the Arctic and Antarctic regions.

The contributed chapters cover spatial and temporal trends of polychlorinated biphenyls (PCBs), chlorinated pesticides, polychlorinated naphthalenes (PCNs), polychlorinated dibenzodioxins/furans (PCDD/DFs), polybrominated diphenyl ethers (PBDEs), hexabromocyclododecanes (HBCDs), perfluorinated compounds (PFCs), synthetic musks, polynuclear aromatic hydrocarbons (PAHs), and octyl- and nonylphenols, in environmental and biological matrices such as the atmosphere, water, soil, sediment, bivalve mollusks, fish, marine mammals, birds, terrestrial mammals, and human breast milk, originating from Australia, Brazil, China, Estonia, Ghana, Hong Kong, India, Italy, Japan, Korea, Norway, Poland, Sweden, the United States, and the Arctic and Antarctic regions.

This book will provide information that will help students; researchers; legislators; environmental agencies; federal, state, and local authorities; and laymen to understand the history, present status, and possible future trends of persistent chemicals in the global environment and to facilitate the development of strategies/practices aimed at protecting the global environment for future generations.

Bommanna G. Loganathan
Paul K.S. Lam

For MATLAB® and Simulink® product information, please contact:

The MathWorks, Inc.
3 Apple Hill Drive
Natick, MA, 01760-2098 USA
Tel: 508-647-7000
Fax: 508-647-7001
E-mail: info@mathworks.com
Web: www.mathworks.com

Acknowledgments

We would like to thank each author and the reviewers for their invaluable contributions and patience throughout the publication process. The senior editor, Bommanna Loganathan, appreciates the Murray State University Committee on Institutional Studies and Research (MSU-CISR) for awarding the 2010 Presidential Research Fellowship for support in this endeavor. We gratefully acknowledge Arpana Gambiraopet and Archana Gambiraopet for their technical assistance. We would also like to thank Ashley Ireland, reference librarian, MSU, for her timely help with environmental research materials. We greatly appreciate Carl Woods, at the MSU Science Resource Center, for his patience and meticulous work on the cover design for this book. Our thanks also to Dr. Judy Ratliff and Dr. Kevin Miller, MSU Department of Chemistry, and Cassidy Palmer, Center for Teaching Learning and Technology, for their help during various stages of preparation of this book. This book would not have been possible without the contributions of Dr. Lance Wobus, Richard Tressider and Kathryn Younce, Taylor & Francis Group, LLC, and Suganthi Thirunavukarasu, SPi Global, India.

Editors

Bommanna G. Loganathan, PhD, is professor of environmental/analytical chemistry at the Murray State University (MSU), Murray, Kentucky. His current research involves investigations on the distribution, environmental transformation, and fate of persistent organic/organometallic pollutants in the environment and their effects on wildlife and human health. A major focus of his research is to evaluate the status and trends of classical as well as emerging pollutants in man-made freshwater lakes, such as Kentucky Lake, in comparison with natural lakes and marine ecosystems and to assess the effects of these compounds on human natural killer cells' ability to kill cancer cells using in vitro assays. He is the author/coauthor of over 100 publications in peer-reviewed journals and book chapters. He also serves as an associate editor for the *Journal of Environmental Monitoring and Restoration.*

Dr. Loganathan received MS and PhD degrees in marine biology from Annamalai University, India. He was awarded the Japanese Government's Monbusho Research Fellowship (1986–1990) to conduct research on temporal trends of persistent organic pollutants at Ehime University, Matsuyama, Japan, and received a second PhD degree in ecotoxicology and environmental chemistry in 1990. Dr. Loganathan was a postdoctoral research associate at the Great Lakes Laboratory, State University of New York (SUNY) College at Buffalo, Buffalo, New York (1990–1993) and a research assistant professor at the Skidaway Institute of Oceanography, University System of Georgia, Savannah, Georgia (1993–1997). He joined MSU in 1997 and holds a joint appointment with the Department of Chemistry and the Watershed Studies Institute. He also has served as an adjunct associate professor at South Carolina State University (SCSU) since 2000 and teaches summer courses in environmental chemistry at SCSU's Environmental Field Station located at the Savannah River Site, Aiken, South Carolina. Dr. Loganathan is the recipient of several research and teaching awards, including the discretionary research award for outstanding environmental research at the Great Lakes Laboratory, SUNY College at Buffalo, New York; the 2007 MSU Alumni Association's Distinguished Researcher Award; the 2010 MSU Service Learning Mentor of the Year Award; 2010 MSU Sandra Flynn Professor of the Year Award; and the 2010 MSU Presidential Research fellowship Award.

Paul K.S. Lam is chair professor of biology at the City University of Hong Kong. He is currently serving as the vice president (student affairs).

Professor Lam has extensive research experience in marine environmental research. He is particularly interested in the responses of organisms to toxic chemicals and algal toxins, as well as the risk assessment of these compounds. He is the author/coauthor of over 240 publications in internationally refereed journals. He has been responsible for over 30 major government consultancy projects in the environmental field, including assessment of risks to Hong Kong cetaceans and waterbirds due to toxic contaminants.

Professor Lam coedited *Inland Waters of Tropical Asia and Australia: Conservation and Management* for the International Association of Theoretical and Applied Limnology (1994) and *Persistent Organic Pollutants in Asia: Sources, Distributions, Transport, and Fate* published by Elsevier Science (2007). He was a guest editor for two special issues of the *Marine Pollution Bulletin* (2005 and 2007). He is also the coauthor of a book entitled *Introduction to Ecotoxicology* published by Blackwell Science. He was an associate editor of the "Environmental Toxicology and Risk Assessment" section of the international journal, *Chemosphere* (2004–2008). He is also a member of the editorial advisory board of *Environmental Science and Technology* (2010-present).

Contributors

Annalisa Abballe
Department of the Environment and Primary
    Prevention
National Institute for Health
Rome, Italy

Clark R. Alexander
Skidaway Institute of Oceanography
Savannah, Georgia

Anthony E. Archibong
Department of Physiology
Meharry Medical College
Nashville, Tennessee

Marie Aune
Department of Research and Development
National Food Administration
Uppsala, Sweden

Michelle L. Berger
Center for Marine Studies
Marine Environmental Research Institute
Blue Hill, Maine

Benjamin O. Botwe
Department of Oceanography and Fisheries
University of Ghana
Accra, Ghana

Paromita Chakraborty
Guangzhou Institute of Geochemistry
Chinese Academy of Sciences
Guangzhou, China
and
Department of Natural Sciences
Savannah State University
Savannah, Georgia

Simonetta Corsolini
Department of Environmental Science
    "G. Sarfatti"
University of Siena
Siena, Italy

Per Ola Darnerud
Department of Research and Development
National Food Administration
Uppsala, Sweden

Elena De Felip
Department of the Environment and Primary
    Prevention
National Institute for Health
Rome, Italy

Gilberto Fillmann
Institute of Oceanography
Federal University of Rio Grande
Rio Grande, Brazil

Geir Wing Gabrielsen
Fram Centre
Norwegian Polar Institute
Tromsø, Norway

Toshitaka Gamo
Department of Chemical Oceanography
The University of Tokyo
Kashiwa, Japan

Anders Glynn
Department of Research and Development
National Food Administration
Uppsala, Sweden

Ying Guo
New York State Department of Health and
    Department of Environmental Health
    Sciences
State University of New York
Albany, New York

Zhongmao Guo
Department of Physiology
Meharry Medical College
Nashville, Tennessee

Keerthi S. Guruge
National Institute of Animal Health
Tsukuba, Japan

Nobuyasu Hanari
National Metrology Institute of Japan
National Institute of Advanced Industrial
    Science and Technology
Tsukuba, Japan

Lisa Bjørnsdatter Helgason
Fram Centre
Norwegian Polar Institute
Tromsø, Norway

Seongjin Hong
Department of Environmental and Marine
    Sciences
Hanyang University
Ansan, South Korea

Darryl B. Hood
Department of Neuroscience & Pharmacology
Meharry Medical College
Nashville, Tennessee

Yuichi Horii
Center for Environmental Science in Saitama
Kazo, Japan

Anna Maria Ingelido
Department of the Environment and Primary
    Prevention
National Institute for Health
Rome, Italy

Ling Jin
Department of Biology and Chemistry
City University of Hong Kong
Hong Kong, People's Republic of China

and

National Research Centre for Environmental
    Toxicology (EnTox)
The University of Queensland
Brisbane, Queensland, Australia

Kurunthachalam Kannan
New York State Department of Health and
    Department of Environmental Health
    Sciences
State University of New York
Albany, New York

Masahide Kawano
Department of Environment Conservation
Ehime University
Matsuyama, Japan

Karen Y. Kwok
Department of Biology and Chemistry
City University of Hong Kong
Hong Kong, People's Republic of China

James C.W. Lam
Department of Biology and Chemistry
City University of Hong Kong
Hong Kong, People's Republic of China

Paul K.S. Lam
Department of Biology and Chemistry
City University of Hong Kong
Hong Kong, People's Republic of China

Richard F. Lee
Skidaway Institute of Oceanography
Savannah, Georgia

Juliana Leonel
Institute of Oceanography
Federal University of Rio Grande
Rio Grande, Brazil

Sanna Lignell
Department of Research and Development
National Food Administration
Uppsala, Sweden

Wenbin Liu
Research Center for Eco-Environmental
    Sciences
Chinese Academy of Sciences
Beijing, People's Republic of China

Bommanna G. Loganathan
Department of Chemistry and Watershed
    Studies Institute
Murray State University
Murray, Kentucky

Yonglong Lu
Research Center for Eco-Environmental
    Sciences
Chinese Academy of Sciences
Beijing, People's Republic of China

Shigeki Masunaga
Graduate School of Environment and
    Information Services
Yokohama National University
Yokohama, Japan

Munro R. Mortimer
Department of Environment and Resource
    Management
State of Queensland
Indooroopilly, Queensland, Australia

Jochen F. Mueller
National Research Centre for Environmental
    Toxicology
The University of Queensland
Coopers Plains, Queensland, Australia

Sayaka Murata
Department of Chemistry
Kumamoto University
Kumamoto, Japan

Margaret B. Murphy
Department of Biology and Chemistry
City University of Hong Kong
Hong Kong, People's Republic of China

Haruhiko Nakata
Department of Chemistry
Kumamoto University
Kumamoto, Japan

Alicja Niewiadowska
Department of Pharmacology and Toxicology
National Veterinary Research Institute
Pulawy, Poland

William J. Ntow
Department of Plant Sciences
University of California
Salinas, California

Jing Pan
National Research Center for Geoanalysis
Chinese Academy of Geological Sciences
Beijing, People's Republic of China

Gert Petrick
Leibniz-Institute of Marine Sciences
University of Kiel
Kiel, Germany

Aramandla Ramesh
Department of Biochemistry and Cancer
    Biology
Meharry Medical College
Nashville, Tennessee

Ott Roots
Estonian Environmental Research Institute
and
Estonian Marine Institute
University of Tartu
Tallinn, Estonia

Baki B. Sadi
Health Canada
Radiation Protection Bureau
Ottawa, Ontario, Canada

Kjetil Sagerup
Fram Centre
Norwegian Polar Institute
Tromsø, Norway

Kenneth Sajwan
Department of National Sciences and
    Mathematics
Savannah State University
Savannah, Georgia

Hiroshi Sasaki
Department of Chemistry
Kumamoto University
Kumamoto, Japan

Kurunthachalam Senthil Kumar
Department of National Sciences and
    Mathematics
Savannah State University
Savannah, Georgia

Jose L. Sericano
Geochemical and Environmental Research
    Group
Texas A&M University
College Station, Texas

Susan D. Shaw
Center for Marine Studies
Marine Environmental Research Institute
Blue Hill, Maine

Kyung-Hoon Shin
Department of Environmental and Marine
    Sciences
Hanyang University
Ansan, South Korea

Ryu-ichi Shinohara
Department of Chemistry
Graduate School of Science and Technology
Kumamoto University
Kumamoto, Japan

Sachi Taniyasu
Research Institute for Environmental
    Management Technology
National Institute of Advanced Industrial
    Science and Technology
Tsukuba, Japan

Vu Duc Thao
Institute for Environmental Science and
    Technology
Hanoi University of Technology
Hanoi, Vietnam

Leisa-Maree Leontjew Toms
Faculty of Science and Technology
Queensland University of Technology
Brisbane, Queensland, Australia

Anna Törnkvist
Department of Research and Development
National Food Administration
Uppsala, Sweden

Allen D. Uhler
NewFields Environmental Forensics Practice
Rockland, Massachusetts

Terry L. Wade
Texas A&M University
College Station, Texas

Ying Xing
Research Center for Eco-Environmental
    Sciences
and
National Science Library
Chinese Academy of Sciences
Beijing, People's Republic of China

Nobuyoshi Yamashita
Research Institute for Environmental
    Management Technology
National Institute of Advanced Industrial
    Science and Technology
Tsukuba, Japan

Yongliang Yang
National Research Center for Geoanalysis
Chinese Academy of Geological Sciences
Beijing, People's Republic of China

Leo W.Y. Yeung
Department of Chemistry
University of Toronto
Toronto, Ontario, Canada

Gan Zhang
Guangzhou Institute of Geochemistry
Chinese Academy of Sciences
Guangzhou, China

Vladimir Zitko
St. Andrews, New Brunswick, Canada

Jan Zmudzki
Department of Pharmacology and Toxicology
National Veterinary Research Institute
Pulawy, Poland

Yasuyuki Zushi
Graduate School of Environment and
    Information Services
Yokohama National University
Yokohama, Japan

Part I

Classical and Emerging Persistent Organic Chemicals: Perspectives

1 Global Contamination Trends of Persistent Organic Chemicals
An Overview

*Bommanna G. Loganathan**

CONTENTS

1.1 Introduction 3
1.2 Historical Background of Organohalogen Compounds 4
    1.2.1 Organochlorine Compounds 4
        1.2.1.1 Polychlorinated Biphenyls 4
        1.2.1.2 Chlorinated Pesticides 7
        1.2.1.3 Dioxins and Furans 7
    1.2.2 Brominated Compounds 8
    1.2.3 Perfluorinated Compounds 9
1.3 Other Persistent Chemicals of Concern 9
    1.3.1 Polychlorinated Naphthalenes 9
    1.3.2 Pharmaceutical and Personal Care Products 10
    1.3.3 Bisphenol A 11
    1.3.4 Polycyclic Aromatic Hydrocarbons 11
1.4 Value of Trend Monitoring Studies 12
1.5 Factors Affecting Spatial and Temporal Trends of Persistent Organic Chemicals 13
    1.5.1 Spatial Viewpoint 13
    1.5.2 Biological Viewpoint 14
    1.5.3 Chemical Characteristics Viewpoint 15
1.6 Spatial and Temporal Trends in Specific Ecosystems 16
    1.6.1 Terrestrial Ecosystem 16
    1.6.2 Inland and Coastal Aquatic Ecosystems 17
    1.6.3 Open Ocean Ecosystems 18
1.7 Summary and Conclusions 19
References 23

1.1 INTRODUCTION

Over 10 million organic compounds have been discovered or synthesized and an estimated 100,000 new compounds are discovered or prepared in the laboratory each year [1]. These compounds are used in foods, flavors, fragrances, medicines, toiletries, glues, plastics, paints, and other industrial and consumer products. Several of these chemicals protect humans and animals from deadly diseases, improve

* E-mail: bommanna.loganathan@murraystate.edu (Chapter corresponding author).

agricultural food production to meet the demands of increasing populations, and improve the overall quality of our lives. A few classes of chemicals, although of benefit to mankind, lead to environmental contamination and change the quality of the environment on a global scale due to widespread usage, which can adversely affect plants and animals worldwide. A typical example of such environmental damage and harmful biological effects is that caused by persistent human-made chemicals, particularly organohalogen compounds. Organohalogens are organic compounds which contain chlorine, bromine, and/or fluorine atoms, the corresponding molecules of which are called chlorinated, brominated, and/or fluorinated compounds respectively. In addition to organohalogen compounds, there are certain other synthetic organic chemicals that are currently used in industry, and consumer products, many of which are reported as stable under extreme environmental conditions and cause negative effects on wildlife and humans. For example, chemicals such as bisphenol A (BPA), used in liners of canned food containers, can withstand the high temperatures used during sterilization of food.

The term "persistent chemicals" hereby refers to any synthetic organic chemical species that is stable in the environment for long periods of time and causes unintended effects on the environment and health of wildlife and humans. Persistence is often measured as a half-life in the environment. According to Rodan (2002), persistence screening values for the Stockholm Convention are based on 2 months in water or 6 months in soil or sediments, with a 2 day screening criterion for air transport [2]. Table 1.1a through d shows structures of selected classical and emerging man-made persistent chemicals that are/were heavily produced and used in industry, agriculture, public health, pharmaceuticals, and as by-products. This chapter reviews the spatial and temporal trends of classic organochlorine compounds as well as other emerging pollutants (including brominated and fluorinated compounds, pharmaceuticals, and chemicals used in consumer products) in the world and the future implications of these trends. Existing information was compiled and interpreted to understand the clearance rates in various geographical locations and biota. The conclusions derived are based on trend analyses performed during the past two decades and on information obtained from published data.

1.2 HISTORICAL BACKGROUND OF ORGANOHALOGEN COMPOUNDS

1.2.1 ORGANOCHLORINE COMPOUNDS

1.2.1.1 Polychlorinated Biphenyls

Chlorinated hydrocarbons were synthesized as early as 1825, when Michael Faraday first reported, before the Royal Society of London the formation of "benzene hexachloride" by the reaction of benzene with chlorine in presence of sunlight. Polychlorinated biphenyls (PCBs) were first synthesized in 1881 by Schmidt and Schultz [3] and their commercial production began in 1929, when industrial applications were realized. Over a million tons of PCBs were produced from 1929 through the mid-1970s [4]. PCBs were primarily produced by developed nations. Commercial PCB formulations were sold under a variety of trade names. For example, in the United States and Great Britain, Aroclor was the most common trade name for PCBs. PCB mixtures were named according to their chlorine content. Aroclor 1254 contains 54% chlorine by weight, and Aroclor 1260 contains 60%. The PCB mixture formulations were different depending on the country of origin and were produced in Germany (Clofen), France (Phenoclor and Pyralene), Japan (Kanechlor), Italy (Fenclor), Russia (Sovol), and Czechoslovakia (Delor). PCBs consist of potentially 209 congeners produced by chlorinating biphenyls with 1–10 chlorines (Table 1.1a). PCBs were widely used as lubricants, heat transfer and dielectric fluids, in capacitors, plasticizers, protective coatings, and copy paper, etc. [5].

Properties of PCBs vary widely depending on the number and position of chlorine atoms in the biphenyl rings (Table 1.1a). For example, molar mass ranges from 188 to 498; K_{ow} (octanol–water partition coefficient) $\sim10^{6.5}$, K_{AW} (air–water partition coefficient) $\sim10^{-2.3}$, and K_{OA} (octanol–air partition coefficient) $\sim10^{8.8}$ [6]. The unique physical and chemical properties such as heat and chemical stability, hydrophobicity, compatibility with organic acids, and resistance to biodegradation

TABLE 1.1
Classical and Emerging Persistent Organic Chemicals

Compound	Molecular Weight	Chemical Structure

(a) Industrial chemicals

PCBs　　　　188–498

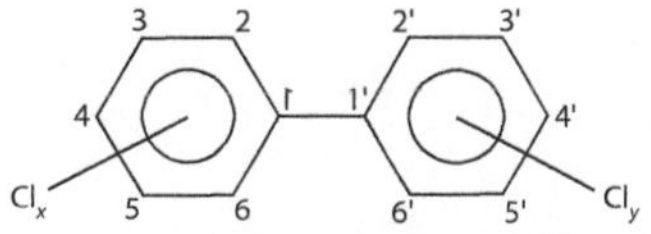

$x + y = 1$–8
Generalized structure of PCBs

PBDEs　　　　249–959

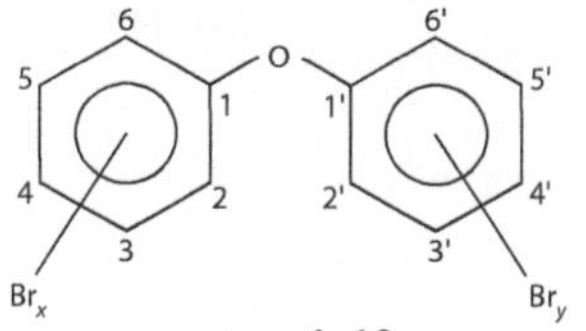

$x + y = 1$–10
Generalized structure of PBDE

PFCs　　　　478

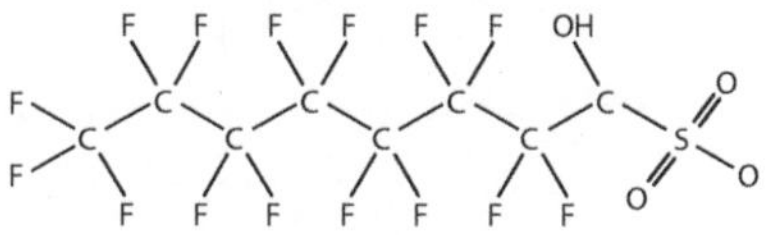

Perfluorooctane sulfonate (PFOS)

BPA　　　　228

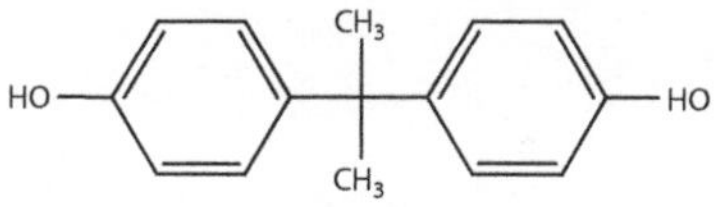

Bisphenol A

(b) Agricultural and public health

DDT (*p,p′*-DDT)　　　　355

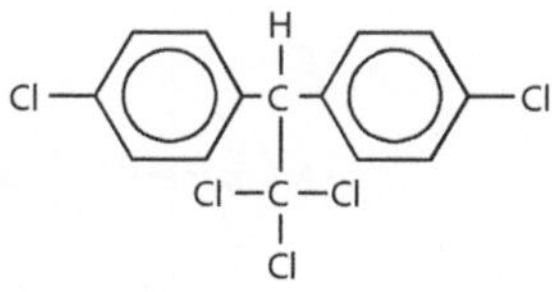

HCHs　　　　291

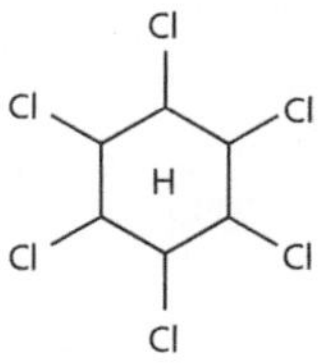

Chlordane　　　　410

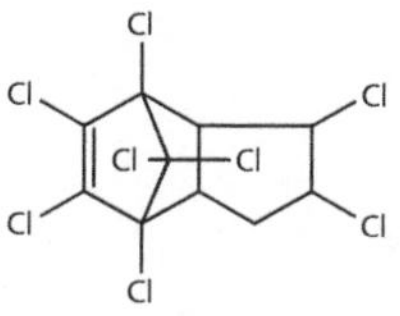

(*continued*)

TABLE 1.1 (continued)
Classical and Emerging Persistent Organic Chemicals

Compound	Molecular Weight	Chemical Structure

(c) By-products

PCDDs   218–531

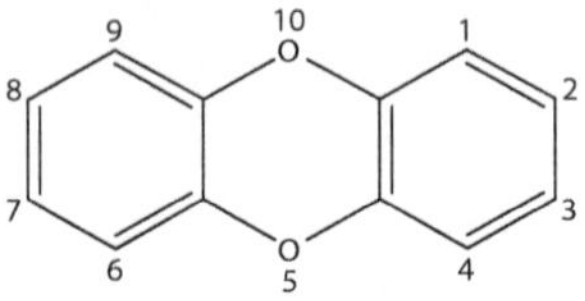

Dioxins

PCDFs   202–444

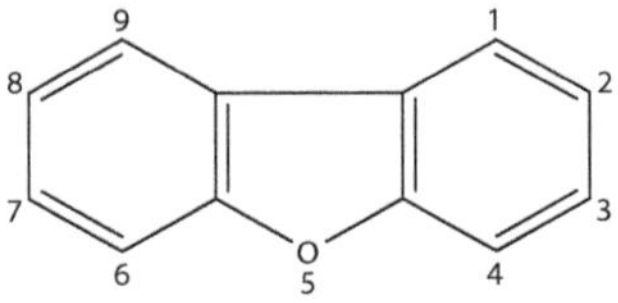

Dibenzofurans

PCNs   162–404

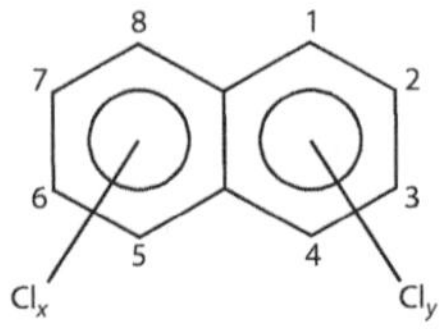

$x + y = 1–8$
Polychlorinated naphthalenes

PAHs   252

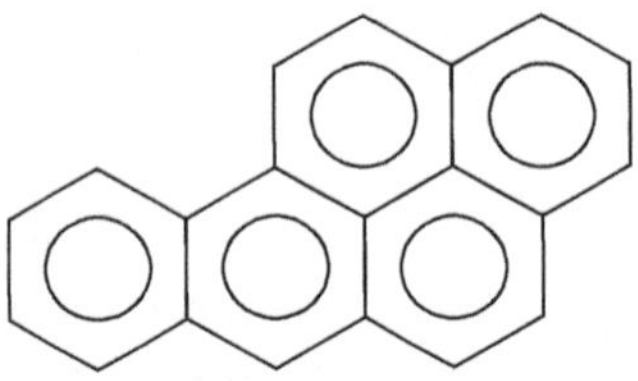

Benzo(a)pyrene

(d) Pharmaceutical and personal care products

Triclosan   289.5

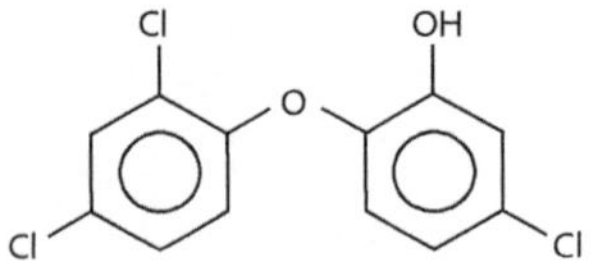

Azithromycin   749

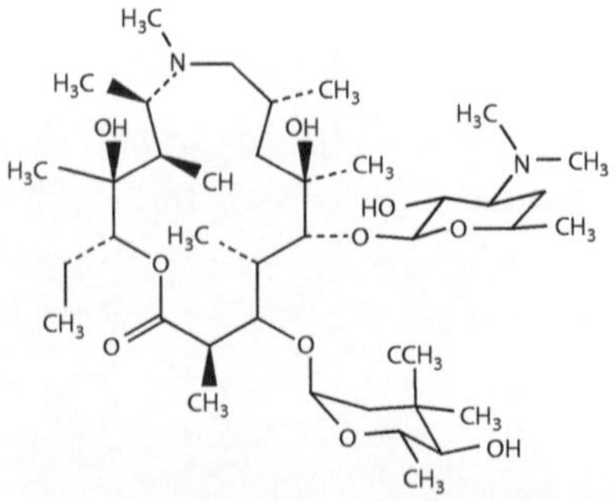

TABLE 1.1 (continued)
Classical and Emerging Persistent Organic Chemicals

Compound	Molecular Weight	Chemical Structure
HHCB	258	

make these compounds highly useful in industries. The same properties also make PCBs stable in the environment, and pervade every component of the global ecosystem including the Arctic and Antarctic environments and biota [7–9]. Due to their persistent properties, long-range atmospheric transport, bioaccumulation and biomagnification (in the food chain) potential, and toxic effects on wildlife and humans, PCBs were banned from production in most developed countries during the early to–mid-1970s.

1.2.1.2 Chlorinated Pesticides

DDT (dichlorodiphenyl trichloroethane or 1,1,1-trichloro-2, 2-bis (4-chlorophenyl) ethane) was first synthesized by Zeilder in 1874 by the reaction of chloral and chlorobenzene in the presence of sulfuric acid (Table 1.1b). Their insecticidal value was discovered just prior to World War II and it was put to field use in the 1940s. Subsequently, this chlorine containing insecticide replaced compounds with arsenic, lead, and copper, which were in use previously as insecticides. DDT was extremely effective against crop and household pests. Applications of DDT contributed to the rapid reduction of malaria, typhoid fever, cholera, and other insect-borne diseases, and led to an astonishing increase in agricultural productivity in many regions of the world. The utility of DDT was so remarkable that it saved 5 million lives and prevented 100 million incidents of serious diseases [10]. The great success in the application of DDT during and in the years following World War II earned Paul Müller the Nobel Prize in Medicine [11]. Other organochlorines such as HCHs (hexachlorocyclohexanes) and CHLs (chlordane compounds) were introduced in 1945 (Table 1.1b). These insecticides also contributed to human welfare as agricultural and domestic pest control agents. During the late 1960s, a number of reports documented that several bird species populations were declining, especially fish eating birds due to egg shell thinning and subsequent failure in hatching. Investigations revealed that DDT and its metabolites were responsible for adverse effects on nontarget species such as wildlife and humans. These negative effects on environment and health resulted in bans or severe restrictions on the production and use of DDTs in most of the developed countries beginning in the late 1960s. However, countries in the tropical regions, particularly developing countries, continue to use these inexpensive pesticides for public health purposes to control malaria and other insect-borne diseases.

1.2.1.3 Dioxins and Furans

Dibenzo-*p*-dioxin was first prepared by Lesimple in the year 1866 from triphenylphosphate and lime, and the structure was determined in 1871 by Hoffmeister [12]. In 1872, German chemists Mertz and Weitz reported preparation of the first chlorinated dibenzo-*p*-dioxin [13]. The most toxic congener, 2,3,7,8-tetrachlorodibenzo-*p*-dioxin (TCDD), was not reported until the mid-1950s [13]. TCDD and other polychlorinated dibenzo-*p*-dioxins (PCDDs) and the corresponding 2,3,7,8-tetrachlorodibenzofuran (TCDF) and polychlorinated dibenzofurans (PCDFs) have never been produced deliberately, but are released into the environment as by-products of the combustion of organic matter in the presence of chlorine. PCDDs and PCDFs consist of 75 possible chlorinated dibenzo-*p*-dioxins and 135 possible chlorinated dibenzofurans with one to eight chlorine substituents (Table 1.1c). Dioxins and furans are formed as derivatives during the production of industrial chemicals such as PCBs, polychlorinated naphthalenes (PCNs), chlorinated phenols, and polyvinyl chlorides, and in chlorine bleaching in paper making and metal smelting, etc. [14–17]. These compounds are also formed during the incineration of municipal and industrial wastes, as well as forest fires, combustion engines, and home fire places [18,19]. Due to anthropogenic and natural processes, dioxins and furans are widely dispersed in the global environment and their presence has been reported in water, air, soil, sediment, and aquatic and terrestrial organisms including human tissues [8,20–23].

Kjeller et al. analyzed lake sediment data for North America and Western Europe as well as archived plants and soils from the United Kingdom and found that levels of PCDDs and PCDFs were very low during 1920–1940, corresponding to a period of significant increase in the manufacturing of synthetic organic chemicals [24]. Temporal trend studies exhibited low levels of dioxins in ancient mummies and 100–400 year old frozen Eskimos compared to those in human tissue today in industrialized nations [25,26]. Todaka et al. measured concentrations of PCDDs, PCDFs, and dioxin-like PCBs in the blood of 195 pregnant women in Sapporo City and found that the levels of these compounds in maternal blood have decreased compared to the past levels in Japan [27]. A similar decreasing trend was also observed in PCDDs/Fs levels of mother's milk from Sweden [28]. A steady state or tendency for a declining trend in PCDD/PCDF and dioxin-like PCB levels was observed in the blood of the adult population in Munich and suburban and rural areas in the southern parts of Germany [29]. PCDDs/PCDFs and dioxin-like PCB concentrations (TEQs) in mother's milk collected from Italy during 1987–2000 also revealed a declining trend on the order of 40%–60% [30]. In contrast, Yang and coworkers observed no declining trend of PCDD/DFs in the blood of urban dwellers in Korea from 2002 to 2007 [31]. Llobet and coworkers analyzed PCDD/PCDFs and PCBs in a variety of foodstuffs from Spain and compared the TEQ values obtained to the TEQ values of similar foodstuffs analyzed for these compounds in the previous years and found a significant decrease in the dietary intake of these compounds in recent years. The authors concluded that the current estimated intake for a Spanish adult male is 1.12 pg WHO-TEQ/kg bodyweight per day, which is lower than most intakes reported in a number of other countries around the world [32]. The present decreasing trend of PCDDs/PCDFs and dioxin-like PCBs may be attributable to the extensive efforts of many countries to reduce the exposure against these compounds. In contrast, brominated dioxins such as polybrominated dibenzo-*p*-dioxins (PBDDs) and polybrominated dibenzofurans (PBDFs) (which are considered less toxic than chlorinated dioxins) are of concern for future environmental and health since these compounds are released into the environment when heating or incinerating of bromine containing compounds such as polybrominated diphenyl ethers (PBDEs) that are still used extensively in household/consumer products [5,33].

1.2.2 BROMINATED COMPOUNDS

The next class of organohalogens of concern are brominated compounds which include PBDEs and PBDDs/PBDFs. PBDEs are made up of two phenyl rings linked by oxygen, and are thus designated as ethers (Table 1.1a). PBDEs are structurally similar to PCBs and are highly resistant to physical, chemical, and biological degradation. These properties make PBDEs excellent flame retardants. PBDEs are added to consumer products so that the products will not catch fire or will burn more slowly if exposed to flame or heat. PBDEs are added to plastics, upholstery, fabrics and foams and are common in products such as computers, televisions, mobile phones, furniture and carpet pads. Approximately 90% of electronic and electrical appliances contain PBDEs, which enable up to 15 times greater escape time in case of fire.

In contrast to PCBs, PBDE formulations are currently being produced and used in household materials. PBDEs are primarily indoor pollutants. PBDEs leach into the environment when household wastes decompose in landfills or are incompletely incinerated. Human health concerns have arisen due to the fact that PBDEs are persistent, bioaccumulative, toxic, and subject to long-range atmospheric transport. PBDE concentrations are rapidly increasing in the global environment and in human blood, breast milk, and liver tissue [28,34,35]. Although PBDEs are widespread in the environment and bioaccumulate in wildlife and humans, a little is known about their toxic effects [36]. During the last three decades, North America has consumed more than half of the global production of PBDEs and 95% of the penta-BDE product, the isomers of which are more persistent and bioaccumulative than those of octa- or deca-BDEs [37]. Accordingly, PBDE burdens in wildlife and humans in North America are the highest in the world; in contrast to Europe, North American body burdens have been increasing over time [38]. Because of environmental and public health concerns, penta- and octa-BDE formulations were banned in Europe in 2004 and subsequently withdrawn from commerce in the United States [39]. In 2007, the state of Maine passed legislation prohibiting the use of PBDE-209 (deca-BDE) in mattresses and residential upholstered furniture sold within its borders and will extend the ban to electronics in 2010. Washington State prohibits the use of deca-BDE in mattresses and is considering a ban on its use in furniture and electronics. Similar legislation is pending in other states in the United States and in Asian countries [34]. Nevertheless, large amounts of deca-BDE have been released into the global environment and this flame retardant is still in high-volume use. Shaw and Kannan roughly estimated that more than 1 million tons of PBDE have been produced globally and the production of deca-BDE continues [34]. Spatial and temporal trend monitoring of PBDEs and brominated dioxins are warranted to prevent exposure and health effects.

1.2.3 PERFLUORINATED COMPOUNDS

Perfluorinated compounds (PFCs) are another group of organic compounds in the persistent organohalogen family (Table 1.1a). PFCs are emerging environmental pollutants [40]. Most fluorinated organohalogens have amphiphilic (ionic and neutral) properties. Because of their thermodynamically strong covalent C–F bonds, these compounds were initially considered non-metabolizable and nontoxic. PFCs may readily convert to one or more additional fluorinated compounds. For example, fluorotelomer alcohols—which are used in paper coatings, paints, lubricants, and carpeting—can readily convert to perfluorooctanoic acid and related perfluorocarboxylates. Despite their persistence and bioaccumulative potentials, the physicochemical and biochemical properties such as vapor pressure, water solubility, liphophilicity, metabolic degradability, and partitioning behavior of PFCs are different from conventional organohalogens such as PCBs. Therefore, prediction of their environmental dynamics cannot be generalized with those of PCBs or PBDEs. PFCs with unique surface modification properties readily bind to surfaces such as blood globulin [40].

The chemical stability and nondegradable nature of PFCs, coupled with their widespread use, have led to global environmental contamination and accumulation of PFCs in aquatic and terrestrial organisms, including humans. Perfluorooctane sulfonate (PFOS; Table 1.1a) and perfluorooctanoic acid (PFOA) have been routinely detected in environmental matrices, wildlife, and human tissues [40–44]. These findings have raised concerns regarding widespread environmental contamination and their possible impacts on ecosystems and on human health. Although PFCs are ubiquitous in the global environment, they bioaccumulate in wildlife and humans and their toxic properties, including mechanisms of toxicity, are still under investigation.

1.3 OTHER PERSISTENT CHEMICALS OF CONCERN

1.3.1 POLYCHLORINATED NAPHTHALENES

PCNs consist of two fused aromatic rings substituted with one to eight chlorine atoms (Table 1.1c). Based on the number and position of the chlorine substitution, a total of 75 congeners are possible with varying physicochemical properties. Since PCNs are structurally similar to PCBs, PCNs have similar properties and are used in similar industrial applications. PCNs were first synthesized in the early 1800s and large amounts of PCNs were manufactured in several countries from the 1910s to the 1980s [45]. The production and use of PCNs were phased out in the 1980s [46]. The estimated global production was 150,000 metric tons [45]. PCNs are formed as combustion by-products from various industrial and waste incineration processes in a similar fashion to dioxins and furans [47]. Commercial technical PCB mixtures also contain PCNs as by-products and contribute one of the sources to the environment [45,48]. Global production of PCNs as a by-product has been estimated at about 169 ton [48]. Environmental sources of PCNs include municipal solid waste incineration and metallurgical processes like copper roasting, and chlor-alkali processes. PCNs have been detected in air, sediment, pine needles, fish, birds, and human tissues [49–52]. Exposure to PCNs causes similar toxic effects as PCDDs/DFs and PCBs do. PCNs bind to aryl hydrocarbon hydroxylase enzymes and bring about Ah-receptor-mediated toxic effects such as hepatotoxicity, neurotoxicity, reproductive toxicity, and immune suppression [53]. Although PCNs were phased out over two decades ago, residues of PCNs are still found in environmental and biological samples [50–52]. Very limited information is available on PCN contamination status as well as, spatial and temporal trends.

1.3.2 Pharmaceutical and Personal Care Products

During the 1970s and 1980s, environmental scientists focused their research on industrial and agricultural pollutants such as DDT, PCBs, and dioxins, and thus overlooked the subtle connection between personal human activities and the subsequent release of residues from personal care products and pharmaceuticals into the natural environment [54]. This is evidenced from a gap in publications on this subject for several years since Hignite and Azarnoff reported findings of aspirin, caffeine, and nicotine in wastewater effluent in 1977 [55]. Watts et al. reported the presence of three pharmaceuticals (erythromycin, tetracycline, and theophylline), BPA, and other endocrine disrupting compounds in river water samples in 1984 [56]. Renewed interest in this subject arose after an authoritative and seminal publication by Daughton and Ternes in 1999 [57]. Since then, the number of publications has increased from 2 publications during the 1980s to over 300 scientific publications per year.

The occurrence of pharmaceutical and personal care product (PPCP) residues in the environment has received considerable attention in recent years as these compounds have been implicated for their negative effect on biota and the ecosystem. PPCPs are emerging environmental pollutants in the aquatic environment, and have been detected in surface water, groundwater, municipal wastewater, fish, and biosolids [58–60]. Earlier studies have linked the presence of PPCPs in the development of antibiotic resistant bacteria, feminization of male fish, and genotoxicity in aquatic organisms [57,56–64].

PPCPs enter the environment through multiple sources. Pharmaceutical drugs given to people and to domestic animals including antibiotics, hormones, pain relievers, tranquilizers, and chemotherapy chemicals given to cancer patients are being excreted and distributed into the environment through septic and sewer system and though the use of manure and sewage sludge in agriculture. Similarly, a large number of personal care products are being released into open waters via washing–bathing and swimming [58,65]. Many drugs are designed to be persistent and lipophilic so that they can retain their chemical structure long enough to perform their therapeutic function. After they are excreted, these chemicals persist in the environment, enter the food chain through bioaccumulation and cause harmful effects to wildlife and humans. For example synthetic polycyclic musks such as 1,3,4,6,7,8-hexahydro-4,6,6,7,8,8-hexamethylcyclopenta(γ)-2-benzopyran (HHCB) and 7-acetyl-1,1,3,4,4,6-hexamethyl-1,2,3,4-tetrahydronaphthalene (AHTN) are used as fragrances in a wide range of consumer products [66]. HHCB and AHTN are used as washing and cleaning agents and in personal care products and have been recognized as important organic contaminants in aquatic environments [67]. Production and use of nitro musks such as musk xylene and musk ketone have been decreased since the late 1980s due to concerns about toxicity [68]. Consequently, the use of polycyclic musks such as HHCB and AHTN has increased in the last decade [69]. The production of HHCB and AHTN in Europe in the late 1990s was approximately 1800 ton/year [70]. In the United States in 2000, 6500 ton of synthetic musks were used [71]. HHCB and AHTN accounted for 95% of the total market volume of polycyclic musks [72]. More studies are warranted to better understand the environmental distribution, behavior, and fate of PPCPs in the global environment. Regulations have been put in place to restrict or ban the use of persistent synthetic fragrances. However, such regulations are often limited to selected countries. Compounds such as HHCB have been detected in marine organisms from open waters suggesting widespread contamination by these compounds.

1.3.3 BISPHENOL A

Bisphenol A (BPA: 2,2-bis(hydroxyphenyl)propane ($C_{15}H_{16}O_2$; molecular weight 228.29) was first synthesized in 1881 by Diannin through condensation of acetone with phenol [73]. Industrial application was not discovered until the late 1930s while large-volume production of BPA began through its use as a plastic component in the 1940s. BPA has also been used in the synthesis of polycarbonate and epoxy resins. Epoxy and polystyrene resins are extensively used in plastic coatings, food packaging, dental fillings, water containers, baby bottles, beverage and food can linings, and optical lenses [74,75]. By the 1960s, BPA was used in all can linings. Currently, BPA is a $6 billion global industry. Approximately 940,000 ton of BPA is produced in the United States annually, according to the National Institute of Health [76]. BPA has been detected in wastewater treatment plant samples, river waters, and sediments as well as indoor air and dust [59,77–80].

Epoxy resins, when exposed to high temperatures, acidic conditions or cleaning detergents are able to leach BPA into the food or food materials within the container [81]. Humans are exposed to BPA via consumption of contaminated foods or beverages. BPA is known as "endocrine disruptor." It can mimic natural hormones and cause changes in physiological functions. BPA can bind to the estrogen receptors and induce estrogen receptor–mediated toxicity [82]. BPA can also bind to thyroid receptors and affect thyroid hormone functions [83]. Recent studies on humans have shown that male workers who were exposed to extremely high levels of BPA experienced consistently higher sexual dysfunction than unexposed workers in the same plastic manufacturing facility [84]. Hai et al. [81] showed that BPA decreases sperm quantity, causes various types of cancer, and affects the brain and cardiovascular system. These harmful effects of BPA cause great concern for the scientific community and general public. More detailed studies are needed to understand the contamination profile, environmental behavior, and fate of BPA. BPA is relatively less persistent than PCBs and DDTs. However, continuous exposure to BPA on a daily basis can pose risk to human and wildlife health. Spatial and temporal trends of BPA in the environment and biota have not been elucidated.

1.3.4 POLYCYCLIC AROMATIC HYDROCARBONS

Polycyclic (polynuclear) aromatic hydrocarbons are fused benzene rings that exhibit benzene-like properties. Polycyclic aromatic hydrocarbons (PAHs) are complex mixtures containing over 100 compounds, including highly carcinogenic benzo(a)pyrene, chrysene, benz(a)anthracene, benzo(b)fluoranthene, and benz(a,h)anthracene. Of these over 100 compounds, 16 PAHs are identified by the United States Environmental Protection Agency (U.S. EPA) as priority pollutants [85]. PAHs are formed during the combustion of organic matter as by-products and are thus not produced intentionally. These compounds are formed during the incomplete burning of coal, gas, oil, garbage, wood, or other organic matter that result from industrial or human activities [86–88]. PAHs are common pollutants in air, water, soil, sediment, and biota and contribute to serious environmental and health problems including cancer [89]. PAHs can enter surface water through the deposition of airborne PAHs, discharge of municipal wastewater, run off from urban stormwater and coal storage areas, wood treatment plants and other industries, oil spills, and petroleum processing. PAHs in soil result from public sewage treatment plant sludge disposal, irrigation with coke oven effluent, leachate from bituminous coal storage sites, use of soil compost and fertilizers, vehicle exhaust, and emissions from the wear of tires and asphalt [90–92]. Because PAHs are nonpolar, hydrophobic, and relatively stable, they tend to accumulate in environmental (sediment) and biological matrices [93]. Environmental contamination by PAHs has become of great concern due to their distribution in soil, water, and sediment; and their bioaccumulation in terrestrial and aquatic plants, invertebrates, and fish. The release of PAHs into the environment through human activities continues to increase. Low-molecular-weight PAHs such as naphthalene tend to get transported from tropical to polar regions via long-range atmospheric transport. Although PAHs are photodegradable, spatial, and temporal trend studies are important considering the continued input of PAHs into the environment by anthropogenic activities and their carcinogenic properties.

1.4   VALUE OF TREND MONITORING STUDIES

Even after several decades of banning organochlorine compounds, such as PCBs and DDT, residues of these compounds have been found in several environmental samples and biota and elicit chronic toxic effects. These persistent chemicals continue to cycle through food chain transfer, atmospheric, and hydrospheric transport. Considering the persistent properties of these man-made chemicals and their long-term effects on humans and wildlife, trend monitoring studies are useful to assess the risks of environmental pollution based on the simple philosophy that increasing environmental levels are more dangerous than decreasing ones. Spatial and temporal trend data are valuable in (1) determining whether government regulations taken to reduce the degree of environmental pollution have had the anticipated effects, (2) elucidating the sources/hot spots of contamination, and (3) providing information during the study on the impact of certain substances on a particular population. If the contamination levels decrease, but the ecological effects remain, there is a chance that another contaminant is solely or partly responsible for the effects.

Given the long-range intercontinental transport of these chemicals, no one government taking action can protect its people or the environment from persistent organochlorines. Therefore, organohalogen compounds have been the subject of an intensive international, national, regional attempt to limit their production and use as well as the control and disposal of old or obsolete chemical stocks. The United Nations Environment Program (UNEP)'s Stockholm Convention on persistent organic pollutants (POPs) is a global treaty established to protect human health and the environment that was adopted in 2001 and entered into force in 2004, requiring countries to take measures to eliminate or reduce the release of persistent chemicals into the environment. UNEP's Governing Council mandates eliminating or reducing emission and even in some cases halting the production and use of 12 POPs. The 12 POPs consist of PCBs, PCDDs, PCDFs, aldrin, chlordane, dieldrin, DDT, endrin, heptachlor, hexachlorobenzene, mirex, and toxaphene [94]. The UNEP's Stockholm Convention on POPs' fourth meeting held in May 2009 added nine new POPs under this Stockholm Convention [95]. The newly added POPs are alpha-hexachlorocyclohexane, beta-hexachlorocyclohexane, chlordecone, hexabromophenyl, hexabromobiphenyl ether and heptabromobiphenyl ether, lindane, pentachlorobenzene, perfluorooctane sulfonic acid and its salts, perfluorooctane sulfonyl fluoride, tetrabromodiphenyl ether, and pentabromodiphenyl ether [95]. The countries that signed the treaty are expected to monitor and report trends on the occurrence and distribution of these chemicals in their national environments. New chemicals added to the

Stockholm Convention in 2009 emphasize the value of continued monitoring and assessment of environmental contaminants.

1.5 FACTORS AFFECTING SPATIAL AND TEMPORAL TRENDS OF PERSISTENT ORGANIC CHEMICALS

Temporal trend studies on persistent organic chemicals have been performed extensively using environmental and biological matrices from different regions of the world including Japan [96–99], the western Mediterranean [100], the Swiss Plateau [101], Italy [30,102], Catalonia, Spain [103], Greenland [104], the United States [105,106], and the Canadian Arctic [107]. Several of these investigations were carried out with the objective of understanding the existing status of pollutant concentrations when compared to those in the past. The diverse factors that determine trends of contaminant levels in the environment were not clearly explained.

Due to limitations on temporal trend data on brominated compounds, fluorinated compounds, and other emerging persistent pollutants, the factors that determine spatial and temporal trends of these compounds are not discussed here. The factors determining the long-term contamination, behavior, and fate of organochlorines such as PCBs, DDTs, and HCHs were described based on the author's systematic studies carried out in Japan and the United States. For detailed sample information and results, refer to Loganathan et al. [96–98,105,108]. The factors considered for trend analysis included spatial (point, nonpoint, and remote from source), biological (short life span with low degradable capacity, long life span with high degradable capacity and long life span with low degradable capacity) and chemical characteristics (lipophilic and stable, less lipophilic and unstable characteristics). Tissue samples of fish, human, and striped dolphin represent point source, nonpoint source, and remote areas of contamination respectively. Fish samples were collected during 1968–1986 from the River Nagaragawa, Japan. This river had a history of serious environmental problems due to the discharge of industrial and agricultural wastes. Human adipose tissues from Japanese males was collected in 1928–1985 from various hospitals in Japan. Blubber samples of striped dolphin caught off Taiji, on the Pacific coast of central Japan during the period 1978–1979 and 1986, were used for trend monitoring studies. The observations are summarized in schematic figures (Figures 1.1 through 1.3) and interpretations were made from spatial, biological, and chemical characteristic viewpoints.

1.5.1 SPATIAL VIEWPOINT

Figure 1.1 shows the schematic representation of the temporal trends of persistent organic chemicals from a spatial viewpoint. In the point source area, such as the River Nagaragawa, an increase in residue levels of PCBs, HCHs, and DDTs was observed during the period of usage and then a rapid decline was noticed after restriction. In the terrestrial environment—nonpoint source contamination sites—organochlorines showed an increasing trend in concentrations from the beginning to end of the usage of these chemicals. During the post-ban period, a steady state or gradual decrease was observed until recent years. In the open ocean environment—remote areas—the residues of these persistent organochlorines showed a very slight decline or a stable condition in the contamination levels, even after several years of banning the chemicals (Figure 1.1).

Based on the results of spatial trend analysis, it may be inferred that in the point source areas, especially in the riverine environment, organochlorine concentrations increased during their usage and declined at a rapid rate when the restrictions were imposed. Contaminant levels in biota in the terrestrial environment and in the remote areas gradually increased with usage, and after government restrictions on their use, a slow rate of decline could be observed. Remote areas such as the open ocean environment reveal slow clearance rates of contamination. Therefore, it takes a long time to see the effectiveness of restriction on the use of persistent chemicals (Figure 1.1). Similar trends were also observed in point source (riverine), nonpoint source (terrestrial), and remote from source (open ocean) in various parts of the world [99,109–112].

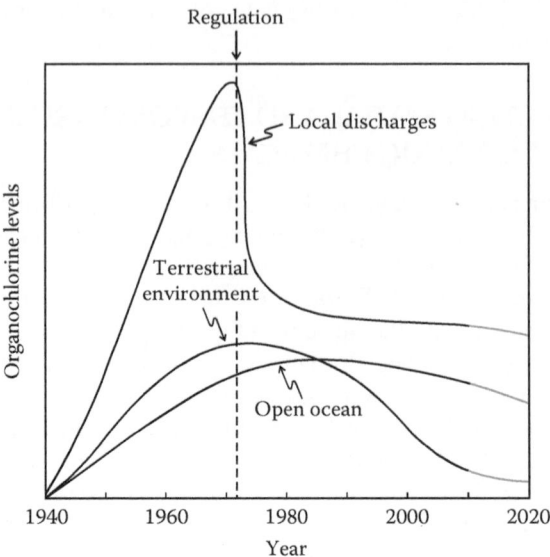

FIGURE 1.1 Schematic representation of the temporal trends of persistent organic chemicals from a spatial viewpoint. Local discharges, terrestrial environment and open ocean represent point source, away from the source and remote from the source respectively.

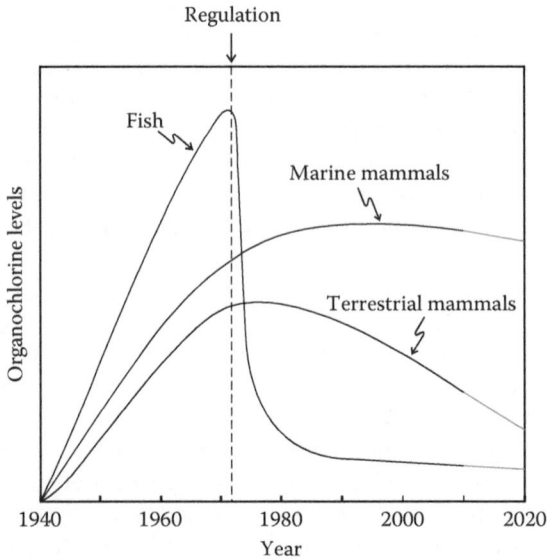

FIGURE 1.2 Schematic representation of the temporal trends of persistent man-made chemicals from a biological viewpoint. Fish, terrestrial mammals and marine mammals represent short life-span with low metabolic degradable capacity, long life-span with high metabolic degradable capacity and long life-span with low metabolic degradable capacity respectively.

1.5.2 Biological Viewpoint

Figure 1.2 shows a schematic representation of the temporal trends of a persistent man-made chemicals from a biological viewpoint. Fish samples collected from point source areas revealed elevated levels of organochlorine residues during the periods of usage of these compounds. The concentrations declined when the production and consumption of these chemicals were stopped. On the other hand, the levels of organochlorines in human adipose tissue gradually increased until the ban on these compounds. A slow rate or steady state was observed during postban periods. Interestingly, organochlorine levels in striped dolphin showed that the contamination of PCBs and DDTs remained unchanged or showed tendency for gradual decline in the open ocean environment.

Therefore, animals that have a short life span and less metabolic capacity to degrade organochlorines such as fish are contaminated rapidly. The clearance rate is faster for these animals when regulations are imposed on the use of these chemicals. However, in organisms with long life spans and high metabolic capacities, such as humans, both the concentrations and clearance rates are slower. Animals with a long life span but lesser capability for metabolizing man-made chemicals recorded extremely slow clearance rates. Hence, it can be surmised that animals that have long life spans and low metabolic capacities (e.g., marine mammals) are susceptible to long-term accumulation and are at greater toxic threat of persistent organic chemicals. This viewpoint can be applied to a global scale, since similar trends were also discernible from published reports from other parts of the world [32,36]. Furthermore, it is worth indicating that animals having a relatively higher content of fat in their bodies show a slower clearance of organochlorines than other animals [111–115].

1.5.3 CHEMICAL CHARACTERISTICS VIEWPOINT

Figure 1.3 shows a schematic representation of temporal trends of environmental contamination by persistent organic chemicals from a chemical characteristic viewpoint. Organochlorine residues that are relatively less lipophilic and unstable than PCBs and DDTs, such as HCHs exhibited a declining trend after withdrawal from production and use in all the biological samples from different ecosystems, whereas highly lipophilic and stable contaminants like PCBs and DDTs continued to exhibit higher concentrations even after restriction on production and use (Figure 1.3). It has been made clear from our results that the less lipophilic and comparatively unstable compounds (e.g., HCHs) contaminate the environment gradually and get cleared rapidly upon restriction. Highly lipophilic and stable compounds contaminate the environment quickly and their clearance rates are very slow. As a result, highly lipophilic and more stable compounds elicit long-term effects on

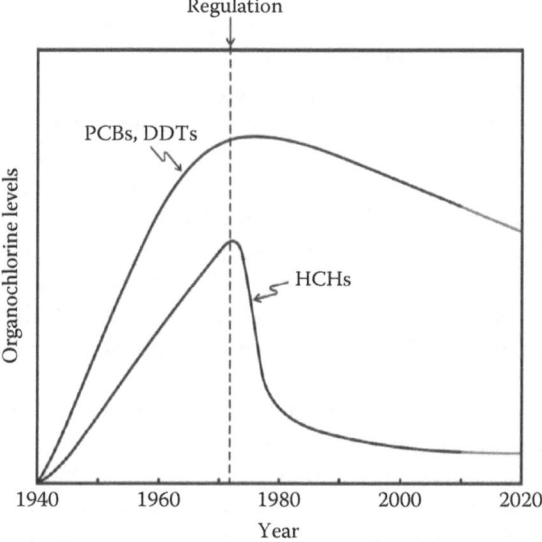

FIGURE 1.3 Schematic representation of temporal trends of environmental contamination by persistent organic chemicals from a chemical characteristic viewpoint. PCBs and DDTs represent lipophilic and stable compounds and HCHs represent less lipophilic and unstable compounds.

humans and wildlife. These observations can be projected for the entire global environment and supported from the results obtained by various authors [111,114–117].

One of the main goals of temporal trend studies is to predict future levels of contamination. Trend monitoring studies are of further use for local governments in making appropriate legislative actions and making policy decisions to combat pollution in the environment. Very limited studies were conducted to develop suitable mathematical models to fit the time trend data for assessing the future trends of contaminant levels. Ritter et al. (2009) reported a multi-individual pharmacokinetic model framework for interpreting time trends of persistent chemicals in human populations [118]. Applications and further development of such models are necessary to make clear the future course of pollutant behavior in various ecosystems. Influences of different variables governing pollutant characteristics should also be taken into considerations in those models. Since chemodynamics of any chemical species depend on various abiotic and biotic factors, development of models must be site-specific and representative of the prevailing conditions. The following section deals with the spatial and temporal trends in specific ecosystems based on published studies.

1.6   SPATIAL AND TEMPORAL TRENDS IN SPECIFIC ECOSYSTEMS

1.6.1   Terrestrial Ecosystem

A variety of sample types have been used to trace the historical trends of organochlorine pollution in terrestrial ecosystems. Human adipose tissue has been one of the most commonly used matrices. The first evidence of accumulation of DDT in human fat was reported by Howell [119]. Kutz et al. reviewed worldwide data on the residue levels of organochlorines in human adipose tissues [120]. In general, the clearance rates of PCBs based on the time trend surveys carried out in the United States [115], Canada [110], China [117], Italy [30], Spain [32], and Japan [99,108] using human adipose fat indicated a very slow declining trend in their residues, implying a continuing exposure of humans to PCBs. This could be attributable to the estimate that two-thirds of PCBs produced on the global scale are still in use or in landfills and thus their environmental levels are unlikely to show any significant decline in the near future [121]. Additionally, in humans, PCBs are eliminated at a slower rate as a result of metabolic degradation [122,123]. PCB pollution has been regarded as a problem for industrialized nations in North America, Europe, and Japan. Studies in tropical Asia revealed moderate-to-high concentrations of PCBs in various classes of foods used for human consumption [124–126]. The observed trends for PCBs in humans from both developing and developed nations suggest the need for suitable PCB pollution abatement strategies.

Unlike PCBs, a rapid reduction in the DDT content of human fat has occurred since restrictions on the use of this compound were imposed [108,115,127]. The clearance rate of DDT has been slower in human tissues than in aquatic ecosystems [128]. During the late 1970s and 1980s, there was a remarkable reduction in the human DDT body burdens in developing countries after which reduction rates were relatively slow. Monitoring studies on contaminant levels in human adipose tissue, blood, and breast milk from the United States [120,129], Canada [130], and Norway [131,132] suggested that DDT levels have been declining since the early 1970s while PCB concentrations have been declining relatively more slowly. This trend may be applicable to most higher trophic predators of terrestrial food chains in developed countries.

In developing Asian and African countries, few long-term studies on human contamination by organochlorine pesticides exist despite the increasing use of these chemicals. Comparison of HCH and DDT concentrations in human fat, reported for India in the 1960s and 1970s, with the present levels showed no significant reduction [133]. Similarly, considerable amounts of DDT were present in human blood samples from Bangladesh [134], Slovakia [135], and South Africa [136]. These results indicate continued human DDT exposure in Africa and tropical Asia. This trend is likely to continue unless suitable alternatives to DDT are developed to control mosquitoes that cause malaria and other diseases. Based on these results, it can be surmised that in humans the clearance rates of organochlorines are slow; PCBs exhibit a tendency to decline over decades. HCH, DDT, and chlordanes continue to decline in human tissues in developed nations.

Global monitoring studies have been conducted by various authors using plant foliage such as leaves, bark, lichens, mosses, pine needles, etc. Pine needles have been used as a fixed site, nondestructive, passive monitoring matrix for the evaluation of distribution and transport of lipophilic air pollutants in the continental atmosphere [51,137,138]. Jones et al. used this approach to determine temporal trends of air in the United Kingdom from 1965 to 1989 [139]. The use of plant foliage may indicate organochlorines derived from long-range atmospheric transport in addition to local/regional sources. For example, an unexpected increase in DDT levels in pine needles from Sweden in 1984 was correlated to the forest spraying of DDT in East Germany during the same period [137]. Therefore, organochlorine trends observed in plant foliage or ambient air at a site not only represent inputs of local contaminants but also the inputs through atmospheric transport from various point source areas.

Soil and peat cores have been used to examine the long-term inputs of organohalogen compounds. These studies concluded that the atmosphere responds rapidly to changes in chemical usage and disposal. A sharp clearance or rapid reduction was seen in atmospheric levels following a local ban on organochlorine usage [140]. However, ice, sediment, peat-, and soil-core analyses may give poor temporal resolution when the contaminants are subjected to post-depositional changes [141].

Another type of biological matrix commonly used in tracing persistent man-made organochlorine trends is bird tissues. Interpretation of bird tissue data is complicated due to the effect of biological variables, including feeding habits and migratory behavior, which results in the organochlorine data from bird tissues being highly variable. For fish-eating waterbirds, the organochlorine trends often represent organochlorine behavior found in aquatic ecosystems rather than in terrestrial environments. When eggs have been used for trend studies, it cannot be assumed that the eggs collected during several years belong to the same population group.

PCB and DDT trends observed using bird tissues in the United Kingdom [142], the United States [143], Canada [132], Italy [133], and Norway [134] indicated a decline in the body burdens of the compounds between the early 1970s and 1980s with the reduction rate for PCBs slower than that for DDT. As found for humans, birds also exhibited a slow clearance rate, which is in contrast to freshwater biota (e.g., fish) inhabiting environments close to the point source of pollution. The major metabolite of DDT and DDE has been reported to degrade slowly due to its longer half-life in birds, but the rate of decline has been slow [144]. For example, the British sparrow hawk population has maintained relatively constant DDE levels for over 20 years [145]. Most studies show clearly that the decline in populations of several bird species in the Northern Hemisphere was due to high DDT burdens. Thus it is important to protect birds in the Southern Hemisphere where there is continued large-scale use of DDT for public health purposes. Trend monitoring studies in tropical regions, particularly in Africa, Latin America, and Asia, are warranted.

1.6.2 Inland and Coastal Aquatic Ecosystems

Shell fish (mussels), fish (including fish products), and waterfowl have been used to determine temporal trends of organochlorines in the aquatic environment. Trend monitoring studies of riverine fish indicated a sharp decline in residue levels following the prohibition of organochlorine use [30,96–151]. Periodic monitoring surveys carried out by the U.S. Fish and Wildlife Service have shown a consistent decline in DDT concentrations in the early 1980s [151]. Similarly, a rapid downward trend has been reported for PCB levels in fish residing in the Hudson River in the United States, since 1975 [152]. Several studies have shown that the concentrations of organochlorines in gill-breathing animals such as shellfish and teleost fish are controlled by equilibrium partitioning [153–155] and residue levels in such organisms are related to the ambient levels in the water. Organochlorines with higher water solubilities, such as HCH, reach rapid equilibrium with water and accumulate rapidly in gill-breathing organisms. While equilibrium partitioning is the primary factor governing organochlorine levels in small fish, intake of these compounds via food is of major importance in higher trophic level predatory fish. Rapid reduction of organochlorine concentrations in fish over time indicates fast clearance of these compounds in rivers. In general, the riverine ecosystem concentrations reflect the effectiveness of restrictions on uses of a chemical. The rapid reduction in persistent organochlorine chemicals in rivers may also reflect the relatively short residence time of water in rivers.

Among lakes in the United States, the Great Lakes are the most frequently investigated fresh water system because of extensive contamination by persistent organochlorine chemicals. Reports indicate declining concentrations of PCBs, DDTs, and other organochlorine compounds in the Great Lakes [128,156–158]. The rate of decrease in concentrations depends on fish species and sampling locations chosen. In general, the declining rate was faster in the 1970s than in the 1980s. DDT declined rapidly from 1970 to 1980, but after that it has declined at a slower rate or remained almost constant. A slow and gradual decline was found for PCBs. Slower clearance rates in recent years may be due to the continuous atmospheric input of these chemicals from other areas that still use these inexpensive organochlorine pesticides. Results of studies on Swedish lakes indicate declining concentrations of organochlorines during the late 1970s and early 1980s [159,160]. These studies have shown that local restrictions on the use of organochlorines have had positive effects on local freshwater environments. The declining rates in lakes have been slower than those in rivers but faster than those in the marine environment [128,149,161]. Very limited information is available on the contamination of lakes in the Southern Hemisphere. As closed aquatic systems are prone to rapid contamination by persistent chemicals, careful management is essential to protect these ecosystems.

Organochlorine concentrations in fish and other higher trophic level aquatic animals indicate that the highest concentrations are found in the higher trophic level predators inhabiting semi-enclosed seas and coastal regions. For example, PCB concentrations in striped dolphins from the Mediterranean Sea have been determined to be greater than 1000 ppm, which is the highest level ever recorded in a marine mammal [162]. Predatory eagles in the Baltic Sea contained over 100 ppm of PCBs and DDT [163]. The Seto Inland Sea of Japan has been shown to receive considerable amounts of PCBs and other organochlorines from the surrounding terrestrial ecosystem [164]. Time trend studies in fish from the Swedish [160], Finnish [165], and Polish coasts [166] of the Baltic Sea as well as the North Sea [167] Norwegian coast [113] and in mussels from the Adriatic Sea [168] indicated decreasing concentrations of PCBs and DDT. However, the clearance rate has been slow in most coastal locations. For more volatile chemicals such as HCH, a steady state is observed [166].

Cycling and flux of organic contaminants in the biota of the Mediterranean suggested that their residence time is long and varies from 3 to 30 years [169]. The concentrations of organochlorines have been observed to be comparatively high in semi-enclosed and coastal areas even in recent years [104]. The heavy utilization of pesticides and PCBs was centered in regions of industrialization and farming, resulting in direct discharges into adjacent coastal areas. Semi-enclosed seas received direct discharges of chemicals from the terrestrial ecosystem via rivers, and, therefore, these ecosystems become rapidly contaminated to a high degree in contrast to open oceans, which are contaminated by atmospheric deposition, transport via coastal water currents, and/or ocean dumping. Because the clearance rates of such chemicals are relatively slow, biota in semi-enclosed seas are at risk for chronic toxic effects. Environmental management and appropriate use of enclosed coastal seas, as well as conservation strategies are essential to improve protection of enclosed seas.

1.6.3 Open Ocean Ecosystems

The ocean is the ultimate sink for terrestrial matter including man-made chemicals. For example, a large portion of PCBs that have escaped into the global environment reside in coastal sediments and open ocean waters, suggesting the role of the marine environment as a reservoir for persistent and semi-volatile organochlorines [170]. The ultimate effect of these toxic chemicals is mainly on marine mammals, which are the top predators of the oceanic food chain and lack the capacity to metabolize/degrade many of these toxic contaminants. The widespread occurrence of *tris* (4-chlorophenyl) methanol in open ocean biota from various regions suggests the transport of polymers, agrochemicals, and synthetic dyes used in the terrestrial environment to the ocean [171].

While there are many reports of organochlorine levels in oceanic samples, only a few dealt with their temporal trends. In Arctic ringed seals, residual levels of PCBs and DDTs declined from the 1970s to the 1980s and then leveled off during the 1980s and early 1990s [172]. Norstrom et al. [173] showed that PCB and DDT levels did not change significantly between 1969–1971 and 1984 in polar bears from the Canadian Arctic. Increasing trends of DDTs from 1994 to 2004 was found in both sexes of seals as well as in male sculpins from Central West Greenland [104]. Loganathan et al. analyzed striped dolphins from the western North Pacific and the results showed no reduction in the concentrations of PCBs and DDT from 1978 to 1986, whereas HCHs and HCB showed a significant decrease during this period [98]. PCB and HCH trends in northern fur seals collected from the Pacific coast of northern Japan suggested a very slow declining trend while that for DDT showed a rapid drop in concentrations after the early 1970s [174]. Dolphins from the Indian Ocean showed a decline in PCB burden over several years, while there has been an increase in DDT concentrations during the 1980s [175]. A uniform distribution of HCH reported in surface waters of the Atlantic Ocean was suggested to be due to the continuous inputs of HCH from Africa and South America [176]. However, the ratios of DDE/Σ DDT in North Atlantic pinnipeds and cetaceans based on the analysis of literature data suggest a reduction in DDT input into the Atlantic [177]. In general, it appears that the open ocean environment exhibits a very slow declining trend in organochlorine concentrations and continues to serve as a sink for semi-volatile organochlorines [178]. Although the magnitude of contamination in various oceans differs considerably according to the meteorological parameters prevailing in the region and the physicochemical properties of the compounds involved, atmospheric transport on regional and global scales is recognized as a major route for the transfer of contaminants to the ocean. More volatile compounds (e.g., HCH) have long atmospheric residence times in temperate regions and short aquatic residence times in the tropics. HCH is highly transportable but relatively little accumulates in marine mammals, while DDT has a shorter residence time in the atmosphere and tends to bioaccumulate in organisms to a greater extent than HCH. Despite the ban on the use of DDT in many Northern Hemisphere countries, DDT was reported to be present in air above northern oceans, suggesting inputs from the tropics. It has been shown that the North Pacific receives the highest atmospheric depositions of HCH and DDT while maximum depositional rates of PCBs and dieldrin occur in the North Atlantic as compared to oceans in the Southern Hemisphere [179].

Marine pollution by organochlorines may not decline significantly unless strict regulations are imposed on their use throughout the world. Marine mammals have long life spans are at the highest trophic level in the oceanic food chain and possess low ability to metabolize some organochlorines; therefore, these animals are considered to be at risk for long-term exposure and negative health effects.

1.7   SUMMARY AND CONCLUSIONS

Spatial and temporal trend investigations of POPs in various environments are useful for a better understanding of the behavior of organochlorines for the purpose of establishing strategies to limit their dispersal. However, trend monitoring studies are often stalled by the unavailability of archived samples or suitable analytical instrumentation. The chemicals in question often were not analyzed in the past, and tissues and environmental samples are not archived for future analysis. Even if archived samples are available, not all background information that influence the chemicals' concentrations was recorded. One of the reasons for this may be due to the lack of complete knowledge of chemical behavior at that time.

A comparison of temporal trend studies in one region with those of another required careful consideration of biological and chemical factors. The issue of organochlorine pollution is complicated by the difficulty in obtaining data of the amount of production and usage of these chemicals in various countries. Biological and chemical precision can be improved by international programs such as Mussel Watch, which monitors trace organic substances on a worldwide comparative basis. Although the present review provides a generalized assessment, it is representative of the existing situation for organochlorine trends in the global environment.

Figures 1.4 through 1.6 summarize organohalogen trends in the global environment. As mentioned in an earlier section, semi-enclosed seas, the open ocean, and tropical regions are the areas of special attention for future investigations pertaining to organochlorine dynamics and their effect on the environment and human health. People in developing countries are exposed to high levels of organochlorines from food and air as a result of the continued use of these chemicals. Since many developed nations import foodstuffs from developing countries, the populations in developed countries are exposed to organochlorines through their food. This has been shown by the steady state or very slow decline of organochlorine burden in the populations of both developing

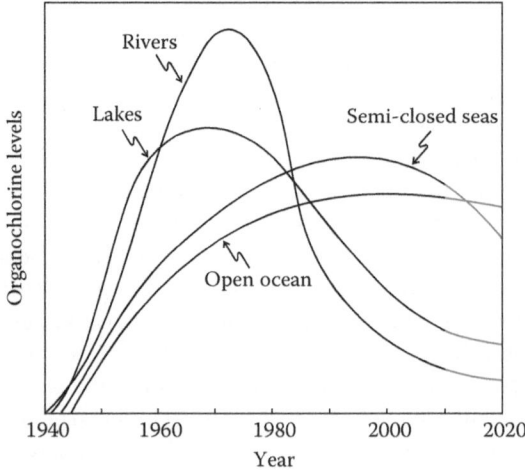

FIGURE 1.4   Schematic representation of the organochlorine contamination trends and clearance rates in aquatic biota.

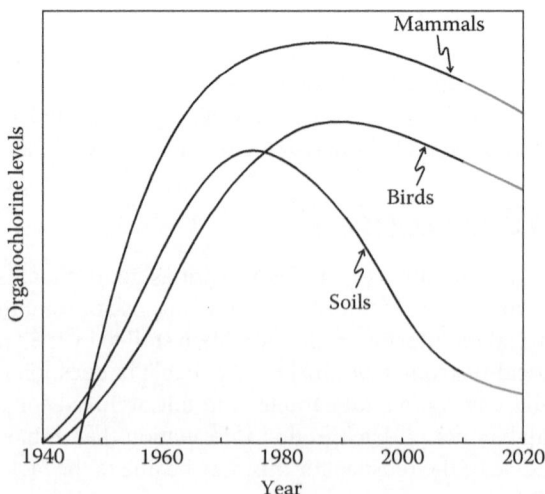

FIGURE 1.5   Schematic representation of the organochlorine contamination trends and clearance rates in terrestrial biota.

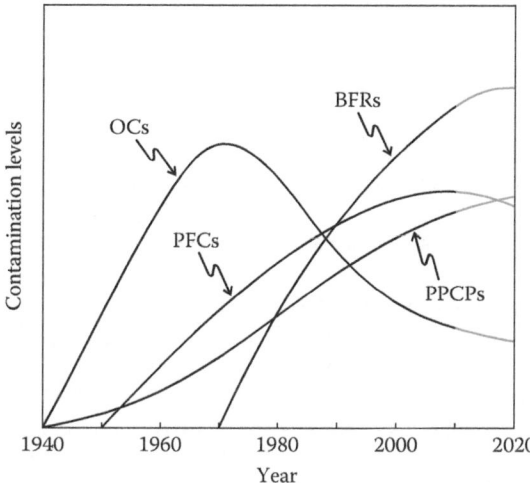

FIGURE 1.6 Schematic representation of global environmental contamination trends of organohalogen compounds. OCs, organochlorines; PFCs, perfluorinated compounds; BFRs, brominated flame retardants; PPCPs, pharmaceutical and personal care products.

and developed nations. Furthermore, developing countries are being used as dumping grounds for hazardous chemicals including highly toxic pesticides and brominated flame retardants in e-wastes because many of the countries have few policies to check the inflow of hazardous chemicals [180–184]. Despite this, no comprehensive contaminant monitoring programs, including long-term trend studies and toxicological investigations, have been carried out in those countries likely because of the economic and political implications of such studies [183,184]. The international Mussel Watch Program is an example of a legitimate tool for pollution monitoring in the coastal environment of the developing countries.

The exposure pathway for chlorinated organic compounds in humans is mainly from outdoor environments via consumption of contaminated food (food chain accumulation). However, for the emerging environmental pollutants such as BFRs, PFCs, and PPCPs, the human exposure pathway is predominantly from indoor contamination. Given the widespread use of PBDEs in household items, indoor contamination may be a significant source of human exposure. Similarly, PFCs are used in a variety of consumer products and human exposure from the indoor environment and consumption of PFC-contaminated foodstuffs are inevitable. Pharmaceuticals and personal care products enter the environment via wastewater from homes and indirectly affect humans by affecting aquatic organisms such as fish and developing antibiotic resistant bacteria. The human exposure and health effects caused by these emerging environmental pollutants will continue for a long period of time, even after the baning their production, as has been evidenced for organochlorines [185]. Potential "compounds of concern" for environmental and health issues is compiled in Table 1.2.

Figures 1.5 and 1.6 show a schematic representation of time perspectives of environmental contamination and exposure to humans. Chlorinated compounds such as PCBs and pesticides contaminate the environment and biota very rapidly during the periods of their use in agriculture and/or for public health purposes. The contamination levels declined after a ban or severe restriction was placed on the production and use of these compounds in most developed countries. In contrast, brominated and fluorinated compounds are being produced in large quantities and used globally (by both developing and developed countries). These compounds are heavily used in indoor appliances and materials. Unlike organochlorines (exposure via food chain), human exposure pathways for PBDEs and PFCs are direct. Considerable data have been amassed on

TABLE 1.2

"Compounds of Concern" for Potential Future Environmental and Health Issues

Abbreviation	Compound Name	CAS Number
BPA	Bisphenol A (4,4-dihydroxy-2,2-diphenyl propane)	80-05-7
BPAF	Bisphenol AF (hexafluorobisphenol A) (1,1,1,3,3,3-hexafluoro-2,2-bis(4-hydroxyphenyl)propane)	1478-61-1
BEHTBP	Bis(2-ethylhexyl)tetrabromophthalate	26040-51-7
BTBPE	1,2-Bis(2,4,6-tribromophenoxy)ethane	37853-59-1
BTBPIE	1,2-Bis(tetrabromophthalimido)ethane	32588-76-4
DBHC-TCTD or HCDBCO	5,6-Dibromo-1,10,11,12,13,13-hexachloro-11-tricyclo[8.2.1.02,9]tridecene	51936-55-1
DBDPE	Decabromodiphenylethane	84852-53-9
DEHTBP or TBPH	Di(ethylhexyl)tetrabromophthalate	26040-51-7
DP	Dechlorane Plus, Bis(hexachlorocyclopentadieno)cyclo-octane	13560-89-9
EH-TBB or TBB	2-Ethylhexyl-2,3,4,5-tetrabromobenzoate	183658-27-7
HBB	Hexabromobenzene	87-82-1
HBCDD or HBCD	Hexabromocyclododecane, major isomers are α, β, γ-HBCDD	3194-55-6
PBEB	Pentabromoethylbenzene	85-22-3
PBT	Pentabromotoluene	87-83-2
SCCP	Short-chain chlorinated paraffins	85535-84-8 and 71011-12-6
TBBPA	Tetrabromobisphenol A	79-94-7
TBBPA-DAE	Tetrabromobisphenol A diallyl ether	25327-89-3
TBBPA-DBPE	Tetrabromobisphenol A bis(2,3-dibromopropyl) ether	21850-44-2
TBECH	1,2-Dibromo-4-(1,2-dibromoethyl)cyclohexane	3322-93-8
TBP-AE or ATT	2,4,6-tribromophenyl allyl ether	3278-89-5
TCEP	Tris(2-chloroethyl)phosphate	115-96-8
TDCPP or TDCP	Tris(1,3-dichloroisopropyl)phosphate	13674-87-8
POSF	Perfluorooctane sulfonyl fluoride (1,1,2,2,3,3,4,4,5,5,6,6,7,7,8,8,8-heptadecafluorooctane-1-sulfonyl fluoride)	307-35-7
PFOS	Perfluorooctane sulfonic acid (1,1,2,2,3,3,4,4,5,5,6,6,7,7,8,8,8-heptadecafluorooctane-1-sulfonic acid)	1763-23-1
PFOS K	Perfluorooctane sulfonate potassium salt (1,1,2,2,3,3,4,4,5,5,6,6,7,7,8,8,8-heptadecafluorooctane-1-sulfonate potassium salt)	2795-39-3
N-EtFOSA	*N*-ethyl-perfluorooctanesulfonamide (1,1,2,2,3,3,4,4,5,5,6,6,7,7,8,8,8-heptadecafluorooctane-1-sulfonamide)	754-91-6
PFBA	Perfluorobutanoic acid (2,2,3,3,4,4,4-heptafluorobutanoic acid)	375-22-4
PFPeA	Perfluroropentanoic acid (2,2,3,3,4,4,5,5,5-nonafluoropentanoic acid)	2706-90-3
PFHxA	Perfluorohexanoic acid (2,2,3,3,4,4,5,5,6,6,6-undecafluorohexanoic acid)	307-24-4
PFHpA	Perfluroroheptanoic acid (2,2,3,3,4,4,5,5,6,6,7,7,7-tridecafluoroheptanoic acid)	375-85-9
PFOA	Perfluorooctanoic acid (2,2,3,3,4,4,5,5,6,6,7,7,8,8,8-pentadecafluorooctanoic acid)	335-67-1
PFNA	Perfluorononanoic acid (2,2,3,3,4,4,5,5,6,6,7,7,8,8,9,9,9-heptadecafluorononanoic acid)	375-95-1
PFDA	Perfluorodecanoic acid (2,2,3,3,4,4,5,5,6,6,7,7,8,8,9,9,10,10,10-nonadecafluorodecanoic acid)	335-76-2
PFUnDA	Perfluoroundecanoic acid (2,2,3,3,4,4,5,5,6,6,7,7,8,8,9,9,10,10,11,11,11-henicosafluoroundecanoic acid)	2058-94-8

TABLE 1.2 (continued)
"Compounds of Concern" for Potential Future Environmental and Health Issues

Abbreviation	Compound Name	CAS Number
PFDoDA	Perfluorododecanoic acid	307-55-1
	(2,2,3,3,4,4,5,5,6,6,7,7,8,8,9,9,10,10,11,11,12,12,12-tricosafluorododecanoic acid)	
8:2 FTOH	8:2 fluorotelomer alcohol	678-39-7
	(3,3,4,4,5,5,6,6,7,7,8,8,9,9,10,10,10-heptadecafluoro-1-decanol)	
HHCB	1,3,4,6,7,8-hexahydro-4,6,6,7,8,8-hexamethylcyclopenta[g]-2-benzopyrane	1222-05-5
AHTN	7-acetyl-1,1,3,4,4,6-hexamethyltetrahydeonaphthalene	1506-02-1
UV-326	2-(3-*tert*-butyl-2-hydroxy-5-methylphenyl)-5-chlorobenzotriazole	3896-11-5
UV-327	2,4-di-*tert*-butyl-6-(5-chloro-2H-benzotriazol-2-yl) phenol	3864-99-1
UV-328	2-(2H-benzotriazol-2-yl)-4,6-di-tert-pentylphenol	25973-55-1

Prepared based on Yin et al. [186], DiGangi et al. [187], Matsushima et al. [188], Yamashita et al. [189], Nakata et al. [190].

the presence of PBDEs and PFCs in indoor environmental media (air, water, dust, lint, clothing, food packaging materials, etc.) and human tissues (blood, breast milk, liver, fetus, etc.) have been amassed. Evidence of PPCP contamination and ecotoxicological effects on fish and other aquatic organisms and indirect effect on humans via antibiotic resistant bacteria is mounting. Because of their recalcitrant properties, bioaccumulation, and their toxic potentials it can be predicted that the environmental contamination, human exposure, and health effects of PBDEs and PFCs will continue to increase for several decades in both developed and developing countries (Figure 1.6). These factors need consideration in the effort to protect human and wildlife health from the long-term effects caused by these persistent organic chemicals.

REFERENCES

1. Bettelheim, F.A., Brown, W.H., and March, J. 2001. *Introduction to General, Organic and Biochemistry.* Harcourt College Publishers, Philadelphia, PA, 713pp.
2. Rodan, B.D. 2002. Genesis of the global persistent organic pollutants treaty. In *The Foundation for Global Action on Persistent Organic Pollutants: A United States Perspective.* Office of Research and Development, Washington, DC. EPA/600/P-01/003F.
3. Schmidt, H. and Schultz, G. 1881. Uber benzidin (a-di-amidophenyl). *Am. Chem. Liebigs.* 207: 320.
4. Schmidt, C. 2010. How PCBS are like grasshoppers. *Environ. Sci. Technol.* 44: 2752.
5. Kodavanti, P.R., Senthilkumar, K., and Loganathan, B.G. 2008. Organohalogen pollutants and human health. In: *Encyclopedia of Public Health* (H.K. Heggenhougen and S. Quah, eds.), vol. 4. Academic Press, San Diego, CA, pp. 686–693.
6. Klecka, G., Boethling, B., Franklin, J. et al. (eds.). 2000. *Evaluation of Persistence and Long-Range Transport of Organic Chemicals in the Environment.* SETAC Special Publication Series, Pensacola, FL, 400pp.
7. Fraser, B. 2010. Researchers find little known PCB "pretty much everywhere." *Environ. Sci. Technol.* 44: 2753–2754.
8. Loganathan, B.G. and Kannan, K. 1994. Global organochlorine contamination: An overview. *AMBIO* 23: 187–191.
9. Trowbridge, A.G. and Swackhamer, D.L. 2001. Biomagnification of toxic PCB congeners in the Lake Michigan foodweb. In: *Persistent, Bioaccumulative and Toxic Chemicals I: Fate and Exposure* (R.L. Lipnick, J. Hermens, K. Jones, and D. Muir, eds.), ACS Monograph Series vol. 772. American Chemical Society, Washington, DC, 308pp.
10. Magill, F.N. (ed.). 1995. *Great Events from History II: Ecology and Environment Series.* The United States Bans DDT, vol. 3. Salem Press, Pasadena, CA, pp. 1966–1973.

11. Klassen, C.D. (ed.). 1996. *Casarett and Doull's Toxicology. The Basic Science of Poisons*, 5th edn. McGraw Hill, New York, pp. 671–673.
12. Evers, B., Hawkins, S., Ravenscroft, M., Rounsaville, J.F., and Schultz, G. (eds.) 1989. In: *Ullmann's Encyclopedia of Industrial Chemistry*, vol. A1, 25th edn. VCH, Weinheim, Germany, 130pp.
13. Rappe, C. 1978. 2,3,7,8-Tetrachlorodibenzo-p-dioxin (TCDD)—Introduction. In: *Dioxin: Toxicological and Chemical Aspects* (F. Cattabeni, A. Cavallaro, and G. Galli, eds.). SP Medical and Scientific Books, New York, pp. 9–11.
14. Hutzinger, O., Choudhry, G.G., Chittim, B.G., and Johnston, L.E. 1985. Formation of polychlorinated dibenzofuran and dioxins during combustion, electrical equipment fires and PCB incineration. *Environ. Health Perspect.* 60: 3–9.
15. Hryhorczuk, D.O., Orris, P., Kominski, J.R., Melius, I., Burton, W., and Hinkamp, D.L. 1986. PCB, PCDF, PCDD exposure following a transformer fire—Chicago. *Chemosphere* 15: 1297–1303.
16. ATSDR, 2001. Toxicological report for pentachlorophenol (update). Agency for Toxic Substances and Disease Registry Publication No. PB/2001/109106/AS. U.S. Department of Health and Human Services, Atlanta, GA.
17. Masunaga, S., Takasuga, T., and Nakanishi, J. 2001. Dioxin and dioxin like impurities in some Japanese agrochemical formulations. *Chemosphere* 44: 873–885.
18. Loganathan, B.G., Kanna, K., Sajwan, K.S., Chetty, C.S., Giesy, J.P., and Owen, D.A. 1997. Polychlorinated dibenzo-*p*-dioxins, dibenzofurans and polychlorinated biphenyls in street dusts and soil samples from Savannah, Georgia. *Organohalogen Compd.* 32: 192–197.
19. Feil, V.J. and Larsen, G.L. 2001. Dioxins in food from animal sources. In: *Persistent, Bioaccumulative and Toxic Chemicals I: Fate and Exposure* (R.L. Lipnick, J.L.M. Hermens, K.C. Jones, and D.C.G. Muir, eds.), ACS Symposium Series vol. 772. American Chemical Society, Washington, DC, pp. 245–251.
20. Safe, S. 1990. Polychlorinated dibenzo-*p*-dioxins (PCDDs),dibenzofurans (PCDFs) and related compounds: Environmental and mechanistic considerations which support the development of toxic equivalency factors (TEFs). *Crit. Rev. Toxicol.* 21: 51–88.
21. Giesy, J. and Kannan, K. 1998. Dioxin-like and non-dioxin like toxic effects of polychlorinated biphenyls (PCBs): Implications for risk assessment. *Crit. Rev. Toxicol.* 28: 511–569.
22. Masunaga, S., Yao, Y., Ogura, I., Nakai, S., Kanai, Y., Yamamura, M., and Nakanishi, J. 2001. Identifying sources and mass balance of dioxin pollution in Lake Shinji Basin, Japan. *Environ. Sci. Technol.* 35: 1967–1973.
23. Ogura, I., Masunaga, S., and Nakanishi, J. 2001. Congener specific characterization of PCDDs/PCDFs in atmospheric deposition: Comparison of profiles between deposition, source and environmental sink. *Chemosphere* 45: 173–181.
24. Kjeller, L.-O., Jones, K.C., Johnston, A.E., and Rappe, C. 1991. Increase in polychlorinated dibenzo-p-dioxin and furan content of soils and vegetation sites since 1840s. *Environ. Sci. Technol.* 25: 1619–1627.
25. Webster, T. and Commoner, B. 1994. Overview: The dioxin debate. In: *Dioxin and Health* (A. Schecter, ed.). Plenum Press, New York, pp. 1–6.
26. Petreas, M., She, J., McKinney, M., Visita, P., Winkler, J., Mok, M., and Hooper, K. 2001. *Dioxin Body Burden in California Populations* (R.L. Lipnick, J.L.M. Hermens, K.C. Jones, and D.C.G. Muir, eds.), ACS Symposium Series vol. 772. American Chemical Society, Washington, DC, pp. 252–265.
27. Todaka, T., Hori, T., Hirakawa, H., Kajiwara, J., Yasutake, K., Onozuka, D., Washino, N., Konishi, K., Sasaki, S., Yoshioka, E., Yuasa, M., Kishi, R., Iida, T., and Furue, M. 2008. Concentrations and congener profiles of non-dioxin-like polychlorinated biphenyls in blood collected from 195 pregnant women in Sapporo City, Japan. *Organohalogen Compd.* 70: 1597–1600.
28. Lignell, S., Aune, M., Darnerud, P.O., Cnattingius, S., and Glynn, A. 2009. Persistent organochlorine and organobromine compounds in mother's milk from Sweden 1996–2006: Compound-specific temporal trends. *Environ. Res.* 109: 760–767.
29. Michael, A., Sigrun, B., Kerstin, B., Jürgen, W., Richard, M., and Hermann, F. 2008. PCDD/PCDF and dioxin-like polychlorinated biphenyls in blood of an adult population. *Organohalogen Compd.* 70: 1044–1047.
30. Abballe, A., Ballard, T., Dellatte, E. et al. 2008. Persistent environmental contaminants in human milk: Concentrations and time trends in Italy. *Chemosphere* 73: 5220–5227.
31. Yang, J., Kim, H., Ho, M., Shin, D., Chang, Y., and Lim, Y. 2008. Trend analysis of PCDD/F levels in the blood of urban dwellers in Korea. *Organohalogen Compd.* 70: 1586–1589.
32. Llobet, J.M., Marti-Cid, R., Castell, V., and Domingo, J.L. 2008. Significant decreasing trend in human dietary exposure to PCDD/PCDFs and PCBs in Catalina, Spain. *Toxicol. Lett.* 178: 117–126.

33. Loganathan, B.G., Kannan, K., Watanabe, I., Kawano, M., Irvine, K., Kumar, S., and Sikka, H.C. 1995. Isomer-specific determination and toxic evaluation of polychlorinated biphenyls, polychlorinated/brominated dibenzo-*p*-dioxins and dibenzofurans, polybrominated biphenyl ethers, and extractable organic halogen in carp from Buffalo River, New York. *Environ. Sci. Technol.* 29: 1832–1838.

34. Shaw, S.D. and Kannan, K. 2009. Polybrominated diphenyl ethers in marine ecosystems of the American Continents: Foresight from current knowledge. *Rev. Environ. Health* 24: 157–229.

35. Vorkamp, K., Riget, F.F., Glasius, M., Muir, D.C.G., and Dietz, R. 2008. Levels and trends of persistent organic pollutants in ringed seals (*Phoca hispida*) from central west Greenland, with particular focus on polybrominated diphenyl ethers (PBDEs). *Environ. Int.* 34: 499–508.

36. Kodavanti, P. 2005. Neurotoxicity of persistent organic pollutants: Possible modes of action and further considerations. *Dose Response* 3: 273–305.

37. Hale, R.C., Kin, S.L., Harvey, E. et al. 2008. Antarctic research bases: Local sources of polybrominated diphenyl ether(PBDE) flame retardants. *Environ. Sci. Technol.* 42: 1452–1457.

38. Hites, R.A. 2004. Polybrominated diphenyl ethers in the environment and in people: A meta analysis of concentrations. *Environ. Sci. Technol.* 38: 945–956.

39. Betts, K. 2008. New thinking on flame retardants. *Environ. Health Perspect.* 116: A210–A213.

40. Kannan, K., Corsolini, S., Falandyz, J. et al. 2004. Perfluorooctanesulfonate and related fluorochemicals in human blood from several countries. *Environ. Sci. Technol.* 38: 4489–4495.

41. Giesy, J.P. and Kannan, K. 2001. Global distribution of perfluorooctane sulfonate in wildlife. *Environ. Sci. Technol.* 35: 1339–1342.

42. Kannan, K., Corsolini, S., Falandysz, J., Oehme, G., Focardi, S., and Giesy, J.P. 2002. Perfluorooctane sulfonate and related fluorinated hydrocarbons in marine mammals, fishes and birds from coasts of the Baltic and Mediterranean Seas. *Environ. Sci. Technol.* 36: 3210–3216.

43. Taniyasu, S., Kannan, K., Horii, Y., Hanari, N., and Yamashita, N. 2003. A survey of perfluoroocatane sulfonate and related perfluorinated organic compounds in water, fish, birds, and humans from Japan. *Environ. Sci. Technol.* 37: 2634–2639.

44. Mak, Y.L., Taniyasu, S., Yeung, L.W.Y., Lu, G.H., Jin, L., Yang, Y.L., Lam, P.K.S., Kannan, K., and Yamashita, N. 2009. Perfluorinated compounds in tap water from China and several other countries. *Environ. Sci. Technol.* 43: 4824–4829.

45. Kucklick, J.R. and Helm, P.A. 2006. Advances in the environmental analysis of polychlorinated naphthalenes and toxaphene. *Anal. Bioanal. Chem.* 386: 819–836.

46. Falandysz, J. 1998. Polychlorinated naphthalenes: An environmental update. *Environ. Pollut.* 101: 77–90.

47. Bidleman, T.F., Helm, P.A., Braune, B.M., and Gabrielsen, G.W. 2010. Polychlorinated naphthalenes in polar environments—A review. *Sci. Total Environ.* 48: 2919–2935.

48. Yamashita, N., Kannan, K., Imagawa, T., Miyazaki, A., and Giesy, J.P. 2000. Concentrations and profiles of polychlorinated naphthalene congeners in eighteen technical polychlorinated biphenyl preparations. *Environ. Sci. Technol.* 34: 4236–4241.

49. Harner, T., Lee, R.G.M., and Jones, K. 2000. Polychlorinated naphthalenes in the atmosphere of the United Kingdom. *Environ. Sci. Technol.* 34: 3137–3142.

50. Kannan, K., Imagawa, T., Blankenship, A.L., and Giesy, J.P. 1998. Isomer specific analysis and toxic evaluation of polychlorinated naphthalenes in soil, sediment, and biota collected near the site of a former chlor-alkali plant. *Environ. Sci. Technol.* 32: 2507–2514.

51. Loganathan, B.G., Senthilkumar, K., Seaford, K.D., Sajwan, K.S., Hanari, N., and Yamashita, N. 2008. Distribution of persistent organohalogen compounds in pine needles from selected locations in Kentucky and Georgia, USA. *Arch. Environ. Contam. Toxicol.* 54: 422–439.

52. Kannan, K., Hilscherova, K., Imagawa, T., Yamashita, N., Williams, L.L., and Giesy, J.P. 2001. Polychlorinated naphthalenes, -biphenyls, -dibenzo-*p*-dioxins and -dibenzofurans in double-crested cormorants and herring gulls from Michigan waters of Great Lakes. *Environ. Sci. Technol.* 35: 441–447.

53. Eljarrat, E. and Barcelo, D. 2003. Priority list for persistent organic pollutants and emerging contaminants based on their relative toxic potency in environmental samples. *Trends Anal. Chem.* 22: 655–665.

54. Jones-Lepp, T., Alvarez, D., and Loganathan, B.G. 2011. On the frontier: Analytical chemistry and the occurrence of illicit drugs in surface waters in the USA. In: *Illicit Drugs in the Environment: Occurrence, Analysis and Fate Using Mass Spectrometry*, ed. S. Castiglioni, E. Zuccato, and R. Fennelli, 171–188. John Wiley & Sons, Inc., New York.

55. Hignite, C. and Azarnoff, D. 1977. Drugs and drug metabolites as environmental contaminants: Chlorophenoxyisobutyrate and salicylic acid in sewage water effluents. *Life Sci.* 20: 337–341.

56. Watts, C.D., Crathorn, B., Fielding, M., and Steel, C.P. 1984. Identification of non-volatile organics in water using field desorption mass spectrometry and high performance liquid chromatography. In: *Analysis of Organic Micropollutants in Water* (G. Angeletti and A. Bjorseth, eds.). *Proceedings of the Third European Symposium*, September 19–21, 1983, Oslo, Norway, pp. 120–131.

57. Daughton, C. and Ternes, T. 1999. Pharmaceutical and personal care products in the environment: Agents of subtle change? *Environ. Health Perspect.* 107: 907–938.

58. Loganathan, B.G., Phillips, M., Mowery, H., and Jones-Lepp, T.L. 2009. Contamination profiles and mass loadings of macrolide antibiotics and illicit drugs from a small urban wastewater treatment plant. *Chemosphere* 75: 70–77.

59. Kolpin, D. Furlong, E., Meyer, M., Thurman, E., Zaugg, S., Barber, L., and Button, H. 2002. Pharmaceutical hormones, and other wastewater contaminants in US streams, 1999–2000: A national reconnaissance. *Environ. Sci. Technol.* 36: 1202–1211.

60. Jones-Lepp, T. and Stevens, R. 2007. Pharmaceuticals in biosolids—The interface between analytical chemistry and regulation. *Anal. Bioanal. Chem.* 387: 1173–1183.

61. Schwartz, T., Kohnen, W., Jansen, B., and Obst, U. 2003. Detection of antibiotic resistant bacteria and their resistance genes in wastewater and surface water and drinking water biofilms. *FEMS Microbiol. Ecol.* 43: 325–335.

62. Jobling, S., Williams, R., Johnson, A. et al. 2006. Predicted exposure to steroid estrogens in UK rivers correlate with widespread sexual disruption in wild fish populations. *Environ. Health Perspect.* 114 (Suppl. 1): 32–39.

63. Nash, J.P., Kime, D.E., Van der Ven, L.T.M., Wester, P.W., Brion, E., Maack, G., Scabbschmidt-Alner, P., and Tyler, C.R. 2004. Long-term exposure to environmental concentrations of the pharmaceutical ethynylestradiol cause reproductive failure in fish. *Environ. Health Perspect.* 112: 1725–1733.

64. Kostich, M. and Lazorchak, J. 2008. Risk to aquatic organisms posed by human pharmaceutical use. *Sci. Total Environ.* 389: 329–339.

65. Grange, A.H. and Sovocool, G.W. 2003. Ion composition and elucidation (ICE) of ions from trace level of pharmaceuticals and disinfection byproducts in water supplies. U.S. EPA Report. U.S. EPA National Exposure Research Laboratory, Las Vegas, NV.

66. Horii, Y., Reiner, J.L., Loganathan, B.G., Kumar, K.S., Sajwan, K., and Kannan, K. 2007. Occurrence and fate of polycyclic musks in wastewater treatment plants in Kentucky and Georgia, USA. *Chemosphere* 68: 2011–2020.

67. Herberer, T. 2002. Occurrence, fate, and assessment of polycyclic musk residues in the aquatic environment or urban areas—A review. *Acta Hydrochem. Hydrobiol.* 30: 227–243.

68. Gattermann, R., Biselli, S., Hübnerfuss, H., Rimkus, G.G., Hecker, M., and Karbe, L. 2002. Synthetic musk in the environment. Part 1. Species dependent bioaccumulation of polycyclic and nitro musk fragrances in freshwater fish and mussels. *Arch. Environ. Contam. Toxicol.* 42: 437–446.

69. Sommer, C. 2004. *The Role of Musk and Musk Compounds in the Fragrance Industry*. Springer, New York.

70. Rimkus, G.G., 1999. Polycyclic musk fragrances in the aquatic environment. *Toxicol. Lett.* 111: 37–56.

71. Peck, A.M., Linebaugh, E.K., and Hornbuckle, K.C. 2006. Synthetic musk fragrances in Lake Erie and Lake Ontario sediment cores. *Environ. Sci. Technol.* 40: 5629–5635.

72. Balk, F. and Ford, R.A. 1999. Environmental risk assessment for the polycyclic musks, AHTN and HHCB: II effect assessment and risk characterization. *Toxicol. Lett.* 111: 81–94.

73. Dianin, D. 1891. Bisphenol A. *Zh. Russ. Fiz.-Khim. O-va.* 23: 492.

74. Weisbons, W.V., Nagel, S.C., and van Seal, P.S. 2006. Large effects from small exposures: III Endocrine mechanisms mediating effects of bisphenol A at levels of human exposure. *Endocrinology* 147: 856–869.

75. Padmanaban, V., Siefert, K., Ranson, S., Johnson, T., Pinkerton, J., Anderson, I., Tao, L., and Kannan, K. 2008. Maternal bisphenol A levels at delivery: A looming problem. *J. Perinatol.* 28: 258–263.

76. Voith, M. 2009. Can conundrum. *Chem. Eng. News* 87: 28–29.

77. Amaral, L., Barcelo, D., Dinix, M. et al. 2006. Characterization of selected endocrine disrupting compounds in a Portuguese wastewater treatment plant. *Environ. Monit. Assess.* 118: 75–87.

78. Boyd, G.R., Palmeri, J.M., Zhang, S., and Grimm, D.A. 2004. Pharmaceutical and personal care products (PPCPs), and endocrine disrupting chemicals (EDCs), in stormwater canals and Bayou St. John in New Orleans, Louisiana, USA. *Sci. Total Environ.* 333: 137–148.

79. Khim, J.S., Lee, K.T., Villeneuve, D.I., and Kannan, K. 2001. *In vitro* bioassay determination of dioxin-like and estrogenic activity in sediment and water from Ulsan Bay and its vicinity, Korea. *Arch. Environ. Contam. Toxicol.* 40: 151–160.

80. Rudel, R.A., Brody, J.G., Spengler, J.D., Vallario, J., Geno, P.W., Sun, G., and Yau, B. 2001. Identification of selected hormonally active agents and animal mammary carcinogens in commercial and residential air and dust samples. *J. Air Water Manag. Assoc.* 51: 499–513.

81. Hai, G., Huanshun, Y., Lin, C., Lusheng, Z., and Shiyun, A. 2010. Electrochemical determination of bisphenol A at $Mg-Al-CO_3$ layered double hydroxide modified glassy carbon electrode. *Electrochim. Acta* 55: 603–610.

82. Matthews, J.B., Twomey, K., and Zacharewski, T.R. 2001. *In vitro* and *in vivo* interaction of bisphenol A and its metabolite bisphenol A glucuronide, with estrogen receptors alpha and beta. *Chem. Res. Toxicol.* 14: 149–157.

83. Zneller, R.T., Bansal, R., and Parris, C. 2005. Bisphenol A an environmental contaminant that acts as a thyroid hormone receptor antagonist *in vitro,* increases serum thyroxine and alters RC3/neurogranin expression in the developing rat brain. *Endocrinology* 146: 607–612.

84. Erickson, B.E. 2009. BPA linked to male sexual dysfunction. *Chem. Eng. News* 87: 27.

85. Maruya, K.A., Loganathan, B.G., Kannan, K., McCumber-Khan, S., and Lee, R.F. 1997. Organic and organometallic compounds in estuarine sediments from the Gulf of Mexico (1993–1994). *Estuaries* 20: 700–709.

86. Ohura, T., Amagai, T., Fusaya, M., and Matsushita, H. 2004. Polycyclic aromatic hydrocarbons in indoor and outdoor environments and factors affecting their concentrations. *Environ. Sci. Technol.* 38: 77–83.

87. Bzdusek, P.A. and Christensen, E.R. 2004. Source apportionment of sediment PAHs in Lake Calumet, Chicago: Application of factor analysis with nonnegative constraints. *Environ. Sci. Technol.* 38: 97–103.

88. Terzi, E. and Samara, C. 2004. Gas-particle partitioning of polycyclic aromatic hydrocarbons in urban, adjacent coastal and continental background sites of western Greece. *Environ. Sci. Technol.* 38: 4973–4978.

89. Jacob, J. 1996. The significance of polycyclic aromatic hydrocarbons as environmental carcinogens. *Pure Appl. Chem.* 68: 301–308.

90. Li, A., Jang, J.K., and Scheff, P.A. 2003. Application of EPA CMB8.2 Model for source apportionment of sediment PAHs in Lake Calumet, Chicago. *Environ. Sci. Technol.* 37: 2958–2965.

91. Thorsen, W.A., Cope, W.G., and Shea, D. 2004. Bioavailability of PAHs: Effects of soot carbon and PAHs source. *Environ. Sci. Technol.* 38: 2029–2037.

92. Sverdrup, L.E., Nielson, T., and Krogh, P.H. 2002. Soil ecotoxicity of polycyclic aromatic hydrocarbons in relation to soil sorption, lipophilicity and water solubility. *Environ. Sci. Technol.* 36: 2429–2435.

93. Lapviboonsuk, J. and Loganathan, B.G. 2007. Polynuclear aromatic hydrocarbons in sediments and mussel tissue from the lower Tennessee River and Kentucky Lake. *J. Ky. Acad. Sci.* 68: 186–197.

94. Buccini, J. 1999. Progress in developing a United Nations convention on persistent organic pollutants (POPs). *Organohalogen Compd.* 43: 459–460.

95. http://chm.pops.int

96. Loganathan, B.G., Tanabe, S., Tatsukawa, R., and Goto, M. 1989. Temporal trends of organochlorine contamination in lizard goby, *Rhinogobius flumineus* from the River Nagaragawa, Japan. *Environ. Pollut.* 62: 237–251.

97. Loganathan, B.G., Tanabe, S., Tatsukawa, R., Ogawa, K., and Goto, M. 1989. Temporal changes of morphologic abnormalities and parasite infestation in fishes from the River Nagaragawa, Japan. *Nippon Suis. Gak. (Bull. Jpn. Soc. Sci. Fish.)* 55: 769–774.

98. Loganathan, B.G., Tanabe, S., Tanaka, H., Miyazaki, N., Amano, M., and Tatsukawa, R. 1990. Comparison of persistent residues in striped dolphin *Stenella coeruleoalba* from western North Pacific in 1978–1986. *Mar. Pollut. Bull.* 21: 435–439.

99. Kunisue, T., Muraoka, M., Ohtake, M. et al. 2006. Contamination status of persistent organochlorines in human breast milk from Japan: Recent levels and temporal trends. *Chemosphere* 64: 1601–1608.

100. Deudero, S., Box, A., March, D. et al. 2007. Organic compounds temporal trends at some invertebrate species from Balearics, western Mediterranean. *Chemosphere* 68: 1650–1659.

101. Zennegg, M., Kohler, M., Hartman, P. et al. 2007. The historical record of PCB and PCDD/F deposition in Greifensee, a lake of the Swiss plateau, between 1848 and 1999. *Chemosphere* 67: 1754–1761.

102. Valle, M.D., Marcomini, A., Sweetman, A.J., and Jones, K.C. 2005. Temporal trends in the sources of PCDD/Fs to and around the Venice lagoon. *Environ. Int.* 31: 1040–1046.

103. Abad, E., Caixach, J., Rivera, J., Gustem, L., Massague, G., and Puig, O. 2004. Temporal trends of PCDDs/PCDFs in ambient air in Catalonia (Spain). *Sci. Total Environ.* 334–335: 279–285.

104. Riget, F., Dietz, R., Vorkamp, K., Johansen, P., and Muir, D. 2004. Levels and spatial and temporal trends of contaminants in Greenland biota: An updated review. *Sci. Total Environ.* 331: 29–52.

105. Alexander, C.R., Smith, R., Loganathan, B.G., Ertel, J., Windom, H.L., and Lee, R.F. 1999. Pollution history of the Savannah River Estuary and comparisons with Baltic Sea pollution history. *Limnologica* 29: 267–273.

106. Marvin, C.H., Painter, S., Charlton, M.N., Fox, M.E., and Lina Thiessen, P.A. 2004. Trends in spatial and temporal levels of persistent organic pollutants in Lake Erie sediments. *Chemosphere* 54: 33–40.

107. Hung, H., Blanchard, P., Halsall, C.J. et al. 2005. Temporal and spatial variabilities of atmospheric polychlorinated biphenyls (PCBs), organochlorine (OC) pesticides and polycyclic aromatic hydrocarbons (PAHs) in the Canadian Arctic: Results from decade of monitoring. *Sci. Total Environ.* 342: 119–144.

108. Loganathan, B.G., Tanabe, S., Hidaka, Y., Kawano, M., Hidaka, H., and Tatsukawa, R. 1993. Temporal trends of persistent organochlorine residues in human adipose tissues from Japan, 1928–1985. *Environ. Pollut.* 81: 31–39.

109. Novak, M.A., Reilly, A.A., and Jackling, S.J. 1988. Long-term monitoring of polychlorinated biphenyls in the Hudson River (New York) using Caddis fly larvae and other macroinvertebrates. *Arch. Environ. Contam. Toxicol.* 17: 699–710.

110. Boer, J. 1989. Organochlorine compounds and bromodiphenyl ethers in livers of Atlantic cod (*Gadus morhua*) from the North Sea. *Chemosphere* 18: 2131–2140.

111. Muir, D.C.G. and Norstrom, R.J. 2000. Geographical differences and time trends of persistent organic pollutants in the Arctic. *Toxicol. Lett.* 112–113: 93–101.

112. Corsolini, S. 2009. Industrial contaminants in Antarctic biota. *J. Chromatogr. A.* 1216: 598–612.

113. Skare, J.U., Stenersen, U., Kveseth, N., and Polder, A. 1985. Time trends of organochlorine chemical residues in seven sedentary marine fish species from a Norwegian Fjord during the period 1972–1982. *Arch. Environ. Contam. Toxicol.* 14: 33–41.

114. Kannan, K., Falandysz, J., Yamashita, N., Tanabe, S., and Tatsukawa, R. 1992. Temporal trends of organochlorine concentrations in cod liver oil from the southern Baltic proper, 1971–1989. *Ambio.* 24: 358–363.

115. Robinson, P.E., Mack, G.A., Remmers, J. et al. 1990. Trends of PCBs, hexachlorobenzene, and β-benzene hexachloride levels in the adipose tissue of the US population. *Environ. Res.* 53: 175–192.

116. Pearce, P.A., Elliott, J.E., Peakall, D.B., and Norstrom, R.J. 1989. Organochlorine contaminants in eggs of seabirds in the northwest Atlantic 1968–1984. *Environ. Pollut.* 56: 217–235.

117. Xing, Y., Lu, Y., Dawson, R.W. et al. 2005. A spatial temporal assessment of pollution from PCBs in China. *Chemosphere* 60: 731–739.

118. Ritter, R., Scheringer, M., Macleod, M., Schenker, U., and Hungerbuhler, K. 2009. A multi-individual pharmacokinetic model framework for interpreting time trends of persistent chemicals in human populations: Application to a postban situation. *Environ. Health Perspect.* 117: 1280–1286.

119. Howell, D.E. 1948. A case of DDT storage in human fat. *Proc. Oklahoma Acad. Sci.* 29: 31–32.

120. Kutz, F.W., Wood, P.H., and Bottimore, D.P. 1991. Organochlorine pesticides and polychlorinated biphenyls in human adipose tissue. *Rev. Environ. Contam. Toxicol.* 120: 1–82.

121. Tanabe, S. 1988. PCB problems in the future: Foresight from current knowledge. *Environ. Pollut.* 50: 5–28.

122. Yakushili, T. 1988. Contamination, clearance and transfer of PCBs from human milk. *Rev. Environ. Contam. Toxicol.* 101: 139–164.

123. Bühler, F., Schmid, P., and Schlatter, Ch. 1988. Kinetics of PCB elimination in man. *Chemosphere* 17: 1717–1726.

124. Tanabe, S., Kannan, K., Tabucanon, M.S., Siriwong, C., Ambe, Y., and Tatsukawa, R. 1991. Organochlorine pesticide and polychlorinated biphenyl residues in foodstuffs from Bangkok, Thailand. *Environ. Pollut.* 72: 191–203.

125. Kannan, K., Tanabe, S., Ramesh, A., Subramanian, A.N., and Tatsukawa, R. 1992. Persistent organochlorine residues in foodstuffs from India and their implications on human dietary exposure. *J. Agric. Food Chem.* 40: 518–524.

126. Kannan, K., Tanabe, S., Quynh, H.T., Hue, N.D., and Tatsukawa, R. 1992. Residue pattern and dietary intake of persistent organochlorine compounds in foodstuffs from Vietnam. *Arch. Environ. Contam. Toxicol.* 22: 367–374.

127. Mes, J. 1990. Trends in the levels of some chlorinated hydrocarbon residues in adipose tissue of Canadians. *Environ. Pollut.* 65: 269–278.

128. Kannan, K., Ridal, J., and Struger, J. 2006. Pesticides in the Great Lakes. *Hdb. Env. Chem.* 5: 151–199.

129. Hovinga, M.E., Sowers, M.F., and Humphrey, H.E.B. 1992. Historical changes in serum PCB and DDT levels in an environmentally exposed cohort. *Arch. Environ. Contam. Toxicol.* 22: 362–366.

130. Frank, R., Rasper, J., Smout, M.S., and Braun, H.E. 1988. Organochlorine residues in adipose tissues, blood and milk from Ontario residents, 1976–1985. *Can. J. Public Health* 79: 150–158.

131. Skaare, J.U., Tuvneg, J.M., and Sanade, H.A. 1989. Organochlorine pesticides and polychlorinated biphenyls in material adipose tissue, blood milk and cord blood from mothers and their infants living in Norway. *Arch. Environ. Contam. Toxicol.* 17: 55–63.

132. Noren, K. 1983. Organochlorine contaminants in Swedish human milk from the Stockholm region. *Acta. Paediatr. Scand.* 72: 259–264.

133. Ramachandran, M., Banerjee, B.D., Gulati, M., Grover, A., Zaidi, S.S.A., and Hussain, Q.C. 1984. DDT and HCH residues in the body fat and blood samples from some Delhi hospitals. *Indian J. Med. Res.* 80: 590–593.

134. Mamun, M.I.R., Zamir, R., Nahar, N., Mosihuzzaman, M., Linderholm, L., Athanasiadou, M., and Bergman, A. 2007. Traditional organochlorine pollutants in blood from humans living in the Bangladesh capital area. *Organohalogen Compd.* 69: 241–245.

135. Jana, C., Kamil, C., Sona, W., Anton, K., and Milena, D. 2008. Dioxins, PCBs and organochlorine pesticides in the blood serum of Slovak residents. *Organohalogen Compd.* 70: 1177–1180.

136. Rollin, H.B., Sandanger, T.M., and Odland, J.O. 2008. DDTs, and other persistent organic pollutants in plasma of delivering women from selected areas of South Africa—Results of pilot study. *Organohalogen Compd.* 70: 1345–1348.

137. Eriksson, G., Jensen, S., Kylin, H., and Strachan, W. 1989. The pine needle as a monitor of atmospheric pollution. *Nature* 341: 42–44.

138. Jones, K.C. 1991. Contaminant trends in soils and crops. *Environ. Pollut.* 69: 311–325.

139. Jones, K.C., Sanders, G., Wild, S.R., Brunett, V., and Johnston, A.E. 1992. Evidence for a decline of PCBs and PAHs in rural vegetation and air in the United Kingdom. *Nature* 356: 137–140.

140. Rapaport, R.A. and Eisenreich, S.J. 1988. Historical atmospheric inputs of high molecular weight chlorinated hydrocarbons to eastern North America. *Environ. Sci. Technol.* 22: 931–941.

141. Sanders, G., Jones, K.C., Taylor, J.H., and Dürr, H. 1992. Historical inputs of polychlorinated biphenyls and other organochlorines to a dated lacustrine sediment core in rural England. *Environ. Sci. Technol.* 26: 1815–1821.

142. Newton, I., Haas, M.B., and Freestone, P. 1990. Trends in organochlorine and mercury levels in gannet eggs. *Environ. Pollut.* 63: 1–12.

143. UNEP. 1987. *Environmental Data Report*. United Nations Environment Program. Basil Blackwell Ltd., Oxford, U.K., 352pp.

144. Stickel, W.H., Stickel, L.F., Dyrland, R.A., and Hughes, D.L. 1984. DDE in birds; lethal residues and loss rates. *Arch. Environ. Contam. Toxicol.* 13: 1–6.

145. Walker, C.H. and Stanley, P.I. 1987. Organochlorine insecticide residues in predatory birds: Long-term trends and bioaccumulation. In: *Pesticide Science and Biotechnology* (R. Greenbalgh and I.R. Roberts, eds.). Blackwell Scientific Publications, London, U.K., pp. 367–370.

146. Elliott, J.E., Norstrom, R.J., and Keith, J.A. 1988. Organochlorines and eggshell thinning in northern gannets (*Sula bassanus*) from eastern Canada. 1968–1984. *Environ. Pollut.* 52: 81–102.

147. Fasola, M., Vecchio, I., Caccialanza, G., Gandini, C., and Kitsos, M. 1987. Trends of organochlorine residues in eggs of birds from Italy. 1977–1985. *Environ. Pollut.* 48: 25–36.

148. Fimreite, N., Brevik, E.M., and Torp, R. 1982. Mercury and organochlorines in eggs from a Norwegian gannet colony. *Bull. Environ. Contam. Toxicol.* 28: 58–60.

149. Sajwan, K.S., Senthilkumar, K., Nune, S., Fowler, A., Richardson, J.P., and Loganathan, B.G. 2008. Persistent organochlorine pesticides, polychlorinated biphenyls, polybrominated diphenylethers in fish from coastal waters off Savannah, GA, USA. *Toxicol. Environ. Chem.* 90: 81–96.

150. Fukushima, M., Kawai, S., Yamamoto, O., Oda, K., and Morioka, T. 1988. Organochlorine pollution in Yodo River basin, Japan: Historical trend and present status. In: *Pollution in the Urban Environment. POLMET 88* (P. Hills, R. Keen, K.C. Lam, C.T. Leung, M.A. Oswell, M. Stokes, and E. Turner, eds.), vol. 2. Vincent Blue Copy Co. Ltd., Hong Kong, pp. 395–400.

151. Schmitt, C.J., Zajicek, J.L., and Peterman, P.H. 1990. National contaminant biomonitoring program: Residues of organochlorine chemicals in US freshwater fish. 1976–1984. *Arch. Environ. Contam. Toxicol.* 19: 748–781.

152. Thomann, R.V., Mueller, J.A., Winfield, R.P., and Huang, C.R. 1991. Model of fate and accumulation of PCB homologues in Hudson estuary. *J. Environ. Eng.* 117: 161–178.

153. Hamelink, J.L., Waybrant, R.C., and Ball, R.C. 1971. A proposal: Exchange equilibria control the degree chlorinated hydrocarbons are biologically magnified in lentic environments. *Trans. Am. Fish. Soc.* 100: 207–214.

154. Clayton, J.R. Jr., Pavlou, S.P., and Breitner, N.F. 1977. Polychlorinated biphenyls in coastal marine zooplankton: Bioaccumulation by equilibrium partitioning. *Environ. Sci. Technol.* 11: 676–682.

155. Schneider, R. 1982. Polychlorinated biphenyls (PCBs) in cod tissues from the western Baltic: Significance of equilibrium partitioning and lipid composition in the bioaccumulation of lipophilic pollutants in gill-breathing animals. *Meeresforschung* 29: 69–79.

156. Amant, S.J.R., Pariso, M.E., and Sheffy, T.B. 1984. Polychlorinated biphenyls in seven species of Lake Michigan fish. 1971–1981. In: *Toxic Contaminants in the Great Lakes* (J.O. Nriagu and M.S. Simmons, eds.). John Wiley & Sons, New York, pp. 311–319.

157. Baumann, P.C. and Whittle, D.M. 1988. The status of selected organics in the Laurentian, Great Lakes: An overview of DDT, PCBs, dioxins, furans and aromatic hydrocarbons. *Aquat. Toxicol.* 11: 241–257.

158. Borgmann, U. and Whittle, D.M. 1991. Contaminant concentration trends in Lake Ontario lake trout (*Salvelinus namaycush*): 1977 to 1988. *J. Great Lakes Res.* 17: 368–381.

159. Anderson, O., Linder, C.E., Olsson, M., Reutergardh, L., Uvemo, U.B., and Wideqvist, U. 1988. Spatial differences and temporal trends of organochlorine compounds in biota of the northwestern hemisphere. *Arch. Environ. Contam. Toxicol.* 17: 755–765.

160. Olsson, M., Reutergardh, L., Uvemo, U.B., and Wideqvist, U. 1988. Spatial differences and temporal trends of organochlorine compounds in biota of the northwestern hemisphere. *Arch. Environ. Contam. Toxicol.* 17: 755–765.

161. Summers, J.K., Macauley, J.M., Heitmuller, P.T., Engle, V.D., Brooks, G.T., Barrow, M., and Adams, A.M. 1994. Annual statistical summary: EMAP—Estuaries Louisianian Province 1992. EPA/620/R-94/002. United States Environmental Protection Agency, Office of Research and Development, Gulf Breeze, FL.

162. Borrell, A. and Aguilar, A. 1991. Pollution by PCBs in striped dolphins affected by the western Mediterranean epizootic. In: *Proceedings of Mediterranean Striped Dolphin Mortality International Workshop.* (X. Pastor and M. Simmonds, eds.). Green Peace International, Madrid, Spain, pp. 121–127.

163. Falandysz, J., Jakuezun, B., and Mizera, T. 1988. Metals and organochlorines in four white-tailed eagles. *Mar. Pollut. Bull.* 19: 521–526.

164. Tanabe, S., Kannan, N., Fukushima, M., Okamoto, T., Wakimoto, T., and Tatsukawa, R. 1989. Persistent organochlorines in Japanese coastal waters: An introspective summary from a far east developed nation. *Mar. Pollut. Bull.* 20: 344–352.

165. Haahu, H. Levels and trends of organochlorines in cod and herring in the northern Baltic. *Mar. Pollut. Bull.* 19: 29–32.

166. Kannan, K., Falandysz, J., Yamashita, N., Tanabe, S., and Tatsukawa, R. 1992. Temporal trends of organochlorine concentrations in cod-liver oil from the southern Baltic proper. 1971–1989. *Mar. Pollut. Bull.* 24: 358–363.

167. De Boer, J. 1989. Organochlorine compounds and bromodiphenylethers in livers of Atlantic cod (*Gadus morhus*) from the North Sea. 1977–1987. *Chemosphere* 18: 2131–2140.

168. Picer, M. and Picer, N. 1991. Levels and long-term trends of some high molecular chlorinated hydrocarbons in mussels collected from the western Istrian coastal waters—Northern Adriatic. *Chemosphere* 23: 742–759.

169. Burns, K.A. and Villeneuve, J.P. 1987. Chlorinated hydrocarbons in the open Mediterranean ecosystem and implications for mass balance calculations. *Mar. Chem.* 20: 337–359.

170. Tatsukawa, R. and Tanabe, S. 1990. Fate and bioaccumulation of persistent organochlorine compounds in the marine environment. In: *Oceanic Process in Marine Pollution* (D.J. Baumgartner and I.W. Duedall, eds.), vol. 6. Krieger Publishing Co., Malabar, FL, pp. 39–52.

171. Jarman, W.M., Simon, M., Norstrom, R.J., Burns, S.A., Bacon, C.A., Simonelt, B.R.T., and Risebrough, R.W. 1992. Global distribution of tris (4-chlorophenyl) methanol in high trophic level birds and mammals. *Environ. Sci. Technol.* 26: 1770–1774.

172. Muir, D., Braune, B., DeMarch, B. et al. 1999. Spatial and temporal trends and effects of contaminants in the Canadian Arctic marine ecosystem: A review. *Sci. Total Environ.* 230: 83–144.

173. Norstrom, R.J., Simon, M., Muir, D.C.G., and Schweinsburg, R.E. 1988. Organochlorine contaminants in Arctic marine food chains: Identification, geographical distribution and temporal trends in polar bears. *Environ. Sci. Technol.* 22: 1063–1071.

174. Sung, J.K., Tanabe, S., Choi, D.Y., Tatsukawa, R., Baba, N., Kiyota, M., and Yoshida, K. 1991. Variations of organochlorine residue levels with age and time in northern fur seals (*Callorhinus ursinus*) from Pacific coast of Japan since 1971. In: *Twelfth Annual Meeting of the Society of Environmental Toxicology and Chemistry*, Seattle, WA. Abstract, p. 2.

175. De Kock, A.C. and Lord, D.A. 1989. Predicting the fate and effects of chlorinated hydrocarbons in a coastal marine system—Southeast Indian Ocean. *Int. J. Environ. Anal. Chem.* 36: 133–138.

176. Fischer, R.C., Krämer, W., and Ballschmiter, K. 1991. Hexachlorocyclohexane isomers as markers in the water flow of the Atlantic Ocean. *Chemosphere* 23: 889–900.

177. Aguilar, A. 1984. Relationship of DDE/S DDt in marine mammals to the chronology of DDT input into the ecosystem. *Can. J. Fish. Aquat. Sci.* 41: 840–844.

178. Loganathan, B.G. and Kannan, K. 1991. Time perspectives of organochlorine contamination in the global environment. *Mar. Pollut. Bull.* 21: 582–584.

179. Law, R.J., Allchin, C.R., and Harwood, J. 1989. Concentrations of organochlorine compounds in the blubber of seals from eastern and north-eastern England. 1988. *Mar. Pollut. Bull.* 20: 110–115.

180. Anon. 1988. Report of the ICES advisory committee on marine pollution, 1987. *ICES Coop. Res. Rep.* 150.

181. Vir, A.K. 1989. Toxic trade with Africa. *Environ. Sci. Technol.* 23: 23–25.

182. French, H.F. 1990. A most deadly trade. *World Watch* 3: 11–17.

183. Jain, V. 1992. Disposing of pesticides. *Environ. Sci. Technol.* 26: 226–228.

184. Mrinalini, N. 1984. Pesticide hazards: A growing global problem. *Chem. India* 35: 347–352.

185. Loganathan, B.G., Tanabe, S., Tanaka, H., Watanabe, S., Miyazaki, N., Amano, M., and Tatsukawa, R. 1990. Comparison of organochlorine residue levels in the striped dolphin from western North Pacific. 1978–1979 and 1986. *Mar. Pollut. Bull.* 21: 435–439.

186. Yin, J., Meng, Z., Zhu, Y., Song, M., and Wang, H. 2011. Dummy molecularly imprinted polymer for selective screening of trace bisphenols in river water. *Anal. Methods* 3: 173–180.

187. DiGangi, J., Blum, A., Bergman, A., Lucas, D., Schecter, A., Scheringer, M., Shaw, S.D., and Webster, T.F. 2010. San Antonio statement on brominated and chlorinated flame retardants. *Environ. Health Perspect.* 118: A516–A518.

188. Matsushima, A., Liu, X., Okada, H., Shimohigashi, M., and Shimohigashi, Y. 2010. Bisphenol AF is a full agonist for the estrogen receptor ERα but a highly specific antagonist for ERβ. *Environ. Health Perspect.* 118: 1267–1272.

189. Yamashita, N., Yeung, L.W.Y., Taniyasu, S., Kwok, K.Y., Petrick, G., Gamo, T., Guruge, K.S., Lam, P.K.S., and Loganathan, B.G. 2011. Global distribution of PFOS and related chemicals. In: *Global Contamination Trends of Persistent Organic Chemicals* (B.G. Loganathan and P.K.S. Lam, eds.). CRC Press, Boca Raton, FL.

190. Nakata, H., Shinohara, R., Murata, S., and Sasaki, H. 2011. Synthetic musks and benzotriazole UV stabilizers in marine ecosystems. In: *Global Contamination Trends of Persistent Organic Chemicals* (B.G. Loganathan and P.K.S. Lam, eds.). CRC Press, Boca Raton, FL.

2 Spatial and Temporal Trends of Polybrominated Diphenyl Ethers

Ying Guo, Susan D. Shaw, and Kurunthachalam Kannan*

CONTENTS

2.1 Introduction 33
    2.1.1 Chemical and Physical Properties and Uses...34
    2.1.2 Sources and Environmental Fate 36
    2.1.3 Toxicity 37
2.2 PBDEs in the Global Environment..39
    2.2.1 Human Samples: Spatial and Temporal Trends..39
    2.2.2 Sediment, Soil, and House Dust: Spatial and Temporal Trends........................42
        2.2.2.1 Surface Sediment and Sediment Core 42
        2.2.2.2 Soil 43
        2.2.2.3 House Dust..43
    2.2.3 Wildlife: Spatial and Temporal Trends...46
        2.2.3.1 Mussels..46
        2.2.3.2 Fish Species 47
        2.2.3.3 Birds and Bird Eggs 48
        2.2.3.4 Mammals 50
2.3 E-Waste...52
2.4 Summary and Perspectives..53
References...55

2.1 INTRODUCTION

Polybrominated diphenyl ethers (PBDEs) are a class of brominated flame retardants that are used in numerous polymer-based commercial and household products such as textiles, furniture, and electronic equipment, to reduce the likelihood of ignition, slow the burn rate should products catch fire, and meet fire safety standards [1,2]. PBDEs can be substituted with up to 10 bromine (Br) atoms and, depending on the locations and number of Br atoms, there are 209 possible congeners. PBDEs and their hydroxylated metabolites, HO-PBDEs, are structurally similar to thyroxine (T_4), and laboratory studies indicate that PBDE exposures interfere with early neurodevelopment [3]. Since the 1990s, PBDEs have been recognized as a global problem, as they have been detected worldwide in all environmental matrices examined (air, water, soil, sediment, sludge, dust, mussels, fish, mammals, and human samples) [4–7]. The less brominated congeners have been found in remote areas (e.g., the Arctic) distant from their known use or production [5] and in the open oceans [8]. PBDEs

---

* E-mail: kkannan@wadsworth.org (Chapter corresponding author).

are lipophilic and readily biomagnify in food webs; top-level marine predators accumulate higher PBDE concentrations than do terrestrial biota [5]. The oceans are the global sinks for many hydrophobic persistent organic pollutants (POPs) such as polychlorinated biphenyls (PCBs) and PBDEs [9–11]. The detection of PBDEs in diverse deep-sea organisms including sperm whales [8], cephalopods [12], and deep-sea fish species [10,11,13,14] confirms that these contaminants have reached the deep ocean environment.

For the past three decades, North America has dominated the global market for PBDEs, consuming 95% of the penta-BDE and 44% of the deca-BDE formulation [15]. Accordingly, PBDE concentrations in biota and humans from North America are generally the highest in the world and are still increasing [4]. The distribution of PBDEs in the European, Asian, and Arctic environments has been the subject of numerous studies [4–7,16].

PBDEs are similar to PCBs with regard to structure, physicochemical properties, and the volume of global production and use. Based on the estimate of global production of PBDEs in 2001 (67,400 metric tons [MT] [15]) and the production duration of over 30 years, it can be roughly estimated that over 1 million MT of PBDEs have been produced globally, and the production of deca-BDE is still ongoing. For PCBs, the total global production from 1929 to the late 1970s was estimated to be 1.2 million MT [17]. Tanabe predicted that the majority of the PCBs (66%) would remain stockpiled in equipment (transformers and capacitors) long after PCBs were banned from production [17]. A similar scenario can be expected for PBDEs. Despite recent restrictions on penta- and octa-PBDE mixtures, large amounts of polymer-based products, building materials, and plastics containing PBDEs are still in use and will be disposed of after their lifetimes, creating second-tier outdoor reservoirs (e.g., landfills, wastewater treatment plants, electronic waste [E-waste] recycling facilities, and stockpiles of hazardous wastes) for the future dispersal of PBDEs to surface waters and the oceans. Moreover, deca-BDE remains the highest-volume global PBDE in use today and the discharge of BDE-209 to aquatic and marine ecosystems has been increasing. In this chapter, we describe the temporal and spatial distribution of PBDEs on a global scale.

2.1.1   Chemical and Physical Properties and Uses

PBDEs are brominated aromatic compounds consisting of two phenyl rings linked by an ether bond with variable bromine substitutions (Figure 2.1). Mixtures of PBDEs are added to a variety of polymers such as polystyrene foams, high-impact polystyrene, and epoxy resins that are applied to consumer products including electronic equipment (circuit boards, computers, and monitors), textiles, commercial and residential construction materials, insulation, mattresses and foam cushions in furniture, baby products, and automobile and aircraft interiors to increase their fire resistance [1,18]. As additive flame retardants, PBDEs are not chemically bound (e.g., covalently bound) but are physically blended with polymers; thus, over time, PBDEs have leached out of products, accumulated in indoor air and dust, and eventually entered the natural environment [19].

PBDEs have been produced since the 1960s as three commercial mixtures (penta-, octa-, and deca-BDE) that vary in degree of bromination (Table 2.1). PBDEs are structurally similar to PCBs and, analogous to PCBs, 209 distinct PBDE isomers are possible; however, each commercial mixture contains only a limited number of congeners from each homologue group. The major constituents of penta-BDE, namely, BDE-47, -99, and -100, with minor contributions from BDE-153, -154, and -85, are highly lipophilic, persistent and bioaccumulative, and subject to long-range transport [1,20]. The major congener found in octa-BDE is hepta-BDE-183 (Table 2.1); other constituents are nona-BDE-203 and several octa- and nona-BDEs, whereas deca-BDE contains primarily (97%) BDE-209 and low levels of nona-BDEs [1].

Penta-BDE mixtures were mainly used in polyester and flexible polyurethane foam formulations at amounts that could be as high as 30% by weight in the finished material. The main use of octa-BDE was in a variety of thermoplastic resins, in particular acrylonitrile-butadiene-styrene (ABS) plastics, which can contain up to 12% by weight octa-BDE. Deca-BDE, the most widely used PBDE

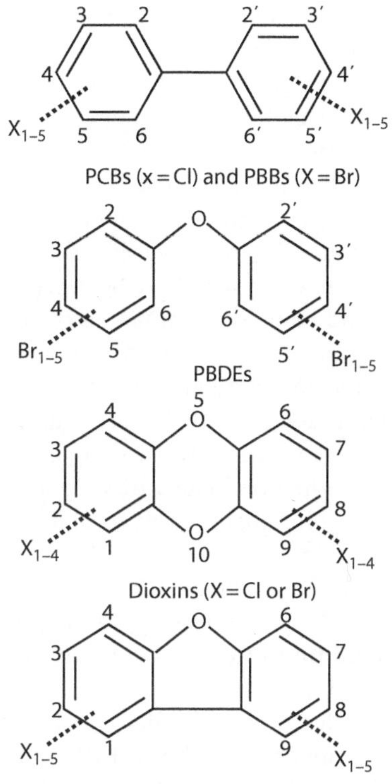

FIGURE 2.1 PBDEs are structurally similar to PBBs, PCBs, dioxins, and furans.

TABLE 2.1
Approximate Composition of Commercial PBDE Mixtures

PBDE Mixtures	Congener Composition (% of Total)
Penta-	24%–38% tetra-BDEs, 50%–60% penta-BDEs, 4%–8% hexa-BDEs; BDE 47 (25%–37%), BDE 99 (35%–50%)
Octa-	10%–12% hexa-BDEs, 44% hepta-BDEs, 31%–35% octa-BDEs, 10%–11% nona-BDEs <1% deca-BDEs; BDE-183 (40%), BDE-197 (21%)
Deca-	<3% nona-BDEs, 97%–98% deca-BDE

globally, is added to various plastic polymers such as polyvinyl chloride, polycarbonates, and high-impact polystyrene, as well as back coating for textiles (commercial furniture, automobile fabrics, and carpets).

Over the past three decades, North America has consumed more than half of the global production of PBDEs and 95% of the penta-BDE product, the constituents of which are more persistent and bioaccumulative than those of the octa- or deca-BDEs [21]. Accordingly, PBDE burdens in humans and biota from North America are the highest in the world, and in contrast to Europe, North American burdens have continued to increase over time [4]. In the United States and Canada, PBDE concentrations are increasing exponentially in fish, birds, and marine mammals, with concentrations in species in some areas doubling as rapidly as every 3–4 years [4,22–25]. In urbanized regions, levels of tetra- to hexa-brominated PBDEs are surpassing those of the PCBs as the top contaminant in air, water, and sewage sludge, and are rivaling PCBs in many fish and marine mammal populations [20].

Largely because of environmental and public health concerns, the penta- and octa-BDE mixtures were banned in Europe in 2004 and were also subsequently withdrawn from commerce in the United States in 2004 [26]. In May 2009, the Stockholm Convention included these PBDEs (penta- and octa-BDE) in the list of POPs that by definition are environmentally persistent, bioaccumulative, and toxic to humans and the environment, and subject to long-range transport. Despite regulations, large amounts of PBDEs are present in long-lived, in-service, and discarded polymer products and electronics, ensuring continuous releases to the environment for decades. Moreover, deca-BDE remains a high-production-volume flame retardant worldwide. Substantial releases of deca-BDE from industrial sources directly to the environment have been reported [19]. Furthermore, concerns have been raised about the abiotic and biotic debromination of BDE-209, the major constituent of deca-BDE, to more bioaccumulative and potentially toxic congeners [20,27–30]. Comparatively few data exist regarding the environmental distribution of the highly brominated PBDEs, which are either present in or derived from octa- and deca-BDE formulations. These hydrophobic congeners bind to particles and are concentrated in soils, sewage sludge, and indoor dust, especially in and around urban areas [19]. BDE-209 is the dominant PBDE found in marine sediments [31]; thus, the deep oceans are vast sinks for ongoing inputs of deca-BDE and its debromination products to marine ecosystems.

Restrictions have been imposed on the production and use of deca-BDE in Europe and in some U.S. states [26]. In 2007, the state of Maine enacted legislation prohibiting the use of deca-BDE in mattresses and residential upholstered furniture sold within its borders, and will extend the ban to electronics in 2010. Washington state prohibits the use of deca-BDE in mattresses and is considering a ban on its use in furniture and electronics. Similar legislation is pending in other U.S. states and in some Asian countries. In December 2009, the USEPA announced the phase-out of production of deca-BDE by the end of 2012 and uses by 2013. Nevertheless, large amounts of deca-BDE have already been released to the global environment and this flame retardant is still in high-volume use. The global market demand for deca-BDE in 2001 was ~56,100 MT [15].

PBDEs have captured the attention of scientists and policy makers, because levels in the environment and humans have increased rapidly since these chemicals came into use. The continuous rise in PBDE concentrations in blood and breast milk both from the United States and Canada throughout the 1990s and 2000s, coupled with the fact that North American body burdens exceed those of Europeans and others by factors of 10 or more, has served to focus considerable attention on North American exposure to PBDEs.

2.1.2 Sources and Environmental Fate

Commercial production of PBDEs began in 1976 [32]. Commercial formulations are mixtures of PBDE congeners with penta-, octa-, and deca-BDE; their Br content is about 71%, 79%, and 83%, respectively. Prior to the end of production in 2004, ~95% of penta-BDE was used as an additive flame retardant in flexible polyurethane foam materials; these were used as seat cushioning and backing material for domestic furniture, in bedding mattresses, and in cushioning for automobile seats. Octa-BDE was an additive flame retardant in the production of the ABS-based plastics used in the manufacture of housings of office equipment, business machines, and computer casings. Deca-BDE was used in a wide range of plastics having many applications. High-impact polystyrene (used in housings of televisions, audio/video equipment, mobile phones, remote controls, and computer monitors), polyethylene, polypropylene, polybutylene terephthalate, and unsaturated polyesters are common plastics treated with deca-BDE.

PBDEs are basically additive flame retardants and are not bonded chemically to the matrix. This means that the potential exists for PBDEs to escape the matrix through volatilization to the air. BDE-47 has a high vapor pressure and has the greatest tendency to volatilize from plastics. Environmental releases can also occur during the disposal of electronic wastes via leaching and volatilization.

The sources of PBDE input to the environment include emissions to air and surface waters from plants manufacturing the technical mixtures and facilities incorporating these flame retardants into polymers, as well as release during the life cycle of treated consumer products (degradation, recycling, and disposal) [20]. E-waste recycling facilities in Asia have recently been identified as major sources of PBDEs [33]. Airborne emissions of less-brominated PBDE congeners (up to hexa-substituted) are expected to exist in both the vapor and particulate phases and therefore be subject to long-range atmospheric transport [34].

In the United States municipal sewage treatment plants (STPs) and landfills are additional point sources of PBDEs entering the environment [35]. High levels of PBDEs have been found in STP effluent and sludge and may be widely dispersed through the use of sludge as fertilizer on agricultural lands [20]. Because of their low vapor pressure and high hydrophobicity, PBDEs are strongly adsorbed into soils, sediments, and suspended organic material in the water column, thus facilitating their transfer to aquatic organisms. PBDEs enter coastal waters through municipal and industrial wastewater outfalls, landfill leachate, and atmospheric deposition from multiple sources [5,36]. They readily travel through the movement of air, water, particles, and biota, and are surpassing PCBs as the top contaminant in many coastal ecosystems of North America [24]. The highest concentrations of PBDEs are often found near industrial or municipal outfalls [6]. PBDEs have been detected at low concentrations in North American air, water, and sediment, but at much higher levels in aquatic biota in areas receiving discharges [21]. Whereas sediment concentrations of PCBs are determined by environmental processes (e.g., sediment accumulation and mixing rates), those of more recently introduced PBDEs are strongly controlled by proximity to sources and use patterns. In particular, the entry of PBDEs into domestic dust facilitates the transport of these compounds to the coastal oceans by municipal wastewater systems [37]. Because PBDEs, especially the less environmentally mobile, highly brominated congeners (e.g., BDE-209), are persistent in sediments, they are available for foraging by benthic organisms [10] through which they reenter marine food webs. High-molecular-weight PBDEs exhibit a propensity for the breakdown and decomposition in soils by UV light. For example, photolysis of BDE-47 and -99 in air is a major atmospheric removal process for these congeners. In contrast, photolysis of BDE-209 in air is a minor atmospheric removal process. Atmospheric wet and dry deposition of BDE-209 is the most significant atmospheric removal pathway. Higher-brominated congeners can undergo metabolic debromination in fish, mammals, and birds to form lower-brominated congeners. Soils and sediments are also environmental sinks for PBDEs.

2.1.3 TOXICITY

Detailed reviews of the toxicology of PBDEs, including mechanisms of toxicity, can be found in the literature [18,38–41]. Laboratory studies conducted over the past two decades have shown that PBDEs have the potential to disrupt the endocrine system at multiple target sites in amphibians, birds, fish, mice, and rats, resulting in effects on thyroid, ovarian, and androgen function. Many of these reports concern disruption of thyroid hormone homeostasis by PBDEs (both in vivo and in vitro) and characterized by a reduction in T_4 concentrations. Possible mechanisms include interference in the transport of T_4 via competitive binding to thyroid transport proteins (transthyretin, TTR) and thyroid hormone receptors, induction of thyroid hormone metabolic activity, and interference with the hypothalamus-pituitary-thyroid axis [42]. Recent studies have demonstrated effects of PBDEs on estrogen- and androgen-mediated processes as well, notably anti-androgenic effects shown both in vivo and in vitro [40].

The greatest concern for the potential health effects of PBDEs arises from reports of toxicity due to developmental exposure [18,43]. In animal models, toxic effects following prenatal or neonatal exposure to PBDEs include effects on liver enzymes [44], endocrine disruption (altered thyroid hormone levels) [45], reproductive damage [38,46,47], neurotoxic effects [48,49], and immunotoxicity [50]. In offspring of mice prenatally exposed to BDE-209, immunotoxic effects included a dose-dependent increase in titers of respiratory synctytial virus in the lungs, increased levels of interferon-Y in the bronchoalveolar lavage fluids, and, at higher doses (10,000 µg/g deca-BDE), an increased incidence of pneumonia was observed [51]. Experiments conducted by Eriksson and coworkers in mice developmentally exposed either to penta- or higher BDEs [48,49,52–55] and in rats exposed to BDE-209 [56,57] during the period of brain growth spurt have shown neurotoxic effects including impairment of spontaneous behavior, cholinergic transmitter susceptibility, and habituation capability. Deficits in learning and memory were observed to last into adulthood and worsen with age. Similarly, a study by Rice et al. demonstrated that dietary exposure of neonatal mice to deca-BDE resulted in developmental delays, changes in spontaneous locomotor activity, and a dose-related reduction in serum T_4 concentrations [58]. These results underline the critical role of thyroid hormones during brain development and suggest that the neurodevelopmental effects of PBDEs are related to perturbations in thyroid hormone homeostasis in the neonate [18]. The USEPA's Integrated Risk Information System recently updated a toxicological review on deca-BDE and published an oral reference dose (RfD) of 0.007 mg/kg day based on neurobehavioral effects [59].

Despite the widespread and escalating PBDE exposure among North Americans, data on the potential adverse health effects of PBDEs in humans are still limited. A study by Turyk et al. reported that PBDEs were associated with increased T_4 levels and thyroglobulin antibodies in the blood of adult male consumers of Great Lakes sport fish [60]. Effects were observed at PBDE levels comparable to those found in the general U.S. population and were independent of PCB exposure and sport fish consumption. Another recent study reported elevated serum levels of thyroid-stimulating hormone (TSH) in workers exposed to PBDEs by working in or living near an E-waste recycling center in China [61]. Elevated TSH levels may be a compensation for the reduction of circulating thyroid hormones and are indicative of stress on the thyroid system. However, an earlier study by Julander et al. at an E-waste recycling facility in Sweden did not find a significant change in thyroid hormone levels among PBDE-exposed workers [62]. In a study of mother–boy pairs from Denmark and Finland, elevated PBDE levels in breast milk were correlated with cryptorchidism in the children [63]. PBDE levels associated with cryptorchidism were also positively correlated with serum luteinizing hormone (LH) concentrations in the infants, which suggested a possible compensatory mechanism to achieve normal testosterone levels and is consistent with the anti-androgenic effects of PBDEs observed in experimental animals. Elevated levels of PBDEs in breast milk of pregnant Taiwanese women were significantly associated with an adverse birth outcome regarding weight, length, and chest circumference of their infants [64]. In both latter studies, effects were observed at levels much lower than average PBDE levels in the adult U.S. population. A pilot study conducted by Japanese researchers reported that elevated blood levels of BDE-153 were correlated with decreased sperm count and decreased testes size [65]. The importance of house dust as a major exposure route for PBDEs in humans has been highlighted [66–68], and a recent study reported a relationship between altered hormone levels in American men and PBDE levels in house dust [69]. Findings included significant inverse associations between PBDEs in house dust and serum concentrations of the free androgen index, LH, and follicle-stimulating hormone (FSH) and positive associations between PBDEs and sex hormone binding globulin (SHBG) and as well as T_4. The positive relationship between PBDEs and free T_4 is consistent with the findings of Turyk et al. in adult male fish consumers [60], but inconsistent with developmental exposure studies in animals that have consistently reported a decrease in T_4 related to PBDE exposure [39,40]. The authors suggest that there are important differences in PBDE effects on thyroid signaling and T_4 levels in developmentally exposed animals compared with adult human exposure. However, the relationships between PBDEs and enhanced thyroid hormone levels reported in studies of marine mammals suggest that, in addition to species and maturation differences, possible interactions of contaminant mixtures may be involved in thyroid alterations. In the United States, congenital hypothyroidism has been on the rise for the past 20 years. Studies assessing the relationship between in utero exposure to compounds that alter thyroid hormone levels, such as PBDEs, and congenital hypothyroidism are needed.

The carcinogenic potential of PBDEs has not been adequately addressed in animals or humans, as few studies have been conducted. A study by Hardell et al. reported an association between BDE-47 concentrations and an increased risk for non-Hodgkin's lymphoma (NHL) [70]. In the highest risk, highest exposure group, BDE-47 was also significantly correlated with elevated titers to Epstein Barr IgG, a human herpes virus that has been associated with some subgroups of NHL. In a long-term feeding study, a significant increase in the incidence of follicular-cell hyperplasia of the thyroid gland was observed in mice (male and female) exposed to high doses of BDE-209 [71]. The incidence of thyroid cancer has been increasing in the United States during the past several decades, especially among women, and it is hypothesized that part of the observed increase in thyroid cancer rates may be related to the increasing population exposure to PBDEs and other thyroid hormone–disrupting compounds [42].

2.2 PBDES IN THE GLOBAL ENVIRONMENT

PBDEs are ubiquitous in the environment and have been detected in environmental and biological samples collected from all over the world including the remote marine environment and deep sea (>1000 m depth) organisms. PBDEs are lipophilic and hydrophobic compounds and readily bioaccumulate into terrestrial and aquatic food webs. This tendency has resulted in extensive accumulation of PBDEs in a wide variety of birds, fish, insects, and aquatic and terrestrial mammals reflective of dietary exposures through the food chain. The following sections describe global trends of PBDEs in environmental and biological specimens.

2.2.1 HUMAN SAMPLES: SPATIAL AND TEMPORAL TRENDS

Biomonitoring of PBDEs in human specimens such as breast milk, adipose fat, and serum, has been widely used to assess exposure levels and body burdens. Since the first time-trend study published in the late 1990s that showed an exponential increase in PBDE levels in Swedish human breast milk [72,73], numerous reports of PBDEs in human samples, mainly breast milk and blood, have been published.

Studies have shown that PBDE levels in human samples from North America are much higher (10–100 times) than the levels reported for Asia and Europe [4,6,74,75]. Compilation and meta-analysis of all human biomonitoring data for the Americas, Europe, and Asia from the studies published after 2000 showed median PBDE levels of 40, 4, and 3.5 ng/g lipid wt, respectively (Figure 2.2),

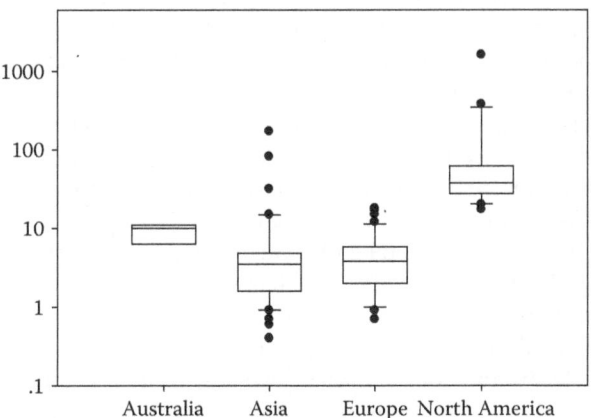

FIGURE 2.2 Reported PBDE levels in human tissues from several continents (ng/g lipid wt). The horizontal lines indicate the 10th, 25th, 50th, 75th, and 90th percentiles of the distribution. Outliers are shown as dots. Median values (mean value was used when median value was not available) from references: Australia (including New Zealand) [76,77,209]; Asia [64,80,81,85,86,93,94,210–226]; Europe [72,86,87,90,92,99,227–252]; and North America [86,102,106,253–265].

with no significant difference in concentrations between Asia and Europe ($p > 0.05$). Mean PBDE concentrations in breast milk collected from Brisbane, Australia, in 2003 and 2007 were 10.2 and 10 ng/g lipid wt, respectively [76,77].

In occupationally exposed populations (Table 2.2), PBDE levels in samples of foam and carpet installation workers in North America [78] and E-waste recyclers in China [79–81] were on the order of 100 ng/g lipid wt. PBDE levels in occupationally exposed workers in Europe were one order of magnitude lower (~10 ng/g lipid wt) than the levels reported for Chinese and North American populations with occupational exposures [82–84].

Several studies have reported PBDE levels in breast milk from European countries, North America, eastern China, and Japan; and limited data are available for South America, Africa, and Russia (Figure 2.3). PBDE levels reported for breast milk from North America are among the highest globally, with concentrations ranging from 10 to 100 ng/g lipid wt, while the levels reported for other countries are generally below 10 ng/g lipid wt. In Asia, the highest concentration of PBDEs was reported in breast milk from Laizhou Bay, Shandong, China (82 ng/g lipid wt) (not shown in Figure 2.3) [85]; this very high value was similar to the values reported for the United States. Among European countries, higher levels of PBDEs have been reported in northern and western European countries such as Ireland [86], the United Kingdom [87], Denmark [88] and Finland [89] than in eastern and central European countries such as Bulgaria [86], Hungary [86], and Turkey [90].

Temporal trends of PBDE concentrations in breast milk collected from several European countries, for data reported through 2009, are shown in Figure 2.4. In general, PBDE concentrations in Europe peaked around 1999 and then declined. During 1972–1997, PBDE levels in Swedish breast milk increased with a doubling time of 5 years [91]. A similar trend was reported by Fängström et al. who showed peak values in 2001 for BDE-153 [72]. PBDE concentrations in serum of Norwegians collected from 1977 to 2003 indicated that PBDE concentrations reached a steady state in 2000 [92]. The temporal trends of PBDEs in breast milk from Europe indicate that the regulation of the usage of PBDEs in the early 2000s has been effective in reducing human exposures.

Time-trend studies of PBDEs in human samples from the Americas and Asia are limited. Concentrations of PBDEs in breast milk collected from Osaka, Japan, during 1973–2000 [93] showed that the concentrations reached its peak in 1998 (2.3 ng/g lipid wt) and then stabilized, as has been reported for Swedish mothers' milk [91]. PBDE concentrations in human adipose tissues collected between 1970 and 2000 from Tokyo, Japan [94], showed a 40-fold increase in concentrations (from 0.03 to 1.3 ng/g lipid wt) during that period.

Human exposure to PBDEs via internal and external routes has been reviewed by Frederiksen et al. [95]. Some studies, mainly from Europe, suggested that PBDE exposures via dietary sources are important (contributing >70% of total PBDE intake) [96–98]. Fish consumption was strongly associated with PBDE levels in human serum [99]. In China, PBDE levels were high in breast milk collected from Shandong [85], a region with high fish consumption rates (Table 2.2). However, studies in North America have shown that house dust ingestion is the major source of PBDE exposures [67,68,100–102] (contributing >80% of the total PBDE exposure doses). In a recent report of the USEPA, it was estimated that the U.S. adult intake dose of total PBDEs via house dust was 6.4 ng/kg bw/day, accounting for 90% of the total daily intake [103]. Daily intake of PBDEs in U.S. nursing infants was estimated to be 90 ng/kg bw/day, which is 10- to 20-fold higher than the doses estimated for adults [68]. Studies have shown that PBDE levels in children (serum) were two to five times higher than the levels found in their parents/adults [104,105]. It was estimated that the total intake of PBDEs was 47.2 ng/kg bw/day for children of ages 1–5, which is higher than the intake values estimated for adults (7.1 ng/kg bw/day) in the United States [103]. House dust is a major source of PBDE exposure in North American population (due to strict flammability standards that require use of flame retardants in several household products), whereas diet can be a major source of PBDE exposures in Asian general populations.

BDE-47, -99, and -153 were the more predominant PBDE congeners found in human samples. A few studies have shown the predominance of BDE-153 over BDE-47 in human samples. In breast

TABLE 2.2
Concentrations of PBDEs in Tissues of Occupationally Exposed Populations (ng/g Lipid wt)

Country	Location	Year	N	Sample Type	#28	#47	#99	#100	#153	#154	#183	#209	Median	Reference
United States	Baltimore	1999–2001	12	Serum of foam worker	3.2	77.8	26.1	20.5	33.1	3.3	0.5	nd	160	[78]
United States	Santa Ana	1999–2001	3	Serum of carpet installer	4.6	100	23.5	18.3	22.0	1.6	<0.4	nd	178	[78]
China	Guiyu	2005	21	Serum of worker with e-waste	2.5	9.5	2.7	1.3	18	2.3	5.5	310	600	[80]
China	Zhejiang	2007	19	Kidney of worker with e-waste	48	54	16.4	3.1	15.8	17.0	4.5	191	377	[79]
China	Zhejiang	2007	55	Liver of worker with e-waste	35	50	8.7	5.0	19.1	17.4	9.2	118	311	[79]
China	Zhejiang	2007	7	Lung of worker with e-waste	37	39	12.3	6.6	6.8	39.5	6.8	270	442	[79]
Korea[a]	Seoul	2001	13	Serum of incinerator worker	0.4	6.3	2.6	1.3	4.4	0.3	4.1	na	19.3	[223]
Korea	Seoul	2001	30	Blood of incinerator worker	0.3	5.8	2.4	1.0	5.1	0.3	1.7	na	17.7	[224]

Note: nd, not detectable; na, no analysis.
[a] Mean value.

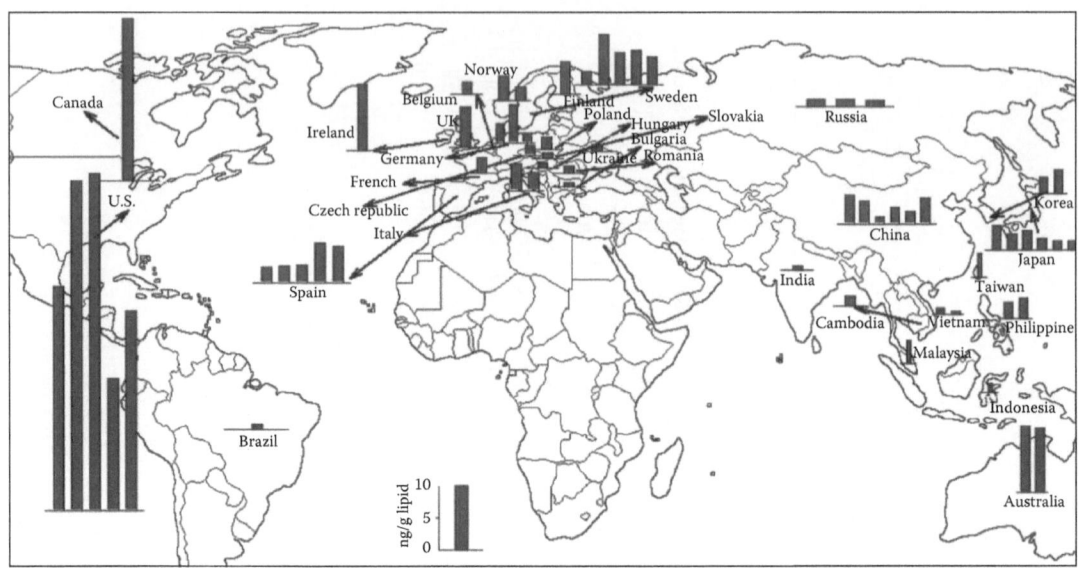

FIGURE 2.3 Global distribution of PBDE levels (after 2000) in human milk (ng/g lipid wt). Median values (mean value was used when median value was not available) from references: Brazil [86]; Canada [253]; the United States [106,256–259]; Cambodia [210]; China [211–215]; India [210]; Japan [93,210,215,219–221]; Korea [210,215]; Indonesia [225]; Malaysia [210]; Philippines [86,210]; Taiwan [64]; Vietnam [210,215]; Belgium [86,230]; Czech Republic [231]; French [234]; Finland [86]; Germany [86,235,236]; Hungary [86]; Ireland [86]; Italy [86,237]; Norway [240,241]; Poland [242]; Romania [86]; Russia [243,244]; Slovakia [86]; Spain [86,246,247,249]; Sweden [72,250]; United Kingdom [87]; Ukraine [86]; and Australia [76,77].

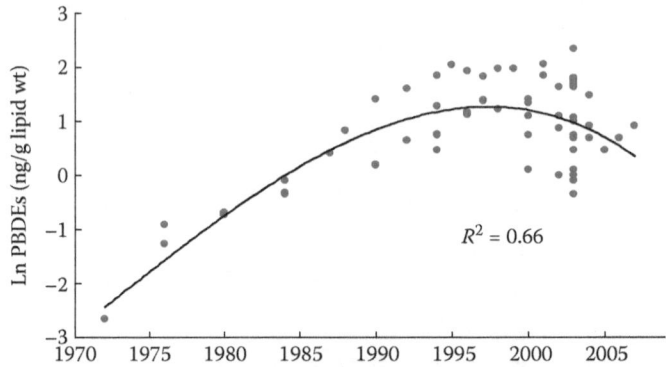

FIGURE 2.4 Temporal trends of PBDE levels in breast milk from several European countries (ng/g lipid wt). Median values (mean value was used when median value was not available) from references: [72,73,86–89,91,230,231,234–237,240–244,246,247,250,266–268].

milk collected from the Pacific Northwest of the United States [106], BDE-153 was the most abundant congener in 3 of 40 samples analyzed. The authors suggested that the dominance of BDE-153 reflected exposure sources (occupational exposures) and the long half-life of BDE-153.

2.2.2 SEDIMENT, SOIL, AND HOUSE DUST: SPATIAL AND TEMPORAL TRENDS

2.2.2.1 Surface Sediment and Sediment Core

Sediment has been identified as an ultimate sink, as well as a potential source, of PBDEs. In North America, PBDEs have been reported in sediments of the Great Lakes [107–110] and in coastal marine sediments, by the National Oceanic and Atmospheric Administration's (NOAA) Mussel Watch Project [111]. Total PBDE concentrations in surface sediments of the Great Lakes ranged from 10 to 236 ng/g dry wt, with BDE-209 contributing to major proportions (from 8.9 to 230 ng/g [112]). The highest PBDE concentration was found in Lake Ontario sediment. The NOAA's Mussel Watch Project collected surface sediments throughout the U.S. coast [111] and showed that high levels of PBDEs were found in industrialized and in densely populated areas, such as Marina del Rey in California (88 ng/g dry wt). High PBDE concentrations have been reported in sediments collected from urban and industrial areas in Australia (Figure 2.5) [113] and in the Pearl River Delta, China [114].

Sediment cores have been used to reconstruct the history of contamination by persistent pollutants. Analysis of PBDEs in sediment cores collected from the Great Lakes [107–110] provided information on the doubling time, inventory, surface flux, and loading rates of PBDEs in sediment [112]. Total concentrations of PBDEs (excluding BDE-209) and BDE-209 increased annually since the 1980s, where the doubling time of BDE-209 in sediments from Lakes Michigan, Huron, Erie, and Ontario was 19, 10, 10, and 13 years, respectively [112]. Studies of sediment cores collected globally showed that PBDE levels have been increasing annually, but at different rates (i.e., doubling times) (Figure 2.5). BDE-209, as the major component (>96%) of deca-BDE technical mixture [115], was also shown to increase in sediment cores on a global scale. As shown in Figure 2.6, the highest concentrations of BDE-209 were found in sediment samples collected from Asia. Sediment cores collected from the Pearl River Delta in China [116] showed that BDE-209 concentrations remained similar until 1990, and then increased notably, with a doubling time of 3–6 years. This increasing tendency reflected the high market demand for deca-BDE mixture after 1990 in China.

BDE-209 was the predominant PBDE congener in nearly all surface sediment and core sediment samples, on a global scale, except for samples collected from or near some point sources of pollution [117]. The high abundance of BDE-209 in sediment suggests its high production and usage, and high K_{ow} value octanol-water partition coefficients, (log $K_{ow} \sim 10$), resulting in preferential partitioning to the sinking sediment particles. BDE-47, -99, -153, and -154 have also been reported in sediments frequently.

2.2.2.2 Soil

High concentrations of PBDE have been reported in soils collected near E-waste recycling facilities and waste dump sites, and in soils treated with sludge as an amendment. BDE-209 was the most abundant PBDE congener reported for soils. In the United States, PBDE concentrations in floodplain soils from Michigan ranged from 0.02 to 55.1 ng/g dry wt [118]. In China, a low PBDE concentration (0.01 ng/g dry wt) was detected in surface soils from the Tibetan Plateau [119] and high concentrations (~2700 ng/g) were found in soils collected near E-waste dismantling sites in Guangdong [120]. Concentrations of PBDEs in soils from rural, urban, and suburban areas in the United Kingdom were 0.2, 1.8, and 0.3 ng/g dry wt, respectively [121]. Sellström et al. determined PBDEs in Swedish soils following the application of sewage sludge and found that PBDE concentrations were 100–1000-fold higher in sludge-applied soils than in reference soils [122]. A median PBDE level in background soils in the United Kingdom and Norway was reported to be 1.4 ng/g dry wt [123].

2.2.2.3 House Dust

Concentrations of PBDEs in house dust are much higher than in soil or sediment. House dust collected from the United Kingdom and North America contained the highest PBDE levels, on the order of several thousands of nanograms per gram, followed by dust samples from Eastern Asia, Australia, and European countries (other than United Kingdom) (Figure 2.7). High PBDE concentrations were detected in dust collected from car interiors (median: 28–720 μg/g [124,125]), television sets (300 μg/g [126] and 72 μg/g [127]), houses with several computers (6 μg/g [127]), and an E-waste recycling area [128].

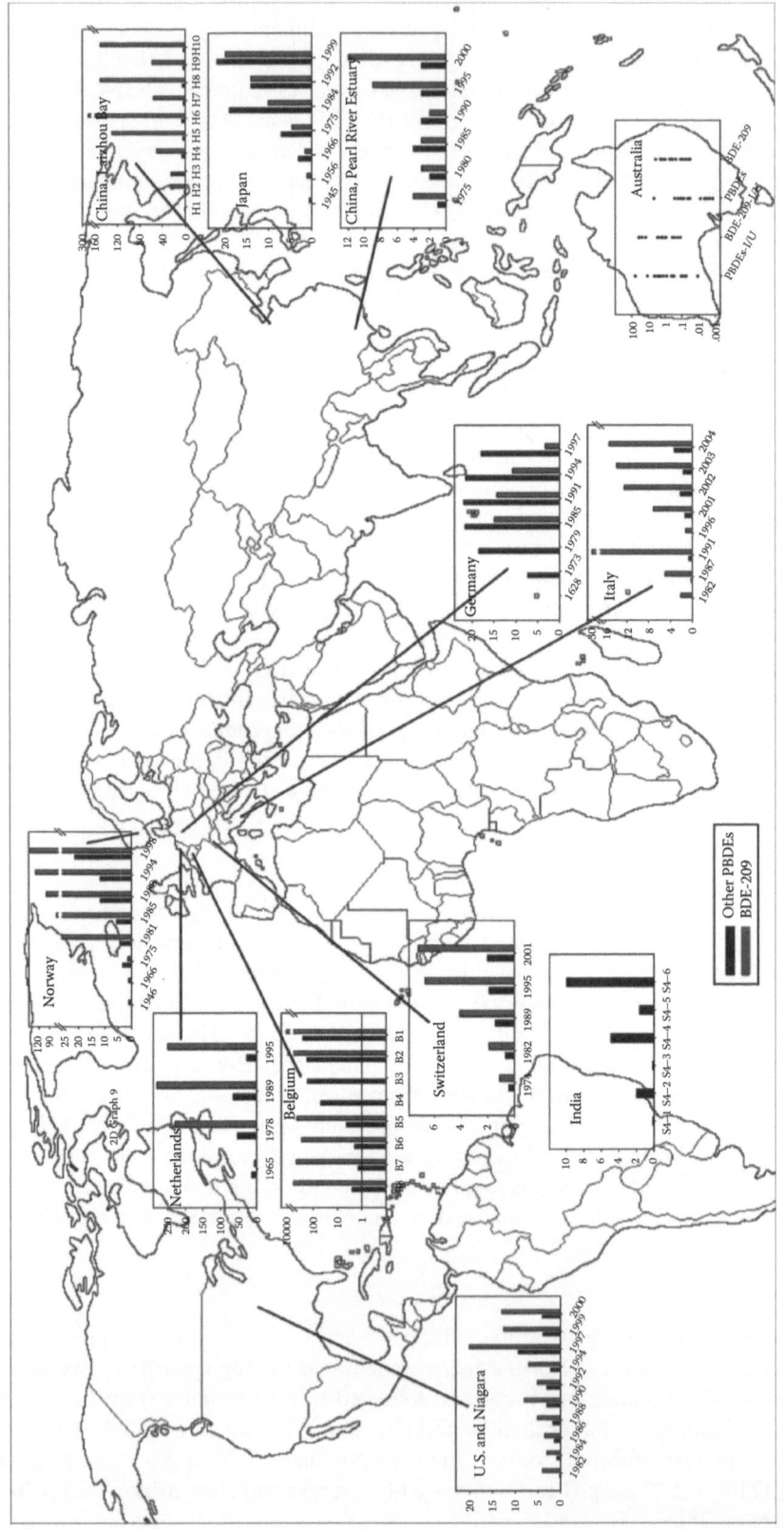

FIGURE 2.5 Global distribution and temporal trends of PBDEs in sediment core (ng/g dry weight). Data: the United States, Niagara River (surface sediment in different year) [269]; Norway, the Netherlands, and Germany (data on total organic carbon basis) [270]; Switzerland [271]; Belgium [272]; India [273]; Italy [274]; China, Laizhou Bay [275]; China, Pearl River Estuary (data estimated from figure) [116]; Japan, Tokyo Bay (data estimated from figure) [276]; Australia, surface sediments by salinity and land-use type. I/U = Industry/Urban. Other sediments including agricultural and remote samples [113].

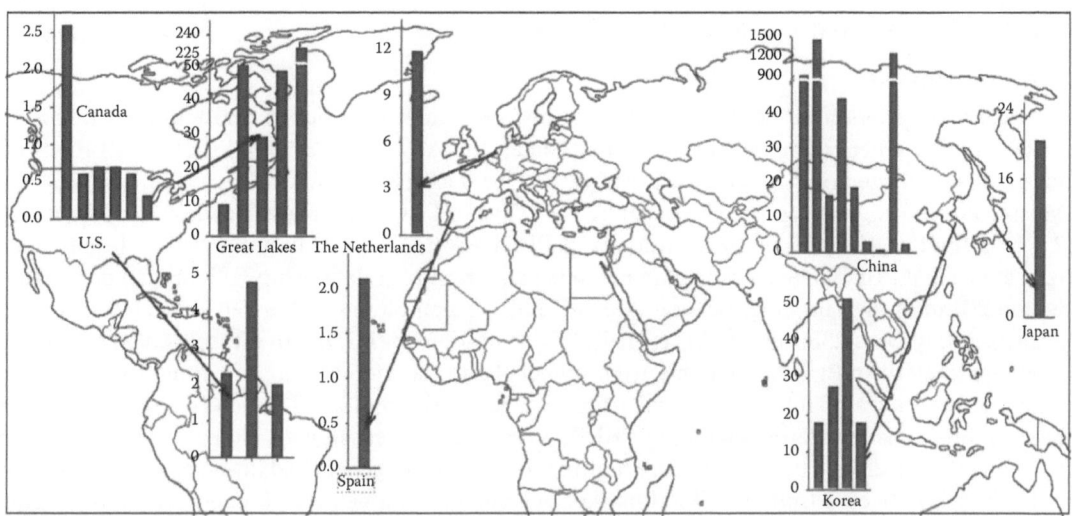

FIGURE 2.6 Global distribution of BDE-209 concentrations in surface sediment (ng/g dry weight). Median values (mean value was used when median value was not available) from references: Canada [132]; the United States [118]; Great Lakes [112]; the Netherlands [277]; Spain [278]; China [114,128,135,137]; Japan [276]; Korea [36,279,280].

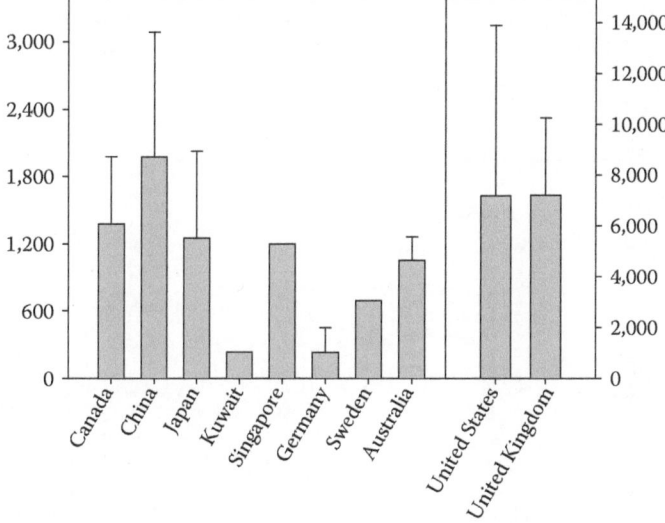

FIGURE 2.7 PBDE concentrations in house dust and outdoor dust from several countries (ng)/g dry weight. Median values (mean value was used when median value was not available) from references: Canada [281,282]; China [127]; Japan [283]; Kuwait [284]; Singapore [285]; Germany [286,287]; Sweden [251]; Australia [77,287]; the United States [68,100,125,282,287]; and the United Kingdom [124,282,287].

House dust ingestion has been suggested as an important source of human expose to PBDEs. Allen et al. reported that the concentrations of PBDEs in dust samples from living rooms (Boston, Unites States) were higher than concentrations found in bedrooms and vacuum cleaner bags, with a geometric mean of 13,700, 6,260, and 4,270 ng/g dry wt, respectively [100]. Another study estimated the emission of PBDEs from U.S. houses and reported that a total of 4100 kg of PBDEs were emitted from U.S. houses every year [124]. Significance of dust as a major source of human exposure in the United States has been discussed in detail [67,68].

2.2.3 Wildlife: Spatial and Temporal Trends

2.2.3.1 Mussels

Mussels and oysters have been widely used to assess spatial distribution and temporal trends of PBDEs in coastal environments in the United States and Asia [111,129,130]. PBDE levels reported for mussels collected from several locations, on a global basis, are shown in Figure 2.8. The NOAA's Mussel Watch Program reported a concentration range of PBDEs in mussels and oysters, from 1 to 270 ng/g lipid wt [111]. High concentrations of PBDEs were detected in urbanized and industrialized areas, and PBDE concentrations in mussels had a positive relationship with human population within 20 km of the sampling location. Mean PBDE concentrations of mussels collected from San Francisco Bay were 2380 ng/g lipid wt [131]. Concentrations of PBDEs in mussels collected near a municipal outfall in British Columbia were higher than in a reference area (with no local source), with median values of 1000 and <10 ng/g lipid wt, respectively [132]. This study also showed that biota-sediment accumulation factors of PBDEs in mussels increased with increasing log K_{ow} of up to ~7 and then declined, that is, the more highly brominated congeners had low accumulation potential. Mussels collected from a seafood market in Canada contained PBDE concentrations ranging from 4 to 57 ng/g lipid wt [133], lower than in mussels sampled from Lake Winnipeg, Canada (110–161 ng/g lipid wt [134]).

Mussels collected during 2003–2004 in coastal waters of Cambodia, China, Hong Kong, India, Indonesia, Japan, Korea, Malaysia, the Philippines, and Vietnam were analyzed for PBDEs [130]. PBDEs were detected in all mussels, indicating the widespread contamination in the Asian coastal environment (0.66–440 ng/g lipid wt). Relatively higher PBDE concentrations were found in mussels collected from the Philippines, Korea, and China than in Indonesia and Japan. PBDE levels in mussels from India, Cambodia, Malaysia, and Vietnam were lower than in other Asian countries, with concentrations below 10 ng/g lipid wt. Monitoring of PBDEs in mussels from China revealed

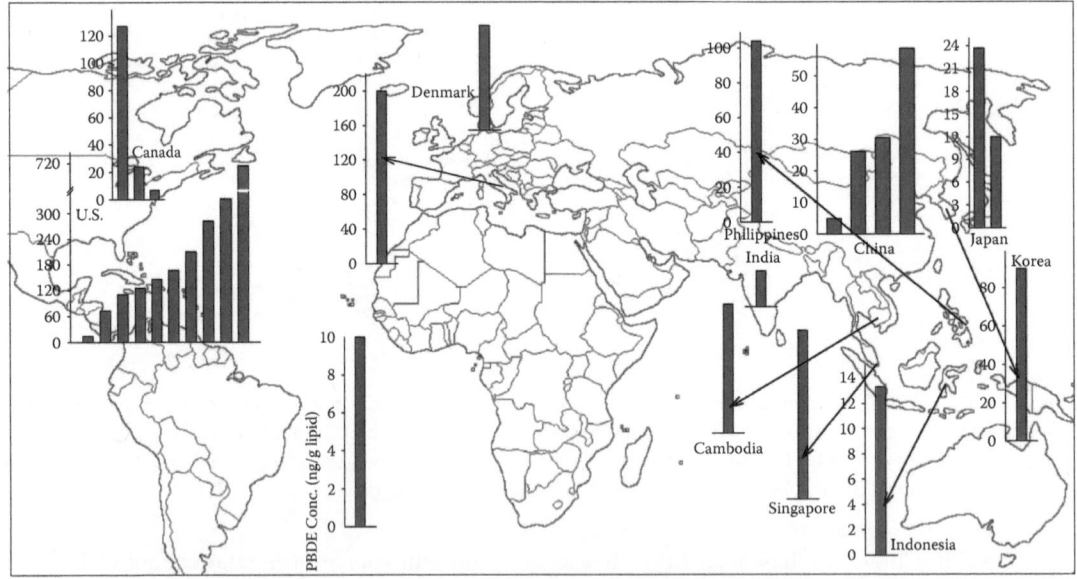

FIGURE 2.8 Global distribution of PBDEs in mussels (ng/g lipid wt). Columns of concentrations lower than 10 ng/g lipid are shown without scaling on left and concentrations are compared with column as high as 10 ng/g. Median values (mean value was used when median value was not available) from references: the United States [111]; Canada [133,134]; Cambodia [130]; China [130,135,136,138]; Japan [130,288]; Korea [130,289]; Vietnam [130]; India [130]; Indonesia [130]; Philippines [130]; Malaysia [130]; Singapore [290]; Italy [139]; and Denmark [291].

that samples collected from South China contained higher concentrations than in samples from North China [135–138].

PBDE concentrations reported for mussels from European countries were generally below 10 ng/g lipid wt. High concentrations of PBDEs were detected in mussels collected in 2005 from Lake Maggiore, Italy [139] with values ranging from 40 to 447 ng/g lipid wt (mean: 200 ng/g lipid wt).

BDE-47 was the predominant congener found in mussels, followed by BDE-99. BDE-209 was predominant in mussels collected from heavily contaminated locations in Canada [132], China [135–138], Korea and the Philippines [130], and Lake Maggiore, Italy [139].

2.2.3.2 Fish Species

PBDEs have been measured in various fish species globally. A compilation of data reported for PBDEs in fish from several continents is shown in Figure 2.9. Fish from North America contained approximately one order of magnitude higher PBDE concentrations than fish from Europe, Asia, and Australia. Median PBDE concentrations reported for marine wild fish (summary of data from several studies) from Asia, Australia, Europe, North America, and the North Pacific Ocean were 20, 30, 30, 150, and 5 ng/g lipid wt, respectively.

Skipjack tuna has been used as a bioindicator of global PBDE contamination in oceans [140]. Skipjack tuna were collected from offshore waters of Japan, Taiwan, the Philippines, Indonesia, Seychelles, and Brazil, as well as from the Japan Sea, East China Sea, South China Sea, Indian Ocean, and the North Pacific Ocean. PBDEs were found in all fillet samples of skipjack tuna, ranging in concentrations from <0.1 to 53 ng/g lipid wt. Relatively higher PBDE concentrations (>20 ng/g lipid wt) were found in tuna from the Japan Sea, East China Sea, off-Taiwan, and the South China Sea than in tuna from the Bay of Bengal, off-Indonesia, off-Seychelles, off-Philippines, and off-Brazil (<15 ng/g lipid wt). PBDE levels in fish from the Northern hemisphere are higher than the levels in the Southern hemisphere.

Several studies have reported the differences in PBDE concentrations in farmed versus wild-caught fish [141–144]. This was important because more than half of the salmon consumed globally

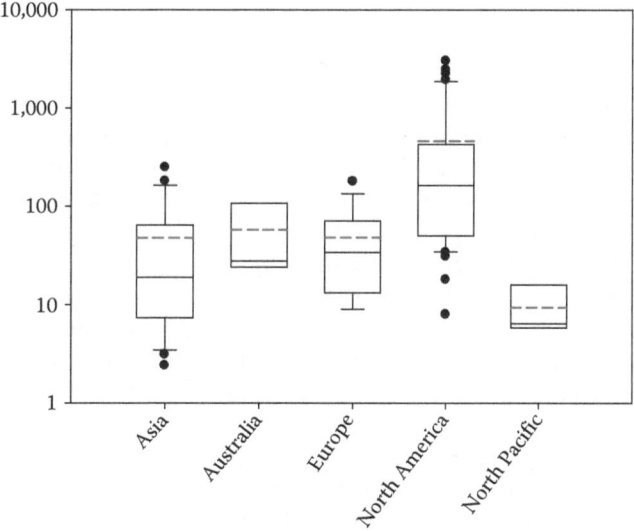

FIGURE 2.9 Summary of PBDE concentrations in marine wild fish species (ng/g lipid wt) from several geographical regions. The horizontal lines indicate the 10th, 25th, 50th, 75th, and 90th percentiles of the distribution. Outliers are shown as dots. The short dash lines indicate the mean values. Median values (mean value was used when median value was not available) from references: Asia [140,144,292,293]; Australia [294,295]; Europe [4,142,149,296–301]; North America [23,30,152,203,297,302–311]; and North Pacific [140].

is farmed. A comprehensive study of PBDEs in farmed and wild salmon from several global locations across Europe, North America, and Chile showed significantly higher PBDE concentrations in farmed salmon than in wild salmon [141]. PBDE levels in whitefish fillets from Swiss Lakes (36–165 ng/g lipid wt) were significantly higher than the levels found in farmed rainbow trout (12–24 ng/g lipid wt) from Swiss fish farms [142]. PBDEs were determined in farmed salmon sampled from Norway, the United States, and Canada and in wild salmon from Alaska [143]. No significant location-specific or farm-specific differences in PBDE concentrations were observed. PBDEs were determined in muscle tissues of seven freshwater fish species ($n=390$), three farmed seawater, and three wild seawater fish species, collected from 11 coastal cities in Guangdong, South China [144]. Significantly higher PBDE levels were found in seawater-farmed fish (median: 13.6 ng/g lipid wt) and freshwater-farmed fish (median: 10.1 ng/g lipid wt) than in wild fish (median: 4.5 ng/g lipid wt). Among the 13 fish species analyzed, PBDE concentrations in carnivorous fish (median: 13.8 ng/g lipid wt) were significantly higher than in herbivorous fish (median: 11.2 ng/g lipid wt) and detritivorous fish (median: 4.1 ng/g lipid wt). The higher PBDE concentrations in farmed fish than in wild fish have been attributed to higher PBDE levels in fish feed [141,145–148]. Median concentrations of PBDEs in farmed and wild bluefin tuna collected from the Mediterranean Sea were reported as 37 and 66 ng/g lipid wt, respectively [149].

Fish collected near point sources of pollution contained significantly higher PBDE concentrations than in reference locations. For example, PBDE levels in five fish species collected from the Nanyang River (contaminated by e-waste recycling operations) in Guiyu, China, ranged from 550 to 6400 ng/g lipid wt [150]. Fish collected from Qingyuan (E-waste recycling area), China, contained PBDE concentrations ranging from 50 to 850 ng/g lipid wt [151]. High PBDE levels (around 2000 ng/g lipid wt) were detected in fish collected from Hadley Lake (near a suspected PBDE manufacturing site) in the United States [152,153]. BDE congeners -28, -47, -49, -99, -100, and -153 were commonly found in fish species, with BDE-47 being the most abundant, followed by BDE-99 and BDE-100.

2.2.3.3 Birds and Bird Eggs

PBDE concentrations in eggs of black guillemot sampled from Greenland ranged from non-detectable to 25 ng/g lipid wt [154], and in eggs of three species of water birds from South Africa, mean levels were below 20 ng/g lipid wt [155]. Significantly high PBDE concentrations were found in eggs of water birds collected from North America. For example, mean PBDE concentrations were between 4670 and 5870 ng/g lipid wt in bird eggs sampled from San Francisco Bay and Gray's Harbor, United States [156], while the mean concentration of PBDEs in eggs of herring gulls sampled from the Great Lakes was 6700 ng/g lipid wt [157]. Concentrations of PBDEs in bird eggs collected from Asian countries were lower than the levels reported for North American birds. The total concentrations of PBDEs in bird eggs collected from South China were 30–1000 ng/g lipid wt [158] and in North China were below 100 ng/g lipid wt [159].

In 2004, for the first time, BDE-183 and BDE-209 were quantified in higher trophic level, peregrine falcon eggs collected from Sweden; total mean PBDE concentrations in eggs of peregrine falcons were 3800 and 4450 ng/g lipid wt for samples from south and north Sweden, respectively [160]. Mean PBDE concentrations in peregrine falcon eggs collected from the northeastern United States ranged from 885 ng/g lipid wt (in Maine) to 2190 ng/g lipid wt (in New Hampshire [161]) peregrine falcons from California had PBDE levels ranging from 80 to 53,100 ng/g lipid wt, with a median value of 4530 ng/g lipid wt [162]. Significantly high PBDE levels were reported in tissues of common kestrels collected from north China (Beijing); with respective mean values of 12,300 and 12,200 ng/g lipid wt, in muscle and liver [160]. Concentrations of PBDEs in little owl eggs from Belgium [163] and in eggs of eiber, tawny owl, and European shag from Norway [164–166] were below 250 ng/g lipid wt. PBDEs were determined in Great Tit eggs collected from 22 sampling sites, including urban, rural, and remote areas, in 14 European countries [167]. Total PBDE concentrations were below 250 ng/g lipid wt, ranging from 4.0 to 136 ng/g lipid wt. Overall, this study showed that total PBDE concentrations were significantly higher in urban areas than in rural and remote areas.

The temporal trends of PBDEs in bird eggs reported from studies in North America, Norway, and South Greenland are summarized in Table 2.3. These studies showed that PBDE levels in bird eggs increased from the 1980s to the early 2000s. The doubling time of total PBDEs in water bird eggs was ~5.7 years during the period from 1979 to 2002 in Canada and was ~3 years during the period from 1981 to 2000 for the Great Lakes. In the study of PBDEs in peregrine falcon eggs sampled from the northeastern United States [161], BDE-209 concentrations increased with a doubling time of 5 years; however, no differences in total PBDE concentrations were found between 1996 and 2006. Concentrations of PBDEs decreased by 62% in tawny owl eggs collected from central Norway during 1986–2004 [165].

PBDE congener profiles in birds are highly influenced by diet. BDE-47 is the most abundant congener in fish-eating water birds, followed by BDE-100 and BDE-99, and then by BDE-153 and BDE-154. Terrestrial birds sampled in urban areas contained relatively higher abundances of highly brominated PBDE congeners such as BDE-209, -153, -154, and -183 than less highly brominated congeners such as BDE-47. In peregrine falcon eggs from the northeastern United States [161], BDE-153 was the dominant congener. BDE-209 was the dominant congener in peregrine eggs from urban areas of California [162] and in birds collected from Beijing, China [160]. The relatively high abundance of more highly brominated PBDE congeners in terrestrial birds from urban areas suggests widespread exposure to deca-BDE mixture in the urban environment.

TABLE 2.3
Temporal Trends of PBDEs Reported for Bird Eggs in Several Global Locations

Species	Location	Total PBDEs Range (ng/g lipid wt)	Temporal Trends	Reference
Double-crested cormorant	British Columbia, Canada	50–5,260	Exponentially increased with doubling time of 5.7 years for total PBDEs 1979–2002	[25]
Great blue herons	Vancouver, Canada	200–7,690	Exponentially increased with doubling time of 5.3 years for total PBDEs 1983–2000	[25]
Herring gull	Great Lakes	90–16,800	Sum of BDE-47, -99, and -100 exponentially increased, respectively, with doubling time of 2.8, 2.6, and 3.1 for Lake Ontario, Lake Michigan and Huron, 1981–2000	[157]
Caspian tern	San Francisco Bay	2,590–6,760	No trend, 2000–2003	[156]
Forster's tern	San Francisco Bay	2,160–9,420	No trend, 2000–2003	[156]
Peregrine falcons	California, United States	80–53,100	More than tripled each decade of total PBDEs in 1986–2007	[162]
Peregrine falcons	South Greenland	300–12,900	BDE-99 and BDE-100 levels increased ~10% per year in 1986–2003	[312]
Peregrine falcons	Northeastern United States	74.5–6,610[a]	BDE-209 with a doubling time of 5 years and no trends for other PBDEs in 1996–2006	[161]
Tawny owls	Central Norway	9.8–5,273	Total PBDEs decreased 62% in 1986–2004	[165]

[a] Values based on wet weight.

2.2.3.4 Mammals

2.2.3.4.1 Marine Mammals

Marine mammals accumulate high concentrations of POPs such as PBDEs through the food chain magnification because of marine mammals' long life span and higher trophic position in the marine food chain. Polar bears (*Ursus maritimus*) are the most carnivorous members of the bear family and are apex predators of the Arctic marine ecosystem. They feed primarily on marine mammals such as ringed and bearded seals and occasionally on larger prey such as walruses and beluga whales.

PBDE concentrations and profiles were investigated in various samples, such as liver, blood, and adipose fat of polar bears from the Arctic regions [168–172]. Mean concentrations of PBDEs were 10–16 ng/g lipid wt in livers of polar bears from the Beaufort Sea and Chukchi Sea, Alaska [169], 10-fold lower than the values reported for plasma of Norwegian polar bears [172]. The mean concentration of PBDEs in fat tissues of polar bears from Alaska [168] was ~6.5 ng/g lipid wt. Concentrations of PBDEs in fat tissues of polar bears from Alaskan Beaufort Sea coast [170] and East Greenland [171] were 80 and 70 ng/g lipid wt, respectively. PBDE concentrations were not significantly different between the genders [168,169], and a decrease in concentrations with increasing age was noted for Alaskan polar bears [169].

Tissues of seals, sea lions, dolphins, and whales have also been used as biomonitors of pollution by PBDEs in marine ecosystems. PBDE concentrations ranged from below 1 ng/g lipid wt in Northern fur seals from the Pacific Ocean [173] to 236,000 ng/g lipid wt in California sea lions [174]. The geographical distribution of PBDEs in marine mammals from Asia was investigated by Kajiwara et al. [175]. Concentrations of PBDEs reported for small cetaceans collected from several Asian countries (6–6000 ng/g lipid wt) were in the following order: Hong Kong > Japan > Philippines > India. PBDE concentrations in marine mammals from North America have been reviewed by Shaw and Kannan [75]. High PBDE concentrations were reported in marine mammals collected during 1994–2006 from California [174]. Mean PBDE concentrations were 28,700, 2,200, and 360 ng/g lipid wt in California sea lions, Pacific harbor seals, and Northern elephant seals, respectively. BDE congeners -28, -47, -99, -100, -153, and -154 were the frequently detected congeners in marine mammals with BDE-47 as the most abundant congener, followed by BDE-99 and BDE-100. BDE-209 was seldom detected in marine mammals.

2.2.3.4.2 Terrestrial Mammals

Very few studies have reported on PBDEs in terrestrial mammals. PBDEs were determined in muscle, liver, and adipose fat samples of red foxes from Belgium [176]. Median concentrations of PBDEs (excluding BDE-209) were 2.2, 2.4, and 3.4 ng/g lipid wt in fox adipose fat, liver, and muscle, respectively. BDE-209 was found in 15%–40% of the tissue samples analyzed, with concentrations ranging from <3.7 to 760 ng/g lipid wt. BDE-209 was the dominant congener in some samples, accounting for up to 70% of the total PBDE concentrations. PBDEs were determined in serum of pet cats [177]. PBDEs were found in all cats, with mean PBDE concentrations of 4.3 and 10.5 ng/mL for young and old cats, respectively. BDE-47, -99, -207, and -209 were the dominant congeners. PBDEs were determined in various tissues of foraging hens from an E-waste recycling area in South China [178]. The highest PBDE concentrations were found in muscle (18,000 ng/g lipid wt), followed by fat, intestine, heart, liver, oviduct, gizzard, blood, skin, and ovum (125 ng/g lipid wt). BDE-209 was found in all samples (33–18,000 ng/g lipid wt) and was the dominant congener. PBDEs were also found in captive giant panda and red panda from China [176]. Total PBDE concentrations in panda ranged from 16.4 to 2160 ng/g lipid wt. BDE-209 was the most abundant congener, followed by BDE-206, -207, -203, -47, and -153.

2.2.3.4.3 Temporal Trends of PBDEs in Marine Mammals

Several studies have reported temporal trends of PBDEs in marine mammals (Table 2.4). No clear trend was found for PBDE concentrations in California sea lions collected during 1993–2006 from

TABLE 2.4
Temporal Trends of PBDEs Reported for Marine Mammals from Several Global Locations

Species	Location	Total PBDEs Range (ng/g lipid wt)	Temporal Trends	Reference
California sea lion	California, United States	0.6–24	No clear trend, 1993–2003	[179]
California sea lion	California, United States	0.06–236	No clear trend, 1994–2006	[174]
Northern elephant seal		0.04–2.0	No clear trend, 1994–2006	
Harbor seal		0.3–7.2	No clear trend, 1994–2006	
Harbor seal	NW Atlantic Ocean	0.08–26	No clear trend, 1991–2005	[181]
White-sided dolphin	Massachusetts, United States	—	No clear trend, 1993–2000	[182]
Indo-Pacific dolphin	Hong Kong, China	0.3–51	No clear trend, 1997–2008	[313]
Harbor seal	San Francisco Bay	0.09–8.3	Increased between 1988 and 2000	[314]
Male ringed seal	Arctic Canada	—	Exponentially increased, an order of magnitude between 1981 and 2000	[22]
Beluga seal	St. Lawrence estuary	0.2–1.1	Exponentially increased with a doubling time years between 1988 and 1999	[183]
Bottlenose dolphin	Florida coast, United States	0.03–4.5	Exponentially increased with a doubling time of 3–4 years between 1993 and 2004	[23]
Guiana dolphin	Southeast Brazil	0.01–1.6	Yearly increased between 1994 and 2006	[184]
Striped dolphin	Japan	0.01–0.09	Continuously increased from the late 1978 to 2003	[315]
Melon-head whales	Pacific coast of Japan	0.02–0.5	25 times higher in 2006 than in 1982 for adult male	[316]
Finless porpoise	South China Sea	0.08–1.0	An order of magnitude higher in 1990 than in 2001	[13]
Northern fur seal	Japan	0.0003–1.1	Reached peak around 1991–1994, and then decreased to about 50% around 1997–1998	[173]
Harbor porpoise	United Kingdom	nd–15.7	Peaked around 1998 and then reduced 67.6% to 2008	[185]

Note: nd, not detectable.

the California coast [174,179], in sea otter livers collected during 1992–2002 from the California coast [180], in harbor seals collected during 1991–2005 from California [174] and the Northwest Atlantic Ocean [181], and in white-sided dolphin collected during 1993–2000 from Massachusetts, United States [182]. Several other studies demonstrated exponential increase in PBDE concentrations in marine mammals. Total PBDE concentrations increased by 10-fold in ringed seals collected during 1981–2000 from the Canadian Arctic [22], and increased exponentially with a doubling time of ~3 years in beluga whales collected during 1988–1999 from St. Lawrence Estuary [183] and in bottlenose dolphin collected during 1993–2004 from Florida coast, United States [23]. PBDE concentrations in Guiana dolphin collected during 1993–2004 from Southeast Brazil showed increasing concentrations with time [184]. A few studies showed decreasing PBDE concentrations in marine mammals after 2000. PBDE concentrations in archived blubber tissues of Northern fur seals from Sanriku, Japan, increased during 1972–1994 sampling (about 150 times in 1994 to that in 1972), and then decreased by 50% during 1997–1998 sampling [173]. A recent study reported temporal trends of PBDEs in harbor porpoise collected during 1992–2008 from the U.K. coast [185]. The available data indicate that PBDE concentrations in marine mammals increased from the 1970s to the mid-1990s, all over the world. A decreasing trend of PBDEs in marine mammals from the United Kingdom and Japan after the late 1990s may be due to the earlier restriction on the usage of penta- and octa-PBDE formulations. Further studies are needed to evaluate the trends of PBDEs in North America after restrictions have been imposed on the production of PBDEs since the mid-2000s. Moreover, time-trend studies are needed in Asia (especially China and India), as the economies of several countries are increasing considerably and there has been a heavy demand for PBDEs in Asia in recent years. Furthermore, because no regulations have been imposed on PBDE production and usage, it is expected that the environmental levels will continue to increase in some Asian countries.

2.3   E-WASTE

Waste electrical and electronic equipment (EEE), known as E-waste, has been recognized as the fastest-growing solid-waste stream. E-waste is the term used to describe old, end-of-life EEE, such as computers, laptops, television sets, DVD players, mobile phones, etc. The world's production of E-waste was estimated at 20–50 million MT per year, accounting for more than 5% of all municipal solid waste, and was growing steadily [186]. At present, most of the E-waste is produced in developed countries in Europe, the United States, and Australia. It was estimated that the total production of E-waste was 5.5 million MT per year for the 15-member-state European Union (EU) [187] and 2.4 million MT in the United States in 2005 [103]. China, Eastern Europe, and Latin America are expected to become the major E-waste producers in the next 10 years [188]. A study predicted that the volume of obsolete personal computers generated in developing countries will exceed the volume in developed countries by 2016–2018 [189].

E-waste is discarded in landfills, or incinerated, or recycled. Only less than 10% of the E-waste is recycled [190]. It was estimated that 50%–80% of the E-waste generated in the United States are exported to Asia, mainly China [191], and 60%–75% of the E-waste generated in the EU is exported to Asia and Africa [192]. China plays a key role in the global recycling of E-waste, because 90% of the E-waste exported to Asia is destined to China [191], in addition to the huge domestic production, which is estimated at 2.5 million MT in 2005 [193]. Significant quantities of E-waste are also exported to India, Pakistan, Vietnam, the Philippines, Malaysia, Nigeria, and Ghana [188].

E-waste disposal has led to serious environmental problems in developing countries. E-waste contains over 60% (by weight) of metals, such as iron, copper, aluminum, gold, and others, and 2.7% of organic pollutants [194], most of which are highly toxic, especially when incinerated or recycled in an uncontrolled environment. PBDEs are used to reduce or prevent flammability of the plastic housing of EEE and circuit boards. Average concentrations of penta-, octa-, and deca-BDE in small-size E-waste materials in Switzerland were reported to be 34, 530, and 510 μg/g, respectively [195]. Mean concentrations of total PBDEs (excluding BDE-209) and BDE-209 in electronic shredder residues (printed circuit board and plastic materials discarded after recovery of metals) collected from the E-waste recycling area in eastern China were 42 and 3.3 μg/g, respectively [196]. In the United States, it has been estimated that 1224 MT of deca-PBDE and 1272 MT of octa-PBDE were present in 2.4 million MT of E-waste in 2005 [103].

PBDE concentrations of dust collected from inside television sets (300 μg/g) [126] and dust from houses with several computers (6 μg/g) contained very high PBDE concentrations [127], suggesting off-gassing of PBDEs from the housing of electronic equipment. It was estimated that ~19 MT of BDE-209 were released into the air from the production and manufacturing of deca-PBDEs, and 45 kg of BDE-209 were released from the incineration of E-waste every year in the United States [103].

TABLE 2.5
PBDE Levels in Environmental Samples from E-Waste Recycling Facilities in China

Sample	Location	PBDEs (Median)[a]	BDE-209	Reference[a]	Reference
Air	Guiyu, Guangdong	11,742 pg/m^3 (day time)	2164 pg/m^3	376 pg/m^3	[317]
Water	Qingyuan, Guangdong	24.4 ng/L	0.60 ng/L		[151]
Soil	Guiyu, Guangdong	48.2–3570 ng/g	37.3–1270 ng/g		[318]
Soil	Qingyuan, Guangdong	39.1–2689 ng/g	29.9–1539 ng/g		[121]
Soil	Taizhou, Zhejiang	1910 ng/g	1800 ng/g	0.30 ng/g	[197]
Sediment	Guiyu, Guangdong	156–9357 ng/g	30–36 ng/g		[117]
Dust	Taizhou, Zhejiang	30,700 ng/g	29,800 ng/g		[196]
Water birds	Qingyuan, Guangdong	37–2200 ng/g lipid wt	nd–532 ng/g lipid wt		[319]
Forage hen	Taizhou, Zhejiang	125–17,976 ng/g lipid wt	33–17,796 ng/g lipid wt		[178]
Fish	Guiyu, Guangdong	551–6377 ng/g lipid wt	nd		[150]
Blood	Guangdong	126.1 ng/g lipid wt	83.5 ng/g lipid wt	10.7 ng/g lipid wt	[320]
Serum	Guiyu, Guangdong	600 ng/g lipid wt	310 ng/g lipid wt	170 ng/g lipid wt	[80]
Human viscera	Zhejiang	311–442 ng/g lipid wt	118–270 ng/g lipid wt		[79]

Note: nd, not detectable.
[a] Total concentration of PBDEs, mean value was used when median value was not available.

Plastics are the second most predominant components by weight, representing ~21% of the EEE [194]. It was estimated that brominated flame retardants constitute 5%–30% of the plastic by weight [197], and 10% of the flame retardants are PBDEs [191]; thus, EEE contain ~0.1%–0.6% of PBDEs. As mentioned above, the total PBDEs accounted for ~0.1% of the total weight of small-size E-waste in Switzerland [195]. The global release of PBDEs through E-waste disposal is estimated to be at least 20,000 MT per year (based on 20–50 million MT of E-waste produced per year). Asia has received ~18,000 MT of PBDEs per year, with China accounting for 14,000 MT annually. It was further estimated that only 2 MT of total PBDEs were discharged from the Pearl River Delta into its coastal waters in South China, where most of the country's E-waste recycling facilities are located [198]. The results suggest the existence of large PBDE inventories in China and in many other countries. The E-waste recycling operations in China have spread from Guangdong to many other provinces including Hunan, Zhejiang, Shanghai, Tianjin, Fujian, and Shandong [199].

PBDE concentrations are high in various environmental matrices in E-waste recycling facilities in China (Table 2.5). Reports indicate that in Guiyu, the world's largest E-waste recycling area, 80% of the families of a total population of 160,000 are involved in E-waste recycling operations [191]. High PBDE concentrations in the environment, human exposures to it, and health concerns caused by PBDEs in E-waste recycling operations need further investigations.

2.4 SUMMARY AND PERSPECTIVES

PBDEs are ubiquitous and detectable in the global environment, even in remote marine environments. Several reviews have appeared on PBDEs in various environmental matrices in East Asia [200], Europe [6,74], and the Arctic marine environment [201], in wildlife [202], in marine biota from the Canadian Arctic [203], in marine environment of the United States [204], and in the marine ecosystem of the American continents [75]. Despite an exponential increase in the number of publications on PBDEs in the past 10 years, data from South America, Africa, and Oceania countries are still limited.

Based on the available information, the following generalizations can be made. PBDEs are produced to retard fires and thus can have a direct benefit. Yet the demonstrated persistence, bio-accumulation, and toxic potential of these compounds in animals and in humans are of increasing concern. Since their introduction in the 1970s, PBDEs have been increasing in abiotic and biotic matrices in coastal marine environments and are beginning to rival PCBs as the predominant contaminants in water and sediments in urban and near-urban source areas. Whereas a penta-BDE signature characterizes the accumulation patterns in fish and piscivorous wildlife, exposure to octa- and deca-BDE mixtures is indicated by the patterns of terrestrial species and in abiotic matrices. In certain aquatic/marine food webs, bioaccumulation and magnification of constituents of all three PBDE commercial mixtures are evident, presenting an increasing health risk to marine animals. The synergistic interactions and effects due to exposure to multiple contaminants (especially the mixture of PCBs and PBDEs) in marine organisms need further investigation.

PBDE concentrations are high in the environment and biota collected from North America. PBDE levels in human samples from North America were one order of magnitude higher than in Asia or Europe. The PBDE concentrations are higher in the environmental and biological specimens from the Northern hemisphere than in the Southern hemisphere. Soils and sediments appear to be the final environmental sinks for PBDEs. PBDE concentrations are high in house dust from North America, and dust ingestion is the main route for human exposure to PBDEs in the United States. High concentrations of PBDEs in dust from North America are related to the strict flammability standards that require the use of flame retardants in many household products.

Diet may be a major source of human exposure to PBDEs in Asian countries. The temporal trends of PBDEs in human breast milk, archived animal tissues, sediment cores, and bird eggs indicate that PBDEs were introduced in the 1970s, and the concentrations increased until the late 1990s; PBDE concentrations have continued to increase in some locations until recently, whereas in other regions such as Europe, PBDE levels were stable in the 2000s, reflecting an earlier restrictions on the production and usage. The decreasing trends of PBDEs in human breast milk from some European countries and in marine animals from the United Kingdom demonstrate that the regulations on usage of PBDEs are having a clear effect.

BDE-47 is the most abundant congener in biota samples, followed by BDE-99, -100, -153, and -154. BDE-209 is the most abundant congener in abiotic samples, such as sediment, soil, and dust. Despite its large molecular size, BDE-209 is detectable in mussel, fish, birds, mammals, and human samples; and in biota from point source areas, BDE-209 dominated the total PBDE concentrations. E-waste disposal will continue to contribute PBDE pollution on a global scale and further studies are needed to evaluate the restrictions that have been placed on PBDEs in several countries.

Similar to our experience with PCBs (which have been banned from production and use in most developed countries during the past three decades), it is likely that regulatory actions aimed at controlling PBDEs by banning manufacturing and new uses will be insufficient to stop this trend immediately. PBDE congeners from the penta-, octa-, and deca-BDE mixtures will continue to be released into the environment during the use, disposal, and recycling of existing fire-retardant-containing products for years to come. This situation is especially troublesome when one considers that large amounts of electronics and furniture, manufactured when PBDEs were most heavily used, are likely being recycled and disposed of at present or will be in the near future. Similarly, Tanabe [17] predicted that most of the PCBs (66%) were stockpiled in products (transformers and capacitors) long after PCBs were banned from production in the 1970s.

As flame retardants, PBDEs may save several lives by reducing fire. However, the increasing levels of PBDEs in the environment coupled with its potential toxicity to biota present an emerging risk for environment and human health. In May 2009, the commercial penta- and octa-PBDE mixtures have been added to the list of POPs by the United Nations Stockholm Convention. Commercial penta- and octa-PBDE mixtures were banned in the Europe and the United States, although deca-PBDE is still currently being used globally. BDE-209 can be debrominated by both organisms (fish and mammal) and in the environment (photodegradation and biodegradation) [205] and can be the source of other BDE congeners that are persistent and bioaccumulative. Bioavailability and mechanism of biotransformation BDE 209 in human bodies need further investigation.

Global regulations and strategies for the management of E-wastes are needed to control the cycling of toxic contaminants present [206]. Education is the key, and the public should be made aware of health impacts of toxic contaminants present in E-waste. For example, it has been reported that 67% of people in the United States are not aware of E-waste disposal restrictions or policies [207] and more than 90% of Chinese citizens are reluctant to pay for the recycling of their E-waste [208]. Concerted action is needed not only to ban the production and use of PBDEs but also to find ways of reducing the existing indoor reservoirs and managing the end of life of PBDE-containing products. PBDE discharges continue to increase, and these compounds are loading into all compartments of the environment. The evidence suggests that if PBDEs were banned today, it would take decades after the end of discharge for marine sediment to bury them [75].

REFERENCES

1. Alaee, M., Arias, P., Sjodin, A., and Bergman, Å. 2003. An overview of commercially used brominated flame-retardants, their applications, their patterns in different countries/regions and possible modes of release. *Environment International*. 29: 683–689.
2. Lorber, M. and Cleverly, D. 2010. *An Exposure Assessment of Brominated Diphenyl Ethers*. National Center for Environmental Assessment Office of Research and Development, U.S. Environmental Protection Agency, Washington, DC.
3. Darnerud, P. O. 2003. Toxic effects of brominated flame retardants in man and in wildlife. *Environment International*. 29: 841–853.
4. Hites, R. A. 2004. Polybrominated diphenyl ethers in the environment and in people: A meta-analysis of concentrations. *Environmental Science & Technology*. 38: 945–956.
5. de Wit, C. A. 2002. An overview of brominated flame retardants in the environment. *Chemosphere*. 46: 583–624.
6. Law, R. J., Allchin, C. R., de Boer, J., Covaci, A., Herzke, D., Lepom, P., Morris, S., Tronczynski, J., and de Wit, C. A. 2006. Levels and trends of brominated flame retardants in the European environment. *Chemosphere*. 64: 187–208.
7. Tanabe, S. 2008. Temporal trends of brominated flame retardants in coastal waters of Japan and South China: Retrospective monitoring study using archived samples from es-Bank, Ehime University, Japan. *Marine Pollution Bulletin*. 57: 267–274.
8. de Boer, J., Wester, P. G., Klammer, J. J. C., Lewis, W. E., and Boon, J. P. 1998. Do flame retardants threaten ocean life? *Nature*. 394: 28–29.
9. Loganathan, B. G. and Kannan, K. 1994. Global organochlorine contamination trends: An overview. *Ambio*. 23: 187–191.
10. Covaci, A., Losada, S., Roosens, L., Vetter, W., Santos, F. J., Neels, H., Storelli, A., and Storelli, M. M. 2008. Anthropogenic and naturally occurring organobrominated compounds in two deep-sea fish species from the Mediterranean Sea. *Environmental Science & Technology*. 42: 8654–8660.
11. Toyoshima, S., Isobe, T., Ramu, K., Miyasaka, H., Omori, K., Takahashi, S., Nishida, S., and Tanabe, S. 2009. Organochlorines and brominated flame retardants in deep-sea ecosystem of Sagami Bay. In *Interdisciplinary Studies on Environmental Chemistry-Environmental Research in Asia*; Obayashi, Y., Isobe, T., Subramanian, A., Suzuki, S., and Tanabe, S. (Eds.). TERRAPUB, Japan, pp. 83–90.
12. Urger, M. A., Harvey, E., Vadas, G. G., and Vecchione, M. 2008. Persistent pollutions in nine species of deep-sea cephalopods. *Marine Pollution Bulletin*. 56: 1498–1500.
13. Ramu, K., Kajiwara, N., Lam, P. K. S., Jefferson, T. A., Zhou, K., and Tanabe, S. 2006. Temporal variation and biomagnification of organohalogen compounds in finless porpoises (*Neophocaena phocaenoides*) from the South China Sea. *Environmental Pollution*. 144: 516–523.
14. Oshihoi, T., Isobe, T., Takahashi, S., Kubodera, T., and Tanabe, S. 2009. Contamination status of organohalogen compounds in deep-sea fishes in northwest Pacific Ocean off-Tohoku, Japan. In *Interdisciplinary Studies on Environmental Chemistry-Environmental research in Asia*; Obayashi, Y., Isobe, T., Subramanian, A., Suzuki, S., and Tanabe, S. (Eds.). TERRAPUB, Japan, pp. 67–72.

15. Bromine Science and Environmental Forum (BSAF). 2003. *Major Brominated Flame Retardants Volume Estimated: Total Markets Demand by Region in 2001.* http://www.bsef-site.com/docs/BFR_vols_2001. doc (accessed June 2010).

16. Jenssen, B. M., Sormo, E. G., Baek, K., Bytingsvik, J., Gaustad, H., Ruus, A., and Skaare, J. U. 2007. Brominated flame retardants in north-east Atlantic marine ecosystems. *Environmental Health Perspectives.* 115: 35–41.

17. Tanabe, S. 1988. PCB problems in the future: Foresight from current knowledge. *Environmental Pollution.* 50: 5–28.

18. Birnbaum, L. S. and Staskal, D. F. 2004. Brominated flame retardants: Cause for concern? *Environmental Health Perspectives.* 112: 9–17.

19. Hale, R. C., La Guardia, M. J., Harvey, E., Gaylor, M. O., and Mainor, T. M. 2006. Brominated flame retardant concentrations and trends in abiotic media. *Chemosphere.* 64: 181–186.

20. Hale, R. C., Alaee, M., Manchester-Neesvig, J. B., Stapleton, H. M., and Ikonomou, M. G. 2003. Polybrominated diphenyl ether flame retardants in the North American environment. *Environment International.* 29: 771–779.

21. Hale, R. C., Kim, S. L., Harvey, E., La Guardia, M. J., Mainor, T. M., Bush, E. O., and Jacobs, E. M. 2008. Antarctic research bases: Local sources of polybrominated diphenyl ether (PBDE) flame retardants. *Environmental Science & Technology.* 42: 1452–1457.

22. Ikonomou, M. G., Rayne, S., and Addison, R. F. 2002. Exponential increases of the brominated flame retardants, polybrominated diphenyl ethers, in the Canadian Arctic from 1981 to 2000. *Environmental Science & Technology.* 36: 1886–1892.

23. Johnson-Restrepo, B., Kannan, K., Addink, R., and Adams, D. H. 2005. Polybrominated diphenyl ethers and polychlorinated biphenyls in a marine foodweb of coastal Florida. *Environmental Science & Technology.* 39: 8243–8250.

24. Ross, P. S., Couillard, C. M., Ikonomou, M. G., Johannessen, S. C., Lebeuf, M., Macdonald, R. W., and Tomy, G. T. 2009. Large and growing environmental reservoirs of Deca-BDE present an emerging health risk for fish and marine mammals. *Marine Pollution Bulletin.* 58: 7–10.

25. Elliott, J. E., Wilson, L. K., and Wakeford, B. 2005. Polybrominated diphenyl ether trends in eggs of marine and freshwater birds from British Columbia, Canada, 1979–2002. *Environmental Science & Technology.* 39: 5584–5591.

26. Betts, K. 2008. New thinking on flame retardants. *Environmental Health and Perspectives.* 116: A202–A208.

27. Stapleton, H. M. and Dodder, N. G. 2008. Photodegradation of decabromodiphenyl ether in natural sunlight. *Environmental Toxicology & Chemistry.* 27: 306–312.

28. Stapleton, H. M., Alaee, M., Letcher, R. J., and Baker, J. E. 2004. Debromination of the flame retardant decabromodiphenyl ether by juvenile carp (*Cyprinus carpio*) following dietary exposure. *Environmental Science & Technology.* 38: 112–119.

29. La Guardia, M. J., Hale, R. C., and Harvey, E. 2007. Evidence of debromination of decabromodiphenyl ether (BDE-209) in biota from a wastewater receiving stream. *Environmental Science & Technology.* 41: 6663–6670.

30. Shaw, S. D., Berger, M. L., Brenner, D., Kannan, K., Lohmann, N., and Papke, O. 2009. Bioaccumulation of polybrominated diphenyl ethers and hexabromocyclododecane in the northwest Atlantic marine food web. *Science of the Total Environment.* 407: 3323–3329.

31. de Boer, J., Wester, P. G., van der Horst, A., and Leonards, P. E. G. 2003. Polybrominated diphenyl ethers in influents, suspended particulate matter, sediments, sewage treatment plant and effluents and biota from the Netherlands. *Environmental Pollution.* 122: 63–74.

32. IPCS (International Program on Chemical Safety). 1994. *Environmental Health Criteria 162: Brominated Diphenyl Ethers.* World Health Organization, Geneva, Switzerland. http://www.inchem.org/documents/ehc/ehc/ehc162.htm (accessed June 2010).

33. Wang, X. M., Ding, X., Mai, B. X., Xie, Z. Q., Xiang, C. H., Sun, L. G., Sheng, G. Y., Fu, J. M., and Zeng, E. Y. 2005. Polybrominated diphenyl ethers in airborne particulates collected during a research expedition from the Bohai Sea to the Arctic. *Environmental Science & Technology.* 39: 7803–7809.

34. Moon, H. B., Kannan, K., Lee, S. J., and Choi, M. 2007. Atmospheric deposition of polybrominated diphenyl ethers (PBDEs) in coastal areas in Korea. *Chemosphere.* 66: 585–593.

35. Hale, R. C., La Guardia, M. J., Harvey, E. P., Mainor, T. M., Duff, W. H., and Gaylor, M. O. 2001. Polybrominated diphenyl ether flame retardants in Virginia freshwater fishes (USA). *Environmental Science & Technology.* 35: 4585–4591.

36. Moon, H.-B., Kannan, K., Choi, M., and Choi, H.-G. 2007. Polybrominated diphenyl ethers (PBDEs) in marine sediments from industrialized bays of Korea. *Marine Pollution Bulletin*. 54: 1402–1412.

37. Johannessen, S. C., Macdonald, R. W., Wright, C. A., Burd, B., Shaw, D. P., and van Roodselaar, A. 2008. Joined by geochemistry, divided by history: PCBs and PBDEs in Strait of Georgia sediment. *Marine Environmental Research*. 108: 158–167.

38. Talsness, C. E. 2008. Overview of toxicological aspects of polybrominated diphenyl ethers: A flame-retardant additive in several consumer products. *Environmental Research*. 108: 158–167.

39. Darnerud, P. O. 2008. Brominated flame retardants as possible endocrine disrupters. *International Journal of Andrology*. 31: 152–160.

40. Legler, J. 2008. New insights into the endocrine disrupting effects of brominated flame retardants. *Chemosphere*. 73: 216–222.

41. Hamers, T., Kamstra, J. H., Sonneveld, E., Murk, A. J., Kester, M. H. A., Andersson, P. L., Leglder, J., and Brouwer, A. 2006. *In vitro* profiling of the endocrine-disrupting potency of brominated flame retardants. *Toxicological Sciences*. 92: 157–173.

42. Zhang, Y., Guo, G. L., Han, X., Zhu, C., Kilfoy, B. A., Zhu, Y., Boyle, P., and Zheng, T. 2008. Do polybrominated diphenyl ethers (PBDE) increase the risk of thyroid cancer? *Bioscience Hypotheses*. 1: 195–199.

43. Costa, L. G. and Giordano, G. 2007. Developmental neurotoxicity of polybrominated diphenyl ether (PBDE) flame retardants. *Neurotoxicology*. 28: 1047–1067.

44. Lundgren, M., Darnerud, P. O., Molin, Y., Lilienthal, H., Blomberg, J., and Ilbäck, N.-G. 2007. Viral infection and PBDE exposure interact on CYP gene expression and enzyme activities in the mouse liver. *Toxicology*. 242: 100–108.

45. Kuriyama, S. N., Wanner, A., Fidalgo-Neto, A. A., Talsness, C. E., Koerner, W., and Chahoud, I. 2007. Developmental exposure to low-dose PBDE-99: Tissue distribution and thyroid hormone levels. *Toxicology*. 242: 80–90.

46. Lilienthal, H., Hack, A., Roth-Härer, A., Grande, S. W., and Talsness, C. E. 2006. Effects of developmental exposure to 2,2′,4,4′,5-pentabromodiphenyl ether (PBDE-99) on sex steroids, sexual development and sexually dimorphic behavior in rats. *Environmental Health and Perspectives*. 114: 194–201.

47. McDonald, T. A. 2005. Polybrominated diphenyl ether levels among United States residents: Daily intake and risk of harm to the developing brain and reproductive organs. *Integrated Environmental Assessment and Management*. 1: 343–354.

48. Eriksson, P., Jakobsson, E., and Fredriksson, A. 2001. Brominated flame retardants: A novel class of developmental neurotoxicants in our environment? *Environmental Health Perspectives*. 109: 903–908.

49. Eriksson, P., Viberg, H., Jakobsson, E., Orn, U., and Fredriksson, A. 2002. A brominated flame retardant, 2,2′,4,4′,5-pentabromodiphenyl ether: Uptake, retention, and induction of neurobehavioral alterations in mice during a critical phase of neonatal brain development. *Toxicological Sciences*. 67: 98–103.

50. Martin, P. A., Mayne, G. J., Bursian, S. J., Tomy, G., Palace, V., Pekarik, C., and Smits, J. 2007. Immunotoxicity of the commercial polybrominated diphenyl ether mixture DE-71 in ranch mink (*Mustela vison*). *Environmental Toxicology & Chemistry*. 26: 988–997.

51. Watanabe, W., Shimizu, T., Hino, A., and Kurokawa, M. 2008. Effects of decabrominated diphenyl ether (DBDE) on developmental immunotoxicity in offspring mice. *Environmental Toxicology and Pharmacology*. 26: 315–319.

52. Viberg, H., Fredriksson, A., Jakobsson, E., Orn, U., and Eriksson, P. 2003. Neurobehavioral derangements in adult mice receiving decabrominated diphenyl ether (PBDE 209) during a defined period of neonatal brain development. *Toxicological Sciences*. 76: 112–120.

53. Viberg, H., Fredriksson, A., and Eriksson, P. 2003. Neonatal exposure to polybrominated diphenyl ether (PBDE 153) disrupts spontaneous behaviour, impairs learning and memory, and decreases hippocampal cholinergic receptors in adult mice. *Toxicology and Applied Pharmacology*. 192: 95–106.

54. Viberg, H., Fredriksson, A., and Eriksson, P. 2004. Neonatal exposure to the brominated flame-retardant, 2,2′,4,4′,5-pentabromodiphenyl ether, decreases cholinergic nicotinic receptors in hippocampus and affects spontaneous behaviour in the adult mouse. *Environmental Toxicology and Pharmacology*. 17: 61–65.

55. Viberg, H., Johansson, N., Fredriksson, A., Eriksson, J., Marsh, G., and Eriksson, P. 2006. Neonatal exposure to higher brominated diphenyl ethers, hepta-, octa-, or nonabromodiphenyl ether, impairs spontaneous behavior and learning and memory functions of adult mice. *Toxicological Sciences*. 92: 211–218.

56. Viberg, H., Fredriksson, A., and Eriksson, P. 2007. Changes in spontaneous behaviour and altered response to nicotine in the adult rat, after neonatal exposure to the brominated flame retardant, decabrominated diphenyl ether (PBDE 209). *Neurotoxicology*. 28: 136–142.

57. Johansson, N., Viberg, H., Fredriksson, A., and Eriksson, P. 2008. Neonatal exposure to deca-brominated diphenyl ether (PBDE 209) causes dose-response changes in spontaneous behaviour and cholinergic susceptibility in adult mice. *Neurotoxicology*. 29: 911–919.

58. Rice, D. C., Reeve, E. A., Herlihy, A., Zoeller, R. T., Thompson, W. D., and Markowski, V. P. 2007. Developmental delays and locomotor activity in the C57BL6/J mouse following neonatal exposure to the fully-brominated PBDE, decabromodiphenyl ether. *Neurotoxicology and Teratology*. 29: 511–520.

59. US Environmental Protection Agency-Integrated Risk Information System (IRIS). 2008. *Toxicological Review of Decabromodiphenyl Ether (BDE-209) (CAS No. 1163-19-5)*. http://www.epa.gov/IRIS/toxreviewa/0035tr.pdf (accessed June 2010).

60. Turyk, M. E., Persky, V. W., Imm, P., Knobeloch, L., Chatterton, R., and Anderson, H. A. 2008. Hormone disruption by PBDEs in adult male sport fish consumers. *Environmental Health Perspectives*. 116: 1635–1641.

61. Yuan, J., Chen, L., Chen, D.-H., Guo, H., Bi, X.-H., Ju, Y., Jiang, P. et al. 2008. Elevated serum polybrominated diphenyl ethers and thyroid-stimulating hormone associated with lymphocytic micronuclei in Chinese workers from an E-waste dismantling site. *Environmental Science & Technology*. 42: 2195–2200.

62. Julander, A., Karlsson, M., Hagstrom, K., Ohlson, C. G., Engwall, M., Bryngelsson, I. L., Westberg, H., and van Bavel, B. 2005. Polybrominated diphenyl ethers—Plasma levels and thyroid status of workers at an electronic recycling facility. *International Archives of Occupational and Environmental Health*. 78: 584–592.

63. Main, K. M., Kiviranta, H., Virtanen, H. E., Sundqvist, E., Tuomisto, J. T., Tuomisto, J., Vartiainen, T., Skakkebek, N. E., and Toppari, J. 2007. Flame retardants in placenta and breast milk and cryptorchidism in newborn boys. *Environmental Health and Perspectives*. 115: 1519–1526.

64. Chao, H.-R., Wang, S.-L., Lee, W.-J., Wang, Y.-F., and Päpke, O. 2007. Levels of polybrominated diphenyl ethers (PBDEs) in breast milk from central Taiwan and their relation to infant birth outcome and maternal menstruation effects. *Environment International*. 33: 239–245.

65. Akutsu, K., Takatori, S., Nozawa, S., Yoshiike, M., Nakazawa, H., Hayakawa, K., Makino, T., and Iwamoto, T. 2008. Polybrominated diphenyl ethers in human serum and sperm quality. *Bulletin of Environmental Contamination and Toxicology*. 80: 345–350.

66. Sjödin, A., Wong, L.-Y., Jones, R. S., Park, A., Zhang, Y.-L., Hodge, C., DiPietro, E. et al. 2008. Serum concentrations of polybrominated diphenyl ethers (PBDEs) and polybrominated diphenyl (PBB) in the United States population: 2003–2004. *Environmental Science & Technology*. 42: 1377–1384.

67. Lorber, M. 2008. Exposure of Americans to polybrominated diphenyl ethers. *Journal of Exposure Science and Environmental Epidemiology*. 18: 2–19.

68. Johnson-Restrepo, B. and Kannan, K. 2009. An assessment of sources and pathways of human exposure to polybrominated diphenyl ethers in the United States. *Chemosphere*. 76: 542–548.

69. Meeker, J. D., Johnson, P. I., Camann, D., and Hauser, R. 2009. Polybrominated diphenyl ether (PBDE) concentrations in house dust are related to hormone levels in men. *Science of the Total Environment*. 407: 3425–3429.

70. Hardell, L., Eriksson, M., Lindström, G., van Bavel, B., Linde, A., Carlberg, M., and Liljegren, G. 2001. Case-control study on concentration of organohalogen compounds and titers of antibodies to Epstein-Barr virus antigens in the etiology of non-Hodgkin lymphoma. *Leukemia and Lymphoma*. 42: 619–629.

71. National Toxicology Program (US). 1986. *Toxicology and Carcinogenesis Studies of Decabromodiphenyl Oxide (CAS No. 1163-19-5) in F344/N Rats and B6C3F1 Mice (Feed Studies)*. http://ntp.niehs.nih.gov/ntp/htdocs/LT_rpts/tr309.pdf (accessed June 2010).

72. Fängström, B., Athanassiadis, L., Odsjö, T., Norén, K., and Bergman, Å. 2008. Temporal trends of polybrominated diphenyl ethers and hexabromocyclododecane in milk from Stockholm mothers, 1980–2004. *Molecular Nutrition and Food Research*. 52: 187–193.

73. Meironyté, D., Norén, K., and Bergman, A. 1999. Analysis of polybrominated diphenyl ethers in Swedish human milk. A time-related trend study, 1972–1997. *Journal of Toxicology and Environmental Health-Part A*. 58: 329–341.

74. Law, R. J., Herzke, D., Harrad, S., Morris, S., Bersuder, P., and Allchin, C. R. 2008. Levels and trends of HBCD and BDEs in the European and Asian environments, with some information for other BFRs. *Chemosphere*. 73: 223–241.

75. Shaw, S. D. and Kannan, K. 2009. Polybrominated diphenyl ethers in marine ecosystems of the American continents: Foresight from current knowledge. *Reviews on Environmental Health*. 24: 157–229.

76. Toms, L. M. L., Harden, F. A., Symons, R. K., Burniston, D., Furst, P., and Muller, J. F. 2007. Polybrominated diphenyl ethers (PBDEs) in human milk from Australia. *Chemosphere*. 68: 797–803.

77. Toms, L. M. L., Hearn, L., Kennedy, K., Harden, F., Bartkow, M., Temme, C. and Mueller, J. F. 2009. Concentrations of polybrominated diphenyl ethers (PBDEs) in matched samples of human milk, dust and indoor air. *Environment International*. 35: 864–869.

78. Stapleton, H. M., Sjodin, A., Jones, R. S., Niehoser, S., Zhang, Y., and Patterson, D. G. 2008. Serum levels of polybrominated diphenyl ethers (PBDEs) in foam recyclers and carpet installers working in the United States. *Environmental Science & Technology*. 42: 3453–3458.

79. Zhao, G. F., Wang, Z. J., Zhou, H. D., and Zhao, Q. 2009. Burdens of PBBs, PBDEs, and PCBs in tissues of the cancer patients in the e-waste disassembly sites in Zhejiang, China. *Science of the Total Environment*. 407: 4831–4837.

80. Bi, X. H., Thomas, G. O., Jones, K. C., Qu, W. Y., Sheng, G. Y., Martin, F. L., and Fu, J. M. 2007. Exposure of electronics dismantling workers to polybrominated diphenyl ethers, polychlorinated biphenyls, and organochlorine pesticides in South China. *Environmental Science & Technology*. 41: 5647–5653.

81. Wu, K. S., Xu, X. J., Liu, J. X., Guo, Y. Y., Li, Y., and Huo, X. 2010. Polybrominated diphenyl ethers in umbilical cord blood and relevant factors in neonates from Guiyu, China. *Environmental Science & Technology*. 44: 813–819.

82. Thomsen, C., Lundanes, E., and Becher, G. 2001. Brominated flame retardants in plasma samples from three different occupational groups in Norway. *Journal of Environmental Monitoring*. 3: 366–370.

83. Sjödin, A., Hagmar, L., Klasson-Wehler, E., Kronholm-Diab, K., Jakobsson, E., and Bergman, A. 1999. Flame retardant exposure: Polybrominated diphenyl ethers in blood from Swedish workers. *Environmental Health Perspectives*. 107: 643–648.

84. Jakobsson, K., Thuresson, K., Rylander, L., Sjödin, A., Hagmar, L., and Bergman, A. 2002. Exposure to polybrominated diphenyl ethers and tetrabromobisphenol A among computer technicians. *Chemosphere*. 46: 709–716.

85. Jin, J., Wang, Y., Yang, C. Q., Hu, J. C., Liu, W. Z., Cui, J., and Tang, X. Y. 2009. Polybrominated diphenyl ethers in the serum and breast milk of the resident population from production area, China. *Environment International*. 35: 1048–1052.

86. Kotz, A., Malisch, R., Kypke, K., and Oehme, M. 2005. PBDE, PBDD/F and mixed chlorinated-brominated PXDD/F in pooled human milk samples from different countries. *Organohalogen Compounds*. 67: 1540–1544.

87. Kalantzi, O. L., Martin, F. L., Thomas, G. O., Alcock, R. E., Tang, H. R., Drury, S. C., Carmichael, P. L., Nicholson, J. K., and Jones, K. C. 2004. Different levels of polybrominated diphenyl ethers (PBDEs) and chlorinated compounds in breast milk from two UK regions. *Environmental Health Perspectives*. 112: 1085–1091.

88. Fängström, B., Strid, A., Athanassiadis, L., Grandjean, P., Weihe, P., and Bergman, Å. 2004. A retrospective time trend study of PBDEs and PCBs in human milk from the Faroe Islands. *Organohalogen Compounds*. 66: 2795–2799.

89. Strandman, T., Koistinen, J., and Vartiainen, T. 2000. Polybrominated diphenyl ethers (PBDEs) in placenta and human milk. *Organohalogen Compounds*. 47: 61–64.

90. Erdoğrul, Ö., Covaci, A., Kurtul, N., and Schepens, P. 2004. Levels of organohalogenated persistent pollutants in human milk from Kahramanmaras region, Turkey. *Environment International*. 30: 659–666.

91. Norén, K. and Meironyté, D. 2000. Certain organochlorine and organobromine contaminants in Swedish human milk in perspective of past 20–30 years. *Chemosphere*. 40: 1111–1123.

92. Thomsen, C., Liane, V. H., and Becher, G. 2007. Automated solid-phase extraction for the determination of polybrominated diphenyl ethers and polychlorinated biphenyls in serum—Application on archived Norwegian samples from 1977 to 2003. *Journal of Chromatography B-Analytical Technologies in the Biomedical and Life Sciences*. 846: 252–263.

93. Akutsu, K., Kitagawa, M., Nakazawa, H., Makino, T., Iwazaki, K., Oda, H., and Hori, S. 2003. Time-trend (1973–2000) of polybrominated diphenyl ethers in Japanese mother's milk. *Chemosphere*. 53: 645–654.

94. Choi, J. W., Fujimaki, S., Kitamura, K., Hashimoto, S., Ito, H., Suzuki, N., Sakai, S., and Morita, M. 2003. Polybrominated dibenzo-p-dioxins, dibenzofurans, and diphenyl ethers in Japanese human adipose tissue. *Environmental Science & Technology*. 37: 817–821.

95. Frederiksen, M., Vorkamp, K., Thomsen, M., and Knudsen, L. E. 2009. Human internal and external exposure to PBDEs—A review of levels and sources. *International Journal of Hygiene and Environmental Health*. 212: 109–134.

96. Harrad, S., Wijesekera, R., Hunter, S., Halliwell, C., and Baker, R. 2004. Preliminary assessment of U.K. human dietary and inhalation exposure to polybrominated diphenyl ethers. *Environmental Science & Technology*. 38: 2345–2350.

97. Meng, X.-Z., Zeng, E. Y., Yu, L.-P., Guo, Y., and Mai, B.-X. 2007. Assessment of human exposure to polybrominated diphenyl ethers in China via fish consumption and inhalation. *Environmental Science & Technology*. 41: 4882–4887.

98. Wilford, B. H., Harner, T., Zhu, J., Shoeib, M., and Jones, K. C. 2004. Passive sampling survey of polybrominated diphenyl ether flame retardants in indoor and outdoor air in Ottawa, Canada: Implications for sources and exposure. *Environmental Science & Technology*. 38: 5312–5318.

99. Thomsen, C., Knutsen, H. K., Liane, V. H., Froshaug, M., Kvalem, H. E., Haugen, M., Meltzer, H. M., Alexander, J., and Becher, G. 2008. Consumption of fish from a contaminated lake strongly affects the concentrations of polybrominated diphenyl ethers and hexabromocyclododecane in serum. *Molecular Nutrition and Food Research*. 52: 228–237.

100. Allen, J. G., McClean, M. D., Stapleton, H. M., and Webster, T. F. 2008. Critical factors in assessing exposure to PBDEs via house dust. *Environment International*. 34: 1085–1091.

101. Hazrati, S. and Harrad, S. 2006. Causes of variability in concentrations of polychlorinated biphenyls and polybrominated diphenyl ethers in indoor air. *Environmental Science & Technology*. 40: 7584–7589.

102. Zota, A. R., Rudel, R. A., Morello-Frosch, R. A., and Brody, J. G. 2008. Elevated house dust and serum concentrations of PBDEs in California: Unintended consequences of furniture flammability standards? *Environmental Science & Technology*. 42: 8158–8164.

103. Lorber, M. and Cleverly, D. 2010. *An Exposure Assessment of Brominated Diphenyl Ethers*. National Center for Environmental Assessment Office of Research and Development, U.S. Environmental Protection Agency, Washington, DC.

104. Fischer, D., Hooper, K., Athanasiadou, M., Athanassiadis, I., and Bergman, A. 2006. Children show highest levels of polybrominated diphenyl ethers in a California family of four: A case study. *Environmental Health Perspectives*. 114: 1581–1584.

105. Zuurbier, M., Leijs, M., Schoeters, G., Tusscher, G. T., and Koppe, J. G. 2006. Children's exposure to polybrominated diphenyl ethers. *Acta Paediatrica*. 95: 65–70.

106. She, J. W., Holden, A., Sharp, M., Tanner, M., Williams-Derry, C., and Hooper, K. 2007. Polybrominated diphenyl ethers (PBDEs) and polychlorinated biphenyls (PCBs) in breast milk from the Pacific Northwest. *Chemosphere*. 67: S307–S317.

107. Song, W. L., Ford, J. C., Li, A., Mills, W. J., Buckley, D. R., and Rockne, K. J. 2004. Polybrominated diphenyl ethers in the sediments of the Great Lakes. 1. Lake Superior. *Environmental Science & Technology*. 38: 3286–3293.

108. Song, W. L., Li, A., Ford, J. C., Sturchio, N. C., Rockne, K. J., Buckley, D. R., and Mills, W. J. 2005. Polybrominated diphenyl ethers in the sediments of the great lakes. 2. Lakes Michigan and Huron. *Environmental Science & Technology*. 39: 3474–3479.

109. Song, W. L., Ford, J. C., Li, A., Sturchio, N. C., Rockne, K. J., Buckley, D. R., and Mills, W. J. 2005. Polybrominated diphenyl ethers in the sediments of the Great Lakes. 3. Lakes Ontario and Erie. *Environmental Science & Technology*. 39: 5600–5605.

110. Zhu, L. Y. and Hites, R. A. 2005. Brominated flame retardants in sediment cores from lakes Michigan and Erie. *Environmental Science & Technology*. 39: 3488–3494.

111. Kimbrough, K. L., Johnson, W. E., Lauenstein, G. G., Christensen, J. D., and Apeti, D. A. 2009. *Mussel Watch Program. Assessment of Polybrominated Diphenyl Ethers (PBDEs) in Sediment and Bivalves of the U.S. Coastal Zone*. http://ccma.nos.noaa.gov/about/coast/nsandt/pdf/PBDEreport/PBDEreport.pdf (accessed May 2010).

112. Hites, R. A. 2006. Brominated flame retardants in the Great Lakes. In *The Handbook of Environmental Chemistry*; Vol. 5N/2006. Springer, Berlin/Heidelberg, Germany, pp. 355–390.

113. Toms, L. M. L., Mortimer, M., Symons, R. K., Paepke, O., and Mueller, J. F. 2008. Polybrominated diphenyl ethers (PBDEs) in sediment by salinity and land-use type from Australia. *Environment International*. 34: 58–66.

114. Mai, B. X., Chen, S. J., Luo, X. J., Chen, L. G., Yang, Q. S., Sheng, G. Y., Peng, P. A., Fu, J. M., and Zeng, E. Y. 2005. Distribution of polybrominated diphenyl ethers in sediments of the Pearl River Delta and adjacent South China Sea. *Environmental Science & Technology*. 39: 3521–3527.

115. La-Guardia, M. J., Hale, R. C., and Harvey, E. 2006. Detailed polybrominated diphenyl ether (PBDE) congener composition of the widely used penta-, octa-, and deca-PBDE technical flame-retardant mixtures. *Environmental Science & Technology*. 40: 6247–6254.

116. Chen, S. J., Luo, X. J., Lin, Z., Luo, Y., Li, K. C., Peng, X. Z., Mai, B. X., Ran, Y., and Zeng, E. Y. 2007. Time trends of polybrominated diphenyl ethers in sediment cores from the Pearl River Estuary, South China. *Environmental Science & Technology*. 41: 5595–5600.

117. Luo, Q., Cai, Z. W., and Wong, M. H. 2007. Polybrominated diphenyl ethers in fish and sediment from river polluted by electronic waste. *Science of the Total Environment.* 383: 115–127.
118. Yun, S. H., Addink, R., McCabe, J. M., Ostaszewski, A., Mackenzie-Taylor, D., Taylor, A. B., and Kannan, K. 2008. Polybrominated diphenyl ethers and polybrominated biphenyls in sediment and flood-plain soils of the Saginaw River watershed, Michigan, USA. *Archives of Environmental Contamination and Toxicology.* 55: 1–10.
119. Wang, P., Zhang, Q. H., Wang, Y. W., Wang, T., Li, X. M., Li, Y. M., Ding, L., and Jiang, G. B. 2009. Altitude dependence of polychlorinated biphenyls (PCBs) and polybrominated diphenyl ethers (PBDEs) in surface soil from Tibetan Plateau, China. *Chemosphere.* 76: 1498–1504.
120. Luo, Y., Luo, X. J., Lin, Z., Chen, S. J., Liu, J., Mai, B. X., and Yang, Z. Y. 2009. Polybrominated diphenyl ethers in road and farmland soils from an e-waste recycling region in Southern China: Concentrations, source profiles, and potential dispersion and deposition. *Science of the Total Environment.* 407: 1105–1113.
121. Harrad, S. and Hunter, S. 2006. Concentrations of polybrominated diphenyl ethers in air and soil on a rural-urban transect across a major UK conurbation. *Environmental Science & Technology.* 40: 4548–4553.
122. Sellström, U., De Wit, C. A., Lundgren, N., and Tysklind, M. 2005. Effect of sewage-sludge application on concentrations of higher-brominated diphenyl ethers in soils and earthworms. *Environmental Science & Technology.* 39: 9064–9070.
123. Hassanin, A., Breivik, K., Meijer, S. N., Steinnes, E., Thomas, G. O., and Jones, K. C. 2004. PBDEs in European background soils: Levels and factors controlling their distribution. *Environmental Science & Technology.* 38: 738–745.
124. Harrad, S., Ibarra, C., Abdallah, M. A., and Boon, R. 2008. Concentrations of brominated flame retardants in dust from United Kingdom cars, homes, and offices: Causes of variability and implications for human exposure. *Environment International.* 34: 1170–1175.
125. Batterman, S. A., Chernyak, S., Jia, C. R., Godwin, C., and Charles, S. 2009. Concentrations and emissions of polybrominated diphenyl ethers from US houses and garages. *Environmental Science & Technology.* 43: 2693–2700.
126. Takigami, H., Suzuki, G., Hirai, Y., and Sakai, S. 2008. Transfer of brominated flame retardants from components into dust inside television cabinets. *Chemosphere.* 73: 161–169.
127. Huang, Y. M., Chen, L. G., Peng, X. C., Xu, Z. C., and Ye, Z. X. 2010. PBDEs in indoor dust in South-Central China: Characteristics and implications. *Chemosphere.* 78: 169–174.
128. Shi, T., Chen, S. J., Luo, X. J., Zhang, X. L., Tang, C. M., Luo, Y., Ma, Y. J., Wu, J. P., Peng, X. Z., and Mai, B. X. 2009. Occurrence of brominated flame retardants other than polybrominated diphenyl ethers in environmental and biota samples from southern China. *Chemosphere.* 74: 910–916.
129. Monirith, I., Ueno, D., Takahashi, S., Nakata, H., Sudaryanto, A., Subramanian, A., Karuppiah, S. et al. 2003. Asia-Pacific mussel watch: Monitoring contamination of persistent organochlorine compounds in coastal waters of Asian countries. *Marine Pollution Bulletin.* 46: 281–300.
130. Ramu, K., Kajiwara, N., Sudaryanto, A., Isobe, T., Takahashi, S., Subramanian, A., Ueno, D. et al. 2007. Asian mussel watch program: Contamination status of polybrominated diphenyl ethers and organochlorines in coastal waters of Asian countries. *Environmental Science & Technology.* 41: 4580–4586.
131. Oros, D. R., Hoover, D., Rodigari, F., Crane, D., and Sericano, J. 2005. Levels and distribution of polybrominated diphenyl ethers in water, surface sediments, and bivalves from the San Francisco Estuary. *Environmental Science & Technology.* 39: 33–41.
132. deBruyn, A. M. H., Meloche, L. M., and Lowe, C. J. 2009. Patterns of bioaccumulation of polybrominated diphenyl ether and polychlorinated biphenyl congeners in marine mussels. *Environmental Science & Technology.* 43: 3700–3704.
133. Tittlemier, S. A., Forsyth, D., Breakell, K., Verigin, V., Ryan, J. J., and Hayward, S. 2004. Polybrominated diphenyl ethers in retail fish and shellfish samples purchased from Canadian markets. *Journal of Agricultural and Food Chemistry.* 52: 7740–7745.
134. Law, K., Halldorson, T., Danell, R., Stern, G., Gewurtz, S., Alaee, M., Marvin, C., Whittle, M., and Tomy, G. 2006. Bioaccumulation and trophic transfer of some brominated flame retardants in a Lake Winnipeg (Canada) food web. *Environmental Toxicology & Chemistry.* 25: 2177–2186.
135. Wang, Z., Ma, X., Lin, Z., Na, G., and Yao, Z. 2009. Congener specific distributions of polybrominated diphenyl ethers (PBDEs) in sediment and mussel (*Mytilus edulis*) of the Bo Sea, China. *Chemosphere.* 74: 896–901.
136. Pan, J., Yang, Y.-L., Xu, Q., Chen, D.-Z., and Xi, D.-L. 2007. PCBs, PCNs and PBDEs in sediments and mussels from Qingdao coastal sea in the frame of current circulations and influence of sewage sludge. *Chemosphere.* 66: 1971–1982.

137. Liu, Y., Zheng, G. J., Yu, H., Martin, M., Richardson, B. J., Lam, M. H. W., and Lam, P. K. S. 2005. Polybrominated diphenyl ethers (PBDEs) in sediments and mussel tissues from Hong Kong marine waters. *Marine Pollution Bulletin*. 50: 1173–1184.

138. Guo, J.-Y., Zeng, E. Y., Mai, B.-X., Luo, X.-J., and Wu, F.-C. 2007. Polybrominated diphenyl ethers in seafood products of South China. *Journal of Agricultural and Food Chemistry*. 55: 9152–9158.

139. Binelli, A., Guzzella, L., and Roscioli, C. 2008. Levels and congener profiles of polybrominated diphenyl ethers (PBDEs) in Zebra mussels (D-polymorpha) from Lake Maggiore (Italy). *Environmental Pollution*. 153: 610–617.

140. Ueno, D., Kajiwara, N., Tanaka, H., Subramanian, A., Fillmann, G., Lam, P. K. S., Zheng, G. J. et al. 2004. Global pollution monitoring of polybrominated diphenyl ethers using Skipjack Tuna as a bioindicator. *Environmental Science & Technology*. 38: 2312–2316.

141. Hites, R. A., Foran, J. A., Schwager, S. J., Knuth, B. A., Hamilton, M. C., and Carpenter, D. O. 2004. Global assessment of polybrominated diphenyl ethers in farmed and wild salmon. *Environmental Science & Technology*. 38: 4945–4949.

142. Zennegg, M., Kohler, M., Gerecke, A. C., and Schmid, P. 2003. Polybrominated diphenyl ethers in whitefish from Swiss lakes and farmed rainbow trout. *Chemosphere*. 51: 545–553.

143. Shaw, S. D., Berger, M. L., Brenner, D., Carpenter, D. O., Tao, L., Hong, C. S., and Kannan, K. 2008. Polybrominated diphenyl ethers (PBDEs) in farmed and wild salmon marketed in the Northeastern United States. *Chemosphere*. 71: 1422–1431.

144. Meng, X.-Z., Zeng, E. Y., Yu, L.-P., Mai, B.-X., Luo, X.-J., and Ran, Y. 2007. Persistent halogenated hydrocarbons in consumer fish of China: Regional and global implications for human exposure. *Environmental Science & Technology*. 41: 1821–1827.

145. Montory, M. and Barra, R. 2006. Preliminary data on polybrominated diphenyl ethers (PBDEs) in farmed fish tissues (*Salmo salar*) and fish feed in Southern Chile. *Chemosphere*. 63: 1252–1260.

146. Minh, N. H., Minh, T. B., Kajiwara, N., Kunisue, K., Iiwata, H., Viet, P. H., Tu, N. P. C., Tuyen, B. C., and Tanabe, S. 2006. Contamination by polybrominated diphenyl ethers and persistent organochlorines in catfish and feed from Mekong River Delta, Vietnam. *Environmental Toxicology & Chemistry*. 25: 2700–2709.

147. van Beusekom, O. C., Eljarrat, E., Barceló, D., and Koelmans, A. A. 2006. Dynamic modeling of food-chain accumulation of brominated flame retardants in fish from the Ebro River Basin, Spain. *Environmental Toxicology & Chemistry*. 25: 2553–2560.

148. Guo, Y., Yu, H.-Y., Zhang, B.-Z., and Zeng, E. Y. 2009. Persistent halogenated hydrocarbons in fish feeds manufactured in South China. *Journal of Agricultural and Food Chemistry*. 57: 3674–3680.

149. Pena-Abaurrea, M., Weijs, L., Ramos, L., Borghesi, N., Corsolini, S., Neels, H., Blust, R., and Covaci, A. 2009. Anthropogenic and naturally-produced organobrominated compounds in bluefin tuna from the Mediterranean Sea. *Chemosphere*. 76: 1477–1482.

150. Luo, Q., Wong, M. H., and Cai, Z. W. 2007. Determination of polybrominated diphenyl ethers in freshwater fishes from a river polluted by e-wastes. *Talanta*. 72: 1644–1649.

151. Wu, J.-P., Luo, X.-J., Zhang, Y., Luo, Y., Chen, S.-J., Mai, B.-X., and Yang, Z.-Y. 2008. Bioaccumulation of polybrominated diphenyl ethers (PBDEs) and polychlorinated biphenyls (PCBs) in wild aquatic species from an electronic waste (e-waste) recycling site in South China. *Environment International*. 34: 1109–1113.

152. Dodder, N. G., Strandberg, B., and Hites, R. A. 2000. Concentrations and spatial variation of polybrominated diphenyl ethers in fish and air from the Northeastern United States. *Organohalogen Compounds*. 47: 69–72.

153. Dodder, N. G., Strandberg, B., and Hites, R. A. 2002. Concentrations and spatial variations of polybrominated diphenyl ethers and several organochlorine compounds in fishes from the Northeastern United States. *Environmental Science & Technology*. 36: 146–151.

154. Vorkamp, K., Christensen, J. H., Glasius, M., and Riget, F. F. 2004. Persistent halogenated compounds in black guillemots (*Cepphus grylle*) from Greenland—Levels, compound patterns and spatial trends. *Marine Pollution Bulletin*. 48: 111–121.

155. Polder, A., Venter, B., Skaare, J. U., and Bouwman, H. 2008. Polybrominated diphenyl ethers and HBCD in bird eggs of South Africa. *Chemosphere*. 73: 148–154.

156. She, J. W., Holden, A., Adelsbach, T. L., Tanner, M., Schwarzbach, S. E., Yee, J. L., and Hooper, K. 2008. Concentrations and time trends of polybrominated diphenyl ethers (PBDEs) and polychlorinated biphenyls (PCBs) in aquatic bird eggs from San Francisco Bay, CA 2000–2003. *Chemosphere*. 73: S201–S209.

157. Norstrom, R. J., Simon, M., Moisey, J., Wakeford, B., and Weseloh, D. V. C. 2002. Geographical distribution (2000) and temporal trends (1981–2000) of brominated diphenyl ethers in Great Lakes Herring Gull eggs. *Environmental Science & Technology*. 36: 4783–4789.

158. Lam, J. C. W., Kajiwara, N., Ramu, K., Tanabe, S., and Lam, P. K. S. 2007. Assessment of polybrominated diphenyl ethers in eggs of waterbirds from South China. *Environmental Pollution*. 148: 258–267.

159. Gao, F., Luo, X.-J., Yang, Z.-F., Wang, X.-M., and Mai, B.-X. 2009. Brominated flame retardants, polychlorinated diphenyls, and organochlorine pesticides in bird eggs from the Yellow River Delta, North China. *Environmental Science & Technology*. 43: 6956–6962.

160. Chen, D., Mai, B., Song, J., Sun, Q., Luo, Y., Luo, X., Zeng, E. Y., and Hale, R. C. 2007. Polybrominated diphenyl ethers in birds of prey from Northern China. *Environmental Science & Technology*. 41: 1828–1833.

161. Chen, D., La Guardia, M. J., Harvey, E., Amaral, M., Wohlfort, K., and Hale, R. C. 2008. Polybrominated diphenyl ethers in Peregrine Falcon (*Falco peregrinus*) eggs from the Northeastern US. *Environmental Science & Technology*. 42: 7594–7600.

162. Park, J. S., Holden, A., Chu, V., Kim, M., Rhee, A., Patel, P., Shi, Y. T. et al. 2009. Time-trends and congener profiles of PBDEs and PCBs in California Peregrine Falcons (*Falco peregrinus*). *Environmental Science & Technology*. 43: 8744–8751.

163. Jaspers, V., Covaci, A., Maervoet, J., Dauwe, T., Voorspoels, S., Schepens, P., and Eens, M. 2005. Brominated flame retardants and organochlorine pollutants in eggs of little owls (*Athene noctua*) from Belgium. *Environmental Pollution*. 136: 81–88.

164. Herzke, D., Nygard, T., Berger, U., Huber, S., and Rov, N. 2009. Perfluorinated and other persistent halogenated organic compounds in European shag (*Phalacrocorax aristotelis*) and common eider (*Somateria mollissima*) from Norway: A suburban to remote pollutant gradient. *Science of the Total Environment*. 408: 340–348.

165. Bustnes, J. O., Yoccoz, N. G., Bangjord, G., Polder, A., and Skaare, J. U. 2007. Temporal trends (1986–2004) of organochlorines and brominated flame retardants in tawny owl eggs from northern Europe. *Environmental Science & Technology*. 41: 8491–8497.

166. Murvoll, K. M., Skaare, J. U., Anderssen, E., and Jenssen, B. M. 2006. Exposure and effects of persistent organic pollutants in European shag (*Phalacrocorax aristotelis*) hatchlings from the coast of Norway. *Environmental Toxicology & Chemistry*. 25: 190–198.

167. Van den Steen, E., Pinxten, R., Jaspers, V. L. B., Covaci, A., Barba, E., Carere, C., Cichon, M. et al. 2009. Brominated flame retardants and organochlorines in the European environment using great tit eggs as a biomonitoring tool. *Environment International*. 35: 310–317.

168. Muir, D. C. G., Backus, S., Derocher, A. E., Dietz, R., Evans, T. J., Gabrielsen, G. W., Nagy, J. et al. 2006. Brominated flame retardants in polar bears (*Ursus maritimus*) from Alaska, the Canadian Arctic, East Greenland, and Svalbard. *Environmental Science & Technology*. 40: 449–455.

169. Kannan, K., Yun, S. H., and Evans, T. J. 2005. Chlorinated, brominated, and perfluorinated contaminants in livers of polar bears from Alaska. *Environmental Science & Technology*. 39: 9057–9063.

170. Bentzen, T. W., Muir, D. C. G., Amstrup, S. C., and O'Hara, T. M. 2008. Organohalogen concentrations in blood and adipose tissue of Southern Beaufort Sea polar bears. *Science of the Total Environment*. 406: 352–367.

171. Dietz, R., Rigét, F. F., Sonne, C., Letcher, R. J., Backus, S., Born, E. W., Kirkegaard, M., and Muir, D. C. G. 2007. Age and seasonal variability of polybrominated diphenyl ethers in free-ranging East Greenland polar bears (*Ursus maritimus*). *Environmental Pollution*. 146: 166–173.

172. Verreault, J., Gabrielsen, G. V., Chu, S. G., Muir, D. C. G., Andersen, M., Hamaed, A., and Letcher, R. J. 2005. Flame retardants and methoxylated and hydroxylated polybrominated diphenyl ethers in two Norwegian Arctic top predators: Glaucous gulls and polar bears. *Environmental Science & Technology*. 39: 6021–6028.

173. Kajiwara, N., Ueno, D., Takahashi, A., Baba, N., and Tanabe, S. 2004. Polybrominated diphenyl ethers and organochlorines in archived northern fur seal samples from the Pacific coast of Japan, 1972–1998. *Environmental Science & Technology*. 38: 3804–3809.

174. Meng, X.-Z., Blasius, M. E., Gossett, R. W., and Maruya, K. A. 2009. Polybrominated diphenyl ethers in pinnipeds stranded along the southern California coast. *Environmental Pollution*. 157: 2731–2736.

175. Kajiwara, N., Kamikawa, S., Ramu, K., Ueno, D., Yamada, T. K., Subramanian, A., Lam, P. K. S. et al. 2006. Geographical distribution of polybrominated diphenyl ethers (PBDEs) and organochlorines in small cetaceans from Asia waters. *Chemosphere*. 64: 287–295.

176. Voorspoels, S., Covaci, A., Lepom, P., Escutenaire, S., and Schepens, P. 2006. Remarkable findings concerning PBDEs in the terrestrial top-predator red fox (*Vulpes vulpes*). *Environmental Science & Technology*. 40: 2937–2943.

177. Dye, J. A., Venier, M., Zhu, L., Ward, C. R., Hites, R. A., and Birnbaum, L. S. 2007. Elevated PBDE levels in pet cats: Sentinels for humans? *Environmental Science & Technology*. 41: 6350–6356.

178. Liang, S. X., Zhao, Q., Qin, Z. F., Zhao, X. R., Yang, Z. Z., and Xu, X. B. 2008. Levels and distribution of polybrominated diphenyl ethers in various tissues of foraging hens from an electronic waste recycling area in South China. *Environmental Toxicology & Chemistry*. 27: 1279–1283.

179. Stapleton, H. M., Dodder, N. G., Kucklick, J. R., Reddy, C. M., Schantz, M. M., Becker, P. R., Gulland, F., Porter, B. J., and Wise, S. A. 2006. Determination of HBCD, PBDEs and MeO-BDEs in California sea lions (*Zalophus californianus*) stranded between 1993 and 2003. *Marine Pollution Bulletin*. 52: 522–531.

180. Kannan, K., Perrotta, E., Thomas, N. J., and Aldous, K. M. 2007. A comparative analysis of polybrominated diphenyl ethers and polychlorinated biphenyls in southern sea otters that died of infectious diseases and noninfectious causes. *Archives of Environmental Contamination and Toxicology*. 53: 293–302.

181. Shaw, S. D., Brenner, D., Berge, M. L., Fang, F., Hong, C. S., Addink, R., and Hilker, D. 2008. Bioaccumulation of polybrominated diphenyl ethers in harbor seals from the northwest Atlantic. *Chemosphere*. 73: 1773–1780.

182. Tuerk, K. J. S., Kucklick, J. R., Becker, P. R., Stapleton, H. M., and Baker, J. E. 2005. Persistent organic pollutants in two dolphin species with focus on toxaphene and polybrominated diphenyl ethers. *Environmental Science & Technology*. 39: 692–698.

183. Lebeuf, M., Gouteux, B., Measures, L., and Trottier, S. 2004. Levels and temporal trends (1988–1999) of polybrominated diphenyl ethers in Beluga whales (*Delphinapterus leucas*) from the St. Lawrence estuary, Canada. *Environmental Science & Technology*. 38: 2971–2977.

184. Dorneles, P. R., Lailson-Brito, J., Dirtu, A. C., Weijs, L., Azevedo, A. F., Torres, J. P. M., Malm, O. et al. 2010. Anthropogenic and naturally-produced organobrominated compounds in marine mammals from Brazil. *Environment International*. 36: 60–67.

185. Law, R. J., Barry, J., Bersuder, P., Barber, J. L., Deaville, R., Reid, R. J., and Jepson, P. D. 2010. Levels and trends of brominated diphenyl ethers in blubber of harbor porpoises (*Phocoena phocoena*) from the U.K., 1992–2008. *Environmental Science & Technology*. 44: 4447–4451.

186. UNEP. 2006. Call for global action on E-waste. United Nations Environment Program, Nairobi.

187. Huisman, J. and Magalini, F. 2007. Where are WEEE now? Lessons from WEEE: Will EPR work for the US? In *Proceeding of the 2007 IEEE International Symposium on Electronic & the Environment*, Orlando, pp. 149–154.

188. Robinson, B. H. 2009. E-waste: An assessment of global production and environmental impacts. *Science of the Total Environment*. 408: 183–191.

189. Yu, J.-L., Williams, E., Ju, M.-T., and Yang, Y. 2010. Forecasting global generation of obsolete personal computers. *Environmental Science & Technology*. 44: 3232–3237.

190. Ladou, J. and Lovegrove, S. 2008. Export of electronics equipment waste. *International Journal of Occupational Medicine and Environmental Health*. 14: 1–10.

191. The Basel Action Network (BAN) and Silicon Valley Toxic Coalition (SVTC). 2002. *Exporting Harm: The High-Tech Trashing of Asia*. http://www.ban.org/E-waste/technotrashfinalcomp.pdf (accessed May 2010).

192. SwedWatch. 2009. *Out of Control: E-Waste Trade Flows from the EU to Developing Countries*. http://makeitfair.org/the-facts/reports (accessed May 2010).

193. Liu, X., Tanaka, M., and Matsui, Y. 2006. Generation amount prediction and material flow analysis of electronic waste: A case study in Beijing, China. *Waste Management and Research*. 24: 434–445.

194. Widmer, R., Oswald-Krapf, H., Sinha-Khetriwal, D., Schnellmann, M., and Böni, H. 2005. Global perspectives on e-waste. *Environmental Impact Assessment Review*. 25: 436–458.

195. Morf, L. S., Tremp, J., Gloor, R., Huber, Y., Stengele, M., and Zennegg, M. 2005. Brominated flame retardants in waste electrical and electronic equipment: Substance flows in a recycling plant. *Environmental Science & Technology*. 39: 8691–8699.

196. Ma, J., Addink, R., Yun, S. H., Cheng, J. P., Wang, W. H., and Kannan, K. 2009. Polybrominated dibenzo-p-dioxins/dibenzofurans and polybrominated diphenyl ethers in soil, vegetation, workshop-floor dust, and electronic shredder residue from an electronic waste recycling facility and in soils from a chemical industrial complex in eastern China. *Environmental Science & Technology*. 43: 7350–7356.

197. Darnerud, P. O., Eriksen, G. S., Jóhannesson, T., Larsen, P. B., and Viluksela, M. 2001. Polybrominated diphenyl ethers: Occurrence, dietary exposure, and toxicology. *Environmental Health Perspectives*. 109 (Suppl 1): 49–68.

198. Guan, Y.-F., Wang, J.-Z., Ni, H.-G., Luo, X.-J., Mai, B.-X., and Zeng, E. Y. 2007. Riverine inputs of polybrominated diphenyl ethers from the Pearl River Delta (China) to the coastal ocean. *Environmental Science & Technology*. 41: 6007–6013.

199. Hicks, C., Dietmar, R., and Eugster, M. 2005. The recycling and disposal of electrical and electronic waste in China—Legislative and market responses. *Environmental Impact Assessment Review*. 25: 459–471.

200. Wang, Y. W., Jiang, G. B., Lam, P. K. S., and Li, A. 2007. Polybrominated diphenyl ether in the East Asian environment: A critical review. *Environment International*. 33: 963–973.
201. de Wit, C. A., Herzke, D., and Vorkamp, K. 2010. Brominated flame retardants in the Arctic environment—Trends and new candidates. *Science of the Total Environment*. 15: 2885–2918.
202. Law, R. J., Alaee, M., Allchin, C. R., Boon, J. P., Lebeuf, M., Lepom, P., and Stern, G. A. 2003. Levels and trends of polybrominated diphenyl ethers and other brominated flame retardants in wildlife. *Environment International*. 29: 757–770.
203. Braune, B. M., Outridge, P. M., Fisk, A. T., Muir, D. C. G., Helm, P. A., Hobbs, K., Hoekstra, P. F. et al. 2005. Persistent organic pollutants and mercury in marine biota of the Canadian Arctic: An overview of spatial and temporal trends. *Science of the Total Environment*. 351–352: 4–56.
204. Yogui, G. T. and Sericano, J. L. 2009. Polybrominated diphenyl ether flame retardants in the US marine environment: A review. *Environment International*. 35: 655–666.
205. Environment Canada. 2010. *State of the Science Report on the Bioaccumulation and Transformation of Decabromodiphenyl Ether*. http://www.ec.gc.ca/CEPARegistry/subs_list/decaBDE/SR.pdf (accessed May 2010).
206. Ni, H.-G. and Zeng, E. Y. 2009. Law enforcement and global collaboration are the keys to containing E-waste Tsunami in China. *Environmental Science & Technology*. 43: 3991–1994.
207. Ogunseitan, O. A., Schoenung, J. M., Saphores, J.-D. M., and Shapiro, A. A. 2009. The electronics revolution: From E-wonderland to E-wasteland. *Science*. 326: 670–671.
208. Liu, X., Tanaka, M., and Matsui, Y. 2006. Electrical and electronic waste management in China: Progress and the barriers to overcome. *Waste Management and Research*. 24: 92–101.
209. Harrad, S. and Porter, L. 2007. Concentrations of polybrominated diphenyl ethers in blood serum from New Zealand. *Chemosphere*. 66: 2019–2023.
210. Sudaryanto, A., Kajiwara, N., Tsydenova, O., Iwata, H., Adibroto, T. A., Yu, H., Chung, K.-H., Subramanian, A., Prudente, M., Tana, T. S., and Tanabe, S. 2005. Global contamination of PBDEs in human milk from Asia. *Organohalogen Compounds*. 67: 1315–1318.
211. Sudaryanto, A., Kajiwara, N., Tsydenova, O. V., Isobe, T., Yu, H. X., Takahashi, S., and Tanabe, S. 2008. Levels and congener specific profiles of PBDEs in human breast milk from China: Implication on exposure sources and pathways. *Chemosphere*. 73: 1661–1668.
212. Bi, X. H., Qu, W. Y., Sheng, G. Y., Zhang, W. B., Mai, B. X., Chen, D. J., Yu, L., and Fu, J. M. 2006. Polybrominated diphenyl ethers in South China maternal and fetal blood and breast milk. *Environmental Pollution*. 144: 1024–1030.
213. Li, J., Yu, H., Zhao, Y., Zhang, G., and Wu, Y. 2008. Levels of polybrominated diphenyl ethers (PBDEs) in breast milk from Beijing, China. *Chemosphere*. 73: 182–186.
214. Zhu, L.-Y., Ma, B.-L., Li, J.-G., Wu, Y.-N., and Gong, J. 2009. Distribution of polybrominated diphenyl ethers in breast milk from North China: Implication of exposure pathways. *Chemosphere*. 74: 1429–1434.
215. Haraguchi, K., Koizumi, A., Inoue, K., Harada, K. H., Hitomi, T., Minata, M., Tanabe, M. et al., 2009. Levels and regional trends of persistent organochlorines and polybrominated diphenyl ethers in Asian breast milk demonstrate POPs signatures unique to individual countries. *Environment International*. 35: 1072–1079.
216. Zhu, L. Y., Ma, B. L., and Hites, R. A. 2009. Brominated flame retardants in serum from the general population in Northern China. *Environmental Science & Technology*. 43: 6963–6968.
217. Chen, C., Chen, J. W., Zhao, H. X., Xie, Q., Yin, Z. Q., and Ge, L. K. 2010. Levels and patterns of polybrominated diphenyl ethers in children's plasma from Dalian, China. *Environment International*. 36: 163–167.
218. Hedley, A. J., Hui, L. L., Kypke, K., Malisch, R., van Leeuwen, F. X. R., Moy, G., Wong, T. W., and Nelson, E. A. S. 2010. Residues of persistent organic pollutants (POPs) in human milk in Hong Kong. *Chemosphere*. 79: 259–265.
219. Eslami, B., Koizumi, A., Ohta, S., Inoue, K., Aozasa, O., Harada, K., Yoshinaga, T. et al. 2006. Large-scale evaluation of the current level of polybrominated diphenyl ethers (PBDEs) in breast milk from 13 regions of Japan. *Chemosphere*. 63: 554–561.
220. Inoue, K., Harada, K., Takenaka, K., Uehara, S., Kono, M., Shimizu, T., Takasuga, T., Senthilkumar, K., Yamashita, F., and Koizumi, A. 2006. Levels and concentration ratios of polychlorinated biphenyls and polybrominated diphenyl ethers in serum and breast milk in Japanese mothers. *Environmental Health Perspectives*. 114: 1179–1185.
221. Kawashiro, Y., Fukata, H., Omori-Inoue, M., Kubonoya, K., Tomomi, J., Takigami, H., Sakai, C.-I., and Mori, C. 2008. Perinatal exposure to brominated flame retardants and chlorinated biphenyls in Japan. *Endocrine Journal*. 5: 1071–1084.

222. Kumsue, T., Takayanagi, N., Isobe, T., Takahashi, S., Nose, M., Yamada, T., Komori, H., Arita, N., Ueda, N., and Tanabe, S. 2007. Polybrominated diphenyl ethers and persistent organochlorines in Japanese human adipose tissues. *Environment International*. 33: 1048–1056.

223. Kim, B. H., Ikonomou, M. G., Lee, S. J., Kim, H. S., and Chang, Y. S. 2005. Concentrations of polybrominated diphenyl ethers, polychlorinated dibenzo-p-dioxins and dibenzofurans, and polychlorinated biphenyls in human blood samples from Korea. *Science of the Total Environment*. 336: 45–56.

224. Lee, S. J., Ikonomou, M. G., Park, H., Baek, S. Y., and Chang, Y. S. 2007. Polybrominated diphenyl ethers in blood from Korean incinerator workers and general population. *Chemosphere*. 67: 489–497.

225. Sudaryanto, A., Kajiwara, N., Takahashi, S., Muawanah, and Tanabe, S. 2008. Geographical distribution and accumulation features of PBDEs in human breast milk from Indonesia. *Environmental Pollution*. 151: 130–138.

226. Li, Q. Q., Loganath, A., Chong, Y. S., and Obbard, J. P. 2005. Determination and occurrence of polybrominated diphenyl ethers in maternal adipose tissue from inhabitants of Singapore. *Journal of Chromatography B-Analytical Technologies in the Biomedical and Life Sciences*. 819: 253–257.

227. Covaci, A., de Boer, J., Ryan, J. J., Voorspoels, S., and Schepens, P. 2002. Distribution of organobrominated and organochlorinated contaminants in Belgian human adipose tissue. *Environmental Research*. 88: 210–218.

228. Covaci, A., Voorspoels, S., Roosens, L., Jacobs, W., Blust, R., and Neels, H. 2008. Polybrominated diphenyl ethers (PBDEs) and polychlorinated biphenyls (PCBs) in human liver and adipose tissue samples from Belgium. *Chemosphere*. 73: 170–175.

229. Covaci, A. and Voorspoels, S. 2005. Optimization of the determination of polybrominated diphenyl ethers in human serum using solid-phase extraction and gas chromatography-electron capture negative ionization mass spectrometry. *Journal of Chromatography B—Analytical Technologies in the Biomedical and Life Sciences*. 827: 216–223.

230. Colles, A., Koppen, G., Hanot, V., Nelen, V., Dewolf, M. C., Noël, E., Malisch, R. et al. 2008. Fourth WHO-coordinated survey of human milk for persistent organic pollutants (POPs): Belgian results. *Chemosphere*. 73: 907–914.

231. Kazda, R., Hajšlová, J., Poustka, J., and Čajka, T. 2004. Determination of polybrominated diphenyl ethers in human milk samples in the Czech Republic comparative study of negative chemical ionisation mass spectrometry and time-of-flight high-resolution mass spectrometry. *Analytica Chimica Acta*. 520: 237–243.

232. Fängström, B., Hovander, L., Bignert, A., Athanassiadis, I., Linderholm, L., Grandjean, P., Weihe, P., and Bergmant, Å. 2005. Concentrations of polybrominated diphenyl ethers, polychlorinated biphenyls, and polychlorobiphenylols in serum from pregnant faroese women and their children 7 years later. *Environmental Science & Technology*. 39: 9457–9463.

233. Leijs, M. M., van Teunenbroek, J., Olie, K., Koppe, J. G., ten Tusscher, G. W., van Aalderen, W. M. C., and de Voogt, P. 2008. Assessment of current serum levels of PCDD/Fs, dl-PCBs and PBDEs in a Dutch cohort with known perinatal PCDD/F exposure. *Chemosphere*. 73: 176–181.

234. Antignac, J. P., Cariou, R., Zalko, D., Berrebi, A., Cravedi, J. P., Maume, D., Marchand, P. et al. 2009. Exposure assessment of French women and their newborn to brominated flame retardants: Determination of tri- to deca- polybromodiphenylethers (PBDE) in maternal adipose tissue, serum, breast milk and cord serum. *Environmental Pollution*. 157: 164–173.

235. Fürst, P. 2006. Dioxins, polychlorinated biphenyls and other organohalogen compounds in human milk—Levels, correlations, trends and exposure through breastfeeding. *Molecular Nutrition and Food Research*. 50: 922–933.

236. Raab, U., Preiss, U., Albrecht, M., Shahin, N., Parlar, H., and Fromme, H. 2008. Concentrations of polybrominated diphenyl ethers, organochlorine compounds and nitro musks in mother's milk from Germany (Bavaria). *Chemosphere*. 72: 87–94.

237. Ingelido, A. M., Ballard, T., Dellatte, E., di Domenico, A., Ferri, F., Fulgenzi, A. R., Herrmann, T. et al. 2007. Polychlorinated biphenyls (PCBs) and polybrominated diphenyl ethers (PBDEs) in milk from Italian women living in Rome and Venice. *Chemosphere*. 67: S301–S306.

238. Meijer, L., Weiss, J., Van Velzen, M., Brouwer, A., Bergman, A., and Sauerf, P. J. J. 2008. Serum concentrations of neutral and phenolic organohalogens in pregnant women and some of their infants in the Netherlands. *Environmental Science & Technology*. 42: 3428–3433.

239. Meneses, M., Wingfors, H., Schuhmacher, M., Domingo, J. L., Lindström, G., and Bavel, B. 1999. Polybrominated diphenyl ethers detected in human adipose tissue from Spain. *Chemosphere*. 39: 2271–2278.

240. Polder, A., Thomsen, C., Lindstrom, G., Loken, K. B., and Skaare, J. U. 2008. Levels and temporal trends of chlorinated pesticides, polychlorinated biphenyls and brominated flame retardants in individual human breast milk samples from Northern and Southern Norway. *Chemosphere*. 73: 14–23.

241. Thomsen, C., Stigum, H., Froshaug, M., Broadwell, S. L., Becher, G., and Eggesbo, M. 2010. Determinants of brominated flame retardants in breast milk from a large scale Norwegian study. *Environment International*. 36: 68–74.

242. Jaraczewska, K., Lulek, J., Covaci, A., Voorspoels, S., Kaluba-Skotarczak, A., Drews, K., and Schepens, P. 2006. Distribution of polychlorinated biphenyls, organochlorine pesticides and polybrominated diphenyl ethers in human umbilical cord serum, maternal serum and milk from Wielkopolska region, Poland. *Science of the Total Environment*. 372: 20–31.

243. Polder, A., Gabrielsen, G. W., Odland, J. Ø., Savinova, T. N., Tkachev, A., Løken, K. B., and Skaare, J. U. 2008. Spatial and temporal changes of chlorinated pesticides, PCBs, dioxins (PCDDs/PCDFs) and brominated flame retardants in human breast milk from Northern Russia. *Science of the Total Environment*. 391: 41–54.

244. Tsydenova, O. V., Sudaryanto, A., Kajiwara, N., Kunisue, T., Batoev, V. B., and Tanabe, S. 2007. Organohalogen compounds in human breast milk from Republic of Buryatia, Russia. *Environmental Pollution*. 146: 225–232.

245. Fernandez, M. F., Araque, P., Kiviranta, H., Molina-Molina, J. M., Rantakokko, P., Laine, O., Vartiainen, T., and Olea, N. 2007. PBDEs and PBBs in the adipose tissue of women from Spain. *Chemosphere*. 66: 377–383.

246. Schuhmacher, M., Kiviranta, H., Ruokojärvi, P., Nadal, M., and Domingo, J. L. 2009. Concentrations of PCDD/Fs, PCBs and PBDEs in breast milk of women from Catalonia, Spain: A follow-up study. *Environment International*. 35: 607–613.

247. Schuhmacher, M., Kiviranta, H., Vartiainen, T., and Domingo, J. L. 2007. Concentrations of polychlorinated biphenyls (PCBs) and polybrominated diphenyl ethers (PBDEs) in milk of women from Catalonia, Spain. *Chemosphere*. 67: S295–S300.

248. Ramos, J. J., Gomara, B., Fernandez, M. A., and Gonzalez, M. J. 2007. A simple and fast method for the simultaneous determination of polychlorinated biphenyls and polybrominated diphenyl ethers in small volumes of human serum. *Journal of Chromatography A*. 1152: 124–129.

249. Gómara, B., Herrero, L., Ramos, J. J., Mateo, J. R., Fernandez, M. A., Garcia, J. F., and González, M. J. 2007. Distribution of polybrominated diphenyl ethers in human umbilical cord serum, paternal serum, maternal serum, placentas, and breast milk from Madrid population, Spain. *Environmental Science & Technology*. 41: 6961–6968.

250. Guvenius, D. M., Aronsson, A., Ekman-Ordeberg, G., Bergman, Å., and Norén, K. 2003. Human prenatal and postnatal exposure to polybrominated diphenyl ethers, polychlorinated biphenyls, polychlorobiphenylols, and pentachlorophenol. *Environmental Health Perspectives*. 111: 1235–1241.

251. Karlsson, M., Julander, A., van Bavel, B., and Hardell, L. 2007. Levels of brominated flame retardants in blood in relation to levels in household air and dust. *Environment International*. 33: 62–69.

252. Thomas, G. O., Wilkinson, M., Hodson, S., and Jones, K. C. 2006. Organohalogen chemicals in human blood from the United Kingdom. *Environmental Pollution*. 141: 30–41.

253. Ryan, J. J., Patry, B., Mills, P., and Beaudoin, G. 2002. Recent trends in levels of brominated diphenyl ethers (BDEs) in human milk from Canada. *Organohalogen Compounds*. 58: 173–176.

254. Doucet, J., Tague, B., Arnold, D. L., Cooke, G. M., Hayward, S., and Goodyer, C. G. 2009. Persistent organic pollutant residues in human fetal liver and placenta from Greater Montreal, Quebec: A Longitudinal Study from 1998 through 2006. *Environmental Health Perspectives*. 117: 605–610.

255. Sandanger, T. M., Sinotte, M., Dumas, P., Marchand, M., Sandau, C. D., Pereg, D., Berube, S., Brisson, J., and Ayotte, P. 2007. Plasma concentrations of selected organobromine compounds and polychlorinated biphenyls in postmenopausal women of Quebec, Canada. *Environmental Health Perspectives*. 115: 1429–1434.

256. Schecter, A., Pavuk, M., Päpke, O., Ryan, J. J., Birnbaum, L., and Rosen, R. 2003. Polybrominated diphenyl ethers (PBDEs) in U.S. mothers' milk. *Environmental Health Perspectives*. 111: 1723–1729.

257. Daniels, J. L., Pan, I. J., Jones, R., Anderson, S., Patterson, D. G., Needham, L. L., and Sjödin, A. 2010. Individual characteristics associated with PBDE levels in US human milk samples. *Environmental Health Perspectives*. 118: 155–160.

258. Johnson-Restrepo, B., Addink, R., Wong, C., Arcaro, K., and Kannan, K. 2007. Polybrominated diphenyl ethers and organochlorine pesticides in human breast milk from Massachusetts, USA. *Journal of Environmental Monitoring*. 9: 1205–1212.

259. Wu, N., Herrmann, T., Paepke, O., Tickner, J., Hale, R., Harvey, E., La Guardia, M., McClean, M. D., and Webster, T. F. 2007. Human exposure to PBDEs: Associations of PBDE body burdens with food consumption and house dust concentrations. *Environmental Science & Technology*. 41: 1584–1589.

260. Mazdai, A., Dodder, N. G., Abernathy, M. P., Hites, R. A., and Bigsby, R. M. 2003. Polybrominated diphenyl ethers in maternal and fetal blood samples. *Environmental Health Perspectives*. 111: 1249–1252.

261. Qiu, X. H., Bigsby, R. M., and Hites, R. A. 2009. Hydroxylated metabolites of polybrominated diphenyl ethers in human blood samples from the United States. *Environmental Health Perspectives*. 117: 93–98.

262. Johnson-Restrepo, B., Kannan, K., Rapaport, D. P., and Rodan, B. D. 2005. Polybrominated diphenyl ethers and polychlorinated biphenyls in human adipose tissue from New York. *Environmental Science & Technology*. 39: 5177–5182.

263. Anderson, H. A., Imm, P., Knobeloch, L., Turyk, M., Mathew, J., Buelow, C., and Persky, V. 2008. Polybrominated diphenyl ethers (PBDE) in serum: Findings from a US cohort of consumers of sport-caught fish. *Chemosphere*. 73: 187–194.

264. Miller, M. F., Chernyak, S. M., Batterman, S., and Loch-Caruso, R. 2009. Polybrominated diphenyl ethers in human gestational membranes from women in southeast Michigan. *Environmental Science & Technology*. 43: 3042–3046.

265. Sjödin, A., Jones, R. S., Focant, J., Lapeza, C., Wang, R. Y., McGahee III, E. E., Zhang, Y. et al. 2004. Retrospective time-trend study of brominated diphenyl ether and polybrominated and polychlorinated biphenyl levels in human serum from the United States. *Environmental Health and Perspectives*. 112: 654–658.

266. Baumann, B., Hijman, W., van Beuzekom, S., Hoogerbrugge, R., Houweling, D., and Zeilmaker, M. 2003. PBDEs in human milk from the Dutch 1998 monitoring program. *Organohalogen Compounds*. 61: 187–190.

267. Darnerud, P. O., Aune, M., Atuma, S., Becker, W., Bjerselius, R., Cnattingius, S., and Glynn, A. 2002. Time trend of polybrominated diphenyl ether (PBDE) levels in breast milk from Uppsala, Sweden, 1996–2000. *Organohalogen Compounds*. 58: 233–236.

268. Lind, Y., Darnerud, P. O., Atuma, S., Aune, M., Becker, W., Bjerselius, R., Cnattingius, S., and Glynn, A. 2003. Polybrominated diphenyl ethers in breast milk from Uppsala County, Sweden. *Environmental Research*. 93: 186–194.

269. Marvin, C., Williams, D., Kuntz, K., Klawunn, P., Backus, S., Kolic, T., Lucaciu, C., MacPherson, K., and Reiner, E. 2007. Temporal trends in polychlorinated dibenzo-p-dioxins and dibenzofurans, dioxin-like PCBs, and polybrominated diphenyl ethers in Niagara river suspended sediments. *Chemosphere*. 67: 1808–1815.

270. Zegers, B. N., Lewis, W. E., Booij, K., Smittenberg, R. H., Boer, W., de Boer, J., and Boon, J. P. 2003. Levels of polybrominated diphenyl ether flame retardants in sediment core from Western Europe. *Environmental Science & Technology*. 37: 3803–3807.

271. Kohler, M., Zennegg, M., Bogdal, C., Gerecke, A. C., Schmid, P., Heeb, N. V., Sturm, M., Vonmont, H., Kohler, H. P. E., and Giger, W. 2008. Temporal trends, congener patterns, and sources of octa-, nona-, and deca-bromodiphenyl ethers (PBDE) and hexabromocyclododecanes (HBCD) in Swiss lake sediments. *Environmental Science & Technology*. 42: 6378–6384.

272. Covaci, A., Gheorghe, A., Voorspoels, S., Maervoet, J., Steen Redeker, E., Blust, R., and Schepens, P. 2005. Polybrominated diphenyl ethers, polychlorinated biphenyls and organochlorine pesticides in sediment cores from the Western Scheldt river (Belgium): Analytical aspects and depth profiles. *Environment International*. 31: 367–375.

273. Binelli, A., Sarkar, S. K., Chatterjee, M., Riva, C., Parolini, M., Bhattacharya, B. D., Bhattacharya, A. K., and Satpathy, K. K. 2007. Concentration of polybrominated diphenyl ethers (PBDEs) in sediment cores of Sundarban mangrove wetland, northeastern part of Bay of Bengal (India). *Marine Pollution Bulletin*. 54: 1220–1229.

274. Guzzella, L., Roscioli, C., and Binelli, A. 2008. Contamination by polybrominated diphenyl ethers of sediments from the Lake Maggiore basin (Italy and Switzerland). *Chemosphere*. 73: 1684–1691.

275. Jin, J., Liu, W. Z., Wang, Y., and Tang, X. Y. 2008. Levels and distribution of polybrominated diphenyl ethers in plant, shellfish and sediment samples from Laizhou Bay in China. *Chemosphere*. 71: 1043–1050.

276. Minh, N. H., Isobe, T., Ueno, D., Matsumoto, K., Mine, M., Kajiwara, N., Takahashi, S., and Tanabe, S. 2007. Spatial distribution and vertical profile of polybrominated diphenyl ethers and hexabromocyclododecanes in sediment core from Tokyo Bay, Japan. *Environmental Pollution*. 148: 409–417.

277. Klamer, H. J. C., Leonards, P. E. G., Lamoree, M. H., Villerius, L. A., Akerman, J. E., and Bakker, J. F. 2005. A chemical and toxicological profile of Dutch North Sea surface sediments. *Chemosphere*. 58: 1579–1587.

278. Labandeira, A., Eljarrat, E., and Barceló, D. 2007. Congener distribution of polybrominated diphenyl ethers in feral carp (*Cyprinus carpio*) from the Llobregat River, Spain. *Environmental Pollution.* 146: 188–195.

279. Ramu, K., Isobe, T., Takahashi, S., Kim, E.-Y., Min, B.-Y., We, S.-U., and Tanabe, S. 2010. Spatial distribution of polybrominated diphenyl ethers and hexabromocyclododecanes in sediments from coastal waters of Korea. *Chemosphere.* 79: 713–719.

280. Moon, H.-B., Kannan, K., Lee, S.-J., and Choi, M. 2007. Polybrominated diphenyl ethers (PBDEs) in sediment and bivalves from Korean coastal waters. *Chemosphere.* 66: 243–251.

281. Wilford, B. H., Shoeib, M., Harner, T., Zhu, J., and Jones, K. C. 2005. Polybrominated diphenyl ethers in indoor dust in Ottawa, Canada: Implications for sources and exposure. *Environmental Science & Technology.* 39: 7027–7035.

282. Harrad, S., Ibarra, C., Diamond, M., Melymuk, L., Robson, M., Douwes, J., Roosens, L., Dirtu, A. C., and Covaci, A. 2008. Polybrominated diphenyl ethers in domestic indoor dust from Canada, New Zealand, United Kingdom and United States. *Environment International.* 34: 232–238.

283. Suzuki, G., Takigami, H., Takahashi, S., and Sakai, S. 2006. PBDEs and PBDD/Fs in house and office dust from Japan. *Organohalogen Compounds.* 68: 1843–1846.

284. Gevao, B., Al-Bahloul, M., Al-Ghadban, A. N., Al-Omair, A., Ali, L., Zafar, J., and Helaleh, M. 2006. House dust as a source of human exposure to polybrominated diphenyl ethers in Kuwait. *Chemosphere.* 64: 603–608.

285. Tan, J., Cheng, S. M., Loganath, A., Chong, Y. S., and Obbard, J. P. 2007. Polybrominated diphenyl ethers in house dust in Singapore. *Chemosphere.* 66: 985–992.

286. Fromme, H., Körner, W., Shahin, N., Wanner, A., Albrecht, M., Boehmer, S., Parlar, H., Mayer, R., Liebl, B., and Bolte, G. 2009. Human exposure to polybrominated diphenyl ethers (PBDE), as evidenced by data from a duplicate diet study, indoor air, house dust, and biomonitoring in Germany. *Environment International.* 35: 1125–1135.

287. Sjödin, A., Päpke, O., McGahee, E., Focant, J. F., Jones, R. S., Pless-Mulloli, T., Toms, L. M. L. et al. 2008. Concentration of polybrominated diphenyl ethers (PBDEs) in household dust from various countries. *Chemosphere.* 73: S131–S136.

288. Ueno, D., Isobe, T., Ramu, K., Tanabe, S., Alaee, M., Marvin, C., Inoue, K. et al. 2010. Spatial distribution of hexabromocyclododecanes (HBCDs), polybrominated diphenyl ethers (PBDEs) and organochlorines in bivalves from Japanese coastal waters. *Chemosphere.* 78: 1213–1219.

289. Hong, S. H., Munschy, C., Kannan, N., Tixier, C., Tronczynski, J., Heas-Moisan, K., and Shim, W. J. 2009. PCDD/F, PBDE, and nonylphenol contamination in a semi-enclosed bay (Masan Bay, South Korea) and a Mediterranean lagoon (Thau, France). *Chemosphere.* 77: 854–862.

290. Bayen, S., Thomas, G. O., Lee, H. K., and Obbard, J. P. 2003. Occurrence of polychlorinated biphenyls and polybrominated diphenyl ethers in green mussels (*Perna viridis*) from Singapore, Southeast Asia. *Environmental Toxicology & Chemistry.* 22: 2432–2437.

291. Christensen, J. H., Glasius, M., Pecseli, M., Platz, J., and Pritzl, G. 2002. Polybrominated diphenyl ethers (PBDEs) in marine fish and blue mussels from southern Greenland. *Chemosphere.* 47: 631–638.

292. Xiang, C.-H., Luo, X.-J., Chen, S.-J., Yu, M., Mai, B.-X., and Zeng, E. Y. 2007. Polybrominated diphenyl ethers in the biota and sediments of the Pearl River Estuary, South China. *Environmental Toxicology & Chemistry.* 26: 616–623.

293. Akutsu, K., Obana, H., Okihashi, M., Kitagawa, M., Nakazawa, H., Matsuki, Y., Makino, T., Oda, H., and Hori, S. 2001. GC/MS analysis of polybrominated diphenyl ethers in fish collected from the Inland Sea of Seto, Japan. *Chemosphere.* 44: 1325–1333.

294. Losada, S., Roach, A., Roosens, L., Santos, F. J., Galceran, M. T., Vetter, W., Neels, H., and Covaci, A. 2009. Biomagnification of anthropogenic and naturally-produced organobrominated compounds in a marine food web from Sydney Harbour, Australia. *Environment International.* 35: 1142–1149.

295. Hermanussen, S., Matthews, V., Paepke, O., Limpus, C. J., and Gaus, C. 2008. Flame retardants (PBDEs) in marine turtles, dugongs and seafood from Queensland, Australia. *Marine Pollution Bulletin.* 57: 409–418.

296. Burreau, S., Broman, D., and Zebühr, Y. 1999. Biomagnification quantification of PBDEs in fish using stable nitrogen isotopes. *Organohalogen Compounds.* 40: 363–366.

297. Asplund, L., Hornung, M., Peterson, R. E., Turesson, K., and Bergman, Å. 1999. Levels of polybrominated diphenyl ethers (PBDEs) in fish from the Great Lakes and Baltic Sea. *Organohalogen Compounds.* 40: 351–354.

298. Vives, I., Grimalt, J. O., Lacorte, S., Guillamon, M., and Barcelo, D. 2004. Polybrominated diphenyl ether flame retardants in fish from lakes in European high mountains and Greenland. *Environmental Science & Technology.* 38: 2338–2344.

299. Vorkamp, K., Christensen, J. H., and Riget, F. 2004. Polybrominated diphenyl ethers and organochlorine compounds in biota from the marine environment of West Greenland. *The Science of the Total Environment.* 331: 143–155.

300. Schmid, P., Kohler, M., Gujer, E., Zennegg, M., and Lanfranchi, M. 2007. Persistent organic pollutants, brominated flame retardants and synthetic musks in fish from remote alpine lakes in Switzerland. *Chemosphere.* 67: S16–S21.

301. Erdogrul, Ö., Covaci, A., and Schepens, P. 2005. Levels of organochlorine pesticides, polychlorinated biphenyls and polybrominated diphenyl ethers in fish species from Kahramanmaras, Turkey. *Environment International.* 31: 703–711.

302. Holden, A., She, J. W., Tanner, M., Lunder, S., Sharp, R., and Hooper, K. 2003. PBDEs in the San Francisco Bay area: Measurements in fish. *Organohalogen Compounds.* 61: 255–258.

303. Brown, F. R., Winkler, J., Visita, P., Dhaliwal, J., and Petreas, M. 2006. Levels of PBDEs, PCDDs, PCDFs, and coplanar PCBs in edible fish from California coastal waters. *Chemosphere.* 64: 276–286.

304. Rice, C. P., Chernyak, S. M., Begnoche, L., Quintal, R., and Hickey, J. 2002. Comparisons of PBDE composition and concentration in fish collected from the Detroit River, MI and Des Plaines River, IL. *Chemosphere.* 49: 731–737.

305. Boon, J. P., Lewis, W. E., Tjoen-A-Choy, M. R., Allchin, C. R., Law, R. J., de Boer, J., ten Hallers-Tjabbes, C. C., and Zegers, B. N. 2002. Levels of polybrominated diphenyl ether (PBDE) flame retardants in animals representing different trophic levels of the North Sea food web. *Environmental Science & Technology.* 36: 4025–4032.

306. Easton, M. D. L., Luszniak, D., and Von der Geest, E. 2002. Preliminary examination of contaminant loadings in framed salmon, wild salmon and commercial salmon feed. *Chemosphere.* 46: 1053–1074.

307. Manchester-Neesvig, J. B., Valters, K., and Sonzogni, W. C. 2001. Comparison of polybrominated diphenyl ethers (PBDEs) and polychlorinated biphenyls (PCBs) in Lake Michigan salmonids. *Environmental Science & Technology.* 35: 1072–1077.

308. Luross, J. M., Alaee, M., Sergeant, D. B., Cannon, C. M., Whittle, D. M., Solomon, K. R., and Muir, D. C. G. 2002. Spatial distribution of polybrominated diphenyl ethers and polybrominated biphenyls in lake trout from the Laurentian Great Lakes. *Chemosphere.* 46: 665–672.

309. Zhu, L. Y. and Hites, R. A. 2004. Temporal trends and spatial distributions of brominated flame retardants in archived fishes from the Great Lakes. *Environmental Science & Technology.* 38: 2779–2784.

310. Ikonomou, M. G., Rayne, S., Fischer, M., Fernandez, M. P., and Cretney, W. 2002. Occurrence and congener profiles of polybrominated diphenyl ethers (PBDEs) in environmental samples from coastal British Columbia, Canada. *Chemosphere.* 46: 649–663.

311. Ikonomou, M. G., Fernandez, M. P., and Hickman, Z. L. 2006. Spatio-temporal and species-specific variation in PBDE levels/patterns in British Columbia's coastal waters. *Environmental Pollution.* 140: 355–363.

312. Vorkamp, K., Thomsen, M., Falk, K., Leslie, H., Moller, S., and Sorensen, P. B. 2005. Temporal development of brominated flame retardants in peregrine falcon (*Falco peregrinus*) eggs from South Greenland (1986–2003). *Environmental Science & Technology.* 39: 8199–8206.

313. Lam, J. C. W., Lau, R. K. F., Murphy, M. B., and Lam, P. K. S. 2009. Temporal trends of hexabromocyclododecanes (HBCDs) and polybrominated diphenyl ethers (PBDEs) and detection of two novel flame retardants in marine mammals from Hong Kong, South China. *Environmental Science & Technology.* 43: 6944–6949.

314. She, J. W., Petreas, M., Winkler, J., Visita, P., McKinney, M., and Kopec, D. 2002. PBDEs in the San Francisco Bay Area: Measurements in harbor seal blubber and human breast adipose tissue. *Chemosphere.* 46: 697–707.

315. Isobe, T., Ochi, Y., Ramu, K., Yamamoto, T., Tajima, Y., Yamada, T. K., Amano, M., Miyazaki, N., Takahashi, S., and Tanabe, S. 2009. Organohalogen contaminants in striped dolphins (*Stenella coeruleoalba*) from Japan: Present contamination status, body distribution and temporal trends (1978–2003). *Marine Pollution Bulletin.* 58: 396–401.

316. Kajiwara, N., Kamikawa, S., Amano, M., Hayano, A., Yamada, T. K., Miyazaki, N., and Tanabe, S. 2008. Polybrominated diphenyl ethers (PBDEs) and organochlorines in melon-headed whales, *Peponocephala electra*, mass stranded along the Japanese coasts: Maternal transfer and temporal trend. *Environmental Pollution.* 156: 106–114.

317. Chen, D. H., Bi, X. H., Zhao, J. P., Chen, L. G., Tan, J. H., Mai, B. X., Sheng, G. Y., Fu, J. M., and Wong, M. H. 2009. Pollution characterization and diurnal variation of PBDEs in the atmosphere of an E-waste dismantling region. *Environmental Pollution.* 157: 1051–1057.

318. Leung, A. O. W., Luksemburg, W. J., Wong, A. S., and Wong, M. H. 2007. Spatial distribution of poly-brominated diphenyl ethers and polychlorinated dibenzo-p-dioxins and dibenzofurans in soil and com-busted residue at Guiyu, an electronic waste recycling site in Southeast China. *Environmental Science & Technology*. 41: 2730–2737.

319. Luo, X. J., Zhang, X. L., Liu, J., Wu, J. P., Luo, Y., Chen, S. J., Mai, B. X., and Yang, Z. Y. 2009. Persistent halogenated compounds in waterbirds from an E-waste recycling region in South China. *Environmental Science & Technology*. 43: 306–311.

320. Qu, W. Y., Bi, X. H., Sheng, G. Y., Lu, S. Y., Fu, H., Yuan, J., and Li, L. P. 2007. Exposure to polybromi-nated diphenyl ethers among workers at an electronic waste dismantling region in Guangdong, China. *Environment International*. 33: 1029–1034.

3 Temporal Trends of Perfluorinated Compounds from Various Environmental Matrices

Yasuyuki Zushi and Shigeki Masunaga*

CONTENTS

3.1 Perfluorinated Compounds 73
3.2 Temporal Trends of PFC Production and Emission Amounts in the World 74
    3.2.1 Temporal Trends of PFOA Production and Emission 74
    3.2.2 Temporal Trends of PFOSF/PFOS Production and Emission 74
    3.2.3 Temporal Trends of PFC Transaction Volume from Products Register System 75
3.3 Temporal Trends of PFC Pollution from Various Environmental Matrices 76
    3.3.1 Temporal Trends in Japan 76
    3.3.2 Temporal Trends in Europe 79
    3.3.3 Temporal Trends in North America 80
    3.3.4 Temporal Trends in the Arctic 82
    3.3.5 Other Areas 82
3.4 Perspective of PFC Pollution in the Future 82
Acknowledgments 83
References 83

3.1 PERFLUORINATED COMPOUNDS

Perfluorooctane sulfonate (PFOS) has been affirmed as a new persistent organic pollutant (POPs) after the Stockholm Convention on POPs held in May 2009. Global contamination by PFOS was first recognized in 2001 in a study conducted by Giesy and Kannan [1], which monitored PFOS globally. 3M Ltd., the world's major producer of PFOS and its homologue perfluorooctanoate (PFOA), decided to start phasing out the production of PFOS, its derivatives (i.e., Perfluorooctane sulfonamides [FOSAs], perfluorooctane sulfonamidoethanols [FOSEs]), and its synthetic starting material (i.e., perfluorooctyl sulfonyl fluoride [PFOSF]) beginning in 2000. The company planned to completely phase out production by 2003 [2] due to the high risk that PFOS and its derivatives posed for human health and ecosystems. Perfluorinated compounds (PFCs) have been used industrially and commercially since the 1950s in various applications, such as textiles, upholstery manufacturing processes, fire-fighting foams, and hydraulic fluids. Additionally, PFOS is heavily used as an emulsifying agent in the production of polytetrafluoroethylene (PTFE) and in photography. PFCs are regarded as valuable compounds because of their unique properties such as chemical stability,

---

* E-mail: masunaga@ynu.ac.jp (Chapter corresponding author).

interfacial activity, and leveling property. These characteristics arise from the coexistence of the perfluorinated alkyl chain (which induces lipophobicity and hydrophobicity) and the hydrophilic group (which induces hydrophilicity). Acrylate polymers with perfluoroalkyl side chains, which need PFCs as their building block for production, exhibit the property of water–oil repellency, and they are used for upholstery, textiles, and so on. Until the end of the twentieth century, the production of PFOS and its related compounds increased steadily due to high demand [3]. With the turn of the century, however, the production and use of PFOS have decreased significantly due to the Stockholm Convention regulations set by various countries and the EU [4–6], due to voluntary phase-out programs enacted by various companies [2]. Additionally, voluntary regulations concerning production, use, and import of PFOA and other higher-chain-length homologues [7–9] and of fluorotelomer alcohols (FTOHs) [10], which are used in the production of fluoropolymers and degrade to PFOA, have been also introduced.

In the following section, the global emission volumes of PFCs prior to the enactment of these regulations are summarized in relation to today's post-regulation emission volumes, referring several reports of their production. In the next section, temporal trends of PFCs constructed by various historically archived samples in the world are reviewed. The perspective of PFC pollution for the future is discussed throughout this chapter.

3.2 TEMPORAL TRENDS OF PFC PRODUCTION AND EMISSION AMOUNTS IN THE WORLD

3.2.1 TEMPORAL TRENDS OF PFOA PRODUCTION AND EMISSION

PTFE, known as Teflon® nowadays, was accidentally invented in 1938. In 1951, PFOA was introduced as the emulsifier in the polymerization process of tetrafluoroethylene to produce PTFE. Production amounts of ammonium perfluorooctanoate (APFO) since 1951 have been estimated by Prevedouros et al. [11]. They found that the production amount of APFO gradually increased with time: Annual productions were 5–25 ton from 1951 to 1964, 30–50 ton from 1965 to 1979, 100–150 ton from 1980 to 1994, and 200–300 ton from 1995 to 2002. They also estimated the PFOA and APFO emissions based on the production amounts. Taking into consideration the fact that fluoropolymer manufacturing contributed the most PFOA and APFO emissions to the environment, the global PFOA/APFO emission trend can be extrapolated from the emission trend of APFO in the fluoropolymer manufacturing sector. It was determined that annual emissions of APFO from fluoropolymer manufacturing also gradually increased: 3–15 ton during 1951–1964, 20–30 ton during 1965–1979, 50–100 ton during 1980–1994, and 100–200 ton during 1995–2004. Furthermore, it was reported that the estimated global emissions of APFO from the fluoropolymer manufacturing sector started to decrease in 2001—the emission amounts were approximately 40 ton in 2006. The global APFO emissions trend is shown in Figure 3.1. Global APFO emissions started to increase in the early 1950s, peaked in 2000, and then decreased with a half-value period of approximately 4 years.

The stewardship program, which started in early 2006 and was attended by eight of the major producers of PFCA [7], aims to reduce global emissions of PFCAs with carbon chains composed of more than eight carbons and its precursors by 95% below the levels found in 2000 by 2010, and 100% by 2015. Thus, we expect to see a further reduction in APFO (PFOA) emissions starting from 2006. The progress of the program is displayed on the USEPA's Web site [7].

3.2.2 TEMPORAL TRENDS OF PFOSF/PFOS PRODUCTION AND EMISSION

PFOSF and its derivatives (FOSEs, FOSAs, and PFOS) are primarily produced by electrochemical fluorination (ECF) and are generically called PFOSF-based products. The ECF process was scaled up to manufacturing pilot scale by 3M Ltd. in 1949 [12]. Although information concerning their initial production operations of PFOSF-based products was scarce, estimated production

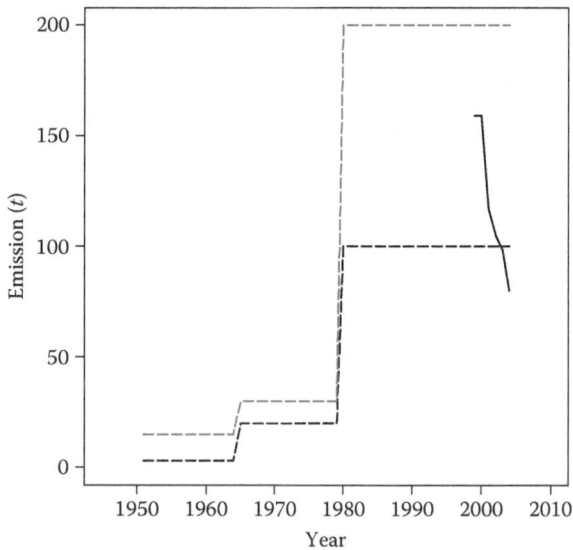

FIGURE 3.1 Global historical trends of APFO emissions from fluoropolymer manufacturing. Minimum/maximum estimation of APFO (---) and estimation based on other reported information by Prevedouros et al. (—). (Based on Prevedouros, K. et al., *Environ. Sci. Technol.*, 40, 32, 2006.)

amounts of PFOSF were reported starting in 1957 [13]. According to the report, the estimated global PFOSF production amount gradually increased from 1957 until 2002, after which it decreased, due to the phase-out of PFOSF production by 3M Ltd. Specifically, 3,905 ton of PFOSF were produced during 1957–1975, 12,859 ton during 1976–1984, 11,907 ton during 1985–1989, 13,608 ton during 1990–1994, 22,881 ton during 1995–2002, and 1,460 ton during 2003–2010. Also, global PFOS emissions from PFOSF manufacturing to water environments were estimated as 24 ton during 1957–1975, 80 ton during 1976–1984, 74 ton during 1985–1989, 85 ton during 1990–1994, 105 ton during 1995–2002, and 5 ton during 2003–2010. Other reports estimated amounts of PFOSF production amounts increased beginning in 1970 until 2000, after which levels decreased by 3–9 times [11,14,15]. Although the trend of PFOS production was not described in detail, the majority of PFOS emissions was estimated based on the percentage of PFOS as PFOSF product impurity and the amount of PFOS in the direct release of aqueous fire fighting foams (AFFFs) [3]. PFOS oceanic mass balance calculations revealed that estimated amounts of PFOS emission corresponds to percentage of impurity of this compound in PFOSF. The trends of global PFOSF production and global PFOS emission are shown in Figure 3.2. The PFOS emission trend was based on the trend of PFOSF production. There was a gradual increase in PFOS emissions during the late 1950s until 2000, after which there was a rapid decrease.

3.2.3 TEMPORAL TRENDS OF PFC TRANSACTION VOLUME FROM PRODUCTS REGISTER SYSTEM

KemI conducted a survey on perfluorocarboxylate (PFCA) and PFOS/perfluoroalkyl sulfonate (PFAS) use in Sweden [16]. A rapid decrease in the use of PFOS and PFAS-related substances after 2000 was reported; approximately 23 tons were used in 2000 while only 0.7 ton was used in 2004. This result corresponds with the PFOSF production and PFOS emission trends previously described. As for PFCA, a clear decreasing trend was not observed despite the rapidly decreasing behavior in APFO emissions shown in Figure 3.1, indicating that the recent usage of APFO (PFOA) does not correlate to their production. Otherwise, the data from the products register that KemI used represents the "total" quantities of PFCA used, and thus it might be difficult to localize

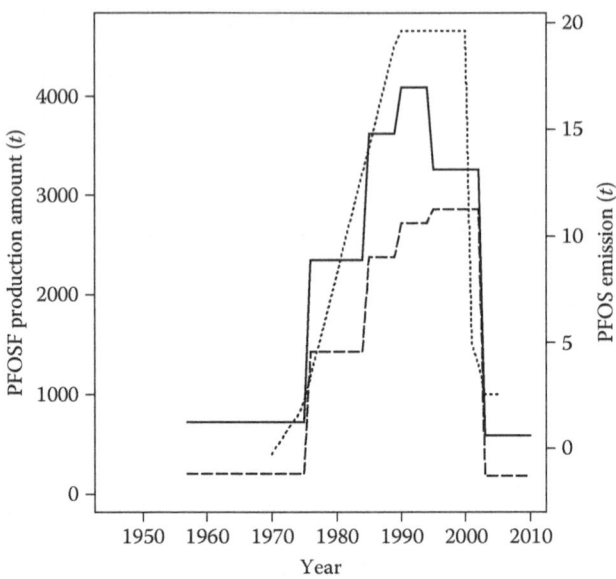

FIGURE 3.2 Global historical trends of PFOSF production and PFOS emissions. Shown are PFOSF production and PFOS emission amounts estimated by Armitage et al. (---), PFOSF production amounts estimated by Paul et al. (···) (see Ref. [14]), and PFOS emission amounts estimated by Armitage et al. (—). (Estimated by Armitage, J.M. et al., *Environ. Sci. Technol.*, 43, 9274, 2009.)

any correspondence between the production of PFCA and the emission of APFO. As a result, more detailed information concerning PFCA usage, production, and emission is needed.

Likewise, information concerning temporal trends in the production and/or emission of PFOS and PFOA derivatives/precursors, except for PFOSF, was also scarce.

3.3 TEMPORAL TRENDS OF PFC POLLUTION FROM VARIOUS ENVIRONMENTAL MATRICES

3.3.1 TEMPORAL TRENDS IN JAPAN

Several studies concerning the temporal trends of PFC pollution were conducted in Japan. Harada et al. reported the temporal trends of PFOS and PFOA emissions using archived human serum samples in the Kyoto prefecture during 1983–1999 at 4 year intervals [17]. Concentrations of PFOS and PFOA in all serum samples were greater than the limit of quantification (LOQ), and a clear increasing trend in PFOA concentration was observed in male and female serum from 1983 to 1999. PFOS, on the other hand, did not significantly increase, and reached a plateau in the same time period. Other districts in Japan were also studied, and increasing trends in PFOA concentration from the 1990s to 2003 (Yokote, Akita) and from the late 1970s to 2003 (Taiwa, Miyagi) were observed; the PFOS trend was varied among districts [18]. The temporal trends of PFOS and PFOA concentrations after 2003 were gathered using serum samples obtained in Osaka, Sendai, and Takayama in Japan were reported [19]. A large fluoropolymer manufacturer, Daikin Co., operates in Osaka and they decreased their PFOA emissions by 82% (compared to 2000 levels) in 2006 and by 89% in 2007. This decrease in PFOA emissions was reflected in the concentration of PFOA measured in the human serum collected from Osaka; we did not observe a clear decrease in PFOA concentration in the samples from Sendai and Takayama. The concentrations of PFOS decreased in all districts during 2004–2008; a 22.3%–66.7% decrease from 2003 to 2004 levels [17] in 2008 was observed, concluding that the phase-out of PFOS production by 3M Ltd. around 2000–2002 was beginning to have an effect.

The temporal trends of PFOS, FOSA, and PFCAs of several chain lengths, using the liver tissue samples of melon-headed whales mass-stranded along the Japanese coast in 1982–2006 (1982, 2001–2002, 2006) have been reported [20]. Melon-headed whales are widely found in tropical and warm temperate oceans; thus, the results do not measure PFC pollution in Japan only. A significant increase was observed between 1985 and 2001–2002 in PFC concentrations in all the liver samples of melon-headed whales. No significant changes in concentration between 2001–2002 and 2006 were observed for PFOS, perfluoroundecanoate (PFUnDA), or perfluorododecanoate (PFDoDA). However, perfluorononanoate (PFNA), and perfluorodecanoate (PFDA) concentrations significantly increased during 2001–2002 to 2006, and the FOSA concentration decreased during the same period. The unique trend of FOSA was observed in other reports as will become apparent below.

Sediment core from Tokyo Bay was also analyzed to obtain the temporal trend of PFC pollution in Japan [21]. Particulate matter from the atmosphere and matter drained into the Tokyo Bay basin by nearby rivers accumulate at the bottom of Tokyo Bay at a rate of approximately 1 cm/year. Thus, the columnar core of the sediment preserves information on heavy metals, POPs, and other compounds over time. The sediment core sample was sliced, separated at 2 cm intervals in the vertical direction, and analyzed after homogenizing and freeze-drying treatments. The date of each layer of sediment sample was determined using a radioisotope dating technique that measures the concentration of ^{210}Pb in the sample [22]. The sediment date determined by ^{210}Pb dating was erroneously reported in our previous paper [21] and is corrected here, as described below.

The concentration of PFCs in sediment samples were corrected using the concentration of organic carbon (OC) in each sliced sediment sample because the partitioning of PFC in the sediment was highly affected by the concentration of OC in the sample [23]. The temporal trends of PFC concentration during the mid-1960s–2004 at 2–3 year intervals were obtained and the results are shown in Figure 3.3. A locally weighted scatterplot smoother (LOESS) line was applied using "R" software to extrapolate the generalized trends of PFC. A detailed description of the methodology behind LOESS and its application to time series analysis for pollutants is described by Fryer and Nicholson [24]. The important parameter of smoother span, which gives the proportion of points in the plot influencing the smooth at each value, was set depending on the ratio of "the number year for which data are available" to the "year length of the data." This setting of smoother span was used to adjust for the differences in year lengths of the data, and the number of year for data are available in various studies concerning temporal trend of PFC, and allowed us to better compare the PFC trend obtained at various time intervals. The results of this comparison will be discussed below. Increasing trends in PFOA, PFNA, PFDA, PFUnDA, PFDoDA, and perfluorotridecanoate (PFTrDA) concentrations over the last three decades were observed in our study. Decreasing trends in PFOS, N-methylperfluoro-1-octanesulfonamidoacetate (NMeFOSAA), and N-ethylperfluoro-1-octanesulfonamidoacetate (NEtFOSAA) concentrations were observed after an initial increase during the late 1970s to the mid-1990s (Figure 3.3). The rapid decrease in the concentrations of NMeFOSAA and NEtFOSAA, which degraded to PFOS [25,26], compared to the more gradual decrease in PFOS concentration is noteworthy. The increases in PFOA concentrations in human serum and in PFNA and PFDA concentrations in the livers of melon-headed whales, on the contrary to the nonincrease of PFOS in those samples during the late 1990s–2000 as mentioned above were comparable with this results. Additionally, FOSA, which is one of the compounds involved in the degradation process of PFOSF-based compounds like NMeFOSAA and NEtFOSAA to PFOS, significantly decreased in those studies as well as the NMeFOSAA and NEtFOSAA themselves which were found in our studies. The PFOS concentrations in the late 1990s/early 2000s in our study did not correlate with the PFOS emission amounts estimated from PFOSF production amounts (Figure 3.2), but the concentrations of PFOS precursors like NMeFOSAA and NEtFOSAA do. Concentrations of fluorotelomer carboxylates (FTCAs) and fluorotelomer unsaturated carboxylates (FTUCAs), which are PFCA precursors [27,28], were below the LOQ in all sediment samples and so, their temporal behaviors could not be discerned.

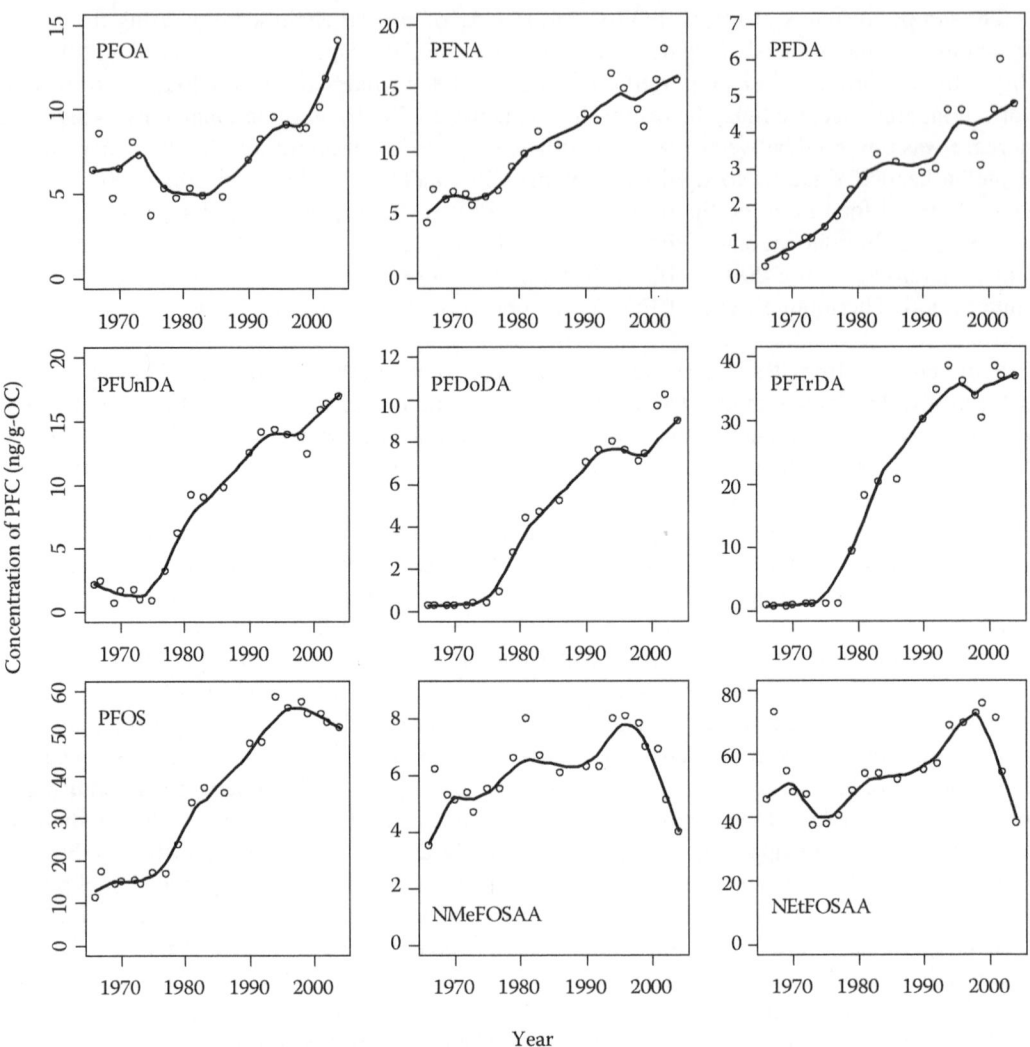

FIGURE 3.3 Temporal trends of PFC concentrations (ng/g-OC dry wt.) in sediment core from Tokyo Bay. The smooth, solid lines represent the LOESS lines. Smoother span was set depending on the ratio of "the number year for which data are available" to "year length of the data."

The analysis of the branched-chain isomer of PFCA provides interesting findings on PFC temporal trend and their sources. In Japan, branched-chain PFCA isomers were industrially produced starting from the mid-1970s. The branched-chain isomers are made from different synthetic materials (isopropyl iodide) than the materials for linear PFCAs production (ethyl iodide) [29]. We observed from the sediment core samples a gradual decrease over time of the ratio of linear-to-branched chain isomers of PFTrDA during the late 1980s–2004, indicating an increased production over time of branched PFTrDA isomers. In light of the shift in the production of various types of PFCs, such as from linear chain isomers of PFTrDA to branched-chain isomers of PFTrDA, development of analysis for various types of PFCs is required for understanding the complex PFC emission trend observed from various environmental samples. The sediment core of Tokyo Bay was also measured in another study for PFCs with the similar temporal trend of PFCAs and NEtFOSAA obtained in our study [30].

An increasing trend in PFCA concentration, decreasing trend in the concentrations of PFOS precursors, and a steady plateau state for PFOS concentrations have been identified in Japan from several reports. The trends for PFCA and PFOS concentrations did not correspond with those of APFO (Figure 3.1) and PFOS (Figure 3.2) concentrations. More studies that include the most recent data on PFC emissions and environmental measurement in Japan are needed.

3.3.2 Temporal Trends in Europe

Many studies similar to those completed in Japan were conducted in Europe. Holmström et al. reported the temporal trend of PFOS concentration in Guillemot Eggs obtained from Stora Karlsö (Sweden) during 1968–2003 at 2–3 year intervals. PFOA was also measured and concentrations were below the detection limit in all samples. We observed a rapid increase in PFOS concentrations from 1968 to the late 1990s, followed by a rapid decrease thereafter (Figure 3.4). The rapid decrease in PFOS concentrations was not observed in Japan, but its existence in Europe has been also con- firmed by another study. In this study, the concentrations of PFOS in historical samples of human milk (1996–2004 at 1 year intervals) from Sweden [31] and in human serum samples (1977–2006

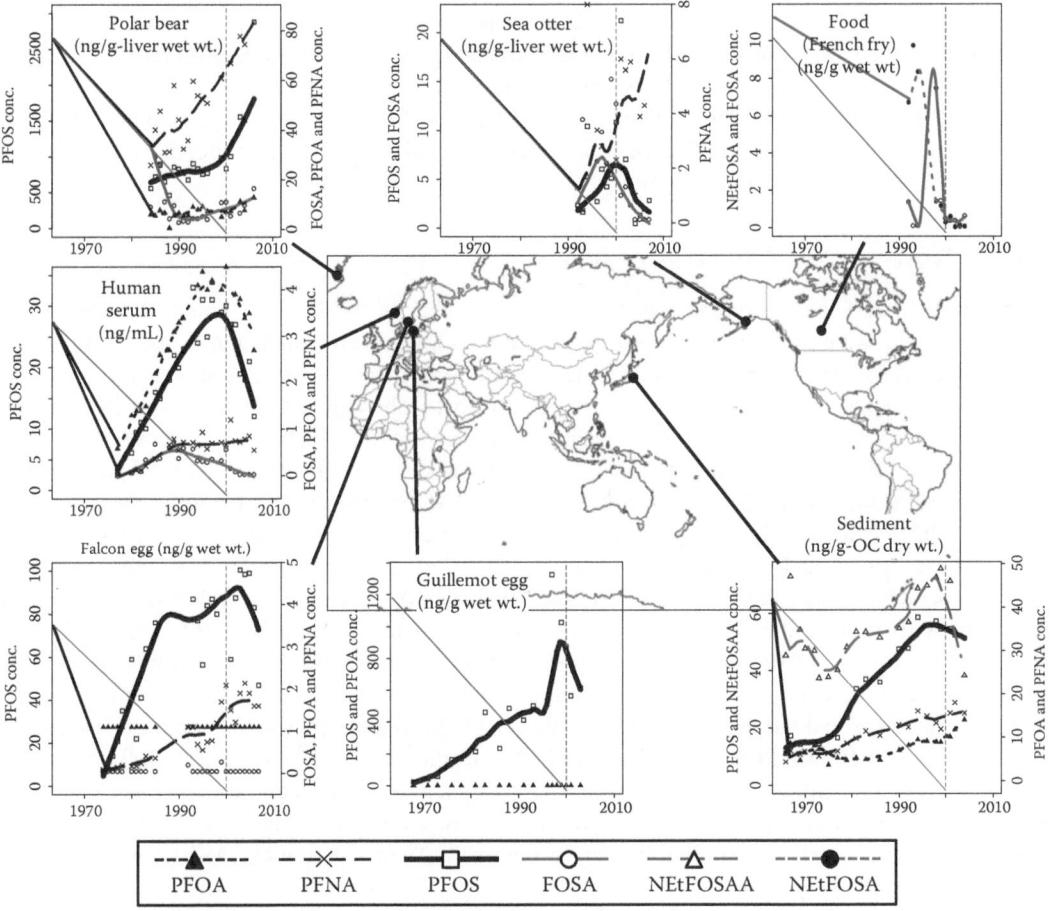

FIGURE 3.4 Temporal trends of PFC concentrations from environmental samples collected from various parts of the world. The smooth, solid lines represent LOESS lines. Smoother span was set after correcting for the differences among all the samples in year length of data and in the number year for which data are avail- able, ensuring a uniform degree of smoothness among the studies.

at approximately 1 year intervals) from Norway [32], peaked in the late 1990s/early 2000s and then rapidly decreased. These results correlate with the PFOS emission trends estimated from the PFOSF production data (Figure 3.2).

The trend in the concentration of PFOS at the end of the twentieth century lagged 3–5 years behind the PFOS emissions trend determined from the historical samples of Falcon eggs (1974–2007 at approximately 1 year intervals) from Sweden [33] (see Figure 3.4) and of Harbor seal livers (1999–2008 at 1 year intervals) from Germany [34]. The decreasing trend of PFOS probably will also be observed in Japan, if the extended monitoring until in the late 2000s is conducted. Initial reports from Osaka, Japan indicate a 22.3%–66.7% decrease from 2003–2004 [17] to 2008 [19].

Looking at PFCAs, we observed a rapid decrease in PFOA concentrations only in the Norwegian human serum samples from the late 1990s, though the other PFCAs were at a plateau [32]. In the late 2000s, PFCA concentrations in Falcon eggs [33] and Harbor seal livers [34] started to decrease. This decrease happened after the decrease in APFO inventory, which showed a rapid decrease starting from the early 2000s (Figure 3.1). Thus, the stewardship program on PFCA emissions starting in 2006 must have had an effect on PFCA decrease in the late 2000s.

Several patterns in FOSA concentrations were also obtained in Europe. The concentration of FOSA peaked in 1999 in the samples of Harbor seals liver [34] and around the early 1990s in the samples of Norwegian human serum [32].

The decrease in PFOS concentrations in Europe was observed in Norwegian serum samples and corresponded well with the trends in the world inventory of PFOSF/PFOS (Figure 3.1). However, some studies have indicated a time delay between the decreasing trends in PFOS concentrations and in the PFOSF/PFOS inventory as well as the PFOS trend in Japan. PFCA decrease in Europe after a few years in the decrease of APFO production was observed and once again indicates the reflection of the effect of the stewardship program.

3.3.3 TEMPORAL TRENDS IN NORTH AMERICA

Many studies were conducted in North America as well as in Europe, though studies in PFC monitoring with long time duration were few. Several studies were conducted in the United States on the changes in PFC concentrations in the human body using serum samples taken from large populations during the time period in which 3M Ltd. was phasing out production of PFOS-based products. The concentrations of PFOS, perfluorohexanesulfonate (PFHxS), PFOA, PFNA, FOSA, NMeFOSAA, NEtFOSAA, and of five other PFCs in the serum of the general U.S. population (≥12 years old) were measured [35]. The results of the 2094 samples collected from the participants in the 2003–2004 National Health and Nutrition Examination Survey (NHANES) and the results of 1562 samples gathered from the participants in the 1999–2000 NHANES were compared [36]. Perfluorooctanesulfonamide, NMeFOSAA, and NEtFOSAA were detected at lower frequencies in the samples from the 2003–2004 NHANES than in the samples from the 1999–2000 NHANES. However, we could not directly compare the concentrations for the aforementioned three compounds because the concentrations in the samples were all near the limit of detection (LOD) level and the LOD threshold values differ between the 1999–2000 and 2003–2004 samples. Instead, the concentrations of PFOS, PFHxS, PFOA, and PFNA in the 1999–2000 and 2003–2004 NHANES were examined, and the results showed that the geometric mean concentrations (GMC) of PFOS, PFHxS, and PFOA in the 2003–2004 survey samples were lower than those in the 1999–2000 survey samples. This clearly shows that the phase-out of PFOS production by 3M Ltd. was influential and effective, especially for PFOS.

The decrease in PFOA during 1999–2004 was not widely observed in Japan and Europe, except in the Norwegian human serum samples [32]. The GMC of PFNA, however, in the 2003–2004 samples was approximately two times higher than that in the 1999–2000 samples. With the introduction of the stewardship program in 2006, we expect to see a decrease in PFNA concentrations starting 2006.

A PFC analysis of 645 adult human plasma samples from the American Red Cross between 2000 and 2006 was also conducted with the age, sex, and location properly adjusted [37]. Among the six PFC compounds (PFOS, PFHxS, PFOA, PFBS, NMeFOSAA, and NEtFOSAA), PFOS, PFOA, and PFHxS were detected with the highest frequency. We observed a decrease in the concentrations of PFOS, PFOA, and PFHxS that was similar to that observed in the NHANES samples.

The temporal trends of PFCA, PFAS, and FOSA concentrations in archived samples of lake trout from Lake Ontario (1979–2004 at approximately 5 year intervals) has been previously reported [38]. A significant increase in PFOS concentrations in lake trout (whole fish homogenates) from 1979 to 1993 followed by a decrease in 1998 was observed. The levels of PFOS in 2004 returned to those in the 1980s–1990s. A similar trend was observed for most of the PFCAs though no significant changes in PFOA were observed during the time period of interest. As was the case for the Norwegian human serum samples, a peak in concentration in 1993 for FOSA was observed. Although the 5 year intervals made understanding the specifics of PFC pollution over time more difficult compared to results obtained from samples gathered at shorter time intervals, the results in the study of the lake trout do not correlate well with the PFOS and APFO emission estimates (Figures 3.1 and 3.2). Unlike what we saw for the sediment core samples of Tokyo Bay, an increasing trend in the ratio of linear-to-branched chain isomers of PFUnDA and PFTrDA was observed in the lake trout [21]. The authors suggest that an alternative source of the branched isomers might have been introduced in the Great Lakes environment.

Historical samples of southern sea otters during 1992–2002 were obtained for the analysis of PFOS and PFOA [39]. The samples were collected at approximately 1 year intervals. An increasing trend in PFOA concentrations over the entire monitoring period and an initial increase in PFOS concentrations followed by a slight decrease after 1998 were found. These different trends between PFOS and PFOA concentrations in the sea otters are similar to what was observed in most of the studies in Japan and Europe.

Historical samples of Northern Sea otters taken over a longer time period were collected during 1992–2007 at approximately 1 year intervals [40]. The concentrations of PFOS, FOSA, and PFNA were measured. An overall increasing trend for PFNA concentrations was observed, whereas the concentrations of PFOS and FOSA rapidly decreased starting 2000 (Figure 3.4). The increase in PFNA concentrations from 1999–2000 to 2003–2004 was also observed in the human serum samples from the NHANES as mentioned above [35] however the continued increase in PFNA concentrations after the introduction of the stewardship program in 2006 was unexpected. The results for PFNA from the Northern Sea otters were different than those we obtained in Europe.

The concentrations of perfluorooctanesulfonamides including FOSA and NEtFOSA were measured in food sold in Canada and were obtained from samples collected from four different grocery stores during 1992–2004 at mostly 1 year intervals [41]. The concentrations in the foods were mostly below LOQ, and the historically sequential data of the PFC could not be obtained except for chicken nuggets, pizza, and French fries. The temporal trends for FOSA and NEtFOSA concentrations obtained from the analysis of French fries are shown in Figure 3.4. A decrease in NEtFOSA concentrations during 1992–2004 was observed. The timing of the start of the decrease in NEtFOSA concentration seems to be somewhat early (Figure 3.4) and it is probably due to the lack of data around then. Although the timing was somewhat early, such a rapid decrease in NEtFOSA concentrations was observed in NEtFOSAA concentrations (NEtFOSAA is the precursor to NEtFOSA) from 1995 to 1997 in the sediment core in Tokyo Bay [21]. Although the concentrations of FOSA were mostly below LOQ even in French fries, the concentration of FOSA peaked in the mid-late 1990s, just as was the case for the Northern Sea otters in the United States [40] and the harbor seals in Germany [34].

In summary, a decrease in PFOS concentration from around 2000 was observed in North America, which corresponds well with the trend in the world inventory of PFOSF/PFOS (Figure 3.1). Such a strong and direct correlation was not observed in some of the studies in Europe and Japan. With the exception of the human serum study [35,37], most studies reported an increasing trend in PFOA concentration. To better understand the details of PFC concentration behavior, more studies using samples collected in North America at shorter time intervals are needed, especially for PFOA.

3.3.4 TEMPORAL TRENDS IN THE ARCTIC

Several studies on the temporal trends of PFC concentrations were conducted in the Arctic. The trends of the concentrations of PFOS, FOSA, several PFCAs, FTCAs, and FTUCAs from the Arviat ringed seal (1992–2005) and the ringed seal in Resolute Bay (1972–2005) were reported [42]. An increase in PFOS and FOSA concentrations from 1992 to 1998 followed by a decrease from 1998 to 2005 was observed in the historical samples of Arviat ringed seals. An increase in the concentrations of PFCAs from 1992 to 2003 followed by a slight decrease was observed. A similar trend was observed from the ringed seals samples from Resolute Bay. These results correspond well with the trend of the inventory of PFOSF/PFOS and with that of the APFO inventory, albeit with a slight time delay (Figures 3.1 and 3.2). Although most of the FTCA/FTUCA concentrations were below the LOQ in the ringed seals samples, a peak in the concentration of 10:2 FTUCA was observed in 2003.

Two studies have shown prolonged increases in PFOS and PFCA concentrations in the samples of polar bear livers from the North American Arctic (1972–2002 at variable intervals) [15] and from East Greenland (1984–2006 at approximately 1 year intervals) [43]. An increase in the concentration of PFOS from 1983 to 1993 followed by a leveling off until 2003 was observed in the eggs of herring gulls in northern Norway (1983, 1993, and 2003). In short, the concentration of PFOS did not decrease at all from 1983 to 2003. The behavior of PFOS concentrations in the Arctic zone [44] compared with other areas is remarkably different. In addition, an increase in the concentration of FOSA was also observed in two previous studies [43,44]. Even after regulations were enacted, PFC concentrations are still increasing, as described in a report by Butt et al. [45].

The increasing trends in PFC concentrations in the Arctic did not correlate with the emission trends of PFOS and APFO (Figures 3.1 and 3.2). The PFC concentration trends in the Arctic show a time delay in the response to the production/emission trend of PFCs as a result of the slow oceanic transport of PFCs [13,43].

Studies on the temporal trends of PFC concentrations in ice cap core [46] and sediment core [47] samples from Arctic area were also conducted. The comparison of those results with the results of other studies may not be meaningful due to differences in monitoring time interval, thus, the details were not mentioned here.

3.3.5 OTHER AREAS

A few studies identifying temporal trends of PFC concentrations were conducted in other regions outside of Japan, Europe, the United States, and the Arctic. PFC levels in the livers of Baikal seals between 1992 and 2005 were determined [48]. The concentrations of PFOS, PFNA, and PFDA in 2005 were higher than in 1992. Also, the PFOS and PFOA levels in human serum samples from Busan (1994, 2000, 2008) and Seoul (1994, 2007), Korea did not consistently decrease with decreasing production/emission of PFCs [19].

China has been increasing their production of PFOS-based products since 2003 and further increase in PFOS production/emission is expected [49]. More studies examining the temporal trend of PFC concentrations and careful monitoring of PFC concentrations in developing countries and other areas will thus be extremely insightful.

3.4 PERSPECTIVE OF PFC POLLUTION IN THE FUTURE

Temporal trends of PFC concentrations in various regions have been reviewed in the previous sections. The PFC concentrations determined by these different studies were measured at 1–3 year intervals over long periods and are exhibited in Figure 3.4. In Figure 3.4, the LOESS line is shown to better identify the historical and regional trends of PFC pollution among the studies. Smoother spans of the lines were corrected to standardize the degree of smoothing for each study.

A rapid decrease in PFOS levels around the year 2000 was observed only in Europe and North America. This decrease corresponds with the emissions trend of PFOS. The time lags of PFOS decreases were also observed from the results in Japan and one of the results in Europe as shown in Figure 3.4. The rapid decrease in PFOS concentrations in Europe and North America is considered to be due to the early introduction of restrictions by an EU directive [6] and due to the drastic phase-out initiative of PFOS-based products by 3M in the United States. In contrast, the continued increase in PFOS levels observed in the Arctic is speculated to be due to the slow ocean transport of PFOS.

Rapid decreases in NEtFOSA levels in food sold in Canada and in NErFOSAA levels in sediment cores in Japan were observed. A somewhat different situation with PFOS in application and emission of these precursors was considered. In light of the appearance of the rapid response of these precursors ahead of the response of PFOS against PFOSF/PFOS emissions, these precursors would be useful as a predictor of PFOS pollution trend. Thus, the careful monitoring of those precursors is recommended.

The peak of the concentrations of FOSA occurring in the mid-1990s was observed in many studies. However, their temporal trend during the 1980s–2000s was highly varied, even within the same region.

In many parts of the world, PFOA and PFNA (PFCAs) concentrations are increasing. Pollution caused by PFCAs is expected to significantly drop after the introduction of the stewardship program in 2006, which aims to reduce PFCA emissions. Signs of the stewardship program having an effect on the environment have appeared in the temporal trends of PFCA concentrations in the falcon eggs. We expect to see more of the same evidence in the near future, except in the Arctic and in developing countries.

Fluorinated compounds have been used over the past 50 years and have, in recent years, come to be recognized as environmental pollutants. Thus, they are now one of the new persistent pollutants having comparable long history of use and emission. Unlike the case for chlorinated compounds [50], the concentration levels in some PFC homologues will continue to increase. Fortunately, several regulations have been enacted recently and we expect to see an overall decrease in the concentrations of most PFC homologues within a decade, except in the Arctic regions and in developing countries.

As for the remaining issue, the shift of different PFC types such as linear-to-branched isomer, PFOS/PFOA to short chain of PFC, etc., should be paid attention. Also, the deterioration of PFC pollution in developing countries due to the increased demand of PFC is an important issue.

ACKNOWLEDGMENTS

This study was supported by JSPS research fellowships for Young Scientists (ID: 213467).

REFERENCES

1. Giesy, J.P. and Kannan, K. 2001. Global distribution of perfluorooctane sulfonate in wildlife. *Environ. Sci. Technol.* 35: 1339–1342.
2. 3M, 2000. Phase-out plan for POSF-based products. U.S. EPA public Docket AR226-0600.
3. Paul, A.G., Jones, K.C., and Sweetman, A.J. 2009. A first global production, emission, and environmental inventory for perfluorooctane sulfonate. *Environ. Sci. Technol.* 43: 386–392.
4. Government of Canada 2009. Regulations adding perfluorooctane sulfonate and its salts to virtual elimination list. *Canada Gaz.* 143: http://gazette.gc.ca/rp-pr/p2/2009/2009-2002-02-04/html/sor-dors15-eng.html
5. USEPA 2002. Perfluoroalkyl sulfonates; significant new use rule; final and supplemental proposed rule. *Fed. Regist.* 67: http://www.epa.gov/fedrgstr/EPA-TOX/2002/March/Day-11/t5747.pdf

6. EU Directive 2006. Directive 2006/122/EC of the European Parliament and of the Council. http://eur-lex. europa.eu/LexUriServ/LexUriServ.do?uri=OJ:L:2006:372:0032:0034:EN:PDF

7. USEPA 2009. 2010/15 PFOA Stewardship Program. http://www.epa.gov/opptintr/pfoa/pubs/stewardship/ index.html

8. DuPont, 2005. DuPont global PFOA strategy—Comprehensive source reduction. U.S. EPA public docket AR226-1914.

9. Asahi Glass Co. Ltd. 2008. New Asahi Glass "Fluon PTFE E-SERIES" fluorinated resin will be free of PFOA. *Focus Surfactants* 2008: 3–3.

10. Environment Canada. 2006. Ecological screening assessment report on perfluorooctane sulfonate, its salts and its precursors that contain the $C_8F_{17}SO_2$ or $C_8F_{17}SO_3$, or $C_8F_{17}SO_2N$ moiety. Canadian Environmental Protection Act, 1999 (CEPA-1999), pp. 81.

11. Prevedouros, K., Cousins, I.T., Buck, R.C., and Korzeniowski, S.H. 2006. Sources, fate and transport of perfluorocarboxylates. *Environ. Sci. Technol.* 40: 32–44.

12. 3M, 1999. Fluorochemical use, distribution and release overview. U.S. EPA public docket AR226-0550.

13. Armitage, J.M., Schenker, U., Scheringer, M., Martin, J.W., MacLeod, M., and Cousins, I.T. 2009. Modeling the global fate and transport of perfluorooctane sulfonate (PFOS) and precursor compounds in relation to temporal trends in wildlife exposure. *Environ. Sci. Technol.* 43: 9274–9280.

14. Paul, C.R., Clare, L.M., and Tom, H. 2009. Perfluorooctane sulphonate and perfluorooctanoic acid in drinking and environmental waters. *Phil. Trans. Math. Phys. Eng. Sci.* 367: 4119–4136.

15. Smithwick, M., Norstrom, R.J., Mabury, S.A., Solomon, K., Evans, T.J., Stirling, I., Taylor, M.K., and Muir, D.C.G. 2006. Temporal trends of perfluoroalkyl contaminants in polar bears (*Ursus maritimus*) from two locations in the North American Arctic, 1972–2002. *Environ. Sci. Technol.* 40: 1139.

16. KemI 2006. Perfluorinated substances and their uses in Sweden. http://www.kemi.se/upload/Trycksaker/ Pdf/Rapporter/Report7_06.pdf

17. Harada, K., Koizumi, A., Saito, N., Inoue, K., Yoshinaga, T., Date, C., Fujii, S., Hachiya, N., Hirosawa, I., Koda, S., Kusaka, Y., Murata, K., Omae, K., Shimbo, S., Takenaka, K., Takeshita, T., Todoriki, H., Wada, Y., Watanabe, T., and Ikeda, M. 2007. Historical and geographical aspects of the increasing perfluorooctanoate and perfluorooctane sulfonate contamination in human serum in Japan. *Chemosphere* 66: 293–301.

18. Harada, K. and Koizumi, A. 2009. Environmental and biological monitoring of persistent fluorinated compounds in Japan and their toxicities. *Environ. Health Prev. Med.* 14: 7–19.

19. Harada, K.H., Yang, H.-R., Moon, C.-S., Hung, N.N., Hitomi, T., Inoue, K., Niisoe, T., Watanabe, T., Kamiyama, S., Takenaka, K., Kim, M.-Y., Watanabe, K., Takasuga, T., and Koizumi, A. 2010. Levels of perfluorooctane sulfonate and perfluorooctanoic acid in female serum samples from Japan in 2008, Korea in 1994–2008 and Vietnam in 2007–2008. *Chemosphere* 79: 314–319.

20. Hart, K., Kannan, K., Isobe, T., Takahashi, S., Yamada, T.K., Miyazaki, N., and Tanabe, S. 2008. Time trends and transplacental transfer of perfluorinated compounds in melon-headed whales stranded along the Japanese coast in 1982, 2001/2002, and 2006. *Environ. Sci. Technol.* 42: 7132–7137.

21. Zushi, Y., Tamada, M., Kanai, Y., and Masunaga, S. 2010. Time trends of perfluorinated compounds from the sediment core of Tokyo Bay, Japan (1950s–2004). *Environ. Pollut.* 158: 756–763.

22. Kanai, Y. 1993. Well-type Ge semi-conductor detector for measuring small amount of environmental samples. *Radioisotopes* 42: 169–172 (in Japanese).

23. Higgins, C.P. and Luthy, R.G. 2006. Sorption of perfluorinated surfactants on sediments. *Environ. Sci. Technol.* 40: 7251–7256.

24. Fryer, R.J. and Nicholson, M.D. 1999. Using smoothers for comprehensive assessments of contaminant time series in marine biota. *ICES J. Mar. Sci.* 56: 779–790.

25. Rhoads, K.R., Janssen, E.M.L., Luthy, R.G., and Criddle, C.S. 2008. Aerobic biotransformation and fate of N-ethyl perfluorooctane sulfonamidoethanol (N-EtFOSE) in activated sludge. *Environ. Sci. Technol.* 42: 2873–2878.

26. Higgins, C.P., Mcleod, P.B., Macmanus-Spencer, L.A., and Luthy, R.G. 2007. Bioaccumulation of perfluorochemicals in sediments by the aquatic oligochaete *Lumbriculds variegatus*. *Environ. Sci. Technol.* 41: 4600–4606.

27. Gauthier, S.A. and Mabury, S.A. 2005. Aqueous photolysis of 8:2 fluorotelomer alcohol. *Environ. Toxicol. Chem.* 24: 1837–1846.

28. Dinglasan, M.J.A., Ye, Y., Edwards, E.A., and Mabury, S.A. 2004. Fluorotelomer alcohol biodegradation yields poly- and perfluorinated acids. *Environ. Sci. Technol.* 38: 2857–2864.

29. Katsushima, A., Hisamoto, I., Nagai, M., Fukui, T., and Kato, T. 1976. Method for water and oil repellant treatment. Japanese Patent No. 0831272. (In Japanese), Japan.

30. Ahrens, L., Yamashita, N., Yeung, L.W.Y., Taniyasu, S., Horii, Y., Lam, P.K.S., and Ebinghaus, R. 2009. Partitioning behavior of per- and polyfluoroalkyl compounds between pore water and sediment in two sediment cores from Tokyo Bay, Japan. *Environ. Sci. Technol.* 43: 6969–6975.

31. Karrman, A., Ericson, I., van Bavel, B., Darnerud, P.O., Aune, M., Glynn, A., Lignell, S., and Lindstrom, G. 2007. Exposure of perfluorinated chemicals through lactation: Levels of matched human milk and serum and a temporal trend, 1996–2004, in Sweden. *Environ. Health Perspect.* 115: 226.

32. Haug, L.S., Thomsen, C., and Becher, G. 2009. Time trends and the influence of age and gender on serum concentrations of perfluorinated compounds in archived human samples. *Environ. Sci. Technol.* 43: 2131–2136.

33. Holmström, K.E., Johansson, A.-K., Bignert, A., Lindberg, P., and Berger, U. 2010. Temporal trends of perfluorinated surfactants in Swedish peregrine falcon eggs (*Falco peregrinus*), 1974–2007. *Environ. Sci. Technol.* 44: 4083–4088.

34. Ahrens, L., Siebert, U., and Ebinghaus, R. 2009. Temporal trends of polyfluoroalkyl compounds in harbor seals (*Phoca vitulina*) from the German Bight, 1999–2008. *Chemosphere* 76: 151–158.

35. Calafat, A.M., Wong, L.Y., Kuklenyik, Z., Reidy, J.A., and Needham, L.L. 2007. Polyfluoroalkyl chemicals in the U.S. population: Data from the National Health and Nutrition Examination Survey (NHANES) 2003–2004 and comparisons to NHANES 1999–2000. *Environ. Health Perspect.* 115: 1596–1602.

36. Calafat, A.M., Kuklenyik, Z., Reidy, J.A., Caudill, S.P., Tully, J.S., and Needham, L.L. 2007. Serum concentrations of 11 polyfluoroalkyl compounds in the US population: Data from the National Health and Nutrition Examination Survey (NHANES) 1999–2000. *Environ. Sci. Technol.* 41: 2237–2242.

37. Olsen, G.W., Mair, D.C., Church, T.R., Ellefson, M.E., Reagen, W.K., Boyd, T.M., Herron, R.M., Medhdizadehkashi, Z., Nobiletti, J.B., Rios, J.A., Butenhoff, J.L., and Zobel, L.R. 2008. Decline in perfluorooctanesulfonate and other polyfluoroalkyl chemicals in American Red Cross adult blood donors, 2000–2006. *Environ. Sci. Technol.* 42: 4989–4995.

38. Furdui, V.I., Helm, P.A., Crozier, P.W., Lucaciu, C., Reiner, E.J., Marvin, C.H., Whittle, D.M., Mabury, S.A., and Tomy, G.T. 2008. Temporal trends of perfluoroalkyl compounds with isomer analysis in lake trout from Lake Ontario (1979–2004). *Environ. Sci. Technol.* 42: 4739–4744.

39. Kannan, K., Perrotta, E., and Thomas, N.J. 2006. Association between perfluorinated compounds and pathological conditions in southern sea otters. *Environ. Sci. Technol.* 40: 4943–4948.

40. Hart, K., Gill, V., and Kannan, K. 2009. Temporal trends (1992–2007) of perfluorinated chemicals in Northern sea otters (*Enhydra lutris kenyoni*) from South-Central Alaska. *Arch. Environ. Contam. Toxicol.* 56: 607–614.

41. Tittlemier, S.A., Pepper, K., and Edwards, L. 2006. Concentrations of perfluorooctanesulfonamides in Canadian total diet study composite food samples collected between 1992 and 2004. *J. Agric. Food Chem.* 54: 8385–8389.

42. Butt, C.M., Muir, D.C.G., Stirling, I., Kwan, M., and Mabury, S.A. 2007. Rapid response of Arctic ringed seals to changes in perfluoroalkyl production. *Environ. Sci. Technol.* 41: 42–49.

43. Dietz, R., Bossi, R., Riget, F.F., Sonne, C., and Born, E.W. 2008. Increasing perfluoroalkyl contaminants in east Greenland polar bears (*Ursus maritimus*): A new toxic threat to the Arctic bears. *Environ. Sci. Technol.* 42: 2701–2707.

44. Verreault, J., Berger, U., and Gabrielsen, G.W. 2007. Trends of perfluorinated alkyl substances in herring gull eggs from two coastal colonies in northern Norway: 1983–2003. *Environ. Sci. Technol.* 41: 6671–6677.

45. Butt, C.M., Berger, U., Bossi, R., and Tomy, G.T. 2010. Levels and trends of poly- and perfluorinated compounds in the arctic environment. *Sci. Total Environ.* 408(15): 2936–2965.

46. Young, C.J., Furdui, V.I., Franklin, J., Koerner, R.M., Muir, D.C.G., and Mabury, S.A. 2007. Perfluorinated acids in arctic snow: New evidence for atmospheric formation. *Environ. Sci. Technol.* 41: 3455–3461.

47. Stock, N.L., Furdui, V.I., Muir, D.C.G., and Mabury, S.A. 2007. Perfluoroalkyl contaminants in the Canadian Arctic: Evidence of atmospheric transport and local contamination. *Environ. Sci. Technol.* 41: 3529–3536.

48. Ishibashi, H., Iwata, H., Kim, E.-Y., Tao, L., Kannan, K., Amano, M., Miyazaki, N., Tanabe, S., Batoev, V.B., and Petrov, E.A. 2008. Contamination and effects of perfluorochemicals in Baikal seal (*Pusa sibirica*). 1. Residue level, tissue distribution, and temporal trend. *Environ. Sci. Technol.* 42: 2295–2301.

49. UNEP 2009. Report of the conference of the parties of the Stockholm convention on persistent organic pollutants on the work of its fourth meeting. http://www.pops.int/documents/meetings/poprc/submissions/submission_PFOS_2008.htm

50. Kodavanti, P.R.S., Kumar, K.S., and Loganathan, B.G. 2008. Organohalogen pollutants and human health. *Int. Encyclopedia Public Health* 4: 686–693.

4 Synthetic Musks and Benzotriazole UV Stabilizers in Marine Ecosystems

Haruhiko Nakata, Ryu-ichi Shinohara,
Sayaka Murata, and Hiroshi Sasaki*

CONTENTS

4.1 Introduction 87
4.2 Synthetic Musks..88
4.3 Benzotriazole UV Stabilizers 91
4.4 Summary and Future Trends 95
Acknowledgments..95
References...95

4.1 INTRODUCTION

There is increasing public concern for the occurrence and contamination of persistent emerging chemicals, such as synthetic musks and benzotriazole ultraviolet stabilizers (BUVSs), in the environment. Synthetic musk fragrances have been used in a wide variety of personal care products such as perfumes, skin creams, deodorants, soaps, detergents, and household cleaners. Among several types of musk fragrances, polycyclic musks are dominant in terms of production volume; the industrial use of HHCB (1,3,4,6,7,8-hexahydro-4,6,6,7,8,8-hexamethylcyclopenta[g]-2-benzopyrane; CAS #: 1222-05-5) and AHTN (7-acetyl-1,1,3,4,4,6-hexamethyltetrahydronaphthalene; CAS #: 1506-02-1) was 1473 and 385 ton, respectively, in Europe in 1998 [1] (Figure 4.1). HHCB has been listed as a high-production-volume chemical by the U.S. EPA, and more than 400 ton of this compound are produced in or imported into the United States every year [2].

Recently, it was reported that high concentrations of synthetic musks were detected in the effluents and sewage sludge of wastewater treatment plants (WWTPs) in European countries and in the United States [3–6]. As a result, HHCB and AHTN have been found in air, sediments, and organisms in the aquatic ecosystems [7–10]. This indicates that discharge from WWTPs may be a possible source of synthetic musks, and that these compounds have the potential for bioaccumulation in aquatic ecosystems due to their high lipophilic properties (log K_{ow} of HHCB and AHTN are 5.9 and 5.7, respectively [11]). However, little information is available on detailed bioaccumulation profiles of synthetic musks in the marine food chains, especially the occurrence and concentrations in higher trophic animals, such as seabirds and marine mammals.

BUVSs are known to absorb the full spectrum of UV light, UV-A and UV-B [12]. BUVSs have been used for various industrial and consumer products such as building materials, automobile components, wax, paint, adhesive agent, film, shoes, glasses, and some sports equipment. In Japan,

---

* E-mail: nakata@aster.sci.kumamuto-u.ac.jp (Chapter corresponding author).

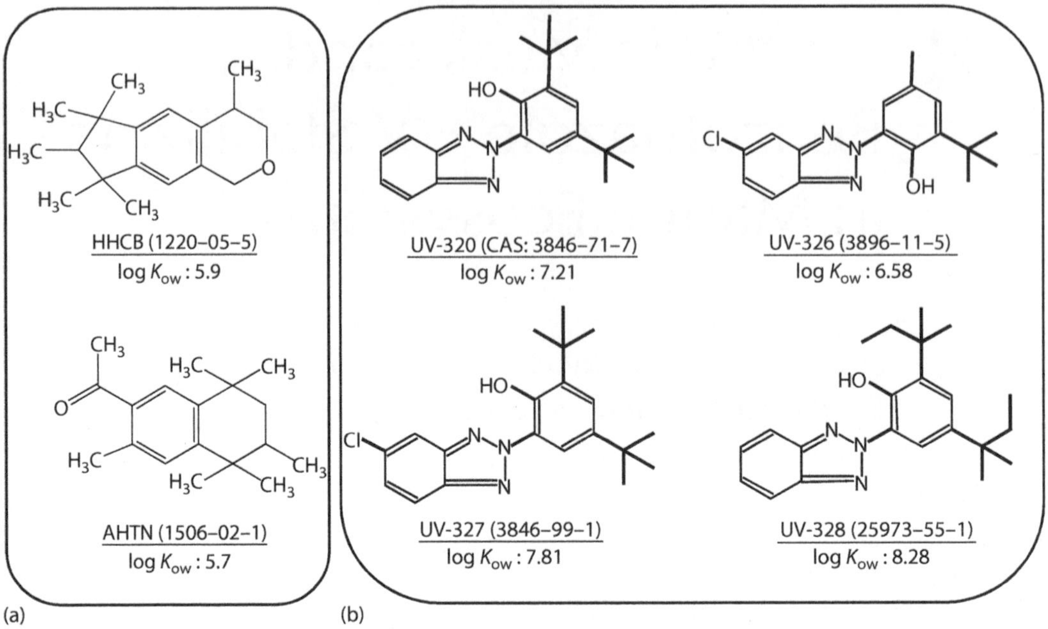

FIGURE 4.1 Chemical structures of (a) polycyclic musks and (b) benzotriazole UV stabilizers analyzed in this study.

2-(3-t-butyl-2-hydroxy-5-methylphenyl)-5-chlorobenzotriazole (UV-326, CAS#: 3896-11-5), 2,4-di-t-butyl-6-(5-chloro2H-benzotriazol-2-yl) phenol (UV-327; 3864-99-1), and 2-(2H-benzotriazol-2yl)-4,6-di-t-pentylphenol (UV-328; 25973-55-1) have been registered and used (Figure 4.1). 2-(3,5-di-t-butyl-2-hydroxyphenyl)-benzotriazole (UV-320; CAS#: 3846-71-7) has been manufactured, but the Japanese government banned the production, use, and import since 2007 due to its persistent, bioaccumulative, and toxic properties [13].

In the 1970s, BUVSs were detected in wastewater, river water, and sediments in the vicinity of a manufacturing plant of these compounds in Rhode Island [14]. Severe contamination by BUVSs was found in all matrices analyzed with the highest concentrations of UV-327 and UV-328 in river sediments reported as 300 and 100 μg/g, respectively. BUVSs were detected in clams purchased from a seafood store at concentrations of several ng/g to several tens of ng/g wet wt. basis [15]. However, details of bioaccumulation potencies of BUVSs in the aquatic ecosystems were not known.

During the past 5 years, our research group has investigated the occurrence, concentrations, and bioaccumulation profiles of emerging persistent contaminants, synthetic musks, and BUVSs in the marine ecosystem. We have analyzed tidal flat and shallow water species such as lugworm, lamp shell, oyster, clam, gastropod, crab, shrimp, mudskippers, shallow water teleost and cartilaginous fish, seabird, and marine mammals collected from the Ariake Sea, Japan between 1999 and 2009 (Figure 4.2). In this chapter, the results of these bioaccumulation profiles of synthetic musks and BUVSs are shown and discussed.

4.2 SYNTHETIC MUSKS

Gas chromatography-mass spectrometry (GC-MS) chromatograms and mass spectra of HHCB in standard and finless porpoise are shown in Figure 4.3. The retention time and mass spectra of HHCB in the sample agreed well with those in standard mixture. While the blank contained HHCB, the abundance of this under selected ion monitoring (SIM) (using sample peaks m/z 243, 258, and 213) was approximately two orders of magnitude less than that in the samples. These results strongly suggest the occurrence of HHCB in the blubber of marine mammals.

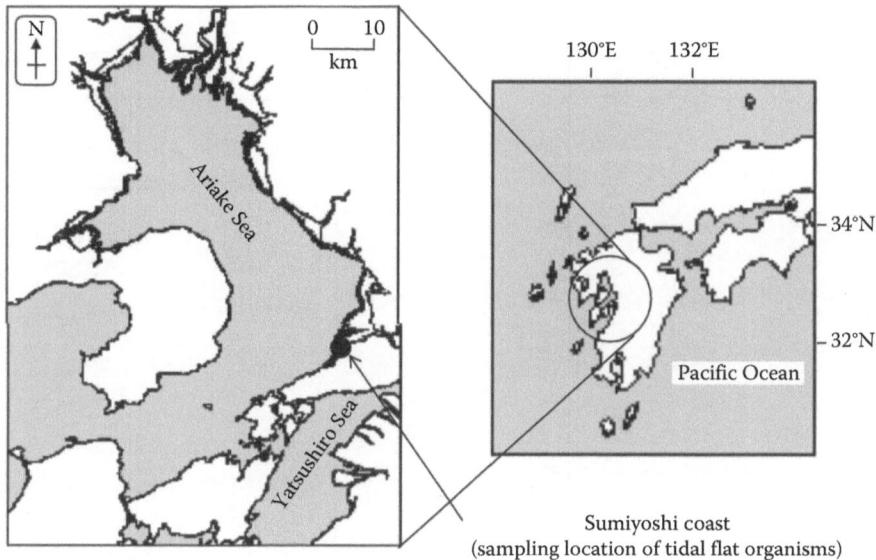

FIGURE 4.2 Map showing the sampling location of marine organisms in the Ariake Sea, western Japan.

HHCB and AHTN were detected in most marine organisms. In general, HHCB concentrations in samples collected in tidal flat areas were higher than those in the shallow water areas (Figure 4.4). The highest concentrations of HHCB were observed in tidal flat gastropods, at the average concentration of 1530 ng/g (lipid wt.). Elevated concentrations of HHCB were found in oyster and clam samples where the concentrations were more than 700 ng/g lipid wt. basis. These levels were comparable with those in mussels from the eastern coast of Canada (1650 ng/g lipid wt.) [9], and significantly higher than those in blue mussels from the North Sea in 1991–1992 (<30–110 ng/g) [4].

In shallow water organisms, large variations of HHCB concentrations were found among fish species. The concentrations in mullets, eagle rays, and hammerhead sharks were generally high at 100 ng/g lipid wt. (Figure 4.4). On the other hand, solefish and sweetrips showed lower concentrations of HHCB (<0.8 ng/g wet wt.), which may be due to species-specific accumulation of HHCB in fish. Interestingly, HHCB was detected in the blubber of finless porpoises and in the livers of seabirds, mallards, and black-headed gulls. Since these animals are the top predators of marine ecosystems, these results suggest that HHCB is persistent and bioaccumulative throughout the marine food chain.

A negative correlation between HHCB concentrations and the trophic level of the marine food chain was found (Figure 4.4). As described earlier, the highest concentrations of HHCB were found in gastropods, followed by clams and fish in tidal flat areas. The mean concentration of HHCB in herbivorous mudskippers (160 ng/g lipid wt.) was more than twofold greater than those in omnivorous mudskippers (67 ng/g), although the former species is at a lower trophic level. In addition, mean concentrations of HHCB in seabirds (19 ng/g lipid wt.) were comparable to those in mullets (14 ng/g) and flatheads (20 ng/g). These results are in contrast to the bioaccumulation pattern of persistent organochlorines such as polychlorinated biphenyls (PCBs). PCB concentrations in mussels, crustaceans, and fish from the tidal flat area of the Ariake Sea have been reported to be 590, 1100, and 2900 ng/g lipid wt. respectively, showing a positive correlation between the concentration and the trophic status of an organism in the tidal flat food web [15]. The difference in the bioaccumulation pattern between HHCB and PCBs is probably due to differing retention and metabolism of these compounds in those organisms.

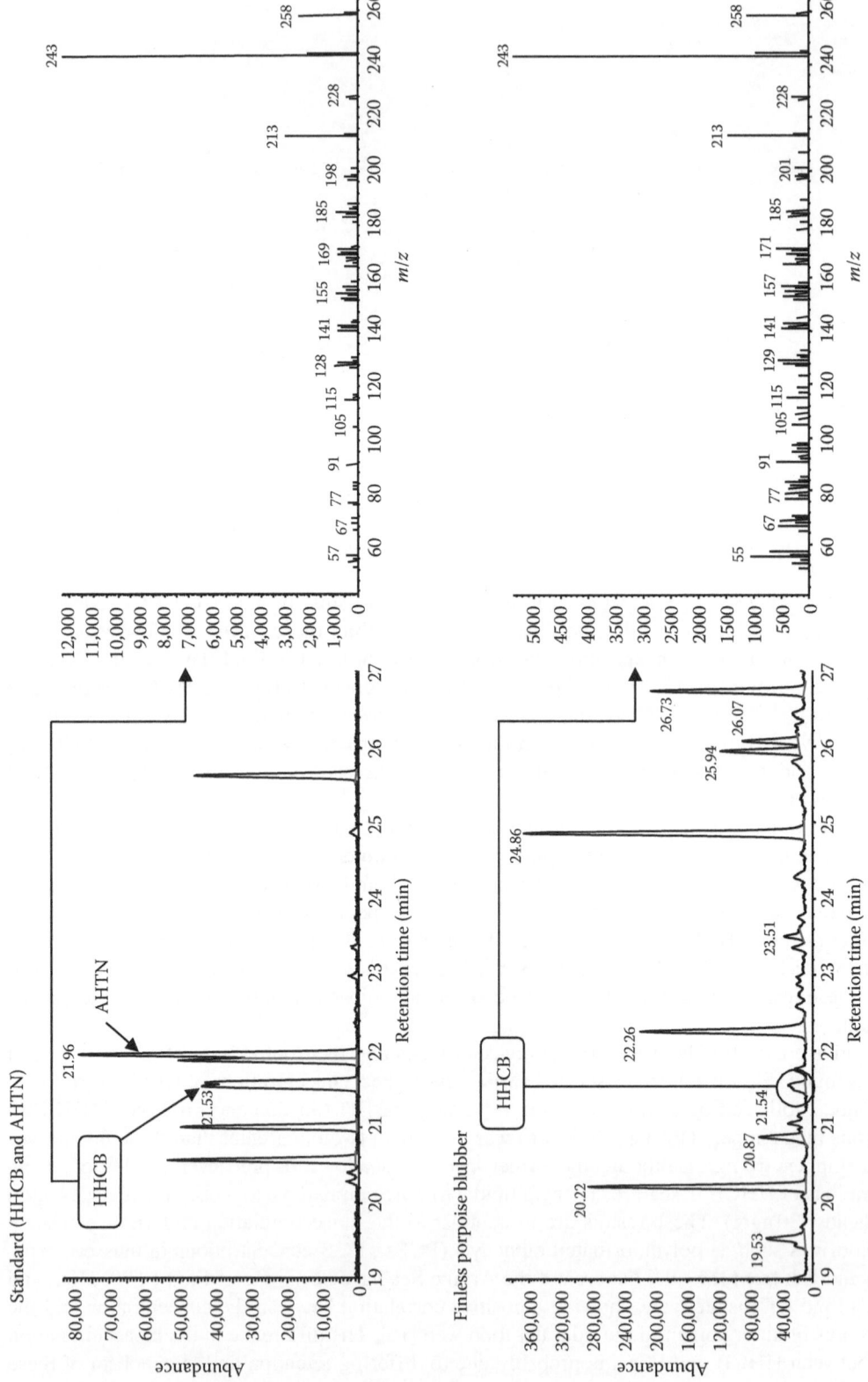

FIGURE 4.3 GC-MS chromatograms (SCAN mode) and mass spectra of HHCB in standard and blubber tissue of finless porpoise collected from the Ariake Sea.

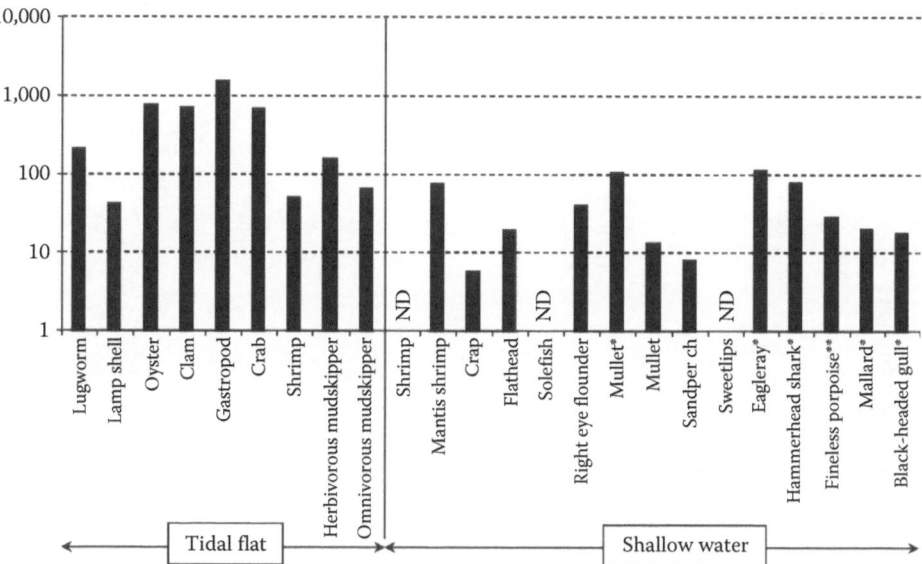

FIGURE 4.4   Concentrations of HHCB (ng/g lipid wt.) in tidal flat and shallow water species collected from the Ariake Sea, Japan. *, liver; **, blubber.

It was reported that HHCB concentrations in seawater of the Ariake Sea were 0.76 and 0.45 ng/L in summer and winter respectively (mean: 0.61 ng/L [16]). When this value is applied to the calculation of the bioconcentration factor (BCF), the BCF of HHCB in small fish (flathead, righteye flounder, sandperch) is 1459, which is comparable to the BCF investigated under the laboratory conditions (1584) [11]. Further, the BCF of HHCB in finless porpoise was estimated. The body burden of HHCB was calculated from the concentration and the weight of the blubber tissue of the animal. It was reported that blubber weight occupied 28.8% of the whole body weight of an adult finless porpoise [17], and it is well known that blubber of marine mammals plays an important role as a reservoir of persistent organic pollutants (POPs) [18]. Based on such information, HHCB concentration in finless porpoises (whole body weight basis) was calculated using the following formula:

$$\text{Concentrations (ng/g wet wt.)} = \frac{\text{total HHCB burden in blubber}}{\text{body weight}}$$

As a result, the concentration of HHCB in finless porpoise was 7.5 ng/g and the BCF of HHCB between seawater and marine mammal was calculated to be 12,295. The BCF values of HHCB in fish and finless porpoise are comparable with that of organochlorine pesticides, hexachlorocyclohexane (fish BCF: 1,000, marine mammal BCF: 37,000), in the western North Pacific ecosystem [19]. These observations may imply that HHCB is a potential POP in the marine environment because of the persistent and bioaccumulative properties through aquatic food chain. Further studies on the distribution, fate, and ecotoxicological aspects of HHCB are needed to assess the exposure and risk in the environment.

4.3   BENZOTRIAZOLE UV STABILIZERS

BUVSs were detected in all samples analyzed. The total ion chromatogram and mass spectra of UV-328 in a standard and in a liver sample of hammerhead shark are shown in Figure 4.5. The retention time of UV-328 in shark agreed well with that in the standard. The dominant fragment ion of the

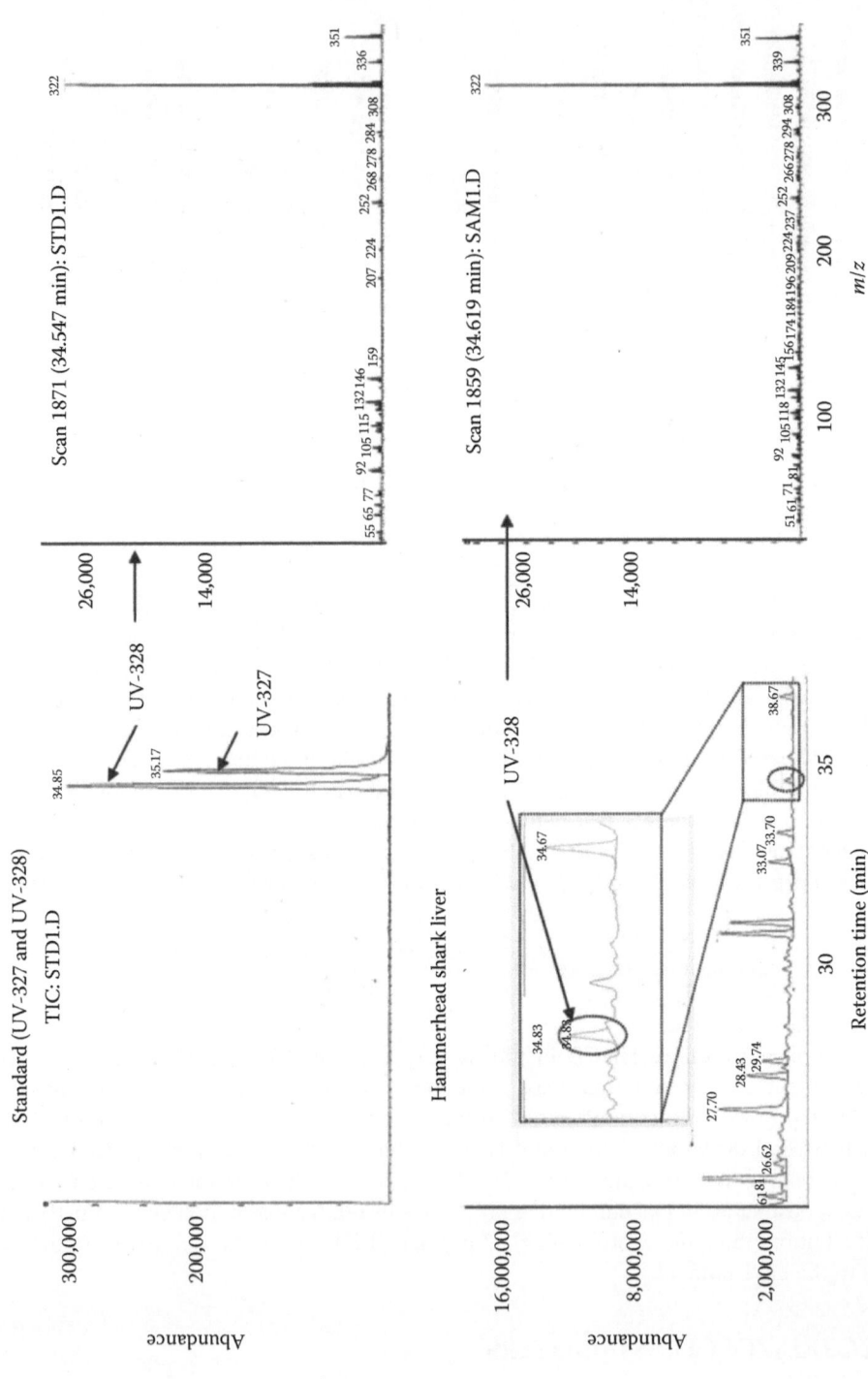

FIGURE 4.5 GC-MS chromatograms (SCAN mode) and mass spectra of UV-328 in standard and liver tissue of hammerhead shark collected from the Ariake Sea.

standard UV-328 under SCAN mode (electron impact ionization GC-MS) was m/z 322, followed by m/z 351 and were similar to those found in the sample. These results clearly indicate the occurrence and accumulation of UV-328 in liver tissue of the hammerhead shark. UV-326 was also identified based on the retention time and mass spectra of the standard solution and oyster sample [20].

The concentrations of BUVSs in tidal flat and shallow water species from the Ariake Sea are shown in Figure 4.6. UV-326, UV-327, and UV-328 were the dominant compounds at concentrations ranging from several to 460 ng/g lipid wt. basis. In contrast, concentrations of UV-320 in samples were low due to the low usage of this compound in Japan. As described previously, the domestic production, use, and import of UV-320 was been banned in 2007 because of its persistent, bioaccumulative, and toxic properties [13].

In general, concentrations of BUVSs in tidal flat organisms were greater than those in shallow water species, which was similar to those of the synthetic musk HHCB (Figure 4.6). The average concentrations of UV-320 and UV-326 in tidal flat species were approximately 10–20-fold higher than those in shallow water organisms. The tidal flat clam showed the highest concentrations of UV-320 and UV-326 at 74 and 219 ng/g (lipid wt.) respectively. Elevated concentrations of UV-326 were also found in oysters and gastropods in the tidal flat area. These results suggest the presence of BUVSs in sediment resulting in the accumulation of these compounds in benthic organisms. On the other hand, the low concentrations of UV-326 in shallow water species might be explained by low BCF of this compound as compared with other BUVSs. The BCFs of UV-326 in carp are 54–109, 196–802, 548–892 at the exposure concentrations of 0.5, 0.05, and 0.005 mg/L respectively [21], which in turn are approximately one order of magnitude lower than those of UV-320 (1,380–10,000).

UV-327 was the most frequently detected compound in marine organisms (Figure 4.6). The average concentrations of UV-327 in tidal flat organisms were only twofold higher than those in shallow water species. The tidal flat clam, crab, and herbivorous mudskipper contained high concentrations of UV-327, greater than 100 ng/g (lipid wt.), followed by gastropods and oysters. In shallow water fish, concentrations of UV-327 were three- to fourfold higher in the liver than in the carcass. These results are consistent with the concentration profiles of UV-328 in mullet, suggesting the preferential accumulation and decreased biodegradation of this compound in the liver of some fish species. Interestingly, seabirds accumulate UV-327 in the liver at average concentrations of 90 ng/g in a spot-billed duck and 59 ng/g in mallards (Figure 4.6). UV-327 and UV-328 were detected in the blubber tissue of finless porpoises. The occurrence of UV-327 in seabirds and marine mammals suggests the significant bioaccumulation in higher trophic species through the aquatic food chain.

It was reported that UV-327 concentrations in seawater from four coastal areas of Tokyo Bay were less than 0.5 ng/L [22]. The Ministry of Environment of Japan investigated UV-327 concentrations in river, lake, and coastal water samples ($n=44$) and reported the geometric mean concentration of 0.12 ng/L [23]. When this value was applied to the calculation of BCF the BCF of UV-327 in small fish was 3250, which is comparable to the values examined under laboratory conditions (3400–9000) [21]. The concentrations of UV-327 and the BCF in finless porpoise were also calculated at values of 4.0 ng/g and 33,400 respectively. These results suggest that UV-327 appears to be persistent and bioaccumulative in the environment and further studies are needed to assess the ecotoxicological risks of this compound in the aquatic ecosystem.

Concentrations of UV-328 in biota were variable and species specific (Figure 4.6). The highest concentration was found in the tidal flat gastropod at 460 ng/g (lipid wt.), followed by the mullet (120 ng/g in whole body and 250 ng/g in liver) and the hammerhead shark (130 ng/g in liver) from shallow water areas. The oysters and clams in the tidal flats contained high concentrations of UV-328, at >100 ng/g. The log K_{ow} of UV-328 was the highest (8.28) among benzotriazole UV stabilizers analyzed (Figure 4.1), but the BCFs of UV-328 were relatively low, 570–1400 and 620–2700 at the exposure concentrations of 0.1, 0.01 for 60 days respectively [21]. The large variations in UV-328 concentrations observed in this study might be due to differences in biodegradation of this compound in marine organisms.

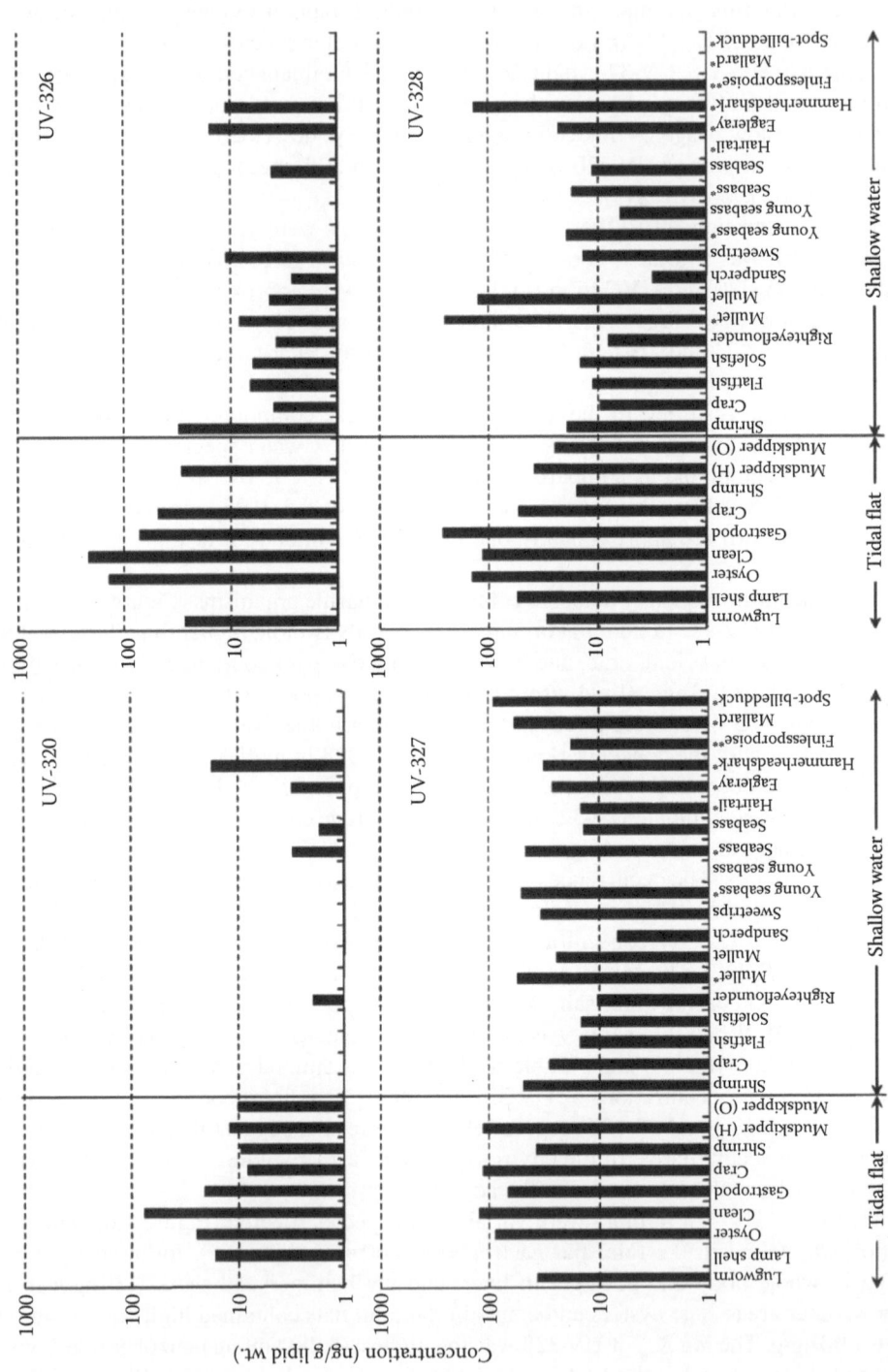

FIGURE 4.6 Concentrations of benzotriazole UV stabilizers in tidal flat and shallow water species collected from the Ariake Sea, Japan. *, liver; **, blubber.

4.4 SUMMARY AND FUTURE TRENDS

The bioaccumulation profiles of synthetic musks and BUVSs in the aquatic ecosystem were investigated by analyzing marine organisms at various trophic levels, such as lugworm, clam, crustacean, fish, marine mammal, and seabird samples collected from the Ariake Sea, Japan. A polycyclic musk, HHCB, and BUVSs, such as UV-326, UV-327, and UV-328, were detected in most samples analyzed, which suggest a ubiquitous contamination and distribution of these compounds in Japanese coastal waters. The high concentrations of HHCB and BUVSs were found in clams and oysters in the tidal flat areas, implying their significant accumulations in sediment. Interestingly, HHCB and UV-327 were also detected in higher trophic species, such as sharks, marine mammals, and seabirds. The occurrence of musks and BUVSs in higher trophic species strongly suggests their persistent and bioaccumulative properties in the marine food chains.

As for future trends on the study on emerging contaminants in the environment, we have recently investigated geographical distribution and temporal trend of synthetic musks and BUVSs in the environment. The concentrations of musks and BUVSs in blue and green mussels collected from 10 Asian countries and regions such as Cambodia, China, Hong Kong, India, Indonesia, Japan, Korea, Malaysia, the Philippines, and Vietnam were examined and we found that HHCB and BUVSs were detected in most samples analyzed at the concentrations of several to several tens of ng/g wet wt. [24]. HHCB and UV-326 were the dominant compounds, and the concentrations were high in mussels from Korea, Japan, and Hong Kong. Musks and BUVSs were also detected in mussels from the western coast of the United States, which suggests the widespread contamination of these emerging pollutants in the Asia-Pacific regions [25].

The temporal trends of BUVSs concentrations were examined by analyzing sediment core samples collected from Tokyo Bay, Japan. UV-326 was the most abundant in sediments, at the highest concentration of 37 ng/g dry wt. basis [26]. UV-326, UV-327, and UV-328 were detected in sediment layers around the 1970s, which suggests that BUVSs have been used in Japan for the past 30 years. Further investigations on the global contamination, source identification, and human exposure to emerging pollutants, such as musks and BUVSs, are necessary to elucidate their risks and hazards in the environment.

ACKNOWLEDGMENTS

This study was partly supported by a grant-in-aid for scientific research from the Japan Society for the Promotion of Science (JSPS) (Grant No. 20510067), and the Global Environmental Research Fund (RF-0904) of the Ministry of the Environment, Japan.

REFERENCES

1. Sommer, C. 2004. The role of musk compounds in the fragrance industry. In: *Synthetic Musk Fragrances in the Environment*, Rimkus, G. G. (ed.). Springer-Verlag, Berlin, Germany, pp. 1–16 (ISBN: 3-540-43706-1).
2. USEPA. 2003. Toxic Substances Control Act inventory, High Production Volume (HPV) Chemical List Database. Available at http://www.epa.gov/chemrtk/opptsrch.htm (accessed on March 18, 2009).
3. Kallenborn, R., Gatermann, R., Planting, S., Rimkus, G. G., Lund, M., Schlabach, M., and Burkow, I. C. 1999. Gas chromatographic determination of synthetic musk compounds in Norwegian air samples. *J. Chromatogr. A*, 846: 295–306.
4. Rimkus, G. 1999. Polycyclic musk fragrances in the aquatic environment. *Toxicol. Lett.*, 111: 37–56.
5. Gatermann, R., Hühnerfuss, H., Rimkus, G., Wolf, M., and Franke, S. 1995. The distribution of nitrobenzene and other nitro aromatic compounds in the North Sea. *Mar. Pollut. Bull.*, 30: 221–227.
6. Reiner, J., Berset, J. D., and Kannan, K. 2007. Mass flow of polycyclic musks in two wastewater treatment plants. *Arch. Environ. Contam. Toxicol.*, 52: 451–457.
7. Peck, A. M. and Hornbuckle, K. C. 2004. Synthetic musk fragrances in Lake Michigan. *Environ. Sci. Technol.*, 38: 367–372.

8. Fromme, H., Otto, T., and Pitz, K. 2001. Polycyclic musk fragrances in different environmental compartments in Berlin (Germany). *Water Res.*, 35: 121–128.

9. Gatermann, R., Hellou, J., Huhnerfuss, H., Rimkus, G., and Zitko, V. 1999. Polycyclic and nitro musks in the environment: A comparison between Canadian and European aquatic biota. *Chemosphere*, 38: 3431–3441.

10. Nakata, H., Sasaki, H., Takemura, A., Yoshioka, M., Tanabe, S., and Kannan, K. 2007. Bioaccumulation, temporal trend, and geographical distribution of synthetic musks in the marine environment. *Environ. Sci. Technol.*, 41: 2216–2222.

11. Balk, F. and Ford, R. A. 1999. Environmental risk assessment for the polycyclic musks AHTN and HHCB in the EU I. Fate and exposure assessment. *Toxicol. Lett.*, 111: 57–79.

12. Tenkazai.com. 2007. Market trend of resin additives 'Light stabilizer'. Available at http://www.tenkazai.com/market.html, accessed on March 18, 2009.

13. Ministry of Economy, Trade and Industry, Japan. 2006. 2-(3,5-di-*t*-butyl-2-hydroxyphenyl)-benzotriazole. Available at www.meti.go.jp/committee/materials/downloadfiles/g60705a03j.pdf, accessed on March 18, 2009.

14. Jungclaus, G. A., Lopez-Avila, V., and Hite, R. A. 1978. Organic compounds in an industrial wastewater: A case study of their environmental impact. *Environ. Sci. Technol.*, 12: 88–96.

15. Nakata, H., Sakai, Y., and Miyawaki, T. 2002. Growth-dependent and species specific accumulation of polychlorinated biphenyls (PCBs) in tidal flat organisms collected from the Ariake Sea, Japan. *Arch. Environ. Contam. Toxicol.*, 42: 222–228.

16. Nakata, H., Sakanashi, Y., and Takikawa, K. 2008. Synthetic musks in air and seawater of the Ariake Sea. *Proceedings of the 17th Symposium on Environmental Chemistry*, Kobe, June 11–13, pp. 646–647 (in Japanese).

17. Iwata, H., Tanabe, S., Mizuno, T., and Tatsukawa, R. 1995. High accumulation of toxic butyltins in marine mammals from Japanese coastal waters. *Environ. Sci. Technol.*, 29: 2959–2962.

18. Tanabe, S., Tatsukawa, R., Tanaka, H., Maruyama, K., Miyazaki, N., and Fujiyama, T. 1981. Distribution and total burdens of chlorinated hydrocarbons in bodies of striped dolphins (*Stenella coeruleoalba*). *Agric. Biol. Chem.*, 45: 2569–2578.

19. Tanabe, S., Tanaka, H., and Tatsukawa, R. 1984. Polychlorobiphenyls DDT, and hexachlorocyclohexane isomers in the Western North Pacific ecosystem. *Arch. Environ. Contam. Toxicol.*, 13: 731–738.

20. Nakata, H., Murata, S., and Filatreau, J. 2009. Occurrence and concentrations of benzotrizole UV stabilizers in marine organisms and sediments from the Ariake Sea, Japan. *Environ. Sci. Technol.*, 43: 6920–6926.

21. National Institute of Technology and Evaluation, Japan. Chemical Collaboration Knowledge database (J-CHECK). Available at http://www.safe.nite.go.jp/jcheck/Top.do;jsessionid=E9FD388CF12981F80CEC776C7C897DA1, accessed on March 18, 2009.

22. Konuma, S., Ogawa, A., Masunaga, S., and Nakamura, Y. 2009. UV filters in seawater and sediments: Partitioning and horizontal distributions on the coastal shoreline in and around the Tokyo Bay. *Proceedings of the 18th Symposium on Environmental Chemistry*, Tsukuba, June 9–11, pp. 448–449 (in Japanese).

23. Ministry of Environment, Japan. 2006. Available at www.env.go.jp/chemi/kurohon/2006/shosai/03_0.pdf, accessed on March 18, 2009. Report on investigation results of chemical exposure (Bakuro-ryou chosa Kekka Hokokusyo).

24. Nakata, H., Murata, S., Shinohara, R., Filatreau, J., Isobe, T., Takahashi, S., and Tanabe, S. 2009. Occurrence and concentrations of persistent personal care products, organic UV filters, in the marine environment. *Interdisciplinary Studies on Environmental Chemistry-Environmental Research in Asia*, Obayashi, Y., Isobe, T., Suzuki, S., and Tanabe, S. (eds.). TERRAPUB Publishing, Tokyo, Japan, pp. 239–246.

25. Nakata, H., Shinohara, R., Nakazawa, Y., Watanabe, M., Isobe, T., Tanabe, S., Kannan, K., and Ueno, D. 2010. Monitoring of synthetic musks and UV stabilizers concentrations in bivalves collected from Asian coastal and US West coast waters. *Proceedings of the 19th Symposium on Environmental Chemistry*, Nagoya, June 21-23, pp. 562–563 (in Japanese).

26. Nakata, H., Shinohara, R., Isobe, T., Tanabe, S., and Watanabe, M. 2010. Benzotriazole UV stabilizers in sediment core samples from Tokyo Bay, and evaluation of its potential sources. *Proceedings of the 19th Symposium on Environmental Chemistry*, Nagoya, June 21-23, pp. 68–69 (in Japanese).

5 Global Environmental Distribution and Human Health Effects of Polycyclic Aromatic Hydrocarbons

Aramandla Ramesh,* Anthony E. Archibong, Darryl B. Hood, Zhongmao Guo, and Bommanna G. Loganathan

CONTENTS

5.1 Introduction 98
    5.1.1 Properties...99
    5.1.2 Sources...100
5.2 Environmental Occurrence and Concentration Levels.......................................101
    5.2.1 PAHs in Air 101
    5.2.2 PAHs in Water 102
    5.2.3 PAHs in Soil 103
    5.2.4 PAHs in Sediment...104
    5.2.5 Relative Contribution of Different Sources to PAH Pollution.................106
5.3 Global Distribution Trends and Current Status 107
5.4 Human Exposure and Intake Estimates 109
5.5 Human Health Effects 110
    5.5.1 Neurotoxicity 111
    5.5.2 Reproductive Toxicity..112
        5.5.2.1 Male Reproduction...112
        5.5.2.2 Female Reproduction 113
    5.5.3 Uterine Tumors 114
    5.5.4 Osteoporosis 114
    5.5.5 Cancer...114
        5.5.5.1 Dermal 114
        5.5.5.2 Esophageal...114
        5.5.5.3 Colorectal..115
        5.5.5.4 Breast 115
        5.5.5.5 Lung 115
    5.5.6 Immunotoxicity 116
    5.5.7 Obesity 116
    5.5.8 Atherosclerosis..116

* E-mail: aramesh@mmc.edu (Chapter corresponding author).

5.6   Conclusions and Future Research Directions 117
Acknowledgments 117
References 117

5.1   INTRODUCTION

The British Petroleum (BP) oil spill in the Gulf of Mexico, the coal ash spill in Eastern Tennessee, and the wildfires near Flagstaff, Arizona, and in California share one thing in common. These environmental disasters spew a family of toxicants called polycyclic aromatic hydrocarbons (PAHs) into the environment. PAHs are a family of ubiquitous environmental contaminants that consist of more than 100 chemicals. The structures of some commonly encountered PAHs are shown in Figure 5.1.

PAHs pose the greatest risk to the environment [1]. Based on calculations made in NAS [1], it has been estimated that as of May 2010, approximately 270–3300 metric ton of PAHs may have been released from the *Deepwater* rig spill [2]. The 100–1000 metric ton of PAHs released from the World Trade Twin Towers' collapse in 2001 [3] pales in comparison to the Gulf oil spill. Being water soluble and semi-volatile, these chemicals are mobilized into the environment in a rapid manner. The U.S. EPA has designated 16 PAH compounds as priority pollutants for environmental monitoring purposes (EPA 16 PAHs). These PAHs are naphthalene, acenaphthylene, acenaphthene, fluorene, phenanthrene, anthracene, fluoranthene, pyrene, benzo[*a*]anthracene, chrysene, benzo[*b*]fluoranthene, benzo[*k*]fluoranthene, benzo[*a*]pyrene, dibenz[*a,h*]anthracene, benzo[*g,h,i*] perylene, and indeno[*1,2,3-cd*]pyrene [4]. Chronic exposures to even low concentrations of these chemicals cause long-lasting damage such as cancer, infertility, and neurotoxicity to humans and wildlife [4–6].

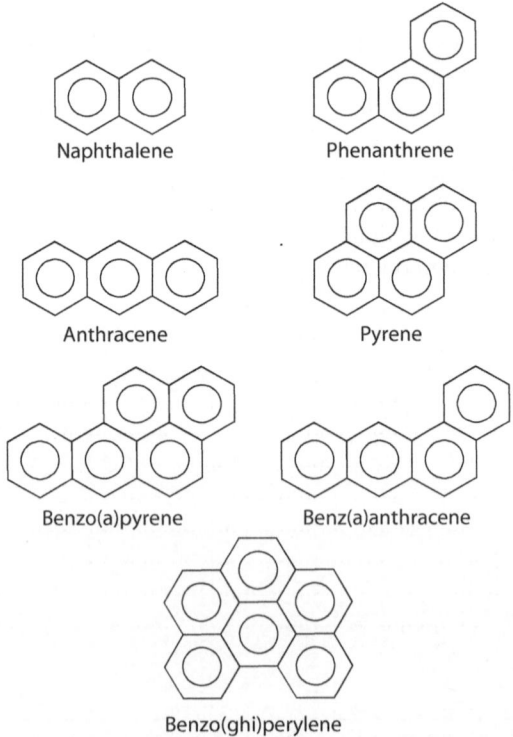

FIGURE 5.1   Some representative PAH compounds.

5.1.1 PROPERTIES

PAHs are compounds with two or more fused aromatic rings. They are sparingly soluble in water, but are highly lipophilic. These compounds possess low vapor pressure. They are adsorbed on particles in the air, soil, water, and sediment. PAHs can undergo photodecomposition when exposed to ultraviolet light from solar radiation. These compounds are not released into the environment as single compounds, but often occur as mixtures, the composition of which depends on the release from various point sources. In the environment, they coexist with amines, chlorinated hydrocarbons, and metals. In the atmosphere, PAHs can react with pollutants such as ozone, nitrogen oxides, chlorine, and sulfur dioxide, yielding diones, nitro- and dinitro-PAHs, chlorinated PAHs, and sulfonic acids, respectively. These compounds in soil and sediments are also subjected to microbial degradation [7,8].

PAHs have been categorized into two groups, the peri- and cata-condensed. Peri-condensed PAHs are defined as those whose lines connect the ring centers, and form cycles. They are subdivided into two classes: alternants, which are formed exclusively by six-membered rings, and non-alternants that include some five-membered rings. Cata-condensed PAHs are defined as those systems whose lines do not form cycles, and can be classified as branched or not branched, the former of which is thermodynamically more stable and chemically less reactive than their nonbranched counterparts of the same size. The categorization of PAHs on the basis of their structures is shown in Figure 5.2.

The structure of PAHs (relative positions of carbon atoms in space of the fused rings) determine their biological activity; the "K" region is defined as the external corner of a phenanthrene moiety; the "L" region consists of a pair of opposed open anthracenic point atoms; the "bay" region is defined as an open inner corner of a phenanthrene moiety; the distal bay region also known as the "M" region; and the peri position, which corresponds to the carbon atom opposite the bay region and adjacent to the angular ring. The organization of carbon atoms as a bay region imparts a high degree of biochemical reactivity to some PAHs and their metabolites. The structure of benz(a) anthracene, a PAH molecule detailing the various regions, is depicted in Figure 5.3.

Compound	Formula	Graph	Characteristics
Peri-condensed			
Pyrene	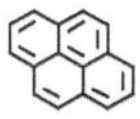		Cycle Alternant
Fluoranthene	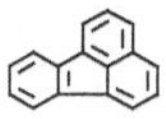		Cycle Non-alternant
Cata-condensed			
Chrysene	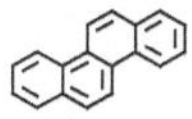		Linear Nonbranched Alternant
Triphenylene	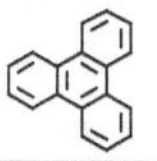		Linear Branched Alternant

FIGURE 5.2 Main structural categorizations of PAHs. (Reprinted from Ramesh, A. et al., *Int. J. Toxicol.*, 23, 301, 2004. With permission.)

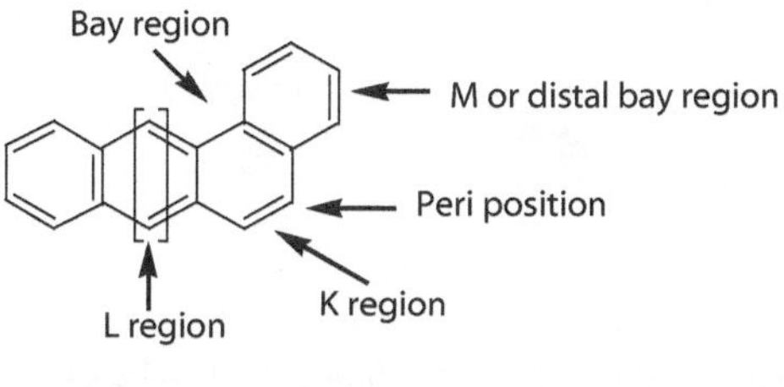

FIGURE 5.3 Regions related to biological activity in PAHs. (Reprinted from Ramesh, A. et al., *Int. J. Toxicol.*, 23, 301, 2004. With permission.)

The toxic and carcinogenic characteristics of PAHs prompted several countries to include them in their environmental monitoring programs. As a result, the PAH pollution from point sources (oil shipping and refinery operations, sewage and industrial effluents) has successfully been curtailed, whereas PAH pollution from nonpoint sources (atmospheric deposition and surface runoffs) is difficult to abate and still constitutes major inputs into the environment.

5.1.2 Sources

PAMs are released into the environment during volcanic eruptions; forest fires; burning of coal, wood, and incineration of municipal refuse; and expulsion of fumes from manufacturing industries such as coke, aluminum, graphite-electrode, carbon-electrode, and petroleum. Lifestyle/domestic activities such as smoking, incense, candle and mosquito coil burning, and cooking also contribute significantly to the contamination of the environment by PAHs, which have the propensity to accumulate to toxic levels in the body within a short period of time. By far, the greatest emissions of PAHs result from human and industrial activities [5,6].

Some representative PAH sources are depicted in Figure 5.4. Most of the PAHs found in the environment arise from three principal sources: pyrogenic, petrolytic, and diagenic. Pyrogenic PAHs are formed from incomplete, high-temperature combustion of organic matter. These PAHs are believed to result from breakdown of organic matter to lower-molecular-weight compounds during pyrolysis, followed by the rapid rearrangement into nonalkylated PAH structures [10,11]. The petrogenic PAHs derive from biological materials in sediments over the geologic timescale, resulting in the formation of fossil fuels enriched with PAHs [11,12]. These PAHs have predominantly alkylated structures. Diagenic PAHs are derived from biogenic material through anaerobic processes. They are present in older sediments and predate industrial activity [13]. The origin and composition of these different categories of PAHs is a matter of debate [14]. Attempts to identify the origin of PAHs (petrogenic or pyrogenic) in field studies are hampered by the distribution of these compounds regardless of source, in the environment; hence the PAH signatures from one source can be obscured by PAHs from another source [15].

FIGURE 5.4 Sources and routes of exposure to environmental PAHs. (Reprinted from Davila, D.R. et al., *Toxicol. Ecotoxicol. News*, 4, 5, 1997. With permission from Taylor & Francis.)

5.2 ENVIRONMENTAL OCCURRENCE AND CONCENTRATION LEVELS

Levels of PAH in the hydrosphere and geosphere are affected by both wet and dry deposition as well as deposition of contaminated refuse. However, PAH levels in the biosphere are impacted by exposure to polluted air and biomagnifications through the food chain [16]. Partitioning between water and air, water and sediment, and water and biota are key factors for the environmental persistence of PAHs. The affinity of PAHs for organic materials in soil and sediments and lipids in biota are high enough to cause accumulation in the above-mentioned matrices with long-lasting consequences for environmental health. The photo- and microbial degradation of PAHs notwithstanding, considerable amounts of these compounds are present in the environment.

From a regulatory and academic standpoint, the ecotoxicology and human health-related issues of PAHs are of paramount importance. Some PAHs have been implicated as causative agents for lung, breast, esophageal, pancreatic, gastric, colorectal, bladder, skin, prostate, and cervical cancers in animal models and humans. In addition to carcinogenicity, PAHs have also been reported to cause hemato-, cardio-, renal-, neuro-, immuno-, reproductive-, and developmental-toxicities in laboratory animals and humans (reviewed in [5,6,17–22]).

5.2.1 PAHs in Air

PAHs in the atmosphere occur in gaseous as well as particulate phases. In urban areas, emissions from gasoline and diesel vehicles contribute to a major proportion of particulate PAHs. In rural areas biomass burning and coal (residential heating and cooking) contribute to particulate PAHs. The concentrations of PAHs in air from different geographical regions are shown in Figure 5.5. Greater levels of PAHs were registered in Asia, Africa, and Latin America compared to Europe, North America (except Mexico), and Australia. The low population densities, strict environmental regulations, and compliance and enforcement aspects have led to a decrease in PAH emissions into the atmosphere in developed nations. On the contrary, increase in population growth, reliance on biofuel, and lack of atmospheric PAH reduction measures (from traffic and industrial emissions) catapulted the PAH emissions in developing countries. This trend is likely to continue for some more years.

The PAH levels in air show seasonal variation. Studies (cited in [45]) have shown that heavy usage of biofuel during the fall and winter seasons in developing countries to meet the residential energy requirements led to high levels of emissions into the atmosphere compared to the spring and summer seasons. While forest fires occur in summer, their relative contribution to atmospheric PAH levels is not very high.

It is not uncommon to see that emissions from one country could affect the PAH levels of neighboring countries. For example, while coal consumption in Japan and Korea has not changed much since 1990, that of China has increased since 2000. Since China consumes 34% of the world's coal, PAH emissions from coal combustions in China and North Korea are likely to influence the PAH levels in South Korea and Japan considerably [45]. PAH emissions from China travel thousands of kilometers to other Asian countries and the Pacific Ocean. Lang et al. [46] showed that of the PAH outflow from China (8092 ton total), 227, 71, 746, and 131 ton reached North Korea, South Korea, Russia-Mongolia region, and Japan, respectively. Furthermore, these authors [46] also reported that 1.4 ton of the above-mentioned emissions reached North America through Trans-Pacific atmospheric transport.

Biomass burning is a common practice in developing countries in Asia and Africa. Field burning after harvest and biofuel burning in rural areas for indoor heating and cooking purposes in Asia and Africa release greater amounts of PAHs into the atmosphere. Asian megacities are the other significant contributors to high levels of local air pollution by PAHs. The increasing urbanization and vehicular emissions contribute to high PAH levels. Thus, while emissions from regional sources dominate the local PAH concentrations, transport of low-molecular-weight PAHs from Asia, Africa, and Latin America to temperate and polar regions is significant.

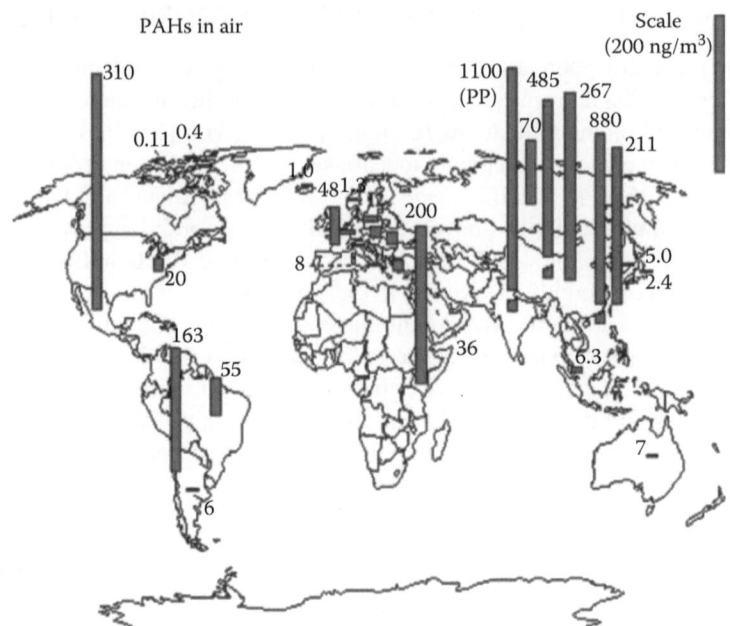

FIGURE 5.5 Spatial variations in atmospheric PAH concentrations (ng/m^3) measured in different parts of the world. The data presented in this map are taken from published literature. The pertinent reference for each country was given in parentheses. The countries represented are Argentina [23], Australia [24], Belgium [25], Brazil [26], Canadian Arctic [27], Chile [28], China [29], Czech Republic [30], Germany [31], Greece [32], Hungary [33], Iceland [34], Ireland [34], India [35], Italy [36], Japan [37], Kenya [38], Malaysia [39], Mexico [40], Norway [34], Russia [34], South Korea [41], Taiwan [42], Turkey [43], and the United States [44]. The values represent the sum total of major PAHs. Most of the data reported here represent the PAH concentrations in the particulate phase, while for few locations, composite data (particulate + vapor phase) are shown. The measurements were made during different years. Due to space constraints, the name of the sampling location and year was not marked on the top of the bar. Some measurements were made in urban and some in rural locations. Also, measurements were made during different seasons.

5.2.2 PAHs in Water

PAHs in water occur in dissolved, particulate, and suspension phases. Atmospheric deposition (wet and dry) and land runoff are major sources of PAH pollution in the aquatic environment. Wastewater and refinery effluents discharge PAHs into urban bodies of water while refuse burning and dry and wet depositions contribute to PAH pollution in rural bodies of water. A coal ash spill from a coal-fired electric power plant in Harriman, Tennessee, in 2008 released 9.4 million cubic yards of coal ash into the surrounding environment, a large amount of which was let into the Emory River [47]. Environmental disasters like this contribute to a significant loading of three to four ring PAHs that constitute a larger percent of coal ash [48] into the aquatic environment, affecting the water quality. The emission rates for PAHs discharging into coastal waters could reach up to 100 Mton/year [49].

The concentrations of PAHs in bodies of water from different geographical regions are presented in Figure 5.6. The geographical distribution trend of PAH concentrations in water (Figure 5.6) mirrors that of air (Figure 5.5). The spatial variations clearly show the contribution of developing nations toward PAH pollution. The PAH distributions in these countries are symptomatic of industrial discharges and petroleum sources. In developed countries such as the United States, regulations like the Clean Water Act of 1972 mandate that all discharges into bodies of water must be regulated by local/federal government agencies. The absence or lack of implementation of such legislation in developing countries leads to deterioration of surface water quality. As mentioned earlier, increasing

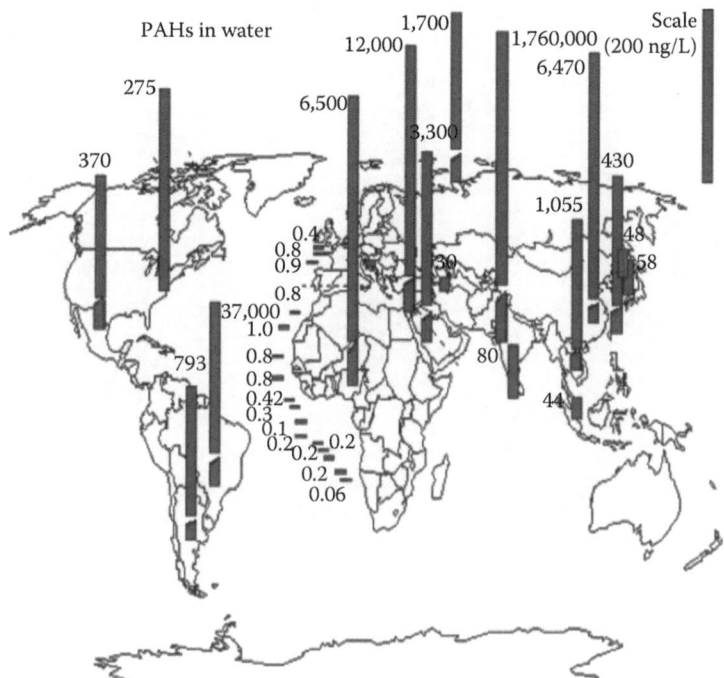

FIGURE 5.6 Spatial variations in water PAH concentrations (ng/L) measured in different parts of the world. The data presented in this map are taken from published literature. The pertinent reference for each country was given in parentheses. The countries/territories represented are Argentina [50], Baltic Sea [51], Brazil [52], China [53], Egypt [54], India [55], Iran [56], Italy [57], Japan [58], Korea [59], Mexico [60], Nigeria [61], North-South Atlantic transect [62], Saudi Arabia [63], Singapore [64], Sri Lanka [65], Taiwan [66], Thailand [67], and the United States [68]. The values represent the sum total of major PAHs. The data reported here represent the PAH concentrations in dissolved phase and from subsurface waters. The measurements were made during different years and during different seasons. Due to space constraints, the name of the sampling location and year was not marked on the top of the bar. Most of the samples collected were from marine, estuarine or riverine locations.

urbanization and population growth will contribute to unabated input of PAHs to freshwater and coastal marine environments in Asia, Africa, and Latin America in the coming years.

Riverine inputs discharge PAHs into estuarine waters. As a result, inshore waters register greater levels of PAHs compared to offshore waters. Due to their high water solubility, low-molecular-weight PAHs such as naphthalene are dissolved in water and could be transported to greater distances such as offshore, temperate, and polar regions through movement of water masses. The high-molecular-weight PAHs are sparingly soluble in water and tend to associate with particulate matter. Hence, high-molecular-weight PAHs in marine waters are indicative of prominent local inputs.

5.2.3 PAHs in Soil

The concentrations of PAHs in the soil are several folds greater than those in the air. Both wet and dry depositions contribute to PAH levels in soil. The partitioning and accumulation of PAHs in soil depends on their physicochemical properties (molecular weight, solubility, electrochemical stability, polarity of soil matrix—i.e., moieties in soil organic matter containing nitrogen and oxygen), aromaticity, carbon type, and content in soil [14,69,70]. Soil is a complex matrix which consists of aqueous phase colloids, nonaqueous phase liquids, soot carbon, and organic carbon. In general, PAHs show a strong affinity for colloidal and organic carbon. However, PAHs released from point source areas near manufactured gas plant sites show preferential accumulation in soil that contains nonaqueous

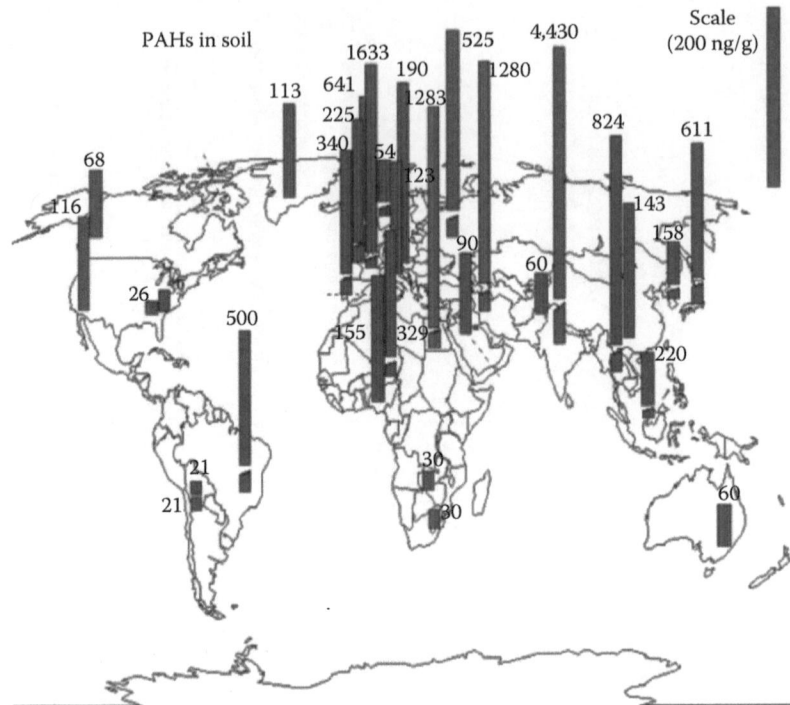

FIGURE 5.7 Spatial variations in soil PAH concentrations (ng/g dry wt.) measured in different parts of the world. The data presented in this map are taken from published literature. The pertinent reference for each country was given in parentheses. The countries/territories represented are Australia [71], Bahrain [72], Bolivia [73], Brazil [74], Canada [75], Chile [73], China [76], Egypt [77], France [78], Greenland [79], India [80], Iran [81], Italy [82], Japan [83], Malaysia [39], Niger [84], Nigeria [85], Norway [71], Pakistan [71], Russia [86], Serbia [87], South Africa [71], South Korea [71], Spain [88], Switzerland [89], Thailand [90], the United Kingdom [71], United States—South Carolina, Georgia [91], and Texas [92]—and Zambia [71]. The values represent the sum total of major PAHs. Most of the data reported here represent the PAH concentrations in rural and urban road side soils. The measurements were made during different years. Due to space constraints, the names of the sampling locations and years were not marked on the top of the bar.

phase liquids [14]. The different phases have specific partitioning characteristics that affect the chemical fate and bioavailability of PAHs. For instance, PAHs with less than three rings are relatively more amenable to microbial degradation compared to PAHs with more than three rings [8].

The concentrations of PAHs in bodies of water from different geographical regions are depicted in Figure 5.7. As in the case of air and water, countries in Asia and Africa registered high levels of PAHs. However, countries from the western hemisphere also showed comparable levels to some developing countries. The distribution profiles also permit the formulation of the following generalizations. Locations that lie in close proximity to long-term emission sources (discharge of industrial effluents, petroleum refining, and transport activities) that also receive high atmospheric depositional inputs and through accidental release of raw and refined petroleum products contribute to localized loadings of PAHs into the soil. As observed in the case of PAH levels in air, a positive correlation between population density and soil PAH concentrations [73] suggests that soil PAH levels are inextricably linked to human activities.

5.2.4 PAHs IN SEDIMENT

As detailed in the situation with the soil, PAH distribution and accumulation in sediments is governed by particulate carbon content and sediment composition (grain size, sand, silt, colloid, etc.).

The partitioning of PAHs between water and sediment plays a major role in PAH bioavailability. If the PAH loadings are of recent occurrence, they are rapidly desorbed and bioavailable. If the loadings are of long-term duration, they tend to accumulate and can be observed in aged sediments. Pyrogenic sources contribute to PAH inputs in the aquatic environment to a greater extent. Low-molecular-weight PAHs released during pyrolysis are transported rapidly into the sediment environment from where global distribution ensues. PAHs with higher molecular weights that are generated from pyrolysis are resistant to microbial degradation and are more likely to sediment. Petrogenic PAHs behave in a similar manner, but their contribution to sediment PAH levels are minor [14].

The concentrations of PAHs in sediments from different geographical regions are mapped in Figure 5.8. On a broader scale, the global PAH levels in sediment reflect those in the water and soil. Countries in Asia, Africa, and Latin America show high levels than in the rest of the regions. The twentieth century has seen rapid industrialization in some of the countries in Asia and Africa. A similar scenario occurred in the eighteenth and nineteenth centuries in Europe and the Americas. As a result of this industrial revolution, population growth, and per capita consumption of energy, more PAHs are released into the terrestrial environment in the above-mentioned geographical regions that eventually reached bodies of water, partitioned into the sediments, and served as sediment PAH depots.

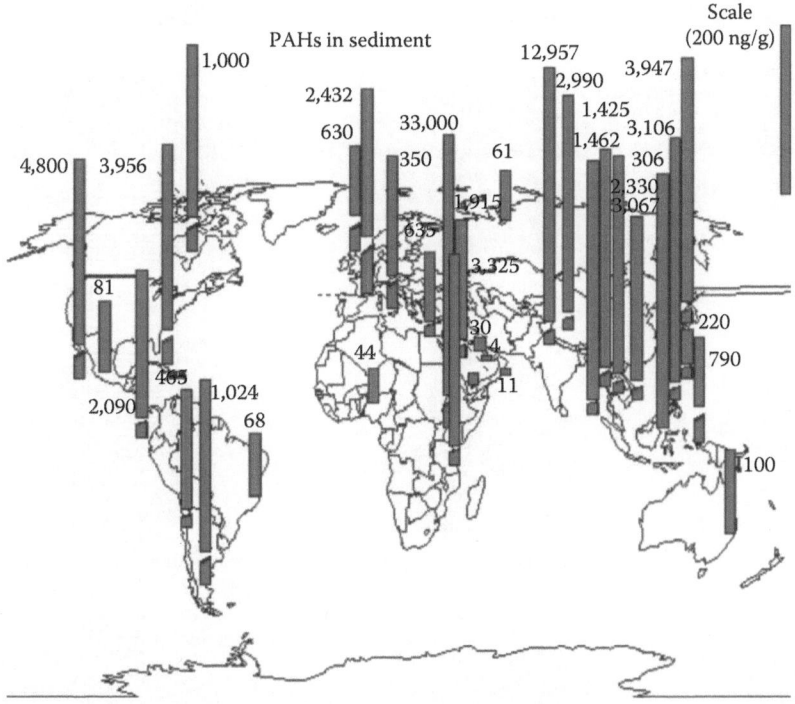

FIGURE 5.8 Spatial variations in sediment PAH concentrations (ng/g dry wt.) measured in different parts of the world. The data presented in this map are taken from published literature. The pertinent reference for each country was given in parentheses. The countries/territories represented are Argentina [50], Australia [93], Bahrain [94], Brazil [95], Cambodia [96], Canada [97], Chile [98], China [99], Colombia [100], Egypt [101], France [102], Guam [103], India [96], Indonesia [96], Italy [104], Japan [96], Kenya [105], Laos [96], Malaysia [96], Mexico [106], New Zealand [107], Nigeria [108], Oman [94], the Philippines [96], Qatar [94], Russia [109], South China Sea [110], Thailand [96], Vietnam [96], the United Arab Emirates [94], the United Kingdom [111], the United States [112], and Yemen [113]. The values represent the sum total of major PAHs. The measurements were made during different years and during different seasons. Due to space constraints, the names of the sampling locations and years were not marked on the top of the bar. Most of the samples collected were from marine, estuarine or riverine locations.

5.2.5 RELATIVE CONTRIBUTION OF DIFFERENT SOURCES TO PAH POLLUTION

The relative contributions of various combustion sources to the environmental PAHs on a global basis and for several representative countries are shown Figure 5.9. In this figure, the global PAH source profiles are shown on the basis of both PAH16 (the 16 PAHs prioritized by U.S. EPA) and BaPeq (benzo(*a*)pyrene; BaP toxic equivalents) emission basis. On a global scale, biomass burning (both biofuel combustion and wildfires) contributed to 73.7% of the total global PAH emissions. No less important are consumer products, traffic oil combustion, and domestic coal combustion, which together contributed to 15.4% of global emissions. Among the other sources, major industrial activities contributed less than 10% to the total global PAH emissions [114].

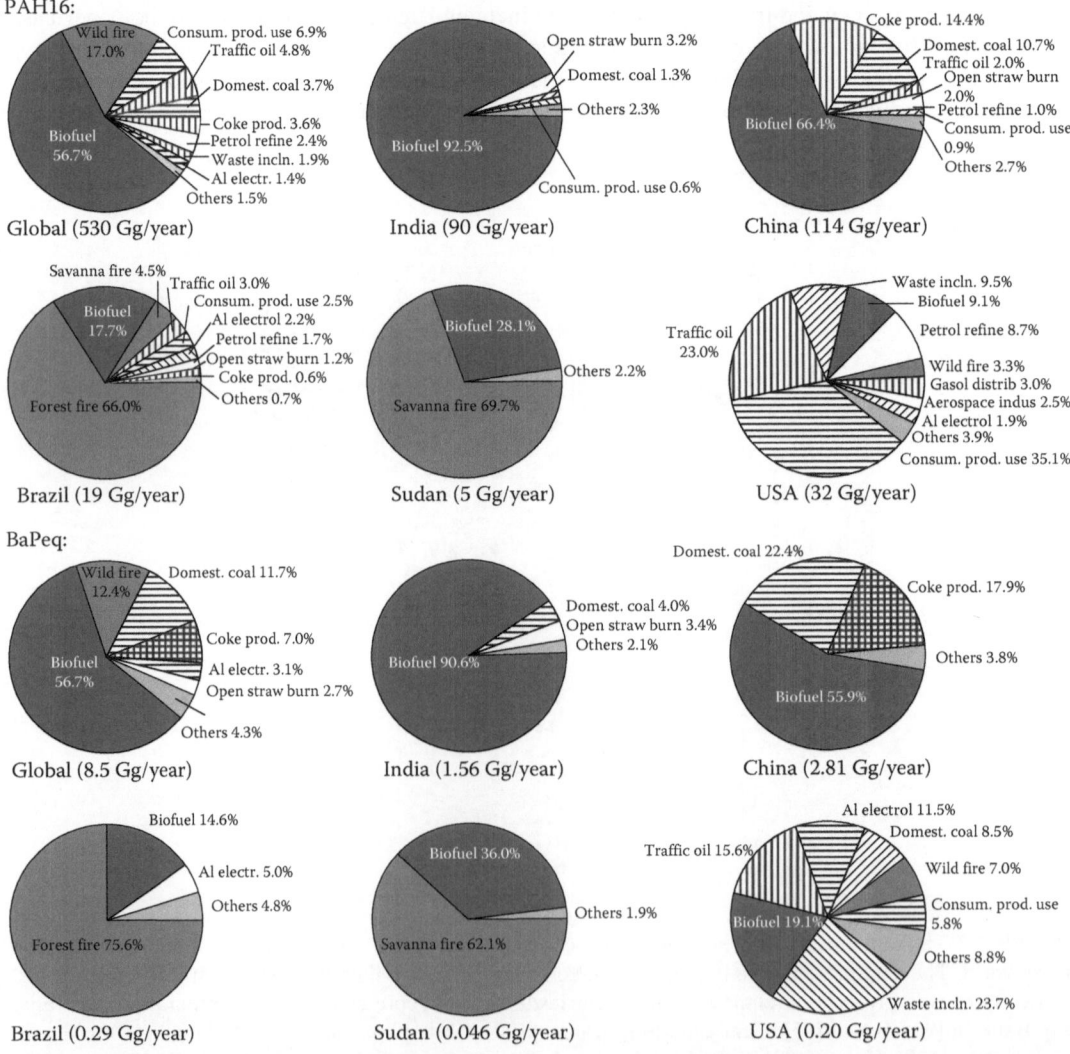

FIGURE 5.9 Relative contributions of various sources to PAH16 and BaP (BaPeq) equivalents emission in the world. Data were shown for representative countries such as China, India, the United States, Brazil, and Sudan. The corresponding total atmospheric emissions of PAH16 (16 PAHs listed as the U.S. EPA priority pollutants) and BaPeq (BaP toxic equivalents) in 2004 are given in the parentheses. (Reprinted from *Atmos. Environ.*, 43, Zhang, Y. and Tao, S., Global atmospheric emission inventory of polycyclic aromatic hydrocarbons (PAHs) for 2004, 812–819, Copyright (2009), with permission from Elsevier.)

As exemplified in Figure 5.9, factors such as energy structure, status of development, population density, and vegetation cover determine the relative contribution of different PAH sources in different countries. India offers one such example where biofuel contributed to over 90% of the total PAH emissions because of overreliance on biomass burning and petroleum for domestic energy in this country. China is rich in coal deposits and hence coal is extensively used for coke production and domestic cooking and heating [115]. However, compared to India, China's biomass consumption is lower. In other countries in the tropical belt (e.g., Brazil and Sudan), forest fires and savanna fires exceeded biofuel sources in PAH emissions due to the high vegetation cover and low population density in these countries. On the other hand, in developed countries, represented by the United States, consumer product use and traffic oil combustion represented major PAH emission sources, followed by waste incineration, biofuel combustion, and petroleum refining activities. With regards to the PAH source profiles of other countries, they were similar to the countries represented in Figure 5.7.

5.3 GLOBAL DISTRIBUTION TRENDS AND CURRENT STATUS

The global distribution maps for PAHs in diverse environmental matrices such as air, soil, water, and sediment are shown in Figures 5.5 through 5.8, respectively. As much as possible, data presented in these maps were drawn from published literature that covered rural or semi-urban to urban residential areas, far removed from point sources such as factories, hazardous waste sites, and petrochemical industries. Furthermore, the readers should be cognizant of the fact that data on global distribution of PAHs cited in the aforementioned figures are by no means an exhaustive compilation of published literature for a specific environmental matrix or a specific location. While data were available from certain geographical locations for the last 20 years, in others the data were dated and current distribution profiles are not available. Hence, we used data only from the last decade for all the locations to minimize the impact of temporal variations of PAH residue levels on spatial heterogeneity, which otherwise could confound the conclusions.

These spatial distribution maps are presented to provide a better perspective of the ecotoxicological significance of PAHs. Furthermore, they also furnish clues on source identification and environmental partitioning [71]. Also, the data from a specific location could be used by concerned governmental agencies for developing further surveillance programs to assure that industries are in compliance with environmental policies promulgated in regard to emissions [88].

As represented in the residue profiles of PAHs in all the environmental media, developing countries in Asia, Africa, and Latin America bear the brunt of high levels of contamination compared to developed countries. Among Asian countries, China and India account for the highest PAH emissions. In this context, it is interesting to note that a positive correlation exists between population density and atmospheric and soil PAH concentrations [71,116].

A review of the global distribution of individual PAH compounds in the atmosphere and precipitation revealed the preponderance of three- and four-ringed compounds (phenanthrene, fluoranthene, and pyrene; [49]). These distribution patterns are indicative of a majority contribution from combustion and petroleum sources into the atmosphere. A receptor modeling study was performed by Singh et al. [117] to map the relative contribution of PAHs to the urban atmosphere, using India as a prototypical example. Their studies revealed that the combustion sources (natural gas, wood, coal/coke, and biomass) contributed 19%–97% of various PAHs, vehicular emissions 0%–70%, diesel-based sources 0%–81%, and other miscellaneous sources 0%–20% of different PAHs. The contributions of major pyrolytic and petrogenic sources to total PAHs were estimated to be 56% and 42%, respectively. On the other hand, the levels of BaP were found to be higher in the atmosphere of China relative to other locations in the world. As mentioned previously, coal is the predominant source of fossil fuels in China, whose utilization for power generation, home heating, and domestic cooking, in conjunction with the population density, accelerated economic growth, and burgeoning manufacturing activities may have contributed to elevated levels of this PAH compounds in air.

Because PAHs are semi-volatile compounds, they travel through "long-range atmospheric transport" [118], which facilitates global distribution. After a cursory glance through the global distribution profiles of PAHs, one may wonder what lies in store for the far flung and pristine environments of the Arctic and Antarctic regions, given the fact that most of the persistent organic pollutants (POPs) including PAHs are transported to these regions over great distances from their sources of production. The POPs evaporate slowly at normal environmental temperatures from their locations (soil or water). As the vapor pressure of chemicals increases exponentially with increases in ambient temperature, they evaporate in tropical and semitropical areas. A contrasting trend is seen in colder regions of the world where the colder temperatures favor the condensation and adsorption of gaseous compounds onto atmospheric particles, which are eventually deposited onto the Earth's surface. Thus, polar regions are sinks for the POPs that are not deposited at lower latitudes because of their low volatility. Since temperatures are extremely low in these regions, these chemicals remain in these environments for a protracted period of time [119]. From a mobility standpoint, PAHs that have more than four rings (B[a]P and benzo[ghi]perylene) show low mobility while those with four rings (pyrene and benz(a)anthracene) show relatively low mobility. The PAHs with three rings (anthracene and phenanthrene) show relatively high mobility while their counterparts with two rings (naphthalene) show very high mobility [118]. Environmental monitoring studies have been conducted in the Arctic region to encourage countries that are in close proximity to promulgate necessary restrictions on the release of POPs of concern for the aforementioned region. Hung et al. [120] conducted spatial and temporal PAH distribution monitoring studies in four Canadian and two Russian arctic sites. These studies showed that air PAH concentrations were dominated by lighter PAHs such as fluorene, phenanthrene, fluoranthene, and pyrene. These findings are in agreement with the report of Halsall et al. [121] who showed through mathematical modeling that lighter PAHs, such as fluorene and phenanthrene, could reach the Arctic region while heavier compounds such as B(a)P are likely to be removed from the atmosphere before reaching the Arctic, either by reaction with hydroxy radicals or deposition along the transport pathway.

While low-molecular-weight PAHs (MW < 202) were known to be rapidly transported to polar regions through a long-range atmospheric transport process, the higher-molecular-weight PAHs (MW > 202) with four to six aromatic rings, such as B(a)P, B(e)P, fluoranthene, pyrene, benzo(a) anthracene, chrysene, benzo(b+ k)fluoranthene, indene(1,2,3-c,d)pyrene, dibenzo (a,h)-anthracene, and benzo(g,h,i)perylene, are transported to polar regions through water, oil spillages, petroleum seeps, or from combustion processes in countries close to the polar regions, as shown for Antarctica by Martins et al. [122].

As a result of their chemical behavior, PAHs bioaccumulate through the food chain and were reported to be sequestered in adipose and other tissues of marine birds and mammals [123,124] in the polar regions. The Inuit communities who rely on marine mammals for a considerable portion of their dietary requirements are therefore vulnerable to PAH exposure [123].

In addition to air mass movement, climatic factors influencing PAH concentrations in environmental media also influence the turnover of organic matter in soil, sediment, and water. The composition of various PAH sources and the changes these compounds undergo in environmental media such as air and soil results in characteristic PAH spectra (fingerprints) that are specific for climatically defined ecosystems [86].

Change in emission source and/or transport pathways could influence the qualitative and distribution profiles of PAHs. The distribution of PAHs in sediment cores serves as a good indicator of temporal trends and effectiveness of government regulations in curbing pollution from point sources. PAH pollution in the Savannah River estuary in Georgia, United States, offers a good example in this regard [125]. This estuary receives both nonpoint source and point source inputs. The Port of Savannah handles coal, ferrous metals, fuel, oil, and raw and processed chemicals. Analyses of sediment cores from a sampling station further down the river from the city of Savanna revealed a gradual increase in PAH concentrations from a depth of 37 (1951) to 9 cm (1984), followed by a decrease from 9 cm to the surface (1993). On the other hand, another sampling station close to industrial locations registered a large increase in PAH concentrations from a depth of 175 (1948) to 135 cm (1958), followed by a decrease in concentration from 135 cm to the surface. The distribution profiles of PAHs in the sediment cores suggest a significant input of PAHs into the estuary between 1948 and 1958, keeping in pace with the growth of both population and industry around the estuary. The decrease in PAH concentrations after 1976 could be attributed to the passage of pollution control laws.

We did not dwell into the details of global PAH isomer-specific distribution; however, interested readers may refer to the works of Wilcke and coworkers (cited in [86]) about fingerprint identification for selected areas in Europe, North America, and countries in the tropical belt. Furthermore, the effect of global climate change on altering the transport pathways of chemical contaminants [126] including PAHs cannot be overemphasized.

5.4 HUMAN EXPOSURE AND INTAKE ESTIMATES

Humans are exposed to PAHs through several routes such as air, water, food, skin contact, and occupational settings. For a large section of the general population not occupationally exposed to PAHs and atmospheric pollution, food ingestion is the major route of exposure compared to inhalation [127]. Studies conducted on human exposure to B(a)P revealed that the range and magnitude of dietary exposures (2–500 ng/day) were larger than for inhalation (10–50 ng/day [128]). Diet makes a substantial contribution (more than 70% in nonsmokers) to the nonoccupational exposure to PAHs [129,130].

PAH contamination of food arises from a number of sources including environment, food processing techniques, and methods used for analysis. Unprocessed food consists of vegetables, fruits, grains, vegetable oils, dairy products, and seafood. For plants (leafy vegetables and tubers), uptake through atmosphere and soil are prime sources of PAH contamination. Vegetables with broad leaves such as lettuce have a larger surface area that is ideal for deposition of airborne particles containing PAHs [131]. The accumulation of PAHs in foods of animal origin, especially livestock, is due to the consumption of contaminated pasture and vegetation [132]. PAHs in fish and shellfish are a result of contamination of fresh and coastal waters. Processed food (through smoking [133]) and cooked food (charcoal cooked [134]) also contribute substantially to the intake of PAHs. The type of cooking, cooking temperature, time, amounts of fat, and type of oil influence the formation of PAHs [135,136]. Drying techniques used for cereal preservation such as combustion gas heating and smoking increase PAH concentrations [137]. Grilled vegetables contain higher PAH concentrations than raw vegetables [138]. Similarly, the variations in refining processes contribute to the differences in PAH concentrations in oils of plant origin [139].

Concentrations of PAHs found in products of plant and animal origin have extensively been reviewed in Ramesh et al. [5]. Global dietary intake of PAHs is shown in Table 5.1. The intake ranged from 0.02 to 3.6 µg/person/day. The variety of food items considered for intake estimation, analytical methods, and the PAHs used for computing dietary intake varied. Among the variety of foods that contributed to high PAH intake are red meat, fats, oils, cereals, and vegetables. Intake of PAHs through water was reported to vary from 0.0002 to 0.12 µg/day, (assuming a daily consumption of 2 L of water/person/day [20]), which is significantly below the estimated intakes from food. Runoffs from road traffic [145], oil fields [146], and former industrial sites [147] contaminate groundwater. Additionally, the use of tar-coated water pipes was reported to release PAHs into drinking water [148].

Among other sources of exposure, PAH intake from ambient air was reported to be in the range of 0.02–3 µg/day [20]. However, the levels vary depending on the area people live, local traffic, industrial pollution, type of heating, and personal habits like smoking etc. Occupational exposures to PAHs vary with the setting; 9.6–450 µg/m^3 in aluminum smelters, petroleum refineries, and copper mines. High occupational exposure to PAHs occurs in aluminum production, coal gasification, coke production, iron and steel foundries, tar distillation, shale oil extraction, wood impregnation, roofing, road paving, carbon black production, carbon electrode production, restaurant cooking, diesel engine servicing, fire fighting, aviation fuel handling, chimney sweeping, and calcium carbide production [149].

TABLE 5.1
Estimated Dietary Intakes of PAH for Various Countries[a]

Country	Estimated Intake (μg/Person/Day)
Australia[b]	0.03
Austria	0.39
Brazil	2.90
China	3.56
Czech Republic	0.19
Denmark	0.02
Finland	2.34
France	0.09
Germany	0.19
Greece	0.005
India[c]	11.0
Italy[b]	0.2
Japan[b]	0.09
Netherlands	0.52
New Zealand[b]	0.16
Nigeria[b,d]	6.0
Norway	0.02
South Korea[d]	1.10
Spain	0.11
Sweden	0.08
United Kingdom	0.06
USA	0.05

Sources: Data obtained from Benford, D. et al., whqlibdoc.who.int/publica-tions/2006/9241660554_PAH_eng.pdf, 2006, downloaded on June 20, 2010; Pandey, M.K. et al., *J. Am. Oil Chem. Soc.*, 81, 1131, 2004; Akpan, V. et al., *Bull. Environ. Contam. Toxicol.*, 53, 246, 1994; Suzuki, K. and Yoshinaga, J., *Int. Arch. Occup. Environ. Health*, 81, 115, 2007; Moon, H.-B. et al., *Arch. Environ. Contam. Toxicol.*, 58, 214, 2010.

[a] Data obtained from Benford et al. [140], except for India [141], Nigeria [142], Japan [143], and South Korea [144].

[b] Data pertain to benzo(a)pyrene intake only.

[c] Data were collected from intake of cooking oils.

[d] Data were collected from consumption of fish and other seafood items only.

Aspects pertaining to absorption, distribution, metabolism, excretion, and biotransformation of PAHs; and also genomic and oxidative DNA damage caused by PAHs to target tissues have been the subject of numerous review articles, book chapters, and monographs. Interested readers may refer to the reviews in Ramesh et al. [5], WHO [6], ATSDR [17], Penning [150], Luch [151], Shimada [152], and Shimada and Guengerich [153].

5.5 HUMAN HEALTH EFFECTS

Several of the PAH compounds have been reported to cause toxicity and cancer in a plethora of animal species including humans. Most of the published literature on effects of PAHs on human health have mainly focused on their potential to cause cancers of the skin, lung, breast, scrotum, bladder,

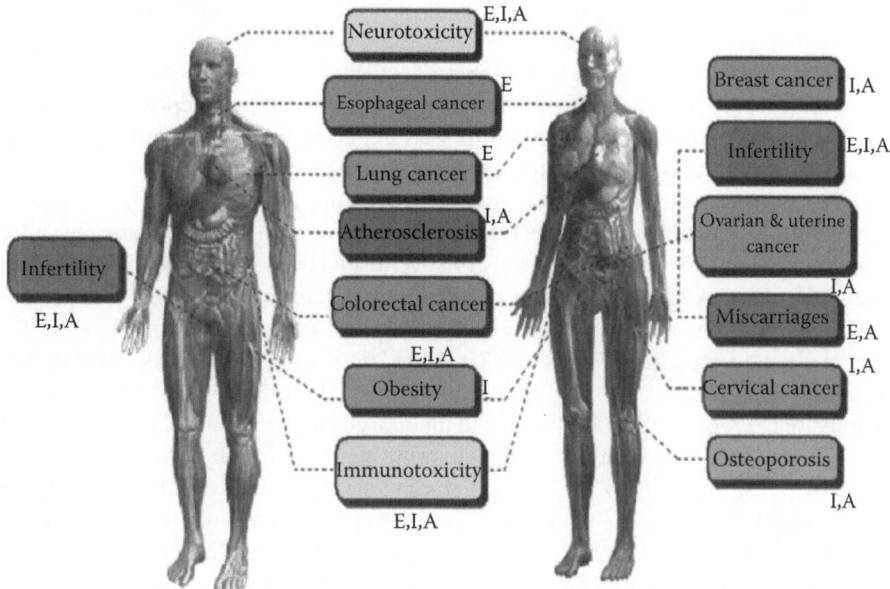

FIGURE 5.10 Schematic representation of various systems targeted by PAHs and the resulting toxic effects. "E" represents that the denoted "health effect" was supported by epidemiology studies; "I" represents that the denoted "health effect" was supported by in vitro studies; "A" represents that the denoted "health effect" was supported by in vivo (animal model) studies. (Images licensed from Zygote Media Group, Inc.)

and colon, etc. [154,155]. Excess incidences of these cancers have been associated with a wide spectrum of occupational settings such as aluminum production, coal gasification, coke production, iron and steel foundries, tar distillation, shale oil extraction, wood impregnation, roofing, road paving, carbon black production, carbon electrode production, chimney sweeping, calcium carbide production, diesel engine repair, and fire fighting [156].

In addition to cancer, the ability of PAHs to cause reproductive, developmental, and neurotoxicities has come to the center stage in recent years, thanks to the efforts of several research groups working with animal models and humans (reviewed in Hood et al. [157–159] and Ramesh and Archibong [160]). The potential targets of PAH action that potentially result in toxicity or cancer in humans is schematically illustrated in Figure 5.10.

All the health effects mentioned here were chosen based on evidence from studies using either animal models (in vivo) or isolated cell preparations/organelles (in vitro) or epidemiological studies or a combination of any two or all of these models.

Even though other reports are available pertaining to health effects of B(a)P on organ systems, they are not reported in this figure because of lack of convincing evidence. Some of the estimates were imprecise (arising out of small sample size or statistically insignificant results) and may have led to spurious associations, which may have emerged by more of a chance than of the actual exposure–causal relationship.

The various human health effects caused by PAHs are detailed next.

5.5.1 NEUROTOXICITY

Acute and subacute PAH exposures in laboratory animals were shown to cause neurobehavioral deficits [161–165]. On the other hand, several epidemiological studies were in support of the view that exposures to PAHs will cause behavioral deficits. Children born during the early 1980s in the Czech Republic had learning disorders and were presumed to be exposed in utero to elevated levels of environmental PAHs that were the result of mining and combustion of coal [166]. Coke production plant workers in Poland developed neurotic syndromes with vegetative dysregulation and a loss of short-term memory, the prevalence of which depended on the level of exposure to a PAH (B[a]P [167]). Multiple myeloma and brain cancer incidences were reported among firefighters exposed to PAHs [168]. Neurological symptoms were reported from a community that was chronically exposed to B(a)P, benz(*a*)anthracene, chrysene, naphthalene, fluorine, and pyrene dumped at nearby NPL hazardous waste sites in Texas from the 1960s until the 1970s [169]. Similarly, residents in close proximity to a combustion Superfund site in Louisiana displayed neurophysiological and neuropsychological impairments. High levels of several PAHs including B(a)P were found in soil, sludge, and water samples along with other solvents and metals at this site that was previously used for processing used motor oil and chemical wastes during the mid-1970s and early 1980s [170]. Paternal occupational exposure to PAHs was associated with an increased risk of primitive neuroectodermal tumors in children from Italy, France, and Spain [171]. Similarly, an association between paternal exposure to creosote (rich in PAHs) and diagnosed cases of neuroblastoma in children were observed [172]. In a study of children of nonsmoking native African American and Dominican-American women residing in New York City, Perera et al. [173] observed that prenatal exposure to PAHs affected children's IQs. The same research group [174] also found that the gene–PAH interaction profoundly affects child mental development. Using two parallel prospective cohort studies of African-American and Dominican mothers in New York City and of Caucasian mothers and children in Krakow, Poland, Wang et al. [175] reported polymorphisms in genes involved in PAH metabolic activation or detoxification. A recent study from China also reports PAH neurotoxicity in occupational workers [176]. Benzo(a)pyrene has been reported to perturb neurobehavioral function (significant decrease in learning and memory) in coke oven workers from China. Also, a correlation was observed between behavioral parameters, acetylcholine levels, and acetylcholine esterase activity in these occupational cohorts [176].

5.5.2 REPRODUCTIVE TOXICITY

5.5.2.1 Male Reproduction

Benzo(a)pyrene has been reported to affect sperm morphology (Figure 5.11a and b). This toxicant is also known to affect important motions characteristic of human spermatozoa such as hyperactivation as well as premature acrosome reaction [179]. Thus, caregivers should be aware of the correlation between B(a)P exposure and abnormal semen parameters and question patients about previous exposures to a PAH/B(a)P sources particularly during counseling of occupationally exposed workers suffering from idiopathic male infertility. A correlation between urinary concentrations of PAH metabolites and increased idiopathic male infertility was reported by Xia et al. [180]. The presence of aryl hydrocarbon receptor (AhR) and aryl hydrocarbon receptor nuclear translocator (ARNT) in spermatozoa [181] suggests that exposure to PAHs stimulates the AhR that in turn increases the metabolism of PAHs to reactive metabolites that can influence sperm function. Microarray analysis and Q-PCR studies also have shown that mRNA of CYP1A1 and CYP1B1 are detectable in spermatozoa [182], confirming the ability of germ cells to produce these CYP enzymes to metabolize environmental toxicants such as PAHs.

Zenzes and coworkers [183] were the first team to report B(a)P-DNA adducts in the spermatozoa of smokers. This team [184] also detected B(a)P-DNA adducts in preimplantation embryos derived from the fertilization of a nonsmoker's mature ova with spermatozoa from smoking fathers. These studies bring forth an interesting argument that developmental problems found in children are male mediated. Subsequent studies stress that paternal transmission of DNA which has been modified upon exposure to chemical carcinogens will have profound multigenerational effects. In in vitro studies, B(a)P has been shown to cause DNA strand breaks in human spermatozoa [185]. Polyaromatic hydrocarbon-DNA adduct concentrations in human spermatozoa were used as a biomarker for DNA damage. Samples collected from an infertility clinic in Italy revealed a significant

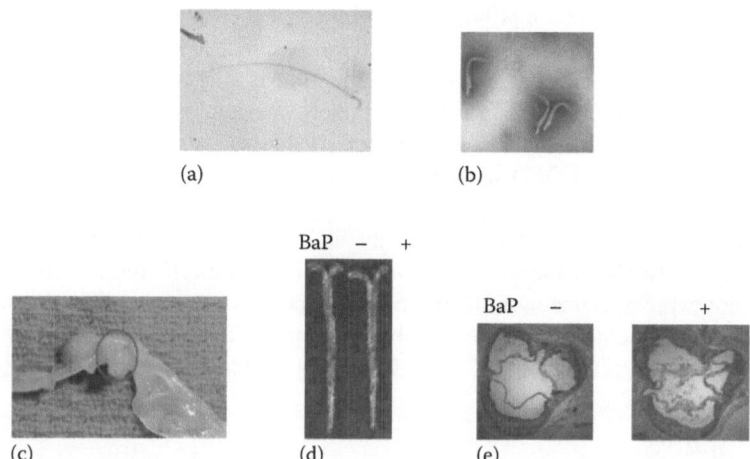

FIGURE 5.11 Photomicrographs showing (a) a normal spermatozoan taken from unexposed F-344 rat; (b) spermatozoa with decapitated heads taken from a F-344 rat exposed to subchronic exposure concentration of 75 μg benzo(a)pyrene/m³. (Reprinted from *Exp. Toxicol. Pathol.*, 60, Ramesh, A. et al., Alteration of fertility endpoints in adult male F-344 rats by subchronic exposure to inhaled benzo(a)pyrene, 269–280, Copyright 2008, with permission from Elsevier.) (c) Representative tumor in the distal colon of Apc^Min mouse treated with 100 μg B(a)P/kg bw for 60 days via oral gavage in saturated fat. (Reprinted from Harris, D.L. et al., *Toxicol. Pathol.*, 37, 938, 2009. With permission from Sage Publications.) (d) En face preparations of distal aortas taken from apolipoprotein-E–deficient (ApoE⁻/⁻ control mice). (e) Cross sections of distal aortas taken from ApoE⁻/⁻ mice treated with BaP. Stained areas are atherosclerotic lesions. (Reprinted from *Atherosclerosis* 207, Yang, H., Zhou, L., Wang, Z. et al., Overexpression of antioxidant enzymes in ApoE-deficient mice suppresses benzo(a)pyrene-accelerated atherosclerosis, 51–58, Copyright 2009, with permission from Elsevier.)

relationship between occupational exposure to PAHs and PAH-DNA adduct concentrations in spermatozoa [186]. Hsu et al. [187] reported a relationship between PAH exposure and sperm DNA damage in coke oven workers. The small sample size notwithstanding, these studies indicate the possibility that PAHs may impact the integrity of DNA in spermatozoa. Sperm samples collected from infertile men showed a correlation between PAH exposure (assessed by urinary PAH metabolites) and sperm DNA damage [188]. Interestingly, some of these patients who were presumed to have been exposed to PAHs also showed a polymorphism in the x-ray repair cross complementing the group 1 (XRCC1) gene involved in base excision repair of DNA. The combined effect of XRCC1 polymorphisms and PAH exposure seem to exacerbate the damage to DNA of spermatozoa in these patients. In this context it is imperative to mention the studies conducted earlier in the Czech Republic wherein men exposed to polluted air rich in PAHs showed abnormal chromatin/fragmented DNA in spermatozoa [189]. These men were found to carry a homozygous null glutathione-S-transferase 1 (GSTM1) gene. Also, a statistically significant relationship was observed between GSTM1 null genotype and increased DNA fragmentation index in these individuals. Studies conducted by Paracchini et al. [190] in infertile men in Italy also exhibited the same trend observed by Rubes et al. [191] with regard to GSTM1 deficiency. The GSTM1 deletions in these men showed a significant increase in PAH-DNA adduct concentrations in spermatozoa. The above-mentioned studies document key pieces of evidence for the presence of gene–environment interaction in PAH-induced toxicity of the male reproductive system.

5.5.2.2 Female Reproduction

There is a significant association between smoking and reduced fertility among female smokers. [192,193]. If a single acute exposure to a high dose of B(a)P could sustain reactive metabolites and adducts up to a month after exposure [194], the likelihood of the damage caused in subacute (short term) and subchronic (long term) exposures in humans is of concern from the standpoint of toxicity and carcinogenesis. Evidence subscribing to this viewpoint is furnished by the studies of Zenzes et al. [195] who report the presence of B(a)P-DNA adducts in granulosa cells, and oocytes of female cigarette smokers. Cigarette smoke is a rich source of PAHs that puts these women at an increased risk since transmission of altered DNA to preimplantation embryos increases the likelihood of contributing to childhood cancers.

Human ovarian tissue samples collected from hospitals also showed the presence of B(a)P-DNA adducts [196], an indication that mature ova from the sample donors could carry defective genes that may or may not express themselves during intrauterine residence but could do so post partum. Besides the ovary, B(a)P metabolites have also been reported to exert inhibitory effect on human cervical cells by enhancing cell death [197] and B(a)P-DNA adduct formation [198].

5.5.3 Uterine Tumors

McGarry et al. [199] reported that B(a)P alters cell adhesion proteins in human uterine cells. Qin et al. [200] found significantly higher concentrations of PAHs in adipose tissue of patients with uterine leiomyomas (noncancerous tumors of the uterus) compared to normal subjects.

5.5.4 Osteoporosis

The repression of estrogen biosynthesis by PAHs can result in the reduction of the circulating concentrations of this ovarian steroid below the physiological threshold essential for maintaining bone health. Much of the data available to document an association between bone damage and PAH exposure is from animal studies only. The PAH compound 3-methylcholanthrene (3-MC) was reported to inhibit the proliferation and differentiation of osteoblasts (bone forming cells; [201]). Investigations by Naruse et al. [202] also revealed that 3-MC affects osteoclast supporting cells which is consistent with the findings of Naruse et al. [201,202], Voronov et al. [203] who observed that B(a)P inhibits rabbit osteoclast (bone destroying cells) differentiation.

So far, only two studies document an association between bone damage and PAH exposure in vivo. Exposure to 3-MC was found to cause a delay in the ossification of the forelimb, hindlimb, and cervical and thoracic vertebrae in fetal mice [201]. Also, exposure to B(a)P resulted in a loss of bone mass and bone strength in ovariectomized adult rats [204]. There is a clear data gap in the knowledge regarding the mechanisms of bone damage caused by PAHs in an in vivo model. There is every reason to believe that exposure to toxicants such as PAHs contribute to an early onset of menopause, one of the consequences of which is osteoporosis.

5.5.5 Cancer

5.5.5.1 Dermal

Though animal bioassay studies have revealed considerable dermal toxicity and increased cancer risk upon exposure to PAHs [205], no conclusive evidence has been obtained from human biomonitoring studies conducted on road pavers [206]. Recent studies have shown that dermal application of coal tar and ultraviolet radiation to treat psoriasis caused mutagenic and temporary genotoxic effects in children [207].

5.5.5.2 Esophageal

An association between ingestion of PAHs and esophageal cancer (esophageal squamous cell carcinoma) has been reported from Iran [208] and China [209]. The Iranian study detected 1-hydroxypyrene glucuronide (a stable metabolite of PAHs) in 83% of the subjects. The Chinese studies detected greater concentrations of PAH-DNA adducts [209] and elevated AhR gene expressions [210] in esophageal cell samples from patients compared to healthy control subjects. While increase in environmental pollution by PAHs has been attributed to esophageal cancer in Iran, high levels of carcinogenic PAHs in food [211] and indoor air pollution (domestic heating with coal) have been implicated as the causative factors for esophageal cancer in China.

5.5.5.3   Colorectal

A review of literature conducted by Ramesh et al. [5] indicates considerable contamination of food-stuffs, especially fatty ones including red meat by PAHs. Therefore, it is assumed that regular intake of PAH-contaminated diet can significantly increase colorectal cancer risk.

Findings from a recent clinic-based case-control study further lend credence to the view point that dietary intake of PAHs is associated with colorectal cancer risk [212]. This study created a B(a)P (as a surrogate for PAHs) database using a wide range of food items and linked it to a food frequency questionnaire (FFQ). The FFQ has detailed questions on meat-cooking methods, doneness levels, frequency of consumption, and portion size. This questionnaire was used to gather information approximately 1 year before the patients underwent sigmoidoscopy/colonoscopy. Increased risk of colorectal adenomas with high intake of B(a)P was found from total diet and from meat only. Using the same study design and a sample size of about 4000 adenoma cases, Sinha et al. [213] also showed that consumption of well-done red meat was associated with increased risks for adenoma of the descending colon and sigmoid colon for B(a)P.

Studies conducted by our research group [177] in a mouse model also established that dietary fat, especially saturated fat, potentiates the development of colon tumors caused by B(a)P (Figure 5.11c). An increased prevalence ($P < .05$) of adenomas in colon of mice that ingested B(a)P through saturated dietary fat compared to unsaturated fat and unexposed controls was noticed. Interestingly, we have also observed adenomas with high-grade dysplasia in B(a)P + saturated fat group and these observations were of frequent occurrence at a dose level of 100 µg B(a)P/kg. Hence, it is conceivable that colorectal cancers are promoted by increased intake of PAHs through saturated dietary fat that in turn influences the biotransformation and metabolic processing of these toxic chemicals.

5.5.5.4   Breast

Studies in animal models and human cell cultures have shown that PAHs induce breast tumors. Benzo(a)pyrene has been shown to inhibit cell proliferation and alter the expression of BRCA-1 gene. The metabolism of B(a)P to its reactive metabolite 7r, 8t-dihydroxy-9t,10t-epoxy-7,8,9,10-tetrahydrobenzo[a]pyrene (BPDE) has been reported to downregulate BRCA-1 [214]. Though an association between breast tumors and the levels of PAH-DNA adducts have been suspected to be indicative of breast cancer, a hospital-based case-control study using tumor tissues from patients and control subjects showed no such association [215]. A population-based case-control study showed that data from PAH exposure suggested that PAH compounds did not cause a major change in overall p53 mutations. This finding opens the possibility that PAH exposure may be associated with specific breast tumor p53 mutation [216]. As part of the Long Island Breast Cancer Study Project (LIBCSP), DNA samples from 1100 women with and without breast cancer were analyzed. A review of the epidemiologic evidence using data from the LIBCSP revealed that breast cancer risk increased with variants in GST genotypes, oxidative stress, DNA repair, and apoptosis genes. The magnitude of risk tends to increase when these parameters are considered individually rather than jointly [217]. From the foregone studies, it can be safely inferred that genetic damage resulting from individual exposure and susceptibility to PAH may play a role in the causation and progression of sporadic breast cancer.

5.5.5.5   Lung

The detection of greater levels of PAHs in lung tissue of smokers led to the belief that PAHs in cigarette smoke are likely to contribute to the development of lung cancer in humans [218]. Occupational exposure to PAHs via inhalation has also been implicated in developing lung cancer. The exposure history and smoking habits of a cohort of 16,000 persons working in the aluminum smelter industry were studied to estimate the exposure–response function. A clear tendency for increased risk with increased cumulative exposure to PAH (measured as BaP) was seen in at least 680 cases. While chronic occupational exposure increases the likelihood of risk for lung cancer, the combination of risk due to occupational PAH and smoking did not appear to be higher than occupational exposure alone [219]. Wide variations in PAH metabolizing enzyme activities were recorded in human lung tissues [220]. As observed in the case of breast cancer, the polymorphic variability in drug bioactivation/detoxification capacity [220] and DNA repair gene expression [221] can play an important role in individual susceptibility to PAH-induced lung cancer.

5.5.6 IMMUNOTOXICITY

PAHs have been reported to exert immunotoxicity in experimental animals and humans [9]. Since immunotoxicity contributes to an enhanced potential for cancer development, studies have been carried out in occupational workers. These toxicants were found to increase the percentage of some T-cells, monocytes, and serum immunoglobulin levels in road-paving workers [222]. The immunosuppressions induced by PAHs in road pavers were less than the ones observed by Szczeklik et al. [223] in coke-oven workers whose serum immunoglobulin levels were much reduced. Both of the above-mentioned occupational cohorts were reported to exhibit suppression of T-lymphocyte proliferation upon exposure to PAHs [224].

5.5.7 OBESITY

The report by Hutcheon et al. [225] has shown a positive correlation between human plasma B(a) P levels and body mass index. The circulating B(a)P partitions into adipose tissue which serves as a storage depot for B(a)P and other PAHs [226]. Exposure of mouse primary adipocytes to B(a)P (0–100 μM conc.) was found to inhibit epinephrine-induced lipolysis in a dose-dependent manner [227]. Studies by these authors [227] also show a decreased response to epinephrine-induced lipolysis in C57BL/6J mice to subacute doses (14 days exposure) of 400 μg/kg B(a)P. Also, a significant weight gain (43%) and increased fat mass was observed in these mice. The excess weight gain was not corrected in these animals even after withdrawal from B(a)P dosing without changes in food uptake. The decrease in lipolysis was attributed to inhibition of β-adrenergic and ACTH receptor signaling by B(a)P. Given the fact that B(a)P is a ubiquitous environmental toxicant and food pollutant [5], the propensity of humans to relish fast foods will always contribute to an enhanced intake of B(a)P that can disrupt homeostatic control of adipogenesis and energy balance, hence contributing to obesity in susceptible populations.

5.5.8 ATHEROSCLEROSIS

In as much as PAHs are important contaminants of barbecued red meats and other fatty diets, the implications of B(a)P residing in adipose tissue calls attention to health implications of this chemical from the standpoint of cardiovascular toxicity. Chronic exposure (12–24 weeks) of Apo-E knockout mice to 5 mg/kg B(a)P has been reported to accelerate atherosclerotic plaques [228]. Furthermore, these authors have observed a correlation between the stages of plaques and the levels of B(a)P reactive metabolites and DNA adducts. Benzo(a)pyrene-accelerated atherosclerosis (Figure 5.11d and e) has been shown to be mediated through altered expression of antioxidant enzymes and enhanced generation of reactive oxygen species in this mouse model [178]. Furthermore, enhanced expression of AhR target gene expression observed in mouse aortic endothelial cells indicates that the upregulation of AhR and its target genes play a key role in B(a)P-induced atherogenesis [229].

5.6 CONCLUSIONS AND FUTURE RESEARCH DIRECTIONS

PAHs are seldom released into the environment as individual compounds. They coexist with a myriad of compounds such as heavy metals, heterocyclic amines, and their own chlorinated, methylated, and alkylated analogs which are environmentally more persistent than the native PAHs. Like any complex chemical mixtures, PAHs also exhibit antagonism, synergism, or additivity in their interactions. Therefore, PAH mixture toxicity studies are warranted in animal models to develop robust "toxic equivalency factors" from the standpoint of cancer/toxicity risk assessment.

The concentrations/doses and the duration for which humans and wildlife are exposed to PAHs will have a bearing on the health of the ecosystem and people living in communities that are in close proximity to abandoned industrial sites, ammunition depots, petroleum refineries, thus making the inhabitants vulnerable to PAH emissions emanated from these point sources. Given their generation from industrial outfalls, PAHs are sequestered in soil, sediment, and soot deposits, and these pollutants therefore are concentrated in urban and industrial areas. From a risk reduction standpoint, human exposure to diffuse sources of PAH pollution such as automobile emissions, residential wood burning, and refuse burning is of equal importance. Federal and state cleanup limits are currently based on risk estimates at or below ambient urban background concentrations. A comprehensive geospatial mapping of urban background concentrations and human monitoring are essential to the formulation of effective regulatory policies to minimize risks from PAH exposures.

ACKNOWLEDGMENTS

The authors gratefully acknowledge the research funding from the National Institutes of Health through grant numbers 1RO1CA142845-O1A1, 1RO3CA130112-01 (Ramesh), U54RR026140 (Archibong), 5-G12-RR03032 (Archibong and Ramesh), R56ES017448 (Hood), U54NS041071 (Hood), S11ES014156 (Hood, Ramesh and Archibong), 1R01ES014471-01A1, and 1R01 HL089382 (Guo), which provide support for two decades of research in their laboratories on various health effects of PAHs. We also appreciate the assistance rendered by Mounika Aramandla in mapping the global distribution of PAHs.

REFERENCES

1. NAS. 2003. *Oil in the Sea III: Inputs, Fates, and Effects*. Washington, DC: National Academies Press, 280pp.
2. Chameides, B. 2010. Oil spill reduction. http://www.nicholas.duke.edu/thegreengrok/-gulfspill-pahs/mobile_view. Downloaded on June 25, 2010.
3. Pleil, J.D., Vette, A.F., and Johnson, B.A. 2004. Air levels of carcinogenic polycyclic aromatic hydrocarbons after the World Trade Center disaster. *Proc. Natl Acad. Sci. USA* 101: 11685–11688.
4. Varanasi, U. 1989. *Metabolism of Polycyclic Aromatic Hydrocarbons in the Aquatic Environment*. Boca Raton, FL: CRC Press.
5. Ramesh, A., Walker, S.A., Hood, D.B. et al. 2004. Bioavailability and risk assessment of orally ingested polycyclic aromatic hydrocarbons (review). *Int. J. Toxicol.* 23: 301–333.
6. WHO. 2010. Some non-heterocyclic polycyclic aromatic hydrocarbons and some related exposures. *IARC Monographs on the Evaluation of Carcinogenic Risks to Humans*, Vol. 92. Lyon, France: International Agency for Research on Cancer.
7. WHO. 2000. Polycyclic aromatic hydrocarbons (PAHs). In: *Air Quality Guidelines*. Copenhagen, Denmark: WHO Regional Office for Europe.
8. Kanaly, R.A. and Harayama, S. 2000. Biodegradation of high-molecular-weight polycyclic aromatic hydrocarbons by bacteria. *J. Bact.* 182: 2059–2067.
9. Davila, D.R., Mounho, B.J., and Burchiel, S.W. 1997. Toxicity of polycyclic aromatic hydrocarbons to the human immune system: Models and mechanisms. *Toxicol. Ecotoxicol. News* 4: 5–9.
10. Neff, JM. 1979. *Polycyclic Aromatic Hydrocarbons in the Aquatic Environment*. London, U.K.: Applied Science Publishers Ltd.

11. Meyer, P.A. and Ishiwatari, R. 1993. Lacustrine organic geochemistry—An overview of indicators of organic matter sources and diagenesis in lake sediments. *Org. Geochem.* 20: 867–900.

12. Boehm, P.D., Page, D.S., Burns, W.A. et al. 2001. Resolving the origin of the petrogenic hydrocarbon background in Prince William Sound, Alaska. *Environ. Sci. Technol.* 35: 471–479.

13. Gschwend, P.M., Chen, P.H., and Hites, R.A. 1983. On the formation of perylene in recent sediments: Kinetic models. *Geochim. Cosmochim. Acta* 47: 2115–2119.

14. Burgess, R.M., Ahrens, M.J., Hickey, C.W. et al. 2003. Geochemistry of PAHs in aquatic environments: Source, persistence and distribution. In: *PAHs: An Ecotoxicological Perspective*, ed. P.E.T. Douben. Hoboken, NJ: John Wiley & Sons Ltd., pp. 35–45.

15. Wang, Z., Fingas, M., Shu, Y.Y. et al. 1999. Quantitative characterization of PAHs in burn residue and soot samples and differentiation of pyrogenic PAHs from petrogenic PAH-the 1994 mobile burn study. *Environ. Sci. Technol.* 33: 3100–3109.

16. Douben, P.E.T. 2003. PAHs: *An Ecotoxicological Perspective*. Hoboken, NJ: John Wiley & Sons Ltd.

17. ATSDR. 1995. Toxicological profile for polycyclic aromatic hydrocarbons (PAHs). *Agency for Toxic Substances and Disease Registry, US Department of Health and Human Services*, Atlanta, GA: US Public Health Service, 271pp.

18. Ramesh, A., Inyang, F., Lunstra, D.D. et al. 2008. Alteration of fertility endpoints in adult male F-344 rats by subchronic exposure to inhaled benzo(a)pyrene. *Exp. Toxicol. Pathol.* 60: 269–280.

19. WHO. 1999. Selected non-heterocyclic polycyclic aromatic hydrocarbons. *Environmental Health Criteria*. No. 202. Geneva, Switzerland: World Health Organization, p. 905.

20. WHO. 2003. Selected nitro- and nitro-oxy-polycyclic aromatic hydrocarbons. *Environmental Health Criteria*. Geneva, Switzerland: World Health Organization.

21. Archibong, A.E., Inyang, F., Ramesh, A. et al. 2002. Alteration of pregnancy related hormones and fetal survival in F-344 rats exposed by inhalation to benzo(a)pyrene. *Reprod. Toxicol.* 16: 801–808.

22. Inyang, F., Ramesh, A., Kopsombut, P. et al. 2003. Disruption of testicular steroidogenesis and epididymal function by inhaled benzo(a)pyrene. *Reprod. Toxicol.* 17: 527–537.

23. Ares, J. and Zavatti, J. 1993. Comparative analysis of emissions and diffusion of air PAHs at a coastal arid site (Patagonia, Argentina). *Bull. Environ. Contam. Toxicol.* 50: 333–339.

24. Lim, M.C.H., Ayoko, G.A., and Morawska, L. 2005. Characterization of elemental and polycyclic aromatic hydrocarbon compositions of the urban air in Brisbane. *Atmos. Environ.* 39: 463–476.

25. Ravindra, K., Bencs, L., Wauters, E. et al. 2006. Seasonal and site-specific variation in vapour and aerosol phase PAHs over Flanders (Belgium) and their relation with anthropogenic activities. *Atmos. Environ.* 40: 771–785.

26. Vasconcellos, P.C., Zacarias, D., Pires, M.A.F. et al. 2003. Measurements of polycyclic aromatic hydrocarbons in airborne particles from the metropolitan area of São Paulo city, Brazil. *Atmos. Environ.* 37: 3009–3018.

27. Hung, H., Blanchard, P., Halsall, C.J. et al. 2005. Temporal and spatial variabilities of atmospheric polychlorinated biphenyls (PCBs), organochlorine (OC) pesticides and polycyclic aromatic hydrocarbons (PAHs) in the Canadian Arctic: Results from a decade of monitoring. *Sci. Total Environ.* 342: 119–144.

28. Adonis, M., Martinez, V., Riqueline, R. et al. 2003. Susceptibility and exposure biomarkers in people exposed to PAHs from diesel exhaust. *Toxicol. Lett.* 144: 3–15.

29. Liu, S., Tao, S., and Liu, W. 2008. Seasonal and spatial occurrence and distribution of atmospheric polycyclic aromatic hydrocarbons (PAHs) in rural and urban areas of the north Chinese plain. *Environ. Pollut.* 156: 651–656.

30. Beneš, I., Novák, J., and Pinto, J.P. 2006. Air pollution in Teplice and Prachatice in 1995 and 2003. In *Environmental Health in Central and Eastern Europe*, ed. K.C. Donnelly and L.H. Cizmas. New York: Springer, pp. 13–22.

31. Fertmann, R., Tesseraux, I., Schümann, M. et al. 2002. Evaluation of ambient air concentrations of polycyclic aromatic hydrocarbons in Germany from 1990 to 1998. *J. Exp. Anal. Environ. Epidemiol.* 12: 115–123.

32. Mantis, J., Chaloulakou, A., and Samara, C. 2005. PM10-bound polycyclic aromatic hydrocarbons (PAHs) in the greater area of Athens, Greece. *Chemosphere* 59: 593–604.

33. Gelencsér, A., Barcza, T., Kiss, G.Y. et al. 1998. Distribution of *n*-alkanes and PAHs in atmospheric aerosols. *Atmos. Res.* 46: 223–231.

34. Jaward, F.M., Farrar, N.J., Harner, T. et al. 2004. Passive air sampling of polycyclic aromatic hydrocarbons and polychlorinated naphthalenes across Europe. *Environ. Toxicol. Chem.* 23: 1355–1364.

35. Sharma, H., Jain, V.K., and Khan, Z.H. 2007. Characterization and source identification of polycyclic aromatic hydrocarbons (PAHs) in the urban environment of Delhi. *Chemosphere* 66: 302–310.

36. Menichini, E., Iacovella, N., Monfredini, F. et al. 2007. Atmospheric pollution by PAHs, PCDD/Fs and PCBs simultaneously collected at a regional background site in central Italy and at an urban site in Rome. *Chemosphere* 69: 422–434.

37. Tham, Y.W.F., Takeda, K., and Sakugawa, H. 2008. Polycyclic aromatic hydrocarbons (PAHs) associated with atmospheric particles in Higashi Hiroshima, Japan: Influence of meteorological conditions and seasonal variations. *Atmos. Res.* 88: 224–233.

38. Muendo, M., Hanai, Y., Kameda, Y. et al. 2006. Polycyclic aromatic hydrocarbons in urban air: Concentration levels, patterns, and source analysis in Nairobi, Kenya. *Environ. Forensics* 7: 147–157.

39. Omar, N.Y.M.J., Abas, M.R.B., Ketuly, K.A. et al. 2002. Concentrations of PAHs in atmospheric particles (PM-10) and roadside soil particles collected in Kuala Lumpur, Malaysia. *Atmos. Environ.* 36: 247–254.

40. Marr, L.C., Grogan, H.W., Molina, L.T. et al. 2004. Vehicle traffic as a source of polycyclic aromatic hydrocarbons exposure in the Mexico City Metropolitan Area. *Environ. Sci. Technol.* 38: 2584–2592.

41. Lee, J.Y., Kim, Y.P., Kang, C.H. et al. 2006. Temporal trend and long range transport of particulate PAHs at Gosan in Northeast Asia between 2001 and 2004. *J. Geophys. Res.* 111: D11303.

42. Fang, M.-D., Hsieh, P.-C., Ko, F.-C. et al. 2007. Sources and distribution of polycyclic aromatic hydrocarbons in the sediments of Kaoping river and submarine canyon system, Taiwan. *Mar. Pollut. Bull.* 54: 1179–1189.

43. Bozlaker, A., Muezzinoglu, A., and Odabasi, M. 2008. Atmospheric concentrations, dry deposition and air-soil exchange of polycyclic aromatic hydrocarbons (PAHs) in an industrial region in Turkey. *J. Hazard. Mater.* 153: 1093–1102.

44. Dachs, J., Glenn, T.R. IV, Gigliotti, C.L. et al. 2002. Processes driving the short-term variability of PAHs in the Baltimore and northern Chesapeake Bay atmosphere, USA. *Atmos. Environ.* 36: 2281–2295.

45. Lee, J.Y. and Kim, Y.P. 2007. Source apportionment of the particulate PAHs in Seoul, Korea: Impact of long range transport to a mega city. *Atmos. Chem. Phys.* 7: 3587–3596.

46. Lang, C., Tao, S., Liu, W. et al. 2008. Atmospheric transport and outflow of polycyclic aromatic hydrocarbons from China. *Environ. Sci. Technol.* 42: 5196–5201.

47. Stokes, S.C., Hood, D.B., Zokovitch, J. et al. 2010. Blueprint for communicating risk and preventing environmental injustice. *J. Health Care Poor Underserved* 21: 35–52.

48. Arditsoglou, A., Petaloti, C., and Terzi, E. 2004. Size distribution of trace elements and polycyclic aromatic hydrocarbons in fly ashes generated in Greek lignite-fired power plants. *Sci. Total Environ.* 323: 153–167.

49. Latimer, J.S. and Zheng, J. (eds.). 2003. *The Sources, Transport, and Fate of PAHs in the Marine Environment.* Chichester, U.K.: Wiley.

50. Arias, A.H., Vazquez-Botello, A., Tombesi, N. et al. 2010. Presence, distribution, and origins of polycyclic aromatic hydrocarbons (PAHs) in sediments from Bahía Blanca estuary, Argentina. *Environ. Monit. Assess.* 160: 301–314.

51. Pikkarainen, A.-L. and Lemponen, P. 2005. Petroleum hydrocarbon concentrations in Baltic Sea subsurface water. *Boreal Environ. Res.* 10: 125–134.

52. Dórea, H.S., Bispo, J.R.L., Aragão, K.A.S. et al. 2007. Analysis of BTEX, PAHs and metals in the oilfield produced water in the State of Sergipe, Brazil. *Microchem. J.* 85: 234–238.

53. Guo, W., He, M., and Yang, Z. 2007. Distribution of polycyclic aromatic hydrocarbons in water, suspended particulate matter and sediment from Daliao River watershed, China. *Chemosphere* 68: 93–104.

54. Said, T.O. and Agroudy, N.E. 2006. Assessment of PAHs in water and fish tissues from Great Bitter and El Temsah lakes, Suez Canal, as chemical markers of pollution sources. *Chem. Ecol.* 22: 159–173.

55. Reddy, M.S., Basha, S., and Joshi, H.V. 2005. Seasonal distribution and contamination levels of total PHCs, PAHs and heavy metals in coastal waters of the Alang–Sosiya ship scrapping yard, Gulf of Cambay, India. *Chemosphere* 61: 1587–1593.

56. Habibi, M.H. and Hadjmohammadi, M.R. 2008. Determination of some polycyclic aromatic hydrocarbons in the Caspian Seawater by HPLC following preconcentration with solid-phase extraction. *Iran. J. Chem. Chem. Eng.* 27: 91–96.

57. Manodori, L., Gambaro, A., and Piazza, R. et al. 2006. PCBs and PAHs in sea-surface microlayer and sub-surface water samples of the Venice Lagoon (Italy). *Mar. Pollut. Bull.* 52: 184–192.

58. Iwasaki, K., Ozaki, N., Kojima, K. et al. 2009. Estimation of river discharge loadings of PAHs in a suburban river in Hiroshima Prefecture, Japan. *J. Water. Environ. Technol.* 7: 109–120.

59. Hashmi, I., Kim, J–G., Kim, K.S. et al. 2005. Hydrocarbons (PAHs) levels from two industrial zones (Sihwa and Banwal) located in An-san city of the Korean Peninsula and their influence on lake. *J. Appl. Sci. Environ. Manage* 9: 63–69.

60. Armenta-Arteaga, G. and Elizaide-Gonzalez, M.P. 2003. Contamination by PAHs, PCBs, PCPs and heavy metals in the mecoafcin lake estuarine water and sediments after oil spilling. *J. Soils Sedim.* 3: 35–40.

61. Anyakora, C. and Coker, H. 2006. Determination of polynuclear aromatic hydrocarbons (PAHs) in selected water bodies in the Niger Delta. *Afr. J. Biotechnol.* 5: 2024–2031.

62. Nizzetto, L., Lohmann, R., Giola, R. et al. 2008. PAHs in air and seawater along a North-South Atlantic transect: Trends, processes and possible sources. *Environ. Sci. Technol.* 42: 1580–1585.

63. Al-Farawati, R.K., El-Maradny, A., and Niaz, G.R. 2009. Fecal sterols and PAHs in sewage polluted marine environment along the eastern Red Sea coast, south of Jeddah, Saudi Arabia. *Ind. J. Mar. Sci.* 38: 404–410.

64. He, J. and Balasubramanian, R. 2010. The exchange of SVOCs across the air-sea interface in Singapore's coastal environment. *Atmos. Chem. Phys.* 10: 1837–1852.

65. Pathiratne, K.A.S., De Silva, O.C.P., Hehemann, D. et al. 2007. Occurrence and distribution of polycyclic aromatic hydrocarbons (PAHs) in Bolgoda and Beira Lakes, Sri Lanka. *Bull. Environ. Contam. Toxicol.* 79: 135–140.

66. Doong, R.A. and Lin, Y.T. 2004. Characterization and distribution of polycyclic aromatic hydrocarbon contaminations in surface sediment and water from Gao-ping River, Taiwan. *Water Res.* 38: 1733–1744.

67. Wattayakorn, G. 2003. Polycyclic aromatic hydrocarbons in the Chao Phraya estuary, Thailand. *J. Sci. Res. Chula. Univ.* 28, Special Issue I: 15–27.

68. Hwang, H.-M. and Foster, G.D. 2006. Characterization of polycyclic aromatic hydrocarbons in urban stormwater runoff flowing into the tidal Anacostia River, Washington, DC, USA. *Environ. Pollut.* 140: 416–426.

69. Harvey, R.G. 1997. *Polycyclic Aromatic Hydrocarbons.* New York: Wiley-VCH.

70. Zander, M. 1983. Physical and chemical properties of polycyclic aromatic hydrocarbons. In: *Handbook of Polycyclic Aromatic Hydrocarbons,* ed. A. Bjørseth. New York: Marcel Dekker, Inc., pp. 1–26.

71. Nam, J.J., Gustafssono, O., Kurt-Karakus, P. et al. 2008. Relationships between organic matter, black carbon and persistent organic pollutants in European background soils: Implications for sources and environmental fate. *Environ. Pollut.* 156: 809–817.

72. Al-Haddad, A. 2005. Benzo[a]pyrene concentrations in topsoils of North Bahrain. *Toxicol. Environ. Chem.* 87: 159–165.

73. Nam, J.J., Sweetman, A.J., and Jones, K.C. 2008. Polynuclear aromatic hydrocarbons (PAHs) in global background soils. *J. Environ. Monit.* 11: 45–48.

74. Bourotte, C., Forti, M.C., Lucas, Y. et al. 2009. Comparison of polycyclic aromatic hydrocarbon (PAHs) concentrations in urban and natural forest soils in the Atlantic Forest (São Paulo State). *Ann. Acad. Bras. Cienc.* 81: 127–136.

75. Choi, S.-D., Shunthirasingham, C., and Daly, G.L. 2009. Levels of polycyclic aromatic hydrocarbons in Canadian mountain air and soil are controlled by proximity to roads. *Environ. Pollut.* 157: 3199–3206.

76. Li, Y.T., Li, F.B., and Chen, J.J. 2008. The concentrations, distribution and sources of PAHs in agricultural soils and vegetables from Shunde, Guangdong, China. *Environ. Monit. Assess.* 139: 61–76.

77. Zohair, A., Salim, A., and Soyibo, A.A. 2006. Residues of polycyclic aromatic hydrocarbons (PAHs), polychlorinated biphenyls (PCBs) and organochlorine pesticides in organically-farmed vegetables. *Chemosphere* 63: 541–553.

78. Motelay-Massei, A., Ollivon, D., and Garban, B. 2004. Distribution and spatial trends of PAHs and PCBs in soils in the Seine River basin, France. *Chemosphere* 55: 555–565.

79. Riget, F.F., Christensen, J., and Johansen, P., eds. 2003. *AMAP Green-land and the Faroe Islands. Vol. 2: The Environment of Greenland. DANCEA Report.* Copenhagen, Denmark: Ministry of the Environment, 193pp.

80. Ray, S., Khillare, P.S., Agarwal, T. et al. 2008. Assessment of PAHs in soil around the international airport, Delhi, India. *J. Hazard. Mater.* 156: 9–16.

81. Samimi, S.V., Akbari Rad, R., and Ghanizade, F. 2009. Polycyclic aromatic hydrocarbon contamination levels in collected samples from the vicinity of a highway. *Iran. J. Environ. Health. Sci. Eng.* 6: 47–52.

82. Orecchio, S. 2010. Assessment of polycyclic aromatic hydrocarbons (PAHs) in soil of a natural reserve (Isola delle Femmine) (Italy) located in front of a plant for the production of cement. *J. Hazard. Mater.* 173: 358–368.

83. Yang, Y., Zhang, X.X., and Korenaga, T. 2002. Distribution of polynuclear aromatic hydrocarbons (PAHs) in the soil of Tokushima, Japan. *Water Air Soil Pollut.* 138: 51–60.

84. Abbas, A.O. and Brack, W. 2005. Polycyclic aromatic hydrocarbons in Niger Delta soil: Contamination sources and profiles. *Int. J. Environ. Sci. Technol.* 2: 343–352.

85. Olajire, A.A., Alade, A.O., Adeniyi, A.A. et al. 2007. Distribution of polycyclic aromatic hydrocarbons in surface soils and water from the vicinity of Agbabu bitumen field of southwestern Nigeria. *J. Environ. Sci. Health A* 42: 1043–1049.

86. Wilcke, W., Krauss, M., Safronov, G. et al. 2005. Polycyclic aromatic hydrocarbons (PAHs) in soils of the Moscow region-concentrations, temporal trends, and small-scale distribution. *J. Environ. Qual.* 34: 1581–1590.

87. Crnkovic, D., Ristić, M., Jovanović, A. et al. 2007. Levels of PAH in the soils of Belgrade and its environs. *Environ. Monit. Assess.* 125: 75–83.

88. Nadal, M., Schuhmacher M., and Domingo, J.L. 2007. Levels of metals, PCBs, PCNs and PAHs in soils of a highly industrialized chemical/petrochemical area: Temporal trend. *Chemosphere* 66: 267–276.

89. Bucheli, T.D., Blum, F., and Desaules, A. et al. 2004. Polycyclic aromatic hydrocarbons, black carbon, and molecular markers in soils of Switzerland. *Chemosphere* 56: 1061–1076.

90. Amagai, T., Takahashi, Y., and Matsushita, H. 1999. A survey on polycyclic aromatic hydrocarbon concentrations in soil in Chiang-Mai, Thailand. *Environ. Int.* 25: 563–572.

91. Kannan, K., Battula, S., Loganathan, B.G. et al. 2003. Trace organic contaminants, including toxaphene and trifluralin, in cotton field soils from Georgia and South Carolina, USA. *Arch. Environ. Contam. Toxicol.* 45: 30–36.

92. De La Torre-Roche, R.J., Lee, W.-Y., and Campos-Díaz, S.I. 2009. Soil-borne polycyclic aromatic hydrocarbons in El Paso, Texas: Analysis of a potential problem in the United States/Mexico border region. *J. Hazard. Mater.* 163: 946–958.

93. McCready, S., Sleeà, D.J., Birch, G.F. et al. 2000. The distribution of polycyclic aromatic hydrocarbons in surcial sediments of Sydney harbour, Australia. *Mar. Pollut. Bull.* 40: 999–1006.

94. Tolosa, I., de Mora, S.J., and Fowler, S.W. 2005. Aliphatic and aromatic hydrocarbons in marine biota and coastal sediments from the Gulf and the Gulf of Oman. *Mar. Pollut. Bull.* 54: 1619–1633.

95. Meire, R.O., Azeredo, A., Pereira, M.D. et al. 2008. Polycyclic aromatic hydrocarbons assessment in sediment of national parks in southeast Brazil. *Chemosphere* 73 (Suppl. 1): S180–S185.

96. Saha, M., Togo, A., Mizukawa, K. et al. 2009. Sources of sedimentary PAHs in tropical Asian waters: Differentiation between pyrogenic and petrogenic sources by alkyl homolog abundance. *Mar. Pollut. Bull.* 58: 189–200.

97. Bolton, J.L., Stehr, C.M., Boyd, D.T. et al. 2003. Organic and trace metal contaminants in sediments and English sole tissues from Vancouver Harbour, Canada. *Mar. Environ. Res.* 57: 19–36.

98. Quiroz, R., Popp, P., Urrutia, R. et al. 2005. PAH fluxes in the Laja lake of south central Chile Andes over the last 50 years: Evidence from a dated sediment core. *Sci. Total Environ.* 349: 150–160.

99. Shen, Q., Wang, K.Y., and Zhang, W. 2009. Characterization and sources of PAHs in an urban river system in Beijing, China. *Environ. Geochem. Health* 31: 453–462.

100. Johnson-Restrepo, B., Olivero-Verbel, J., Lu, S. et al. 2008. Polycyclic aromatic hydrocarbons and their hydroxylated metabolites in fish bile and sediments from coastal waters of Colombia. *Environ. Pollut.* 151: 452–459.

101. El Deeb, K.Z., Said, T.O., El Naggar, M.H. et al. 2007. Distribution and sources of aliphatic and polycyclic aromatic hydrocarbons in surface sediments, fish and bivalves of Abu Qir Bay (Egyptian Mediterranean Sea). *Bull. Environ. Contam. Toxicol.* 78: 373–379.

102. Mille, G., Asia, L., Guiliano, M. et al. 2007. Hydrocarbons in coastal sediments from the Mediterranean Sea (Gulf of Fos area, France). *Mar. Pollut. Bull.* 54: 566–575.

103. Denton, G.R.W., Concepcion, L.P., and Wood, H.R. 2006. Polycyclic aromatic hydrocarbons (PAHs) in small island coastal environments: A case study from harbours in Guam, Micronesia. *Mar. Pollut. Bull.* 52: 1090–1117.

104. De Luca, G., Furesi, A., Micera, G. et al. 2005. Nature, distribution and origin of polycyclic aromatic hydrocarbons (PAHs) in the sediments of Olbia harbor (Northern Sardinia, Italy). *Mar. Pollut. Bull.* 50: 1223–1232.

105. Kwach, B.O. and Lalah, J.O. 2009. High concentrations of polycyclic aromatic hydrocarbons found in water and sediments of car wash and Kisat areas of Winam Gulf, Lake Victoria-Kenya. *Bull. Environ. Contam. Toxicol.* 83: 727–733.

106. Macías-Zamora, J.V., Mendoza-Vega, E., and Villaescusa-Celaya, J.A. 2002. PAHs composition of surface marine sediments: A comparison to potential local sources in Todos Santos Bay, B.C., Mexico. *Chemosphere* 46: 459–468.

107. Ahrens, M.J. and Depree, C.V. 2004. Inhomogeneous distribution of polycyclic aromatic hydrocarbons in different size and density fractions of contaminated sediment from Auckland Harbour, New Zealand: An opportunity for mitigation. *Mar. Pollut. Bull.* 48: 341–350.

108. Olajirea, A.A., Altenburger, R., Kuster, E. et al. 2005. Chemical and ecotoxicological assessment of polycyclic aromatic hydrocarbon—Contaminated sediments of the Niger Delta, Southern Nigeria. *Sci. Total Environ.* 340: 123–136.

109. Savinov, V.M., Savinova, T.N., Carroll, J. et al. 2000. Polycyclic aromatic hydrocarbons (PAHs) in sediments of the White Sea, Russia. *Mar. Pollut. Bull.* 40: 807–818.

110. Qiu, Y.-W., Zhang, G., Liu, G.-Q., Guo, L.-L., Li, X.-D., and Wai, O. 2009. Polycyclic aromatic hydrocarbons (PAHs) in the water column and sediment core of Deep Bay, South China. *Estuar. Coast Shelf Sci.* 83: 60–66.

111. Vane, C.H., Harrison, I., and Kim, A.W. 2007. Assessment of polyaromatic hydrocarbons (PAHs) and polychlorinated biphenyls (PCBs) in surface sediments of the Inner Clyde Estuary, UK. *Mar. Pollut. Bull.* 54: 1287–1306.

112. Kannan, K., Johnson-Restrepo, B., Sharons, Y. et al. 2005. Spatial and temporal distribution of polycyclic aromatic hydrocarbons in sediments from Michigan inland lakes. *Environ. Sci. Technol.* 39: 4700–4706.

113. Mostafa, A.R., Wade, T.L., Sweet, S.T. et al. 2009. Distribution and characteristics of polycyclic aromatic hydrocarbons (PAHs) in sediments of Hadhramout coastal area, Gulf of Aden, Yemen. *J. Mar. Syst.* 78: 1–8.

114. Zhang, Y. and Tao, S. 2009. Global atmospheric emission inventory of polycyclic aromatic hydrocarbons (PAHs) for 2004. *Atmos. Environ.* 43: 812–819.

115. International Energy Agency. 2006. *Energy Statistics and Balances of Non-OECD Countries, 2003–2004.* Paris, France: Organization for Economic Co-operation and Development.

116. Hafner, W.D., Carlson, D.L., and Hites, R.A. 2005. Influence of local human population on atmospheric polycyclic aromatic hydrocarbon concentrations. *Environ. Sci. Technol.* 39: 7374–7379.

117. Singh, K.P., Malik, A., Kumar, R. et al. 2008. Receptor modeling for source apportionment of polycyclic aromatic hydrocarbons in urban atmosphere. *Environ. Monit. Assess.* 136: 183–196.

118. Wania, F. and Mackay. D. 1996. Tracking the distribution of persistent organic pollutants. *Environ. Sci. Technol.* 30: 390A–396A.

119. Baird, C. and Cann, M. 2008. *Environmental Chemistry.* New York: W.H. Freeman and Company.

120. Hung, H., Kallenborn, R., Brevik, K. et al. 2010. Atmospheric monitoring of organic pollutants in the Arctic under the Arctic monitoring of organic pollutants in the Arctic monitoring and assessment programme (AMAP): 1993–2006. *Sci. Total Environ.* 408: 2854–2873.

121. Halsall, C.J., Sweetman, A.J., Barrie, L.A. et al. 2001. Modelling the behaviour of PAHs during atmospheric transport from the UK to the Arctic. *Atmos. Environ.* 35: 255–267.

122. Martins, C.C., Bícego, M.C., Rose, N.L. et al. 2010. Historical record of polycyclic aromatic hydrocarbons (PAHs) and spheroidal carbonaceous particles (SCPs) in marine sediment cores from Admiralty Bay, King George Island, Antarctica. *Environ. Pollut.* 158: 192–200.

123. Bard, S.M. 1999. Global transport of anthropogenic contaminants and the consequences for the Arctic marine ecosystem. *Mar. Pollut. Bull.* 38: 356–379.

124. Taniguchi, S., Montone, R.C., Bícego, M.C. et al. 2009. Chlorinated pesticides, polychlorinated biphenyls and polycyclic aromatic hydrocarbons in the fat tissue of seabirds from King George Island, Antarctica. *Mar. Pollut. Bull.* 58: 129–133.

125. Alexander, C., Smith, R., Loganathan, B. et al. 1999. Pollution history of the Savannah river estuary and comparisons with Baltic sea pollution history. *Limnologica.* 29: 267–273.

126. Macdonald, R.W., Harner, T., and Fyfe, J. 2005. Recent climate change in the Arctic and its impact on contaminant pathways and interpretation of temporal trend data. *Sci. Total Environ.* 342: 5–86.

127. Butler, J.P., Post, G.B., Lioy, P.J. et al. 1993. Assessment of carcinogenic risk from personal exposure to benzo(a)pyrene in the total human environmental exposure study (THEES). *J. Air Waste Manage. Assoc.* 43: 970–977.

128. Lioy, P.L., Waldman, J.M., and Greenberg, A. 1988. The Total Human Environmental Exposure Study (THEES) to benzo(a)pyrene: Comparison of the inhalation and food pathways. *Arch. Environ. Health* 43: 304–312.

129. Beckman Sundh, O., Thuvander, A., and Andersson, C. 1998. Preview of PAHs in food: Potential health effects and contents in food, Report 8, Swedish National Food Administration, Uppsala, Sweden.

130. Phillips, D.H. 1999. Polycyclic aromatic hydrocarbons in the diet. *Mutat. Res.* 443: 139–147.

131. Wickstrom, K., Pyysalo, H., Plaami-Heikkila, S. et al. 1986. Polycyclic aromatic compounds (PAC) in leaf lettuce. *Z. Lebensm. Unters. Forsch.* 183: 182–185.

132. Crepineau, C., Rychen, G., and Feidt, C. 2003. Contamination of pastures by polycyclic aromatic hydrocarbons (PAHs) in the vicinity of a highway. *J. Agric. Food Chem.* 51: 4841–4845.

133. Yakibu, H.Y., Martins, M.S., and Takahashi, M.Y. 1993. Levels of benzo(a)-pyrene and other polycyclic aromatic hydrocarbons in liquid smoke flavour and some smoked foods. *Food Addit. Contam.* 10: 399–405.
134. Knize, M.G., Salmon, C.P., Pais, P. et al. 1999. Food heating and the formation of heterocyclic aromatic amine and polycyclic aromatic hydrocarbon mutagens/carcinogens. *Adv. Exp. Med. Biol.* 459: 179–193.
135. Vainiotalo, S. and Matveinen, K. 1993. Cooking fumes as a hygienic problem in the food and catering industries. *Am. Ind. Hyg. Assoc. J.* 54: 376–382.
136. Perez, C., Lopez de Carain, A., and Bello, J. 2002. Modulation of mutagenic activity in meat samples after deep-frying in vegetable oils. *Mutagenesis* 17: 63–66.
137. Klein, H., Speer, K., and Schmidt, E.H.F. 1993. Polycyclic aromatic hydrocarbons in raw coffee and roasted coffee. *Bundesgesundheitsblatt* 36: 98–100.
138. Tateno, T., Nagumo, Y., and Suenaga, S. 1990. Polycyclic aromatic hydrocarbons produced from grilled vegetables. *J. Food Hyg. Soc. Jpn.* 31: 271–276.
139. Guillén, M.D. and Sopelana, P. 2003. Polycyclic aromatic hydrocarbons in diverse foods. In: *Food Safety: Contaminants and Toxins*, ed. J.P.F. D'Mello. Oxford, U.K.: CAB International, pp. 175–197.
140. Benford, D., Agudo, B., Carrington, C. et al. 2006. Polycyclic aromatic hydrocarbons. whqlibdoc.who.int/publications/2006/9241660554_PAH_eng.pdf. Downloaded on June 20, 2010.
141. Pandey, M.K., Mishra, K.K., Khanna, S.K. et al. 2004. Detection of polycyclic aromatic hydrocarbons in commonly consumed edible oils and their likely intake in the Indian population. *J. Am. Oil Chem. Soc.* 81: 1131–1136.
142. Akpan, V., Lodovici, M., and Dolara, P. 1994. Polycyclic aromatic hydrocarbons in fresh and smoked fish samples from three Nigerian cities. *Bull. Environ. Contam. Toxicol.* 53: 246–253.
143. Suzuki, K. and Yoshinaga, J. 2007. Inhalation and dietary exposure to polycyclic aromatic hydrocarbons and urinary 1-hydroxypyrene in non-smoking university students. *Int. Arch. Occup. Environ. Health* 81: 115–121.
144. Moon, H.-B., Kim, H.-S., Choi, M. et al. 2010. Intake and potential health risk of polycyclic aromatic hydrocarbons associated with seafood consumption in Korea from 2005 to 2007. *Arch. Environ. Contam. Toxicol.* 58: 214–221.
145. Ishimaru, T., Inouye, H., and Morioka, T. 1990. Risk assessment of drinking water in a reservoir contaminated by PAH's originated from road traffic. *Sci. Total Environ.* 93: 125–130.
146. Literathy, P., Quinn, M., and Al-Rashed, M. 2003. Pollution potential of oil-contaminated soil on groundwater resources in Kuwait. *Water Sci. Technol.* 47: 259–65.
147. Boyce, C.P. and Gary, M.R. 2003. Developing risk-based target concentrations for carcinogenic polycyclic aromatic hydrocarbon compounds assuming human consumption of aquatic biota. *J. Toxicol. Environ. Health B* 6: 497–520.
148. Maier, M., Maier, D., and Lloyd, B.J. 2002. Factors influencing the mobilization of polycyclic aromatic hydrocarbons (PAHs) from the coal-tar lining of water mains. *Water Res.* 34: 773–786.
149. Boffetta, P., Jourenkova, N., and Gustavsson, P. 1997. Cancer risk from occupational exposure to polycyclic aromatic hydrocarbons. *Cancer Causes Control* 8: 444–472.
150. Penning, T.M. 1993. Dihydrodiol dehydrogenase and its role in polycyclic aromatic hydrocarbon metabolism. *Chem.-Biol. Interact.* 89: 1–34.
151. Luch, A. (ed.). 2005. *The Carcinogenic Effects of Polycyclic Aromatic Hydrocarbons*. London, U.K.: Imperial College Press.
152. Shimada, T. 2006. Xenobiotic-metabolizing enzymes involved in activation and detoxification of carcinogenic polycyclic aromatic hydrocarbons. *Drug Metab. Pharmacokinet.* 21: 257–276.
153. Shimada, T. and Guengerich, F.P. 2006. Inhibition of human cytochrome P450 1A1-, 1A2-, and 1B1-mediated activation of procarcinogens to genotoxic metabolites by polycyclic aromatic hydrocarbons. *Chem. Res. Toxicol.* 19: 288–294.
154. Menzie, C.A., Potocki, B.B., and Santodonato, J. 1992. Exposure to carcinogenic PAHs in the environment. *Environ. Sci. Technol.* 26: 1278–83.
155. Mastrangelo, G., Fadda, E., and Marzia, V. 1996. Polycyclic aromatic hydrocarbons and cancer in man. *Environ. Health Perspect.* 104: 1166–1170.
156. McGregor, D. 2007. Risk of cancer of the colon and rectum in firemen. Chemical substances and research projects. Report R-516. Quebec, Canada.
157. Hood, D.B., Ramesh, A., and Aschner, M. 2009. Polycyclic aromatic hydrocarbons: Exposure from emission products and terrorist attacks on US targets: Implications for developmental central nervous system toxicity. In: *Handbook of Toxicology of Chemical Warfare Agents*, ed. R.C. Gupta. London, U.K.: Academic Press, pp. 229–243.

158. Hood, D.B., Campbell, D., and Levitt, P. 2011. An emerging gene-environment interaction model: Autism spectrum disorder phenotypes resulting from exposure to environmental contaminants during gestation. In: *Developmental Neurotoxicology Research: Principles, Models, Techniques, Strategies and Mechanisms*, Vol. 1, ed. C. Wang, W. Slikker Jr. pp. 543–562. London, U.K.: John Wiley & Sons.

159. Hood, D.B., Ramesh, A., Chirwa, S., Khoshbouei, H., and Archibong, A. 2011. Developmental toxicity of polycyclic aromatic hydrocarbons. In: *Reproductive and Developmental Toxicology*, ed. R.C. Gupta. London, U.K.: Academic Press, pp. 593–606.

160. Ramesh, A. and Archibong, A. 2011. Reproductive toxicity of polycyclic aromatic hydrocarbons: Occupational relevance. In: *Reproductive and Developmental Toxicology*, ed. R.C. Gupta, London, U.K.: Academic Press, pp. 577–591.

161. Wormley, D.D., Ramesh, A., and Hood, D.B. 2004. Environmental contaminant-mixture effects on CNS development, plasticity, and behavior. *Toxicol. Appl. Pharmacol.* 197: 49–65.

162. Brown, L.A., Khoshbouei, H., Goodwin, S.J. et al. 2007. Downregulation of early ionotrophic glutamate receptor subunit developmental expression as a mechanism for observed plasticity deficits following gestational exposure to benzo(a)pyrene. *Neurotoxicology* 28: 965–978.

163. McCallister, N.M., Maguire, M., Ramesh, A. et al. 2008. Prenatal exposure to benzo(a)pyrene impairs later-life cortical neuronal function. *Neurotoxicology* 29: 846–854.

164. Bouayed, J., Desor, F., Rammal, H. et al. 2009. Effects of lactational exposure to benzo[alpha]pyrene (B[alpha]P) on postnatal neurodevelopment, neuronal receptor gene expression and behavior in mice. *Toxicology* 259: 97–106.

165. Dutta, K., Ghosh, D., Nazmi, A. et al. 2010. A common carcinogen benzo[a]pyrene causes neuronal death in mouse via microglial activation. *PLoS One* 5: e9984.

166. Otto, D., Skalik, I., and Bahboli, R. 1997. Neurobehavioral performance of Czech school children born in years of maximal air pollution. *Neurotoxicology* 18: 903.

167. Majchrzak, R., Sroczyński, J., and Chełmecka, E. 1990. Evaluation of the nervous system in workers in the furnace and coal divisions of the coke-producing plants. *Med. Pr.* 41: 108–113.

168. Golden, A.L., Markovitz, S.B., and Landrigan, P.J., 1995. The risk of cancer in firefighters. *Occup. Med.* 10: 803–820.

169. Dayal, H., Gupta, S., Trieff, N. et al. 1995. Symptom clusters in a community with chronic exposure to chemicals in two superfund sites. *Arch. Environ. Health* 50: 108–111.

170. Kilburn, K.H. and Warshaw, R.H. 1995. Neurotoxic effects from residential exposure to chemicals from an oil reprocessing facility and superfund site. *Neurotoxicol. Teratol.* 17: 89–102.

171. Cordier, S., Lefeuvre, B., Filippini, G. et al. 1997. Parenteral occupation, occupational exposure to solvents and polycyclic aromatic hydrocarbons and risk of childhood brain tumors (Italy, France, Spain). *Cancer Causes Control* 8: 688–697.

172. Kerr, M.A., Nasca, P.C. and Mundt, K.A. 2000. Parenteral occupational exposures and risk of neuroblastoma: A case control study (United States). *Cancer Causes Control* 11: 635–643.

173. Perera, F.P., Li, Z., Whyatt, R. et al. 2009. Prenatal airborne polycyclic aromatic hydrocarbon exposure and child IQ at age 5 years. *Pediatrics* 124: 195–202.

174. Wang, S., Zheng, T., Chanock, S. et al. 2010. Methods for detecting interactions between genetic polymorphisms and prenatal environment exposure with a mother-child design. *Genet. Epidemiol.* 34: 125–132.

175. Wang, S., Chanock, S., and Tang, D. 2010. Effect of gene-environment interactions on mental development in African American, Dominican, and Caucasian mothers and newborns. *Ann. Hum. Genet.* 74: 46–56.

176. Niu, Q., Zhang, H., Li, X. et al. 2010. Benzo[a]pyrene-induced neurobehavioral function and neurotransmitter alterations in coke oven workers. *Occup. Environ. Med.* 67: 444–448.

177. Harris, D.L., Washington, M.K., Hood, D.B. et al. 2009. Dietary fat-influenced development of colon neoplasia in Apc^{Min} mouse exposed to benzo(a)pyrene. *Toxicol. Pathol.* 37: 938–946.

178. Yang, H., Zhou, L., Wang, Z. et al. 2009. Overexpression of antioxidant enzymes in ApoE-deficient mice suppresses benzo(a)pyrene-accelerated atherosclerosis. *Atherosclerosis* 207: 51–58.

179. Mukhopadhyay, D., Nandi, P., Varghese, A.C. et al. 2010. The in vitro effect of benzo(a)pyrene on human sperm hyperactivation and acrosome reaction. *Fertil. Steril.* 94: 595–598.

180. Xia, Y., Zhu, P., Han, Y. et al. 2009. Urinary metabolites of polycyclic aromatic hydrocarbons in relation to idiopathic male infertility. *Hum. Reprod.* 1: 1–8.

181. Khorram, O., Han, G., and Magee, T. 2010. Cigarette smoke inhibits endometrial epithelial cell proliferation through a nitric oxide-mediated pathway. *Fertil. Steril.* 93: 257–263.

182. Linschooten, J.O., van Schooten, F.J., and Baumgartner, A. 2009. Use of spermatozoam RNA profiles to study gene-environment interactions in human germ cells. *Mutat. Res.* 667: 70–76.

183. Zenzes, M.T., Bielecki, R., and Reed, T.E. 1999. Detection of benzo(a)pyrene diol epoxide-DNA adducts in sperm of men exposed to cigarette smoke. *Fertil. Steril.* 72: 330–335.

184. Zenzes, M.T., Puy, L.A., Bielecki. R. et al. 1999. Detection of benzo(a)pyrene diol epoxide-DNA adducts in embryos from smoking couples: Evidence for transmission by spermatozoa. *Mol. Hum. Reprod.* 5: 125–131.

185. Russo, A., Troncoso, N. Sanchez, F. et al. 2006. Propolis protects human spermatozoa from DNA damage caused by benzo(a)pyrene and exogenous reactive oxygen species. *Life Sci.* 78: 1401–1406.

186. Gaspari, L., Chang, S.S., Santella, R.M. et al. 2003. Polycyclic aromatic hydrocarbon-DNA adducts in human sperm as a marker of DNA damage and infertility. *Mutat. Res.* 535: 155–160.

187. Hsu, P.C., Chen, I.Y., and Pan, C.H. 2006. Sperm DNA damage correlates with polycyclic aromatic hydrocarbon biomarker in coke-oven workers. *Int. Arch. Occup. Environ. Health* 79: 349–356.

188. Ji, G., Gu, A., and Zhu, P. 2010. Joint effects of XRCC1 polymorphisms and polycyclic aromatic hydrocarbons exposure on sperm DNA damage and male infertility. *Toxicol. Sci.* 116: 92–98.

189. Rubes, J., Selevan, S.G., and Sram, R.J. 2007. GSTM1 genotype influences the susceptibility of men to sperm DNA damage associated with exposure to air pollution. *Mutat. Res.* 625: 20–28.

190. Paracchini, V., Chang, S.-S., Santella, R.M. et al. 2005. GSTM1 deletion modifies the levels of polycyclic aromatic hydrocarbon-DNA adducts in human sperm. *Mutat. Res.* 586: 97–101.

191. Rubes, J., Rybar, R., Prinosilova, P. et al. 2010. Genetic polymorphisms influence the susceptibility of men to sperm DNA damage associated with exposure to air pollution. *Mutat. Res.* 683: 9–15.

192. Zenzes, M.T. 2000. Smoking and reproduction: Gene damage to human gametes and embryos. *Hum. Reprod. Update* 6: 122–131.

193. Soares, S.R., Simon, C., Remohí, J. et al. 2007. Cigarette smoking affects uterine receptiveness. *Hum. Reprod.* 22: 543–547.

194. Ramesh, A., Archibong, A.E., and Niaz, M.S. 2010. Ovarian susceptibility to benzo(a)pyrene: Tissue burden of metabolites and DNA adducts in F-344 rats. *J. Toxicol. Environ. Health* 73: 1611–1625.

195. Zenzes, M.T., Puy, L.A., and Bielecki, R. 1998. Immunodetection of benzo(a)pyrene adducts in ovarian cells of women exposed to cigarette smoke. *Mol. Hum. Reprod.* 4: 159–165.

196. Shamsuddin, A.K. and Gan, R. 1988. Immunocytochemical localization of benzo(a)pyrene-DNA adducts in human tissue. *Hum. Pathol.* 19: 309–315.

197. Rorke, E.A., Sizemore, N., Mukhtar, H. et al. 1998. Polycyclic aromatic hydrocarbons enhance terminal cell death of human ectocervical cells. *Int. J. Oncol.* 13: 557–563.

198. Melikian, A.A., Sun, P., and Prokopczyk, B. 1999. Identification of benzo(a)pyrene metabolites in cervical mucus and DNA adducts in cervical tissues in humans by gas chromatography-mass spectrometry. *Cancer Lett.* 146: 127–134.

199. McGarry, M.A., Charles, G.D., Medrano, T. et al. 2002. Benzo(a)pyrene, but not 2,3,7,8-tetrachloro-dibenzo-p-dioxin alters cell adhesion proteins in human uterine RL95-2 cells. *Biochem. Biophys. Res. Commun.* 294: 101–107.

200. Qin, Y.Y., Leung, C.K.M., Leung, A.O.W. et al. 2009. Persistent organic pollutants and heavy metals in adipose tissues of patients with uterine leiomyomas and the association of these pollutants with seafood diet, BMI, and age. *Environ. Sci. Pollut. Res. Int.* 17: 229–240.

201. Naruse, M., Ishihara, Y., Miyagawa-Tomita, S. et al. 2002. 3-Methylcholanthrene, which binds to the aryl hydrocarbon receptor, inhibits proliferation and differentiation of osteoblasts in vitro and ossification *in vivo. Endocrinology* 143: 3575–3581.

202. Naruse, M., Otsuka, E., Naruse, M. et al. 2004. Inhibition of osteoclast formation by 3-methylcholanthrene, a ligand for arylhydrocarbon receptor: Suppression of osteoclast differentiation factor in osteogenic cells. *Biochem. Pharmacol.* 67: 119–127.

203. Voronov, I., Heersche, J.N.M., Casper, R.F. et al. 2005. Inhibition of osteoclast differentiation by polycyclic aryl hydrocarbons is dependent on cell density and RANKL concentration. *Biochem. Pharmacol.* 70: 300–307.

204. Lee, L.L., Lee, J.S.C., Waldman, S.D. et al. 2002. Polycyclic aromatic hydrocarbons present in cigarette smoke cause bone loss in an ovariectomized rat model. *Bone* 30: 917–923.

205. Knafla, A., Phillipps, K.A., Brecher, R.W. et al. 2006. Development of a dermal cancer slope factor for benzo[a]pyrene. *Regul. Toxicol. Pharmacol.* 45: 159–168.

206. Väänänen, V., Meila, M., and Kalliokoski, P. 2005. Dermal exposure to polycyclic aromatic hydrocarbons among road pavers. *Ann. Occup. Hyg.* 49: 167–178.

207. Borska, L., Andrys, C., Krejsek, J. et al. 2010. Genotoxic and apoptotic effects of Goeckerman therapy for psoriasis. *Int. J. Dermatol.* 49: 289–294.
208. Kamangar, F., Strickland, P.T., Pourshams, A. et al. 2005. High exposure to polycyclic aromatic hydrocarbons may contribute to high risk of esophageal cancer in northeastern Iran. *Anticancer Res.* 25: 425–428.
209. Van Gijssel, H.E., Schild, L.J., Watt, D.L. et al. 2004. Polycyclic aromatic hydrocarbon-DNA adducts determined by semiquantitative immunohistochemistry in human esophageal biopsies taken in 1985. *Mutat. Res.* 547: 55–62.
210. Roth, M.J., Wei, W.Q., Baer, J. et al. 2009. Aryl hydrocarbon receptor expression is associated with a family history of upper gastrointestinal tract cancer in a high-risk population exposed to aromatic hydrocarbons. *Cancer Epidemiol. Biomarkers Prev.* 18: 2391–2396.
211. Roth, M.J., Strickland, K.L., Wang, G.Q. et al. 1998. High levels of carcinogenic polycyclic aromatic hydrocarbons present within food from Linxian, China may contribute to that region's high incidence of oesophageal cancer. *Eur. J. Cancer* 34: 757–758.
212. Sinha, R., Kulldorff, M., Gunter, M.J. et al. 2005. Dietary benzo(a)pyrene intake and risk of colorectal adenoma. *Cancer Epidemiol. Biomarkers Prev.* 14: 2030–2034.
213. Sinha, R., Peters, U., Cross, A.J. et al. 2005. Meat, meat cooking methods and preservation and risk for colorectal adenoma. *Cancer Res.* 65: 8034–8041.
214. Jeffy, B.D., Schultz, E.U., Selmin, O. et al. 1999. Inhibition of *BRCA-1* expression by benzo[*a*]pyrene and its diol epoxide. *Mol. Carcinog.* 26: 100–118.
215. Rundle, A., Tang, D., Hibshoosh, H. et al. 2000. The relationship between genetic damage from polycyclic aromatic hydrocarbons in breast tissue and breast cancer. *Carcinogenesis.* 21: 1281–1289.
216. Mordukhovich, I., Rossner, P. Jr., Terry, M.B. et al. 2010. Associations between polycyclic aromatic hydrocarbon-related exposures and p53 mutations in breast tumors. *Environ. Health Perspect.* 118: 511–518.
217. Gammon, M.D. and Santella, R.M. 2008. PAH, genetic susceptibility and breast cancer risk: An update from the Long Island Breast Cancer Study Project. *Eur. J. Cancer* 44: 636–640.
218. Goldman, R., Enewold, L., Pellizzari, E. et al. 2001. Smoking increases carcinogenic polycyclic aromatic hydrocarbons in human lung tissue. *Cancer Res.* 61: 6367–6371.
219. Armstrong, B.G. and Gibbs, G. 2009. Exposure–response relationship between lung cancer and polycyclic aromatic hydrocarbons (PAHs). *Occup. Environ. Med.* 66: 740–746.
220. Elovaara, E., Mikkola, J., and Stockmann-Juvala, H. 2007. Polycyclic aromatic hydrocarbon (PAH) metabolizing enzyme activities in human lung, and their inducibility by exposure to naphthalene, phenanthrene, pyrene, chrysene, and benzo(a)pyrene as shown in the rat lung and liver. *Arch. Toxicol.* 81: 169–182.
221. Shen, M., Berndt, S.I., Rothman, N. et al. 2005. Polymorphisms in the DNA base excision repair genes APEX1 and XRCC1 and lung cancer risk in Xuan Wei, China. *Anticancer Res.* 25: 537–542.
222. Karakaya, A., Yücesoy, B., Turhan, A. et al. 1999. Investigation of some immunological functions in a group of asphalt workers exposed to polycyclic aromatic hydrocarbons. *Toxicology* 135: 43–47.
223. Szczeklik, A., Szczeklik, J., Galuszka, Z. et al. 1994. Humoral immunosuppression in men exposed to polycyclic aromatic hydrocarbons and related carcinogens in polluted environments. *Environ. Health Perspect.* 102: 302–304.
224. Karakaya, A., Ates, I., and Yucesoy, B. 2004. Effects of occupational polycyclic aromatic hydrocarbon exposure on T-lymphocyte functions and natural killer cell activity in asphalt and coke oven workers. *Hum. Exp. Toxicol.* 23: 317–322.
225. Hutcheon, D.E., Kantrowitz, J., Van Gelder, R.N. et al. 1983. Factors affecting plasma benzo[a]pyrene levels in environmental studies. *Environ. Res.* 32: 104–110.
226. Moir, D., Viau, A., Chu, I. et al. 1998. Pharmacokinetics of benzo[a]pyrene in the rat. *J. Toxicol. Environ. Health A* 53: 507–530.
227. Irigaray, P., Ogier, V., Jacquenet, S. et al. 2006. Benzo[a]pyrene impairs beta-adrenergic stimulation of adipose tissue lipolysis and causes weight gain in mice. A novel molecular mechanism of toxicity for a common food pollutant. *FEBS J.* 273: 1362–1372.
228. Curfs, D.M., Lutgens, E., Gijbels, M.J. et al. 2004. Chronic exposure to the carcinogenic compound benzo[a]pyrene induces larger and phenotypically different atherosclerotic plaques in ApoE-knockout mice. *Am. J. Pathol.* 164: 101–108.
229. Wang, Z., Yang, H., Ramesh, A. et al. 2009. Overexpression of Cu/Zn-superoxide dismutase and/or catalase accelerates benzo(a)pyrene detoxification by upregulation of the aryl hydrocarbon receptor in mouse endothelial cells. *Free Radic. Biol. Med.* 47: 1221–1229.

Part II

Persistent Organic Chemicals in the Asia-Pacific Region

6 Organohalogen Contamination Profiles and Trends in Australia

*Munro R. Mortimer, Leisa-Maree Leontjew Toms,
and Jochen F. Mueller**

CONTENTS

6.1 Introduction 129
    6.1.1 History of Persistent Organic Chemicals Use in Agriculture and Industries
        in Australia 130
    6.1.2 Environmental Levels, Environmental Impact, and Ban/Restriction of
        POCs in Australia 132
        6.1.2.1 Dioxins and Dioxin-Like Chemicals 132
        6.1.2.2 PBDEs... 144
6.2 Trend Monitoring Studies in Australia... 151
    6.2.1 OCP Trends in Australia... 152
        6.2.1.1 OCP Trends in the Australian Marine Environment................... 152
        6.2.1.2 OCP Trends in Human Milk Samples in Australia 155
        6.2.1.3 OCP Trends in Historic Butter Samples in Australia 156
    6.2.2 Trends in Concentrations of Dioxin-Like Chemicals in Australia.......... 158
6.3 Conclusions.. 159
References... 159

6.1 INTRODUCTION

Australia is an "island continent" spanning some 4000 km from east to west, 3700 km from north to south, with a total area of 7.69 million km^2 bounded by the Indian, Pacific, and Southern Oceans. Approximately one-third of the land mass lies to the north of the Tropic of Capricorn, and the continent spans a diverse range of climatic zones. Australia's average elevation is only 330 m, reaching little more than 2200 m at its highest point. Approximately 70% of the landscape is arid or semi-arid, with a large portion of the continent center unsuitable for settlement and too dry to sow pasture or crops. However, much of this area has sufficient native plant cover to support extensive grazing, provided ground water is available. The most fertile areas are close to the coast, particularly in the high rainfall strips along the east coast, and the southwest corner. The highly productive coastal region comprises about 6% of Australia and contains all its major cities.

Australia has a current population of over 22 million persons, of whom 90% are urbanized, and some 80% live within 100 km of the coast (see Figure 6.1). The population has tripled since 1960. Government is by federal parliamentary democracy and the Federal Parliament itself is located in Canberra. Each state and territory has a government to make laws over matters not controlled by the Federal Government.

* E-mail: j.mueller@uq.edu.au (Chapter corresponding author).

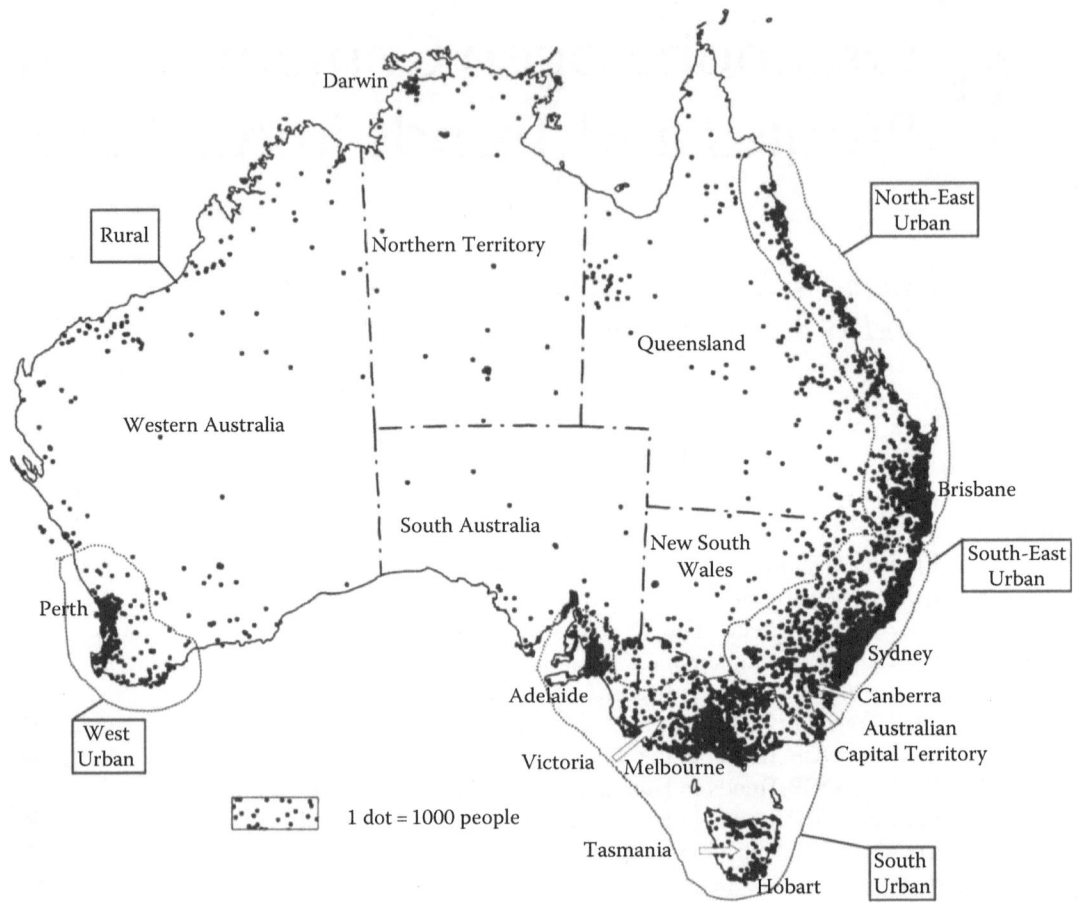

FIGURE 6.1 Australia showing population distribution (Australian Bureau of Statistics map from 2006 census), capital cities, state and territory boundaries, together with sampling locations discussed in the text, and superimposed with the major population regions used in national studies of contaminant concentrations in blood and breast milk.

Australia is rich in natural resources and is a major exporter of agricultural products and minerals such as iron ore, coal, and liquefied natural gas. Major imports comprise manufactured goods, crude oil, and petroleum. GDP by sector is agriculture 4%, industry 2.48%, and services 71.2% [1]. See Table 6.1 for attribution of land uses in Australia.

6.1.1 History of Persistent Organic Chemicals Use in Agriculture and Industries in Australia

Australia has a history of intensive organochlorine insecticide, hexachlorobenzene (HCB), and industrial chemical polychlorinated biphenyl (PCB) use, but with a gradual phasing out of persistent chlorohydrocarbon pesticides through restrictions on agricultural use and availability of less persistent and bioaccumulative alternatives. However, as reported by Connell et al. [2], reliable data on usage are scant. In 1972, the fungicide HCB was withdrawn from use and prohibited but was still produced as an industrial by-product in chemical manufacturing [3]. The use of organochlorines for termite control was discontinued and replaced by alternative chemicals and physical barriers in the 1980s. Importation of PCBs and equipment containing PCBs was banned in 1975, although use in existing equipment has been allowed to continue. Australia is a signatory to the Stockholm

TABLE 6.1
Percentages of Land Uses in
Australia

Nature conservation	6.5
Other protected areas including indigenous uses	13.4
Minimal use	15.7
Livestock grazing	56
Forestry	2
Dryland agriculture	5.2
Irrigated agriculture	0.3
Built environment	0.3
Water bodies	0.6

Source: Australian Government online
sources.

TABLE 6.2
Australia's Action on the 12 POPs Listed under the Stockholm Convention prior
to May 2009

Aldrin	Final registration cancelled 1992 and importation prohibited
Dieldrin	Final registration cancelled 1988 and importation prohibited
DDT	Final registration cancelled 1987 and importation prohibited
Endrin	Final registration cancelled 1987 and importation prohibited
Chordane	Final registration cancelled 1997 and importation prohibited
Hexachlorobenzene	Final registration cancelled 1980 and importation prohibited. National strategy for HCB waste provides for management and destruction of stockpile
Mirex	Final registration cancelled 2007 and importation prohibited
Toxaphene	Final registration cancelled 1987
Heptachlor	Final registration cancelled 1997 and importation prohibited
PCBs	Importation of PCB is banned unless explicit permission is granted by the Minister for Customs. National strategy proposes to remove and destroy all PCBs by end of 2009
PCDD and PCDFs	No federal emission standards, but most states have some regulations. Reporting under National Pollution Inventory

Source: DEWHA, Australia's action on the 12 POPs listed under the convention, available at: www.
environment.gov.au/settlements/chemicals/international/pop.html#action, 2010.

Convention on persistent organic pollutants (POPs) and its actions on the 12 POPs listed under the Convention, prior to the listing of nine additional POPs in May 2009, are listed in Table 6.2. At the date of writing this chapter, this list from the Australian Government Web site had not been updated with respect to the additional POPs.

Polybrominated diphenyl ethers (PBDEs) are imported into Australia in raw chemical form and are also already incorporated into manufactured products. In 2003–2004, the responsible federal government agency estimated that 180 ton of deca-BDE product, 20 ton of penta-BDE product, and less than 10 ton of octa-BDE product were imported in raw chemical form into Australia [5]. The agency also reported a decrease in the use of approximately 90% of octa-BDE and approximately 70% of penta-BDE was seen in 2003–2004 compared to 1998–1999 [5]. The amount of BFRs imported into Australia in manufactured products remains unknown. In Australia, NICNAS removed octa-BDE products from the Australian Inventory of Chemical Substances on February

6, 2007 which means that it is not allowed to be manufactured or imported. Further, penta-BDE is under an interim ban while the assessment is underway, and deca-BDE is also under review [5].

6.1.2 ENVIRONMENTAL LEVELS, ENVIRONMENTAL IMPACT, AND BAN/ RESTRICTION OF POCs IN AUSTRALIA

Richardson [6] observed that no systematic national assessment of persistent organic chemicals (POCs) in Australia was conducted prior to 1995. Although there was a substantial quantity of data available in the mid-1990s from studies in specific areas of concern, as listed in the literature review of POPs in the Southern Hemisphere by Connell et al. [2], there were many information gaps. The conclusion of Connell et al. [2] was that persistent chlorohydrocarbon levels in various media in river and estuarine waters in Australia were generally acceptable relative to national guidelines, except possibly for elevated PCB levels around Sydney. Using the previous review data, augmented by more recent data, Connell et al. [7] concluded that the bans on persistent chlorohydrocarbons listed in Table 6.2 had resulted in a marked decrease in environmental levels since the 1980s, citing a steady decline in pesticide levels in marine and estuarine urban areas of Australia.

In the case of dioxins and PBDEs, there was a series of coordinated nationwide sampling-based assessments. Dioxin-like compounds were covered by the National Dioxins Program [8] and PBDEs by a series of consultancies all of which were commissioned by the Federal Government [9–11] targeting specific environmental media (aquatic sediments, indoor environments, and human blood). An important and significant aspect of these studies was that the sampling strategies used were designed to avoid potential local hot spots and sources but to capture background concentrations of contamination.

6.1.2.1 Dioxins and Dioxin-Like Chemicals

6.1.2.1.1 Polychlorinated Dibenzo-p-dioxins/Furans and PCBs in Australian Aquatic Environment

For assessment and reporting consistency across the various components of studies conducted under the National Dioxins Program [8], Australia was geographically subdivided into three regions (Northern, South-Eastern, and South-Western) common across study components. Similarly, land-use types were categorized as "urban," "industrial," "agricultural," or "remote." These subdivisions and categorizations are illustrated in Figure 6.2 with respect of the aquatic component of the study.

In the aquatic environment study conducted by Mueller et al. [12] as a component of the National Dioxins Program [8], dioxin-like chemicals were found in all aquatic sediments analyzed (cores from 62 locations), with middle bound concentrations ranging from 0.002 to 520 pg TEQ g^{-1} dm. Highest concentrations were found in the sediments sampled from the Parramatta River estuary (100 and 520 pg TEQ g^{-1} dm) and the western section of Port Jackson (78 and 130 pg TEQ g^{-1} dm), which are in close proximity to historical manufacturing point sources around Homebush Bay. In addition, elevated concentrations were also found in other estuarine waters of Sydney (Botany Bay) as well as the estuaries in or near Brisbane, Melbourne, Hobart, Perth, and Wollongong.

Considering all sediment samples, the median concentrations were 0.2, 2.3, and 0.12 pg TEQ g^{-1} dm in sediments from freshwater, estuarine, and marine locations respectively. Homologue and congener profiles for the polychlorinated dibenzo-p-dioxins/furans (PCDDs/PCDFs) were strongly dominated by OCDD with 1,2,3,4,6,7,8-heptachloro dibenzodioxin usually being the congener with the second highest concentration.

For most sediment samples, PCDD/PCDF dominated the mixture of dioxin-like chemicals present, accounting for more than 80% of the total TEQ. However, a range of samples such as those from the Brisbane River (south-east Queensland), the Torrens River (Adelaide), or from south-west

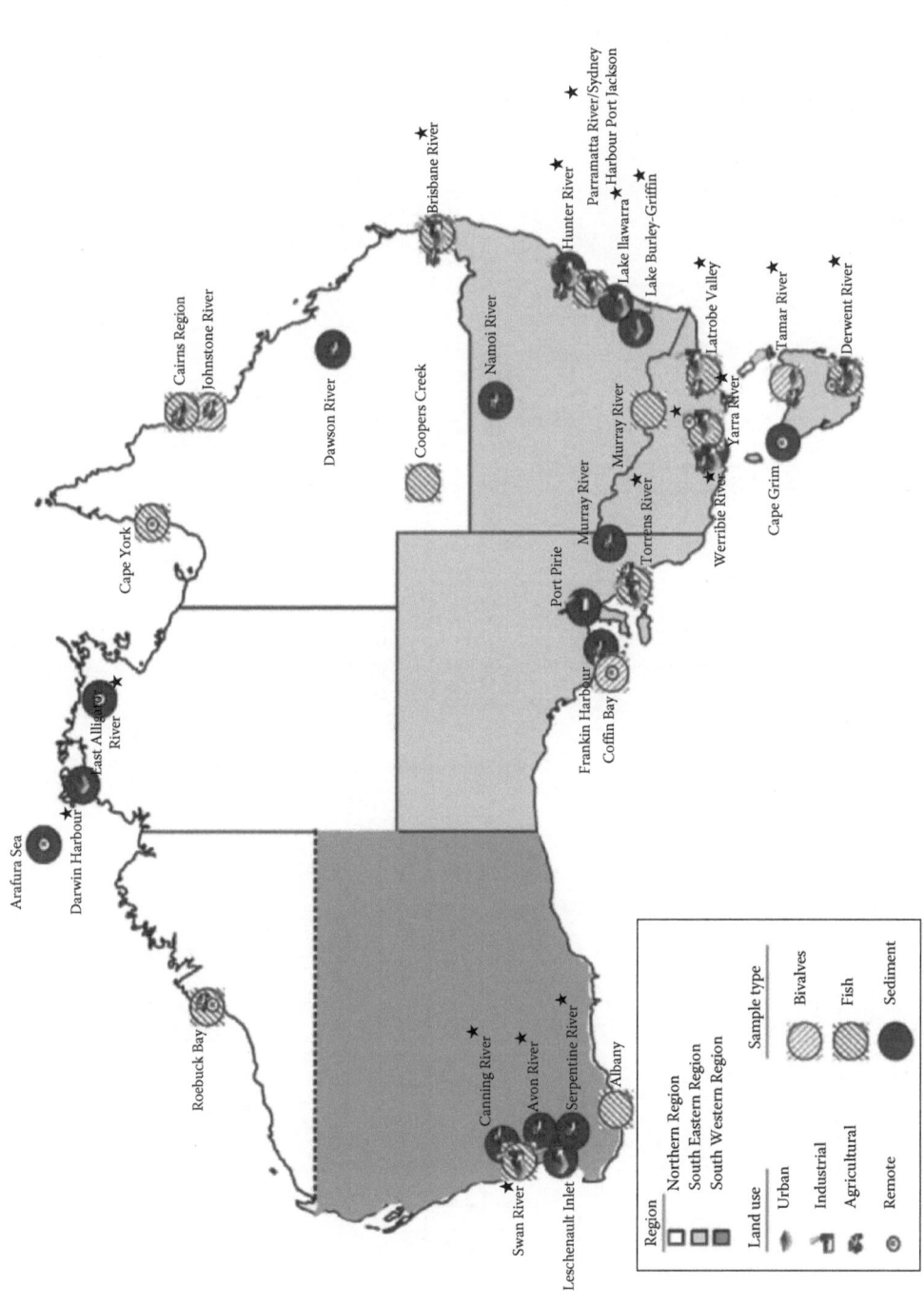

FIGURE 6.2 Geographical regions from which samples were collected for the aquatic component of the National Dioxins Program, together with regional stratification and land-use types. Sampling locations which are asterisked were those used to determine PBDE concentrations. (From Mueller, J. et al., Dioxins in aquatic environments in Australia, National Dioxins Technical Report No. 5, Australian Government Department of the Environment and Heritage, Canberra, Australia, 2004.)

Western Australia showed contributions of PCB exceeding 50%. This suggests local sources of PCB have influenced the compound profiles at those sampling locations. For PCBs, congener 118 dominated the profile whereas the most toxic PCB contributed less than 1% to the congener profile. The overall results for sediments on a water-type basis (fresh/estuarine/marine) are illustrated in Figure 6.3, and for associated land-use–type basis in Figure 6.4.

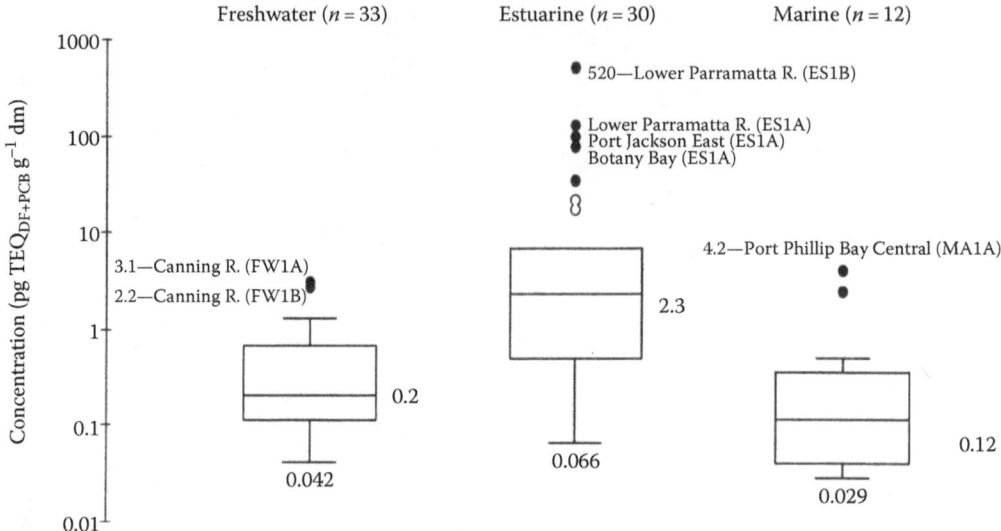

FIGURE 6.3 Concentrations of PCDD/PCDFs and PCBs in sediments by water type. (From Mueller, J. et al., Dioxins in aquatic environments in Australia, National Dioxins Technical Report No. 5, Australian Government Department of the Environment and Heritage, Canberra, Australia, 2004.)

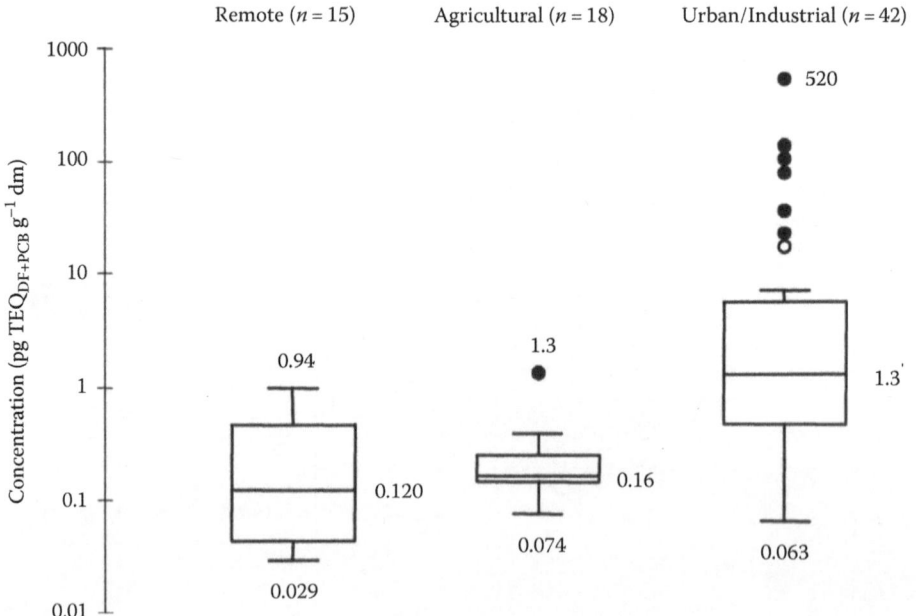

FIGURE 6.4 Concentrations of PCDD/PCDFs and PCBs in sediments by associated land-use types. (From Mueller, J. et al., Dioxins in aquatic environments in Australia, National Dioxins Technical Report No. 5, Australian Government Department of the Environment and Heritage, Canberra, Australia, 2004.)

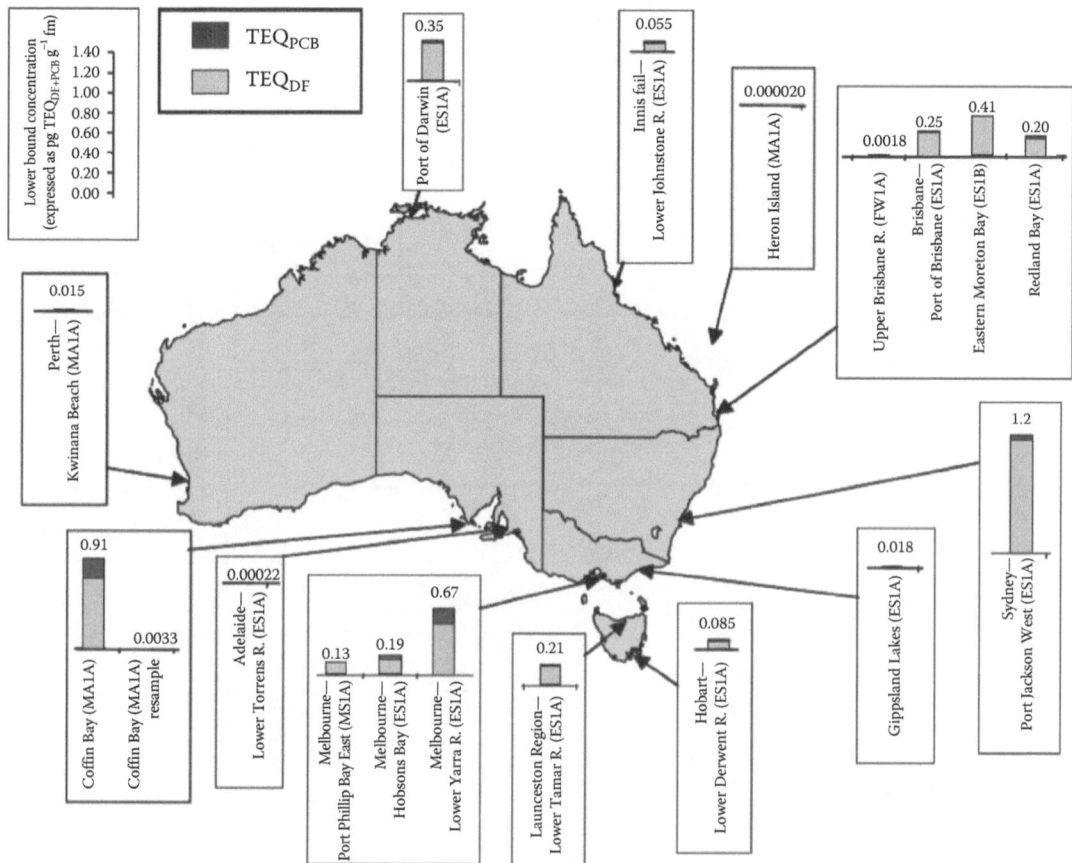

FIGURE 6.5 Concentrations of PCDD/PCDFs and PCBs in bivalves. (From Mueller, J. et al., Dioxins in aquatic environments in Australia, National Dioxins Technical Report No. 5, Australian Government Department of the Environment and Heritage, Canberra, Australia, 2004.)

The middle bound concentrations of dioxin-like chemicals in 18 bivalves samples ranged from 0.0043 pg TEQ g^{-1} fm to about 1.2 pg TEQ g^{-1} fm when expressed using fish toxic equivalent factors, with the greatest concentrations in samples from Port Jackson and the Yarra estuary (see Figure 6.5).

Dioxin-like chemicals were also analyzed in 23 fish samples from around the country and middle bound concentrations ranged from 0.0053 to about 0.49 pg TEQ$_{FISH}$ g^{-1} fm (Figure 6.6). The level of dioxin-like chemicals was highest in a fish sample obtained from the Sydney/Port Jackson area.

The bivalve results followed a similar pattern to the sediment results confirming the existence of areas with elevated environmental exposure levels of dioxin-like chemicals (Figure 6.7). However, the fish were relatively unaffected, with consistently low levels of dioxin-like chemicals found.

Statistical analysis showed significant differences between TEQ values of dioxin-like chemicals across locations with different associated land uses. Urban/industrial locations have significantly greater TEQ levels than remote and agricultural regions.

6.1.2.1.2 PCDD/PCDFs and PCBs in Australian Soils

In the soils study conducted by Mueller et al. [13] as a component of the National Dioxins Program [8], dioxin-like chemicals were found in most of the 116 Australian soils sampled, with middle bound concentrations ranging from the limit of detection (0.05 pg TEQ g^{-1} dm) to 23 TEQ g^{-1} dm. Median concentrations expressed as TEQ for dioxin-like chemicals in soils across all land-use types in the

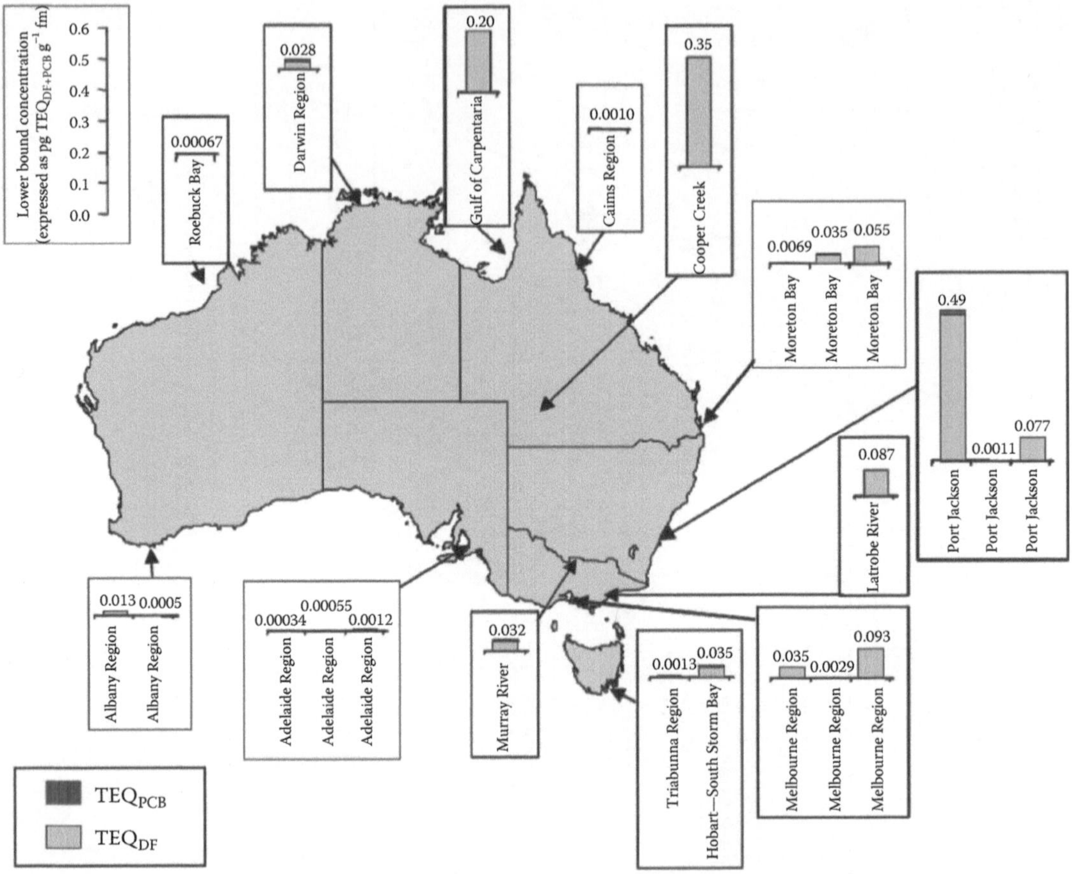

Lower bound concentration (expressed as pg TEQ_{DF+PCB} g^{-1} fm)
0.6
0.5
0.4
0.3
0.2
0.1
0.0
0.00067
Roebuck Bay
0.028
Darwin Region
0.20
Gulf of Carpentaria
0.0010
Cairns Region
0.35
Cooper Creek
0.0069
0.035 0.055
Moreton Bay
Moreton Bay
Moreton Bay
0.49
0.0011
0.077
Port Jackson
Port Jackson
Port Jackson
0.087
Latrobe River
0.013 0.0005
Albany Region
Albany Region
0.00034
0.00055
0.0012
Adelaide Region
Adelaide Region
Adelaide Region
0.032
Murray River
0.0013
0.035
Triabunna Region
Hobart—South Storm Bay
0.035
0.0029
0.093
Melbourne Region
Melbourne Region
Melbourne Region
TEQ_{PCB}
TEQ_{DF}

FIGURE 6.6 Concentrations of dioxin-like chemicals in fish. (From Mueller, J. et al., Dioxins in aquatic environments in Australia, National Dioxins Technical Report No. 5, Australian Government Department of the Environment and Heritage, Canberra, Australia, 2004.)

Concentration in biota (pg TEQ g^{-1} lipid)
1000
100
10
1
0.1
0.1
1
10
100
1,000
10,000
Estuary
Freshwater
Marine
Coffin Bay
Low Yarra
Pt Jackson
Moreton Bay
Pt Phil. Hobsons Bay
Torrens
Coffin Bay

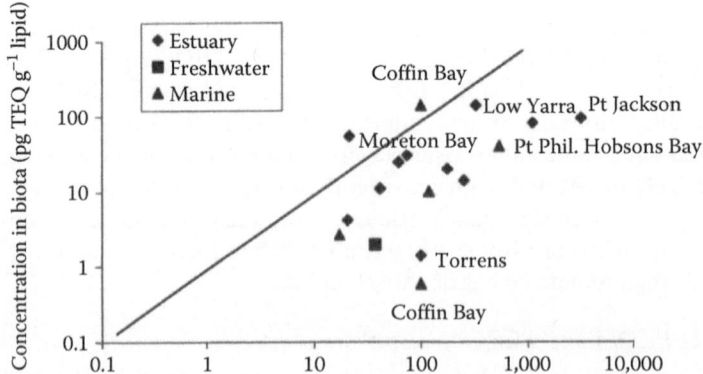

FIGURE 6.7 Concentrations of dioxin-like chemicals in biota versus concentrations in sediments. (From Mueller, J. et al., Dioxins in aquatic environments in Australia, National Dioxins Technical Report No. 5, Australian Government Department of the Environment and Heritage, Canberra, Australia, 2004.)

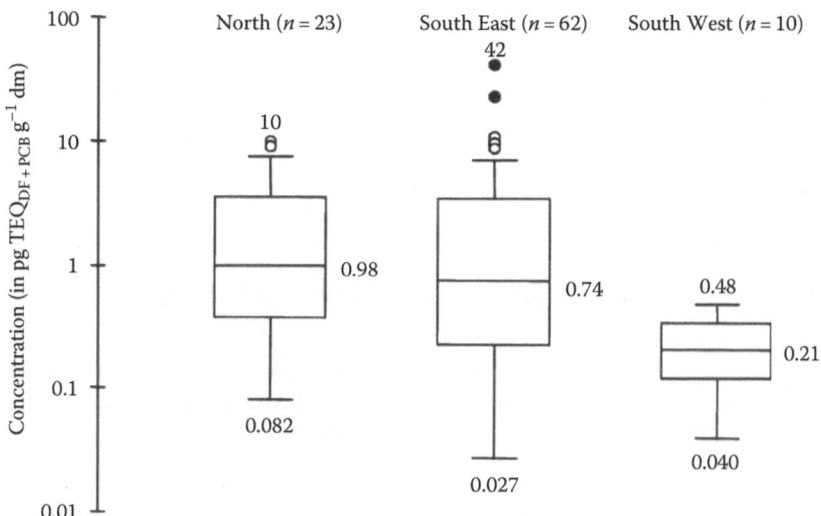

FIGURE 6.8 Concentrations of dioxin-like chemicals in soil samples from North, South-East and South-West regions. (From Mueller, J. et al., Dioxins in soils in Australia, National Dioxins Technical Report No. 6, Australian Government Department of the Environment and Heritage, Canberra, Australia, 2004.)

North and South-East study regions were similar, but the median concentration in the South-West study region was less ($p < .05$). The distribution of concentrations by region is shown in Figure 6.8. The geographic regions allocated in the soils component of the National Dioxins Program is the same as that for the aquatic component (see Figure 6.2). The greatest concentrations were found in soils collected near centers of population within the south-east coastal area of Australia, whereas concentrations were consistently low from locations in Western Australia and inland areas across all regions.

Levels of dioxin-like chemicals in soils from urban and industrial locations were substantially higher than agricultural land use and remote locations (Figure 6.9).

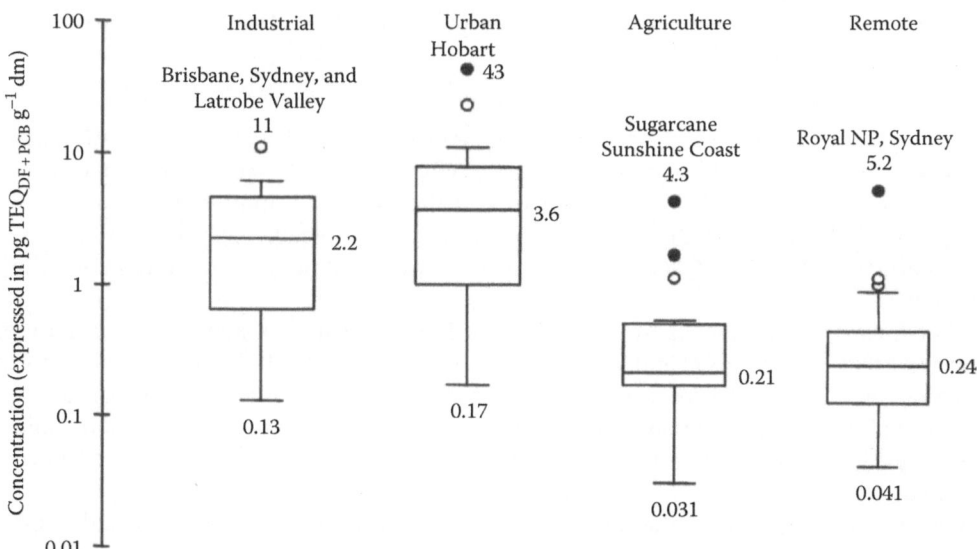

FIGURE 6.9 Concentrations of dioxin-like chemicals in soil samples from different land-use types. (From Mueller, J. et al., Dioxins in soils in Australia, National Dioxins Technical Report No. 6, Australian Government Department of the Environment and Heritage, Canberra, Australia, 2004.)

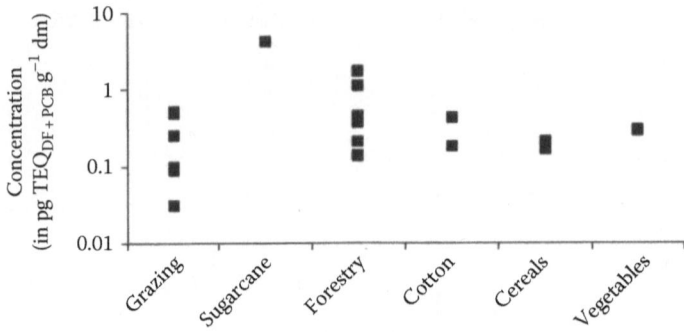

FIGURE 6.10 Concentrations of dioxin-like chemicals in soil samples from specific types of agriculture. (From Mueller, J. et al., Dioxins in soils in Australia, National Dioxins Technical Report No. 6, Australian Government Department of the Environment and Heritage, Canberra, Australia, 2004.)

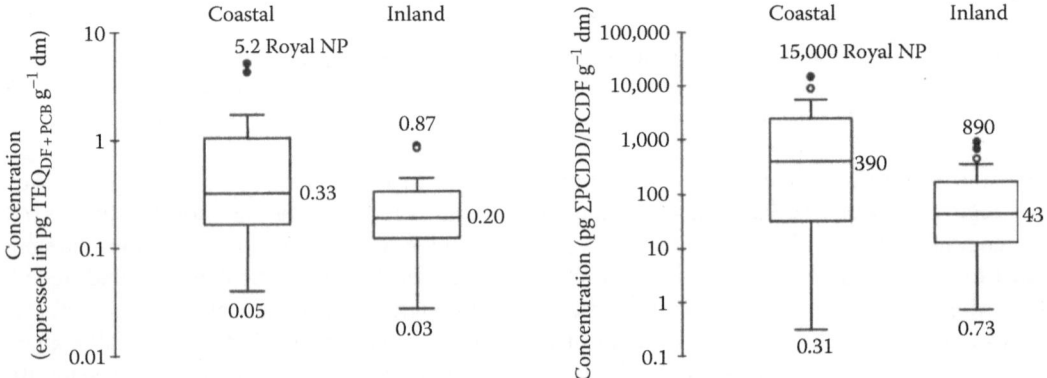

FIGURE 6.11 Concentrations of dioxin-like chemicals in soil samples from coastal and inland areas, excluding sites within urban and industrial land-use categories. (From Mueller, J. et al., Dioxins in soils in Australia, National Dioxins Technical Report No. 6, Australian Government Department of the Environment and Heritage, Canberra, Australia, 2004.)

Across agricultural land uses, concentrations were similar with the exception of sugarcane growing, in which the concentrations were greater (Figure 6.10). However, other evidence covered in detail by Prange et al. [14,15] suggests that this is a geographic rather than a land-use related phenomenon. These studies [14,15] have shown a widespread contamination of soils with higher chlorinated PCDDs in Queensland's coastal environment but low contamination of inland environments.

Similarly, as illustrated in Figure 6.11, the data from the national study [13] showed greater concentrations of dioxin-like chemicals at coastal relative to inland locations (using only data from remote and agricultural locations, to remove the influence of urban and industrial sources since almost all of Australia's urban centers are located in the coastal region whereas the inland is sparsely populated). This relationship did not hold when locations within 200 km of urban centers were excluded.

Homologue and congener profiles for PCDD/PCDFs in soils were strongly dominated by OCDD. Similarly, the tetra-heptachlorinated 2,3,7,8-chlorine substituted profiles were dominated by the highest chlorinated PCDD, 1,2,3,4,6,7,8-heptachloro dibenzodioxin. On average more than 80% of the TEQ across soil samples was contributed by 2,3,7,8-PCDD/PCDF, with a greater contribution in industrial and urban locations.

6.1.2.1.3 PCDD/PCDFs and PCBs in Australian Fauna

The fauna study conducted by Correll et al. [16], a component of the National Dioxins Program [8], involved the collection of several hundred fauna samples (primarily roadkill for terrestrial animals and stranded animals for marine mammals) with emphasis on spatial and biological diversity. The collected specimens were pooled into 66 samples covering all states as well as the Northern Territory. The median and range of TEQ concentrations across fauna classes are listed in Table 6.3.

Dioxin-like chemicals were present in all samples and TEQs ranged from the limit of detection to 3900 pg TEQ g^{-1}. Overall the survey found greatest concentrations in birds of prey (sparrow-hawks, goshawks, falcons, eagles, etc.) with a maximum concentration of 3900 pg TEQ g^{-1} lipid (middle bound). Piscivorous marine mammals also had high levels with a dolphin from the Port River in South Australia having a level of 590 pg TEQ g^{-1} lipid. In contrast, concentrations were generally low in herbivorous animals such as macropods, a galah, and a dugong (marine mammal that feeds exclusively on seagrass).

Concentrations of dioxin-like chemicals in the 22 macropod samples (Figure 6.12) were relatively low with a median concentration of 0.71 pg g^{-1} lipid. The highest concentration of dioxin-like chemicals in macropods (TEQ of 25 pg g^{-1} lipid) was in a pooled sample of three kangaroos collected from the Para Wirra National Park located 25 km northeast of Adelaide. The TEQs of the other marsupials (possum, koala, and bandicoot) were low and comparable to that of the macropods.

As illustrated in Figure 6.13, PCBs contributed significantly to the TEQ load of birds and marsupials, and was most dominant in marine mammals. The PCB profile of the marine mammals was similar across all species examined, suggesting that these contaminants have a common origin or pathway. In general, the concentrations of dioxin-like compounds were low by overseas fauna standards, including the TEQs in the marine mammals.

The authors of the fauna study commented that since the sample material was collected during a period of severe drought affecting much of southern Australia in 2002, the terrestrial animals would have been in poor condition, and it is likely that they would have used stored body fat during the drought. The effects of the drought would not have been confined to primary feeders such as kangaroo, but would have been perpetrated throughout the food chain. Under such circumstances,

TABLE 6.3
TEQ Levels of Dioxin-Like Chemicals Found in Pooled Samples of Australian Fauna

Fauna Class	No. of Pooled Samples	Minimum TEQ[a]	Median TEQ	Maximum TEQ
Bird	19	0.64	300	3900
Dingo	2	1.7	2.0	2.3
Macropod	22	0.14	0.71	25
Marine mammal	13	1.1	28	590
Monotreme	5	9.3	23	60
Other marsupial	4	0.95	2.0	13
Reptile	1		0.65	

Source: Correll, R. et al., Dioxins in fauna in Australia, National Dioxins Program Technical Report No. 7, Australian Government Department of the Environment and Heritage, Canberra, Australia, 2004.

a Units are pg g^{-1} lipid.

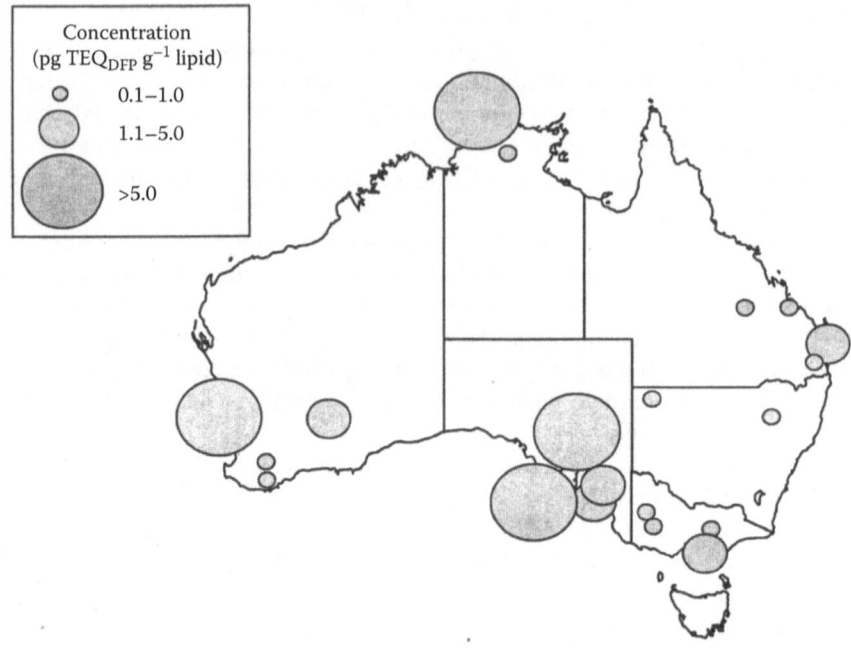

FIGURE 6.12 Concentrations of dioxin-like chemicals in macropods. (From Correll, R. et al., Dioxins in fauna in Australia, National Dioxins Technical Report No. 7, Australian Government Department of the Environment and Heritage, Canberra, Australia, 2004.)

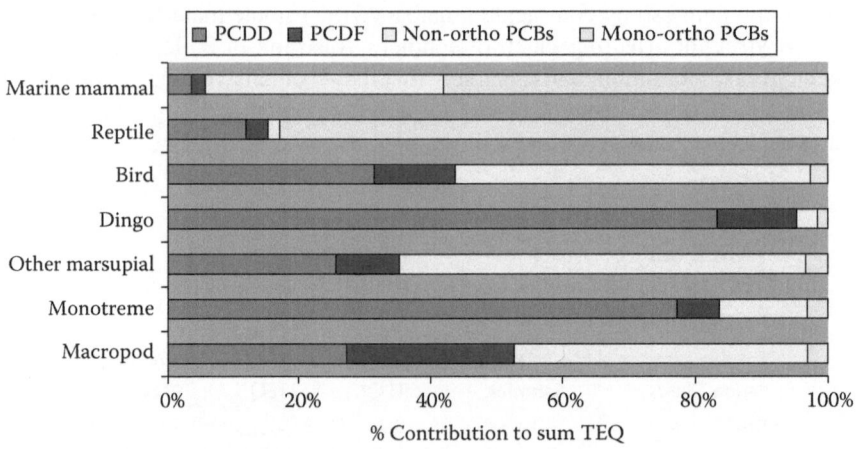

FIGURE 6.13 Comparison of the percent contribution of PCDD/PCDFs and dioxin-like PCBs to the total TEQ across each faunal class. (From Correll, R. et al., Dioxins in fauna in Australia, National Dioxins Technical Report No. 7, Australian Government Department of the Environment and Heritage, Canberra, Australia, 2004.)

it is likely that stored lipid would have been removed preferentially to the dioxin-like compounds, so those compounds would have been concentrated in the fatty tissue of the animal. If this was the case, the results presented in this study would if anything be an overestimate of the general levels in Australian terrestrial fauna.

There was no evidence from the national survey of higher concentrations of dioxin-like chemicals in fauna in the industrial/urban or agricultural areas. The higher concentrations found were in

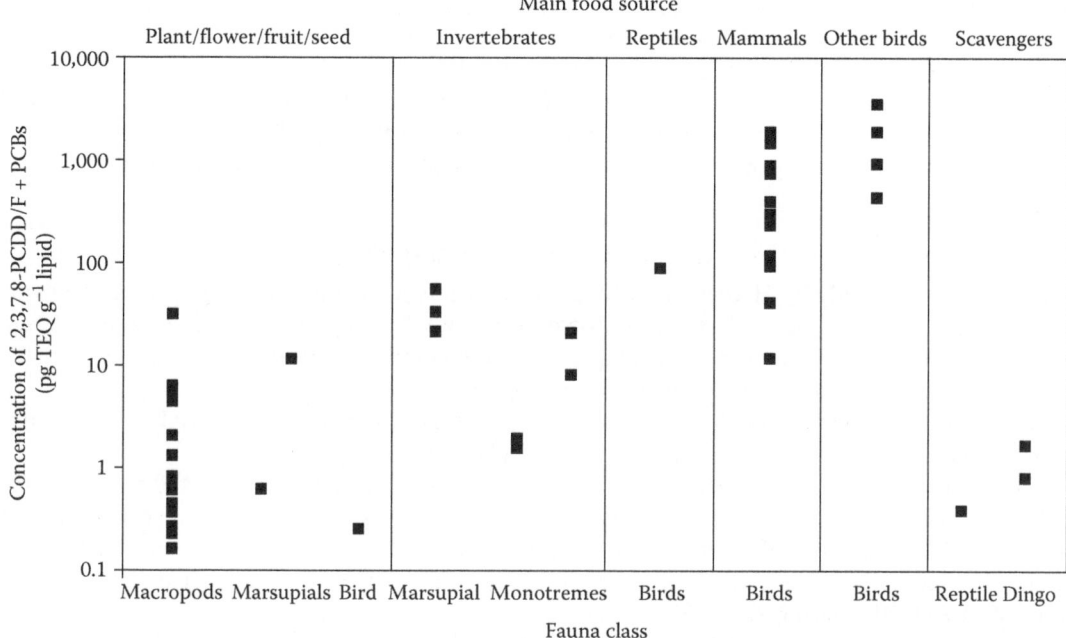

FIGURE 6.14 Concentrations of dioxin-like chemicals across fauna classes. (From Correll, R. et al., Dioxins in fauna in Australia, National Dioxins Technical Report No. 7, Australian Government Department of the Environment and Heritage, Canberra, Australia, 2004.)

remote locations. There was significant evidence of a trend for increasing TEQ with increasing trophic level (Figure 6.14) consistent with food chain biomagnification. However, the scavengers (dingo and reptile) did not follow this trend, but the sample numbers for these is very small.

6.1.2.1.4 PCDD/PCDFs and PCBs in Ambient Air in Australia

The study by Gras et al. [17], conducted as a component of the National Dioxins Program [8], involved the use of high-volume samplers on a month by month basis at 10 stations for a period of 12 months. A major finding was a very clear, strong, seasonal cycle in PCDD/PCDF concentrations, both as mass concentrations and TEQ, with a winter concentration maximum in all of the major population centers studied, from Perth, through Adelaide, Melbourne, Sydney, and as far north as Brisbane. The seasonal variation in PCDD/PCDF concentration, correlation with aerosol non-seasalt potassium (nssK), a tracer for biomass burning, and association of PCDD/PCDF with nssK by factor analysis, all pointed to a residential wood smoke source for the strong winter enhancement. Congener patterns showed strong resemblances to congener concentration patterns found in a study of wood smoke emissions from Australian residential wood heaters [18], and to homologue patterns reported for emissions from Christchurch and Masterton, New Zealand, during winter [19]. Seasonal cycles in TEQ_{DF} were also observed in rural Queensland and rural Victoria, although these were weaker than in the major urban locations. Despite the winter enhancement in PCDD/PCDFs, overall annual mean concentrations in the major Australian cities were very low by world standards. For example, for the two Perth locations, annual mean concentrations, expressed as middle bound TEQs, were approximately 14 fg TEQDF m^{-3}, and Brisbane 9, Sydney 15, Melbourne 17, and Adelaide 15 fg TEQDF m^{-3}, respectively.

A potentially significant source of airshed contamination by combustion products such as PCDD/PCDFs in northern Australia is the burning off of vegetation during the dry season, and a dry season to wet season difference was found in Darwin with TEQ_{DF}, concentrations in the dry season around four times those found during the wet season. However, mean concentrations in Darwin were very low (less than 3 fg TEQ_{DF} m^{-3} annual mean); hence, the impact of top-end dry season burning on PCDD/PCDF concentrations appears to be relatively minor (during this study).

Extremely low PCDD/PCDF concentrations were observed in clean marine air and also in rural locations removed from the major urban centers (typically less than 2 fg TEQ_{DF} m^{-3}). This indicates a very clean regional background, with the major sources being local and associated with the urban population.

6.1.2.1.5   PCDD/PCDFs and PCBs in Human Milk in Australia

The human milk study by Harden et al. [20], conducted as a component of the National Dioxins Program [8], involved the analysis of 20 pooled samples of breast milk obtained from different regions of Australia. Of these pooled samples, 17 were composites from milk collected over the 2002–2003 period, and the other 3 were composites from archived breast milk samples collected from Melbourne women in 1993 during a previous study. The latter sample material enabled a limited historical comparison between contaminant levels for the two periods.

Dioxin-like chemicals were detected in all pooled samples. For samples collected during 2002–2003, the mean and median concentrations were 9.0 and 8.9 pg TEQ g^{-1} lipid, respectively. The average lipid content in 2002–2003 pooled samples was 3.7% ± 0.5%. No systematic differences were observed in the levels of dioxin-like chemicals in breast milk samples collected from different regions of Australia during 2002–2003.

For samples collected in 1993, the mean and median levels were 16.0 and 16.4 pg TEQ g^{-1} lipid respectively. Lipid content in 1993 pooled samples gave an average concentration of 3.9% ± 0.7%.

The comparison of the samples collected from Melbourne women in 1993 with those collected for the 2002–2003 study showed clearly that the levels of dioxin-like chemicals in human milk decreased over the 10-year time period. It should, however, be noted that comparison of the two sample populations is complicated because details of maternal parity and infant age at the date of collection was not made available for the older samples. Despite these limitations, a clear decrease in the levels of these compounds over time was observed. The concentration of dioxin-like chemicals in pooled human milk samples decreased by almost a factor of 2 from 1993 to 2002–2003, from 16 ± 1.4 to 9.1 ± 1.3 pg g^{-1} lipid. PCDD/PCDFs as well as PCBs decreased consistently across congeners in this period (Figures 6.15 through 6.18). This decrease is consistent with the worldwide trends for the levels of dioxin-like compounds to have decreased over the 10-year period from 1993 to 2003 by approximately 60%.

6.1.2.1.6   PCDD/PCDFs and PCBs in Human Blood in Australia

The human blood study, conducted by Harden et al. [21] as a component of the National Dioxins Program [8], involved the analysis of 9090 samples of blood collected in 2002–2003 and aggregated into 96 pooled samples, selected to represent regional, age, and gender stratification. The regional stratification ("rural," "west urban," "south urban," "south-east urban," and "north-east urban") with mapped human population distributions is shown in Figure 6.1. The age stratification was <16, 16–30, 31–45, 46–60, and >60 years. The source of blood samples was de-identified surplus pathology material provided by a pathology service with an extensive national network.

Dioxin-like compounds were detected in all strata. Overall, the levels in the Australian population were very low by international standards. The mean and median levels expressed as upper bound TEQ values for all pooled samples were 10.9 and 8.3 pg TEQ g^{-1} lipid respectively. For males and females, the mean levels were 10.4 and 11.5 pg TEQ g^{-1} lipid, respectively. The total PCDD and PCDF TEQ concentrations by region, age, and gender are shown in Figure 6.19.

A relationship of increasing dioxin-like compound levels in blood with increasing age was observed (Figure 6.20) and could be described by the following equation:

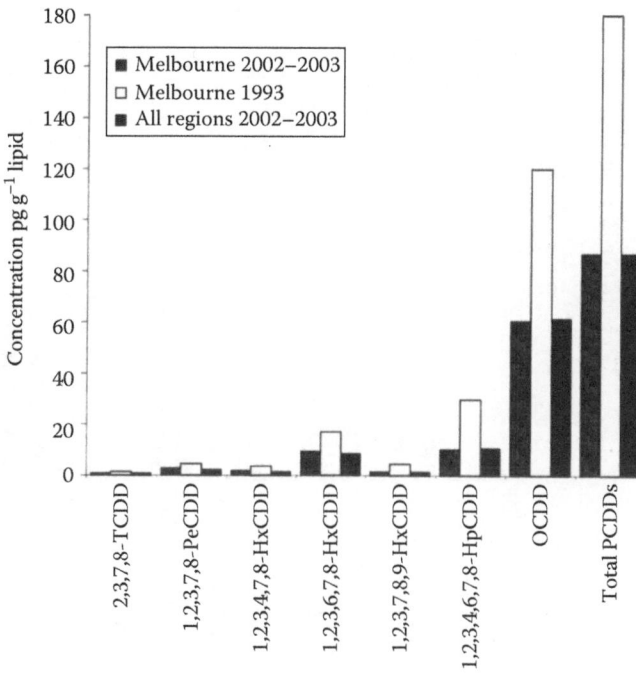

FIGURE 6.15 Comparison of concentrations of PCDD congeners in Melbourne 2002/2003, Melbourne 1993, and all regions (including Melbourne) pooled milk samples. Values are unweighted means. (From Harden, F. et al., Dioxins in the Australian population: Levels in human milk, National Dioxins Program Technical Report No. 10, Australian Government Department of the Environment and Heritage, Canberra, Australia, 2004.)

$$\text{Levels in blood expressed as pg TEQ g}^{-1}\text{ lipid} = 3.3\ \exp^{0.0251\text{age}}\ (r^2 = 0.87)$$

This relationship was found to hold from approximately 25 years of age until at least the eighth decade.

No systematic differences were observed in the levels of dioxin-like compounds in blood samples collected from males and females. However, slightly higher levels of dioxin-like compounds were observed in females in the >60 years age group. This result could not be explained on the basis of differences in the mean age between males and females in this group.

The levels of dioxin-like compounds across the five regions were remarkably similar within each age range. General trends were noted:

- The levels of dioxin-like compounds across all regions and within each age range appeared to be very similar.
- Despite the similarity in levels, for all strata except the <16-year-old females, the samples from the Southeast region exhibited slightly higher levels of dioxin-like compounds.

Since the samples were de-identified, any assessment of the length of time an individual had resided in a particular area prior to their sample being collected or the recording of either food intake or possible exposure to environmental contaminants in that region could not be assessed.

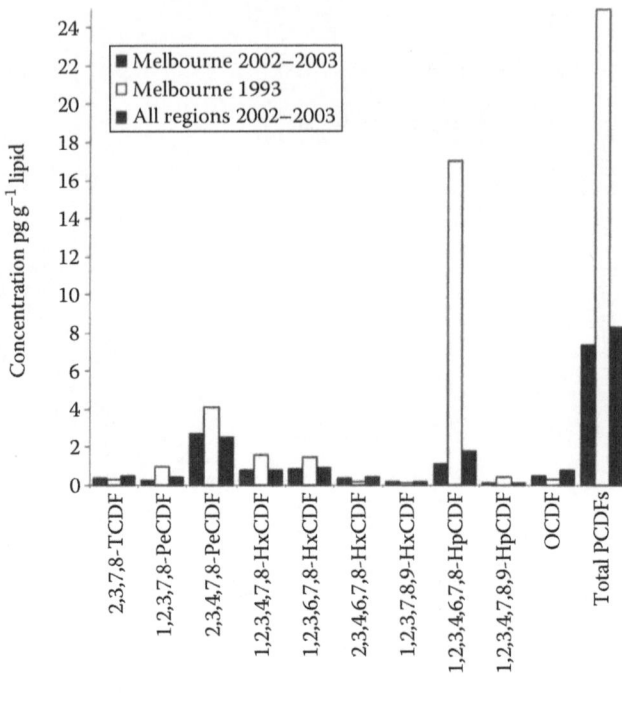

FIGURE 6.16 Comparison of concentrations of PCDF congeners in Melbourne 2002/2003, Melbourne 1993, and all regions (including Melbourne) pooled milk samples. Values are unweighted means. (From Harden, F. et al., Dioxins in the Australian population: Levels in human milk, National Dioxins Program Technical Report No. 10, Australian Government Department of the Environment and Heritage, Canberra, Australia, 2004.)

For dioxins and furans, the higher chlorinated PCDD and OCDD dominated the overall congener profile in all samples. OCDD contributes an average of approximately 80% to the total of all detected congeners. 1,2,3,7,8-pentachlorodibenzodioxin and 3,3′,4,4′,5-pentachlorobiphenyl were the single most relevant components in the congener profile, each contributing approximately 20% to the overall TEQ value. 2,3,7,8-TCDD was not detected in 57 of the 96 pooled samples but was consistently close to the detection limit. Overall, the PCDD/PCDF congener profile was dominated by higher chlorinated PCDDs whereas concentrations of higher chlorinated PCDFs (Cl6 or greater) were almost exclusively below the limit of detection in all samples.

6.1.2.2 PBDEs

6.1.2.2.1 PBDEs in Aquatic Environments in Australia

The aquatic environment study by Toms et al. [9] involved the re-analysis (for BFRs) of sediment samples collected in 2002–2003 to ascertain background concentrations of dioxin-like compounds as part of the National Dioxins Program [8]. In addition, six sediment samples from up- and downstream of the outfall of sewage treatment plants (STPs) were collected in 2005 to assess contamination from this potential point source.

Sediment samples were analyzed from 39 locations from all states and territories of Australia (Figure 6.2). At seven locations, two samples were analyzed representing similar sites within the same location. In total, samples from 46 sites were analyzed. The sampling locations were chosen to be representative of various land uses (with numbers of locations of each type in parenthesis)—remote (5), remote/agricultural (2), agricultural (7), urban (11), urban/industrial (9), industrial/urban/

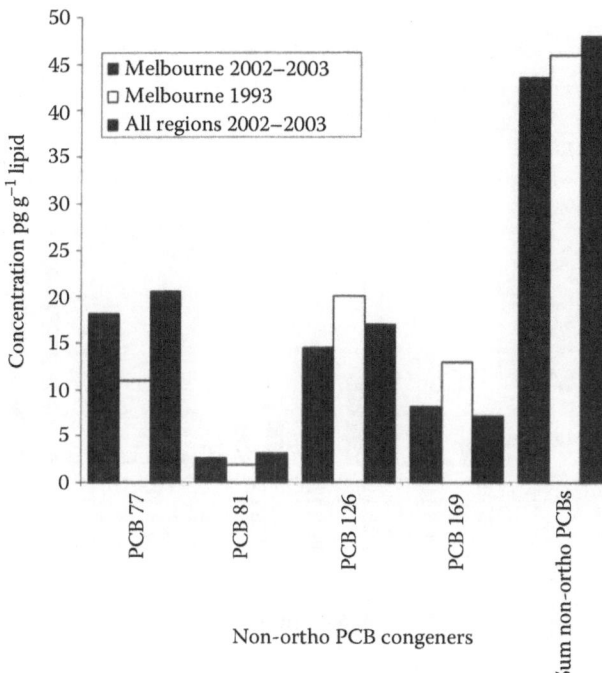

FIGURE 6.17 Comparison of concentrations of non-ortho PCB congeners in Melbourne 2002/2003, Melbourne 1993, and all regions (including Melbourne) pooled milk samples. Values are unweighted means. (From Harden, F. et al., Dioxins in the Australian population: Levels in human milk, National Dioxins Program Technical Report No. 10, Australian Government Department of the Environment and Heritage, Canberra, Australia, 2004.)

agricultural (1), industrial (7), and STPs (4), and a range of salinities—freshwater (20), marine (1), and estuarine (25).

PBDEs were detected in samples from 35 of 46 sites and the ΣPBDE concentration (excluding the LOD (limit of detection)) ranged from non-detectable to 60,900 pg g^{-1} dry weight (dw) with an overall mean (± standard deviation) and median of 4707 ± 12,580 and 305 pg g^{-1} dw, respectively. As expected, the sites with the highest concentrations were the estuaries with the highest degree of urbanization and industrialization. Marine and freshwater locations on the whole had lower PBDE concentrations than estuarine locations. Overall, there was a trend with land use which showed the concentrations of ΣPBDEs to be higher in the industrial/urban areas and followed in descending order of ΣPBDE concentration by industrial, STPs, urban, remote, agricultural, agricultural/remote, and agricultural/urban/industrial areas (Figure 6.21). It should be noted that those sediment samples from remote, remote/agricultural, agricultural, and agricultural/urban/industrial land uses had nondetectable or low concentrations of PBDEs.

In 86% of sediment samples, the congener profile was dominated by BDE-209 (excluding samples where PBDEs were not detected). The profile of the samples obtained near the outfall of STPs was dominated by BDE-209; however, it differed slightly from other samples with contributions from congeners BDE-17, -47, -49, -99, -206, and -207 (Figure 6.22). This suggests the sources of PBDEs in the outfall from STPs differed from that in other aquatic environment locations.

Overall, with the exception of the samples collected from Port Phillip Bay (Melbourne), the concentrations of PBDEs in Australian sediment were relatively low when compared to studies on PBDEs in sediments in industrialized countries from the northern hemisphere.

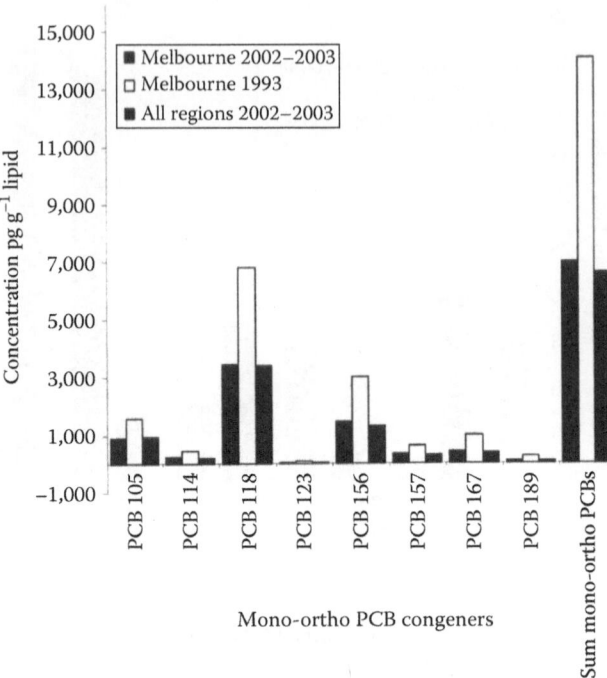

FIGURE 6.18 Comparison of concentrations of mono-ortho PCB congeners in Melbourne 2002/2003, Melbourne 1993, and all regions (including Melbourne) pooled milk samples. Values are unweighted means. (From Harden, F. et al., Dioxins in the Australian population: Levels in human milk, National Dioxins Program Technical Report No. 10, Australian Government Department of the Environment and Heritage, Canberra, Australia, 2004.)

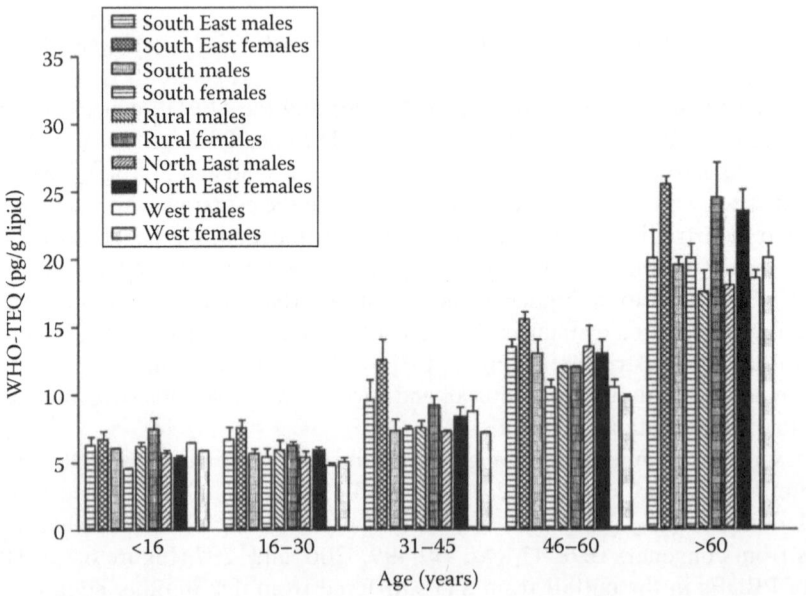

FIGURE 6.19 Average upper bound concentrations of dioxin-like chemicals in pooled blood samples from the Australian population. (From Harden, F. et al., Dioxins in the Australian population: Levels in human milk, National Dioxins Program Technical Report No. 9, Australian Government Department of the Environment and Heritage, Canberra, Australia, 2004.)

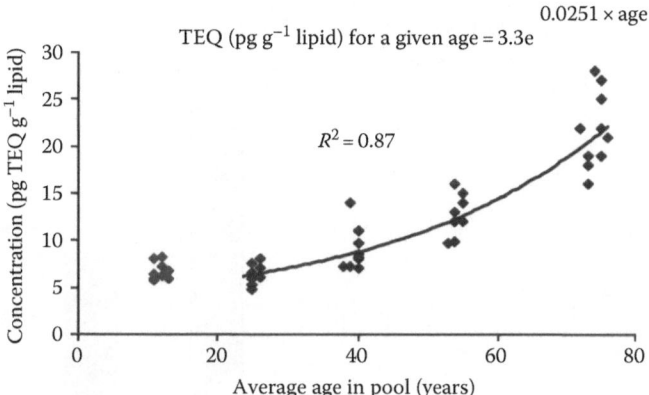

FIGURE 6.20 Relationship between age and the level of dioxin-like chemicals in pooled blood samples from the Australian population. (From Harden, F. et al., Dioxins in the Australian population: Levels in human milk, National Dioxins Program Technical Report No. 9, Australian Government Department of the Environment and Heritage, Canberra, Australia, 2004.)

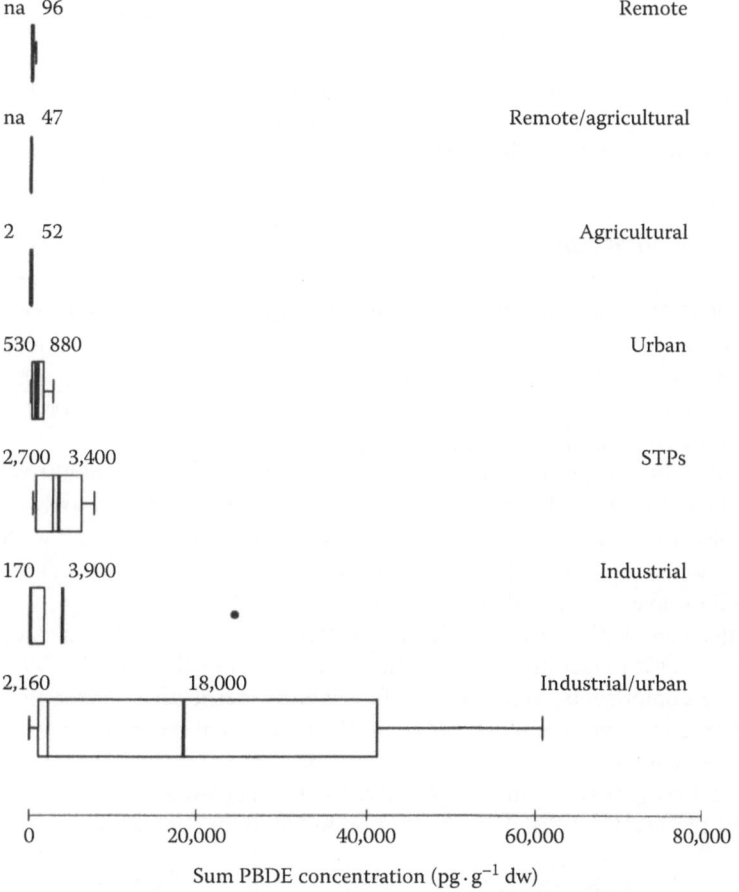

FIGURE 6.21 ΣPBDE concentrations in aquatic sediments by land-use types expressed as pg g^{-1} dry weight. (From Toms, L. et al., *Assessment of Concentration of Polybrominated Diphenyl Ether Flame Retardants in Aquatic Environments in Australia*, Australian Government Department of the Environment and Heritage, Canberra, Australia, 2006.)

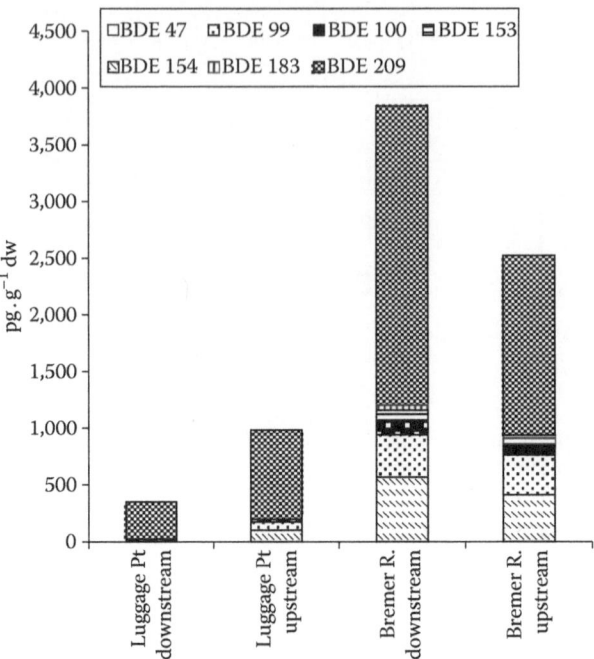

FIGURE 6.22 BDE congeners near sewage outfalls in Queensland showing domination by BDE-209. (From Toms, L. et al., *Assessment of Concentration of Polybrominated Diphenyl Ether Flame Retardants in Aquatic Environments in Australia*, Australian Government Department of the Environment and Heritage, Canberra, Australia, 2006.)

6.1.2.2.2 PBDEs in Human Blood in Australia

The human blood study conducted by Tome et al. [11] was structured similarly to the blood study which assessed PCDD/PCDFs and PCBs in blood in Australia [20], and involved the use of de-identified blood samples provided by a pathology service with an extensive national network, and aggregated on the basis of regional, age, and gender stratification into 85 pooled samples. The regions and human population distributions covered were the same as those for the PCDD/PCDFs and PCBs in the blood studies shown in Figure 6.1.

In the PCDD/PCDFs and PCB blood study [20], only blood collected in 2002–2003 was available, but for the PBDE study [11], the archived 2002–2003 blood sample resource was augmented with additional blood sampled in 2004–2005 from persons in one region only (Northeast). This enabled an age stratification on the basis of 0–4 and 5–15 years (2004–2005 samples only), <16 years (2002–2003 samples only), and 16–30, 31–45, 46–60, and >60 years.

PBDEs were detected in all age and gender strata with 24 out of 35 BDE congeners detected. The concentration of ΣPBDEs ranged from 6.4 to 80 ng g^{-1} lipid. Typically BDE-47, -99, -100, -153, -207, and -209 were key components, although the detectability, respective concentration, and overall contribution of the latter two were more variable. The percentage contribution of congener groups is illustrated in Figure 6.23.

An inverse relationship between mean age and ΣPBDE concentration was observed (Figure 6.24) which enables the prediction of PBDE concentration in blood for ages >2.4

$$y = 28.45 \times \exp^{(-0.006461x)} + 80.79 \times \exp^{(-0.2030x)} - 5.53$$

where
  y is the predicted ΣPBDE concentration (ng g^{-1} lipid)
  x is age

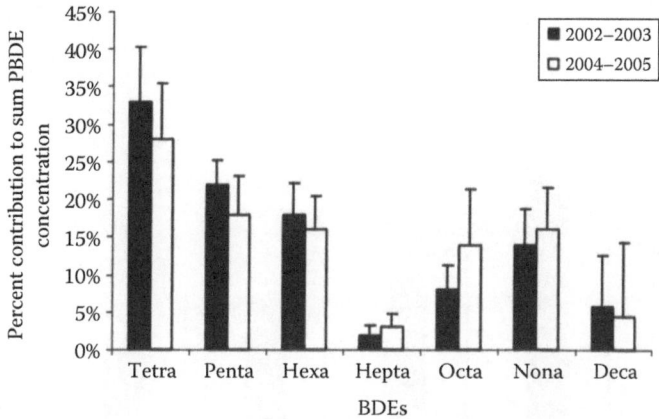

FIGURE 6.23 Percent contribution of congener groups to the sum PBDE concentrations in pooled blood samples (gender, age, and region combined) by year of collection. (From Toms, L. et al., *Assessment of Concentration of Polybrominated Diphenyl Ether Flame Retardants in the Australian Population: Levels in Blood*, Australian Government Department of the Environment and Heritage, Canberra, Australia, 2006.)

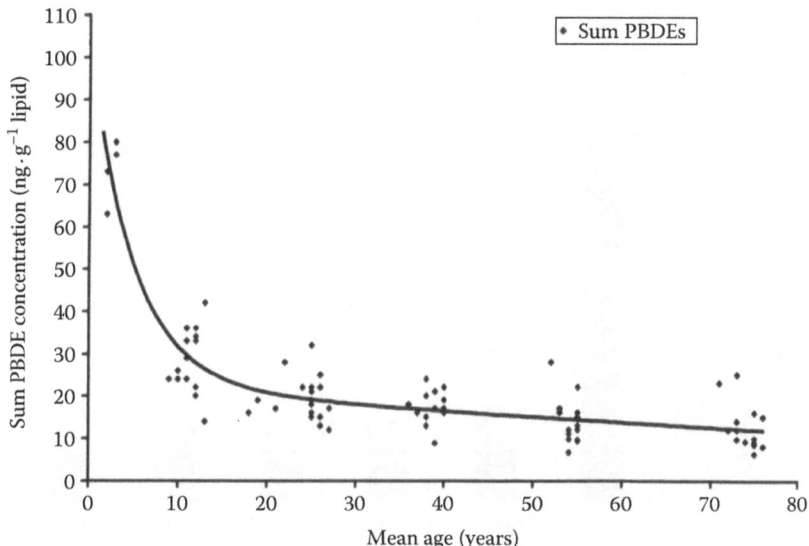

FIGURE 6.24 Sum of PBDE concentrations (ng g^{-1} lipid) in each pooled blood sample by the respective mean age (years) of donors in each pool. (From Toms, L. et al., *Assessment of Concentration of Polybrominated Diphenyl Ether Flame Retardants in the Australian Population: Levels in Blood,* Australian Government Department of the Environment and Heritage, Canberra, Australia, 2006.)

When age group data from both the 2002–2003 and 2004–2005 samples were investigated, an exponential decrease in the concentrations of PBDEs from the youngest age group was seen (Figures 6.25 and 6.26). The concentrations observed in the 0–4 years age group from the 2004–2005 pools were twice as high as the 5–15 years age group and four times higher than the >16 years age group. Toms et al. [11] suggest that the elevated concentrations of PBDEs in the youngest population along with the decreasing levels by age are likely to be related to factors including history of exposure, differences in exposure pathways (i.e., relatively high exposure of infants through breast milk and other pathways related to child behaviors), and the half-lives of PBDEs in humans enabling excess body burden from childhood to be depleted through degradation as well as growth dilution.

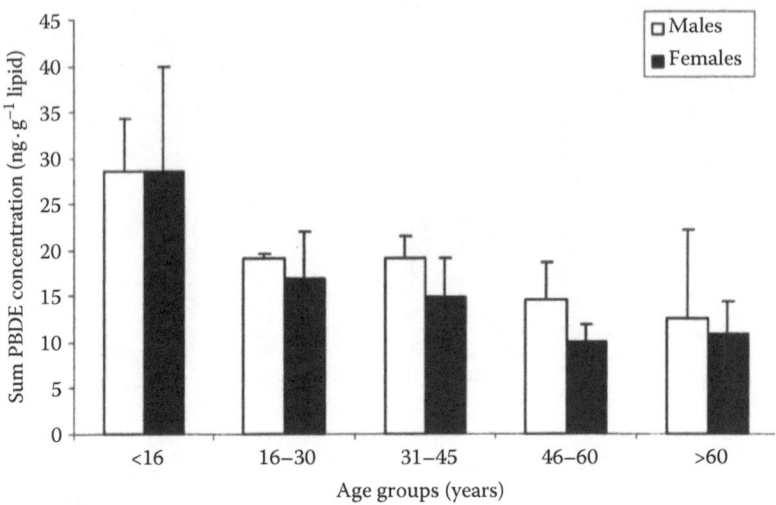

FIGURE 6.25 Mean and standard deviation for sum PBDE concentrations (ng g^{-1} lipid) in pooled blood samples by gender and age for the 2002–2003 samples. (From Toms, L. et al., *Assessment of Concentration of Polybrominated Diphenyl Ether Flame Retardants in the Australian Population: Levels in Blood,* Australian Government Department of the Environment and Heritage, Canberra, Australia, 2006.)

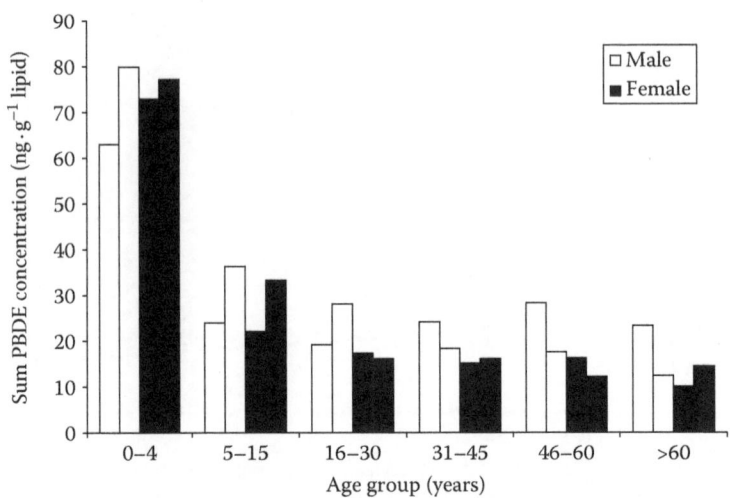

FIGURE 6.26 Mean sum PBDE concentrations (ng g^{-1} lipid) by gender and age for the 2004–2005 pooled blood samples. (From Toms, L. et al., *Assessment of Concentration of Polybrominated Diphenyl Ether Flame Retardants in the Australian Population: Levels in Blood,* Australian Government Department of the Environment and Heritage, Canberra, Australia, 2006.)

Concentrations were slightly higher in males than females (Figures 6.25 and 6.26) and were similar across all regions of Australia within each of the designated age ranges (Figure 6.27).

6.1.2.2.3 PBDEs in Indoor Environments in Australia

The study by Toms et al. [10] was an indoor air study of limited geographic extent that assessed PBDE concentrations in buildings of different ages (<2 and >5 years) and characteristics (carpeting and air-conditioning). Samples collected comprised 9 indoor air samples, 2 outdoor air samples, 9 dust samples, and 10 surface wipes from the populated southeast region of the Australian state of Queensland (Figure 6.28).

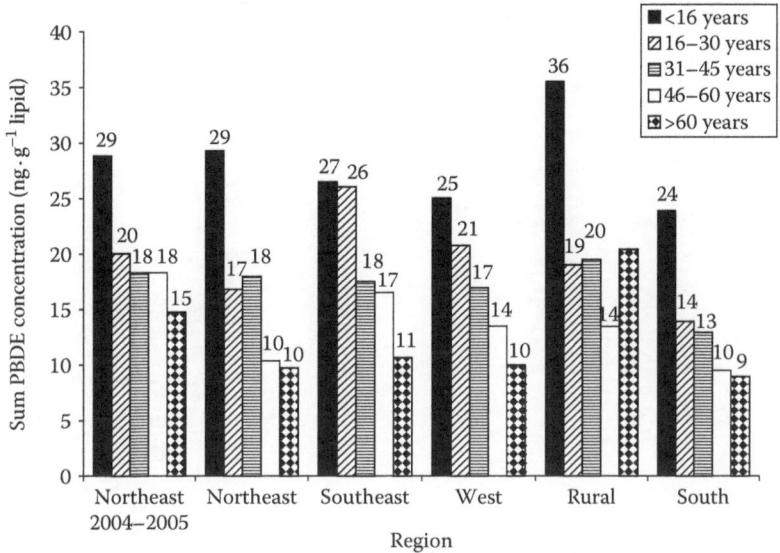

FIGURE 6.27 Mean sum PBDE concentrations (ng g⁻¹ lipid) in pooled blood samples by region and age (combined gender). (From Toms, L. et al., *Assessment of Concentration of Polybrominated Diphenyl Ether Flame Retardants in the Australian Population: Levels in Blood,* Australian Government Department of the Environment and Heritage, Canberra, Australia, 2006.)

Due to the small sample size used the results cannot be assumed to be representative of all indoor environments in Australia, but patterns were found in the data indicating that building characteristics may affect exposure to these chemicals in indoor air in Australia.

PBDEs were detected in all samples of air and dust, and in 90% of surface wipe samples with 24 out of 26 congeners detected in total. Concentrations of PBDEs were greater in indoor air than in outdoor air. For indoor air, the concentration of ΣPBDEs ranged from 0.5 to 179 pg m⁻³ for homes and 15–487 pg m⁻³ for offices. The mean (± standard deviation) and median concentrations of ΣPBDEs in homes were 50 ± 70 and 19 pg m⁻³ and in offices 173 ± 272 and 18 pg m⁻³ respectively (excluding LOD). However, the small sample sizes mean that a comparison of concentrations in homes and offices and across building ages lacks statistical power.

As shown in Figure 6.29, ΣPBDE concentration ranged from 87 to 3070 ng g⁻¹ dust in dust samples with mean and median concentrations 897 ± 944 and 591 ng g⁻¹ dust respectively (± standard deviation, excluding LOD).

6.2 TREND MONITORING STUDIES IN AUSTRALIA

There is very limited information available concerning trends in POC concentrations in Australia. Concentrations of PCDD/PCDFs in environmental media such as soil, water, and fauna were not assessed outside of known hot spots prior to the commencement of the National Dioxins Program, and although there is a substantial number of published reports on concentrations of organochlorine pesticides (OCPs) and PCBs in various media, as noted by Connell et al. [22], these studies were not designed to evaluate changes over time. Connell et al. [2] reported that most of the residue data for organochlorines in Australia up to 1996 involved statewide monitoring of waterways (e.g., Queensland), university and CSIRO research projects, and investigations by state agencies in major coastal areas of urbanization and a few intensive agricultural regions. Additionally, the study found that POCs such as dioxins and furans received some attention in relation to airsheds (e.g., Sydney incinerator emissions) and monitoring for coastal outfalls (e.g., Ninety Mile Beach), and

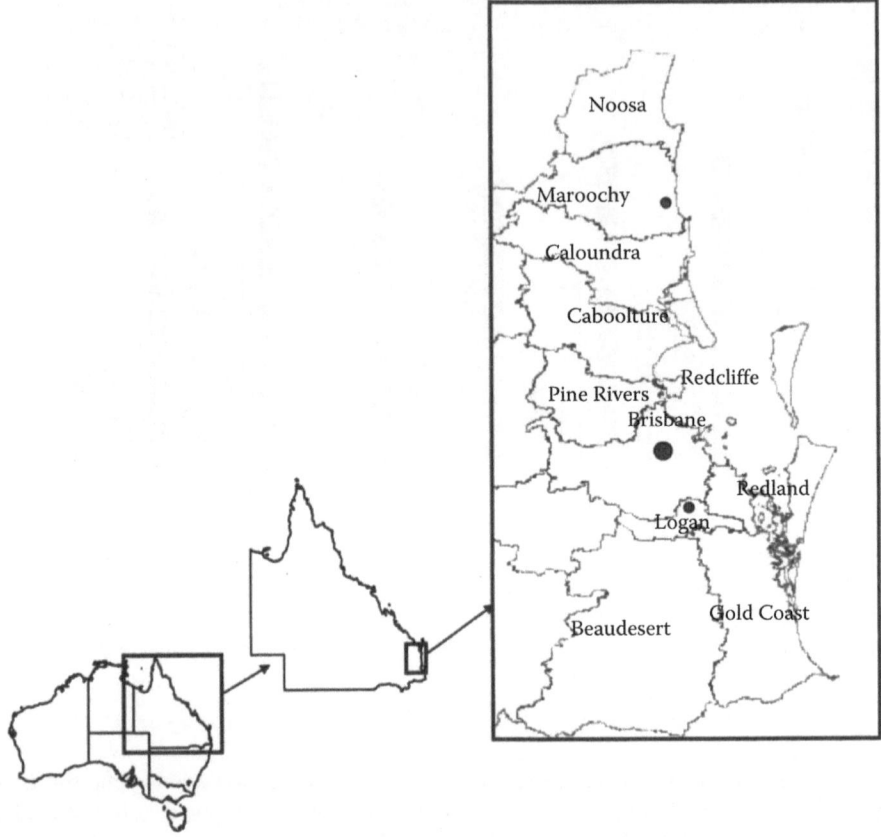

FIGURE 6.28 Sampling locations for the study of PBDEs in indoor air and dust in southeast Queensland, Australia. (From Toms, L. et al., *Assessment of Concentration of Polybrominated Diphenyl Ether Flame Retardants in Indoor Environment in Australia,* Australian Government Department of the Environment and Heritage, Canberra, Australia, 2006.)

that unpublished data existed for technical investigations of industrial contamination, for example, Homebush Bay, Sydney, and stack emission testing.

6.2.1   OCP Trends in Australia

Connell et al. [2] highlighted that useful and reliable data on the use and release of OCPs in Australia were almost entirely lacking, quantitative data concerning individual agricultural chemicals manufactured in or imported into Australia were either unavailable or inaccessible, presumably for reasons of commercial confidentiality, and that although agricultural uses were centrally registered, including approved rates of application, there had been very few documented studies of the actual amounts of chemicals applied.

A summary of the Australian OCP data in the review by Connell et al. [22] is provided in Tables 6.4 through 6.6.

6.2.1.1   OCP Trends in the Australian Marine Environment

A review of OCP contaminant concentrations in Australia was commissioned by the Federal Government in 1999 and this included an assessment of changes in concentrations in the Australian marine environment subsequent to the banning of most of these compounds in the 1970s and 1980s (Table 6.2). In a summary of the marine environment component of this OCP contaminants review,

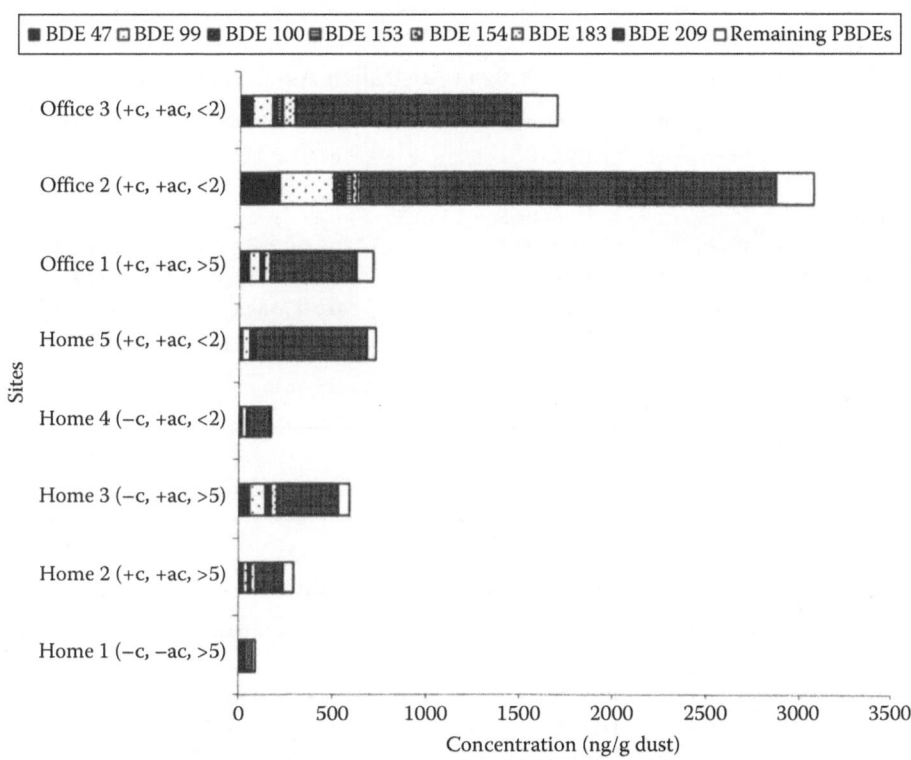

FIGURE 6.29 Concentrations of PBDE congeners in indoor dust (ng g⁻¹ dust) by site (where + is with, − is without, c is carpet, ac is air conditioning, <2 is less than 2 years old, and >5 is greater than 5 years old). (From Toms, L. et al., *Assessment of Concentration of Polybrominated Diphenyl Ether Flame Retardants in Indoor Environment in Australia,* Australian Government Department of the Environment and Heritage, Canberra, Australia, 2006.)

TABLE 6.4
Concentrations of OCPs in Australian Waters

Compound	Environment	Year(s)	Number	Concentration, ng L⁻¹	Source
DDT	Fresh water estuarine	1979–1991	500	<1–42,900	[23]
HCH	Rivers	1974–1980	>411	<1–1100	[24]
Dieldrin	Rivers	1964–1989	ca. 600	<1–1100	[24]
Heptachlor and heptachlor epoxide	Rivers	1982	—	20	[24]

Source: Connell, D.W. et al., *Crit. Rev. Environ. Sci. Technol.*, 29, 47, 1999.

published by Connell et al. [7], it was noted that most work on OCPs in Australian waters had been conducted around the sewage outfalls in the Sydney area citing the publications: [6,7,29,37–41]. Of particular note in this near offshore area was the concentration of HCB (mean 871 μg kg⁻¹) in the sediments [29]. The source of this HCB was largely from chemical manufacturing and storage of HCB-containing wastes at Botany Bay, Sydney.

There were sufficient data available from the studies in the Sydney area for Connell and Miller [7] to show as a general trend that OCPs had decreased markedly in Australian coastal waters and sediments since the 1980s, and similarly to show that for the waters of the Brisbane River (Brisbane

TABLE 6.5
Concentrations of OCPs and PCBs in Australian Aquatic Sediments

Compound	Environment	Year(s)	Number	Concentration	Sources
DDT	Fresh water agricultural	1972–1991	—	0.002 to 11.3 mg g^{-1}	[24]
PCB	Rivers	1976–1998	17 > 120	0.49–790 ng g^{-1} dw	[25]
HCH	Rivers	1979–1991	—	0.18–17 ng g^{-1} dw	[24,26]
Dieldrin	Rivers	1973–1985	—	0.08–410 ng g^{-1}	[24,27]
HCB	Marine	1990–1993	6	<0.6–871 ng g^{-1}	[28,29]
Chlordane	River	1979	—	1.5 mean ng g^{-1}	[24]

Source: Connell, D.W. et al., *Crit. Rev. Env. Sci. Technol.*, 29, 47, 1999.

TABLE 6.6
Concentrations of OCPs and PCBs in Australian Aquatic Biota

Compound	Environment	Year(s)	Number	Concentration	Sources
DDT	Invertebrates	1969–1994	—	<0.001–4.4 mg g^{-1}	[24]
DDT	Fish	1969–1994	—	<Detection limit to 40.3 mg g^{-1}	[24]
PCB	Invertebrates	1970–1994	—	<Detection limit to 930 ng g^{-1}	[25,30]
PCB	Fish	1970–1994	—	<Detection limit to 5000 ng g^{-1}	[25]
HCH	Fish		215	0.06–130 ng g^{-1}	[24,31–33]
HCH	Invertebrates	1969–1976	104	0.01–2.0 ng g^{-1}	[33,34]
Dieldrin	Fish	1979–1994	570	0.09–6000 ng g^{-1}	[24,29,31,33,35]
HCB	Invertebrates	1974–1993	>20	<Detection limit to 0.002 ng g^{-1}	[24,31]
HCB	Fish	1973–1994	860	0.001–3.0 ng g^{-1}	[24,29,31]
Heptachlor and heptachlor epoxide	Invertebrates	1977	—	<Detection limit to 0.03 ng kg^{-1}	[24]
Heptachlor and heptachlor epoxide	Fish	1979–1994	>660	<Detection limit to 4.8 ng g^{-1}	[24]
Chlordane	Invertebrates	1977–1993	—	<Detection limit to 0.5 mg g^{-1}	[24,31]
Chlordane	Fish	1988–1993	940	<Detection limit to 1.7 mg g^{-1}	[29,31]
Aldrin	Various	1969–1992	—	0.001–33 mg g^{-1}	[34,36]
Endrin	Invertebrates	1977	—	<1–40 ng g^{-1}	[24]

Source: Data from Connell, D.W. et al., *Crit. Rev. Env. Sci. Technol.*, 29, 47, 1999.

is the capital city of Queensland) the concentrations had declined from a maximum of 1.7 μg L^{-1} in the 1970s to non-detectable in 1986–1987 (using grab samples of water).

This trend in the Brisbane River has recently been confirmed by a comparison of OCP concentration data from an estuarine site within Brisbane City measured using passive samplers (semipermeable membrane devices) deployed on two sampling occasions 11 years apart. The site was first monitored in the summer of 1997 over a 1-month deployment and again in the summer of 2008 using similar techniques and including performance reference compounds. The results showed that over the decade elapsed since the initial assessment, concentrations of dieldrin (the OCP at the greatest concentration) had decreased from 3.9 to 1.4 ng L^{-1}, DDE (dichlorodiphenyldichloroethylene) from 0.084 to 0.015 ng L^{-1}, and DDD (dichlorodiphenyldichloroethane) from 0.13 to 0.014 ng L^{-1} [42].

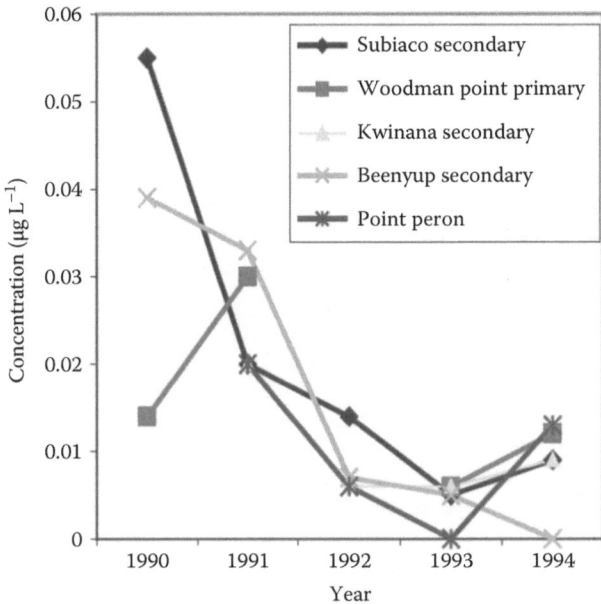

FIGURE 6.30 Dieldrin concentrations in effluent from several secondary sewage treatment plants in Western Australia (1990–1994). (From *Mar. Pollut. Bull.*, 45, Connell, D.W., Miller, G.J., and S.M. Anderson, Chlorohydrocarbon pesticides in the Australian marine environment after banning in the period from the 1970s to 1980s, 78–83, Copyright 2002, with permission from Elsevier.)

A similar temporal decline in OCP concentrations in the tissues of estuarine crabs from a range of locations along the Queensland coast is apparent when comparing the data from the 1996–1997 study by Mortimer [43] with that of 2005–2006 by Negri et al. [44]. Over that period, body burden concentrations of key OCPs declined by at least an order of magnitude. In crab tissues collected during the earlier study, total dichlorodiphenyltrichloroethane (DDT) concentrations ranged from <30 to 3200 µg kg^{-1} lipid and declined to a range of <5 to 240 µg kg^{-1} lipid by 2005–2006; dieldrin concentrations declined from a previous range of <30 to 5500 µg kg^{-1} lipid to a range of <5 to 98 µg kg^{-1} lipid by 2005–2006; and heptachlor epoxide which was detected previously in crabs from most locations became rarely detected even with the improved level of detection in the more recent study. However, direct comparison between the two studies should be made only with caution, since the locations from which crabs were sampled were not common to both studies.

Connell et al. [7] were able to plot a substantial decline in dieldrin concentrations in effluent from several sewage treatment plants in Western Australia during the 1990s (Figure 6.30), which they attributed to a response to declining levels in the catchments served by the treatment plants. They also found a substantial downward trend in dietary intake of total DDT (Figure 6.31) estimated from Australian Market Basket Surveys (NH&MRC and NFA, 1971–1996). These surveys are assessments of consumers' dietary exposure (intake) to pesticide residues, contaminants, and other substances, conducted by Food Standards Australia approximately every 2 years (see: www.foodstandards.gov.au/scienceandeducation/monitoringandsurveillance/australiantotaldiets1914.cfm).

6.2.1.2 OCP Trends in Human Milk Samples in Australia

Human milk was used by Mueller et al. [45] as a surrogate for the assessment of body burden and exposure to OCPs. In total, 157 samples of human milk from 12 regions of Australia covering both rural and urban locations were collected during 2002–2003 and analyzed as 17 pooled samples. A further 24 samples archived from a collection in Melbourne were also used, although an absence

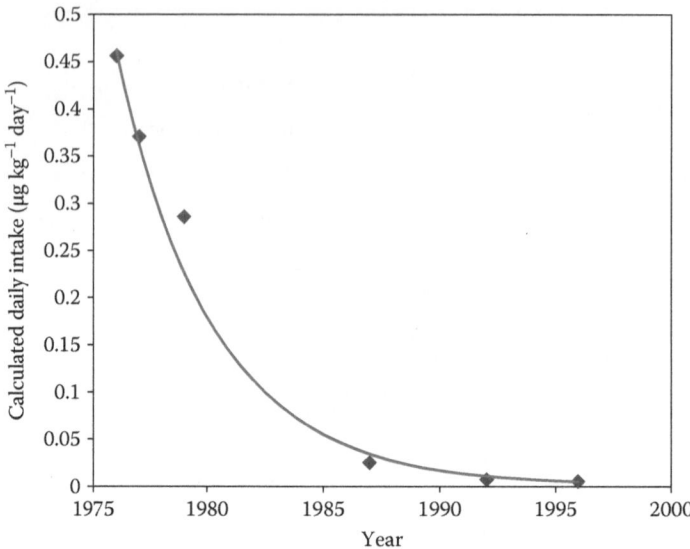

FIGURE 6.31 Trends in estimated dietary intake of total DDT for Australian adult males. (From *Mar. Pollut. Bull.*, 45, Connell, D.W., Miller, G.J., and S.M. Anderson, Chlorohydrocarbon pesticides in the Australian marine environment after banning in the period from the 1970s to 1980s, 78–83, Copyright 2002, with permission from Elsevier.)

of information concerning the demographics of the donors for those "historical" samples limited the interpretation of data.

OCP compounds were detected in all pooled human milk samples used in the study ranging from about 200 ng g^{-1} in a pool collected in Tasmania to approximately 1600 ng g^{-1} in a pool collected in Melbourne. Overall, the mean and median concentrations of sum OCPs were 460 ± 340 and 385 ng g^{-1} lipid, respectively. Overall, DDE was the dominating OCP detected, contributing between 53% and 88% to the sum OCPs. The other key component was β-hexachlorocyclohexane (HCH) (contribution between 3% and 42%). In addition, there were consistent detections of HCB and dieldrin (on average 5% of sum OCP), heptachlorepoxide, oxychlordane, trans-chlordane, and the parent DDT (2%–3% average contribution). Other OCPs—α- and γ-HCH isomers, DDD, pp-DDT, and mirex—were consistently present as very minor components in the OCP profile (each <0.2% contribution to sum OCP). Some regional groupings were apparent in that human milk from large metropolitan centers had the greatest concentrations of DDTs, HCHs, and mirex, whereas in more rural areas the tendency was toward nonachlor, dieldrin, and chlordanes.

A comparison of the 1993 and 2002–2003 Melbourne samples showed little, if any, decrease in the concentrations of most OCPs. Only HCB, heptachlorepoxide and dieldrin showed consistent and notable decreases (58%, 50%, and 23% respectively).

For DDT and dieldrin in human milk, a larger dataset was available by drawing on results from other studies and is presented in Figure 6.32. This shows a substantial decrease in DDT concentrations from the early 1970s (several µg g^{-1} lipid) to well below 1 µg g^{-1} lipid in the 1990s when restrictions on usage were introduced. Similarly, concentrations of dieldrin have decreased from several hundred µg g^{-1} lipid in the 1970s and 1980s to a mean concentration in the range 10–20 µg g^{-1} lipid in the 1990s.

6.2.1.3 OCP Trends in Historic Butter Samples in Australia

Analysis of sealed butter samples was used by Müller et al. [53] to construct plots of historic contamination trends by OCPs. The source of the samples was tinned butter archived at the Australian War Memorial, Canberra, and rubbish tip sites near stations in the Australian Antarctic Territory.

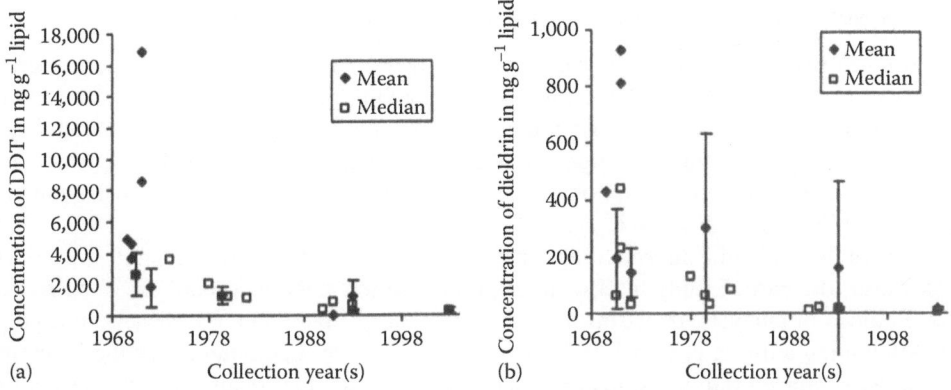

FIGURE 6.32 Historic trends in the concentration of DDTs (a) and dieldrin (b) in human milk samples from Australia. Data for 1969–1970 and 1970–1971 (all Western Australia) from Stacey and Thomas [46]; for 1970 Newton and Greene (in Smith [47]); for 1971 urban and rural Queensland from Miller and Fox [48]; for 1971–1972 (four metropolitan areas) from Siyali [49]; for 1974, 1978, 1980, 1982, 1990 and 1991 (all Western Australia) from Stevens et al. [50]; 1990–1991 from Quinsey et al. [51]; dieldrin for 1994 from Sim et al. [52] and for 1993 and 2002/2003 from this study. (From *Chemosphere*, 70, Mueller, J.F., Harden, F., Toms, L., Symons, R., and Furst, P., Persistent organochlorine pesticides in human milk samples from Australia, 712–720, Copyright 2008, with permission from Elsevier.)

All samples had been sealed since manufacture and hence were an archive of OCPs in milk contaminant levels at that time. The samples covered a period of approximately 60 years.

DDTs and HCHs as well as dioxin-like chemicals were detected in all samples. HCH concentrations ranged from 530 ng g^{-1} lipid in the late 1950s to <2 ng g^{-1} lipid in the 1980s. Similarly, the concentrations of DDTs ranged form several µg g^{-1} lipid in the 1950s and 1960s to about 20 ng g^{-1} lipid in 2000.

As illustrated in Figure 6.33, the HCH profile was dominated by α-HCH in the 1950s and 1960s, which is consistent with the use of technical mixtures of HCH in Australia at the time. By contrast, γ-HCH was dominant in the 1987 sample. Similarly, for DDTs, the samples from the 1950s to the 1960s were dominated by the parent compound p,p′-DDT widely used at that time, but there was a shift back to p,p′-DDE dominance in the 1987 sample consistent with the ban of the parent compound in Australia, and subsequent aging and metabolism of residual DDT in the environment.

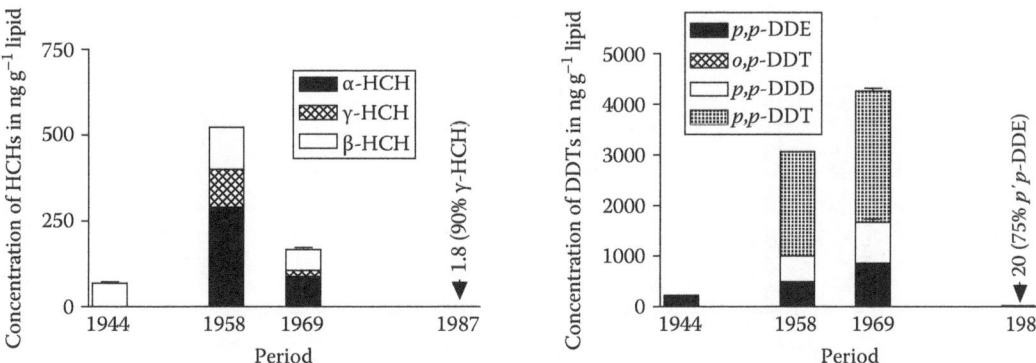

FIGURE 6.33 Concentration and isomer distribution of HCHs (left), and concentrations of isomers and metabolites of DDT (right) in historical butter samples from Australia. (From Mueller, J.F. et al., *Organohalogen Compd.*, 64, 203, 2003.)

6.2.2 TRENDS IN CONCENTRATIONS OF DIOXIN-LIKE CHEMICALS IN AUSTRALIA

As noted by Connell et al. [22], opportunities to measure contaminant concentration trends over time are very limited because available data concerning POC concentrations in the Australian environment over a suitable time scale does not come from studies designed to evaluate changes over time. Another difficulty in interpretation is that critical variables such as lipid or organic carbon in samples are generally not recorded. Some such historical data for PCBs in the Australian environment are listed in Tables 6.5 and 6.6.

PCBs were detected in estuarine crabs in 4 out of 11 rivers in the recent study by Negri et al. [44]. However, the earlier study by Mortimer [43] using the same methodology of collection and similar methods of analysis, including the measurement of lipid content in tissues, did not report PCBs. Attempting a direct comparison between the data from these studies for the assessment of a trend in PCB concentrations is unsound because the collection locations were not common to both studies.

The only historical material available to the National Dioxins Program studies was a set of 10 archived soil samples covering some 80 years from a single location and in very small quantities. The measured concentrations suggested confounding by artifacts of storage and sampling, and were not able to indicate a trend. However, the historic butter samples available to Müller et al. [53] (see Section 6.2.1.3) were more useful, showing a clear trend of reduced PCB concentrations from the 1950s to 1987 (Figure 6.34), and in PCDD/PCDF concentrations (Figure 6.35).

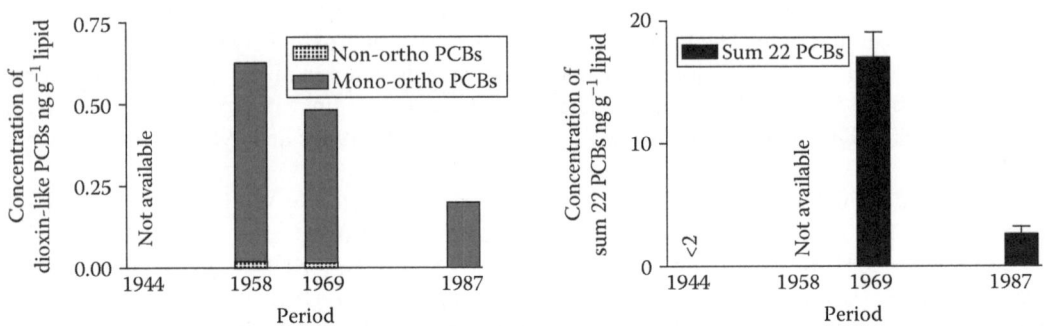

FIGURE 6.34 Concentration and isomer distribution of non-ortho and mono-ortho PCBs (left), and concentrations of sum of 22 PCBs (right) in historical butter samples from Australia. (From Mueller, J.F. et al., *Organohalogen Compd.*, 64, 203, 2003.)

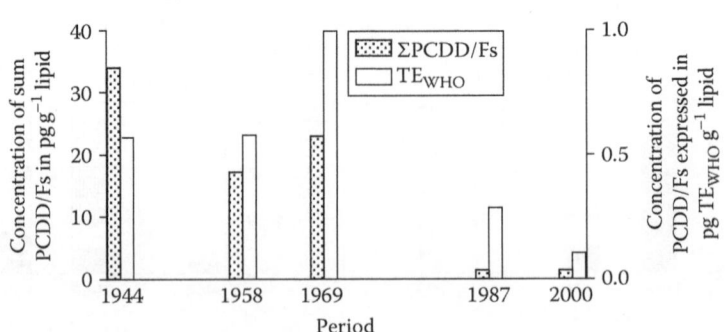

FIGURE 6.35 Concentration of PCDD/PCDFs expressed as sum 2,3,7,8 PCDD/PCDFs ($Y1$ axis) and TE_{WHO} ($Y2$ axis) in historical butter samples from Australia. (From Mueller, J.F. et al., *Organohalogen Compd.*, 64, 203, 2003.)

6.3 CONCLUSIONS

Australia has a history of POC use in both agricultural and industrial applications. Residual contamination by organochlorine pesticides such as DDTs and dieldrin is widespread, particularly in urban/industrial areas and estuaries close to urban centers, as evidenced in recently measured levels in soils, sediments, fauna, and humans. However, levels are low by comparison with most other developed nations, particularly those in the Northern Hemisphere.

Although there is a limited set of data available, there are clear indications of a downward trend in OCP contaminants in the Australian environment, consistent with the bans placed on these chemicals in the 1970s and 1980s.

For dioxin-like chemicals, a sound national baseline was established in a range of media as a consequence of studies undertaken as components of the National Dioxins Program in the early 2000s. Human milk studies show a decrease in concentrations of dioxin-like chemicals over time, and human blood studies an increase in concentrations with age of donor. However, as yet, apart from a downward trend for PCBs in a small set of historic butter samples and the human milk studies, there is insufficient data available at this time to establish any temporal trends for these chemicals in the Australian environment. Generally, PCDD/PCDFs dominate total TEQ in Australian samples, although in a few cases the contribution from PCBs exceeds 50%. Overall, urban/industrial locations have a significantly greater TEQ level than remote and agricultural areas.

In fauna, PCBs contribute significantly to the TEQ level in birds and marsupials,and is dominant in marine mammals. There is significant evidence for a trend of increasing TEQ with increasing trophic level, consistent with food chain biomagnification.

A national baseline for PBDEs was established by studies in the mid 2000s. The occurrence of environmental PBDEs in Australia is associated with urban/industrial areas. BDE-209 dominates the congener profile. Human blood studies show a decline with age.

Ambient air sampling shows a strong seasonal concentration cycle for PCDD/PCDFs in major population centers associated with smoke from residential wood burning, and associated concentrations are greater than in non-urban areas.

REFERENCES

1. CIA. 2010. *World Factbook, Australia*. Central Intelligence Agency of the United States of America. Available at: www.cia.gov/library/publications/the-world-factbook//geos/as.html (accessed on April 6, 2011).
2. Connell, D.W., Miller, G.J., Mortimer, M.R., Shaw, G.R., and Anderson, S.M. 1996. Persistent organic pollutants (POPs) in the Southern Hemisphere. A consultancy report prepared for the Department of Environment, Sport and Territories, Environment Protection Agency, Environment Standards Branch, Canberra, Australia.
3. Selinger, B. 1995. HCB waste background and issues paper. Ben Selinger, Anutech Pty Ltd. (cited in Connell et al., 1996 [2]). Available at: www.environment.gov.au/settlements/publications/chemicals/scheduled-waste/hcbplan.html (accessed on April 6, 2011).
4. DEWHA. 2010 Australia's action on the 12 POPs listed under the Convention. Available at: www.environment.gov.au/settlements/chemicals/international/pop.html#action (accessed on April 6, 2011).
5. NICNAS. 2007. Interim Public Health Risk Assessment Report on certain PBDE congeners contained in commercial preparations of pentabromodiphenyl ether and octabromodiphenyl ether. Available at: http://www.nicnas.gov.au/Publications/CAR/other/PBDE_PDF.pdf (accessed on July 30, 2008).
6. Richardson, B.J. 1995. The problem of chlorinated compounds in Australia's marine environment. In: *The State of the Marine Environment Report for Australia. Technical Annex: 2* (Technical Paper). Zann, L.P. and Sutton, D.C. (Eds.), p. 47. Ocean Rescue 2000. Department of the Environment, Sport and Territories, Canberra, Australia. Available at: www.environment.gov.au/archive/coasts/publications/somer/annex2/richardson.html (accessed on April 6, 2011).
7. Connell, D.W., Miller, G.J., and Anderson, S.M. 2002. Chlorohydrocarbon pesticides in the Australian marine environment after banning in the period from the 1970s to 1980s. *Marine Pollution Bulletin* 45: 78–83.

8. DEH. 2001–2004. National Dioxins Program. Australian Government Department of the Environment and Heritage, Canberra, Australia. Available at: www.environment.gov.au/settlements/chemicals/dioxins/index.html (accessed on April 6, 2011).

9. Toms, L., Mueller, J., Mortimer, M., Symons, M., Stevenson, R., and Gaus, C. 2006. *Assessment of Concentrations of Polybrominated Diphenyl Ether Flame Retardants in Aquatic Environments in Australia.* Australian Government Department of the Environment and Heritage, Canberra, Australia.

10. Toms, L., Mueller, J., Bartkow, M., and Symons, R. 2006. *Assessment of Concentrations of Polybrominated Diphenyl Ether Flame Retardants in Indoor Environments in Australia.* Australian Government Department of the Environment and Heritage, Canberra, Australia.

11. Toms, L., Harden, F., Hobson, P., Papke, O., Ryan, J., and Mueller, J. 2006. *Assessment of Concentrations of Polybrominated Diphenyl Ether Flame Retardants in the Australian Population: Levels in Blood.* Australian Government Department of the Environment and Heritage, Canberra, Australia.

12. Mueller, J., Muller, R., Goudkamp, K. et al. 2004. Dioxins in aquatic environments in Australia. National Dioxins Technical Report No. 5, Australian Government Department of the Environment and Heritage, Canberra, Australia.

13. Mueller, J., Muller, R., Goudkamp, K. et al. 2004. Dioxins in soils in Australia. National Dioxins Technical Report No. 6, Australian Government Department of the Environment and Heritage, Canberra, Australia.

14. Prange, J.A., Gaus, C., Papke, O., and Müller, J.F. 2001. PCDDs in geologically old samples from Queensland, Australia. *Organohalogen Compounds* 50: 534–537.

15. Prange, J.A., Gaus, C., Papke, O., and Müller, J.F. 2002. Investigations into the PCDD contamination of topsoil, river sediments and kaolinite clay in Queensland, Australia. *Chemosphere* 46: 1335–1342.

16. Correll, R., Müller, J., Ellis, D. et al. 2004. Dioxins in fauna in Australia, National Dioxins Program Technical Report No. 7. Australian Government Department of the Environment and Heritage, Canberra, Australia.

17. Gras, J., Müller, J., Graham, B., Symons, R., Carras, J., and G. Cook. 2004. Dioxins in ambient air in Australia. National Dioxins Program Technical Report No. 4, Australian Government Department of the Environment and Heritage, Canberra, Australia.

18. Gras, J., Meyer, C., Weeks, I. et al. 2002. Characterisation of emissions from solid-fuel burning appliances (wood-heaters, open fireplaces). Final report CSIRO Atmospheric Research, Aspendale, Victoria, Australia, 266 pp. Also available as Gras J.L. 2002. Emissions from domestic solid fuel appliances. Technical Report No. 5, Environment Australia, Canberra, Australia. Available at: http://www.ea.gov.au/atmosphere/airtoxics/report5/index.html (accessed on April 6, 2011).

19. Buckland, S.J., Ellis, H.K., and Salter, R.T. 1999. *Organochlorines in New Zealand: Ambient Concentrations of Selected Organochlorines in Air.* Ministry for the Environment, Wellington, New Zealand.

20. Harden, F., Müller, J., Toms, L. et al. 2004. Dioxins in the Australian population: Levels in human milk. National Dioxins Program Technical Report No. 10, Australian Government Department of the Environment and Heritage, Canberra, Australia.

21. Harden, F., Müller, J., Toms, L. et al. 2004. Dioxins in the Australian population: Levels in blood. National Dioxins Program Technical Report No. 9, Australian Government Department of the Environment and Heritage, Canberra, Australia.

22. Connell, D.W., Miller, G.J., Mortimer, M.R., Shaw G.R., and Anderson, S.M. 1999. Persistent lipophilic contaminants and other chemical residues in the Southern Hemisphere. *Critical Reviews in Environmental Science & Technology* 29: 47–82.

23. Garman, D.E.J. 1983. Water quality issues in Australia. Water 2000: Consultants Report No. 7, Australian Government Publishing Service, Canberra, Australia.

24. Cullen, M.C. 1990. Aspects of the behaviour of chlorinated hydrocarbon pesticides in terrestrial and aquatic environments. Dissertation for Master of Philosophy, Division of Australian Environmental Studies, Griffith University, Brisbane, Australia.

25. Shaw, G.R. and Connell, D.W. 1980. Polychlorinated biphenyls in the Brisbane River estuary, Australia. *Marine Pollution Bulletin* 11: 356.

26. Iwata, H., Tanabe, S., Sakai, N., Nishimura, A., and Tatsukawa, R. 1994. Geographical distribution of persistent organochlorines in air, water and sediments from Asia and Oceania, and their implications for global redistribution from lower latitudes. *Environmental Pollution* 85(1): 15–33.

27. Connell, D.W. 1986. Ecotoxicology of lipophilic pollutants in Australian inland waters. In *Limnology in Australia.* de Deckker, P. and Williams, W.D. (Eds.), 573 pp. CSIRO, Clayton, Victoria, Australia.

28. EPA. 1994. *Sydney Deepwater Outfalls. Final Report Series. Volume 4. Trace Metals and Organochlorines in the Marine Environment.* Environmental Monitoring Program. Environmental Protection Authority, New South Wales, Australia.

29. Mortimer, M.R. and Connell, D.W. 1995. A model of the environmental fate of chlorohydrocarbon contaminants associated with Sydney sewage discharges. *Chemosphere* 30(11): 2021–2038.

30. Phillips, D.J.H., Richardson, B.J., Murray, A.P., and Fabris, J.G. 1992. Trace metals, organochlorines and hydrocarbons in Port Phillip Bay, Victoria: A historical review. *Marine Pollution Bulletin* 25: 200–217.

31. Miskiewicz, A.G. and Gibbs, P.J. 1994. Organochlorine pesticides and hexachlorobenzene in tissues of fish and invertebrates caught near a sewage outfall. *Environmental Pollution* 84(3): 269–277.

32. Nicholson, G.J., Theodoropoulos, T., and Fabris, G.J. 1994. Hydrocarbons, pesticides, PAH in Port Phillip Bay (Victoria) sand flathead. *Marine Pollution Bulletin* 28(2): 115–120.

33. Olafson, R.W. 1978. Effect of agricultural activity on levels of organochlorine pesticides in hard corals, fish and molluscs from the Great Barrier Reef. *Marine Environmental Research* 1(2): 87–107.

34. Thomson, J.M. and Davie, J.D.S. 1974. Pesticide residues in the fauna of the Brisbane River estuary. *Search* 5: 152.

35. Cullen, M.C. and Connell, D.W. 1992. Bioaccumulation of chlorohydrocarbon pesticides by fish in the natural environment. *Chemosphere* 25(11): 1579–1587.

36. FPMEC. 1992. Organochlorine termiticides: Residues in fisheries products: Position Paper. Fisheries Pollution and Marine Environment Committee, Canberra, Australia.

37. Thompson, G.B., Chapman, J.C., and Richardson, B.J. 1992. Disposal of hazardous wastes in Australia: Implications for marine pollution. *Marine Pollution Bulletin* 25: 155–162.

38. Lincoln Smith, M.P. and Mann, R.A. 1989. Bioaccumulation in nearshore marine organisms. I. Organochlorine compounds and trace metals in rocky reef animals near the Malabar Ocean outfall. State Pollution Control Commission, Sydney, Australia.

39. Lincoln Smith, M.P. and Mann, R.A. 1989. Bioaccumulation in nearshore marine organisms. II. Organochlorine compounds in the Red Morwong, *Cheilodactylus fuscus*, around Sydney's three major sewage ocean outfalls. State Pollution Control Commission, Sydney, Australia.

40. ANZECC. 1991. Persistent chlorinated organic compounds in the marine environment. Public Information Paper. Australian and New Zealand Environment and Conservation Council, Canberra, Australia.

41. EPA NSW. 1995. The State of the Environment Report 1995. Environmental Protection Authority, Sydney, Australia.

42. Mueller, J.F., Mortimer, M., Shaw, G., Connell, D., and O'Brien, J. 2010. Exposure trends for OCs and PAHs in the Brisbane River Estuary and Moreton Bay. In: *6th International Conference on Marine Pollution and Ecotoxicology*, Hong Kong, May 31–June 3, 2010. (Special Issue *Marine Pollution Bulletin*—Accepted.)

43. Mortimer, M.R. 2000. Pesticide and trace metal concentration in Queensland estuarine crabs. *Marine Pollution Bulletin* 41(7–12): 359–366.

44. Negri, A.P., Mortimer, M., Carter, S., and Müller, J.F. 2009. Persistent organochlorines and metals in estuarine mud crabs of the Great Barrier Reef. *Marine Pollution Bulletin* 58(5):769–773.

45. Mueller, J.F., Harden, F., Toms, L., Symons, R., and Furst, P. 2008. Persistent organochlorine pesticides in human milk samples from Australia. *Chemosphere* 70: 712–720.

46. Stacey, C. and Thomas, B. 1975. Organochlorine pesticide residues in human milk, Western Australia 1970–1971. *Pesticides Monitoring Journal* 9: 64–66.

47. Smith, D. 1999. Worldwide trends in DDT levels in human breast milk. *International Journal of Epidemiology* 28: 179–188.

48. Miller, G. and Fox, J. 1973. Chlorinated hydrocarbon pesticide residues in Queensland human milks. *The Medical Journal of Australia* 2: 261–264.

49. Siyali, D.S. 1973. Polychlorinated biphenyls, hexachlorobenzene and other organochlorine pesticides in human milk. *The Medical Journal of Australia* 2: 815–818.

50. Stevens, M.F., Ebell, G.F. and Psaila-Savona, P. 1993. Organochlorine pesticides in western Australian nursing mothers. *The Medical Journal of Australia* 158: 238–241.

51. Quinsey, P., Donohue, D., and Ahokas, J. 1995. Persistence of organochlorines in breast milk of women in Victoria, Australia. *Food and Chemical Technology* 33: 49–56.

52. Sim, M., Forbes, A., McNeil, J., and Roberts, G. 1998. Termite control and other determinants of high body burdens of cyclodiene insecticides. *Archives of Environmental Health* 53: 114–121.

53. Müller, J.F., Jacobs, M., Covaci, A., and Papke, O. 2003. Organohalogen compounds in historic butter. *Organohalogen Compounds* 64: 203–206.

7 Environmental Contamination Status of Polychlorinated Biphenyls in China

Ying Xing,* Yonglong Lu, and Wenbin Liu

CONTENTS

7.1   Introduction 163
7.2   Estimation of Total PCB-Containing Pollutants and Their Distribution in China 164
7.3   PCB Levels in the Environment at the National Scale 164
    7.3.1   PCB Pollution Status in Surface Soils 165
    7.3.2   PCB Pollution Status in Surface Water Bodies 167
    7.3.3   PCB Pollution Status in Surface Sediments 168
    7.3.4   PCB Pollution Status in Atmosphere 171
7.4   Conclusions 173
Acknowledgments 174
References 174

7.1   INTRODUCTION

In China, polychlorinated biphenyl (PCB) compounds were manufactured from 1965 to 1974 by four major producers: Xi'an Chemical Plant, Shanghai Electric Chemical Plant, Suzhou Solvent Plant, and Shanghai Sanzao Chemical Plant. Total amounts of PCBs produced were approximately 10,000 MT, of which 1,000 MT were pentachlorobiphenyl and 9,000 MT were trichlorobiphenyl. Pentachlorobiphenyl was used as a paint additive and trichlorobiphenyl was used as dielectric fluid in capacitors. In the 1950s–1980s, uninformed of the risks, China imported from other countries a great deal of electric equipment containing PCBs [1].

Following the prohibition in 1974 of the production and use of PCBs in China, most of the outdated PCB-containing equipment was removed from use and stored in industrial and rural storage areas, either dispersed or centralized. Most storage was temporary, in poor condition, and without proper management and supervision, which resulted in equipment damage and leakage. Some of the PCB-containing equipment was transferred both legally and illegally into e-waste dismantling sites where the PCB-containing transformers and capacitors were dismantled and incinerated openly and without any measures for environmental protection, which caused serious point-source PCB pollution. In the 1980s, for example, e-waste dismantling operations emerged and boomed in the Wentai area of Zhejiang Province on the southeast coast of China. Unfortunately, this was not the only example of such pollution in China.

Due to the rising concerns of governors, scientists, and stakeholders on PCB pollution in China there has been an increase in related research. There is still, however, a knowledge gap in our

---

* E-mail: xingy@mail.las.ac.cn (Chapter corresponding author).

understanding of the extent and spatial character of PCB pollution in China. This work provides a comprehensive profile of PCB levels in China based on existing data, estimates the quantity of PCB-containing pollutants, and analyzes the major sources across the country.

7.2 ESTIMATION OF TOTAL PCB-CONTAINING POLLUTANTS AND THEIR DISTRIBUTION IN CHINA

It is difficult to obtain accurate information about the status of disposal sites due to their long history and insufficient records and documents. Jin et al. [2] reported that 750,000 capacitors containing PCBs had been produced historically in China. Accordingly, the total weight of PCB-containing capacitors has reached 37.5 kt based on the average weight of one capacitor being 50 kg. An estimated 50 kt of PCB-containing equipment was calculated to exist in China including imported transformers. Types of PCB-containing pollutants with some level of contaminants were presumed to be 1000 kt in the whole of China. Since the quantity of PCB equipment is related to power generation, PCB pollutants distributing in different regions can be calculated approximately based on the power generation of each province in 1974. The burden of PCB pollutants in six regions was estimated and plotted on a map of China (Figure 7.1). The eastern regions, especially the eastern coastal areas, had more PCB-containing pollutants than the central and western regions.

7.3 PCB LEVELS IN THE ENVIRONMENT AT THE NATIONAL SCALE

While the number of studies on PCBs in China has increased recently as concern has been raised over the risks that persistent organic pollutants (POPs) pose to humans and the environment, there has so far not been a systematic national survey of PCB pollution status conducted by the government. Only a few research groups have conducted large-scale investigations of PCB levels in soil [3], sediment [4,5] and air [6,7]. Other studies have been conducted mainly by local researchers acting individually. The current sites of PCB investigations of soil, water, sediment, and air are presented in Figure 7.2. Research on PCBs in soil and air has increased significantly compared to research on water and sediment.

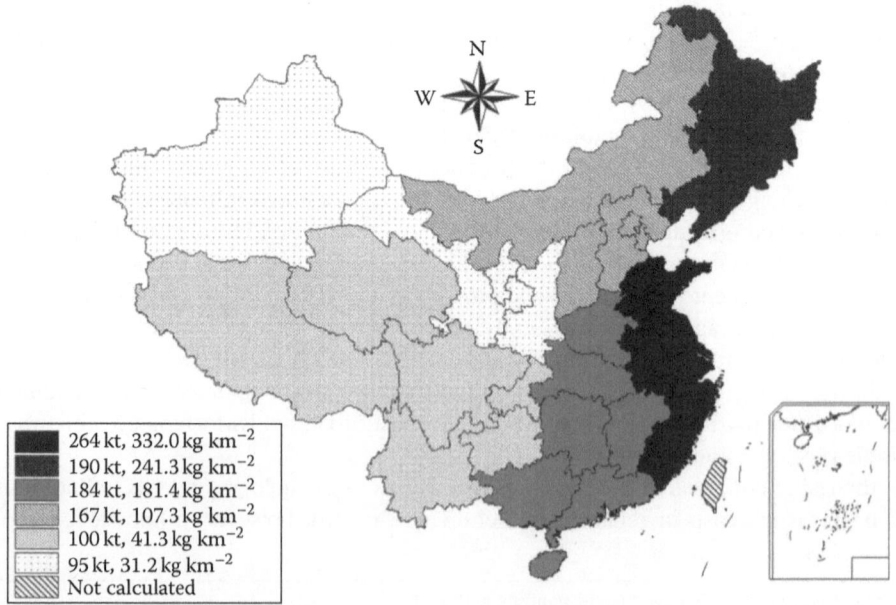

Legend:
- 264 kt, 332.0 kg km^{-2}
- 190 kt, 241.3 kg km^{-2}
- 184 kt, 181.4 kg km^{-2}
- 167 kt, 107.3 kg km^{-2}
- 100 kt, 41.3 kg km^{-2}
- 95 kt, 31.2 kg km^{-2}
- Not calculated

FIGURE 7.1 Distribution of PCB pollutants estimated in China.

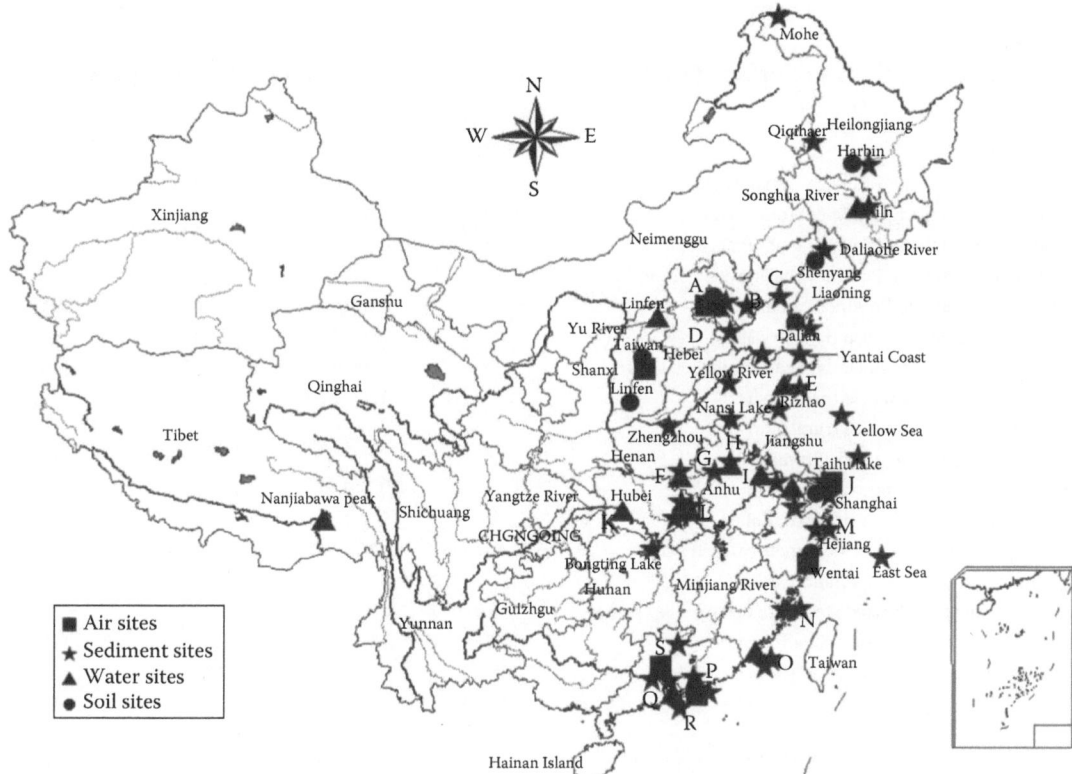

FIGURE 7.2   Sample sites of PCB investigation in China. A, Beijing City (Tonghui River, Yongding River-Guanting Reservoir, Huairou); B, Luanhe River; C, Jinzhou Bay; D, Huaihe River; E, Jiaozhou Bay, Qingdao, Shandong Province; F, Xinyang, Henan Province; G, Fuyang, Anhui Province; H, Huainan, Anhui Province; I, Nanjing, Jiangsu Province; J, Yangtze River Estuary, Huangpu River, Shanghai City; K, Three Gorge Reservoir; L, Wuhan, Yangtze River, Dong Lake, Ya-Er Lake, Hubei Province; M, Hangzhou, Qiantangjiang River, Zhejiang Province; N, Minjiang Estuary, Fuzhou, Fujian Province; O, Jiulongjiang Estuary, Xiamen, Fujian Province; P, Daya Bay, Guangdong Province; Q, Guangzhou, Pearl River, West River, East River, Guangdong Province; R, Pearl River Estuary, Shiziyang, Hongkong, Macao; S, Shaoguan, Beijiang, Pearl River, Guangdong Province.

7.3.1   PCB POLLUTION STATUS IN SURFACE SOILS

PCB pollution status in soils was seldom investigated before 2005 with significant increases after that time. Related surface soil PCB data are listed in Table 7.1.

In China, the background value of PCB concentration in soil was 0.625–3.501 ng g^{-1} at a site located on the Tibetan Plateau unaffected by PCB pollution [20] and 0.14 ng g^{-1} in Tibet in other research [3]. By comparison, agricultural soil sampled in suburban Beijing had a concentration of 0.18 ng g^{-1} [11].

Heavy PCB pollution in soil was always associated with pollution incidents [22]. Polluted soils in some PCB-containing equipment storage locations had levels as high as 4.54 mg g^{-1}, with levels in surrounding areas and farmland as high as 2930 ng g^{-1} [1]. At Wentai, Zhejiang Province, the irregular trading or dismantling of PCB-containing equipment had been a local industry that boomed for a long time, which caused PCB pollution of the environment and health risks to the residents. PCB concentrations were reported as follows. Chu et al. [11] reported that PCB concentrations were as high as 788 ng g^{-1} dry wt. There, PCB levels in a rice field adjacent to a pollution incident that occurred in 1989 reached 1101.4 ng g^{-1} in 1993. Following the ban on arbitrary

TABLE 7.1

PCB Concentrations in Surface Soil from Various Sites in China (ng g^{-1} Dry Weight)

Location	Average Concentration	Concentration Range	Time of Sample	Reference
Harbin, Heilongjiang Province (urban sites)	2.24	0.53–6.17	2006	[8]
Harbin, Heilongjiang Province (rural sites)	1.59	0.45–2.42	2006	[8]
Harbin, Heilongjiang Province (suburban sites)	0.64	0.51–1.08	2006	[8]
Harbin, Heilongjiang Province (background site)	0.3		2006	[8]
Shenyang, Liaoning Province (urban area)		6.4–15.2	1992	[9]
Dalian, Liaoning Province (industry sites)	3.09	2.36–3.92	2007	[10]
Dalian, Liaoning Province (business/residence sites)	2.86	1.75–4.77	2007	[10]
Dalian, Liaoning Province (garden sites)	2.2	1.87–2.53	2007	[10]
Dalian, Liaoning Province (rural site)	1.34		2007	[10]
Huairou, Beijing City (agricultural soil)	0.18		1990–1992	[11]
Linfen, Shanxi Province (urban sites)	1	0.2–3.4	2006	[12]
Linfen, Shanxi Province (industrial plant sites)	2.4	0.5–14.8	2006	[12]
Taiyuan, Shanxi Province (symmetrical grid sites)	0.64	<4.7	2006	[13]
Zhejiang Province (nonpollution areas)	45.4	7.5–263	2003	[14]
WenTai, Zhejiang Province (e-waste disassembly sites)	146.79	27.77–738.96	2007	[15]
WenTai, Zhejiang Province (control site)	16.26	12.56–20.81	2007	[15]
WT, southeast China (pollution site)	788		1990–1992	[11]
WenTai, Zhejiang Province (rice field beside pollution site)	1101.4		1993	[16]
WenTai, Zhejiang Province (rice field beside pollution site)	744.1	568.15–920.08	1997	[16]
WenTai, Zhejiang Province (rice field beside pollution site)	4.68	2.01–7.34	1999	[16]
WenTai, Zhejiang Province (pollution site)	4,211	2,458–5,644	2003	[17]
Ya-Er Lake, Wuhan, Hubei Province (polluted lake)	7.7	7.1–8.2	1991–1994	[18]
Wujing, Shanghai City (chemical industrial sites)	48.91	0.5–586.85	2005	[19]
Nanjiabawa Peak, Tibet (not influenced by industry)		0.625–3.501		[20]
Tibet (background and rural sites)	0.19	0.05–0.42	2005	[21]
Tibet (background and rural sites)	0.14		2005	[3]
All of China (all the sites)	0.52	0.14–1.84	2005	[3]
All of China (background and rural sites)	0.42	0.14–1.14	2005	[3]

disposal of used electrical appliances, pollution levels decreased to 568.15–920.08 ng g^{-1} in 1997 and were down to 2.01–7.34 ng g^{-1} by 1999 [16]. However, there is still considerable environmental risks from PCBs that must be addressed because of the persistence of high average PCB concentration (4211 ng g^{-1}), with an upper limit of 5644 ng g^{-1} in 2003 [14]. At the same site, other research has measured a mean PCB level of 146.79 ng g^{-1}, ranging from 27.77 to 738.96 ng g^{-1} in 2007 [15].

Ren et al. [3] conducted a systematic investigation of PCB pollution in soil at a national level in 2005. Fifty-two sites including urban, rural, and background localities were sampled and analyzed, with the sample localities and the relevant PCBs concentrations presented in Figure 7.3. Nearly all provinces showed occurrences of PCBs with the average concentration of total PCBs being 0.52 ng g^{-1} and ranging from 0.14 to 1.84 ng g^{-1}, which meant the PCB concentrations were not very high. Other published PCB data listed in Table 7.1 gave the same trend except for the heavily polluted areas.

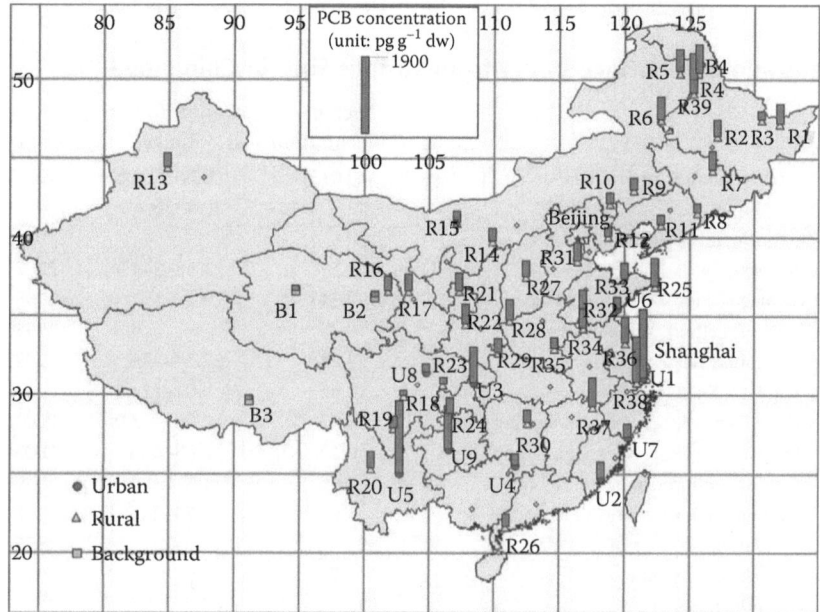

FIGURE 7.3   Sample sites and concentrations of Σ_{51}PCBs in surface soils in China (ng g^{-1}). (From Ren, N.Q. et al., *Environ. Sci. Technol.*, 41(11), 3871, 2007. With permission. Copyright © 2007 American Chemical Society.)

From research conducted at several city locations such as Harbin in Heilongjiang Province [8] and Dalian in Liaoning Province [10], PCB concentrations decreased with distance away from urban sites to suburban and rural sites, with the lowest concentrations at the background sites. This indicates that cities are a major source of PCBs, with industrial zones as the main urban PCB source. For example, in the Wujing industrial area of Shanghai, the greatest concentration was found at a PVC workshop in a chlor-alkali plant [19]. In Linfen, Shanxi Province, PCB levels were higher at industrial plant sites than other sites [12]. It has been suggested that the occurrence of PCBs at remote background sites may be due to atmospheric transport.

7.3.2 PCB Pollution Status in Surface Water Bodies

Some important Chinese rivers such as the Yangtze and Pearl Rivers have been investigated by different researchers from the 1980s to 2005 (Table 7.2). Based on this data, it can be concluded that PCB pollution was not serious in the Yangtze and Pearl Rivers, with concentrations being lower than the Chinese National Environmental Quality Standard of Surface Water (20 ng L^{-1}). However, PCB pollution status was severe at Minjiang and Jiulongjiang (both in Fujian Province, southeast China), Daya Bay (on the coast of Guangdong Province), Shenyang industrial sewage water (Liaoning Province), Tonghui River (near the Beijing), and Taihu Lake (in an industrialized area).

Higher PCB concentrations usually occurred in estuarine environments such as the Minjiang and Jiulongjiang estuaries. Estuaries have been subjected to pollution from upstream leaching, emission of municipal sewage and harbor and shipping pollution and are thought to be sinks for a range of chemicals, especially for POPs [34,35]. Taihu Lake is one of the most developed regions in China and its water quality has been greatly degraded after decades of intensive industrial development.

TABLE 7.2
PCB Concentrations in Surface Water from Various Sites in China (ng L^{-1})

Water and Location	Average Concentration	Concentration Range	Time of Sample	Reference
Songhua River (Jilin section), Jilin Province	0.013	0.003–0.088	1983	[23]
Shenyang, Liaoning Province (water in park)	43.9	30.9–56.9	1992	[9]
Shenyang, Liaoning Province (water in industrial sewage)	128.8		1992	[9]
Tonghui River, Beijing City	105.5	31.58–344.9	2002	[24]
Huaihe River (Xinyang section), Henan Province	0.24		1998	[25]
Huaihe River (Huainan section), Anhui Province	0.28		1998	[25]
Yu river (DaTong section), Shanxi Province	28.22	22.54–33.9	1998	[26]
Jiaozhou Bay (Qingdao), Shandong Province	10.82	9.01–12.63		[27]
Yangtze River (Three Gorge Reservoir), Chongqing City	0.004	0.001–0.006	2005	[28]
Yangtze River (Three Gorge Reservoir), Chongqing City	0.19	0.08–0.51	2008	[29]
Yangtze River (Nanjing section), Jiangsu Province	2.96	1.8–5.0	1999	[30]
Yangtze River (Nanjing section), Jiangsu Province	1.91	1.74–2.00	1998	[31]
Taihu Lake, Jiangsu Province	631		1999	[32]
Dong Lake, Wuhan, Hubei Province	2.7		1994	[33]
Minjiang (Estuary), Fujian Province	985.2	203.9–2,473	1999	[34]
Jiulongjiang (Estuary), Fujian Province	355	0.36–1,505	1999	[35]
Jiulongjiang (Estuary), Fujian Province	17.5	0.1–34.8	1998	[36]
Daya Bay, Guangdong Province	313.6	91.1–1,355.3	1999	[37]
Pearl River (Estuary), Guangdong Province	0.77	0.12–0.47	2005	[38]
Pearl River (Estuary), Guangdong Province	3.92	2.47–6.75	2000	[39]
Pearl River (Estuary), Guangdong Province	1.87	0.48–3.93	2000	[40]
Nanjiabawa Peak, Tibet	0.0093			[20]

7.3.3 PCB POLLUTION STATUS IN SURFACE SEDIMENTS

The PCB pollution status of sediments has been investigated more than in any other environmental medium. The concentration and range in surface sediments in the main rivers, estuaries, coastal areas, and lakes are listed in Tables 7.3 through 7.6 respectively.

Based on the published data, the Songhuajiang River in northeast China and the Pearl River in southeast China had the highest PCB concentrations. PCB levels beside a drainage outlet of the Songhuajiang River reached 337 ng g^{-1} [41]. The Guangzhou section and East River branch of the Pearl River system had high PCB concentrations as well. The PCB concentration at Fangcun in the Guangzhou section was 485.5 ng g^{-1} [46]. Two sites along the East River had high levels of 270 and 240 ng g^{-1} [49]. The high PCB levels in Guangzhou may result from high levels of industrial emissions and municipal sewage. Concentrations in sediments in old industrial parks (e.g., Zhichang, Yuancun, and Huangpu in the Guangzhou sector) were also higher than those at new industrial parks or in adjacent residential areas [46,47]. China's two largest rivers, the Yangtze and Yellow Rivers, did not show prominent PCB pollution based on the current records, though they were seriously polluted with many other substances.

PCB concentrations in the Pearl River Estuary, including the Hong Kong and Macao areas, were higher than Yangtze, Yellow, Minjiang, and Jiulongjiang estuaries. PCB levels at Hong Kong were 461 ng g^{-1} [58], at Macao Inner Harbor 338.5 ng g^{-1} [46] and at Victoria Harbor 97.9 ng g^{-1} [61]. This indicates serious pollution in the Pearl River Estuary. Fu et al. [66] also reported that besides Guangzhou, Hong Kong was another PCB pollution source to the Pearl River Estuary based on the PCB distribution profile. PCBs in Pearl River Delta sediments seemed to result from trace

TABLE 7.3
PCB Concentrations in Sediment in Rivers of China (ng g⁻¹ Dry Weight)

Location	Average Concentration	Concentration Range	Time of Sample	Reference
Heilongjiang River (Mohe section), Heilongjiang Province	21		Early of 1990s	[4]
Nenjiang River (Qiqihaer section), Heilongjiang Province	25		Early of 1990s	[4]
Songhuajiang River (Haerbin section), Heilongjiang Province	20		Early of 1990s	[4]
Jilin section, Songhuajiang River	0.62	0.12–1.04	1983	[23]
Songhuajiang River (Jilin section), Jilin Province	89.3	0.6–337		[41]
Daliaohe River, Liaoning Province	2.3	1.9–2.7	1989	[5]
Yongding River-Guanting Reservoir, Beijing City	5.1	0.81–9.72	1999	[42]
Tonghuihe River, Beijing City	3.29	0.78–8.47	2002	[24]
Haihe River, Tianjin City	3.2	2.9–3.5	1985	[5]
Luanhe River, Hebei Province	1.4		1991	[5]
Huaihe River (Bangbu section), Anhui Province	15		Early of 1990s	[4]
Huaihe River (Xinyang section), Henan Province	8.24			[43]
Huaihe River (Huainan section), Anhui Province	6.34			[43]
Yellow River (Zhengzhou section), Henan Province	17		Early of 1990s	[4]
Yellow River (Mid- and down-stream sections)	3.1		2004	[44]
Hanshui River (Hanyang section, Branch of Yangtze River), Hubei Province	18.5		Early of 1990s	[4]
Yangtze River (Wuhan section), Hubei Province	20		Early of 1990s	[4]
Yangtze River (Wuhan section), Hubei Province	9.2	1.2–45.1	2005	[45]
Yangtze River (Nanjing section), Jiangsu Province	15		Early of 1990s	[4]
Huangpujiang River (Branch of Yangtze River), Shanghai City	19.9		1988	[5]
Qiantangjiang River (Hangzhou section), Zhejiang Province	22		Early of 1990s	[4]
Qiantangjiang River (Hangzhou section), Zhejiang Province	12.8		1984	[5]
Minjiang River (Fuzhou section), Fujiang Province	25		Early of 1990s	[4]
Pearl River (Guangzhou section), Guangdong Province	23		Early of 1990s	[4]
Pearl River (Guangzhou section), Guangdong Province	195.3	48.3–485.5	1997	[46]
Pearl River (Guangzhou section), Guangdong Province	31.52	12.88–65.31	2000	[47]
Pearl River (Shiziyang), Guangdong Province	22.2	15.98–30.27	1997	[46]
East River (branch of Pearl River), Guangdong Province	13.8	8.0–22.0	1996	[48]
East River (branch of Pearl River), Guangdong Province	131	48–270	2007	[49]
West River (branch of Pearl River), Guangdong Province	13.2	11.04–14.92	1997	[46]
Beijiang River (Shaoguan section), Guangdong Province	12.5		Early of 1990s	[4]

TABLE 7.4
PCB Concentrations in Sediment in Estuaries in China (ng g^{-1} Dry Weight)

Location	Average Concentration	Concentration Range	Time of Sample	Reference
Yellow Estuary	1.3	0.7–2.4	1985	[5]
Yellow Estuary	0.03	0.01–0.04	2004	[50]
Yangtze Estuary	7.1	3.0–9.5	1987	[5]
Yangtze Estuary	18.12	10.5–28.6	2000	[51]
Yangtze Estuary	2.7	0.19–18.95	2001	[52]
Yangtze Estuary	2.3	0.92–9.69	2004	[53]
Yangtze Estuary	0.02	0.01–0.04	2004	[50]
Minjiang Estuary	6.63	4.69–7.27	1996	[54]
Minjiang Estuary	34.49	15.13–57.93	1999	[55]
Jiulongjiang Estuary	0.8	0.45–1.15	1993	[5]
Jiulongjiang Estuary	1.74	0.05–7.24	1993	[56]
Jiulongjiang Estuary	0.18	0.16–0.19	1994	[57]
Jiulongjiang Estuary	0.16	ND-0.32	1998	[36]
Jiulongjiang Estuary	8.37	4.1–14.29	1999	[54]
Pearl River Estuary	19.3	3.2–81	1992	[56]
Pearl River Estuary	2.49		1994	[54]
Pearl River Estuary		46–461	1995	[58]
Pearl River Estuary	0.67	0.18–1.8	1996	[59]
Pearl River Estuary	11.5	10.16–12.51	1997	[46]
Pearl River Estuary	28.9	26.7–32	1997	[60]
Pearl River Estuary (Macao Harbor)	338.5		1997	[46]
Pearl River Estuary	17.2	0.5–97.9	1998	[61]
Pearl River Estuary	21.3	0.21–63.76	<1999	[62]
Pearl River Estuary	14.5	7.32–36.2	2000	[63]
Pearl River Estuary	16.5	11.13–23.23	2000	[39]
Pearl River Estuary	3.86	0.1–25.1		[64]
Pearl River Estuary	1.24	0.23–4.7	2004	[65]

discharges into storm water, sewage, industrial waste, agricultural runoff, and some atmospheric deposition.

Some typical bays and coastlines along eastern China have been studied, and the results are listed in Table 7.5.

Data from Dalian Bay show high PCB concentrations, while other sites did not show any pollution. This may be attributed to the dilution effect of the ocean. However, the Yellow Sea receives a vast amount of fresh water from both Chinese and Korean rivers, and PCBs from these inputs are accumulating gradually in sediments [73].

Lakes are subject to pollution by PCBs since they are usually the sinks for various pollutants from upstream. Current data on Chinese lakes is scarce. However, it can be seen that PCBs in Ya'er Lake, located in Hubei Province, suffered severe pollution from a nearby chemical manufacturing plant in the 1970s.

In addition to PCBs in the sediments of main water bodies, heavy PCB pollution (691 ng g^{-1}) occurs on the southeast coast of China due to illegal spilling from and improper dismantling of PCB-containing equipment [11]. The Ministry of Environmental Protection of China [1] reported

TABLE 7.5
PCB Concentrations in Sediment at Coasts and in Bays in China (ng g^{-1} Dry Weight)

Location	Average Concentration	Concentration Range	Time of Sample	Reference
Dalian Bay, Liaoning Province	19.1	1.02–153.1	1996	[67]
Dalian Bay, Liaoning Province	2.14	0.04–3.23	1999	[68]
Jinzhou Bay, Liaoning Province	5.84	0.6–32.6	1996	[67]
Yantai Coast, Shandong Province	0.35		2002	[69]
Rizhao Coast, Shandong Province	0.27		2002	[69]
Qingdao Coast, Shandong Province	4.86	0.65–32.9	1999	[70]
Xiamen Island Coast, Fujian Province	0.05	0.03–0.08	1995	[71]
Daya Bay, Guangdong Province	8.83	0.85–27.4	1999	[37]
East Sea	0.81			[72]
Yellow Sea	0.92			[72]
Southern Yellow Sea	1.72	0.52–5.85	2004	[73]
Southern Yellow Sea	1.78		2003	[73]

TABLE 7.6
PCB Concentrations in Sediment in Lakes in China (ng g^{-1} Dry Weight)

Location	Average Concentration	Concentration Range	Time of Sample	Reference
Nansi Lake, Shandong Province	0.63		2002	[69]
Ya-er Lake, Wuhan, Hubei Province	1,503	<5,970	1991–1994	[18]
Eight Lakes in Wuhan City, Hubei Province	16.7	0.9–46.1	2005	[74]
Dongting Lake, Hunan Province	22.5		Early 1990s	[4]
Taihu Lake, Jiangsu Province	6.71	0.89–29.7	2002	[75]
Taihu Lake, Jiangsu Province	2.51	0.37–4	2000	[76]

that concentrations reached 116–183 µg g^{-1} in sediments around closed PCB storage sites lacking proper control measures.

7.3.4   PCB Pollution Status in Atmosphere

The first available public record of atmospheric PCB levels was reported by Jing et al. [9]. However, the first systematic research in mainland China was not carried out until 2005 by Jaward et al. [6]. Zhang et al. [7] conducted another large-scale survey of atmospheric PCB concentration across the entire country. Some local investigations were also conducted with all the results being presented in Table 7.7, Figures 7.4 and 7.5.

The highest average atmospheric PCB concentration was 100,030 pg m^{-3}, ranging from 74,400 to 131,000 pg m^{-3} in Shenyang City (Liaoning Province), which was surveyed in 1992. Shenyang City (located in northeast China) has been one of the biggest industrial cities with numerous factories using PCB equipment, and thus significant PCBs have been lost into ambient environments from the 1980s to the investigated time [9].

Other high PCB levels were found on the southeast coast of China, which is one of the most developed areas in the country, in areas such as Shanghai City and the Zhejiang Province. The higher

TABLE 7.7
PCB Concentrations in Atmosphere from Various Sites in China (pg m^{-3})

Location	Average Concentration	Concentration Range	Time of Sample	Reference
Shenyang, Liaoning Province (urban sites)	100,030	74,400–131,000	1992	[9]
Beijing City (urban center)	44.00	22–65	2007	[77]
Taiyuan, Shanxi Province (urban sites)	47.50	16–190	2006	[13]
Taizhou, Zhejiang Province (PCBs containing equipments had been dismantled in the 1980s)	7,220.00	4,232–11,352	2005	[78]
Wujing, Shanghai City (chlor-alkali chemical factory)	535.81	40.75–1,869.0	2004–2005	[79]
Wujing, Shanghai City (coking and chemical factory)	1,736.98	90.89–3,742.2	2004–2005	[79]
Wujing, Shanghai City (coal-fired power plant)	3,778.86	ND-14,149.2	2004–2005	[79]
Wujing, Shanghai City (chlor-alkali plant, the biggest in China)	396.11	32.1–1,870	2005–2006	[19]
Wujing, Shanghai City (coke-oven plant)	2,680.00	9.22–14,150	2005–2006	[19]
Wujing, Shanghai City (thermoelectric plant)	1,816.80	34.8–7,450	2005–2006	[19]
Guangzhou, Guangdong Province (urban center)	1,961.00	1,383–2,720	2004	[80]
Guangzhou, Guangdong Province (city background site)	921.70	307.2–1,544	2004	[80]
Guangzhou, Guangdong Province (eastern of an industrial zone)	363.10	172.6–696	2004	[80]
Guangzhou, Guangdong Province (western of an industrial zone)	497.40	350–1,100	2004	[80]
Hong Kong, Pearl River Delta (residential area)	4.33	1.91–11.3	2004–2005	[81]
Hong Kong, Pearl Estuary (urban residential site with mixed commerce and industry)	2.97	1.06–7.94	2004–2005	[81]
Hong Kong, Pearl Estuary (rural remote background site)	1.57	0.8–5.56	2004–2005	[81]
All of China (urban sites)	18.82	5.44–65.80	2004	[6]
All of China (rural/background sites)	13.51	4.8–29.95	2004	[6]
All the China (urban sites)	350.00		2005	[7]
All the China (rural sites)	230.00		2005	[7]
All the China (background sites)	77.00		2005	[7]

average atmospheric concentration was 7220 pg m^{-3} in Taizhou, Zhejiang Province, a historically polluted area due to improper disassembly of PCB-containing capacitors and transformers [78]. High concentrations also occurred in a 50-year-old industrial area in Shanghai City. At a coal-fired power plant, the average concentration was 3778.86 pg m^{-3}, ranging 9.22–14,150 pg m^{-3}. A survey of coking and chlor-alkali chemical factories also showed high concentrations [19,79]. Atmospheric research conducted at Guangzhou, Pearl River Estuary, also showed high PCB levels [80].

Jaward et al. [6] presented a similar distribution (Figure 7.4). Higher PCB levels were observed in the three most developed and populated zones along the east coast, i.e., Tianjin-Tsingtao, the Yangtze River Delta, and the Pearl River Delta, with the highest PCB concentrations in the Tianjin urban area (65.80 pg m^{-3}).

Jaward [6] also showed that PCB concentrations at urban sites were higher than those at rural or background sites. This conclusion was also shown by Zhang et al. [7] and Chen et al. [80]. Zhang [7] presented a map of national PCB distribution based on 97 samples (Figure 7.5), which showed that urban sites had the highest level of pollution, followed by rural sites and then background sites. The research by Chen [80] showed the highest PCB levels occurred in the urban center of Guangzhou.

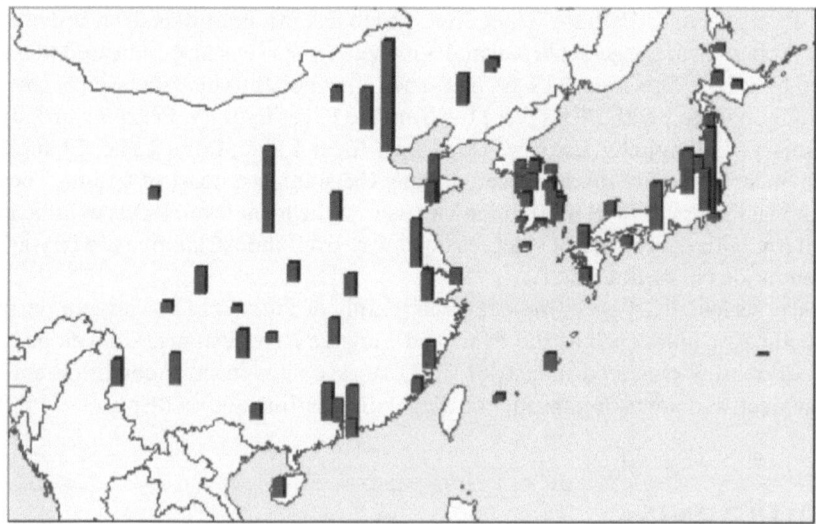

FIGURE 7.4 Sample sites and concentrations of Σ_{29}PCBs in air in some areas of Asia (pg m^{-3}). *Note*: The value of the highest bar on the figure is 65.8 ng sample^{-1}. (From Jaward, T.M. et al., *Environ. Sci. Technol.*, 39(22), 8638, 2005. With permission. Copyright 2005 American Chemical Society.)

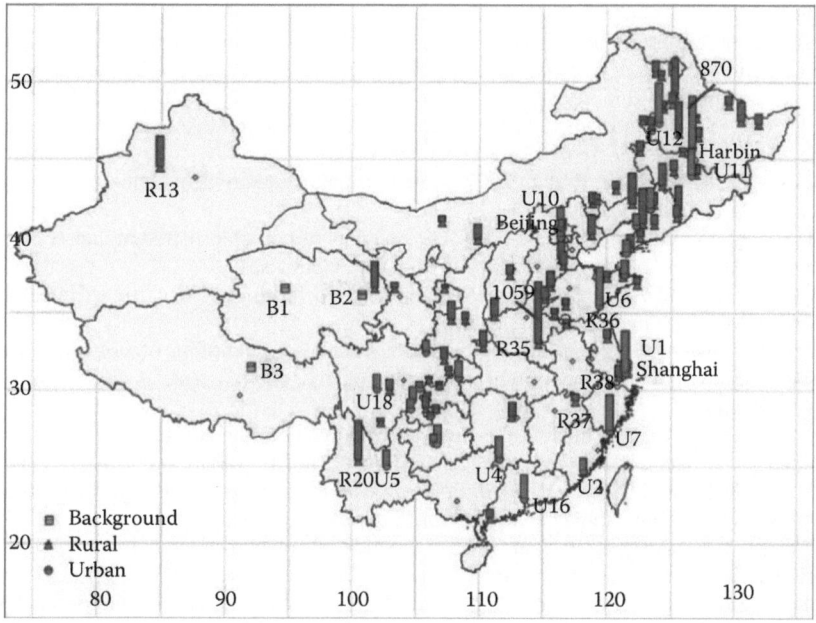

FIGURE 7.5 Sample sites and concentrations of Σ_{60}PCBs in air in China (pg m^{-3}). *Note:* The value of the highest bar (Site R35) is indicated on the figure as 1050 pg m^{-3}. (From Zhang, Z. et al., *Environ. Sci. Technol.*, 42(17), 6514, 2008. With permission. Copyright 2008 American Chemical Society.)

7.4 CONCLUSIONS

Based on the analysis of soil, water, sediment, and atmosphere it is concluded that at the national scale PCB concentrations are relatively low. The main reason for this may be that PCB output was lower due to the shorter history of PCB production in China compared to developed countries.

However, PCBs occur in both the Tibet area, which has minimal industrial influence and in the middle of the Yellow Sea. Large-scale soil and atmospheric surveys also indicated that all provinces were polluted by PCBs. This means PCBs pose a risk that needs to be addressed across the country.

Areas heavily polluted with PCBs were the Yangtze River Estuary, Pearl River/Estuary system, Minjiang Estuary, Jiulongjiang Estuary, Daya Bay, Ya'er Lake, Taihu Lake, Shanghai City, and Zhejiang Province. These are mainly located along the southeast coast of China. These areas correspond to known historical PCB production and use. At the local level, PCBs exhibit a point source pollution pattern, with concentrations decreasing away from industrial areas to city areas, with the lowest concentrations in rural areas.

In summary, serious PCB pollution occurred mainly at four types of sites: Important estuaries located in southeast China, such as the Pearl and Yangtze River estuaries, developed cities with a long history of densely clustered industries, PCB storage sites in poor condition and with inadequate management, and areas of improper or illegal dismantling and/or disposal of PCB-containing equipment.

ACKNOWLEDGMENTS

This research was supported by National Basic Research Program of China (Grant No. 2007CB407307) and National Natural Science Foundation of China (Grant No. 40601089).

REFERENCES

1. China SEPA. 2003. Building the capacity of the People's Republic of China to implement the Stockholm Convention on POPs and develop a National Implementation Plan. *Global Environment Facility Project Brief* (Project Number: GF/CPR/02/010). http://gefonline.org/projectDetailsSQL.cfm?projID=1412
2. Jin, C. Y., Yang, G. X., and Peng, W. S. 1990. PCBs pollution and disposal with environmental sound measure. *Envir. Protect. Sci.* 16 (3):31–35 (in Chinese).
3. Ren, N. Q., Que, M. X., Li, Y. F. et al. 2007. Polychlorinated biphenyls in Chinese surface soils. *Environ. Sci. Technol.* 41 (11):3871–3876.
4. Chen, J. S., Gao, X. M., Qi, M. et al. 1999. The contents of polychlorinated biphenyl in river sediments in eastern China. *Acta Sci. Circumst.* 19 (6):614–618 (in Chinese).
5. Wu, Y., Zhang, J., and Zhou, Q. 1999. Persistent organochlorine residues in sediments from Chinese river/estuary systems. *Environ. Pollut.* 105 (1):143–150.
6. Jaward, T. M., Zhang, G., Nam, J. J. et al. 2005. Passive air sampling of polychlorinated biphenyls, organochlorine compounds, and polybrominated diphenyl ethers across Asia. *Environ. Sci. Technol.* 39 (22):8638–8645.
7. Zhang, Z., Liu, L. Y., Li, Y. F. et al. 2008. Analysis of polychlorinated biphenyls in concurrently sampled Chinese air and surface soil. *Environ. Sci. Technol.* 42 (17):6514–6518.
8. Ma, W. L., Li, Y. F., Sun, D. Z. et al. 2009. Polycyclic aromatic hydrocarbons and polychlorinated biphenyls in topsoils of Harbin, China. *Arch. Environ. Contam. Toxicol.* 57 (4):670–678.
9. Jing, Z. Y., Li, Y. H., Feng, X. B. et al. 1992. The study of polychlorinated biphenyls for their loss, contamination and protection strategy in Shenyang City. *Congkan Environ. Sci.* 13 (5):1–28 (in Chinese).
10. Wang, D. G., Yang, M., Jia, H. L. et al. 2008. Levels, distributions and profiles of polychlorinated biphenyls in surface soils of Dalian, China. *Chemosphere* 73 (1):38–42.
11. Chu, S. G., Yang, C., Xu, X. B. et al. 1995. Polychlorinated biphenyl congener residues in sediment and soil from pollution area. *China Environ. Sci.* 15 (3):199–203 (in Chinese).
12. Fu, S., Cheng, H., Liu, Y. et al. 2008. Polychlorinated biphenyls residues in the soil in Linfen, China. *Bull. Environ. Contam. Toxicol.* 81 (6):594–598.
13. Fu, S., Cheng, H. X., Liu, Y. H. et al. 2009. Spatial character of polychlorinated biphenyls from soil and respirable particulate matter in Taiyuan, China. *Chemosphere* 74 (11):1477–1484.
14. Gao, J., Luo, Y. M., Li, Q. B. et al. 2006. Distribution patterns of polychlorinated biphenyls in soils collected from Zhejiang Province, East China. *Environ. Geochem. Health* 28 (1–2):79–87.
15. Zhao, G. F., Wang, Z. J., Dong, M. H. et al. 2008. PBBs, PBDEs, and PCBs levels in hair of residents around e-waste disassembly sites in Zhejiang Province, China, and their potential sources. *Sci. Tot. Environ.* 397 (1–3):46–57.

16. Bi, X. H., Chu, S. G., Meng, Q. Y. et al. 2002. Movement and retention of polychlorinated biphenyls in a paddy field of WenTai area in China. *Agr. Ecosyst. Environ.* 89 (3):241–252.

17. Gao, L., Zhao, X., Zheng, M. et al. 2005. Distribution of polychlorinated biphenyls in different depths of soil from a polluted area in the People's Republic of China. *Bull. Environ. Contam. Toxicol.* 74 (5):962–967.

18. Wu, W. Z., Schramm, K. W., Henkelmann, B. et al. 1997. PCDD/F-s, PCBs, HCHs and HCB in sediments and soils of Ya-Er lake area in China: Results on residual levels and correlation to the organic carbon and particle size. *Chemosphere* 34 (1):191–202.

19. Ma, J., Cheng, J. P., Xie, H. Y. et al. 2007. Seasonal and spatial character of PCBs in a chemical industrial zone of Shanghai, China. *Environ. Geochem. Health* 29 (6):503–511.

20. Sun, W. X., Chen, R. L., Sun, A. Q. et al. 1986. Pollution of organochlorines at Nanjiabawa peak. *Environ. Sci.* 7 (6):64–69 (in Chinese).

21. Wang, P., Zhang, Q. H., Wang, Y. W. et al. 2009. Altitude dependence of polychlorinated biphenyls (PCBs) and polybrominated diphenyl ethers (PBDEs) in surface soil from Tibetan Plateau, China. *Chemosphere* 76 (11):1498–1504.

22. Xing, X., Lu, Y. L., Dawson, R. W. et al. 2005. A spatial temporal assessment of pollution from PCBs in China. *Chemosphere* 60 (6):731–739.

23. Li, M. X., Yue, G. C., Gao, F. M. et al. 1989. Transport and distribution of PCBs and organochlorine pesticides in the second Songhua River. *Environ. Chem.* 8 (2):49–54 (in Chinese).

24. Zhang, Z. L., Huang, J., Yu, G. et al. 2004. Occurrence of PAHs, PCBs and organochlorine pesticides in the Tonghui River of Beijing, China. *Environ. Pollut.* 130 (2):249–261.

25. Wang, Y., Wang, Z. J., Liu, J. A. et al. 1999. Monitoring toxic and organic pollutants in the Huaihe River using Triolein-SPMD. *Environ. Monitor. China* 15 (4):8–11 (in Chinese).

26. Bi, X. H., Chu, S. G., and Xu, X. B. 2001. Polycyclic aromatic hydrocarbons and polychlorinated biphenyl contamination in DaTong City, China. *Bull. Environ. Contam. Toxicol.* 67 (1):141–148.

27. Chu, S. G., Xu, X. B., and Tong, Y. P. 1995. Transport and distribution of polychlorinated biphenyls in a polluted area. *Acta Sci. Circumst.* 15 (4):423–432 (in Chinese).

28. Chen, J. A., Luo, J. H., Qiu, Z. Q. et al. 2008. PCDDs/PCDFs and PCBs in water samples from the Three Gorge Reservoir. *Chemosphere* 70 (9):1545–1551.

29. Wang, J. X., Bi, Y. H., Pfister, G. et al. 2009. Determination of PAH, PCB, and OCP in water from the Three Gorges Reservoir accumulated by semipermeable membrane devices (SPMD). *Chemosphere* 75 (8):1119–1127.

30. Sun, C., Dong, Y., Xu, S. et al. 2002. Trace analysis of dissolved polychlorinated organic compounds in the water of the Yangtse River (Nanjing, China). *Environ. Pollut.* 117 (1):9–14.

31. Jiang, X., Xu, S. F., Martens, D. et al. 2000. Polychlorinated organic contaminants in waters, suspended solids and sediments of the Nangjing section, Yangtze River. *China Environ. Sci.* 20 (3):193–197 (in Chinese).

32. Wang, H., Wang, C. X., Wu, W. Z. et al. 2003. Persistent organic pollutants in water and surface sediments of Taihu Lake, China and risk assessment. *Chemosphere* 50:557–562.

33. Xi, Z. Q., Chu, S. G., Xu, X. B. et al. 1998. Determination of polychlorinated biphenyls in Donghu Lake. *Oceanol. Limnol. Sin.* 29 (4):436–440 (in Chinese).

34. Zhang, Z. L., Hong, H. S., Zhou, J. L. et al. 2003. Fate and assessment of persistent organic pollutants in water and sediment from Minjiang River Estuary, Southeast China. *Chemosphere* 52:1423–1430.

35. Zhang, Z. L., Chen, W. Q., Khalid, M. et al. 2000. Study on PCBs in water of Jiulong River Estuary. *YunNan Environ. Sci.* 19:124–127 (in Chinese).

36. Zhou, J. L., Hong, H., Zhang, Z. et al. 2000. Multi-phase distribution of organic micropollutants in Xiamen Harbour, China. *Water Res.* 34 (7):2132–2150.

37. Zhou, J. L., Maskaoui, K., Qiu, Y. W. et al. 2001. Polychlorinated biphenyl congeners and organochlorine insecticides in the water column and sediments of Daya Bay, China. *Environ. Pollut.* 113: 373–384.

38. Guan, Y. F., Wang, J. Z., Ni, H. G. et al. 2009. Organochlorine pesticides and polychlorinated biphenyls in riverine runoff of the Pearl River Delta, China: Assessment of mass loading, input source and environmental fate. *Environ. Pollut.* 157 (2):618–624.

39. Nie, X. P., Lan, C. Y., Wei, T. L. et al. 2005. Distribution of polychlorinated biphenyls in the water, sediment and fish from the Pearl River estuary, China. *Mar. Pollut. Bull.* 50 (5):537–546.

40. Nie, X. P., Lan, C. Y., Luan, T. G. et al. 2002. Rapid determination of PCBs in seawater of Pearl River Delta by solid-phase microextraction. *Mar. Environ. Sci.* 21 (2):65–68 (in Chinese).

41. Liu, J. A., Wang, W. H., and Wang, Z. J.1998. Sorts and contents of persistent pollutants in the sediment samples of the second Songhua River. *China Environ. Sci.* 18 (6):518–520 (in Chinese).

42. Ma, M., Wang, Z. J., and Sodergren, A. 2001. Contamination of PCBs and organochlorinated pesticides in the sediment samples of Guanting Reservoir and Yongding River. *Environ. Chem.* 20 (3):238–243 (in Chinese).

43. Wang, Z. J., Wang, Y., and Ma, M. 2001. Assessing the ecological risk of sediment-associated polychlorinated biphenyls in Huaihe River. *China Environ. Sci.* 21 (3):262–265 (in Chinese).

44. He, M. C., Sun, Y., Li, X. R. et al. 2006. Distribution patterns of nitrobenzenes and polychlorinated biphenyls in water, suspended particulate matter and sediment from mid- and down-stream of the Yellow River (China). *Chemosphere* 65 (3):365–374.

45. Yang, Z. F., Shen, Z. Y., Gao, F. et al. 2009. Occurrence and possible sources of polychlorinated biphenyls in surface sediments from the Wuhan reach of the Yangtze River, China. *Chemosphere* 74 (11):1522–1530.

46. Kang, Y. H., Sheng, G. Y., Fu, J. M. et al. 2000. Preliminary study on the distribution and characterization of polychlorinated biphenyls in some of surface sediments from Pearl River Delta. *Environ. Chem.* 19 (3):262–269 (in Chinese).

47. Nie, X. P., Lan, C. Y., Luan, T. G. et al. 2001. Polychlorinated biphenyls in the waters, sediments and benthic organisms from Guangzhou reach of Pearl River. *China Environ. Sci.* 21 (5):417–421 (in Chinese).

48. Ho, K. C. and Hui, K. C. C. 2001. Chemical contamination of the East River (Dongjiang) and its implication on sustainable development in the Pearl River Delta. *Environ. Int.* 26:303–308.

49. Ren, M., Peng, P. A., Chen, D. Y. et al. 2009. Patterns and sources of PCDD/Fs and dioxin-like PCBs in surface sediments from the East River, China. *J. Hazard. Mater.* 170 (1):473–478.

50. Hui, Y. M., Zheng, M. H., Liu, Z. T. et al. 2009. PCDD/Fs and dioxin-like PCBs in sediments from Yellow Estuary and Yangtze Estuary, China. *Bull. Environ. Contam. Toxicol.* 83 (4):614–619.

51. Chen, M. R., Yu, L. Z., Xu, S. Y. et al. 2003. Spatial distribution of PCBs in the sediments of Changjiang Estuary tidal-flat. *Mar. Environ. Sci.* 22 (2):20–23 (in Chinese).

52. Yang, Y., Liu M., and Hou, L. J. 2002, Distribution of polychlorinated organic compound in Yangtze Estuary and its correlation with TOC and particle size. *Shanghai Environ. Sci.* 21 (9):530–532 (in Chinese).

53. Shen, M., Yu, Y. J., Zheng, G. J. et al. 2006. Polychlorinated biphenyls and polybrominated diphenyl ethers in surface sediments from the Yangtze River Delta. *Mar. Pollut. Bull.* 52 (10):1299–1304.

54. Yuan, D. X., Yang, D. N., Wade, T. L. et al. 2001. Status of persistent organic pollutants in the sediment from several estuaries in China. *Environ. Pollut.* 114:101–111.

55. Zhang, Z. L., Hong, H. S., and Yu, G. 2002. Preliminary study on persistent organic pollutants (POPs)—PCBs in multi-phase matrices in Minjiang River Estuary. *Acta Sci. Circumst.* 22 (6):788–791 (in Chinese).

56. Hong, H., Xu, L., Zhang, L. et al. 1995. Environmental fate and chemistry of organic pollutants in the sediment of Xiamen and Victoria Harbours. *Mar. Pollut. Bull.* 31:229–236.

57. Chen, W. Q., Zhang, L. P., Xu, L. et al. 1996. Vertical distribution characteristics of organochlorinated pesticides and polychlorinated biphenyls in sediments of Xiamen Bay. *Mar. Sci.* 2:56–60 (in Chinese).

58. Zhou, H. Y., Cheung, R. Y. H., and Wong, M. H. 1999. Residues of organochlorines in sediments and tilapia collected from inland water systems of Hong Kong. *Arch. Environ. Contam. Toxicol.* 36:424–431.

59. Hong, H. S., Chen, W. Q., Xu, L. et al. 1999. Distribution and fate of organochlorine pollutants in the Pearl River Estuary. *Mar. Pollut. Bull.* 39:376–382.

60. Mai, B. X., Zeng, E. Y., Luo, X. J. et al. 2005. Abundances, depositional fluxes, and homologue patterns of polychlorinated biphenyls in dated sediment cores from the Pearl River Delta, China. *Environ. Sci. Technol.* 39 (1):49–56.

61. Richardson, B. J. and Zheng, G. J. 1999. Chlorinated hydrocarbon contaminations in Hong Kong surficial sediments. *Chemosphere* 39 (6):913–923.

62. Wong, C. K. C., Yeung, H. Y., Cheung, R. Y. H. et al. 2000. Ecotoxicological assessment of persistent organic and heavy metal contamination in Hong Kong coastal sediment. *Arch. Environ. Contam. Toxicol.* 38:486–493.

63. Nie, X. P., Lan, C. Y., An, T. C. et al. 2006. Distributions and congener patterns of PCBs in fish from major aquaculture areas in the Pearl River Delta, South China. *Hum. Ecol. Risk Assess.* 12 (2):363–373.

64. Tam, N. F. Y. and Yao, M. W. Y. 2002. Concentrations of PCBs in coastal mangrove sediments of Hong Kong. *Mar. Pollut. Bull.* 44:642–651.

65. Terauchi, H., Takahashi, S., Lam, P. K. S. et al. 2009. Polybrominated, polychlorinated and monobromopolychlorinated dibenzo-p-dioxins/dibenzofurans and dioxin-like polychlorinated biphenyls in marine surface sediments from Hong Kong and Korea. *Environ. Pollut.* 157 (3):724–730.

66. Fu, J. M., Mai, B. X., Sheng, G. Y. et al. 2003. Persistent organic pollutants in environment of the Pearl River Delta, China: An overview. *Chemosphere* 52:1411–1422.

67. Li, H., Fu, Y. Z., Zhou, C. G. et al. 1998. Distribution characteristics of organic chlorine pesticide and PCB in the surface sediments in Dalian Bay and Jinzhou Bay. *Mar. Environ. Sci.* 17 (2):73–76 (in Chinese).

68. Liu, X. M., Xu, X. R., Zhang, X. T. et al. 2001. Organochlorine pesticides and PCBs in Dalian Bay. *Mar. Environ. Sci.* 20 (4):40–44 (in Chinese).

69. Yang, Y. L., Pan, J., Li, H. L. et al. 2003. Polychlorinated biphenyls in sediments from the Nansi Lake and coastal seas of Yantai and Rizhao, Shandong. *Bull. Mineral. Petrol. Geochem.* 22 (2):108–113 (in Chinese).

70. Yang, Y. L., Pan, J., Li, Y. et al. 2003. Horizontal and vertical distributions of PCBs in sediments and mussel pollution in Qingdao coastal sea. *China Environ. Sci.* 23 (5):515–520 (in Chinese).

71. Chen, W. Q., Zhang, L. P., Xu, L. et al. 1996. Concentrations and distributions of HCHs, DDTs and PCBs in surface sediments of sea area between Xiamen and Jinmen. *J. Xiamen Univ. (Nat. Sci.)* 35 (6):936–940 (in Chinese).

72. Tanabe, S., Yasuhara, Y., and Tatsukawa, R. 1985. Persistent organochlorines in the East China Sea— Their distribution, behavior and fate inherent in marginal sea. In: *Pollution in the Urban Environment*, eds. Chan, M. W. H., Hoare, R. W. M., Holmes, P. R., Law, R. J. S., and Reed, S. B. London: Elsevier Applied Science Publishers. *POLMET* 85:670–678.

73. Zhang, P., Song, J. M., Liu, Z. G. et al. 2007. PCBs and its coupling with eco-environments in southern Yellow Sea surface sediments. *Mar. Pollut. Bull.* 54 (8):1105–1115.

74. Yang, Z. F., Shen, Z. Y., Gao, F. et al. 2009. Polychlorinated biphenyls in urban lake sediments from Wuhan, Central China: Occurrence, composition, and sedimentary record. *J. Environ. Qual.* 38 (4):1441–1448.

75. Zhang, Q. H. and Jiang, G. B. 2005. Polychlorinated dibenzo-p-dioxins/furans and polychlorinated biphenyls in sediments and aquatic organisms from the Taihu Lake, China. *Chemosphere* 61 (3):314–322.

76. Yuan, X. Y., Wang, Y., Sun, C. et al. 2004. Characteristics and environment effects of polychlorinated biphenyls in sediments from Taihu Lake. *Resour. Environ. Yang. Basin* 13 (3):272–276 (in Chinese).

77. Li, Y. M., Zhang, Q. H., Ji, D. S. et al. 2009. Levels and vertical distributions of PCBs, PBDEs, and OCPs in the atmospheric boundary layer: Observation from the Beijing 325-m meteorological tower. *Environ. Sci. Technol.* 43 (4):1030–1035.

78. Li, Y. M., Jiang, G. B., Wang, Y. W. et al. 2008. Concentrations, profiles and gas-particle partitioning of PCDD/Fs, PCBs and PBDEs in the ambient air of an E-waste dismantling area, southeast China. *Chinese Sci. Bull.* 53 (4):521–528.

79. Cheng, J. P., Wu, Q., Xie, H. Y. et al. 2007. Polychlorinated biphenyls (PCBs) in PM10 surrounding a chemical industrial zone in Shanghai, China. *Bull. Environ. Contam. Toxicol.* 79 (4):448–453.

80. Chen, L. G., Peng, X. C., Huang, Y. M. et al. 2009. Polychlorinated biphenyls in the atmosphere of an urban city: Levels, distribution, and emissions. *Arch. Environ. Contam. Toxicol.* 57 (3):437–446.

81. Choi, M. P. K., Ho, S. K. M., So, B. K. L. et al. 2008. PCDD/F and dioxin-like PCB in Hong Kong air in relation to their regional transport in the Pearl River Delta region. *Chemosphere* 71 (2):211–218.

8 Organochlorine Pesticides, Polychlorinated Biphenyls, and Polybrominated Diphenyl Ethers in the Indian Atmosphere

Paromita Chakraborty* and Gan Zhang

CONTENTS

8.1 Introduction 179
8.2 Materials and Methods 181
    8.2.1 Sampling Sites 181
    8.2.2 Brief Description of Wetland Sites 181
    8.2.3 Sample Collection.. 181
    8.2.4 Sample Extraction and Analysis.. 185
    8.2.5 Quality Control/Quality Assurance... 186
8.3 Results and Discussion 186
    8.3.1 Back Trajectory Analysis... 188
    8.3.2 Spatial Distribution and Potential Sources of OCPs 188
        8.3.2.1 Spatial Distribution of HCHs.. 188
        8.3.2.2 Spatial Distribution of DDTs 192
        8.3.2.3 Spatial Distribution and Sources of Chlordane 193
        8.3.2.4 Spatial Distribution and Sources of Endosulfan....................................... 195
        8.3.2.5 Spatial Distribution and Sources of HCB 196
        8.3.2.6 Spatial Distribution and Sources of PCBs... 196
        8.3.2.7 Spatial Distribution and Sources of PBDEs... 198
8.4 Overview... 199
Acknowledgments..200
References...200

8.1 INTRODUCTION

Persistent organic pollutants (POPs) include a wide range of xenobiotic chemicals like organo-chlorine pesticides (OCPs), hexachlorocyclohexane (HCHs), dichlorodiphenyltrichloroethane and related compounds (DDTs), and chlordanes. Polychlorinated biphenyls (PCBs), a family of synthetic chlorinated organic compounds, are widespread and have been used widely as flame retardants,

---

* E-mail: parochakraborty@gmail.com (Chapter corresponding author).

coolants, and lubricants in electrical equipment [1]. Polybrominated diphenyl ethers (PBDEs) are flame retardants used in a variety of materials, such as electronic appliances, furniture, and textiles [2]. The atmosphere is recognized as the major transport route for POPs in the environment and the behavior of PCBs in the atmosphere is well understood [3]. Atmospheric transport from the emission source via deposition is the main route for delivery of POPs to aquatic and terrestrial ecosystems. POPs are transboundary pollutants that undergo long-range atmospheric transport (LRAT) from the source of emission to remote regions [4,5]. POPs emitted from primary and secondary point sources in tropical regions undergo volatilization and are transferred through atmospheric movement in the immediate vicinity of POP sources and via LRAT [4–6]. They accumulate in more remote/pristine regions of the globe far away from the source of emission. A coordinated international regulatory framework is now focused on the reduction of POP emissions to the atmosphere [7,8]. India, which is a tropical country under the influence of tropical monsoons yielding heavy rains and fed by plenty of rivers depositing fertile silt, has agriculture as the main source of livelihood. In India, pesticides are used not only for agriculture but also for public health purposes. For nearly four decades after their introduction, India was dominated by OCPs because of their cost benefits and bioefficacy potential. Even in the 1990s, more than 70% of the gross tonnage of pesticides used in agricultural applications in India consisted of formulations that are banned or severely restricted throughout the world [9]. The tropical monsoonal climate often leads to high upwelling of surrounding water bodies, thereby causing heavy damage to the coastal population, which leads to the usage of OCPs like DDT to combat the vector-borne diseases. Though a complete ban on DDTs was imposed in many developed nations, in India it has only been banned from agricultural usage which HCH remains in use (still used for agriculture). India has been permitted to use DDT up to 10,000 t/year for its vector control practices, under the Stockholm Convention, until an alternative can be found [7]. Though stockpiles of obsolete POPs exist, posing a major threat to the environment, there is no known government program to monitor them. In India, the annual consumption of pesticides is approximately 85,000 metric tons, of which DDTs, HCHs, and malathion accounted for 70% [9]. Most biomonitoring studies and food products have been reported with higher levels of DDT and γ-HCH (lindane) than other parts of the world [10–12]. At present, India generates approximately 150,000 t/year of WEEE (i.e., quantity of electrical and electronic waste). Mumbai, Delhi, and Bangalore, respectively, produce 11,017, 9,730, and 4,648 t of electronic waste (e-waste) annually.

There is currently a paucity of reliable environmental data on the levels of most POP chemicals in the Indian atmosphere from which to assess the effectiveness of international efforts to minimize the release of these chemicals to the environment. The recent efforts of the global atmospheric passive sampling (GAPS) study provided substantial information about the global distribution of POPs [13]. A PUF-PAS (polyurethane foam disk passive air sampler) sampling campaign in Asia including China, Japan, Korea, and Singapore has been also conducted [14]. However, neither of these studies reported data for South Asia, in particular India, where data on POPs in the atmosphere has been sparse. In order to address these needs, a project was designed using the cost-effective and simple passive air sampling (PAS) technique that made use of polyurethane foam (PUF) disks as a method of sample collection. PUF disks of 14 cm diameter and 1.2 cm thickness have been tested as passive samplers previously and are known to sample POPs at a rate of a few cubic meters of air per day when calibrated against an active sampler [15]. With appropriate instrumental detection limits and low blank values, this allows for the detection of many classes of POPs after several weeks of exposure in ambient air. Passive air samplers using polyurethane foam disk (PUF-PAS) are one of the most improved techniques in recent years for monitoring of POPs. The feasibility of coordinated sampling using PUF-PAS at regional scales has been demonstrated in the Great Lakes region of North America, and in South America [16] and Europe [17]. PUF-PAS was therefore deployed along the east and west coasts at 19 locations in India during 2006 to elucidate the temporal and spatial trends and identify possible sources of organochlorine pesticides, polychlorinated biphenyls, and polybrominated diphenyl ethers in the industrialized coastal metropolitan and large cities (as urban sites) as well as rural sites and wetland (or background) sites.

8.2   MATERIALS AND METHODS

8.2.1   Sampling Sites

The Indian coastline is about 7517 km in length, of which about 5423 km is the mainland and 2094 km consists of the Andaman, Nicobar, and Lakshadweep Islands. The Eastern Coastal Plain of India is a wide stretch of land lying between the Eastern Ghats and the Bay of Bengal. The Western Coastal Plain is a narrow strip of land sandwiched between the Western Ghats and the Arabian Sea. The Indian Ocean surrounds the southern portion of India. Nineteen locations were selected on the basis of

1. Urban sites, including highly populated and industrialized cities and commercially growing large cities
2. Rural sites influenced by localized agricultural activities
3. Wetland regions on the east, west, and southeast coasts (refer to Figure 8.1)

Wetlands are located in the estuarine regions and are presumed to be pristine if not affected by the surrounding ambient media or human interference.

8.2.2   Brief Description of Wetland Sites

The coastal zone of India is endowed with the presence of extensive and diverse mangrove wetlands. Sunderban (SB), the largest mangrove in the world, is situated at the mouth of the Bay of Bengal in the Ganga–Brahmaputra delta. Sunderban covers an area of around 10,000 km². Of this, 4262 km² is in India and the rest is in Bangladesh. Of the 60 varieties of mangroves and mangrove associates that are found in India, the Sunderban accounts for 50, many of which are rare. Known for its biodiversity, the region has been identified as a "world heritage site" by the International Union for the Conservation of Nature. The ecological significance of the Sunderban is immense. Apart from serving as a shield against natural calamities, it checks atmospheric pollution and has a seemingly unlimited capacity to absorb pollutants from both air and water.

Pichavaram mangrove wetlands are located in the northernmost end of the Cauvery delta. It is an estuarine type of mangrove situated at the confluence of the Uppanar, a tributary of the Coleroon River. Fishing villages, croplands, and aquaculture ponds surround the area. Two major rivers in this area, the Vellar River and the Coleroon River drain into the Bay of Bengal. The area between the two rivers is identified as brackish water with mangrove vegetation. Nearly 77% of the Gujarat mangroves are confined to the Gulf of Kutch. The bioclimate is hot, with a rather cold season, subdesertic and with very strong average annual thermal amplitude of about 12°C. Further details of individual sites and other sampling parameters are given in Table 8.1.

The principles behind PAS for POPs have been discussed in detail previously [15,17]. Gas-phase compounds partition into the sampling medium during the uptake phase and approach equilibrium if the exposure time is long enough, which is in turn a function of temperature. Depending on the deployment conditions, the uptake rate may be affected by wind speed (i.e., air-side resistance can limit uptake rates).

8.2.3   Sample Collection

The design and deployment of the PUF-PAS has been described in detail elsewhere [14]. PUF disks were precleaned using an accelerated solvent extractor (ASE 300, Dionex, Sunnyvale, CA) at the State Key Laboratory of Organic Geochemistry in Guangzhou, China, and transferred to the sampling locations in sealed, solvent-cleaned amber glass jars. The samplers were assembled at the deployment sites to avoid contamination during transit. PUF-PASs were deployed at each site for

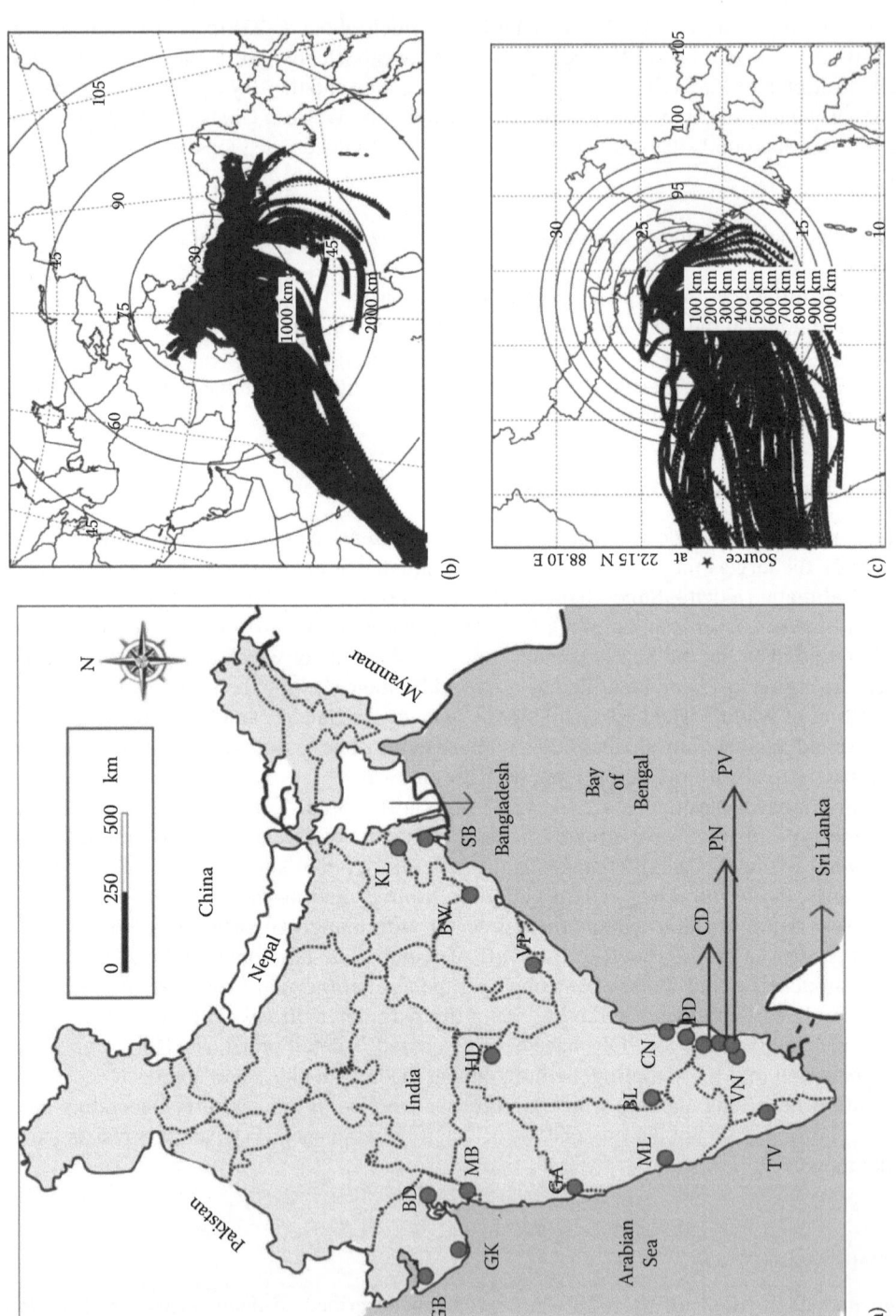

FIGURE 8.1 (a) Sampling sites in India. (b) Representative back trajectories for India during July 30 to September 26, 2006. (c) 5-Day back trajectory analysis ending at Sunderban during sampling period.

TABLE 8.1
Details of the Sampling Location, Duration of Sampling, and the Mean Ambient Temperature during the Sampling Period

ID	Sampling Sites	Latitude	Longitude	Description	Duration	Mean Temperature (°C)
GB	Gujarat-Bhuj	23°26′N	69°66′E	Arid region located at extreme west at a very close proximity to Arabian Sea	August 1–September 18, 2006	28.5
GK	Gulf of Kutch	22°15′N	69°00′E	A seasonally marshy wetland region located in the Thar desert	August 6–September 23, 2006	28
BD	Baroda	22°18′N	73°12′E	A flourishing industrial city in the west coast of India with the establishment of large-scale petrochemical, pharmaceutical, textile and machinery industries	August 2–September 20, 2006	27
MB	Mumbai	18°58′N	72°49′E	The most populous, metropolitan, industrialized, port city in the west coast of India has a deep natural harbor and the port handles over a significant amount of cargo	August 4–September 21, 2006	26.5
GA	Goa	15°24′N	73°47′E	A coastal city on the Western Ghats range with large-scale mining and fishing units apart from tourism as the primary industry	August 5–September 22, 2006	26.5
ML	Mangalore	12°52′N	74°52′E	A coastal city with Western Ghats to its east and its economy is dominated by agricultural processing and port related activities, and is a home to automobile leaf spring industry	August 5–September 22, 2006	
BL	Bangalore	12°58′N	69°66′E	A cosmopolitan city located on the Deccan Plateau evolved into manufacturing hub for public sector heavy industries, IT hub of India and is referred as "Silicon Valley of India"	30.7.06–September 24.9.06	24.5

(*continued*)

TABLE 8.1 (continued)
Details of the Sampling Location, Duration of Sampling, and the Mean Ambient Temperature during the Sampling Period

ID	Sampling Sites	Latitude	Longitude	Description	Duration	Mean Temperature (°C)
TV	Thiruvananthapuram	08°28′N	69°66′E	A city sandwiched between Western Ghats and Arabian Sea at the extreme southern end of India. Large-scale industrial establishments are low compared to other cities of south India	August 6–September 23, 2006	27
KL	Kolkata	22°34′N	88°22′E	A metropolitan city linearly spread along the bank of Hugly and is home to many industrial units ranging from electronics to jute	August 2–September 19, 2006	29
SB	Sunderban	22°15′N	88°10′E	Largest mangrove of the world covering an area of about 1 mha, of which 60% is located in Bangladesh and the remaining western portion, comprising 40%, lies in India. The Indian part of the Sunderbans is located in the western part of the Ganges–Brahmaputra delta	August 9–September 26, 2006	29
BW	Bhubaneswar	20°13′N	85°49′E	A city on the eastern coast of India is a bustling centre for commerce and is a developing IT hub of India	August 3–September 20, 2006	28
VP	Visakhapatnam	17°41′N	83°17′E	A port city located on the eastern shore of India nestled among the Eastern Ghat Hill ranges and facing the Bay of Bengal in the east. It is home to several state owned heavy industries, has one of country's the largest ports and its oldest shipyard and the only natural harbor on the eastern coast of India	August 4–September 24, 2006	26.5

TABLE 8.1 (continued)
Details of the Sampling Location, Duration of Sampling, and the Mean Ambient Temperature during the Sampling Period

ID	Sampling Sites	Latitude	Longitude	Description	Duration	Mean Temperature (°C)
HD	Hyderabad	17°22′N	78°28′E	A metropolitan city located at Deccan plateau in the Telangana region with rapid growth of IT, IT enabled services, and pharmaceutical industries	August 11–October 1, 2006	29.5
CN	Chennai	13°04′N	80°16′E	A metropolitan coastal city of south India lying on the thermal equator. It is a large commercial and industrial centre with a base of automobile industry, software services and hardware manufactures	July 31– September 20, 2006	29
PD	Pondicherry	11°55′N	79°49′E	The Union territory in the east coast of south India	August 1–September 20, 2006	29.5
CD	Cuddalore	11°45′N	79°45′E	A port town with agriculture as the mainstay and marine fishing being a lucrative trade	July 31– September 17, 2006	29.5
PN	Portonovo	11°28′N	79°46′E	A fishing village on the southeast coast of India	August 2–September 21, 2006	29.5
PV	Pichavaram	11°27′N	79°47′E	A tropical mangrove belt in southeast India	August 2–September 21, 2006	29.5
VN	Vedharanyam	10°22′N	79°50′E	A tropical mangrove belt in southeast India	August 30–September 19, 2006	29.5

about 6 weeks between July 30 and September 26, 2006. At the end of the deployment period, the PUF disks were retrieved, resealed, and returned to Guangzhou where they were stored frozen until extraction and analysis.

8.2.4 Sample Extraction and Analysis

A mixture of surrogate standards of 2,4,5,6-tetrachloro-*m*-xylene (TCmX), decachlorobiphenyl (PCB209), $^{13}C_{12}$-PCB138, and $^{13}C_{12}$-PCB180 was added to each of the samples prior to extraction. Activated copper granules were added to the collection flask to remove potential elemental sulfur. Target organic compounds adsorbed in the PUF disks were Soxhlet extracted with dichloromethane (DCM) for 18 h. The extract was concentrated, solvent-exchanged with hexane and purified on an 8 mm i.d. alumina/silica column packed, from the bottom to top, with neutral alumina (6 cm, 3% deactivated), neutral silica gel (10 cm, 3% deactivated), 50% sulfuric acid silica (10 cm), and anhydrous sodium sulfate. The column was eluted with 50 mL of DCM/hexane (2:3) to yield the organochlorine fraction. The fraction was concentrated to 25 mL under a gentle high purity nitrogen stream after 25 mL of dodecane was added as a solvent keeper. A known quantity of pentachloro-nitrobenzene (PCNB) was added as an internal standard prior to gas chromatography-mass spectrometry (GC-MS) analysis.

PCBs and OCP analysis was carried out on a Thermo Finigan-TRACE GC-MS system with a CP-Sil 8 CB capillary column (50 m, 0.25 mm, 0.25 mm), operating under single ion monitoring (SIM) mode. Helium was used as the carrier gas at 1.2 mL min^{-1} under constant-flow mode. The oven temperature began at 60°C for 1 min and increased to 290°C (10 min hold time) at a rate of 4°C min^{-1}. Splitless/split injection of a 1 mL sample was performed with a 5 min solvent delay time. The injector temperature was at 250°C.

The inlet degradation of DDT was checked daily and controlled within 15% before injecting and analyzing field samples. PBDEs were analyzed separately on a Fisons MD 800 GC MS instrument with a negative chemical ionization source in SIM mode using ammonia as the reagent gas. Details of the instruments, GC temperature programs, and monitored ions have been reported elsewhere [17]. A total of 10 OC compounds (HCH, HCB, o,p'-DDT, p,p'-DDE [DDE, dichlorodiphenylethylene], p,p'-DDT, and p,p'-DDD [DDD, dichlorodiphenylethane], cis- and trans-chlordane) 28 PCB congeners (PCB-8, -28, -37, -44, -49, -60, -66, -70, -74, -77, -82, -87, -99, -101, -105, -114, -118, -126, -128, -138, -153, -158, -166, -169, -179, -180, -18, -187) and 9 PBDEs (PBDE-28, -47, -66, -85, -99, -100, -153, -154, -183) regularly detected in samples, were quantified using an internal standard method. The present data focuses on the International Council for the Exploration of the Seas (ICES) congeners (PCB-28, -52, -90/101, -118, -138, -153/132, and -180), Σ_{28}PCBs, o,p'-DDT, p,p'-DDE, p,p'-DDD, p,p'-DDT, α-, β-, γ-, and δ-HCH, HCB, chlordanes, endosulfan and Σ_9PBDEs.

8.2.5 QUALITY CONTROL/QUALITY ASSURANCE

All chemical standards were purchased from AccuStandard Co., New Haven, CT. All analytical procedures were monitored using strict quality assurance and control measures. Laboratory and field (i.e., samplers sent to/from field sites unopened) blanks consisting of preextracted PUF disks were extracted and analyzed in the same way as the samples. Analytical blanks consisted of six field and three laboratory blanks. There were no significant differences (t-test, $p > 0.05$) between analyte concentrations in the laboratory and field blanks, indicating that contamination was negligible during transport, storage, and analysis. Method detection limits (MDLs) were derived from the blanks and quantified as three times the standard deviation of the mean blank concentrations. In addition, peaks were only integrated when the signal-to-noise ratio was >3 otherwise they were considered nondetects. Recoveries were between 67% and 135% for all the compounds studied. Reported values were recovery and field blank corrected.

8.3 RESULTS AND DISCUSSION

The amounts of POPs sequestered over the sampling period were converted to estimated air concentrations using typical sampling rates of ca. 3–4 m^3 of air per day (in this case using the average value of 3.5 m^3 day^{-1}), as derived from previous calibration studies against active samplers [18]. The sampling period used in this study was 6 weeks so it can be assumed that over the study period the PUF disk was operating in the "kinetic" phase [15] although uptake of hexachlorobenzene (HCB; log $K_{oa} = 7.4$) during the 6-week sampling period is expected to have approached/reached equilibrium via saturation in the PUF [15]. Hence, for HCB we used a sampling rate of 2.6 m^3 day^{-1} based on the recent calibration data [18].

Atmospheric concentrations of POPs estimated in this study showed varying levels of contamination among the sampling sites (Table 8.2). Levels of Σ_{28}PCBs and Σ_9PBDEs varied by ~3 orders of

TABLE 8.2

Concentrations of OCPs, PBDEs, and PCBs in Air Samples Collected from Sites in India (in pg m^{-3})

	Urban Sites ($n=9$)			Rural Sites ($n=6$)			Wetlands ($n=3$)			Sunderban ($n=1$)
	Min	Max	Mean	Min	Max	Mean	Min	Max	Mean	
PCB28	18	140	76	31	175	66	9	47	22	87
PCB52	1	66	31	0.06	137	27	2	23	9	6
PCB99/101	BDL	56	16	BDL	116	44	3	18	9	15
PCB118	BDL	19	7	BDL	2	BDL	BDL	1	BDL	BDL
PCB153	0.38	16	8	0.46	3	2	2	17	7	1
PCB138	0.38	77	22	0.29	9	2	1	9	7	1
PCB180	0.09	8	2	0.07	1	1	BDL	15	6	1
Σ_{28}PCBs	216	1077	662	279	805	464	120	320	238	544
α-HCH	22	1691	451	12	167	53	20	31	25	394
β-HCH	BDL	149	36	1	28	16	11	27	17	141
γ-HCH	135	3562	909	31	437	174	34	100	61	537
δ-HCH	8	214	75	4	42	16	BDL	7	3	47
ΣHCH	210	5404	1471	71	645	259	66	144	105	1120
HCB	32	290	136	28	199	79	29	45	38	199
α/γ HCH	0.11	4	0.15	0.17	0.47	0	0.31	1	BDL	1
β/α HCH	BDL	0.53	0.15	BDL	2.00	0.70	0.36	1.00	1.00	0.36
o,p'-DDT	32	768	326	BDL	307	100	BDL	227	101	1667
p,p'-DDT	2	220	83	3	55	19	9	39	23	122
p,p'-DDE	62	2061	541	11	135	53	6	15	12	609
p,p'-DDD	2	56	32	3	39	18	BDL	16	9	118
ΣDDT	194	2952	981	30	488	190	16	298	145	2517
p,p'-DDD+p,p'-DDE/ p,p'-DDT	2	72	14	2	8	5	1	1	1	6
o,p'-DDT/p,p'-DDT	2	19	7	BDL	20	9	BDL	6	3	14
CC	7	340	62	3	39	15	3	12	7	36
TC	16	230	89	9	55	28	3	11	7	81
ΣChlordane	29	921	251	15	249	108	9	28	18	188
ENDO-I	3	680	264	1	369	127	BDL	12	5	19
ENDO-II	19	143	76	BDL	160	53	BDL	10	6	30
ΣEndosulfan	22	761	341	1	465	180	BDL	22	11	49
PBDE-28	BDL	6	1	BDL	BDL	BDL	BDL	BDL	BDL	0.44
PBDE-47	BDL	72	12	BDL	3	1	BDL	1	1	2
PBDE-66	BDL	3	1	BDL	1	BDL	BDL	BDL	BDL	0.15
PBDE-100	BDL	7	1	BDL	BDL	BDL	BDL	BDL	BDL	0.11
PBDE-99	BDL	27	4	BDL	1	BDL	BDL	1	BDL	1
PBDE-85	2	61	13	2	8	5	1	4	2	5
PBDE-154	BDL	4	1	BDL	1	BDL	BDL	BDL	BDL	0.07
PBDE-153	BDL	1	BDL	BDL	BDL	BDL	BDL	BDL	BDL	0.03
PBDE-183	BDL	1	BDL	BDL	BDL	BDL	BDL	BDL	BDL	0.08
Σ_9PBDE	4	181	69	3	13	8	1	6	3	8

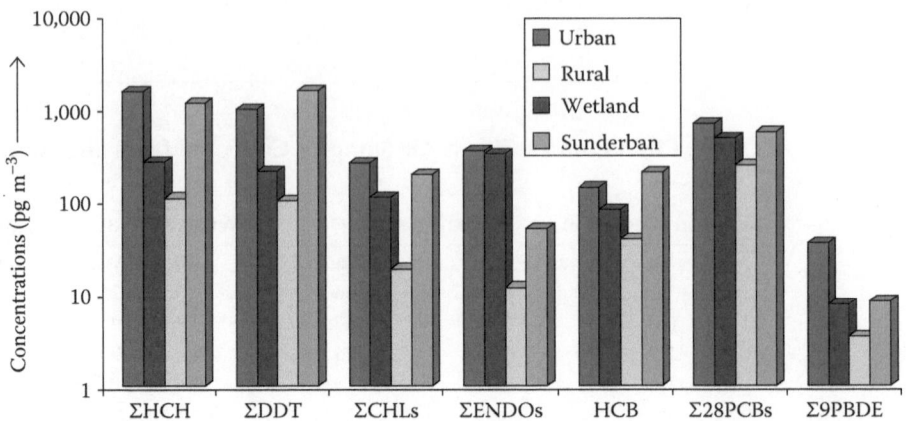

FIGURE 8.2 Distribution of the average concentrations of HCHs, DDTs, CHLs, ENDOs, HCB, PCBs, and PBDEs in India in log scale.

magnitude along the data range, showing an urban–rural gradient with the highest values at the core urban sites and the lowest values in the remote wetland sites (Figure 8.2). Among the wetland locations Sunderban wetlands had significantly higher concentrations. PCBs tend to have an elevated urban signal compared to many other POPs [17,19,20]. Indian Sunderban, located between the city of Kolkata and Bangladesh, was found to have elevated levels of POPs and thus has been reported separately.

8.3.1 Back Trajectory Analysis

In order to assess the possible sources of POPs in the air samples, the HYSPLIT model (Hybrid Single-Particle Lagrangian Integrated Trajectory, Version 4.7), a comprehensive modeling system developed by the National Oceanic and Atmospheric Administration (NOAA) Air Resource Laboratory, Maryland was used. Five-day back trajectories ending in India at 0600 UTC, that is, 11:30 local during July 30–September 26, 2006 was calculated (Figure 8.1b). In order to classify the air masses, the trajectories ended at the height of 500 m AGL (above ground level), a level of about half the height of the mean day time planet boundary layer (PBL), to represent general transport conditions in the PBL. To identify the possible source route of POPs at Sunderban, a similar 5-day back trajectory was also calculated. The air mass during the sampling period encircled most of a 500 km grid along the eastern part of India before ending at Sunderban (Figure 8.1c). Backward trajectory analysis therefore shows that the POPs detected in the wetland region of Indian Sunderban were influenced by large, regional-scale air mass movement at the time of sampling, especially from eastern parts of India where high levels of these compounds have been observed during this study.

8.3.2 Spatial Distribution and Potential Sources of OCPs

8.3.2.1 Spatial Distribution of HCHs

Technical HCH and lindane are two formulations of HCH and the pattern of HCH isomers in the environment is largely dependent on the formulation used. ΣHCH values in the air samples ranged between 66 and 5404 pg m^{-3}, while α-HCH levels varied between 12 and 1691 pg m^{-3}. Iwata et al. reported considerably higher concentrations of HCH over 10,000 pg m^{-3} in the air of the Bay of Bengal and Arabian Sea [21]. γ-HCH, the main component of lindane, was once a globally used pesticide. Spatial trends for γ-HCH in India are similar to those observed for α-HCH but with much higher concentrations contributing to 61% of total HCH concentration (Figure 8.3). The highest concentrations of α- and γ-HCH were measured at CN, 1691 and 3562 pg m^{-3} respectively. The urban sites and wetland region of Sunderban also showed higher concentrations of both of these isomers.

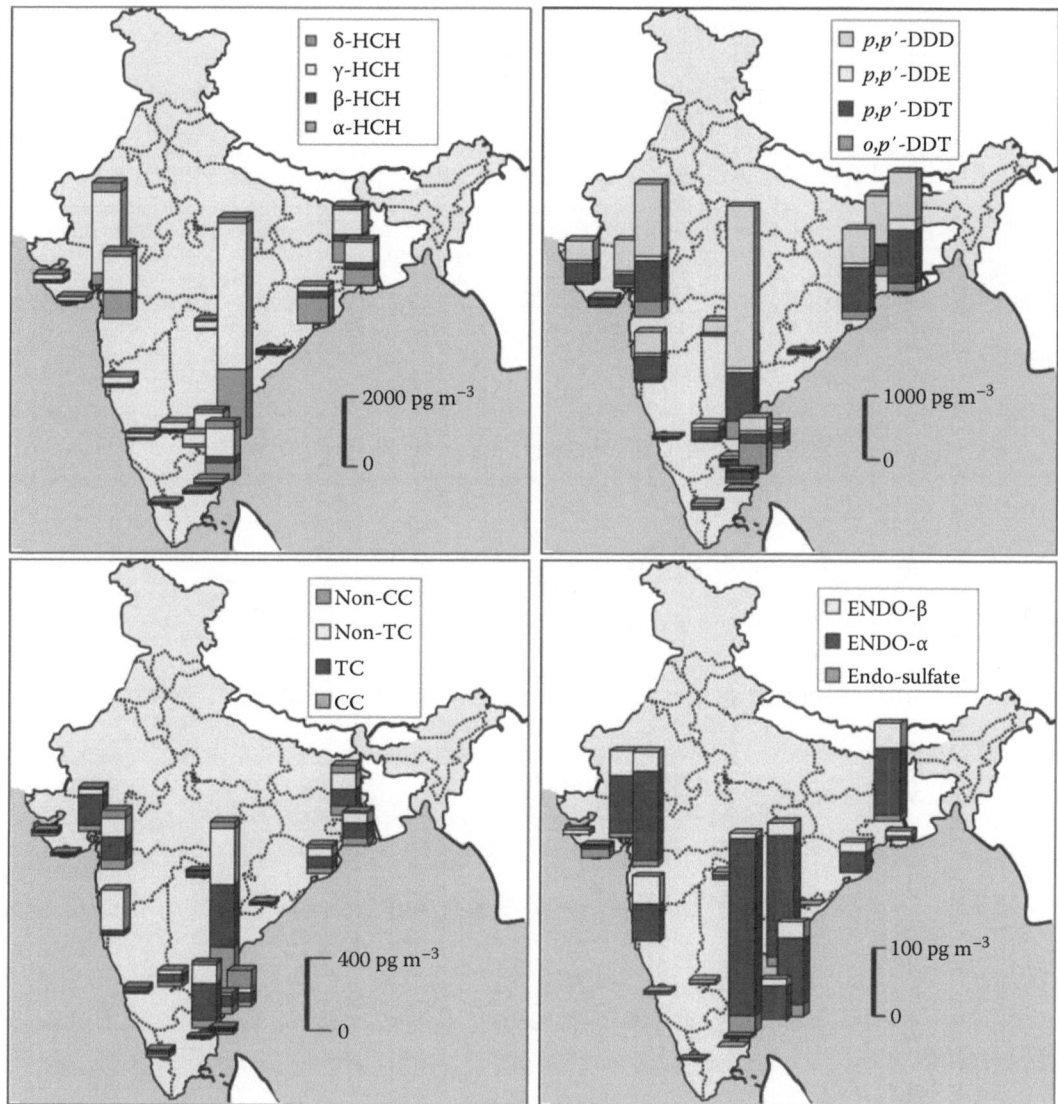

FIGURE 8.3 Spatial distribution of OCPs in air samples in India.

Detected β-HCH and δ-HCH were below the detection limits (BDL)-149 pg m^{-3} and BDL-214 pg m^{-3} respectively. Elevated concentrations of α-HCH have previously been attributed to the ongoing use of technical HCH (α = 60%–70%; β = 5%–12%; γ = 10%–12%; δ = 6%–10%; ε = 3%–4%) [22,23] but the isomeric composition in the present study was α = 30%; β = 4%; γ = 60%; δ = 6%, indicating current use of lindane in India. A plausible reason for the highest concentration of γ-HCH occurring at CN is its use in the surrounding agricultural fields. The sampling time coincided with the main cropping season of rice paddies followed by black and green gram. The application of HCH (10% dust) is the possible source of the higher concentrations encountered in air due to HCH release into the atmosphere through evaporation/volatilization, aerial drift, and suspension in aerosols [24].

Concentrations of ΣHCHs decreased at GA, MB, and KL when compared to previous records [25] but not at CN and BL and the range is higher than those reported by other PUF-PAS studies across the globe (Table 8.3). In the early 1990s, a higher percentage of α-HCH followed by γ-, β-, and

TABLE 8.3

A Comparative Study of the Present Study with Various Studies across the Globe Using PUF Disks as Passive Air Samplers

Sampling Sites	Year	ΣPCBs	α-HCH	γ-HCH	HCB	o,p'-DDT	p,p'DDT	p,p'-DDE	ENDO-I	TC/CC	ΣPBDEs
India[a]	Present study										
Kolkata (urban)		689	513	625	164	262	124	608	372	2	9
Mumbai (urban)		813	637	912	177	524	188	925	498	3	4
Baroda (urban)		688	300	1975	133	135	39	395	317	5	14
Chennai (urban)		644	1691	3562	290	620	220	2061	680	1	228
Bangalore (urban)		1077	42	185	54	211	11	62	3	2	7
Pondicherry (urban)		216	167	437	199	129	55	89	369	3	78
Cuddalore (rural)		444	58	231	97	41	27	48	185	1	10
Portonova (rural)		805	410	686	144	99	387	161	992	5	3
Pichavaram		274	23	50	41	227	39	15	4	1	3
Vedaranyam		120	20	34	29	BDL	9	6	0	2	7
Gulf of Kutch		320	31	100	45	77	21	14	12	2	2
Sunderban		544	394	537	199	1667	122	609	19	2	9
Urban-rural trend, Toronto [20]	2000										
Gage		547	55.2	52.2	X	21.7	27	60	370	1.1	
Junction		343	40.8	29.1	X	13	14.4	33	408	1.1	
Downsview		167	41.4	25.6	X	8	10.5	30	405	1	
R.hill (A)		104	51.7	22.3	X	13.7	17.4	39	465	0.9	
Aurora		104	39.8	23.9	X	9.7	11.7	27	254	0.9	
Egbert		116	61.5	40.2	X	40	52.9	305	817	0.9	
Europe [17]	2002	20–1700	<14–100	9–390	11–50		0.6–190		<0.4–2.5		0.5–250
The Great Lakes Basin [64]	2002–2003	14–900	15–73	13–100					40–1090		<3–37
Global Pilot Study[b] [47]	2002–2004										
Alert			41	8					52	0.3	
Fox lake yukon			106	18					61	0.3	
Whistler mountain,			30	6.1					116	0.5	
Bermuda			3.1	0.32					22	0.6	
Cape grim			1.4	0.78					223	0.6	

	Year										
Chile			5	7						36	0.5
Toronto (urban)			60	79						104	1.2
GAPS study^c [13]	2004–2005										
Australasia		BDL-2826	BDL-2	BDL-2			BDL-190	BDL-24	17-66	BDL-1.3	BDL
Asia		20-497	4-145	10-68			BDL-131	BDL-58	9-101	0.3-1.1	BDL-17
Europe		5.7-644	0.8-36	39,134			BDL-36.3	BDL-192	21-1760	BDL-0.3	BDL-4.7
Africa		BDL-252	BDL-117	0.4-67			BDL	BDL-2	9-564	BDL-1	BDL-6
North America		6-196	1.3-102	0.6-17			BDL	BDL-232	7-129	0.3-0.9	BDL-24
Central America		BDL-32	BDL-1	2-5.5			BDL	BDL	22-39	0.7-0.8	1.0-6.0
South America		BDL-10	BDL-2	0.4-9			BDL	BDL-4	29-14,600	BDL-0.4	BDL-8
Polar region		BDL-58	BDL-48	BDL-9			BDL	BDL-6	BDL-60	BDL-0.6	BDL-5.3
Asia [14]	2004										
China		21-336			10.4-462	4.7-472	2-928	2.8-380			<0.13-340
Singapore		5.0-31			9.5-24.5	1.3-53	<1.9-16	<1.5-10			10.0-29
Japan		7-247			14-95	1.4-70	4.4-146		1.62-544		5.0-71
South Korea		12-84			26-136	<0.3-14	<1.9-20	<15-25			2.0-27
Pearl River Delta, South China [18]	2005										
PRD			X	X	27-216	90-1200	90-2300	50-1700		0.4-1	
Mainland			X	X	40-216	90-1200	120-2300	110-700		0.4-0.8	
Hong Kong			X	X	27-135	110-250	90-1400	50-1700		0.7-1	
China (summer) [33]					48-3790	BDL-6620	BDL-2730	24-3650	BDL-170	0.07-1.6	

a Concentration of present study.
b Mean concentration.
c Mean concentration range of all the sampling sites in each continent.

δ-isomers in atmospheric air was reported in sites similar to PV and CD of the present study [24,26]. Increased levels of HCHs in the breast milk samples collected from Chennai (site-CN) compared to Porto Novo (site-PN) was found due to the current use of a significant amount of HCHs in Chennai for vector control by the health department or in home gardens by the public [27]. γ-HCH concentrations were increased at PV versus previously reported data [25] but the α-HCH concentration was reduced. Urban sites showed higher concentration of both isomers. Faster disappearance of HCHs from the Indian soils did not necessarily mean they were highly degraded. Approximately 95% of their residues were evaporated into the atmosphere [28]. In dumping sites, increased levels of HCHs in soil likely cause disequilibrium in the air–soil interface, thus favoring their emission from the soil.

8.3.2.1.1 Sources of HCHs

HCH technical mixtures have been produced and largely used in India until very recently [9], leading to increased HCH levels in various environmental media. Use of α-HCH in India increased constantly before 1970 and from 1980 to 1990, peaking at 36,000 t year^{-1} in 1990, and remaining at approximately 17,000 t year^{-1} since 1991. Use of γ-HCH in India reached 7700 t year^{-1} in 1990 and has remained at approximately 600 t year^{-1} since 1991 [29]. The Government of India is now encouraging the usage of lindane (γ-HCH) for pest control after banning the technical mixture containing all four isomers in 1997 [9]. The ratio of α- to the γ-isomer can be used to identify the source of HCH contamination and may be also used as an indicator of LRAT. The α:γ HCH values in the present study were only higher than the technical HCH ratio (3–7) [21] at site BW (4.35). At all other locations the ratio was below unity. The range is comparable to Toronto (1–2) [20] and GAPS data (1–10) [13]. Such a low ratio implies the recent use of lindane whereas in earlier studies α:γ-HCH ratios of 4.8–9.6 in the Bay of Bengal and Arabian Sea, and 0.65–2.4 for the Eastern Indian Ocean were observed and account for the relatively high use of the technical HCH mixture in the past [30]. The levels of HCH isomers at Sunderban (α = 394; β = 141; γ = 537; δ = 47) reflects the ongoing use of technical HCH in addition to the sporadic use of lindane in India.

8.3.2.2 Spatial Distribution of DDTs

The spatial distribution of DDTs (ΣDDT, sum of p,p'-DDT, o,p'-DDT, p,p'-DDE, and p,p'-DDD), derived from this study ranged between 16 and 2952 pg m^{-3} with the highest level at CN (Figure 8.3). The probable reason for the higher concentration of DDT at CN may be due to the indiscriminate spraying of DDT in areas of South India and the Andaman and Nicobar Islands before and after the Asian tsunami period to prevent vector-borne diseases [31]. The other sites on the east coast at a proximal distance from CN in South India (PD, CD, PN) showed a marked rise in DDT concentrations compared to previous studies [24], possibly due to their usage in vector control programs. High ΣDDT levels have been reported in blubber tissues of cetacean specimens from southeast coast of India [27,32] and in the mother's milk samples at CN [27]. The atmospheric concentration in the present study is much higher than those observed in Europe and other Asian countries like Singapore, Japan, and Korea using identical PUF-PAS [17], but is lower than those measured in China [14] (Table 8.3). The concentration range is lower than the summer concentration of the Pearl River Delta area in southern China [18] and much lower than mainland China [33]. Present levels in India are much lower than those observed in early 1990s in most of the sites similar to those reported in this study [25].

The mean concentration of p,p'-DDT values ranged between 2 and 387 pg m^{-3} and contributed 10% of ΣDDT concentrations, a range that is higher than the reported data for Europe (0.6–190 pg m^{-3}) [17] but less than those reported for China during a similar sampling time (Table 8.3) as well as Hong Kong (90–1400) [18]. The o,p'-DDT values ranged from BDL to 1667 pg m^{-3}, contributing 41% of ΣDDT concentrations. The highest level was observed at Sunderban (1667 pg m^{-3}) followed by BW (768 pg m^{-3}), CN (620 pg m^{-3}), and MB (524 pg m^{-3}). p,p'-DDE (a metabolite of p,p'-DDT) values ranged between 6 and 2061 pg m^{-3} contributing 44% of ΣDDT. The highest level of p,p'-DDE was measured at CN (2061 pg m^{-3}) followed by KL (608 pg m^{-3}), Sunderban

(609 pg m^{-3}), and MB (925 pg m^{-3}) on the west coast. Sarkar et al. reported p,p'-DDE as the most detected metabolite from the Arabian Sea along the west coast of India [34]. Sediment samples near major cities like Chennai (site-CN), Pondicherry (site-PD), and Cuddalore (site-CD) also contained high DDT levels [35]. In this study, the concentration of the metabolite of DDT (i.e., p,p'- DDE) was found to be more than that of the parent compounds (i.e., p,p'- and o,p'-DDT), which is consistent with a previous study [24]. Iwata et al. observed higher concentrations of the parent compound (i.e., p,p'-DDT) than its metabolites at a similar site Parangipettai (site-PV) [25]. Moreover, Ramesh et al. also estimated a higher concentration of p,p'-DDT than its metabolites in Portonovo (Parangipetti, site-PV) [26].

The mangrove regions have been marked with lower ΣDDT levels, especially GK (125 pg m^{-3}) in the extreme west coast and at PV (286 pg m^{-3}) and VN (15 pg m^{-3}) in the southeast coast with exceptionally higher concentration at Sunderban (2517 pg m^{-3}) located adjacent to the mouth of Bay of Bengal. High DDT concentrations have been reported in the mother's milk samples collected from the dumping ground of Kolkata (site-KL) [27] but surface sediment sample along the stretch of Hoogly estuary and Sunderban (the locations adjacent to our study areas) were found with lower concentration of DDTs [36,37]. This level of DDT at Sunderban may be due to an atmospheric efflux of DDT to Sunderban from Kolkata and its surrounding regions and also possibly from Bangladesh where it is still used [38].

8.3.2.2.1 Sources of DDT

The use of DDT in agriculture has been banned in India since 1989. DDT use is restricted to health programs such as the National Malaria Program (NAMP) which used 3750 t of DDT in 2001 in rural and periurban areas [9].

The higher ratio of o,p'-DDT/p,p'-DDT (0–20) (Figure 8.4) is unusual since no dicofol use has been reported in India unlike China where it is a major source of DDT [39]. A higher o,p'-DDT/p,p'-DDT ratio indicates a fresh input of DDT whereas the high p,p'-DDE + p,p'-DDD/p,p'-DDT ratio may be due to transformation of p,p'-DDT to its metabolites (p,p'-DDE + p,p'-DDD), thereby reflecting an aged DDT source such as previously treated agricultural soils or DDT application to prevent vector-borne diseases. India faces coastal hazards, especially due to heavy monsoonal rain almost every year hence DDT usage is prevalent. For example, around 10,000 L of concentrated DDT was sprayed over the dumping ground and 1,000 L all over Mumbai within 30 days after the large-scale flooding in 2005 (site-MB) [40] and after the 2004 Asian tsunami [31]. The vapor pressure of o,p'-DDT is 7.5 times greater than p,p'-DDT [41] and p,p'-DDT is formed more rapidly in subtropical soils [42]; therefore the amounts of p,p'-DDT volatilized from the soil surface may be relatively small compared to those of o,p'-DDT, although some p,p'-DDT may volatilize from soil surfaces, especially right after application. After DDT application ceases, much of the DDT may be converted to p,p'-DDE. Higher concentrations of p,p'-DDE have been interpreted as a result of its conversion to p,p'-DDE by UV radiation after prolonged exposure in the environment during atmospheric transport [43]. Thus, with the lower concentration of p,p'-DDT, concurrently higher levels of p,p'-DDE have been observed in all the sites except Porto Nova (site PV) where recent use of the technical DDT is suspected. The higher o,p'-DDT/p,p'-DDT ratio at the other sites is possibly due to the faster degradation of the parent compound in the tropical atmosphere. Hence the ΣDDT concentration in India results from the combined effect of past and present use.

8.3.2.3 Spatial Distribution and Sources of Chlordane

Chlordane isomers (*trans*- and *cis*-) were detected at all sites (Figure 8.6). The chlordane concentrations (*trans*-, *cis*-, *trans*-nonachlor, and *cis*-nonachlor) derived from this study (Table 8.2) ranged from 9 to 921 pg m^{-3}, showing higher values in Chennai, Mumbai, and Kolkata compared to other sites. The values are comparatively higher than the previous study in Europe [17], Asia [14], and the GAPS study [13] global background locations (0.7–338 pg m^{-3}) (Table 8.3).

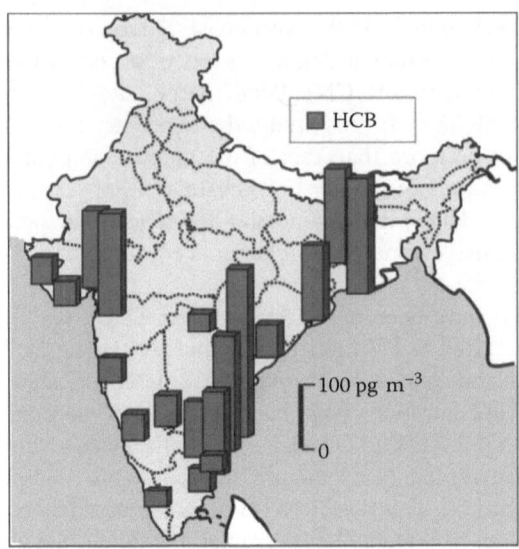

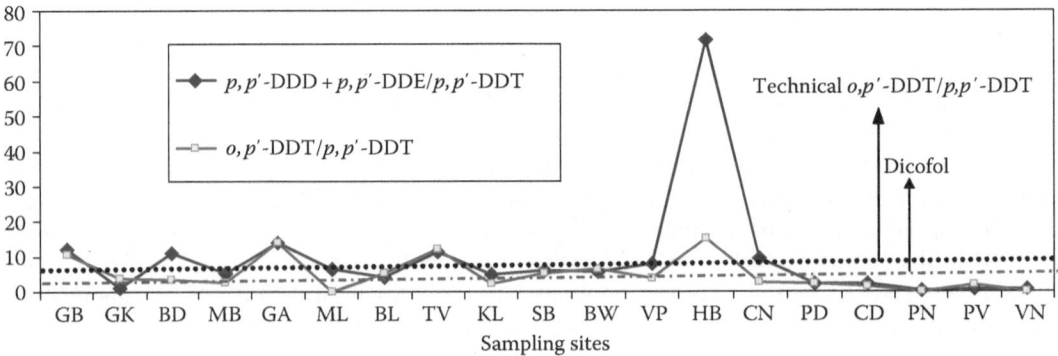

FIGURE 8.4   Variation of o,p'-DDT/p,p'-DDT and p,p'-DDE+ p,p'-DDD/p,p'-DDT in India.

Technical chlordane generally has a TC/CC ratio of 1.2. After a calibration of the vapor pressure difference between TC and CC, the estimated TC/CC ratios in the air as predicted by technical chlordane evaporation (at 25°C) will be 1.6 for the documented world formulation and 1.0 for the local product. The TC/CC ratios of the present study sites ranged from 1 to 5. This high ratio implies that India may use/have used chlordane of domestic formulation [16]. The higher chlordane concentrations in most of the sampling sites indicate fresher chlordane input in the atmosphere, thereby displaying higher TC/CC ratios and significant correlation between TC and CC ($R^2 = 0.7887$, $p < 0.001$) (Figure 8.5). Levels of chlordane found in urban air in tropical Southeast Asia, specifically the Arabian Sea and Bay of Bengal (0.3–14 pg m^{-3}) were considerably higher than levels in open ocean air from the same region [25]. This strongly suggests the ongoing usage of chlordane in India. The highest levels of all of the chlordane isomers were measured at CN (Chennai). The higher chlordane concentrations in most of the sampling sites indicate fresher chlordane input in the atmosphere, thereby displaying higher TC/CC ratios. Previous studies [25] reported much higher concentrations of chlordane in India than this study due to the high amount of past usage. The fresher pattern for chlordane observed at urban sites suggest that the urban burden might be on account of the nondegraded *trans*-chlordane present in heptachlor [44,45].

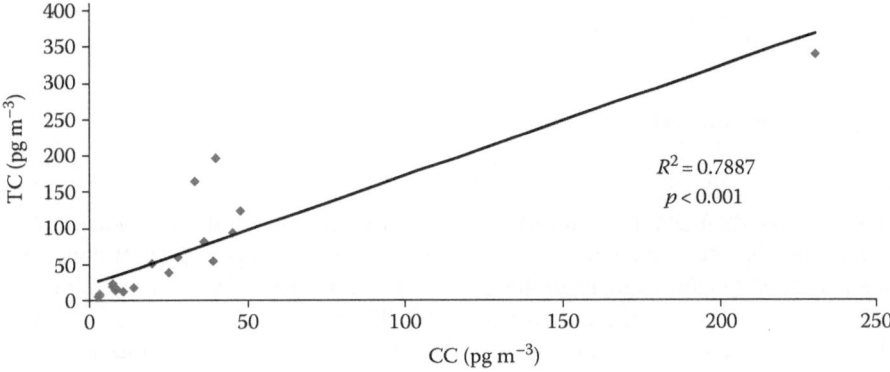

FIGURE 8.5   Correlation between TC and CC.

8.3.2.4   Spatial Distribution and Sources of Endosulfan

Agricultural use of endosulfan in India was estimated to be 5200 metric tons in 1994–1995 [46]. Endosulfan is a broad-spectrum insecticide and an acaricide. It has been widely used in agriculture and forestry throughout the world. India is one of the major producers of endosulfan. Since 1996–1997 it has produced an average of 8,206 MTPA totaling 41,033 MT between 1995 and 2000. Endosulfan consists of two isomers, α and β, in a ratio of 7:3. In this study, the sum of endosulfan isomers and its metabolite (endosulfan sulfate) ranged from 275 to 1122 pg m^{-3} (Figure 8.3), with the highest concentration of α-isomer at Portonovo (site PN). Except for South America (29–14,600 pg m^{-3}) the level is comparable to Europe (21–1760 pg m^{-3}) and higher than rest of the world [13]. The range is also consistent with a global pilot study of PUF sampling for POPs (40–1090 pg m^{-3}) [47]. Endosulfan alone accounts for over 10% of the total insecticide consumption in India, where one of the important crops on which it is used is cotton. It is also used in the cultivation of vegetables and crops like coffee, jute, wheat, and tea and also in cashew plantations. Aerial spraying of endosulfan has been banned recently after a tragic account on a cashew plantation at Kasargod, south India, where endosulfan was sprayed for 24 years, three times every year [48]. The plausible route of high concentrations of endosulfan in the atmosphere is by volatilization from plant and water surfaces. The ratio of the α and β isomers (Figure 8.6) in the present study is quite similar to that of the technical grade and the concentration of endosulfan sulfate has been

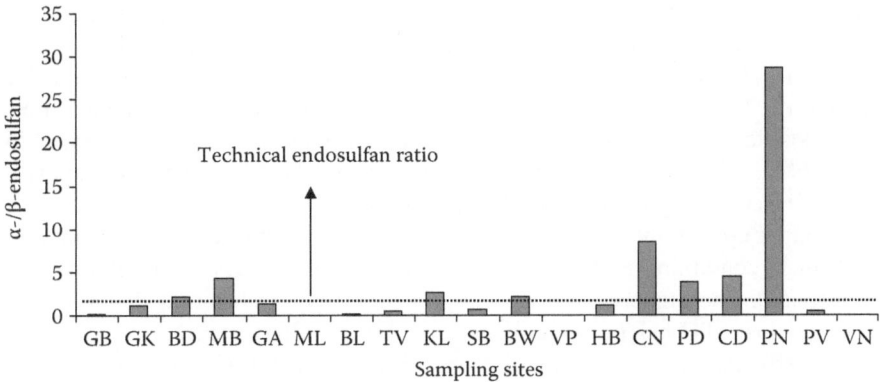

FIGURE 8.6   Ratio of α,β-endo isomers.

found to be very low, thereby implying the ongoing use of technical endosulfan mainly for cashew plantations and tea cultivation by other methods than aerial spraying.

8.3.2.5 Spatial Distribution and Sources of HCB

Concentrations of HCB measured in the present study (28–290 pg m^{-3}) were uniformly distributed (Figure 8.3) in India with the exception of comparatively lower levels measured at the wetland sites excluding Sunderban. Average ambient HCB concentrations in the atmosphere of the temperate northern hemisphere have been estimated to be ~50 pg m^{-3} [49,50]. HCB concentrations in other Asian countries ranged from 10 to 460 pg m^{-3} in China, 10 to 24 pg m^{-3} in Singapore, 14 to 95 pg m^{-3} in Japan, and 25 to140 pg m^{-3} in Korea [14]. The level of HCB in India is comparable to that of China [14]. HCB concentrations in the European atmosphere were quite uniform ranging between 11 and 50 pg m^{-3} [17]. Sources of HCB include several pesticides, which contain chlorine, production of solvents such as perchloroethylene, metal industries, combustion processes, municipal waste incineration and cement production [49]. Recent studies have yielded a profile of global HCB emissions [14] and indicate that no single source type dominates in air, where (largely past) pesticide applications to soil, manufacturing, and combustion processes all influence ambient levels. The main source of input of HCB in Indian coastal cities may be rapid industrial growth. This can be interpreted as evidence of the extremely high volatility and atmospheric persistence of HCB.

An elevated level of surface sediment concentrations of HCB was recorded at Babughat, a site located to the west of the metropolitan megacity Kolkata (site-KL) [36]. High concentrations of HCB in biota have been reported near the industrialized city of Mumbai (site-MB) on the western coast of India [51]. HCB is not only used as a fungicide but is also generated as a byproduct during the production and usage of several agrochemical and industrial chemicals. Furthermore, HCB has also been released into the environment by waste incineration [52] and in a variety of reactions where it persists because of its thermodynamic stability [53]. Hence, the present results may be reflecting the levels of HCB generated in the industrial and highly populous coastal cities of India. The residues of HCB at the rest of the sites might reflect its limited sources and the volatile nature of this compound. Breivik et al. rightly pointed out that HCB emissions can be expected to decrease as developing countries improve their chemical and metal production and handling practices [53].

8.3.2.6 Spatial Distribution and Sources of PCBs

PAS-derived Σ_{28}PCB air concentrations in India range from 120 to 1077 pg m^{-3} (Table 8.1). This indicates a much higher concentration in India than the levels obtained in the recent Asian PAS survey values ranging from ~20 to 340 pg m^{-3} for China, ~5 to 30 pg m^{-3} for Singapore, 7 to 250 pg m^{-3} for Japan, and from 12 to 80 pg m^{-3} for Korea [14]. However, the concentration range is comparatively less than that reported by the European PAS survey where the concentration ranged between 20 and 1700 pg m^{-3} [17] and the GAPS study recorded much higher concentration at Philippines with a range of 0.12–2800 pg m^{-3} [13]. The highest PCB levels in this study were measured at urban sites or at sites close to urban centers (Figure 8.7). A predominant gradient in the maximum concentration of Σ_{28}PCB has been observed between urban–rural–wetland locations with an exceptionally higher level in the wetland region of Sunderban (544 pg m^{-3}).

Current global air concentrations of PCBs are mainly the result of emissions from existing and disposed equipment containing PCBs [51]. A growing source of PCBs for developing countries is electronic waste [54]. The highest level measured in this study occurred at BL (1077 pg m^{-3}), the information technology (IT) hub of India. At present, this city alone generates about 8000 t of computer waste annually and, in the absence of proper disposal, they find their way to scrap dealers. Many informal scrap dealers help in the release of toxic substances in the atmosphere during the process of salvaging the precious metals from the discarded computers.

Elevated levels of PCBs were also observed at MB (813 pg m^{-3}) and BD (688 pg m^{-3}).

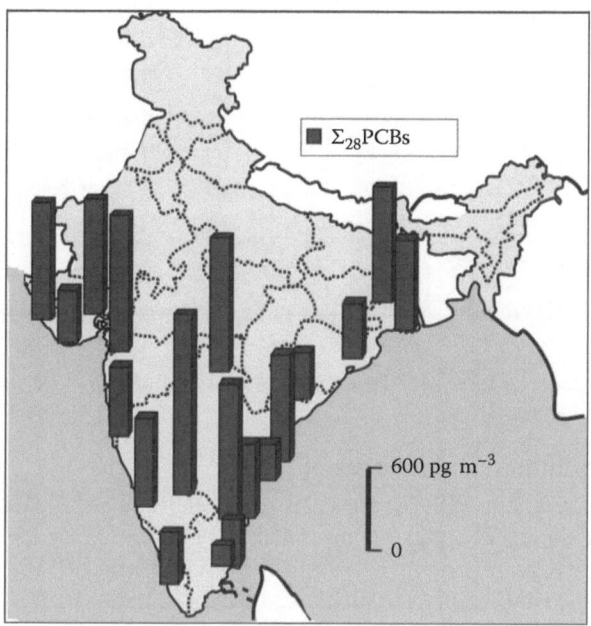

FIGURE 8.7 Spatial distributions of Σ_{28}PCBs (pg m^{-3}).

India is the world's largest shipbreaking nation in terms of volume: 38% of the global shipbreaking activities take place in India [55]. The shipbreaking industry is mostly confined to Alang and Sosiya, two villages in the Bhavnagar district of Gujarat (close to the site-BD) and Mumbai (site-MB). The shipbreaking yard of Mumbai is within the city limits. A typical merchant ship to be dismantled for scrap contains between 250 and 800 kg of PCBs, which is found principally in the paint as well as left on the scrap metal in the vessel machinery that is rerolled or remelted and could be another source of PCB pollution [56]. The plausible reason for an elevated level at the coastal sites of BD and MB could be due to improper management of shipbreaking activities in addition to electronic waste. Elevated levels of PCBs were detected at PN (805 pg m^{-3}) possibly related to leakage from the adjacent industrialized area. The industry produces polyvinylchloride (PVC), which requires polymerization of vinyl chloride (VC) monomer and during the process there is a probable chance of releasing PCBs to the environment. Sources of PCBs in the urban sites also include offgassing from PCB-treated construction materials in older homes and leakage from closed systems such as older equipment (e.g., transformers that contain large quantities of PCB fluids [51]). With the exception of Sunderban (544 pg m^{-3}), PCB levels in the wetlands of GK (320), PV (274), and VN (120) were at least two orders of magnitude less than those from the urban locations, emphasizing the diffusion of atmospheric emissions of PCBs from urban locations. Hence further in-depth research is essential in Sunderban.

It is noteworthy that the PCB congener profile in the present study showed a higher abundance of the heavier congeners at the urban sites, thereby reflecting the potential presence of primary sources, while a lower molecular weight composition enriched in more volatile PCB congeners was observed at the more remote locations (Figure 8.8). As the distance increases from the urban locations, the likely source areas of PCBs, there is a compositional variation in the congener pattern with heavier congeners at the urbanized/industrialized areas and atmospherically driven lighter congeners at remote locations [20]. Pentachlorinated congeners contributed about 43% of the total PCBs in the present study. Almost 38% of the 43% of pentachlorinated congeners has been found to be present in the urban sites, 16% in the rural sites, and 9% in the wetland locations excluding Sunderban as it alone contributed 37%, which is close to the urban level of contamination.

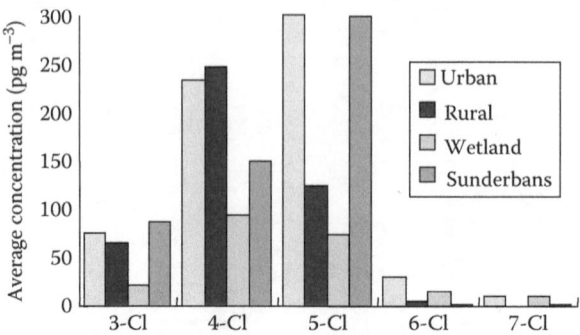

FIGURE 8.8 Average concentration of PCB congeners at urban ($n=9$), rural ($n=6$), wetland/mangroves ($n=3$) sites and at Sunderban ($n=1$).

8.3.2.7 Spatial Distribution and Sources of PBDEs

Polybrominated diphenyl ethers (PBDEs) are a class of recalcitrant and bioaccumulative halogenated compounds that have emerged as a major environmental pollutant. Since PBDEs are not chemically bound to products they may continuously leak to the environment [2]. Figure 8.8 shows the congener-specific spatial distribution pattern for the Σ_9PBDEs in India. PBDE production and use has been a relatively recent phenomenon, with the resultant environmental emission peaking much later than for the organochlorine compounds discussed earlier [57,58]. Major commercial products principally contain penta, octa-, and deca-BDE mixtures. The penta-product contains a mixture of tetra- to hexa-BDEs including BDE47, BDE99, BDE100, BDE153, and BDE154 as well as trace amounts of BDE17 and BDE28. The octa-product consists primarily of BDE183 followed by BDE153 and BDE 154, whereas the deca-product is mostly composed of BDE209 (>97%) [59]. The concentration range of PBDEs in the present study was found to be much higher than Singapore (~6–30 pg m^{-3}) and Korea (12–80 pg m^{-3}) [14] but close to the range in Europe (0.5–250 pg m^{-3}) [17]. The highest concentration in this study was observed at Metropolitan Chennai with elevated levels at PD (504 pg m^{-3}) and BD (409 pg m^{-3}) probably arising due to the usage of Bromkal 70-5DE. Chennai is the third largest commercial and industrial center in India and is considered the country's automobile capital. Chennai has emerged as an electronic manufacturing hub with multinational corporations setting up their production facilities and is highly loaded with e-waste due to rising development in the field of information technology. Industries using PBDEs in their production processes (i.e., car producers, textile and electronic industries) have been found to be potential sources of PBDEs [60]. PD is a coastal site adjacent to CN and the location of large-scale electronic industries. BD witnessed a sudden spurt in industrial activity and the city and the surrounding areas are today humming with industrial activities. BDE-47 (2,2′,4,4′-tetraBDE) and BDE-85 (2,2′,3,4,4′-pentaBDE) dominated the congener pattern (Figure 8.9) contributing ca. 66% to the overall Σ_9PBDE burden. The other congeners detected were BDE 28 (2,4,4′-triBDE), BDE-66 (2,3′,4,4′-tetraBDE), BDE-99 (2,2′,4,4′,5-pentaBDE), BDE-100 (2,2′,4,4′,6-pentaBDE), BDE-153 (2,2′,4,4′,5,5′-hexaBDE), BDE-154 (2,2′,4,4′,5,6′-hexaBDE), and BDE-183 (2,2′,3,4,4′,5′,6-heptaBDE). BDE 209 levels were BDL in all the samples; the most likely explanation for the "absence" of BDE 209 in this study may be related to its association with aerosol particles at ambient temperatures. The PAS used in the present study is designed to sample gas-phase chemicals. BDE-209 may also undergo photolysis in the tropical atmosphere into lower congeners. Degradation of deca-BDE as a solid film by sunlight produced no PBDE congeners with less than five bromine atoms [61]. Recent studies also report the photochemical decomposition of BDE-209 into lower molecular weight congeners under laboratory conditions [62]. In India, the concentration of Σ_9PBDEs at other cities, remote sites, and wetland regions were comparatively very low or almost not detected. CN, BD, and PD showed similar range of concentration gradient for BDE-47 and -99, possibly related to the growth of e-waste, which is an emerging problem for developing nations [54].

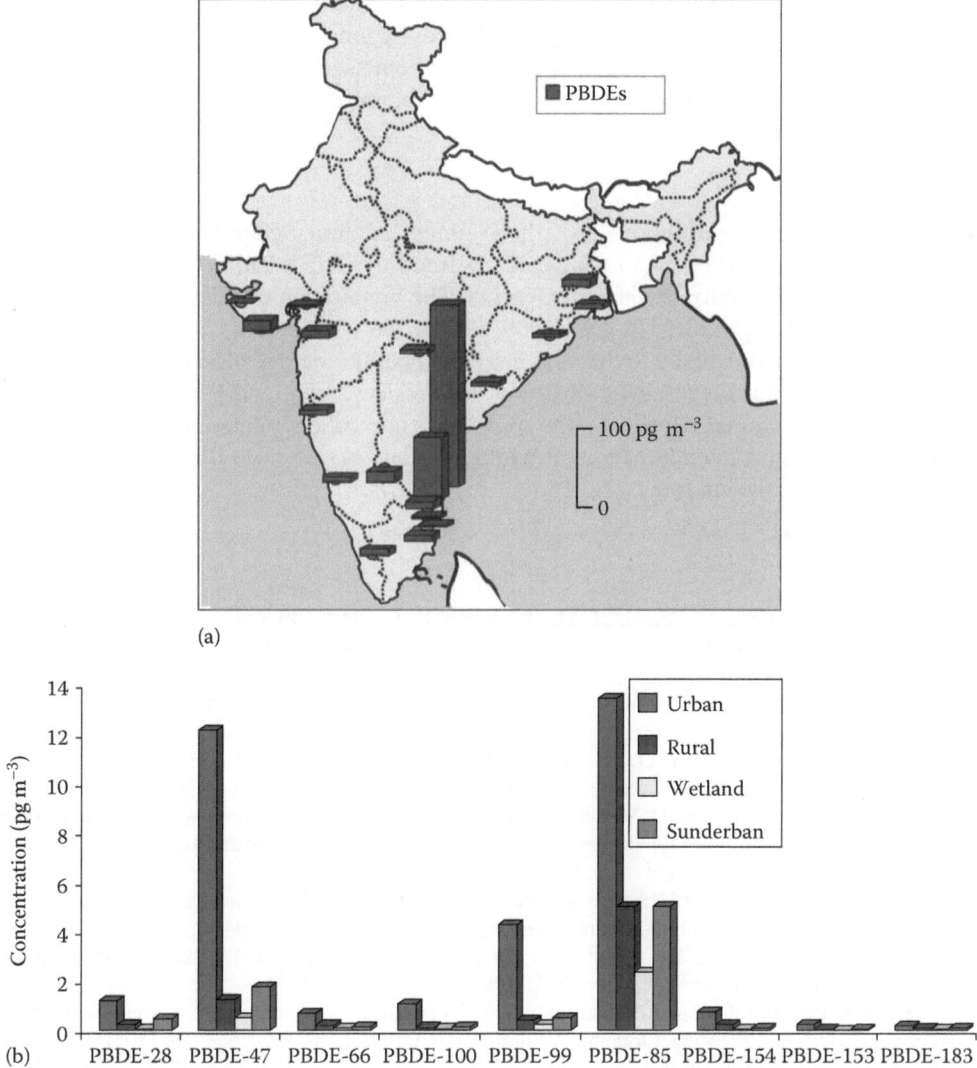

FIGURE 8.9 (a) Spatial distribution of Σ_9PBDEs (pg m^{-3}). (b) Congener-specific distribution of Σ_9PBDEs (pg m^{-3}) in India.

8.4 OVERVIEW

This study gives a snapshot of the atmospheric levels and possible sources of POPs in India. In general, the geographical pattern of all the compounds studied reflected regional emission sources. The highest values were found in urban regions while the lowest values were from the remote wetlands excluding Sunderban. The higher levels of OCPs and PCBs at Sunderban indicate that additional research in this unique mangrove area is essential. In general, the geographical pattern of all the compounds studied reflected suspected regional emission patterns and localized hot spots. The massive decline in the levels of DDTs and HCHs in the coastal cities from previous records reflect the ban on the technical mixture of these compounds for agricultural purpose except for the use of DDT in vector control programs. The tropical atmospheric conditions in India lead to rapid breakdown of most POPs. The destruction of PCBs by hydroxy radicals during their residence in the higher tropical atmosphere has been found to be most efficient [63]. This is evident in the case of PBDEs and DDTs. Hence the POPs reported in this chapter, which are released in the tropical environment of India, not only cause local/regional contamination problems but may also contribute to pollution in areas far away from the point sources by long-range atmospheric transport. This is powerful evidence that there are continuing primary emissions of some POPs in India and that natural resources like Sunderban located far from point sources are being contaminated as a result.

ACKNOWLEDGMENTS

This chapter is a joint contribution from the National Natural Scientific Foundation of China (NSFC 40590391, 40673076), the Global COE awarded to CMES Ehime University, Japan, and the Savannah State University, Savannah, Georgia. The preparation of this chapter was supported by U.S.-DOE and EPA contract No. DE-FG09-96SR18558. The authors would like to thank the students at Annamalai University, India, and Dr. P. Sampathkumar for their help in field sampling, as well as Professor Shinsuke Tanabe of Ehime University, Japan, and Professor Kevin Jones of Lancaster University, United Kingdom, who were the prime collaborators for this chapter and supported the successful completion of this research. The authors are also thankful to Dr. Margaret Murphy for reviewing the chapter.

REFERENCES

1. Breivik, K., Sweetman, A., Pacyna, J. M. and Jones, K. C. 2002. Towards a global historical emission inventory for selected PCB congeners—A mass balance approach: 2. Emissions. *Sci. Total Environ.* 290: 199–224.
2. de Wit C. A. 2002. An overview of brominated flame retardants in the environment. *Chemosphere* 46(5): 583–624.
3. Bidleman, T. F. 1988. Atmospheric processes. *Environ. Sci. Technol.* 22(4): 361–367.
4. Klecka, G., Boethling, B., Franklin, J., Graham, G., Grady, L., Howard, P., Kannan, K., Larson, R., Mackay, D., Muir, D. and vandeMeent, D. 2000. *Evaluation of Persistence and Long-Range Transport of Organic Chemicals in the Environment.* SETAC Special Publication Series, Society of Environmental Toxicology and Chemistry, Pensacola, FL, pp. 1–5.
5. Beyer, A., Mackay, D., Matthies, M., Wania, F. and Webster, E. 2000. Assessing long-range transport potential of persistent organic pollutants. *Environ. Sci. Technol.* 34(4): 699–703.
6. Wania, F. 2003. Assessing the potential of persistent organic chemicals for long-range transport and accumulation in polar regions. *Environ. Sci. Technol.* 37(7): 1344–1351.
7. UNEP. 1998. Preparation of an Internationally Legally Binding Instrument for Implementing International Action on Certain Persistent Organic Pollutants, United Nations Environment Programme, Nairobi, Kenya, UNEP/POPs/Inc.1/6.
8. UNEP. 2001. Regionally based assessment of persistent toxic substances: Central and northeast Asia region; United Nations Environment Programme, Nairobi, Kenya, No. 10.
9. Gupta, P. K. 2004. Pesticide exposure—Indian scene. *Toxicology* 198(1–3): 83–90.
10. Tanabe, S., Gondaira, F., Subramanian, A., Ramesh, A., Mohan, D., Kumaran, P., Venugopalan, V. K. and Tatsukawa, R. 1990. Specific pattern of persistent organochlorine residues in human breast milk from south India. *J. Agric. Food Chem.* 38(3): 899–903.
11. Tanabe, S. and Kunisue, T. 2007. Persistent organic pollutants in human breast milk from Asian countries. *Environ. Pollut.* 146(2): 400–413.
12. Kannan, K., Tanabe, S., Ramesh, A., Subramanian, A. and Tatsukawa, R. 1992. Persistent organochlorine residues in foodstuffs from India and their implications on human dietary exposure. *J. Agric. Food Chem.* 40(3): 518–524.
13. Pozo, K., Harner, T., Wania, F., Muir, D. C. G., Jones, K. C. and Barrie, L. A. 2006.Toward a global network for persistent organic pollutants in air: Results from the GAPS study. *Environ. Sci. Technol.* 40(16): 4867–4873.
14. Jaward, F. M., Zhang, G., Nam, J. J., Sweetman, A. J., Obbard, J. P., Kobara, Y. and Jones, K. C. 2005. Passive air sampling of polychlorinated biphenyls, Organochlorine compounds, and polybrominated diphenyl ethers across asia. *Environ. Sci. Technol.* 39(22): 8638–8645.
15. Shoeib, M. and Harner, T. 2002. Characterization and comparison of three passive air samplers for persistent organic pollutants. *Environ. Sci. Technol.* 36(19): 4142–4151.

16. Pozo, K., Harner, T., Shoeib, M., Urrutia, R., Barra, R., Parra, O. and Focardi, S. 2004. Passive-sampler derived air concentrations of persistent organic pollutants on a north-south transect in Chile. *Environ. Sci. Technol.* 38(24): 6529–6537.

17. Jaward, F. M., Farrar, N. J., Harner, T., Sweetman, A. J. and Jones, K. C. 2004. Passive air sampling of PCBs, PBDEs, and organochlorine pesticides across Europe. *Environ. Sci. Technol.* 38(1): 34–41.

18. Wang, X.-P., Yao, T.-D., Cong, Z.-Y., Yan, X.-L., Kang, S.-C. and Zhang, Y. 2006. Gradient distribution of persistent organic contaminants along northern slope of central-Himalayas, China. *Sci. Total Environ.* 372(1): 193–202.

19. Hafner, W. D. and Hites, R. A. 2003. Potential sources of pesticides, PCBs, and PAHs to the atmosphere of the Great Lakes. *Environ. Sci. Technol.* 37(17): 3764–3773.

20. Harner, T., Shoeib, M., Diamond, M., Stern, G. and Rosenberg, B. 2004. Using passive air samplers to assess urban-rural trends for persistent organic pollutants. 1. Polychlorinated biphenyls and organochlorine pesticides. *Environ. Sci. Technol.* 38(17): 4474–4483.

21. Iwata, H., Tanabe, S., Sakai, N. and Tatsukawa, R. 1993. Distribution of persistent organochlorines in the oceanic air and surface seawater and the role of ocean on their global transport and fate. *Environ. Sci. Technol.* 27(6): 1080–1098.

22. Kutz, F. W., Wood, P. and Bottimore, D. P. 1991. Organochlorine pesticides and polychlorinate biphenyls in human adipose tissue. *Rev. Environ. Contam. Toxicol.* 120: 1–82.

23. Iwata, H., Tanabe, S. and Tatsukawa, R. 1993. A new view on the divergence of HCH isomer composition in oceanic air. *Mar. Pollut. Bull.* 26: 302–305.

24. Babu Rajendran, R., Venugopalan, V. K. and Ramesh, R. 1999. Pesticide residues in air from coastal environment, South India. *Chemosphere* 39(10): 1699–1706.

25. Iwata, H., Tanabe, S., Sakai, N., Nishimura, A. and Tatsukawa, R. 1994. Geographical distribution of persistent organochlorines in air, water and sediments from Asia and Oceania, and their implications for global redistribution from lower latitudes. *Environ. Pollut.*. 85 (1): 15–33.

26. Ramesh, A., Tanabe, S., Tatsukawa, R., Subramanian, A. N., Palanichamy, S., Mohan, D. and Venugopalan, V. K. 1989. Seasonal variations of organochlorine insecticide residues in air from Porto Novo, South India. *Environ. Pollut.* 62(2–3): 213–222.

27. Subramanian, A., Ohtake, M., Kunisue, T., Tanabe, S. 2007. High levels of organochlorines in mothers' milk from Chennai (Madras) city, India. *Chemosphere* 68(5): 928–939.

28. Takeoka, H., Ramesh, A., Iwata, H., Tanabe, S., Subramanian, A. N., Mohan, D., Magendran, A. and Tatsukawa, R. 1991. Fate of the insecticide HCH in the tropical coastal area of South India. *Mar. Pollut. Bull.* 22(6): 290–297.

29. Macdonald, R. W., Barrie, L. A., Bidleman, T. F., Diamond, M. L., Gregor, D. J., Semkin, R. G., Strachan, W. M. J. et al. 2000. Contaminants in the Canadian Arctic: 5 years of progress in understanding sources, occurrence and pathways. *Sci. Total Environ.* 254(2–3): 93–234.

30. Wurl, O., Potter, J. R., Obbard, J. P. and Durville, C. 2006. Persistent organic pollutants in the equatorial atmosphere over the open Indian Ocean. *Environ. Sci. Technol.* 40(5): 1454–1461.

31. Sengupta, A. K. 2005. *Tsunami Situation Report*; www.searo.who.int/LinkFiles/India_ind-14jan05.pdf

32. Karuppiah, S., Subramanian, A. and Obbard, J. P. 2005. Organochlorine residues in odontocete species from the southeast coast of India. *Chemosphere* 60(7): 891–897.

33. Liu, X., Zhang, G., Li, J., Yu, L.-L., Xu, Y., Li, X.-D., Kobara, Y. and Jones, K. C. 2009. Seasonal patterns and current sources of DDTs, chlordanes, hexachlorobenzene, and endosulfan in the atmosphere of 37 Chinese cities. *Environ. Sci. Technol.* 43(5): 1316–1321.

34. Sarkar, A., Nagarajan, R., Chaphadkar, S., Pal, S. and Singbal, S. Y. S. 1997. Contamination of organochlorine pesticides in sediments from the Arabian Sea along the west coast of India. *Water Res.* 31(2): 195–200.

35. Rajendran, R. B., Imagawa, T., Tao, H. and Ramesh, R. 2005. Distribution of PCBs, HCHs and DDTs, and their ecotoxicological implications in Bay of Bengal, India. *Environ. Int.* 31(4): 503–512.

36. Bhattacharya, B., Sarkar, S. K. and Mukherjee, N. 2003. Organochlorine pesticide residues in sediments of a tropical mangrove estuary, India: Implications for monitoring. *Environ. Int.* 29(5): 587–592.

37. Guzzella, L., Roscioli, C., Vigano, L., Saha, M., Sarkar, S. K., Bhattacharya, A. 2005. Evaluation of the concentration of HCH, DDT, HCB, PCB and PAH in the sediments along the lower stretch of Hugli estuary, West Bengal, northeast India. *Environ. Int.* 31(4): 523–534.

38. Santillo, D, Johnston, P. and Stringer, R. A. 1997. Catalogue of gross contamination: Organochlorine production and exposure in India. *Pest. News* 36: 4–6.

39. Harner, T., Bartkow, M., Holoubek, I., Klanova, J., Wania, F., Gioia, R., Moeckel, C., Sweetman, A. J. and Jones, K. C. 2006. Passive air sampling for persistent organic pollutants: Introductory remarks to the special issue. *Environ. Pollut.* 144(2): 361–364.

40. Black rain claims 2,500 buffaloes. Available at: www.rediff.com/news/2005/jul/29jcm.htm
41. Spencer, W. and Cliath, M. M. 1972. Volatility of DDT and related compounds. *J. Agric. Food Chem.* 20(3): 645–649.
42. Talekar, N. S., Sun, L. T., Lee, E. M. and Chen, J. S. 1977. Persistence of some insecticides in subtropical soil. *J. Agric. Food Chem.* 25(2): 348–352.
43. Atlas, E. and Giam, C. S. 1988. Ambient concentration and precipitation scavenging of atmospheric organic pollutants. *Water Air Soil Pollut.* 38(1): 19–36.
44. Bidleman, T. F., Jantunen, L. M. M., Helm, P. A., Brorstrom-Lunden, E. and Juntto, S. 2002. Chlordane enantiomers and temporal trends of chlordane isomers in arctic air. *Environ. Sci. Technol.* 36 (4): 539–544.
45. Jantunen, L. M. M., Bidleman, T. F., Harner, T. and Parkhurst, W. J. 2000. Toxaphene, chlordane, and other organochlorine pesticides in Alabama air. *Environ. Sci. Technol.* 34(24): 5097–5105.
46. Indian Ministry of Chemicals & Fertilisers. *Information compiled by Plant Protection, Adviser, Department of Agriculture & Corporation*, Ministry of Agriculture and Corporation, New Delhi, 2000.
47. Harner, T., Pozo, K., Gouin, T., Macdonald, A.-M., Hung, H., Cainey, J., Peters, A. 2006. Global pilot study for persistent organic pollutants (POPs) using PUF disk passive air samplers. *Environ. Pollut.* 144(2): 445–452.
48. Effects of endosulfan on human beings. Available at http://www.indiaenvironmentportal.org.in/files/Effect_of_endosulfan.pdf
49. Bailey, R. E. 2001. Global hexachlorobenzene emissions. *Chemosphere* 43(2): 167–182.
50. Barber, J. L., Sweetman, A. J., van Wijk, D. and Jones, K. C. 2005. Hexachlorobenzene in the global environment: Emissions, levels, distribution, trends and processes. *Sci. Total Environ.* 349(1–3): 1–44.
51. Monirith, I., Ueno, D., Takahashi, S., Nakata, H., Sudaryanto, A., Subramanian, A., Karuppiah, S. et al. 2003. Asia-Pacific mussel watch: Monitoring contamination of persistent organochlorine compounds in coastal waters of Asian countries. *Mar. Pollut. Bull.* 46(3): 281–300.
52. van-Birgelen, A. 1998. Hexachlorobenzene as a possible major contributor to the dioxin activity of human milk. *Environ. Health Perspect.* 106: 683–688.
53. Breivik, K., Alcock, R., Li, Y., Bailey, R. E., Fiedler, H. and Pacyna, J. M. 2004. Primary sources of selected POPs: Regional and global scale emission. *Environ Pollut.* 128: 3–16.
54. Wong, M. H., Wu, S. C., Deng, W. J., Yu, X. Z., Luo, Q., Leung, A. O. W., Wong, C. S. C., Luksemburg, W. J. and Wong, A. S. 2007. Export of toxic chemicals—A review of the case of uncontrolled electronic-waste recycling. *Environ. Pollut.* 149(2): 131–140.
55. International Metal Workers' Federation Status of Shipbreaking Workers in India—A Survey; IMF-FNV project in India 2004–2007. Available at http://www.imfmetal.org/files/06042810465779/Shipbreaking_survey.pdf
56. Hess, R., Rushworth, D., Hynes, M. V., and Peters, J. E. 2001. *Disposal Options for Ships*, Rand, Santa Monica, CA.
57. Prevedouros, K., Jones, K. C., Sweetman, A. J. 2004. Estimation of the production, consumption, and atmospheric emissions of pentabrominated diphenyl ether in Europe between 1970 and 2000. *Environ. Sci. Technol.* 38(12): 3224–3231.
58. Prevedouros, K., Jones, K. C. and Sweetman, A. J. 2004. European-scale modeling of concentrations and distribution of polybrominated diphenyl ethers in the pentabromodiphenyl ether product. *Environ. Sci. Technol.* 38(22): 5993–6001.
59. Hale, R. C., Alaee, M., Manchester-Neesvig, J. B., Stapleton, H. M. and Ikonomou, M. G. 2003. Polybrominated diphenyl ether flame retardants in the North American environment. *Environ. Int.* 29(6): 771–779.
60. ter Schure, A. 2000. Describing the flows of synthetic musks and brominated flame retardants in the environment: A new ecotoxicological problem? Introductory Paper No. 120, Department of Ecology, Lund University, Lund, Sweden.
61. Watanabe, I. and Sakai, S.-I. 2003. Environmental release and behavior of brominated flame retardants. *Environ. Int.* 29(6): 665–682.
62. Bezares-Cruz, J., Jafvert, C. T. and Hua, I. 2004. Solar photodecomposition of decabromodiphenyl ether: Products and quantum yield. *Environ. Sci. Technol.* 38(15): 4149–4156.
63. Mandalakis, M., Berresheim, H., Stephanou, E. G. 2003. Direct evidence for destruction of polychloro-biphenyls by OH radicals in the subtropical troposphere. *Environ. Sci. Technol.* 37(3): 542–547.
64. Gouin, T., Jantunen, L., Harner, T. Blanchard, P. and Bidleman, T. 2007, Spatial and temporal trends of chiral organochlorine signatures in Great Lakes air using passive air samplers. *Environ. Sci. Technol.* 41(11): 3877–3883.

9 Historical Trends of Dioxin Sources and Contamination in Japan

*Shigeki Masunaga**

CONTENTS

9.1 Introduction 203
9.2 Trends of Dioxin Emission in Japan...204
    9.2.1 Dioxin Impurities in Japanese Agrochemicals...204
    9.2.2 Trends of Dioxin Emissions from Agrochemical Use...................................205
    9.2.3 Municipal Solid Waste Incineration 205
    9.2.4 Industrial Waste Incineration..206
    9.2.5 Trends of Dioxin Emission from Various Sources 206
9.3 Source Apportionment of Dioxin Pollution...206
    9.3.1 Changes in the Dioxin Concentration in Tokyo Bay Sediment Core Samples.........207
    9.3.2 Source Identification of Dioxins in Tokyo Bay Sediment 207
    9.3.3 Source Apportionment of Dioxin Pollution in Tokyo Bay 208
9.4 Trends of Human Dioxin Exposure in Japan...210
    9.4.1 Estimated Dioxin Intake by Total Diet Studies 210
    9.4.2 Dioxin Body Burden: Dioxins in Mother's Milk..211
9.5 Summary 212
Acknowledgments..213
References...213

9.1 INTRODUCTION

Dioxin contamination associated with waste incinerators became an important environmental health issue in the 1990s in Japan. The media reported the contamination of human breast milk with dioxins in 1991. Subsequent reports indicated high concentrations of dioxins in human blood (1996) and vegetables (1999), and high levels of newborn mortality (1997) in some areas near solid waste incinerators. Although some of those reports were refuted, concerned citizens requested that the Japanese government take measures to prevent future dioxin contamination. Japan has the largest number of waste incinerators in the world and potential dioxin emissions from these incinerators are of concern. As a result, the Japanese government gradually introduced stricter emission rules for incinerators from 1997 to 2002 and passed the Law Concerning Special Measures against Dioxins in 1999.

The studies relating to dioxin sources of pollution afterword, however, have clearly indicated that incinerators were not necessarily the sole contributors of dioxin pollution in Japan. This chapter deals with the trends and sources of dioxin pollution in the country, and it is hoped that these

---

* E-mail: masunaga@ynu.ac.jp (Chapter corresponding author).

results may be extrapolated to other countries in order to catalyze future measures against dioxin pollution.

9.2 TRENDS OF DIOXIN EMISSION IN JAPAN

9.2.1 DIOXIN IMPURITIES IN JAPANESE AGROCHEMICALS

Dioxins can be formed as by-products during chemical syntheses. The occurrence of polychlorinated dibenzo-p-dioxins (PCDDs) and polychlorinated dibenzofurans (PCDFs) in pentachlorophenol (PCP) was first reported in the 1970s [1,2]. The contamination of Agent Orange, a herbicide used in the Vietnam War, with 2,3,7,8-tetrachlorodibenzo-p-dioxin (2,3,7,8-TCDD) was also heavily publicized [3]. Additionally, a accidental release of dioxins from a chemical plant occurred in Seveso, Italy, in 1976 [4]. Thus, the formation of dioxins as by-products of chemical manufacturing was well established as one of the major sources of dioxins. However, as described in the previous section, only waste incineration and industrial combustion were considered to be the major targets for countermeasures against dioxin pollution in the 1990s in Japan. A group of researchers (including the author) started to look into the real sources of dioxin in detail.

Old, unused, agrochemical formulations were collected from farmers because we could not obtain original samples from the agrochemical companies. The dioxin contents of the samples were analyzed [5]. The results indicated that some Japanese agrochemicals had high concentrations of dioxin impurities. PCP and 2,4,6-trichlorophenyl-4′-nitrophenyl ether (chloronitrophen, CNP), which were used extensively as rice field herbicides in the 1960s and 1970s had especially high dioxin contents. Some other agrochemicals had intermediate levels of dioxin however those chemicals were not heavily used. When we first presented these results at a conference in January 1999 [6], the manufacturer of CNP denied our results insisting that CNP did not contain toxic dioxins (2,3,7,8-chlorine substituted PCDDs/PCDFs). The company claimed that they had not used chemical

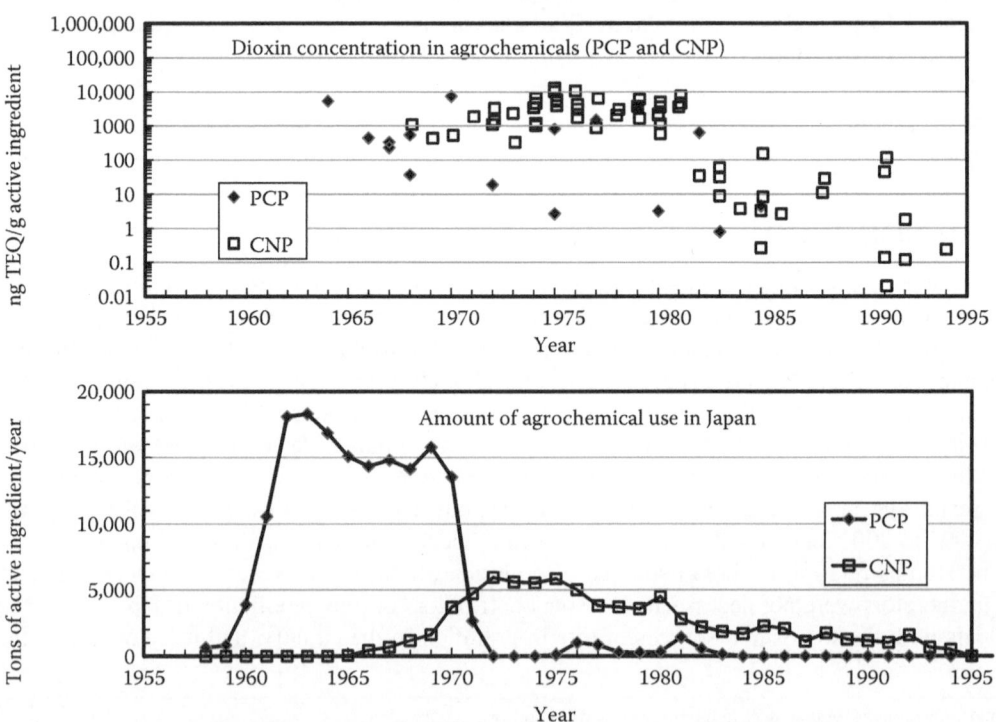

FIGURE 9.1   Time trend of dioxin concentration in agrochemicals (PCP and CPN) and their use in Japan.

synthesis pathways which could form 2,3,7,8-PCDDs/PCDFs as by-products. The company, however, admitted that CNP contained high levels of 2,3,7,8-chlorine substituted dioxin in July 1999 [7,8]. Figure 9.1 shows a summary of all the available data on dioxin concentrations in Japanese PCP and CNP formulations [5,9,10]. There did not appear to be a time trend in the dioxin impurities in PCP, while the dioxin levels in CNP decreased substantially in the early 1980s. The manufacturing company claimed that they changed their production processes in 1982 and 1983 to reduce the impurities in 2,4,6-trichlorophenol, one of two chemicals used to synthesize CNP. By changing the synthetic processes through the introduction of distillation and crystallization steps, the impurities in 2,4,6-trichlorophenol, which might have been tetrachlorophenol and other trichlorophenol isomers, might have decreased substantially and, as a result, contributed to the reduction of 2,3,7,8-chlorine-substituted dioxins in CNP. Figure 9.1 also shows the trends in PCP and CNP use [11].

9.2.2 TRENDS OF DIOXIN EMISSIONS FROM AGROCHEMICAL USE

Using the dioxin impurity contents and government statistics on the use of agrochemicals, a time trend of dioxin emissions and the overall emissions from past agrochemical usage were estimated and are shown in Table 9.1 and Figure 9.2. These results showed that PCP and CNP were the two largest sources of dioxin emissions.

To compare the relative importance of chlorinated agrochemicals as sources of dioxin, a rough estimate of dioxin emissions from incineration sources has been presented in the following sections.

9.2.3 MUNICIPAL SOLID WASTE INCINERATION

Dioxin concentrations in the flue gas of municipal solid waste incineration (MSWI) facilities were measured nationwide by the Ministry of Health and Welfare (MHW) [12]. Arithmetic means for different treatment systems, namely batch, mechanical batch, and semicontinuous and continuous systems, were calculated. These values, together with the time trend of waste incinerated and the capacity of different systems, were used to estimate the trend of dioxin emissions from MSWI. The results are shown in Figure 9.2. The total dioxin toxic equivalent (TEQ) emitted as flue gas between 1955 and 1994 was estimated to be around 72 kg TEQ. This was probably a conservative estimate because fewer air pollution control devices were installed during the former part of the time period.

TABLE 9.1
Estimated Dioxin Emission from Agrochemical Use in Japan

Agro-Chemicals[a]	Period of Use (Year)	Total Amount of Use (Ton Active Ingredient)	Average Dioxin Concentration[b] (ng TEQ/g Active Ingredient)[c]	Estimated Total Dioxin Emission (kg TEQ)[c]
PCP	1955–1983	164,000	1,500 ($n=14$)	250
CNP	1965–1994	78,000	Before 1981: 3,600 ($n=45$)	210
			After 1982: 22 ($n=19$)	
PCNB	1956–1996	28,000	1.52 ($n=3$)	0.043
2,4,5-T	1965–1974	160	1,040 ($n=29$)	0.98
NIP	1963–1982	7,500	1.53 ($n=1$)	0.012

[a] PCP, pentachlorophenol; CNP, chloronitrophen (2,4-dichlorophenyl-3'-methoxy-4'-nitrophenylether); PCNB, pentachloronitrobenzene; 2,4,5-T, 2,4,5-trichlorophenoxyacetic acid; NIP, nitrophen (2,4-dichlorophenyl-4-nitoro-phenylether).

[b] Average dioxin concentration (TEQ) from Japanese agrochemicals except for 2,4,5-T. Data for 2,4,5-T is the average 2,3,7,8-TCDD concentration from the world.

[c] Dioxin TEQ was calculated using WHO-TEF (1998).

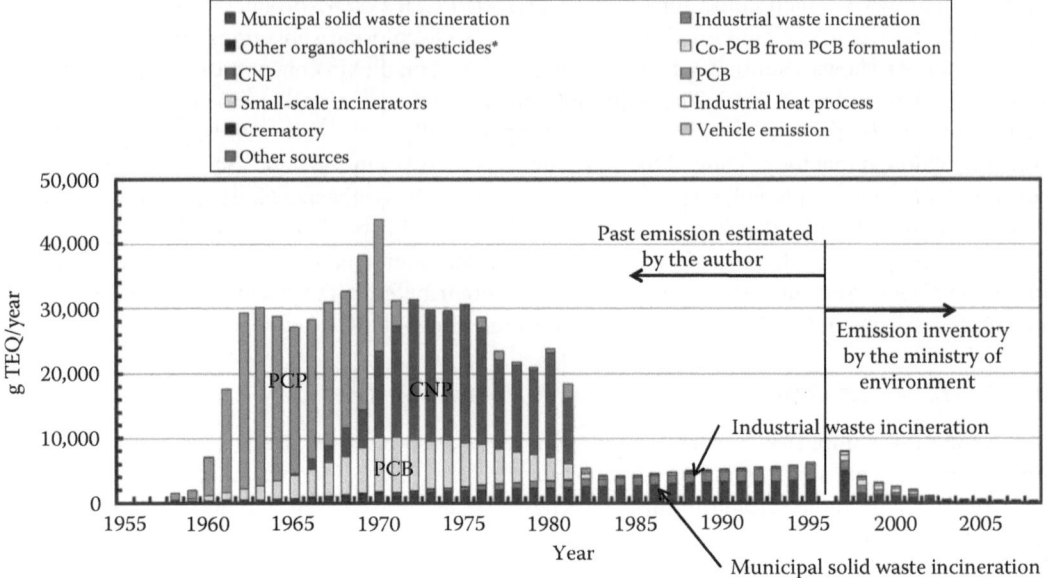

FIGURE 9.2 Time trend of dioxin emission from various sources in Japan. The amount of dioxin in terms of TEQ was calculated using WHO-TEF (1998). * Other organochlorine pesticides include 2,4-dichlorophenyl-4′-nitrophenyl ether (Nitrophen, NIP), 2,4,5-trichlorophenoxyacetic acid (2,4,5-T), 2,4-dichlorophenoxyacetic acid (2,4-D), 2-methyl-4-chloro-phenoxyacetic acid (MCP), tetrachloro-*iso*-phthalonitrile (Chlorothalonil, TPN), and pentachloronitrobenzene (PCNB).

9.2.4 INDUSTRIAL WASTE INCINERATION

There are no reliable historical statistics on industrial waste incineration (IWI). Thus, we used the MHW's survey of industrial waste incineration and trends from a number of IWI facilities to estimate the amount of industrial waste incinerated in the past [13]. The Environment Agency's report on emission factors in sludge and other industrial waste incineration was used to estimate emissions [14]. The total TEQ emissions between 1955 and 1994 was calculated to be around 55 kg TEQ. It should be noted that this value was a rough estimate since only flue gas emissions were included (emission as ash was not counted).

9.2.5 TRENDS OF DIOXIN EMISSION FROM VARIOUS SOURCES

Trends of dioxin emissions from chemical manufacturing, use, and incineration are summarized in Figure 9.2. The emission data from 1950 to 1994 was estimated by the author [6,15] and the emission data in recent years (1996–2008) was estimated by the Ministry of Environment (MOE) [16]. The MOE started extensive monitoring of dioxins in flue gas and industrial wastewater and collecting reports from industries under the Law Concerning Special Measures against Dioxins. Thus, the recent estimate by the MOE should be relatively precise.

The estimated past dioxin emission data from agrochemical usage clearly indicates that the contribution of agrochemicals to the total dioxin burden was quite large in Japan compared to those from other sources, including incineration.

9.3 SOURCE APPORTIONMENT OF DIOXIN POLLUTION

Sediment core samples taken from the center of Tokyo Bay were analyzed to determine if agrochemicals were the major source of dioxin emission in Japan. Because Tokyo Bay receives runoff and waste water from the Tokyo metropolitan area and rivers draining the surrounding agricultural lands, its sediment should reflect the impact of human activities quite accurately.

9.3.1 CHANGES IN THE DIOXIN CONCENTRATION IN TOKYO BAY SEDIMENT CORE SAMPLES

The sediment core sample was collected in September 1993 at the center of the Tokyo Bay (Figure 9.3) and dated [17]. Additionally, some surface sediment samples of the bay and some soil samples in the basin were later collected in 1995. The dioxins in those samples were analyzed congener specifically [18,19]. The change in the dioxin concentration in the core is shown in Figure 9.4. The total concentration of PCDDs/PCDFs increased rapidly during the 1960s, reached its maximum around the early 1970s, and decreased gradually thereafter. This trend is consistent with the estimated emission trend shown in Figure 9.2. The congener demonstrating the greatest increase in concentration during the 1960s was octachlorodibenzo-p-dioxin, which was measured in high concentrations in PCP. The concentration of TeCDD increased in the early 1970s and corresponded to the large amount of 1,3,6,8-TeCDD and 1,3,7,9-TeCDD emissions from the use of CNP.

9.3.2 SOURCE IDENTIFICATION OF DIOXINS IN TOKYO BAY SEDIMENT

The congener specific dioxin concentrations in the sediment core, surface sediments, and some soil samples were used as input data for factor analysis. Three factors were extracted and are shown in Table 9.2. The origins of the factors were interpreted by analyzing the characteristic congeners in the factors. Factor 1 correlated strongly with most of the tetrachlorodibenzofurans (TeCDFs), pentachlorodibenzofurans (PeCDFs), and hexachlorodibenzofurans (HxCDFs), half

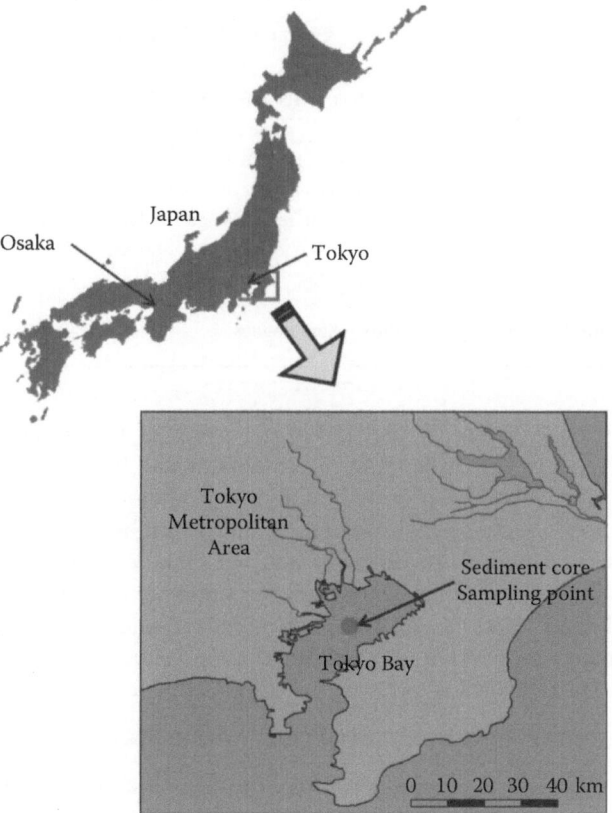

FIGURE 9.3 Location of the sediment core sampling point in Tokyo Bay.

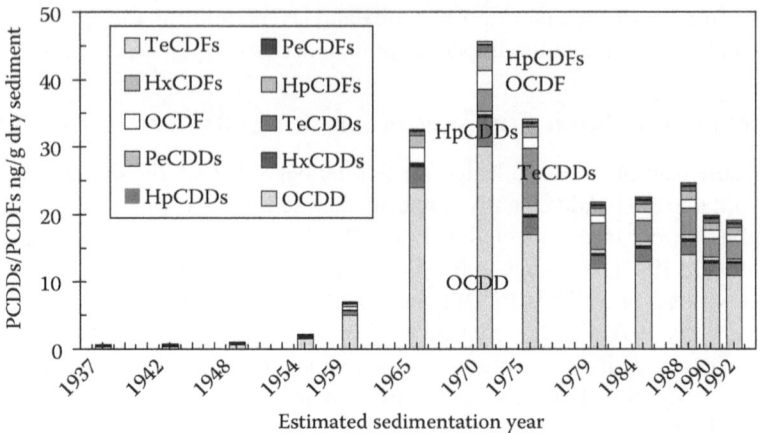

FIGURE 9.4 Dioxin concentration in the Tokyo Bay sediment core. Concentration presented by homologue on a mass basis (not TEQ). TeCDFs, tetrachlorodibenzofurans; PeCDFs, pentachlorodibenzofurans; HxCDFs, hexachlorodibenzofurans; HpCDFs, heptachlorodibenzofurans; OCDF, octachlorodibenzofuran; TeCDDs, tetrachlorodibenzo-*p*-dioxins; PeCDDs, pentachlorodibenzo-*p*-dioxins; HxCDDs, hexachlorodibenzo-*p*-dioxins; HpCDDs, heptachlorodibenzo-*p*-dioxins; OCDD, octachlorodibenzo-*p*-dioxin.

TABLE 9.2
Result of Factor Analysis on the Dioxin Congener Concentration in Sediment and Soil Samples from the Tokyo Bay Basin (after Varimax Rotation)

Factor	Contribution	Cumulative Contribution	Characteristic Congeners (Factor Loadings > 0.7)	Interpretation
PC1	0.50	0.50	Most of TeCDFs, half of PeCDDs, most of PeCDFs, some of HxCDDs, most of HxCDFs	Combustion
PC2	0.25	0.75	Most of HxCDDs, some of HxCDF, HpCDDs, most of HpCDFs, OCDD, OCDF	PCP
PC3	0.18	0.93	Some of TeCDDs and PeCDDs, especially 1,3,6,8- and 1,3,7,9-substituted TeCDDs and PeCDDs, 2,4,6,8-TeCDF	CNP

Refer to Figure 9.4 for the abbreviations of dioxin homologues.

---

of the pentachlorodibenzo-*p*-dioxins (PeCDDs) and some of the hexachlorodibenzo-*p*-dioxins (HxCDDs). This correlation with lower chlorinated dioxins and furans indicated that Factor 1 was best represented as being caused by an incineration source. Factor 2 correlated strongly with highly chlorinated dioxins and furans, and was therefore interpreted as an impurity of PCP [5]. Factor 3 was strongly correlated with some TeCDDs and PeCDDs, especially 1,3,6,9-TeCDD, 1,3,7,9-TeCDD, and 2,4,6,8-TeCDF [5]. Those congeners are known to be present in CNP. Factor 3 was therefore interpreted as an impurity of CNP. Thus, incineration, impurities of PCP and CNP were indicated to be the major sources of dioxin pollution in the Tokyo Bay basin.

9.3.3 SOURCE APPORTIONMENT OF DIOXIN POLLUTION IN TOKYO BAY

Three sources were identified as the major cause of dioxin pollution in the Tokyo Bay. To estimate the relative quantitative contributions of the three sources, the average dioxin congener profiles were obtained and submitted, along with the dioxin congener profiles in the Tokyo Bay sediment, to multiple regression analysis (same as the chemical mass balance method) [20]. The calculation was conducted for each homologue since PCDDs and PCDFs with different numbers of chlorine substitution should demonstrate different chemical/physical properties and environmental fate (degradation, transportation, and/or partition in/between environmental media may be different). The proposed calculation method was better than the ordinary method which input all of the congener data into multiple regression analyses simultaneously [21]. Adoption of the proposed method was made possible because the dioxin congeners were analyzed by specific congeners whenever possible.

Results obtained for each homologue were summed up and are presented in Figure 9.5 in terms of PCDD/PCDF mass basis and dioxin-TEQ. The results showed that the amount of dioxins that were estimated to originate from PCP impurities increased rapidly in the 1960s. This corresponded well with the past use of PCP as an herbicide (Figure 9.1). On the other hand, amount of dioxin estimated to originate from CNP impurity increased in the early 1970s (Figure 9.5), which corresponded well with the past use of CNP (Figure 9.1). The amount of dioxin from CNP was significant in terms of dioxin mass basis (Figure 9.5, upper graph), but was not significant in terms of dioxin TEQ (Figure 9.5, lower graph), this was probably because CNP contained high concentrations of 1,3,6,8- and 1,3,7,9-TeCDDs, which were nontoxic dioxin congeners. However, CNP is not necessarily less important when compared to PCP because the lowly chlorinated 2,3,7,8-chlorine substituted dioxins in CNP are more bioaccumulative through the food chain than the highly chlorinated 2,3,7,8-chlorine substituted dioxins which are abundant in PCP [22].

Although the general trends in the use of dioxin containing agrochemicals and dioxin source apportionment in the core were in good agreement, the relative estimated dioxin emissions from PCP and CNP (Table 9.2) and those recorded in the sediment core did not agree as well. Even though the estimated dioxin emissions (based on TEQ) from PCP and CNP were almost the same, the amount reserved in the core from PCP was much larger than that from CNP. The cause of this discrepancy may have resulted from (1) the uncertainty of the average dioxin content in agrochemicals,

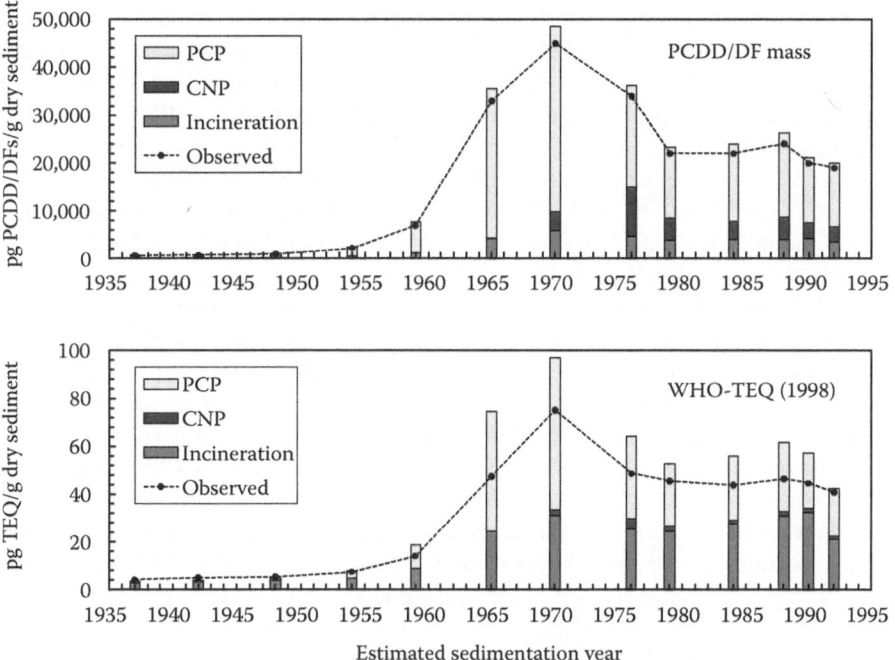

FIGURE 9.5 Source apportionment of dioxins in Tokyo Bay sediment core. Upper graph: dioxin mass. Lower graph: PCDDs/PCDFs (excluding coplanar PCBs) in WHO-TEQ (1998). Those figures do not include the contributions of coplanar PCBs.

(2) the difference between the relative use ratio between PCP and CNP in the Tokyo Bay basin and the Japanese average, and/or (3) the larger percentage of runoff from PCP originated dioxins than from CNP due to differences between the periods of use.

In the previous section, the dioxin emissions from waste incineration were estimated to have increased continuously from the 1960s to the mid-1990s. The amount of dioxin apportioned to incineration or heat processes, however, increased greatly during the 1960s but then leveled off after the 1970s and did not agree well with the previous estimation. It is assumed that dioxin emissions increased during the 1960s with the high economic growth in that period. Even though the amount of waste incinerated continued to increase even after the 1960s, the introduction of better flue gas treatment systems might have reduced the concentration and probably counterbalanced any increase.

A similar study was carried out in the sediment core of Lake Shinji, which is located in a rural area of Japan [23]. The results obtained were nearly the same as those for Tokyo Bay and showed that incineration and two agrochemicals (PCP and CNP) had been the most important dioxin sources in Japan.

9.4   TRENDS OF HUMAN DIOXIN EXPOSURE IN JAPAN

The general trends of dioxin pollution in conservative environmental media in Japan were discussed in the previous sections. These trends indicated that the emission and pollution levels were highest during the 1960s and 1970s. In this section, the trends of exposure will be examined.

9.4.1   ESTIMATED DIOXIN INTAKE BY TOTAL DIET STUDIES

The National Nutrition Survey are being conducted in Japan since the 1940s [24]. Food samples for the total diet study were bought and prepared according to the results of the annual survey. The samples were assumed to represent the average food intake of the Japanese population. The dioxin concentration in both the ongoing total diet study samples and in the archived total diet samples were determined by the Ministry of Health, Labor and Welfare in Japan [25]. The results of the archived samples from the Kansai Area of Japan are summarized in Figure 9.6. It was found that the total dioxin daily intake decreased over the past 20 years, especially from 1977 to 1992. This trend was similar for both PCDDs/DFs and coplanar PCBs. The contributions of different food categories to total TEQ, however, were quite different for PCDDs/DFs and coplanar PCBs.

In the case of PCDDs/DFs, the contributions of three food categories, namely fish and shellfish, meat and eggs, and milk and dairy products, were similar during 1977 and 1988. After the 1990s, the contribution of two food categories, meat and eggs, and milk and dairy products, decreased while that of fish and shellfish remained the same. These phenomena can be explained by the finding of the author's group described in the previous sections. Agrochemicals, namely PCP and CNP, contaminated with PCDD/DF impurities, but not with coplanar PCBs, were extensively used in Japanese agriculture during the 1960s and 1970s. The feed harvested in or around the impacted field must have also been contaminated with PCDDs/DFs. For example, rice straw was used as cattle feed in the 1960s and 1970s [26]. The cessation of PCP use in the early 1970s and the reduction of toxic dioxin content in CNP in the early 1980s, in addition to the change of feed sources from domestic to import, might have been the major causes of the reduction of dioxin intake from those food categories. Fortunately rice, which was the major crop in which PCP and CNP were used, was not an important exposure pathway even in the 1970s and 1980s because dioxin did not accumulate in rice grains.

In the case of coplanar PCBs, fish and shellfish were the major route of human intake for the entire study period. The sole source of coplanar PCBs must have been the PCB formulations used in various processes and products. The production and use of PCB formulations were completely banned in 1972 in Japan.

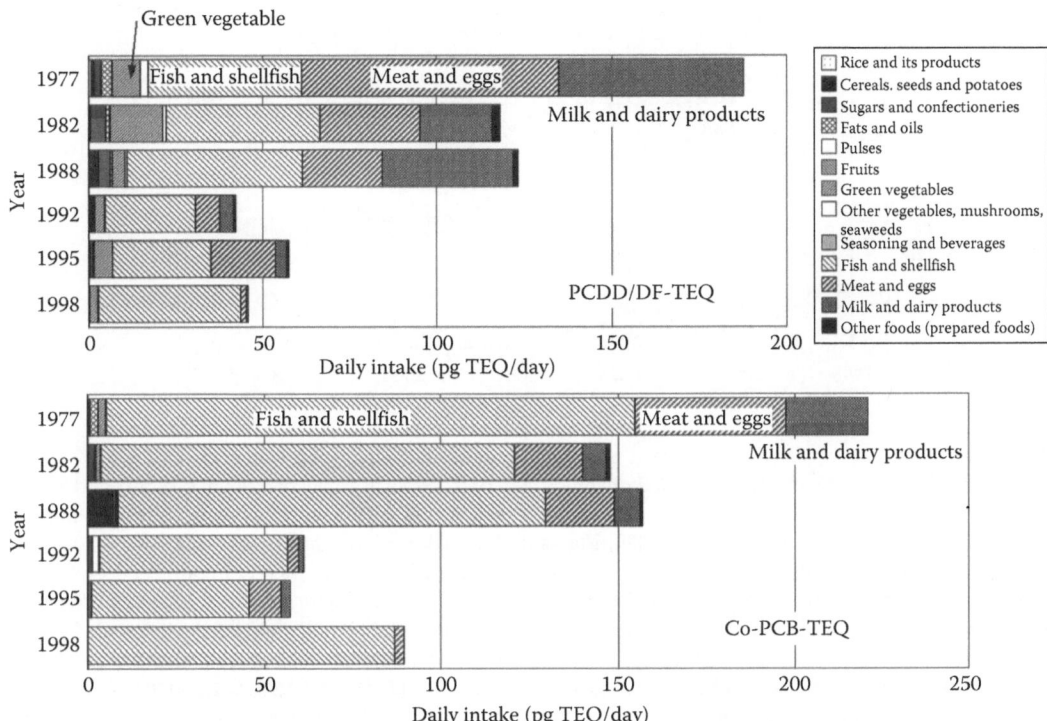

FIGURE 9.6 Trends in daily dioxin intake from food in Kansai, Japan. Data from the total diet study by the Ministry of Health, Labor and Welfare. TEQ was calculated using WHO-TEF (1998). (Data from Ministry of Health, Labour and Welfare, Policy on the dioxin in food, http://www.mhlw.go.jp/topics/bukyoku/iyaku/syoku-anzen/dioxin/, accessed June 1, 2010, in Japanese.)

The recent trend of daily dioxin exposure in Tokyo, Japan is shown in Figure 9.7. This figure is based on the results of the total diet study and ambient dioxin monitoring carried out every year by the Tokyo metropolitan government [27,28]. Because of the introduction of monitoring and the strict regulation of incinerators, a reduction in dioxin exposure through air and soil was observed around 2000. However, this contributed little to the reduction of total exposure because dioxin exposure through air and soil was small compared to the exposure through food. On the other hand, the reduction in exposure through food was not as obvious. This may have been because 70%–80% of the dioxin exposure occurred through seafood consumption, and the dioxin concentration in seafood did not decrease in spite of control measures. These situations indicated that once the coastal area and rivers were contaminated, the reduction of exposure through food, especially from fish and shellfish, was very difficult.

9.4.2 DIOXIN BODY BURDEN: DIOXINS IN MOTHER'S MILK

In the previous sections, it was shown that the estimated trend of dioxin emissions in Japan (Figure 9.2) was consistent with the dioxin contamination recorded in the Tokyo Bay sediment core and also in the archived food samples. Now, historical trends of dioxin body burden in humans will be investigated. Hori et al. [29] analyzed pooled archived human milk samples collected in the Osaka area in Japan. Their results showed that the dioxin concentration had dropped from 60 pg TEQ/g lipid in the early 1970s to about 25 pg TEQ/g lipid in the late 1990s (Figure 9.8). The decreasing trend shown in the dioxin intake study was supported by this human milk study. Even though the PCDD/DF concentration decreased by about 50%, the decrease in the coplanar PCB concentration was

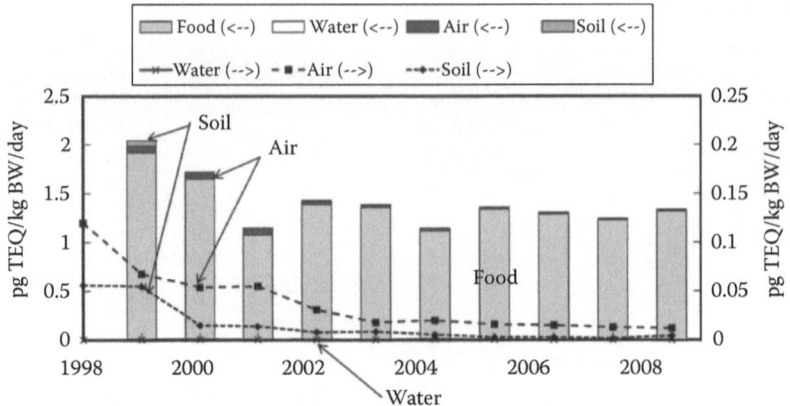

FIGURE 9.7 Recent trends in daily dioxin exposure through food, drinking water, air and soil in Tokyo, Japan. (Data from Sasamoto, T. et al., *Chemosphere* 64, 634, 2006; Bureau of Social Welfare and Public Health, Tokyo Metropolitan Government, Dioxin exposure survey, 2010, http://www.fukushihoken.metro.tokyo.jp/kankyo/kankyo_eisei/chosa/dxn_chemi/taisaku/dxn_bakuro/index.html, accessed June 1, 2010, in Japanese.) Dioxin exposures through various media were calculated as follows: (1) Food: Total diet study. (2) Drinking water: Average dioxin concentration in tap water and the amount of water intake (2 L/day/person). (3) Air: Average dioxin concentration in air and the amount of air inhaled (15 m^3/day/person). (4) Soil: Average concentration in playground soil and the amount of soil ingested (0.1 g/day/person) or contacted through derma (2.5 g/day/person). The percentage absorbed (25% and 1%) and contact frequency (1 and 0.17/day) for soil ingested and soil contacted through derma, respectively. The average body weight of a Japanese adult is assumed to be 50 kg. TEQ was calculated using WHO-TEF (2006). Refer to the left scale for bar graphs. Refer to the right enlarged scale for line graphs.

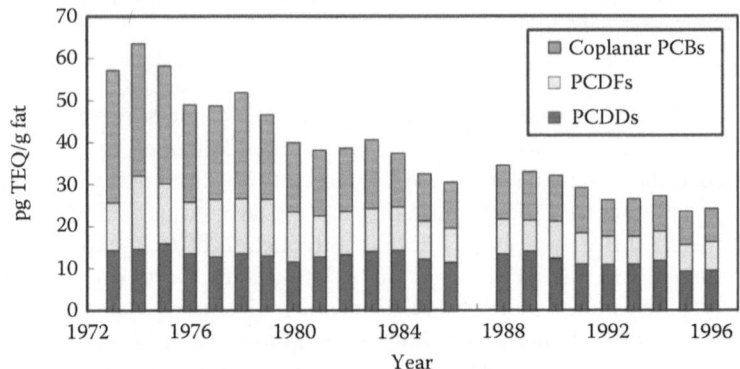

FIGURE 9.8 Trends in dioxin concentrations in archived human milk samples in Japan. (Data from Hori, S. et al., *Organohalogen Compd.*, 44, 141, 1999.) TEQ was calculated using WHO-TEF (1998).

more remarkable and dropped by 75% in the same period. In the 1990s, the intake of dioxin TEQ from coplanar PCBs was somewhat larger than that from PCDDs/DFs (Figure 9.6). The dioxin TEQ from PCDDs/DFs, however, was over twice as high than the TEQ from coplanar PCBs in human milk. This difference may be due to the longer half-lives of PCDD/DF congeners in the human body.

9.5 SUMMARY

In this chapter, the past and present status of dioxin contamination in Japan was investigated. As a whole, the trends of estimated dioxin emissions by the author, the dioxin contamination recorded in the Tokyo Bay sediment core, the dioxin intake calculated from the archived food samples from the total diet study and the dioxin concentrations in archived human milk samples were all in good agreement. They collectively show that dioxin contamination and exposure have been decreasing in Japan since the early 1970s. The use of agrochemicals containing high concentrations of PCDDs/DFs as impurities, in addition to the use of PCB formulations, was found to be the major determinant of these trends in Japan. The contribution of dioxins that originated from waste incineration and other heat processes seemed to have less of an impact on the decreasing trends. However, because the impact of the historical use of agrochemicals and PCB formulations decreased, the contribution of combustion processes has become comparatively larger.

These observations showed that before the government started to take active measures to reduce dioxin emission around 1997, the dioxin exposure was already decreasing due to the cessation of the use of certain chlorinated organic chemicals and the introduction of flue gas treatment intended to reduce pollutants other than dioxins.

ACKNOWLEDGMENTS

The author would like to express his sincere thanks to his colleagues, Dr. Junko Nakanishi, Dr. Takeo Sakurai, Dr. Kikuo Yoshida, Dr. Isamu Ogura, Dr. Yuan Yao, Dr. Youn-Seok Kang, and Hiroshi Hamada, who carried out this dioxin study together as part of the JST CREST project.

REFERENCES

1. Buser, H. R. and Bosshardt H. P. 1976. Determination of polychlorinated dibenzo-*p*-dioxins and dibenzofurans in commercial pentachlorophenols by combined gas chromatography-mass spectrometry. *Journal of the AOAC* 59: 562–569.
2. Rappe, C., Gara, A., and Buser, H. R. 1978. Identification of polychlorinated dibenzofurans (PCDFs) in commercial chlorophenol formulations. *Chemosphere* 7: 981–991.
3. Institute of Medicine. 1994. *Veterans and Agent Orange: Health Effects of Herbicides Used in Vietnam.* Washington, DC: The National Academies Press.
4. Ramondetta M. and Repossi, A. 1998. *Seveso 20 Years After: From Dioxin to the Oak Wood.* Fondazione Lombardia per l'Ambiente: Milano, Italy, Dossier No. 33.
5. Masunaga, S., Takasuga, T., and Nakanishi, J. 2001. Dioxin and dioxin-like PCB impurities in some Japanese agrochemical formulations. *Chemosphere* 44: 873–885.
6. Masunaga, S. 1999. Toward a time trend analysis of dioxin emission and exposure. *Proceedings of the 2nd International Workshop on Risk Evaluation and Management of Chemicals*, Yokohama, Japan, January 28, 1999, Vol. 1, pp. 1–10.
7. Mitsui Chemicals. July 8, 1999. Press release: Analytical result of dioxins in CNP and recall, storage of CNP herbicide (in Japanese). http://jp.mitsuichem.com/ (accessed in July, 2000).
8. Ministry of Agriculture, Forestry and Fisheries. July 8, 1999. Press release: Reconfirmation of dioxins contained in agrochemicals (in Japanese). http://www.maff.go.jp/work/990714.pdf (accessed in July, 2000).
9. Seike, N., Otani, T., Ueji, M., Takasuga, T., and Tsuzuki, N. 2003. Temporal change of polychlorinated dibenzo-*p*-dioxins, dibenzofurans and dioxin-like polychlorinated biphenyls source in paddy soils. *Journal of Environmental Chemistry* 13: 117–131, Tokyo, Japan (in Japanese).
10. Mitsui Chemicals. 2002. Press release: Analytical results of dioxins in CNP herbicide (in Japanese). http://jp.mitsuichem.com/release/2002/pdf/020412a.pdf (accessed on March 30, 2011).
11. Ministry of Agriculture, Forestry and Fisheries. 1980–1997. *Noyaku Yoran (Agrochemical Digest)*, Nihon Shokubutu Boueki Kyokai (Japan Plant Protection Association) (in Japanese).
12. Ministry of Health and Welfare. 1997/1998. On dioxin concentration in flue gas from municipal solid waste incinerators (in Japanese). http://www.env.go.jp/recycle/kosei_press/h971017a.html; http://www.env.go.jp/recycle/kosei_press/h970624a.html; http://www.env.go.jp/recycle/kosei_press/h970411a.html (accessed in March, 2011).
13. Ministry of Health and Welfare. 1997. Present status of treatment and disposal of industrial waste (in Japanese). http://www.env.go.jp/recycle/kosei_press/h970910b.html (accessed in March, 2011).

14. Environment Agency. 1998. Report of advisory board on reduction of dioxin emission (in Japanese). http://www.env.go.jp/press/press.php?serial=2192 (accessed in March, 2011).

15. Weber, R. et al. 2008. Dioxin- and POP-contaminated sites—Contemporary and future relevance and challenges. *Environmental Science & Pollution Research* 15: 363–393.

16. Government of Japan. 2009. Information Brochure Dioxins. http://www.env.go.jp/en/chemi/dioxins/brochure2009.pdf (accessed on June 1, 2010).

17. Sanada, Y. et al. 1999. Estimation of sedimentation processes in Tokyo Bay using radionuclides and anthropogenic molecular markers. *Chikyukagaku (Geochemistry)* 33: 123–138 (in Japanese).

18. Sakurai, T. et al. 2000. Polychlorinated dibenzo-*p*-dioxins and dibenzofurans in sediment, soil, fish, shellfish, and crab samples from Tokyo Bay area, Japan. *Chemosphere* 40: 627–640.

19. Yao, Y., Masunaga, S., Takada, H., and Nakanishi, J. 2002. Identification of polychlorinated dibenzo-*p*-dioxin, dibenzofuran, and coplanar polychlorinated biphenyl sources in Tokyo Bay, Japan. *Environmental Toxicology & Chemistry* 21: 991–998.

20. Masunaga, S., Yao, Y., Takada, H., Sakurai, T., and Nakanishi, J. 2001. Source apportionment of dioxin pollution recorded in Tokyo Bay sediment core based on congener composition. *Chikyukagaku (Geochemistry)* 35: 159–168 (in Japanese).

21. Masunaga, S., Yao, Y., Ogura, I., Sakurai, T., and Nakanishi, J. 2003. Source and behavior analyses of dioxins based on congener-specific information and their application to Tokyo Bay basin. *Chemosphere* 53: 315–324.

22. Naito, W., Jin, J., Kang, Y. S., Yamamuro, M., Masunaga S., and Nakanishi, J. 2003. Dynamics of PCDDs/DFs and coplanar-PCBs in an aquatic food chain of Tokyo Bay. *Chemosphere* 53: 347–362.

23. Masunaga, S., Yao, Y., Ogura, I., Nakai, S., Kanai, Y., Yamamuro, M., and Nakanishi, J. 2001. Identifying sources and mass balance of dioxin pollution in Lake Shinji Basin, Japan. *Environmental Science & Technology* 35: 1967–1973.

24. National Institute of Health and Nutrition. National Nutrition Survey. http://www.nih.go.jp/eiken/chosa/kokumin_eiyou/index.html (accessed June 1, 2010) (in Japanese).

25. Ministry of Health, Labour and Welfare. Policy on the dioxin in food. http://www.mhlw.go.jp/topics/bukyoku/iyaku/syoku-anzen/dioxin/ (accessed June 1, 2010) (in Japanese).

26. Kameda, Y., Masunaga, S., Hamada, H., and Nakanishi, J. 2003. Historical trends of dioxin and agrochemicals in rice straw and their impact on daily PCDD/Fs intake via foods. *Journal of Environmental Chemistry* 13: 369–383 (in Japanese).

27. Sasamoto, T., Ushio, F., Kikutani, N., Saitoh, Y., Yamaki, Y., Hashimoto, T., Horii, S., Nakagawa, J., and Ibe, A. 2006. Estimation of 1999–2004 dietary daily intake of PCDDs, PCDFs and dioxin-like PCBs by a total diet study in metropolitan Tokyo, Japan. *Chemosphere* 64: 634–641.

28. Bureau of Social Welfare and Public Health, Tokyo Metropolitan Government. Dioxin exposure survey. http://www.fukushihoken.metro.tokyo.jp/kankyo/kankyo_eisei/chosa/dxn_chemi/taisaku/dxn_bakuro/index.html (accessed in June 1, 2010) (in Japanese).

29. Hori, S., Konishi, Y., and Kuwabara, K. 1999. Decrease of PCDDs, PCDFs and co-PCBs levels in human milk from Osaka (1973–1996). *Organohalogen Compounds* 44: 141–144.

10 Polychlorinated Naphthalenes
Use and Contamination Trends in Japan and China

Sachi Taniyasu,* Yuichi Horii, Nobuyasu Hanari,
Nobuyoshi Yamashita, Jing Pan, Yongliang Yang,
and Bommanna G. Loganathan

CONTENTS

10.1 Introduction 216
    10.1.1 Chemical Structure and Physicochemical Properties........................... 216
    10.1.2 Production and Uses 218
    10.1.3 Sources.. 219
    10.1.4 Toxicity 220
    10.1.5 PCN Regulation in Japan... 221
10.2 Environmental Occurrence and Concentration Levels.. 222
    10.2.1 Abiotic Environment.. 222
        10.2.1.1 Water.. 222
        10.2.1.2 Sediment 226
        10.2.1.3 Gas 226
        10.2.1.4 Ash 227
        10.2.1.5 Ambient Air 228
    10.2.2 Biotic Environment.. 230
        10.2.2.1 Animal 231
        10.2.2.2 Human... 231
10.3 Isomer-Specific Analysis of PCNs... 232
    10.3.1 PCN Standards 232
    10.3.2 Extraction.. 233
    10.3.3 Clean-Up and Separation 233
    10.3.4 Quantification 234
    10.3.5 Interlaboratory Comparison 234
    10.3.6 Advanced Applications... 235
        10.3.6.1 Comprehensive Two-Dimensional Gas Chromatography (GC×GC) 235
        10.3.6.2 Congener-Specific Carbon Isotopic Analysis of PCNs....... 235
10.4 Historical Reconstruction of PCN Pollution Using Sediment Core in Japan...... 236
    10.4.1 Tokyo Bay 238
    10.4.2 Lake Kitaura 239
    10.4.3 CN Homologue and Congener Patterns.. 239
    10.4.4 Flux... 241

* E-mail: s-taniyasu@aist.go.jp (Chapter corresponding author).

10.4.5 Occurrence of PCNs in the Preindustrial Period..242
10.4.6 Novel Halogenated Polycyclic Aromatic Hydrocarbons 242
10.5 Accidental Contamination of PCNs in Industrial Products in Japan 242
10.5.1 Background...242
10.5.2 Isomer-Specific Analysis of PCNs in Contaminated Commercial Products 243
10.5.3 Results and Discussions...245
10.5.3.1 Risk of PCN in Contaminated Commercial Products to Our Daily Life 246
10.6 Polychlorinated Naphthalenes in China 249
10.6.1 Introduction 249
10.6.2 Research on PCNs in China 249
10.6.2.1 Sources...249
10.6.2.2 Levels of PCNs 250
10.6.2.3 Sediments...250
10.6.2.4 Soils...250
10.6.2.5 Sewage Sludge 250
10.6.2.6 Biota..251
10.6.2.7 Fly Ash...251
10.6.2.8 Human Exposure 251
10.6.3 Conclusions..252
References..252

10.1 INTRODUCTION

10.1.1 Chemical Structure and Physicochemical Properties

Polychlorinated naphthalenes (PCNs) are industrial chemicals and known as persistent organic pollutants [1–4], consisting of two fused aromatic rings substituted with one to eight chlorine atoms, forming a total of 75 possible congeners with different physical and chemical properties (Figure 10.1 and Table 10.1). PCNs are structurally similar to polychlorinated biphenyls (PCBs).

They were first synthesized in 1833, and commercial production began early in the twentieth century [5]. Since the physicochemical properties of PCNs are similar to PCBs with their hydrophobic, high thermal stability, and inertness, they have been used in similar industrial application in the electrical industry as dielectric fluids in transformers and capacitors and as cable insulators [2,4,6]. Physicochemical properties for some PCN congeners have been measured but not all 75 [7–10]. PCNs are water insoluble in principle, with a span of aqueous solubility ranging from 2.87 mg/dm³ for CN1 to 0.08 µg/dm³ for CN75 [11,12]. Physicochemical properties such as octanol–water partition coefficients (K_{OW}), octanol–air partition coefficients (K_{OA}), and air–water partition coefficients (K_{AW}) are important in order to characterize environmental fate. Puzyn and Falandysz calculated log K_{OW} and log K_{OA} for all 75 PCN congeners by quantitative structure–property relationships (QSPR). The predicted log K_{OW} values of CNs are 3.93–3.97 for mono-CNs, 4.20–4.67 for di-CNs, 4.59–5.50 for tri-CNs, 5.14–6.10 for tetra-CNs, 5.67–6.49 for penta-CNs, 6.02–6.68 for hexa-CNs, 6.48–6.57 for hepta-CNs, and 6.43 for octa-CN [13]. The log $K_{OW} > 5$ is one of the

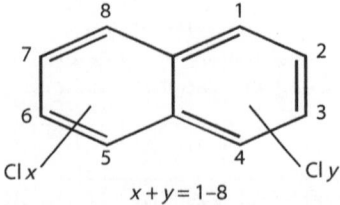

FIGURE 10.1 Naphthalene structure and chlorine substitution position numbered.

TABLE 10.1
Chlorine Substitute Position and IUPAC Number for PCN

No.	Substitution Position	No.	Substitution Position
Mono-CNs		**Tetra-CNs**	
1	1-CN	39	1,2,6,7-tetra-CN
2	2-CN	40	1,2,6,8-tetra-CN
Di-CNs		41	1,2,7,8-tetra-CN
3	1,2-di-CN	42	1,3,5,7-tetra-CN
4	1,3-di-CN	43	1,3,5,8-tetra-CN
5	1,4-di-CN	44	1,3,6,7-tetra-CN
6	1,5-di-CN	45	1,3,6,8-tetra-CN
7	1,6-di-CN	46	1,4,5,8-tetra-CN
8	1,7-di-CN	47	1,4,6,7-tetra-CN
9	1,8-di-CN	48	2,3,6,7-tetra-CN
10	2,3-di-CN	**Penta-CNs**	
11	2,6-di-CN	49	1,2,3,4,5-penta-CN
12	2,7-di-CN	50	1,2,3,4,6-penta-CN
		51	1,2,3,5,6-penta-CN
13	1,2,3-tri-CN	52	1,2,3,5,7-penta-CN
14	1,2,4-tri-CN	53	1,2,3,5,8-penta-CN
15	1,2,5-tri-CN	54	1,2,3,6,7-penta-CN
16	1,2,6-tri-CN	55	1,2,3,6,8-penta-CN
17	1,2,7-tri-CN	56	1,2,3,7,8-penta-CN
18	1,2,8-tri-CN	57	1,2,4,5,6-penta-CN
19	1,3,5-tri-CN	58	1,2,4,5,7-penta-CN
20	1,3,6-tri-CN	59	1,2,4,5,8-penta-CN
21	1,3,7-tri-CN	60	1,2,4,6,7-penta-CN
22	1,3,8-tri-CN	61	1,2,4,6,8-penta-CN
23	1,4,5-tri-CN	62	1,2,4,7,8-penta-CN
24	1,4,6-tri-CN	**Hexa-CNs**	
25	1,6,7-tri-CN	63	1,2,3,4,5,6-hexa-CN
26	2,3,6-tri-CN	64	1,2,3,4,5,7-hexa-CN
		65	1,2,3,4,5,8-hexa-CN
27	1,2,3,4-tetra-CN	66	1,2,3,4,6,7-hexa-CN
28	1,2,3,5-tetra-CN	67	1,2,3,5,6,7-hexa-CN
29	1,2,3,6-tetra-CN	68	1,2,3,5,6,8-hexa-CN
30	1,2,3,7-tetra-CN	69	1,2,3,5,7,8-hexa-CN
31	1,2,3,8-tetra-CN	70	1,2,3,6,7,8-hexa-CN
32	1,2,4,5-tetra-CN	71	1,2,4,5,6,8-hexa-CN
33	1,2,4,6-tetra-CN	72	1,2,4,5,7,8-hexa-CN
34	1,2,4,7-tetra-CN	**Hepta-CNs**	
35	1,2,4,8-tetra-CN	73	1,2,3,4,5,6,7-hepta-CN
36	1,2,5,6-tetra-CN	74	1,2,3,4,5,6,8-hepta-CN
37	1,2,5,7-tetra-CN	**Octa-CN**	
38	1,2,5,8-tetra-CN	75	1,2,3,4,5,6,7,8-octa-CN

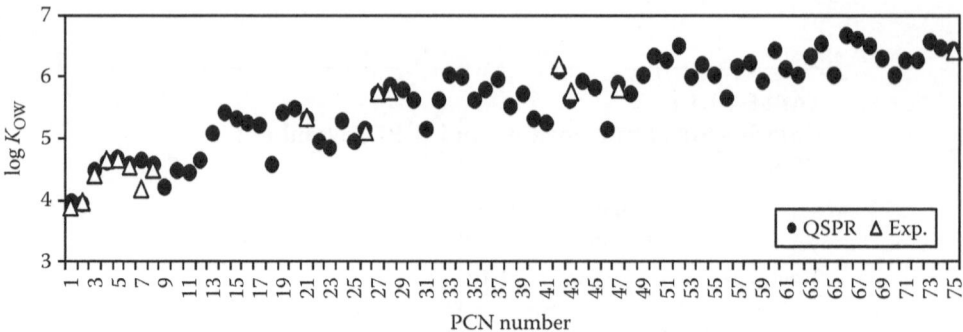

FIGURE 10.2 QSPR predicted and experimentally measured log K_{OW} values. (QSPR data from Puzyn, T. and Falandysz, J., *J. Phys. Chem. Ref. Data*, 36, 203, 2007; Exp. data from Opperhulzen, A. et al., *Chemosphere*, 14, 1871, 1985.)

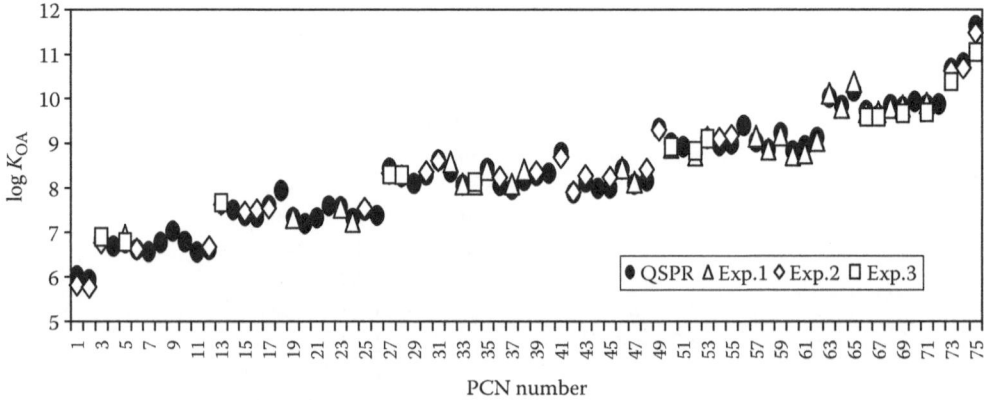

FIGURE 10.3 QSPR predicted and experimentally measured log K_{OA} values. (QSPR data from Puzyn, T. and Falandysz, J., *J. Phys. Chem. Ref. Data*, 36, 203, 2007; Exp. 1 data from Harner, T. and Bidleman, T.F., *Environ. Sci. Technol.*, 32, 1494, 1998; Exp. 2 data from Su, Y.S. et al., *J. Chem. Eng. Data*, 47, 449, 2002; Exp. 3 data from Wania, I.R. et al., *Anal. Chem.*, 74, 3476, 2002.)

criteria of bioaccumulation for screening persistent organic pollutants (POPs) under the Stockholm Convention and some of the di-CNs to octa-CN have been identified as bioaccumulative congeners. The predicted log K_{OA} values of CNs are 5.93–6.02 for mono-CNs, 6.55–7.02 for di-CNs, 7.19–7.94 for tri-CNs, 7.88–8.79 for tetra-CNs, 8.79–9.40 for penta-CNs, 9.62–10.17 for hexa-CNs, 9.62–10.17 for hepta-CNs, and 11.64 for octa-CN. These values are comparable to that obtained for PCBs with similar degrees of chlorination [13,14]. Figures 10.2 and 10.3 show log K_{OW} and log K_{OA} values for both predicted values by QSPR and experimentally measured values.

10.1.2 Production and Uses

PCNs were first patented as flame retardants and dielectric fluids for capacitors in the early 1900s [15]. Melted naphthalene, chlorine, and catalyst (FeCl$_3$ or SbCl$_5$) are the substrates involved in the synthesis of PCNs on a technical scale [4]. PCNs were used in a variety of industrial applications such as flame retardants, cable insulation, engine oil additives, wood preservatives, electroplating masking compounds, and feedstock for dye production [1,2,5]. PCNs are no longer commercially produced but when they were, they were sold as mixtures of several congeners with different product names. PCNs were originally synthesized in 1833 and large amounts of PCNs were manufactured in several countries between 1910 and 1980 [1,2,5]. Technical PCN mixtures with chlorine contents ranging from 22 to 70 wt%, known as Halowaxes™, have been produced by Koppers Company in the United States since the 1920s [2,16]. N-Oil and N-Wax product (Halochem), Basileum products (Desowag-Bayer, Germany), Nirben wax products (I.G.Farbenindustrie/BayerLeverkusen, Germany), Seekay wax Products (ICI Runcorn, Great Britain), and Clonacire wax products (Prodelec, Paris, France) were also major technical PCN mixtures. Some minor PCN technical mixtures include Cerifal Material Product (Caffaro, Italy), Wako-PCN(Wako Chemicals, Japan), and Woskol Product (Zakady Azotowe, Poland) [17]. Until the 1970s PCNs were high volume chemicals. The manufacture and use of technical PCN mixtures were largely phased out in the 1980s [1,18], but up until then they were high volume chemicals. In fact, the global production was estimated to be approximately 150,000 metric tons or 10% of the global PCB production [1]. The total PCNs production was quantified as 4004.6 metric tons from 1940 to 1976 [19,20]. Today, manufacture of PCNs is thought to have ended, although 18 ton of PCN-containing products were illegally imported from the United Kingdom into Japan [21]. This will be discussed later in Section 10.5.

10.1.3 SOURCES

PCNs have legacy sources due to evaporation from PCB products used in various applications and from contaminated soil sources. PCNs are also formed as a combustion by-product from various industrial and waste incineration processes in a similar fashion to polychlorinated dibenzo-p-dioxins (PCDDs), polychlorinated dibenzofurans (PCDFs), and PCBs [15,19,22–24].

Chlorine content depends on technical PCN mixture compositions (Figure 10.4). The technical formulations show a wide range of patterns from nearly pure mono-CNs (Halowax 1031) to nearly

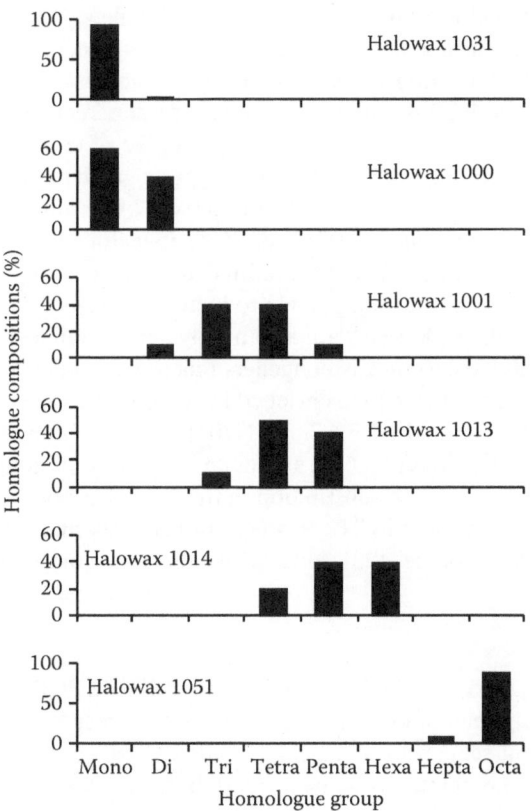

FIGURE 10.4 Homologue composition of technical PCN mixtures of selected Halowaxes. (Data from Falandysz, J. et al., *J. Environ. Sci. Health A*, 41, 2237, 2006.)

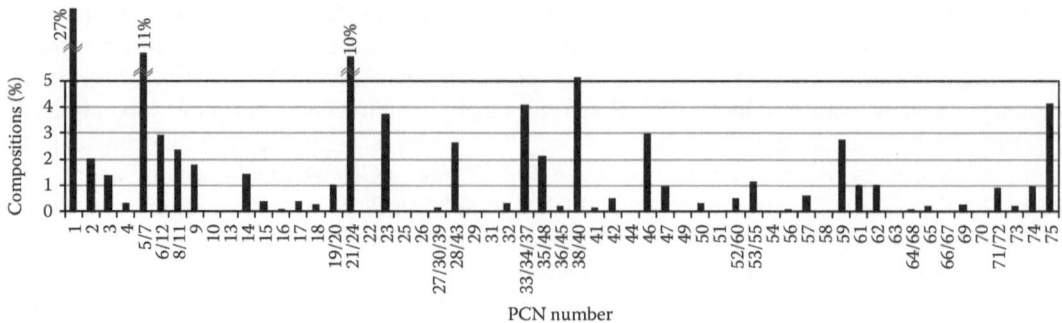

FIGURE 10.5 PCN congener composition of Equi-Halowax. (Data from Falandysz, J. et al., *J. Environ. Sci. Health A*, 35, 281, 2000.)

pure octa-CNs (Halowax 1051). The most abundant CN congeners in the Halowax mixtures depend on the homologue group, type of formulation, and quality control [18]. PCN congener composition in Equi-Halowax (an equivalent mixture of Halowax 1031, 1000, 1001, 1099, 1013, 1014, and 1051) is shown in Figure 10.5. Some of the PCN congeners such as 1,4,5-(CN23), 1,4,6-(CN24), 1,2,4,7-(CN34), 1,2,5,8-(CN40), 1,4,5,8-(CN46), 1,2,4,5,8-(CN59), 1,2,4,5,6,8-/1,2,4,5,7,8-(CN71/ CN72) have been detected in high concentration in several Halowax mixtures [16,18,23]. It was indicated that the congeners most abundant in Halowax mixtures are usually chlorinated at the vicinal α-positions (1,4,5,8-) of the naphthalene [16,23].

Commercial technical PCB mixtures also contain PCN by-products and is one source of contamination to the environment (technical biphenyl contains naphthalene [1,25,26]). Yamashita et al. and Taniyasu et al. reported concentrations and profiles of tri- through octa-PCN congeners in technical PCB mixtures such as Aroclors, Kanechlors, Clophens, Phenoclors, Sovol, Chlorofen, and Delor. In general, highly chlorinated PCB mixtures contained greater percentages of the more chlorinated naphthalene congeners. On the other hand, the average degree of chlorination of each of the CNs was greater than that for the corresponding CB. This suggests greater reactivity of naphthalene than of biphenyl [26]. PCNs were found in all the PCB mixtures as by-products at concentrations ranging from 5.2 to 730 µg/g and contained up to 0.073% of PCNs in technical mixtures of Sovol PCB from the former USSR. Total worldwide production is estimated at 169 ton of by-product in the production of PCBs. This amount is <0.1% of the total global production of PCBs [25]. PCN congeners detected with high concentrations in the Halowaxes are also highly detected in technical PCB mixtures.

Many studies have identified certain CN congeners that are absent or occur at low levels in commercial PCN and PCB mixtures, but yet are enriched in combustion sources such as 1,2,3,6-(CN29), 1,3,6,7-(CN44), 1,2,3,6,7-(CN54), and 1,2,3,6,7,8-(CN70) [15,18,22]. Many PCN congeners with substitution in more than two β-positions (2,3,6,7-) are found in combustion sources such as fly ash, while there were scarce PCN congeners with substitution in the vicinal α-positions (1,4,5,8-). Meanwhile, there are more α-position substituted PCN congeners in Halowax and more β-position substituted PCN congeners in combustion sources [23]. This will be discussed more in Section 10.2.1.

10.1.4 TOXICITY

The mechanism of toxic action for PCNs is similar to PCDDs, PCDFs, and PCBs due to their structural similarities. PCNs can bind to the aryl hydrocarbon receptor (AhR). The AhR can then link up with many different sequences along the strands of DNA and thus influence the synthesis of at least 20 proteins. 2,3,7,8-Tetrachlorodibenzo-p-dioxin (TCDD) is one of the compounds that is strongly attracted to the AhR and is, therefore, toxic at low doses [27]. The dioxin-like toxicity of a single compound is expressed as relative potency (REP). A compound that has REP can cause AhR-mediated effects by comparing the amount of the chemical required to cause the same level

TABLE 10.2
REPs of Individual PCNs

CN No.	Substitution Position	REP
5	1,4-di-CN	3.5×10^{-8} [a]
40	1,2,6,8-tetra-CN	1.65×10^{-5} [b]
54	1,2,3,6,7-penta-CN	1.7×10^{-4} [a]
56	1,2,3,7,8-penta-CN	4.6×10^{-5} [a]
57	1,2,4,5,6-penta-CN	3.5×10^{-6} [a]
63	1,2,3,4,5,6-hexa-CN	2.0×10^{-3} [a]
64	1,2,3,4,5,7-hexa-CN	2.0×10^{-5} [a]
66	1,2,3,4,6,7-hexa-CN	4.0×10^{-3} [c]
67	1,2,3,5,6,7-hexa-CN	1.0×10^{-3} [c]
66/67	1,2,3,4,6,7/1,2,3,5,6,7-hexa-CN	1.3×10^{-3} [c]
68	1,2,3,5,6,8-hexa-CN	1.5×10^{-4} [c]
69	1,2,3,5,7,8-hexa-CN	2.0×10^{-3} [a]
70	1,2,3,6,7,8-hexa-CN	9.9×10^{-3} [a]
73	1,2,3,4,5,6,7-hepta-CN	6.9×10^{-4} [a]

[a] Data from Villeneuve, D.L. et al., *Arch. Environ. Sci. Technol.*, 39, 273, 2000.

[b] Data from Kannan, K. et al., *Environ. Sci. Technol.*, 32, 2507, 1998.

[c] Data from Blankenship, A.L. et al., *Environ. Sci. Technol.*, 34, 3153, 2000.

of response as TCDD [28]. REP values have been determined for some of the PCN congeners by induction of aryl hydrocarbon hydroxylase (AHH), ethoxyresorufin-O-deethylase (EROD), and luciferase using H4IIE rat hepatoma cell [29–31].

REPs determined in several studies are summarized in Table 10.2 [2,30,31]. The most potent RPEs were found for 1,2,3,4,5,6-(CN63), 1,2,3,5,7,8-(CN69), and 1,2,3,6,7,8-(CN70), with values of 2.0×10^{-3}, 2.0×10^{-3}, and 9.9×10^{-3} respectively. These results were similar to those for some PCBs; however, experimental data on the toxicity of individual PCNs are rather limited. REPs for all 75 PCN congeners are available by QSAR estimation. REP values estimated for individual CNs were based on the H4IIE-luc assay and ranged from 4.0×10^{-8} to 1.8×10^{-3}. 1,2,3,4,6,7-(CN66), 1,2,3,5,6,7-(CN67), 1,2,3,6,7,8-(CN70), and 1,2,3,4,5,6,7-(CN73) exhibited the greatest REP values of 2.9×10^{-3}, 1.7×10^{-3}, 7.1×10^{-4}, and 1.8×10^{-3} [28]. REPs are used as toxic equivalency factors (TEFs), which are fractional potencies that relate a compound's potency to that of TCDD. The toxic equivalents (TEQs; estimated by multiplying concentrations with TEFs) were evaluated in several matrixes.

10.1.5 PCN REGULATION IN JAPAN

The production and use of PCNs were largely phased out in the United States and in Europe in the 1980s due to their toxicity and environmental persistence [19]. PCNs were also banned for importation, manufacture, and use starting on August 14, 1979 in Japan under the Act on the Evaluation of Chemical Substances and Regulation of Their Manufacture, etc. [19,32]. This law was enacted in 1973 in Japan because serious environmental pollution of PCBs had emerged in the late 1960s. This law regulates manufacture and importation of persistent, bioaccumulative, and toxic chemical substances such as PCBs in order to prevent future environmental pollution. PCNs (limited to those with no less than three chlorine elements) were listed in the Class I Specified Chemical Substances under this law (similar to PCBs). Permission is required for the manufacture and/or importation (de facto prohibition) and also any uses other than those specified by the Class I Specified Chemical Substances are prohibited.

10.2 ENVIRONMENTAL OCCURRENCE AND CONCENTRATION LEVELS

Since PCNs are ubiquitous global pollutants, PCN occurrences in the abiotic and biotic environmental matrices have been found. Similar to that of other countries, it was indicated that environmental releases of PCNs in Japan were analogous to those of PCBs by referring to reports of environmental contamination. Furthermore, these PCN profiles seem to be a combination of technical PCN preparations (PCN formulations), PCNs originating from PCB products, and incineration processes in the presence of chlorine. PCNs are generally present at low levels (pg/L and pg/Nm^3) in water and in ambient air samples from background sites, while high PCN concentration levels (ng/g) have been detected in sediment and ash samples from urbanized/industrialized sites in Japan. Significant high PCN concentrations have been reported in human adipose tissue samples examined (249 ng/g on a fat weight basis) from Osaka, where several types of chemical manufacturing plants are located [33]; though uses of PCN itself were restricted beginning in 1979, leading to a gradual decrease in the environment. In addition, patterns of PCN congeners in more contaminated samples seem to resemble that of Halowax (see Section 10.1.3 for the detailed description of Halowax profiles), while less contaminated samples do not resemble any single Halowax. Even though few studies are available in the literature describing PCN contaminations in Japan. It is important to indicate the environmental occurrence and concentration levels on PCNs in this chapter. Some comprehensive data on concentrations of total PCNs in the environment are summarized in Table 10.3. Apart from data in the 1980s and early 1990s, PCN profiles in environmental matrices have become available. From the data in these practical reports, typical trends of PCN contaminations in the environment are described as below.

10.2.1 ABIOTIC ENVIRONMENT

In the abiotic environment, PCN contaminations were found in several areas including water, sediment, flue gas, ash, and ambient air. Of these, flue gas and ash samples indicated much higher contaminations over other abiotic environmental matrices. In water and sediment samples, not only the concentration levels but also the time trend on PCN contaminations seem to be clarified by referring to reports (especially the report from the Japan Ministry of the Environment). Fingerprints of PCNs in the abiotic environment can also be performed by using PCN profiles in some environmental samples.

10.2.1.1 Water

In water samples in Japan, PCN concentration levels have been reported in river, sea, and tap water. From the results in a recent report [34] the mean value of total PCNs, including mono- to octa-CNs, were determined as 64 pg/L in raw water supply, 40 pg/L in filtrated water, and 39 pg/L in drinking water by a GC/HRMS method. Moreover, Kosaka et al. [34] reported that tri- and tetra-CNs were the predominant congeners in tap water (Figure 10.6), and then suggested that more volatile mono- and di-CNs may be formed by chlorination of the water supply. Nakano et al. [35] investigated profiles of mono- to hexa-CNs in particles in rain, and it was shown that 1,3,7-/1,4,6-tri-CNs and 1,2,3,4,6,7-/1,2,3,5,6,7-hexa-CNs were the predominant isomers. According to the report on environmental survey and monitoring of chemicals from the Japan Ministry of the Environment [36–38], total PCNs decreased from 100 to 450 ng/L in 1977, to 8 to 40 ng/L in 1979, to 0.0052 to 0.094 ng/L in 2002, in water media in Japan. This time trend seems to depend on the restriction of PCN uses and can be supported in the literature as follows. Yoshida and Takeshita [39] reported the total PCNs (2–20 ng/L, a part of tri- to hexa-CN congeners) measured by using a GC-ECD method

TABLE 10.3
Concentrations of Polychlorinated Naphthalenes in the Environment in Japan

Matrix and Site	Measure	Concentration	References
Abiotic			
Water		*ng/L*	
Water media (type and site were unknown)	Range (*n* = 143 or 148) Sum of PCNs (unknown)	100–450	[36]
Water media (type and site were unknown)	Range (*n* = 75) Sum of PCNs (unknown)	8–40	[37]
Sea and river water (Hokkaido and western in Japan)	Range (*n* = 24[a]) Sum of PCNs (mono- to octa-[b])	0.0052–0.094	[42]
Sea water (around Tokyo and Osaka Bay)	Range (*n* = 21[a]) Sum of PCNs (tri- to hexa-)	2–20	[39]
Tap water (Tsukuba)	Range (*n* = 3[a]) Sum of PCNs (mono- to di-[b])	0.03–0.59	[40]
Tap water including raw, filtrated, and drinking water (urbanized sites)	Range (*n* = 16[a]) Sum of PCNs (mono- to octa-[b])	0.019–0.095	[34]
Rain water (only particle shown)	Sum of PCNs (mono- to hexa-)	Only profile	[35]
Sediment		*ng/g dw*	
Sediment media (type and site were unknown)	Range (*n* = 138) Sum of PCNs (unknown)	5–670	[36]
Sediment media (type and site were unknown)	Range (*n* = 75) Sum of PCNs (unknown)	20–1000	[37]
River, port and offshore sediment (Hokkaido and western in Japan)	Range (*n* = 24[a]) Sum of PCNs (mono- to octa-[b])	0.02–4.1	[38]
River sediment (Tokyo and Niigata)	Range (*n* = 3[a]): ng/g Sum of PCNs (tri- to hexa-[b,d])	n.d.–100[c]	[41]
Bay and Cannal sediment (Tokyo and Fukuoka)	Range (*n* = 2[a]): ng/g Sum of PCNs (tri- to hexa-[b,d])	7.3–27.4[c]	[41]
River, lake and offshore sediment (Okayama)	Range (*n* = 12[a]) Sum of PCNs (mono- to octa-[b])	n.d.–1.6 (figure information)	
Sediment media (type and site were unknown)	Range (*n* = 5): ng/g Sum of PCNs (mono- to octa-[b])	0.78–74[c]	[42]
Core sediment (Tokyo Bay)	Depth (2–90 cm): ng/g Sum of PCNs (tri- to octa-[b,d])	0.1–4.4[c] (figure information)	[23]
Core sediment (Tokyo Bay)	Depth (2–90 cm) Sum of PCNs (tri- to octa-[b,d])	0.1–4.4 (figure information)	[25]
Core sediment (River Kurashiki)	Depth (0–60 cm) Sum of PCNs (mono- to octa-[b])	1.0–8.0 (figure information)	
Core sediment (Lake Kitaura)	Depth (0–292 cm) Sum of PCNs (tri- to octa-[b,d])	0.0024–0.733	[43,44]
Core sediment (Lake Kitaura)	Depth (0–292 cm) Sum of PCNs (tri- to octa-[b,d])	0.002–1.293	[45]
Gas		*ng/Nm^3*	
Flue gas from incineration with oxidative conditions (pilot-scale two-stage municipal solid waste incinerator)	Range (*n* = 6[a]) Sum of PCNs (di- to octa-[b])	180–6000	[46]
Flue gas (municipal solid waste incinerator: MSWI)	Range (*n* = 3[a]): ng/m^3 Sum of PCNs (mono- to octa-[b])	4300–15,000[c]	[47]

(continued)

TABLE 10.3 (continued)
Concentrations of Polychlorinated Naphthalenes in the Environment in Japan

Matrix and Site	Measure	Concentration	References
Flue gas at kiln exit (plant-scale incinerator: MSWI type)	Range (n = 1) Sum of PCNs (mono- to octa-)	83,000	[19]
Flue gas at bag filter entry and exit, and at final exit (plant-scale incinerator: MSWI type)	Range (n = 3a) Sum of PCNs (mono- to octa-)	2.7–23	[19]
Flue gas at kiln exit from co-incineration of rubber belt containing PCNs (plant-scale incinerator: MSWI type)	Range (n = 2a) Sum of PCNs (mono- to octa-)	63,000–66,000	[19]
Flue gas at bag filter entry and exit, and at final exit from co-incineration of rubber belt containing PCNs (plant-scale incinerator: MSWI type)	Range (n = 6a) Sum of PCNs (mono- to octa-)	3.3–100	[19]
Flue gas (incinerator type was unknown)	Range (n = 54, mean value) Sum of 5 PCNs (tetra- to hepta-b)	17	[48]
Ash		*ng/g*	
Bottom and fly ash (small-scale and municipal waste incinerators)	Range (n=12a): ng/g dw Sum of PCNs (unknown)	0.74–610c	[49]
Fly ash (municipal waste incinerators)	Range (n = 12a) Sum of PCNs (tri- to octa-b)	10–1200 (figure information)	[24]
Fly ash (MSWI)	Range (n = 3a) Sum of PCNs (mono- to octa-b,d)	370–1400	[47]
Electron precipitator ash (MSWI)	Range (n = 2a) Sum of PCNs (mono- to octa-b,d)	0.33–2.1	[47]
Fly ash (incinerator type was unknown)	Range (n = 3) Sum of PCNs (mono- to octa-b)	90–700	[42]
Bottom and fly ash (plant-scale incinerator: MSWI type)	Range (n = 2a) Sum of PCNs (mono- to octa-)	0.96–1.7	[19]
Bottom and fly ash from co-incineration of rubber belt containing PCNs (plant-scale incinerator: MSWI type)	Range (n = 4a) Sum of PCNs (mono- to octa-)	0.17–1.2	[19]
Air		*ng/m³*	
Ambient air (all over the Japan except for Okinawa)	Range (n = 42a) Sum of PCNs (mono- to octa-)	0.011–0.86	[50]
Ambient air (all over the Japan except for Okinawa)	Range (n = 33a) Sum of PCNs (mono- to octa-b)	n.d.–0.55	[51]
Ambient air (site was unknown)	Range (unknown) Sum of PCNs (mono- to octa-)	100 (figure information)	[35]
Ambient air (western in Japan)	Range (n = 2a) Sum of PCNs (mono- to octa-b,d)	0.382–0.395	[47]
Ambient air (Shizuoka; middle part in Japan)	Range (n = 20a,e, mean value) Sum of PCNs (tetra- to octa-b)	0.0022	[52]
Air particulate (Shizuoka; middle part in Japan)	Range (n = 20a,e, mean value) Sum of PCNs (tetra- to octa-b)	0.002453	[52]

TABLE 10.3 (continued)
Concentrations of Polychlorinated Naphthalenes in the Environment in Japan

Matrix and Site	Measure	Concentration	References
Pine needles (Tokyo Bay)	Range ($n = 5^a$): ng/g ww Sum of PCNs (tri- to octa-[b])	0.63–1.75	[53]
Pine needles (Tokyo Bay)	Range ($n = 10^a$): ng/g ww Sum of PCNs (tri- to octa-[b,d])	0.25–2.1	[54–57] [69,70]
Others			
Food (all over the Japan. species was unknown)	Range ($n = 50^a$): ng/g fresh weight Sum of PCNs (mono- to octa-[b])	n.d. (0.005)–0.30[c]	[51]
Waste and related samples (rubber, refuse fuel and automobile residue)	Range ($n = 31^a$): ng/g Sum of PCNs (mono- to octa-[b])	3.1–130[c]	[42]
Printer belt and Neoprene FB which is rubber containing PCNs	Range ($n = 22^a$): ng/g Sum of PCNs (mono- to octa-[b])	0.89–36,000,000[c]	[42]
Biotic			
Animal		*ng/g ww*	
Fish (species and site were unknown)	Range ($n = 39$) Sum of PCNs (unknown)	n.d.–350	[36]
Fish (species and site were unknown)	Range ($n = 63$ or 66) Sum of PCNs (unknown)	2–130 or n.d.–130	[37]
Aquatic organism (all over the Japan except for Okinawa, species was unknown)	Range ($n = 30^a$) Sum of PCNs (mono- to octa-[b])	0.012–2.0	[51]
Mussel (all over the Japan except for Hokkaido and Okinawa)	Range ($n = 31^a$) Sum of PCNs (mono- to octa-[b])	0.019 (n.q.)–1.2	[58]
Marine fish (all over the Japan)	Range ($n = 80^a$) Sum of PCNs (mono- to octa-[b])	n.d.–2.7	[58]
Bird (northern in Japan)	Range ($n = 10^a$) Sum of PCNs (mono- to octa-[b])	0.011 (n.q.)–0.027	[58]
Marine fish marketed (all over the Japan except for northern part and Okinawa) (11 species)	Range ($n = 19^a$): ppm Sum of PCNs (tri- to hexa-)	0.00017–0.39[c]	[59]
Pig (species and site were unknown)	Range ($n = 3$): ng/g lipid weight Sum of PCNs (61 tri- to octa-[b,d])	0.16[c]	[60]
Chicken (species and site were unknown)	Range ($n = 5$): ng/g lipid weight Sum of PCNs (61 tri- to octa-[b,d])	0.4[c]	[60]
Human			
Adipose tissue (different age and gender samples, site was unknown)	Range ($n = 10^a$): ppm Sum of PCNs (tri- to hexa-)	0.0028–0.0169[c]	[61]
Adipose tissue (PCB polluted)	Range (unknown): mg/kg Sum of PCNs (unknown)	2.6[c]	[62]
Adipose tissue (Osaka and Ehime Prefecture)	Range ($n = 21^a$): ng/g fat weight Sum of PCNs (unknown)	2.02–249[c]	[33]

[a]   The number means sample number including different sites and treatment.

[b]   Concentrations of each congener are available in the literature.

[c]   Unit is ununiformity within categories of environmental matrices.

[d]   Concentrations of each isomer are available in the literature.

[e]   Seven days of samples were collected for every week from December 2, 2004 up to February 3, 2005 and for 1 week every month from February 3 to December 28, 2005.

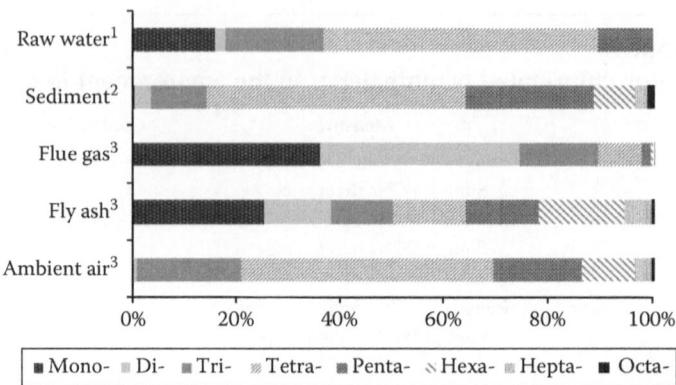

FIGURE 10.6 Profiles of PCN congeners in the abiotic environmental matrices. (1, Data from Kosaka, K. et al., *J. Jpn. Water Works Assoc.*, 77, 2, 2008; 2, Data from Japan Ministry of Economy, Trade and Industry, Chemicals in the environment, Report on environmental survey and monitoring of chemicals (in Japanese), 2002; 3, Data from Takasuga, T. et al., *Arch. Environ. Contam. Toxicol.*, 46, 419, 2004.) The n.d. (not detected) value was treated as 0, and half values were used in case of n.q. (not quantified). The average value of median from each sampling site was used in the profile of sediment.

in most sea water samples examined around Tokyo and Osaka Bay. Subsequent work by Shiraishi et al. [40] determined the concentration (0.03 to 0.44 ng/L for mono-CNs, and not detected (n.d.) to 0.15 ng/L for di-CNs) in tap water by using a GC/MS method. Since the total PCNs are at low levels (pg/L) in the most recent water samples, it is difficult to describe PCN profiles in detail.

10.2.1.2 Sediment

In sediment samples in Japan, PCN concentrations have been reported in river and/or sea and core sediments. Total PCNs in sediment samples from urbanized/industrialized sites are apparently higher until the 1980s when uses of PCNs were restricted under government regulations. In addition to the water samples, the total PCNs decreased from 5 to 670 ng/g on a dry weight (dw) basis in 1977 to 20 to 1000 ng/g dw in 1979, and 0.02 to 4.2 ng/g dw in 2002 in sediment media in Japan, according to the report from the Japan Ministry of the Environment [36–38]. Moreover, Imagawa and Yamashita [41] measured PCN concentrations by using a GC/MS method in sediments collected using grab samplers, and reported that the sum of tri- to hexa-CNs ranged from n.d. to 100 ng/g in river sediments collected around Tokyo and Niigata Prefecture (where it faced the Sea of Japan), and 7.3–27.4 ng/g in marine sediments collected around Tokyo and Dokai (where it faced both the Sea of Japan and the East China Sea; western part) Bay. In core sediment samples, the maximum PCN concentrations were found in certain layers that corresponded to 1980 through 1985 [25,43–45]. Therefore, a similar time trend can be observed in core sediment samples from Tokyo Bay and Lake Kitaura (located near Tokyo), compared to the data from the Japan Ministry of the Environment. In the PCN profiles there was abundant tetra- and penta-CNs in the sediment samples examined (Figure 10.6). Section 10.4 discusses the pattern of PCN congeners in greater detail. Except for the above mentioned, sediment data were also reported [45].

10.2.1.3 Gas

In Japan, the number of incinerators equipped with small- and industrial-scale incineration systems is comparatively larger than that of the United States or the and EU. Since many reports of PCN concentrations and environmental releases of PCNs originating from incineration, such as flue gas and ash samples, are available, it is important to note or analyze PCN data from the literature in order to clarify the atmospheric pollution in Japan. In flue gas samples, PCN concentrations have been reported, some of which describe the relationship of total PCNs and PCN profiles between flue gas and ash samples [19,47]. Takasuga et al. [47] measured PCN concentrations in flue gas samples collected at a municipal solid waste incinerator (MSWI) by using a GC/HRMS method, and reported that the total PCNs, including mono- to octa-CNs, were 15,000 ng/m^3 for the start-up stage, 4,300 ng/m^3 for the steady stage, and 15,000 ng/m^3 for the shutdown stage. Noma et al. [19] measured PCN concentrations in flue gas samples collected at a plant-scale MSWI by using a GC/ HRMS method, and reported that total the PCNs, including mono- to octa-CNs, ranged from 2.7 to 23 ng/Nm^3 at bag entry/exit gas and at final exit gas, and 83,000 ng/Nm^3 at kiln exit gas. Moreover, it was reported that the total PCNs ranged from 3.3 to 100 ng/Nm^3 at bag entry/exit gas and at final exit gas from coincineration of rubber belts containing PCNs (see Section 10.5 for the detailed description of rubber belts), and from 63,000 to 66,000 ng/Nm^3 at kiln exit gas from the coincineration of rubber belts containing PCNs [19]. As a result, adding wastes containing PCNs to MSWI will not influence the environmental release of PCNs. Ohura et al. [48] have determined the mean concentrations (3.8 ng/Nm^3 for 1,2,3,4-tetra-CN, 7.2 ng/Nm^3 for 1,3,5,7-tetra-CN, 3.2 ng/Nm^3 for 1,2,3,5,7-penta-CN, 1.7 ng/Nm^3 for 1,2,3,5,6,7-hexa-CN, and 1.1 ng/Nm^3 for 1,2,3,4,5,6,7-hepta-CN) in flue gas by using a GC/MS method. Furthermore, Sakai et al. [46] reported that the sum of di- to octa-CNs measured by using a GC/HRMS method in flue gas samples collected at a pilot-scale two-stage MSWI in oxidative conditions ranged from 180 to 250 ng/Nm^3 at O$_2$ concentration 9.6%, 260 to 6000 ng/Nm^3 at O$_2$ concentration 3.2%, and 160 to 4300 ng/Nm^3 at O$_2$ concentration 2.7%. For PCN profiles, mono- to octa-CNs were present in flue gas samples, and more volatile mono- to tri-CNs were the predominant congeners compared to ash samples (Figure 10.6). Takasuga et al. [47] also reported that 1- and 2-mono-CNs, 1,4-/1,6-di-CNs, 1,3,6-tri-CN, 1,3,6,8-/1,2,5,6-tetra-CNs, 1,2,3,5,7-/1,2,4,6,7-penta-CNs, 1,2,3,4,6,7-/1,2,3,5,6,7-hexa-CNs, and 1,2,3,4,5,6,7-hepta-CN were the predominant isomers in the flue gas samples. On the other hand, formation of toxic PCN isomers in the ash samples was larger than that of the flue gas compared to PCN profiles of both flue gas and ash samples [47].

10.2.1.4 Ash

For ash samples, it is necessary to investigate the incineration process of PCN and/or combustion of consumer goods containing PCN in order to clarify the emission pathway of PCNs as well as for the flue gas samples. As a result, some reports describing the incineration process may be characterized as environmental releases of PCNs in the atmosphere. Kawano et al. [49] reported that PCN concentrations measured by using a GC/HRMS method in bottom and fly ash samples collected at small-scale and/or MSWI ranged from 0.74 to 610 ng/g on a dw basis. Takasuga et al. [47] measured PCN concentrations in fly ash and electron precipitator (EP) ash samples collected at a MSWI by using a GC/HRMS method, and reported that the total PCNs, including mono- to octa-CNs, ranged from 370 to 1400 ng/g in fly ash samples collected in each of three stages (start-up, steady, and shutdown stages) and 0.33 to 2.1 ng/g in EP ash samples collected under the original and dechlorinated conditions. Noma et al. [19] measured PCN concentrations in bottom and fly ash samples collected at a plant-scale MSWI by using a GC/HRMS method, and reported that the total PCNs, including mono- to octa-CNs, ranged from 0.96 to 1.7 ng/g in bottom and fly ash samples and 0.17 to 1.2 ng/g in bottom and fly ash samples from coincineration of rubber belts containing PCNs. Mono- to octa-CNs were present in ash samples, and tri- to hexa-CNs were the most predominant congeners (Figure 10.6). Also, PCN profiles showed that both destruction and synthesis simultaneously occurred during the incineration process [19] and therefore PCN profiles of both ash and flue gas samples were different from those of the Halowax series.

On the pattern of PCN isomers, 2,3,6,7-tetra-CN and 1,2,3,6,7-penta-CN can be characterized as a combustion related source because their PCN isomers have shown abundance in ash samples compared to technical PCN and PCB preparations (PCN an PCB formulations) (Figure 10.7) [23,35,41,63,64]. Imagawa [23] also reported the difference of PCN isomer profiles between two types of incinerator (stoker and fluidized bed types—the major types of incineration in Japan). The 1,2,8-tri-CN, 1,2,4,7,8-penta-CN, and 1,2,3,4,5,6-hexa-CN in fluidized bed type incinerator seem

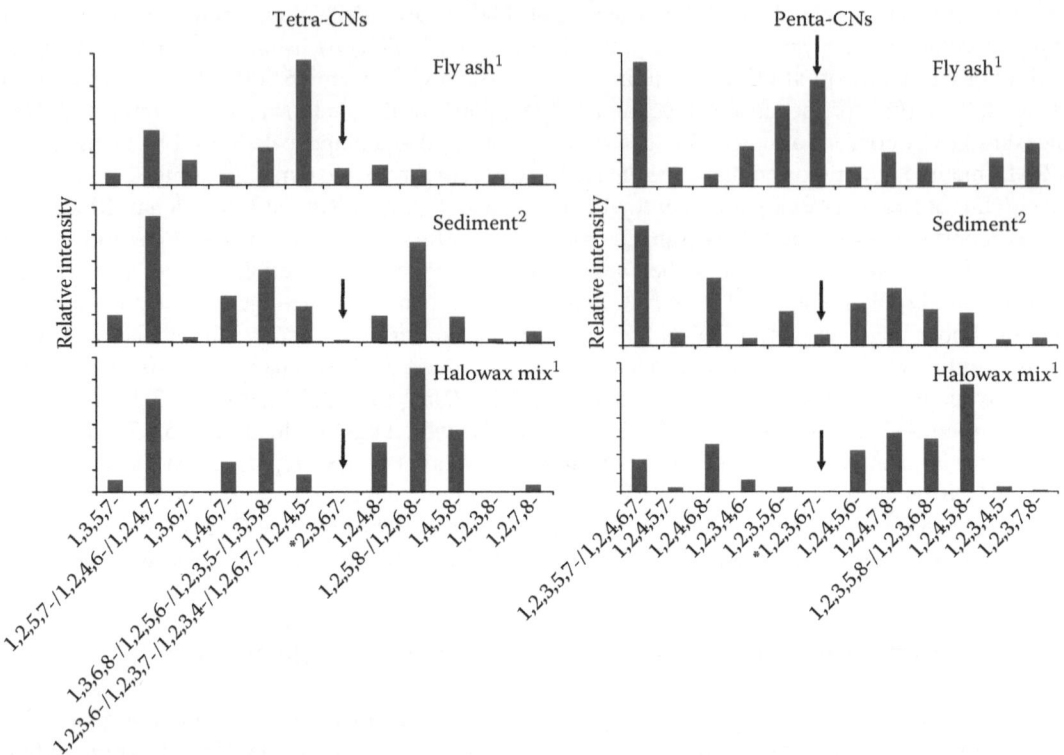

FIGURE 10.7 Tetra- and penta-CN profiles in fly ash, core sediment, and Halowax mixtures. (1. Data of fly ash and Halowax mixture from Takasuga, T. et al., *Arch. Environ. Contam. Toxicol.*, 46, 419, 2004.; 2. Data of core sediment from Horii, Y. et al., *J. Environ. Sci. Health A*, 39, 587, 2004.) A part of the PCN isomers were combined because of comparison of PCN patterns.

to be larger than those in stoker type one (Figure 10.8). Thus, fingerprint analysis of PCNs in ash samples can be performed by using certain PCN isomers, which were characterized as a combustion related source (Figure 10.8).

10.2.1.5 Ambient Air

According to the report on environmental survey and monitoring of chemicals from the Japan Ministry of the Environment [50,51], PCN concentrations have been reported in ambient air samples. Total PCNs ranged from 0.011 to 0.86 ng/m^3 in 1999 and n.d. to 0.55 ng/m^3 in 2003. Takasuga et al. [47] measured PCN concentrations in ambient air samples collected using a high-volume air sampler and GC/HRMS, and reported that the total PCNs, including mono- to octa-CNs, ranged from 0.382 to 0.395 ng/m^3 in the ambient air sampled in each of the seasons (summer and winter) in western Japan. Ohura et al. [52] have determined mean concentrations (0.43 pg/m^3 for 1,2,3,4-tetra-CN, 0.14 pg/m^3 for 1,3,5,7-tetra-CN, 0.37 pg/m^3 for penta-CNs, 0.13 pg/m^3 for hexa-CNs, and 0.21 pg/m^3 for hepta-CNs, and 0.92 pg/m^3 for octa-CN) in ambient air by using a GC/HRMS method. With regard to seasonal trends, the ambient concentrations of high chlorinated PCNs (hexa- to octa-CNs) seemed to be higher during the cold season [47,52]. In PCN profiles, tri- and tetra-CNs show much higher abundance than hexa- to octa-CNs (Figure 10.6). This is similar to profiles observed in atmospheric samples from the Arctic [65], Great Lakes [66], and U.K. [67].

It was known that pine needles can be used as an indicator of atmospheric pollution [68], and numerous reports that discussed the level of environmental pollutants in ambient air and air deposition are available. Some studies in Japan have used pine needles as an indicator of PCN atmospheric pollution [53,69,70]. Hanari et al. [54–56] measured PCN concentrations in pine needle samples

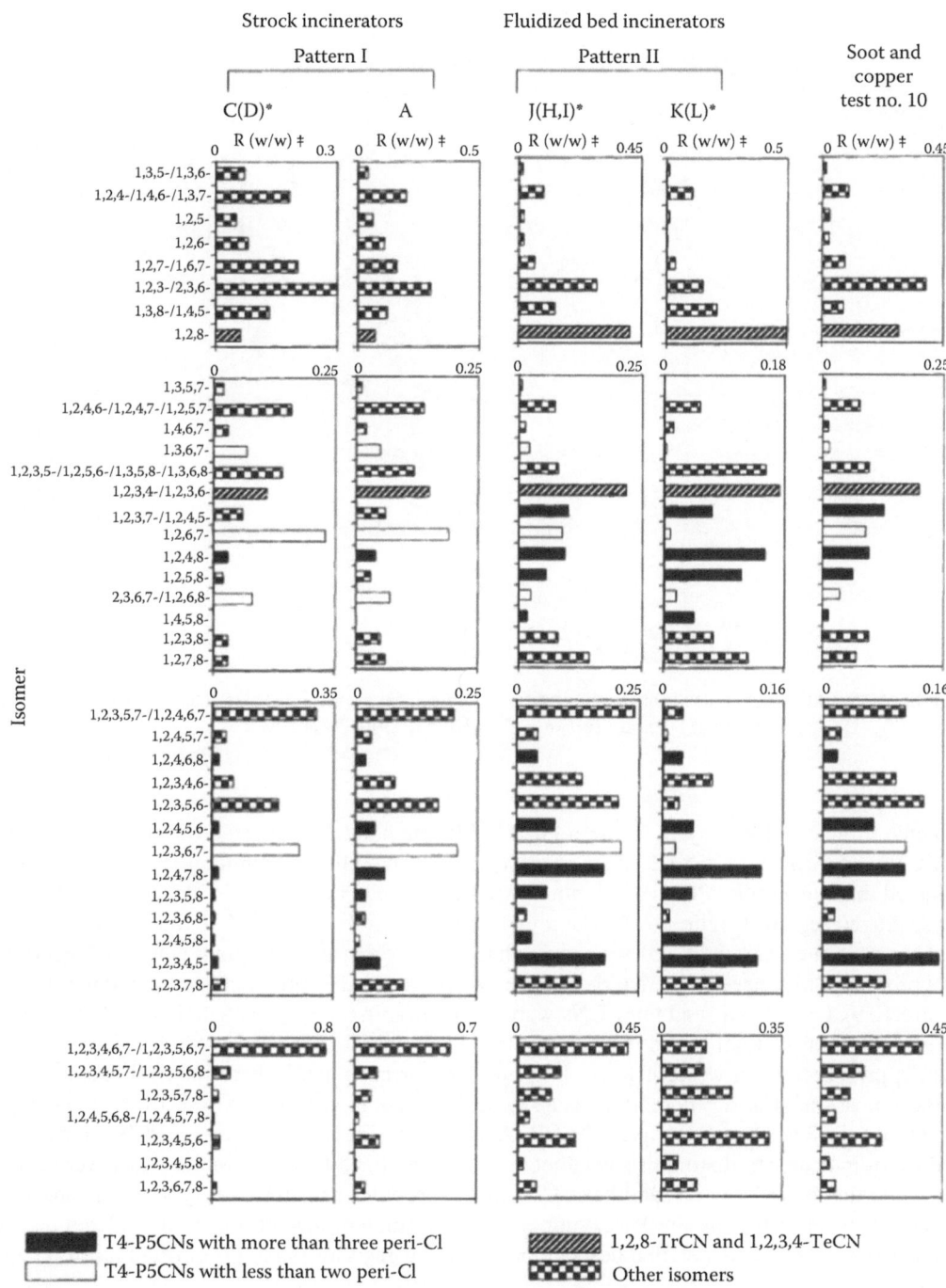

FIGURE 10.8 Isomer pattern of PCNs in fly ash samples collected from municipal incinerators. (Reprinted from Report on the National Institute for Resources and Environment, 29, Fig. 6-3, Imagawa, T., studies on formation of polychloromepthelenes from waste incineration (in Japanese), 49, Copyright 2009, With permission from Elsevier.)

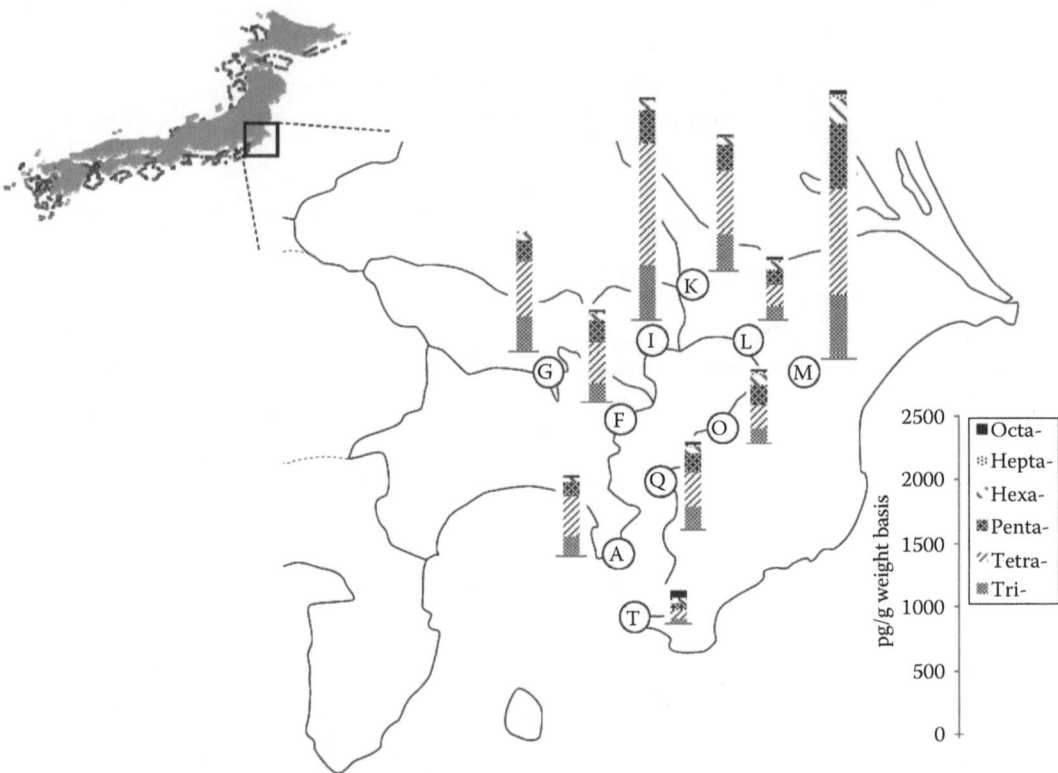

FIGURE 10.9 PCN concentrations, including tri- to octa-CNs, in pine needles from sites selected around Tokyo Bay. (Data from Hanari, N. et al., *Bunseki Kagaku*, 53, 1399, 2004.)

collected around Tokyo Bay by using a GC/HRMS method, and reported that the sum of tri- to octa-CNs ranged from 0.25 to 2.1 ng/g on a wet weight (ww) basis. Sampling sites of pine needles indicated that higher total PCNs corresponded to sites located in the heart of densely populated and industrialized regions (Figure 10.9). The area that indicated the lowest PCN concentration was the southernmost site examined and was somewhat distant from densely populated sites. Moreover, PCN profiles of pine needles were largely different for the southernmost site compared to other sites (Figure 10.9). Overall, tri- and tetra-CNs were the predominant congeners, collectively accounting for 54%–80% of the total PCNs in pine needles for samples examined from Tokyo Bay as well as in the atmosphere from western Japan [47] and the northern hemisphere [65–67] (except for the southernmost site). Furthermore, there is a similarity in the fingerprint analysis of tri- to hexa-CNs in pine needles and fly ash samples [23,57]. As shown in Figure 10.8, certain PCN isomers were used as an indicator to distinguish whether stoker or fluidized bed type incinerators were used. It was found that patterns of tri- and hexa-CNs in pine needles and in stoker type incinerators were similar (Figure 10.10). This similarity supports the transition of incinerator from fluidized bed type to stoker type in Japan. A detailed description of other categories [45,53] is omitted in this chapter.

10.2.2 BIOTIC ENVIRONMENT

PCNs are usually found in all types of samples from the biotic environment, and contaminations were observed in both animal and human samples in Japan as well as water and sediment samples. The time trend on PCN contaminations can be clarified by referring to reports on environmental

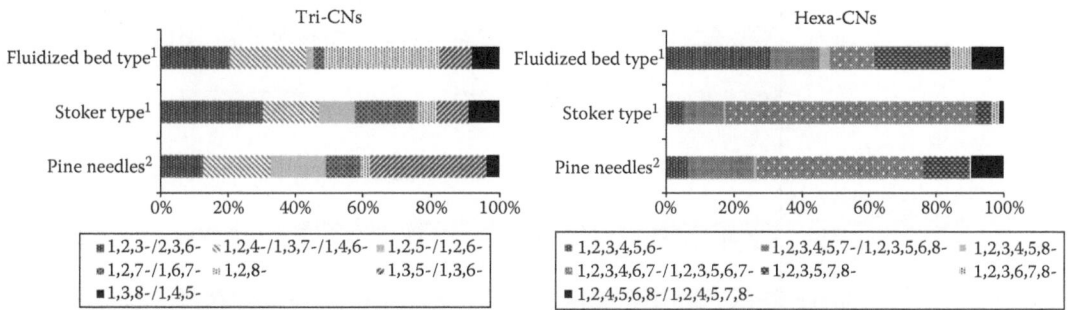

FIGURE 10.10 Tri- and hexa-CN profiles in fly ashes from two types of incinerators (fluidized bed and stoker) and pine needles from sites selected around Tokyo Bay. (1, Data from Imagawa, T., Studies on formation of polychloronaphthalenes from waste incineration (in Japanese), Report on the National Institute for Resources and Environment, 29, 2000; Imagawa, T. and Yamashita, N., *J. Environ. Chem.*, 6, 495, 1996; 2, Data from Hanari, N. et al., *Bunseki Kagaku*, 53, 1399, 2004.)

survey and monitoring of chemicals arising from the Japan Ministry of the Environment [36,37,58]. For PCN profiles, congener patterns in birds and humans as well as higher trophic levels can be compared to those in biological samples at lower trophic levels.

10.2.2.1 Animal

In animal samples in Japan, PCN concentrations have been reported in mussel, fish, bird, and domestic animals as well as water and sediment samples. The reports of the Japan Ministry of the Environment indicated total PCNs and PCN contamination trends over time. In fish samples, concentration levels decreased from n.d. to 350 ng/g on a ww basis in 1977, 2 (or n.d.) to 130 ng/g ww in 1979, and n.d. to 2.7 ng/g on ww in 2007 (the report from [36,37,58]). Takeshita and Yoshida [59] reported the total PCNs (17–390 ppb, a part of tri- to hexa-CN congeners) measured by using a GC-ECD method in marine fish samples examined and indicated that the total PCNs in marine fishes were 300–1000 times lower than the total PCBs. Total PCNs from animals of different species in the report of the Japan Ministry of the environment [51,58] were also reported (aquatic organism; 0.012–2.0 ng/g ww in 2003, mussel; 0.019 [this value was not quantified; n.q.] to 1.2 ng/g ww in 2007, bird; 0.011 [n.q.]–0.027 ng/g ww in 2007]). In domestic animal samples, Guruge et al. [60] measured PCN concentrations in pig and chicken fats by using a GC/HRMS method, and reported that the sum of tri- to octa-CNs were 0.16 ng/g lipid weight in pig fat and 0.4 ng/g lipid weight in chicken fat. In PCN profiles, there were predominant congeners such as tri- and tetra-CNs in mussel samples, while there were relatively abundant penta- and hexa-CNs in fish, bird, and domestic animals (Figure 10.11). Guruge et al. [60] also reported 1,2,3,7,8-/1,2,4,6,7-penta-CNs as the abundant isomers in penta-CNs. PCN profiles, including congeners and isomers, in samples from the biotic environment seem to vary in the food web as well as PCBs (Figure 10.11).

10.2.2.2 Human

In human samples in Japan, PCN concentrations have been reported in only adipose tissue samples. Takeshita and Yoshida [61] measured PCN concentrations in adipose tissues by using a GC/MS method and reported that total PCNs (a part of tri- to hexa-CN congeners) ranged from 0.0028 to 0.0109 ppm in different age and gender samples. Also, it was indicated that the total PCNs in adipose tissue were 500–1000 times lower than the total PCBs. Kawano et al. [33] measured PCN concentrations in adipose tissues by using a GC/HRMS method and reported that the total PCNs ranged from 3.32 to 10.3 ng/g on fat weight basis in Ehime Prefecture and 2.02–249 ng/g on fat weight in Osaka. Moreover, it was indicated that 1,2,5,8-/1,2,6,8-tetra-CNs, 1,2,3,6,7- and

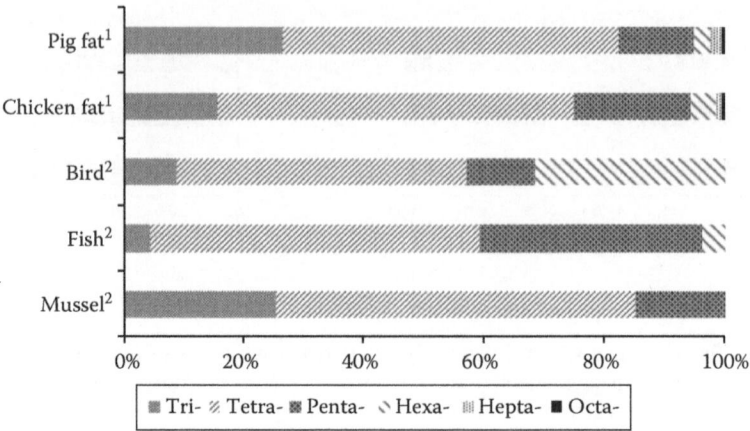

FIGURE 10.11 Profiles of tri- to octa-CN congeners in examined Japanese animals. (1, Data from Guruge, K.S. et al., *J. Environ. Monitor.*, 6, 753, 2004; 2, Data from the Japan Ministry of the Environment 2007, Chemicals in the environment, Report on environmental survey and monitoring of chemicals (in Japanese).) The n.d. (not detected) value was treated as 0, and the half value was used in the case of n.q. (not quantified) value. The median value of total samples examined was used in bird, fish, and mussel samples.

1,2,3,5,7-/1,2,4,6,7-penta-CNs, and 1,2,3,4,6,7-/1,2,3,5,6,7-hexa-CNs were the predominant isomers in human adipose tissue samples [33]. Interestingly, Haglund et al. [62] reported that the total PCNs in Yusho (the site of the accidental poisoning of PCBs in Japan in the 1968) rice oil was 2.6 mg/kg. In addition, PCN profiles have been discussed for the Yusho rice oil and adipose tissue of the victims of the Yusho poisoning, and notable amounts of five tetra-CN isomers were found. More detailed descriptions of PCN profiles in human samples is difficult because of limited sample numbers and information in the literature.

10.3 ISOMER-SPECIFIC ANALYSIS OF PCNS

Early analytical methods for PCNs were reviewed by Brinkman and Reymer, and recent advances in environmental analysis of PCNs were highlighted by Kucklick and Helm [5] and van Leeuwen and de Boer [71]. PCNs are structurally similar to dioxins and PCBs and have similar physical/chemical properties [72]. Therefore, similar analytical methods for dioxins can be applied for PCNs. Method improvements have been made for PCNs over the past decade in isomer-specific quantification, peak resolution, and separation from interferences. A historical timeline for the publication of major milestones in method development for the environmental analysis of PCNs was summarized by Kucklick and Helm [5] (Figure 10.12). In this section, each step of the environmental analysis of PCNs such as extraction, clean-up and separation, and quantification are explained briefly; then, some recent advanced techniques in PCN analysis are highlighted.

10.3.1 PCN STANDARDS

Today, about 30 individual PCN congeners out of the 75 possible are commercially available, whereas all individual congeners are available for PCB. Equivalent amounts of technical mixtures with different chlorination degrees (e.g., Halowax 1000, 1001, 1014, and 1051) were often used as the standard for congener identification. Isotopically labeled PCNs, including CN27, 42, 52, 64, 67, 73, and 75, became commercially available beginning in 2002 (Figure 10.12) and would enable one to perform an isotope dilution method for environmental analysis of PCNs. ^{13}C-Labeled

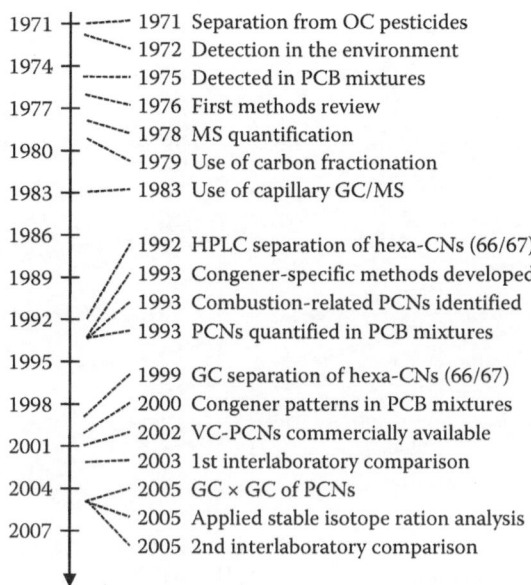

1971 — ------ 1971 Separation from OC pesticides
   ------. 1972 Detection in the environment
1974 — ------- 1975 Detected in PCB mixtures
1977 — ------. 1976 First methods review
   ------. 1978 MS quantification
1980 — ------. 1979 Use of carbon fractionation
1983 — ------ 1983 Use of capillary GC/MS

1986 — ------ 1992 HPLC separation of hexa-CNs (66/67)
1989 — 1993 Congener-specific methods developed
   1993 Combustion-related PCNs identified
1992 — 1993 PCNs quantified in PCB mixtures

1995 — 1999 GC separation of hexa-CNs (66/67)
1998 — 2000 Congener patterns in PCB mixtures
   2002 VC-PCNs commercially available
2001 — ------ 2003 1st interlaboratory comparison
2004 — 2005 GC × GC of PCNs
   2005 Applied stable isotope ration analysis
2007 — 2005 2nd interlaboratory comparison

FIGURE 10.12 Historical timeline for the publication of major milestones in method development for the environmental analysis of PCNs. (With bind permission from Springer Science+Business Media: *Anal. Bioanal. Chem.*, Advances in the environmental analysis of polychlorinated naphthalenes and toxaphene, 386, 2006, 819–836, Kucklick, J.R. and Helm, P.A.)

mono- to tri-CN congeners are not yet available Further advances in labeled standard availability will improve quality control of environmental analysis of PCNs, particularly for low chlorinated CNs found heavily in air and water.

10.3.2 EXTRACTION

Determination of target chemicals typically starts with extracting them from sample matrices. Soxhlet extraction is the classical method used to isolate PCNs from solid matrices and is still used widely due to the simplicity of the method. Materials are either extracted wet after homogenizing with anhydrous sodium sulfate or after drying the samples. Pressurized liquid extraction (e.g., accelerated solvent extraction [ASE®]) employed at elevated temperatures and pressures is becoming more common for extracting POPs from solid and semisolid matrices due to the reduced extraction time and solvent use. Other choices of extraction techniques include supercritical fluid extraction, but it has yet to find a broad application.

10.3.3 CLEAN-UP AND SEPARATION

The clean-up of crude extracts of environmental samples can be achieved by the removal of bulk lipid, polar compounds, and organic matter using column chromatographic techniques such as gel permeation chromatography (GPC), silica-gel, Florisil, and alumina. Removal of sulfur using activated copper or $AgNO_3$-silica gel is one of the most important steps of clean-up since sulfur easily damages the stationary phase of the GC capillary column, which causes poor peak resolution and sensitivity of PCNs. Multilayered silica-gel columns (destructive method), which are typical for dioxin analysis, efficiently remove interfering compounds including lipids and sulfur. However, some OCPs (e.g., dieldrin and endrin) degrade under the strong acidic conditions [71]. Further removal of interferences is achieved by using porous graphitic carbon or PYE columns, because

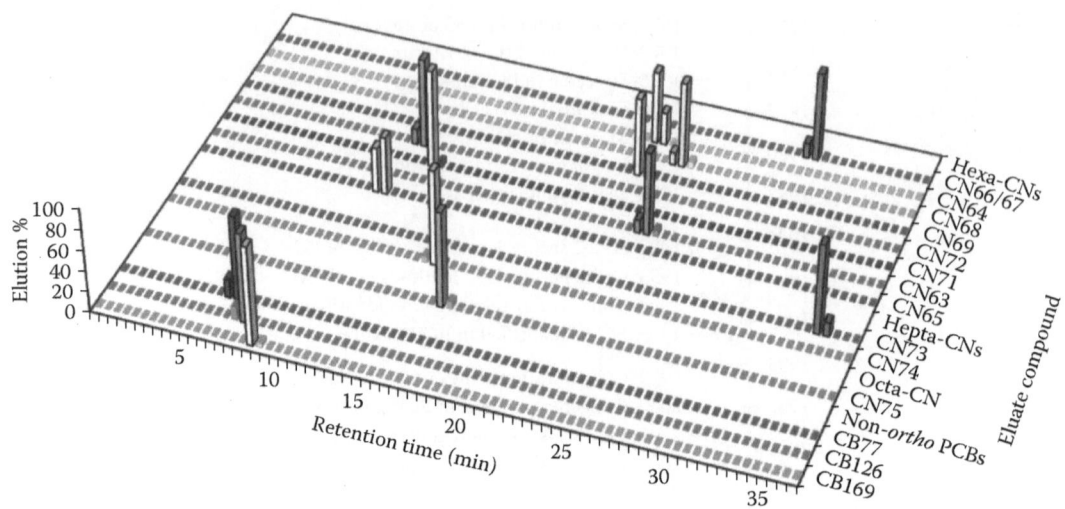

FIGURE 10.13 Elution profiles of individual PCN congeners (hexa to octa-CN) and non-*ortho* PCBs on HPLC equipped with PYE column. (Modified from Hanari, N. et al., *J. Environ. Monitor.*, 6, 305, 2004.)

the molecular planarity allows selective separation of PCNs and interferences. Carbon columns are used for the separation of OCPs and bulk PCBs from dioxin-like compounds, including PCNs.

The double column HPLC clean-up was developed to separate PCNs from interferences and to solve coelution of hexa-CN congeners [26,45,55]. First, planar compounds such as PCDDs, PCDFs, PCNs, and non-*ortho* PCBs are fractionated from the bulk of PCBs using a Hypercarb HPLC column and eluted by back flushed toluene. This fraction is further subdivided into four fractions using a PYE HPLC column (elution profiles of individual CN congeners are shown in Figure 10.13). This enables the isolation of hexa-CN congeners, such as pairs of CN64 and 68 and CN71 and 72, which would otherwise coelute in common GC columns.

10.3.4 QUANTIFICATION

Instrumental analysis of PCNs is most commonly undertaken using capillary GC with either low-resolution or high-resolution mass spectrometers as the detectors (GC/MS, GC/HRMS) [71]. Jarnberg et al. [73] determined the retention behavior of a PCN standard mixture on six capillary columns, including Ultra1, Ultra2, HT-5, CP-Sil-88, SB-octyl 50, and SB-smectic. Ultra1 and Ultra2 showed better separation in 44 out of 75 possible congeners. Congener-specific analysis of PCNs (mono- to octa-CN) in seven Halowax mixtures using HRGC/HRMS coupled with an Ultra2 capillary column provided detailed concentration and composition analysis of individual CN congeners in each Halowax mixture [74–76]. Imagawa and Yamashita [77] used α- and β-cyclodextrin coated columns to separate hexa-CNs. Helm et al. [78] reported that the use of the liquid phase of a proprietary cyclodextrin material (Rt-βDEXcst) enables successful separation of all isomers of penta- and hexa-CNs, found in Halowax 1014 and 1051. The 5% phenyl-dimethylpolysiloxane stationary phase is the standard choice for PCN analysis as with other halogenated analyses [5].

10.3.5 INTERLABORATORY COMPARISON

Two interlaboratory studies were carried out to investigate consistency in reported environmental data of PCNs. In Phase I [79], Halowax solutions were quantified for total PCNs and for several individual congeners by nine different labs worldwide. In Phase II, two test materials, SRM 1649a Urban Dust and SRM 1944 New York/New Jersey Waterway Sediment, and Halowax 1014 as a control material, were distributed to seven labs. Five labs submitted the results of the quantification. Although a small number of labs participated in the interlaboratory comparison in Phase II, the results showed a very good agreement for both Halowax 1014 and SRM materials, with RSDs, generally, being 10% or lower.

10.3.6 ADVANCED APPLICATIONS

10.3.6.1 Comprehensive Two-Dimensional Gas Chromatography (GC × GC)

In the past decade, GC × GC emerged as a powerful analytical technique and is an excellent choice when the composition of complex samples has to be unraveled. Advances in environmental toxicant analysis using multidimensional gas chromatography and GC × GC were highlighted by Marriott et al. [80] and Adahchour et al. [81].

Group separation of organohalogen compounds such as PCDDs, PCDFs, PCBs, PCNs, OCPs, and polychlorinated terphenyls were performed using GC × GC equipped with several combinations of capillary columns [82]. The DB-1 × LC-50 set provides a successful group separation, based on planarity—planar compounds such as the dioxins and PCNs are more retained than nonplanar analytes and are thus separated. The DB-1 × 007-65HT column set is the most efficient column combination for in-class separation of OCPs and PCNs. A combination of nonpolar and shape selective columns in a GC × GC system consisting of DB-5MS × LC-50 enabled separation of all possible 22 isomeric tetrachloronaphthalenes in the technical CN Halowax mixtures into 18 peaks, covering 15 single separated compounds and 7 coelutings (Figure 10.14) [83].

10.3.6.2 Congener-Specific Carbon Isotopic Analysis of PCNs

Stable isotopic compositions of carbon have been used in the study of biogeochemical processes over the last few decades. Variations in the isotopic ratio of $^{13}C/^{12}C$ in environmental matrices have been used in the understanding of environmental biogeochemistry of natural and synthetic hydrocarbons. In recent years, stable carbon isotopic analysis is becoming increasingly popular in understanding the sources and fate of anthropogenic organic contaminants in the environment. Compound-specific carbon isotopic analysis (CSIA) of anthropogenic organic contaminants such as PCBs and PAHs, using GC-isotope ratio mass spectrometry (GC-IRMS) is emerging as a powerful analytical tool to trace the origin and fate of organic pollutants [84–86]. For example, technical PCB mixtures from Eastern European countries (Delors, Sovol, Trichlorodiphenyl, and Chlorofen) had distinct $\delta^{13}C$ values, suggesting possible variations in the raw materials used or production processes of technical PCB mixtures in different geographical regions [87].

While several studies on CSIA of PCBs were conducted, only a little information is available for PCNs. Horii et al. [87] determined $\delta^{13}C$ values of 15 individual congeners in Halowaxes using 2D GC-IRMS equipped with moving capillary stream switching (heart cut system). The $\delta^{13}C$ values of CN congeners ranged from −26.3‰ (CN5/6/7/12, HW 1000) to −21.7‰ (CN73/74, HW 1051) (Figure 10.15) and was comparable to that found for technical PCB mixtures [84,87]. Similarly, in PCBs, $\delta^{13}C$ values decreased with increasing chlorination in the Halowax series. Furthermore, $\delta^{13}C$ values for individual CN congeners increased with increasing chlorination of technical mixtures (Figure 10.15). For instance, $\delta^{13}C$ values of CN46 in Halowaxes 1001, 1013, and 1014 were −25.1‰, −23.9‰, and −23.0‰ respectively, suggesting a ^{13}C depletion with decreasing chlorination for the same congeners in each Halowax. These values would provide baseline data for future studies to understand the origin and fate of these complex mixtures in the environment. The research group also determined isotopic ratios of PCNs in commercial goods contaminated with PCNs using the same technique. This will be discussed later in Section 10.5.

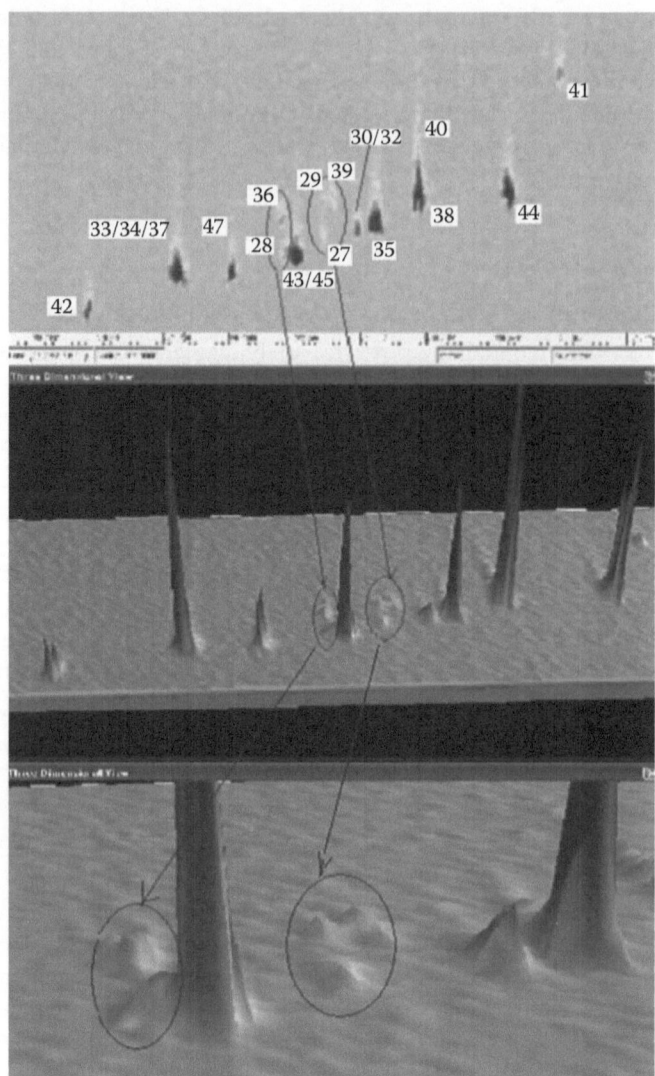

FIGURE 10.14 GC×GC chromatograms of tetrachlorinated naphthalene congeners in Halowax 1099, separated with DB-5MS as the first column and LC-50 as the second column. (From Lukaszewicz, E. et al., *J. Environ. Sci. Health A*, 42, 1612, 2007, Figure 1. With permission.)

10.4 HISTORICAL RECONSTRUCTION OF PCN POLLUTION USING SEDIMENT CORE IN JAPAN

PCNs persist in ecosystems for many years and sediments can be a major sink for PCNs in the water, owing to their low water solubility and their propensity for adsorption onto solid particles (log K_{OW}: 3.9–6.4) [72]. Vertical profiles of residues in dated sediment cores have been used as historical records of pollution [88]. Several studies have examined vertical profiles of POPs such as PCDD, PCDFs, PCBs, and PAHs in sediment cores and found that the depositional histories of these compounds are often preserved in the sediment bed [89]. Therefore, the vertical profiles of POPs in dated sediment cores can reflect production and usage trends of chemicals and emission from industrial activities.

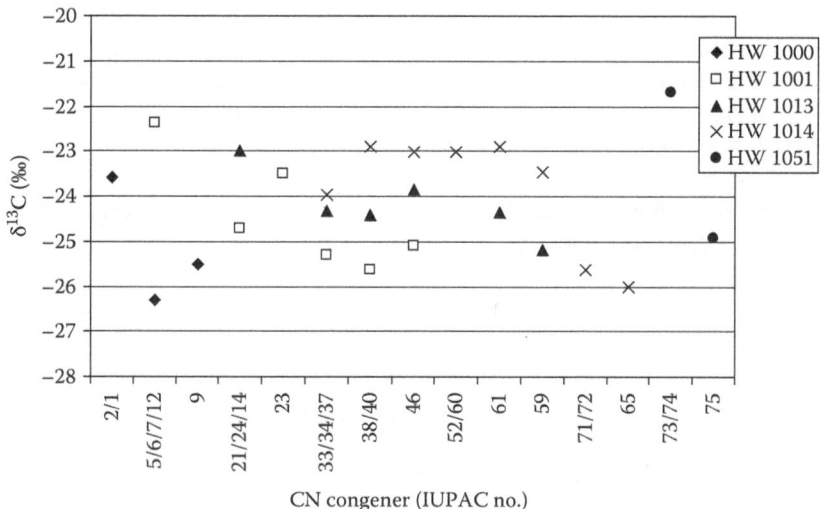

FIGURE 10.15 Carbon isotopic ratios of PCN congeners in Halowax series. (Data from Horii, Y. et al., *Environ. Sci. Technol.*, 39, 4206, 2005.)

To reconstruct historical trends of pollution, it is important to take undisturbed sediment core samples. Sediment cores are collected using a corer attached to an acrylic tube or by inserting an acrylic tube directly on the surface, bottom of a lake, or sea by a diver (Figure 10.16). A piston corer is also used to take relatively long sediment core samples up to several meters (Figure 10.16). The sediment core collected is then sliced into 1–2 cm segments using a stainless steel slicer, depending on the sedimentation rate or the required time resolution (Figure 10.16). Profiles of radioisotopes

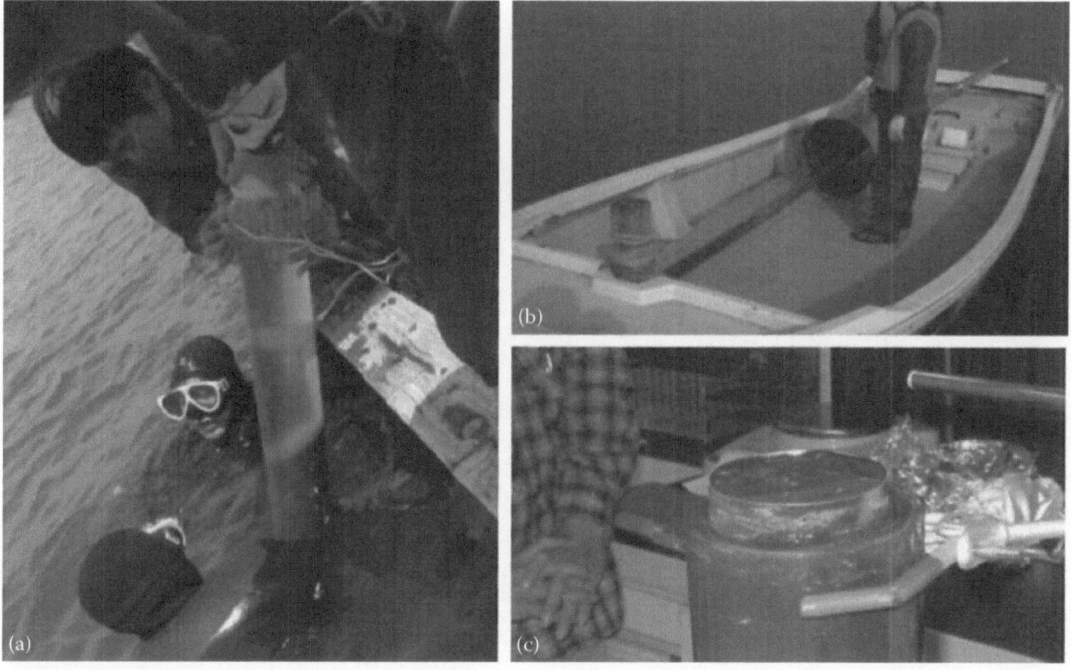

FIGURE 10.16 Sediment core collection using (a) acrylic tube (120 cm long × 11 cm i.d.) and (b) piston corer (4 m long × 7 cm i.d.), and (c) sediment core slice.

such as ^{210}Pb, ^{137}Cs, and ^{14}C were used for estimation of sedimentation rate in the core. Volcanic ash stratigraphy is also a useful tool to determine the age of sediment layers if the sediment core contains a known volcanic ash event. The time interval of each slice is estimated, based on sedimentation rate and depth.

While several studies have reported historical profiles of dioxins and PCBs, not much data were available for PCNs. In this section, two studies on dated sediment cores collected from Tokyo Bay and Lake Kitaura, Japan are introduced. The occurrence, profiles, and fluxes of PCNs in Japanese sediment cores were compared with data worldwide.

10.4.1 Tokyo Bay

Yamashita et al. [90] investigated PCNs in 90 cm-dated sediment core from Tokyo Bay and reconstructed the historical pollution level of PCNs over the past 100 years (Figure 10.17). Tokyo Bay is surrounded by major cities such as Tokyo and Yokohama and is heavily contaminated by industrial and urban wastes. The vertical profile of PCNs was characterized by lesser concentrations in deeper layers (before the 1940s, <0.5 ng/g dw), followed by an exponential increase to a subsurface peak (year 1980, 4.4 ng/g dw), and a gradual decrease to the surface (1990s, 1.8 ng/g dw). They also investigated the time trend of dioxin and PCB concentrations in the sediment core. Increases in concentrations of PCNs in sediment were observed as early as the 1960s (25–30 cm), which predates 20 years those for PCBs. The total production of technical PCNs in Japan from 1940 to 1976 was 4005 ton [20], and technical PCNs were banned after 1979 in Japan. PCNs have been used in numerous applications in various industrial products and commercial goods (see Section 10.5), while from the 1930s these compounds were accordingly substituted with PCBs, due to the similar physical/chemical properties [72]. Despite a ban on the production of PCNs in 1976, the sediment core profile showed the peak concentrations to be in the early 1980s. This may suggest a time lag between peak

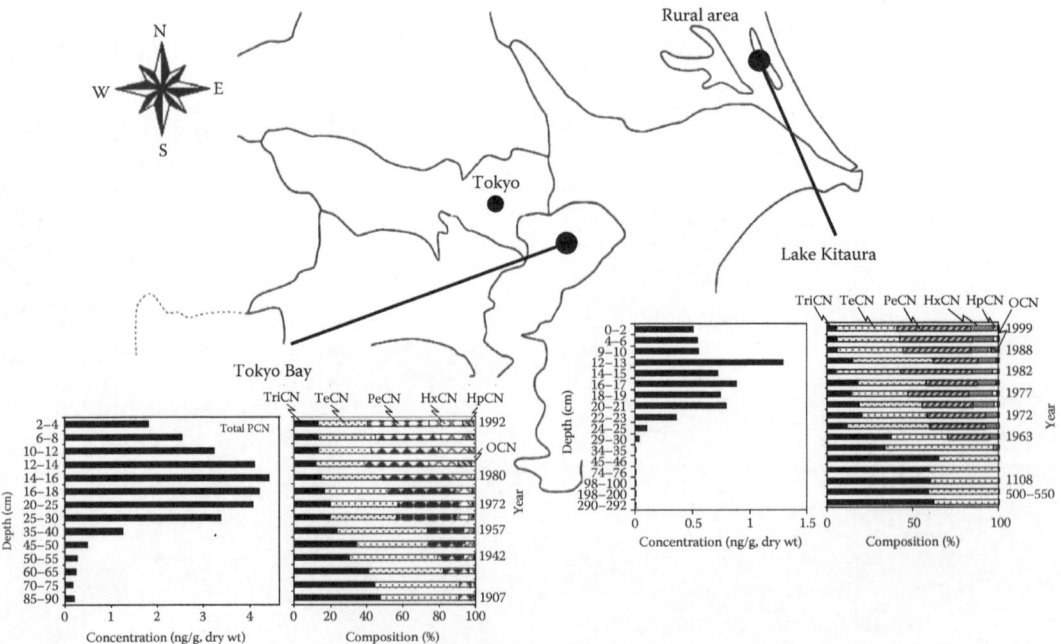

FIGURE 10.17 Vertical profiles of concentrations and homologue compositions of PCNs in dated sediment core from Tokyo Bay and Lake Kitaura, Japan. (From Yamashita, N. et al., *Environ. Sci. Technol.*, 34, 35647, 2000, Figure 4; Horii, Y. et al., *J. Environ. Sci. Health A*, 39, 600, 2004, Figure 3. With permission.)

periods of production and usage and the deposition to coastal marine environment by particulate transport and deposition.

In terms of toxicity, several PCN congeners were found to elicit dioxin-like toxic potencies in vitro bioassay [30,31]. Kannan et al. [91] reported vertical profiles of dioxin-like and estrogenic activities associated with a sediment core from Tokyo Bay. In vitro luciferase assay with recombinant rat hepatoma cell (H4IIE-luc) was used to measure dioxin-like activities in florisil fractions of sediment core extracts and the predicted and estimated that the dioxin-like activities in the fractions were compared at selected depth of the core. The greatest PCN-TEQ concentrations at 0.44 pg/g dw were observed at a depth of 12–14 cm (early 1980s). CN 66/67 accounted for 51%–86% of the total PCN-TEQs. The PCN-TEQs calculated were less than the concentrations required to yield a significant response in the H4IIE-luc assay suggesting that PCNs alone were not responsible for the AhR-mediated activity associated with sediment core extracts from Tokyo Bay.

10.4.2 Lake Kitaura

Horii et al. [45] investigated concentrations and fluxes of PCNs in sediment core from Lake Kitaura, Japan, in the past 15 centuries (Figure 10.17). Lake Kitaura is located in the rural region in the Ibaraki Prefecture about 60 km away from Tokyo Bay. The Lake Kitaura surface area is 35.2 km^2 and the mean depth is 4.5 m. Three meter-long sediment core samples were collected in December 2000, using a piston corer (Picture 10.1b). Total CN concentrations of the Lake Kitaura sediments dated from 2000 to before 500 ranged from 2.0 to 1300 pg/g dw. The vertical profile of PCN concentrations is very similar to that of Tokyo Bay in that both subsurface peaks were found at sediment layers corresponding to the early 1980s, while concentration levels in Lake Kitaura were four times lower than those for Tokyo Bay. The surface sediment layers (0–10 cm) contained PCNs ranging in concentration from 510 to 560 pg/g dw, which is within the range of 30–7700 pg/g dw documented for the "background" or "semibackground" sites elsewhere in the world [1,79]. At the polluted sites studied in the United States, Sweden, and Germany, because of inputs due to discharge from the nearby specific point pollution sources such as the chlor-alkali industry, the total PCN content of the surface sediment was up to 23,000 ng/g dw [2,92,93].

10.4.3 CN Homologue and Congener Patterns

The pattern or profile of CN homologue groups of the sediment core varied depending on the sediment depth (Figures 10.17 and 10.18). Tetra-CNs (~48%), penta-CNs (~43%), and tri-CNs (~37%) highly dominated the upper layers of the cores from Tokyo Bay and Lake Kitaura (1960s–1990s). This profile is most similar to Halowax 1001 and 1099, which are technical PCNs containing predominant proportions of tri- and tetra-CNs.

Major sources of PCNs in the environment were technical PCNs, MSWI, and technical PCB mixtures. Due to their relative abundance in products of combustion (flue gas, fly ash) and absence or scarcity, respectively, in technical PCN and PCB mixtures, the congeners CN26, CN29, CN44, CN48, CN54, and CN70—are considered as possible combustion tracers or combustion marker congeners [19,26,94,95]. All six congeners were quantified in the sediment core in Lake Kitaura, and specific vertical profiles for some of them are shown in Figure 10.19. The vertical profiles of combustion marker congeners are very interesting because their concentration increases or remains stable from the subsurface layer of 14–15 cm (dated on 1981–1982) to the top. This finding highly suggests that the origin of those PCNs are from sources related to combustion. When compared to the combustion marker CN congeners, the vertical profile of the congeners, which are typical to the technical Halowax mixtures, that is, CN39 and CN44, was different and a maximum concentration occurred at the layer of 12–13 cm (Figure 10.19). This profile is similar to that of major congeners in Halowax mixtures (e.g., CN38 and CN59). Generally, the same picture is drawn for the Tokyo Bay sediment core.

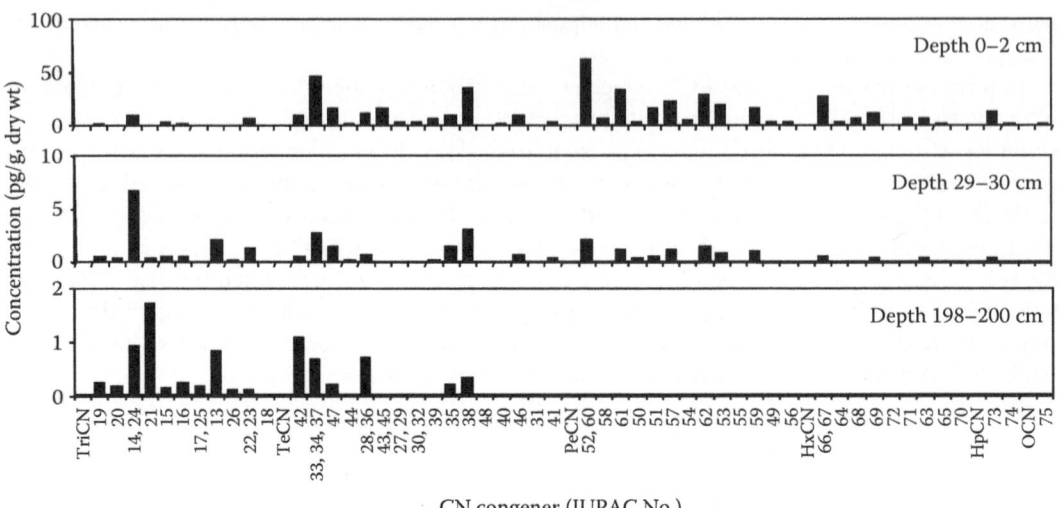

FIGURE 10.18 Patterns of CN congeners in selected layers of sediment core from Lake Kitaura. (From Horii, Y. et al., *J. Environ. Sci. Health A*, 39, 601, 2004, Figure 4. With permission.)

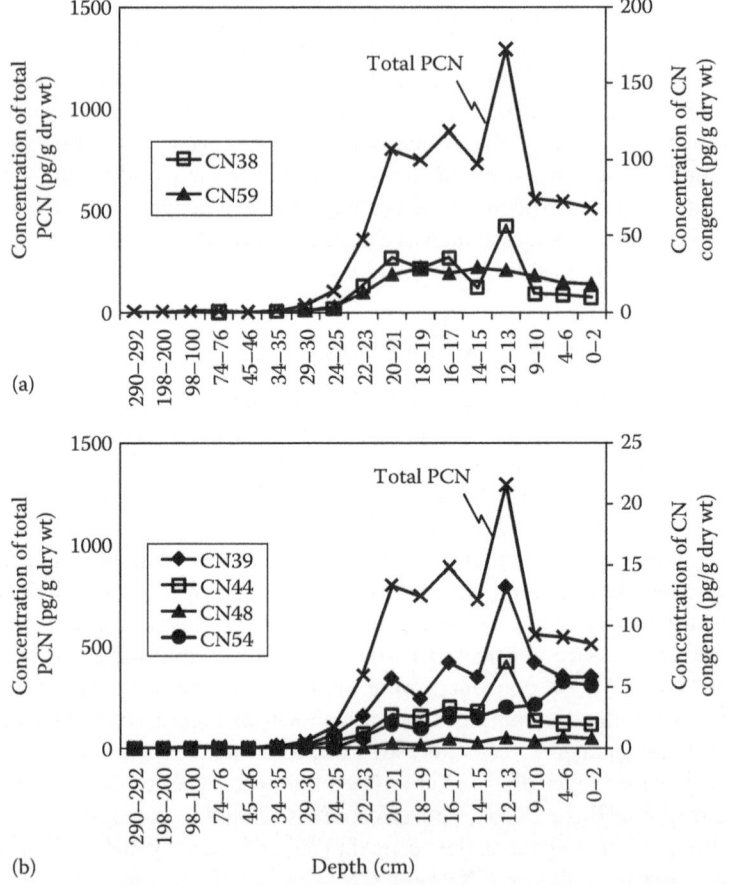

FIGURE 10.19 Vertical profiles of total CNs and marker CN congeners: (a) combustion sources (e.g., waste incineration) and (b) technical CN mixture (e.g., Halowax). (From Horii, Y. et al., *J. Environ. Sci. Health A*, 39, 601, 2004, Figure 5. With permission.)

10.4.4 Flux

Annual flux of PCN is calculated based on the concentration of total PCNs, sedimentation rate, and bulk densities. There were only a few studies carried out on the historical or actual flux of PCNs in the global environment including the sediment. The available estimates are summarized in Table 10.4. In Lake Kitaura, the annual PCN fluxes ranged from 0.0007 to 1.1 $\mu g/m^2$. In the preindustrial period (up to 1926) alleged annual flux of PCNs was at a low value of 0.0007–0.003 $\mu g/m^2$, and then sharply increased to 0.5–1.1 $\mu g/m^2$ between 1971 and 1975. After 1985, the annual values of CN flux decreased and by 1997–2000 it was 0.17 $\mu g/m^2$. In Tokyo Bay, the annual fluxes of PCNs were two orders of magnitude higher than those for Lake Kitaura and were 2–3 times higher than those for Lake Esthwaite, which is located in a semirural area in the United Kingdom [96]. When compared to Lake Kitaura, comparable or slightly higher PCN fluxes were found in the Gulf of Bothnia, which is located in the northern part of the Baltic Sea [97]. In comparison with other POPs, PCN fluxes in Tokyo Bay were of the same order of magnitude as the fluxes of PCDFs (~23 $\mu g/m^2$/year), and were >10-fold lower than those for PCBs (~690 $\mu g/m^2$/year) [90].

TABLE 10.4
Fluxes Estimation ($\mu g/m^2$/year) of CNs in Water Bodies Worldwide

Lake Kitaura, Japan[a]		Tokyo Bay, Japan[b]		Lake Esthwaite, United Kingdom[c]		Gulf of Bothnia, Baltic Sea[d]	
Period	Flux	Period	Flux	Period	Flux	Period	Flux
2000–1997	0.17						
1995–1992	0.24			1995	2.7		
		1993–1991	9.1			1992–1991	0.49–0.93
1989–1987	0.35			1989	4.3		
1985–1984	0.95	1985–1983	16				
1982–1981	0.6	1983–1981	21	1981	4.2		
1980–1978	0.83						
		1979–1976	22				
1977–1976	0.77						
1975–1973	1.1						
1972–1971	0.5			1972	4.8		
		1975–1970	20				
1970–1968	0.14						
1963–1962	0.055			1962	12		
		1950–1945	2.5	1948	2.2		
1926–1917	0.0026			1927	1.4		
		1905–1910	1	1909	0.7		
				1889	0.5		
				1870	0.6		
1828–500	0.0007–0.003						

[a] Data from Horii, Y. et al., *J. Environ. Sci. Health* A, 39, 587, 2004 (sum of tri to octa-CNs).
[b] Data from Yamashita, N. et al., *Environ. Sci. Technol.*, 34, 3560, 2000 (sum of tri- to octa-CNs).
[c] Data from Gevao, B. et al., *Environ. Sci. Technol.*, 34, 33, 2000 (sum of tetra- to hepta-CNs).
[d] Data from Lundgren, K. et al., *Environ. Pollut.*, 126, 93, 2003. (sum of tetra to hepta-CNs).

10.4.5   OCCURRENCE OF PCNs IN THE PREINDUSTRIAL PERIOD

PCNs dominated by tri- and tetra-CNs were found in deep sediment layers, that is, the preindustrial period (pre-1900s). When moving downward from the sediment core in all three of those studies [45,90,96], tri- and tetra-CNs were the relatively more abundant PCN constituents. For example, the tri-CN contribution in the deepest layers was 34%–65% for Lake Kitaura (dated <1967), 48% for Tokyo Bay (dated ~1907), and 38%–46% for Lake Esthwaite (dated 1870–1909). In the oldest sediment layers of Lake Kitaura (dated <550) only tri- and tetra-CNs at an extremely small concentration could be quantified (Figure 10.18).

There are several hypotheses that might explain the presence of low molecular weight congeners such as tri-CNs, which are the only congeners quantified in the deepest sediment layers. One explanation is that because of the better water solubility and weak adsorbtivity of tri-CNs, and to lesser degree the tetra-CNs as well, those compounds might preferentially diffuse downward through the pore water. Unfortunately, mono- and di-CNs, which are much more water soluble than the remaining CN congeners, were not determined in the sediment cores discussed above. The second possibility is the breakdown of hexa- and octa-CN to lower molecular weight PCNs due to microbial activity in anaerobic conditions, but no experimental data exist to support such a suggestion. Other causes could be the natural formation of lower molecular weight PCNs from unknown natural sources, or the combustion of wood in preindustrial period.

10.4.6   NOVEL HALOGENATED POLYCYCLIC AROMATIC HYDROCARBONS

Chlorinated and brominated polycyclic aromatic hydrocarbons (Cl-/Br-PAHs) are a class of halogenated aromatic hydrocarbons that are structurally similar to PCNs, PCBs, PCDDs, and PCDFs. In terms of toxicity, AhR-mediated activities of several Cl-/Br-PAHs have been reported using a YCM3 cell and H4IIE-luc [98,99]. Relative potencies of several Cl-/Br-PAHs are of the same order of magnitude as the RePs of mono-ortho PCBs. Recent syntheses of individual Cl-/Br-PAHs (3–5 aromatic rings) has made the congener-specific analysis possible and Cl-/Br-PAHs were found in several environmental matrices [100,101]. Horii et al. [102] reported the concentrations and profiles of 20 ClPAHs in sediment core that was collected in July 2004 from Tokyo Bay (N35°34′58″, E139°54′57″) near a site studied [90,91].

ClPAHs were found throughout the sediment core at total ClPAHs ranging from 36 to 1210 pg/g dw (Figure 10.20). The highest concentration of ClPAHs was found in the 14–16 cm section of the core, corresponding to the mid-1990s. The concentrations of ClPAH increased with decreasing depth of up to 14–16 cm (1994–1995), then gradually decreased up to the surface layer, corresponding to the year 2004 (673 pg/g dw). Among individual ClPAHs, 6-monochlorobenzo[a]pyrene (29%) was the predominant compound, followed in abundance by 1-monochloropyrene (19%) and 3-monochlorofluoranthene (8.7%). In general, ClPAH profiles found in the Tokyo Bay sediment core were similar to the profiles in other environmental matrices such as incineration ash or urban air particles [101,103]. The annual fluxes of ClPAHs ranged from 0.29 to 5.7 μg/m². The highest flux of ClPAHs was approximately 30-fold lower than the highest flux of total PCDD/Fs, and 120-fold lower than that of PCBs in 1979–1981.

10.5   ACCIDENTAL CONTAMINATION OF PCNS IN INDUSTRIAL PRODUCTS IN JAPAN

10.5.1   BACKGROUND

PCNs are widely used industrial chemicals produced under different trade names, such as Halowax (United States), Nibren wax (Germany), Seekay wax (United Kingdom), Clonacire wax (France), and Cerifal (Italy), from the 1910s to the 1980s [1]. According to the Japanese Chemical Substance

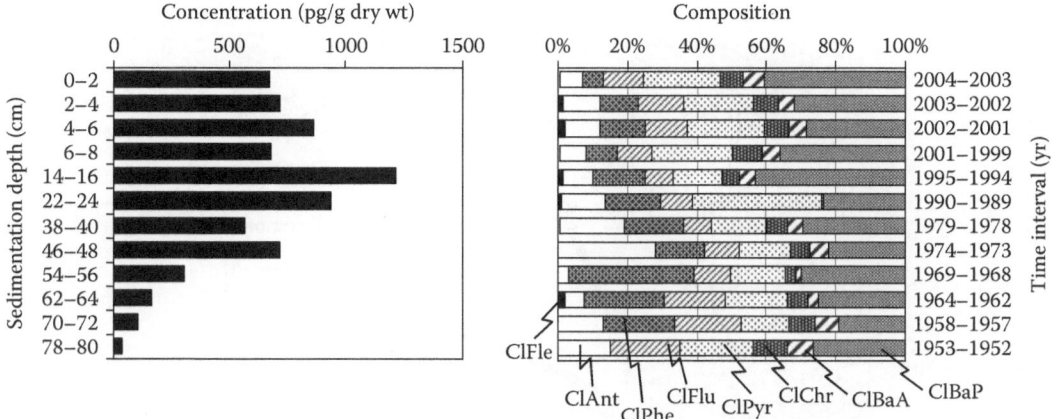

FIGURE 10.20 Vertical profiles in concentrations (pg/g dw) and composition (%) for ClPAHs, along with time interval represented, in Tokyo Bay sediment core. chlorinated fluorene (ClFle), chlorinated anthracene (ClAnt), chlorinated phenanthrene (ClPhe), chlorinated fluoranthene (ClFlu), chlorinated pyrene (ClPyr), chlorinated chrysene (ClChr), chlorinated benz[*a*]anthracene (ClBaA), chlorinated benzo[*a*]pyrene (ClBaP). (From Horii, Y. et al., *Arch. Environ. Contam. Toxicol.*, 57, 656, 2009, Figure 1. With permission.)

Control Law, PCNs were classified as Class I hazardous chemicals in 1979 in a similar fashion to the contaminants of PCBs in that they can pose the highest risk to humans [20].

On January 16, 2002, the Ministry of Economy, Trade and Industry (METI) of Japan announced that more than 18 metric tons of PCNs had been imported by a Japanese company (Showa DDE Co.) from the United Kingdom from 1998 to 2000 without abidance to the rule of regulation and manufacture of chemical substances in Japan [21]. The PCNs imported illegally were used to make "Neoprene FB," a synthetic rubber product by polymerization of chloroprene from 1999 to 2001. The total amount of Neoprene FB manufactured was reported as 259 metric tons of which more than 207 metric tons were exported to the international market. The rest of them were manufactured as "rubber compounds" and were used in weather-resistant products such as adhesives, shoe soles, paints, and many kinds of products in Japan. Only less than 800 kg of Neoprene FB was recalled by end of February while more than 29 metric tons was supposed to be used in domestic commercial products. It is known that Neoprene FB and the rubber compounds have been used by 42 domestic companies to make a wide variety of commercial products in Japan as of April 30, 2002 (Figure 10.21).

More unexpected trouble was reported in February 2002. About 54 metric tons of rubber compound for adhesive made from Neoprene FB (which contains PCNs) had been imported by Sumitomo 3 M Co. from Canada after December, 1995. This rubber compound was used to manufacture three series of aerosol adhesive bombs. Total amount of product was reported to be more than 210,000 bombs and a recall was announced at the end of February under the direction of the Ministry of Economy, Trade, and Industry (METI). The amount of PCNs in the adhesive material that was imported was declared to be about 40 kg by Sumitomo 3 M (Figure 10.22).

Despite these ignorant illegal mistakes of PCN usage, they can provide useful information to understand what will happen "IF" future toxic chemicals are released into commercial produce in our social system. In this chapter, we tried to describe these incidents from both the domestic and international points of view.

10.5.2 Isomer-Specific Analysis of PCNs in Contaminated Commercial Products

After the incident occurred, METI encountered a difficulty because there was no information on isomer-specific data of PCNs for any material or product related to Neoprene FB. As a result, it is essential to conduct isomer-specific risk assessments for PCNs as some of them have been estimated

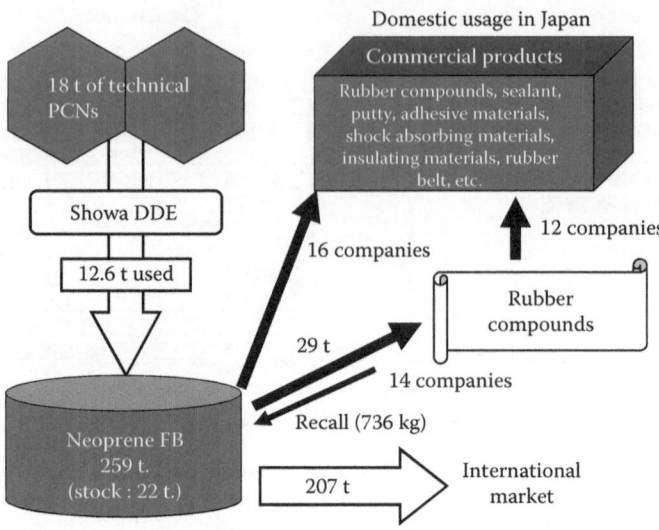

FIGURE 10.21 Flowchart of distribution and usage of imported PCN mixtures—Case A.

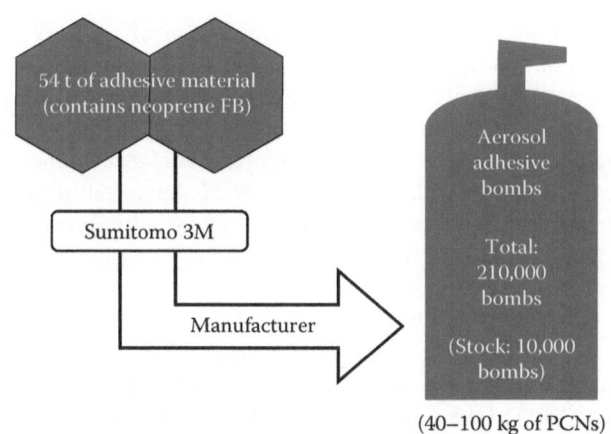

FIGURE 10.22 Flowchart of distribution and usage for imported PCN mixtures—Case B.

to have toxicity equivalency factors (TEFs) similar to that of dioxin-like compounds. The National Institute for Advanced Industrial Science and Technology (AIST) was requested to conduct a follow-up study of the incident from the scientific side and applied isomer specific analytical methods to determine PCN residues in the above industrial products. Aerosol adhesive bombs and its controls (similar to those manufactured by Sumitomo 3 M) were commercially purchased in March 2002. Neoprene FB and related rubber compounds were provided upon the request to cooperate for this national follow-up project. Samples were dissolved into methylene chloride and toluene before quantification. An exact amount of solution was passed through the acidic silica multilayer glass column with 200 mL of hexane. The eluate was concentrated and applied to a double-column HPLC (graphitic carbon and pyrenil silica HPLC) clean-up, which enabled high resolution separation of each PCN isomer.

HRGC-HRMS (JMS-700D, JEOL Co.) was used to obtain isomer-specific data of PCNs. A pair of single ion monitors with DB-1701 capillary column (J&W, 30 m, 0.25 mm i.d., 0.25 µm of film) was carried out to measure the exact m/z of each PCN congener as follows: T3CNs: 229.9457, 231.9427, T4CNs: 263.9067, 265.9038, P5CNs: 299.8648, 301.8618, H6CNs: 333.8258, 335.8229,

H7CNs: 367.7868, 369.7839, O8CN: 401.7479, 403.7449. ^{13}C-labled PCDD and coplanar PCBs were added into the sample solutions before clean-up, as internal and recovery standards. Detailed analytical procedures are reported elsewhere [21].

10.5.3 RESULTS AND DISCUSSIONS

Isomer-specific data of PCNs in contaminated commercial products in Japan are presented in Figure 10.23 and Table 10.5. Isomer composition of Halowax (HW1101) and imported PCNs were similar to each other and also comparable to Neoprene FB related rubber compounds. The amount of PCNs in Neoprene FB was 4.5% and was close to the value (about 4%) that was declared by manufacturer. Two different lots of aerosol adhesive bomb 333 showed similar isomer compositions of PCNs to that of Neoprene FB. It is clear that the control bombs (made with chloroprene) from different companies contain no PCN. Rubber coated sheets, which are used as rubber bands for office automation instruments, contained much lower PCN values than Neoprene FB, but

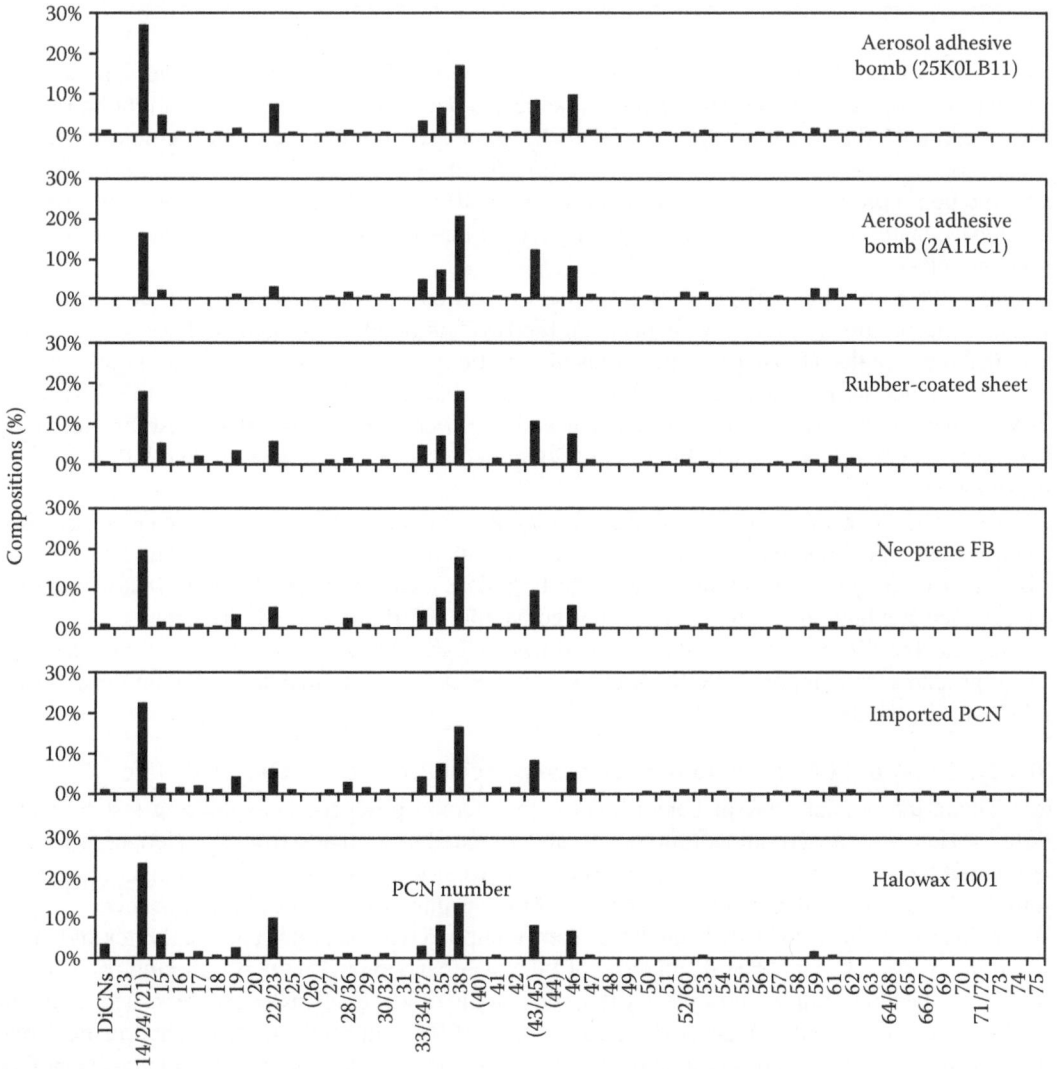

FIGURE 10.23 Isomer composition of di-, tri-, tetra-, penta-, and hexa-CNs in Neoprene FB related materials. (Data From Yamashita, N. et al., *J. Environ. Sci. Health A*, 38, 1745, 2003.)

TABLE 10.5
PCN Concentrations in Neoprene FB–Related Commercial Products

	Halowax 1001 %	Imported PCN %	Neoprene FB µg/g	Rubber Coated Sheet µg/g	Aerosol Adhesive Bomb 333 (Serial No.)		Control µg/mL
					(#2A1LC1) µg/mL	(#25K0LB1) µg/mL	
Di-CNs	3.3	1.4	494	8.5	4.2	14	—
Tri-CNs	42	47	45,860	351	282	510	—
Tetra-CNs	43	59	24,980	560	721	608	—
Penta-CNs	4.6	7.2	3,630	75	156	64	—
Hexa-CNs	—	0.081	46	—	3.2	1.0	—
Hepta-CNs	—	—	—	—	—	—	—
Octa-CNs	—	—	—	—	—	—	—
Total PCNs	93	115	45,010	995	1166	1200	—

still significant concentrations (at several hundred ppm) of PCNs were still found. The average concentration of PCNs in two lots of aerosol adhesive bomb was 1180 ug/mL. It might be reasonable to estimate that nearly 500 mg of PCNs exist in each bomb (each bomb contains 430 mL of adhesive solution). On account of the 210,000 bombs produced, the total amount of PCN in adhesive bomb products could be estimated at about 100 kg. Although declared amount of PCNs in adhesive material used was about 40 kg, difference between the two is acceptable as a variation of manufacturers.

Table 10.6 shows that TEQs of PCN isomers based on dioxin-like toxicity. In contrast to the similar concentrations of total PCNs between the two lots of adhesive bombs, there was a significant difference in the TEQs. The same kinds of variations in TEQs was also found in the imported PCN and Neoprene FB related compounds. It is possible to explain that most of the dioxin-like PCN isomers are constituted of penta, hexa, and hepta chlorinated and that these are relatively difficult to detect without the combination of double-column HPLC separation and HRGC-HRMS measurement.

As a result, it is worth to mention that a significant amount of PCNs was supposed to still exist in a wide variety of commercial products but yet no information was available about human exposure. It is thus necessary to conduct an isomer-specific risk-assessment of PCN-contaminated commercial products not only in Japan but also exported Neoprene FB and related products.

Analytical results of commercial products showed clearly that the estimated use of contaminated rubber products from imported PCN could be used to make a quantitative model of human exposure to PCN in the incident.

10.5.3.1 Risk of PCN in Contaminated Commercial Products to Our Daily Life

To evaluate the possible risk of contaminated commercial products, a weathering test of rubber coated sheets was carried out as follows. A rubber coated sheet made from the Neoprene FB and designated for a timing belt was cut into 10 mm² pieces. It was assumed that the temperature of the timing belt was about 40°C when used in an operating printer. The pieces of the rubber coated sheet were held at 40°C in electrically heated chambers equipped with a passive chemical filter and maintained up to 168 h. Isomer specific analysis of PCN was carried out on the weathering test samples in a similar fashion. Results show clearly that the amount of low chlorinated isomers such as di- and tri-CNs were dramatically decreased, and loss rates of 98% and 66% were reached respectively. On the other side, the loss of more chlorinated homologue groups such as tetra-, penta-, hexa-CNs was very small. The initial content (77%) of higher chlorinated homologue groups remained after termination of the 168 h test. TEQ value on the test sheets showed the same results as the higher

TABLE 10.6
2,3,7,8-TCDD Like TEQs of Individual Isomers of PCNs in Neoprene FB Related Commercial Products

PCN No.	Structure	TEF	Halowax 1001 ng-TEQ/g	Imported PCN ng-TEQ/g	Neoprene FB ng-TEQ/g	Rubber-Coated Sheet ng-TEQ/g	Aerosol Adhesive Bomb 333 (Serial No.)	
							(#2A1LC1) ng-TEQ/mL	(#25K0LB1) ng-TEQ/mL
54	1,2,3,6,7	0.00017	—	967	—	—	—	—
56	1,2,3,7,8	0.000046	—	—	2	—	0.099	0.034
57	1,2,4,5,6	3.5E-06	10	27	1	0.021	0.042	0.019
63	1,2,3,4,5,6	0.002	—	—	12	—	0.57	0.27
64	1,2,3,4,5,7	0.00002	—	2	—	—	0.006	0.003
66/67	1,2,3,4,6,7/1,2,3,5,6,7	0.0013	—	201	2	—	0.26	—
68	1,2,3,5,6,8	0.00015	—	18	1	—	0.044	0.019
69	1,2,3,5,7,8	0.002	—	710	25	—	1.5	0.51
70	(1,2,3,6,7,8)	0.0099	—	—	—	—	—	—
73	1,2,3,4,5,6,7	0.00069	—	—	—	—	—	—
Total TEQ			10	1925	43	0.021	2.5	0.85

Source: Yamashita, N. et al., *J. Environ. Sci. Health A*, 38, 1745, 2003. With permission.

chlorinated homologue groups because most of the toxic isomers were penta-, hexa-, and hepta-CNs in the contaminated rubber sheets (Table 10.6).

Considering these results, the reconstruction of two PCN incidents for quantitative risk assessment was carried out (Figures 10.24 and 10.25). Case A (contaminated rubber sheets) can be described as follows. About 1.3 t of PCN in 29 t of Neoprene FB were in use in the domestic market of Japan.

About 1.2 ton of low chlorinated PCNs are easily released from commercial products such as the timing belt of operating printer and the resulting daily exposure to human. 0.1 ton of highly chlorinated isomers may remain in these products and discarded into the garbage. This contains

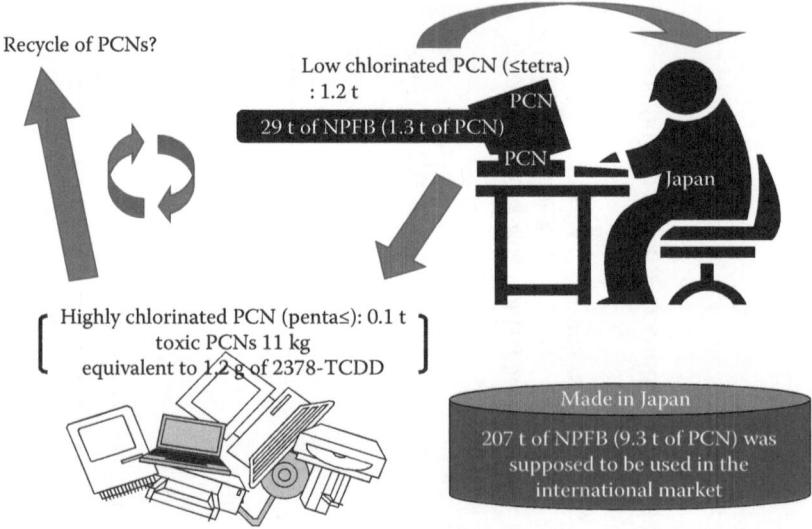

FIGURE 10.24 Risk model of accidental contamination of PCNs in industrial products in Japan—Case A.

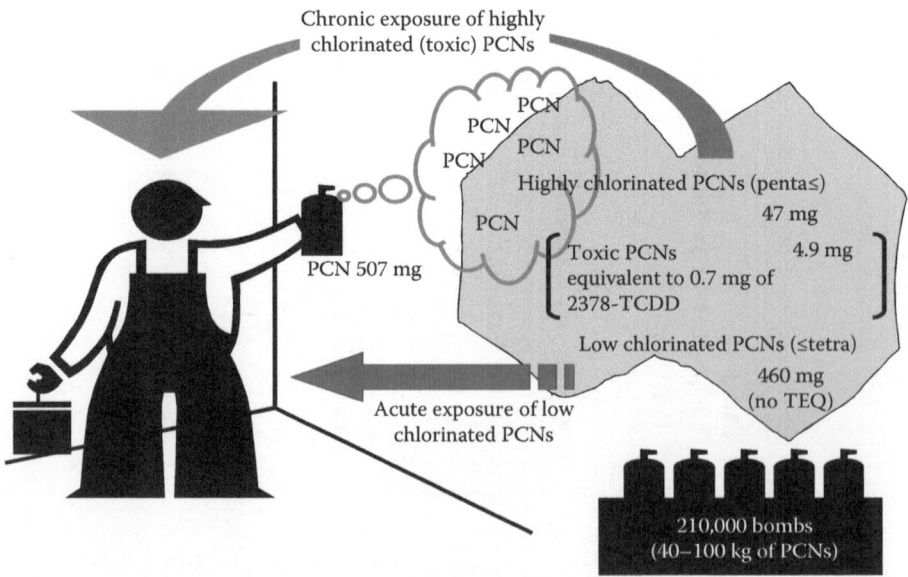

FIGURE 10.25 Risk model of accidental contamination of PCNs in industrial products in Japan—Case B.

11 kg of toxic PCNs which equates to 1.2 g of 2,3,7,8-TCDD. Recycling of contaminated commercial products may also cause human exposure to PCNs.

For Case B (about 3 M AAB), about 507 mg of PCN was contained in each bomb. Analysis of one bomb used for indoor environments resulted in an acute exposure of 460 mg of low chlorinated isomers. The remaining 47 mg of highly chlorinated isomer may expose humans for longer periods to chronic exposure. This contains 4.9 mg of toxic PCN and it is equivalent to 0.7 mg of 2,3,7,8-TCDD.

For now, no incidents of accidental use of PCN contaminated commercial products from the above two events have been reported. Human exposure to hazardous chemicals are complicated and it is difficult to identify one reason or source and its weakness of any modeling effort to know adverse effect of environmental chemicals. Findings into the accidental use of PCNs in commercial products in Japan, described here, showed that the approach of analytical chemistry in using accurate instrumental techniques is useful to reveal the human exposure, hiding our daily life.

10.6 POLYCHLORINATED NAPHTHALENES IN CHINA

10.6.1 INTRODUCTION

PCBs and PCDD/Fs in the environment and in humans in China have been studied extensively. However, due to the lack of regulations and monitoring capacity, PCNs have not yet been regularly monitored in China. Data about the occurrence and the pollution status of PCNs were extremely scarce in China and still lack the necessary base-line data. In this part, sources and levels of PCNs in the environmental media of China are summarized and are based mainly on available scientific literature.

10.6.2 RESEARCH ON PCNS IN CHINA

10.6.2.1 Sources

There were no PCN technical formulating products in China and the information on the application of PCN products in China is also rare. The main PCN sources in the environment in China have been from the importation of PCN products from abroad and from various products containing PCNs such as impurities or unintentionally produced by-products.

Preliminary investigations showed that solid waste incineration [104] and atmospheric transportation [105] may have been important PCN sources. Guo et al. [106] indicated that the industrial input might be the most significant source of PCNs in sewage sludge, and thermal processes such as waste incineration and coal combustion may be other sources of PCN contamination. Ba et al. [107] selected typical secondary copper, aluminum, zinc, and lead plants to investigate the emissions of PCNs in secondary nonferrous productions in China. Secondary nonferrous production was addressed as one of the potential sources of the unintentionally produced persistent organic pollutants (UP-POPs) due to raw material impurities. Although there are inventories of dioxin emissions from secondary nonferrous metallurgical facilities, release inventories of PCNs are scarce.

Pan et al. [108] reported that there were two pathways of input of PCNs in sediments from Qingdao coastal sea, one is the river discharge of sewage sludge and the other is atmospheric precipitation. The historical trends of contributions from different sources of PCNs in the sediment core in Jiaozhou Bay, China were studied by Pan et al. [109]. PCNs began to occur in the 1950s before the industrialization period of China, probably coming from wood and coal combustions. The beginning of the increase of isomers PCN71/72 and PCN59 (representatives of the technical product Halowax 1014) since the early 1950s in the sediment core may reflect the industrialization processes in the surrounding regions as PCN pollution was transported through the Yellow Sea current system to the Jiaozhou Bay area. The impact of this PCN source on the sediments reached its maximum in the late 1990s and has diminished since then, reflecting the ban of PCNs production and usage. The rapid increase of indicatory isomers PCN66/67 and PCN52/60, representatives of the PCNs in fly ash formed by municipal solid waste incinerators (MSWI), since the late 1970s in the core, may indicate that the contribution from MSWI becomes ever more important. Enormous amounts of solid waste were produced as a result of the fast economic growth in China, and this trend of contribution from MSW incinerations is now showing no sign of slowing down. The increase of isomer PCN14 in the sediment core, representative of the contribution from coal burning process, may also reflect the growth of the economy as well as the population in China. The indicatory isomer reached its maximum in the mid-1980s and then decreased, reflecting the change in the fuel structure in China.

10.6.2.2 Levels of PCNs

Due to the lack of regulations and monitoring capacity, only very limited knowledge on the pollution of PCNs in China is available. This section briefly reviewed the existing status of PCNs in China. The amount of available data is inadequate to precisely evaluate the pollution situation. Most of the data were collected from public scientific works and therefore are not systematic.

10.6.2.3 Sediments

Levels of PCNs (31 congeners) in the surface sediments collected from Qingdao coastal sea, Shandong Peninsula were reported by Yang et al. [110]. The total concentrations of PCNs in sediments ranged from 212 to 1209 pg/g dw with the mean of 585 pg/g dw. In general, the level of PCNs in the sediments of China was relatively low compared to those reported in industrialized countries. Influence of current circulation and sewage sludge on spatial distributions of PCNs in the sediments was also investigated. The maximum concentrations were all found near the Haibo River mouth, which was affected by sewage sludge input from the river.

Pan et al. [109] measured concentrations of PCNs (36 congeners) in a sediment core collected from Jiaozhou Bay, China. PCNs were dominated by tri-CNs and tetra-CNs, accounting for over 80% of the total PCNs, similar to the composition of technical Halowax products in which tri-CNs and tetra-CNs are typically predominant. The total PCN concentrations in the core ranged from 3.86 to 56.4 pg/g dw (mean: 29.8 pg/g dw), which was lower than those of sediments reported in the Baltic river [111], Tokyo Bay [90], and Switzerland [112]. PCN-TEQs in all the sections of the sediment core from Jiaozhou Bay were lower than 0.1 pg TEQ/g dw, with relatively minor contributions to the total TEQs. The maximum PCN sedimentation fluxes occurred in the mid-1970s inside Jiaozhou Bay, reflecting the influence of PCN pollutions in countries and regions around the Yellow Sea.

10.6.2.4 Soils

Although soil is believed to act as a significant repository for POPs, data on the pollution situation of PCNs in the soils of China is still not clear because only one report has been conducted. The concentrations of PCNs in five top soils collected from Wolong area of Southwest China ranged from 13.0 to 29.0 pg/g dw [105]. The pollution level was substantially lower than the soil collected from the United Kingdom (300–9000 pg/g dw) and Spain (32–180 pg/g dw) [64,113]. The mean concentration (21.4 pg/g dw) in soils from the Wolong area was about 6 orders of magnitude less than those reported for soils (17.9 µg/g dw) collected near a chlor-alkali plant from Japan [2]. The homologue of PCNs was dominated by tri-CNs. The PCN-TEQ in the soils were all less than 0.001 pg TEQ/g dw, mainly contributed from PCNs 66/67, 73, and 69 [105].

10.6.2.5 Sewage Sludge

The concentrations of PCNs in sewage sludge from wastewater treatment plants give an indication of the general exposure to and uses of these compounds. In China, data on PCNs in sewage sludge are scarce. PCNs in sewage sludge collected from eight different provinces of China were measured [106]. The homologue distribution patterns of PCNs in the sewage sludge samples were generally similar, with di- and tri-CNs as the predominant components. The total PCN (70 congeners) concentrations were in the range of 1.48–28.21 ng/g dw and the PCN-TEQ ranged from 0.11 to 2.45 pg TEQ/g dw. The results showed that PCNs levels in the sewage sludge in China were much lower than those in the sludge from other industrialized countries [114,115].

10.6.2.6 Biota

Available data on PCNs in biota were obtained using seafood or poultry samples in China. Most of the investigations were carried out in eastern and southeastern China.

Yang et al. [110] investigated concentrations of 31 PCN isomers in mussels (*Mytilus edulis Linne*) collected from one Chinese coastal city (Qingdao) and found 9.0 ng/g lipid. The mussels were enriched significantly in PCNs relative to the sediments. The PCN-TEQ in mussels was 0.063 pg TEQ/g lipid, lower than that of PCBs. Contributions to the total TEQs were mainly from PCN66 and PCN73. PCNs in five seafood categories (fish, bivalves, shrimp, crab, and cephalopods) collected in two Chinese coastal cities (Guangzhou and Zhoushan) were reported [116]. The predominant congeners were tetra-CNs and penta-CNs in Guangzhou seafood samples, accounting for more than 60% of the total PCNs. The concentrations of total PCNs in seafood were 93.8–1300 pg/g lipid depending on sampling locations and species. PCN concentrations in fish from Guangzhou and Zhoushan were 545 and 137 pg/g lipid respectively, which were similar to concentrations in fish samples collected from the Arctic (81.3–915 pg/g lipid) [117].

Nine pooled samples, belonging to two food categories (fish and duck), were collected from the coastal waters of Qingdao and Chongming Island of Shanghai [118]. The mean concentrations of total PCNs in fish were 225 and 640 pg/g lipid respectively. In duck meat, the mean concentration was 43.8 pg/g lipid in Shanghai. The PCN-TEQ ranged from 0.001 to 0.21 pg TEQ/g lipid for fish and duck meat. The concentrations of total PCNs in fish were comparable with those reported in China [116], the Arctic area [117], and the lower than those in fish samples collected in Japan (190–3300 pg/g lipid) [60], the United States (19–31,400 pg/g ww and 27 ng/g lipid), [2] and Poland (26 ng/g lipid) [111].

Two yak samples were collected from the Wolong area of Southwest China [105]. The PCN-TEQ in the yak meat and yak fat were 122.8 and 224.1 pg/g lipid respectively, which is unlikely to cause adverse health effect on the yak-meat consuming people. The pollution levels of PCNs in biota reported by China were generally low in the region.

10.6.2.7 Fly Ash

Zhang et al. [104] determined the concentrations of PCNs in fly ash collected from three types of cement kilns (vertical shaft kiln, wet-process rotary kiln, and dry-process rotary kiln) and two types of waste incinerators. Different types of cement kiln fly ash presented similar PCNs homologue patterns. The predominant homologues were tetra-CNs. PCN 66/67, which has dioxin-like toxicity, was the most abundant congener in all fly ash. The results showed that the total PCN-TEQ in cement kiln fly ash, which were in the range of 0.47–2.8 ng TEQ/kg, were much lower than that of fly ash from waste incinerators.

10.6.2.8 Human Exposure

Exposure to PCNs from consumption of fish in two Chinese coastal cities (Guangzhou and Zhoushan) was estimated to be 16.6 and 19.6 pg/kg body weight/day (bw/day), assuming that the non-detectable levels were at the level of the limit of quantifications (LOQs) [116]. Expressed in terms of WHO-TEQ, the daily intake of PCNs was less than 0.01 pg TEQ/kg bw/day for the Chinese coastal population. The dietary intake of dioxin-like compounds (including PCDD/Fs, coplanar PCBs, and PCNs) was at the lower end of the tolerable daily intake (TDI), 1–4 pg TEQ/kg bw for the general population.

Incidents of consumption of contaminated rice oil in two populations in the Taiwan area (called Yusho or Yucheng, i.e., oil disease) of China with symptoms similar to those described for PCN exposure have been extensively studied. Furthermore, PCNs have also been reported in whole blood (1,2,3,4,6,7/1,2,3,5,6,7-hexa-CN: 8,590–30,400 ng/kg lipid) and adipose tissue specimens [119,120] of the Yucheng victims, who were exposed to PCBs and their heat-degradation products, mainly PCDFs, from the ingestion of contaminated rice oil in 1978–1979.

10.6.3 CONCLUSIONS

The sources, environmental levels, and current situation of PCNs in China were reviewed in this section. Because there is such a limited amount of published data, the occurrence and the current pollution status of PCNs in China is still not clear. The attention caused by PCNs is not as high as other POPs in China; however there is an immediate need for the Chinese government and research institutions to set up a PCN standardized method and to popularize the method, so as to conduct nationwide inventory surveys on PCNs. As one of the components of the Stockholm Convention, improving monitoring capability is a principal and crucial step to get a thorough understanding of PCNs for the Chinese government.

REFERENCES

1. Falandysz, J. 1998. Polychlorinated naphthalenes: An environmental update. *Environ. Pollut.* 101: 77–90.
2. Kannan, K., Imagawa, T., Blankenship, A.L., and Giesy, J.P. 1998. Isomer-specific analysis and toxic evaluation of polychlorinated naphthalenes in soil, sediment, and biota collected near the site of a former chlor-alkali plant. *Environ. Sci. Technol.* 32: 2507–2514.
3. Falandysz, J. 2003. Chloronaphthalenes as food-chain contaminants: A review. *Food Addit. Contam.* 20: 995–1014.
4. Falandysz, J., Puzyn, T., Szymanowska, B., Kawano, M., Markuszewski, M., Kaliszan, R., Skurski, P., Blazejowski, J., and Wakimoto, T. 2001. Thermodynamic and physico-chemical descriptors of chloronaphthalenes: An attempt to select features explaining environmental behaviour and specific toxic effects of these compounds. *Pol. J. Environ. Stud.* 10: 217–235.
5. Kucklick, J.R. and Helm, P.A. 2006. Advances in the environmental analysis of polychlorinated naphthalenes and toxaphene. *Anal. Bioanal. Chem.* 386: 819–836.
6. Domingo, J.L. 2004. Polychlorinated naphthalenes in animal aquatic species and human exposure through the diet: A review. *J. Chromatogr. A.* 1054: 327–334.
7. Su, Y.S., Lei, Y.D., Daly, G.L., and Wania, F. 2002. Determination of octanol-air partition coefficient (KOA) values for chlorobenzenes and polychlorinated naphthalenes from gas chromatographic retention times. *J. Chem. Eng. Data* 47: 449–455.
8. Harner, T. and Bidleman, T.F. 1998. Octanol-air partition coefficient for describing particle/gas partitioning of aromatic compounds in urban air. *Environ. Sci. Technol.* 32: 1494–1502.
9. Wania, I.R., Lei, Y.D., and Harner, T. 2002. Estimating octanol-air partition coefficients of nonpolar semivolatile organic compounds from gas chromatographic retention times. *Anal. Chem.* 74: 3476–3483.
10. Opperhulzen, A., Vandervelde, E.W., Gobas, F.A.P.C., Liem, D.A.K., Vandersteen, J.M.D., and Hutzinger, O. 1985. Relationship between bioconcentration in fish and steric factors of hydrophobic chemicals. *Chemosphere* 14: 1871–1896.
11. Opperhuizen, A. 1987. Relationships between octan-1-ol water partition-coefficients, aqueous activity-coefficients and reversed phase HPLC capacity factors of alkylbenzenes, chlorobenzenes, chloronaphthalenes and chlorobiphenyls. *Toxicol. Environ. Chem.* 15: 249–264.
12. Puzyn, T. and Falandysz, J. 2005. Octanol/water partition coefficients of chloronaphthalenes. *J. Environ. Sci. Heal. A.* 40: 1651–1663.
13. Puzyn, T. and Falandysz, J. 2007. QSPR modeling of partition coefficients and Henry's law constants for 75 chloronaphthalene congeners by means of six chemometric approaches—A comparative study. *J. Phys. Chem. Ref. Data* 36: 203–214.
14. Wania, F. and Mackay, D. 1996. Tracking the distribution of persistent organic pollutants. *Environ. Sci. Technol.* 30: A390–A396.
15. Bidleman, T.F., Helm, P.A., Braune, B.M., and Gabrielsen, G.W. 2010. Polychlorinated naphthalenes in polar environments—A review. *Sci. Total Environ.* 48: 2919–2935.

16. Falandysz, J., Kawano, M., Ueda, M., Matsuda, M., Kannan, K., Giesy, J.P., and Wakimoto, T. 2000. Composition of chloronaphthalene congeners in technical chloronaphthalene formulations of the Halowax series. *J. Environ. Sci. Health A*. 35: 281–298.

17. Falandysz, J., Chudzynski, K., Takekuma, M., Yamamoto, T., Noma, Y., Hanari, N., and Yamashita, N. 2008. Multivariate analysis of identity of imported technical PCN formulation. *J. Environ. Sci. Health A* 43: 1381–1390.

18. Falandysz, J., Nose, K., Ishikawa, Y., Lukaszewicz, E., Yamashita, N., and Noma, Y. 2006. HRGC/HRMS analysis of chloronaphthalenes in several batches of Halowax 1000, 1001, 1013, 1014 and 1099. *J. Environ. Sci. Health A* 41: 2237–2255.

19. Noma, Y., Yamamoto, T., Giraud, R., and Sakai, S. 2006. Behavior of PCNs, PCDDs, PCDFs, and dioxin-like PCBs in the thermal destruction of wastes containing PCNs. *Chemosphere* 62: 1183–1195.

20. Japan Ministry of Economy, Trade and Industry, Chemical council, Safety guideline section. 1979. The regulation of polychlorinated naphthalenes and hexa-chlorobenzenes (in Japanese).

21. Yamashita, N., Taniyasu, S., Hanari, N., Horii, Y., and Falandysz, J. 2003. Polychlorinated naphthalene contamination of some recently manufactured industrial products and commercial goods in Japan. *J. Environ. Sci. Health A* 38: 1745–1759.

22. Muir, D.C.G. and de Wit, C.A. 2010. Trends of legacy and new persistent organic pollutants in the circumpolar arctic: Overview, conclusions, and recommendations. *Sci. Total Environ.* 406: 3044–3051.

23. Imagawa, T. 2000. Studies on formation of polychloronaphtalenes from waste incineration (in Japanese). Report on the National Institute for Resources and Environment, 29. Available at: http://www.aist.go.jp/NIRE/publica/hokoku/h29/h29.htm

24. Imagawa, T. and Lee, C.W. 2001. Correlation of polychlorinated naphthalenes with polychlorinated dibenzofurans formed from waste incineration. *Chemosphere* 44: 1511–1520.

25. Yamashita, N., Kannan, K., Imagawa, T., Miyazaki, A., and Giesy, J.P. 2000. Concentrations and profiles of polychlorinated naphthalene congeners in eighteen technical polychlorinated biphenyl preparations. *Environ. Sci. Technol.* 34: 4236–4241.

26. Taniyasu, S., Kannan, K., Holoubek, I., Ansorgova, A., Horii, Y., Hanari, N., Yamashita, N., and Aldous, K.M. 2003. Isomer-specific analysis of chlorinated biphenyls, naphthalenes and dibenzofurans in Delor: Polychlorinated biphenyl preparations from the former Czechoslovakia. *Environ. Pollut.* 126: 169–178.

27. Eljarrat, E. and Barcelo, D. 2003. Priority lists for persistent organic pollutants and emerging contaminants based on their relative toxic potency in environmental samples. *Trac-Trend. Anal. Chem.* 22: 655–665.

28. Puzyn, T., Falandysz, J., Jones, P.D., and Giesy, J.P. 2007. Quantitative structure–activity relationships for the prediction of relative in vitro potencies (REPs) for chloronaphthalenes. *J. Environ. Sci. Health A* 42: 573–590.

29. Hanberg, A., Waern, F., Asplund, L., Haglund, E., and Safe, S. 1990. Swedish dioxin survey: Determination of 2,3,7,8-TCDD toxic equivalent factors for some polychlorinated-biphenyls and naphthalenes using biological tests. *Chemosphere* 20: 1161–1164.

30. Villeneuve, D.L., Kannan, K., Khim, J.S., Falandysz, J., Nikiforov, V.A., Blankenship, A.L., and Giesy, J.P. 2000. Relative potencies of individual polychlorinated naphthalenes to induce dioxin-like responses in fish and mammalian in vitro bioassays. *Arch. Environ. Contam. Toxicol.* 39: 273–281.

31. Blankenship, A.L., Kannan, K., Villalobos, S.A., Villeneuve, D.L., Falandysz, J., Imagawa, T., Jakobsson, E., and Giesy, J.P. 2000. Relative potencies of individual polychlorinated naphthalenes and halowax mixtures to induce Ah receptor-mediated responses. *Environ. Sci. Technol.* 34: 3153–3158.

32. Japan Ministry of Economy, Trade and Industry. 1973. Japanese Law concerning the examination and regulation of manufacture of chemical substances.

33. Kawano, M., Ueda, M., Falandysz, J., Matsuda, M., and Wakimoto, T. 2000. Polychlorinated naphthalenes (PCNs) in human adipose tissue in Japan. *Organohalogen Compd.* 47: 159–162.

34. Kosaka, K., Itoh, M., Nakazawa, Y., and Mori, K. 2008. Occurrence of polychlorinated naphthalenes and chlorinated polycyclic aromatic hydrocarbons in water supply system (in Japanese). *J. Jpn. Water Works Assoc.* 77: 2–12.

35. Nakano, T., Matsumura, C., and Fujimori, K. 2000. Isomer specific analysis of polychlorinated naphthalenes for environmental sample. *Organohalogen Compd.* 47: 178–181.

36. Japan Ministry of the Environment. 1977. Chemicals in the environment. Report on environmental survey and monitoring of chemicals (in Japanese).

37. Japan Ministry of the Environment. 1979. Chemicals in the environment. Report on environmental survey and monitoring of chemicals (in Japanese).

38. Japan Ministry of the Environment. 2002. Chemicals in the environment. Report on environmental survey and monitoring of chemicals (in Japanese). Available at: http://www.env.go.jp/chemi/kurohon/http2002/index.html

39. Yoshida, H. and Takeshita, R. 1979. Studies on environmental contamination by polychlorinated naphthalenes (PCN). VI. Contamination of sea water by PCN (in Japanese). *Eisei Kagaku* 25: 334–337.

40. Shiraishi, H., Pilkington, N.H., Otsuki, A., and Fuwa, K. 1985. Occurrence of chlorinated polynuclear aromatic-hydrocarbons in tap water. *Environ. Sci. Technol.* 19: 585–590.

41. Imagawa, T. and Yamashita, N. 1996. Estimation of emission sources of polychlorinated naphthalenes using finger-print method for isomer composition (in Japanese). *J. Environ. Chem.* 6: 495–501.

42. Yamamoto, T., Noma, Y., Hirai, Y., Nose, K., and Sakai, S. 2005. Congener-specific analysis of polychlorinated napthalenes in the waste samples. *Organohalogen Comp.* 67: 708–711.

43. Horii, Y., Okada, M., Amano, K., Hanari, N., Taniyasu, S., and Yamashita, N. 2002. Analysis of individual polychlorinated naphthalene congeners and dioxin-like compounds in a dated sediment core from Lake Kitaura, Japan. *Bunseki Kagaku* 51: 1009–1018.

44. Horii, Y., Okada, M., and Yamashita, N. 2002. Historical records of polychlorinated naphthalenes, -biphenyls, -dibenzo-p-dioxins, -dibenzofurans in a dated sediment core from Lake Kitaura, Japan. *Organohalogen Compd.* 57: 29–32.

45. Horii, Y., Falandysz, J., Hanari, N., Rostkowski, P., Puzyn, T., Okada, M., Amano, K., Naya, T., Taniyasu, S., and Yamashita, N. 2004. Concentrations and fluxes of chloronaphthalenes in sediment from Lake Kitaura in Japan in past 15 centuries. *J. Environ. Sci. Health A* 39: 587–609.

46. Sakai, S., Hiraoka, M., Takeda, N., and Shiozaki, K. 1996. Behavior of coplanar PCBs and PCNs in oxidative conditions of municipal waste incineration. *Chemosphere* 32: 79–88.

47. Takasuga, T., Inoue, T., Ohi, E., and Kumar, K.S. 2004. Formation of polychlorinated naphthalenes, dibenzo-p-dioxins, dibenzofurans, biphenyls, and organochlorine pesticides in thermal processes and their occurrence in ambient air. *Arch. Environ. Contam. Toxicol.* 46: 419–431.

48. Ohura, T., Kitazawa, A., Amagai, T., and Shinomiya, M. 2007. Relationships between chlorinated polycyclic aromatic hydrocarbons and dioxins in urban air and incinerators. *Organohalogen Compd.* 69: 2902–2905.

49. Kawano, M., Ueda, M., Matsui, M., Kashima, Y., Matsuda, M., and Wakimoto, T. 1998. Extractable organic halogens (EOX: Cl, Br and I), polychlorinated naphthalenes and polychlorinated dibenzo-p-dioxines and dibenzofurans in ashes from incinerators located in Japan. *Organohalogen Compd.* 36: 221–224.

50. Japan Ministry of the Environment. 1999. Chemicals in the environment. Report on environmental survey and monitoring of chemicals (in Japanese). Available at: http://www.env.go.jp/chemi/kurohon/http1999/index.html

51. Japan Ministry of the Environment. 2003. Chemicals in the environment. Report on environmental survey and monitoring of chemicals (in Japanese). Available at: http://www.env.go.jp/chemi/kurohon/http2003/sec2_1_1.html

52. Ohura, T., Fujima, S., Amagai, T., and Shinomiya, M. 2008. Chlorinated polycyclic aromatic hydrocarbons in the atmosphere: Seasonal levels, gas-particle partitioning, and origin. *Environ. Sci. Technol.* 42: 3296–3302.

53. Hanari, N., Horii, Y., Bochentin, I., Orlikowska, A., Puzyn, T., Falandysz, J., and Yamashita, N. 2003. PCNs in pine needles around the Tokyo bay, Japan. *Organohalogen Compd.* 62: 356–359.

54. Hanari, N., Horii, Y., Taniyasu, S., and Yamashita, N. 2003. Analysis of polychlorinated naphthalenes and dioxin-like compounds in pine needle leaf by high-resolution GC high-resolution MS. *Bunseki Kagaku* 52: 127–138.

55. Hanari, N., Horii, Y., Okazawa, T., Falandysz, J., Bochentin, I., Orlikowska, A., Puzyn, T., Wyrzykowska, B., and Yamashita, N. 2004. Dioxin-like compounds in pine needles around Tokyo Bay, Japan in 1999. *J. Environ. Monitor.* 6: 305–312.

56. Hanari, N., Horii, Y., Taniyasu, S., Falandysz, J., Bochentin, I., Orlikowska, A., Puzyn, T., and Yamashita, N. 2004. Isomer specific analysis of polychlorinated naphthalenes in pine trees (*Pinus thunbergi* Parl.) and (*Pinus densiflora* Sieb. et Zucc) needles around Tokyo Bay, Japan. *Pol. J. Environ. Stud.* 13: 139–151.

57. Hanari, N., Orlikowska, A., Bochentin, I., Wyrzykowska, B., Falandysz, J., Horii, Y., Taniyasu, S., Okazawa, T., and Yamashita, N. 2004. Source identification of polychlorinated naphthalenes, dioxins and related compounds in pine needles from Tokyo Bay, Japan and Poland (in Japanese). *Bunseki Kagaku* 53: 1399–1409.

58. Japan Ministry of the Environment. 2007. Chemicals in the environment. Report on environmental survey and monitoring of chemicals (in Japanese). Available at: http://www.env.go.jp/chemi/kurohon/2007/index.html

59. Takeshita, R. and Yoshida, H. 1979. Studies on environmental contamination by polychlorinated naphthalenes (PCN). IV. Contamination of marine fishes by PCN (in Japanese). *Eisei Kagaku* 25: 29–33.

60. Guruge, K.S., Seike, N., Yamanaka, N., and Miyazaki, S. 2004. Accumulation of polychlorinated naphthalenes in domestic animal related samples. *J. Environ. Monitor.* 6: 753–757.

61. Takeshita, R. and Yoshida, H. 1979. Studies on environmental contamination by polychlorinated naphthalenes (PCN). III. Contamination of human body by PCN (in Japanese). *Eisei Kagaku* 25: 24–28.

62. Haglund, P., Jakobsson, E., and Masuda, Y. 1995. Isomerspecific analysis of polychlorinated naphthalenes in Kanechlor KC 400, Yusho rice oil and adipose tissue of a Yusho victim. *Organohalogen Compd.* 26: 405–410.

63. Imagawa, T., Kannan, K., Yamashita, N., Miyazaki, A., and Giesy, J.P. 2000. Profiles of specific isomers of polychlorinated naphthalenes in Tokyo Bay sediment core sample. *Organohalogen Compd.* 47: 155–158.

64. Meijer, S.N., Harner, T., Helm, P.A., Halsall, C.J., Johnston, A.E., and Jones, K.C. 2001. Polychlorinated naphthalenes in UK soils: Time trends, markers of source, and equilibrium status. *Environ. Sci. Technol.* 35: 4205–4213.

65. Harner, T., Kylin, H., Bidleman, T.F., Halsall, C., and Strachan, W.M.J. 1998. Polychlorinated naphthalenes and coplanar polychlorinated biphenyls in arctic air. *Environ. Sci. Technol.* 32: 3257–3265.

66. Helm, P.A., Bidleman, T.F., Jantunen, L.M.M., and Ridal, J. 2000. Polychlorinated naphthalenes in Great Lakes air: Source and ambient air profiles. *Organohalogen Compd.* 47: 17–20.

67. Harner, T., Lee, R.G.M., and Jones, K.C. 2000. Polychlorinated naphthalenes in the atmosphere of the United Kingdom. *Environ. Sci. Technol.* 34: 3137–3142.

68. Eriksson, G., Jensen, S., Kylin, H., and Strachan, W. 1989. The pine needle as a monitor of atmospheric-pollution. *Nature* 341: 42–44.

69. Loganathan, B.G., Kumar, K.S., Seaford, K.D., Sajwan, K.S., Hanari, N., and Yamashita, N. 2008. Distribution of persistent organohalogen compounds in pine needles from selected locations in Kentucky and Georgia, USA. *Arch. Environ. Contam. Toxicol.* 54: 422–439.

70. Orlikowska, A., Hanari, N., Wyrzykowska, B., Bochentin, I., Horii, Y., Yamashita, N., and Falandysz, J. 2009. Airborne chloronaphthalenes in Scots pine needles of Poland. *Chemosphere* 75: 1196–1205.

71. van Leeuwen, S.P.J. and de Boer, J. 2008. Advances in the gas chromatographic determination of persistent organic pollutants in the aquatic environment. *J. Chromatogr. A* 1186: 161–182.

72. Crookes, M.J. and Howwe, P.D. 1993. *Environmental Hazard Assessment. Halogenated Naphthalenes*, London, U.K.: Department of the Environment.

73. Jarnberg, U., Asplund, L., and Jakobsson, E. 1994. Gas-chromatographic retention behavior of polychlorinated naphthalenes on nonpolar, polarizable, polar and smectic capillary columns. *J. Chromatogr. A* 683: 385–396.

74. Noma, Y., Yamamoto, T., Falandysz, J., Lukaszewicz, E., Gutfranska, A., and Sakai, S. 2004. By-side impurities in chloronaphthalene mixtures of the halowax series: All 12 chlorobenzenes. *J. Environ. Sci. Health A.* 39: 2011–2022.

75. Noma, Y., Yamamoto, T., Falandysz, J., Gutfranska, A., Lukaszewicz, E., and Sakai, S. 2004. By-side impurities in chloronaphthalene mixtures of the Halowax series: All 19 chlorophenols. *J. Environ. Sci. Health A* 39: 2023–2034.

76. Noma, Y., Ishikawa, Y., Falandysz, J., Jecek, L., Gulkowska, A., Miyaji, K., and Sakai, S. 2004. By-side impurities in chloronaphthalene mixtures of the Halowax series: All 209 chlorobiphenyls. *J. Environ. Sci. Health A* 39: 2035–2058.

77. Imagawa, T. and Yamashita, N. 1997. Gas chromatographic isolation of 1,2,3,4,5,7-, 1,2,3,5,6,8-, 1,2,4,5,6,8- and 1,2,4,5,7,8-hexachloronaphthalene. *Chemosphere* 35: 1195–1198.

78. Helm, P.A., Jantunen, L.M.M., Bidleman, T.F., and Dorman, F.L. 1999. Complete separation of isomeric penta- and hexachloronaphthalenes by capillary gas chromatography. *J. High Resolut. Chromatogr.* 22: 639–643.

79. Harner, T. and Kucklick, J. 2003. Interlaboratory study for the polychlorinated naphthalenes (PCNs): Phase 1 results. *Chemosphere* 51: 555–562.

80. Marriott, P.J., Haglund, P., and Ong, R.C.Y. 2003. A review of environmental toxicant analysis by using multidimensional gas chromatography and comprehensive GC. *Clin. Chim. Acta.* 328: 1–19.

81. Adahchour, M., Beens, J., and Brinkman, U.A.T. 2008. Recent developments in the application of comprehensive two-dimensional gas chromatography. *J. Chromatogr. A* 1186: 67–108.

82. Korytar, P., Leonards, P.E.G., de Boer, J., and Brinkman, U.A.T. 2005. Group separation of organohalogenated compounds by means of comprehensive two-dimensional gas chromatography. *J. Chromatogr. A* 1086: 29–44.

83. Lukaszewicz, E., Ieda, T., Horii, Y., Yamashita, N., and Falandysz, J. 2007. Comprehensive two-dimensional GC (GC×GC) qMS analysis of tetrachloronaphthalenes in Halowax formulations. *J. Environ. Sci. Health A* 42: 1607–1614.

84. Jarman, W.M., Hilkert, A., Bacon, C.E., Collister, J.W., Ballschmiter, K., and Risebrough, R.W. 1998. Compound-specific carbon isotopic analysis of Aroclors, Clophens, Kaneclors, and Phenoclors. *Environ. Sci. Technol.* 32: 833–836.

85. Drenzek, N.J., Eglinton, T.I., May, J.M., Wu, Q.Z., Sowers, K.R., and Reddy, C.M. 2001. The absence and application of stable carbon isotopic fractionation during the reductive dechlorination of polychlorinated biphenyls. *Environ. Sci. Technol.* 35: 3310–3313.

86. Omalley, V.P., Abrajano, T.A., and Hellou, J. 1994. Determination of the $^{13}C/^{12}C$ ratios of individual PAH from environmental-samples—Can PAH sources be apportioned. *Org. Geochem.* 21: 809–822.

87. Horii, Y., Kannan, K., Petrick, G., Gamo, T., Falandysz, J., and Yamashita, N. 2005. Congener-specific carbon isotopic analysis of technical PCB and PCN mixtures using two-dimensional gas chromatography—Isotope ratio mass spectrometry. *Environ. Sci. Technol.* 39: 4206–4212.

88. Hites, R.A., La Flamme, R.E., and Farrington, Y.W. 1977. Sedimentary polycyclic aromatic hydrocarbons: The terrestrial record. *Science* 198: 829–831.

89. Eisenreich, S.J., Capel, P.D., Robbins, J.A., and Bourbonniere, R. 1989. Accumulation and diagenesis of chlorinated hydrocarbons in lacustrine sediments. *Environ. Sci. Technol.* 23: 1116–1126.

90. Yamashita, N., Kannan, K., Imagawa, T., Villeneuve, D.L., Hashimoto, S., Miyazaki, A., and Giesy, J.P. 2000. Vertical profile of polychlorinated dibenzo-p-dioxins, dibenzofurans, naphthalenes, biphenyls, polycyclic aromatic hydrocarbons, and alkylphenols in a sediment core from Tokyo Bay, Japan. *Environ. Sci. Technol.* 34: 3560–3567.

91. Kannan, K., Villeneuve, D.L., Yamashita, N., Imagawa, T., Hashimoto, S., Miyazaki, A., and Giesy, J.P. 2000. Vertical profiles of dioxin-like and estrogenic activities associated with a sediment core from Tokyo Bay, Japan. *Environ. Sci. Technol.* 34: 3568–3573.

92. Falandysz, J., Strandberg, L., Bergqvist, P.A., Strandberg, B., and Rappe, C. 1997. Spatial distribution and bioaccumulation of polychlorinated naphthalenes (PCNs) in mussel and fish from the Gulf of Gdansk, Baltic Sea. *Sci. Total Environ.* 203: 93–104.

93. Brack, W., Kind, T., Schrader, S., Moder, M., and Schuurmann, G. 2003. Polychlorinated naphthalenes in sediments from the industrial region of Bitterfeld. *Environ. Pollut.* 121: 81–85.

94. Noma, Y., Yamamoto, T., and Sakai, S.I. 2004. Congener-specific composition of polychlorinated naphthalenes, coplanar PCBs, dibenzo-p-dioxins, and dibenzofurans in the halowax series. *Environ. Sci. Technol.* 38: 1675–1680.

95. Helm, P.A. and Bidleman, T.F. 2003. Current combustion-related sources contribute to polychlorinated naphthalene and dioxin-like polychlorinated biphenyl levels and profiles in air in Toronto, Canada. *Environ. Sci. Technol.* 37: 1075–1082.

96. Gevao, B., Harner, T., and Jones, K.C. 2000. Sedimentary record of polychlorinated naphthalene concentrations and deposition fluxes in a dated Lake Core. *Environ. Sci. Technol.* 34: 33–38.

97. Lundgren, K., Tysklind, M., Ishaq, R., Broman, D., and van Bavel, B. 2003. Flux estimates and sedimentation of polychlorinated naphthalenes in the northern part of the Baltic Sea. *Environ. Pollut.* 126: 93–105.

98. Horii, Y., Khim, J.S., Higley, E.B., Giesy, J.P., Ohura, T., and Kannan, K. 2009. Relative potencies of individual chlorinated and brominated polycyclic aromatic hydrocarbons for induction of aryl hydrocarbon receptor-mediated responses. *Environ. Sci. Technol.* 43: 2159–2165.

99. Ohura, T., Morita, M., Makino, M., Amagai, T., and Shimoi, K. 2007. Aryl hydrocarbon receptor-mediated effects of chlorinated polycyclic aromatic hydrocarbons. *Chem. Res. Toxicol.* 20: 1237–1241.

100. Ohura, T. 2007. Environmental behavior, sources, and effects of chlorinated polycyclic aromatic hydrocarbons. *Sci. World J.* 7: 372–380.

101. Horii, Y., Ok, G., Ohura, T., and Kannan, K. 2008. Occurrence and profiles of chlorinated and brominated polycyclic aromatic hydrocarbons in waste incinerators. *Environ. Sci. Technol.* 42: 1904–1909.

102. Horii, Y., Ohura, T., Yamashita, N., and Kannan, K. 2009. Chlorinated polycyclic aromatic hydrocarbons in sediments from industrial areas in Japan and the United States. *Arch. Environ. Contam. Toxicol.* 57: 651–660.

103. Ohura, T., Sawada, K.I., Amagai, T., and Shinomiya, M. 2009. Discovery of novel halogenated polycyclic aromatic hydrocarbons in urban particulate matters: Occurrence, photostability, and AhR activity. *Environ. Sci. Technol.* 43: 2269–2275.

104. Zhang, J., Ni, Y.W., Zhang, H.J., Zhang, X.P., Zhang, Q., and Chen, J.P. 2009. Patterns of PCDD/Fs, PCBs and PCNs homologues in fly ash from cement kilns (in Chinese). *Environ. Sci.* 30: 568–573.

105. Yang, Y.L., Pan, J., Zhu, X.H., Liu, X.D., Chen, D.Z., and Yamashita, N. 2009. Altitude gradient distributions of dioxin-like compounds and PCNs in soils from Wolong area and human health risk assessment for yak consumption (in Chinese). *Environ. Chem.* 28: 276–283.

106. Guo, L., Zhang, B., Xiao, K., Zhang, Q.H., and Zheng, M.H. 2008. Levels and distributions of polychlorinated naphthalenes in sewage sludge of urban wastewater treatment plants. *Chin. Sci. Bull.* 53: 508–513.

107. Ba, T., Zheng, M.H., Zhang, B., Liu, W.B., Su, G.J., Liu, G.R., and Xiao, K. 2010. Estimation and congener-specific characterization of polychlorinated naphthalene emissions from secondary nonferrous metallurgical facilities in China. *Environ. Sci. Technol.* 44: 2441–2446.

108. Pan, J., Yang, Y.L., Xu, Q., Chen, D.Z., and Xi, D.L. 2007. PCBs, PCNs and PBDEs in sediments and mussels from Qingdao coastal sea in the frame of current circulations and influence of sewage sludge. *Chemosphere* 66: 1971–1982.

109. Pan, J., Miyake, Y., Yang, Y.L., Yeung, L.W.Y., WU, X.L., Taniyasu, S., and Yamashita, N. 2009. The vertical trends of polychlorinated naphthalenes in a dated sediment core from Qingdao coastal sea, China. *Organohalogen Compd.* 71: 2496–2500.

110. Yang, Y.L., Pan, J., Li, Y., Yin, X.C., and Shi, L. 2004. Persistent organic pollutants PCNs and PBDEs in sediments from coastal waters of Qingdao, Shandong Peninsula. *Chin. Sci. Bull.* 49: 98–106.

111. Falandysz, J., Strandberg, L., Bergqvist, P.A., Kulp, S.E., Strandberg, B., and Rappe, C. 1996. Polychlorinated naphthalenes in sediment and biota from the Gdansk Basin, Baltic Sea. *Environ. Sci. Technol.* 30: 3266–3274.

112. Bogdal, C., Kohler, M., Schmid, P., Sturm, M., Grieder, E., Scheringer, M., and Hungerbühler, K. 2006. Polychlorinated naphthalenes: Congener specific analysis and source identification in a dated sediment core from Lake Thun, Switzerland. *Organohalogen Compd.* 68: 300–303.

113. Schuhmacher, M., Nadal, M., and Domingo, J.L. 2004. Levels of PCDD/Fs, PCBs, and PCNs in soils and vegetation in an area with chemical and petrochemical industries. *Environ. Sci. Technol.* 38: 1960–1969.

114. Stevens, J.L., Northcott, G.L., Stern, G.A., Tomy, G.T., and Jones, K.C. 2003. PAHs, PCBs, PCNs, organochlorine pesticides, synthetic musks, and polychlorinated n-alkanes in UK sewage sludge: Survey results and implications. *Environ. Sci. Technol.* 37: 462–467.

115. Nylund, K., Asplund, L., Jansson, B., Jonsson, P., Litzen, K., and Sellstrom, U. 1992. Analysis of some polyhalogenated organic pollutants in sediment and sewage-sludge. *Chemosphere* 24: 1721–1730.

116. Jiang, Q.T., Hanari, N., Miyake, Y., Okazawa, T., Lau, R.K.F., Chen, K., Wyrzykowska, B., So, M.K., Yamashita, N., and Lam, P.K.S. 2007. Health risk assessment for polychlorinated biphenyls, polychlorinated dibenzo-p-dioxins and dibenzofurans, and polychlorinated naphthalenes in seafood from Guangzhou and Zhoushan, China. *Environ. Pollut.* 148: 31–39.

117. Corsolini, S., Kannan, K., Imagawa, T., Focardi, S., and Giesy, J.P. 2002. Polychloronaphthalenes and other dioxin-like compounds in Arctic and Antarctic marine food webs. *Environ. Sci. Technol.* 36: 3490–3496.

118. Yang, Y.L., Pan, J., Zhu, X.H., Liu, X.D., Lu, G.H., Li, Q., and X.D., L. 2009. Studies on coplanar-PCBs and PCNs in edible fish and duck in Qingdao and Chongming Island (in Chinese). *Res. Environ. Sci.* 22: 187–193.

119. IPCS, UNEP/ILO/WHO. 2001. Concise international chemical assessment document No. 34: Chlorinated naphthalenes. Available at: http://www.who.int/ipcs/publications/cicad/en/cicad34.pdf

120. Ryan, J.J. and Masuda, Y. 1944. Polychlorinated naphthalenes in the rice oil poisonings. *Organohalogen Compd.* 21: 251–254.

11 Sources, Distributions, and Temporal Trends of Nonylphenols in South Korea

Seongjin Hong* and Kyung-Hoon Shin

CONTENTS

11.1 Introduction 259
11.2 Use and Control of Legislation for NPs in South Korea 260
11.3 Sources, Distributions, and Temporal Trends of NPs in Aquatic Environments
in South Korea 261
    11.3.1 Industrialized Area: Lake Shihwa..262
        11.3.1.1 Description of Lake Shihwa 262
        11.3.1.2 Sources and Distributions 263
        11.3.1.3 Temporal Trends 263
        11.3.1.4 NPs in Biota: Polychaetes 265
    11.3.2 Wastewater Treatment Plant Outfall: Masan Bay......................................266
        11.3.2.1 Description of Masan Bay 266
        11.3.2.2 Sources and Distributions 268
        11.3.2.3 Temporal Trends 268
        11.3.2.4 NPs in Biota: Mussels 270
    11.3.3 Ocean Waste Disposal: Dumpsite "Byung"...270
        11.3.3.1 Description of Dumpsite "Byung"..270
        11.3.3.2 Sources, Distributions, and Temporal Trends............................271
        11.3.3.3 Korean Government Plans for Waste Disposal at Sea................273
    11.3.4 Urbanized Area: Han River...273
        11.3.4.1 Description of Han River..273
        11.3.4.2 Sources and Distributions 273
        11.3.4.3 Comparison with Other Rivers of East Asian Countries............273
11.4 Comparison of Contamination Levels in Each Aquatic Environment and
Environmental Quality Guidelines..274
11.5 Summary 275
Acknowledgments..275
References..275

11.1 INTRODUCTION

Alkylphenols (APs) are degradation products of alkylphenol ethoxylates (APEOs), which have been used as detergents, wetting agents, dispersing agents, and emulsifiers in various commercial, industrial, and household surfactants [1,2]. APEOs enter the natural environment mainly through

---

* E-mail: hongseongjin@gmail.com (Chapter corresponding author).

TABLE 11.1

Water and Sediment Quality Guidelines for Nonylphenol (NP) of the United States and Canada

| Aquatic Life | EPA NP Criteria[a] Water (μg L^{-1}) | | Canadian NP Guidelines[b] | |
	Acute	Chronic	Water (μg L^{-1})	Sediment (μg g^{-1} dw)[c]
Freshwater	28	6.6	1	1.4
Marine	7	1.7	0.7	1

[a] Refer from U.S. EPA [15].
[b] Refer from CCME [5].
[c] Expressed on a TEQ basis and assumes 1% TOC.

direct urban or industrial inputs and sewage treatment plant (STP) effluents [3,4]. Once APs and their ethoxylates have entered the aquatic environment, they are degraded by photooxidation and/or microbial activity in the water column and adsorbed onto suspended particles; they then settle down and accumulate in the sediments and aquatic organisms according to their physicochemical properties [5–8]. The breakdown product, nonylphenol (NP), has more severe aquatic toxicity than the original surfactant and is a more persistent and lipophilic chemical [9–11]. NPs are also recognized as endocrine-disrupting chemicals, which can cause estrogenic effects in fish and other aquatic organisms [12]. Because of their hazardous characteristics, NPs have been controlled by worldwide environmental agencies over the last few years. The European Union (EU) has classified NPs as a "Priority Hazardous Substance" that has been banned for all uses since 2003 [13], and the United Nations Environment Programme (UNEP) has included NPs in the list of 28 "Persistent Toxic Substances" [14]. The Canadian Council of Ministers of the Environment (CCME) suggested water and sediment quality guidelines for NP and its ethoxylates based on the acute and chronic toxicity and an equilibrium partitioning approach for the protection of aquatic life [5]. The U.S. Environmental Protection Agency (U.S. EPA) has also developed water quality criteria (WQC) for NPs [15]. Canadian water quality guidelines (WQG) for NPs were lower than U.S. chronic WQC, as shown in Table 11.1. Sediment quality guidelines (SQG) for NPs in Canada are for guidance only.

11.2   USE AND CONTROL OF LEGISLATION FOR NPS IN SOUTH KOREA

During the last five decades, approximately 500,000 ton of APEOs were produced annually worldwide [16]. APEOs most commonly consist of nonylphenol ethoxylates (NPEOs, 80%), octylphenol ethoxylates (OPEOs, 15%), and other phenolic ethoxylates [17,18]. Because of their harmful effects, some developed countries have banned the use of APEO products [19,20], but they are still used in many developing countries because APEOs are inexpensive and have superior cleaning properties [21]. In South Korea, all NPs have been imported as products (X-100, NX, or NP) containing up to 25% APEOs, and the import volume was 11,216 ton in 2004. The most common uses for NPs in South Korea were in household detergents and cleaning agents (60%). They are also utilized in manufacturing processes of paints and epoxy resins (12%), copper laminates (9%), ink binders (5%), agricultural pesticides (2%), and other uses (12%) [22,23].

Figure 11.1 shows amounts of NP and octylphenol (OP) chemicals emitted and transported into the environment and the consumption volumes of NP chemicals in South Korea from 2004 to 2007 [24,25]. The amounts of NP and OP emitted gradually decreased after 2004, and the temporal trend may be related to the banning of NP use for household detergent since 2002, while transport amounts of APs were shown to have increased, suggesting that APs are not readily degradable in

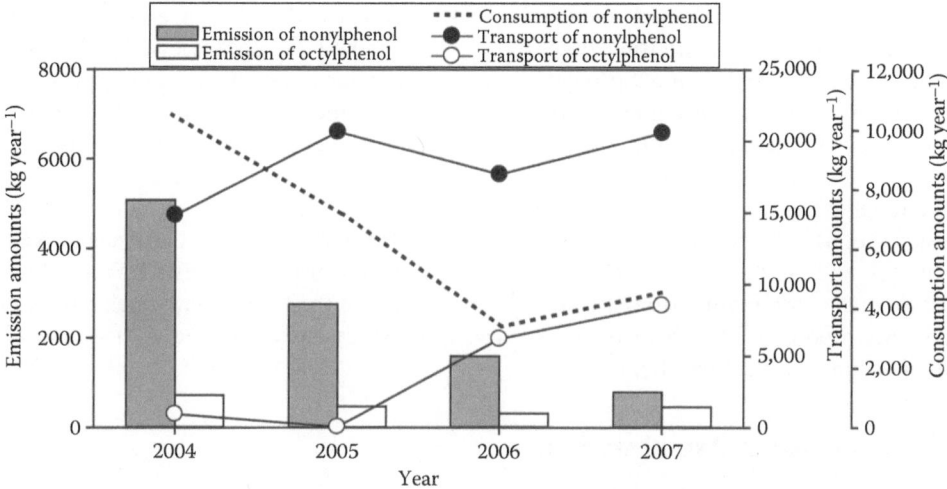

FIGURE 11.1   Amounts of NP and OP chemicals emitted and transported into the environment and consumption volumes of NP chemicals in South Korea from 2004 to 2007. (Adapted from National Institute of Environmental Research (NIER), Toxics release inventory open information system, http://ncis.nier.go.kr/triopen, KMOE, Korea, 2007; NIER, Investigation of emission characteristics for endocrine disrupting chemicals, NIER annual report, KMOE, Korea, 2007.)

the environment. The Korean government named NPs as priority chemicals in 2001, banned the use of NPs in kitchen cleaners in 2002, and phased out all domestic applications in 2007 [22,23]. Furthermore, NPs were designated as restricted chemicals in 2007, which restricts the manufacture, import, sale, storage, transportation, and use of these chemicals. South Korea has recently enforced a ban on the use of NPs in paints and ink binders in 2010, almost completely phasing out the use of NPs, but some uses remain.

11.3   SOURCES, DISTRIBUTIONS, AND TEMPORAL TRENDS OF NPs IN AQUATIC ENVIRONMENTS IN SOUTH KOREA

NPs enter aquatic environments directly or as breakdown products of NPEOs [4]. NPEOs are degraded to nonylphenol diethoxylate (NP2EO) via sequential elimination of ethoxy chains, which then accumulate in aquatic environments and/or may be further degraded to nonylphenol monoethoxylate (NP1EO) and NP by microbial activity. Long-chain NPEOs (more than two ethoxy chains) are more rapidly degraded than short-chain NPEOs (two or fewer ethoxy chains) in aquatic environments [9,26–28]. NPs are degraded by photooxidation and/or microbial activity in the water column and adsorbed onto suspended particles; they then settle down and accumulate in sediments and aquatic organisms according to their physicochemical characteristics [5–8]. NP released to the atmosphere is likely to be degraded by reaction with hydroxyl radicals, with a calculated half-life of 0.3 days, and the fraction of chemical adsorbed onto aerosol particles is also small [29]. The potential for transport of NP in the atmospheric environment is low, so NP is unlikely to be transported far from its emission source [29,30].

Studies on sources, distributions, and temporal trends of NPs in aquatic environments of South Korea have progressed in various ways since the late 1990s. A total of 19 SCI (Science Citation Index) articles have previously been published, focusing mainly on industrialized regions such as Lake Shihwa (six articles) and Masan Bay (six articles). According to the published papers, input sources of NPs into aquatic environments of South Korea can be divided into four types, including discharges from (1) industrialized areas, (2) wastewater treatment plant (WWTP) outfall, (3) ocean waste disposal, and (4) urbanized areas. In this chapter, we have classified the four sources of NPs and reviewed distributions, concentration levels, and temporal trends of NPs in aquatic environments of South Korea based on the published case studies. We searched for peer-reviewed SCI papers using "Science Direct," "Scopus," and "Springer Link." Domestic articles and research reports were not included in this review because they are not available worldwide and are written in Korean.

Sources, levels, and temporal trends of NPs in water, sediment, and biota are discussed in relation to Lake Shihwa and adjoining industrial complexes in Section 11.3.1. Section 11.3.2 presents spatial and temporal trends of NP pollution in sediments associated with the WWTP outfall as a point source in Masan Bay. Section 11.3.3 discusses the temporal trends and influxes of NPs in the dumpsite "Byung" in the East Sea (Sea of Japan) and the government plans for waste disposal at sea. NP chemicals originating from the urbanized area around the Han River, which passes through the Seoul Metropolitan City, are also discussed as a case study and compared with similar rivers in East Asian countries (Section 11.3.4).

11.3.1 Industrialized Area: Lake Shihwa

11.3.1.1 Description of Lake Shihwa

Lake Shihwa is an artificial lake located on the west coast of the Korean Peninsula (Figure 11.2a). The Shihwa dike (12.7 km) was constructed from 1987 to 1994 to supply freshwater to industrial and agricultural regions, resulting is the complete separation of Lake Shihwa from the Yellow Sea. The water and sediments of Lake Shihwa have been rapidly contaminated because of the lack of waste-water treatment facilities and increasing pollutant loads from the Shihwa and Banweol industrial complexes and cities [21,31]. Consequently, the Korean government abandoned its original plan to change Lake Shihwa into a freshwater lake and constructed a water gate at the end of the Shihwa dike in 1999 for the purpose of seawater replacement from the Yellow Sea [32,33]. Although water quality has improved in the vicinity of the gate, various organic pollutants such as organochlorine pesticides

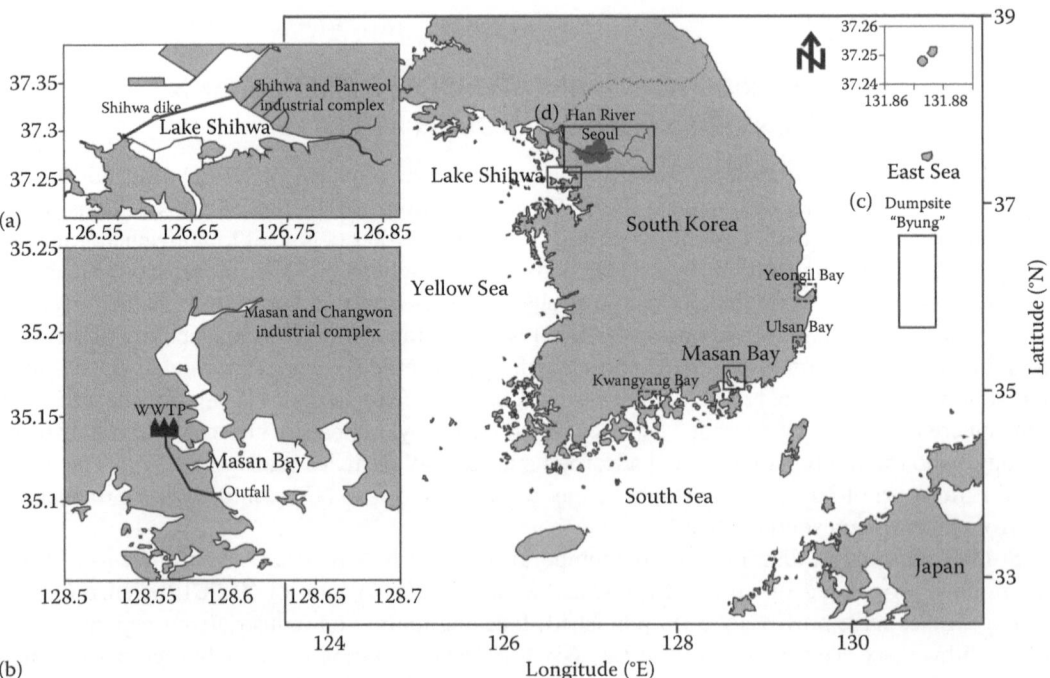

FIGURE 11.2 Map of (a) Lake Shihwa, (b) Masan Bay, (c) waste dumpsite "Byung," (d) Han River, and other industrialized bays (dotted square) of South Korea.

(OCPs), polycyclic aromatic hydrocarbons (PAHs), perfluorinated alkyl compounds (PFAs), and NPs have been detected in sediments from the inner region of Lake Shihwa [21,31,34,35].

11.3.1.2 Sources and Distributions

The main sources of NP pollution in the lake were wastewater from local industrial complexes via surrounding creeks and directly discharged untreated sewage from cities because of a lack of adequate WWTPs. NP concentrations in water, sediment, and biota in Lake Shihwa and surrounding creeks are presented in Figures 11.3a and 11.4a and Table 11.2. High NP concentrations were found in creek waters adjacent to industrial complex areas, ranging from 700 to 41,300 ng L^{-1} (average = 8,186 ng L^{-1}) in 2000, from 131 to 15,821 ng L^{-1} (average = 3,241 ng L^{-1}) in 2002, and from 312 to 16,598 ng L^{-1} (average = 4,283 ng L^{-1}) in 2008 [21,33,36]. Considerably higher NP levels were detected in the creeks of the industrial complex areas compared to the municipal area creeks, indicating that the major source of NP was the industrial area [33]. In all previous reports, concentration levels of NPs in water from creeks were higher than those from Lake Shihwa. Consequently, the effluent from those contaminated creeks affects the water quality of the lake. Although there have been many efforts to improve water quality in the Lake Shihwa environment, high concentrations of NPs are still present in the surrounding creeks.

11.3.1.3 Temporal Trends

In Figure 11.4a, pollution trends of NPs in surface sediments of Lake Shihwa are shown over the last 10 years. The current concentrations of NPs detected in Lake Shihwa sediments are relatively low compared with the concentrations noted in previous reports. This trend is attributed to the many environmental improvement efforts that have been enacted during the intervening years, such as a ban by the Korean government on household use of NPEOs since 2002, phasing out of all domestic applications by 2007, and the water exchange that was initiated through the installation and operation of a water gate at the end of the Lake Shihwa dike in 1999 [36].

Historical pollution trends of Lake Shihwa were recorded in core sediments over the past 30 years, as shown in Figure 11.5. The average sedimentation rate in the core samples from inner regions of Lake Shihwa was 2.41 cm year^{-1}, as estimated using ^{210}Pb dating. NP concentrations gradually increased after 1980, peaked in the late 1980s, and then decreased sharply until 1990. These trends may be related to increases in NP use caused by industrialization and urbanization associated with the development of the Banweol industrial complex from 1977 to 1987 and the construction and operation of the WWTP in 1990. NP concentrations also peaked in the early 1990s, indicating that untreated wastewater was directly discharged into Lake Shihwa because the adjacent cities lacked adequate WWTPs. High organic carbon (OC) content and low δ^{13}C (carbon stable isotope ratio) values could indicate large influxes of land-derived organic matter during the corresponding periods. The δ^{13}C value is a well-known indicator of land-derived organic matter in sediments, with −26 to −28% indicating terrestrial-derived and −19 to −21‰ indicating marine phytoplankton-derived organics respectively [37–40]. The increased sediment NP concentrations detected from the mid- to late-1990s may reflect the isolation of Lake Shihwa until 1999.

NP concentrations declined after 2000 on account of environmental improvement efforts such as construction of a water gate and intensive water quality control in the surrounding creeks. Despite these efforts, high concentrations of NP chemicals, the highest OC contents and lowest values of δ^{13}C were detected in the mid-2000s. These trends may reflect the insufficiency of the WWTP in proportion to the expanding local population coinciding with the construction of massive apartment complexes near Lake Shihwa after 2004. However, the amounts of NP deposited recently have declined rapidly, reflecting reduced NP inflows caused by expansion of the WWTP facilities in the surrounding areas and a ban on the domestic use of NP chemicals. High OC contents and low δ^{13}C values in the sediments corresponded to high NP concentrations at the same sediment core depths. A strong correlation between low δ^{13}C values and high NP concentrations and OC contents suggests that NP should be transported with land-derived organic matter to Lake Shihwa [36].

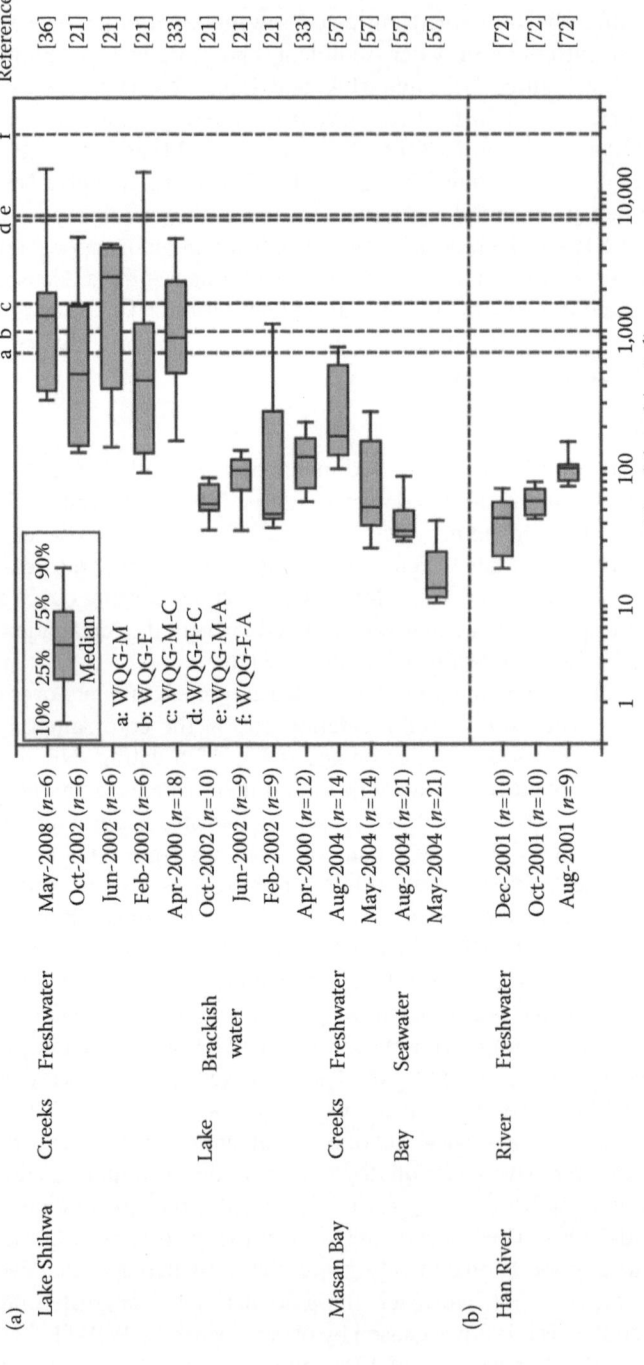

FIGURE 11.3 Box plots for the measured concentrations of NP in waters from (a) industrialized area and (b) urbanized area of South Korea (M: marine; F: freshwater; C: chronic; A: acute; WQG: water quality guideline of Canada; WQC: water quality criteria of the United States). (Adapted from Canadian Council of Ministers of the Environment (CCME), *Canadian sediment quality guidelines for the protection of aquatic life: Nonylphenol and its ethoxylates*, in *Canadian Environmental Quality Guidelines 1999*, Canadian Council of Ministers of the Environment, Winnipeg, Canada, 2001; United States Environmental Protection Agency (U.S. EPA), *Aquatic life ambient water quality criteria-Nonylphenol*, EPA-822-R-05-005, Office of Science and Technology. Washington, DC, 2005.)

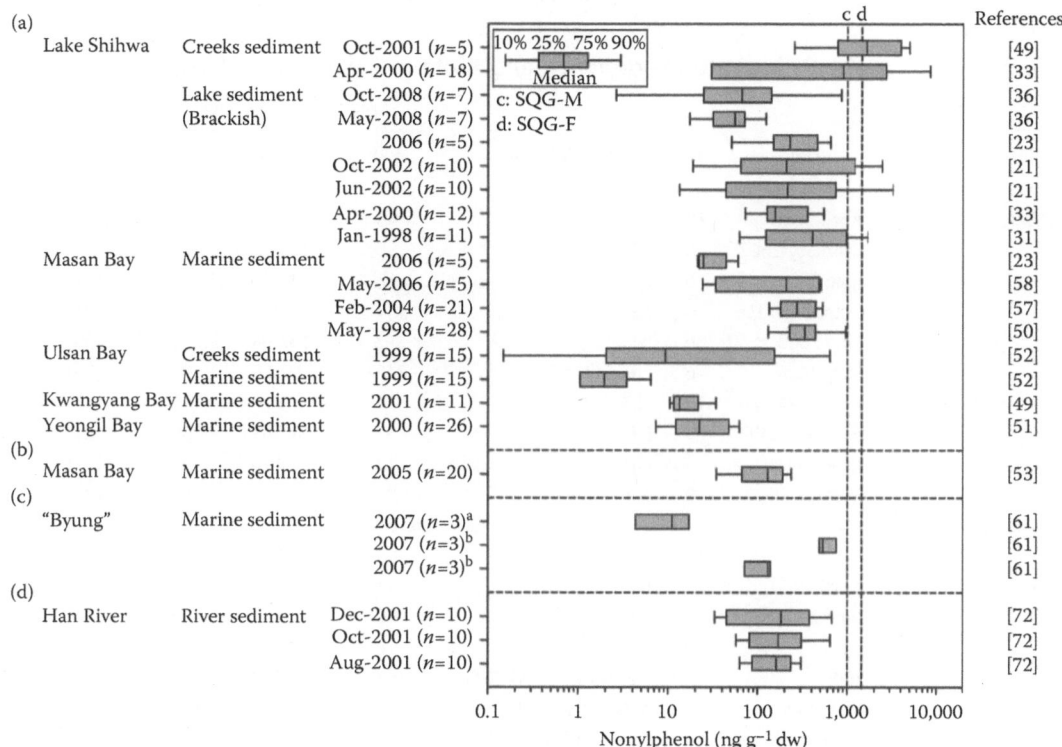

FIGURE 11.4   Box plots for the measured concentrations of NP in surface sediments from (a) industrialized area, (b) WWTP outfall, (c) waste dumpsite, and (d) urbanized area of South Korea ([a]non-dumpsite [0–6 cm depth]; [b]dumpsite [0–6 cm depth]; M: marine; F: freshwater; SQG: sediment quality guideline of Canada [normalized to 1% TOC]). (Adapted from Canadian Council of Ministers of the Environment (CCME), Canadian sediment quality guidelines for the protection of aquatic life: Nonylphenol and its ethoxylates, in *Canadian Environmental Quality Guidelines 1999*, Canadian Council of Ministers of the Environment, Winnipeg, Canada, 2001.)

The data obtained from the core sediments are in good agreement with previously reported data on surface sediments collected in January 1998 [31], April 2000 [33], and 2006 [23] within a 5 km radius of the core sampling site. This correlation between NP concentrations during the study period and previously reported data suggests that NP can be persistent under the strongly hypoxic conditions found in the Lake Shihwa sediments.

11.3.1.4   NPs in Biota: Polychaetes

Polychaetes (*Neanthes succinea*) have been used as pollution indicator species in hypoxic and polluted sediments [41], which have been found frequently in Lake Shihwa during the spring bloom season. The NP concentrations in polychaetes collected from Lake Shihwa sediments ranged from 1,156 to 2,757 ng g^{-1} dry weight (dw) (average = 1,916 ng g^{-1} dw), appearing higher than those previously reported for other organisms in other polluted regions. NP concentrations in mussels collected from Masan Bay in South Korea ranged from 50 to 290 ng g^{-1} dw [42], and NP concentrations in snails collected along the coast of Taiwan ranged from 130 to 1,560 ng g^{-1} dw [43]. NP concentrations in fish from the Kalamazoo River in Michigan ranged from <3.3 (method detection limit) to 29.1 ng g^{-1} wet weight (ww) [44]. David et al. [45] reviewed the NP concentrations in various marine organisms and found that they were higher in bivalves (mussels and oysters) and gastropods (snails) than in fishes. These differences in bioaccumulation of NPs may be related to the lipid contents, biological cycles, trophic levels, and feeding behaviors of these organisms. The bioaccumulated NP concentrations in the polychaetes strongly correlated with NP concentrations in surface sediments ($r^2 = 0.90$)

TABLE 11.2
Concentrations of NPs in Water, Sediments, and Biota in Lake Shihwa Environment

Samples	Year	n	Compounds	NPs Concentrations			Reference
				Min	Max	Mean	
Water				ng L^{-1}			
Creeks	April 2000	18	NP	46	41,300	3,580	[33]
	February 2002	6	NP	90	15,821	3,010	[21]
	June 2002	6	NP	118	4,324	2,309	[21]
	October 2002	6	NP	131	5,239	1,339	[21]
	May 2008	6	NP	312	16,598	3,632	[36]
	May 2008	6	NP1EO+NP2EO	746	9,606	3,709	[36]
Lake Shihwa	April 2000	12	NP	37	770	183	[33]
	February 2002	9	NP	35	1,533	280	[21]
	June 2002	9	NP	17	140	92	[21]
	October 2002	10	NP	31	87	61	[21]
Sediment				ng g^{-1} dw			
Lake Shihwa	January 1998	11	NP	20	1,820	616	[31]
	April 2000	12	NP	11	624	235	[33]
	June 2002	10	NP	10	5,054	826	[21]
	October 2002	10	NP	16	2,513	772	[21]
	2006	5	NP+NP1EO+NP2EO	103	702	353	[23]
	May 2008	7	NP+NP1EO+NP2EO	24	352	159	[36]
	August 2008	7	NP+NP1EO+NP2EO	<2	1,046	210	[36]
Biota (polychaete)				ng g^{-1} dw			
Lake Shihwa	May 2008	4	NP	1,156	2,757	1,917	[36]

from the same sites. This relationship may be explained by the polychaete feeding behavior, because polychaetes are nonselective deposit feeders that directly ingest sedimentary organic matter [46]. The average BAF (bioaccumulation factor) values for NP in polychaetes were estimated to be 40 ± 5 using the NP concentrations in the biota and sediment. Polychaetes are important food sources for fish and other predators, and the bioaccumulated NP in polychaetes may transfer to higher trophic levels in the marine ecosystem [47,48]. The establishment of a fishery in the inner regions of Lake Shihwa has been prohibited since 1998 because of severe pollution, but many species of fish move in and out of the lake through the water gate. Therefore, it is still possible that bioaccumulation of NPs in Lake Shihwa poses risks to the marine ecosystem and eventually to human health.

In the case of Lake Shihwa, NPs released from the industrial complex are revealed as the main contamination source in the aquatic environment. Many reports indicate that Masan Bay, Ulsan Bay, Kwangyang Bay, and Yeongil Bay in South Korea are also affected by industrial complexes like Lake Shihwa (Figure 11.4a) [49–52]. Therefore, there is a need to control NP sources for the health and management of the aquatic ecosystem.

11.3.2 WASTEWATER TREATMENT PLANT OUTFALL: MASAN BAY

11.3.2.1 Description of Masan Bay

Masan Bay, located on the south coast of South Korea, is a long and narrow inlet of semi-enclosed bay (Figure 11.2b). Because of its geological features, Masan Bay has shown a slow rate of water exchange and a trapping effect on contaminants discharged from surrounding industrial complexes and cities. Masan and Changwon industrial complexes, including petrochemical, heavy metal, electrical, and plastic industries, as well as heavily populated cities, are located adjacent to the bay

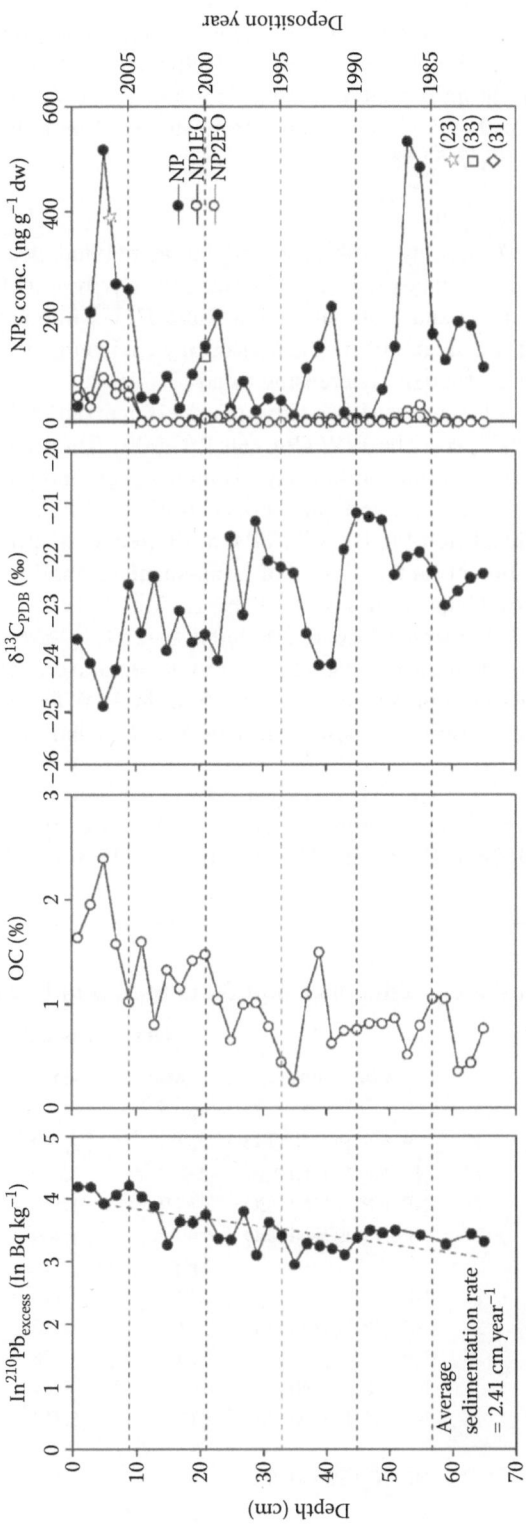

FIGURE 11.5 Vertical profiles of $^{210}Pb_{excess}$ activity, OC, $\delta^{13}C$, and NPs from sediment cores in Lake Shihwa (dotted line: deposition year by ^{210}Pb dating). (From Hong, S. et al., *Mar. Pollut. Bull.*, 60, 308, 2010. With permission.)

[50,53]. Masan Bay has been listed as a special management coastal zone (SMCZ) by the Korean government since 1983, in association with increased contaminant levels. There are some reports of the distribution and characteristics of organic contaminants such as OCPs, polybrominated diphenyl ethers (PBDEs), polychlorinated biphenyls (PCBs), polychlorinated dibenzo-p-dioxin/dibenzo-furans (PCDD/Fs), and NPs in Masan Bay environments [53–56]. These studies showed that Masan Bay is heavily polluted by toxic chemicals originating from surrounding industrial complexes and cities, similar to Lake Shihwa [37,53,57,58].

11.3.2.2 Sources and Distributions

Spatial distributions of NPs in Masan Bay exhibit generally greater abundance in the inner bay than in the outer bay. NP concentrations measured in the Masan Bay environment during the past few years are presented in Figures 11.3a and 11.4a and b and Table 11.3. Although there is insufficient data to understand the temporal trends of NPs in the Masan Bay environment, the concentrations of NPs in surface sediments show a gradually decreasing trend.

On the other hand, a WWTP is located at Masan City, and its effluent outlet is recognized as a point source of NPs into Masan Bay [53,57]. The WWTP treats 260,000t d^{-1} of industrial and domestic wastewater and discharges its effluent into Masan Bay via an underground pipeline [53,59] (Figure 11.2b). Li et al. [57] and Moon et al. [53] investigated concentrations of NPs, PCDD/Fs, dioxin-like PCBs (dl-PCBs), PBDEs, and fecal sterol in the WWTP outfall area to identify the impact of effluents. They collected surface sediment samples based on a fan-shaped sampling scheme from around the WWTP outfall (Figure 11.6). Concentrations of NP, NP1EO, and NP2EO in surface sediments were 68–600, 23–3,491, and 27–2,800 ng g^{-1} dw in 2004 and 18–269, 18–576, and 4–258 ng g^{-1} dw in 2005 respectively. There were clear decreasing trends for toxic organic contaminants such as NPs, PCDD/Fs, dl-PCBs, and PBDEs with increasing distance from the WWTP outfall, suggesting that the WWTP discharge is the major source of these toxic substances, including NPs [53,57].

11.3.2.3 Temporal Trends

Temporal trends of NPs were recorded in sediment cores collected near the outfall of the WWTP. Vertical profiles of OC, PCDD/Fs, dl-PCBs, and NPs are shown in Figure 11.7 [60]. NPs increased

TABLE 11.3
Concentrations of NPs in Water, Sediments, and Biota in Masan Bay Environment

Samples	Year	*n*	Compounds	NPs Concentrations			Reference
				Min	Max	Mean	
Water				ng L^{-1}			
Creeks	May 2004	14	NP+NP1EO+NP2EO	432	2,500	1,229	[57]
	August 2004	14	NP+NP1EO+NP2EO	334	3,630	1,433	[57]
Masan Bay	May 2004	21	NP+NP1EO+NP2EO	15.9	1,880	158	[57]
	August 2004	21	NP+NP1EO+NP2EO	40.8	36,400	1,869	[57]
Sediment				ng g^{-1} dw			
Masan Bay	May 1998	28	NP	113	3,890	510	[50]
	February 2004	21	NP+NP1EO+NP2EO	131	2,810	581	[57]
	2006	5	NP	24	506	248	[58]
	2006	5	NP+NP1EO+NP2EO	49	124	71	[23]
Around WWTP outfall	October 2004	14	NP+NP1EO+NP2EO	117	6,817	1,327	[57]
	February 2005	20	NP+NP1EO+NP2EO	40	1,208	411	[53]
Biota (mussel)				ng g^{-1} dw			
Masan Bay	May 2006	5	NP	51	289	140	[57]

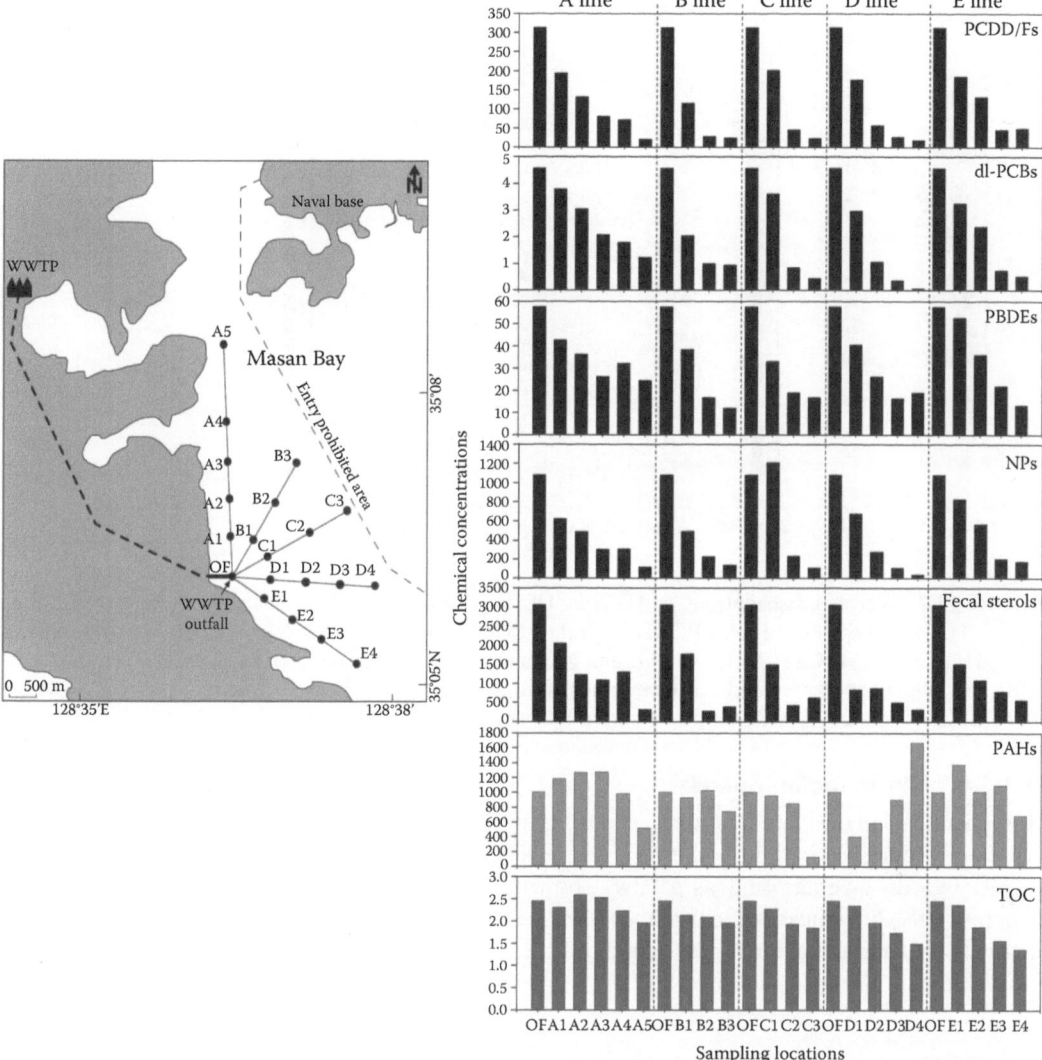

FIGURE 11.6   Distributions of organic contaminants and TOC in marine sediments collected at the locations of each transect line from the WWTP outfall in Masan Bay (ng TEQ/kg dw for PCDD/Fs and dl-PCBs; μg kg⁻¹ dw for PBDEs, NPs, fecal sterols and PAHs; % for TOC). (From *Chemosphere*, 73, Moon, H.-B., Yoon, S.-P., Jung, R.-H., and Choi, M., Wastewater treatment plants (WWTPs) as a source of sediment contamination by toxic organic pollutants and fecal sterols in a semi-enclosed bay in Korea, 880–889, Copyright 2008, with permission from Elsevier.)

gradually from the early 1990s, with the highest concentration detected using ^{210}Pb dating at 8–10 cm depth, corresponding approximately to the year 2000 and subsequently decreased. There were similar trends in the PCDD/F and dl-PCB concentration profiles. These trends are closely related to the operation and efficiency of the WWTP. The WWTP was established in 1994 and was upgraded to use an activated sludge treatment method in the early 2000s [59,60]. Spatial and vertical profiles of NPs around the WWTP in Masan Bay sediments suggested that discharges from the WWTP apparently contributed to NP contamination in this region. Therefore, an adequate capacity of WWTPs needs to be established, as well as continuous monitoring for quality improvement of the aquatic environment.

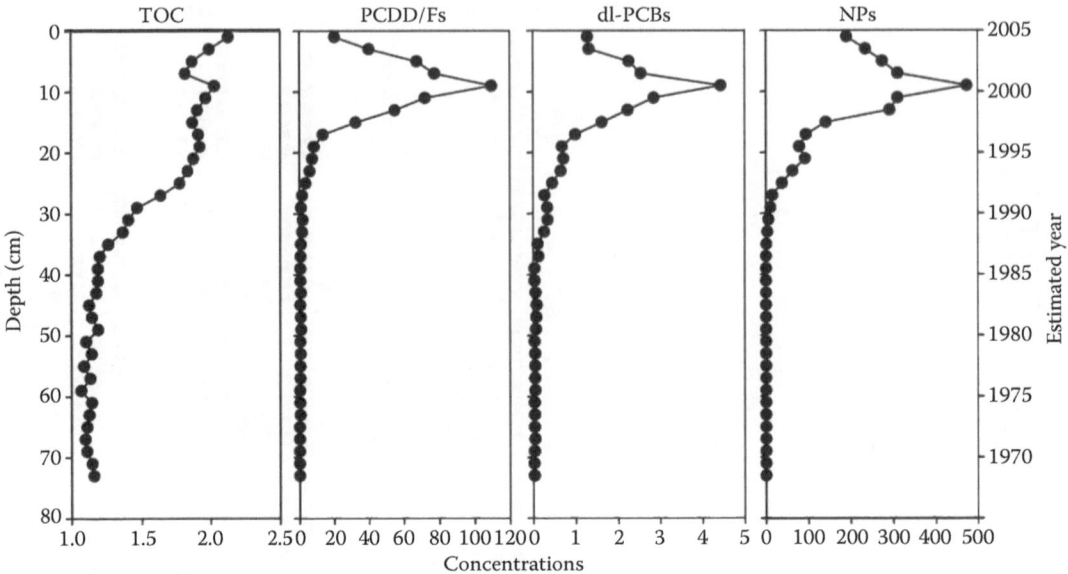

FIGURE 11.7 Vertical distributions of TOC, PCDD/Fs, dl-PCBs, and NPs in sediment cores from Masan Bay (% TOC; ng TEQ kg^{-1} dw for PCDD/Fs and dl-PCBs; and μg kg^{-1} dw NPs). (From *Chemosphere*, 75, Moon, H.-B., Choi, M., Choi, H.-G., Ok, G., and Kannan, K., Historical trends of PCDDs, PCDFs, dioxin-like PCBs and nonylphenols in dated sediment cores from a semi-enclosed bay in Korea: Tracking the sources 565–571, Copyright 2009, with permission from Elsevier.)

11.3.2.4 NPs in Biota: Mussels

Most chemicals in the water column could be assimilated by various aquatic organisms and concentrated or metabolized in their bodies [57]. NP levels in mussels found in Masan Bay ranged from 51 to 289 ng g^{-1} dw (average = 140 ng g^{-1} dw), showing high correlation with water ($r^2 = 0.98$) and sediment ($r^2 = 0.90$) NP concentrations at the same sites. Wang et al. [42] calculated the bioconcentration factors (BCF) and the average was found to be 2,990 obtained from NP concentrations in mussels and water.

11.3.3 OCEAN WASTE DISPOSAL: DUMPSITE "BYUNG"

11.3.3.1 Description of Dumpsite "Byung"

The Korean government have designated some areas in the waters off the coast of Gunsan, Ulsan, and Pohang as maritime dumping zones since 1988. The dumpsite "Byung" in the East Sea is located 125 km off the coast of Pohang (area: 3700 km^2; depth: 200–2000 m) (Figure 11.2c). Approximately 48 million ton of waste (excrements, livestock wastewater, domestic and industrial wastewater treatment sludge, etc.) have been discharged into the dumpsite "Byung" in the past 18 years [61]. Dumped waste is diluted in surface water bodies by advection and/or diffusion, but much of the waste settles directly down through the water column and is deposited on the sea floor where it accumulates as sediment [62–64].

This waste contains various organic pollutants such as PCDD/Fs, PCBs, OCPs, PAHs, and NPs, depending on their different sources. There are many reports concerning organic pollutants in WWTP and STP sludge samples. High concentrations of NPs have been detected in the effluents of many municipal STPs and in industrial and domestic sludge samples that are aerobically or anaerobically stabilized [1,10,65–68]. Sewage sludge samples included not only high concentrations of NPs, but also heavy metals, PAHs, PCBs, and PBDEs [67,69–71]. As a result, large amounts of waste dumping into the ocean could cause adverse effects on living organisms in the marine environment.

11.3.3.2   Sources, Distributions, and Temporal Trends

Vertical profiles of NPs in the dumpsite "Byung" in Korea's East Sea are shown in Figure 11.8. A small amount of NPs were found outside the dumpsite, while high levels of NPs were detected at the dumpsite itself (Figures 11.4c and 11.8). NPs were mainly detected at depths above 10 cm. The highest NP concentrations were at 2–4 cm in the dumpsite sediment core. According to ^{210}Pb dating, NPs increased gradually from the late 1980s and early 1990s, peaked after the year 2000, and subsequently decreased. This profile can be associated with the amount of waste disposed into the dumpsite "Byung" since 1988, which increased from 1989 to 2005 and has decreased since 2005 (Figure 11.9a). Vertical profiles of OC, total nitrogen (TN), δ^{13}C and stable nitrogen isotope ratio (δ^{15}N) in sediments from the dumpsite and non-dumpsite area are presented in Figure 11.8. Maximum amounts of OC and TN appeared in the

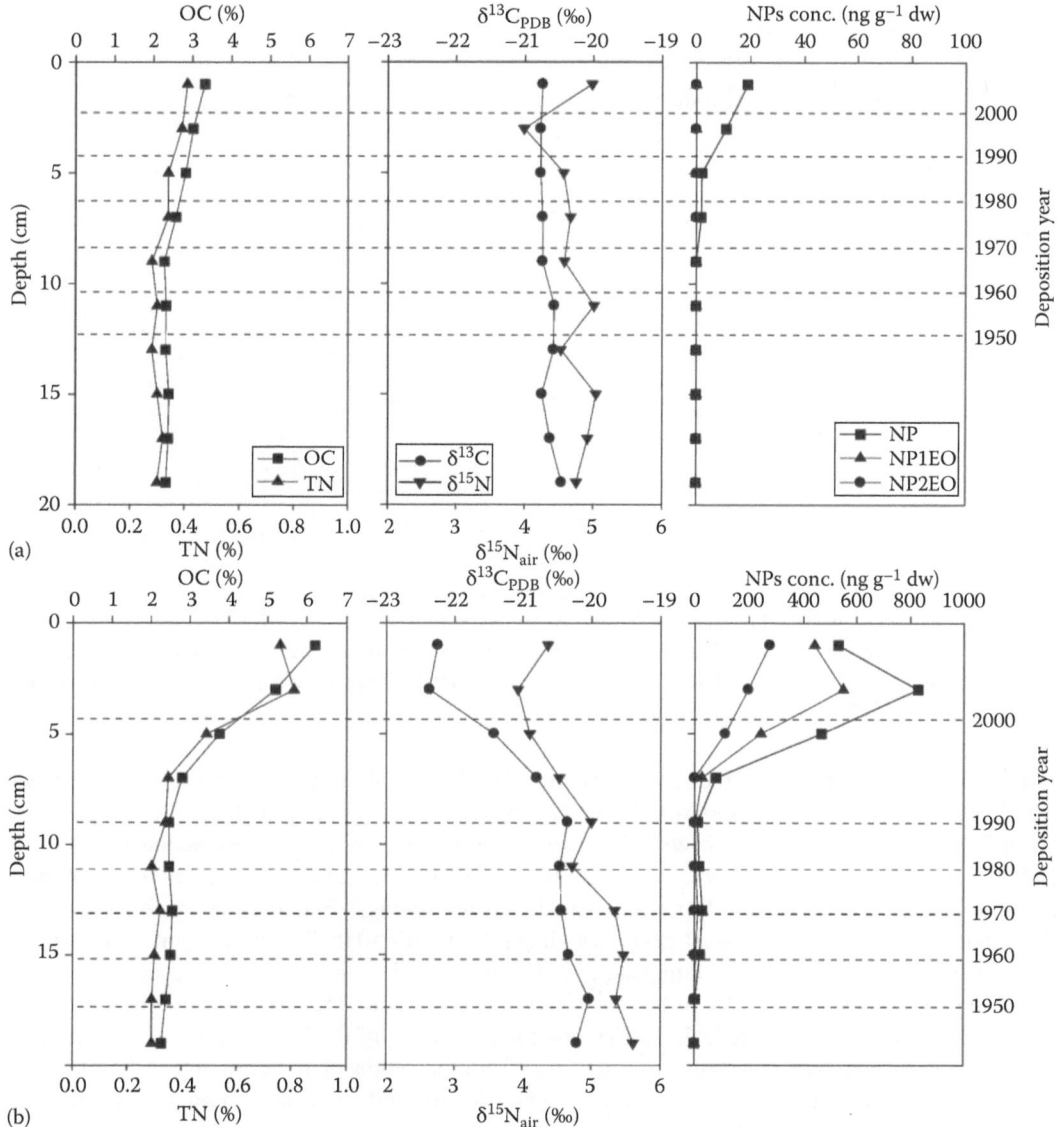

FIGURE 11.8   Vertical profiles of NPs, OC, TN, δ^{13}C, and δ^{15}N from (a) non-dumpsite sediments and (b) dumpsite sediments ("Byung") in the East Sea (dotted lines: deposition year by ^{210}Pb dating). (From Hong, S. and Shin, K.-H., *Mar. Pollut. Bull.*, 58, 1566, 2009. With permission.)

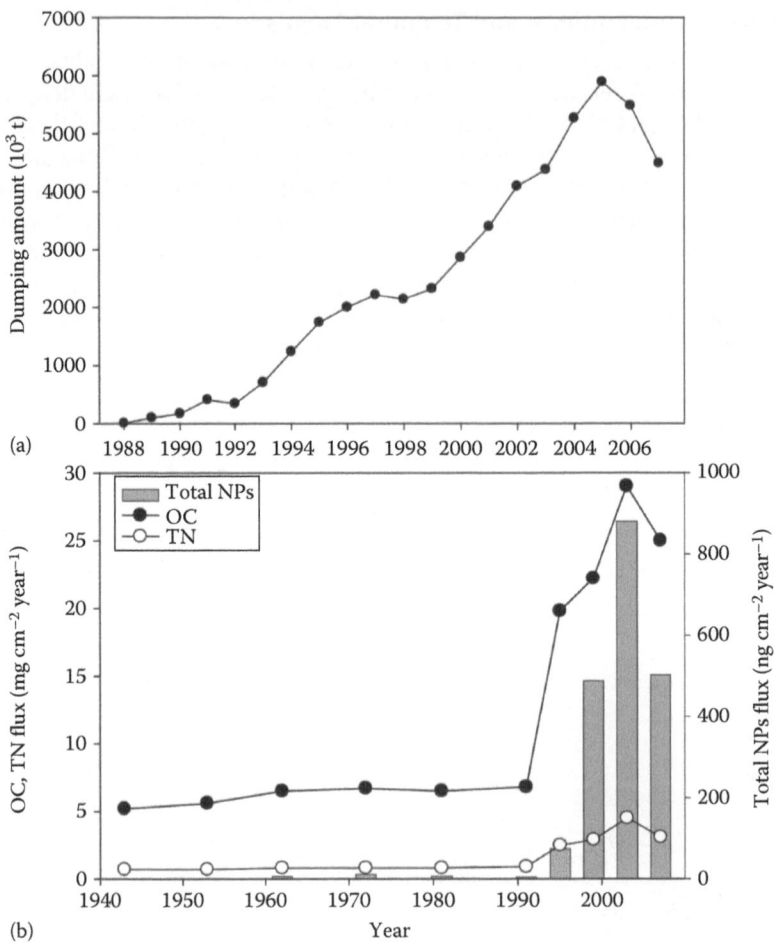

(a)

(b)

FIGURE 11.9 (a) Amount of waste disposal into the dumpsite "Byung" in the East Sea since 1988 and (b) deposition fluxes of OC, TN, and total NPs (NP + NP1EO + NP2EO) (Flux = C_i ρ_i γ_i, where C_i is the measured concentration of total NP and the content of OC and TN in the sediment layer i. ρ_i (g cm^{-3}) and γ_i (cm year^{-1}) are the dry mass and sedimentation rate, respectively). (From Hong, S. and Shin, K.-H., *Mar. Pollut. Bull.*, 58, 1566, 2009. With permission.)

surface slice (0–2 cm depth) of the sediment cores at each site. OC and TN contents showed decreasing trends with increasing core depth. OC and TN contents in the upper layers from the dumpsite were about 2–3 times higher compared with those from the non-dumpsite area. Deposition fluxes of OC and TN showed rapid increases with NPs from the late 1980s and early 1990s, which was profoundly linked to the amount of waste disposal into the dumpsite "Byung" (Figure 11.9a and b). The vertical distribution of $\delta^{13}C$ showed constant values for all depths (0–20 cm) in the non-dumpsite core, whereas lighter values were present in the dumpsite at 0–10 cm depth. The lightest values occurred at 2–4 cm depth and relatively heavier values at 0–2 cm depth. The $\delta^{13}C$ values are well-known indicators of land waste-derived organic matter [62,63,71]. The end-member value of $\delta^{13}C$ not affected by ocean dumping may be −20.29‰, which was the average value from the depths of 10–20 cm at the dumpsite. The upper depths of the dumpsite sediment cores, which were affected by ocean dumping, had $\delta^{13}C$ values (maximum ~2‰) lighter than the end-member value. These results showed a good correlation ($r^2 > 0.90$) between $\delta^{13}C$ values and NP concentrations in core sediments from the dumpsite "Byung." This relationship indicates that most of the nonylphenolic chemicals originated from dumped wastes, and the benthic environment in the dumpsite "Byung" had been affected by an excess of ocean dumping.

11.3.3.3 Korean Government Plans for Waste Disposal at Sea

The Korean government has considered counterproposals about waste disposal at sea, enacted in 2006 in order to protect the ocean environment and marine creatures. The relatively low concentrations of NPs in the surface sediments (0–2 cm depth) of the dumpsite might result from the ocean disposal management system, which has been based on the guidelines of discharge permits and on the reduction of ocean disposal in terms of recent regulations of the Korean government in relation to international regulations (London Convention 1972 and London Protocol 1996). The government further amended its policy on waste disposal at sea and banned ocean dumping of construction sludge and water treatment sludge from 2006 and 2007, respectively. Ocean disposal of sewage treatment sludge and livestock sludge is scheduled to be prohibited in 2012. The Korean government has also enforced supervision of illegal dumping and continuous monitoring in the dumping area in anticipation of the restoration of marine environments.

11.3.4 URBANIZED AREA: HAN RIVER

11.3.4.1 Description of Han River

The Han River, which is one of the largest rivers in South Korea, passes through the Seoul metropolitan area and flows westward to the Yellow Sea (Figure 11.2d). The total length of the river is about 514 km, with an area of 26,000 km^2 [72]. Its downstream supplies drinking water to approximately 15 million people living in Seoul City, which is the largest urban area of South Korea, with many apartment complexes and business districts. Because of the rapid development of the economy and industries around the central part of South Korea in the 1970s, the water quality of the Han River deteriorated dramatically [73]. There are four STPs in Seoul City which treat mainly domestic wastewater and discharge the effluent directly into the Han River [74]. However, the capacity of the plants is insufficient to receive the large volume of wastewater produced during the rainy seasons [72], so untreated wastewater is discharged directly into the river. Hence, discharge of treated or untreated urban effluent into the Han River gives rise to critical concerns about aquatic environmental management.

11.3.4.2 Sources and Distributions

There is an insufficient amount of published data available to understand the distribution of NPs in the Han River ecosystem. Only one paper has reported the distribution, behavior, and seasonal variation of NPs in the Han River environment [72]. Concentrations of NPs in water, suspended solids and sediments in 2001 are presented in Figures 11.3b and 11.4d and Table 11.4, in which concentrations were much lower than previously reported acute and chronic toxicity levels (Table 11.1). According to the paper, the main input sources of NPs in the Han River were effluents containing large amounts of sewage and wastewater from Seoul City. An increasing trend in the concentrations was shown in the water, suspended solids and sediments along the river. Seasonal variation of NPs in the Han River system depended principally on microbial activities in accordance with water temperature changes. Several papers have reported that NPEOs are rapidly degraded to NP2EO, NP1EO and NP at high water temperatures by microbial activity, while the degradation rate becomes very slow at low water temperatures [72,75,76]. As a result, NP concentrations associated with the water and suspended particles were higher in summer than in winter, but there is little seasonal difference in sediments because of the different degradation mechanisms between water and sediments [72]. NPEO degradation rates in sediments are strongly influenced by oxygen concentrations, temperature, OC content, electron acceptors (such as sulfate, nitrate and iron (III)) and microbial community activity [11,26,28,77]. Further investigation is needed to establish such a correlation.

11.3.4.3 Comparison with Other Rivers of East Asian Countries

Distribution patterns, sources and environmental levels of NPs in the Han River are compared with other rivers of urbanized areas in East Asian countries, including the Sumidagawa River (Japan) and

TABLE 11.4
Concentrations of NPs in Water, Suspended Solids, and Sediments in Rivers of East Asian Countries

Samples	Year	n	Compounds	NP Concentrations			Reference
				Min	Max	Mean	
Water				ng L^{-1}			
Han River, South Korea	August 2001	9	NP	71	188	103	[72]
	October 2001	10	NP	42	84	60	[72]
	December 2001	10	NP	17	72	43	[72]
Sumidagawa River, Japan	1997	12	NP	80	1,080		[18]
Yellow River, China	November 2004	12	NP	42	376	101	[78]
Suspended solid				µg g^{-1} dw			
Han River, South Korea	August 2001	9	NP	3.8	17	8.7	[72]
	October 2001	10	NP	1.1	9.3	3.6	[72]
	December 2001	10	NP	1.0	6.9	3.1	[72]
Sumidagawa River, Japan	1997	12	NP	2	7	3.54	[18]
Yellow River, China	November 2004	12	NP	0.05	2.84	0.47	[78]
Sediment				ng g^{-1} dw			
Han River, South Korea	August 2001	10	NP	11	624	251	[72]
	October 2001	10	NP	56	357	157	[72]
	December 2001	10	NP	35	550	192	[72]
Sumidagawa River, Japan	1997–1999	11	NP	520	13,000		[18]
Yellow River, China	November 2004	12	NP	39	863	155	[78]

the Yellow River (China) (Table 11.4). The Sumidagawa River flows through Tokyo City, and the main sources of NPs were drainage from 10 municipal STPs as well as effluent from the city [18]. NPs in the Lanzhou Reach of the Yellow River originated from municipal and industrial wastewater [78]. The concentration of NPs in the Han River environment was similar to levels in the Yellow River but lower than those in the Sumidagawa River. A seasonal trend, in which the NP concentration in river water was higher in warmer seasons than in colder seasons, was a common feature observed in all three rivers. Spatial distribution patterns of NPs in the three rivers were generally higher downstream than upstream, characteristic of wastewater discharged from pollution sources along the three rivers. In the Han River, about 60% of NPs were detected in the particulate phase compared with about 20% in the Sumidagawa River and over 45% in the Yellow River. This indicated that effective removal of suspended solids occurs during sewage treatment processes on the Sumidagawa River of Japan [72].

The Han River ecosystem is a very important resource for drinking water, as well as for industrial and agricultural uses. However, there have been NPs, persistent organic pollutants (OCPs, PCDD/Fs, PCBs), pharmaceutical residues, personal care products and other endocrine-disrupting chemicals detected in the Han River environment [74,79,80]. Water quality management needs to be planned for the Han River.

11.4 COMPARISON OF CONTAMINATION LEVELS IN EACH AQUATIC ENVIRONMENT AND ENVIRONMENTAL QUALITY GUIDELINES

The concentrations of NPs in water and sediments from South Korea were compared with guidelines from the United States and Canada for risk assessment in relation to NP exposure. Acute and chronic criteria for NPs according to the U.S. EPA are 28 and 6.6 µg L^{-1} in freshwater, and 7 and 1.7 µg L^{-1} in seawater respectively [15]. The WQG for NPs in freshwater and seawater in Canada are

1 and $0.7\,\mu g\,L^{-1}$ [5]. Box plots for the measured concentrations of NPs in waters from industrialized areas (Lake Shihwa and Masan Bay) and an urbanized area (Han River) of South Korea are shown in Figure 11.3. The concentrations of NPs in seawater were mainly below the WQG and WQC concentrations. NP concentrations in all freshwater and seawater samples of South Korea did not exceed the acute WQC concentrations. However, some freshwater samples from creeks around Lake Shihwa contained high NP levels that exceeded the WQG and chronic WQC concentrations. The industrialized area waters showed higher NP levels than those in the urbanized areas. In Figure 11.4, the box plot shows NP concentration ranges from surface sediments in four sources from published reports. Industrialized areas (Lake Shihwa and Masan Bay) had greater NP contamination than other pollution source regions and relatively higher levels in freshwater sediments than in brackish and marine sediments of industrialized areas. This is because more sources of NPs are in creeks than in lakes and/or bays, and it indicates that the influx of NPs originating from industrial complexes is the main NP pollution source in aquatic environments of South Korea. SQG for NPs are currently proposed in Canada, with values of 1.4 and $1\,\mu g\,g^{-1}$ dw (normalized to 1% total organic carbon [TOC]) in freshwater and marine sediments respectively. Concentrations of NP in sediments from South Korea are generally lower than the SQG of Canada, but some hotspot regions exceed the guidelines.

11.5 SUMMARY

This chapter reviews sources, levels and temporal trends of NPs in aquatic environments of South Korea. There are four main sources of NPs in aquatic ecosystems, including discharges from industrialized areas, WWTP outfalls, urbanized areas and land waste disposal at sea. NP concentration levels are generally lower than those that cause acute and chronic toxicity to biota. However, NPs in some waters and sediments were detected at levels above guidelines set forth by the United States and Canada, suggesting adverse health effects on aquatic organisms. NP concentrations in some aquatic environments also indicate bioaccumulation potential in aquatic biota. Temporal trends of NPs in the aquatic environment have gradually decreased following recent actions such as banning the use of NP chemicals and expansion of WWTPs. However, more investigation should be carried out to establish environmental quality guidelines for regulating NPs in South Korea. Therefore, there is a need to implement regulations to control NPs and protect the aquatic ecosystems of South Korea.

ACKNOWLEDGMENTS

The authors would like to thank Prof. Jong Seong Khim (Korea University, Seoul, Korea) and Prof. Hyo-Bang Moon (Hanyang University, Ansan, Korea) for their helpful comments and suggestions.

REFERENCES

1. Giger, W., Brunner, P.H., and Schaffner, C. 1984. 4-Nonylphenol in sewage sludge: Accumulation of toxic metabolites from nonionic surfactants. *Science* 225: 623–625.
2. Renner, R. 1997. European bans on surfactant trigger transatlantic debate. *Environ. Sci. Technol.* 31: 316A–320A.
3. Ahel, M., Giger, W., and Koch, M. 1994. Behaviour of alkylphenol polyethoxylate surfactants in the aquatic environment I. Occurrence and transformation in sewage treatment. *Water Res.* 28: 1131–1142.
4. Ying, G.-G., Williams, B., and Kookana, R. 2002. Environmental fate of alkylphenols and alkylphenol ethoxylates—A review. *Environ. Int.* 28: 215–226.
5. Canadian Council of Ministers of the Environment (CCME). 2001. Canadian sediment quality guidelines for the protection of aquatic life: Nonylphenol and its ethoxylates. In: *Canadian Environmental Quality Guidelines 1999.* Canadian Council of Ministers of the Environment. Winnipeg, Canada.
6. Correa-Reyes, G., Viana, M.T., Marquez-Rocha, F.J., Licea, A.F., Ponce, E., and Vazquez-Duhalt, R. 2007. Nonylphenol algal bioaccumulation and its effect through the trophic chain. *Chemosphere* 68: 662–670.
7. Ferrara, F., Ademollo, N., Delise, M., Fabietti, F., and Funari, E. 2008. Alkylphenols and their ethoxylates in seafood from the Tyrrhenian Sea. *Chemosphere* 72: 1279–1285.

8. Soares, A., Guieysse, B., Jefferson, B., Cartmell, E., and Lester, J.N. 2008. Nonylphenol in the environment: A critical review on occurrence, fate, toxicity and treatment in wastewaters. *Environ. Int.* 34: 1033–1049.

9. John, D.M. and White, G.F. 1998. Mechanism for biotransformation of nonylphenol polyethoxylates to xenoestrogens in *Pseudomonas putida. J. Bacteriol.* 180: 4332–4338.

10. Loyo-Rosales, J., Rice, C., and Torrents, A. 2007. Fate of octyl- and nonylphenol ethoxylates and some carboxylated derivatives in three American wastewater treatment plants. *Environ. Sci. Technol.* 41: 6815–6821.

11. Lu, J., Jin, Q., He, Y., Wu, J., Zhang, W., and Zhao, J. 2008. Anaerobic degradation behavior of nonylphenol polyethoxylates in sludge. *Chemosphere* 71: 345–351.

12. Routledge, E.J. and Sumpter, J.P. 1997. Structure features of alkylphenolic chemicals associated with estrogenic activity. *J. Biol. Chem.* 272: 3280–3288.

13. Directive 2003/53/EC. 2003. Amending for the 26th time the Council directive 76/769/EEC relating to restrictions on the marketing and use of certain dangerous substances and preparations (nonylphenol, nonylphenol ethoxylate and cement). European Parliament and the Council of the European Union. Luxembourg.

14. United Nations Environment Programme (UNEP). 2003. Regionally based assessment of persistent toxic substances. UNEP Chemicals. Geneva, Switzerland.

15. United States Environmental Protection Agency (U.S. EPA). 2005. Aquatic life ambient water quality criteria-Nonylphenol. EPA-822-R-05-005. Office of Science and Technology. Washington, DC.

16. Hawrelak, M., Bennett, E., and Metealfe, C. 1999. The environmental fate of the primary degradation products of alkylphenol ethoxylate surfactants in recycled paper sludge. *Chemosphere* 39: 745–752.

17. Naylor, G.C., Mierure, J.P., Weeks, J.A., Castaldi, F.J., and Romano, R.R. 1992. Alkylphenol ethoxylates in the environment. *J. Am. Oil Chemists Soc.* 69: 695–703.

18. Isobe, T., Nishiyama, H., Nakashma, A., and Takada, H. 2001. Distribution and behavior of nonylphenol, octylphenol, and nonylphenol monoethoxylate in Tokyo metropolitan area: Their association with aquatic particles and sedimentary distributions. *Environ. Sci. Technol.* 35: 1041–1049.

19. Harris, R.M., Waring, R.H., Kirk, C.J., and Hughes, P.J. 2000. Sulfation of "estrogenic" alkylphenols and 175-estradiol by human platelet phenol sulfotransferases. *J. Biol. Chem.* 275: 159–166.

20. Pedersen, K.H., Pedersen, S.N., Pedersen, K.L., Korsgaard, B., and Bjerregaard, P. 2003. Estrogenic effect of dietary 4-*tert*-octylphenol in rainbow trout (*Oncorhynchus mykiss*). *Aquat. Toxicol.* 62: 295–303.

21. Li, Z., Li, D., Oh, J.-R., and Je, J.-G. 2004. Seasonal and spatial distribution of nonylphenol in Shihwa Lake, Korea. *Chemosphere* 56: 611–618.

22. Korea Ministry of Environment (KMOE). 2006. A designation of nonylphenolic compounds as the restricted chemical in Korea. Environmental Policy Office. Department of Hazardous Chemical Management. Seoul, Korea.

23. Choi, M., Moon, H.-B., Yu, J., Kim, S.-S., Pait, A.S., and Choi, H.-G. 2009. Nationwide monitoring of nonylphenolic compounds and coprostanol in sediments from Korean coastal waters. *Mar. Pollut. Bull.* 58: 1086–1092.

24. National Institute of Environmental Research (NIER). 2007. Toxics release inventory open information system (http://ncis.nier.go.kr/triopen). KMOE, Korea.

25. NIER. 2007. Investigation of emission characteristics for endocrine disrupting chemicals. NIER annual report. KMOE, Korea.

26. Ferguson, P.L., Bopp, R.F., Chillrud, S.N., Aller, R.C., and Brownawell, B.J. 2003. Biogeochemistry of nonylphenol ethoxylates in urban estuarine sediments. *Environ. Sci. Technol.* 37: 3499–3506.

27. Liu, X., Tani, A., Kimbara, K., and Kawai, F. 2006. Metabolic pathway of xenoestrogenic short ethoxy chain-nonylphenol to nonylphenol by aerobic bacteria, *Ensifer* sp. Strain AS08 and *Pseudomonas* sp. Strain AS90. *Appl. Microbiol. Biotechnol.* 72: 552–559.

28. Lu, J., He, Y., Wu, J., and Jin, Q. 2009. Aerobic and anaerobic biodegradation of nonylphenol ethoxylates in estuary sediment of Yangtze River, China. *Environ. Geol.* 57: 1–8.

29. European Communities. 2002. European Union risk assessment report-4-nonylphenol (branched) and nonylphenol. EUR20387EN. Official Publications of the European Communities. Luxembourg.

30. Fenner, K., Kooijman, C., Scheringer, M., and Hungerbuhler, K. 2002. Including transformation products into the risk assessment for chemicals: The case of nonylphenol ethoxylate usage in Switzerland. *Environ. Sci. Technol.* 36: 1147–1154.

31. Khim, J.S., Villeneuve, D.L., Kannan, L. et al. 1999. Alkylphenols, polycyclic aromatic hydrocarbons, and organochlorines in sediment from Lake Shihwa, Korea: Instrumental and bioanalytical characterization. *Environ. Toxicol. Chem.* 18: 2424–2432.

32. Han, M.W. and Park, Y.C. 1999. The development of anoxia in the artificial Lake Shihwa, Korea, as a consequence of intertidal reclamation. *Mar. Pollut. Bull.* 38: 1194–1199.

33. Li, D., Kim, M., Oh, J.-R., and Park, J. 2004. Distribution characteristics of nonylphenols in the artificial Lake Shihwa, and surrounding creeks in Korea. *Chemosphere* 56: 783–790.

34. Koh, C.-H., Khim, J.S., Kannan, K., Villeneuve, D.L., Johnson. B.G., and Giesy, J.P. 2005. Instrumental and bioanalytical measures of dioxin-like and estrogenic compounds and activities associated with sediment from the Korean coast. *Ecotox. Environ. Safe.* 61: 366–379.

35. Rostkowski, P., Yamashita, N., So, M.K. et al. 2006. Perfluorinated compounds in streams of the Shihwa Industrial zone and Lake Shihwa, South Korea. *Environ. Toxicol. Chem.* 25: 2374–2380.

36. Hong, S., Won, E.J., Ju, H.-J., Kim, M.-S., and Shin, K.-H. 2010. Current nonylphenol pollution and the past 30 years record in an artificial Lake Shihwa, Korea. *Mar. Pollut. Bull.* 60: 308–313.

37. Gearing, P., Plucker, F.E., and Parker, P.L. 1977. Organic carbon stable isotope ratios of continental margin sediments. *Mar. Chem.* 5: 251–266.

38. Fry, B. and Sherr, E.B. 1984. $\delta^{13}C$ measurements as indicators of carbon flow in marine and freshwater ecosystem. *Contrib. Mar. Sci.* 27: 13–47.

39. Emerson, S. and Hedges, J.I. 1988. Processes controlling the organic carbon content of open ocean sediments. *Paleoceanography* 3: 621–634.

40. Van Dover, C.L., Grassle, J.F., Fry, B., Garritt, R.H., and Starczak, V.R. 1992. Stable isotope evidence for entry of sewage-derived organic material into a deep-sea food web. *Nature* 360: 153–156.

41. Giangrande, A., Margherita, L., and Musco, L. 2005. Polychaetes as environmental indicators revisited. *Mar. Pollut. Bull.* 50: 1153–1162.

42. Wang, J., Dong, M., Shim, W.J., Kannan, N., and Li, D. 2007. Improved cleanup technique for gas chromatographic–mass spectrometric determination of alkylphenols from biota extract. *J. Chromatogr. A* 1171: 15–21.

43. Cheng, C.-Y., Liu, L.-L., and Ding, W.-H. 2006. Occurrence and seasonal variation of alkylphenols in marine organisms from the coast of Taiwan. *Chemosphere.* 65: 2152–2159.

44. Kannan, K., Keith, T.L., Naylor, C.G., Staple, C.A., Snyder, S.A., and Giesy, J.P. 2003. Nonylphenol and nonylphenol ethoxylates in fish, sediment, and water from the Kalamazoo River, Michigan. *Arch. Environ. Contam. Toxicol.* 44: 77–82.

45. David, A., Fenet, H., and Gomez, E. 2009. Alkylphenols in marine environments: distribution monitoring strategies and detection considerations. *Mar. Pollut. Bull.* 58: 953–960.

46. Jørgensen, A., Giessing, A.M.B., Rasmussen, L.J., and Andersen, O. 2008. Biotransformation of polycyclic aromatic hydrocarbons in marine polychaetes. *Mar. Environ. Res.* 65: 171–186.

47. Pruell, R.J., Taplin, B.K., McGovern, D.G., McKinney, R., and Norton, S.B. 2000. Organic contaminant distributions in sediments, polychaetes (*Nereis virens*) and American lobster (*Homarus americanus*) from a laboratory food chain experiment. *Mar. Environ. Res.* 49: 19–36.

48. Magni, P., Rajagopal, S., Velde, G. et al. 2008. Sediment features, macrozoobenthic assemblages and trophic relationships ($\delta^{13}C$ and $\delta^{15}N$ analysis) following a dystrophic event with anoxia and sulphide development in the Santa Giusta lagoon (western Sardinia, Italy). *Mar. Pollut. Bull.* 57: 125–136.

49. Kho, C.-H., Khim, J.S., Villeneuve, D.L., Kannan, K., Johnson, B.G., and Giesy, J.P. 2005. Instrumental and bioanalytical measures of dioxin-like and estrogenic compounds and activities associated with sediment from the Korean coast. *Ecotox. Environ. Safe.* 61: 366–379.

50. Khim, J.S., Kannan, K., Villeneuve, D.L., Kho, C.H., and Giesy, J.P. 1999. Characterization and distribution of trace organic contaminants in sediment from Masan Bay, Korea. 1. Instrumental analysis. *Environ. Sci. Technol.* 33: 4199–4205.

51. Kho, C.H., Khim, J.S., Villeneuve, D.L., Kannan, K., and Giesy, J.P. 2006. Characterization of trace organic contaminants in marine sediment from Yeongil Bay, Korea: 1. Instrumental analyses. *Environ. Pollut.* 142: 39–47.

52. Khim, J.S., Lee, K.T., Villeneuve, D.L., Kannan, K., Giesy, J.P., and Kho, C.H. 2001. *In vitro* bioassay determination of dioxin-like and estrogenic activity in sediment and water from Ulsan Bay and its vicinity, Korea. *Arch. Environ. Con. Tox.* 40: 151–160.

53. Moon, H.-B., Yoon, S.-P., Jung, R.-H., and Choi, M. 2008. Wastewater treatment plants (WWTPs) as a source of sediment contamination by toxic organic pollutants and fecal sterols in a semi-enclosed bay in Korea. *Chemosphere* 73: 880–889.

54. Yim, U.H., Hong, S.H., Shim, W.J., and Chang, M. 2005. Spatio-temporal distribution and characteristics of PAHs in sediments from Masan Bay, Korea. *Mar. Pollut. Bull.* 50: 319–326.

55. Hong, S.H., Yim, U.H., Shim, W.J., Oh, J.R., and Lee, I.S. 2003. Horizontal and vertical distribution of PCBs and chlorinated pesticides in sediments from Masan Bay, Korea. *Mar. Pollut. Bull.* 46: 244–253.

56. Moon, H.-B., Choi, H.-G., Lee, P.-Y., and Ok, G. 2008. Congener-specific characterization and sources of PCDD/Fs and dioxin-like PCBs in marine sediments from industrialized bays of Korea. *Environ. Toxicol. Chem.* 27: 323–333.

57. Li, D., Dong, M., Shim, W.J., Yim, U.H., Hong, S.H., and Kannan, N. 2008. Distribution characteristics of nonylphenolic chemicals in Masan Bay environments, Korea. *Chemosphere* 71: 1162–1172.

58. Hong, S.H., Munschy, C., Kannan, N. et al. 2009. PCDD/F, PBDE, and nonylphenol contamination in a semi-enclosed bay (Masan Bay, South Korea) and a Mediterranean lagoon (Thau, France). *Chemosphere* 77: 854–862.

59. Korea Ministry of Maritime and Fisheries (KMMF) 2007. Study on fundamental plan of total maximum daily loading (TMDL) management for coastal pollution. Ministry of Maritime and Fisheries. Seoul, Korea.

60. Moon, H.-B., Choi, M., Choi, H.-G., Ok, G., and Kannan. K. 2009. Historical trends of PCDDs, PCDFs, dioxin-like PCBs and nonylphenols in dated sediment cores from a semi-enclosed bay in Korea: Tracking the sources. *Chemosphere* 75: 565–571.

61. Hong, S. and Shin, K.-H. 2009. Alkylphenols in the core sediment of a waste dumpsite in the East Sea (Sea of Japan), Korea. *Mar. Pollut. Bull.* 58: 1566–1571.

62. Burnett, W.C. and Schaeffer, O.A. 1980. Effect of ocean dumping on $^{13}C/^{12}C$ ratios in marine sediments from the New York Bight. *Estuar. Coast. Mar. Sci.* 11: 605–611.

63. Sweeney, R.E., Kalil, E.K., and Kaplan, I.R. 1980. Characterisation of domestic and industrial sewage in southern California coastal sediments using nitrogen carbon, sulfur and uranium tracers. *Mar. Environ. Res.* 3: 225–243.

64. Takada, H., Farrington, J.W., Bothner, M.H., Johnson, C.G., and Tripp, B.W. 1994. Transport of sludge-derived organic pollutants to deep-sea sediments at deep water dump site 106. *Environ. Sci. Technol.* 28: 1062–1072.

65. La Guardia, M.J., Hale, R.C., Harvey, E., and Mainor, T.M. 2001. Alkylphenol ethoxylates degradation products in land-applied sewage sludge (biosolids). *Nature* 35: 4798–4804.

66. Nakada, N., Tanishima, T., Shinohara, H., Kiri, K., and Takada, H. 2006. Pharmaceutical chemicals and endocrine disrupters in municipal wastewater in Tokyo and their removal during activated sludge treatment. *Water Res.* 40: 3297–3303.

67. Santos, J.L., Gonzalez, M.D.M., Aparcio, I., and Alonso, E. 2007. Monitoring of di-(2-ethylhexyl) phthalate, nonylphenol, nonylphenol ethoxylates, and polychlorinated biphenyls in anaerobic and aerobic sewage sludge by gas chromatography-mass spectrometry. *Int. J. Environ. Anal. Chem.* 87: 1033–1042.

68. Pothitou, P. and Voutsa, D. 2008. Endocrine disrupting compounds in municipal and industrial wastewater treatment plants in Northern Greece. *Chemosphere* 73: 1716–1723.

69. Wild, S.R., Waterhouse, K.S., McGrath, S.P., and Jones, K.C. 1990. Organic contaminants in an agricultural soil with a known history of sewage sludge amendments: Polynuclear aromatic hydrocarbons. *Environ. Sci. Technol.* 24: 1706–1711.

70. Knoth, W., Mann, W., Meyer, R., and Nebhuth, J. 2007. Polybrominated diphenyl ether in sewage sludge in Germany. *Chemosphere* 67: 1831–1837.

71. Rogers, K.M. 2003. Stable carbon and nitrogen isotope signatures indicate recovery of marine biota from sewage pollution at Moa Point, New Zealand. *Mar. Pollut. Bull.* 46: 821–827.

72. Li, D., Kim, M., Shim, W.J., Yim, U.H., Oh, J.-R., and Kwon, Y.-J. 2004. Seasonal flux of nonylphenol in Han River, Korea. *Chemosphere* 56: 1–6.

73. Kim, J.-W., Ki, S.J., Moon, J. et al. 2008. Mass load-based pollution management of the Han River and its tributaries, Korea. *Environ. Manage.* 41: 12–19.

74. Choi, K., Kim, Y., Park, J., Park, C.K., Kim, M.Y., and Kim, H.S. 2008. Seasonal variations of several pharmaceutical residues in surface water and sewage treatment plants of Han River, Korea. *Sci. Total Environ.* 405: 120–128.

75. Staples, C.A., Williams, J.B., Blessing, R.L., and Varineau, P.T. 1999. Measuring the biodegradability of non-ylphenol ether carboxylates, octylphenol ether carboxylates, and nonylphenol. *Chemosphere* 38: 2029–2039.

76. Manzano, M.A., Perales, J.A., Sales, D., and Quiroga, J.M. 1999. The effect of temperature on the bio-degradation of a nonylphenol polyethoxylate in river water. *Water Res.* 33: 2593–2600.

77. Ferguson, P.L. and Brownawell, B.J. 2003. Degradation of nonylphenol ethoxylates in estuarine sediment under aerobic and anaerobic conditions. *Environ. Toxicol. Chem.* 22: 1189–1199.

78. Xu, J., Wang, P., Guo, W., Dong, J., Wang, L., and Dai, S. 2006. Seasonal and spatial distribution of nonylphenol in Lanzhou Reach of Yellow River in China. *Chemosphere* 65: 1445–1451.

79. Kim, K.-S., Lee, S.C., Kim, K.-H. et al. 2009. Survey on organochlorine pesticides, PCDD/Fs, dioxin-like PCBs and HCB in sediments from the Han River, Korea. *Chemosphere* 75: 580–587.

80. Yoon, Y., Ryu, J., Oh, J., Choi, B.-G., and Snyder, S.A. 2010. Occurrence of endocrine disrupting compounds, pharmaceuticals, and personal care products in the Han River (Seoul, South Korea). *Sci. Total Environ.* 408: 636–643.

12 Spatial and Temporal Trends of Persistent Organic Chemicals in Vietnam Soils

Masahide Kawano* and Vu Duc Thao

CONTENTS

12.1 Introduction 279
12.2 Production of DDTs, HCHs, and PCBs in Vietnam...281
    12.2.1 DDTs and HCHs..281
    12.2.2 PCBs...281
12.3 Spatial Distribution of DDTs, HCHs, and PCBs in Vietnam 282
    12.3.1 Levels of DDTs and HCHs 282
    12.3.2 Levels of PCBs 285
    12.3.3 Comparison to Other Recent Data Reported on POPs in Soil 285
12.4 Temporal Trends and Half-Lives of DDTs, HCHs, and PCBs 293
    12.4.1 Temporal Trends and Half-Lives of POPs...293
    12.4.2 Factors Affecting Half-Lives of POPs in Soil 294
        12.4.2.1 Non-Biological Processes 294
        12.4.2.2 Biological Processes 297
12.5 Conclusions..297
Acknowledgments..298
References...298

12.1 INTRODUCTION

Global environmental contamination and toxic effects of persistent organic pollutants (POPs) have been one of the most serious environmental issues and has received considerable attention during the past five decades. POPs are controlled and managed by international treaties, namely the Stockholm Convention on Persistent Organic Pollutants, due to their persistence in the environment, their abilities of long-range transport around the globe, their bioaccumulation properties in wildlife and in humans, and their harmful effects to organisms including humans [1]. This chapter deals with contamination profiles and temporal trends of POPs in Vietnam.

    Vietnam is located at the center of the Southeast Asian region and it is bordered by the People's Republic of China (PRC) to the north, Laos to the northeast, Cambodia to the southwest, and the South China Sea (Figure 12.1). Vietnam's coastline spans about 3400 km along the Gulf of Tonkin, South China Sea, and the Gulf of Thailand. Urbanization, industrialization, and intensive farming are having an impact on the country's environment. These factors have contributed to water pollution, air pollution, and noise pollution, especially in urban and industrial centers such as Ho Chi Minh City

* E-mail: mkawano@agr.ehime-u.ac.jp (Chapter corresponding author).

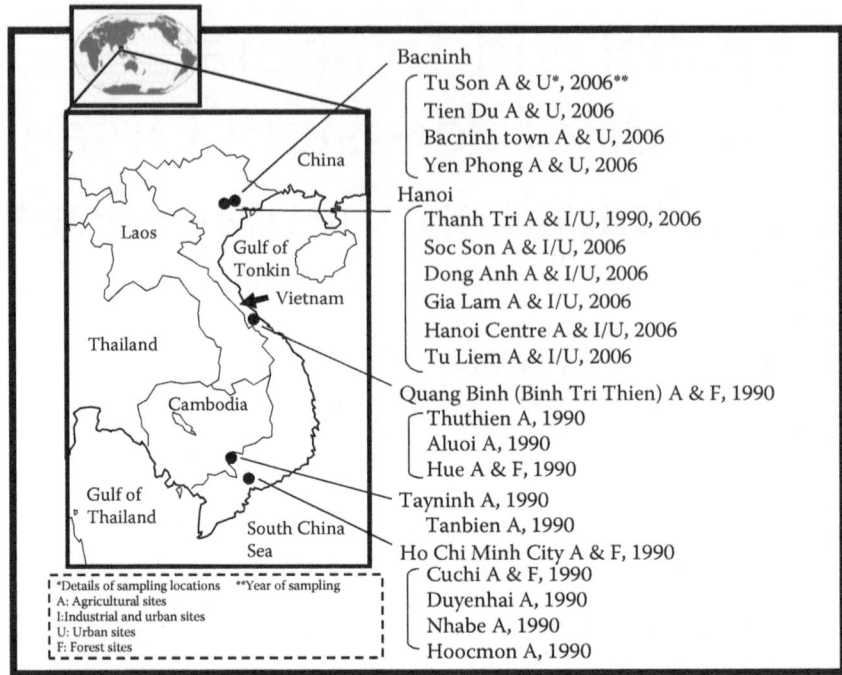

FIGURE 12.1 Map showing sampling locations.

and Hanoi [2]. Land use practices have also led to severe deforestation, soil erosion, sedimentation of rivers, flooding in the deltas, declining fish yields, and pollution of the coastal marine environment. Furthermore, the lingering effect of Agent Orange (used during the Second Indochina War (1954–1975)) in the form of POPs continues to contribute to the incidences of various diseases and birth defects [2]. The above factors as well as the rapid development of agriculture and industrialization have made Vietnam an important subject for extensive studies dealing with environmental pollution during the past decade. A comprehensive review of the research work dealing with organochlorine (OC) pollutants in Vietnam soils and temporal trends are presented in the following sections.

Human health can be affected directly through ingestion of mineral, chemical, and biological components of soils, and can also be influenced in more indirect ways when soils interact with the atmosphere, biosphere, and hydrosphere [3]. In this context, soil is one of the most important reservoirs of POPs in the environment. OC compounds such as DDT compounds (DDTs), *alpha-*, *beta-* and *gamma-*hexachlorocyclohexane isomers (HCHs), and polychlorinated biphenyl (PCBs), are still being used in tropical countries even after bans on production and use in developed countries such as the United States, Europe or several other European countries, and Japan. Among the OCs, DDT formulation is still in use for malaria control in several tropical countries. Since these compounds are transported to far off places via the atmosphere, tropical countries serve as point source for countries in other parts of the world [4]. Understanding the contamination status and time trend of POP residues in tropical areas may be useful to evaluate the ecotoxicological effects of POPs over the globe in the future.

The vast majority of investigations on OC contamination in soil have been performed in temperate regions, predominately in developed countries such as the United States [5], Canada [6], and several European countries [7]. However, approximately one half of the earth's population and roughly one third of its land area are present in the tropics. From a global point of view, tropical regions are thought to be a significant source area of POPs, owing to a long history of widespread use of persistent chemicals. Vietnam is a tropical country and has extensive literature dealing with POPs levels, especially during the last decade. Although well-designed studies on temporal trends of POPs are limited, sufficient data dealing with contamination in soil of Vietnam are available for evaluation of temporal trends and clearance rates (Figure 12.1). To illustrate the environmental fate of POPs, much effort was focused on the analysis of these compounds in soil samples and the investigation of their behavior in the environment [8–13]. In this chapter, spatial distribution and time trend residue of POPs, namely, DDTs, HCHs and PCBs in the tropical soil of Vietnam have been discussed in detail.

12.2 PRODUCTION OF DDTs, HCHs, AND PCBs IN VIETNAM

As seen in other developing countries, the statistical data on production, import/export, and use of DDTs, HCHs, and PCBs in Vietnam are not available. Here, we discuss the available statistical data reported so far on DDTs, HCHs, and PCBs in Vietnam.

12.2.1 DDTs AND HCHs

In 1990 and 1991, the Food and Agriculture Organization (FAO), International Program for Integrated Pest Control in rice in South and Southeast Asia reported that pesticides were applied more frequently to rice in southern Vietnam (an average of 5.3 applications per crop) than elsewhere in Asia. In northern Vietnam, the average was 1.0 application per crop. Comparatively, China used 3.5 applications, the Philippines 2.0 applications, and India 2.4 applications per crop [14].

The Ministry of Health in Vietnam indicated that DDT importation was banned in 1992 and stockpiles were allowed to clear the system resulting in use termination approximately before 1995. Deltamethrin, vectron icon, and other pyrethroids are now used as substitutes for DDT. It was also reported that the Ministry of Health still has approximately 2 tons of obsolete DDT stockpiled in its possession [15]. It had been marketed with the commercial names Neocid, Pentachlorine, and Chlorophenothane [15].

Before 1985, Vietnam used pesticides imported (6500–9000 ton/year) from the former Soviet Union. Since Vietnam had one of the highest rates of malaria disease in the world, it has kept data on DDT imports from source countries such as Russia and the Netherlands and the respective quantities from years 1957–1990 [15]. DDT use went from 315 tons/year in 1961 to 22 ton by 1974 due to supply cutbacks by the USSR and increasing restrictions on DDT [15]. Consumption apparently increased again as values from 1962, 1963, and 1981 were reported to be at 1000 tons/year [15]. A paper presented by Vietnam at the 1999 UNEP POPs Workshop stated that DDT was banned in 1992, but remained in limited use for "health care" purposed until 1995. Although Vietnam officially reported a stoppage of DDT use for malaria in 1995 [15], it was still used in Vietnam until 1994 as a chemical agent used to control mosquitoes that spread malaria. In 1994, the DDT used in Vietnam was reported to contain 2.92 tons of active ingredient [16]. The amount of DDT used for malaria control in another report had been estimated around 20,000 tons or more [17]. The data on HCH use in Vietnam is not available. Global technical DDT and HCH usages have been estimated to be approximately 2 million [18] and 10 million tons [19] respectively.

12.2.2 PCBs

Vietnam never manufactured PCBs. However, it is possible that imported PCBs from other countries might have been used in Vietnam for many years in industrial fluids such as hydraulic fluids, lubricating oils, and as plasticizers [15]. It is estimated that one major use has likely been as a dielectric fluid in transformers and in electrical capacitors [15]. It has been documented that PCBs were imported from China [15]. In the past, Vietnam also received supplies of PCB oil from Russia and Romania [15]. It has been documented that 27,000–30,000 tons of PCB oil have been imported from Russia, China, and Romania [15]. Recently, an inventory of PCB stockpiles reported oil containing PCBs as well as oil not identified to contain PCBs throughout the north, central, and south economic areas as well as remaining provinces of Vietnam [15]. It has been revealed that there was a great deal of oil containing PCBs (9,639,116 kg) present in storage on Vietnamese land [15]. The number of PCB-containing capacitors and transformers accounted for 5,453 and 74,987 case respectively [15]. These stockpiles have a potential adverse effect on the environment and human health due to their leakage and release from containers and storage houses into the Vietnamese environment [20]. The global PCB production has been estimated at 1.1–1.324 million tons [21,22].

12.3 SPATIAL DISTRIBUTION OF DDTs, HCHs, AND PCBs IN VIETNAM

Analyses of POPs in Vietnam soils have been performed since the early 1990s. In recent years in particular, the soil samples collected from Hanoi and Bacninh (northern Vietnam) were analyzed. The following sections describe the spatial distribution in northern and other areas of Vietnam, and the resulting data was compared to levels reported in soils from other areas of the world.

12.3.1 LEVELS OF DDTs AND HCHs

Figures 12.2 through 12.4 describe the levels of DDTs, HCHs, and PCBs in soil samples collected from agricultural and industrial areas within the Hanoi and Bacninh Provinces in 2006. Figures 12.5 through 12.7 present analytical data of soil samples collected in 1990. The analytical results showed that the levels of DDTs and HCHs found in soil samples from agricultural areas were higher than those from industrial areas. DDT concentrations were higher than HCH concentrations in both the agricultural areas, although the difference in levels in samples from Hanoi and Bacninh were not remarkably different. Besides pesticide, technical DDT has been used in malaria control for many decades in Vietnam [15,16]. Consequently, public health use of DDTs might have contributed to elevated levels of DDT residues in soil samples.

Hanoi is an industrialized city with an estimated population of 6.5 million people, and it covers an area of 3344.7 km². It is the capital and second largest city of Vietnam [23,24]. The city comprises of a downtown area, 10 urban districts, and 18 rural districts [24]. On the other hand, the Bacninh Province, which is located east of Hanoi, has an estimated population of just over 1.03 million people [25]. The total natural area of Bacninh consists of about 804 km² of which agricultural land accounts for 64.7%, forestry land accounts for 0.7%, specialized and residential area accounts

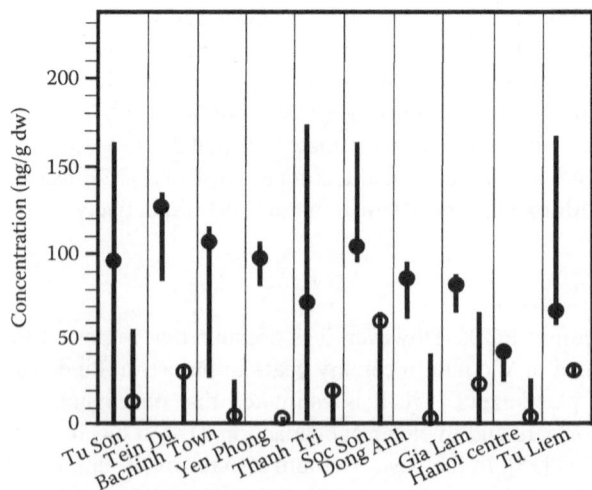

FIGURE 12.2 ΣDDT concentrations in soil samples collected in agricultural and industrial/urban areas in 2006.

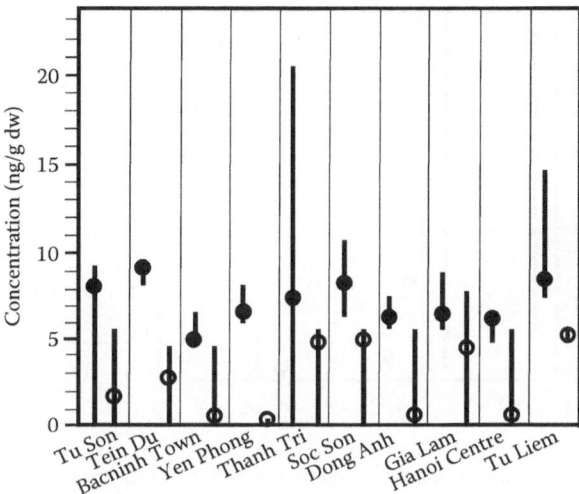

FIGURE 12.3 ΣHCH concentrations in soil samples collected in agricultural and industrial/urban areas in 2006.

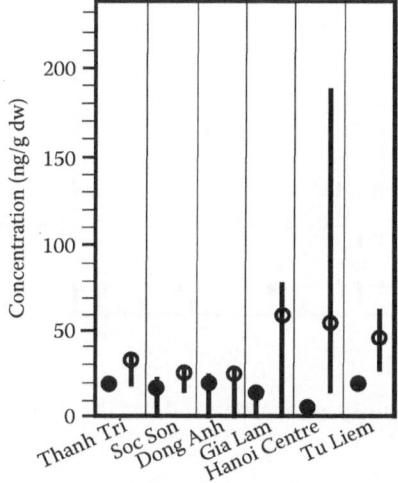

FIGURE 12.4 ΣPCB concentrations in soil samples collected in agricultural and industrial/urban areas in 2006.

for 23.5%, and the remaining as unused land [25]. Recently, modern farming practices have led to elevated crop production, which required the application of necessary pesticides. Bacninh is one of the agricultural cities adjacent to Hanoi [25].

In order to understand the differences between the levels of POPs in agricultural and industrial/urban soils of Hanoi and Bacninh, a statistical investigation, namely the Mann-Whitney U-test, was performed (Table 12.1). Mean and median concentrations were calculated using the available data. A half value of detection limits was used for concentrations below the detection limits. A statistically significant difference ($p < 0.01$) was observed between total DDTs and HCHs measured from the agricultural and industrial sites in Hanoi and agricultural and urban sites in Bacninh. DDT concentrations detected from agricultural sites were higher than those from industrial and urban sites. On the contrary, the concentration of PCBs from industrial sites was higher than those from agricultural sites.

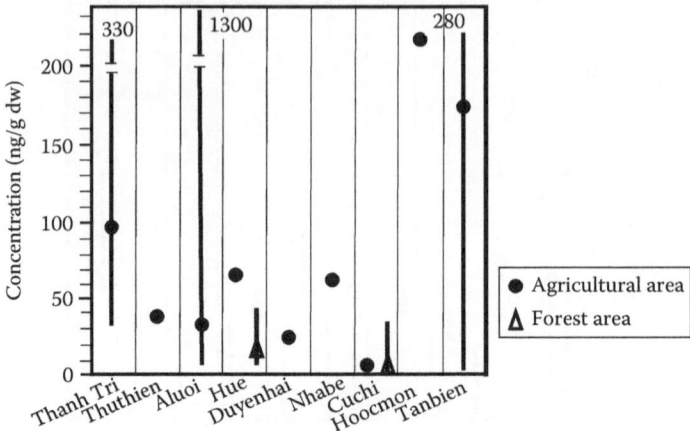

FIGURE 12.5 ΣDDT concentrations in soil samples collected in agricultural and forest areas in 1990.

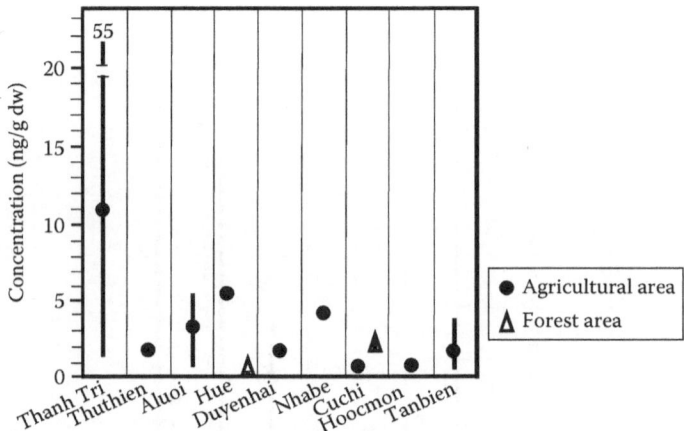

FIGURE 12.6 ΣHCH concentrations in soil samples collected in agricultural and forest areas in 1990.

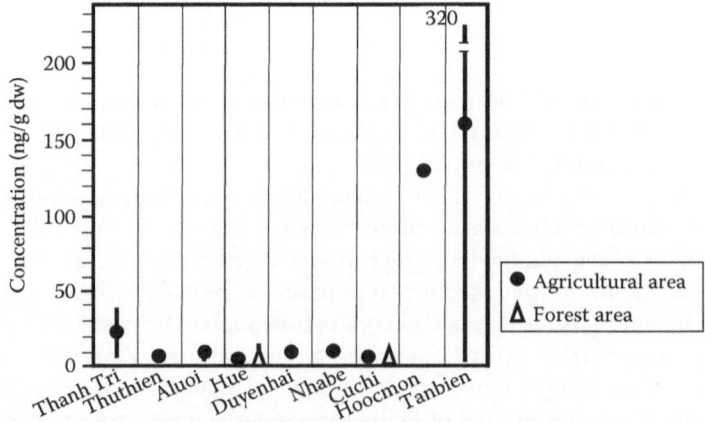

FIGURE 12.7 ΣPCB concentrations in soil samples collected in agricultural and forest areas in 1990.

TABLE 12.1
Concentration Differences of POPs between Agricultural and Industrial/ Urban Soil Samples from Hanoi and Bacninh

POPs	Sampling Sites
Hanoi	
DDTs	Agricultural* >> industrial
HCHs	Agricultural* >> industrial
PCBs	Agricultural << industrial*
Bacninh	
DDTs	Agricultural* >> urban
HCHs	Agricultural* >> urban

Note: Mann–Whitney U-tests.
* Significant difference ($p < 0.01$).

12.3.2 LEVELS OF PCBs

It is well known that commercial PCB mixtures were used in a wide variety of industrial applications. Their commercial utility was based largely on their chemical stability, which included low flammability and their desirable physical properties, including electrical insulating properties [21]. In Vietnam, PCBs were used in closed systems such as capacitors and transformers [20]. However, there exists no information regarding the use of PCBs in open systems. It has been reported that landfills are one contamination source of PCBs [26].

The analytical data of PCBs in Hanoi and Bacninh soil samples from industrial sites were relatively higher when compared with the samples from agricultural sites of the both cities. It could be due to careless disposal practices, accidents, or leakage from various industrial facilities and chemical waste disposal sites. In Vietnam, as shown in Figure 12.7, relatively high PCB contamination was discovered in the battleground soil collected from Tanbien, which is a rural area of Tay Ninh, Vietnam. This suggests that military activities could be another important source of PCB contamination [9]. PCB contamination was also found to be heavy in the soil samples collected from a former Soviet army base located in western Poland near the border of former East Germany [27].

12.3.3 COMPARISON TO OTHER RECENT DATA REPORTED ON POPs IN SOIL

Table 12.2 shows a summary of POPs data in soils from Vietnam as well as data reported on and later than 2000. The concentrations in soils of Vietnam exhibit the following order: DDTs > PCBs > HCHs. It was found that the concentrations of DDTs ranged from 0.02 to 1300 ng/g, PCB concentrations ranged from 0.61 to 9960 ng/g, and HCHs ranged from 0.05 to 71 ng/g for 480 soil samples collected from geographically different areas of Vietnam [12]. These results indicated that higher concentrations of DDTs in soil samples from agricultural areas. In contrast, higher levels of PCBs were found in soil samples from industrial areas.

Careful considerations must be given to compare these data due to the differences in analytical techniques used to quantitate PCBs. A few reported the sum of measured values with only seven regulatory congeners while others reported the sum as the complete list of all congeners. Accuracy and precision are very important for POPs analysis. High level of accuracy can be achieved by using certified reference materials and standards.

TABLE 12.2

Concentrations of DDTs, HCHs, and PCBs in Soils from Asian, African, and Other Geographic Areas (ng/g Dry Weight)

Location	Survey Year	Land Use/Location	ΣDDTs	ΣHCHs	ΣPCBs	References
Asia						
China Tibetan Plateau	2005	Agricultural sites (n=6)	NA	NA	139–423[a]; 288[b], 281[c]	[28]
		Not cultivated sites (n=9)	NA	NA	47–236[a]; 127[b], 122[c]	
China Loess Plateau, Jiaodong Peninsula and Beijing Suburb	2007	Orchard sites (n=99)	0.456–252.5[a]; 2.38[b], 18.8[c]	NA	NA	[29]
China Beijing		Urban park sites	5.942–1039[a]; 64.87[b], 162.0[c], 65.64[d]	0.249–197.0[a]; 1.768[b], 10.54[c], 2.129[d]	NA	[30]
China Beijing	2007	Agricultural sites (n=104)	ND—117[a,l]; 0.40[b], 0.67[c]	ND—5.56[a,m]; 0.40[b], 0.67[c]	NA	[31]
China Guangzhou	2004	Agricultural sites around urban area (n=32)	7.61–662.9[a,j]; 67.32[b], 64.57[d]	0.21–103.92[a,m]; 4.44[b], 6.20[d]	NA	[32]
China Watershed of the Pearl River Delta	2005	Agricultural sites (n=55)	3.58–831[a,j]; 82.1[c]	0.19–42.3[a,m]; 4.42[c]		[33]
China Guanting Reservoir, Beijing	2003	Agricultural sites (n=56)	ND—76[j]; ND[b], 9.5[c]	ND—7.3[a,m]; ND[b], 0.66[c]	NA	[34]
China Hong Kong	2000	Woodland (n=36 including grassland)	0.32[c]	5.72[c]	NA	[35]
		Grassland (n=36 including woodland)	0.24[c]	6.32[c]	NA	[35]
		Farmland (n=7)	1.85[c]	6.53[c]	NA	[35]
		Wetland (n=2)	0.43[c]	7.64[c]	NA	[35]
		Reclamation (n=1)	0.13[c]	6.56[c]	NA	[35]
China Guanting Reservoir Hebei/Beijing	2003	Farmland (n=30)	<1.2–94.07[a,l]; 5.11[b]	<0.8–8.96[a,m]; 0.56[b]	NA	[36]
China Beijing	2003	Agricultural sites (n=17)	7.2–2910[a,l]	2.0–760.3[a,m]	NA	[37]
China		Apple orchard	381.3[d]	32[d]		

Location	Year	Site				Ref.
Shanghai	2001	Agricultural sites ($n=2$)	$84^{u,c}$	$2.4^{o,c}$	NA	[38]
Haining	2001	Agricultural sites ($n=2$)	$83^{u,c}$	$0.71^{o,c}$	NA	[38]
Lake Tai	2000	Agricultural sites ($n=2$)	$7.8^{u,c}$	$0.68^{o,c}$	NA	[38]
Taiwan Across the whole land	1994	Agricultural sites ($n=204$)	$<2.67\text{–}149^{a,h}$ 5.82^c	$<0.14\text{–}24.9^{a,n}$ 1.10^c	NA	[39]
India Delhi, Haryana, Uttar Pradesh	2002	Agricultural sites ($n=16$)	NA	$0.20\text{–}212.20^{\#}$ $14.57^b, 44.81^c$	NA	[40]
Vietnam Hanoi	2006	Agricultural sites ($n=31$)	$<0.02\text{–}171.83^a$ $86.26^b, 92.86^c$	$<0.05\text{–}20.57^{a,m}$ $7.39^b, 8.30^c$	$<0.02\text{–}24.37^a$ $18.36^b, 18.77^c$	[8,9,11]
	2006	Industrial sites ($n=29$)	$<0.02\text{–}67.82^a$ $22.43^b, 21.22^c$	$<0.05\text{–}7.76^a$ $4.86^b, 3.25^c$	$<0.02\text{–}190.42^a$ $35.74^b, 41.90^c$	[8,9,11]
Bacninh	2006	Agricultural sites ($n=24$)	$<0.02\text{–}161.46^a$ $103.61^b, 111.4^c$	$<0.05\text{–}9.58^a$ $8.22^b, 7.70^c$	NA	[10,11]
	2006	Urban sites ($n=16$)	<0.02 $<0.02^b, 12.1^c$	$<0.05\text{–}5.86^a$ $<0.05^b, 1.74^c$	NA	[10,11]
Middle and South America						
Costa Rica ($n=15$) Olivia-Paraiso	Not indicated	Recreation area	$10\text{–}14,000^d$ 180^d	NA	NA	[41]
Costa Rica ($n=16$) Estrada	Not indicated	Recreation area	$10\text{–}13,000^d$ 90^d	NA	NA	[41]
Mexico ($n=20$) Nuevo Nicapa	Not indicated	Recreation area	$10\text{–}19,000^d$ 70^d	NA	NA	[41]
Honduras ($n=15$) Ceiba Grande	Not indicated	Recreation area	$<LOD\text{—}1,300^d$ 110^d	NA	NA	[41]
Honduras ($n=16$) Feo	Not indicated	Recreation area	$<LOD\text{—}900^d$ 70^d	NA	NA	[41]
Nicaragua ($n=4$) Miriam Tinoco	Not indicated	Recreation area	$50\text{–}110^a$ 70^d	NA	NA	[41]
Nicaragua ($n=3$) Nuevo Amanecer	Not indicated	Recreation area	$10\text{–}60^d$ 20^d	NA	NA	[41]
El Salvador ($n=10$) San Luis	Not indicated	Recreation area	$<LOD\text{—}200^d$ 40^d	NA	NA	[41]

(continued)

TABLE 12.2 (continued)
Concentrations of DDTs, HCHs, and PCBs in Soils from Asian, African, and Other Geographic Areas (ng/g Dry Weight)

Location	Survey Year	Land Use/Location	ΣDDTs	ΣHCHs	ΣPCBs	References
Guatemala (n=22) Sehix A.V.	Not indicated	Recreation area	<LOD—300[a] 30[j]	NA	NA	[41]
Panama (n=32) Bisira	Not indicated	Recreation area	<LOD <LOD[d]	NA	NA	[41]
Brazil	1997	DDT-sprayed sites	344[h]	NA	NA	[42]
Brazil	1995–2005	Indoor	2.0–55.4[a,i]	NA	NA	[43]
Madeira River Basin, Amazon Brazil	Forest		1.6–13.3[a,i]	NA	NA	
Northeastern part of Sao Paulo State Brazil	2005	Environmental reserves (n=18)	0.12–11.01[l]	0.05–0.92[a,m]	0.02–0.25[a,7]	[44]
Tapajos River Basin, Amazon	1994	Urban sites (n=3)	281–1224[j] 546[b], 684[c]	NA	NA	[45]
Madeira River Basin, Amazon	1997	Urban sites (n=3)	4–123[j] 116[b], 81[c]	NA	NA	[45]
Mexico	2005	Rural sites (n=4)	NA[k] 0.17[c], ND[d]	NA ND[c], ND[d]	NA	[46]
		Urban sites (n=9)	NA[k] 45[c], 4.7[d]	NA 0.043[c], 0.032[d]	NA	[46]
		Agricultural sites (n=16)	NA[k] 10[c], 1.5[d]	NP 0.044[c], 0.029[d]	NA	[46]
Peru Andes	2004	Mountain sites (n=10)	0.02–1.65[a,k] 0.51[c]	<0.01[a,m] <0.01[c]	<0.01–0.44[a] 0.08[c]	[47]
Africa Republic of Mali (West Africa)	2003 DDT was not applied during 10 years	Agricultural sites (n=15)	<6–132[a,j]	NA	NA	[48]
North America Canada	2002	Forest	1.0–18[a]	<0.005–0.068[b]	NA	[49]

Location	Year	Site				Ref
			5.0^b, 6.81^c	0.23^b, 0.026^c		
USA Ohio, Pennsylvania Indiana and Illinois	1995–1996	Agricultural site and home gardens (n=40)	<0.5–11,846^a 9.63^d	NA	NA	[50]
USA Maine	1993	Forest, DDT sprayed sites	270–1898^{a,j} 999^b, 1007^c	NA	NA	[51]
USA	1993	Forest, DDT unsprayed sites	0–11^{a,j} 2^b, 6.5^c	NA	NA	[51]
USA South Carolina	1999	Agricultural sites (n=16)	0.11–45^{a,j} 11^c	<0.1^o <0.1^c	<0.5^{100}	[52]
Georgia	1999	Agricultural sites (n=16)	0.34–34^j 13^c	<0.1^o <0.1^c	<0.5^{100}	[52]
United States (n=27)	2003	Rural and remote sites	NA	NA	0.255–24.6^{a,209} 3.09^c	[53]
Far East Europe Turkey (n=18) Gölbaşı, Ankara	1997	Rural, industrial	NA	NA	530–464,000^{a,e}	[54]
Turkey (n=1) Antalya	1988	Uncultivated	NA	NA	0.344^{29}	[54]
Turkey (n=1) Aliaga, Izmir	2001	Industrial	NA	NA	640^7	[54]
Turkey (n=48) Aliaga, Izmir	2004–2006	Rural, industrial	NA	NA	0.23–805^{a,41}	[55]
Turkey (n=11) Golbasi, Ankara	2007	Rural, industrial	NA	NA	ND—10,000^{a,f}	[54]
Turkey (n=30) Gölbaşı, Ankara	2008	Rural, industrial	NA	NA	ND—84^{a,f}	[54]
Turkey (n=20) Iskenderun, Hatay	2008	Industrial	NA	NA	17^{41}	[54]

(continued)

TABLE 12.2 (continued)
Concentrations of DDTs, HCHs, and PCBs in Soils from Asian, African, and Other Geographic Areas (ng/g Dry Weight)

Location	Survey Year	Land Use/Location	ΣDDTs	ΣHCHs	ΣPCBs	References
East, Middle and South Europe						
Estonia	2005–2006	Industrial sites (n=11)	<0.01–15.8[a,k]; 1.3[b], 4.3[c]	<0.01–0.05[a,n]; <0.01[b], <0.01[c]	0.43–81[a,26]; 4.0[c], 13.4[c]	[56]
Czech Republic	Since 1992	Arable soil (n=39)	4.00–1018.3[a]; 34.2[b], 113.7[c]	0.65–4.00[a,m]; 4.00[b], 3.54[c]	3.50–42.1[a]; 3.60[b], 6.86[c]	[57]
		Grassland soil (n=22)	2.04–28.2[a]; 9.34[b], 12.19[b]	0.38–4.00[a,m]; 0.88[b], 1.18[c]	2.01–29.21[a]; 6.31[b], 9.34[b,c]	[57]
		Hilly forest soil (n=9)	2.20–954.9[a]; 8.50[b], 119.5[c]	0.55–2.34[a,m]; 1.09[b], 1.18[c]	3.42–13.37[a]; 8.40[b], 8.87[c]	[57]
		Mountain forest soil (n=9)	8.80–1908.3[a]; 69.2[b], 315.2[c]	0.26–1.66[a,m]; 0.65[b], 0.86[c]	7.90–36.18[a]; 22.64[b], 22.76[c]	[57]
Lithuania (n=5)	2006	Background and industrial sites	0.3–7.7[a]; 1.92[c]	0.4–1.1[a,m]; 0.8[c]	0.6–24.[a,7]; 5.94[c]	[58]
Central and Southern Europe	2005	Background sites	NA	NA	1.3–2.3[a,7]; 1.8[b], 1.8[c]	[59]
	2005	Residential, rural and urban sites	NA	NA	1.8–20.1[7]; 5.9[b], 6.8[c]	[59]
	2005	Industrial sites	NA	NA	0.7–68.4[a,7]; 23.3[b], 27.2[c]	[59]
Italy Central Italian Alps	9007	Background sites		0.04–1.93[a,m]; 0.27[b], 0.38[c]	0.37–5.52[a]; 1.47[b], 1.63[c]	[60]
Spain	2005	Petrochemical industry sites (n=7)	NA	NA	314–17,895[a]; 728[b], 4673[c]	[61]
		Chemical industry sites (n=9)	NA	NA	257–14,997[a]; 4139[b], 4633[c]	[61]
		Urban/residential sites (n=5)	NA	NA	185–10,543[a]; 4679[b], 4435[c]	[61]
		Unpolluted sites (n=5)	NA	NA	ND–2105[a]; 334[b], 766[c]	[61]

Great Britain ($n=200$)	Rural sites		NA	NA	0.274–80.6[a,33] 2.52[b], 5.03[c]	[62]
Germany, Austria, Switzerland and Slovenia sides of Alps	Forest sites		0.4–28.8[a,m]	0.3–8.8[a,n]		[63]
Italy	Mountain sites ($n=19$)	2003	0.18–11[k]	<0.01–1.88[a,m]	0.61–8.9[a]	[47]
Alps			2.2[c]	0.51[c]	3.6[c]	
Others						
Across the world ($n=191$)	Background sites	1998	NA	NA	0.026–97[a,27] 5.4[c]	[64]
Antarctica	Pristine sites ($n=9$)	2005	0.51–3.68[j]	0.49–1.34[a,m]	0.51–1.82[a,7]	[65]
James Ross Island						

Notes

1. Concentration
[a] Minimum concentration – Maximum concentration
[b] Median
[c] Mean
[d] Geometric mean

2. PCBs
[e] The concentration as Aroclor 1260
[f] The sum of all PCBs given by the sum of Aroclor 1016 and 1260.
A superscript number indicates the number of PCB congeners analyzed.

3. DDTs
[g] p,p'-DDT
[h] Sum of p,p'-DDT and p,p'-DDE;
[i] Sum of o,p'-DDT, p,p'-DDT, o,p'-DDE, p,p'-DDE and p,p'-DDD
[j] Sum of p,p'-DDT, p,p'-DDE and p,p'-DDD
[k] Sum of o,p'-DDT, p,p'-DDT, o,p'-DDE, p,p'-DDE, o,p'-DDD and p,p'-DDD
[l] Sum of o,p'-DDT, p,p'-DDT, p,p'-DDE and p,p'-DDD

4. HCHs
[m] Sum of *alpha-*, *beta-*, *gamma-* and *delta*-isomers
[n] *gamma*-HCH
[o] Sum of *alpha-*, *beta-*, and *gamma*-isomers
[p] Sum of *alpha-* and *gamma*-isomers

In countries in Asia as well as other countries, DDTs and HCHs data that were analyzed during the 2000s were limited (Table 12.2). These data were cited from the limited reports that have been published; however, it did roughly indicate the status of contamination in recent years in the respective countries. Judging from several reports, China had concentration levels of DDTs and HCHs in almost the same ranges, namely on the order of 10 ng/g for DDTs and 1 ng/g for HCHs. It may be true that the mass of DDT application in these areas was much higher than that of HCH. DDT was used as an insecticide (to control mosquitoes that spread malaria and other diseases) in residential areas as well as to control pests in agricultural fields to increase crop yield for a long time in tropical countries. Some of these heavily contaminated samples are listed in Table 12.2. On the other hand, HCHs were mainly used for pest control in agricultural activities. Another reason for higher concentrations of DDTs when comparing with HCHs was attributed to the lower vapor pressure of DDTs versus HCHs [66–68]. This results in longer-term residues of DDTs in the soil. PCB data of recent soil samples collected during or after 2000 are also limited in developing countries as well as developed countries, compared to the POPs data reported on soil samples collected in developed countries in the 1980s and 1990s. The limited data show that relatively high concentrations of PCBs detected from some of the industrial areas are the same as those from the Vietnamese soil samples. As seen in Table 12.2, higher concentrations were found in soil samples collected from industrial areas of developed countries in Europe. The data is not enough to discuss the contamination status of developed countries located over tropical and subtropical areas. It is necessary to have more data in order to discuss in more detail.

The residue composition of individual DDTs provides some environmental information such as the degradation conditions after application of pesticide DDT to the field. Consequently, under aerobic conditions, a high rate of formation of p,p'-DDE from p,p'-DDT takes place, whereas the conversion of p,p'-DDT into p,p'-DDD takes place rapidly under anaerobic conditions [69]. It is estimated that pesticide DDT would be applied to a paddy field and according to the above context, the constituent of its agent, p,p'-DDT, would be converted into p,p'-DDD under anaerobic conditions, however the remainder, p,p'-DDT, would be converted to p,p'-DDE under aerobic conditions during dry season [9,69]. Therefore, a wide variation in percent composition of DDT and its metabolites was observed in Vietnam soil samples as well as the other tropical countries. Thao et al. reported that the percent composition was found to be on the order of p,p'-DDE $> p,p'$-DDD $> p,p'$-DDT $> o,p'$-DDT in paddy field samples collected in 1990 and 1991 from northern and south western Vietnam [9]. Field conditions and climate are thought to be important factors in the transformation of p,p'-DDT to its metabolites, however soil properties, the mode of application, and the time elapsing after application may influence the residues of insecticides and their metabolites in soil [9].

In comparison with the analytical data obtained from the samples collected in 1990 [9], the p,p'-DDE found in samples collected in 2006 were relatively high, suggesting that the p,p'-DDE might be more stable than p,p'-DDD or p,p'-DDT [10]. This implies that the half-life of p,p'-DDE is longer than both the latter compounds. The ratio of metabolites of DDTs/total DDTs was 0.75–0.99 with a mean of 0.79, suggesting a significant degradation of pesticide DDT. These observations suggested that there have been no recent applications of pesticide DDT in this region [10].

With regard to HCHs, in general, the order of concentrations determined in the soil samples collected from northern Vietnam in 2006 were found to be in the following order: *beta*-HCH $>$ *alpha*-HCH $>$ *gamma*-HCH $>$ *delta*-HCH. This order was different from the samples collected in northern Vietnam in 1990. Namely, the compositions of *alpha*-HCH found in the samples collected in 1990 were higher than the samples collected in 2006, indicating easy dissipation of *alpha*-HCH in the soil versus *beta*-HCH. In other words, *beta*-HCH was more stable in soil when compared to the other HCH isomers [67,68].

PCB formulations are normally composed of many congeners and isomers. Soil samples analyzed from Vietnam contained several PCB congener residues. The residue patterns fundamentally depend on the PCB formulations that were used in industries and industrial materials. PCBs detected in soil samples from northern Vietnam might be due to leaks from industrial sources [9].

As mentioned earlier, relatively high concentrations of PCBs were detected in paddy fields far from industrial areas and residential areas in the far interior province of Tay Ninh [9]. With regards to the sources of this PCB contamination, the only possible explanation is the so-called "Ranch Hand" operation, which lasted from 1961 to 1971. During the Second Indochina War (Vietnam War), the U.S. Army may have used PCB formulations as lubricants or in other application during their army activities [9]. In a similar manner, other army bases may have been important contamination sources of PCBs throughout Vietnam. It is difficult to compare the PCB data from the samples collected in 1990 and 2006 due to measuring different congeners [9,11].

12.4 TEMPORAL TRENDS AND HALF-LIVES OF DDTs, HCHs, AND PCBs

Soil is one of the most important reservoirs for persistent organic chemicals (POC). Physicochemical properties of these compounds include hydrophobicity, semi-volatility, and the tendency to adsorb to organic matter. As a result, these compounds tend to accumulate in soils. Consequently, the contents of the soil materials control the residence time of POC in the soil environment. The organisms living in the soil are also important factors for POP residues. Climate conditions such as high temperature, etc. are also important. In general, the organic content of tropical soils is relatively low, and in the dry season the atmospheric temperature is very high during the daytime. Hence, POPs might dissipate due to their evaporation from the soil surface. However, some amounts may remain bound in the soil due to physicochemical interactions with organic matters and/or inorganic materials in soil.

Persistency of chemicals in environmental media is normally expressed as time trend residue and half-life. The half-life of these compounds plays an important role since it is possible to estimate the fate of chemicals in the environment and predicts future trends. Studies concerning the time trend residues of POPs were performed by some environmentalists by using various historical environmental samples, long-term surveys and predicting models [7,51,56,70–77]. From these results, the declining trends of POPs in environmental samples were confirmed. This result was useful in estimating the present status of contamination, and evaluating the future trends [78,79]. In this study, we calculated the half-life values of POPs in Vietnamese soil.

12.4.1 Temporal Trends and Half-Lives of POPs

As shown in Figures 12.2 and 12.5, agricultural soils were collected during 1990 [9] and 2006 [13] in Thanh Tri, Hanoi, at an interval of 16 years. The median concentrations of DDTs, HCHs, and PCBs were 97.0, 11.0, and 22.5 ng/g from the samples collected in 1990, and 71.4, 7.37, and 17.3 ng/g from the samples collected in 2006. In the same manner, the arithmetic mean concentrations of those compounds were 139, 19.5, and 22.4 ng/g from the 1990 samples and 81.1, 9.33, and 18.7 ng/g for the 2006 samples. The results calculated for half-lives of DDTs, HCHs, and PCBs were 36, 27, and 42 years by median values, and 21, 15, and 61 years by arithmetical mean concentrations.

Half-lives of DDTs and HCHs in Bacninh soils were reported by Toan et al. [13]. The mean concentrations of DDTs in soil samples from Bacninh in 1992, 1995, 1998, 2001, and 2006 were 278.06, 254.73, 156.46, 124.85, and 68.92 ng/g respectively [13]. In the same manner, the mean concentrations of HCHs in the soil samples from Bacninh collected in 1992, 1995, 1998, 2001, and 2006 were determined to be 33.86, 29.96, 17.46, 11.85, and 5.06 ng/g, respectively. The results are illustrated in Figure 12.8. The estimated half-lives of DDTs and HCHs were calculated by the concentration data mentioned above as 6.7 and 4.9 years, respectively.

A few half-life data reported so far are shown in Table 12.3. It seems that the reported half-lives of DDTs [81–83,86,87,94] and HCHs [80,94] in soils of tropical conditions are considerably lower than the values reported by Toan et al. [13].

Half-life values calculated from the environmental soil samples fluctuate considerably due to many factors. The values were depended on the analytical results measured. The analytical error

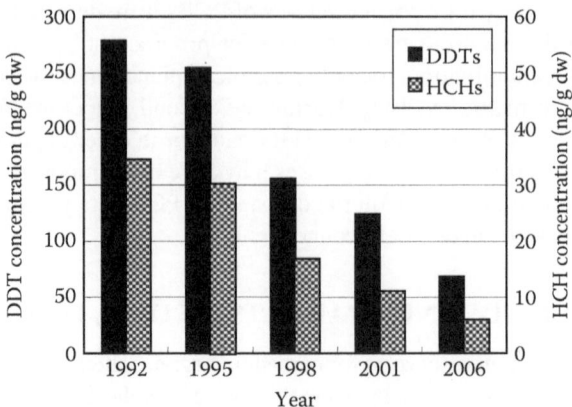

FIGURE 12.8 Time trends of DDTs and HCHs from 1992 to 2006 in agricultural fields.

could be reduced by using state-of-the-art analytical equipment and methods. Certified reference materials (CRM) or standard reference materials (SRM) for chemical analysis are necessary to get reasonably accurate and precise data for the calculation of half-life. The results of half-life calculations may be affected if the continuous input of POPs is conducted via multimedia. In this case, the calculated results of apparent half-lives could be higher compared to the figures reported from the other areas.

Furthermore, the figures could be variable due to the quality and quantity of constituents in soil, especially organic matter (OM). It is thought that the chemicals, which were applied or released into the field/ground, were dispersed onto the soil surface. Some of them are present on the soil surface are free from clay and OM. These could penetrate into the OM of the soil surface. There may be bound residue due to higher octanol–water partition coefficients. From this context, states of POP residues in soil could be explained as two types, namely "shallow" and "deep." POPs that are found in the shallow phase of the soil could be easily released from the soil. On the other hand, the residue in the "deep" phase could be difficult to release from the soil due to strong binding with organic materials. These conditions affect the residue's half-life in soils. It may also imply that POPs dissipation in soil could take place as two processes. Just after application, DDTs are present in the shallow phase, suggesting short half-lives, then later in the deep soil phase, suggesting longer half-lives.

12.4.2 Factors Affecting Half-Lives of POPs in Soil

POPs may be lost from the soil due to several factors. It is thought that without degradation, residues may be lost to the atmosphere by volatilization, or in solution or suspension by leaching, run off, and erosion. Loss may also occur by degradation under biological and non-biological processes. To understand the critical effects on the environment, it is necessary to consider that the contaminants are transferred to other environmental compartments or removed from the environment by degrading processes.

12.4.2.1 Non-Biological Processes

12.4.2.1.1 Volatilization

Dissipation of POPs from soil as vapor to atmosphere is one of the significant routes for loss. Volatilization of POPs from the soil surface depends on the vapor pressure of the compounds [99]. The vapor pressure of p,p'-DDT is measured as 1.5×10^{-7} mmHg at 20°C [100]. It has been reported that volatilization is a major pathway for loss of applied DDT in tropical areas [101].

In addition to vapor pressure, soil moisture may also be an important factor in actual vaporization rates [102,103]. Volatilization may be reduced in dry soils when insufficient water is present to

TABLE 12.3
Half-Lives of DDTs, HCHs, and PCBs in Soil

Location	Remarks on Soil Type and the Others	POPs	Half-Life	Reference
Asia				
Malaysia	Agricultural soil, sandy loam	*gamma*-HCH	1.0–1.7 years	[80]
Indonesia	Loam	p,p'-DDT	9 weeks	[81]
	Sandy loam	p,p'-DDT	25 weeks	[81]
Pakistan	Clay loam	p,p'-DDT	144–313 days	[82]
	Clay loam	p,p'-DDE	191 days	[82]
India	Not indicated	p,p'-DDT	319, 343 days	[83]
India, Tamil Nadu		DDT	35 days	[84]
		HCH	45 days	[84]
India, Delhi	Sandy loam	DDTs	188 days	[85]
		HCHs	37.5 days	[85]
China	Subtropical soil	p,p'-DDT	75 weeks	[86]
	Tropical soil	p,p'-DDT	31 weeks	[86]
Philippine	Sandy clay loam	p,p'-DDT	261, 210 days	[87]
		p,p'-DDE	151 days	[87]
Vietnam		DDTs	6.7 years	[11]
		HCHs	4.9 years	[11]
Middle and North America				
USA, Hawaii	Sandy clay loam	p,p'-DDT	13, 24 weeks	[88]
Canada	Well-drained and aerobic sandy	p,p'-DDT	20–50 years	[89]
USA, Arizona	Agricultural soil	DDTs	7 years	[90]
USA, Arizona	Agricultural soil	DDTs	>20 years	[91]
USA, Maine	Forest soil	DDTs	20–30 years	[51]
Brazil	Sandy loam	p,p'-DDT	>200 weeks	[92]
	Sandy loam	p,p'-DDE	88 weeks	[92]
Argentina, Balcarce	Non-cultivated and loam soil	*gamma*-HCH	8.6–9.4 weeks	[93]
Africa				
Kenya		p,p'-DDT	64.5–245.6 days	[94]
		p,p'-DDE	145 days	[94]
		gamma-HCH	5–8 days	[94]
Kenya		p,p'-DDT	110 days	[95]
Tanzania	Tropical (hot and wet climate)	p,p'-DDT	174 days	[96]
	Tropical (cool and wet climate)	p,p'-DDT	335 days	[96]
Europe				
England	Sandy loam	PCB 28	193 days	[97]
	Sandy loam	PCB 52	267 days	[97]
	Sandy loam	PCB 101	462 days	[97]
	Sandy loam	PCB 138	495 days	[97]
	Sandy loam	PCB 180	8664 days	[97]
United Kingdom	Five different types of soils	PCBs	About 5 years	[72]
The Netherlands		p,p'-DDT	About 20 years	[98]
		gamma-HCH	3–4 years	[98]

form a layer over soil particles [104]. It was investigated that co-distillation processes would be taking place when sufficient water would be available [103,105]. It can be easily estimated that actual volatilization rates may be reduced by the association of POPs with organic soil matter [106]. In tropical climates, however, organics can be easily mineralized due to high temperatures throughout the year. Accordingly, the content of OM in tropical soil is generally lower than that in temperate soil. However, Harner et al. concluded from their study on pesticide residues in soil of temperate areas that DDT and HCH residues were not proportional to organic carbon content in soil, indicating that residue concentrations were a reflection of pesticide application history and dissipation rates rather than air–soil equilibria [107]. Furthermore, Wong et al. have suggested that DDTs in the air of Mexico were the result of a combination of ongoing regional usage and reemission of old DDT residues from soils [46,108]. These findings suggested that one of the dissipation routes of DDT residue in soil was evaporation from soil to atmosphere.

12.4.2.1.2 Photodegradation
Photolytic degradation on the soil surface did not significantly contribute to the dissipation of POPs in the environment. It was commonly understood that photolytic degradation was not an important mechanism of pesticide loss from soil because the compounds are not expected to undergo direct photolysis due to the lack of absorption in the environmental UV spectrum (<290 nm). Furthermore, it was thought that only a small amount of POPs were present on the soil surface compared to the amount of residue in the total soil profile. However, Chu et al. have recently reported the accelerating and quenching of PCB photolysis in the presence of OM such as humic materials and surfactants [109]. In tropical environments, with the exception of artificial surfactant contamination, the organic content in soil is very low due to the high atmospheric temperature throughout the year, so organic materials will easily decompose to inorganic constituents by soil organisms. Accordingly, organic content of tropical soil were at commonly low levels.

12.4.2.1.3 Leaching, Run-Off, and Erosion
Due to low solubility of DDTs, leaching will not be a major loss route. Jury et al. had obtained the results of DDT mobility testing for soil columns with a soil chemical screening model (SCSM) [110]. The results of the model indicated that DDTs did not leach into deeper layers. Negligible vertical movement of DDTs was found in laboratory [111,112] and field studies [113]. These results supported the hypothesis that leaching is a minor factor in the total DDT losses from soil. In the case of lindane (*gamma*-HCH), which is more soluble [67,68] dissolved into soil solution [111,117,118]. Champion and Olsen estimated that the chlorine atoms in the ethylene group of DDT possess a partial negative charge that appeared mainly responsible for the adsorption of DDT onto OM and colloidal clay particles [116]. Physical and chemical adsorption would take place between DDTs and SOM/soil colloid [114,115]. It was thought that the solubility and, hence, movement of residues might be enhanced by association with water-soluble OM [119]. This implies that water-mediated downward transport depends on the amount of rainwater and irrigation water percolating through the soil profile. The profiles of POPs will be determined depending on the different compound's physicochemical properties. However, Moeckel et al. described the occurrence of downward transport and hence limited surface-air exchange of more volatile POPs as they are removed from the top layers [120]. Furthermore, such soils may be able to retain higher amounts of these compounds than just addressed due to the capacity of their surface soil layers in the middle- and high-latitude areas [120]. It is possible that some downward movement of POPs adsorbed onto soil particulate matter through soil pores and cracks might also take place.

The total amount of rainfall or irrigating water received, the intensity (water flux), and frequency of received water all appear to effect movement of POPs in soils. Consequently, run-off and erosion are likely to be more important in the movement of POPs, especially under tropical climate conditions and even more so in the tropical monsoon climate most commonly experienced in northern Vietnam. The climate in northern Vietnam is cool and dry between November and April and hot and rainy from May to October. The temperature varies considerably between these two seasons with a difference of roughly 12°C. Typhoons sometime attack with heavy rainfall. Consequently, POPs will be washed out by rainfall during the rainy season. In general, in tropical areas, this process may be critical for the movement of POPs from terrestrial to aquatic environments.

12.4.2.2 Biological Processes

Soil microorganisms play an important role in the degradation of pesticides [121]. Several reports have documented degradation of DDT and HCH in soils [122–124]. Different microorganisms may contribute to the reactions that can be associated with aerobic and anaerobic conditions. Under anaerobic soil conditions, the conversion of DDT to DDD occurred, suggesting reductive DDT degradation by anaerobic microbes [69,125]. Boul et al. reported that two agricultural practices, namely, irrigation and superphosphates application, enhanced degradation of p,p'-DDT via p,p'-DDD, while, on the other hand, p,p'-DDE was still a significant proportion of the soil residues [74]. Since DDT is reductively dechlorinated to DDD and dehydrochlorinated to DDE, and the metabolite DDE is relatively stable. Accordingly, the metabolites/parent or metabolites/total DDTs ratios, in general, gradually increases after application as a function of time elapsed, although there may be a large variability with soil conditions including soil types [9,126,127]. The same tendency was found in the *gamma*-HCH/total HCHs ratio because *gamma*-HCH dissipates more rapidly in moisture soil [128].

In soil, lindane broke down rapidly and produced their respective dechlorinated metabolites [125]. Recently, Middeldorp et al. investigated the reductive dechlorination of *beta*-HCH in a number of contaminated soil samples under methanogenic conditions [129]. They have revealed that *gamma*-HCH is more unstable than the other HCH-isomers and is also unstable in the presence of common constituents in comparison to other isomers.

An incubation experiment with the addition of glucose to the soils contaminated with DDTs and PCBs was carried out in order to understand the relationships between their degradation and the potential microbial activities [130]. These results showed that the concentrations of DDT, DDD, and DDE decreased whereas the PCBs concentrations were not affected at all by microbial activities. This suggests that an increase in soil microbial degradation caused by addition of the organic substances can contribute to a complex set of environmental conditions, particularly in the SOM [130].

The biodegradation of PCBs, which is possible with some aerobic bacteria, depends on the degree of chlorination and the position of the chlorine substitution. Reductive dechlorination can also take place under anaerobic conditions [131]. Degradation decreases with increasing chlorination [132]. Under anaerobic conditions of sediment, some PCB congeners can be degraded to lower chlorine substituted congeners through a dechlorination processes [131,133].

This fact strongly indicates that during the wet season of tropical areas and in paddy fields with irrigated water conditions, some constituents of PCB formulation will degrade due to anaerobic microbes.

12.5 CONCLUSIONS

Data on production, import/export, and use of persistent chemicals in Vietnam are limited. Therefore, it is difficult to discuss any relationships between mass of use/production and the concentrations in Vietnamese soil. However, the levels of POPs in Vietnam are comparable to the levels in other tropical countries reported so far. With respect to PCBs contamination in Vietnam, it appears that industrial areas and damping sites might be considered heavily polluted. Although the Vietnam War took place several decades ago, the legacy of it still remains in the land and on the seashore. Any former army bases, battlefields, naval ports, and similar areas might be considered contaminated heavily by PCBs as well as dioxins. People living near these sites may be exposed to elevated levels of POPs and may be at higher risk for toxic effects.

It is well known that POP dissipation in tropical areas is enhanced by atmospheric temperature, humidity, precipitation, and volatilization, implying shorter half-lives than the temperate areas. Among the data for tropical countries, the half-lives of POPs in Vietnamese soils seem to be longer compared to the data from other tropical countries reported so far. Very limited studies have been conducted in Vietnam with the focus on temporal trends of classical as well as emerging new pollutants such as polybrominated diphenyl ethers (PBDEs), perfluorinated compounds (PFCs), etc. With regard to the temporal trends of POC in Vietnam, the available data demonstrated a declining trend in OC insecticides contamination such as DDTs and HCHs, while PCBs exhibit a relatively slower declining trend. Systematic studies dealing with time trends of organohalogen pollutants are needed in order to predict the future trends of contamination.

ACKNOWLEDGMENTS

This work was supported in part by Grand-in-Aid for Scientific Research (B) of the Ministry of Education, Culture, Sports, Science and Technology, Japan (Project No. 22310021). I wish to thank Dr. Loganathan for critical reading of this chapter.

REFERENCES

1. United Nations Environment Programme (UNEP). Persistent organic pollutants. http://www.unep.org/ (accessed March 18, 2011).
2. Library of Congress—Federal Research Division. Country Profile: Vietnam. 2005. http://memory.loc.gov/ammem/ndex.html/ (accessed March 18, 2011).
3. Abrahams, P. W. 2002. Soils: Their implications to human health. *The Science of the Total Environment* 291: 1–32.
4. Loganathan, B. G. and Kannan, K. 1994. Global organochlorine contamination trends: An overview. *Ambio* 23: 187–91.
5. Woodruff, L. G., Canon, W. F., Eberl, D. D., Smith, D. B., Kilburn, J. E., Horton, J. D., Garrett, R. G., and Klassen, R. A. 2009. Continental-scale patterns in soil geochemistry and mineralogy: Results from two transects across the United States and Canada. *Applied Geochemistry* 24: 1369–1381.
6. Kurt-Karakus, P. B., Bildeman, T. F., Staebler, R. M., and Jones, K. C. 2006. Measurement of DDT fluxes from a historically treated agricultural soil in Canada. *Environmental Science & Technology* 40: 4578–4585.
7. Meijer, S. N., Halsall, C. J., Harner, T., Peters, A. J., Ockenden, W. A., Johnston, A. E., and Jones, K. C. 2001. Organochlorine pesticide residues in archived UK soil. *Environmental Science & Technology* 35: 1989–1995.
8. Thao, V. D., Kawano, M., Matsuda, M., Wakimoto, T., Tatsukawa, R., Cau, H. D., and Quynh, H. T. 1993. Chlorinated hydrocarbon insecticide and polychlorinated biphenyl residues in soils from southern provinces of Vietnam. *International Journal of Environmental Analytical Chemistry* 50: 147–159.
9. Thao, V. D., Kawano, M., and Tatsukawa, R. 1993. Persistent organochlorine residues in soils from tropical and sub-tropical Asian countries. *Environmental Pollution* 81: 61–71.
10. Toan, V. D., Thao, V. D., Walder, J., Schmutz, H. R., and Ha, C. T. 2007. Contamination by selected organochlorine pesticides (OCPs) in surface soils in Hanoi, Vietnam. *Bulletin of Environmental Contamination and Toxicology* 78: 195–200.
11. Toan, V. D., Thao, V. D., Walder, J., Schmutz, H. R., and Ha, C. T. 2007. Level and distribution of polychlorinated biphenyls (PCBs) in surface soils from Hanoi, Vietnam. *Bulletin of Environmental Contamination and Toxicology* 78: 211–216.
12. Thao, V. D., Toan, V. D., and Kawano, M. 2008. Variation with time of persistent organochlorine residues in soils from Vietnam. *International Symposium on Environmental Sciences.* Abstract, Matsuyama, Japan, p. 14.
13. Toan, V. D., Thao, V. D., Walder, J., and Ha, C. T. 2009. Residue, temporal trend and half-life of selected organochlorine pesticides (OCPs) in surface soils from Bacninh, Vietnam. *Bulletin of Environmental Contamination and Toxicology* 82: 516–521.
14. Tenenbaum, D., 1996. The value of Vietnam. *Environmental Health Perspectives* 104: 1280–1285.
15. Global Environment Facility, World Bank, accession, implementation and enforcement of the newly-signed Stockholm Convention on POPs. Project No. 1450, http://gefonline.org/ (accessed August 11, 2010).

16. Zaim, M. and Jambulingam, P. *Global Insecticide Use for Vector-Borne Disease Control.* Department of Control of Neglected Tropical Diseases (NTD), World Health Organization Pesticide Evaluation Scheme (WHOPES), Geneva, Switzerland, 2007.

17. Hien, N. V. 1999. *Proceedings of the Regional Workshop on the Management of Persistent Organic Pollutants.* United Nations Environment Programme, Hanoi, Vietnam, March 16–19, pp. 241–249.

18. Agency for Toxic Substances and Disease Registry (ATSDR), Public Health Service, U.S. Department of Health and Human Services, Atlanta, Toxicological profile for DDT, DDE and DDD. http://www.atsdr.cdc.gov/taxprofiles/tp35.html (accessed March 11, 2011).

19. Li, Y. F. 1999. Global technical hexachlorocyclohexane usage and its contamination consequence in the environment: From 1948 to 1997. *The Science of the Total Environment* 232: 121–158.

20. Vietnam Environmental Protection Agency (VEPA). PCB inventory in Vietnam—Practicalities and challenges. http://www.chem.unep.ch/pops/pcb.activities/ (accessed August 11, 2010).

21. Erickson, M. D. *Analytical Chemistry of PCBs.* New York: CRC Press, Inc., Lewis Publishers, 1997.

22. Breivik, K., Sweetman, A., Pacyna, J. M., and Jones, K. C. 2002. Towards a global historical emission inventory for selected PCB congeners—A mass balance approach. 1. Global production and consumption. *The Science of the Total Environment* 290: 181–198.

23. Encyclopaedia Brittanica. Inc. Hanoi. http://www.britannica.com/EBchecked/topic/254479/Hanoi (accessed March 14, 2011).

24. Hanoi Department of International and Communication. "Overall Databases of Hanoi. http://www.english.hanoi.gov.vn/ 2010 (accessed March 14, 2011).

25. Vietnam Investment Network Corporation. Invest in Vietnam. http://www.investinvietnam.vn/ (accessed March 14, 2011).

26. Minh, T. B., Minh, N. H., Iwata, H., Takahashi, S., Viet, P. H., Tuyen, B. C., and Tanabe, S. 2007. Persistent organic pollutants in Vietnam: Levels, patterns, trends, and human health implications. *Developments in Environmental Science* edited by Li, S., Tanabe, S., Jiang, Cr., 7: 515–555.

27. Kawano, M., Falandysz, J., and Wakimoto, T. 1997. Polychlorinated biphenyls (PCBs) concentrations of soil in a former Soviet army base in Poland. *Organohalogen Compounds* 32: 172–177.

28. Wang, P., Zhang, Q., Wang, Y., Wang, T., Li, X., Li, Y., Ding, L., and Jiang, G. 2009. Altitude dependence of polychlorinated biphenyls (PCBs) and polybrominated diphenyl ethers (PBDEs) in surface soil from Tibetan Plateau, China. *Chemosphere* 76: 1498–1504.

29. Yang, S., Zhang, A., Ma, Y., Yang, S., and Yang, Z. 2009. Residues of persistent organic pollutants in orchard topsoil of major fruit producing regions in China. *Journal of Food, Agriculture & Environment* 17: 829–834.

30. Li, X.-H., Wang, W., Wang, J., Cao, X.-L., Wang, X.-F., Liu, J.-C., Liu, X.-F., Xu, X.-B., and Jiang, X.-N. 2008. Contamination of soils with organochlorine pesticides in urban parks in Beijing, China. *Chemosphere* 70: 1660–1668.

31. Hu, W., Lu, Y., Wang, G., Wang, T., Luo, W., Shi, Y., Zhang, X., and Jiao, W. 2009. Organochlorine pesticides in soils around watersheds of Beijing reservoirs: A case study in Guanting and Miyun reservoirs. *Bulletin of Environmental Toxicology* 82: 694–700.

32. Gao, F., Jia, J., and Wang, X. 2008. Occurrence and ordination of dichlorodiphenyltrichloroethane and hexachlorocyclohexane in agricultural soils from Guangzhou, China. *Archives of Environmental Contamination and Toxicology* 54: 155–166.

33. Ma, X., Ran, Y., Gong, J., and Zou, M. 2008. Concentrations and inventories of polycyclic aromatic hydrocarbons and organochlorine pesticides in watershed soils in the Pearl River Delta, China. *Environmental Monitoring and Assessment* 145: 453–464.

34. Wang, T., Lu, Y., Shi, Y., Giesy, J. P., and Luo, W. 2007. Organochlorine pesticides in soils around Guanting Reservoir, China. *Environmental Geochemistry and Health* 29: 491–501.

35. Zhang, H. B., Luo, Y. M., Zhao, Q. G., Wong, M. H., and Zhang, G. L. 2006. Residues of organochlorine pesticides in Hong Kong soils. *Chemosphere* 63: 633–641.

36. Hong, Z., Yonglong, L., Dawson, R. W., Yajuan, S., and Tieyu, W. 2005. Classification and ordination of DDT and HCH in soil samples from the Guanting Reservoir, China. *Chemosphere* 60: 762–769.

37. Shi, Y., Meng, F., Guo, F., Lu, Y., Wang, T., and Zhang, H. 2005. Residues of organic chlorinated pesticides in agricultural soils of Beijing, China. *Archives of Environmental Contamination and Toxicology* 49: 37–44.

38. Nakata, H., Hirakawa, Y., Kawazoe, M., Nakabo, T., Arizono, K., Abe, S.-I., Kitano, T., Shimada, H., Watanabe, I., Li, W., and Ding, X. 2005. Concentrations and compositions of organochlorine contaminants in sediments, soils, crustaceans, fishes and birds collected from Lake Tai, Hangzhou Bay and Shanghai city region, China. *Environmental Pollution* 133: 415–429.

39. Lin, H. T., Wong, S. S., and Li, G. C. 1997. Residual level of chlorinated hydrocarbon insecticides in the field soil of Taiwan. *Plant Protection Bulletin* 39: 173–180.

40. Prakash, O., Suar, M., Raina, V., Dogra, C., Pal, R., and Lal, R. 2004. Residues of hexachlorocyclohexane isomers in soil and water samples from Delhi and adjoining areas. *Current Science* 87: 73–77.

41. Perez-Maldonado, N., Trejo, A., Ruepert, C., Jovel, R. D. C., Mendez, M. P., Ferrari, M., Saballos-Sobalvarro, E., Alexander, C., Yanez-Estrada, L., Lopez, D., Henao, S., Pinto, E. R., and Diaz-Barriga, F. 2010. Assessment of DDT levels in selected environmental media and biological samples from Mexico and Central America. *Chemosphere* 78: 1244–1249.

42. Vieira, E. D. R., Torres, J. P. M., and Malm, O. 2001. DDT environmental persistence from its use in a vector control program: A case study. *Environmental Research Section A* 86: 174–182.

43. Torres, J. P. M., Lailson-Brito, J., Saldanha, G., Dorneles, P. E., Silva, C. E. A., Malm, O., Guimaraes, J. R., Azeredo, A., Bastos, W. R., Da Silva, V. M. F., Martin, A. R., Claudio, L., and Markowitz, S. 2009. Persistent toxic substances in the Brazilian Amazon: Contamination of man and the environment. *Journal of Brazilian Chemical Society* 20: 1175–1179.

44. Rissato, S., Galhiane, M. S., Ximenes, V. F., De Andrade, R. M. B., Talamoni, J. L. B., Libanio, M., De Almeida, M. V., Apon, B. M., and Cavalari, A. A. 2006. Organochlorine pesticides and polychlorinated biphenyls in soil and water samples in the Northeastern part of Sao Paulo State, Brazil. *Chemosphere* 65: 1949–1958.

45. Torres, J. P. M., Pfeiffer, W. C., Markowitz, S., Pause, R., Malm, O., and Japenga, J. 2002. Dichlorodiphenyltrichloroethane in soil, river sediment, and fish in the Amazon in Brazil. *Environmental Research Section A* 88: 134–139.

46. Wong, F., Alegria, H. A., and Bidleman, T. F. 2010. Organochlorine pesticides in soils of Mexico and the potential for soil-air exchange. *Environmental Pollution* 158: 749–755.

47. Tremolada, P., Villa, S., Bazzarin, P., Bizzotto, E., Comolli, R., and Vighi, M. 2008. POPs in mountain soils from the Alps and Andes: Suggestions for a 'Precipitation effect' on altitudinal gradients. *Water, Air and Soil Pollution* 188: 93–109.

48. Dem, S. B. and Cobb, J. M. 2007. Pesticide residues in soil and water from four cotton growing areas of Mali, West Africa. *Journal of Agricultural, Food, and Environmental Sciences* 1: 1–9.

49. Wong, F., Robson, M., Diamond, M. L., and Truong, J. 2009. Concentrations and chiral signatures of POPs in soils and sediments: A comparative urban versus rural study in Canada and UK. *Chemosphere* 74: 404–411.

50. Aigner, E. J., Leone, A. D., and Falconer, R. L. 1998. Concentrations and enantiomeric ratios of organo-chlorine pesticides in soils from the U.S. corn belt. *Environmental Science & Technology* 32: 1162–1168.

51. Dimond, J. B. and Owen, R. B. 1996. Long-term residue of DDT compounds in forest soils in Maine. *Environmental Pollution* 92: 227–230.

52. Kannan, K., Battula, S., Loganathan, B. G., Hong, C.-S., Lam, W. H., Villeneuve, D. L., Sajwan, K., Giesy, J. P., and Aldous, K. M. 2003. Trace organic contaminants, including toxaphene and trifluralin, in cotton field soils from Georgia and South Carolina, USA. *Archives of Environmental Contamination and Toxicology* 45: 30–36.

53. Environmental Protection Agency (EPA). *Pilot Survey of Levels of PCDD/PCDDFs, PCBs, and Mercury in Rural Soils of the United States, EPA/600/R-05/048F.* United States Environmental Protection Agency, National Center for Environmental Assessment, Office of Research and Development, Washington, DC., 2007.

54. Gedik, K. and Imamoglu, I. 2010. An assessment of the spatial distribution of polychlorinated biphenyl contamination in Turkey. *Clean* 38: 117–128.

55. Bozlaker, A., Odabasi, M., and Muezzinoglu, A. 2008. Dry deposition and soil-air gas exchange of poly-chlorinated biphenyls (PCBs) in an industrial area. *Environmental Pollution* 156: 784–793.

56. Kumar, K. S., Priya, M., Sajwan, K., Kolli, R., and Roots, O. 2009. Residues of persistent organic pollut-ants in Estonia soil (1964–2006). *Estonian Journal of Earth Sciences* 58: 109–123.

57. Holoubek, I., Dusek, L., Sanka, M., Hofman, J., Cupr, P., Jarkovsky, J., Zbiral, J., and Klanova, J. 2009. Soil burdens of persistent organic pollutants—Their levels, fate and risk. Part 1. Variation of concentra-tion ranges according to different soil uses and locations. *Environmental Pollution* 157: 3207–3217.

58. Milukaite, A., Klanova, J., Rimselyte, I., and Kvietkus, K. 2008. Persistent organic pollutants in Lithuania: Assessment of air and soil contamination. *Lithuanian Journal of Physics* 48: 357–366.

59. Ruzikova, P., Klanova, J., Cupr, P., Lammnl, G., and Holoubek, I. 2008. An assessment of air-soil exchange of polychlorinated biphenyls and organochlorine pesticides across Central and Southern Europe. *Environmental Science & Technology* 42: 179–185.

60. Tremolada, P., Parolini, M., Binelli, A., Ballabio, C., Comolli R., and Provini, A. 2009. Preferential reten-tion of POPs on the northern aspect of mountains. *Environmental Pollution* 157: 3298–3307.

61. Nadal, M., Schuhmacher, M., and Domingo, J. L. 2007. Levels of metals, PCBs, PCNs, and PAHs in soils of a highly industrialized chemical/petrochemical area: Temporal trend. *Chemosphere* 66: 267–276.

62. Heywood, E., Wright, J., Wienburg, C. L., Black, H. I. J., Long, S. M., Osborn, D., and Spurgeon, D. J. 2006. Factors influencing the national distribution of polycyclic aromatic hydrocarbons and polychlorinated biphenyls in British soils. *Environmental Science & Technology* 40: 7629–7635.

63. Kirchner, M., Faus-Kessler, T., Jakobi, G., Levy, W., Henkelmann, B., Bernhoft, S., Simoncic, P., Uhl, M., Weiss, P., and Schramm, K.-W. 2009. Vertical distribution of organochlorine pesticides in humus along Alpine altitudinal profiles in relation to ambiental parameters. *Environmental Pollution* 157: 3238–3247.

64. Meijer, S. N., Ockenden, W. A., Sweetman, A., Breivik, K., Grimalt, J. O., and Jones, K. C. 2003. Global distribution and budget of PCBs and HCB in background surface soils: Implication for sources and environmental processes. *Environmental Science & Technology* 37: 667–672.

65. Klanova, J., Matykiewiczova, N., Macka, Z., Prosek, P., Laska, K., and Klan, P. 2008. Persistent organic pollutants in soils and sediments from James Ross Island, Antarctica. *Environmental Pollution* 152: 416–423.

66. WHO, Environmental Health Criteria 9, DDT and its derivatives. Geneva, Switzerland, 1979.

67. WHO, Environmental Health Criteria 123, *Alpha-* and *beta*-hexachlorcyclohexane. Geneva, Switzerland, 1991.

68. WHO, Environmental Health Criteria 124, Lindane. Geneva, Switzerland, 1991.

69. Menzie, C. M. 1972. Fate of pesticides in the environment. *Annual Review of Entomology* 17: 199–222.

70. Loganathan, B. C., Tanabe, S., Tanaka, H., Watanabe, S., Miyazaki, N., Amano, M., and Tatsukawa, R. 1990. Comparison of organochlorine residue levels in the striped dolphin from western North Pacific, 1978–79 and 1986. *Marine Pollution Bulletin* 21: 435–439.

71. Kannan, K., Falandysz, J., Yamashita, N., Tanabe, S., and Tatsukawa, R. 1992. Temporal trends of organochlorine concentrations in cod-liver oil from the southern Baltic proper, 1971–1989. *Marine Pollution Bulletin* 24: 358–363.

72. Alcock, R. E., Johnston, A. E., McGranth, S. P., Berrow, M. L., and Jones, K. C. 1993. Long-term changes in the polychlorinated biphenyls content of United Kingdom soils. *Environmental Science & Technology* 27: 1918–1923.

73. Loganathan, B. G., Tanabe, S., Hidaka, Y., Kawano, M., Hidaka, H., and Tatsukawa, R. 1993. Temporal trends of persistent organochlorine residues in human adipose tissue from Japan, 1928–1985. *Environmental Pollution* 81: 31–39.

74. Boul, H. L., Garnham, M. L., Hucker, D., Baird, D., and Aislabie, J. 1994. The influence of agricultural practices on the levels of DDT and its residues in soil. *Environmental Science & Technology* 28: 1397–1402.

75. Neary, D. G., Bush, P. B., and Michael, J. L. 2009. Fate, dissipation and environmental effects of pesticides in southern forests: A review of a decade of research progress. *Environmental Toxicology and Chemistry* 12: 411–428.

76. Haraguchi, K., Koizumi, A., Inoue, K., Harada, K. H., Hitomi, T., Minata, M., Tanabe, M., Kato, Y., Nishimura, E., Yamamoto, Y., Watanabe, T., Takenaka, K., Uehara, S., Yang, H. R., Kim, M. Y., Moon, C. S., Kim, H. S., Wang, P., Liu, A., and Hung, N. N. 2009. Levels and regional trends of persistent organochlorines and polybrominated diphenyl ethers in Asian breast milk demonstrate POPs signatures unique to individual countries. *Environment International* 35: 1072–1079.

77. Wiener, J. G. and Sandheinrich, M. B. 2010. Contaminants in the upper Mississipi River: Historical trends, responses to regulatory controls, and emerging concerns. *Hydrobiologia* 640: 49–70.

78. Loganathan, B. C. and Kannan, K. 1991. Time perspectives of organochlorine contaminants in the global environment. *Marine Pollution Bulletin* 22: 582–584.

79. Kodavanti, P. R. S., Kumar, K. S., and Loganathan, B. G. 2008. Organohalogen pollutants and human health. *International Encyclopedia of Public Health* 1: 686–693.

80. Cheah, U.-B., Kirkwood, R. C., and Lum, K. Y. 1998. Degradation of four commonly used pesticides in Malaysian agricultural soils. *Journal of Agricultural and Food Chemistry* 46: 1217–1223.

81. Sjoeib, F., Anwar, M. S., and Tungguldihardjo, M. S. 1994. Behaviour of DDT and DDE in Indonesian tropical environments. *Journal of Environmental Science and Health Part B* 29: 17–24.

82. Hussain, A., Maqbool, U., and Asi, M. 1994. Studies on dissipation and degradation of ^{14}C-DDT and ^{14}C-DDE in Pakistani soils under field conditions. *Journal of Environmental Science and Health Part B* 29: 1–15.

83. Agarwal, H. C., Singh, D. K., and Sharma, V. B. 1994. Persistence, metabolism and binding of *p,p′*-DDT in soil in Delhi, India. *Journal of Environmental Science and Health Part B* 29: 73–86.

84. Rajukkannu, K., Basha, A. A., and Habeebullah, B. 1985. Degradation and persistence of DDT, HCH carbaryl and malathion in soils. *Indian Journal of Environmental Health* 27: 237–243.

85. Kaushik, C. P. 1991. Persistence and metabolism of HCH and DDT in soil under subtropical conditions. *Soil Biology & Biochemistry* 23: 131–134.

86. Xu, B., Jiany, G., Yongxi, Z., and Haibo, L. 1994. Behaviour of DDT in Chinese tropical soils. *Journal of Environmental Science and Health Part B* 29: 37–46.

87. Varca, L. M. and Magallona, E. D. 1994. Dissipation and degradation of DDT and DDE in Philippine soil under field conditions. *Journal of Environmental Science and Health Part B* 29: 25–35.

88. Helling, C. S., Engelke, B. F., and Doherty, M. A. 1994. DDT dissipation in Hawaiian in-situ soil columns. *Journal of Environmental Science and Health Part B* 29: 103–119.

89. Crowe, A. S. and Smith, J. E. 2007. Distribution and persistence of DDT in soil at a sand dune-marsh environment: Point Pelee, Ontario, Canada. *Canadian Journal of Soil Science* 87: 315–327.

90. Ware, G. W., Estesen, B. J., Buck, N. A., and Cahill, W. P. 1978. DDT moratorium in Arizona, Agricultural residues after seven years. *Pesticide Monitoring Journal* 12: 1–3.

91. Ware, G. W., Estesen, B. J., and Cahill, W. P. 1974. DDT moratorium in Arizona: Agricultural residues after four years. *Pesticide Monitoring Journal* 8: 98–101.

92. Andrea, M. M., Mello, M. H. S. H., Luchini, L. C., Tomita, R. Y., Mesquita, T. B., and Musumeci, M. R. 1994. Dissipation and degradation of DDT, DDE and parathion in Brazilian soils. *Journal of Environmental Science and Health Part B* 29: 121–132.

93. Pereyra, M. A., Mantecon, J. D., and Barassi, C. A. 1998. Persistence of lindane and heptachlor in a representative soil from Balcarce, Argentina, under natural environmental conditions. *Journal of Environmental Biology* 19: 1–7.

94. Lalah, J. O., Kaig Wara, P. N., Getenga, Z., Mghenyi, J. M., and Wandiga, S. O. 2001. The major environmental factors that influence rapid disappearance of pesticides from tropical soils in Kenya. *Toxicological and Environmental Chemistry* 81: 161–197.

95. Sleicher, C. A. and Hopcraft, J. 1984. Persistence of pesticides in surface soil and relation to sublimation. *Environmental Science & Technology* 18: 514–518.

96. Stephens, J., Maeda, D. N., Ngowi, A. V., Moshi, A. O., Mushy, P., and Mausa, E. 1994. Dissipation and degradation of ^{14}C p,p'-DDT and ^{14}C p,p'-DDE in Tanzanian soils under field conditions. *Journal of Environmental Science and Health Part B* 29: 65–71.

97. Ayris, S. and Harrad, S. 1999. The fate and persistence of polychlorinated biphenyls in soil. *Journal of Environmental Monitoring* 1: 395–401.

98. Martijn, A., Bakker, H., and Schreuder, R. H. 1993. Soil persistent of DDT, dieldrin, and lindane over a long period. *Bulletin of Environmental Contamination and Toxicology* 51: 178–184.

99. Farmer, W. J., Igue, K., Spencer, W. F., and Martin, J. P. 1972. Volatility of organochlorine insecticides from soil: I. Effects of concentration, temperature, air flow rates, and vapor pressure. *Soil Society of America Journal* 36: 443–447.

100. Spencer, W. F. and Cliath, M. M. 1972. Volatility of DDT and related compounds. *Journal of Agricultural and Food Chemistry* 20: 645–649.

101. FAO/IAEA, 1994. Appraisal of overall programme accomplishments. *Journal of Environmental Science and Health Part B* 29: 205–226.

102. Taylor, A. 1978. Post-application volatilization of pesticides under field conditions. *Journal of Air Pollution Control Association* 28: 922–927.

103. Voutsas, E., Vavva, C., Magoulas, K., and Tassios, D. 2005. Estimation of the volatilization of organic compounds from soil surfaces. *Chemosphere* 58: 751–758.

104. Nash, R. G. 1983. Comparative volatilization and dissipation rates of several pesticides from soil. *Journal of Agricultural and Food Chemistry* 31: 210–217.

105. Guenzi, W. D. and Beard, W. E. 1970. Volatilization of lindane and DDT from soils. *Soil Society of America Journal* 34: 443–447.

106. Porter, L. K. and Beard, W. E. 1968. Retention and volatilization of lindane and DDT in the presence of organic colloids isolated from soils and Leonardite. *Journal of Agricultural and Food Chemistry* 16: 344–347.

107. Harner, T., Wideman, J. L., Jantunen, L. M. M., Bidleman, T. F., and Parkhurst, W. J. 1999. Residues of organochlorine pesticides in Alabama soils. *Environmental Pollution* 106: 323–332.

108. Wong, F., Alegria, H. A., Jantunen, L. M., Bidleman, T. F., Salvador-Figueroa, M., Gold-Bouchot, G., Ceja-Moreno, V., Waliszewski, S. W., and Infanzon, R. 2008. Organochlorine pesticides in soils and air of southern Mexico: Chemical profiles and potential for soil emissions. *Atmospheric Environment* 42: 7737–7745.

109. Chu, W., Chan, K. H., Kwan, C. Y., and Jafvert, C. T. 2005. Acceleration and quenching of the photolysis of PCB in the presence of surfactant and humic materials. *Environmental Science & Technology* 39: 9211–9216.

110. Jury, W. A., Farmer, W. J., and Spencer, W. F. 1984. Behavior assessment model for trace organics in soil: II. Chemical classification and parameter sensitivity. *Journal of Environmental Quality* 13: 567–572.

111. Bowman, M. C., Schechter, M. S., and Carter, R. L. 1965. Behavior of chlorinated insecticides in a broad spectrum of soil types. *Journal of Agricultural and Food Chemistry* 13: 360–365.

112. Riekerk, H. and Gessel, S. P. 1968. The movement of DDT in forest soil. *Soil Society of America Journal* 32: 595–596.

113. Guenzi, W. D., Beard, W. E., and Viets, F. G. Jr. 1971. Influence of soil treatment on persistence of six chlorinated hydrocarbon insecticides in the field. *Soil Society of America Journal* 35: 910–913.

114. Bailey, G. W. and White, J. L. 1970. Factors influencing the adsorption, desorption, and movement of pesticides in soil. *Residue Reviews* 32: 29–92.

115. Cornelissen, G., Hassell, K. A., van Noort, P. C. M., Kraaij, R., van Ekeren, P. J., Dijkema, C., de Jager, P. A., and Govers, H. A. J. 2000. Slow desorption of PCBs and chlorobenzenes from soils and sediments: Relations with sorbent and sorbate characteristics. *Environmental Pollution* 108: 69–80.

116. Champion, D. F. and Olsen, S. R. 1971. Adsorption of DDT on solid particles. *Soil Society of America Journal* 35: 887–891.

117. Cheah, U. B., Kirkwood, R. C., and Lum, K. Y. 1997. Adsorption, desorption and mobility of four commonly used pesticides in Malaysian agricultural soils. *Pesticide Science* 50: 53–63.

118. Concha-Grana, E., Turnes-Carou, M. I., Muniategui-Lorenzo, S., Lopez-Mahia, P., Parada-Rodriguez, D., and Fernandez-Fernandez, E. 2006. Evaluation of HCH isomers and metabolites in soils, leachates, river water and sediments of a highly contaminated area. *Chemosphere* 64: 588–595.

119. Ballard, T. M. 1971. Role of humic carrier substances in DDT movement through forest soil. *Soil Society of America Journal* 35: 145–147.

120. Moeckel, C., Nizzetto, L., Guardo, A. D., Steinnes, E., Freppaz, M., Filippa, G., Camporini, P., Benner, J., and Jones, K. C. 2008. Persistent organic pollutants in Boreal and Montane soil profiles: Distribution, evidence of processes and implications for global cycling. *Environmental Science & Technology* 42: 8374–8380.

121. Alexander, M. 1965. Persistence and biological reactions of pesticides in soils. *Soil Science Society of America Journal* 29: 1–7.

122. Guenzi, W. D. and Beard, W. E. 1967. Anaerobic biodegradation of DDT to DDD in soil. *Science* 156: 1116–1117.

123. Matsumura, F. 1973. Degradation of pesticides residues in the environment. In: *Environmental Pollution by Pesticides*. Ed. Edwards, C. A. Plenum Publishing Co., London, U.K., pp. 494–513.

124. Edwards, C. A. 1973. Pesticide residues in soil and water. In: *Environmental Pollution by Pesticides*. Ed. Edwards, C. A. Plenum Publishing Co., London, U.K., pp. 409–458.

125. Menzie, C. M. 1969. *Metabolism of Pesticides*. Bureau of Sport Fisheries and Wildlife, Special Scientific Report Wildlife, No.127, Washington, DC.

126. Miglioranza, K. S. B., Gonzalez Sagrario, M. D. L. A., Aizpun de Moreno, J. E., Moreno, V. J., Escalante, A. H., and Osterrieth, M. L. 2002. Agricultural soil as a potential source of input of organochlorine pesticides into a nearby pond. *Environmental Science and Pollution Research* 9: 250–256.

127. Ramesh, A., Tanabe, S., Murase, H., Subramanian, A. N., and Tatsukawa, R. 1991. Distribution and behaviour of persistent organochlorine insecticides in paddy soil and sediments in the tropical environment: A case study in South India. *Environmental Pollution* 74: 293–307.

128. Wada, H., Senoo, K., and Takai, Y. 1989. Rapid degradation of gamma-HCH in upland soil after multiple application. *Soil Science and Plant Nutrition* 35: 71–77.

129. Middeldorp, P. J. M., van Doesburg, W., Schraa, G., and Stams, A. J. M. 2005. Reductive dechlorination of hexachlorocyclohexane (HCH) isomers in soil under anaerobic conditions. *Biodegradation* 16: 283–290.

130. Muhlbachova, G. 2008. Potential of the soil microbial biomass C to tolerate and degrade persistent organic pollutants. *Soil and Water Research* 3: 12–20.

131. Alder, A. C., Haggblom, M. M., Oppenheimer, S. R., and Young, L. Y. 1993. Reductive dechlorination of polychlorinated biphenyls in anaerobic sediments. *Environmental Science & Technology* 27: 530–538.

132. WHO, Environmental Health Criteria 140, *Polychlorinated Biphenyls and Terphenyls*. Geneva, Switzerland, 1992.

133. Brown, J. F. Jr., Bedard, D. L., Brennan, M. J., Carnahan, J. C., Feng, H., and Wagner, R. E. 1987. Polychlorinated biphenyls dechlorination in aquatic sediments. *Science* 236: 709–712.

Part III

Persistent Organic Chemicals in Europe and Africa

13 Contamination Profiles and Possible Trends of Organohalogen Compounds in the Estonian Environment and Biota

Ott Roots,* Vladimir Zitko, Kurunthachalam Senthil Kumar,
Kenneth Sajwan, and Bommanna G. Loganathan

CONTENTS

13.1 Introduction 307
13.2 History of Persistent Organic Pollutants in Estonia 308
13.3 Sources of Contamination of POPs and Levels 309
13.4 Chemical Analysis 311
    13.4.1 Air Analysis 311
    13.4.2 Soil Analysis 312
    13.4.3 Biota Analysis 313
13.5 Atmospheric Evaluation of POPs 314
    13.5.1 POPs Levels in Air 314
13.6 Persistent Organic Pollutants Levels in Soil 320
    13.6.1 Soil Sampling Locations and Contamination Levels 320
13.7 Persistent Organic Pollutants in Biota 323
    13.7.1 Contamination Levels in Fish 323
13.8 Temporal Trends 326
13.9 Conclusions and Recommendations 330
References 331

13.1 INTRODUCTION

The Estonian Republic is one of the smallest countries in the European Union, with a total land area of 45,227 km^2 and a population of 1.34 million in 2009. Geographically, Estonia is located in northeastern Europe (latitude 57°30′N–59°49′N and longitude 21°46′E–28°13′E) on the east coast of the Baltic Sea. Estonia borders Latvia (339 km) and the Russian Federation (294 km), with a total border length of 633 km. Much of the Estonian land consists of forest and agriculture lands. In particular, half of the land surface is covered by forests (ca. 47%), one-third is agriculture (cropland 28% and pastures 7%), around 2% is under settlements, and the rest of the territory is covered by mires and

* E-mail: ott.roots@klab.ee (Chapter corresponding author).

bogs. There are about 1450 natural and man-made lakes in Estonia (6.1% of the country's territory). The largest industrial and energy enterprises are located in the northeast region of Estonia (Baltic and Estonian thermal power plants, Kunda Nordic Cement, Viru Keemia Grupp AS, etc.).

13.2 HISTORY OF PERSISTENT ORGANIC POLLUTANTS IN ESTONIA

Although Estonia has been bestowed with a significant area of forest and agriculture, organochlorine pesticides (OCPs) were never produced in Estonia [1]. The pesticides applied to the cropland and forests were imported from other countries. The import of OCPs was banned in Estonia beginning in October 21, 1967. Earlier, in 1957, 226 ton of pesticides were used, mainly DDT, hexachloran and, to a lesser degree seed dressing products [1]. After the ban, the quantity of the pesticide used declined sharply (Table 13.1). In addition, major pollution problems in Estonia during the late 1990s were associated with the Soviet Army. Despite the withdrawal of Russian Army from Estonia on August 31, 1994, the damage and pollution remained [2]. In Estonia, the concentration of persistent organic pollutants (POPs) in the ecological system of the Baltic Sea has been studied since 1974. Estonian contribution to mass loadings of toxic contaminants into the Baltic Sea has been reported elsewhere [3,4–22]. The POPs were found to be present in air, water, plankton, fish, algae, molluscs, and seals [4,5,10,23–26]. Since 1994, the analyses of hazardous substances originating from the Estonian environment were conducted as part of the Estonian National Environmental Monitoring Programme [27]. Currently, the movement of some POPs (e.g., PCBs) from southern and southwestern sources outside of Estonia is highly significant [5,15,28,29].

In this chapter, contamination levels of various persistent organic pollutants in environmental and biological samples collected from in and around Estonia are presented. The target compounds included are polychlorinated biphenyls (PCBs), polychlorinated naphthalenes (PCNs),

TABLE 13.1
Chlorinated Plant Protection Products Use (Ton) in Estonia 1960–1970

Plant, Protection, Product	1960	1961	1962	1963	1964	1965	1966	1967	1968	1969	1970
DDT[a] (DDT 10% P[b])	0	154	172	180	196	235	109	81	73	22	0
Hexachlorane* (Hexachlorane 25% P)	0	189	211	204	202	186	200	144	129	17	0
Keltane* (Dicofol 20% E[c])	0	0	0	0	0	0	0	0	1.3	1.2	2.7
Polychloro camphene* (Toksafen 50% E)	0	0	0	0	0	0	0	2.0	0.8	0.4	0
Sulfonate-ester* (Clor-fenson 30% WRP[d])	0	0	3	0	15	8.1	1.5	3.8	3.0	0.10	0.70
Hexachlorobenzol* (Hexachlorobenzol 30% WRP)	0	0	0	0	0	0	0	0	0	0	2.0

Source: Based on Roots, O. and Sweetman, A., *Oil Shale*, 24(3), 483, 2007. With permission.

[a] Active substance.

[b] P, paste.

[c] E, concentrate of the emulsion.

[d] WRP, water-retaining power.

polychlorinated *n*-alkanes (PCAs), polychlorinated terphenyls (PCTs), polybrominated diphenyl ethers (PBDEs), and perfluorinated chemicals (PFCs). In addition, persistent chemicals such as dibenzo-*p*-dioxins, dibenzofurans, and polycyclic aromatic hydrocarbons (PAHs) formed during combustion or incinerations of waste materials were also investigated.

13.3 SOURCES OF CONTAMINATION OF POPs AND LEVELS

Exposure of the average population in Estonia to chlorinated compounds seems lower than in most of western Europe, and current pesticide use is very low [30]. Recently, Roots and Sweetman [15] reported the occurrence of several persistent organic pollutants in air samples from two Estonian monitoring stations. The occurrence of PCBs and PBDEs in Estonian food and fish was also reported [21,31,32]. Recent studies also revealed the presence of PCDD/Fs, DL-PCBs, and PBDEs in various samples [8,9,11,12,14,21,22,31–37].

Concentrations of total PCBs in oil shale and fly ash were 8.6–9.0 and 1.1–2.4 µg/kg, respectively [9]. However, when compared to Finland, Russia, and western European countries, Estonian PCB levels were very low according to passive air samplers deployed at 23 background locations along a broadly west–east transect in eight northern European countries (Ireland, United Kingdom, Denmark, Norway, Sweden, Finland, Estonia, and Russia) for 8–11 weeks during August–October 2004 [15,37–40]. PCB, DDT, PBDEs, and PAH concentrations at the Lahemaa station in Estonia were lower than those in Ireland, Norway, Finland, and Sweden. The highest concentrations were recorded in Russia, Denmark, and the United Kingdom [40]. Olsson et al. [41] documented the lowest PCB concentrations along the Estonian coast and in northern Bothnian Bay. The highest levels of PCBs were measured at urban sites in Russia, France, Italy, Sweden, the United Kingdom, Eastern Europe, Croatia, Hungary, and Estonia (at the Kohtla-Järve site the sum of 29 PCBs was 790 pg/m^3, 99 ng/sample). For most samples, the levels for the sum of 29 PCBs ranged from 2.5 to 280 ng/sample [38,39]. Concentrations of PCBs in soils at the Kohtla-Järve areas (Northeast Estonia) were 0.43–81 ng/g dw [42,43]. These results suggest that part of the PCBs in soil arose from oil shale burning. Effluents from the Järve biological sewage purification station at Kohtla-Järve were analyzed for PCBs in 2006 [15,44]. The PCB concentration in two of three sewage water samples (95 and 103 µg/L), taken from the effluents discharged to the Gulf of Finland from the Järve purification station, exceeded the Estonian PCB concentration limit for sewage waters (50 µg/L). Consequently, PCBs derived from oil shale burning contribute to contamination of the neighborhoods.

The Estonian thermal power station is the world's largest that burns low-grade local oil shale. In preparation of a potential measuring campaign, filter ash and raw oil shale dust samples were analyzed. If dioxins were formed during the combustion of oil shale, then they should be found in such filter ash samples. The analyses revealed that the PCDD/F concentrations were near the lower end of the range covered by PCDD/F filter dust contents found in samples from German hard coal and brown coal combustion plants (0.3–21 ng I-TEQ/kg). The observed filter dust concentrations corresponded with flue gas concentrations of well below 0.1 ng I-TEQ/m^3. Therefore, the results obtained for the oil shale samples indicate neither a considerable input of polychlorinated organic compounds nor their formation and emission during the combustion process.

At the beginning of the 2000s, Denmark, a partner of Estonia, was interested in detailed air pollution data from the Estonian oil shale region. In March 2003, dioxin, PAHs, and naphthalene emissions were measured from a shale oil–producing plant located near the city of Narva in Estonia. The Danish Environment Assistance to Eastern Europe (DANCEE) sponsored such a project and dk-TEKNIK ENERGY & ENVIRONMENT (now FORCE Technology) was responsible for measurements, which were conducted in cooperation with the Estonian Environmental Research Centre in Tallinn. The analytical results for PAHs were handled and presented according to a regulation in the Danish Air Emission Guidelines [45]. This regulation distinguishes between naphthalene and the rest of the 16 EPA (Environmental Protection Agency) PAHs originally selected by the US

TABLE 13.2
PAH and Naphthalene from Oil Plant

Parameter	Unit	Danish Emission Limit Value	Measured Values
Naphthalene	Mg/m^3 (s,d, 6% O$_2$)	300	7.5
PAH	Mg B[a]P – TEQ/m^3 (n,t, 6% O$_2$)	5.0	1.5

Source: Data from Schleicher, O. et al., *Organohalogen Compd.*, 66, 1665, 2004a.

EPA (United States Environmental Protection Agency) and is now widely used internationally in connection with the characterization and assessment of PAH mixtures. The naphthalene and other PAH concentrations were much lower than the Danish emission limit values (Table 13.2). All the dioxin emission concentrations from the oil shale plant were below the EU emission limit value for MSWI at 0.1 ng I-TEQ/m^3. The total emission of dioxins was estimated as 0.2 mg/year for air, and as 700 mg/year including ash [16,17]. The total annual dioxin emission with the ash is based on these measurements and is considered to be very close to zero. The total annual dioxin emission from the oil-shale-fired by Estonian and Baltic power plants into the air is estimated at 160–300 mg I-TEQ [9,16,17,46], which is 10 times lower than previous estimates [47]. Overall, the total annual dioxin emissions with the ash is considered to be zero. However, due to unstable combustion conditions depending up on seasons, it could be higher and needs continuous monitoring [9,16,17,46].

With the financial support of the European Commission, the DG Environment, a project "Dioxin in Candidate Countries" was carried out in the beginning of the 2000s [48]. Overall, 30 industrial facilities located in 10 of 13 candidate countries were proposed for PCDD/F emission measurements. Among these were a number of potentially relevant emission sources such as cement works (Estonia, Cyprus, Lithuania), etc. Results of the measurements carried out in Estonia are shown in Table 13.3. Based on these results, it is unlikely that the oil shale combustion plant in Estonia is a relevant emission source for dioxins and furans [9,49]. The project has been discussed with experts [45] and it has been estimated that the oil shale power plants are not the main sources on a European level, but may be for Estonia.

Based on the available data, it seems that the general widespread exposure to POPs from the Estonian environment is lower than in other European countries [50,51]. The lower exposure of the general population is indicated by lower concentrations of PCDD/Fs and PCBs in milk fat (Table 13.4). Principal component analysis (PCA) of cancer incidences may indicate interesting differences among Nordic, Baltic countries and Russia [11,52]. It remains to be seen whether the suggestions indicated by PCA could be substantiated and related to some environmental or sociological factors. When cancer incidence rates in females and males are considered together, it is Hodgkin's disease and colon cancer that are responsible for the "east versus west" differences. For females, the difference between Estonia, Latvia, and Russia, and between Finland and Iceland are likely

TABLE 13.3
Results of the Measurements Carried Out in Estonia

Plant	PCDD/F Concentration (ng I-TEQ/m^3)	Annual PCDD/F Release (mg I-TEQ/Year)	Emission Factor (µg I-TEQ/Ton)
Cement works	0.018	47	0.07

Source: Data from Quass, U. et al., *Organohalogen Compd.*, 66, 878, 2004.

TABLE 13.4
PCDD/Fs and Marker PCBs (Sum of IUPAC 28, 52, 101, 138, 153 and 180) Concentrations in Human Milk in 1993

Country	PCDD/F (pgTEQ/g Fat)		Marker PCBs (ng/g Fat)	
	Urban Area	Rural Area	Urban Area	Rural Area
Austria	10.7	10.9	381	303
Belgium	26.6	20.8	261	276
Czech Republic	18.4	12.1	1069	532
Croatia	13.5	8.4	220	218
Denmark	15.2	—	209	—
Finland	21.5	12	189	134
Estonia	14.4	12.4	103	136
Germany	16.5	—	375	—
Lithuania	13.3	14.4	322	287
The Netherlands	22.4	—	253	—
Norway	10.1	9.3	273	265
Slovakia	15.1	12.6	1015	489
Spain	25.5	19.4	452	461
United Kingdom	17.9	15.2	130	131
St. Petersburg region of Russia	16.4	—	190	—

Sources: World Health Organization (WHO). Levels of PCBs, PCDDs and PCDFs in human milk: Second round of WHO-coordinated exposure study, WHO, Copenhagen, 1996. Environmental Health in Europe, no 3. EUR/ICP EHPMO2 03 05; Tuomisto, J. and Hagmar, L. *Scand. J. Work Environ. Health*, 25(suppl 3), 65, 1999.

caused by different incidences of lymphoid leukemia in these countries. In contrast to the incidence of cancer in females, the incidence of many different cancers contributes to the differences between "west" and "east" for cancer in males [11,52].

13.4 CHEMICAL ANALYSIS

13.4.1 Air Analysis

The determination of PAHs, PCDD/Fs, DL-PCBs, PBDEs, organochlorine pesticides, and PCBs was carried out according to standard operating procedures at Lancaster University, U.K. [38–40]. The samples were analyzed by gas chromatography–mass spectrometry (GC-MS) with an EI source operating in selected ion mode (SIM) for PCBs and organochlorine pesticides. PBDEs were analyzed separately with a Thermo Trace GC-MS system operated by negative chemical ionization in SIM mode with ammonia as the reagent gas. PAHs were analyzed with a Perkin-Elmer HPLC system and LC250 binary pump, LS40 fluorescence detector, and a ISS200 Auto sampler. Appropriate quality assurance/quality control (QA/QC) analysis was performed [40].

The determination of POPs was done according to standard operating procedures at Masaryk University, Czech Republic [53–55]. All samples were extracted with dichloromethane using a Büchi System B-811 automatic extractor. One laboratory blank and one reference material were analyzed with each set of 10 samples. Surrogate recovery standards PCB 30 and PCB 185 for PCB analyses were spiked on each filter prior to extraction. PCB 121 was used as an internal standard for both PCB and OCP analyses. The volume was reduced after extraction under a gentle nitrogen stream at ambient temperature and fractionation was achieved on a silica gel column. A sulfuric-acid-modified silica gel column was used for PCB/OCPs samples. Samples were analyzed using an HP 5890 gas chromatograph with an electron capture detector and a Quadrex-fused silica column 5% phenyl [53,54].

Quality assurance/quality control: Recoveries were determined for all samples by spiking with the surrogate standards prior to extraction. Amounts were similar to detected quantities of analytes in the samples. Recoveries were higher than 76% for all PCBs. Recovery factors were not applied to any of the data. Recovery of analytes measured in the reference material varied from 88% to 103% for PCBs and from 75% to 98% for OCPs. Laboratory blanks were below the detection limits for selected compounds. Field blanks consisted of pre-extracted polyurethane foam disks were taken at each sampling site. These disks were extracted and analyzed in the same way as the samples, and the levels in the field blanks never exceeded 3% of detected quantities of PCB nor 1% of OCPs respectively. Consequently, minimal contamination occurred during the transport, storage, and analysis [53,54].

For some of the HELCOM countries, the level of PCDD/F emissions in 2006 was higher than those of 1990. In particular, Latvia and Lithuania reported higher values of emissions for 2006 in comparison with the emissions for 1990. In 2006, the total annual PCDD/F emissions of HELCOM countries amounted to 1.4 kg TEQ. Among the HELCOM countries, the largest contributions to the total annual PCDD/F belonged to Russia (54%) followed by Poland (32%) and Germany (6%). Annual emissions of dioxins and furans decreased in HELCOM countries during the period from 1990 to 2006 by 22%. The most significant drop of PCDD/F emissions was noted for Denmark (63%), Finland (60%), and Estonia (53%). Some decrease of emissions were also noted for Sweden (37%), Germany (26%), Russia (22%), and Poland (15%) [78].

13.4.2 Soil Analysis

Polychlorinated biphenyl (PCB) and polybrominated diphenyl ether (PBDE) congeners and organochlorine pesticides (OCPs) were measured in soil by approved procedures [22,33,37,42,43,53,54]. Approximately 10 g of dry soil was Soxhlet extracted by a 3:1 v/v dichloromethane and hexane mixture for 17 h [42,43]. The extract was then concentrated to 10 mL by the Rapid Vap Labconco Evaporation System (Model 79100; The Pump Works Inc. Sanford, FL) and the solvent was changed to hexane. The extract was concentrated to 5 mL using a stream of nitrogen gas to evaporate the solvent and then further subjected to silica gel column chromatography to remove interfering organic and polar species and to separate the PCBs from the pesticides. In the first fraction (F1), PCB, 4,4'-DDE, HCB, and *trans*-nonachlor were eluted using 120 mL of ultra-pure hexane. The second fraction (F2), containing most of the chlorinated pesticides and PBDEs, was eluted with 100 mL of 20% methylene chloride in hexane. The F1 was concentrated using a Rapid Vap apparatus to 10 mL, followed by evaporation to 1 mL, and then micro-concentrated to 100 μL by a gentle stream of nitrogen gas. The extract was transferred to an auto sampler vial and 1 μL was injected onto a gas chromatograph equipped with an electron capture detector (GC-ECD) [42,43]. Twenty-six PCB congeners, 11 PBDE congeners, and several chlorinated pesticides (HCB, γ-HCH, aldrin, dieldrin, mirex, heptachlor, heptachlor epoxide, *trans*-nonachlor, *cis*-chlordane, 2,4'-DDE, 4,4'-DDE, 2,4'-DDD, 4,4'-DDD, 2,4'-DDT, and 4,4'-DDT) were measured by a Varian model CP-3380 gas chromatograph (GC) with the Varian model CP-8410 auto injector (Varian, Inc. Palo Alto, CA). The GC was equipped with a DB-5 (60 m × 0.25 mm; 0.25 μm film thickness) capillary column (J&W Scientific, USA) and a 63 Ni electron capture detector. The initial temperature of the column was 90°C with a 1 min hold time and increased at the rate of 5°C/min–150°C, and was ramped at the rate of 2°C/min–280°C and held for 20 min. The injector and detector temperatures were 270°C and 330°C, respectively. Helium (1.5 mL/min) and nitrogen (28.5 mL/min) were used as carrier and makeup gases respectively. The standard reference material SRM 2262, obtained from the

National Institute of Standards and Technology (NIST, USA), was used for the quantification of PCB congeners, and SRM 2261 from the same source was used to quantify chlorinated pesticides.

The PCB and PBDE congeners and organochlorine pesticides were identified in the sample extract by comparing the retention time from the standard mixture and quantified by their response factors. Appropriate QA/QC analysis was performed, including reagent blank (analyte concentrations were below the method detection limit), calibration curve with the r^2 value of 0.99, surrogate (4,4'-dibromooctafluorobiphenyl) recovery, and matrix spike recovery of $100\% \pm 30\%$. The chemical analysis of selected soil samples were carried out at the Murray State University Chemical Services Laboratory [42,43].

The measurement of PCDD/Fs in the 2003 samples was done according to the standard operating procedure of the National Research Centre for Environment and Health (GSF), Institute of Ecological Chemistry, Germany [33]. Prior to extraction, the sample was spiked with $^{13}C_{12}$-labeled internal standards (Cambridge Isotope Laboratories, Woburn, MA). The samples (15–20 g) were extracted by pressurized fluid (n-hexane and acetone 75:25 (v/v)) in a Dionex ASE extractor. The concentrated crude extract was cleaned up by several sequential liquid chromatography steps. The solvents used were of trace analysis quality. The glass columns were 250 mm long with an inner diameter of 24 mm. Elution flow rates were about 0.1 mL/s. All columns were topped with anhydrous sodium sulfate. The series of clean-up columns were as follows: 25 g alumina, basic, super active, 50–200 μm, filled wet with toluene and eluted with 80 mL toluene, then 200 mL n-hexane/dichloromethane (98/2 v/v). A third fraction of 200 mL n-hexane/dichloromethane (50/50 v/v) contained the PCDD/F congeners. This fraction was evaporated (1000–550 mbar, 343 K) to a volume of 2–3 mL.

The identification and quantification was done with a capillary HRGC/HRMS system. For isomer-specific detection, the Rtx2330 (Restek) polar capillary column (60 m, 0.25 mm ID, 0.1 lm df) was used. The PCDD/Fs were determined by mass spectrometry in the EI mode by tracing their M+, (M+2)$^+$ ions or the most intensive ions of the isotope cluster. The measurement was conducted with high-resolution Thermoquest Finnigan MAT 95S instrument with 10,000 resolution [33].

The determination of POPs was done according to standard operating procedures at the Masaryk University, Czech Republic [22,37,53–55].

13.4.3 Biota Analysis

The determination of PCDD/Fs, DL-PCBs, and other-PCBs from 2002 to 2005 was carried out by the National Research Center for Environment and Health (GSF), Institute of Ecological Chemistry, under quality control according to EN 17025. Fish samples were analyzed for PCDD/F and DL-PCBs, except for the fish collected in 2002, which were analyzed for PCDD/Fs only. The fish samples were freeze-dried and homogenized before accelerated solvent extraction. Cleanup consisted of chromatography on sandwich alumina and florisil columns. The analyses were performed by HRGC/HRMS [8,12,13,19,20]. The tetra to octa PCDD/Fs and the tetra to hepta PCBs were identified and quantified.

The evaluated results are presented according to both WHO-TEF1998 and the recently proposed WHO-TEF 2005 [56]. The concentrations of PCDD/Fs and dioxin-like PCBs are upper-bound values [57], which mean that nondetected congeners are included in the sum by setting the concentrations of the nondetected congeners to their limits of quantification [20].

The effect of season on length, weight, and concentration of lipids and dioxins was analyzed using one-way ANOVA. Prior to the analysis, Bartlett's test was used to check the assumption of homoscedasticity, that is, the equal variation of data or equal variance. Linear regression analyses were used to describe the relationships between the population metrics (i.e., fish length, weight, age) and the concentration of lipids and the pollutants [19].

The determination of PCDD/PCDFs, DL-PCBs, other-PCBs and PBDEs, from 2006 to 2009 was done according to standard operation procedures in the chemistry laboratory at the National Public Health Institute (KTL) [58–62].

The determination of PCBs, HCBs, HCH, DDTs, and other POPs, from 1995 to 2009, was done at the Estonian Environmental Research Centre [7,14,18,26,35,44]. Triangular coordinates, calculated by software of Graham and Midgley [63], were used to visualize the patterns of the DDT group. PCA reduced the dimensions of the chlorobiphenyl group. The PCA was performed by PLS_Toolbox 2 [64], running in MATLAB® 5.0 (The Mathworks, Inc., www.mathworks.com).

13.5 ATMOSPHERIC EVALUATION OF POPs

13.5.1 POPs Levels in Air

The atmosphere is an important contributor of anthropogenic matter to the land and marine ecosystems. Long-range atmospheric transport (intercontinental) of these compounds is possible due to their hydrophobicity and high volatility. Due to their slow rates of chemical, photochemical, and biological degradation, the persistent organic pollutants (such as PCBs, HCB, DDT, γ-HCH, PBDEs, PCNs, etc.) [25,29,37–40,53,54,65,66] provide excellent model compounds in which to study atmospheric transport processes. Passive air sampling is a cheap screening method for comparison of contamination at various sites or for verification of information obtained by active samplers. During the last few years in Estonia attention has been drawn to the most toxic and persistent pollutants and to new pollutants such as PCNs and PBDEs, which were analyzed in Estonian air for the first time. Passive air samplers were successfully applied as a tool for POPs monitoring on the global [38–40,65,66] and regional [15,22,29,37,53–55,67] basis. The relationship between the amount of POPs captured on a polyurethane foam filter and their concentrations in the sampled air has not been fully characterized mathematically. Thus, only empirically estimated information (e.g., based on parallel active and passive measurements) is available to evaluate and interpret the results. Sampling rates of 3.5 m^3/day were determined by empirical measurements, resulting in approximately 100 m^3 for a 28-day sampling cycle [53–55]. As a result, in this study we adopted a passive air sampler for air sample collection. The samples were collected using uniform techniques and were analyzed using similar lab methods [5] for comparability.

Passive disk air samplers containing polyurethane foam disks (housed in the rotation chambers) were employed in the global and regional air pollution studies [15,22,37–40,55,65].

The concentration of pollutants above the Baltic Sea increased with higher south-west winds [7,28,32,36,37]. This endangers the Estonian islands of Saaremaa (Vilsandi) and Hiiumaa since these winds are prevailing on the Baltic Sea. This was the reason why, at the beginning of the 1990s, we were increasingly interested in the long-range transportation of POPs to Estonia. According to the order from 21.10.1967, approved by the Government, the import of organochlorine plant protection products was banned in Estonia [1]. In 1990–1994, PUFs (27 × 40 mm of colorless polyester, density 30 kg/m^3) were used as collection media. The project's coordinator was Lund University in Sweden [29,67]. About 1000 m^3 of air was drawn through the PUF by means of an electric pump at a rate of 30–50 L/min (registered by a flowmeter). The PUFs were replaced with new ones every 14 days. The PUFs were analyzed following the method described by Bremle et al. [68].

In October 1990, a field study was initiated by Lund University in order to determine PCBs (in total 51 identified chlorobiphenyls), DDTs, and HCHs in air and precipitation at 16 sampling stations in the vicinity of the Baltic Sea (Tables 13.5 through 13.7). Sampling was carried out continuously over 1 year. The median concentrations in the air samples for all stations were 57 pg/m^3 for PCBs, 1.6 pg/m^3 for DDTs (sum of pp′-DDT and pp′-DDE), and 25 pg/m^3 for the HCHs (sum of α-HCH and γ-HCH). The station in Latvia (Salaspils) showed the highest values of PCBs and DDTs in air with a median of 454 pg/m^3 of PCBs and 12 pg/m^3 of DDTs. Highest median concentrations of HCHs were

TABLE 13.5
Summary of the PCB Concentrations and Calculated Depositions at the Stations

Latitude	Station	Air (pg/m)	Precipitation (ng/L)	Deposition (ng/m²/Day)
54 00'	Dziwnow (Poland)	55 ($n=5$)	1.4 ($n=2$)	2.3 ($n=2$)
54 15'	Swibno (Poland)	69 ($n=6$)	4.4 ($n=4$)	5.0 ($n=4$)
55 25'	Ventes R. (Lithuania)	61 ($n=10$)	2.0 ($n=15$)	3.7 ($n=15$)
56 14'	Öland (Sweden)	76 ($n=21$)	8.3 ($n=15$)	3.5 ($n=15$)
56 17'	Breanäs (Sweden)	79 ($n=21$)	2.8 ($n=12$)	2.8 ($n=12$)
56 50'	Salaspils (Latvia)	454 ($n=20$)	11 ($n=15$)	18 ($n=15$)
58 20'	Vilsandi (Estonia)	79 ($n=9$)	1.5 ($n=9$)	2.2 ($n=9$)
58 21'	Gotska s. (Sweden)	60 ($n=24$)	2.0 ($n=15$)	3.0 ($n=15$)
59 17'	Stockholm (Sweden)	80 ($n=21$)	1.3 ($n=10$)	2.4 ($n=10$)
59 30'	Lahemaa (Estonia)	49 ($n=16$)	0.8 ($n=12$)	1.8 ($n=12$)
63 02'	Vasa (Finland)	32 ($n=27$)	0.9 ($n=12$)	1.2 ($n=12$)
63 03'	Docksta (Sweden)	50 ($n=24$)	1.8 ($n=15$)	2.6 ($n=15$)
63 32'	Norrbyn (Sweden)	48 ($n=24$)	1.8 ($n=17$)	3.2 ($n=14$)
63 36'	Holmögadd (Sweden)	57 ($n=23$)	4.9 ($n=12$)	5.7 ($n=12$)
64 31'	Bjuröklubb (Sweden)	38 ($n=24$)	2.9 ($n=13$)	2.2 ($n=13$)
65 44'	Kalix (Sweden)	47 ($n=24$)	2.4 ($n=14$)	1.5 ($n=14$)
	All stations	57 ($n=299$)	2.3 ($n=192$)	2.7 ($n=192$)

Source: Data from Agrell, G. et al., Atmospheric and river input of PCBs, DDts and HCHs to the Baltic Sea, in *A System Analysis of the Baltic Sea*, F. Wulff, L. Rahm, and P. Larsson, eds., Springer, Berlin, Germany, 2001, pp. 149–175.

found in two Polish stations (Swibno 103 pg/m³ and Dziwnow 72 pg/m³). These stations had high DDTs concentrations too, correspondingly 6 and 9 pg/m³ [29]. The PCB and pesticides measurements in 1993–1994 were made by the Lund University on samples taken near the Gulf of Riga in the five Baltic air-research stations in Estonia (Vilsandi-Saaremaa and Tahkuse, Western Estonia) and Latvia (Salaspils, Salacgriva, and Slitere) (Figure 13.1). The results showed that the air and rain water samples taken at the Estonian stations were relatively cleaner compared with the samples taken in Latvia [25,29,67]. The results were further treated by PCA and it was shown that the PCB in air samples near the Riga City (Latvia) consisted of congeners present in the original industrial PCB mixture. The sources for POPs in the atmosphere over Saaremaa-Vilsandi were thus not situated in the nearby areas (Nordic, 1999). This refers to either the long-range transportation at or the local waste center situated near the Gulf of Riga. Geometric mean concentrations of organochlorine pesticides were α-HCH 5–17 pg/m³, γ-HCH 0.3–4 pg/m³, and HCB 10–38 pg/m³ [67]. According to the data of Agrell et al. [29], at the beginning of the 1990s the rivers and the atmosphere contributed about equally in PCB load, while for the pesticides the atmospheric deposition was about five to seven times more important [29,67].

In 2002, 71 samplers were successfully deployed in 22 countries. The project coordinator was the Lancaster University Environmental Chemistry and Ecotoxicology Group from the United Kingdom. Local volunteers were given guidance on the choice of deployment location. Precleaned and weighed PUF disks (14 cm diameter 1.35 cm thick; density 0.0213 g/cm³) were used. The PUFs were changed after 6 weeks (June 15–July 30, 2002) and analyzed by methods presented by Jaward et al. [38,39] and Gioia et al. [40]. In 2002, samplers were deployed at remote, rural, and urban locations in 22 countries (71 stations), among them were two Estonian stations: Lahemaa-background EMEP station and the Kohtla-Järve industrial (oil shale chemistry) region

TABLE 13.6
Summary of the DDT Concentrations and Calculated Depositions at the Stations

Latitude	Station	Air (pg/m)	Precipitation (ng/L)	Deposition (ng/m²/Day)
54 00′	Dziwnow	9.0 ($n=5$)	0.21 ($n=2$)	0.30 ($n=2$)
54 15′	Swibno	6.3 ($n=6$)	1.24 ($n=3$)	1.0 ($n=3$)
55 25′	Ventes R.	2.3 ($n=10$)	0.18 ($n=15$)	0.38 ($n=15$)
56 14′	Öland	5.1 ($n=21$)	0.71 ($n=13$)	0.38 ($n=13$)
56 17′	Breanäs	3.3 ($n=20$)	0.17 ($n=12$)	0.19 ($n=12$)
56 50′	Salaspils	12 ($n=20$)	0.40 ($n=15$)	0.64 ($n=15$)
58 20′	Vilsandi	6.9 ($n=8$)	0.28 ($n=5$)	0.23 ($n=5$)
58 21′	Gotska s.	2.0 ($n=24$)	0.15 ($n=15$)	0.19 ($n=15$)
59 17′	Stockholms s.	2.0 ($n=21$)	0.09 ($n=10$)	0.12 ($n=10$)
59 30′	Lahemaa	2.0 ($n=16$)	0.06 ($n=12$)	0.1 ($n=12$)
63 02′	Vasa	0.8 ($n=26$)	0.03 ($n=12$)	0.05 ($n=12$)
63 03′	Docksta	1.2 ($n=24$)	0.08 ($n=15$)	0.08 ($n=15$)
63 32′	Norrbyn	0.9 ($n=24$)	0.07 ($n=16$)	0.09 ($n=16$)
63 36′	Holmögadd	1.2 ($n=22$)	0.18 ($n=8$)	0.16 ($n=8$)
64 31′	Bjuröklubb	0.7 ($n=22$)	0.04 ($n=11$)	0.05 ($n=11$)
65 44′	Kalix	0.9 ($n=24$)	0.07 ($n=14$)	0.05 ($n=14$)
	All stations	1.6 ($n=281$)	0.13 ($n=178$)	0.15 ($n=178$)

Source: Data from Agrell, G. et al., Atmospheric and river input of PCBs, DDts and HCHs to the Baltic Sea, in *A System Analysis of the Baltic Sea*, F. Wulff, L. Rahm, and P. Larsson, eds., Springer, Berlin, Germany, 2001, pp. 149–175.

station. This paper presents ambient air data for a range of PCBs, polycyclic aromatic hydrocarbons (PAHs), HCB, HCH, pp′DDE, pp′DDT and PBDEs, in Estonia. Results were presented in three publications [38–40]. Estonia was very interested in the data on PAHs and PBDEs, because these two toxic groups of compounds were measured in Estonian air samples for the first time (Tables 13.8 and 13.9).

The geographical pattern of all compounds reflected the suspected regional emission patterns and highlighted localized hotspots [38–40]. A model monitoring network in the Czech Republic has been functional since 2005 and a passive air sampling survey of the Central and Eastern Europe (CEE) was initiated in 2006. The design of this study was synchronized with the Czech Republic passive air monitoring network (MONET_CZ), which provides continuous data. CEE partners joined the project in March 2006. The RECETOX of Masaryk University in Brno, Czech Republic, has completed a large CEE air and soil sampling campaign within the framework of the project "Pilot study for development of the monitoring network in the Central and Eastern Europe (MONET_CEEC)." Samplers were deployed and samples were collected at remote, rural, and urban locations (58 stations) in eight countries (Czech Republic, Bosnia and Herzegovina, Estonia, Latvia, Lithuania, Romania, Serbia, Slovakia), five of which are Estonian stations—Tallinn (Kopli), Muuga Port, Lahemaa, Kunda, and Kohtla-Järve—among them (http://monet-ceec.eu). This paper presents data on a range of α-HCH, β-HCH, γ-HCH, δ-HCH, and (p,p′DDE; p,p′DDD; p,p′DDT; o,p′DDE; o,p′DDD; o,p′DDT) determined in ambient air and soil samples collected at Estonian stations (Table 13.10). Long-range atmospheric transport (LRAT) clearly dominates OCPs atmospheric input into Estonia [15,22,29,37]. Based on the results of the MONET_CEEC campaign, the Lahemaa background station seems to be an appropriate candidate for continuous background monitoring of persistent organic pollutants (POPs) [53,54].

TABLE 13.7
Summary of the HCH Concentrations and Calculated Depositions at the Stations

Latitude	Station	Air (pg/m)	Precipitation (ng/L)	Deposition (ng/m²/Day)
54 00′	Dziwnow	72 ($n=5$)	0.63 ($n=2$)	1.4 ($n=2$)
54 15′	Swibno	103 ($n=6$)	8.65 ($n=3$)	5.7 ($n=3$)
55 25′	Ventes R.	26 ($n=10$)	1.63 ($n=15$)	3.2 ($n=15$)
56 14′	Öland	20 ($n=21$)	2.5 ($n=13$)	0.98 ($n=13$)
56 17′	Breanäs	45 ($n=21$)	1.8 ($n=12$)	1.9 ($n=12$)
56 50′	Salaspils	39 ($n=20$)	1.3 ($n=15$)	2.5 ($n=15$)
58 20′	Vilsandi	33 ($n=28$)	2.1 ($n=5$)	3.7 ($n=5$)
58 21′	Gotska s.	45 ($n=24$)	1.4 ($n=15$)	2.2 ($n=15$)
59 17′	Stockholms s.	24 ($n=21$)	1.0 ($n=10$)	1.3 ($n=10$)
59 30′	Lahemaa	26 ($n=16$)	0.31 ($n=12$)	0.53 ($n=12$)
63 02′	Vasa	30 ($n=16$)	0.38 ($n=12$)	1.3 ($n=12$)
63 03′	Docksta	18 ($n=24$)	0.92 ($n=15$)	1.7 ($n=15$)
63 32′	Norrbyn	7 ($n=24$)	0.16 ($n=17$)	0.61 ($n=17$)
63 36′	Holmögadd	20 ($n=23$)	1.3 ($n=8$)	0.82 ($n=8$)
64 31′	Bjuröklubb	28 ($n=15$)	0.46 ($n=10$)	0.22 ($n=10$)
65 44′	Kalix	4 ($n=21$)	0.33 ($n=14$)	0.16 ($n=14$)
	All stations	25 ($n=275$)	1.0 ($n=178$)	1.3 ($n=178$)

Source: Data from Agrell, G. et al., Atmospheric and river input of PCBs, DDts and HCHs to the Baltic Sea, in *A System Analysis of the Baltic Sea*, F. Wulff, L. Rahm, and P. Larsson, eds., Springer, Berlin, Germany, 2001, pp. 149–175.

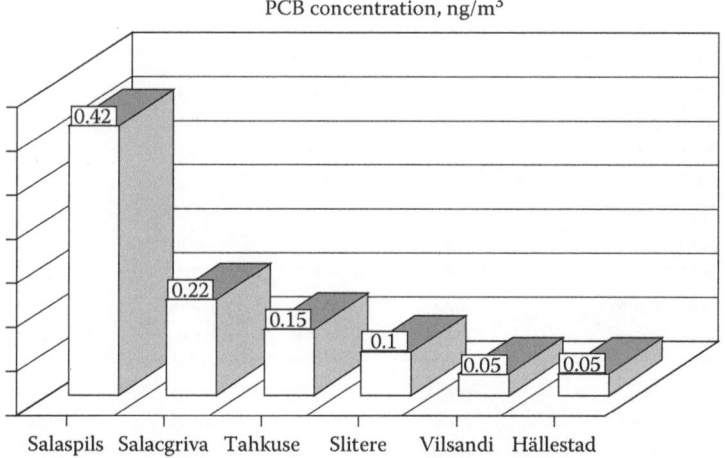

FIGURE 13.1 Atmospheric concentrations of PCB along the route from the town of Riga-Salaspils (Latvia) to the island of Saaremaa-Vilsandi (Estonia) in 1993–1994 (geometric means for the monitoring stations). Levels recorded at a Swedish station (Hällestad), Estonian station (Tahkuse), and Latvian stations (Salacgriva and Slitere) are shown for comparison. (Data from Nordic, Environmental research programme for 1993–1997//Final report and self-evaluation, Nordic Council of Ministers, *TemaNord Environ.*, 548, 135.)

TABLE 13.8
Persistent Organic Pollutants Concentrations Measured in Estonian Lahemaa and Kohtla-Järve Air Monitoring Stations

	Estonia			
	Kohtla-Järve	Lahemaa	Kohtla-Järve	Lahemaa
Chemicals	ng/Sample		pg m³	
PCB 18	10	<0.48	81	nd
PCB 22	5.1	<0.58	41	nd
PCB 28	13	<0.64	100	nd
PCB 31	10	0.55	80	4.4
PCB 41/64	3.6	0.23	29	1.9
PCB 44	4.9	0.35	39	2.8
PCB 49	3.8	0.23	30	1.9
PCB 52	5.9	0.55	47	4.4
PCB 60/56	4.1	0.26	32	2.1
PCB 70	5.7	0.45	45	3.6
PCB 74	3.0	0.20	24	1.6
PCB 87	1.8	0.23	14	1.8
PCB 90/101	4.0	0.81	32	6.4
PCB 95	3.0	0.60	24	4.8
PCB 99	2.2	0.28	18	2.2
PCB 105	1.0	0.14	8.2	1.1
PCB 110	3.6	0.64	29	5.1
PCB 118	2.9	0.40	23	3.1
PCB 123	0.03	<0.03	0.27	nd
PCB 138	2.5	0.68	20	5.4
PCB 141	0.48	0.17	3.8	1.4
PCB 149	2.3	0.83	18	6.6
PCB 151	0.78	0.35	6.2	2.8
PCB 153/132	3.2	0.92	25	7.3
PCB 158	0.25	0.05	2.0	0.40
PCB 170	0.20	<0.03	1.6	nd
PCB 174	0.45	0.19	3.6	1.5
PCB 180	0.63	0.11	5.0	0.90
PCB 183	0.32	0.08	2.6	0.66
PCB 187	0.63	0.22	5.0	1.8
Sum 29 PCBs	99	10	787	83
a-HCH	5.2	2.2	41	17
g-HCH	3.8	1.9	30	15
a-chlordane	0.23	0.11	1.8	0.85
g-chlordane	<0.05	<0.05	nd	nd
HCB	4.0	2.9	31	23
o,p′-DDD	0.63	<0.05	5.0	nd
p,p′-DDE	0.62	0.07	4.9	0.55
p,p′-DDD	0.28	<0.05	2.2	nd
p,p′-DDT	2.1	0.40	17	3.1
p,p′-DDE/p,p′-DDT	0.29	0.17	2.3	1.4
a-HCH/g-HCH	1.4	1.1	11	9.0
PBDE 28	0.14	0.07	1.1	0.55
PBDE 47	1.8	1.5	14	12

TABLE 13.8 (continued)
Persistent Organic Pollutants Concentrations Measured in Estonian Lahemaa and Kohtla-Järve Air Monitoring Stations

	Estonia			
	Kohtla-Järve	Lahemaa	Kohtla-Järve	Lahemaa
Chemicals	ng/Sample		pg m^3	
PBDE 49	0.09	<0.06	0.68	nd
PBDE 75	0.07	<0.06	0.53	nd
PBDE 99	<1.47	3.2	nd	25
PBDE 100	<0.29	0.56	nd	4.4
PBDE 153	<0.09	0.36	nd	2.9
PBDE 154	<0.1	0.26	nd	2.1
Sum PBDEs	3.0	6.0	24	48

Source: Roots, O. and Sweetman, A., *Oil Shale*, 24, 483, 2007. With permission.

TABLE 13.9
Polycyclic Aromatic Hydrocarbons (ng m^3) in Two Estonian Stations

PAHs	Kohtla-Järve	Lahemaa
Naphthalene	1.1	0.73
2-methylnaphthalene	0.80	0.48
1-methylnaphthalene	0.48	0.31
Biphenyl	1.5	0.75
2,6-dimethylnaphthalene	0.33	0.17
Acenaphthylene	0.14	0.02
Acenaphthene	0.20	0.07
2,3,6-trimethylnapthalene	0.29	0.05
Fluorene	0.94	0.17
Phenanthrene	7.0	0.81
Anthracene	0.21	0.02
1-methylphenanthrene	1.1	0.12
Fluoranthene	2.5	0.29
Pyrene	2.2	0.16
Benzo(a)anthracene	0.23	0.01
Chrysene	0.53	0.04
Benzo(b)fluoranthene	0.15	0.01
Benzo(k)fluoranthene	0.12	0.01
Benzo(e)pyrene	0.16	0.01
Benzo(a)pyrene	0.06	<0.005
Perylene	0.01	<0.005
Indeno(123-cd)pyrene	0.04	<0.005
Dibenz(ah)anthracene	0.01	<0.005

Source: Roots, O. and Sweetman, A., *Oil Shale*, 24, 483, 2007. With permission.

TABLE 13.10

Temporal Variations of Organochlorine Pesticides in Ambient Air (ng/Filter) in Five Estonian Stations

Sampling Date/ Compound	Sampling Locations				
	Tallinn	Muuga Port	Lahemaa	Kunda	Kohtla
Sampling date	21.03.06– 19.04.06	21.03.06– 19.04.06	21.03.06– 19.04.06	21.03.06– 19.04.06	21.03.06– 19.04.06
Sum of HCH	19.5	7.3	6.2	13.4	26.4
Sum of DDT	3.3	1.1	1.3	1.8	3.5
Sampling date	19.04.06– 17.05.06	19.04.06– 17.05.06	19.04.06– 17.05.06	19.04.06– 17.05.06	19.04.06– 17.05.06
Sum of HCH	34.7	24.6	10.5	16.7	11.2
Sum of DDT	3.3	2.6	0.7	0.6	2.1
Sampling date	17.05.06– 12.07.06	17.05.06– 12.07.06	17.05.06– 12.07.06	17.05.06– 12.07.06	17.05.06– 12.07.06
Sum of HCH	12	4.6	4.2	1.4	4.5
Sum of DDT	2.3	1.5	0.5	0.9	1.8
Sampling date	12.07.06– 08.08.06	12.07.06– 08.08.06	12.07.06– 08.08.06	12.07.06– 08.08.06	12.07.06– 08.08.06
Sum of HCH	9	63.5	4.2	4.8	9.5
Sum of DDT	2.5	1.8	0.8	0.7	1.4

Sources: Klanova, J. et al., Application of passive sampler for monitoring of POPs in ambient air, Part II: Pilot study for development of the monitoring network in the Central and Eastern Europe (MONET_CEEC) 2006, RECETOX-TOCOEN REPORTS, 319, 1, 2007; Roots, O., *Ecol. Chem.*, 17: 44, 2008; Roots, O. et al., *Environ. Sci. Pollut.* 17, 740, 2010. With permission.

13.6 PERSISTENT ORGANIC POLLUTANTS LEVELS IN SOIL

13.6.1 Soil Sampling Locations and Contamination Levels

Soil samples were collected from the sites of the deployed air sampler stations. In order to reduce the number (and costs) of samples for analysis, soil samples were mainly pooled. A soil plot was divided into 10×10 m subplots [69]. The humus layer was sampled separately with a steel cylinder of known diameter (15 cm). The sample included only the 0–10 cm topsoil layer. About 6–10 small subsamples (approximately 0.3–0.5 kg) from randomly placed subplot points were taken for one composite sample. The aboveground plant and litter material were excluded. All subsamples were placed in one black plastic bag, mixed, labeled, and transported to the laboratory within 24 h. There, the samples were dried (at room temperature on freeze-dried filter paper on trays), thoroughly mixed, and sieved through a 2 mm mesh filter to prepare the material for chemical analyses [42,43,53,69].

Soil samples were collected at Ahja, Eerika, Kohtla-Järve, Muuga Port, Kunda, Lahemaa, and Vilsandi Island and archived. The map of Estonia and location of sampling regions are shown in Figure 13.2. The soils of Lahemaa and Vilsandi Island nature preserves may be classified respectively as calcaric cambisols, whose topsoil textures are composed of calcareous sandy loam, and hyperskeletic-endolithic regosols, where the topsoil consists of coarse carbonate sediments. The samples characterizing the Muuga Port area were taken from coastal, poor, gravelly, sandy salic-epigleyic fluvisols, those of Kunda from fine sandy endogleyic-epidystric luvisol, and of Kohtla-Järve from loamy calcaric-endoruptic anthrosol. Soils of the Eerika and Ahja sampling areas are stagnic albeluvisols, with loamy sand topsoil and loamy subsoil [42,43,70]. The Lahemaa and Kohtla-Järve

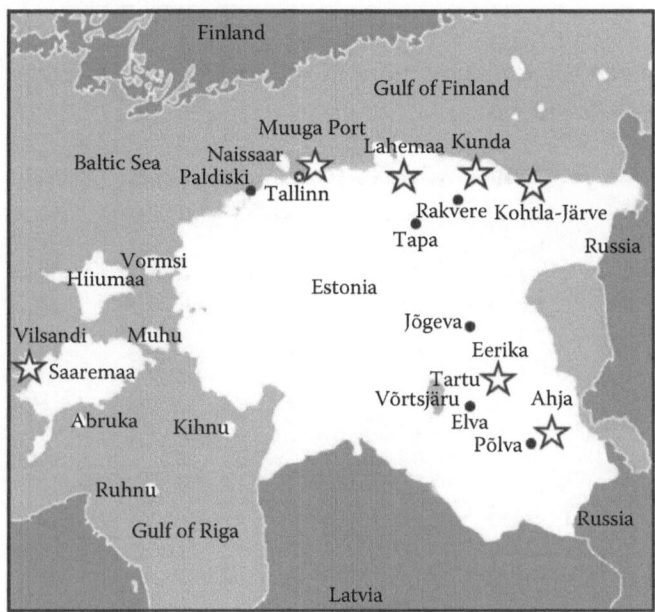

FIGURE 13.2   The Estonian map showing air and soil sampling sites.

sampling areas are situated on till plains with calcareous cover, Vilsandi and Kunda on limestone plains, whereas in Vilsandi the Quaternary cover is very thin and skeletal, but in Kunda relatively thick and sandy. The Muuga area lies on a low, sandy ridge of the North Estonian coastal plain. The Ahja and Eerika areas are situated on a reddish-brown non-calcareous moraine (2.8 m) plain with small variations in elevation. The cultivated massifs of fields are separated by primeval valleys and lowland forest areas [22,42,43,70].

Five sampling sites, including the metropolitan, industrial, urban, suburban, and background sites, were situated in northern Estonia along the shoreline of the Gulf of Finland. The sampling geographical profile was designed in a west to east direction for the region-wide coverage of pollution and wind conditions [22,42,43].

Model monitoring network in the Czech Republic has been functional since 2005 and a passive air sampling survey of the Central and Eastern Europe (CEE) was initiated in 2006. A design of the study was synchronized with the Czech Republic passive air monitoring network (MONET_CZ), which provides continuous data. CEE partners collaborated on the project in March 2006. The RECETOX of Masaryk University in Brno, Czech Republic, completed a large CEE air and soil sampling campaign within the framework of the project entitled "Pilot study for development of the monitoring network in the Central and Eastern Europe (MONET_CEEC)." Samplers were deployed and soil samples were collected at remote, rural, and urban locations (58 stations) in eight countries (Czech Republic, Bosnia and Herzegovina, Estonia, Latvia, Lithuania, Romania, Serbia, Slovakia), five of which are Estonian stations—Tallinn (Kopli), Muuga Port, Lahemaa, Kunda, and Kohtla-Järve—among them. This paper presents data on PCBs (the chlorobiphenyls 28, 52, 101, 118, 153, 138, and 180), HCB, PeCB, α-HCH, β-HCH, γ-HCH, δ-HCH), p,p′DDE, p,p′DDD, p,p′DDT, o,p′DDE, o,p′DDD, o,p′ DDT in air and soil samples collected at Estonian stations [53–55] (Table 13.11).

Various countries banned or restricted p,p′DDT usage through the 1970s–1980s. As expected, the lowest values of DDT were in the north and west of Europe and highest in Russia and Italy. In both cases, low p,p′DDE/p,p′DDT ratios were found, suggestive of a fresh p,p′DDT signal [38,39]. The highest total HCH and DDT concentrations in Estonian monitoring stations were found when the prevailing winds were mainly from the north-eastern or south-eastern directions [37].

TABLE 13.11

PCB and OCP Concentrations in Soil (ng/g) from Different Stations in Estonia

Sampling Site	Tallinn	Muuga Port	Lahemaa	Kunda	Kohtla-Järve
PCB 28	0.2	<LOQ	<LOQ	<LOQ	<LOQ
PCB 52	0.6	0.6	<LOQ	<LOQ	<LOQ
PCB 101	1.1	0.2	<LOQ	<LOQ	0.1
PCB 118	3.6	0.6	<LOQ	<LOQ	0.3
PCB 153	2.4	0.3	0.1	<LOQ	0.2
PCB 138	2.7	0.3	0.1	<LOQ	0.2
PCB 180	1.3	0.1	0.1	<LOQ	<LOQ
Sum of PCBs	12	2.1	0.3	<LOQ	0.8
a-HCH	0.2	0.3	<LOQ	0.3	<LOQ
b-HCH	0.2	0.3	0.1	0.2	0.2
g-HCH	<LOQ	0.5	0.1	<LOQ	0.2
d-HCH	<LOQ	<LOQ	<LOQ	<LOQ	<LOQ
Sum of HCHs	0.4	1.1	0.2	0.5	0.4
o,p'-DDE	0.1	<LOQ	<LOQ	<LOQ	<LOQ
p,p'- DDE	5.7	0.9	0.1	0.3	0.3
o,p'-DDD	0.8	0.5	<LOQ	<LOQ	0.1
p,p'-DDD	4.6	2.1	<LOQ	<LOQ	0.2
o,p'-DDT	<LOQ	0.1	<LOQ	<LOQ	<LOQ
p,p'-DDT	0.8	0.1	<LOQ	<LOQ	<LOQ
Sum of DDTs	11.9	3.6	0.1	0.4	0.5
PeCB	0.1	0.1	<LOQ	<LOQ	0
HCB	0.5	0.1	0.1	<LOQ	0.1

Source: Klanova, J. et al., *RECETOX-TOCOEN REPORTS*, 319, 1, 2007.

Estonia still has no waste incineration facilities which would act as substantial sources of PCDD/Fs. As landfill fires may serve as sources of these compounds, the focus was on the concentrations of PCDD/Fs in soil samples taken in the vicinity of the landfill located at southeast Estonia. Laguja landfill is situated in the rural municipality of Nõo, 35 km from Tartu (Figure 13.2). Concentrations of PCDD/Fs were measured in five soil samples taken in the vicinity of the Laguja landfill in south-east Estonia. Four soil samples were taken in southern, eastern, western, and northern parts not further than 300 m from the landfill, and one sample was taken at the distance of 3 km from the landfill. The PCDD/F concentrations in all soil samples were at background levels (0.13–0.83 pg I-TEQ WHO/g dry weight) (Table 13.12). To maintain this situation, the administrator of the landfill must avoid landfill fires, which are one of the reasons for the generation of PCDD/Fs. The total area of the Laguja landfill is 2 ha. Until 2004, the landfill, which started to operate in 1974, was under the joint administration of AS Ragn Sells and the city government. Now the landfill is closed. If we compare the CDD/F concentrations of the soil samples from Laguja landfill (Table 13.12) with the respective figures obtained in other countries (Table 13.13), it becomes obvious that concentrations in soil samples collected near Laguja landfill are at a background level and that in 1999 only low concentrations were reported from Italy, Spain, and Ireland. However, the database for these countries is very small [71]. As it can be seen from Table 13.13, the concentrations for contaminated soil vary between 332 and 98,000 pg I-TEQ/g per dry weight.

TABLE 13.12
The Moisture Content and Concentrations of Dioxins and Furans

Sample Number	Moisture (%)	Concentration (pg I-TEQ WHO/g per Wet Weight)	Concentration (pg I-TEQ WHO/g per Dry Weight)
Sample 1	18.9	0.29	0.36
Sample 2	17.9	0.22	0.27
Sample 3	17.3	0.11	0.13
Sample 4	18.4	0.21	0.26
Sample 5	26.2	0.61	0.83

Source: Roots, O. et al., *Chemosphere*, 57(5), 337, 2004.

TABLE 13.13
Summary of PCDD/PCDF Concentrations (ngTEQ/kg Dry Matter) in Soils from EU Member States

Country	Any Type	Forest	Pasture	Arable	Rural	Contaminated
Austria		0.01–64	1.6–14			332
Belgium	2.7–8.9				2.1–2.7	
Finland						85,000
Germany	0.1–42	10–30	0.004–30	0.03–25	1.0	30,000
Greece	2.0–45					1,144
Ireland	0.15–8.6	4.8	0.8–13			
Italy	0.057–0.12		0.1–43	1.9–3.1		
Luxembourg	1.8–20	6.0			1.4	
The Netherlands	2.0–55				2.2–17	98,000
Spain	0.63–8.4				0.1–8.4	
Sweden					0.11	11,446
United Kingdom	0.87–87				0.78–20	1,585

Source: Data from Fiedler, H. et al., *Organohalogen Compd.*, 43, 151, 1999.

13.7 PERSISTENT ORGANIC POLLUTANTS IN BIOTA

13.7.1 CONTAMINATION LEVELS IN FISH

The main source of human exposure to persistent organic pollutants is consumption of contaminated food. Therefore, food contamination by potentially hazardous substances is a worldwide public health concern. According to the World Health Organization (WHO), monitoring of persistent organic pollutants (POPs) in food, especially fish, should be carried out worldwide to determine the possible sources of these contaminants in the diet. Bearing in mind the complex hydrographic conditions, and the fact that most Baltic Sea organisms live at the edge of their physiological tolerance range, anthropogenic chemical pollution has to be seen as a further stress factor acting upon the Northern Estonian coastal waters biodiversity. Only during recent decades have we started to gain an advanced understanding of how multiple stressors (e.g., salinity, temperature, hypoxia, and chemical pollution) in combination may affect biota [5,6] and thus biodiversity. For example, accumulation of fat, somatic condition, and growth rate suggested that Baltic herring juveniles have their optimum salinity in 8–12 psu, which is higher than the salinity in most of the nursery grounds in the Baltic Sea [72].

The HELCOM COMBINE Programme is designed to measure levels of contaminants in selected species of biota at specific locations over time in order to detect whether levels are changing in response to the changes in inputs of contaminants to the Baltic Sea. Accordingly, samples of Baltic herring were studied during 1995–2007 to investigate the content of PCBs, HCHs, HCBs, and their distribution in the fish of the Estonian waters of the Baltic Sea. All organochlorine contaminants were measured in muscle tissues of 2–3-year-old female Baltic herring [18]. Baltic herring (*Clupea harengus membras* L.) is the most important fish species in the Baltic and for the Estonian fish processing industry.

Individual fish were collected in the western, middle, and eastern parts of the Gulf of Finland, Gulf of Riga, Open Baltic Sea (Eastern Gotland Basin), Lake Peipsi [8,12,13,18–21], and Estonian River deltas [35] from the catches of commercial fishermen.

All the biota samples were processed prior to analysis. Individual fish were categorized by length, weight, and gender. The otoliths were taken for age determination. Each sample contained from 3 to 60 (for herring and sprat 21–60) specimens; the exact number depended on the size of the fish. The sample number, collection area, fish species, year of sampling, fish number, length (cm), weight (g), sex, maturity of the gonads, age of the fish (years), percent female of the investigated samples and maturity of the gonads, percent dry matter, and percent lipid were reported [8,12,13,18–21,35]. The head, tail fin, and viscera were removed from the fish before analysis. The PCDD/F level in all the fish parts used for human consumption was determined. For perch, the PCDD/F concentration was determined from the back muscle with the skin and scales. Flounder, pike-perch, and perch were collected in the age range of 2–8 years. The total weight and length of all fish were measured to the nearest of 0.1 g and 1 mm respectively. The fish age was determined by the number of hyaline rings with the individual being moved to the next age group on January 1. Herring and sprat were analyzed in the form in which they are sold for human consumption. Fish were cleaned (head, fins, and viscera removed) and pooled. Samples were sealed in polyethylene bags and stored at −20°C until subsequent analysis could be performed.

Baltic herring (*Clupea harengus membras* L.) were selected as a bio-indicator because they can be caught in all parts of the Baltic Sea and are an important commercial species. Fish were caught from the eastern (Kunda, GFE) and western (Muuga, GFW) parts of the Gulf of Finland and Gulf of Riga (Pärnu). The data set contains age, length, weight, and concentrations of lipid (α- and γ-HCH, p,p′-DDE, p,p′-DDD, and p,p′-DDT, HCB), and chlorobiphenyls (CB) 28, 52, 101, 118, 138, 153, and 180, in 1–3 years-old herring [18].

To a certain extent, elevated concentrations of PCBs (Figure 13.3), γ-HCH (Figure 13.4), and HCB (Figure 13.5) in 2001 may have been caused by sampling the Baltic female herring older than 3 years. In 2004–2006, the concentrations of HCB in the muscle tissue of Baltic herring were approximately on the same level. Currently, the movement of some POPs, for example PCBs, from

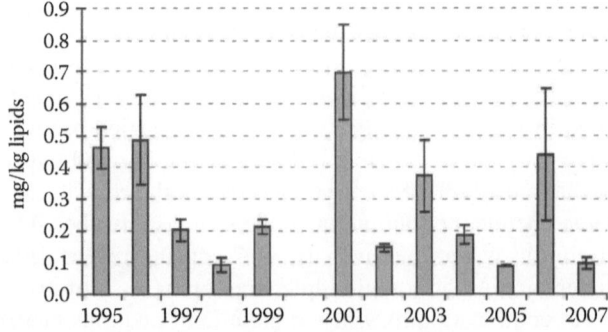

FIGURE 13.3 Concentrations of total PCBs (mg/kg lipid) in the muscle tissue of 2–3-year-old Baltic female herring. (Based on Estonian State Monitoring Programme, http://eelis.ic.envir.ee:88/seireveeb; Lukki, T. et al., *Organohalogen Compd.*, 70, 2098, 2008. With permission.)

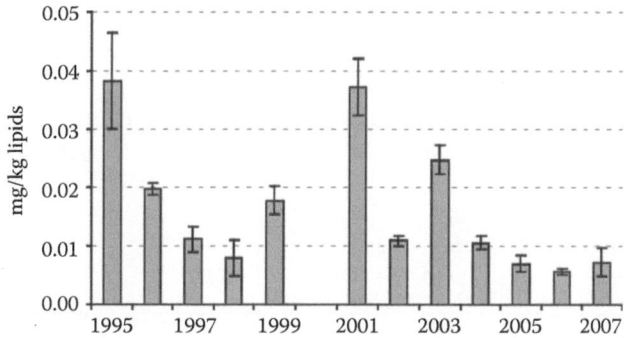

FIGURE 13.4 Concentrations of γ-HCH (mg/kg lipids) in the muscle tissue of 2–3-year-old Baltic female herring. (Based on Estonian State Monitoring Programme, http://eelis.ic.envir.ee:88/seireveeb; Lukki, T. et al., *Organohalogen Compd.*, 70, 2098, 2008. With permission.)

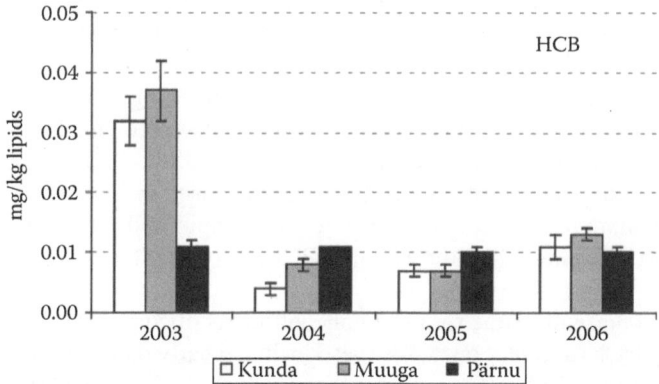

FIGURE 13.5 Concentrations of HCB (mg/kg lipid) in the muscle tissue of 2–3-year-old Baltic female herring. (Based on Estonian State Monitoring Programme, http://eelis.ic.envir.ee:88/seireveeb; Lukki, T. et al., *Organohalogen Compd.*, 70, 2098, 2008. With permission.) Gulf of Finland—eastern (Kunda) and central (Muuga) parts and Gulf of Riga (Pärnu).

remote sources outside Estonia is highly significant. The atmospheric load of pollutants seemed especially interesting since the concentration of PCBs in the atmosphere registered in a previous investigation in the vicinity of Riga (Latvia) was high compared with other regions in the Baltic Sea catchment area [29,67,73,74]. The concentrations of PCBs were high near Riga (0.6 ng/m³, geometric mean), but decreased in a gradient to background values at Saaremaa Island (Estonia) (0.05 ng/m³, geometric mean).

The mean size of herring and the concentrations of lipids and contaminants in years when more than one sample was analyzed were about the same in the three sampling areas, however fish from the western part of the Gulf of Finland were somewhat older and those from the eastern part of the Gulf of Finland were smaller by weight (Table 13.14). For a majority of herring from the eastern part of the Gulf of Finland and from the Gulf of Riga, there was a very good relationship between age and length. However, there was considerable scatter among herring from the western part of the Gulf of Finland. The weights of herring, plotted against age, were considerably more scattered than the length versus age data. There is practically no relationship between lipid concentration and age of herring in the entire data set. The mean contaminant concentrations were similar in all sampling areas except for CB28 and CB52, which were present in much higher concentrations in herring from the Gulf of Riga. The highest concentrations of the CBs 101–153 were in herring from the eastern part of the Gulf of Finland.

TABLE 13.14
Overall Means for Years with >1 Sample

	Gulf of Finland East	Gulf of Finland West	Gulf of Riga
	2003–2006	2002–2006	2002–2006
Age	1.4	1.9	1.4
cm avg	10.5	11.7	11.8
g avg	7.7	10.6	11.2
Lipid	2.08	2.73	2.84

13.8 TEMPORAL TRENDS

The concentrations of all chlorobiphenyls except CB28 and CB52 decreased from 2002 to 2006 in 2-year-old herring from the western part of the Gulf of Finland (Table 13.15). The concentrations of CB28 and CB52 increased in the same time period while those of lipids decreased. The slope of the concentrations of HCB showed a declining trend. A clear pattern of decreasing concentrations is evident in Figures 13.6 through 13.8.

Disregarding the scatter of the concentrations in the intervening years, the decrease in the concentration of p,p'-DDE, HCB, and the HCHs was similar to that predicted by the exponential curves of Bignert et al. [75] in herring from the Baltic Sea (Table 13.16). On the other hand, the concentration of the "industrial" organochlorine compounds increased in herring from the Gulf of Riga. This may indicate a more recent and, possibly, intensified input of these compounds. Currently, the long-range transport of some POPs, for example PCBs, from southern sources outside Estonia is highly significant.

Bignert et al. [75] presented time series of concentrations of compounds measured in this work, in most cases going back to 1978. For PCBs, these included early data obtained on packed columns and the authors devised a method for conversion of these data to values obtained on capillary columns. To avoid potential biases of the conversion, the authors also give time series of concentrations

TABLE 13.15
Gulf of Finland, Western Part, Medians

Year-2002	0.0	1.0	2.0	3.0	4.0	Slope	Intercept	SRSE
cm avg	14	12	11	12	12	−0.4	13	0.25
g avg	17	9.7	8.5	9.7	9.7	−1.5	14	0.95
Lipid	4.2	2.3	1.5	3.3	2.2	−0.3	3.3	1.6
HCB	9.5	34	6.7	7.9	12	−2.1	18	3.3
CB 28	12	6.9	12	5.5	15	0.5	9.1	2.0
CB 52	8.8	12	11	9.1	12	0.4	9.8	0.56
CB 101	31	23	38	15	17	−3.7	32	1.1
CB 118	27	28	35	20	22	−1.8	30	0.61
CB 138	44	44	66	27	35	−3.5	50	1.1
CB 153	43	47	60	26	30	−4.6	51	1.0
CB 180	26	18	16	6.8	2.3	−5.9	26	0.56
a-HCH	7.9	8.6	5.3	3.0	1.9	−1.8	8.9	0.52
g-HCH	14	17	8.2	4.7	4.8	−3.1	16	1.3
p,p'-DDE	64	93	87	36	37	−11	85	1.4
p,p'-DDD	40	75	45	17	19	−10	60	1.7
p,p'-DDT	24	28	17	7.4	15	−4	26	1.6

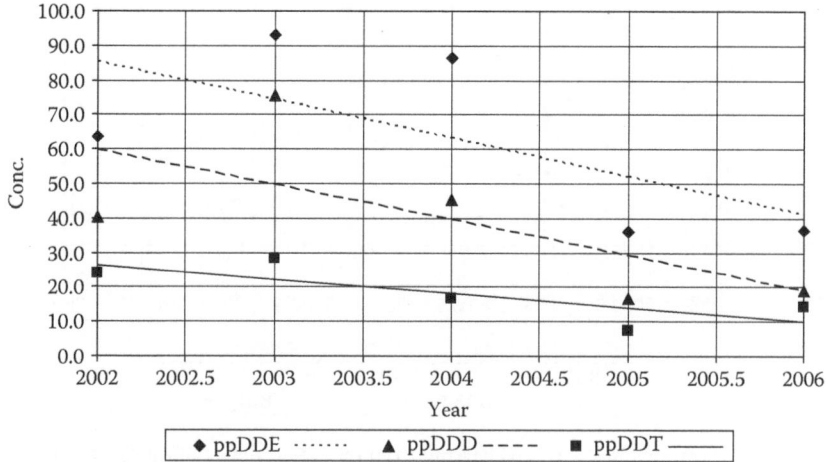

FIGURE 13.6 Concentrations of DDE, DDD, and DDT in 2-year-old herring from the western part of the Gulf of Finland. (Based on Estonian State Monitoring Programme, http://eelis.ic.envir.ee:88/seireveeb. With permission.)

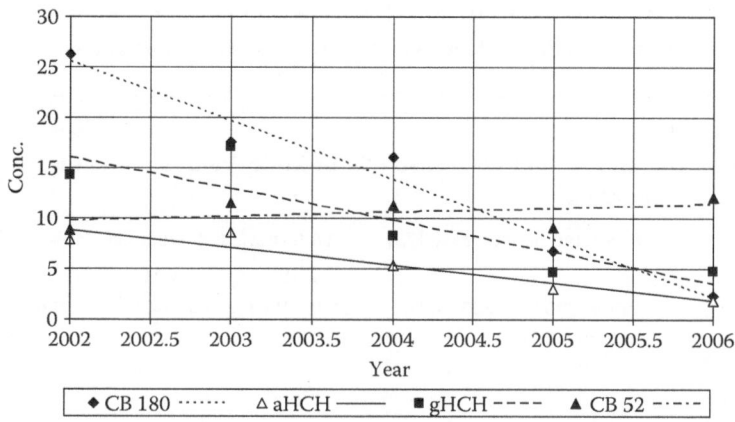

FIGURE 13.7 Concentrations in 2-year-old herring from the western part of the Gulf of Finland. (Based on Estonian State Monitoring Programme, http://eelis.ic.envir.ee:88/seireveeb. With permission.)

of CB153. The present work reports the concentrations of only 7 CB, and this would introduce additional complications in the conversion. Consequently, only the CB153 data were used for comparison with the Bignert et al. data [75] for years slightly overlapping those of the present study (Figure 13.9). It can be seen from Figure 13.10 that there is a slight declining trend in the concentrations of α-HCH. Koistinen et al. [60] reported concentrations of many CBs in small and large herring from Bothnian Bay, Bothnian Sea and the Gulf of Finland in 1999 (Table 13.17). For sake of clarity, these concentrations of CB153 are not included in Figure 13.10. Most of them are considerably higher than those in Figure 13.10 (Table 13.17).

For comparison, the CB profiles of the data of Parmanne et al. [61], Koistinen et al. [60], and Pikkarainen and Parmanne [62], as well as the mean profiles of the current data, and the CB profiles of several PCB preparations by Wyrzykowska et al. [76], Roots et al. [77] are presented in score plots (Figures 13.11 and 13.12). It can be seen from these figures that the profiles of the current CB data from GFW and those reported in the literature are quite similar. On the other hand, those from

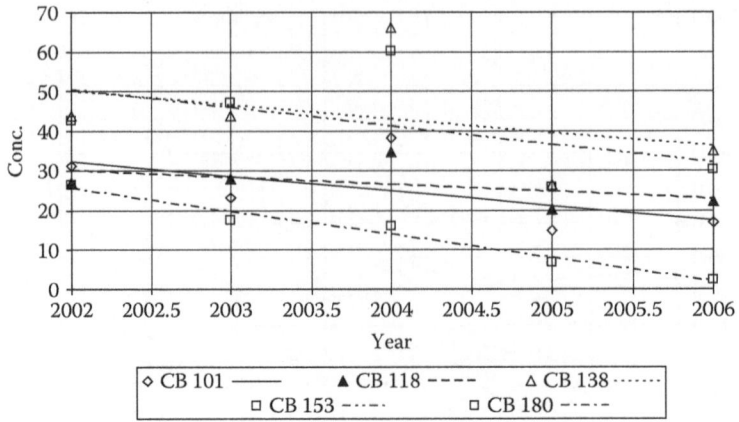

FIGURE 13.8 Concentrations in 2-year-old herring from the western part of the Gulf of Finland. (Based on Estonian State Monitoring Programme, http://eelis.ic.envir.ee:88/seireveeb. With permission.)

GFE and, in particular those from GR contain larger proportions of CB28 and CB52, which are characteristic of the less chlorinated PCB preparations such as Aroclor 1242. This may indicate a more recent input of such preparations into the Gulf of Riga.

Other studies [73,74] showed that Riga and its surroundings are comparatively more polluted areas in Latvia with respect to PCBs. However, the sources are still unknown. POP concentrations

TABLE 13.16
Predicted and Actual Decrease (%) of 2006 Values (Negative Values = Increase)

	Predicted % Decrease [75]		Determined % Decrease (This Work)		
			GFW[b]	GFE	GR
	2006–2001	2006–2001	2006–2001	2006–2001	2006–2002
CB-153[Aa]	26	21	79	92	−26
CB-153[H]	19	16			
DDE[A]	35	29	76	87	38
DDE[H]	35	29			
HCB[A]	34	29	44	39	−20
HCB[H]	14	11			
a-HCH[A]	64	56	66	46	9.0
a-HCH[H]	55	47			
g-HCH	55	48	88	76	38
g-HCH	38	32			
CB-28			56	38	−5897
CB-52			69	41	−1471
CB-101			81	80	−68
CB-118			88	88	−155
CB-138			79	90	−28
CB-180			84	91	41
p,p′-DDD			93	96	36
p,p′-DDT			94	92	3.0

a   [A]Ängskärsclubb; [H]Harufjärden.
b   GFE, GFW, Gulf of Finland east and west, respectively; GR, Gulf of Riga.

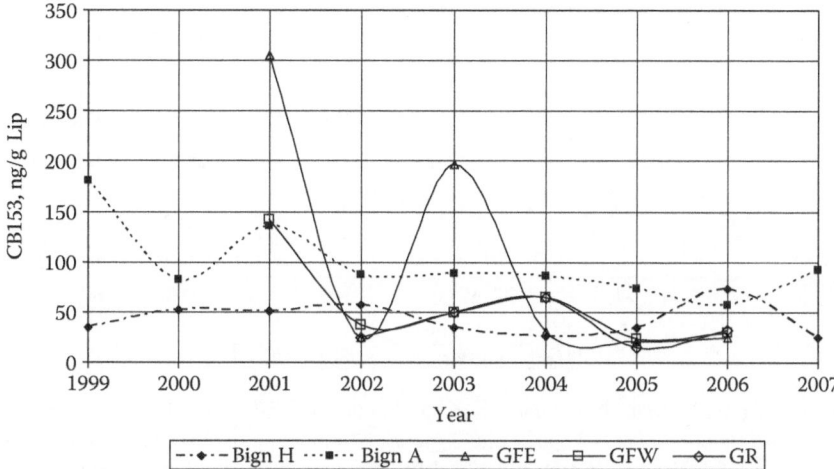

FIGURE 13.9 Trend of the concentration of CB-153. BignH and BignA are concentrations at stations Harufjärden and Ängskärsclubb [75], respectively; GFE, GFW stand for Gulf of Finland east and west, respectively, and GR for Gulf of Riga. (Based on Estonian State Monitoring Programme, http://eelis.ic.envir. ee:88/seireveeb. With permission.)

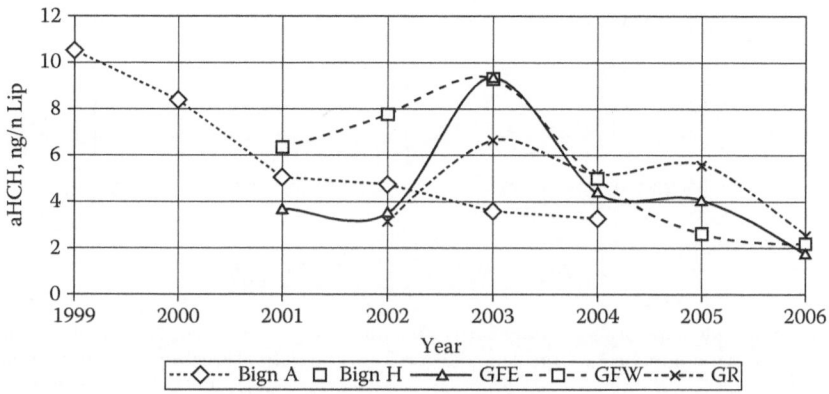

FIGURE 13.10 Trend of the concentration of α-HCH. BignH and BignA are concentrations at stations Harufjärden and Ängskärsclubb [75], respectively; GFE, GFW, stand for Gulf of Finland east and west, respectively, and GR for Gulf of Riga. (Based on Estonian State Monitoring Programme, http://eelis.ic.envir. ee:88/seireweeb; Lukki, T. et al., *Organohalogen Compd.*, 70, 2098, 2008. With permission.)

TABLE 13.17
Concentrations of CB-153 (ng/g Lipid) in 1999

Bothnian Bay		Bothnian Sea		Gulf of Finland	
Small Herring	Large Herring	Small Herring	Large Herring	Small Herring	Large Herring
270	1015	476	685	847	922

Source: Koistinen, J. et al., *Environ. Pollut.*, 154, 172, 2008.

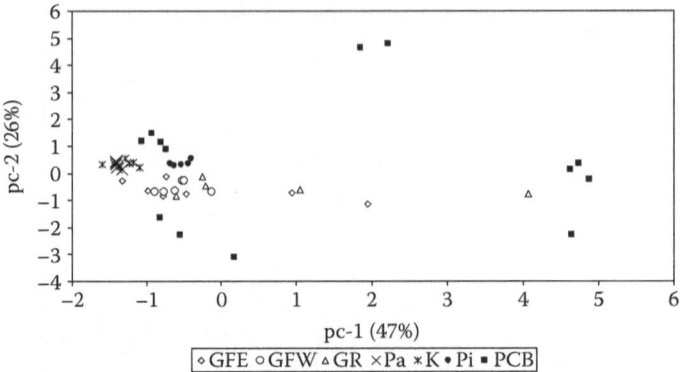

FIGURE 13.11 Score plot: mean or in some cases median concentrations of chlorobiphenyls projected on the plane of the principal components 1 (pc-1) and 2 (pc-2). Numbers on the axes show the proportion of the original variance. GFE, GFW, and GR refer to Gulf of Finland east, west, and Gulf of Riga, respectively, Pa, K, and Pi are data of Parmanne et al. (2006), Koistinen et al. (2008), and Pikkarainnen and Parmanne (2006). (Based on Parmanne, R. et al., *Mar. Pollut. Bullet.*, 52, 149, 2006; Koistinen, J. et al., *Environ. Pollut.*, 154, 172, 2008; Pikkarainnen, A.L. and Parmanne, R. *Mar. Pollut. Bullet.*, 52, 1299, 2006.)

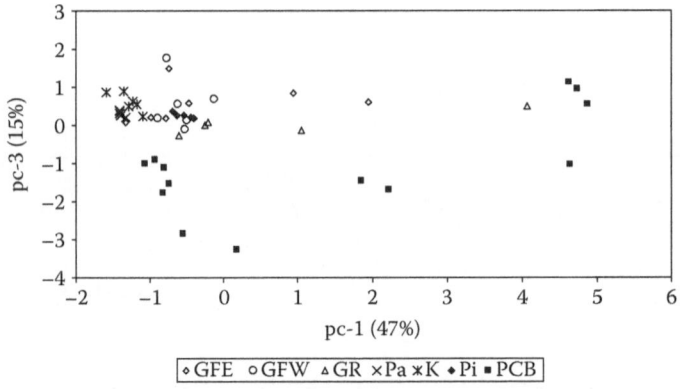

FIGURE 13.12 Score plot: mean or in some cases median concentrations of chlorobiphenyls projected on the plane of the principal components 1 (pc-1) and 3 (pc-3) of the data of this report. For details, see Figure 13.11 and ref. [77].

in other areas of Latvia are in the same range as in samples from Sweden as well as other parts of the Baltic Sea region [73,74].

13.9 CONCLUSIONS AND RECOMMENDATIONS

Organochlorine pesticides were never produced in Estonia [1] and, until 2008, there were no waste incineration facilities and all obsolete POPs residues were destroyed outside of Estonia in 2007 in Germany.

Based on the results for POPs in the air and soil samples from Estonia, the contamination level of these compounds seem to be relatively low [15–17,22,25,29,33,37,42,43,46,48,49,66,67]. The long-range atmospheric transport of some POPs, for example PCB, from southern sources outside Estonia is highly significant. Based on OCPs data of soil samples collected from agricultural-rural (Ahja, Eerika), industrial-urban (Kohtla-Järve, Muuga, Kunda), and reference sites (Lahemaa and Vilsandi), atmospheric input into Estonia is evident. The PCB concentrations were greater in Ahja-1982 and Kohtla-Järve station 1, while Eerika, other Kohtla-Järve, Muuga Port,

Kunda, Lahemaa, and Vilsandi Island concentrations were low. Organochlorine pesticides were slightly higher at Kohtla-Järve stations and PBDEs at Kunda stations. Overall, the predominant contaminants were PCBs (0.43–89 ng/g dw) followed by OCPs (0.21–16 ng/g dw) and PBDEs (<0.01–3.2 ng/g dw).

The occurrence of p,p'-DDT in some samples indicated recent input of DDT of Estonian soils. The PBDEs in Estonian soils were also reported and the levels were low. Considering the current usage of PBDEs, the contamination levels may increase in the future. Persistent organic pollutants concentrations were repeatedly measured at the Estonian Lahemaa background station and for the first time at the Kohtla-Järve industrial region air monitoring stations. The PCN (12 isomers) concentrations in the Estonian two air station samples were under the practical detection limit (0.003–1.4 ng/sample for PCNs) [38,39]. Dioxin, PAHs, and naphthalene emissions were measured in our oil shale region, from Estonian and Baltic Power Plants and a shale oil producing plant located near the city of Narva in Estonia and the values were considered to be very low. Based on the results of air and soil research, the Lahemaa background station seems to be an appropriate candidate for continuous background monitoring of persistent organic pollutants (POPs). Annual emissions of dioxins and furans have decreased in HELCOM countries during the period from 1990 to 2006 by 22% (). The most significant drop in PCDD/F emissions were noted for Denmark (63%), Estonia (53%), and Sweden (35%). Some decrease of emission was also noted for Germany (26%), Russia (22%), Poland (15%), and Finland (60%) [68,78]. Annual emissions of dioxins and furans have decreased in HELCOM countries during the period from 1990 to 2005 by 24%.

In Baltic Sea fish samples (e.g., Baltic herring), the contamination levels vary depending on the time and place of the catch, that is, the population and location, when different regions are compared [23,77]. Besides fish age, other characteristics such as length, sex, weight, fat content, maturity, etc., are to be taken into account [4–6,8,10–14,18,23,26,77]. Disregarding the scatter of the concentrations in the intervening years, the decrease in the concentration of *p,p'*-DDE, PCB, HCB, and HCHs in the herring from the Gulf of Finland is similar to that predicted by the exponential curves published in the studies [18,75,77]. Currently, the movement of some POPs, for example PCBs, from afar from southern outside Estonia is highly significant [15,28,29,77]. For daily intake, it is recommended to increase the share of the perch, the pike-perch, and the flounder as well as the fish deriving from inspected fish farms and imported from the states of the European Union. The results do not eliminate the need to monitor the toxicants in fish in the future because the use of hazardous chemicals in the Baltic Sea region will probably continue.

REFERENCES

1. Müür, J. 1996. *Plant Protection Products Use in Estonia/Estonian Environment 1995* (E. Meikas, ed.). Ministry of Environment of Estonia, Environment Information Center, Tallinn, pp. 66–68.
2. Ratas, R. and Raukas, A. 1997. *Main Outlines of Sustainable Development in Estonia*. Ministry of the Environment Republic of Estonia, Tallinn, 39pp.
3. Roots, O. 2008. Proposal for selection of national priority hazardous substances for Estonian surface water bodies. Ecological Chemistry (Экологическая химия). St. Petersburg. *Thesa* 17: 44–56.
4. Roots, O. and Peikre, E. 1982. Polychlorinated biphenyls and chlororganic pesticides in the ecosystem of the Baltic Sea. In: *Proceedings of the 13th Conference of Baltic Oceanographers*, Helsinki, Vol. 2, pp. 604–616.
5. Roots, O. 1996. *Toxic Chlororganic Compounds in the Ecosystem of the Baltic Sea*. Estonian Environment Information Centre, Tallinn, 144pp (ISBN 9985-9072-0-5).
6. Roots, O. 1999. Did natural changes save the grey seal of the Baltic Sea? Hypothesis or reality. *Toxicological and Environmental Chemistry* 69: 119–131.
7. Roots, O. 2001. Halogenated environmental contaminants in fish from Estonian coastal areas. *Chemosphere* 43(4–7): 623–632.
8. Roots, O., Lahne, R., Simm, M., and Schramm, K.-W. 2003. Dioxins in the Baltic herring and sprat in Estonian coastal waters. *Organohalogen Compounds* 62: 201–203.

9. Roots, O. 2004. Polychlorinated biphenyls (PCB), polychlorinated dibenzo-p-dioxins (PCDD) and dibenzofurans (PCDF) in oil shale and fly ash from oil shale-fired power plant in Estonia. *Oil Shale* 21(4): 333–339.

10. Roots, O. and Peikre, E. 1978. O sodershanii polikhorirovannih biphenilov I khlororganitsheskih pesticidov v ribah Baltiiskogo morja. *Izv. AN ESSR, Serija Khimia* 3: 193–196 (in Russian).

11. Roots, O. and Zitko, V. 2004. Health concerns in the Baltic States, Nordic countries, and Russia. *Proceedings of the Estonian Academy of Sciences. Biology Ecology* 53: 194–207.

12. Roots, O. and Zitko, V. 2004. Chlorinated dibenzo-p-dioxins and dibenzofurans in the Baltic herring and sprat of Estonian coastal waters. *Environmental Science and Pollution* 11: 186–193.

13. Roots, O. and Zitko, V. 2006. The effect of age on the concentration of polychlorinated dibenzo-p-dioxins, dibenzofurans and dioxin-like polychlorinated biphenyls in Baltic herring and sprat. *Fresenius Environmental Bulletin* 15: 207–219.

14. Roots, O., Järv, L., and Simm, M. 2004. DDT and PCB concentrations dependency on the biology and domicile of fish: An example of Perch (*Perca fluviatilis* L.) in Estonian Coastal Sea. *Fresenius Environmental Bulletin* 13: 620–625.

15. Roots, O. and Sweetman, A. 2007. Passive air sampling of persistent organic pollutants in two Estonian air monitoring stations. *Oil Shale* 24: 483–494.

16. Schleicher, O., Jensen, A., Roots, O., Herrmann, T., and Tordik, A. 2004. Dioxin and PAH emissions from a shale oil processing plant in Estonia. *Organohalogen Compounds* 66: 1665–1671.

17. Schleicher, O., Jensen, A., Roots, O., Herrmann, T., and Tordik, A. 2004. Dioxin emission from two oil shale fired power plants in Estonia. *Organohalogen Compounds* 66: 4089–4095.

18. Lukki, T., Roots, O., Simm, M., Talvari, A., and Tuvikene, A. 2008. PCBs, HCHs, and HCB in the Baltic Sea herring of the Estonian coastal sea. *Organohalogen Compounds* 70: 2098–2101.

19. Simm, M., Roots, O., Kotta, J., Lankov, A., Henkelmann, B., Shen, H., and Schramm, K-W. 2006. PCDD/Fs in sprat (*Sprattus sprattus* (L.)) from the Gulf of Finland, the Baltic Sea. *Chemosphere* 65: 1570–1575.

20. Pandelova, M., Henkelmann, B., Roots, O., Simm, M., Järv, L., Benfenati, E., and Schramm, K.-W. 2008. Levels of PCDD/F and dioxin-like PCB in Baltic fish of different age and gender. *Chemosphere* 71: 369–378.

21. Roots, O., Simm, M., Kiviranta, H., and Rantakokko, P. 2008. Persistent organic Pollutants (POPs): Food safety control in Estonia. In: *The Fate of Persistent Organic Pollutants in the Environment* (NATO Science for Peace and Security) (Mehmetli, E. and Koumanova, B., eds.). Springer, Berlin, Germany, pp. 173–185.

22. Roots, O., Holoubek, I., Cupr, P., Klanova, J., Kallis, A., and Kuningas, K. 2008. Air and soil pollution. Part 1. Organo-chlorine pesticides in the north- and north-eastern part of the Estonia. Ecological Chemistry, St. Petersburg University. *Thesa* 17: 88–93 (ISSN 0869-3498).

23. Roots, O. and Lukki, T. 1990. On the hypothesis of uniqueness of the Baltic herring catches on the bases of polychlorinated biphenyls content evaluation. In: *Proceedings of the 17th Conference of the Baltic Oceanographers*, Norrköping, pp. 443–451.

24. Roots, O., Kamenev, V., and Kaup, E. 1990. Polikhlorirovannie biphenili v ekosistemah Antarktiki.—Informatsionni Billetin Sovetskoi Antarktitsheskoi Ekspeditsii (Information Bulletin Soviet Antarctic Expedition), Leningrad. *Gidrometizdat* 113: 95–101 (in Russian).

25. Roots, O. 1995.Organochlorine pesticides and polychlorinated biphenyls in the ecosystem of the Baltic Sea. *Chemosphere* 31: 4085–4097.

26. Roots, O., Zitko, V., and Roose, A. 2005. Persistent organic pollutant patterns in grey seals (*Halichoerus grypus*). *Chemosphere* 60: 914–921.

27. Roots, O. and Saare, L. 1996. Structure and objectives of the Estonian environmental monitoring program. *Environmental Monitoring and Assessment* 40: 289–301.

28. Roots, O. 1992. Interpreting observations on the transport and wet deposition of airborne pollutants over the Baltic Sea and West-Estonian Islands. *AMBIO* 21: 321–322.

29. Agrell, G., Larsson, P., Okla, L., Bremle, G., Johansson, N., Klavins, M., Roots, O., and Zelechowska, A. 2001. Atmospheric and river input of PCBs, DDTs and HCHs to the Baltic Sea. In: *A System Analysis of the Baltic Sea* (Wulff, F., Rahm L., and Larsson, P., eds.). Springer, Berlin, Germany, pp. 149–175.

30. Tuomisto, J. and Hagmar, L. 1999. Environmental health in the east Baltic region—Pesticides and permanent organic compounds. *Scandinavian Journal of Work Environment and Health* 25(suppl 3): 65–71.

31. Roots, O. 2007. PCDDs, PCDFs and dl-PCBs in some selected Estonian and imported food samples. *Fresenius Environmental Bulletin* 16: 1662–1666.

32. Roots, O., Zitko, V., Kiviranta, H., Rantakokko, P., and Ruokojärvi, P. 2010. Polybrominated diphenyl ethers in Baltic herring from Estonian waters, 2006–2008. (Экологическая химия) 19: 14–23.

33. Roots, O., Henkelmann, B., and Schramm, K.-W. 2005. Concentrations of polychlorinated dibenzo-p-dioxins and polychlorinated dibenzofurans in soil in the vicinity of a landfill. *Chemosphere* 2004, 57(5): 337–342 [Corrigendum to Concentrations of polychlorinated dibenzo-p-dioxins and polychlorinated dibenzofurans in soil in the vicinity of a landfill, *Chemosphere* 57 (2004) 337–342—*Chemosphere* 2005, 58: 379].

34. Roots, O., Zitko, V., Schramm, K.-W., Henkelmann, B., and Simm, M. 2005. Polychlorinated dibenzo-p-dioxins and dioxin-like polychlorinated biphenyl pattern in Estonian food. In: *Proceedings of Dioxin 2005/ISPAC 20*, Toronto, Canada, pp. 1495–1497.

35. Roots, O. and Simm, M. 2009. Persistent organic pollutants analysis of river deltas and the complete characterization of Estonian coastal waters. *Ecological Chemistry* 18: 10–21.

36. Roots, O., Zitko, V., Kiviranta, H., Rantakokko, P., and Ruokajärvi, P. 2009. Concentrations and profiles of brominated diphenyl ethers (BDEs) in Baltic and Atlantic herring. *Oceanologia* 51: 515–523.

37. Roots, O., Roose, A., Kull, A., Holoubek, I., Cupr, P., and Klanova, J. 2010. Distribution pattern of PCBs, HCB and PeCB using passive air and soil sampling in Estonia. *Environmental Science and Pollution Research* 17: 740–749.

38. Jaward, F.M., Farrar, N.J., Harner, T., Sweetman, A.J., and Jones, C.K. 2004. Passive air sampling of PCBs, PBDEs, and organochlorine pesticides across Europe. *Environmental Science & Technology* 38: 34–41.

39. Jaward, F.M., Farrar, N.J., Harner, T., Sweetman, A.J., and Jones, C.K. 2004. Passive air sampling of polycyclic aromatic hydrocarbons and polychlorinated naphthalenes across Europe. *Environmental Toxicology and Chemistry* 23: 1355–1364.

40. Gioia, R., Sweetman, A.J., and Jones, K.C. 2007. Coupling passive air sampling with emission estimates and chemical fate modeling for persistent organic pollutants (POPs): A feasibility study for Northern Europe. *Environmental Science & Technology* 41: 2165–2171.

41. Olsson, M., Bignert, A., Aune, M., Haarich, M., Harms, U., Korhonen, M., Poutanen E.-L., Roots, O., and Sapota, G. 2002. Organic contaminants. In: *Environment of the Baltic Sea Area 1994–1998. Baltic Sea Environmental Proceedings*. 82B:133–140.

42. Senthil Kumar, K., Mahalakshmi Priya, A., M., Sajwan, K., Kõlli, R., and Roots, O. 2009. Residues of persistent organic pollutants in Estonian soils (1964–2006). *Estonian Journal of Earth Sciences* 58: 109–123.

43. Sajwan, K., Kumar, K., Roots, O., Kõlli, R., Mowery, H., and Loganathan, B. 2008. Persistent organochlorines and brominated diphenyl ethers in soil collected from rural and urban Estonia. *Organohalogen Compounds* 70: 2474–2477.

44. Roose, A. and Roots, O. 2005. Monitoring of priority hazardous substances in Estonian water bodies and the coastal Baltic Sea. *Boreal Environment Research* 10: 89–102.

45. Status on POPs phase-out in Estonia, Latvia and Lithuania. 2001. Ministry of Environment and Energy, Denmark and Danish Cooperation for Environment in Eastern Europe. *Status Report* 53329: 1–37.

46. Schleicher, O., Roots, O., Jensen, A.A., Herrmann, T., and Tordik, A. 2005. Dioxin emission from two oil shale-fired power plants in Estonia. *Oil Shale* 22: 563–570.

47. Larssen, C., Hansen, E., Jensen, A.A., Olendrynski, K., Kolsut, W., Zurek, J., Kangulewicz, I., Debski, B., Skolkiewicz, J., Holtzer, M., Grochowalski, A., Brante, E., Poltimäe, H., Kallaste, T., and Kapturauskas, J. 2003. Survey of dioxin sources in the Baltic Region. *Environmental Science and Pollution Research* 10: 49–56.

48. Quass, U., Pulles, T., and Kok, H. 2004. The DG environment project "dioxin emissions in candidate countries": Scope, approach and first results. *Organohalogen Compounds* 66: 878–883.

49. Quass, U., Fermann, M., and Bröker, G. 2000. The European dioxin emission inventory, Stage II, Executive Summary, 1: 34–36.

50. World Health Organization (WHO). Levels of PCBs, PCDDs and PCDFs in human milk: Second round of WHO-coordinated exposure study. Copenhagen: WHO, 1996. Environmental Health in Europe, no 3. EUR/ICP EHPMO2 03 05.

51. Mussalo-Rauhamaa, H. and Lindström, G. 1995. PCDD and PCDF levels in human milk in Estonia and Nordic countries. *Organohalogen Compounds* 26: 245–248.

52. Roots, O. 2000. Health concerns in the Baltic countries and environmental quality. *Ecological Chemistry* (Экологическая химия), St. Petersburg, (Thesa), St. Petersburg 9: 54–62 (ISSN 0869-3498).

53. Klanova, J., Kohoutek, J., Hamplova, L., Urbanova, P., and Holoubek, I. 2006. Passive air sampler as a tool for long-term air pollution monitoring: Part 1. Performance assessment for seasonal and spatial variations. *Environmental Pollution* 144: 393–405.

54. Klanova, J., Cupr, P., and Holoubek, I. 2007. Application of passive sampler for monitoring of POPs in ambient air. Part II: Pilot study for development of the monitoring network in the Central and Eastern Europe (MONET_CEEC) 2006. *RECETOX-TOCOEN REPORTS*, 319: 1–177.

55. Kohoutek, J., Holoubek, I., and Klanova, J. 2006. Methodology of passive sampling. TOCOEN, s.r.o. Brno, RECETOX MU Brno. *RECETOX-TOCOEN REPORTS*. 300: 1–14.

56. Van den Berg, M., Birnbaum, L.S., Denison, M. de Vito, Farland, M., Feeley, W., Fiedler, M., Hakansson, H., Hanberg, A., Haws, L., Rose, M., Safe, S., Schrenk, D., Tohyama, C., Tritscher, A., Tuomisto, J., Tysklind, M., Walker, N., and Peterson, R.E. 2006. The 2005 World Health Organization re-evaluation of human and mammalian toxic equivalency factors for dioxins and dioxin-like compounds. *Toxicological Sciences* 93: 223–241.

57. EU. 2006. Council regulation (EC) No. 199/2006 of 3 February 2006 amending Commission Regulation (EC) No. 466/2001 setting maximum levels for certain contaminants in foodstuffs as regards dioxins and dioxin-like PCBs. *Official Journal of the European Communities*.

58. Kiviranta, H., Vartianen, T., Parmanne, R., Hallikainen, A., and Koistinen, J. 2003. PCDD/Fs and PCBs in Baltic herring during the 1990s. *Chemosphere* 50: 1201–1216.

59. Kiviranta, H., Ovaskainen, M.L., and Vartiainen, T. 2004. Market basket study on dietary intake of PCDD/Fs, PCBs, and PBDEs in Finland. *Environment International* 30: 923–932.

60. Koistinen, J., Kiviranta, H., Ruokojärvi, P., Parmanne, R., Verta, M., Hallikainen, A., and Vartiainan, T. 2008. Organohalogen pollutants in herring from the northern Baltic Sea: Concentrations, congener profiles and explanatory factors. *Environmental Pollution* 154: 172–183.

61. Parmanne, R., Hallikainen, A., Isosaar, I.P., Kiviranta, H., Koistinen, J., Laine, O., Rantakokko, P., Vuorinen, P.J., and Vartianen, T. 2006. The dependence of organohalogen compound concentrations on herring age and size in the Bothnian Sea, northern Baltic. *Marine Pollution Bulletin* 52: 149.

62. Pikkarainen, A.-L. and Parmanne, R. 2006. Polychlorinated biphenyls and organochlorine pesticides in Baltic herring 1985–2002. *Marine Pollution Bulletin* 52: 1299–1309.

63. Graham, D.J. and Midgley, N.G. 2000. Graphical representation of particle shape using triangular diagrams: An excel spreadsheet method. *Earth Surface Processes and Landforms* 25: 1473–1477.

64. Wise, B.M. and Gallagher, N.B. 1998. PLS-Tool 2.0, Eigenvector Inc., Manson, WA 98831, USA.

65. Harner, T., Bartkow, M., Holoubek, I., Klanova, J., Wania, F., Gioia, R., and Peters, A. 2006. Global pilot study for persistent organic pollutants (POPs) using PUF disk passive air samplers. *Environmental Pollution* 144: 445–452.

66. Gusev, A., Rozovakaya, O., Shatalov, V., Sokovyh, V., Aas, W., Breivik, K., and Halse, A.K. 2008. Persistent organic pollutants in the environment. *EMEP Status Report* 3: 1–72.

67. Nordic. 1999. Environmental research programme for 1993–1997//Final report and self-evaluation. Nordic Council of Ministers. *TemaNord Environment* 548: 135–140.

68. Bremle, G., Okla, L., and Larsson, P. 1995. Uptake of PCBs in fish in contaminated river system: Bioconcentration factors measured in the field. *Environmental Science & Technology* 29: 2010–2015.

69. Manual for Integrated Monitoring. Part: 7.7. Soil Chemistry. UN ECE Convention on long-range transboundary air pollution. Compiled by ICP IM Programme Centre, Finnish Environmental Institute, 1998.

70. Kõlli, R. 2002. Large-scale digital soil map of Estonia and its application for soil conservation purposes. *ESSC Newsletter* 2: 7–9.

71. Fiedler, H., Golder, D., Coleman, P., King, K., and Petersen, A. 1999. Compilation of EU dioxin exposure and health data: Environmental levels. *Organohalogen Compounds* 43: 151–154.

72. Rajasilta, M., Laine, P., and Paranko, J. 2009. Feeding, somatic condition and growth of the Baltic herring (*Clupea harengus membras*) juveniles in different salinities. In: *Abstract Book, Baltic Sea Science Congress 2009*, August 17–21, 2009, Tallinn University of Technology, pp. 230–230.

73. Valters, K., Olsson, A., Asplund, L., and Bergman, A. 1999. Polychlorinated biphenyls and some pesticides in perch (*Perca fluviatilis*) from inland waters of Latvia. *Chemosphere* 38: 2053–2064.

74. Valters, K. 2001. Assessment of organochlorine contamination in the aquatic environment of Latvia with perch and heron as bioindicators. Doctoral dissertation, Stockholm University, Wallenberg Laboratory, Stockholm, Sweden, 50pp.

75. Bignert, A., Nyberg, E., Asplund, L., Eriksson, U., Wilander, A., and Haglund, P. 2007. Comments Concerning the National Swedish Contaminant Monitoring Programme in Marine Biota, Dnr 721-1692-06Mm. 3-31-2007a. Swedish Museum of Natural History—The Department of Contaminant Research.

76. Wyrzykowska, B., Bochentin, I., Hanari, N., Orlikowska, A., Falandysz, J., Yuichi, H., and Yamashita, N. 2006. Source determination of highly chlorinated biphenyl isomers in pine needles—Comparison to several PCB preparations. *Environmental Pollution* 143: 46–59.

77. Roots, O., Zitko, V., Lukki, T., Simm, M., and Järv, L. Time trends, geography, and composition patterns of organochlorine contaminants in herring from the Gulf of Finland and the Gulf of Riga in the years 2002–2006. *Ecological Chemistry* 19: 75–97.

78. Gusev, A. 2008. Atmospheric depositions of PCDD/Fs on the Batlic Sea. Archive. Indicator fact sheets 2008. Online http://www.helcom.fi/environment2/ifs/ifs2008/en_GB/pcddfemissions/

79. Helsinki Commission (HELCOM) Baltic Marine Environment Protection Commission homepage: http://www.helcom.fi

14 Chlorinated Hydrocarbons in Animal Tissues and Products of Animal Origin from Poland

*Alicja Niewiadowska and Jan Zmudzki**

CONTENTS

14.1 Introduction 337
14.2 Materials and Methods 338
14.3 Results and Discussion 340
    14.3.1 Monitoring 1995–2003 340
    14.3.2 Dietary Intake 342
    14.3.3 Residue Control Plan 1997–2008 343
    14.3.4 Time Trends 345
    14.3.5 Human Milk 349
14.4 Conclusions 351
References 351

14.1 INTRODUCTION

Persistent organic pollutants (POPs) are chemical substances that possess certain toxic properties and, as other pollutants, resist degradation, which makes them particularly harmful for human health and environment. Many of these compounds, due to their environmental persistence, have the ability to bioaccumulate in fatty tissues. Organochlorine pesticides and polychlorinated biphenyls (PCBs) are representatives of a large group of chlorinated hydrocarbons, which are listed as POPs in the Stockholm convention report [1]. Provisions of the convention concern organochlorine pesticides, compounds used in industry, and compounds produced unintentionally. Although the production and use of organochlorine pesticides in agriculture and PCBs in industry has been restricted, or even prohibited in many countries, their residues are still detected in food. Studies of the organochlorine pesticide residues have become the basic element of each monitoring program for food and feed contamination.

The evaluation of food and environmental contaminations by organochlorine pesticides and PCBs began in the 1960s, or early 1970s in many industrialized countries. Results of these studies led to introduction of restrictions and, in later stages, prohibition of production and use of the organochlorine pesticides and PCB.

In Poland, regular monitoring of the pesticide residues in animal tissues, milk, and eggs was introduced in 1969 by the National Veterinary Research Institute in Pulawy (PIWet Pulawy), according to the system developed by Prof. Juszkiewicz [2]. This evaluation was conducted every year by the commission and in cooperation with the Ministry of Agriculture. The veterinary residue program, which originally concerned the organochlorine pesticides, was subsequently expanded to studies

---

* E-mail: zmudzki@piwet.pulawy.pl (Chapter corresponding author).

of other chemical compounds such as PCBs, toxic elements, mycotoxins, and selected groups of veterinary drugs. This research program has been under the constant supervision of experts representing USDA/FSIS and the EU since 1984.

In 2004, with Poland's accession into the European Union (EU), the national veterinary residue control program for live animals and animal products was recognized in accordance with the Council Directive 96/23/EC and approved by the EU [3,4]. The Ministry of Agriculture and Rural Development together with Veterinary Inspection are in charge of the program realization. However, it should be underlined that the PIWet Pulawy has coordinated the monitoring program from the beginning of the studies on pesticide residues in Poland, 40 years ago. For the last 7 years, the PIWet Pulawy has functioned as the National Reference Laboratory. The concept of the monitoring program of the chemical residues and its plan were developed by the PIWet Pulawy, approved by the chief veterinary officer, and accepted by the European Commission. Since 2006, the PIWet Pulawy has also introduced a national control program for dioxin content (PCDDs, PCDFs, dl-PCBs) in products of animal origin and feed [5,6].

In Poland, among the chemicals used in production of pesticide preparations, which have been listed in Stockholm convention, were found DDT, dieldrin, hexachlorobenzene (HCB), hexachlorocyclohexane (HCH), and toxaphene. Additionally, aldrin, endrin, and heptachlor were found in agricultural preparations. Among the above-mentioned substances, DDT and toxaphene were produced in Poland while the other compounds were imported. DDT was produced between 1949 and 1977 (about 75,000 ton) while toxaphene was produced in 1961–1962 (only 40 ton). Pesticides containing chlordan and mirex were not used.

In 1972, DDT-containing pesticides were gradually withdrawn, and in 1975 the use of these chemicals was completely terminated. It is estimated that between 1949 and 1971 about 50,000 ton of DDT were introduced into the environment on Polish territory, with exportation excluded [7]. Small quantities of aldrin and heptachlor were allowed to be used in 1962 and 1966 respectively; however, after 1971, most of the organochlorine pesticides were withdrawn. The use of pesticides containing HCB was permitted for the last time in 1978. In EU countries, the application of γ-HCH as a pesticide, containing less than 99% γ-HCH, was banned from 1981 and a total prohibition of γ-HCH was required starting in 2002. Lindane was completely banned in Poland from 1988 to 1990.

Currently, none of pesticides containing the substances listed in the Stockholm convention are produced, used, imported, or exported in Poland. For over 30 years, the production of pesticides has been based on substances which do not contain persistent organic pollutants.

PCBs have never been produced in Poland on a technical scale. Only small quantities were produced (below <1000 ton). Among them were products with commercial names: Tarnol was produced between 1971 and 1976 and was used as dielectric oil in transformers in electrotechnical industry. Chlorofen was produced between 1965 and 1976 and was used as a lubricant and as hydraulic fluid mainly in coal mining equipment [7,8]. However, there was significant import of PCBs as components for oils used in different equipments and installations. Some electrotechnical devices containing PCBs were also imported.

Poland, PCBs were mainly used in capacitors, imported transformers, and in smaller amounts in other power devices such as oil switches, suppressors, etc. Initial inventory performed in 2002 revealed that about 3000 ton of oils were contaminated with PCBs, while about 1400 capacitors and over 5600 transformers contained PCBs [7]. The total amount of devices containing PCBs is estimated at 15,000 ton.

14.2 MATERIALS AND METHODS

The investigation of organochlorine pesticides and PCB residues in animal tissues and in the food of animal origin was conducted using a classical monitoring and control program according to Directive 96/23. The systems of sampling in these programs assure a good representation of the conducted studies. During the first years of the studies, the sampling system was based mainly on a random system developed by the current administrative division of the country. These principles were also applied during monitoring studies conducted between 1995 and 2003 under the Program of Polish Ministry of Agriculture and Rural Development, which focused on the quality of soils, plants, food, and agricultural products. For the purposes of this program, a total of 1000 to 2000 samples were taken every year. However, starting from 1997, which is from the moment of the implementation of Council Directive 96/23/EC, the rules of the sampling have been dependent on target system, the rate of food production, and the results of the studies from previous years. This system clearly indicates the investigated animal species and the type of samples. The minimal number of tested animals and the minimal number of samples taken from the products of animal origin was calculated on the basis of the data concerning the number of animals slaughtered and the scale of food production during the previous year.

The material for the analysis consisted of adipose tissue samples from different animal species, as well as milk, eggs, and honey. The samples were taken by veterinary inspectors according to the every-year-updated instruction of the chief veterinary officer and were delivered by them along with the certificate of origin. The samples were taken in different places of food production: in slaughterhouses, milk catchment area, on farms, and in animal production units.

Beginning in 1969, the samples were analyzed for the presence of organochlorine pesticide residues, and from 1976 the scope of investigation was expanded to include PCBs. The investigated organochlorine pesticides were as follows: DDT and its metabolites, α-HCH, β-HCH, γ-HCH, HCB, aldrin, dieldrin, endrin, *cis*-chlordan, *trans*-chlordan, oxychlordan, heptachlor, heptachlor epoxide, and metoxychlor. Polychlorinated biphenyls were quantified as total PCBs using commercial mixtures of PCB as standards, the most frequently used being Aroclor 1260. Starting in 2000 and according to international requirements, the content of PCBs was quantified as a sum of seven indicator congeners of PCBs marked with IUPAC numbers 28, 52, 101, 118, 138, 153, and 180. The identification and quantitative analysis of investigated compounds were performed using gas chromatography with an electron capture detector (ECD).

The majority of the analyses were performed by the Department of Pharmacology and Toxicology of PIWet Pulawy. Some analyses were also performed in the Regional Veterinary Inspection Laboratories in Bialystok, Gdansk, Katowice, Poznan, Warsaw, and Wroclaw. All of the laboratories used validated analytical procedures and were accredited according to ISO 17025 by the Polish Centre for Accreditation in 2004.

An important element of conducted analyses is the quality and reliability assurance of the analytical procedures used in laboratories mentioned above. An active quality assurance program for the analysis was established, involving among others the regular participation in proficiency tests and analytical trainings. Starting in 1985 as a requirement of the residues control program, the PIWet Pulawy organizes the proficiency tests which are obligatory for all the laboratories that take part in the control program four times a year. For many years, the PIWet Pulawy, being the National Reference Laboratory, regularly takes part in international proficiency tests organized by the Food and Environmental Research Agency in England in the frame of FAPAS (Food Analysis Performance Assessment Schemes) program as well as in analysis programs organized by the community reference laboratories. The results obtained in these studies are satisfactory. The Z-score values are much lower than warranted criteria, which means that the laboratory delivers reliable results (Z-score values between −2 and +2). The results of all proficiency tests indicate that both the PIWet Pulawy laboratory and the regional veterinary inspection laboratories, which take part in the residue control program, deliver reliable results, and the analytical methods used in these laboratories have satisfactory validation parameters.

14.3 RESULTS AND DISCUSSION

14.3.1 MONITORING 1995–2003

From 1995 to 2003, the total number of analyses performed by the laboratory of PIWet Pulawy, concerning the occurrence and the levels of organochlorine pesticides and PCBs in tissue samples and products of animal origin was 9913. Among the analyzed materials there were 2605 pig adipose tissue samples, 2309 bovine adipose tissue samples, 618 chicken adipose tissue samples, 1145 game animals adipose tissue samples (511 of wild boars, 447 of roe deer and 187 of deer), 944 samples of farmed fish muscle (873 carp and 71 trout), 1735 milk samples, and 145 egg samples. The basic materials were supplemented by 412 adipose tissue samples of other animal species (turkeys, ducks, geese, horses, sheep, and rabbits). The common presence of low concentrations of organochlorine pesticides and PCBs was observed [9,10].

The presence of DDT and its metabolites has been observed in about 95% of all investigated samples. The highest average concentrations of DDTs were observed in game animals, lower in farmed fish and the lowest in the tissues of chickens and horses (Figure 14.1). In tissues of pigs, bovine, poultry, and other species, as well as in eggs, the concentrations of these compounds did not exceed the level of 0.1 mg/kg. The presence of low HCB concentrations was also detected in about 5% of all investigated tissues and products of animal origin, the most frequently in game animals, bovine, and fish. Higher average concentrations of HCB were observed in trout and game animals (Figure 14.2). The isomers of HCH were present in only 14% of samples, the most frequently in fish. Among the various animal products analyzed, the highest average concentrations of the total of HCH isomers were observed in the tissues of carp and trout (Figure 14.3).

Exceedance of the maximum residue levels (MRL) of pesticides were identified in only 1% of the studied samples, including 0.8% of results exceeding limit values for DDT and 0.1% for HCH isomers [11]. Concentrations of DDTs exceeding 1 mg/kg of fat were identified in samples from 51 wild boars, 22 pigs, 3 bovine, and 1 roe deer, and exceeding 0.5 mg/kg of fat in 6 egg samples. In the whole investigated material, there were only 14 exceedances of the limit values for HCH, including γ-HCH concentrations exceeding 0.02 mg/kg in 4 swine and 2 wild boars samples, γ-HCH concentrations above 0.025 mg/kg of fat in 7 milk samples, and β-HCH concentration above 0.1 mg/kg in 1 wild boar sample.

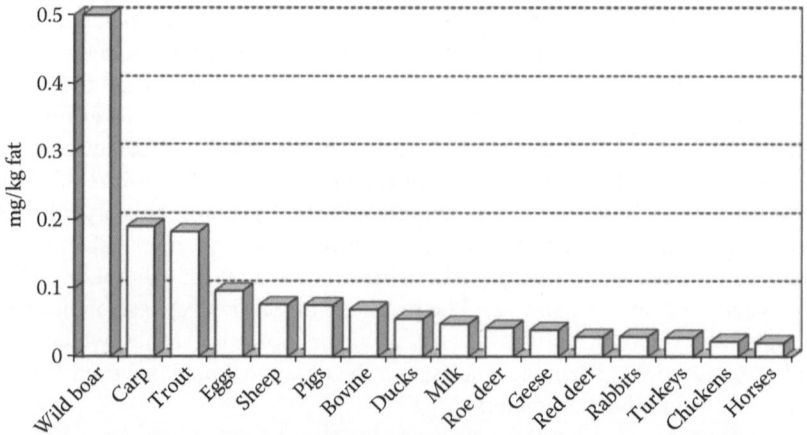

FIGURE 14.1 Levels of total DDT in animal fats, milk, and eggs in Poland, 1995–2003.

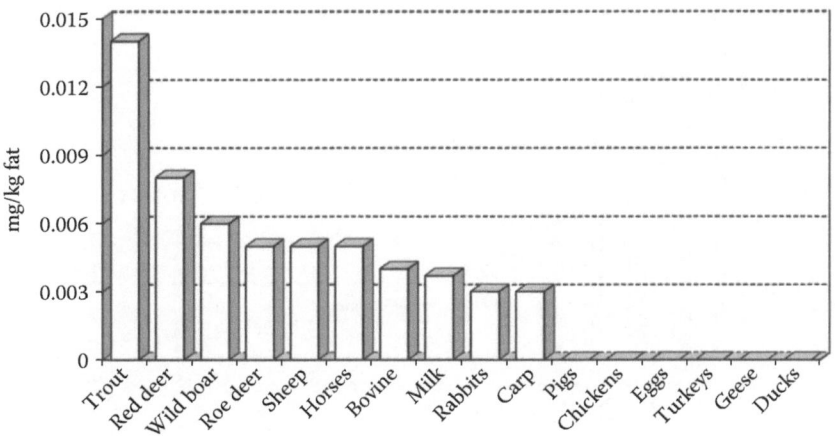

FIGURE 14.2 Levels of HCB in animal fats, milk, and eggs in Poland, 1995–2003.

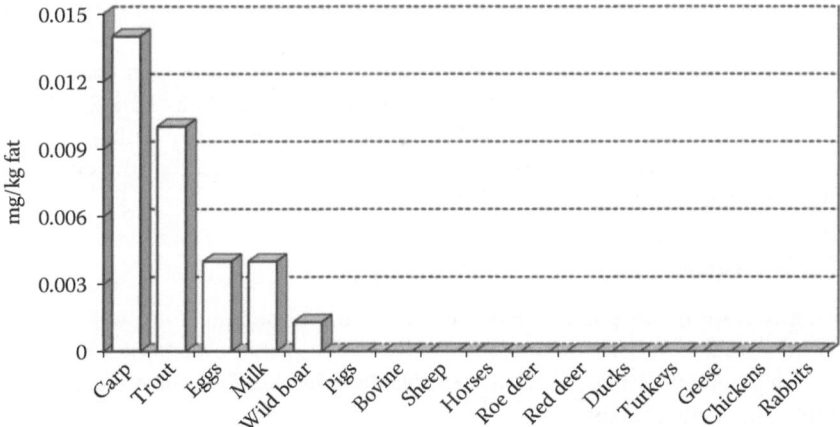

FIGURE 14.3 Levels of HCH isomers in animal fats, milk, and eggs in Poland, 1995–2003.

Starting from 2000, the PCB residues were quantified as IUPAC congener numbers. 28, 52, 101, 118, 138, 153, and 180. In the former years, total PCBs were quantified as Aroclor 1260. In order to compare long-term evaluation of PCB levels in products of animal origin a recounting factor has been established which allows for the recalculation of the sum of the seven indicator PCB congeners into total PCBs.

PCBs were identified in 72% of the samples. The lowest occurrence of PCBs was recorded in swine samples (30%). A comparison of PCB concentrations (as Aroclor 1260) in animal fats is presented in Figure 14.4. Among the studied animal fats, the highest average PCB concentrations were detected in the tissues of farmed fish, while in the remaining tissues the PCB concentrations were at the level of a hundredth or thousandth parts per mg/kg. The most frequently identified PCB congeners were as follows: PCB 153 in 64% of the samples, PCB 138 in 59% of the samples, and PCB 180 in 40% of the samples. The residues of congeners PCB 138 and PCB 153 were identified in over 95% of the tissue samples from game animals and farmed fish and in more than 70% samples of milk and bovine fat. The congener PCB 180 was also frequently identified in tissues of game animals and fish (above 80%). The determined concentrations of PCBs should be considered as very low and are many times lower than acceptable limits in other countries.

The levels of organochlorine pesticides and PCBs did not differ significantly between 1995 and 2003. However, in comparison with the results obtained during our own investigations systematically

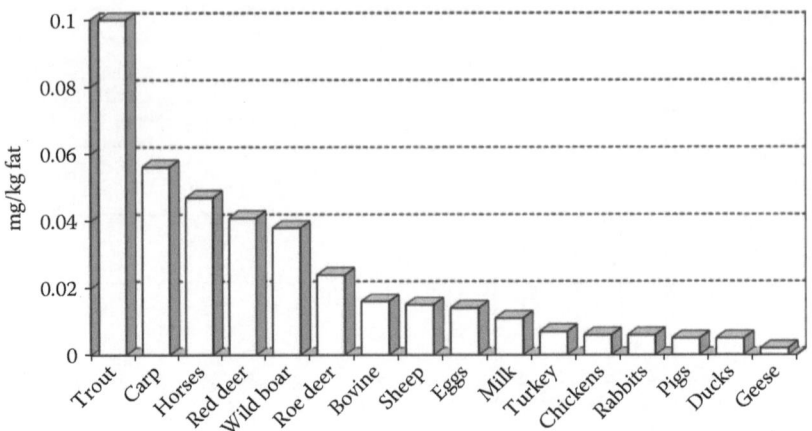

FIGURE 14.4 Levels of total PCBs in animal fats, milk, and eggs in Poland, 1995–2003.

carried out since the 1970s, the considerable declines of the mean levels of these compounds have been observed.

The results of our and other authors' studies confirm the general opinion about low contamination of food of animal origin with persistent chlorinated hydrocarbons in Poland [9,12–21]. The wide studies of domestic and imported food of plant origin, including fruits, vegetables, grains, oils, and processed products for infants and small children indicate the low contamination with organochlorine pesticides [14,22–25].

14.3.2 DIETARY INTAKE

The classical model for the evaluation of chronic exposure of humans to residues of chemical contaminants in food relies on the assessment of daily intake of these compounds with food and its comparison with acceptable values (ADI, acceptable daily intake; TDI, tolerable daily intake; PTDI, provisional tolerable daily intake).

Based on the results of our studies (1995–2005) and the amount of average daily food consumption, the daily intake of organochlorine pesticides and PCBs from food of animal origin by the inhabitants of Poland has been assessed. In order to obtain comparable values representing concentrations of organochlorine pesticides and PCBs in different kinds of food, the weighted average has been calculated taking as a weight the number of samples in consecutive years of studies. In the recalculation of concentrations, it has been assumed that pork contains an average of 35% of fat, beef 15% of fat, meat from poultry, game, and farmed fish no more than 10% of fat, milk—4% of fat and eggs 10% of fat.

Daily intake of DDTs (p,p'-DDT, o,p'-DDT, p,p'-DDE and p,p'-DDD) has been assessed on 5.53 µg/person or 0.092 µg/kg bw assuming that the average body weight is 60 kg. As far as DDT intake in food of animal origin, the highest share was found in meat and meat products (61%) (Figure 14.5). Milk and dairy products provided 33% of daily DDT intake, eggs 6%, and farmed fish 1%.

The HCB daily intake has been assessed on 0.16 µg/person (0.003 µg/kg bw/day) and HCHs daily intake (γ-HCH, α-HCH, and β-HCH) has been assessed on 0.15 µg/person (0.0025 µg/kg bw/day). Milk and dairy products were found to provide the most significant daily intake of HCHs and HCB (91% and 80%, respectively). The meat provided 18% of HCB daily intake and fish and eggs about 1%. Meat, eggs, and fish provided only a few percent of HCHs daily intake.

PCBs daily intake has been assessed on 0.81 µg/person (0.013 µg/kg bw/day). Milk and dairy provided 50% of the PCBs daily intake while meat and meat products contributed 41%. Fish and eggs provided a few % of PCBs daily intake.

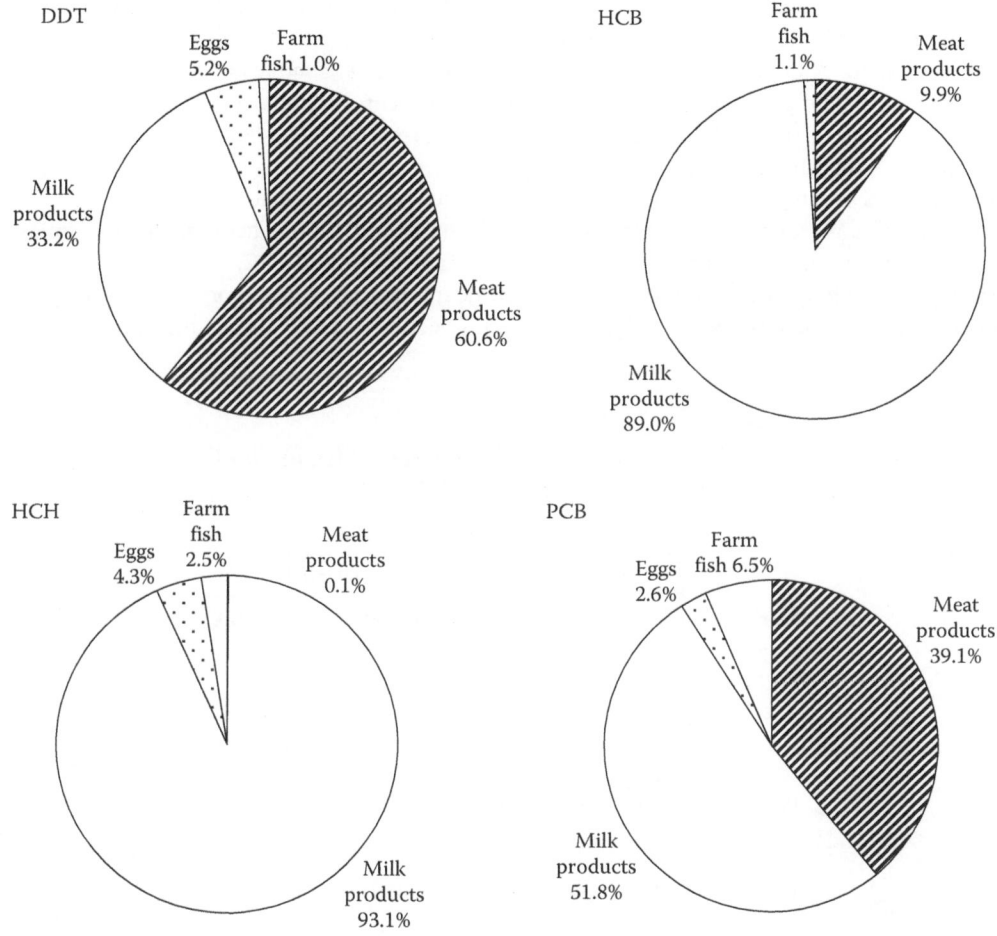

FIGURE 14.5   Contribution of food groups to the dietary intake of chlorinated hydrocarbons in Poland.

For the 90th percentile, the values of daily intake were about twice as high, and for the 95th percentile three times higher. When compared to acceptable daily intake values set by different international organizations and to the data concerning the daily dietary intakes in other countries, these values should be considered as low.

The calculated values are several times lower than the daily intakes assessed in Poland in previous years [25,26]. According to Falandysz, the daily intake of HCB from food in 1970–1996 was between 1.3 and 1.5 μg/person, PCBs daily intake with fish in 1970–1982 was between 2.3 and 5.8 μg/person and PCBs intake for total food was between 8.5 and 11 μg/person and showed decreasing trend [26–28]. In these calculations, the author used values from the average levels in fish from his own studies but in other groups of products he used data from other countries.

14.3.3   RESIDUE CONTROL PLAN 1997–2008

Starting from 1997, the studies on pesticides and PCB residues in tissues and products of animal origin became an element of a national control program for chemical, biological, and veterinary drugs residues in food [4,29,30]. The main purpose of these studies was to eliminate unsafe foods and to protect the health of consumers, as well as to comply with the requirements of international food trade.

Up until 2008, a total number of 19767 samples of tissues and animal products were analyzed including more than 4200 samples of swine and poultry adipose tissue (chickens, turkeys, geese, and ducks), about 3000 samples of bovine adipose tissue, 2500 samples of milk, more than 1000 samples of eggs, wild game adipose tissue (wild boar, roe deer, and red deer) and samples of farmed fish. In the scope of the studies adipose tissue from horses, rabbits, sheep, as well as honey and imported food, mainly marine fish, were also used.

In first years of the program, about 2000 analyses per year were performed, and in the last years the number of analyses per year decreased, but still exceeded 1000. The number of samples analyzed for pesticides and PCB content is presented in Tables 14.1 and 14.2.

Results of the long-term studies indicated significant decreases in the organochlorine pesticide levels in domestic food of animal origin. However, residues of DDT and its metabolites were still

TABLE 14.1
Occurrence of Organochlorine Pesticides in Food of Animal Origin in Poland, 1997–2008

| | | Number of Samples, n (%) | | |
| | | | Detected | |
Year	Number of Samples	None Detected	Below MRL	Above MRL
1997	2076	398 (19.2)	1678 (80.8)	13 (0.6)
				2—pigs
				4—chickens
				4—wild game
				3—milk
1998	2237	328 (14.7)	1909 (85.3)	13 (0.6)
				12—wild game
				1—eggs
1999	2043	330 (16.1)	1713 (83.8)	6 (0.3)
				1—geese
				3—wild game
				2—eggs
2000	1937	330 (17.0)	1607 (83.0)	4 (0.2)
				1—pigs
				3—wild game
2001	2012	605 (30.1)	1407 (69.9)	5 (0.25)
				1—pigs
				1—bovines
				2—wild game
				1—fish
2002	1704	547 (32.1)	1157 (67.9)	0
2003	1518	394 (26.0)	1124 (74.0)	1 (0.06)
				1—wild game
2004	1284	334 (26.0)	950 (74.0)	0
2005	1267	435 (34.3)	832 (65.7)	0
2006	1211	381 (31.5)	830 (68.5)	1 (0.08)
				1—eggs
2007	1251	484 (38.7)	767 (61.3)	3 (0.2)
				2—wild game
				1—fish
2008	1326	648 (48.9)	678 (51.1)	1 (0.08)
				1—wild game

TABLE 14.2
Occurrence of PCBs in Food of Animal Origin in Poland, 1997–2008

Year	Number of Samples	Number of Samples, n (%)	
		None Detected	Detected
1997	2067	1017 (49.2)	1050 (50.8)
1998	2237	1107 (49.5)	1130 (50.5)
1999	1993	1064 (53.4)	929 (46.6)
2000	1936	1347 (69.6)	589 (30.4)
2001	1983	1392 (70.2)	591 (29.8)
2002	1713	736 (43.0)	977 (57.0)
2003	1471	711 (48.3)	760 (51.7)
2004	1284	830 (64.6)	454 (35.4)
2005	1267	820 (64.7)	447 (35.3)
2006	1216	840 (69.1)	376 (30.9)
2007	1274	1003 (78.7)	271 (21.3)
2008	1326	858 (64.7)	468 (35.3)

present in more than 50% of the animal fats investigated. The isomers α-, β-, and γ-HCH and HCB were present in small percentages of examined samples at low levels. In general, the average concentrations of identified compounds were at the level of a hundredth or thousandth parts of mg/kg, which amounts to only few percent within the acceptable limit. Other pesticides were not present. In the studies conducted in 2008, low concentrations of organochlorine pesticides were still observed. The highest number of samples without residues came from swine, poultry, and eggs (Table 14.3).

In only few of the all food samples analyzed maximum residue limit (MRL) of pesticides was exceeded [11]. The most frequently noncompliant result was the concentration of DDTs in adipose tissue of wild boars, which exceeded exceeding 1 mg/kg. The number of noncompliant results in the following years of studies decreased from 13 in 1997 to only 1 in 2008 (Table 14.1). Among the investigated animal fats, the highest levels of DDTs were observed in game animals (higher in wild boars than in roe deer) and in fish.

The obtained results indicate a decline in the occurrence of PCBs in food of animal origin. In 2008, only 35% of the studied samples contained low levels of PCB residues. The lowest occurrence of PCBs was observed in swine and poultry tissues, with below 30% of the samples containing residues (Table 14.3). The most frequently identified congeners were as follows: PCB 153, 138, and 180. In earlier years, the PCBs occurred in more than 50% of samples.

The concentrations of PCB congeners identified in tissues and animal products were at a very low level—hundredth to thousandth parts of mg/kg. Among the studied fats, the highest concentrations of PCBs were observed in marine fish (from import), farmed fish (carp, trout), and game animals. Concentrations of these compounds in swine, bovine, poultry, and other species tissues did not exceed 0.01 mg/kg fat.

14.3.4 TIME TRENDS

Estimation of pesticide and PCB contaminations of animals and food of animal origins has been the subject of studies in the PIWet Pulawy for decades. Results of these long-term, systematic studies show considerable decrease of the levels of organochlorine pesticides and PCB residues in Poland [2,10,21,31–35].

TABLE 14.3
Occurrence of Chlorinated Hydrocarbons in Food of Animal Origin in Poland in 2008

Commodity	No. Samples	Organochlorine Pesticides Number of Samples, n (%)			Polychlorinated Biphenyls Number of Samples, n (%)		
		None Detected	Detected	Above MRL	None Detected	Detected	Above 0.1 mg/kg of Fat
Bovine	167	64 (38.3)	103 (61.7)	0	108 (64.7)	59 (35.3)	0
Pigs	289	180 (62.3)	109 (37.7)	0	216 (74.7)	73 (25.3)	0
Sheep	21	5 (27.6)	16 (72.4)	0	6 (28.6)	15 (71.4)	0
Horses	34	11 (32.4)	23 (67.6)	0	4 (11.8)	30 (88.2)	3 (8.8)
Chickens	183	102 (55.7)	81 (44.3)	0	137 (74.9)	46 (25.1)	0
Turkeys	39	20 (51.3)	19 (48.7)	0	31 (79.5)	8 (20.5)	0
Geese	36	12 (33.3)	24 (66.7)	0	16 (44.4)	20 (55.6)	0
Ducks	27	16 (59.2)	11 (40.8)	0	19 (70.4)	8 (29.6)	0
Farm fish	59	13 (22.0)	46 (78.0)	0	26 (44.1)	33 (55.9)	4 (6.8)
Milk	131	53 (40.4)	78 (59.6)	0	87 (66.4)	44 (33.6)	0
Eggs	133	76 (57.1)	57 (42.9)	0	98 (73.7)	35 (26.3)	0
Rabbits	21	3 (14.3)	18 (85.7)	0	6 (28.6)	15 (71.4)	0
Wild game	89	39 (43.8)	50 (56.2)	1 (1.1)	50 (56.2)	39 (43.8)	7 (7.9)
Honey	16	16 (100)		0	16 (100)		0
Marine fish (import)	81	38 (46.9)	43 (53.1)	0	38 (46.9)	43 (53.1)	23 (28.4)
Total	1326	648 (48.9)	678 (51.1)	1 (0.1)	858 (64.7)	468 (35.3)	37 (2.8)

In the last 40 years, a 40-fold decrease in the average DDT levels has been noted in pig tissues (Figure 14.6). In the first 6 years of research, the average concentrations of DDTs dropped from levels over 2 mg/kg to about 0.5 mg/kg. Results obtained in the next 10 years (1975–1985) showed a further five-fold reduction of these concentrations. The decrease in DDT levels continued in the following years to 0.1 mg/kg, which is 20-fold lower from the permitted values for DDT.

A 20-fold decrease in the concentration of DDTs was also observed in the case of bovine tissues and cow milk (Figures 14.7 and 14.8). In the first 10 years of monitoring, the biggest decline was noted in the case of beef fat, where the levels of DDTs changed from about 1.0 mg/kg to 0.2 mg/kg. The following years showed slower but systematic decreases of the concentration until value reached

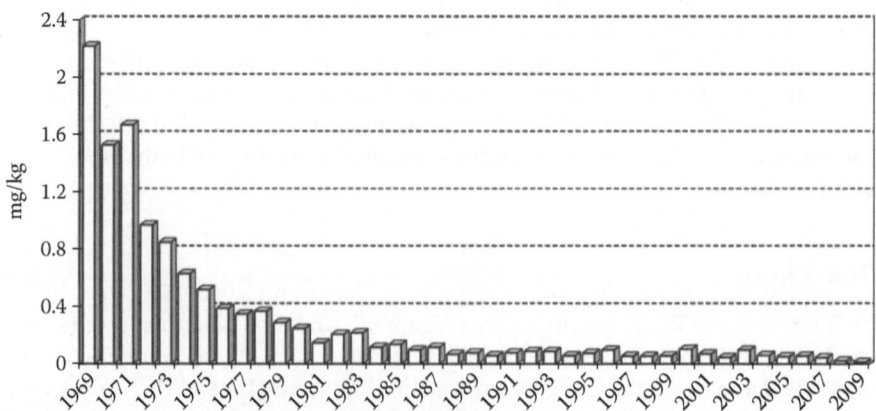

FIGURE 14.6 Levels of total DDT in swine fat in Poland.

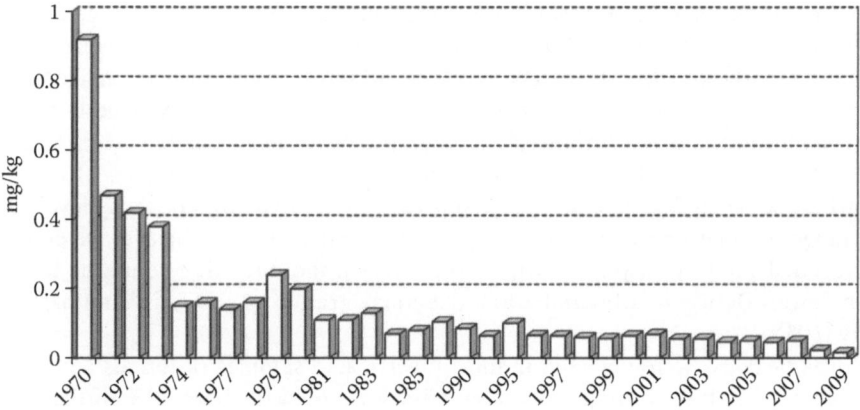

FIGURE 14.7   Residues of total DDT in bovine fat in Poland.

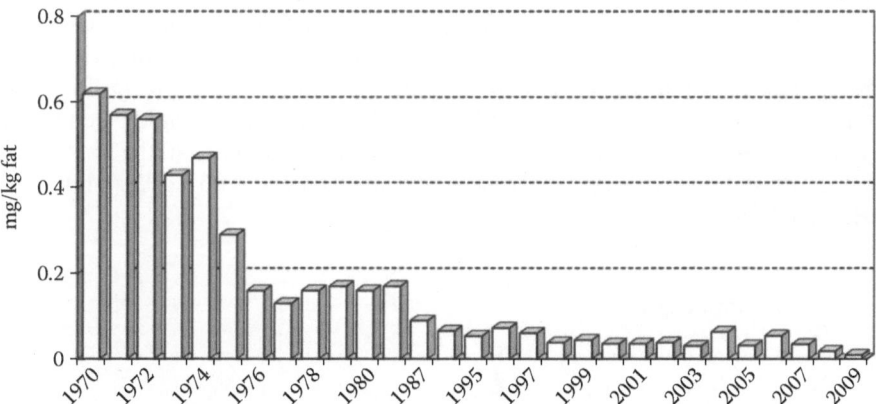

FIGURE 14.8   Residues of total DDT in bovine milk in Poland.

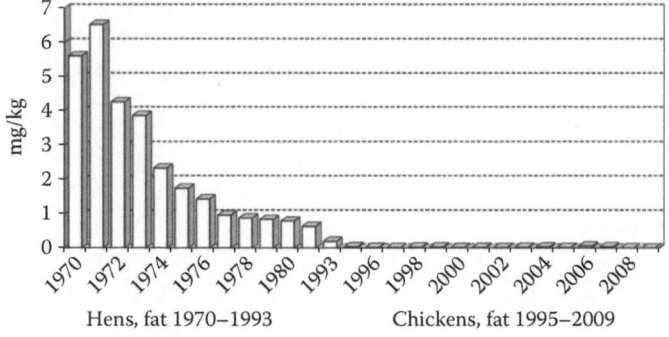

FIGURE 14.9   Residues of total DDT in poultry fat in Poland.

0.1 mg/kg and lower. The levels of DDTs detected in cow milk fat were reduced from 0.6 to 0.2 mg/kg in years 1970–1980. In the following years of the studies, the levels continuously decreased, and in the last years the observed values were well below 0.1 mg/kg [10,32,34,36].

The studies on DDT levels in hens and eggs, conducted for over 20 years, have shown a 30-fold decrease in the average concentrations of DDTs in the case of hen fat, and a 100-fold decrease in the case of eggs (Figure 14.9) [2,10,33,37]. Between 1970 and 1971, the average

DDT levels in hen fat were high, reaching 5–6 mg/kg; until 1977 these average concentrations exceeded the permitted values for DDT, which were in force at that time. Research monitoring conducted in 1993 showed that the DDT levels in adipose tissue from intensive rearing hens amounted to about 10% of the permissible values, while in the case of village hens (free-range farming) the concentrations were higher, reaching 60% of the permissible values. In 1995–2003, the average concentrations of DDTs in investigated adipose tissue samples of chickens were between 0.01 and 0.04 mg/kg. The last years showed further reduction of these levels, which are now in the range of thousandth parts of mg/kg. In case of hen egg fat, the average concentrations of DDTs exceeded 1 mg/kg in years 1970–1980, while in the following years the levels became significantly lower. Between 2000 and 2003, the concentration of DDTs in commercial hen egg fat was only 0.05 mg/kg.

The residues of PCBs were detected in most investigated samples of tissues and animal products. Their presence in bovine tissues and cow milk was noted in about 80% of the samples. For many years, the average PCB concentrations were detected on the level of a hundredth parts of mg/kg.

The studies on PCB levels in animal tissues and milk showed that these changes are not as significant as in the case of DDTs (Figures 14.10 through 14.13) [10,32,36,38]. In Poland, the average PCB concentrations in milk fat amounted to 0.03–0.04 mg/kg between 1976 and 1987, while in the years following it decreased to the level below 0.01 mg/kg. Similar to the cow milk, the concentrations of PCBs in bovine fat reached a level of 0.04 mg/kg in the 1980s. Studies conducted in the following years showed a decline to a level of 0.02 mg/kg. In case of hen fat, the average concentrations of PCBs in Poland in years 1976–1992 were between 0.04 and 0.06 mg/kg, while in broiler fat it did not exceed 0.01 mg/kg.

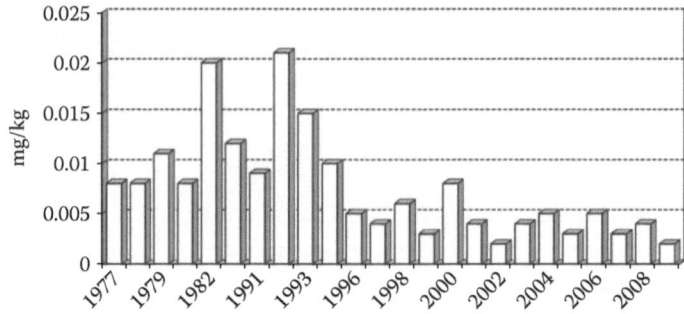

FIGURE 14.10   Residues of PCBs in swine fat in Poland.

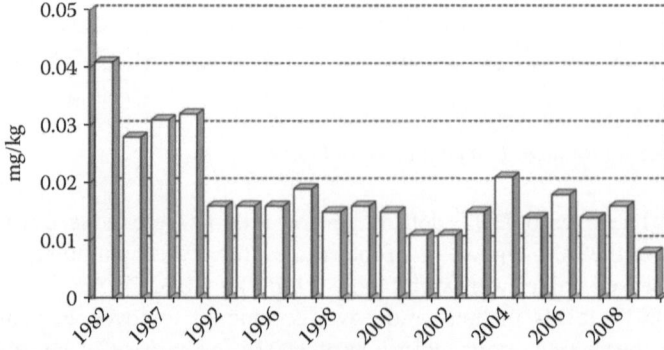

FIGURE 14.11   Residues of PCBs in bovine fat in Poland.

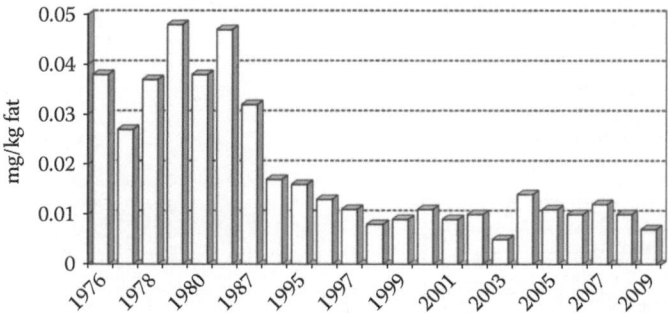

FIGURE 14.12   Residues of PCBs in bovine milk in Poland.

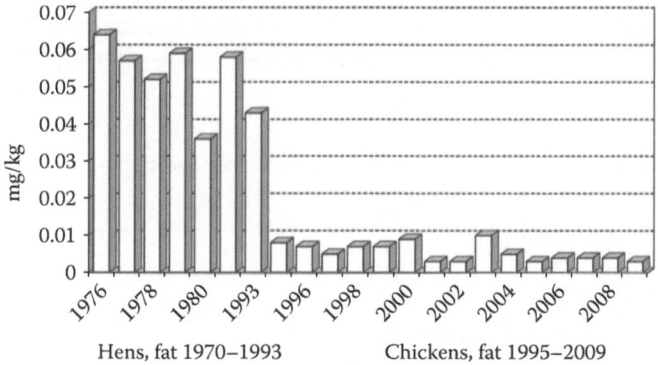

Hens, fat 1970–1993   Chickens, fat 1995–2009

FIGURE 14.13   Residues of PCBs in poultry fat in Poland.

14.3.5 HUMAN MILK

Estimation of human exposure to chemical contaminations can be conducted using biological monitoring, by examination of their content in tissues and body fluids, or with the use of environmental monitoring in which the level of contaminations is assessed in food, water, and air. The main source of exposure to chlorinated aromatic hydrocarbons in human population is food of animal origin, which contributes up to 90% of the total daily intake.

Since 1973, the PIWet Pulawy in cooperation with the Medical University in Lublin, began the monitoring the organochlorine pesticides and PCB content in women's milk and adipose tissue in order to estimate the exposure of the Polish population to chlorinated hydrocarbons [2,36,39,40]. One thousand women's milk samples from different regions of the country were examined until 2003, in a few years' interval. The studies showed significantly higher concentrations of DDTs and PCBs in the investigated samples in comparison to animal tissues. During the period of monitoring, an over 20-fold decrease in the levels of DDTs was noted (Figures 14.14 and 14.15). At the same time, the concentrations of PCBs in women's milk did not show changes as significant as in the case of DDT. The level of PCBs in women's milk samples from industrial regions (Katowice) was higher than in the case of rural regions (Lublin). The obtained results did not differ significantly from the results presented by other authors [22,41–47]. However, comparison of the results was difficult due to the differences in regions of the country, number of investigated samples, methods of samples' collection, limits of detection, calculations of PCBs concentrations, and others.

Food for infants and small children is a subject of special interest. The highest concerns regard the presence of chemical contaminations in women's milk, and fresh and powdered cow milk, which are main, important nutrition factors, and are found as components of a variety of products for infants and children. Our studies (conducted in 2003) of the content of chlorinated hydrocarbons

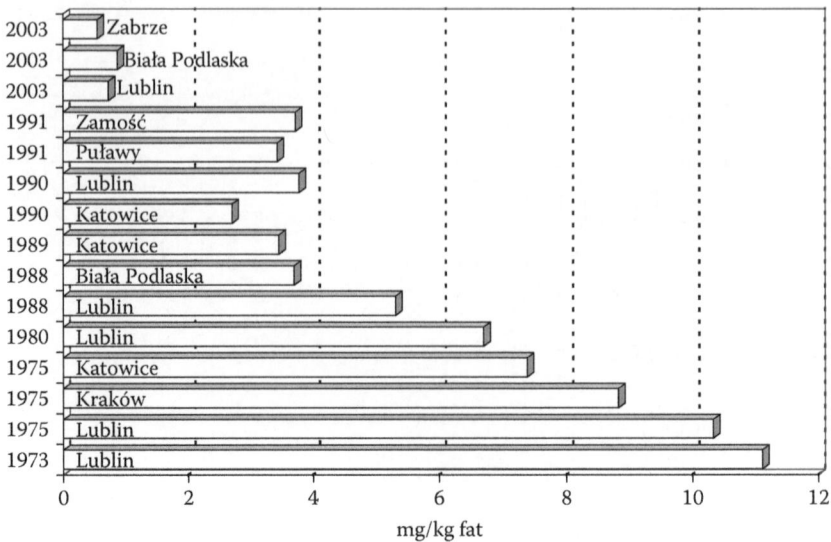

FIGURE 14.14   Residues of total DDT in human milk in Poland (mg/kg of fat).

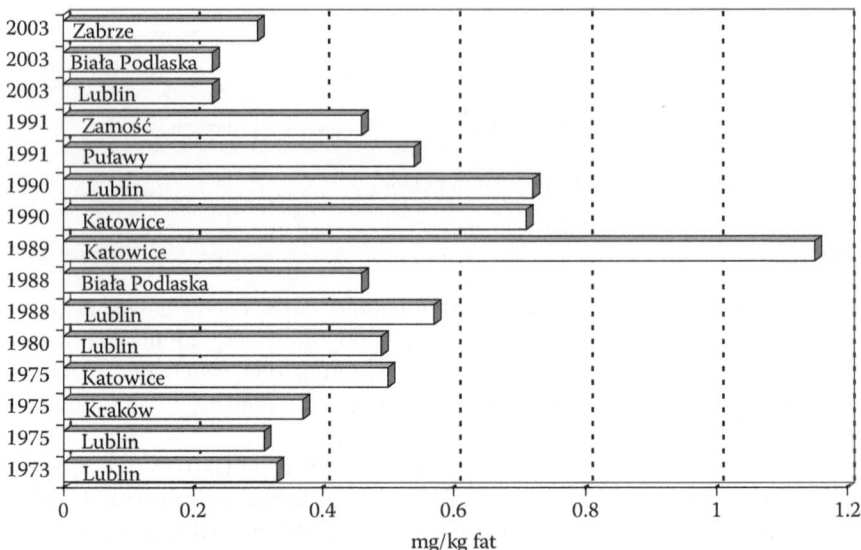

FIGURE 14.15   Residues of PCBs in human milk in Poland (mg/kg of fat).

in women's milk showed the presence of DDTs and PCBs in all 240 investigated samples, HCB in 79% of the samples, and HCHs (mainly β-HCH) in 47% of the analyzed samples. The average concentrations of pesticides, calculated in the milk fat were as follows: DDTs, 0.707 mg/kg, HCB, 0.019 mg/kg, HCHs, 0.012 mg/kg, and PCBs, 0.253 mg/kg. Assuming that the daily requirement of infants for milk is 1 L (an average of 150 mL/kg bw) and taking the average concentrations in women's milk (the average milk fat—3.4%), it can be estimated that the infants in Poland received 24 μg of DDTs, 8.6 μg of PCBs, 0.6 μg of HCB, and 0.4 μg of HCHs with mother's milk. In case of cow milk consumption, the infants and small children received daily about 2 μg of DDTs, 0.4 μg of PCBs, and 0.15 μg of HCB and HCHs. It must be underlined that the amount of organochlorine pesticides received by infants from mother's milk was significantly higher than the amount consumed from the food of animal origin in the population of adults.

14.4   CONCLUSIONS

The results of the long-term studies show a significant decrease in the levels of organochlorine pesticides in the national food of animal origin during the first years of monitoring, when the prohibition of their use has been introduced. Additionally, a continuous, favorable tendency of a constant, slow decrease in the content of pesticides was observed. Although the presence of organochlorine pesticides (50%) and PCBs (35%) residues was still detected in the animal fat samples, their concentrations were mostly on the level of hundredth, or thousandth parts of mg/kg, which constitutes only a few percent of the limit values for these compounds.

The evaluation of the occurrence and the content of organochlorine pesticides (DDTs, HCB, and HCHs) and PCBs in animal tissues and products of animal origin, and the assessment of the daily intake of these contaminants have allowed Polish food of animal origin to be considered safe for consumers.

REFERENCES

1. POPs 2001. Stockholm convention on persistent organic pollutants. http://www.pops.int/documents/convtext/convtext_en.pdf
2. Juszkiewicz, T. and Niewiadowska, A. 1984. Residues of pesticides and polychlorinated biphenyls in animal tissues, milk, eggs and environment in the light of 15-years of their own studies. *Medycyna Wet* 40: 323–327.
3. Council Directive 96/23/EC of 29 April 1996 on measures to monitor certain substances and residues thereof in live animals and animal products and repealing Directives 85/358/EEC, 86/469/EEC and Decision 89/187/EEC and 91/664/EEC. O. J. 1996, L 125, 10-31.
4. Zmudzki, J., Niewiadowska, A., and Wojton, B. 2005. National veterinary residue control program in animal tissues and food of animal origin. *Medycyna Wet* 61: 649–653.
5. Piskorska-Pliszczynska, J., Lizak, R., Maszewski, S. et al. 2009. Survey of persistent organochlorine contaminants (PCDD, PCDF, dl-PCB) in Baltic fish and fish meals. *Bull Vet Inst Pulawy* 53: 825–831.
6. Piskorska-Pliszczynska, J., Malagocki, P., Lizak, R. et al. 2008. Survey of dioxins and dioxin-like compounds in animal feed in Poland. *Organohalogen Compd* 70: 2051–2052.
7. Report of GF/POL/01/004 Project 2003. Enabling activities to facilitate early action on implementation of the Stockholm Convention on Persistent Organic Pollutants. National POPs Profile—Poland, Warsaw.
8. Falandysz, J. and Szymczyk, K. 2001. Data on the manufacture, use, inventory and disposal of polychlorinated biphenyls (PCBs) in Poland. *Polish J Environ Stud* 10: 189–193.
9. Michna, W. et al. 1995–2003. Reports on monitoring investigations on quality of soil, plants, food and agricultural products. Ministry of Agriculture and Rural Development, Warsaw, Poland.
10. Niewiadowska, A. 2007. Exposure assessment of organochlorine insecticides and PCB in food of animal origin. Monograph ISBN 978-83-89946-01-0. PIWet Pulawy.
11. Regulation (EC) No 396/2005 of the European Parliament and of the Council of 23 February 2005 on maximum residue levels of pesticides in or on food and feed of plant and animal origin and amending Council Directive 91/414/EEC. O. J. 2005, L 70, 1–16.
12. Falandysz, J. and Kannan, K. 1992. Organochlorine pesticides and polychlorinated biphenyl residues in slaughtered and game animal fats from the northern part of Poland. *Z Lebensm Unters Forsch* 195: 17–21.
13. Falandysz, J. 2000. Chlordane intake with seafood in Poland. *Roczniki PZH* 51: 229–239.
14. Ludwicki, J.K., Góralczyk, K., and Czaja, K. 1992. Residues of organochlorine insecticides in foods in 1986–1990. *Roczniki PZH* 43: 21–31.
15. Przybycin, J. and Juszkiewicz T. 1993. Residues of polychlorinated biphenyls in game animals. *Medycyna Wet* 49: 318–319.
16. Szymczyk-Kobrzynska, K. and Zalewski, K. 2003. DDT, HCH and PCB residues in fat of red deer (*Cervus Elaphus*) from the region of Warmia and Mazury, 2000–2001. *Polish J Environ Stud* 12: 613–617.
17. Zasadowski, A. 1994. Polychlorinated biphenyls (PCBs) in the adipose tissue of wild boars and roe-deer in the region of Warmia and Mazuria. *Polish J Environ Stud* 3: 43–45.

18. Zasadowski, A. 1995. Residues of polychlorinated biphenyls (PCBs) in the adipose tissue of domestic animals (pigs, cattle) in the region of Warmia and Mazuria. *Polish J Environ Stud* 4: 65–67.

19. Zasadowski, A. 1997. Residues of organochlorine pesticides and polychlorinated biphenyls (PCBs) in adipose fat of geese and turkeys from the region of Warmia and Mazuria. *Polish J Environ Stud* 6: 41–44.

20. Zmudzki, J., Niewiadowska, A., Kowalski, B. et al. 1994. Chemical residues in pork hams. *Medycyna Wet* 50: 623–625.

21. Zmudzki, J., Niewiadowska, A., Szkoda, J. et al. 2001. Toxic contaminants in food of animals origin in Poland. *Medycyna Pracy* 52: 35–40.

22. Goralczyk, K., Czaja, K., and Ludwicki, J.K. 1996. Biological and environmental monitoring of exposure to chlorinated aromatic hydrocarbons. *Roczniki PZH* 47: 25–31.

23. Goralczyk, K., Ludwicki, J.K., Czaja K. et al. 1998. Monitoring of pesticide residues in food products in Poland. *Roczniki PZH* 49: 331–339.

24. Goralczyk, K., Strucinski, P., Hernik, A. et al. 2005. Monitoring and official control of pesticide residues in foodstuffs in Poland in 2004. *Roczniki PZH* 56: 307–316.

25. Ludwicki, J.K., Czaja, K., and Strucinski, P. 1996. An attempt on health risk assessment for the environmental exposure to chlorinated aromatic hydrocarbons. *Roczniki PZH* 47: 33–39.

26. Falandysz, J. 1988. Estimation of dietary intake of polychlorinated biphenyls in Poland. *Roczniki PZH* 39: 366–373.

27. Falandysz, J. 1988. Estimation of the intake of polychlorinated biphenyls with fish consumed in Poland. *Roczniki PZH* 39: 450–453.

28. Falandysz, J., Szymczyk, K., and Gucia, M. 2001. Intake of hexachlorobenzene with food in Poland. *Polish J Environ Stud* 10: 155–160.

29. Niewiadowska, A., Zmudzki, J., and Semeniuk, S. 2005. Pesticide residues in food of animal origin in 1997–2006 in Poland. *Medycyna Wet* 64: 1221–1224.

30. Niewiadowska, A., Zmudzki, J., and Semeniuk, S. 2010. Residues of polychlorinated biphenyls in food of animal origin. *Medycyna Wet* 66: 259–263.

31. Juszkiewicz, T., Niewiadowska, A., Stec, J. et al. 1991. Xenobiotics in animal tissues and food products in Poland. *Bull Vet Inst Pulawy* 34: 1–14.

32. Niewiadowska, A., Zmudzki, J., and Semeniuk, S. 1995. Residues of chlorinated hydrocarbons in milk. *Roczniki PZH* 46: 113–117.

33. Niewiadowska, A., Zmudzki, J., and Semeniuk, S. 1995. Contamination of hens by chlorinated hydrocarbon residues. *Medycyna Wet* 51: 346–348.

34. Niewiadowska, A., Semeniuk, S., and Zmudzki, J. 1996. Residues of chlorinated hydrocarbons in cheeses. *Roczniki PZH* 47: 371–376.

35. Niewiadowska, A. and Zmudzki, J. 1996. Chlorinated hydrocarbons in food of animal origin. *Roczniki PZH* 47: 59–64.

36. Juszkiewicz, T., Niewiadowska, A., Posyniak A. et al. 1983. Residues of organochlorine insecticides and polychlorinated biphenyls in cattle and human milk. *Przeglad Lekarski* 40: 521–523.

37. Niewiadowska, A., Zmudzki, J., and Semeniuk, S. 1996. Residues of organochlorine pesticides and polychlorinated biphenyls (PCB) in hen eggs. *Bromat Chem Toksykol* 29: 79–83.

38. Niewiadowska, A. and Juszkiewicz, T. 1978. Residues of polychlorinated biphenyls in milk. *Bull Vet Inst Pulawy* 22: 30–35.

39. Juszkiewicz, T., Sikorski, R., Niewiadowska, A. et al. 1979. Polychlorinated biphenyl residues in adipose tissue and human milk. *Ginekol Pol* 50: 917–922.

40. Sikorski, R., Paszkowski, T., Radomanski, T. et al. 1990. Human colostrum as a source of organohalogen xenobiotics for a breast-fed neonate. *Reprod Toxicol* 4: 17–20.

41. Czaja, K., Ludwicki, J.K., Góralczyk, K. et al. 1997. Organochlorine pesticides, HCB, and PCBs in human milk in Poland. *Bull Environ Contam Toxicol* 58: 769–775.

42. Falandysz, J., Yamashita, N., Tanabe, S. et al. 1994. Congener data of polychlorinated biphenyl residues in human adipose tissue in Poland. *Sci Total Environ* 149: 113–119.

43. Hernik, A., Goralczyk, K., Czaja, K. et al. 2009. Polybrominated diphenyl ethers (PBDEs), polychlorinated biphenyls (PCBs) and organochlorine pesticides in human milk in Poland. *Organohalogen Compd* 71: 254–256.

44. Jaraczewska, K., Lulek, J., Covaci, A. et al. 2006. Distribution of polychlorinated biphenyls, organochlorine pesticides and polybrominated diphenyl ethers in human umbilical cord serum, maternal serum and milk from Wielkopolska region, Poland. *Sci Total Environ* 372: 20–31.

45. Lulek, J., Polanska, A., Szyrwinska, K. et al. 2002. Levels of selected polychlorinated biphenyls in human milk from Wielkopolska region in Poland. *Fresenius Environ Bull* 11: 102–107.
46. Pietrzak-Fiecko, R. and Smoczynski, S. 2001. Organochlorine insecticides in human milk in Olsztyn in the years 1976, 1986, 1996. *Roczniki PZH* 52: 55–59.
47. Szyrwinska, K. and Lulek, J. 2007. Exposure to specific polychlorinated biphenyls and some chlorinated pesticides via breast milk in Poland. *Chemosphere* 66: 1895–1903.

15 Temporal Trends of Organohalogen Compounds in Mother's Milk from Sweden

Anders Glynn,* Sanna Lignell, Marie Aune,
Per Ola Darnerud, and Anna Törnkvist

CONTENTS

15.1 Introduction 355
15.2 The POPUP Study 1996–2008 356
    15.2.1 Recruitment and Sampling 356
    15.2.2 Analytical Methods and Calculations 357
    15.2.3 Design of the POPUP Study 359
    15.2.4 Representativeness of the POPUP Study 360
15.3 The Stockholm Study 360
15.4 Temporal Trends of PCBs and PCDD/Fs in Sweden....................................... 361
    15.4.1 The POPUP Study 361
    15.4.2 Comparisons with the Stockholm Study 364
    15.4.3 Comparisons with Temporal Trends in Other Matrices 367
15.5 Temporal Trends of Chlorinated Pesticides and Metabolites.......................... 369
    15.5.1 DDT Compounds... 369
    15.5.2 HCB, HCH Compounds, and Chlordane... 370
        15.5.2.1 HCB 370
        15.5.2.2 HCH Compounds... 371
        15.5.2.3 Chlordane Compounds 371
15.6 Temporal Trends of Brominated Flame Retardants 372
15.7 Levels and Temporal Trends in Other Regions of the World 372
    15.7.1 PCBs and PCDD/Fs... 372
    15.7.2 Chlorinated Pesticides/Metabolites 373
    15.7.3 Polybrominated Diphenyl Ethers.. 374
15.8 Conclusions... 375
Acknowledgments... 375
References... 375

15.1 INTRODUCTION

This chapter deals with levels and trends of organohalogen pollutants in mother's milk from Sweden. Man-made organohalogen compounds have been produced for industrial, agricultural, and household use. Some of these chemicals have persisted in the environment, entered

---

* E-mail: anders.glynn@slv.se (Chapter corresponding author).

the biological food webs, bioaccumulated and biomagnified in the food chains, and ultimately have reached humans. Many of these contaminants are present at levels that are of no health risk to humans. However, some chemicals have accumulated in the human body up to levels that are suspected to cause negative health effects. Well-known examples of such compounds are the industrial chemical polychlorinated biphenyls (PCBs), the unintentionally formed polychlorinated dibenzo-p-dioxins and dibenzofurans (PCDD/Fs), chlorinated pesticides (DDT, hexachlorobenzene [HCB], hexachlorocyclohexanes [HCHs], and chlordane), and the brominated flame retardant polybrominated diphenyl ethers (PBDEs). In Sweden and in many other countries, the production and use of PCBs and the chlorinated pesticides have been banned or restricted for many decades [1]. Great efforts have been made to reduce the emissions of PCDD/Fs due to human activities [2].

Temporal trend studies of organohalogen compound levels in mother's milk have in Sweden been used as a follow-up tool of the effects of risk management decisions regarding the emissions of the compounds into the environment. In the Stockholm/Uppsala area, mother's milk has been sampled since 1967 [3,4]. The samples have been banked and the environmental pollutants have been measured in the banked samples when financial support for the expensive analyses became available.

In this chapter, the temporal trend studies of organohalogen compound levels in mother's milk from the Stockholm/Uppsala area are described. Data for the period 1967–2004 have been published by Koidu Norén and coworkers [3,5]. Temporal trends of PCB, PCDD/F, and PBDE levels among women participating in the Persistent Organic Pollutants in Uppsala Primiparas (POPUP) study between 1996 and 2006 have been published by Lignell et al. [4]. New data from 2008 have now been added to the POPUP temporal trends, as well as earlier unpublished data regarding temporal trends of chlorinated pesticides/metabolites. Furthermore, in this chapter, the temporal trends in Swedish mother's milk are compared with temporal trends reported for biota in the Swedish environment (the Baltic Sea) and for food or food-producing animals in Sweden. The results are put into an international perspective by comparing both trends and levels of the studied compounds with data reported from other parts of the world.

In Sweden, the use of the pesticide dichlorodiphenyltrichloroethane (DDT) was banned in the 1970s, but more than 100 metric tons/year were used on arable land in the 1960s. The other chlorinated pesticides HCB, HCH, and chlordane were less extensively used in Sweden before being banned. During the period 1957–1980, 8,000–10,000 metric tons of PCBs were imported. The current major sources of PCDD/F emissions in Sweden are iron and steel manufacturing, concrete production, pulp and paper production, fossil fuel combustion, biofuel combustion, waste incineration, landfill fires, and backyard burning. The estimates of the reductions of PCDD/F emissions and the current quantities of emissions are however uncertain [2]. The use of brominated flame retardants, among others PBDEs, in products on the Swedish market has decreased from between 600 and 400 metric tons/year in 1993–2000 to 100 metric tons in 2006–2007 [6].

15.2 THE POPUP STUDY 1996–2008

15.2.1 RECRUITMENT AND SAMPLING

The recruitment of the POPUP cohort commenced in 1996 with the aim to determine the temporal trends of organohalogen compound exposure of pregnant women and their breast-fed first-born children (Table 15.1). In total, 356 women donated mother's milk from 1996 to 2008.

Between 1996 and 1999, pregnant women, carrying their first child and living in Uppsala County, in late pregnancy (week 32–34) were asked to participate in a study of organohalogen compound exposure of pregnant and nursing women ($N=405$) [7]. Of these women, 211 (52%) agreed to donate mother's milk for chemical analysis. In the temporal trend study, data from women born in the Nordic countries ($N=203$) were used since these women had significantly higher blood serum PCB levels and lower levels of some of the chlorinated pesticides/metabolites than women not born in the Nordic countries [8].

TABLE 15.1
Personal Characteristics of the Women in the POPUP Study[a]

Year (N)	Age (Years)	Pre-Pregnancy BMI (kg/m²)	Weight Gain Pregnancy (% per Week)	Weight Loss (% from Delivery)
1996 (20)	30±5	21.7±2.1	0.61±0.21	9.1±3.5
1997 (67)	28±4	23.3±3.6	0.62±0.20	9.6±3.1
1998 (90)	28±4	22.8±3.5	0.65±0.22	10±3.3
1999 (26)	28±4	22.4±3.0	0.64±0.19	9.4±3.9
2000 (18)	30±4	24.0±5.0	0.60±0.26	8.8±3.3
2001 (11)	29±5	23.5±4.1	0.58±0.36	7.9±2.8
2002 (26)	30±3	22.0±2.1	0.63±0.15	8.8±3.3
2003 (5)	28±3	21.2±1.1	0.61±0.17	6.1±2.7
2004 (32)	29±4	22.7±2.7	0.65±0.23	10±3.0
2006 (30)	30±4	23.1±3.2	0.60±0.23	9.4±2.8
2008 (31)	29±5	23.3±3.7	0.62±0.19	9.3±3.1

[a] Mean±SD. N=number of study participants. No statistically significant temporal trend in personal characteristics were found (simple linear regression, $p > 0.05$).

From year 2000 to 2004, milk was sampled every year, and from 2004 every second year. In this part of the study, 153 primiparous mothers were randomly recruited among women delivering their first child at the Uppsala University Hospital [4]. During each sampling period, 29–32 women (46%–63% of approached women) participated in the study, donating mother's milk sampled during the third week after delivery.

The mother's milk was sampled by the mothers themselves at home during the third week after delivery (day 14–21 postpartum) [9]. A manual mother's milk pump and/or a passive mother's milk sampling cup were used for sampling during normal breast-feeding sessions. The women were instructed to sample milk at the beginning and at the end of sessions. The objective was to sample up to 500 mL milk from each mother during 7 days of sampling. The mother's milk was stored at home by the mother and frozen in acetone-washed glass bottles. Newly sampled milk was poured on top of the frozen milk. After transport to the laboratory in a frozen state, the milk was stored at −20°C.

15.2.2 Analytical Methods and Calculations

The specific compounds that were analyzed in the mother's milk samples from the POPUP study are presented in Table 15.2. PCBs (with a few exceptions, see below) and chlorinated pesticides/metabolites were analyzed by using gas chromatography with dual capillary columns and electron capture detection (GC/ECD) at the National Food Administration (NFA), Sweden, using previously described methods [10,11]. The PBDEs were also analyzed with GC/ECD at the NFA using a method described in Atuma et al. [12], with a few modifications. All samples were fortified with internal standards prior to extraction to correct for analytical losses and to ensure quality control. A number of control samples were analyzed together with the samples to verify the accuracy and precision of the measurements. The laboratory is accredited for analysis of PCBs, chlorinated pesticides, and brominated flame retardants in human milk.

PCDDs and PCDFs were analyzed using isotope dilution and gas chromatography with high-resolution mass spectrometry (GC/HRMS) at the National Institute of Public Health and the Environment (RIVM), the Netherlands, from 1996 to 2004. The method is described by Glynn et al. [9]. In 2006, the analyses were performed by using isotope dilution and GC/HRMS at Umeå

TABLE 15.2
Summary of Compounds That Were Analyzed in the Mother's Milk Samples

Compound	Congeners, Metabolites, etc.
PCBs	
PCBs	28, 52, 101, 105, 114, 118, 138, 153, 156, 157, 167, 170, 180
Non-*ortho* PCBs	77, 126, 169, 81
Chlorinated pesticides[a]	
Hexachlorobenzene (HCB)	
Hexachlorocyclohexane	α-HCH, β-HCH, γ-HCH
Chlordane	Oxychlordane, trans-nonachlor
DDT	*p,p*′-DDE, *p,p*′-DDT, *o,p*′-DDT, *p,p*′-DDD, *o,p*′-DDE
PCDD/Fs	
PCDDs	7 congeners
PCDFs	10 congeners
Brominated flame retardants[b]	
PBDE	28, 47, 66, 99, 100, 138, 153, 154, 183
HBCD	

[a] γ-HCH, *o,p*′-DDT, *p,p*′-DDD, *o,p*′-DDE were not included in the Stockholm study.

[b] PBDE 183 was not included in the Stockholm study. The additional PBDE congeners 85 and 209 were included in the Stockholm study.

University, Sweden [13]. Non-*ortho* PCBs were analyzed using isotope dilution and GC/HRMS at RIVM (1996–1999) and Umeå University (2006), or by using isotope dilution and gas chromatography with low-resolution mass spectrometry (GC/LRMS) at the NFA (2000–2004) [4].

In order to confirm that there were no significant differences in analytical results from the involved laboratories, interlaboratory studies have been performed. In 2000, a calibration study was performed, in which 26 samples were analyzed for non-*ortho* PCBs at both the NFA and RIVM in order to calibrate the results of the two laboratories. CB 126 is the non-*ortho* PCB congener that gives by far the largest contribution to the concentrations of PCB toxic equivalents (TEQs). A comparison of the CB 126 results of the two laboratories showed no systematic errors (Wilcoxon signed rank test, $p = 0.798$). During 2006, a calibration study was performed comparing the PCDD/F, CB 126, and CB 169 results in 10 samples analyzed both at RIVM and Umeå University. Results of non-*ortho* PCB analyses performed at the NFA and Umeå University were also compared. The comparison showed that the differences between the laboratories were small [4]. Correction of analytical results for these slight interlaboratory differences in levels of PCDD/Fs and non-*ortho* PCBs resulted in very small differences in the estimated median total TEQ concentrations (<1%) and in the temporal trends (annual changes between 1996 and 2006) of TEQs (3%–12%) [4]. Uncorrected results were therefore used in the studies of temporal trends.

In cases when levels of individual compounds were below the limit of quantification in more than 25% of the mother's milk samples, statistical analyses of temporal trends were not performed concerning these individual compounds. This was evident for PCB congeners CB 52, CB 77, CB 101, and CB 105, the chlorinated pesticides/metabolites α-HCH, γ-HCH, *o,p*′-DDT, *p,p*′-DDD, and *o,p*′-DDE, the PCDFs 1,2,3,7,8,9-HxCDF, and 1,2,3,4,7,8,9-HpCDF, and the PBDEs BDE 28, BDE 66, BDE 138, BDE 154, and BDE 183 (Table 15.2). Moreover, no temporal trends were determined for the PCB congeners CB 81, CB 114, CB 157, and CB 170 since these congeners were not measured in the early time period of the study. Calculations of ΣPCBs included congeners CB 28,

CB 52, CB 101, CB 105, CB 118, CB 138, CB 153, CB 156, CB 167, and CB 180. In the ΣPBDE calculation, congeners BDE 47, BDE 99, BDE 100, BDE 153, and BDE 154 were included. The levels of PCDD/Fs and dioxin-like PCBs were summed up as toxicity equivalents (TEQs) using the WHO_{2005} toxicity equivalent factors (TEFs) [14]. In cases when comparisons were made with data based on WHO_{1998} TEFs, TEQ levels were calculated using WHO_{1998} TEFs [15]. In cases when analytical results of single compounds were below the limit of quantification in these calculations, the value of 1/2 of the limit of quantification was used. Mother's milk concentrations of POPs were lipid-adjusted, since the studied compounds mainly distribute within the lipid compartment in human tissues [16]. Statistical analyses were performed on logarithmically transformed organohalogen compound data because the distribution of data closely followed a log-normal distribution.

15.2.3 Design of the POPUP Study

Mother's milk is a good matrix for the determination of body burdens of organohalogen compounds among pregnant and nursing women. The milk has a relatively high lipid content, which simplifies the detection of the lipophilic compounds. Moreover, the body burdens of organohalogen compounds among pregnant and nursing women determines the exposure of the developing fetus and nursed infant [17,18]. Humans seem to be susceptible against organohalogen compound exposure early in life [19,20].

One of the important features of the POPUP study is that the samples donated by the participating women are banked for future analyses of currently "unknown" contaminants of mother's milk. Moreover, the POPUP study is designed to have maximum statistical power to detect temporal trends of organohalogen compounds. The omission of women born outside the Nordic area caused a decrease in within-year and between-year variation of organohalogen compound body burdens [8,21,22]. The variation in mother's milk levels of the studied compounds was further restricted by the recruitment of primiparous women only. It is well known that multiparous women may have lower body burdens of organohalogen compounds than primiparous women due to elimination of the compounds during nursing [23,24]. In Sweden, nursing is encouraged and women have paid parental leave for at least 6 months, which is the nursing period recommended by the WHO [25].

The sampling of milk during a short time period (third week after delivery) ensured that variation in organohalogen compounds due to potential time-dependent alterations in organohalogen compound levels during the nursing period were minimized [24].

Earlier studies have shown that age of the mother, body mass index (BMI), weight increase during pregnancy, and weight loss during the period between delivery and milk sampling were significant determinants of the milk levels of PCBs, PCDD/Fs, and several of the chlorinated pesticides [4]. Data on these maternal characteristics were obtained via questionnaires (Table 15.1). This enabled us to adjust the temporal trends of organohalogen compounds for possible temporal trends in these personal characteristics that are strongly associated with organohalogen compounds' body burdens. Moreover, the detailed knowledge about these personal characteristics makes it possible to determine how they influence the observed temporal trends of organohalogen compound levels.

The age dependency of levels of many PCBs, PCDD/Fs, and chlorinated pesticides/metabolites in mother's milk is most probably caused by an age-dependent accumulation of the persistent organohalogenated compounds in combination with a birth cohort effect. This effect is caused by the fact that women in their 40s have experienced higher PCB and PCDD/F exposures during childhood and adolescence than 20-year-old women since the older women grew up during the 1960s–1970s when the levels of these substances were at their peak in the Swedish environment [3,26,27].

The weight increase during pregnancy is negatively associated with the mother's milk levels of many of the studied organohalogen compounds [4]. The body mass increases during pregnancy and this may at least partially cause a dilution effect of the organohalogen compounds accumulated in the adipose tissue before pregnancy [8]. An opposite effect is caused by the weight loss in the period between delivery and mother's milk sampling. This is reflected in the positive association between weight loss after delivery and organohalogen compound levels in mother's milk [4]. The negative association between pre-pregnancy BMI and organohalogen compound levels is also probably due to a dilution effect caused by weight increase before pregnancy [8].

The study design with analyses of individual samples gives information not only about the average levels of organohalogenated compounds in mother's milk, but also about the range of mother's milk levels in the population of primiparous women. The knowledge about the range of mother's milk levels of organohalogen compounds is important for the health risk assessment of the studied compounds in the mothers, fetuses, and nursed infants.

15.2.4   REPRESENTATIVENESS OF THE POPUP STUDY

The participation rate in the POPUP mother's milk study was about 50%. Thus, it may be questioned if the results are representative for the whole Swedish population of women nursing their first child. In the early phase of the study between 1996 and 1999, we looked at possible differences in age and pre-pregnancy BMI between the women who agreed to participate in the study in late pregnancy and the women who declined to participate. Both age of the women and pre-pregnancy BMI are personal characteristics that are significantly associated with the mother's milk levels of many of the studied organohalogen compounds [4]. The pre-pregnancy BMI did not differ between the two groups of women. Those who declined had a median pre-pregnancy BMI of $22.7 \, kg/m^2$ (min-max: $17.8–39.3 \, kg/m^2$) and women agreeing to participate had a median BMI of $22.2 \, kg/m^2$ (min-max: $16.6–43.6 \, kg/m^2$) (Mann–Whitney U test, $p = 0.98$). The women who declined were slightly younger ($N = 58$, median age: 26 years, min-max: 18–39 years) than women agreeing to participate ($N = 325$, median age: 28, min-max: 18–41) (Mann–Whitney U test, $p = 0.003$). Taking into account that an increased age is strongly associated with a higher level of many of the studied compounds in mother's milk [4], the results suggest that women declining to participate in the study on average had slightly lower organohalogen compound levels than women agreeing to participate.

In a comparison of mother's milk levels of PCBs, p,p'-DDE, and PBDE among primiparous women from four different regions of Sweden from south to north, which included a subsample of the Uppsala POPUP women, only slight differences in geometric mean levels were observed between regions (less than twofold) [28]. The variation in levels of the organohalogen compounds was much larger within regions with an almost complete overlap between regions. This suggests that the PCB, p,p'-DDE, and PBDE levels observed in the POPUP cohort is representative of the levels found among primiparous women from the general population in other areas of Sweden.

Food is currently the major source of human exposure to the organohalogenated compounds in Sweden, perhaps with the exception of PBDEs, which may also have consumer products as a source of exposure [29]. Sweden has a well-developed food and consumer product distribution system, reaching all parts of the country. The temporal trends observed in the POPUP study is therefore most likely representative of the general temporal trends in human exposure of the compounds in Sweden.

15.3   THE STOCKHOLM STUDY

In 1970, Westöö and Norén published the first organohalogen compound data for mother's milk from the Stockholm area [30]. Milk was purchased from the Mother's Milk Center in Stockholm, and samples were pooled in the laboratory. The pooled samples were composed of milk from 20 to 210 mothers. Samples were acquired every year or every other year from 1967 to 1980, during the 1980s every fourth year, and between 1990 and 1997 milk was purchased each year or every other year [3]. From 1998 to 2004, milk was sampled every year [5]. From 1972, the samples were banked for future analyses. Between 55% and 80% of the donators were nursing their first child, and after 1985 the donors were nonsmokers and the milk was collected by the mothers during the first 3 months of the nursing period [3]. During the period 1972–1985, the average age of the mothers was 27–28 years, and in 1996–1997 it was 30–31 years.

The organohalogen compounds analyzed in the samples were, among others, PCBs, PCDD/Fs, chlorinated pesticides/metabolites, and PBDEs (Table 15.2) [3,5]. PCBs, PCDD/Fs, chlorinated pesticides/metabolites, and brominated flame retardants were analyzed at Karolinska Institutet, Sweden, and brominated flame retardants at Stockholm University, Sweden. The analyses were performed by GC/ECD and GC/HRMS. The methodology is in general similar to the one used in the POPUP study [3,5].

15.4   TEMPORAL TRENDS OF PCBS AND PCDD/FS IN SWEDEN

15.4.1   THE POPUP STUDY

The raw data shows that the levels of all studied PCB congeners have declined in mother's milk among the POPUP women between 1996 and 2008 (Table 15.3, Figure 15.1). The mean ΣPCB decreased from 190 ng/g lipid to 70 ng/g lipid, an almost threefold decrease. A similar, almost threefold decline was evident for the di-*ortho* congener CB 153 (Figure 15.1), the PCB congener being present at the highest levels. On average, CB 153 contributed 42% to the ΣPCB level (10 congeners). Linear regression showed a strong positive association between the levels of CB 153 and ΣPCB over the whole time period (Figure 15.2), showing that CB 153 is a good indicator of the total PCB level in mother's milk.

As seen in Table 15.3 and Figure 15.1, the mean levels of many of the studied PCDD/F congeners, similarly as in the case of PCBs, decreased in mother's milk from the participating women between 1996 and 2006. For some PCDF congeners, however, a decline in levels is not obvious when looking at the raw data (Table 15.3).

The simple regression analysis showed a similar rate of decline (6%–8% per year) of several of the mono- and di-*ortho* PCBs (Table 15.4). A slower decline was however evident for CB 28, CB 156, and CB 169. The slower decline of CB 169 could partly be due to the levels being closer to the quantification limit than in the case of the other congeners, as shown by a relative high number of samples with CB 169 levels below the quantification limit. In this case, 17% of the samples had levels below the quantification limit between 1996 and 1998 whereas 50% of the samples had levels below the limit of quantification between 2004 and 2006. This most probably resulted in an underestimation of the rate of decline in CB 169 levels in mother's milk in the POPUP study.

For CB 28, several studies have shown that the indoor environment may contribute to human CB 28 exposure due to the presence of high levels of this congener in some building materials such as sealants [31,32]. It may be hypothesized that a continuous CB 28 exposure in the indoor environment among some of the participating individuals may at least partially explain the slower temporal decline of this congener in mother's milk.

The regression analysis of the data revealed a large variation in rates of decline between different PCDD/F congeners (Table 15.4). Among the PCDFs, no significant temporal trend was observed for TCDF, 1,2,3,7,8-PeCDF, and 2,3,4,6,7,8-HxCDF. Over 15% of the mother's milk samples had levels of the two first congeners that were below the LOQ, which could have contributed to the nonsignificant trends of levels of these compounds. Nevertheless, the annual change in levels of PCDF congeners appeared to be slower than for PCDDs. This suggests that levels of PCDFs have declined more slowly in food on the Swedish market than the levels of PCDD. Support for a slower decline of PCDFs is given by a study of PCDD/F levels in serial blood samples from 26 Swedish men [33]. The levels of 5 out of 7 PCDDs had declined between 1987 and 2002, whereas only 2 out of 8 PCDFs had declined significantly during the study period [33]. Moreover, in a temporal trend study of PCDD/Fs in guillemot eggs from the Baltic Sea, PCDFs did not show the same decreasing trend as PCDDs during the most recent 10 years [34]. The slower decline of PCDF in the Swedish environment and in mother's milk may be due to a higher persistence of PCDFs or to a difference in anthropogenic output of PCDDs and PCDFs into the environment. The slower rate of decrease in PCDF levels resulted in a significant increase in the ratio between PCDF and PCDD WHO$_{2005}$ TEQ

TABLE 15.3

Levels of PCBs, PCDDs, and PCDFs in Mother's Milk from Primiparous Women from Uppsala County Participating in the POPUP Study between 1996 and 2008[a]

PCB	1996	1997	1998	1999	2000	2001	2002	2003	2004	2006	2008
	N=20	N=67	N=90	N=26	N=18	N=11	N=26	N=5	N=32	N=30	N=31
CB 28	3.3±2.5	3.9±5.4	2.9±4.3	2.2±1.8	2.3±1.0	1.9±0.9	1.7±1.4	2.2±2.9	1.9±0.7	2.8±5.2	1.3±1.0
CB 118	15±6	15±10	12±5	11±5	11±5	12±6	8.4±2.9	6.8±2.8	7.1±3.0	6.5±2.8	5.7±3.5
CB 126	58±28	50±24	56±25	47±19	47±22	48±24	36±13	35±10	29±14	31±12	
CB 138	39±19	35±16	31±12	30±12	26±8	25±9	25±8	20±5	20±7	20±8	16±11
CB 153	81±37	70±29	61±26	62±26	56±20	53±25	48±16	37±9	38±13	36±17	31±21
CB 156	6.3±3.1	5.4±3.3	4.9±2.5	4.2±2.5	3.7±1.6	3.6±1.8	3.3±1.6	2.3±0.8	3.4±1.4	3.6±1.5	3.1±2.1
CB 167	1.7±0.9	1.4±0.9	1.3±0.8	1.3±0.8	1.8±0.7	1.9±0.9	0.9±0.4	1.0±0.4	1.1±0.5	0.8±0.4	0.6±0.4
CB 169	22±13	21±11	28±12	25±11	26±16	22±12	24±12	16±10	19±10	19±9	
CB 180	39±18	32±13	30±12	28±12	26±11	26±13	23±9	17±5	19±8	19±7	15±11
ΣPCB[b]	188±82	164±70	144±57	141±57	130±44	126±54	113±36	86±24	91±32	91±38	74±50
PCDD/F	N=12	N=40	N=29	N=17	N=14	N=10	N=17		N=15	N=30	
TCDD	1.2±0.5	1.0±0.4	1.1±0.5	1.2±0.5	0.97±0.59	0.92±0.40	0.99±0.27		0.59±0.26	0.65±0.26	
1,2,3,7,8-PeCDD	2.8±1.3	2.4±0.7	3.0±1.0	3.3±1.1	2.3±1.1	2.7±0.7	2.7±0.7		1.7±0.4	1.7±0.6	
1,2,3,4,7,8-HxCDD	1.3±0.7	1.3±0.4	1.4±0.5	1.5±0.5	1.1±0.5	0.96±0.32	1.2±0.3		0.69±0.21	0.75±0.21	
1,2,3,6,7,8-HxCDD	11±5	10±4	9.6±2.8	9.2±3.4	9.3±3.0	7.7±2.7	6.8±2.4		5.3±1.5	5.5±1.9	
1,2,3,7,8,9-HxCDD	2.1±0.6	2.5±1.5	2.1±0.7	1.9±0.6	1.9±0.9	1.8±0.7	1.4±0.5		1.1±0.3	1.2±0.5	
1,2,3,4,6,7,8-HpCDD	20±11	21±12	20±12	12±7	18±12	18±8	9.4±3.4		8.6±3.0	8.8±4.0	

OCDD	97±29	100±55	88±23	74±39	98±44	78±35	60±26	54±19	47±20
TCDF	0.49±0.19	0.40±0.24	0.46±0.17	0.47±0.29	0.44±0.23	0.43±0.24	0.38±0.14	0.44±0.27	0.42±0.23
1,2,3,7,8-PeCDF	0.20±0.08	0.21±0.10	0.18±0.07	0.19±0.11	0.29±0.13	0.29±0.13	0.14±0.05	0.17±0.12	0.24±0.11
2,3,4,7,8-PeCDF	7.5±4.3	6.7±2.9	6.7±3.0	7.1±4.2	7.0±3.5	6.1±2.7	5.3±2.3	5.0±1.6	4.6±1.9
1,2,3,4,7,8-HxCDF	1.6±0.6	1.6±0.5	1.5±0.3	1.5±0.5	1.7±0.5	1.4±0.5	1.1±0.3	1.1±0.3	1.2±0.3
1,2,3,6,7,8-HxCDF	1.4±0.7	1.3±0.4	1.3±0.3	1.3±0.5	1.5±0.6	1.2±0.4	1.1±0.3	0.99±0.23	1.1±0.3
2,3,4,6,7,8-HxCDF	0.76±0.55	0.69±0.28	0.70±0.21	0.61±0.31	0.86±0.59	0.85±0.34	0.49±0.13	0.56±0.14	0.70±0.21
1,2,3,4,6,7,8-HpCDF	2.0±0.7	2.4±1.7	2.0±0.8	1.6±0.7	3.1±2.8	1.9±0.8	1.4±0.6	2.2±1.1	2.0±2.9
OCDF	0.48±0.40	0.43±0.28	0.54±0.21	0.33±0.10	0.60±0.49	0.48±0.33	0.29±0.11	0.61±0.40	0.27±0.24
PCDD WHO$_{2005}$ TEQ	5.6±2.5	5.1±1.4	5.7±1.7	5.9±2.1	4.7±1.9	4.4±1.5	4.7±1.3	3.1±0.8	3.2±1.0
PCDF WHO$_{2005}$ TEQ	2.7±1.5	2.5±1.0	2.4±0.9	2.5±1.4	2.6±1.2	2.3±0.9	1.9±0.8	1.8±0.6	1.8±0.6
PCDD/F WHO$_{2005}$ TEQ	8.4±3.9	7.6±2.3	8.1±2.5	8.4±3.4	7.3±3.0	6.7±2.4	6.6±2.0	4.2±1.3	5.0±1.8
PCB WHO$_{2005}$ TEQ[c]	7.1±3.4	6.3±2.7	7.1±3.0	5.9±2.3	6.2±2.7	5.7±0.9	4.6±1.8	3.9±1.6	4.0±1.5
Total WHO$_{2005}$ TEQ	15±7	14±4	15±5	14±6	13±6	13±5	11±3	9.0±2.6	9.0±3.0

[a] Mean±SD, levels in nanogram per gram milk lipids except in the case of CB 126, CB 169, PCDDs, PCDFs, and TEQs (picogram per gram milk lipids). WHO$_{2005}$ TEQ levels are based on the 2005 WHO TEFs (14).

[b] CB 28, CB 52, CB 101, CB 105, CB 118, CB 138, CB 153, CB 156, CB 167, CB 180.

[c] CB 77, CB 126, CB 169, CB 105, CB 118, CB 156, CB 167.

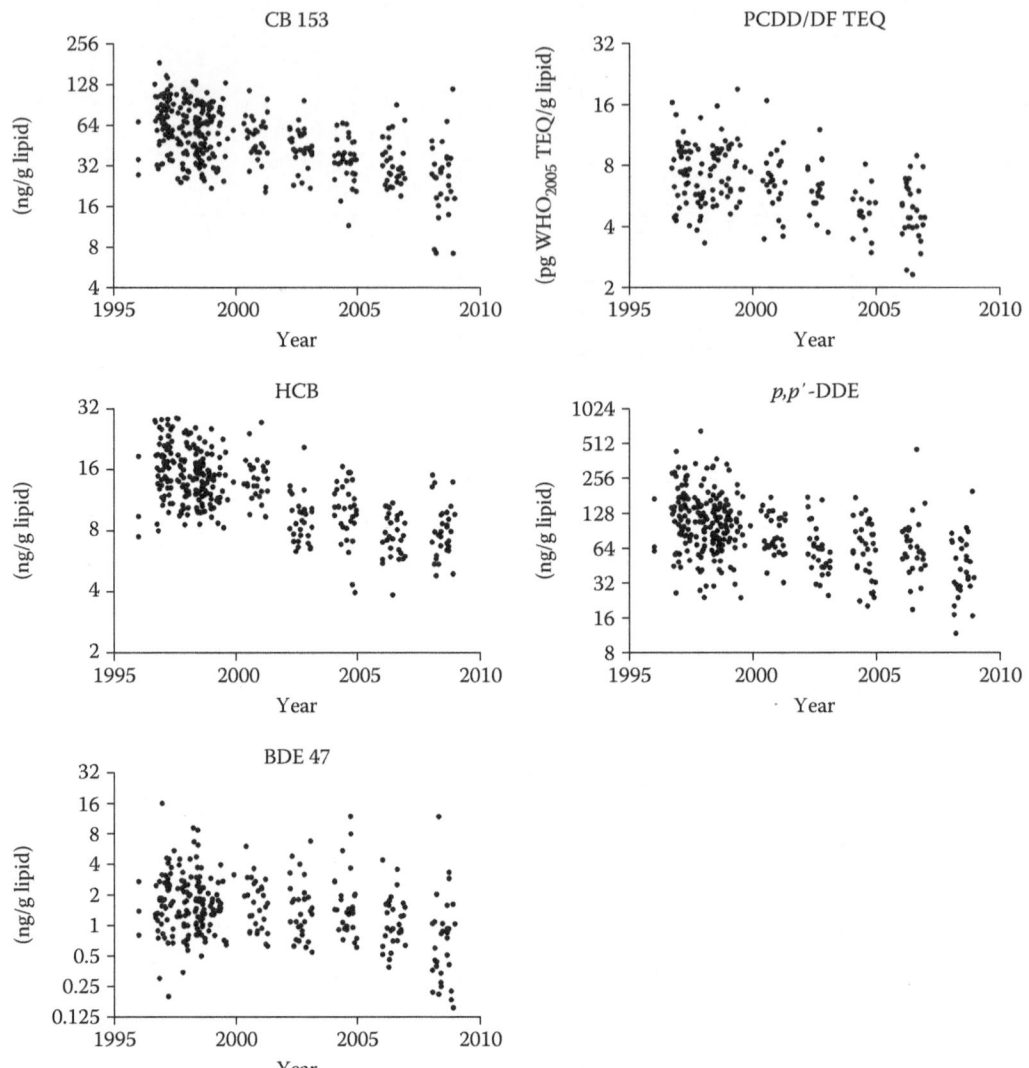

FIGURE 15.1 Levels of organohalogen compounds in mother's milk lipids among primiparous women from the Uppsala area in Sweden between 1996 and 2008.

levels between 1996 (mean ratio: 0.47) and 2006 (mean: 0.57) (linear regression, trend: $p < 0.001$) in the POPUP study mother's milk.

Adjustment of the temporal trends of PCBs and PCDD/Fs for the personal characteristics such as mother's age, pre-pregnancy BMI, and weight change during pregnancy and after delivery resulted in slightly faster declining rates of individual congeners among the POPUP mothers (Table 15.4). The small change in declining rates after adjustment is not surprising since the age, pre-pregnancy BMI, and weight changes did not change significantly during the study period (Table 15.1).

15.4.2 COMPARISONS WITH THE STOCKHOLM STUDY

The Stockholm study of temporal trends of organohalogen compounds in pooled mother's milk samples showed that the levels of CB 153 decreased from about 200 ng/g lipid in the early 1970s to 70 ng/g lipid in the late 1990s [3] (Figure 15.3). The levels of PCB were similar during the time

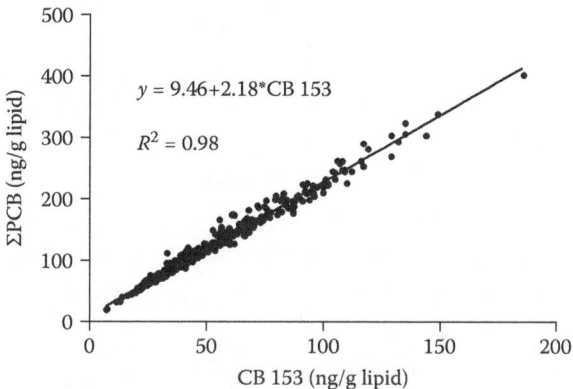

FIGURE 15.2 Linear regression analysis of the association between levels of CB 153 and ΣPCB in mother's milk sampled within the POPUP study.

period of overlap between the Stockholm and POPUP studies in 1996–1997, with mean levels of CB 153 of 70–80 ng/g lipid [3] (Figure 15.3). This is not surprising since marked regional differences in PCB levels could not be detected in mother's milk from primiparas living in geographically separated areas of Sweden [28].

The number of years required for the levels of CB 153 to decline by 50% (half-life) in mother's milk from the Stockholm area was estimated to 17 years [3]. The half-lives for CB 118 and CB 138 were reported to be 11–14 years. The Stockholm half-times were longer than those found in the POPUP study (8–9 years) (Table 15.5). Several factors may contribute to the differences observed, such as lack of age adjustment and recruitment of both primi- and multiparous women in the Stockholm study. The average age of the mothers donating breast milk increased between 1972–1985 (27–28 years) and 1996–1997 (30–31 years) [3]. Consequently, it is possible that the rates of decline of PCBs were underestimated in the Stockholm study. Moreover, in contrast to the Stockholm study, the recruitment of primiparas in the POPUP ensured that nursing-related variation in PCB levels was avoided. Earlier studies have shown that the mother's milk PCB levels decrease both with increasing parity and with time spent breast-feeding [23,24]. The comparisons of half-times between the POPUP and Stockholm studies may also be hampered by the fact that the studies were performed during different time periods (Stockholm study: 1972–1997; POPUP study 1996–2008). However, the steady decline of CB 153 (7% per year) and ΣPCB (9% per year) in guillemot eggs from the Baltic Sea between 1969 and 2006 supports a similar rate of decline in PCB levels in the Swedish environment during the two study periods [34].

The Stockholm study also showed a decline in PCDD/F levels in mother's milk from 1972 to 1997, with mean levels decreasing from about 100 pg PCDD/F WHO$_{1998}$-TEQs/g lipid in the early 1970s to approximately 30 pg TEQ/g lipids in the late 1990s (Figure 15.3) [3]. In the Stockholm study, the half-time for PCDD WHO$_{1998}$ TEQ was estimated at 15 years, which is slower than the half-time estimated in the POPUP study (10 years). This could, similarly as in the case of PCBs, be due to the lack of age adjustment of the temporal trend in the Stockholm study. The half-time of PCDF WHO$_{1998}$ TEQs was estimated at 11 years in the Stockholm study, which is faster than in the POPUP study (15 years). Moreover, the Stockholm study reported faster declines of PCDFs than of PCDDs, whereas the reverse was evident for the POPUP study [3] (Table 15.5). The difference in half-time estimates between the studies may be due to the differences in study design mentioned earlier. Furthermore, the estimated half-times from the Stockholm study is based on only nine pooled samples during the time period 1972–1997 [3], whereas the estimates from the POPUP study was based on PCDD/F levels in 184 individual samples (Table 15.5).

TABLE 15.4
Temporal Trend of PCB, PCDD, PCDF, and TEQ Levels in Mother's Milk from Primiparous Women from Uppsala County Participating in the POPUP Study between 1996 and 2008[a]

Compound[b]	Annual Change (%)	Half-Time (Years)		Half-Time (Years)	%<LOQ
	Simple Regression		Multiple Regression[c]		
PCB 28	-4.7 ± 1.0	14	-4.6 ± 1.0	15	9
PCB 118	-8.2 ± 0.6	8.1	-8.5 ± 0.4	7.8	0
PCB 126	-6.4 ± 0.9	11	-7.0 ± 0.7	9.6	6
PCB 138	-7.0 ± 0.5	9.5	-7.3 ± 0.4	9.1	0
PCB 153	-7.8 ± 0.6	8.5	-8.1 ± 0.4	8.2	0
PCB 156	-5.2 ± 0.6	13	-5.6 ± 0.4	12	0
PCB 167	-6.4 ± 0.9	10	-7.0 ± 0.7	9.5	17
PCB 169	-2.4 ± 1.0	29	-3.3 ± 0.7	21	25
PCB 180	-6.9 ± 0.6	9.7	-7.7 ± 0.3	8.7	0
SumPCB[c]	-6.8 ± 0.5	9.8	-7.7 ± 0.3	8.7	
TCDD	-5.8 ± 0.8	12	-6.9 ± 0.6	9.8	3
1,2,3,7,8-PeCDD	-5.1 ± 0.7	13	-5.8 ± 0.5	11	0
1,2,3,4,7,8-HxCDD	-6.2 ± 0.8	11	-7.1 ± 0.6	9.4	2
1,2,3,6,7,8-HxCDD	-7.2 ± 0.7	9.7	-7.6 ± 0.5	8.8	0
1,2,3,7,8,9-HxCDD	-7.7 ± 0.7	8.9	-7.9 ± 0.7	8.4	0
1,2,3,4,6,7,8-HpCDD	-9.0 ± 1.0	7.3	-9.3 ± 1.0	7.1	0
OCDD	-7.8 ± 0.9	8.6	-8.3 ± 0.8	8.0	0
TCDF	NS		NS		22
1,2,3,7,8-PeCDF	NS		NS		18
2,3,4,7,8-PeCDF	-4.3 ± 0.9	16	-5.3 ± 0.7	13	0
1,2,3,4,7,8-HxCDF	-3.5 ± 0.6	20	-4.0 ± 0.6	17	0
1,2,3,6,7,8-HxCDF	-2.5 ± 0.7	28	-3.1 ± 0.6	22	0
2,3,4,6,7,8-HxCDF	NS		NS		0
1,2,3,4,6,7,8-HpCDF	-3.3 ± 0.9	21	-3.4 ± 0.9	20	0
OCDF	-6.1 ± 1.4	11	-5.9 ± 1.5	12	19
PCDD WHO$_{2005}$ TEQ	-5.9 ± 0.8	11	-6.7 ± 0.5	10	
PCDF WHO$_{2005}$ TEQ	-3.7 ± 0.8	18	-4.6 ± 0.6	15	
PCDD/F WHO$_{2005}$ TEQ	-5.1 ± 0.7	13	-5.9 ± 0.5	11	
Non-*ortho* PCB WHO$_{2005}$ TEQ	-5.9 ± 0.8	11	-6.5 ± 0.7	10	
Mono-*ortho* PCB WHO$_{2005}$ TEQ	-7.0 ± 0.6	9.5	-7.4 ± 0.4	9.1	
Total WHO$_{2005}$ TEQ	-5.5 ± 0.7	12	-6.3 ± 0.5	11	

[a]   Mean $\pm$ SE. NS: No statistically significant temporal trend ($p > 0.05$).

[b]   Sum PCB = B 28, CB 52, CB 101, CB 105, CB 118, CB 138, CB 153, CB 156, CB 167, CB 180. WHO$_{2005}$ TEQ levels are based on the 2005 WHO TEFs (14).

[c]   Age of the mother, pre-pregnancy BMI, weight change during pregnancy, and weight change after delivery were included as covariates in the regression models.

In 1997, the total WHO$_{1998}$TEQ level in the pooled sample of mother's milk in the Stockholm area was reported to be 28 pg/g lipid [3]. In the POPUP study, the mean total WHO$_{1998}$ TEQ level during 1996 and 1997 was 19 pg/g lipid ($N=52$) (Figure 15.3). The lower TEQ level in the POPUP study is probably not due to a difference in PCDD/DF and dioxin-like PCB exposure of the young women from Uppsala and Stockholm. Instead, the differences in average levels may be due to differences in study design and analytical methods.

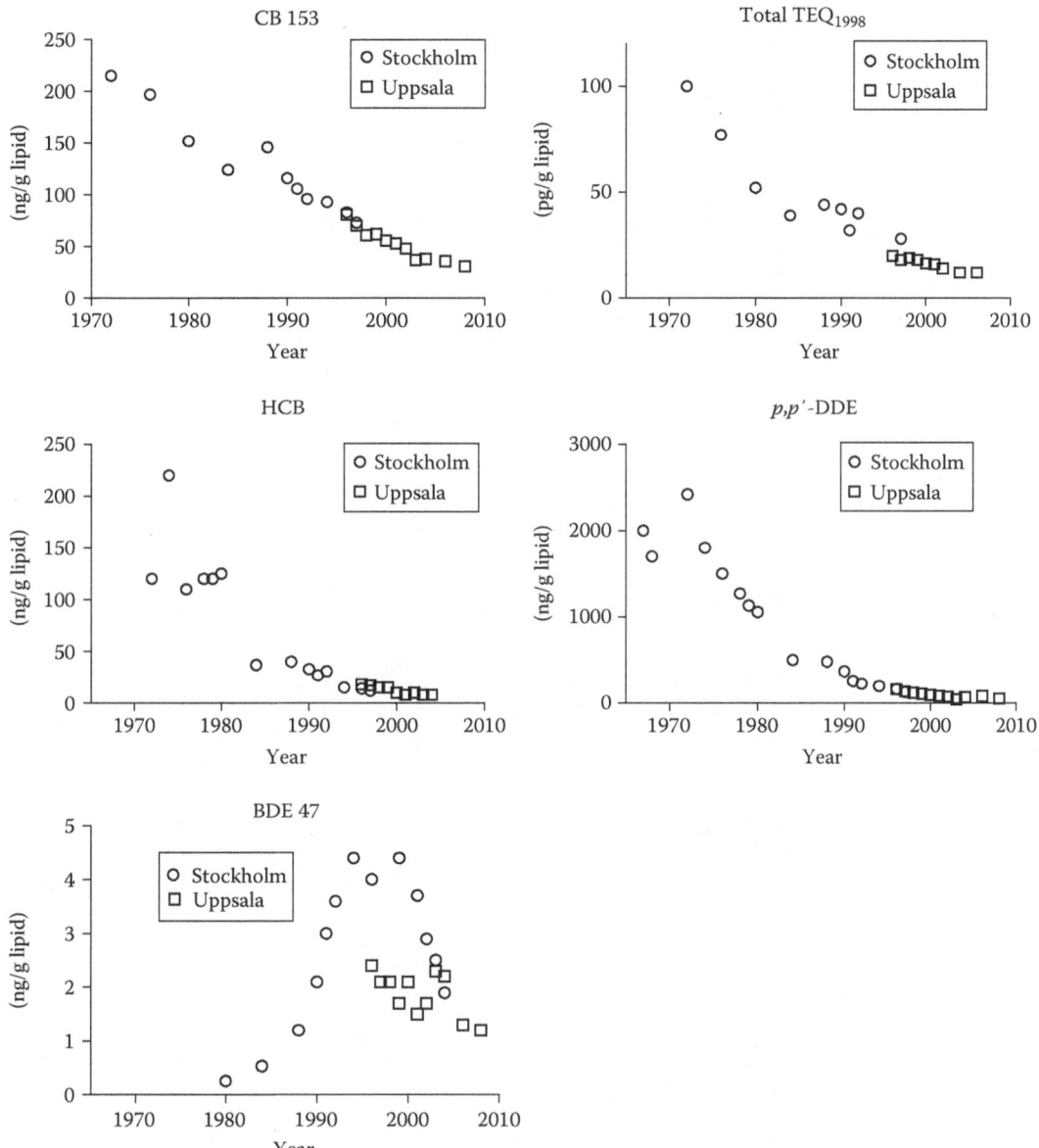

FIGURE 15.3 Temporal trends of average levels of organohalogen compounds in mother's milk among women from the Stockholm area (Stockholm study) [3,5] and Uppsala area (POPUP study).

15.4.3 COMPARISONS WITH TEMPORAL TRENDS IN OTHER MATRICES

Food is currently the major source of PCB exposure and has been so for many decades since the use and production of PCBs were restricted in the early 1970s in Sweden [35,36]. The relatively small difference in half-times for many of the studied PCBs in the POPUP study suggests similar decline rates of different PCB congeners in foods on the Swedish market. This is supported by the similar rate of decline of CB 153 levels in Swedish food-producing animals (pigs and cattle) between 1991 and 2004 (6.5%–9.9% per year) [37] as in mother's milk among the POPUP women (Table 15.4). Moreover, in biota from the Swedish coasts, the PCB levels decreased with 5%–10% per year between 1978 and 2007 [34]. This suggests that PCB levels in food have followed a similar trend

TABLE 15.5

Levels of Chlorinated Pesticides/Metabolites and Brominated Flame Retardants in Mother's Milk from Primiparous Women from Uppsala County Participating in the POPUP Study between 1996 and 2008[a]

Pesticides/ Metabolites	1996 N=20	1997 N=67	1998 N=90	1999 N=26	2000 N=18	2001 N=11	2002 N=26	2003 N=5	2004 N=32	2006 N=30	2008 N=31
HCB	18±6	17±5	15±4	15±4	15±5	15±3	9.6±3.0	8.4±1.8	10±3	7.9±1.9	8.4±2.7
β-HCH	17±12	16±9	13±5	14±15	11±4	11±5	8.2±3.9	6.1±1.0	7.8±3.8	5.7±2.0	4.5±2.1
trans-Nonachlor	9.9±5.8	7.8±4.1	7.4±3.0	6.6±3.0	7.0±3.2	6.9±3.0	6.3±2.7	4.6±1.5	5.4±2.8	5.1±2.7	4.4±3.2
Oxychlordane	5.4±2.8	4.5±2.0	4.3±1.6	3.8±1.4	4.0±1.6	3.8±1.4	3.4±1.4	2.7±0.6	3.2±1.6	2.6±1.1	2.3±1.2
p,p'-DDT	11±11	14±30	7.9±5.4	5.8±2.6	6.3±3.8	5.8±4.0	4.7±1.8	3.6±1.1	5.2±5.6	4.0±1.8	2.3±1.8
p,p'-DDE	160±110	137±95	119±69	108±61	94±39	84±32	74±39	43±13	69±38	82±77	51±36
PBDE	N=17	N=52	N=67	N=22	N=18	N=11	N=25	N=5	N=29	N=30	N=31
BDE 47	2.4±3.6	2.1±1.2	2.1±1.7	1.7±0.8	2.1±1.3	1.5±0.8	1.7±1.1	2.3±2.5	2.2±2.5	1.3±0.9	1.2±2.1
BDE 99	0.49±0.51	0.66±0.68	0.53±0.35	0.41±0.25	0.28±0.18	0.17±0.08	0.22±0.12	0.26±0.21	0.58±0.91	0.25±0.16	0.23±0.26
BDE 100	0.58±1.2	0.36±0.23	0.33±0.23	0.33±0.16	0.33±0.18	0.23±0.09	0.32±0.21	0.55±0.70	0.49±0.55	0.31±0.20	0.23±0.26
BDE 153	0.71±0.97	0.53±0.29	0.53±0.24	0.59±0.30	0.64±0.27	0.57±0.20	0.70±0.23	1.00±0.72	0.87±0.80	0.76±0.30	0.81±0.83
ΣPBDE[b]	4.3±6.25	3.7±2.0	3.6±2.4	3.1±1.3	3.4±1.7	2.6±0.9	3.0±1.5	4.2±4.2	4.2±4.2	2.8±1.3	2.7±3.1

[a] Mean±SD; nanogram per gram lipid.
[b] BDE 47, BDE 99, BDE 100, BDE 153, BDE 154.

as PCB levels in the Swedish environment. Further support for a trend of decreasing PCB levels in food on the Swedish market is the finding of a 39% decline in the per capita intake of CB 153 from fish, meat, dairy products, and eggs between 1999 and 2005 in a Swedish market basket study [38].

Animal feed is the major source of PCB contamination in food-producing animals and the efforts within the European Union to control animal feed contamination may have contributed to the observed decline in human PCB exposure from food in Sweden. Maximum limits of PCDD/Fs and dioxin-like PCBs in animal feed have been in force for almost 10 years [39,40]. This has forced the feed industry to better control their raw materials for feed production [41]. Moreover, the monitoring program for contaminants in animal feed within the EU includes non-dioxin-like PCBs [42].

The slow but steady decline in PCDD/F levels in mother's milk is in contrast to the relatively stable PCDD/F levels in the Baltic Sea environment [34]. For instance, in herring sampled in the Baltic Sea area, no significant temporal trend of PCDD/F levels has been observed during the last 15 years. Fish from the Baltic Sea region is, however, not a major source of PCDD/F exposure among young women in Sweden [35] and a market basket study of PCDD/F levels in food on the Swedish market has suggested significant declines in PCDD/F exposures from food between 1999 and 2005 [38]. The mother's milk levels of the persistent organohalogen compounds are the result of exposure of the women to the compounds during a long time period. Therefore, the temporal trends in mother's milk may not be directly comparable to temporal trends found in biota in the Swedish environment, which probably reflects more recent trends in environmental load of the compounds.

15.5 TEMPORAL TRENDS OF CHLORINATED PESTICIDES AND METABOLITES

p,p'-DDE was the compound with the overall highest mean concentrations of the studied organohalogen compounds among the women in the POPUP study (Table 15.5). Mean concentrations of the other chlorinated pesticides and metabolites were considerably lower. The mean levels of all studied compounds decreased during the study period (Table 15.5, Figure 15.1). Linear regression analyses showed that *p,p'*-DDT and β-HCH decreased at the fastest rate, whereas the slowest rate of decrease was observed for *trans*-nonachlor (Table 15.6).

15.5.1 DDT Compounds

The insecticide DDT has been banned in Sweden since the 1970s, but more than 100 metric tons/year were applied mainly to arable land in the 1960s [36]. *p,p'*-DDT was the predominant DDT isomer in commercial products [43]. The ban of DDT use in Sweden many decades ago is reflected in the relatively high levels of the persistent DDT metabolite *p,p'*-DDE in comparison with levels of *p,p'*-DDT among the women in the POPUP study (Table 15.5). The mean ratio between the *p,p'*-isomers of DDE and DDT among the POPUP women increased from 17 in 1996 to 23 in 2008 (simple linear regression, trend $p < 0.001$, $N = 355$). This was due to a slower annual decrease of *p,p'*-DDE than of *p,p'*-DDT (Table 15.6). Between 1996 and 2008, the mean *p,p'*-DDE level decreased threefold from 160 ng/g lipid to around 50 ng/g lipid, whereas the *p,p'*-DDT level decreased fivefold from about 10 ng/g lipid to 2 ng/g lipid (Table 15.5). In the Stockholm study, *p,p'*-DDT also showed a faster rate of decrease (half-time 4 years) than *p,p'*-DDE in mother's milk (half-time 6 years) [3]. The DDE/DDT ratio was 1.5 in 1967, reflecting the ongoing use of the insecticide, and increased to 9 in 1997 [3]. During the period of 1967–1997 the average *p,p'*-DDT level in Stockholm mother's milk had decreased to only 1% of the initial average level of 1400 ng/g lipid in 1967, whereas the *p,p'*-DDE level had decreased to 5% of the initially highest level of 2500 ng/g lipid in 1972 [3] (Figure 15.3).

A comparison between average *p,p'*-DDE and *p,p'*-DDT levels in mother's milk from the Stockholm area and the Uppsala region during the overlapping time period of the two studies (1996–1997) shows similar levels of the compounds (Figure 15.3). In the Stockholm milk *p,p'*-DDT and *p,p'*-DDE levels were 14 ng/g lipid and 130–160 ng/g lipids, respectively [3], whereas

TABLE 15.6

Temporal trend of Chlorinated Pesticides and Metabolites in Mother's Milk from Primiparous Women from Uppsala County Participating in the POPUP Study between 1996 and 2008[a]

Compound	Annual Change (%)	Half-time (Years)	Half-time (Years)	%<LOQ
	Simple Regression		Multiple Regression[b]	
HCB	−6.9±0.4	9.7	−7.0±0.3	9.5	0
B-HCH	−10±0.5	6.5	−11±0.4	6.1	0
Oxychlordane	−6.7±0.6	10	−7.1±0.4	9.5	0
trans-Nonachlor	−6.1±0.6	10	−6.5±0.4	10	0
p,p'-DDT	−10±0.7	6.3	−11±0.6	6.2	1
p,p'-DDE	−8.5±0.7	7.8	−8.7±0.6	7.6	0
PBDE 47	−6.8±0.9	9.9	−7.0±0.9	9.6	5
PBDE 99	−7.9±0.9	8.4	−7.9±0.9	8.5	230
PBDE 100	−2.8±0.9	25	−2.7±1.0	25	16
PBDE 153[c]	3.4±0.7	21	3.3±0.6	21	2
SumPBDE[d]	−3.1±0.1	22	−3.3±0.7	20	

[a] Mean±SE. All temporal trends were statistically significant ($p \leq 0.05$).

[b] Age of the mother, pre-pregnancy BMI, weight change during pregnancy, and weight change after delivery were included as covariates in the regression models.

[c] Percent increase in levels and the estimated number of years required for the levels to increase by 50% in the population.

[d] BDE 47, BDE 99, BDE 100, BDE 153, BDE 154.

p,p'-DDT levels in Uppsala milk ranged between 11–14 ng/g lipid and *p,p'*-DDE levels between 140 and 160 ng/g lipid (Table 15.5). The human exposure to DDT compounds in Sweden seems to be uniform since no regional differences in *p,p'*-DDE levels in mother's milk were found among primiparous women from four different regions of the country [28].

A study of temporal trends of *p,p'*-DDE in food-producing bovines and pigs 1991–2004 showed that the levels of *p,p'*-DDE decreased in 5 out of 6 Swedish regions studied [37]. The annual decrease in *p,p'*-DDE levels in bovines and pigs on the Swedish market was estimated at 6%–12%. Moreover, the annual decreases in *p,p'*-DDE levels in fish from the Swedish coasts between 1978 and 2007 were estimated at 5%–13% [34]. The average annual decline estimated for mother's milk from the Uppsala area was within the same range (Table 15.6), suggesting that the decline in *p,p'*-DDE levels in Swedish food-producing animals is an important reason behind the declining levels of the compound in mother's milk. The Swedish market basket studies also showed a 30% decline in the per capita intake of *p,p'*-DDE between 1999 and 2005 [38].

15.5.2 HCB, HCH Compounds, and Chlordane

As with DDT, food is probably the major exposure source of hexachlorobenzene (HCB), hexachlorocyclohexane (HCH), and chlordane to the Swedish population. These compounds/compound groups are pesticides/industrial chemicals that have been used less extensively than DDT in Sweden. Similarly, as in the case of DDT, the use has been banned in Sweden for decades [36].

15.5.2.1 HCB

Apart from the ceased industrial and agricultural use of HCB in Sweden, the compound is currently unintentionally produced during combustion processes and as an impurity in the chlorine chemical industry [2,44]. This additional source of environmental HCB pollution may at least partially explain the relatively long half-time of HCB (10 years) in comparison with the DDT compounds and β-HCH in mother's milk from the Uppsala area (Table 15.6, Figure 15.1).

The emissions of HCB to the Swedish environment have declined as illustrated by the average annual decrease in HCB levels of 4%–10% in herring from the Swedish Atlantic and Baltic Sea coast between 1988 and 2007 [34]. A decrease in human exposure to HCB from food in Sweden is supported by the decrease in HCB levels in adipose tissue from food-producing bovines and pigs in Sweden since the early 1990s (4%–16% decline per year) [37,45]. Furthermore, the per capita intake of HCB declined 27% between 1999 and 2005 in Sweden [38].

In the Stockholm study, the half-time of HCB in mother's milk was estimated during a 6 year period between 1974 and 1997, but the average levels (approximately 120 ng/g lipid) did not change much between 1976 and 1980 [3]. HCB as a fungicide was withdrawn from the Swedish market in 1980 [34], and between 1980 and 1984–1985 the average HCB level in the pooled milk samples from Stockholm declined threefold [3] (Figure 15.3). This shows that the human exposure to HCB in Sweden responded rapidly to the restriction of HCB use. The average HCB levels in mother's milk from the Stockholm and Uppsala areas were in the same range during 1996–1997, with an average in the pooled samples from Stockholm of 12–14 ng/g lipid and a mean level of 17–18 ng/g lipid in the POPUP study (Table 15.5, Figure 15.3) [3].

15.5.2.2   HCH Compounds

The half-time of β-HCH decrease in mother's milk among the POPUP women was relatively short, in the range of those found for the DDT compounds (Table 15.6). In the Stockholm study, β-HCH was only studied between 1972 and 1984/85. During this period, the average level decreased from 280 ng/g lipid to 70 ng/g lipid [3]. In 1996, the mean β-HCH was 17 ng/g in mother's milk from the POPUP mothers (Table 15.5), showing that the human exposure to β-HCH has continued to decline since the 1980s. In guillemot eggs from the Baltic Sea, the β-HCH level decreased 9% per year between 1988 and 2007 [34], showing a steady decline of the β-HCH load in the Swedish environment.

β-HCH was a minor part of technical-grade HCH (7%–10%), which also consists of α-HCH (65%–70%), γ-HCH (15%, lindane), and δ-HCH (7%) [46]. In Swedish mother's milk, β-HCH is the dominating HCH isomer, reflecting the relative persistence of this HCH isomer. For instance, in the Stockholm study, the mean levels of α-HCH were more than 10-fold lower than that of β-HCH during the period 1972–1985 [3]. In mother's milk from the Uppsala region, the mean levels of α-HCH and γ-HCH were around 1 ng/g lipid in 1994 [47], whereas the mean β-HCH level among the POPUP women in 1996 was more than 10-fold higher (17 ng/g lipid) (Table 15.5). The relatively low persistence of α-HCH is also illustrated by the relatively large annual decrease in mean levels (14%) in adipose tissue from bovines in Sweden between 1991 and 1997 [45].

15.5.2.3   Chlordane Compounds

Among the POPUP mothers, *trans*-nonachlor and oxychlordane showed relatively low annual decreases in mother's milk levels as HCB (Table 15.6). *trans*-Nonachlor is part of the technical mixture of the pesticide chlordane [48] whereas oxychlordane is the major metabolite of some of the compounds in the chlordane mixture [49]. The low levels of the compounds found in mother's milk in the POPUP study reflect the low use of chlordane in Sweden (Table 15.5). Among the Stockholm women, the levels of the two compounds ranged between 12 and 20 ng/g mother's milk lipids during the period 1972–1989 [3]. No half-time was calculated because of too few sampling points during the study period.

In the Swedish market basket studies, the per capita intake of Σchlordanes (*trans*-nonachlor, α-chlordane, γ-chlordane, and oxychlordane) decreased by 50% between 1999 and 2005 [38]. This supports the possibility that the decrease in chlordane exposures from food is an important factor behind the declined levels of *trans*-nonachlor and oxychlordane in mother's milk in Sweden.

15.6   TEMPORAL TRENDS OF BROMINATED FLAME RETARDANTS

The average levels of ΣPBDEs in mother's milk increased exponentially between 1972 and 1997 in the Stockholm area, from less than 0.1 ng/g lipid to 4 ng/g lipid [3] (Figure 15.3). In the time period around the end of the 1990s, the increase leveled off in the Stockholm study [5]. This was supported by early data from the POPUP study [50]. Later data from the POPUP study suggests that the levels of the tetra- to penta-brominated diphenyl ethers BDE 47, BDE 99, and BDE 100 have started to decline since 1996, whereas the levels of the hexabrominated BDE 153 still increased slowly (Tables 15.5 and 15.6, Figure 15.1). A continuation of the Stockholm study during the period 1998–2004 indicated similar decreasing temporal trends for BDE 47 and BDE 99, and a slow increase in BDE 153 levels [5] (Figure 15.3). The decline in levels of the tetra- to penta-brominated BDEs could be explained by a reduced use of lower brominated PBDEs in Sweden [6]. For BDE 153, a higher persistence could result in a slower response of temporal trends to risk management decisions aiming at decreasing the pollution of the environment [51]. Furthermore, technical PBDE mixtures differ in their composition of BDE 153 and lower brominated PBDEs [52], which may at least in part explain differences in temporal trends in mother's milk between BDE 153 and lower brominated BDEs. An additional source of BDE 153 could be debromination of higher brominated BDEs in the environment [53,54].

Although the temporal trends of PBDEs were similar in the Stockholm and Uppsala areas after 1996, the average PBDE levels in the pooled mother's milk samples from Stockholm were higher than the mean levels in the samples from Uppsala 1996–2004 (Figure 15.3). For instance, average BDE 47 levels in Stockholm decreased from 4 to 2 ng/g lipid during these years, whereas the mean levels decreased from 2.4 to 1.2 ng/g lipid in the Uppsala samples (Table 15.5) [5]. The differences are probably not due to differences in exposure levels in the two areas since both are urban areas and they are geographically very close to each other. Differences in design of the studies and analytical differences may have contributed to the observed differences in PBDE levels between the Stockholm and Uppsala area.

The per capita intake of ΣPBDE from food on the Swedish market was the same in the market basket studies between 1999 and 2005 [38]. This is in agreement with the very slow change of ΣPBDE levels in mother's milk from the Uppsala area between 1996 and 2008 (Table 15.6).

A comparison of the temporal trends of PBDEs in the POPUP and Stockholm studies of mother's milk with temporal trends of PBDEs in biota from the Baltic Sea environment shows a delay in the decline in levels of lower brominated PBDE congeners in humans with approximately 10 years [4,5,55]. In guillemot eggs from the Baltic Sea area, the levels of BDE 47 and BDE 99 peaked around 1985 and decreased rapidly thereafter, whereas the levels of these PBDEs continued to increase in Swedish mother's milk up to the end of the 1990s [4,5,55]. The delay in mother's milk may be caused by the presence of the PBDEs in indoor dust, originating from older consumer products in the home and work environment [29].

15.7   LEVELS AND TEMPORAL TRENDS IN OTHER REGIONS OF THE WORLD

15.7.1   PCBs AND PCDD/Fs

There are surprisingly few published long-term monitoring studies of mother's milk, where organohalogen compounds have been measured in a large number of individual mother's milk samples from the same population. However, studies from Germany, Croatia, and Japan have, similarly to the POPUP study, shown more or less continuously decreasing temporal trends of PCBs (all three countries) and PCDD/Fs (Germany) levels in mother's milk since the beginning of the 1980s up to 2005 [56–60]. In western Germany, however, the marked decrease in PCDD/F seemed to be leveling off in the end of the 1990s and even increased during 1998 and 1999 [57]. It was proposed that this was caused by a general increase in PCDD/F levels in cow's milk in the studied area due to the use of highly contaminated raw materials in animal feed [57]. This illustrates the great importance of animal feed as a source of organohalogen compound contamination of the human food chain. It also points to the possibility to restrict the human exposure by having strict control of the animal feed contamination.

Several studies have tried to determine temporal trends of halogenated compounds in mother's milk by comparing results from more or less comparable studies carried out at different time periods in a country or region. Such studies from Finland, Norway, and Canada have reported decreasing PCDD/F and PCB levels from the early 1980s up to the first years of the twenty-first century [61–65]. Moreover, the fourth WHO survey of PCDD/Fs and PCBs in mother's milk have shown significant decreases in mother's milk levels of the compounds in 15 European countries between 1988 and 2007 [66]. Taken together, the results show that the efforts to phase out the production and use of PCBs and to minimize the emissions of PCDD/Fs in industrialized countries have resulted in an improvement in the mother's milk quality with regard to PCDD/Fs and PCBs.

There may, however, be cases when PCB levels in mother's milk do not follow the more general pattern of decreasing trends. A study of PCB levels in pooled mother's milk samples from the Faroe Islands showed relatively stable levels of more than 10 single PCB congeners over a period of 12 years between 1987 and 1999 [67]. An effort was made to select a homogenous group of women, regarding age, number of deliveries, time period of milk sampling, and dietary habits, in the different pools. The PCB levels were high in the Faroe Island samples, caused by the consumption of PCB-contaminated whale blubber by some of the women [67]. This may have masked a more general decrease in PCB exposure from other foodstuffs.

A comparison of CB 153 levels detected in Swedish mother's milk with those found in other countries show that the current average levels are in the same range as those reported for many other European countries during the last decade (Figure 15.4). The comparisons are however hampered by differences in the design of the studies and use of different analytical laboratories. Moreover, regional differences in levels within a country may exist, as shown in the Czech Republic (Figure 15.4). In this case, high PCB levels in mother's milk have been reported from an area with a history of extensive PCB use in industrial production [68]. High PCB levels in mother's milk have also been reported from an area of Slovakia where PCB was produced [69], and from the Faroe Island among women who ate PCB-contaminated whale blubber [67].

There is a tendency of higher average levels of total WHO TEQ_{1998} levels in mother's milk in densely populated industrial countries than in less populated industrial countries and in developing countries (Figure 15.4). The higher levels in densely populated industrial countries, such as Germany and Belgium, may be due to a higher proportion of the food production located within or in close proximity to the contaminated areas [70,71]. For instance, a study of home-produced eggs from Belgium showed high PCDD/F levels in many of the sampled eggs and a positive correlation between soil and egg PCDD/F levels [72]. Relatively low total WHO_{1998} levels in mother's milk have been found in developing countries where the general PCDD/F and PCB contamination of the human food chain is low (Figure 15.4).

15.7.2 CHLORINATED PESTICIDES/METABOLITES

As in the case of PCBs, many countries in the world have banned or severely restricted the production and use of several persistent chlorinated pesticides [1]. As a result, the levels of pesticides, such as DDT compounds, HCB, HCHs, and chlordane, and some of their metabolites, have declined in mother's milk from European, North American, and Asian countries [56–58,73,74]. The current levels of many chlorinated pesticides/metabolites are considerably lower in industrialized countries in the Northern hemisphere than in countries in subtropical or tropical areas, where the pesticides have been used more recently or where use is still ongoing. For instance, the p,p'-DDE levels in mother's milk from Sweden and other industrialized countries were at least one order of magnitude lower than those found in some countries in the tropics (Figure 15.4). DDT is still recommended by the WHO for use in malaria vector control by spraying inner walls and ceilings of buildings in

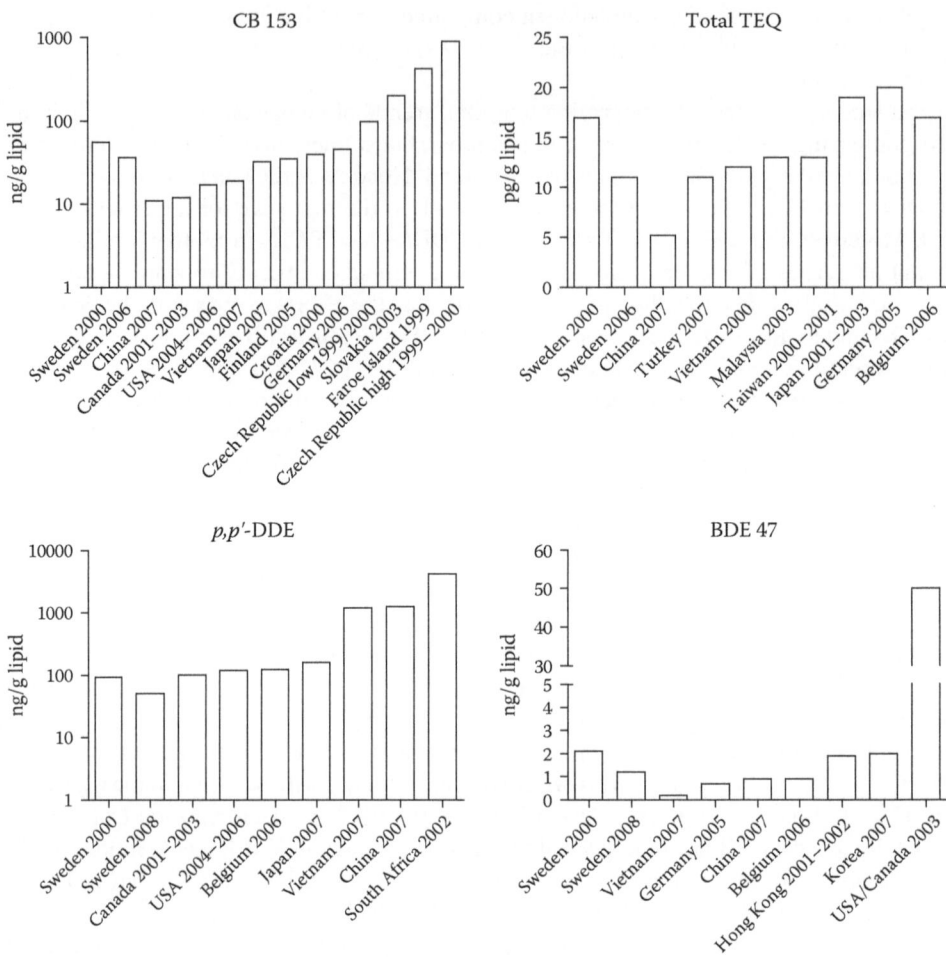

FIGURE 15.4 Average levels of the PCB congener CB 153, total PCDD/F and PCB WHO[1998] TEQ [15], the DDT metabolite *p,p'*-DDE, and of the brominated flame retardant congener BDE 47 in mother's milk from different parts of the world during the first decade of the twenty-first century [56,57,67–69,76,83–94].

affected regions [75]. In areas where DDT is still used in this fashion, high DDT compound levels have been reported in mother's milk [76].

15.7.3 POLYBROMINATED DIPHENYL ETHERS

Temporal trends of PBDEs in human biological samples have only been reported in a few studies, and, similar to those in the Stockholm study, they report increasing levels of PBDEs from the early 1970s to the late 1990s. For instance, an increase in average levels of all studied PBDEs were observed in pooled mother's milk samples from the Faroe Islands between 1987 and 1999 [67]. In Japan, ΣPBDE levels increased in pooled mother's milk samples from Osaka during the period 1973–1988 [77]. A study of PBDE levels in pooled blood serum samples from adult men living in Norway showed that levels of all studied PBDEs increased from the end of the 1970s to the middle/end of the 1990s [78]. In pooled human serum samples from the southeastern and northwestern USA, increasing levels of almost all studied PBDEs from the mid-1980s to the early 2000s were observed [79].

As suggested both in the POPUP and Stockholm study, a decline in levels of tetra- to penta-brominated PBDEs in mother's milk appears to have started in the late 1990s in Sweden (Table 15.6) [5]. In contrast, in fetal liver samples from Montreal, Canada, the levels of almost all tetra- to penta-brominated PBDEs increased significantly between 1998 and 2006 [80]. It is, however, too early to draw the conclusion that current temporal trends of PBDEs differ between regions of the world. The general contamination situation regarding lower brominated PBDEs is much more severe in North America than in other parts of the world due to the extensive use of PBDEs in a variety of commercial products [81] (Figure 15.4). It has been suggested that ingestion and dermal absorption of PBDE-contaminated house dust are the major exposure pathways of the general population in the United States [82].

15.8   CONCLUSIONS

As a consequence of the risk management decisions taken in the early 1970s, the levels of persistent polychlorinated pollutants, such as PCBs, PCDD/Fs, and chlorinated pesticides/metabolites, have slowly declined in mother's milk from Sweden, as shown in the Stockholm study (1967–1997) [3] and the POPUP study (1996–2008). The decreasing levels of the studied organochlorine compounds in Swedish mother's milk during the 1990s and the first decade of the twenty-first century has occurred concomitantly with a decline of the compounds in food on the Swedish market. During the initial period after risk management decisions were taken regarding many of the studied compounds, the decline in levels of some of the compounds in mother's milk was most probably due to a decrease in both direct exposures from products containing the compounds and indirect exposures via food. Later in the study period, the decline of indirect exposure via food was most probably the driving force.

Both the Stockholm and POPUP study have access to banked mother's milk samples. As the PBDEs emerged as a new group of persistent and lipophilic halogenated pollutants in the late 1990s, it was therefore possible to retrospectively study temporal trends of the "new" pollutants in mother's milk. The Stockholm study showed that the levels of tetra- to hexa-brominated PBDEs increased significantly during the 1979–1990s [3]. This sparked risk management efforts to restrict the human exposure to these PBDEs in the late 1990s. Consequently, both the Stockholm study [5] and the POPUP study [50] showed that the increase in levels of tetra- to penta-brominated PBDEs in Swedish mother's milk leveled off around the end of the 1990s. The levels of some of the PBDE congeners have even started to decline slowly during the beginning of the 2000s.

The Swedish studies show that temporal trend studies of persistent and lipophilic pollutants in mother's milk is a good tool for the follow-up of risk management decisions with the aim to limit human exposure to this type of pollutants. The banking of mother's milk samples in the Stockholm and POPUP studies, which hopefully will continue in the future, have secured the possibility for future retrospective investigations of temporal trends of "new" contaminants that may be discovered as potential threats to human health.

ACKNOWLEDGMENTS

We are grateful to the participating women who showed dedication to the project. Appreciation is expressed to the midwives Irma Häggbom, Ragnhild Cnattingius, Margareta Aveskog, Ingela Wessén, Astrid Bengtsson, and Marianne Leimar who assisted in recruitment and sample collection, and to the laboratory personnel Lotta Larsson, Lena Hansson, Ingalill Gadhasson, and Elvy Netzel for technical assistance. The financial support from the Health-Related Environmental Monitoring Programme at the Swedish Environmental Protection Agency is acknowledged.

REFERENCES

1. UNEP. 2003. *On the Reduction and/or Elimination of the Releases of Persistent Organic Pollutants, Master Lists of Actions*, 5th edn. Geneva, Switzerland: United Nations Environment Programme.
2. SNV. 2005. *Survey of Unintentionally Formed Contaminants, Report 5462* (in Swedish). Stockholm, Sweden: Swedish EPA. http://www.naturvardsverket.se/Documents/publikationer/620-5462-5467.pdf

3. Noren, K. and Meironyte, D. 2000. Certain organochlorine and organobromine contaminants in Swedish human milk in perspective of past 20–30 years. *Chemosphere* 40: 1111–1123.

4. Lignell, S., Aune, M., Darnerud, P. O., Cnattingius, S., and Glynn, A. 2009. Persistent organochlorine and organobromine compounds in mother's milk from Sweden 1996–2006: Compound-specific temporal trends. *Environ Res* 109: 760–767.

5. Fangstrom, B., Athanassiadis, I., Odsjo, T., Noren, K., and Bergman, A. 2008. Temporal trends of polybrominated diphenyl ethers and hexabromocyclododecane in milk from Stockholm mothers, 1980–2004. *Mol Nutr Food Res* 52: 187–193.

6. KEMI. 2010 *Flame Retardants*. Stockholm, Sweden: Swedish Chemicals Agency. http://www.kemi.se/templates/Page____3697.aspx

7. Glynn, A., Aune, M., Ankarberg, E., Lignell, S., and Darnerud, P. O. 2007. Polychlorinated dibenzo-p-dioxins (PCDDs) and dibenzofurans (PCDFs), polychlorinated biphenyls (PCBs), chlorinated pesticides and brominated flame retardants in mother's milk from primiparae women in Uppsala County, Sweden—Levels and trends 1996–2006. Stockholm, Sweden: Swedish EPA. http://www.naturvardsverket.se/upload/02_tillstandet_i_miljon/Miljoovervakning/rapporter/halsa/Dioxiner_mm_i_modersmjolk_tidstrend_SLV.pdf

8. Glynn, A., Aune, M., Darnerud, P. O., Cnattingius, S., Bjerselius, R., Becker, W., and Lignell, S. 2007. Determinants of serum concentrations of organochlorine compounds in Swedish pregnant women: A cross-sectional study. *Environ Health* 6: 2.

9. Glynn, A. W., Atuma, S., Aune, M., Darnerud, P. O., and Cnattingius, S. 2001. Polychlorinated biphenyl congeners as markers of toxic equivalents of polychlorinated biphenyls, dibenzo-p-dioxins and dibenzofurans in breast milk. *Environ Res* 86: 217–228.

10. Atuma, S. S. and Aune, M. 1999. Method for the determination of PCB congeners and chlorinated pesticides in human blood serum. *Bull Environ Contam Toxicol* 62: 8–15.

11. Aune, M., Atuma, S., Darnerud, P. O., Glynn, A. W., and Cnattingius, S. 1999. Analysis of organochlorine compounds in human milk. *Organohalogen Compd* 44: 93–96.

12. Atuma, S., Aune, M., Darnerud, P. O., Cnattingius, S., Wernroth, L., and Glynn, A. W. 2000. Polybrominated diphenyl ethers (PBDEs) in human milk from Sweden. In: *Persistent, Bioaccumulative, and Toxic Metals II. Assessment of New Chemicals*, eds. R. L. Lipnick, B. Jansson, D. Mackay, and M. Petreas. Washington, DC: American Chemical Society, pp. 235–242.

13. Danielsson, C., Wiberg, K., Korytar, P., Bergek, S., Brinkman, U. A., and Haglund, P. 2005. Trace analysis of polychlorinated dibenzo-p-dioxins, dibenzofurans and WHO polychlorinated biphenyls in food using comprehensive two-dimensional gas chromatography with electron-capture detection. *J Chromatogr A* 1086: 61–70.

14. Van den Berg, M., Birnbaum, L. S., Denison, M., De Vito, M., Farland, W., Feeley, M., Fiedler, H., Hakansson, H., Hanberg, A., Haws, L., Rose, M., Safe, S., Schrenk, D., Tohyama, C., Tritscher, A., Tuomisto, J., Tysklind, M., Walker, N., and Peterson, R. E. 2006. The 2005 World Health Organization reevaluation of human and Mammalian toxic equivalency factors for dioxins and dioxin-like compounds. *Toxicol Sci* 93: 223–241.

15. Van den Berg, M., Birnbaum, L., Bosveld, A. T., Brunstrom, B., Cook, P., Feeley, M., Giesy, J. P., Hanberg, A., Hasegawa, R., Kennedy, S. W., Kubiak, T., Larsen, J. C., van Leeuwen, F. X., Liem, A. K., Nolt, C., Peterson, R. E., Poellinger, L., Safe, S., Schrenk, D., Tillitt, D., Tysklind, M., Younes, M., Waern, F., and Zacharewski, T. 1998. Toxic equivalency factors (TEFs) for PCBs, PCDDs, PCDFs for humans and wildlife. *Environ Health Perspect* 106: 775–792.

16. Phillips, D. L., Pirkle, J. L., Burse, V. W., Bernert, J. T. Jr., Henderson, L. O., and Needham, L. L. 1989. Chlorinated hydrocarbon levels in human serum: Effects of fasting and feeding. *Arch Environ Contam Toxicol* 18: 495–500.

17. Covaci, A., Jorens, P., Jacquemyn, Y., and Schepens, P. 2002. Distribution of PCBs and organochlorine pesticides in umbilical cord and maternal serum. *Sci Total Environ* 298: 45–53.

18. Mazdai, A., Dodder, N. G., Abernathy, M. P., Hites, R. A., and Bigsby, R. M. 2003. Polybrominated diphenyl ethers in maternal and fetal blood samples. *Environ Health Perspect* 111: 1249–1252.

19. Langer, P. 2008. Persistent organochlorinated pollutants (PCB, DDE, HCB, dioxins, furans) and the thyroid—Review 2008. *Endocr Regul* 42: 79–104.

20. Schantz, S. L., Widholm, J. J., and Rice, D. C. 2003. Effects of PCB exposure on neuropsychological function in children. *Environ Health Perspect* 111: 357–576.

21. Schmid, K., Lederer, P., Goen, T., Schaller, K. H., Strebl, H., Weber, A., Angerer, J., and Lehnert, G. 1997. Internal exposure to hazardous substances of persons from various continents: Investigations on exposure to different organochlorine compounds. *Int Arch Occup Environ Health* 69: 399–406.

22. van der Ven, K., van der Ven, H., Thibold, A., Bauer, O., Kaisi, M., Mbura, J., Mgaya, H. N., Weber, N., Diedrich, K., and Krebs, D. 1992. Chlorinated hydrocarbon content of fetal and maternal body tissues and fluids in full term pregnant women: A comparison of Germany versus Tanzania. *Hum Reprod* 7 (Suppl 1): 95–100.

23. Mes, J., Davies, D. J., Doucet, J., Weber, D., and McMullen, E. 1993. Levels of chlorinated hydrocarbon residues in Canadian human breast milk and their relationship to some characteristics of the donors. *Food Addit Contam* 10: 429–441.

24. Rogan, W. J., Gladen, B. C., McKinney, J. D., Carreras, N., Hardy, P., Thullen, J., Tingelstad, J., and Tully, M. 1986. Polychlorinated biphenyls (PCBs) and dichlorodiphenyl dichloroethene (DDE) in human milk: Effects of maternal factors and previous lactation. *Am J Public Health* 76: 172–177.

25. WHO. 2002. The optimal duration of exclusive breast-feeding. Report of the expert consultation. Geneva, Switzerland: World Health Organization. http://www.who.int/nutrition/publications/optimal_duration_of_exc_bfeeding_report_eng.pdf

26. Bignert, A., Olsson, M., Persson, W., Jensen, S., Zakrisson, S., Litzen, K., Eriksson, U., Haggberg, L., and Alsberg, T. 1998. Temporal trends of organochlorines in Northern Europe, 1967–1995. Relation to global fractionation, leakage from sediments and international measures. *Environ Pollut* 99: 177–198.

27. Bredhult, C., Backlin, B. M., Bignert, A., and Olovsson, M. 2008. Study of the relation between the incidence of uterine leiomyomas and the concentrations of PCB and DDT in Baltic gray seals. *Reprod Toxicol* 25: 247–255.

28. Lignell, S., Glynn, A., Darnerud, P. O., Aune, M., Bergdahl, I., Barregård, L., and Bensryd, I. 2005. Regional differences in levels of persistent organic pollutants in breast milk from primipara women in Uppsala, Göteborg, Lund and Lycksele (Sweden). Uppsala, Sweden: National Food Administration.

29. Stapleton, H. M., Dodder, N. G., Offenberg, J. H., Schantz, M. M., and Wise, S. A. 2005. Polybrominated diphenyl ethers in house dust and clothes dryer lint. *Environ Sci Technol* 39, 925–931.

30. Westöö, G., Norén, K., and Andersson, M. 1997. Levels of organochlorine pesticides and polychlorinated biphenyls in margarine, vegetable oils, and some foods of animal origin on the Swedish market in 1967–1969. *Vår Föda* 2–3: 10–31.

31. Johansson, N., Hanberg, A., Bergek, S., and Tysklind, M. 2001. PCB in sealant is influencing the levels in indoor air. *Organohalogen Compd* 52: 436–440.

32. Johansson, N., Hanberg, A., Wingfors, H. and Tysklind, M. 2003. PCB in building sealant is influencing PCB levels in blood of residents. *Organohalogen Compd* 63: 381–384.

33. Rylander, L., Hagmar, L., Wallin, E., Sjostrom, A. K., and Tysklind, M. 2009. Intra-individual variations and temporal trends in dioxin levels in human blood 1987–2002. *Chemosphere* 76: 1557–1562.

34. Bignert, A., Danielsson, S., Nyberg, E., Asplund, L., Eriksson, U., Berger, U., and Haglund, P. 2009. Comments concerning the National Swedish Contaminant Monitoring Programme in Marine Biota, 2009. Stockholm, Sweden: The Swedish Museum of Natural History. http://www.nrm.se/download/118.6321786f6321122df6365955f80002587/Marina+programmet+80002009.pdf

35. Ankarberg, E., Aune, M., Concha, G., Darnerud, P., Glynn, A., Lignell, S., and Törnkvist, A. 2007. Risk assessment of persistent chlorinated and brominated environmental pollutants in food, NFA Report 9. Uppsala, Sweden: National Food Administration. http://www.slv.se/upload/dokument/rapporter/kemiska/2007_livsmedelsverket_2009_risk_assessment_chlorinated_brominated_pollutants.pdf

36. Bernes, C. 1998. *Persistent Organic Pollutants. A Swedish View of an International Problem.* Stockholm, Uppsala: Swedish Environmental Protection Agency.

37. Glynn, A., Aune, M., Nilsson, I., Darnerud, P. O., Ankarberg, E. H., Bignert, A., and Nordlander, I. 2009. Declining levels of PCB, HCB and *p,p'*-DDE in adipose tissue from food producing bovines and swine in Sweden 1991–2004. *Chemosphere* 74: 1457–1462.

38. Ankarberg, E., Törnkvist, A., Darnerud, P. O., Aune, M., Peterson-Grawé, K., Nordqvist, Y., and Glynn, A. 2006. *Dietary Intake Estimations of Persistent Organic Pollutants (Dioxin, PCB, PBDE, Chlorinated Pesticides and Phenolic Compounds) Based on Swedish Market Basket Data and Levels of Methyl-Mercury in Fish.* Uppsala, Sweden: National Food Administration.

39. EU Commission. 2001. *Setting Maximum Levels of Certain Contaminants in Foodstuffs.* Brussels, Belgium: European Commission.

40. CEU. 2001. *Council Directive 2001/102/EC of November 2001 Amending Directive 1999/29/EC on the Undesirable Substances and Products in Animal Nutrition.* Brussels, Belgium: The Council of the European Union.

41. Commission. 2002. *Limits on Presence of Dioxins in Food and Feed Enter into Force on 1 July. Rapid Press Releases.* Brussels, Belgium: The European Commission.

42. Gallian, B., Boix, A., Di Domenico, A., and Fanelli, R. 2004. Occurrence of non-dioxin-like PCB in food and feed in Europe. *Organohalogen Compd* 66: 3610–3618.

43. WHO. 1979. *DDT and Its Derivates. Environmental Health Criteria 9*. Geneva, Switzerland: World Health Organization.

44. WHO. 1997. *Hexachlorobenzene. Environmental Health Criteria 195*. Geneva, Switzerland: World Health Organization.

45. Glynn, A. W., Wernroth, L., Atuma, S., Linder, C. E., Aune, M., Nilsson, I., and Darnerud, P. O. 2000. PCB and chlorinated pesticide concentrations in swine and bovine adipose tissue in Sweden 1991–1997: Spatial and temporal trends. *Sci Total Environ* 246: 195–206.

46. WHO. 1991. *Lindane. Environmental Health Criteria 124*. Geneva, Switzerland: World Health Organization.

47. Atuma, S. S., Hansson, L., Johnsson, H., Slorach, S., de Wit, C. A., and Lindstrom, G. 1998. Organochlorine pesticides, polychlorinated biphenyls and dioxins in human milk from Swedish mothers. *Food Addit Contam* 15: 142–150.

48. Dearth, M. A. and Hites, R. A. 1991. Complete analysis of technical chlordane using negative ionization mass spectrometry. *Environ Sci Technol* 25: 245–254.

49. Dearth, M. A. and Hites, R. A. 1991. Depuration rates of chlordane compounds from rat fat. *Environ Sci Technol* 25: 1125–1128.

50. Lind, Y., Darnerud, P. O., Atuma, S., Aune, M., Becker, W., Bjerselius, R., Cnattingius, S., and Glynn, A. 2003. Polybrominated diphenyl ethers in breast milk from Uppsala County, Sweden. *Environ Res* 93: 186–194.

51. Geyer, H. J., Schramm, K. W., Darnerud, P. O., Aune, M., Feicht, E. A., Fried, K. W., Henkelmann, B., Lenoier, D., Schmidt, P., and McDonald, T. A. 2004. Terminal elimination half-lives of the brominated flame retardants TBBA, HBCD and lower brominated PBDEs in humans. *Organohalogen Compd* 66: 3820–3825.

52. La Guardia, M. J., Hale, R. C., and Harvey, E. 2006. Detailed polybrominated diphenyl ether (PBDE) congener composition of the widely used penta-, octa-, and deca-PBDE technical flame-retardant mixtures. *Environ Sci Technol* 40: 6247–6254.

53. Lee, L. K. and He, J. 2010. Reductive debromination of polybrominated diphenyl ethers by anaerobic bacteria from soils and sediments. *Appl Environ Microbiol* 76: 794–802.

54. Huang, H., Zhang, S., Christie, P., Wang, S., and Xie, M. 2010. Behavior of decabromodiphenyl ether (BDE-209) in the soil-plant system: Uptake, translocation, and metabolism in plants and dissipation in soil. *Environ Sci Technol* 44: 663–667.

55. Sellstrom, U., Bignert, A., Kierkegaard, A., Haggberg, L., de Wit, C. A., Olsson, M., and Jansson, B. 2003. Temporal trend studies on tetra- and pentabrominated diphenyl ethers and hexabromocyclododecane in guillemot egg from the Baltic Sea. *Environ Sci Technol* 37: 5496–5501.

56. Krauthacker, B., Votava-Raic, A., Herceg Romanic, S., Tjesic-Drinkovic, D., and Reiner, E. 2009. Persistent organochlorine compounds in human milk collected in Croatia over two decades. *Arch Environ Contam Toxicol* 57: 616–622.

57. Furst, P. 2006. Dioxins, polychlorinated biphenyls and other organohalogen compounds in human milk. Levels, correlations, trends and exposure through breastfeeding. *Mol Nutr Food Res* 50: 922–933.

58. Konishi, Y., Kuwabara, K. and Hori, S. 2001. Continuous surveillance of organochlorine compounds in human breast milk from 1972 to 1998 in Osaka, Japan. *Arch Environ Contam Toxicol* 40: 571–578.

59. Raab, U., Preiss, U., Albrecht, M., Shahin, N., Parlar, H., and Fromme, H. 2008. Concentrations of polybrominated diphenyl ethers, organochlorine compounds and nitro musks in mother's milk from Germany (Bavaria). *Chemosphere* 72: 87–94.

60. Zeitz, B. P., Hoopmann, M., Funcke, M., Huppmann, R., Suchenwirth, R., and Gierden, E. 2008. Long-term biomonitoring of polychlorinated biphenyls and organochlorine pesticides in human milk from mothers living in northern Germany. *Int J Hyg Environ Health* 211: 624–638.

61. Becher, G., Skaare, J. U., Polder, A., Sletten, B., Rossland, O. J., Hansen, H. K., and Ptashekas, J. 1995. PCDDs, PCDFs, and PCBs in human milk from different parts of Norway and Lithuania. *J Toxicol Environ Health* 46: 133–148.

62. Craan, A. G. and Haines, D. A. 1998. Twenty-five years of surveillance for contaminants in human breast milk. *Arch Environ Contam Toxicol* 35: 702–710.

63. Johansen, H. R., Becher, G., Polder, A., and Skaare, J. U. 1994. Congener-specific determination of polychlorinated biphenyls and organochlorine pesticides in human milk from Norwegian mothers living in Oslo. *J Toxicol Environ Health* 42: 157–171.

64. Kiviranta, H., Purkunen, R., and Vartiainen, T. 1999. Levels and trends of PCDD/Fs and PCBs in human milk in Finland. *Chemosphere* 38: 311–323.

65. Polder, A., Skaare, J. U., Skjerve, E., Loken, K. B., and Eggesbo, M. 2009. Levels of chlorinated pesticides and polychlorinated biphenyls in Norwegian breast milk (2002–2006), and factors that may predict the level of contamination. *Sci Total Environ* 407: 4584–4590.

66. Tuomisto, J. 2009. *Persistent Organic Pollutants (POPs) in Human Milk, European Environment and Health Information System.* Fact Sheet No. 4.3. Copenhagen, Denmark: World Health Organization. http://www.euro.who.int/document/EHI/enhis_factsheet09_04_03.pdf

67. Fangstrom, B., Strid, A., Grandjean, P., Weihe, P., and Bergman, A. 2005. A retrospective study of PBDEs and PCBs in human milk from the Faroe Islands. *Environ Health* 4: 12.

68. Cerna, M., Bencko, V., Brabec, M., Smid, J., Krskova, A., and Jech, L. 2010. Exposure assessment of breast-fed infants in the Czech Republic to indicator PCBs and selected chlorinated pesticides: Area-related differences. *Chemosphere* 78: 160–168.

69. Yu, Z., Palkovicova, L., Drobna, B., Petrik, J., Kocan, A., Trnovec, T., and Hertz-Picciotto, I. 2007. Comparison of organochlorine compound concentrations in colostrum and mature milk. *Chemosphere* 66: 1012–1018.

70. Schoeters, G. and Hoogenboom, R. 2006. Contamination of free-range chicken eggs with dioxins and dioxin-like polychlorinated biphenyls. *Mol Nutr Food Res* 50: 908–914.

71. Weber, R., Gaus, C., Tysklind, M., Johnston, P., Forter, M., Hollert, H., Heinisch, E., Holoubek, I., Lloyd-Smith, M., Masunaga, S., Moccarelli, P., Santillo, D., Seike, N., Symons, R., Torres, J. P., Verta, M., Varbelow, G., Vijgen, J., Watson, A., Costner, P., Woelz, J., Wycisk, P., and Zennegg, M. 2008. Dioxin- and POP-contaminated sites—Contemporary and future relevance and challenges: Overview on background, aims and scope of the series. *Environ Sci Pollut Res Int* 15: 363–393.

72. Van Overmeire, I., Waegeneers, N., Sioen, I., Bilau, M., De Henauw, S., Goeyens, L., Pussemier, L., and Eppe, G. 2009. PCDD/Fs and dioxin-like PCBs in home-produced eggs from Belgium: Levels, contamination sources and health risks. *Sci Total Environ* 407: 4419–4429.

73. Schade, G. and Heinzow, B. 1998. Organochlorine pesticides and polychlorinated biphenyls in human milk of mothers living in northern Germany: Current extent of contamination, time trend from 1986 to 1997 and factors that influence the levels of contamination. *Sci Total Environ* 215: 31–39.

74. Yu, H. F., Zhao, X. D., Zhao, J. H., Zhu, Z. Q., and Zhao, Z. 2006. Continuous surveillance of organo-chlorine pesticides in human milk from 1983 to 1998 in Beijing, China. *Int J Environ Health Res* 16: 21–26.

75. WHO. 2007. *Global Malaria Programme. The Use of DDT in Malaria Vector Control.* Geneva, Switzerland: World Health Organization.

76. Bouwman, H., Sereda, B., and Meinhardt, H. M. 2006. Simultaneous presence of DDT and pyre-throid residues in human breast milk from malaria endemic area in South Africa. *Environ Pollut* 144: 902–917.

77. Akutsu, K., Kitagawa, M., Nakazawa, H., Makino, T., Iwazaki, K., Oda, H., and Hori, S. 2003. Time-trend (1973–2000) of polybrominated diphenyl ethers in Japanese mother's milk. *Chemosphere* 53: 645–654.

78. Thomsen, C., Lundanes, E., and Becher, G. 2002. Brominated flame retardants in archived serum samples from Norway: A study on temporal trends and the role of age. *Environ Sci Technol* 36: 1414–1418.

79. Sjodin, A., Jones, R. S., Focant, J. F., Lapeza, C., Wang, R. Y., McGahee, E. E., 3rd, Zhang, Y., Turner, W. E., Slazyk, B., Needham, L. L., and Patterson, D. G., Jr. 2004. Retrospective time-trend study of polybrominated diphenyl ether and polybrominated and polychlorinated biphenyl levels in human serum from the United States. *Environ Health Perspect* 112: 654–658.

80. Doucet, J., Tague, B., Arnold, D. L., Cooke, G. M., Hayward, S., and Goodyer, C. G. 2009. Persistent organic pollutant residues in human fetal liver and placenta from Greater Montreal, Quebec: A longitudinal study from 1998 through 2006. *Environ Health Perspect* 117: 605–610.

81. Schecter, A., Papke, O., Tung, K. C., Joseph, J., Harris, T. R., and Dahlgren, J. 2005. Polybrominated diphenyl ether flame retardants in the U.S. population: Current levels, temporal trends, and comparison with dioxins, dibenzofurans, and polychlorinated biphenyls. *J Occup Environ Med* 47: 199–211.

82. Johnson-Restrepo, B. and Kannan, K. 2009. An assessment of sources and pathways of human exposure to polybrominated diphenyl ethers in the United States. *Chemosphere* 76: 542–548.

83. Haraguchi, K., Koizumi, A., Inoue, K., Harada, K. H., Hitomi, T., Minata, M., Tanabe, M., Kato, Y., Nishimura, E., Yamamoto, Y., Watanabe, T., Takenaka, K., Uehara, S., Yang, H. R., Kim, M. Y., Moon, C. S., Kim, H. S., Wang, P., Liu, A., and Hung, N. N. 2009. Levels and regional trends of persistent organo-chlorines and polybrominated diphenyl ethers in Asian breast milk demonstrate POPs signatures unique to individual countries. *Environ Int* 35: 1072–1079.

84. Jarrell, J., Chan, S., Hauser, R., and Hu, H. 2005. Longitudinal assessment of PCBs and chlorinated pesticides in pregnant women from Western Canada. *Environ Health* 4: 10.
85. Pan, I. J., Daniels, J. L., Herring, A. H., Rogan, W. J., Siega-Riz, A. M., Goldman, B. D., and Sjodin, A. 2010. Lactational exposure to polychlorinated biphenyls, dichlorodiphenyltrichloroethane, and dichloro-diphenyldichloroethylene and infant growth: An analysis of the pregnancy, infection, and nutrition babies study. *Paediatr Perinat Epidemiol* 24: 262–271.
86. AMAP. 2009. *AMAP Assessment 2009: Human Health in the Arctic*. Oslo, Norway: Arctic Monitoring and Assessment Programming.
87. Li, J., Zhang, L., Wu, Y., Liu, Y., Zhou, P., Wen, S., Liu, J., Zhao, Y., and Li, X. 2009. A national survey of polychlorinated dioxins, furans (PCDD/Fs) and dioxin-like polychlorinated biphenyls (dl-PCBs) in human milk in China. *Chemosphere* 75: 1236–1242.
88. Cok, I., Donmez, M. K., Uner, M., Demirkaya, E., Henkelmann, B., Shen, H., Kotalik, J., and Schramm, K. W. 2009. Polychlorinated dibenzo-*p*-dioxins, dibenzofurans and polychlorinated biphenyls levels in human breast milk from different regions of Turkey. *Chemosphere* 76: 1563–1571.
89. Tanabe, S. and Minh, T. B. 2010. Dioxins and organohalogen contaminants in the Asia-Pacific region. *Ecotoxicology* 19: 463–478.
90. Wang, S. L., Lin, C. Y., Guo, Y. L., Lin, L. Y., Chou, W. L., and Chang, L. W. 2004. Infant exposure to polychlorinated dibenzo-*p*-dioxins, dibenzofurans and biphenyls (PCDD/Fs, PCBs)-correlation between prenatal and postnatal exposure. *Chemosphere* 54: 1459–1473.
91. Nakamura, T., Nakai, K., Matsumura, T., Suzuki, S., Saito, Y., and Satoh, H. 2008. Determination of dioxins and polychlorinated biphenyls in breast milk, maternal blood and cord blood from residents of Tohoku, Japan. *Sci Total Environ* 394: 39–51.
92. Colles, A., Koppen, G., Hanot, V., Nelen, V., Dewolf, M. C., Noel, E., Malisch, R., Kotz, A., Kypke, K., Biot, P., Vinkx, C., and Schoeters, G. 2008. Fourth WHO-coordinated survey of human milk for persistent organic pollutants (POPs): Belgian results. *Chemosphere* 73: 907–914.
93. Hedley, A. J., Hui, L. L., Kypke, K., Malisch, R., van Leeuwen, F. X., Moy, G., Wong, T. W., and Nelson, E. A. 2010. Residues of persistent organic pollutants (POPs) in human milk in Hong Kong. *Chemosphere* 79: 259–265.
94. She, J., Holden, A., Sharp, M., Tanner, M., Williams-Derry, C. and Hooper, K. 2007. Polybrominated diphenyl ethers (PBDEs) and polychlorinated biphenyls (PCBs) in breast milk from the Pacific Northwest. *Chemosphere* 67: 307–317.

16 Temporal Trends of POPs in Human Milk from Italy

*Annalisa Abballe, Anna Maria Ingelido, and Elena De Felip**

CONTENTS

16.1 Introduction 381
    16.1.1 Persistent Organic Pollutants in Human Milk.. 381
    16.1.2 Earlier Studies on POPs in Human Milk from Italy 382
16.2 Contamination Profiles and Temporal Trends... 384
    16.2.1 Contamination Profiles 385
        16.2.1.1 Dioxins, Furans, and Dioxin-Like PCBs...................................... 385
        16.2.1.2 Non-Dioxin-Like PCBs 385
        16.2.1.3 Polybrominated Diphenyl Ethers 385
    16.2.2 Temporal Trends 386
        16.2.2.1 Dioxins, Furans, and Dioxin-Like PCBs...................................... 387
        16.2.2.2 Non-Dioxin PCBs 387
        16.2.2.3 Polybrominated Diphenyl Ethers 388
16.3 Summary and Conclusions 389
References... 390

16.1 INTRODUCTION

16.1.1 PERSISTENT ORGANIC POLLUTANTS IN HUMAN MILK

Persistent organic pollutants (POPs) have been shown to be present in human milk since the 1950s, when Laug et al. reported the presence of DDT in samples of maternal milk from women living in Washington, DC who had never been professionally exposed to this chemical [1]. With the progressive availability of analytical techniques and instrumentation of high sensitivity and selectivity, other POPs have been detected in human milk, including all those in the Stockholm Convention on POPs, such as polychlorinated-*p*-dibenzodioxins (PCDDs) and polychlorinated dibenzofurans (PCDFs), polychlorinated biphenyls (PCBs), hexachlorobenzene (HCB), a number of organochlorinated pesticides (DDT, aldrin, etc.), polybrominated diphenyl ethers (PBDEs), and the perfluorinated alkylic compounds perfluorooctane sulfonate (PFOS) and perfluorooctanoic acid (PFOA).

Indeed, a number of POPs have been identified as POPs of global concern, and therefore regulatory actions were proposed after being detected in human milk. For example, PBDEs were first found in maternal milk in Germany in the 1980s [2] and afterward in a number of industrialized countries [3–23]. The finding in Sweden that levels of PBDEs in human milk were rapidly increasing, doubling every 5 years [21,24] prompted regulatory authorities to consider progressive phasing out of commercial mixtures of PBDEs at a European Union level.

---

* E-mail: elena.defelip@iss.it (Chapter corresponding author).

Sanitary concerns about the potential risks for breast-fed infants from the presence of POPs in human milk have triggered sanitary authorities and organizations to activate specific monitoring programs on this matrix. Since 1987, the World Health Organization (WHO) has coordinated four international studies to assess the levels and trends of PCDDs, PCDFs, and PCBs in human milk. Together with these pollutants, which have been monitored in all the four studies, others have been progressively included on the list of analytes to be assessed, such as organochlorinated pesticides, PBDEs, hexabromocyclododecanes. After enforcement of the Stockholm Convention on POPs (2004), the WHO studies have been carried out with the double aim to monitor human exposure over time and understand if the Stockholm Convention was actually effective in reducing the release of these chemicals into the environment. Besides the WHO studies, which have the same protocol and design, and therefore ensure the highest degree of comparability, a number of other human biomonitoring studies on human milk have been carried out worldwide. The increasing popularity of these kinds of studies is also due to the fact that, among the biological matrices analyzed for the assessment of POP body burden, human milk is withdrawn by a noninvasive procedure and has the advantage to provide information on the exposure of a subgroup of the adult population and of infant exposure as well.

In Italy, there has been an increasing interest in studies on human milk. Women's concerns about the possible consequences on their infants and children from exposure to POPs have frequently resulted in questions to sanitary authorities on the advisability of breastfeeding, especially in the case of women residing near contaminated areas or industrial complexes. In spite of this, the monitoring of POPs in human milk in Italy has never been carried out on a nationwide, regular basis, but rather on small groups of women in specific areas. The most systematic monitoring in this field has been carried out by the National Institute of Health (ISS), the main sanitary authority involved in the assessment of exposure of the general population to POPs and in the characterization of related sanitary risk.

16.1.2 Earlier Studies on POPs in Human Milk from Italy

A few studies dealing with POPs in human milk are available for Italy and cover the time span 1987–2009. POPs considered for discerning temporal trend are shown in Table 16.1. The target compounds are those of highest toxicological importance, for which a few papers are available to observe a potential temporal trend, namely

- The 17 toxic congeners of the two families of PCDDs and PCDFs.
- Various PCB congeners, specifically the 12 dioxin-like congeners (DL-PCBs), made up of a group of non-*ortho* (or "coplanar") congeners (PCBs 77, 81, 126, and 169) and a group of the mono-*ortho* congeners (PCBs 105, 114, 118, 123, 156, 157, 167, 189) and the most abundant non-dioxin-like congeners (NDL-PCBs), namely, the 6 so-called indicator congeners (PCBs 28, 52, 101, 138, 153, and 180), which make up about 80% of the total PCB content in serum [25].
- The most abundant congeners of polybrominated diphenylethers (PBDEs). Seven congeners (PBDEs 28, 47, 99, 100, 153, 154, and 183) were selected on the basis of their presence in all available studies and of their abundance in commercial mixtures and in human tissues. Other congeners that have been analyzed in some human biomonitoring studies, such as PBDE 209, were not included in our evaluation, because they were not analyzed in all the Italian studies considered for deriving time trends, as well as in most of the available literature from other countries.

A brief description of the studies considered, whose characteristics are reported in Table 16.1, is given in the following text.

Biomonitoring studies conducted during 1987 by Schecter et al. [26] and Larsen et al. [27] analyzed milk samples from four Italian towns at different geographical latitudes. A total of 9 mothers were from Rome (Lazio, central Italy), 9 from Pavia (Lombardia, northern Italy), 27 from Florence

TABLE 16.1

PCDD+PCDF (pg WHO-TEQ/g Lipid Base), DL-PCB (Non-Ortho and Mono-Ortho, pg WHO-TEQ$_{97}$/g Lipid Base), NDL-PCB (Σ_6PCBs, ng/g Lipid Base), and PBDE (Σ_7PBDEs, ng/g Lipid Base) Cumulative Concentrations in Human Milk from Several Areas in Italy in the Period 1987–2009

Year of Sampling	Area	Number of Samples, Pools	PCDD+PCDF; DL-PCB (Non-Ortho and Monoortho)[a]	Σ_6PCB[a]	Σ_7PBDE[a,b]	References
1987	Rome	9, 1 pool	21.0; 15.4[c]	385[d]	—	Schecter et al. [26], Larsen et al. [27]
	Pavia	9, 1 pool	27.0; 13.4	455	—	Schecter et al. [26], Larsen et al. [27]
	Florence	27, 1 pool	27.4; 14.1	420	—	Schecter et al. [26], Larsen et al. [27]
	Milan	14, 1 pool	14.5; 13.5	371	—	Schecter et al. [26], Larsen et al. [27]
1998–2001	Venice[e] LC	10, 1 pool	14.8; 8.34; 11.0	318	2.76[f]	Ingelido et al. [8], Abballe et al. [28]
	Venice MC	13, 1 pool	13.7; 8.81; 10.4	306	2.38	Ingelido et al. [8], Abballe et al. [28]
	Venice HC	6, 1 pool	11.6; 6.46; 6.87	204	1.55	Ingelido et al. [8], Abballe et al. [28]
	Rome	10, 1 pool	9.40; 4.20; 6.81	195	4.06	Ingelido et al. [8], Abballe et al. [28]
2000–2001	Milan	12, 1 pool	11.7; 8.03[g]	—	—	Weiss et al. [29]
	Lombardia (rural area)	12, 1 pool	10.3; 6.12	—	—	Weiss et al. [29]
	Seveso	12, 1 pool	10.7; 6.02	—	—	Weiss et al. [29]
2000–2002	Italy	40, 4 pools	12.7; 16.3[h] (X$_{MED}$)	253 (X$_{MED}$)	—	Malisch and van Leeuwen [39]
—	Siena	47			38.6[i] (X$_{MED}$)	Guerranti et al. [31]
2008–2009	Naples 1	10, 1 pool	6.28; 3.13; 3.02	74.2	1.34	De Felip et al. [30]
	Naples 2	10, 1 pool	6.29; 2.81; 2.71	70.5	1.57	De Felip et al. [30]
	Naples 3	10, 1 pool	6.17; 2.60; 2.39	51.1	1.12	De Felip et al. [30]
	Naples 4	10, 1 pool	8.77; 2.75; 2.63	62.6	1.66	De Felip et al. [30]
	Caserta	5, 1 pool	5.99; 1.82; 2.20	48.7	0.686	De Felip et al. [30]
	Naples/Caserta	7, 1 pool	7.00; 3.00; 2.79	67.7	4.23	De Felip et al. [30]
	Rome	10, 1 pool	7.26; 3.03; 3.95	108	1.47	De Felip et al. [30]

Values rounded off to three figures.

[a] X$_{MED}$, median value.

[b] Sum of the seven PBDE congeners 28, 47, 99, 100, 153, 154, 183, except as otherwise specified.

[c] Data on DL-PCB refer to the sum of non-ortho PCBs only. The mean value of mono-ortho PCBs was 14.48 pg WHO-TEQ/g lb.

[d] Data were estimated multiplying original data (Σ_{16}PCB congeners) by 0.7.

[e] LC, MC, and HC indicate respectively low, medium, and high consumption of local fish and fishery products.

[f] The sum of the seven PBDE congeners 28, 47, 99, 100, 153, 154, 183 was estimated from the mean values reported in the paper for each congener.

[g] Data refer to the sum of non-ortho and mono-ortho PCBs only. Data on mono-ortho PCBs were not available.

[h] Data on DL-PCBs refer to the sum of non-ortho and mono-ortho PCBs. Data on non-ortho and mono-ortho PCBs were not available separately.

[i] Sum of the 12 PBDE congeners 17, 28, 47, 49, 66, 71, 77, 85, 99, 100, 153, 154.

(Toscana, central Italy), and 14 from Milan (Lombardia, northern Italy). The age range was 16–43 years. In the case of Rome, only four mothers were primiparae. Four pools were obtained as summarized in Table 16.1 and analyzed for PCDDs, PCDFs, DL-PCBs, and NDL-PCBs. Milk was collected between the fourth and the ninth week after delivery with the exception of Milan, where milk was collected in the 4th and in the 27th week after delivery.

During 1998–2001, Abballe et al. [28] analyzed PCDDs, PCDFs, DL- and NDL-PCBs, and PBDEs in 29 individual milk samples from primiparous women from Venice (Veneto, northern Italy) and 10 from Rome (Lazio, central Italy). As for Venice, pooling of specimens was carried out in accordance with mothers' fish consumption levels as follows: LC (10 women, low fish consumption, age range 24–38 years), MC (13 women, mean fish consumption, age range 21–38), and HC (six donors, high fish consumption, age range 28–40). The age range was 28–40 years. Sampling was carried out between the fourth and eighth week after delivery.

Weiss et al. [29] investigated levels of dioxin-like compounds in mothers' milk samples collected during 2000–2001 in three sampling sites in the Lombardia Region (northern Italy), the town of Milan, a rural area, and Seveso. In this small town, located approximately 15 km north of Milan, an industrial accident occurred in 1976, which caused the release into the environment of a large quantity of 2, 3, 7, 8-tetrachlorodibenzo-p-dioxin (2, 3, 7, 8-TCDD or "dioxin") resulting in the highest known exposure to 2, 3, 7, 8-TCDD in the residential population.

Twelve mothers were recruited in each sampling site. About one-half of the mothers were primiparae, the remaining were secondiparae. Samples of milk were collected at three different times, at the beginning of lactation (colostrum), after 1 month and after 3 months from delivery. Maternal age range was 25–32 years.

De Felip and di Domenico [30] carried out a study in 2008–2009 in the Campania Region (southern Italy), and involved quite a vast area surrounding (and including) the two towns of Naples and Caserta. It was aimed to assess exposure to PCDDs, PCDFs, DL- and NDL-PCBs, and PBDEs of the general population residing in proximity to illegal waste burning sites, which is quite common in several spots near these towns. The town of Rome was also included in this study as a "reference" area for which data on POPs in human milk was available from other studies.

A total of 62 mothers, 10 from Rome and 52 from the 2 towns of Naples and Caserta, and respective surrounding areas, were recruited. The age range was 18–40 years. Individual milk samples were collected and classified so as to obtain seven pools as summarized in Table 16.1. All mothers were primiparae. Each sample was collected between the fourth and the eighth week after delivery, in accordance to the WHO's protocol. This study reports for the first time data from southern Italy.

Guerranti et al. [31] measured PBDEs in human milk in 47 women (age 23–37 years) living in the town of Siena (Toscana, central Italy) and its surroundings. Milk was collected in the first week after delivery. Since the year of sampling was not specified in the paper, this study was not taken into consideration for deriving temporal trends.

16.2   CONTAMINATION PROFILES AND TEMPORAL TRENDS

The cumulative concentrations of PCDDs, PCDFs, DL- and NDL-PCBs, and PBDEs reported by the studies taken into consideration are reported in Table 16.1. All concentrations are expressed on a lipid weight basis. In general, comparability between the different studies is good with respect to study design and sample type. In fact, in all studies, the analyses were carried out on pooled samples; information on factors known to determine exposure to POPs was gathered by administering questionnaire to each woman a containing, *inter alia*, questions on dietary habits and occupational exposure. There is a good degree of overall inter-study comparability also as to donors' characteristics, such as women's age (age range was mostly 21–40 years) and, to a lesser extent, parity (most of women were primiparae) and the sampling time window (generally comprised between the fourth and the eighth week after delivery).

16.2.1 CONTAMINATION PROFILES

16.2.1.1 Dioxins, Furans, and Dioxin-Like PCBs

The cumulative concentrations of PCDDs, PCDFs, DL-PCBs, NDL-PCBs, and PBDEs detected in Italian human milk in the time period 1987–2009 are shown in Table 16.1. Concentrations of PCDDs, PCDFs, and DL-PCBs are expressed in toxicity equivalent quantity (TEQ) values, obtained by multiplying the analytical concentration of each single congener by the pertinent toxicity equivalency factor (TEF). TEFs used are those defined by WHO in 1997 [32].

Concentrations of PCDDs, PCDFs, and DL-PCBs found by Schecter and Larsen in the four Italian towns sampled in 1987 range from 34 to 48 pg WHO-TEQ/g lb. In this study, data on DL-PCBs refer to coplanar congeners PCB 77, 126, and 169 only. Concentrations of mono-*ortho* congeners for single pools have not been reported by the authors [27], who reported only the mean value for each congener and their sum (14.5 pg WHO-TEQ/g lb). Concentrations of PCDDs, PCDFs, and DL-PCBs ranged between 20.4 and 34.2 pg WHO-TEQ/g lb in the pools sampled in Venice and Rome in 1998–2001 and between 16.4 and 21.3 pg WHO-TEQ/g lb in the three areas of the Lombardia Region sampled in the same period [29].

Concentrations determined in samples collected in the Campania Region and in Rome in the 2008–2009 study were in the range of 10.0–14.2 pg WHO-TEQ/g lb.

Data from the different sampling campaigns suggest a substantial homogeneity in exposure, within the limits represented by the small number of samples and of sampled sites. All sampled locations were not hot spots, with the unique exception of Seveso, where a massive contamination from 2, 3, 7, 8-TCDD had occurred in 1976. Levels assessed in this town about 25 years after the accident show an exposure situation in line with that observed in the other sites from the Lombardia Region sampled during the same period [29].

The study carried out in 2008–2009 involving some areas of the Campania Region, which aimed to assess if illegal burning of waste has lead to an incremental exposure to PCDDs and PCDFs in local residential communities, did not reveal any difference with respect to Rome. In general, levels of PCDDs, PCDFs, and DL-PCBs observed in Italy in the time span considered appear to be in agreement with those reported in the last two decades for other European countries [33–45].

16.2.1.2 Non-Dioxin-Like PCBs

NDL-PCB concentrations in samples collected in Milan, Pavia, Florence, and Rome in 1987 ranged between 371 and 455 ng/g lb [26].

In the 1998–2001 study, the three pools corresponding to women from Venice characterized by different fish consumption, identified as Venice LC, Venice MC, and Venice HC, showed concentrations of 318, 306, and 204 ng/g lb respectively. In the pool from Rome, 195 ng/g lb were found.

In the 2008–2009 study, PCB concentration (Σ_6PCBs) observed in the unique pool from Rome was 108 ng/g lb and levels in the range of 51.1–74.2 ng/g lb were found for Naples.

These values are consistent with data from analogous studies carried out in European countries in the same period [21,35,39,40,43–45].

16.2.1.3 Polybrominated Diphenyl Ethers

Only two Italian studies on PBDEs are available for deriving temporal trends (Table 16.1). Cumulative results of the seven congener PBDEs 28, 47, 99, 100, 153, 154, and 183 in the samples collected in Venice and Rome in 1998–2001 fall in the range 1.55 and 4.06 ng/g lb [28]. In the samples collected in the Campania Region and Rome in 2008–2009, the same congeners concentrations were in the range of 0.686 ng/g and 4.23 ng/g lb. These levels are comparable with data obtained in the same years from other European countries [44,46–48].

In the paper by Guerranti et al. [31] (year of sampling not specified), authors report a concentration of the 12 congeners analyzed (PBDEs 17, 28, 47, 49, 66, 71,77, 85, 99, 100, 153, and 154) of 56.27 ± 90.03 ng/g lb as a mean value (38.65 ng/g lb as the median). The authors correlated and found that the high values observed with dietary exposure to PBDEs in the Siena area were substantially higher than those generally observed in other European countries.

16.2.2 TEMPORAL TRENDS

In order to compare results relative to different time periods, a mean value (weighted by the number of subjects in each pool) was calculated for each sampling campaign and used in discussion. The same approach was used in deriving congener-specific patterns of PCDDs, PCDFs, DL-PCBs, NDL-PCBs, and PBDEs as shown in Figures 16.1 through 16.3 and Figure 16.5.

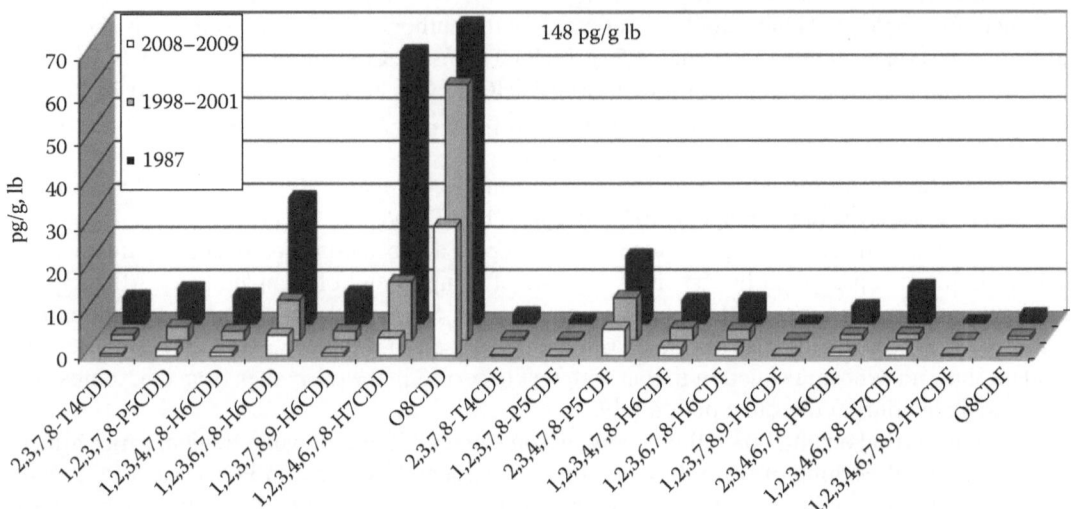

FIGURE 16.1 PCDD and PCDF congener-specific profiles in pools of human milk from Italy, collected in 1987, 1998–2001, and 2008–2009. Mean concentration values are shown for each congener.

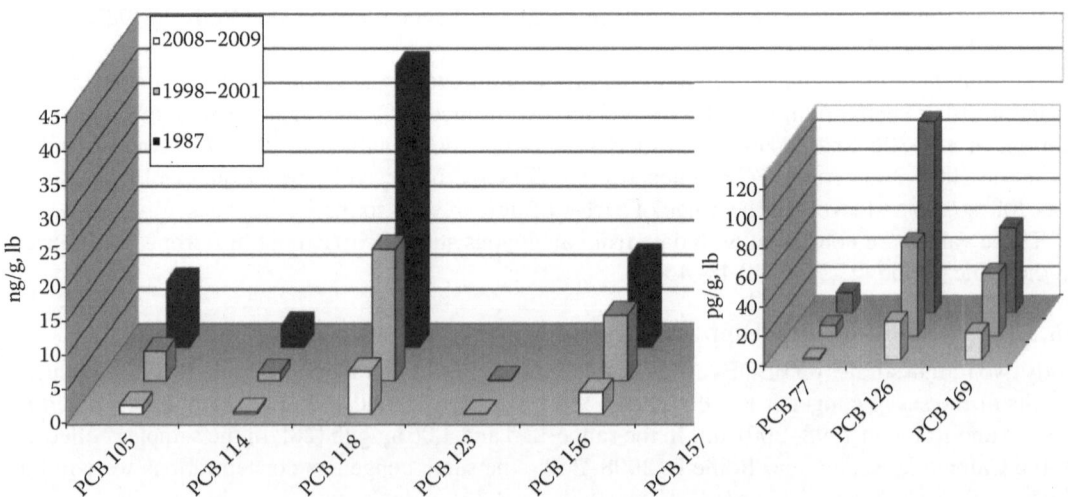

FIGURE 16.2 Congener-specific profiles of DL-PCBs (mono-*ortho* and non-*ortho*-congeners) in pools of human milk from Italy collected in 1987, 1998–2001, and 2008–2009. Mean concentration values are shown for each congener.

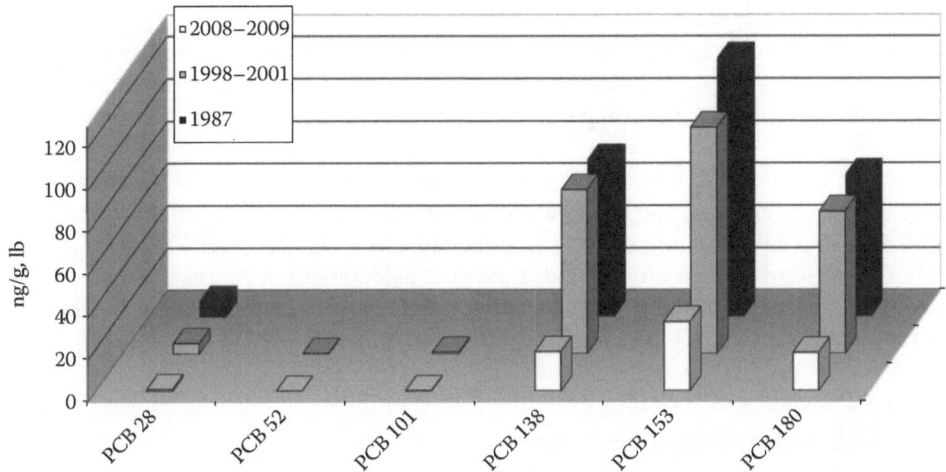

FIGURE 16.3 Congener-specific profiles of the six "indicator" NDL-PCBs in pools of human milk from Italy collected in 1987, 1998–2001, and 2008–2009. Mean concentration values are shown for each congener.

16.2.2.1 Dioxins, Furans, and Dioxin-Like PCBs

The comparison of concentrations of PCDDs, PCDFs, and DL-PCBs found in different studies covering the period 1987–2009 shows a continuous decline in exposure.

In the 1998–2001 study, total TEQs were approximately 40% of those assessed in 1987, while concentrations found in 2008–2009 were about fourfold lower than those observed in 1987. In particular, in the case of Rome, the only geographical location for which data from three studies were available, concentrations found in 1987 were reduced by about 50% in 1998–2001 and by about 70% in 2008–2009. The decrease in human milk concentrations of DL-PCBs, PCDDs, and PCDFs observed in Italy over the period considered appears to be in general agreement with trends observed in other European countries [6,38,39,49–51].

PCDD and PCDF congener-specific patterns are shown in Figure 16.1. The analytical profiles of PCDDs and PCDFs in samples collected in 1987, 1998–2001, and 2008–2009 are very similar and are all characterized by the prevalence of PCDDs on PCDFs. These elements indicate that donors were subjected to a background exposure to these chemicals, mostly determined by dietary intake, known to account for over 90% exposure to these chemicals. The relative contributions of the single congeners to the profiles do not differ appreciably with time. All congeners show a trend in analytical levels negatively associated with time. The concentration of the most toxic congener 2, 3, 7, 8-T_4CDD shows a decrease by 90%, declining from 6.30 pg/g lb in 1987 to 0.60 pg/g lb in 2008–2009. For other congeners, the decrease with time ranges from 36% (1, 2, 3, 7, 8, 9-H_6CDF) to 93% (1, 2, 3, 4, 6, 7, 8-H_7CDD).

Non-*ortho* and mono-*ortho* DL-PCB congener-specific profiles are shown in Figure 16.2. The analytical patterns of both non-*ortho*- and mono-*ortho* DL-PCBs do not show relevant changes from 1987 to 2009, the most important congeners being PCB 126, among non-*ortho* DL-PCBs, and PCB 118, among the mono-*ortho* DL-PCBs, in all the pools analyzed in the three sampling campaigns. This is in agreement with what is generally observed in human biomonitoring studies dealing with these compounds. A declining trend is observed for most DL-PCBs, with decrease in concentrations spanning from about 91% for PCB 77 (from 13.7 to 1.19 pg/g lb) to about 25% for PCB 167 (from 1.20 to 0.90 pg/g lb).

16.2.2.2 Non-Dioxin PCBs

NDL-PCB levels show a temporal trend similar to those observed for PCDDs, PCDFs, and DL-PCBs. Cumulative concentrations of the six indicator congeners progressively decline, passing from the 408 ng/g lb observed in 1987 to 265 ng/g lb in 1998–2001 and 70.7 ng/g lb in 2008–2009. As for Rome, the temporal trend observed for NDL-PCBs overlaps with that observed for PCDDs and PCDFs, showing a mean decrease of about 83%. Additionally, the results appear to be in accordance with the concentration ranges reported by several authors over the same time period [43,44].

NDL-PCB congener profiles of samples collected between 1987 and 2009 are shown in Figure 16.3. They are consistent with the average profile usually observed in human milk, with NDL-PCBs 138, 153, and 180 standing out as the most abundant congeners (PCB 153 > PCB 138 > PCB 180). Decrease in concentrations with time is considerable (on average, about 77%) for all the reported congeners, with the highest decrease (87%, from 7.20 to 0.91 ng/g lb) observed for PCB 28, the least persistent in the environment [52] and human body [53,54] among the indicator congeners.

16.2.2.3   Polybrominated Diphenyl Ethers

Comparison of PBDE concentrations over time may be based only on the two studies available [28,30]. From 1998–2001 to 2008–2009, cumulative levels of the seven congeners decreased by about 40%. The decrease observed for Rome is about 64%, with values declining from 4.06 ng/g lb in 1998 to 1.47 ng/g lb in 2008. This observation suggests that a decrease in exposure is likely to have occurred, within the limits represented by data paucity and by the variability in pattern observed among pools in both studies. This variability appears to be quite relevant in the 2008–2009 study; for example, levels and congener-specific profiles of two samples from the Campania Region are shown (Figure 16.4) for which no significant difference in exposure is known that might account for the observed diversity in profile.

PBDE congener-specific patterns (obtained by reporting each congener as the weighted mean of the analytical determinations carried out on the individual pools) are shown in Figure 16.5. PBDEs 47, 99, 100, 153, and 183 are the most abundant in all the pools analyzed in the two studies, in agreement with the literature data on human milk [4,6,7,10,13,14,16,17,18,22]. A decrease in concentration is observed for most congeners but, while in the 1998–2001 samples profiles are dominated by PBDEs 47, 99, and 153, in samples collected in 2008–2009 the relative contribution of PBDEs 153 and PBDE 183 is definitely more pronounced. Moreover, passing from 1998 to 2009, analytical levels of PBDE 183 and, to a lesser extent, of PBDE 153, increase.

This variation in the relative abundance of the single congeners could be related to the change in the use of PBDE technical mixtures as well as to congener-specific differences in

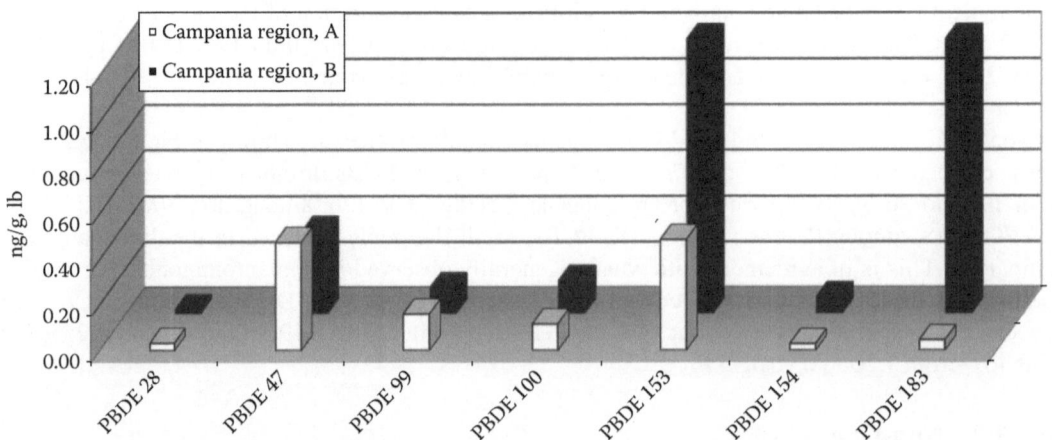

FIGURE 16.4   PBDE congener-specific profiles in pools of human milk from two different locations (A and B) of the Campania Region (2008–2009).

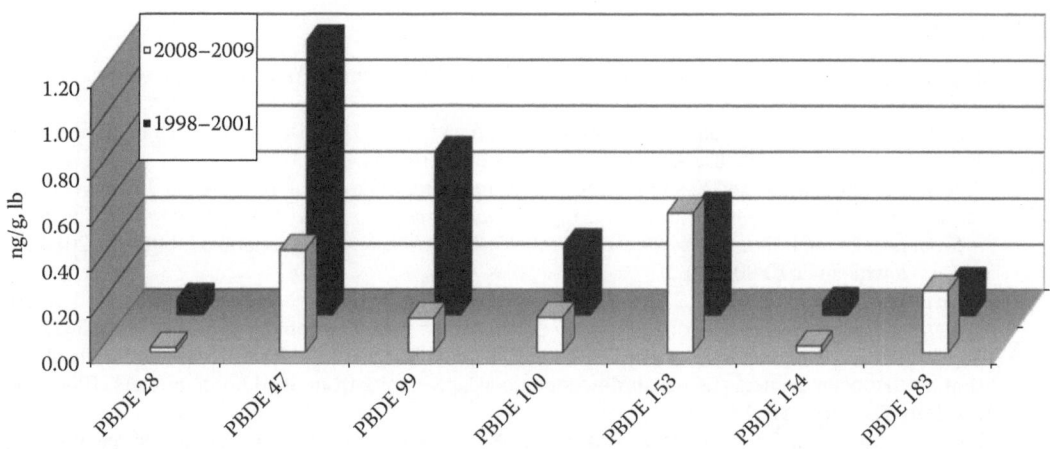

FIGURE 16.5 Congener-specific profiles of PBDEs in pools of human milk from Italy collected in 1998–2001 and 2008–2009. Mean concentration values are shown for each congener.

environmental persistence [55] and half-lives in humans [56,57]. In the past two decades, restrictions on use and successive ban [58–60] of two most widely used commercial mixtures of PBDEs, i.e., the pentabromodiphenyl ether and octabromodiphenyl ether, limited the release into the environment of most of the tetra- to hepta-substituted congeners, in particular of PBDEs 47 and 99 (the major components of the pentabromodiphenyl ether). The third most extensively used commercial mixture, the decabromodiphenyl ether, which is still in use at present, is made primarily from PBDE 209. Although PBDEs 153 and 183, which were both components of the banned commercial mixtures, are not present in the decabromodiphenyl ether, debromination of PBDE 209 could play a role in determining the observed levels of these two congeners. In fact, several studies reported that PBDE 209 undergoes debromination in the environment and biota [61–68] and its degradation products are congeners with a lower degree of bromination, such as PBDEs 153 and 183.

These observations suggest that PBDE temporal trends should be investigated based on congener-specific data. Cumulative PBDE results will however have a bearing on risk assessment, although the number of the toxicologically relevant congeners should be better identified.

16.3 SUMMARY AND CONCLUSIONS

In general, the results of human biomonitoring studies carried out on human milk in Italy in the last two decades show a declining trend in exposure for PCDDs, PCDFs, DL-PCBs, and NDL-PCBs. The limited number of studies available on PBDEs does not allow us to draw sound conclusions, but a declining trend seems to be suggested by the available data.

Although the trends observed appear to confirm the effectiveness of remedial measures undertaken to reduce emissions of POPs into the environment, the small size of the studies and of sampled population does not allow to exclude the presence of population subgroups at higher exposure and unexpected or unpredictable temporal trends.

Monitoring studies on human milk are strongly needed at a national level to monitor exposure to POPs and to respond to a growing demand from population. The continuous increase in the attention paid by sanitary authorities and policy makers to human biomonitoring, together with the pressing demand from population groups to be informed on the levels of contamination "within" their body will prompt the activation of systematic, periodical, nationwide studies to produce reliable and comparable data on human milk.

REFERENCES

1. Laug, E.P. et al. 1951. Occurrence of DDT in human fat and milk. *Arch. Ind. Hyg. Occup. Med.* 3: 2435–2436.
2. Krüger, C. 1988. Biphenyls and polybrominated diphenyl ether: Detection and determination in foods. Thesis, Westfälischen Wilhelms Universitä zu Münster, Germany.
3. Akutsu, K., Kensaku, K., and Yoshimasa, K., 2010. Temporal trend of polybrominated diphenyl ethers in archived breast milk samples from Osaka, Japan. *Paper Presented at the Fifth International Symposium on Brominated Flame Retardants for the BFR 2010*, April 7–9, Kyoto. Available at: http://bfr2010.com/abstract-download/2010/90073.pdf
4. Frederiksen, M., Vorkamp, K., Thomsen, M., and Knudsen, L.E. 2009. Human internal and external exposure to PBDEs—A review of levels and sources. *Int. J. Hyg. Environ. Health* 212: 109–134.
5. Fängström, B., Athanassiadis, I., Odsjö, T., Norén, K., and Bergman, Å. 2008. Temporal trends of polybrominated diphenyl ethers and hexabromocyclododecane in milk from Stockholm mothers, 1980–2004. *Mol. Nutr. Food Res.* 52: 187–193.
6. Glynn, A., Aune, M., Darnerud, P.O., Ankarberg, A., and Törnkvist, A. 2007. Congener specific differences in temporal trends of PCDD, PCDF and PBDE in mother's milk from Uppsala country, Sweden. *Organohalogen Compd.* 69: 1942–1944.
7. Gómara, B., Herrero, L., Ramos, J.J. et al. 2007. Distribution of polybrominated diphenyl ethers in human umbilical cord serum, paternal serum, maternal serum, placentas, and breast milk from Madrid Population, Spain. *Environ. Sci. Technol.* 41 (20): 6961–6968.
8. Ingelido, A.M., Ballard, T.J., Dellatte, E. et al. 2007. Polychlorinated biphenyls (PCBs) and polybrominated diphenyls ethers (PBDEs) in milk from Italian women living in Rome and Venice. *Chemosphere* 67: S301–S306.
9. She, J., Holden, A., Sharp, M., Tanner, M., Williams-Derry, C., and Hooper, K. 2007. Polybrominated diphenyl ethers (PBDEs) and polychlorinated biphenyls (PCBs) in breast milk from the Pacific Northwest. *Chemosphere* 67 (9): S307–S317.
10. Schecter, A.J., Päpke, O., Shah, N., and Harris, T. 2007. Brominated flame retardants in the USA and selected other countries in milk, blood, food, fast food, and air. *Organohalogen Compd.* 69: 694–697.
11. Toms, L.M.L., Harden, F.A., Symons, R.K., Burniston, D., Fürste, P., and Müller, J.F. 2007. Polybrominated diphenyl ethers (PBDEs) in human milk from Australia. *Chemosphere* 68: 797–803.
12. Fängström, B., Strid, A., Grandjean, P., Weihe, P., and Bergman, Å. 2005. A retrospective study of PBDEs and PCBs in human milk from the Faroe Islands. *Environ. Health: A Global Access Sci. Source* 4: 12. Available at: http://www.ehjournal.net/content/4/1/12
13. Kalantzi, O.I., Alcock, R.E., Martin, F.L., Thomas, G.O., and Jones, K.C. 2004. Polybrominated diphenyl ethers (PBDEs) and selected organochlorines in human breast milk samples from the United Kingdom. *Organohalogen Compd.* 61: 9–12.
14. Baumann, B., Hijman, W., Van Beuzekom, S., Hoogerbrugge, R., Houweling, D., and Zeilmaker, M. 2003. PBDEs in human milk from the Dutch 1998 monitoring programme. *Organohalogen Compd.* 61: 187–190.
15. Lind, Y., Darnerud, P.O., Atuma, S. et al. 2003. Polybrominated diphenyl ethers in breast milk from Uppsala County, Sweden. *Environ. Res.* 93: 186–194.
16. Meironyté-Guvenius, D., Aronsson, A., Ekman-Ordebrg, G., Bergman, Å., and Norén, K. 2003. Human prenatal and postnatal exposure to polybrominated diphenyl ethers, polychlorinated biphenyls, polychlorobiphenylols, and pentachlorophenol. *Environ. Health Perspect.* 111: 1235–1241.
17. Pirard, C., De Pauw, E., and Focant, J.F. 2003. Levels of selected PBDEs and PCBs in Belgian human milk. *Organohalogen Compd.* 64: 158–161.
18. Thomsen, C., Frøshaug, M., Leknes, H., and Becher, G. 2003. Brominated flame retardants in breast milk from Norway. *Organohalogen Compd.* 64: 33–36.
19. Solomon, G.M. and Weiss, P.M. 2002. Chemical contaminants in breast milk: Time trends and regional variability. *Environ. Health Perspect.* 110 (6): 339–347.
20. Ryan, J.J., Patry, B., Mills, P., and Beaudoin, G. 2002. Recent trends in levels of brominated diphenyl ethers in human milks from Canada. *Organohalogen Compd.* 58: 173–176.
21. Norén, K. and Meironyté, D. 2000. Certain organochlorine and organobromine contaminants in Swedish human milk in perspective past 20–30 years. *Chemosphere* 40: 1111–1123.
22. Strandman, T., Koistinen, J., and Vartiainen, T. 2000. Polybrominated diphenyl ethers (PBDEs) in placenta and human milk. *Organohalogen Compd.* 47: 61–64.
23. Meironyté, D., Norén, K., and Bergman, Å. 1999. Analysis of polybrominated diphenyl ethers in Swedish human milk. A time-related trend study, 1972–1997. *J. Toxicol. Environ. Health A* 58 (6): 329–341.

24. Meironyté-Guvenius, D. 2002. Organohalogen contaminants in humans with emphasis on polybrominated diphenyl ethers. Thesis, Department of Medical Biochemistry and Biophysics, Karolinska Institute, Stockholm, Sweden.

25. Glynn, A.W., Wolk, A., Aune, M. et al. 2000. Serum concentrations of organochlorines in men: A search for markers of exposure. *Sci. Total Environ.* 263: 197–208.

26. Schecter, A., di Domenico, A., Turrio-Baldassarri, L., and Ryan, J.J. 1992. Dioxin and dibenzofuran levels in the milk of women from four geographical regions in Italy as compared to levels in other countries. *Organohalogen Compd.* 9: 227–230.

27. Larsen, B.R., Turrio-Baldassarri, L., Nilsson, T. et al. 1994. Toxic PCB congeners and organochlorine pesticides in Italian human milk. *Ecotoxicol. Environ. Saf.* 28: 1–13.

28. Abballe, A., Ballard, T.J., Dellatte, E. et al. 2008. Persistent environmental contaminants in human milk: Concentrations and time trends in Italy. *Chemosphere* 73: S220–S227.

29. Weiss, J., Päpke, O., Bignert, A. et al. 2003. Concentrations of dioxins and other organochlorines (PCBs, DDTs, HCHs) in human milk from Seveso, Milan and a Lombardian rural area in Italy: A study performed 25 years after the heavy dioxin exposure in Seveso. *Acta Paediatr.* 92: 467–472.

30. De Felip, E. and di Domenico 2010. Studio epidemiologico sullo stato di salute e sui livelli d'accumulo di contaminanti organici persistenti nel sangue e nel latte materno in gruppi di popolazione a differente rischio d'esposizione nella Regione Campania (SEBIOREC). Scientific report for the Campania Regional Board with the collaboration of: Epidemiological Regional Observatory, Italian National Institute of Health, National Research Council, Regional Tumor Registry, Regional Health Agencies Naples 1, 2, 3, and 4 and Caserta 1 and 2. Italian National Institute for Health (Rome).

31. Guerranti, C., Mariottini, M., and Focardi S. 2008. Dietary exposure to PBDEs and levels in breast milk of women living in central Italy. *Organohalogen Compd.* 70: 2020–2023.

32. Van den Berg, M., Birnbaum, L., Bosveld, A.T. et al. 1998. Toxic equivalency factors (TEFs) for PCBs, PCDDs, PCDFs for humans and wildlife. *Environ. Health Perspect.* 106 (12): 775–792.

33. Deml, E., Mangelsdorf, I., and Greim, H. 1996. Chlorinated dibenzodioxins and dibenzofurans (PCDD/F) in blood and human milk of non-occupationally exposed persons living in the vicinity of a municipal waste incinerator. *Chemosphere* 33: 1941–1950.

34. Päpke, O., Ball, M., Lis, A., and Wuthe, J. 1996. PCDD/PCDFs in humans, follow-up of background data for Germany, 1994. *Chemosphere* 32: 575–582.

35. Kiviranta, H., Purkunen, R., and Vartiainen, T. 1999. Levels and trends of PCDD/Fs and PCBs in human milk in Finland. *Chemosphere* 38: 311–323.

36. Schuhmacher, M., Domingo, J.L., Kiviranta, H., and Vartiainen, T. 1999. PCDD/F concentrations in milk of non-occupationally exposed women living in southern Catalonia, Spain. *Chemosphere* 38: 995–1004.

37. Schuhmacher, M., Domingo, J.L., Kiviranta, H., and Vartiainen, T. 2004. Monitoring dioxins and furans in a population living near a hazardous waste incinerator: Levels in breast milk. *Chemosphere* 57: 43–49.

38. Van Leeuwen, F.X.R. and Malisch, R. 2002. Results of the third round of the WHO coordinated exposure study on the levels of PCBs, PCDDs and PCDFs in human milk. *Organohalogen Compd.* 56: 311–316.

39. Malisch, R. and Van Leeuwen, F.X.R. 2003. Results of the WHO-coordinated exposure study on the levels of PCBs, PCDDs and PCDFs in human milk. *Organohalogen Compd.* 64: 140–143.

40. Roots, O. 2003. Environmental levels of PTS in Estonia. Background document for the Second Technical Workshop of UNEP/GEF Project regional-based assessment of persistent toxic substances—Region III—Europe. TOCOEN Report 243.

41. Takekuma, M., Saito, K., Ogawa, M., Matumoto, R., and Kobayashi, S. 2004. Levels of PCDDs, PCDFs and co-PCBs in human milk in Saitama, Japan, and epidemiological research. *Chemosphere* 54: 127–135.

42. Wittsiepe, J., Fürst, P., Schrey, P. et al. 2004. PCDD/F and dioxin-like PCB in human blood and milk from German mothers. *Organohalogen Compd.* 66: 2831–2837.

43. Costopoulou, D., Vassiliadou, I., Papadopoulos, A., Makropoulos, V., and Leondiadis, L. 2006. Levels of dioxins, furans and PCBs in human serum and milk of people living in Greece. *Chemosphere* 65: 1462–1469.

44. Colles, A., Kopprn, G., Hanot, V. et al. 2008. Fourth WHO-coordinated survey of human milk for persistent organic pollutants (POPs): Belgian results. *Chemosphere* 73: 907–914.

45. Lignell, S., Aune, M., Darnerud, P.O., Cnattingius, S., and Glynn, A. 2009. Persistent organochlorine and organobromine compounds in mother's milk from Sweden 1996–2006: Compound-specific temporal trends. *Environ. Res.* 109: 760–767.

46. Guvenius, D.M., Aronsson, A., Ekman-Ordeberg, G., Bergman, A., and Norèn, K. 2003. Human prenatal and postnatal exposure to polybrominated diphenyl ethers, polychlorinated biphenyls, polychlorobiphenylols and pentachlorophenol. *Environ. Health Perspect.* 111 (9): 1235–1241.

47. Vieth, B., Herrman, T., Mielke, H., Osterann, B., Päpke, O., and Rüdiger, T. 2004. PBDE levels in human milk: The situation in Germany and potential influencing factors—Controlled study. *Organohalogen Compd.* 66: 2643–2648.

48. Jaraczewska, K., Lulek, J., and Covaci, A. 2006. Distribution of polychlorinated biphenyls, organochlorine pesticides and polybrominated diphenyl ethers in human umbilical cord serum, maternal serum and milk from Wielkopolska region, Poland. *Sci. Total Environ.* 372 (1): 20–31.

49. Raaba, U., Schweglera, U., Pressb, U., Albrechtb, M., and Frommea, H. 2007. Bavarian breast milk survey—Pilot study and future developments. *Int. J. Hyg. Environ. Health* 210: 341–344.

50. Wilhelm, M., Ewers, U., Wittsiepe, J. et al. 2007. Human biomonitoring studies in North Rhine-Westphalia, Germany. *Int. J. Hyg. Environ. Health* 210: 307–318.

51. WHO, 2009. Persistent organic pollutants (POPs) in human milk. Available at: http://www.euro.who.int/__data/assets/pdf_file/0003/97032/enhis_factsheet09_4_3.pdf

52. Sinkkonen, S. and Paasivirta, J. 2000. Degradation half-life times of PCDDs, PCDFs and PCBs for environmental fate modeling. *Chemosphere* 40: 943–949.

53. Seegal, R.F., Fitzgerald, E.F., Hills, E.A. et al. 2010. Estimating the half-lives of PCB congeners in former capacitor workers measured over a 28-year interval. *J. Exposure Sci. Environ. Epidemiol.* (online publications 10 March 2010; doi: 10.1038/jes.2010.3) 1–13.

54. Ogura, I. 2004. Half-life of each dioxin and PCB congener in the human body. *Organohalogen Compd.* 66: 3329–3337.

55. Gouin, T. and Harner, T. 2003. Modelling the environmental fate of the polybrominated diphenyl ethers. *Environ. Int.* 29: 717–724.

56. Thuresson, K., Höglund, P., Hagmar, L. Sjödin, A., Bergman, Å., and Jakobsson, K. 2006. Apparent half-lives of hepta- to decabrominated diphenyl ethers in humans as determined in occupationally exposed workers. *Environ. Health Perspect.* 114 (2): 176–181.

57. Geyer, H.J., Schramm, K.W., Darnerud, P.O. et al. 2004. Terminal elimination half-lives of the brominated flame retardants TBBPA, HBCD, and lower brominated PBDEs in humans. *Organohalogen Compd.* 66: 3867–3872.

58. CEE, 1976. Direttiva 76/769/CEE del Consiglio, del 27 luglio 1976, concernente il ravvicinamento delle disposizioni legislative, regolamentari ed amministrative degli Stati Membri relative alle restrizioni in materia di immissione sul mercato e di uso di talune sostanze e preparati pericolosi.

59. EC, 2003. Directive 2003/11/EC of the European parliament and of the council amending for the 24th time council directive 76/769/EEC relating to restrictions on the marketing and use of certain dangerous substances and preparations (pentabromodiphenyl ether, octabromodiphenyl ether).

60. Stockholm Convention on Persistent Organic Pollutants, 2009. The 9 new POPs under the Stockholm Convention. Available at: http://chm.pops.int/Programmes/NewPOPs/The9newPOPs/tabid/672/language/en-US/Default.aspx

61. Lee, L.K. and He, J. 2010. Reductive debromination of polybrominated diphenyl ethers by anaerobic bacteria from soils and sediments. *Appl. Environ. Microbiol.* 76 (3): 794–802.

62. Park, J., Holden, A., Chu, V. et al. 2009. Time-trends and congener profiles of PBDEs and PCBs in California peregrine falcons (*Falco peregrinus*). *Environ. Sci. Technol.* 43: 8744–8751.

63. La Guardia, M.J., Hale, R.C., and Harvey, E. 2007. Evidence of debromination of decabromodiphenyl ether (BDE-209) in biota from a wastewater receiving stream. *Environ. Sci. Technol.* 41 (19): 6663–6670.

64. Van den Steen, E., Covaci, A., Jaspers, V.L.B. et al. 2007. Accumulation, tissue-specific distribution and debromination of decabromodiphenyl ether (BDE 209) in European starlings (*Sturnus vulgaris*). *Environ. Pollut.* 148: 648–653.

65. Stapleton, H.M., Alaee, M., Letcher, R.J., and Baker, J.E. 2004. Debromination of the flame retardant decabromodiphenyl ether by juvenile carp (*Cyprinus carpio*) following dietary exposure. *Environ. Sci. Technol.* 38: 112–119.

66. Stapleton, H.M., Brazil, B., Holbrook, D. et al. 2006. In vivo and in vitro debromination of decabromodiphenyl ether (BDE 209) by Juvenile rainbow trout and common carp. *Environ. Sci. Technol* 40: 4653–4658.

67. Mörck, A., Hakk, H., Örn, U., and Wehler, E.K. 2003. Decabromodiphenyl ether in the rat: Absorption, distribution, metabolism, and excretion. *Drug Metab Dispos.* 31: 900–907.

68. Kierkegaard, A., Balk, L., Tjärnlund, U., de Wit, C.A., and Jansson, B. 1999. Dietary uptake and biological effects of decabromodiphenyl ether in rainbow trout (*Oncorhynchus mykiss*). *Environ. Sci. Technol.* 33: 1612–1617.

17 Contamination Status of Organochlorine Pesticides in Ghana

*William J. Ntow and Benjamin O. Botwe**

CONTENTS

17.1 Introduction 393
    17.1.1 Background Information...393
    17.1.2 History of Persistent Organochlorine Pesticides in Ghana....................................394
    17.1.3 Environmental Impacts of OCPs 395
17.2 Environmental Levels of OCPs in Ghana..396
    17.2.1 Levels of OCPs in Water and Sediment 396
    17.2.2 Levels of OCP in Biological Samples...397
17.3 Methods 404
    17.3.1 Sampling 404
        17.3.1.1 Water and Sediment...404
        17.3.1.2 Vegetables, Human Blood Serum and Breast Milk, and Bivalves 404
    17.3.2 Chemical Analysis..404
        17.3.2.1 Water and Sediment...404
        17.3.2.2 Vegetables, Human Blood Serum and Breast Milk, and Bivalves 405
17.4 Environmental Trends of OCPs in Ghana..406
    17.4.1 Possible Bioindicators for Environmental Monitoring of OCPs in Ghana...............406
    17.4.2 Temporal and Spatial Variations of OCPs in Ghana 406
    17.4.3 Available Alternatives to Banned OCPs in Ghana..406
17.5 Conclusions...407
    17.5.1 Overall Status of OCP Contamination in Ghana...407
    17.5.2 Possible Future Trends of OCPs in Ghana 408
    17.5.3 Possible Future Trends by Emerging Contaminants in Ghana................................408
References...409

17.1 INTRODUCTION

17.1.1 BACKGROUND INFORMATION

Ghana, with a geographical location of $5°36'$ N, $0°10'$ E, is a developing country in West Africa. It is bordered by Cote d'Ivoire to the west, Burkina Faso to the north, Togo to the east, and the Gulf of Guinea to the south (Figure 17.1). It has an area of about $238,500 \, km^2$ and a population of about 23 million. Lake Volta, the world's largest artificial lake, extends through large portions of eastern Ghana. The climate is tropical equatorial with the southern part being hot and humid while the northern part

* E-mail: boseibotwe@yahoo.co.uk (Chapter corresponding author).

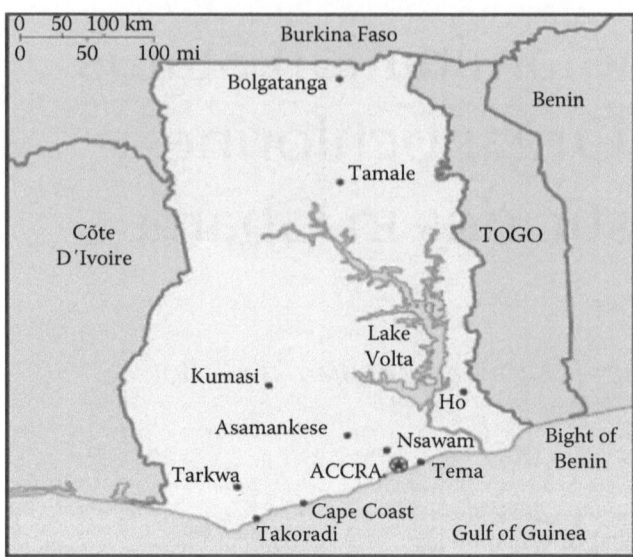

FIGURE 17.1　Map of Ghana where the studies were conducted.

is hot and dry. In the south, there are two rainy seasons (April–June and September–October) and two dry seasons (November–March and July–August) with humidity at about 80%. In the north, the rainy season runs from April to October. The annual rainfall ranges from 1015 to 2300 mm with an average annual temperature of 30°C. Agriculture is the most important economic sector in Ghana, employing about 60% of the national workforce [1]. The majority of this workforce lives in rural areas, where they engage in small-scale farming. Due to diseases and pest attacks in the field, a large portion of agricultural produce is usually lost. The use of pesticides has therefore become necessary to control crop pests and diseases in order to increase crop production to boost the economy. Consequently, many of the farmers in Ghana (about 87%) use chemical pesticides [2,3]. However, most of the farmers lack training in both the choice of chemical pesticides and application techniques and this has resulted in their overuse, misuse, and abuse [3]. Overuse, misuse and abuse of pesticides leads to contamination of the environment as well as agricultural produce with potential adverse health effects on consumers. Inadequate protection of farmers also exposes them to pesticide hazards.

Organochlorine pesticides (OCPs), which include dichlorodiphenyltrichloroethane (DDT), aldrin, dieldrin, hexachlorocyclohexane (HCH; e.g., lindane (γ-HCH)), and hexachlorobenzene (HCB), are of major environmental concern as they are known to be highly toxic and persistent in the physical environment. They also undergo bioaccumulation in fatty tissues and biomagnification through food chains as a result of their persistent and lipophilic nature [4]. Their persistence and adverse effects on human health and the environment have generated interest at local, regional, and global levels [5,6]. Consequently, the use of these OCPs have been severely restricted or banned in many developed countries as well as in some developing countries [7]. Despite their ban, these OCPs continue to be used illegally in some developing countries due to weak import controls, poor pesticide management, and lack of regulatory enforcement [8]. In view of the fact that OCPs are persistent in the environment and their ban in Ghana is quite recent compared to developed countries, it is necessary to monitor them in the environment. This chapter assesses the contamination status of persistent OCPs in Ghana based on selected studies on OCP residue levels in biotic (vegetables, human tissues, and mollusks) and abiotic (water and sediment) samples from Ghana.

17.1.2　HISTORY OF PERSISTENT ORGANOCHLORINE PESTICIDES IN GHANA

DDT and related OCPs were commercially produced worldwide after 1945 and were used extensively for crop protection and disease vector control [9]. DDT in particular gained much popularity

---

TABLE 17.1
List of Some Banned OCPs in Ghana

DDT

Aldrin

Dieldrin

Endrin

Chlordane

HCH (mix isomers)

Heptachlor

Hexachlorobenzene

Toxaphene

Mirex

---

due to its effectiveness against a broad range of insect pests and was used extensively in agriculture and malaria control programs. The introduction of these OCPs on the Ghanaian market may date back to the 1960s and they were used mainly in agriculture and public health programs (especially malaria control). OCPs that have been used for agriculture in Ghana include DDT, aldrin, dieldrin, and lindane. The use of many of these OCPs has been banned in Ghana, although the ban is quite recent compared to developed countries that began the ban in the 1970s [10]. In Ghana, DDT was officially banned in 1985, while lindane was banned in 2002. Table 17.1 shows the list of some banned OCPs in Ghana. Following the ban, Ghana ratified the Stockholm Convention on persistent organic pollutants (POPs), which seeks to protect human health and the environment from POPs. As many as nine of the OCPs, namely DDT, HCB, chlordane, toxaphene, dieldrin, aldrin, endrin, heptachlor, and mirex, are included on the list of POPs targeted by the Stockholm Convention. Furthermore, Ghana has ratified the Rotterdam Convention's Prior Informed Consent (PIC) procedure for certain hazardous chemicals and pesticides in international trade, which include OCPs such as DDT.

17.1.3 Environmental Impacts of OCPs

The use of OCPs has resulted in widespread contamination of the environment [11] as evidenced by their detection in a wide range of environmental media and biota [12]. The mobility and low biodegradability of OCPs in the abiotic and biotic environments have resulted in long-range atmospheric transport and global-scale redistribution [13]. Many OCPs are also lipophilic and therefore they undergo bioaccumulation and biomagnification in organisms throughout various food chains [4]. Exposure to OCPs can result in adverse health effects such as death, cancers, reproductive failure, and immune system suppression. The simultaneous exposure to multiple pollutants can have an additive or synergistic effect in biological systems [14], which may lead to alterations in endocrine and nervous system functions [15–17]. OCP contamination is known to cause deterioration of natural habitats and loss of biodiversity in aquatic systems. OCP contamination of food sources puts the population at risk to developing adverse health effects. Farmers who use pesticides and inhabitants of farming communities where pesticides are used heavily are at a greater risk of potential adverse human health impacts of pesticides. Unfortunately, these impacts may not be easily evident in the exposed populations and ecosystems since these OCP effects are chronic. Knowledge of pesticide residue levels in both biotic and abiotic components of the environment is therefore important in assessing the status of environmental contamination and the potential impacts of OCPs. This chapter assesses the contamination status of persistent OCPs in Ghana based on selected studies of OCP residue levels in biotic (vegetables, human tissues, and mollusks) and abiotic (water and sediment) samples.

17.2 ENVIRONMENTAL LEVELS OF OCPs IN GHANA

17.2.1 LEVELS OF OCPs IN WATER AND SEDIMENT

OCP residues were measured in water and sediment from the Volta Lake, which drains large areas of Ghana [18], as well as stream waters of farming areas [5]. The mean concentrations of the detected OCPs in water and sediment from the Volta Lake are summarized in Table 17.2 and the results for the stream waters are summarized in Table 17.3. Lindane was the most frequently occurring OCP in water, with a mean concentration of 0.008 µg/L. Out of 180 samples of analyzed water, 41 (22.7%) were found to contain detectable levels of lindane. Of all the OCPs, α-endosulfan was present in the highest concentration (mean = 0.036 µg/L) in water, being detected in approximately 16% of the samples analyzed. Residues of β-endosulfan were detected in about 18% of the water samples, while endosulfan sulfate was found in only 10% of the samples. DDT and its metabolite DDE were also found in the sediment. The sediment samples exhibited the greatest number and highest concentrations of OCP residues. Thus, they could be a better indicator of OCP pollution

TABLE 17.2
Organochlorine Pesticide Residues Concentrations (Mean ± SD) in Water (µg/L) and Sediment (µg/kg dw) in the Volta Lake

Organochlorine Pesticide	Water ($n = 180$)	Sediment ($n = 36$)
Lindane	0.008 ± 0.005	2.3 ± 1.4
α-endosulfan	0.036 ± 0.007	0.21 ± 0.05
β-endosulfan	0.024 ± 0.009	0.17 ± 0.04
Endosulfan sulfate	0.023 ± 0.011	0.36 ± 0.04
p,p'-DDT	ND	9 ± 5
p,p'-DDE	ND	52.3 ± 37.8

Source: Data from Ntow, W.J., *Res. Manage.*, 10, 243, 2005.
ND, not detected.

TABLE 17.3
Organochlorine Pesticide Residues Concentrations (Mean ± SD) in Stream Water (ng/L), Sediment (µg/kg dw), and Tomato (µg/kg fw) from Ghana

Organochlorine Pesticide	Water ($n = 50$)	Sediment ($n = 42$)	Tomato ($n = 72$)
HCB	<100	0.9 ± 0.1	<0.1
Lindane	9.5 ± 5.2	3.2 ± 0.6	<0.25
p,p'-DDE	<100	0.46 ± 0.24	<0.1
Heptachlor epoxide	<100	0.63 ± 0.42	1.65 ± 1.48
α-Endosulfan	62.3 ± 7.3	0.19 ± 0.02	<0.5
β-endosulfan	31.4 ± 11.2	0.13 ± 0.01	<0.1
Endosulfan sulfate	30.8 ± 12.3	0.23 ± 0.01	<0.1

Source: Data from Ntow, W.J., *Arch. Environ. Contam. Toxicol.*, 40, 557, 2001; *Environ. Res.*, 106, 17.
<, below detection limit (the compound was identified but could not be quantified).
Dw, dry weight; fw, fresh weight.

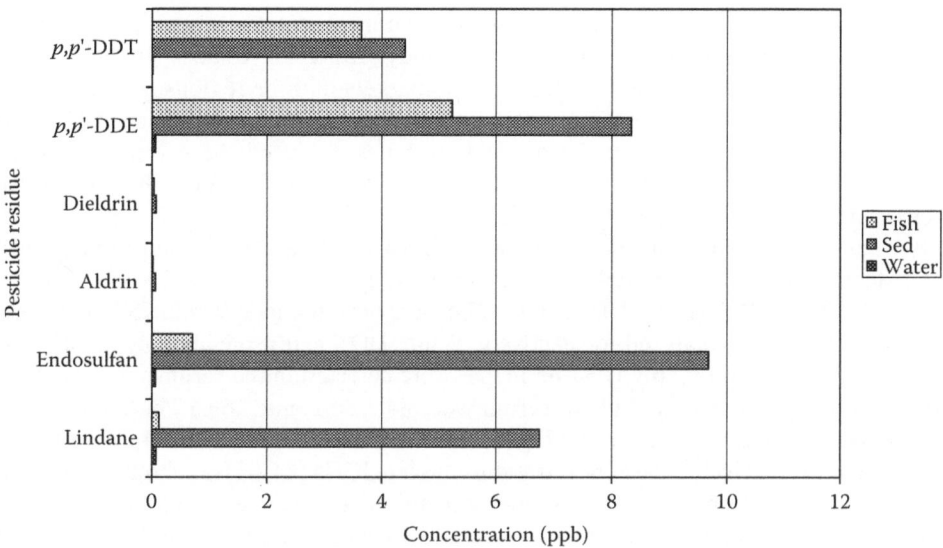

FIGURE 17.2 Pesticide residue concentrations in fish, sediment, and water from Lake Bosomtwi, Ghana. (Data from Darko, G. et al., *Chemosphere*, 72(1), 21, 2008.)

than that of water. All the OCPs found in sediment, except *p,p'*-DDT, appeared in at least 86% of all the samples analyzed, while *p,p'*-DDE was found in all 36 sediment samples (100%), lindane in 33 (91.7%), α-endosulfan in 31 (86.1%), β-endosulfan in 32 (88.9%), endosulfan sulfate in 34 (94.4%), and *p,p'*-DDT in only eight samples (22.2%). The *p,p'*-DDE concentration (mean = 52.3 μg/kg) was highest in the sediment, whereas the β-endosulfan concentration (mean = 0.17 μg/kg) was lowest in the sediment. Darko et al. [19] also detected OCP residues in water and sediments from Lake Bosomtwi in the Ashanti region of Ghana (Figure 17.2).

The results of this study indicated that OCP residues are present in Volta Lake. The OCP residues are believed to have originated mainly from agricultural uses. The natural processes of leaching and runoff are likely to enhance their transfer to the main course of the Volta River, especially during the rainy season. They also may have originated in part from bad fishing activities, with pesticides being used in Ghana to catch fish. The use of OCPs in this manner is an extreme example of the ability of pesticides to enter aquatic ecosystems and be stored in aquatic food resources. The OCP levels found in the present study were comparable to values reported from other aquatic ecosystems in Ghana [5,20] and in other countries [21].

17.2.2 LEVELS OF OCP IN BIOLOGICAL SAMPLES

In a recent study by Botwe et al. (unpublished) [22], residues of DDTs (*p,p'*-DDE + *p,p'*-DDT + *p,p'*-DDD), endosulfan (α-endosulfan + β-endosulfan + endosulfan sulfate), HCHs (α-HCH + β-HCH + γ-HCH + δ-HCH), and methoxychlor were detected in vegetables (tomato, cabbage, pepper, onion, and eggplant) from markets in all 10 regions of Ghana. In most of the vegetables analyzed, methoxychlor residues were found in relatively high concentrations and had the highest frequency of occurrence. The highest mean concentration of methoxychlor (46.95 μg/kg fw) was detected in onion and the lowest concentration (0.85 μg/kg fw) was detected in cabbage. DDTs occurred in all the different vegetable types except peppers. The highest concentration of DDTs (0.45 μg/kg fw) was recorded in cabbage. Low DDT concentrations (0.01–0.03 μg/kg) were detected in tomato, onion, and eggplant. HCHs were detected in all vegetables except eggplant. The highest concentration of HCHs (0.62 μg/kg fw) was detected in cabbage. Generally, eggplant recorded very low concentration of OCP residues. Endosulfan was detected in measurable quantities only in tomato, cabbage, and onion (average concentrations ranged from 0.21 to 3.03 µg/kg fw). The mean concentrations of OCP residues in vegetables are shown in Figure 17.3. In a previous study [23], OCP residues (lindane, endosulfan, and DDT) were also detected in lettuce from urban markets in three regions (Accra, Kumasi, and Tamale) of Ghana at levels ranging from 0.03 to 0.9 mg/kg fw. DDT and lindane (γ-HCH) had the highest levels at 0.9 mg/kg. Most of the pesticide residues exceeded the maximum residue limit (MRL). The OCP residue concentrations in this study were generally higher than the levels found in a recent study [22].

OCP residues have also been detected in human breast milk and the serum of vegetable farmers in Ghana at levels that raise public health concerns [24]. The levels of the detected OCPs are shown in Table 17.4. DDTs (DDT, DDE, and DDD) were detected in 88% and 75% of all the breast milk and serum samples analyzed, respectively. While DDT and it metabolites (DDE and DDD) were detected in breast milk, only DDT and DDE were detected in the serum samples. Total DDT (ΣDDT) levels in breast milk and blood serum were 84.2 ± 6.2 and 7.6 ± 1.7 ng/g fat, respectively. Among the DDTs, DDE occurred at the highest levels of 44.8 ± 4.2 and 7.1 ± 1.2 ng/g fat in breast milk and serum, respectively. The concentrations of total HCHs (ΣHCHs) in breast milk and serum samples were 206 ± 40.6 and 4.3 ± 0.1 ng/g fat, respectively, and were detected in over 50% of each of the sample matrices (either breast milk or serum). Residues of β-HCH were found in both breast milk and serum. ΣDDTs levels were higher than ΣHCHs levels in both breast milk and blood serum. Significantly higher concentrations of DDTs and HCHs were found in breast milk than in blood serum ($p < 0.05$). On the contrary, higher concentrations of dieldrin and HCB were found in serum samples than in breast milk samples, although the differences were not significant ($p > 0.05$). Dieldrin and HCB occurred in at least 60% of breast milk and serum samples analyzed. The mean levels of dieldrin in breast milk and serum samples were 122.8 ± 24.8 and 127.0 ± 27.2 ng/g fat respectively. The mean levels of HCB in breast milk and serum samples were 4.9 ± 0.3 and 5.3 ± 1.9 ng/g fat respectively. The most prevalent OCP residues in both milk and serum were DDT and its metabolites. The levels of ΣDDTs, ΣHCHs, and dieldrin in the breast milk samples were found to correlate positively with ages of the donors of the milk samples (rs = 0.606, 0.770, and 0.540 respectively).

Comparison of OCP residue levels in male and female blood serum for DDTs, HCHs, dieldrin, and HCB showed significantly higher levels ($p < 0.05$) in males than in females for DDTs and dieldrin. For instance, for DDTs, male farmers had a mean concentration of 10.6 ng/g whereas in females it was 7.1 ng/g. For dieldrin, the male farmers had a mean concentration of 134.0 ng/g whereas the female farmers had 114.7 ng/g (data not shown). It is known that adult female excrete lipophilic contaminants such as OCPs via lactation and thus reducing the body burden of such contaminants. However, mean total HCH concentration was higher in females than in males, which is in agreement with other studies [25]. As a result, we could not prove that OCP residue concentrations in human serum are gender dependent. Concentrations of OCP residues in human breast milk can vary with factors such as the age and number of children of the mother [26]. Simple regression analysis between the mothers' age and OCP residue concentrations showed positive correlation for DDTs ($p < 0.001$, rs = 0.606), HCHs ($p < 0.001$, rs = 0.770), and dieldrin ($p < 0.01$, rs = 0.540). This signifies that OCP levels in human breast milk tend to increase with increase of mother's age as found in other studies [27]. There was also an association between OCP residues in breast milk and serum. The presence of OCP residues in milk and blood suggests that they can be transferred from mother to newborn babies and fetus through the milk and placenta, respectively. Estimation of the daily intakes of DDTs and HCHs by infants from human breast milk indicated that OCP residue levels in breast milk in some individual farmers (in the case of DDTs) and all farmers (in the case of HCHs) were above the threshold levels (tolerable daily intake guidelines proposed by Health Canada) for adverse effects [24]. This raises concern about the safety of breast milk and the health of children. Interestingly, the p,p'-DDT/p,p'-DDE ratio for breast milk was estimated at 0.7 and that for blood serum was estimated at 0.07. These ratios suggest a decrease in exposure to DDT in Ghana. The high levels of p,p'-DDE may be due to their high persistence and long-term accumulation in breast milk. In a previous study [5], HCB and p,p'-DDE residues were detected in human

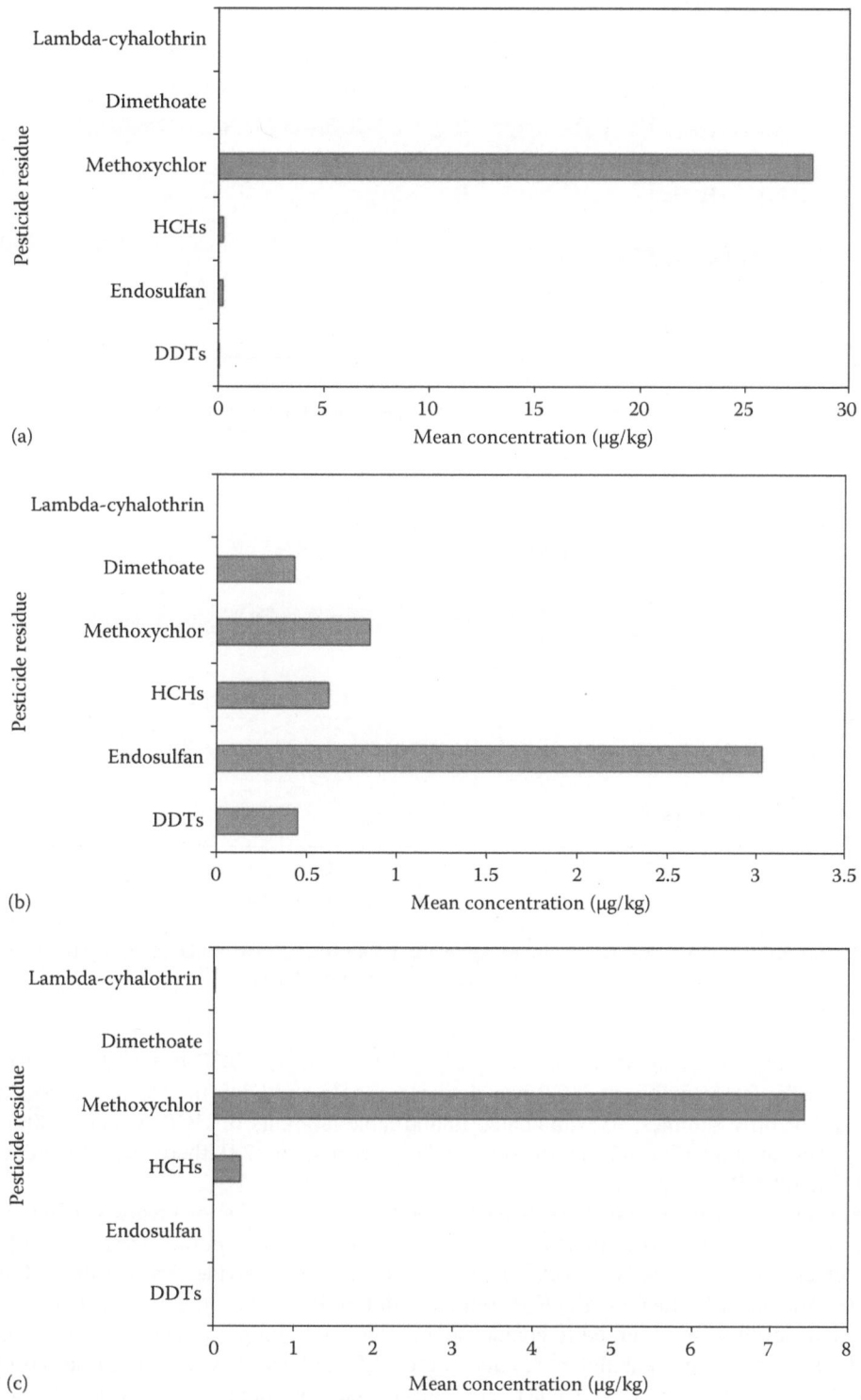

FIGURE 17.3 Levels of pesticide residues in (a) tomato, (b) cabbage, (c) pepper.

(*continued*)

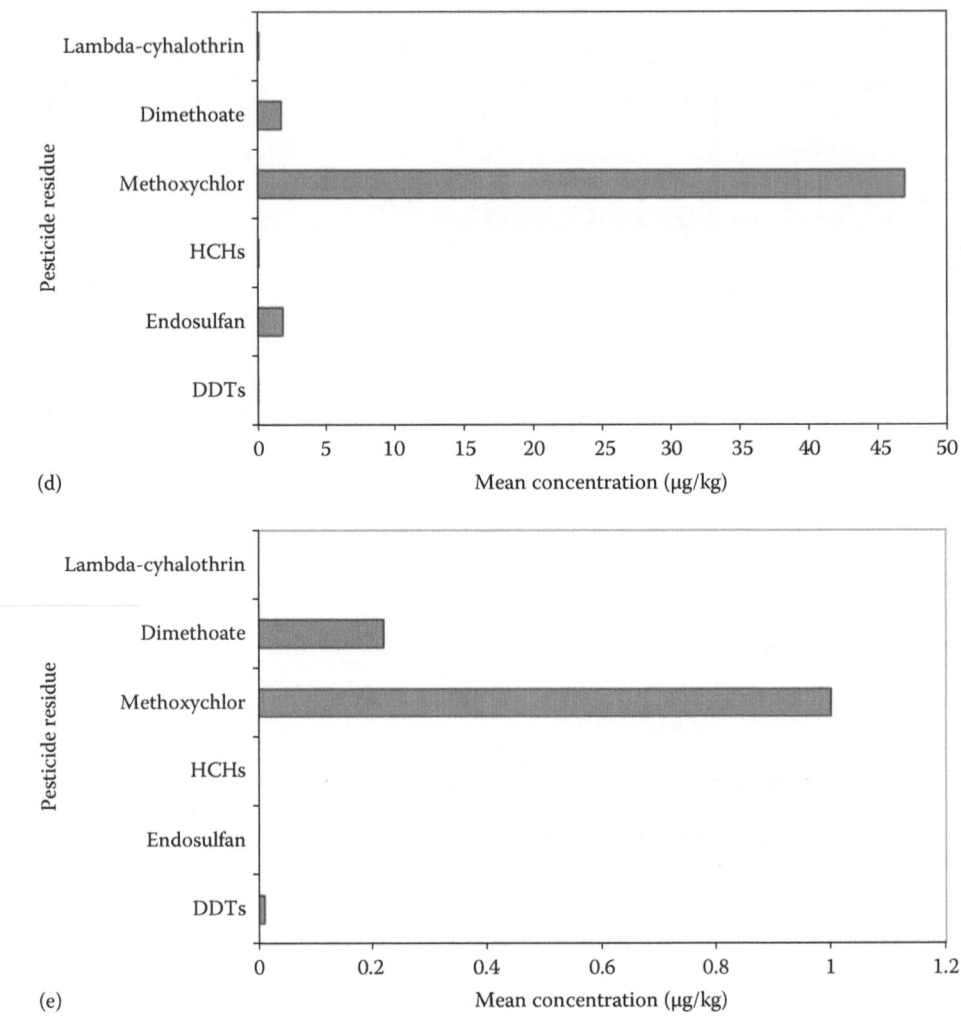

FIGURE 17.3 (continued)   (d) onion, and (e) eggplant. (Data from Botwe, B.O. et al., Pesticide residues in Ghanaian vegetables and their impact on public Health 2009, unpublished.)

breast milk and serum (Figure 17.4). The levels of HCB and p,p'-DDE in blood were 30 ± 10 and $380 \pm 120 \mu g/kg$ fat, respectively. HCB was detected in 55% and DDE in 85% of all blood samples analyzed. For milk samples, 95% indicated quantifiable amounts of HCB, whereas 80% showed DDE. In comparison with levels obtained from the current study [24], there has been a decrease in p,p'-DDE and HCBs.

High levels of OCP residues have been detected in three species of bivalves, namely cockles (*Anadara senilis*), oysters (*Crassostrea tulipa*), and mussels (*Perna perna*) from Ghana [28]. The OCPs detected included HCHs, DDTs, heptachlor, heptachlor epoxide, and dieldrin (Tables 17.5 and 17.6). The results indicate high OCP contamination of the environment. Seasonal variations in OCP residue levels were observed in cockles and oysters, with higher levels in the wet season than in the dry season. The most abundant residues were DDTs and HCHs with concentrations of 73 and 29 ng/g dw, respectively. Ratios of DDT/DDE and heptachlor/heptachlor epoxide reflected recent applications of DDT and heptachlor, while the ratio for aldrin/dieldrin suggested discontinued use of aldrin, at least on the coast of Ghana. Higher DDT/DDE ratios in the wet season was attributed to more recent contamination due to increased application of pesticides for agricultural purposes (pest

TABLE 17.4
Organochlorine Pesticide Residues Concentrations (Mean ± SE, µg/kg fat) in Milk and Blood Serum from Farmers in Ghana

Organochlorine Pesticide	Milk ($n = 109$)	Blood Serum
p,p'-DDE	44.8 ± 4.2	7.1 ± 1.2
p,p'-DDD	8.0 ± 1.0	<LOQ
p,p'-DDT	31.4 ± 4.5	0.5 ± 0.1
α-HCH	192.0 ± 40.4	<LOQ
β-HCH	14.0 ± 2.3	0.2 ± 0.1
δ-HCH	<LOQ	4.1 ± 0.1
Dieldrin	122.8 ± 24.8	127.0 ± 27.2
HCB	4.9 ± 0.3	5.3 ± 1.9
ΣDDTs	84.2 ± 6.2	7.6 ± 1.7
ΣHCHs	206 ± 40.5	4.3 ± 0.1

Source: Data from Ntow, W.J. et al., *Environ. Res.*, 106, 17, 2008.

ΣDDTs, p,p'-DDE + p,p'-DDD + p, p'-DDT.
ΣHCHs, α-HCH + β-HCH + δ-HCH.
ND, not detected.

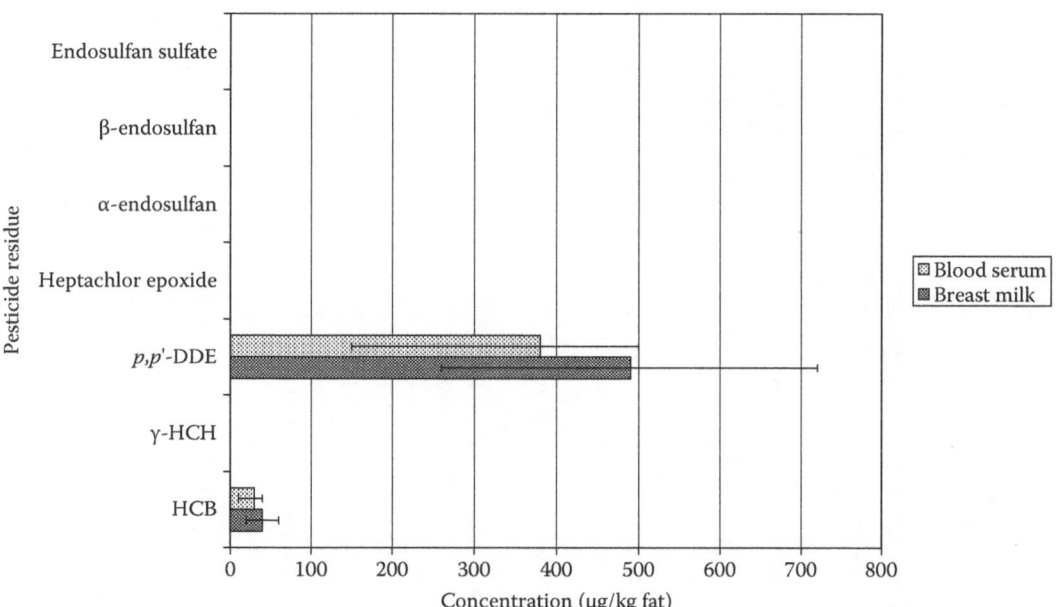

FIGURE 17.4   Pesticide residues in human blood and breast milk from Ghana. (Data from Ntow, W.J., *Arch. Environ. Contam. Toxicol.*, 40, 557, 2001; *Environ. Res.*, 106, 17.)

control) and public health (increased mosquito breeding sites). In another study [19], high levels of OCP residues were found in fish (see Figure 17.2). High levels of OCP residues have also been found in meat [29] and dairy products [30] from Kumasi and Buoho Abattoirs of the Ashanti Region of Ghana. These OCPs included DDTs, lindane, and endosulfan. The OCP residue concentrations were particularly high in meat and dairy products, exceeding the WHO limits in beef fat.

TABLE 17.5
Spatial Trends in Concentrations of Organochlorine Pesticides (ng/g dw) and Lipid Content (mg/g dw) of Bivalves from Benya, Ningo, and Sakumo Lagoons in Ghana in the Dry Season

Location	Benya			Ningo		Sakumo		
Species	Cockles	Oysters	Mussel	Cockles	Oysters	Cockles	Oysters	Mussel
	Median (Range)	Median (Range)	Median	Median (Range)	Median (Range)	Median (Range)	Median (Range)	Median (Range)
ΣHCHs[a]	45 (15–150)	39 (5–118)	47	27 (16–46)	18 (3–39)	41 (25–94)	18 (10–33)	40 (17–75)
Lindane	8.6 (3–14)	5 (1.9–20)	6	5.4 (2.4–26)	3.8 (2–12)	13 (5–22)	4.8 (3–10)	14 (6.7–22)
ΣDDTs[a]	101 (46–220)	88 (13–224)	56	79 (46–116)	69 (8–118)	28 (125–311)	33 (17–92)	80 (31–97)
DDE	22 (5–34)	22 (6–48)	14	25 (21–35)	10 (2–35)	23 (13–32)	8.0 (3–25)	15 (5–28)
DDT/DDE	0.5 (0.2–2)	0.6 (0.2–2)	0.76	0.6 (0.1–2)	0.4 (0.2–1)	0.2 (0.1–1)	0.7 (0.1–1)	0.4 (0.4–1)
ΣCHLs[a]	63 (9–98)	47 (7–107)	92	66 (28–96)	37 (4–49)	3 (24–55)	16 (9–51)	66 (20–222)
Hept/Hept. epo	2.1 (0.2–5)	0.4 (0.2–2)	0.37	2.7 (2.1–3)	1.2 (0.7–5)	1.3 (0.4–3)	1.1 (0.3–3)	3.4 (1.5–4)
Aldrin	2.6 (0.9–14)	4.6 (1–13)	1.4	4.0 (3–31)	4.7 (1–15)	2.6 (1.2–4)	8.3 (1–18)	3.8 (0.9–8)
Dieldrin	20 (9–26)	51 (2–88)	16	30 (12–48)	14 (2–89)	44 (13–88)	3.2 (1–42)	2.4 (4.8–8)
Endrin	6.9 (4–79)	9.8 (2–40)	79	13 (12–15)	9.2 (nd–14)	14 (5–38)	6.2 (2–11)	7 (4.5–19)
α-Endosulfan	8.4 (8–13)	31 (10–93)	9.3	40 (15–98)	18 (3–91)	21 (5–30)	4.4 (2–77)	18 (2–66)
β-Endosulfan	12 (1.9–81)	20 (16–83)	30	4.7 (3–7)	14 (3–47)	51 (2–78)	10 (2–27)	29 (21–66)
Length	41 (37–51)	52 (42–64)	49	51 (45–59)	42 (36–70)	26 (25–32)	46 (38–71)	36 (34–43)
Lipid content	107 (38–160)	192 (75–300)	162	24 (15–114)	81 (66–144)	38 (13–43)	38 (34–90)	53 (49–72)
n	8	9	1	6	10	4	9	5

Source: Data from Otchere, F.A., *Sci. Total Environ.*, 348, 102, 2005.

[a] ΣHCH, sum of α-HCH, γ-HCH, δ-HCH; ΣDDT; ΣDDT, sum of o–p' and p–p' isomers of DDT, DDE and DDD; ΣCHLs, sum of Heptachlor, Heptachlor epoxide and HCB; nd, not detected. *n*, no. of samples.

TABLE 17.6

Spatial Trends in Concentrations of Organochlorine Pesticides (ng/g dw) and Lipid Content (mg/g dw) of Bivalves from Benya, Ningo, and Sakumo Lagoons in Ghana in the Wet Season

Location	Benya			Ningo		Sakumo		
	Cockles	Oysters	Mussel	Cockles	Oysters	Cockles	Oysters	Mussel
Species	Median (Range)	Median	Median (Range)	Median (Range)	Median	Median (Range)	Median (Range)	Median
ΣHCHs[a]	61 (12–187)	18	22 (19–57)	14 (15–39)	18	32 (8–67)	16 (8–51)	—
Lindane	7.5 (2–16)	6.3	14 (6–30)	6.9 (3–8)	5.7	3.7 (1.5–7)	5.9 (3–10)	—
ΣDDTs[a]	70 (21–307)	70	124 (83–214)	152 (56–164)	39	39 (10–82)	46 (16–78)	—
DDE	16 (4–116)	8.9	22 (11–41)	12 (12–14)	11	8.8 (3–15)	9.6 (5–23)	—
DDT/DDE	1.2 (0.5–3)	1.5	1.3 (0.5–2)	1.6 (0.6–4)	0.84	0.9 (0.2–2)	1.0 (0.1–3)	—
ΣCHLs[a]	24 (10–195)	33	18 (8–36)	31 (15–41)	27	19 (16–32)	30 (16–91)	—
Hept./Hept. Epo	0.9 (0.6–1)	1.4	1.2 (0.5–2)	2.7 (2–6)	1.5	3.6 (1–10)	1.8 (0.3–4)	—
Aldrin	4 (0.6–1.1)	4.1	6.9 (3–18)	4.0 (2–5)	1.3	1.7 (0.8–15)	5.3 (2–14)	—
Dieldrin	4.5 (2–10)	8.4	14 (11–60)	2.9 (2.2–5)	22	7.5 (0.7–30)	3.4 (0.3–7)	—
Endrin	11 (3–112)	9.2	14 (6–34)	13 (2.6–17)	8	3.3 (3–15)	4.5 (3–27)	—
α-Endosulfan	14 (7–36)	9.9	26 (13–60)	19 (7–25)	7.6	16 (6–31)	9.4 (nd–17)	—
β-Endosulfan	4.7 (3–78)	13	18 (4–83)	19 (14–24)	8.9	11 (2–20)	15 (5–27)	—
Length	40 (38–49)	27	42 (37–44)	42 (41–49)	71	47 (37–55)	49 (37–70)	—
Lipid content	75 (45–118)	110	90 (60–119)	31 (30–39)	77	35 (13–85)	76 (57–91)	—
n	6	1	7	3	1	10	9	—

Source: Data from Otchere, F.A., *Sci. Total Environ.*, 348, 102, 2005.

[a] ΣHCH, sum of α-HCH, γ-HCH, δ-HCH; ΣDDT, sum of *o–p'* and *p–p'* isomers of DDT, DDE and DDD; ΣCHL, sum of Hept., Hept. epo and HCB; nd, not detected; *n*, no. of samples.

17.3 METHODS

17.3.1 SAMPLING

17.3.1.1 Water and Sediment

One hundred and eighty water samples and thirty-six sediment samples collected from six locations in Volta Lake from February 1995 to January 1996. Water samples were taken from midstream positions of the lake to approximate the mean concentrations of OCP residues at a 0.3 m depth below the water surface using amber 1 L glass bottles previously rinsed with ethyl acetate and heated at 250°C (4 h) before use. After collection, the water samples were stoppered and taken for laboratory analysis. The water samples were stored on ice during transport and kept at 4°C in the laboratory (<3 days) until extraction. Grab samples of waterbed sediment (approximately 1 L) were collected from the bottom of the lake, where fine-textured substrate had accumulated, in glass beakers (or in aluminum foil and wrapped). Composite samples were pooled from three or four subsamples and homogenized. Samples were immediately stored in ice after collection and transported to the laboratory for analysis.

17.3.1.2 Vegetables, Human Blood Serum and Breast Milk, and Bivalves

Identical vegetable items were purchased from local designated markets across all the 10 regions in Ghana. The vegetable items, which included tomatoes, peppers, onions, cabbage, and eggplants, were considered representative of the diet of the Ghanaian population according to a survey in 2005 [3]. At each market, vegetable samples were collected from three randomly selected sellers. Randomization of sellers was accomplished by identifying all sellers of each vegetable item present on the day of sampling and randomly selecting three of them. Three composite samples of identical vegetable items—each containing 3 bulbs of tomato, 3 cabbages, 10 pepper fruits, 3 bulbs of onion, and 3 eggplants—were drawn from each market. Samples were considered representative of all commercially available produce because each local designated market is the sole official fresh produce outlet. The samples (flesh) were given a cold-water wash with a soft brush to remove adhering soil particles, wrapped in aluminum foil according to type, packed in polythene bags, stored in insulated containers with dry ice, and transported to the CSIR Water Research Institute Laboratory in Accra. Samples were kept in a freezer at −4°C until required for extraction, which was carried out within 24 h of their arrival at the laboratory. Blood serum samples were randomly taken from 115 vegetable farmers, made up of 59 females and 56 males, from two agricultural districts (Offinso District and Kassena-Nankana District) in Ghana. The ages of men were between 18 and 74 years (mean=35 years) while the ages of women were between 18 and 53 years (mean=36 years). Blood was drawn with heparin-treated syringes and needles with the assistance of health workers and transferred to 5 mL disposable plastic tubes. The serum was separated by centrifugation (1500 rpm for 5 min) using the Beckman J2-21 Centrifuge and frozen at −4°C in glass vials (prewashed with n-hexane) until analysis. Breast milk samples were collected from lactating mothers by manually expressing the milk into sterilized glass containers by the lactating mothers (these females from whom milk samples were taken also donated blood). The women's ages were between 18 and 40 (mean=28). All samples were stored at −4°C until analysis. The samples were taken from mothers with normal and healthy babies who had reported at the clinic for postnatal observations. Three species of bivalves, namely, cockles (*Anadara senilis*, n=95), oysters (*Crassostrea tulipa*, n=75), and mussels (*Perna perna*, n=30) were collected for the study. Oysters and cockles were collected from three lagoons, namely Benya, Ningo, and Sakumo lagoons, during the 1996 wet season and the 1997 dry season. Mussels were collected from rocky shores adjacent to these lagoons during the same periods.

17.3.2 CHEMICAL ANALYSIS

17.3.2.1 Water and Sediment

The extraction and analyses of water were performed following the Association of Official Analytical Chemists 990.06 and 970.52 methods [31]. Briefly, the 1.0 L unfiltered water samples were extracted sequentially three times with 25 mL *n*-hexane. The extract was dried with anhydrous sodium sulfate and concentrated down to 10 mL by evaporation in a TurboVap (Zymark, Palo Alto, California, USA). A cleanup system, using a chromatographic column packed with florisil, previously activated for 3 h in an oven at 130°C, and anhydrous sulfate (both rinsed with petroleum ether) was used. The extract was transferred to the column. Three fractions were obtained after elution with 6%, 15%, and 50% ethyl ether in petroleum ether. The maximal flux rate of elution was 5 mL/min. Each eluate was evaporated. The extracts were dissolved in 1.5 mL *n*-hexane and diluted to 2 mL with more *n*-hexane. The dissolved extracts were injected into a gas chromatographic system for identification and quantification of the pesticides. Sediment samples were well mixed to obtain a homogeneous sample and then transferred into pans to air-dry at ambient temperature. The air-dried sediment samples were ground in a mortar and sieved through a 2 mm mesh filter before extraction. Briefly, about 5 g of the representative sieved samples were weighed into extraction thimbles and Soxhlet-extracted in methanol and cleaned up in Florisil as described above for water [5]. OCP residue analysis was by gas chromatography coupled with mass spectrometry.

17.3.2.2 Vegetables, Human Blood Serum and Breast Milk, and Bivalves

The frozen vegetable samples were thawed and each cut into four equal segments. Opposite segments were discarded in order to reduce the bulk of the material needed for processing. Following this step, the vegetables were cut into smaller pieces. For the analysis, individuals of identical vegetable items were pooled as one composite. From this composite, three replicate samples were prepared for extraction and subsequent instrumental analysis. This design yielded three residue values for each vegetable item. Sample preparation and extraction followed the procedures described by Ferrer et al. [32]. About 10 g of each vegetable was weighed into a porcelain mortar and ground with 50 g of anhydrous sodium sulfate. The powdered sample was extracted with ethyl acetate. The extract was concentrated down in vacuo at 40°C, and the residue was redissolved in hexane. Sample cleanup followed the procedure of Hsu et al. [33]. The final extract was analyzed by gas chromatography coupled with mass spectrometry.

Blood serum analysis was performed according to method 2 of Lino et al. [34]. Briefly, the extraction of the OC compounds was made in 1 mL of serum with 2 × 5 mL of *n*-hexane–acetone (90 + 10), shaken for 1 min on a vortex mixer and centrifuged at 1250 rpm for 5 min. The extract was transferred to a Florisil SPE cartridge to which 1 cm of sodium sulfate was previously added. Two different eluents were used: E_1 (6 mL *n*-hexane) and E_2 (6 mL *n*-hexane–dichloromethane (5 + 1)). The eluent was concentrated with a stream of nitrogen gas to 1 mL and then quantification and confirmation of results were made using gas chromatography–electron capture detection (GC–ECD) and gas chromatography–mass spectrometry (GC–MS). Complete details of the gravimetric measurement of the lipid percentage of blood serum are described elsewhere [35]. The final extract was analyzed by gas chromatography coupled with mass spectrometry. Breast milk samples were prepared following a previously reported methodology [36]. Briefly, about 2 g of breast milk was weighed in a 100 mL reagent bottle and homogenized with 10 g anhydrous sodium sulfate in a stainless steel blender for 1 min. OCPs residues breast milk samples were extracted with 50 mL of 1:1 acetone and hexane in a shaking incubator at 40°C for at least 12 h. The extraction process was repeated twice. The extract was then concentrated to 5 mL using a rotary evaporator at 80°C. One-fifth of the concentrated extract was used for fat content determination by gravimetric method [26]. The extract was transferred to a Florisil SPE cartridge in which 1 cm of sodium sulfate was previously added for cleanup. The column was washed four times with 8 mL *n*-hexane and the combined eluents were concentrated to 0.5 mL under reduced pressure. The final eluent was diluted to 1.5 mL with *n*-hexane. The final extract was analyzed by gas chromatography coupled with a mass spectrometry.

OCPs were extracted from bivalves with 120 mL (per sample) of a solution containing 90% hexane and 10% acetone for 10 h. Cleanup was done in glass columns filled with 11 g of Florisil (60–100 mesh) and approximately 0.25 g of sodium sulfate anhydride deposited on top. OCPs were separated by two successive elutions: (1) 100 mL of hexane for less polar compounds and (2) 100 mL of hexane–diethyl ether in 1:1 ratio for more polar compounds. The volatile substances were evaporated and the remaining residues recovered in 1.5 mL hexane and sealed in small glass vials for gas chromatographic analysis.

17.4 ENVIRONMENTAL TRENDS OF OCPs IN GHANA

17.4.1 POSSIBLE BIOINDICATORS FOR ENVIRONMENTAL MONITORING OF OCPs IN GHANA

Bioindicators are important tools for the rapid assessment of the status and trends of environmental contaminants. They give early warning of potential human exposure to environmental contaminants. In Ghana, there have not been studies to assess the suitability of specific organisms as bioindicators of OCPs in the environment. In other regions, mollusks (in particular bivalves) have been commonly used as sentinel organisms for rapid assessment of the status of contamination of the environment [37]. Bivalves have been considered as suitable bioindicators of pollutants in the environment due to their ability to accumulate a large number of pollutants over long periods and respond to pollutants at a molecular level [38]. They are unable to degrade most chemicals due to the lack of necessary enzymes, which allows them to reflect environmental contamination. Their wide geographic distribution also facilitates data comparison [39]. Bivalves occur in considerable quantities along the rocky shores of the coast of Ghana and are consumed by the coastal population. Unlike fish, which are mobile, bivalves are sedentary and this affords easy accessibility for collection. A wide range of OCPs and polychlorinated biphenyls (PCBs) have been detected in bivalves from the Ghanaian coastal environment with seasonal variations in the levels of these contaminants [28]. The OCP residue levels in the bivalves are shown in Tables 17.5 and 17.6. Thus, bivalves may be suitable bioindicators for monitoring studies of OCPs in the environment of Ghana. For OCP residue monitoring studies in the human population, breast milk and blood serum may serve as suitable bioindicators [24].

17.4.2 TEMPORAL AND SPATIAL VARIATIONS OF OCPs IN GHANA

Environmental levels of OCPs can vary in time and space. For example, seasonal and spatial variations in OCP levels have been observed in bivalves in Ghana [28]. Generally, levels of OCPs in the wet season were higher than in the dry season. The higher levels of OCP in the wet season were attributed to the impact of agricultural runoff. It has been shown that agricultural runoff can significantly increase the levels of OCPs in the catchment areas [40]. However, a comprehensive assessment of status and trends of OCPs contamination on long time scales has not been conducted in Ghana. For instance, no data exists on environmental levels of OCPs in Ghana prior to the 1960s to provide background levels of OCPs in Ghana. OCP contamination data for the period between 1960 and 1985 when OCPs such as DDT were extensively used in Ghana is also nonexistent. Few existing data only cover the period after the ban, i.e., from 1995. There is therefore the need for a systematic environmental monitoring of OCPs and other persistent organic chemicals in all media of the environment in order to establish long-term environmental trends of these chemicals in Ghana.

17.4.3 AVAILABLE ALTERNATIVES TO BANNED OCPs IN GHANA

Four pesticide products have been approved by the EPA-Ghana in place of DDT. These are Bistar 10 WP (Bifenthrin), Icon 10 CS (Lambda cyhalothrin), Delete 2.5 EC (Deltamethrin), and Vectoguard 40 WP (Pirimiphos methyl). These are organophosphorus pesticides (OPs), which, unlike the OCPs, degrade readily in the environment and biological systems. OPs were developed during the 1970s to replace the OCPs and are widely being used in agriculture and animal production for the control

TABLE 17.7
List of Some Registered Pesticides in Ghana

Trade Name	Active Ingredient	Hazard Class	Crop/Uses
Callidim 400 EC	Dimethoate	II	Insecticide for the control of mealy bugs, mites, thrips, greenflies, and borer larvae in vegetables, pineapples, and ornamentals
Consider 200 SL	Imidacloprid	III	Insecticide for the control of insect pests on vegetables
Cymethoate super EC, Cypadem 43.6% EC	Dimethoate + cypermethrin	II	Insecticide for the control of aphids, caterpillars, whitefly, grasshoppers, bollworms in vegetables and cotton
Cypertex 10 EC	Cypermethrin	III	Insecticide for the control of pests in cotton and vegetables
Decis 25 EC	Deltamethrin	II	Insecticide for the control of insect pests of vegetables
Kilsect 2.5 EC, Kombat 2.5 EC, Kuzithrine 2.5 EC, Lambda Super 2.5 EC, Pawa 2.5 EC, Wreko 2.5 EC, Sunhalothrin 2.5% EC, Bossmate 2.5 EC	Lambda-cyhalothrin	II	Insecticide for the control of Insect pests in vegetables
Dursban 4 E	Chlorpyrifos	II	Insecticide for the control of scale, borers, in cereals, vegetables ornamentals and for public health purposes
Mitox 200 EC, Sumico 20 EC, Sumitox 20 EC	Fenvalerate	II	Insecticide for the control of pests in vegetables
Pyriforce 480 EC	Chlorpyrifos ethyl	II	Contact insecticide for the control of a wide range of insect in agriculture and public health
Suncombi 30% EC	Fenitrothion + fenvalerate	II	Insecticide for the control of insect pests of vegetables and public health purposes

of various insect pests in Ghana presently [41]. Table 17.7 shows a list of some OPs registered in Ghana. The effectiveness of these alternatives is evidenced by, for example, the increased production of cocoa in Ghana in recent times despite the ban of lindane, which used to be the preferred insecticide for cocoa.

17.5 CONCLUSIONS

17.5.1 OVERALL STATUS OF OCP CONTAMINATION IN GHANA

Studies have shown that the Ghanaian environment is contaminated with persistent OCPs as a result of their previous and current use in agriculture. These OCP residues were detected in various environmental and biological samples including water, sediment, vegetables, human blood serum and breast milk, bivalves, and fish. These residues have mainly originated from pesticide application in agriculture. Although the levels of most of the OCPs were relatively low in water and sediment, the levels were high in biological samples. The high OCP levels in biological samples such as human fluids and biota may be due to the persistent and lipophilic nature of OCPs in biological systems, which allow them to reach the top of the food chain through bioaccumulation. DDT and HCH were among the most prevalent OCPs in the environment of Ghana, and they may continue to be prevalent in the environment if they continue to be used. For the general population, the ingestion of contaminated food such as vegetables, fish, meat, and dairy products could be an important source of OCPs exposure. The occurrence of high levels of OCP residues in food sources is therefore of major public health concern. OCP occurrence in breast milk is of particular importance since this puts breast-feeding infants at increased risk of OCPs exposure. The generally low levels of OCPs such as aldrin and dieldrin detected in the various samples indicate their discontinued use.

17.5.2 POSSIBLE FUTURE TRENDS OF OCPs IN GHANA

Residues of DDTs and HCHs are the most prevalent OCP residues found in the Ghanaian environment and these OCPs may continue to be prevalent in the environment due to their persistence. Although DDT is banned in Ghana, the Act (Act 528) establishing this ban is not yet fully implemented. This, therefore, leaves room for noncompliance by pesticide dealers. It is also possible that the stock of obsolete DDTs used for malaria control programs in Ghana may still be illegally distributed and used. The use of DDT to immobilize fish in rivers in Ghana was reported about a decade ago [5], and this practice still continues in some parts of Ghana. Interestingly, the low p,p'-DDT/p,p'-DDE ratios of 0.7 and 0.07 for breast milk and for serum, respectively [24], suggest the exposure to DDT was in the distant past [42]. The observed decrease in exposure to these compounds over the years may be due to the influence of the prohibition of DDT in Ghana. This trend may however change as malaria remains a major public health challenge in many parts of the world, including Africa. In Africa, it accounts for over 70% of all hospital cases and is the leading cause of deaths in children under the age of 5. The WHO estimates some 243 million cases of malaria in 2008, which resulted in some 863,000 deaths. A potential intervention measure is the use of DDT in malaria control programs in malaria endemic regions in Africa, including Ghana. There are therefore plans by the WHO to allow the use of DDT in malaria endemic areas including Ghana for malaria intervention. If this is implemented, it is most likely that environmental levels of DDT in Ghana will increase. Endosulfan is reported to be commonly used in vegetable farming in Ghana [3] and may also be prevalent in the environment. Methoxychlor may also become prevalent in the future as it commonly occurs in vegetables. Methoxychlor, however, is not registered for use in Ghana and therefore its occurrence in the environment may be due to the application of pesticide cocktails in vegetable farming, a practice which is believed to increase pesticide potency [3,43].

17.5.3 POSSIBLE FUTURE TRENDS BY EMERGING CONTAMINANTS IN GHANA

Emerging contaminants in Ghana include endosulfan and organophosphorus pesticides (OPs). Endosulfan is a commonly used OCP on vegetables in Ghana [3] and has been found in vegetables and environmental samples [5,20,23]. For example, at Akumadan, which is a prominent vegetable-farming community in Ghana [5], the pesticide is one of the predominant active ingredients used for controlling various pests such as leaf miners, bollworm, and fruit fly on tomato. Due to much overuse, abuse, and misuse of endosulfan, there is the potential for environmental contamination by this OCP [3]. OPs have become the most widely used pesticides in Ghana for pest control and disease vector eradication over the past decade [44]. It has been reported that OPs are widely being used in agriculture and animal production for the control of various insect pests in Ghana presently [41]. These OPs have replaced the banned OCPs and may now present future environmental concerns. As agricultural production intensifies, pesticide usage in Ghana (mainly OPs) is expected to increase. Other prominent emerging contaminants of concern are polychlorinated biphenyls (PCBs), polycyclic aromatic hydrocarbons (PAHs), and active pharmaceutical ingredients (APIs), which are mainly released into the environment from industrial activities. The major concerns about these emerging contaminants relate to their potential toxicity and bioaccumulation in biological systems and their persistence in the physical environment.

REFERENCES

1. Gerken, A., Suglo, J. V., and Braun, M. (2001) Pesticide policy in Ghana. MoFA/PPRSD, ICP Project, Pesticide Policy Project/GTZ, Accra, Ghana, 185pp.
2. Dinham, B. (2003). Growing vegetables in developing countries for local urban populations and export markets: Problems confronting small-scale producers. *Pest Management Science* 59: 575–582.
3. Ntow, W. J., Gijzen, H. J., Kelderman, P., Drechsel, P. (2006) Farmer perceptions and pesticide use practices in vegetable production in Ghana. *Pest Management Science* 62 (4): 356–365.
4. Patlak, M. (1996). A resting deadline for endocrine disruptors. *Environment Science and Technology* 30: 540–544.
5. Ntow, W. J. (2001). Organochlorine pesticide in water, sediments, crops and human fluids in a farming community in Ghana. *Archive of Environmental Contamination and Toxicology* 40: 557–563; *Environmental Research* 106: 17–26.
6. Cerejeira, M. J., Viana, P., Batista, S., Pereira, T., Silva, E., Valerio, M. J., Silva, A., Ferreira, M., and Silva-Fernandes, A. M. (2003) Pesticides in Portuguese surface and ground waters. *Water Research* 37: 1055–1063.
7. Erdogrul, Ö., Covaci, A., and Schepens, P. (2005). Levels of organochlorine pesticides, polychlorinated biphenyls and polybrominated diphenyl ethers in fish species from Kahramanmaras, Turkey. *Environment International* 31: 703–711.
8. Waichman, A. A., Eve, E., and Nina, N. C. S. (2007). Do farmers understand the information displayed on pesticide product labels? A key question to reduce pesticides exposure and risk of poisoning in the Brazilian Amazon. *Crop Protection* 26: 576–583.
9. Chen, S., Lili, S., Zhengjun, S., and Qiuhui, H. (2007). Determination of organochlorine pesticide residues in rice and human and fish fat by simplified two-dimensional gas chromatography. *Food Chemistry* 104: 1315–1319.
10. Ueno, D., Takahashi, S., Tanaka, H., Subramanian, A. N., Fillmann, G., Nakata, H., Lam, P. K. S., Zheng, J., Muchtar, M., Prudente, M., Chung, K. H., and Tanabe, S. (2003). Global pollution monitoring of PCBs and organochlorine pesticides using Skipjack Tuna as a bioindicator. *Archives of Environmental Contamination and Toxicology* 45: 378–389.
11. Ayas, Z., Ekmekci, G., Ozmen, M., and Yerli, S. V. (2007). Histopathological changes in the livers and kidneys of fish in Sariyar Reservoir, Turkey. *Environmental Toxicology and Pharmacology* 23: 242–249.
12. Tanabe, S. (2000) Asian developing regions: Persistent organic pollutants in the seas. In: Sheppard, C. R. C. (ed.), *Seas at the Millennium: An Environmental Evaluation*. Elsevier Science, Amsterdam, the Netherlands, pp. 447–462.
13. Rodan, B. D., Pennington, D. W., Eckley, N., and Boethling, R. S. (1999). Screening for persistent organic pollutants: Techniques to provide a scientific basis for POPs criteria in international negotiations. *Environmental Science & Technology* 33: 3482–3488.
14. Castillo, L. E., Ruepert, C., and Solis, E. (2000) Pesticide residues in the aquatic environment of banana plantation areas in the North Atlantic Zone of Costa Rica. *Environmental Toxicology and Chemistry* 19: 1942–1950.
15. Cogliano, V. J. (1998). Assessing the cancer risk from environmental PCBs. *Environmental Health Perspective* 106: 317–323.
16. Brouwer, A., Longnecker, M. P., Birnbaum, L. S., Cogliano, J., Kostyniak, P., Moore, J., Schantz, S., and Winneke, G. 1999. Characterization of potential endocrine related health effects at low dose levels of exposure to PCBs. *Environmental Health Perspectives* 107 (Suppl. 4): 639–649.
17. Langer, P., Kocan, A., Tajtakova, M., Petrik, J., Chovancova, J., Drobna, B. et al. (2003). Possible effects of polychlorinated biphenyls and organochlorinated pesticides on the thyroid after long-term exposure to heavy environmental pollution. *Journal of Occupational Environmental Medicine* 45: 526–532.
18. Ntow, W. J. (2005) Pesticide residues in Volta Lake, Ghana. *Lakes & Reservoirs: Research and Management* 10: 243–248.
19. Darko, G., Akoto, O., and Oppong, C. (2008). Persistent organochlorine pesticide residues in water, sediments and fish samples from Lake Bosomtwi, Ghana. *Chemosphere* 72 (1): 21–24.
20. Osafo, S. and Frempong, E. (1998): Lindane and endosulfan residues in water and fish in the Ashanti region of Ghana. *Journal of Ghana Science Association* 1: 135–140.
21. Dabrowski, J. M., Peall, S. K. C., Niekerk, A. V., Reinecke, A. J., Day, J. A., and Schulz, R. (2002). Predicting runoff-induced pesticide input in agricultural sub-catchment surface water: Linking catchment variables and contamination. *Water Research* 36: 4975–4984.

22. Botwe, B. O., Ntow, W. J., Kelderman, P., Drechsel, P., Carboo, D., Nartey, V. K., and Gijzen, H. J. (2009, unpublished). Pesticide residues in Ghanaian vegetables and their impact on public health.

23. Amoah, P., Drechsel, P., Abaidoo, R. C., and Ntow, W. J. (2006). Pesticide and pathogen contamination of vegetables in Ghana's urban markets. *Archives of Environmental Contamination and Toxicology* 50: 1–6.

24. Ntow, W. J., Tagoe, L. M., Drechsel, P., Kelderm, P., Gijzend, H. J., and Nyarko, E. (2008). Accumulation of persistent organochlorine contaminants in milk and serum of farmers from Ghana. *Environmental Research* 106: 17–26.

25. Lino, C. M. and da Silveira, M. I. N. (2006). Evaluation of organochlorine pesticides in serum from students in Coimbra, Portugal: 1997–2001. *Environmental Research* 102: 339–351.

26. Kunisue, T., Someya, M., Monirith, I., Watanabe, M., Tana, T. S., and Tanabe, S. (2004). Occurrence of PCBs, organochlorine insecticides, tris(4-chlorophenyl)methane, and tris(4-chlorophenyl)methanol in human breast milk collected from Cambodia. *Archives of Environmental Contamination and Toxicology* 46: 405–412.

27. Kunisue, T., Someya, M., Kayama, F., Jin, Y., and Tanabe, S. (2004). Persistent organochlorines in human breast milk collected from primiparae in Dalian and Shenyang, China. *Environmental Pollution* 131: 381–392.

28. Otchere, F. A. (2005). Organochlorines (PCBs and pesticides) in the bivalves *Anadara (Senilis) senilis*, *Crassostrea tulipa* and *Perna perna* from the lagoons of Ghana. *Science of the Total Environment* 348: 102–114.

29. Darko, G. and Acquaah, S.O. (2007). Levels of organochlorine pesticides residues in meat. *International Journal of Environmental Science & Technology* 4 (4): 521–524.

30. Darko, G. and Acquaah, S.O. (2008). Levels of organochlorine pesticide residues in dairy products in Kumasi, Ghana. *Chemosphere* 71 (2): 294–298.

31. Rovedatti, M. G., Castañé, P. M., Topalián, M. L., and Salibián, A. (2001) Monitoring of organochlorine and organophosphorus pesticides in the water of the Reconquista River (Buenos Aires, Argentina). *Water Research* 35: 3457–3461.

32. Ferrer, I., García-Reyes, J. F., Mezcua, M., Thurman, E. M., and Fernández-Alba, A. R. (2005). Multi-residue pesticide analysis in fruits and vegetables by liquid chromatography-time-of-flight mass spectrometry. *Journal of Chromatography A* 1082: 81–90.

33. Hsu, R. C., Biggs, I., and Saini, N. K. (1991) Solid-phase extraction cleanup of halogenated organic pesticides. *Journal of Agriculture and Food Chemistry* 39: 1658–1666.

34. Lino, C. M., Azzolini, C. B. F., Nunes, D. S. V., Silva, J. M. R., and Silveira, M. I. N. (1998). Methods for the determination of organochlorine pesticide residues in human serum. *Journal of Chromatography* 716: 147–152.

35. Pauwels, A., Covaci, A., Weyler, J., Delbeke, L., Dhont, M., De Sutter, P., D'Hooghe, T., and Schepens, P. J. C. (2000). Comparison of persistent organic pollutant residues in serum and adipose tissue in a female population in Belgium, 1996–1998. *Archives of Environmental Contamination and Toxicology* 39: 265–270.

36. Poon, B. H. T., Leung, C. K. M., Wong, C. K. C., and Wong, M. H. (2005). Polychlorinated biphenyls and organochlorine pesticides in human adipose tissue and breast milk collected in Hong Kong. *Archives of Environmental Contamination and Toxicology* 49: 274–282.

37. Taleb, Z. M., Benali, I., Gherras, H., Ykhlef-Allal, A., Bachir-Bouiadjra, B., Amiard, J. C., and Boutiba, Z. (2009). Biomonitoring of environmental pollution on the Algerian west coast using caged mussels *Mytilus galloprovincialis. Oceanologia* 51 (1): 63–84.

38. Solè, M. (2000). Assessment of the results of chemical analyses combined with the biological effects of organic pollution on mussels. *Trends in Analytical Chemistry* 19: 1–9.

39. Farrington, J. W., Davis, A. C., Tripp, B. W., Phelps, D. K., and Galloway, W. B. (1987). 'Mussel Watch' measurements of chemical pollutants in bivalves as one indicator of coastal environment quality. In: Boyle, T. P. (ed.), *New Approaches to Monitoring Aquatic Ecosystems*, ASTM STP 940. American Society for Testing and Materials, Philadelphia, PA, pp. 125–139.

40. Ntow, W. J., Drechsel, P., Botwe, B. O., Kelderman, P., and Gijzen, H. J. (2008) The impact of agricultural runoff on the quality of two streams in vegetable farm areas in Ghana. *Journal of Environmental Quality* 37: 696–703.

41. Darko, G. and Akoto, O. (2008) Dietary intake of organophosphorus pesticide residues through vegetables from Kumasi, Ghana. *Food and Chemical Toxicology* 46: 3703–3706.

42. Carreño, J., Rivas, A., Granada, A., Lopez-Espinosa, M. J., Mariscal, M., Olea, N., and Olea-Serrano, F. (2006). Exposure of young men to organochlorine pesticides in Southern Spain. *Environmental Research* 103: 55–61.

43. Danso, G, Drechsel, P., and Fialor, S. C. (2002). Perception of organic agriculture by urban vegetable farmers and consumers in Ghana. *Urban Agriculture Magazine* 6: 23–24.

44. Clarke, E. E. K., Levy, L. S., Spurgeon, A., and Calvert, I. A. (1997). The problems associated pesticide use by irrigation workers in Ghana. *Occupational Medicine* 47: 301–308.

Part IV

Persistent Organic Chemicals
in the Americas

18 Contamination Profiles and Possible Trends of Persistent Organic Compounds in the Brazilian Aquatic Environment

Gilberto Fillmann* and Juliana Leonel

CONTENTS

18.1 Introduction 415
    18.1.1 History of POC Use, Ban, and Restrictions 416
18.2 Contamination Status and Trends... 417
    18.2.1 Sediment 417
    18.2.2 Invertebrates 420
        18.2.2.1 Bivalves.. 420
        18.2.2.2 Other Invertebrates 422
    18.2.3 Vertebrates 422
        18.2.3.1 Teleost Fishes... 422
        18.2.3.2 Cartilaginous Fishes 423
        18.2.3.3 Marine Mammals 423
18.3 Methods 426
    18.3.1 Samples and Sampling Methods.. 426
    18.3.2 Chemical Analysis.. 426
18.4 Conclusions and Recommendations 427
References.. 428

18.1 INTRODUCTION

Brazil in South America is one of the largest countries in the world, covering an area of approximately 8.5 million km^2 and consisting of 190 million inhabitants (about 80% living in the urban areas). Divided into five geographic regions (north, northeast, south, southeast, and central west), Brazil is mostly a tropical country, but extends well into the subtropical zone. Due to the continental dimensions, Brazil possesses a very wide climatic diversity, influenced by its geographical configuration, its significant coastal extension (8512 km), and the dynamics of air masses within its territory [1].

The Brazilian economy is among the 10 largest in the world and is projected to be the 5th largest by 2012. Together with Russia, India, China and Korea, Brazil encompasses over 25% of the world's land coverage and 40% of the world's population, holding a combined gross domestic product (GDP) of $15.435 trillion. The Brazilian GDP itself was about US$ 3 trillion in 2009. Farming and cattle raising account for 27% of the total direct GDP, whereas industry—automobiles, steel, petrochemicals,

* E-mail: docgfill@furg.br (Chapter corresponding author).

electronics, aircraft, and consumer durables—accounted for 28%. The remaining 45% comes from services that comprise various activities, such as trade and repair of motor vehicles; land, sea, and air transportation; financial intermediation; post and telecommunications; retail and repair of personal and household goods; real estate activities; wholesale and commercial representatives; public administration and defense, and social security (education, health and social services) [1].

18.1.1 History of POC Use, Ban, and Restrictions

Information regarding the production, importation, and use of persistent organic compounds (POCs) in Brazil is very scarce, fragmented, and confusing. For example, according to the department that controls the importation in the country (Ministry of Exterior Commerce, Industry, and Development (MDIC)), 11,700 ton of endosulfan was imported from 1989 to 2003 (Table 18.1), whereas according to the Brazilian Environmental Agency (IBAMA) more than 16,000 tons were used in the country during the same period.

Almost all chlorinated pesticides (such as aldrin, DDTs, heptachlor, mirex, lindane, endosulfan) were used in South American countries during 1950s–1970s. There was a declining trend in the 1980s–1990s due to legal restrictions on the production and use [2]. In Brazil, the Ministry of Agriculture (MA) banned the production and commercialization of DDT and HCH for domestic use (parasites extermination) and pest control in pasture in 1971 (Ordinance MA No. 356 and 357/1971, respectively) [3,4]. In 1985, commercialization, use, and distribution of organochlorine compounds were also prohibited for agricultural purposes, but they are still allowed in pest control of tropical and subtropical epidemic diseases (such as malaria), emergency use in agriculture, termite control,

---

TABLE 18.1
Production and Importation of Chlorinated Pesticides and PCBs

Drins: They have been formulated in São Paulo state by Shell Company from 1977 to 1990, where 1250 ton of contaminated soils and approximately 750 kg of pure aldrin, endrin, and dieldrin. Moreover, from 1961 to 1982 a total of 17,000 ton of aldrin and 10,600 ton of dieldrin were imported. Recently, these amounts decreased and aldrin and dieldrin were imported for the last time in 2004 (200 kg) and 2005 (23 kg), respectively.

DDTs: Widely produced and imported; Brazil has imported more than 105,000 ton of DDT between 1959 and 1975, 3200 ton between 1989 and 1991 and only 7 ton in 2001 (the last time DDT was imported). Although DDTs are not been used in Brazil, not even to combat malaria, a significant source to the environment is dicofol production which can have DDT as an impurity.

HCH: 18,400 ton were produced in Brazil from 1955 to 1982, and more 6500 ton were imported in the same period. After 1996 only 14 ton were imported.

Endosulfan: According to the importation data, 6600 ton were imported from 1962 to 1982 and 11,700 ton from 1989 to 2003. However, according to the Brazilian Environmental Agency (IBAMA) more than 16,000 ton were used in the country during the same period (1989–2003). Used in sugarcane, coffee, and cocoa plantations. Its use will be banned in 2013

CHLs: Were not widespread used in Brazil. The only register of importation account for 15 ton from 1989 to 1996.

Mirex: Mirex was used principally to combat ants, but there is no official data about production and/or importation.

HCB: Importation data of HCB shows a decline: 834 ton imported in 1965, 4 ton during 1989–1996, and only 72 kg between 1997 and 2004. Moreover, there are well known highly contaminated sites and stockpiles of HCB in Brazil, e.g., Cubatão, that may be a significant environmental source. In Cubatão, HCB was probably generated from perchloroethylene production and burning of other chlorinated residues.

PCBs: Total PCB burden in Brazil is estimated at around 130,000 ton, which corresponds to approximately 10% of the total worldwide production. PCB used in Brazil was imported.

Sources: Data from PNUMA, Evaluación regional sobre substancias tóxicas persistentes: Informe regional de sudamérica oriental y occidental, Programa de Las Naciones Unidas para el Medio Ambiente, UNEP, Switzerland, 2002; MDIC, Importação Brasileira, 2009, http://aliceweb.mdic.gov.br (accessed June 2010).

and wood preservation [2]. Despite that, it was reported that these compounds have not been used to control epidemic diseases for some years (personal communication: Sérgia de Souza Oliveira, MMA*). Indeed, the last documented time Brazil imported DDTs was in 2002 (7 ton) [5].

There is no information of PCBs production in Brazil. All PCBs used were imported, mostly from General Electric and Westinghouse (USA), under the names of Pyranol and Inerteen, respectively. PCBs-based fluids were formulated from tri- and penta-biphenyls and produced by Monsanto or the Dow Chemical Company. Some PCBs were also imported from Europe, such as Chlophen from Bayer through TUSA (old Trafo Union and now Siemens) and Pyralene from Rhodia (personal communication: Paulo Fernandes, SDM Brazil†). PCBs were commercialized in Brazil as Ascarel and were mainly used by the electro-electronic industry in capacitors and transformers [6]. Although production, commercialization, and use of PCB was prohibited in 1981 (Ordinance MA No. 019/1981), equipment containing PCBs were allowed to remain in use up to the end of their operational life [7]. Approximately 21,000 ton of these dielectric liquids were imported until the prohibition in 1981, but this number is probably underestimated since PCBs were also imported to Brazil inside transformers and capacitors. Despite the incineration of many residues of Ascarel, first in Wales (UK) and after by companies installed in Brazil (Bayer, Rio de Janeiro; CETREL, Bahia; and CINAL, Alagoas), the remaining residues of pure Ascarel and equipment containing Ascarel may still reach about 10,000 ton in Brazil, but the inventory is still underway (personal communication: Paulo Fernandes, SDM Brazil). Very recently 21 ton of Ascarel oil (inside transformers and barrels) were reported to be stored at warehouse A5 of the Rio Grande harbor [8].

Compared to other chlorinated POCs there is almost no data about production, importation and use of polybrominated diphenyl ethers (PBDEs), hexabromocyclododecane (HBCDs) and perfluorinated compounds (PFCs). For brominated flame retardants, however, there is a bill (No. 173/2009) up for voting at the Brazilian Senate that forbids the commercialization of electro-electronic equipments containing more than 0.1% (w/w) of PBDEs and polybrominated biphenyls (PBBs) [9]. Information about PFCs is also scarce, but it is known that PFCs have been used in firefighting foams in the country [10] and in the insecticide sulfluramid (*N*-ethyl perfluorooctane sulfonamide), which is largely used in Brazil for agricultural control of leaf-cutting ants [11].

Despite being one of the most studied group of contaminants in the assessment of POCs in different matrices of Brazilian environments are, in fact, very scarce. Sediments, invertebrates, fish, and mammals are the main addressed matrices, which studies are compiled and summarized below. Although this data was found in the literature, it is likely that more data have been produced and never or just locally been published, making it very difficult to locate and access.

18.2   CONTAMINATION STATUS AND TRENDS

18.2.1   SEDIMENT

Assessments of contamination by POCs in sediment samples from aquatic Brazilian environments are very scarce. So far, only six studies have been found, mostly concentrated in the Southern (Patos Lagoon and adjacent coast (RS) [12,13] and Estuarine Complex of Paranaguá (PR) [14–16]) and Southeastern (Santos–São Vicente Estuary (SP) [17] and Guanabara Bay (RJ) [18]) regions (Figures 18.1 and 18.2). Recently, a very comprehensive study covering water column and superficial sediment of the entire Guanabara Bay (RJ), which included PCB analyses, was organized by CENPES/PETROBRÁS. Unfortunately, this data is still under preparation

---

* Dra. Sérgia de Souza Oliveira, Analista Ambiental, Ministério do Meio Ambiente (MMA), Brasília-DF, Brazil, e-mail: sergia.oliveira@mma.gov.br, tel.: +55 61 2028 1026.
† Paulo Fernandes, Director da SDM do Brazil, Meio Ambiente e Instrumentação Ltda. São Paulo-SP, Brazil, e-mail: paulo@sdmdobrasil.com.br.

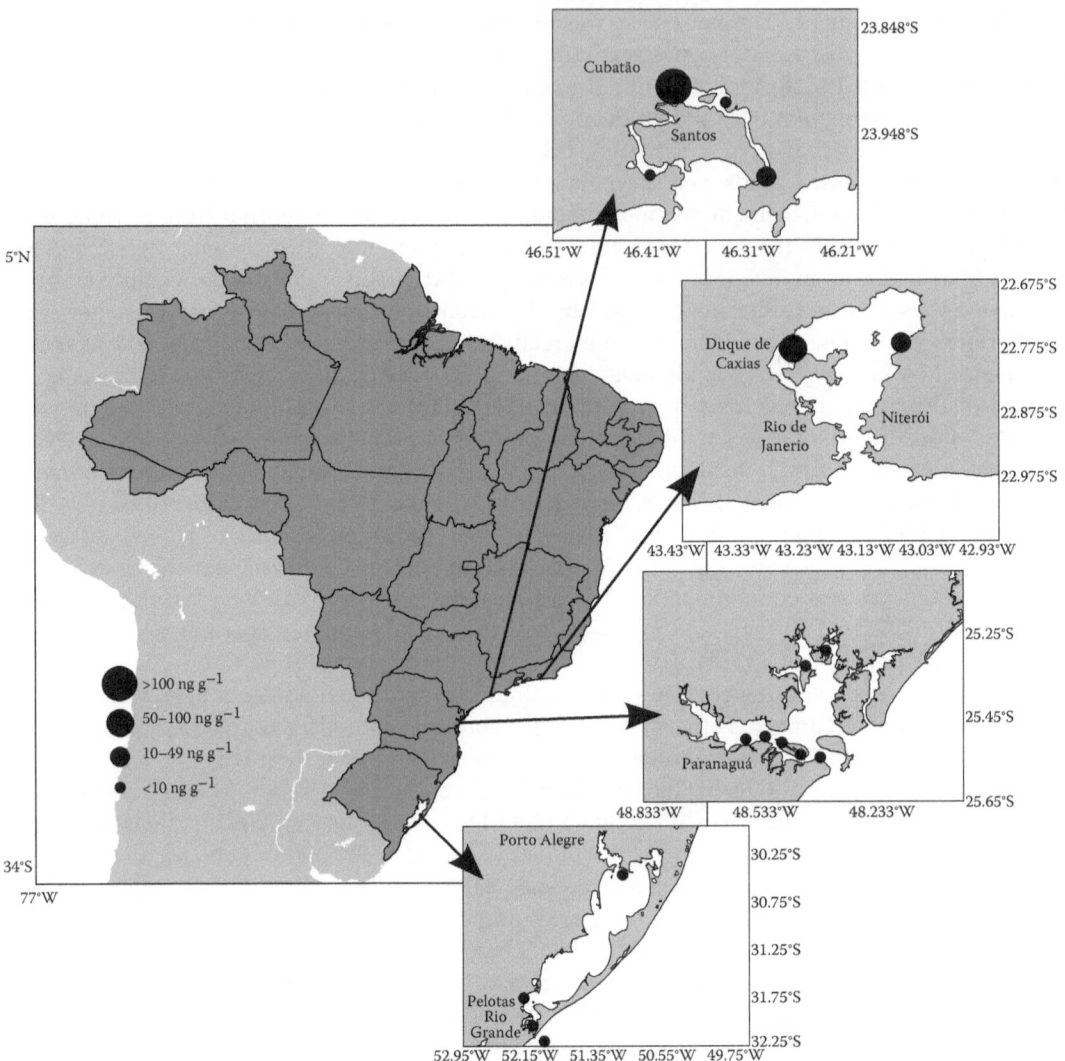

FIGURE 18.1 PCB concentrations (ng g⁻¹ dry weight) in sediment samples from Brazilian coast.

for future publication (personal communication: Maria de Fátima Meniconi, CENPES*). These four areas are very important to the economy of the country since they hold four of the main Brazilian harbors and important industrial and urban activities within their drainage areas [12–18].

PCBs were the dominant compounds in sediments followed by DDTs in concentrations of about one order of magnitude lower. The PCBs/DDTs ratio indicates that industrial sources were more significant than agricultural sources for all four areas, even at the southernmost estate (Rio Grande do Sul (RS)) where the rural activities (agriculture and animal farming) were historically predominant. Among the four studied regions, the higher PCB concentrations were detected in sediment from a very industrial region of Santos–São Vicente Estuary [17]. Levels reached 254 ng g⁻¹ dw (sum of 30 PCB congeners) at Cubatão industrial complex, one of the most important petrochemical

* Dra Maria de Fátima Guadalupe Meniconi, AMA-CENPES/PETROBRÁS, Ilha do Fundão, Rio de Janeiro-RJ, Brazil, e-mail: fatimameniconi@petrobras.com.br, tel.: +55 21 3865 7077.

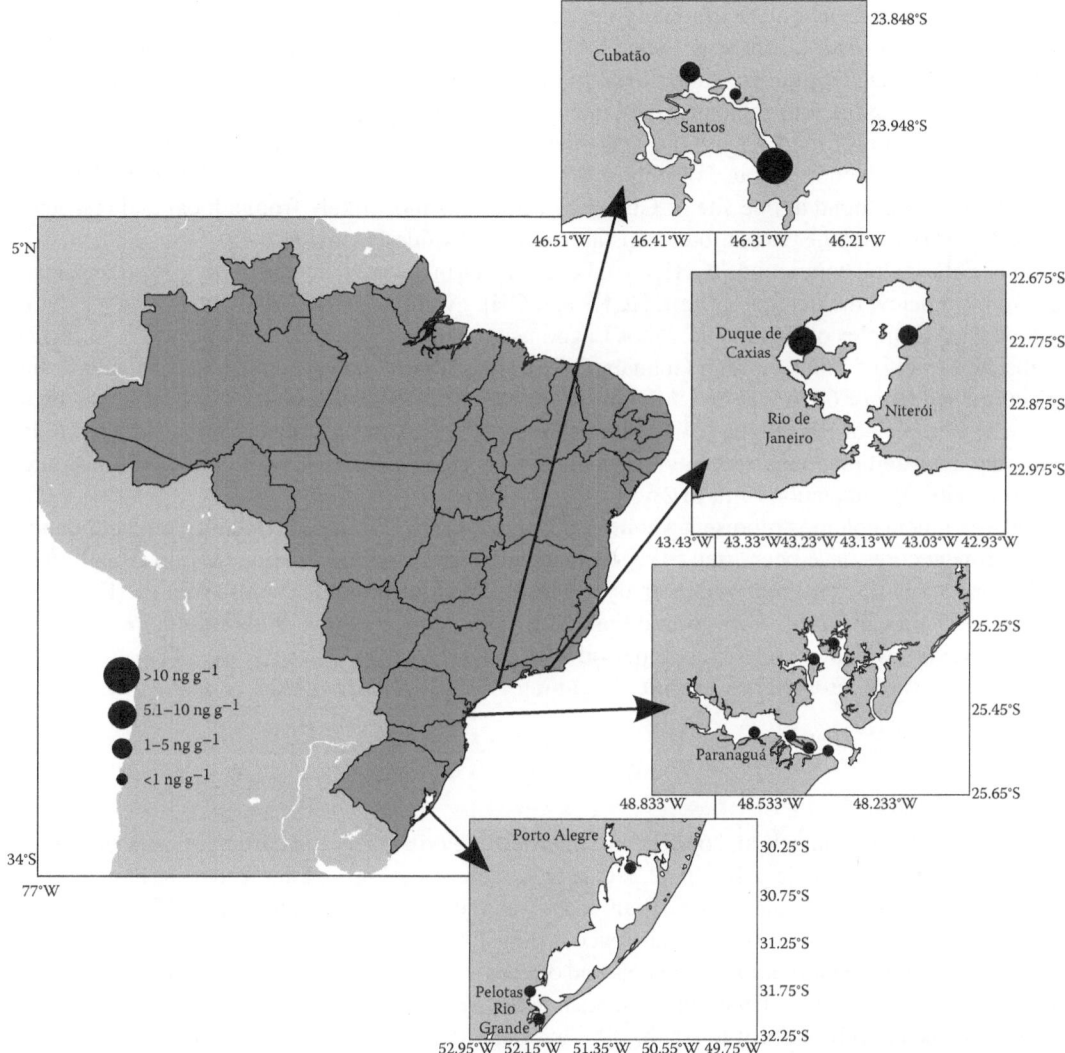

FIGURE 18.2   DDT concentrations (ng g^{-1} dry weight) in sediment samples from Brazilian coast.

and metallurgical industrial centers in Brazil. It is worth mentioning that this value is above the limiting concentration of 200 ng g^{-1} set by the U.S. National Status and Trends Program for a contaminated area [19]. Other data in the Santos–São Vicente region were one or two orders of magnitude lower (1.17–24.5 ng g^{-1} dw). According to Bícego et al. [17], such a high value might be associated to the use of capacitors and transformers in the industrial area. The second highest concentrations were found at the industrial region of Guanabara Bay (184 and 116 ng g^{-1} dw at Duque de Caxias area) (sum of 7 PCBs congeners), although concentration as low as 33 ng g^{-1} dw was found in its adjacent areas [18].

On the other hand, even the highest values detected in the southern Brazil studies were one to two orders of magnitude lower than those from the southeastern region. Values ranged from 0.08 to 8.62 ng g^{-1} dw (sum of 44 PCBs congeners) at the Patos Lagoon region [12,13] and 1.12 to 8.41 ng g^{-1} (sum of 44 PCBs congener) at the Paranaguá Estuarine Complex [14–16]. In spite of harbor and industrial activities and important cities, those anthropogenic inputs at both Paranaguá Estuarine Complex and Patos Lagoon are much more recent and at smaller scale compared to Santos Estuary and Guanabara Bay.

Since the pesticide p,p'-DDT gradually degrades to p,p'-DDE and p,p'-DDD through biological and photochemical transformations under both aerobic and anaerobic conditions, the p,p'-DDT/ (p,p'-DDE + p,p'-DDD) ratio is usually used to know whether DDT input has occurred recently or not [20]. All sites in the four regions except one in Guanabara Bay presented p,p'-DDT/(p,p'-DDE + p,p'-DDD) ratios lower than 1, with most of them lower than 0.5. This is the consequence of legal restrictions on production and use of almost all chlorinated pesticides in the 1980s and 1990s [2]. The value of 2.94, found at one site of Guanabara Bay, was most likely from a local and restricted source of DDT since other sites around this area depicted values as low as 0.45.

Other chlorinated compounds (CHLs, HCB, HCHs, Drins) were detected at very low concentrations or even below the detection limit. HCHs and CHLs were never detected above 1 ng g^{-1} dw in the Estuarine Complex of Paranaguá, Patos Lagoon, and Santos–São Vicente Estuary [12–14,16,17], but reached levels of 2 ng g^{-1} dw in Guanabara Bay [18]. "Drins" compounds (aldrin, dieldrin, and endrin) ranged from 1.68 to 7.43 ng g^{-1} dw in Guanabara Bay [18] and from 0.07 to 0.89 ng g^{-1} dw in Estuarine Complex of Paranaguá [14,16], but were not detected in the Patos Lagoon [12,13]. HCBs were only analyzed in Guanabara Bay and Santos–São Vicente Estuary, with concentrations ranging from below the detection limit to 2 ng g^{-1} dw [17,18].

Since sediment columns represent a continuous sequence of associated contaminants sedimentation, sediment cores have been used to reconstruct a history of contamination [21]. In Brazil, there is only one study of that kind, which analyzed PCB inputs in sediment cores from the Estuarine Complex of Paranaguá [15]. It was found that PCB inputs started in the 1960s and intensified from 1970s to 1990s, which correlates to a period of the stronger use. These values showed a slight decrease afterwards, probably as a result of banning PCBs during the 1980s.

18.2.2 Invertebrates

18.2.2.1 Bivalves

Bivalves, in particular mussels and oysters, are filter-feeding organisms and have been used as bioindicators/biomonitors of aquatic pollution [22]. One of the few studies using filter feeders as bioindicators of contamination by POPs on the Brazilian coast was the Latin American chapter of International Mussel Watch (IMW), which was run just once during 1992 [23,24]. The screening of PCBs, DDTs, and total chlordane contamination in mussels and oysters was done at 12 sites along the Brazilian coast, mostly at urbanized/industrialized areas. At each areas were collected samples from 1 to 3 different points in duplicates. From north to south, the sites were as follows: Bragança (PA), São Luis (MA), Fortaleza (CE), Recife (PE), Mundaú Lagoon-Maceió (AL), Salvador (BA), Vitória (ES), Cabo Frio (RJ), Guanabara Bay (RJ), Santos Bay (SP), Paranaguá Bay (PR), and Patos Lagoon (RS) (Figure 18.3).

DDT levels (sum of p,p'-DDT, o,p'-DDT, p,p'-DDE, o,p'-DDE, p,p'-DDD, and o,p'-DDD) at these sites ranged from below the detection level (<MDL) to 130 ng g^{-1}, with the highest values detected at the northeastern sites (Fortaleza/CE, 130 ng g^{-1} and Recife/PE, 120 ng g^{-1}). DDTs predominated over PCBs only at Bragança (PA), São Luis (MA), and Fortaleza (CE), although the Bragança levels were rather low (DDTs = 3.39 ng g^{-1} and PCBs <LDM). Due to the lack of information, however, it is difficult to identify specific sources for these regions.

Based on the pattern of DDE (51%) >DDD (34%) >DDT (15%) in bivalves collected along the Brazilian coast during the IMW, no recent inputs of DDT have occurred in these Brazilian sites, probably as a consequence of legal restrictions on production and use of almost all chlorinated pesticides in the 1980s–1990s [2].

PCB levels (sum of 23 PCB congeners) ranged from <0.25 (Bragança-PA) to 280 ng g^{-1}. The higher concentrations were detected in regions with intense harbor activities and/or have been receiving industrial and domestic effluents, such as Recife (PE) and Guanabara Bay (RJ) (280 and 210 ng g^{-1}, respectively), followed by Santos (SP) and Vitória (ES) (66 and 90 ng g^{-1}, respectively). For other chlorinated pesticides, the total chlordane (*cis*-chlordane, *trans*-chlordane, *cis*-nonachlor, *trans*-nonachlor, oxychlordane, heptachlor, heptachlor epoxide, metoxychlor) levels were below

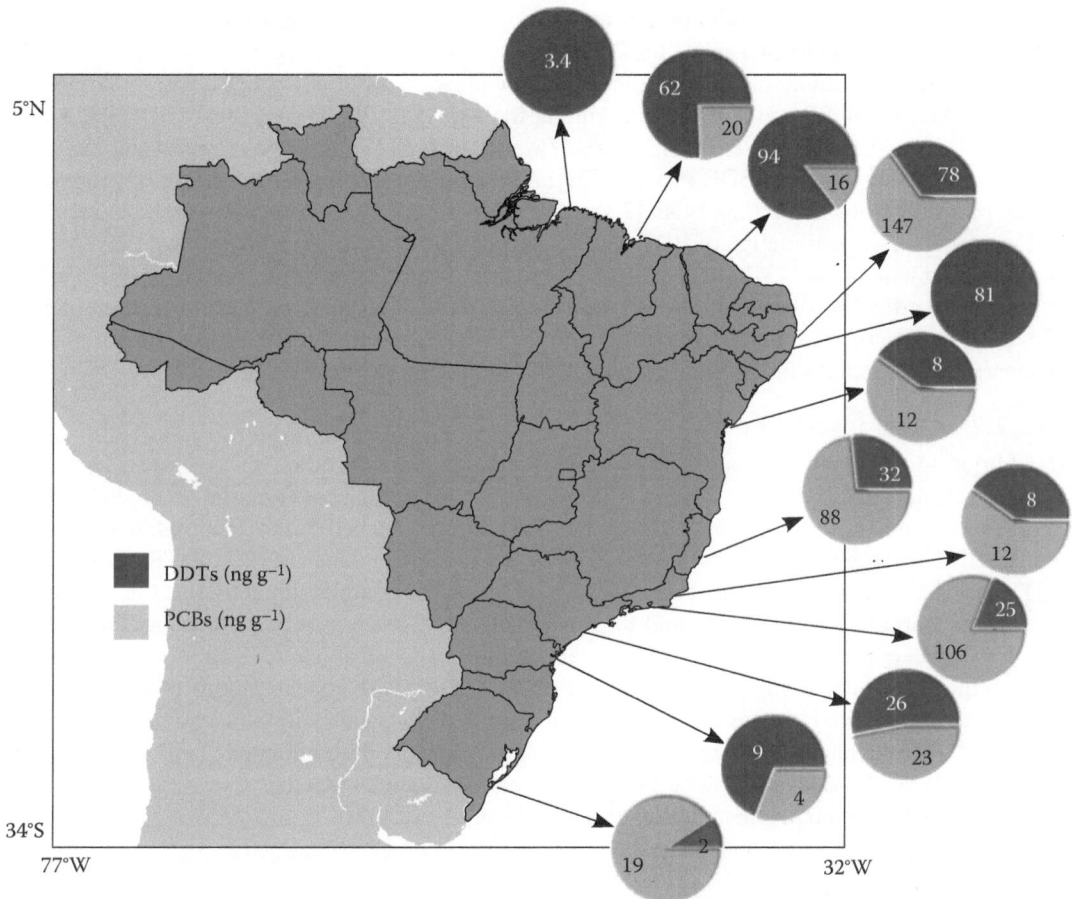

FIGURE 18.3   Mean PCB and DDT concentrations (ng g^{-1} dry weight) from Brazilian coast; data from the Mussel Watch Programme.

10 ng g^{-1}, while HCB levels were at or below the detection limit (≤0.25 ng g^{-1}). Except for Guanabara Bay, where the "Drins" (aldrin, dieldrin, endrin) concentration was 12 ng g^{-1}, Mirex and "Drins" concentrations did not exceed 5 ng g^{-1} at all 12 sites along the coast. HCH (α-, β-, γ-, δ-isomers) concentrations were also low and only at Santos and Guanabara Bay levels were higher than 10 ng g^{-1}.

Despite the contamination, all values in mussels/oysters were below the limits established by the U.S. Food and Drug Administration (USFDA) for fish, which are 500 ng g^{-1} for PCBs, 5000 ng g^{-1} for DDTs, and 300 ng g^{-1} for *cis*-chlordane [25]. Actually, there were no IMW samples for which contaminant concentrations exceed the various national and/or international recommended action limits for these compounds in seafood for human consumption [23].

About 7 years later, Hermanns [12] also analyzed organochlorine compounds in mussels collected from the Patos Lagoon (RS) and did not find any significant temporal trend when comparing with the IMW results. Similarly, Koike [14] could not find any differences in PCBs and chlordane levels detected in oysters collected at Paranaguá Bay, even 14 years after the IMW study, which analyzed levels in mussels instead. DDT levels, nevertheless, showed to be slightly lower in 2006 (3 ng g^{-1}) than in 1992 (8.8 ng g^{-1}). Although using different species of filter-feeder (bivalves) collected a few years before (during 1985–1986) the IMW, Tavares et al. [26] reported similar levels of PCBs and DDTs at Todos os Santos Bay (Salvador (BA); Northeast Brazil).

In addition to organochlorine data, there are two studies of PBDEs (sum of 10 congeners) and PFC contamination in mussels from the adjacent coast of Patos Lagoon [27] and Guanabara Bay

[28], respectively. Levels of both contaminant groups were low, ranging from <LDM to 16 ng g^{-1} for PBDEs, from <0.84 to 14.9 ng g^{-1} for PFOA, and from <0.95 to 4.70 ng g^{-1} for PFOS. PFOA and PFOS were the predominant PFC compounds, but the presence of PFNA, PFDA, and PFUnDA suggest local sources of contamination [28]. Among the analyzed PBDE congeners (BDE 28, 47, 66, 85, 99,100, 138, 153, 154, and 183), only BDE 47 was detected [27]. This may suggest sources originated from the use of penta-BDE mixture [29]. Although the technical penta-BDE mixture has a higher proportion of BDE99 than BDE 47, the predominance of BDE 47 suggests preferential elimination or metabolic degradation of BDE 99. In vivo laboratory exposures have shown that common carp [30] and blue mussel [31] are capable of debrominating BDE 99 to form BDE 47. Therefore, it is possible that the brown mussel has the same capacity, which results in an accumulation of BDE 47 within their tissues and a relative depletion of BDE 99. However, it cannot be ruled out that this congener pattern reflects the original pattern found in the brown mussel diet.

18.2.2.2 Other Invertebrates

In addition to the studies on bivalves, there are very few data on other invertebrates. Gorni and Weber [32] analyzed 37 benthic species collected in 11 different places from the inner shelf area of São Sebastião during 1994–1998, an important tourist resort located in São Paulo State (southeastern Brazil). The largest oil terminal on the Atlantic coast of Latin America (DTCS/TRANSPETRO, PETROBRÁS) is installed at this area, which also has other commercial activities (i.e., fisheries), and receives urban raw discharges and terminal effluents. Chlorinated pesticides and PCB levels, which were predominantly found in crustacean samples, were very low (<18 ng g^{-1}). DDE was the most frequently detected compound (in 12 out of 37 species) with concentrations ranging from 0.7 to 9.7 ng g^{-1}, while HCHs were only detected in 2 out of 37 species with concentration of 5.6 and 17.1 ng g^{-1}. PCBs reached 17.4 ng g^{-1} and the heavier congeners predominated (PCB 138, 153, 170, 180, and 183). All PCB levels were below the National Academy of Science of the USA and USFDA guideline levels for human consumption of fish (500 ng g^{-1}) [25,32].

18.2.3 VERTEBRATES

18.2.3.1 Teleost Fishes

Ueno et al. [33–36] investigated PCBs (sum of 59 PCB congeners), DDTs (sum of p,p'-DDT, p,p'-DDD, and p,p'-DDE), chlordane (CHLs: sum of *trans*-chlordane, *cis*-chlordane, *trans*-nonachlor, *cis*-nonachlor, and oxychlordane), HCHs (sum of α-, β- and γ-HCH), HCB [33], PBDEs (sum of 11 PBDEs congeners) [34], dioxins (total PCDDs), furans (total PCDFs), coplanar PCBs (sum of 12 coplanar PCBs) [35] and HBCDs (sum of α-, β- and γ-HBCD) [36] in liver or muscle of five skipjack tuna (*Katsuwonus pelamis*) collected from the offshore waters of Brazil during 2000. Fish samples were pooled for chemical analysis of PBDEs, PCDFs, and coplanar PCBs and HBCDs. All compounds but PCDDs were above detection limits. Concentrations ranged from 340 to 590 ng g^{-1} lw (lipid weight) for PCBs, 45 to 150 ng g^{-1} lw for DDTs, 32 to 88 ng g^{-1} lw for CHLs, 0.15 to 6.2 ng g^{-1} lw for HCHs, 1.8 to 3.0 ng g^{-1} lw for HCB, 13 ng g^{-1} lw for PBDEs, 4.2 pg g^{-1} lw for PCDFs, 14 ng g^{-1} lw for coplanar PCBs, and 0.28 ng g^{-1} lw for HBCDs. PCBs, DDTs, *cis*-chlordane, and HCB values detected in tuna from offshore waters of Brazil were below the guideline levels of USFDA for human consumption [25].

p,p'-DDE and *trans*-nonachlor were the predominant compounds among DDTs and CHLs respectively, implying no recent inputs of their technical mixtures. However, the predominance of γ-HCH among HCHs congeners (approximately 90%) did not necessarily involve recent use of lindane since total HCHs values were very low (from 0.15 to 6.2 ng g^{-1} lw) [33]. The predominance of PCB and PBDE congeners of lower chlorination/bromination are possibly associated with the higher transportability of these compounds [33,34].

Total concentration of coplanar PCBs (14 ng g^{-1} lw) was three orders of magnitude higher than those of total PCDFs (4.2 pg g^{-1} lw). The estimated toxic equivalents (TEQs) of PCDFs and coplanar

PCBs in skipjack tuna were 0.039 and 0.38 pg TEQs g^{-1} wet weight, respectively. Thus, coplanar PCBs-TEQs accounted for 88% of total-TEQs and more than 80% were attributed to non-*ortho* coplanar PCBs-TEQs [35].

Another study with teleost fishes was conducted with eight species collected in Guanabara Bay from January to March 1999: bluefish (*Pamatomus saltatrix*), yellow croaker (*Micropogonia furnieri*), swordfish (*Trichiurus lepturus*), white mullet (*Mugil liza*), silver mullet (*Mugil curema*) Brazilian sardine (*Sardinella brasiliensis*), fat snook (*Centropomus paralellus*), and common snook (*Centropomus undecimalis*) [37]. Dieldrin was detected at low levels (10 ng g^{-1} ww) only in white mullet, while *p,p'*-DDE was detected in yellow croaker (14 ng g^{-1} ww) and fat snook (16 ng g^{-1} ww). Overall PCB levels (sum of 6 congeners) were higher than those for pesticides, with low means detected in Brazilian sardine (19.2 ng g^{-1} ww) and higher in fat snook (115.8 ng g^{-1} ww).

18.2.3.2 Cartilaginous Fishes

Azevedo and Silva et al. [38] investigated the contamination by PCBs, DDTs, HCB, and γ-HCH in muscle samples of three different species of shark collected during August and September 2001 from the southeastern coast of Brazil: smooth hammerhead (*Sphyrna zygaena*), shortfin mako (*Isurus oxyrinchus*), and bigeye thresher (*Alopias superciliosus*). PCB concentrations (4.54–18.65 ng g^{-1} ww) were higher than DDTs (1.82–2.71 ng g^{-1} ww), HCB (<MDL-0.06 ng g^{-1} ww), and γ-HCH (<MDL-0.12 ng g^{-1} ww) for all studied species. *p,p'*-DDE was the predominant DDT compound, representing between 53% and 84% of total DDTs (sum of *p,p'*-DDE, *p,p'*-DDD, and *p,p'*-DDT). The sum of hexa- and hepta-chlorinated congeners represented 62% and 63% of total PCBs in shortfin mako and bigeye thresher, respectively, while tetra-, hexa- and hepta-chlorinated congeners represented 75% in smooth hammerhead.

All values were below the guideline levels established by USFDA for human consumption [25]. Moreover, the estimated daily intake (EDI) dose of DDTs through the consumption of shark was calculated for general population and fisherman families. The EDI for general population/fisherman families were 0.00064/0.1881, 0.00053/0.1558, and 0.00058/0.1700 μg (kg body weight)$^{-1}$ day^{-1} for smooth hammerhead, shortfin mako, and bigeye thresher, respectively. Thus, levels of EDI were well below the accepted daily intake (ADI) values of 20 μg (kg body weight)$^{-1}$ day^{-1} [38].

18.2.3.3 Marine Mammals

18.2.3.3.1 Cetaceans

The two most analyzed species of cetaceans from the Brazilian coast are *Sotalia guianensis* (estuarine dolphin) and *Pontoporia blainvillei* (franciscana dolphin). The first species (*S. guianensis*) presents a preference for coastal and estuarine brackish waters throughout its distribution, which extends from Santa Catarina State, Southern Brazilian region, toward Central America [39]. Several studies suggest that this species displays long-term residence in bays and estuaries. For example, site fidelity of 86 individuals has been observed in Cananéia Estuary in southeastern Brazil [40]. Therefore, blubber of estuarine dolphins have been frequently used as bioindicators of regional contamination and it is the species with the largest number of studies in different regions of Brazil:

1. Lailson-Brito et al. [41] investigated anthropogenic organic contaminants in carcasses of estuarine dolphin from three regions: Guanabara Bay and Sepetiba Bay, both in Rio de Janeiro State; and Paranaguá Bay in Paraná State.
2. Yogui et al. [42] analyzed samples from a resident group of estuarine dolphins from Cananéia Estuary in São Paulo State.
3. Kajiwara et al. [43] analyzed samples distributed in Cananéia Estuary as well as Paranaguá Bay.

In all of these studies, PCBs and DDTs were the predominant organochlorine compounds, while chlordanes, HCHs, HCB, "drins," and Mirex, were often present but in minor concentrations. The highest average concentrations of PCBs were noticed in dolphins collected in Guanabara Bay (26,000 ng g⁻¹ lw) [40] and Cananéia/Paranaguá region (34,000 ng g⁻¹ lw) [43] whereas the highest average DDT levels were detected in samples from the Cananéia Estuary (36,000 ng g⁻¹) [42]. The high values of PCBs in samples from Guanabara Bay agreed with the industrial characteristics of the region. However, the values from the Cananéia/Paranaguá region were kind of high for what would be expected from an area that is impacted neither by industries nor urbanization. The elevated levels of PCBs in this area could be due to the larger set of PCB isomers and congeners analyzed (101 PCBs against 27 PCBs in other studies). On the contrary, the high DDT levels detected in the Cananéia Estuary could be related to the use of this compound in the area during 1970s–1980s. The predominance of p,p'-DDE (~80%) in all estuarine dolphins reinforces the idea that DDT input to the Cananéia region is not recent.

In addition to organochlorine data, two studies have analyzed PBDEs and PFCs in liver samples of estuarine dolphins from Guanabara Bay only. PBDE levels ranged from 13 to 1620 ng g⁻¹ lw [45] and PFOS from 13 to 902 ng g⁻¹ ww [44]. These values are similar to those reported in cetaceans from the Northern Hemisphere, demonstrating the elevated degree of exposure of high trophic level organisms in southeastern Brazil.

The second most studied cetacean species from the Brazilian coast is *Pontoporia blainvillei* Franciscana dolphin. This species inhabits shallow coastal and estuarine waters from the central coast of Brazil to the central coast of Argentina. Due to its habitat, this species is particularly vulnerable to anthropogenic activities, which is in close proximity to point sources of contamination. The existence of two populations of franciscana dolphin southward and northward of Santa Catarina State is strongly supported by morphological and molecular data [46,47]. The southern population is listed as "vulnerable species" in the International Union for Conservation of Nature and Natural Resources [48].

Leonel et al. [48] analyzed chlorinated pesticides and PCBs in blubber samples of franciscana dolphin collected in southern Brazil (southern populations) during 1994 and 2004. Overall concentrations in mature males (average for the 10 years) were relatively low: 5120 ng g⁻¹ lw for PCBs, 1037 ng g⁻¹ lw for DDTs, 82.5 ng g⁻¹ lw for CHLs, 61.8 ng g⁻¹ lw for Mirex, 56.8 ng g⁻¹ lw for dieldrin, and 30.5 ng g⁻¹ lw for HCB. On the other hand, Kajiwara et al. [43] analyzed the same species from the Northern population (collect along Cananéia-Paranaguá region) and detected similar concentrations of PCBs (5300 ng g⁻¹ lw) and CHLs (64 ng g⁻¹ lw), but higher levels of DDTs (9900 ng g⁻¹ lw) and lower levels of HCB (11 ng g⁻¹ lw) in blubber. PCBs/DDTs ratio in samples from the southern population was approximately 4.9, whereas it was about 0.5 for the northern population. The predominance of DDTs over PCBs in organisms from the Cananéia-Paranaguá region is associated with the intensive use of its technical mixture in the area during 1970s–1980s [42]. The same was observed in estuarine dolphins; organisms from the Cananéia/Paranaguá region showed PCBs/DDTs concentrations lower than 1 (~0.65) [43] and those from the Cananéia Estuary showed values even lower than 0.5 (~0.15) [42].

The occurrence of PFCs and PBDEs was also scanned in franciscana dolphins [48,49]. PBDE levels in franciscana blubber from the southern population (13.38–65.02 ng g⁻¹ lw) were upto one order of magnitude lower than PBDE levels found in franciscana from the northern population (93.62–655.86 ng g⁻¹ lw) [50]. The bigger values in samples from the northern population confirmed that PBDEs, as most other POPs, are found at higher concentrations in environmental matrices collected near industrial or highly urbanized centers, as suggested by Lebeuf et al. [51]. Franciscana samples from the northern population were collected from São Paulo coast, the most developed region of Brazil, where Santos and São Vicente Estuarine System is located. Apart from tourism, the region has one of the most important petrochemical, chemical, and metallurgical industrial centers of Brazil (the Cubatão industrial complex). In addition, the largest commercial harbor of South America (Santos Harbor) is located in this area and represents another potential contamination source. The source of these chemicals in the blubber of franciscana might be linked to the use in fire-preventing foams and materials. Moreover, there could be many other sources that include the coastal development, coupled with sewage, shipping, and industrial discharges.

The PBDEs congener pattern distribution was similar for both populations of franciscana. BDE 47 was the major congener, accounting for 46%–99% (average of 66%) of the total PBDE concentrations in both populations, followed by penta-BDE congeners 99 and 100. Although the lack of difference in composition of PBDEs between the two populations suggests an exposure to similar technical mixtures of PBDEs (probably penta-BDE mixture), BDE 47 accounts for more than 80% of the total PBDEs in the southern population, whereas it accounts for about 60% in the northern population. This might suggest that the northern population was more exposed to the primary sources of penta-BDE technical mixture, while franciscana from the southern population was more exposed to already "aged" technical mixture.

PFCs were only analyzed in franciscana livers from southern populations and the predominant compound detected was the perfluoroalkyl sulfonate (PFOS). Concentrations ranged from 3.6 to 42 ng g^{-1} ww (average of 24 ng g^{-1} ww) [49]. PFDS, PFOSA, and PFUnDA were also detected, but at lower levels (≤1.2 ng g^{-1} ww).

Only one study have analyzed OCs using fresh samples collected from a resident population of bottlenose dolphin (*Tursiops truncatus*) from Patos Lagoon Estuary using biopsy darts [52]. Likewise for franciscana dolphins collected in the same region, PCBs were the predominant compounds (7,900–46,000 ng g^{-1} lw), followed by DDTs (342–5,800 ng g^{-1} lw) and CHLs (21–753 ng g^{-1} lw). HCHs and Dieldrin were also detected, but at lower average concentrations (50 and 5 ng g^{-1} lw, respectively). However, these values were higher than those detected for franciscana dolphin from the same region. Since bottlenose dolphins feed on larger prey, this might have resulted in higher bioaccumulation/biomagnification of chlorinated compounds. In addition *T. truncatus* spends more time inside of the estuary and closer to potential sources of contamination.

Other studies have been carried out for other cetacean species along the Brazilian coast, particularly in southern and southeastern Brazil. Nonetheless, some of these studies were conducted with a small number of samples or with samples from one location. For example, samples of *Stenella frontalis* ($n = 2$) and *Delphinus capensis* ($n = 1$) were collected from the Cananéia/Paranaguá region [43]. Contamination levels were, respectively, 59,000/17,000 ng g^{-1} lw for PCBs (sum of 101 congeners), 36,500/11,000 ng g^{-1} lw for DDTs (sum of *p,p'*-DDE, *p,p'*-DDT, and *p,p'*-DDD), 657/200 ng g^{-1} lw for CHLs (sum oxychlordane, *cis*-chlordane, *trans*-chlordane, *cis*-nonachlor, and *trans*-nonachlor), 275/200 ng g^{-1} lw for dieldrin, 77.5/32 ng g^{-1} lw for HCB, 107/77 ng g^{-1} lw for heptachlor epoxide, and 25.5/24 ng g^{-1} lw for HCHs (sum of γ-, β- and α-isomers). Dorneles et al. [44] have also reported PBDEs in liver samples of five species from the continental shelf: *Stenella frontalis* (96–2440 ng g^{-1} lw), *Pseudorca crassidens* (1210–5960 ng g^{-1} lw), *Tursiops truncatus* (270–1350 ng g^{-1} lw), *Steno bradanensis* (360–1600 ng g^{-1} lw), and *Delphinus delphis* (125–240 ng g^{-1} lw); and four species from the open ocean: *Stenella attenuate* (1215 ng g^{-1} lw), *Stenella longirostris* (150 ng g^{-1} lw), *Stenella coeruleoalba* (210 ng g^{-1} lw), and *Lagenodelphis hosei* (3–28 ng g^{-1} lw). All of these samples were collected between 1994 and 2006 from animals found stranded on beaches of the Rio de Janeiro State, southeastern Brazil. Apparently there is no significant difference in PBDEs load in dolphins from different regions (continental shelf versus open ocean). However, tri-BDE seems to have a higher contribution in *Lagenodelphis hosei* than in other species, especially comparing with those living at the continental shelf. The authors attributed it to the longer transportation capability of more volatile compounds.

Furthermore, there is just one study for mammals from the Northern part of Brazil. DDTs and PCBs concentrations for *Inia geoffrensis* ($n = 4$) collected at the Amazon Basin ranged from 190 to 3176 ng g^{-1} lw and from 151 to 3216 ng g^{-1} lw, respectively. The high values of DDTs appeared to be related to its intensive use to combat malaria, but the PCB values were unexpectedly high since this region is not significantly affected by urban and industrial contamination [53].

18.2.3.3.2 Pinnipedes

Fillmann et al. [54] reported PCBs and chlorinated pesticides (DDTs, CHLs, HCHs, and HCB) contamination in various tissues and organs of eight juvenile of South American fur seals (*Arctocephalus australis*) found stranded on beaches from southern Brazil. Higher concentrations were detected in blubber, although significant levels of organochlorine compounds were also seen in other tissues/organs (such as liver, kidney, skin, adrenal gland, bladder, and heart). The poor nutrition status of these animals has probably allowed for the remobilization of organochlorines stored in lipids to other tissues throughout the body. Concentrations in blubber samples ranged from 940 to 4390 ng g^{-1} lw for PCBs (sum of 55 congeners), 20 to 2000 ng g^{-1} lw for DDTs (sum of *p,p'*-DDE, *o,p'*-DDE, *p,p'*-DDT, *o,p'*-DDT, *p,p'*-DDD, *o,p'*-DDD), 4.9 to 600 ng g^{-1} lw for CHLs (sum of *trans*-chlordane, *cis*-chlorande, *trans*-nonachlor, and *cis*-nonachlor), <0.04 to 5.9 ng g^{-1} lw for HCH (sum of α-, β-, γ-, and δ-isomers), and 0.22 to 4.1 ng g^{-1} lw for HCB.

PCBs and chlorinated pesticides were also analyzed in livers of 19 adult of sub-Antarctic fur seals (*Arctocephalus tropicalis*) found stranded on beaches from southern Brazil [55]. PCBs were the predominant compounds (15,000 ng g^{-1} lw), followed by DDTs (8,400 ng g^{-1} lw), CHLs (530 ng g^{-1} lw), HCHs (400 ng g^{-1} lw), and HCB (50 ng g^{-1} lw). The differences in levels found in both species are probably related to the difference in age of the two groups (juveniles and adults, respectively).

PBDEs were also analyzed in livers of sub-Antarctic fur seals. Among the 10 analyzed PBDEs (BDE 28, 47, 66, 85, 99, 100, 138, 153, 154, and 183), only BDE 47 was detected in concentrations ranging from <MDL to 1100 ng g^{-1} lw.

18.3 METHODS

Despite recent improvements, the small number of studies of POCs in Brazilian environments results from the historical lack of facilities and experts in the subject, as well as lack of resources (money) to cover for the expenses. In addition, there were difficulties in obtaining samples, especially from biological samples such as marine mammals, since the capture of these animals for scientific investigations have several practical as well as ethical and legal issues. The main techniques employed to obtain and to analyze the samples of those few studies listed above for the Brazilian coastal areas are summarized below.

18.3.1 SAMPLES AND SAMPLING METHODS

Generally, fish were captured using a gill net, while mussels are manually collected. Sediment grab samplers were used for surface sediment samples and aluminum tube for core sampling. Marine mammal samples were generally collected from stranded animals found on the beaches or animals incidentally caught in gillnet fisheries and brought to the coast by the fishermen. Just one study has used biopsy darts to collect samples from alive specimen of marine mammals.

18.3.2 CHEMICAL ANALYSIS

Most of the studies were based on single samples per collection site, without any duplicate analyses to confirm the results. In general, similar extraction techniques and solvents were used for OCs and PBDEs analyses in abiotic and biotic matrices. Due to its advantages (simplicity and efficiency), Soxhlet extractions with binary solvent mixtures (acetone: *n*-hexane; *n*-hexane:dichloromethane), were the most used technique for both sediment and biotic samples. After extraction, due to the nonselective nature of the extraction procedures and the complexity of the sample matrices, further purification steps were required before the chromatographic analysis. For sediment samples, the clean-up should ensure sulfur removal, which was reached by addition of active Cu (powder or wire). On the other hand, lipid elimination is a more complex process for biotic samples and several steps were necessary. Most techniques involved and solid phase gel or permeation chromatography. Sometimes, acid treatment was used, especially in samples with high lipid content, such as blubber and liver samples. PCBs and chlorinated pesticides were generally identified and quantified by gas chromatography equipped with an electron capture detector (GC-ECD), while PBDEs were identified and quantified by gas chromatography equipped with a mass spectrometry detector (GC-MSD).

Although the above described techniques are very efficient and plentifully used, they have a few limitations that are important to take into consideration. For example, Soxhlet extraction could be substituted by accelerated solvent extraction (ASE), which produces fewer residues (solvent) and efficiently reduces considerably extraction times [56]. The use of rotary evaporators, combined with nitrogen blowdown, were the most commonly used techniques to reduce the extracts volume during many analytical steps. However, the Kuderna-Danish method is the least sensitive for cross contamination and it also allows for more extracts to be handled at the same time [56]. The Kuderna-Danish method is also preferred over the Turbovap system, since bigger losses were reported for organochlorine pesticides using nitrogen flow. In addition, it is possible to reduce extract volumes to below 100 µl by using a conical Kuderna-Danish receiving flask.

The application of ECD detectors for PCB analyses is straightforward. It is an attractive detector due to its low costs, high sensitivity, and also provides fairly simple and easy-to-interpret chromatograms. However, its selectivity is limited and can be biased since it is sensitive to electronegative interferents, which could coelute with the analytes of interest. Therefore, MS detectors are preferred for accurate determinations (less ambiguous identification), although they present lower sensitivity than ECD when operated in electron impact mode. Operating in negative ionization mode will improve sensitivity but only for higher chlorinated compounds.

GC-MS has been broadly used for analyzing PBDEs, with $[M - Br2]+$ and $[M]+$ being the most commonly monitored ions. They provide not only good selectivity, but also lower sensitivity, especially for the higher PBDE congeners (hepta- to deca-BDE). Therefore, the use of large volume injectors or high-resolution GC-MS (HRMS), which provides a better sensitivity and selectivity, however, HRMS needs a much is recommended higher instrumentation investment with higher operational and maintenance costs.

There were no research groups with capability of analyzing PFCs in Brazil. Thus, all analyses so far have been performed in collaboration with foreign laboratories. Overall analyses were performed using liquid chromatography coupled with a triple-quadrupole mass spectrometer (LC-MS-MS) and operating in electrospray negative ion mode.

To accomplish international standards of quality assurance/quality control (QA/QC), some laboratories have reported analytical routines of analyzing blanks, spiked matrices, duplicate matrices, and certified reference materials. Moreover, it is common practice to use surrogate and internal standards to determine recovery rates and quantification, respectively. Moreover, some of the laboratories periodically participate in intercalibration or proficiency test programs to assure the quality of results.

18.4   CONCLUSIONS AND RECOMMENDATIONS

The most studied POCs along the Brazilian coast were the organochlorine compounds (PCBs and chlorinated pesticides). With few exceptions, the PCBs were the predominant compounds among the analyzed POCs with larger amounts of contaminants being detected in the very industrialized and urbanized regions of Brazil, such as São Paulo and Rio de Janeiro States. A slight decrease in organochlorine pesticides levels was detected during a 10-years long study using dolphins from southern Brazil, whereas no trend was seen for PCBs due probably to the remobilization of PCBs from widespread remaining stocks (equipment and goods containing PCBs). However, it is really difficult to appraise the real situation of POCs contamination along the aquatic environments of Brazil, since there are several places with no studies whatsoever.

Overall, due to the lower usage of organochlorine compounds, PCBs and chlorinated pesticides levels at Brazilian coastal areas were somewhat lower (sometimes 2–3 times) than those reported for matrices from the Northern Hemisphere. Nonetheless, this pattern was not observed for the emerging contaminants, such as PBDEs and PFCs. Actually, PFOS concentrations in males of estuarine dolphins from Guanabara Bay were among the highest already detected for cetaceans. Moreover, the exponential increase of PBDEs levels over the last decade in franciscana dolphin from southern Brazil and the extensive use of these compounds as flame retardants in a large number of products, enhancing the number of diffuse sources, suggests continuous increase of environmental levels in the future. These confirm that "modern" contaminants are widespread and equally distributed in the ecosystems of both hemispheres. However, further studies should be conducted to better evaluate this real threat to the environment.

Hence, the challenge for the next decades is to extend the studies over the Brazilian environments in a more comprehensive manner for better identification of sources (primary and secondary; local, diffuse, and global), pathways, and fate of POCs in the environment.

REFERENCES

1. IBGE, 2010. Brazilian Institute for Geostatistics and Geopolitics, www.ibge.br—accessed May 2010.
2. PNUMA, 2002. Evaluación regional sobre substancias tóxicas persistentes: Informe regional de sudamérica oriental y occidental. Programa de Las Naciones Unidas para el Medio Ambiente, UNEP, Switzerland.
3. Brasil, 1971. Ministério da Agricultura, Portaria n° 356. Diár. Oficial União, Brasília, 15 out. 1971, Seç. 1, p. 8318.
4. Brasil, 1971. Ministério da Agricultura. Portaria n° 357. Diár. Of. União, Brasília, 15 out. 1971, Seç. 1, p. 8318.
5. MDIC, 2009. Importação Brasileira. http://aliceweb.mdic.gov.br (accessed June 2010).
6. Penteado, J. C. P. and Vaz, J. M. 2001. O Legado das Bifenilas Policloradas (PCBs). *Química Nova* 24:390–398.
7. Barreto, H. C., Inomata, O. N. K., and Lara, W. H. 1988. Bifenilas policloradas em óleos minerais usados em transformadores. *Revista do Instituto Adolf Lutz* 48:87–92.
8. Jornal Agora, Edition no. 9679, published at August 14, 2010; www.jornalagora.com.br
9. Brasil, 2009. Projeto de Lei do Senado N° 173, de 2009, proposto pelo Senador João Tenório. Estabelece prazo para que computadores, componentes de computadores e equipamentos de informática em geral, comercializados no Brasil, atendam a requisitos ambientais e de eficiência energética. Poder Executivo, Brasília (DF).
10. Figueredo, R. C. R., Ribeiro, F. A. L., and Sabadini, E. 1999. Ciência de espumas—aplicação na extinção de incêndios. *Química Nova* 22:126–130.
11. Zanuncio, J. C., Zanuncio, T. V., Pereira, J. M., and Oliveira, H. N. 1999. Controle de *Atta laevigata* (Hymenoptera: Formicidae) com a isca landrin-f, em área anteriormente coberta com. *Eucalyptus Ciência Rural* 29:573–576.
12. Hermanns, L. 2004. Diagnóstico da Contaminação por Organoclorados na Lagoa dos Patos (RS) e Costa Adjacente. Dissertação de Mestrado em Oceanografia Física, Química e Geológica, Universidade Federal do Rio Grande, 96 pp.
13. SUPRG, 2010. Superintendência do Porto do Rio Grande, www.portoriogrande.com.br, accessed July, 2010.
14. Koike, 2007. Caracterização do Estado de Contaminação por Organoclorados em água, Sedimento e Ostras do Complexo Estuarino da Baía de Paranaguá (Paraná-Brazil). Dissertação de Mestrado em Oceanografia Física, Química e Geológica, Universidade Federal do Rio Grande, 78 pp.
15. Combi, 2009. Bifenilas Policloradas (PCBs) em Colunas Sedimentares da Baía de Paranaguá, PR. Monografia apresentada ao Curso de Oceanografia da Universidade Federal do Paraná, 49 pp.
16. Fillmann, G., da CostaMachado, E., de Castro Martins, C, and Sá, F. 2007. Poluentes Orgânicos Persistentes nos Sedimentos dos Canais de Acesso aos Portos de Paranaguá e Antonina (PR). In *Dragagens Portuárias no Brasil—Licenciamento e Monitoramento Ambiental*, E. B. Boldrini, C. R. Soares, and E. V. de Paula (Eds.), pp. 264–275. Governo do Estado do Paraná (SEMA/PR); Associação de Defesa do Meio Ambiente e Desenvolvimento de Antonina (ADEMADAN); Faculdades Integradas Espírita (UNIBEM).
17. Bícego, M. C., Taniguchi, S., Yogui, G. T., Montone, R. C., da Silva, D. A. M., Lourenço, R. A., de Castro Martins, C., Sasaki, S. T., Pellizari, V. H., and Weber, R. R. 2006. Assessment of contamination by polychlorinated biphenyls and aliphatic and aromatic hydrocarbons in sediments of the Santos and São Vicente Estuary System, São Paulo, Brazil. *Marine Pollution Bulletin* 52:1804–1816.

18. de Souza, A. S., Torres, J. P. M., Meire, R. O., Neves, R. C., Couri, M. S., and Serejo, C. S. 2008. Organochlorine pesticides (OCs) and polychlorinated biphenyls (PCBs) in sediments and crabs (*Chasmagnathus granulata*, Dana, 1851) from mangroves of Guanabara Bay, Rio de Janeiro State, Brazil. *Chemosphere* 73:S186–S192.

19. NOAA, 1991. Second summary of data on chemical concentrations in sediments from the *National Status and Trends Program*, Technical Memorandum NOS OMA 59.

20. Thomas, J. E., Ou, L. T., and Al-Agely, A. 2008. DDE remediation and degradation. Rev. *Environmental Contamination Toxicology* 154: 55–69.

21. Alexander, C., Smith, R., Loganathan, B., Ertel, J., Windom, H. L., and Lee, R. F. 1999. Pollution history of the savannah river estuary and comparisons with Baltic Sea pollution history. *Limnologica* 29:267–273.

22. Tanabe, S. and Subramanian, A. 2006. Mussels, In *Bioindicators of POPs: Monitoring in Developing Countries*, S. Tanabe and A. Subramanian (Eds.). Trans Pacific Press, Melbourne, pp. 38–63.

23. Sericano, J. L., Wade, T. L., Jackson, T. J., Tripp B. W., Farrington, J. W., Mee, L. D., Readmann, J. W., Villeneuve, J. P., and Golberg, E. D. 1995. Trace organic contamination in the Americas: An overview of the US National Status and Trends and the International "Mussel Watch" programmes. *Marine Pollution Bulletin* 31:214–225.

24. IMW Committee (1994). *International Mussel Watch Project: Initial Implementation Phase, Final Report*, J. W. Farrington and B. W. Tripp (Eds.). Woods Hole Oceanographic Institution, Coastal Research Center, Woods Hole, MA.

25. Macauley, J. M., Summers, J. K., Heitmuller, P. T., Engle, V. D., Brooks, G. T., and Babikow, M. 1992. Statistical Summary: EMAP—Estuaries Lousianian Province—1992. USEPA. Environmental Research Laboratory, Gulf Breeze, FL. EPA/620/R-94/002.

26. Tavares, T. M., Rocha, V. C., Porte, C., Barceló, D., and Albaigés, J. 1988. Application of the mussel watch concept in studies of hydrocarbons, PCBs and DDTs in the Brazilian Bay of Todos os Santos (Bahia). *Marine Pollution Bulletin* 19:575–578.

27. Pieroni, M. C. 2009. Difenis Éter Polibromados (PBDEs) em mexilhão *Perna perna* (Linnaeus, 1758) da costa do Rio Grande do Sul. Monografia apresentada ao Curso de Oceanografia da Universidade Federal do Paraná, 37 pp.

28. Quinete, N., Wub, Q., Zhang, T., Yun, H., Moreira, I., and Kannan, K. 2009. Specific profiles of perfluorinated compounds in surface and drinking waters and accumulation in mussels, fish, and dolphins from southeastern Brazil. *Chemosphere* 77:863–869.

29. de Wit, C. A. 2002. An overview of brominated flame retardants in the environment. *Chemosphere* 46: 583–624.

30. Stapleton, H. M., Letcher, R. J., and Baker, J. E. 2004. Debromination of polybrominated diphenyl ether congeners BDE 99 and BDE 183 in the intestinal tract of the common carp (Cyprinus carpio). *Environmental Science & Technology* 38:1054–1061.

31. Gustafsson, K., Björk, M., Burreau, S., and Gilek, M. 1999. Bioacumulation kinetics of brominated flame retardants (polybrominated diphenyl ethers) in blue mussels (*Mytilus edulis*). *Environmental Toxicology and Chemistry* 18:1218–1224.

32. Gorni, R. and Weber, R. R. 2004. Organochlorine pesticides residues and PCBs in benthic organisms of the inner shelf of the São Sebastião Channel, São Paulo, Brazil. *Brazilian Journal of Oceanography* 52:141–152.

33. Ueno, D., Takahashi S., Tanaka, H., Subramanian, A. N., Fillmann, G., Nakata, H., Lam, P. K. S. et al. 2003. Global pollution monitoring of PCBs and organochlorine pesticides using skipjack tuna as a bioindicator arch. *Environmental Contamination Toxicology* 45:378–389.

34. Ueno, D., Kajiwara, N., Tanaka, H., Subramanian, A. N., Fillmann G., Lam P. K. S., Zheng, J. et al. 2004. Global pollution monitoring of polybrominated diphenyl ethers using skipjack tuna as a bioindicator. *Environmental Science & Technology* 38:2312–2316.

35. Ueno, D., Watanabe, M., Subramanian, A. N., Tanaka, H., Fillmann, G., Lam P. K. S., Zheng, J. et al. 2005. Global pollution monitoring of polychlorinated dibenzo-*p*-dioxins (PCDDs), furans (PCDFs) and coplanar polychlorinated biphenyls (coplanar PCBs) using skipjack tuna as bioindicator. *Environmental Pollution* 136:303–313.

36. Ueno, D., Alaee, M., Marvin, C., Muir, D., Macinnis, G., Reiner, E., Crozier, P. et al. 2006. Distribution and transportability of hexabromocyclododecane in the Asia-Pacific region using skipjack tuna as a bioindicator. *Environmental Pollution* 144:238–247.

37. da Silva, A. M. F., Lemes, V. R. R., Barreto, H. H. C., Oliveira, E. S., Alleluia, I. B., de, Paumgarten, J. R. 2003. Polychlorinated biphenyls and organochlorine pesticides in edible fish species and dolphins from Guanabara Bay, Rio de Janeiro, Brazil. Bull. *Environmental Contamination Toxicology* 70:1151–1157.

38. Azevedo e Silva. C. E., de, Azeredo, A., Dias, A. C. L., Costa, P., Lailson-Brito, J., Malm, O., Guimarães, J. R. D., and Torres, J. P. M. 2009. Organochlorine compounds in sharks from the Brazilian coast. *Marine Pollution Bulletin* 58:290–311.

39. Cunha, H. A., da Silva, V. M. F., Lailson-Brito, J., Santos, M. C. O., Flores, P. A. C., Martin, A. R., Azevedo, A. F., Fragoso, A. B. L., Zanelatto, R. C., and Sole-Cava, A. M. 2005. Riverine and marine ecotypes of Sotalia dolphins are different species. *Marine Biology* 148:449–457.

40. Santos, M. C. O., Acuña, L. B., and Rosso, S. 2001. Insights on site fidelity and calving intervals of the marine tucuxi dolphin (*Sotalia fluviatilis*) in southeastern Brazil. *Journal of the Marine Biological Association of the UK* 81:1049–1052.

41. Lailson-Brito, J., Dorneles, P. R., Azevedo-Silva, C. E., Azevedo, A. F., Vidal, L. G., Zanelatto, R. C., Losinski, C. P. C. et al. 2010. High organochlorine accumulation in blubber of Guiana dolphin, Sotalia guianensis, from Brazilian coast and its use to establish geographical differences among populations. *Environmental Pollution* 158:1800–1808.

42. Yogui, G. T., Santos, M. C. O., and Montone, R. M. 2003. Chlorinated pesticides and polychlorinated biphenyls in marine tucuxi dolphins (*Sotalia fluviatilis*) from the Cananéia estuary, southeastern Brazil. *The Science of the Total Environment* 312:67–78.

43. Kajiwara, N., Matsuoka, S., Iwata, H., Rosas, F. C. W., Fillmann, G., and Readman, J. W. 2004. Contamination by persistent organochlorines in cetaceans incidentally caught along Brazilian Coastal waters. *Archives of Environmental Contamination and Toxicology* 46:124–134.

44. Dorneles, P. R., Lailson-Brito, J., Azevedo, A., Meyer, J., Vidal, L. G., Fragoso, A. B., Torres, J. P., Malm, O., Blust, R., and Das, K. 2008. High accumulation of perfluorooctane sulfonate (PFOS) in marine tucuxi dolphins (*Sotalia guianensis*) from the Brazilian Coast. *Environmental Science & Technology* 42:5368–5373.

45. Dorneles, P. R., Lailson-Brito, J., Dirtu, A. C., Weijs, L., Azevedo, A. F., Torres, J. P. M., Malm, O. et al. 2010. Anthropogenic and naturally-produced organobrominated compounds in marine mammals from Brazil. *Environment International* 36:60–67.

46. Secchi, E. R., Wang, J. Y., Murray, B., Rocha-Campos, C., and White, B.N. 1998. Population differentiation in the Franciscana (*Pontoporia blainvillei*) from two geographic locations in Brazil as determined from mitochondrial DNA control region sequences. *Canadian Journal Zoology* 76:1622–1627.

47. Pinedo, M. C. 1991. Development and variation of the Franciscana, *Pontoporia blainvillei*. PhD thesis, Santa Cruz, University of California, Estados Unidos, 406 pp.

48. Leonel, J., Sericano, J. L., Fillmann, G., Secchi, E, and Montone, R. C. 2010. Long-term trends of polychlorinated biphenyls and chlorinated pesticides in franciscana dolphin (*Pontoporia blainvillei*) from Southern Brazil. *Marine Pollution Bulletin* 60:412–418.

49. Leonel, J., Kannan, K., Tao, L., Fillmann, G., and Montonr, R. C. 2008. A baseline study of perfluorochemicals in Franciscana dolphin and Subantarctic fur seal from coastal waters of Southern Brazil. *Marine Pollution Bulletin* 56:770–797.

50. Leonel, J. 2007. Ocorrência de poluentes orgânicos persistentes em *Pontoporia blainvillei*. PhD thesis, Instituto Oceanográfico, Universidade de São Paulo, 100 pp.

51. Lebeuf, M., Gouteux, B., Measures, L., and Trottier, S. 2004. Levels and temporal trends (1988–1999) of polybrominated diphenyl ethers in beluga whales (*Delphinapterus leucas*) from the St. Lawrence Estuary, Canada. *Environmental Science & Technology* 38:2971–2977.

52. Lago, C. F. 2006. Organoclorados na população do cetáceo *Tursiops truncatus* (Montagu, 1821) do extremo sul do Brasil. *Monografia apresentada ao Curso de Oceanografia da Universidade Federal do Rio Grande*. 77 pp.

53. Lailson-Brito, J., Dorneles, P. R., da Silva, V. M. F., Martin, A. R., Bastos, W. R., Azevedo-Silva, A. E., Azevedo, A. F., Torres, P. M., and Malm, O. 2008. Dolphins as indicators of micropollutant trophic flow in amazon basin. *Oecologia Brasiliensis* 12:531–541.

54. Fillmann, G., Hermanns, L., Fileman, T. W., and Readman, J. W. 2007. Accumulation patterns of organochlorines in juveniles of *Arctocephalus australis* found stranded along the coast of Southern Brazil. *Environmental Pollution* 146:262–267.

55. Leonel, J. 2004. Análise da Contaminação por Organoclorados no Lobo marinho Subantártico (*Arctocephalus tropicalis*). Monografia apresentada ao Curso de Oceanografia da Universidade Federal do Rio Grande, 42 pp.

56. Leeuwen, S. P. J. and de Boer, J. 2008. Advances in the gas chromatographic determination of persistent organic pollutants in the aquatic environment. *Journal of Chromatography A* 1186:161–182.

19 Contamination Profiles and Temporal Trends of Persistent Organic Pollutants in Oysters from the Gulf of Mexico

Jose L. Sericano* and Terry L. Wade

CONTENTS

19.1 Introduction 431
19.2 Bivalve Mollusks as Monitoring Organisms 432
19.3 Mechanism of Bioconcentration 433
    19.3.1 Single-Compartment Model 433
    19.3.2 Accumulation and Depuration of Persistent Organic Pollutants in the Field 435
19.4 Contamination Profiles and Temporal Trends of Persistent Organic Pollutants in the
    Gulf of Mexico 438
    19.4.1 NOAA's National Status and Trends "Mussel Watch" Project 438
        19.4.1.1 Chlorinated Pesticides 438
        19.4.1.2 Polychlorinated Biphenyls 447
        19.4.1.3 Polycyclic Aromatic Hydrocarbons 449
19.5 Case Studies 458
    19.5.1 Galveston Bay, Texas 458
    19.5.2 Tampa Bay, Florida 462
19.6 Conclusions 463
Acknowledgment 464
References 465

19.1 INTRODUCTION

Chemical contamination of the coastal environment has been an issue of concern for scientists and politicians worldwide for the last several decades. Growing urban areas and diversifying industrialization in many countries produce tremendous amounts of municipal and industrial wastes that are introduced into the aquatic system; in addition, significant amounts of chemical contaminants are added from non-point sources. The coastal environment is particularly at risk from the degradative effects of these contaminants that might result in changes that are deleterious, over the long term, to both the integrity of the coastal environment and to human health. In his editorial in *Marine Pollution Bulletin*, Professor Goldberg [1] emphasized that the world ocean continued to receive significant amounts of wastes while the implementation of many proposed global marine monitoring programs was delayed. Without a systematic attempt to assess the marine levels of

* E-mail: jlsericano@geos.tamu.edu (Chapter corresponding author).

known contaminants, Professor Goldberg called for a global marine monitoring program utilizing mussels.

Since the earlier studies in the 1970s [2–5], the use of indigenous oysters has become standard practice to monitor environmental levels of selected contaminants in the coastal environment by various national and international monitoring programs. The use of bivalve mollusks in biomonitoring studies were conducted mostly in the United States (237 reports), followed by Australia, France, Japan, and Mexico [6,7]. The National Pesticide Monitoring Program (NPMP) (1965–1972), the first national monitoring programs in the United States, measured monthly concentrations of 15 selected persistent chlorinated pesticides in nearly 8100 bivalves (clams, mussels, and oysters) samples from a large number of locations along the coasts of 15 states, including the states bordering the Gulf of Mexico [3]. Subsequently, the U.S. Environmental Protection Agency (EPA) initiated the "Mussel Watch" program (1976–1978), which monitored the concentrations of selected contaminants in bivalves from nearly 100 sites; 10 of those locations were located along the northern coast of the Gulf of Mexico [8]. The National Oceanic and Atmospheric Administration (NOAA) started sampling for the Mussel Watch program in 1986 to assess temporal trends in chemical contamination of estuarine and coastal areas or ecosystems of the United States through the application of the Mussel Watch concept. The Mussel Watch project is one of the seven major components of the National Status and Trends (NS&T) Program created a couple of years earlier to address national concerns regarding the quality of the nation's coastal zones. The strategy of the Mussel Watch project, reviewed in various articles [7,9,10], is to collect and analyze annual samples from indigenous bivalve mollusks (clams, mussels, and oysters) populations within a predetermined size range and time of the year. The Mussel Watch project measures concentrations of selected polycyclic aromatic hydrocarbons (PAHs), polychlorinated biphenyl (PCB) congeners, chlorinated pesticides, butyltin compounds, and trace elements in mussels and oysters from over 300 coastal and estuarine locations along the east, Gulf, and west coasts of the United States, including locations in Alaska, Hawaii, and Puerto Rico (for overall details on sampling locations, see O'Connor and Beliaeff [10]). Several publications in scientific journals have presented the findings of the NS&T Mussel Watch project since it started in 1986 [11–21]. These long-term, broad-coverage monitoring programs provided and continue to produce valuable data on selected contaminants to assess the environmental quality of estuarine and coastal waters nationwide against which deleterious changes related to anthropogenic activities (e.g., oil spills) can be evaluated.

This chapter reviews the contamination profiles and temporal trends of persistent organic pollutants in oysters from the Gulf of Mexico by describing the accumulation and elimination kinetics of these contaminants and examining the findings of the NOAA's Mussel Watch program over the last two decades (Personal communication, G.G. Lauenstein, NOAA). Where applicable, the data is compared to historical data produced during the National Pesticide Monitoring and the EPA Mussel Watch programs.

19.2 BIVALVE MOLLUSKS AS MONITORING ORGANISMS

Phillips and Rainbow [22] summarize the advantages, discussed in greater detail by several authors, of using organisms (i.e., bivalve mollusks) over other environmental matrices (e.g., water, sediment) for monitoring chemical contamination in coastal areas. The rationale behind this approach has also been extensively discussed in the open literature [7,22–25]. In general, a significant correlation exists between the pollutant contents in cosmopolitan, sedentary bivalve mollusks within a stable population and the average pollutant concentrations in the surrounding habitat. Bivalve mollusks have a reasonably high tolerance to many types of pollution and bioaccumulate contaminants to reach concentrations many-fold higher than seawater concentrations. Bivalves are relatively metabolically passive. Many bivalve species are often of a reasonable size, providing adequate tissue for analysis, sufficiently long-lived to allow the sampling of more than 1 year-class, easy to sample and hardy enough to survive in the laboratory to allow studies of pollutant uptake or to be transplanted to other areas for experimentation. Finally, bivalve mollusks are commercially valuable seafood and a measure of chemical contamination is of public health interest. Contaminant concentrations in bivalve mollusks are the result of passive chemical uptake and depuration processes.

19.3   MECHANISM OF BIOCONCENTRATION

The relationship between pollutant concentrations in organisms and their aquatic habitat was first explained as a simple partition process across external membrane surfaces [26]. Since then, the dynamic equilibrium between uptake from and depuration to water, together with the balance between ingested and excreted matter has been widely used to explain bioaccumulation data. After the introduction of the *n*-octanol- water partition coefficient (K_{OW}) to assess the potential bioaccumulation of different organic compounds under equilibrium conditions, several studies have reported a correlation between the concentration factors of organic contaminants in tissues and the logarithms of their K_{OW} coefficients [27–29]. The K_{OW} coefficient has been found to be very useful in predicting the environmental partitioning of some lipophilic compounds, which tend to accumulate and be retained in lipid-rich tissues of bivalve mollusks. This preferential accumulation and retention of persistent organic pollutants led to the argument that a multiple compartment model would better explain the mechanism of bioconcentration. Although some studies have focused on the analysis of lipophilic contaminants in gills, hepatopancreas, or gonads [30,31], most monitoring studies concentrate on the analysis of homogenates of the whole body of the mollusks as a single compartment as discussed below.

19.3.1   SINGLE-COMPARTMENT MODEL

Bioconcentration is defined as the balance between uptake and depuration processes, which may proceed by first-order kinetics characterized by rate constants [32]. The bioconcentration factor (BCF) is defined as the proportionality constant relating the concentration of a chemical in water to the concentration in an aquatic organism at steady state equilibrium. The following characteristics describing the kinetics of bioconcentration using a single-compartment model was adapted from Connell [33]. The one-compartment approach is the mathematical expression of the hydrophobicity model, which considers bioconcentration as the partitioning of a chemical between the exposure media and the lipidic pools of an organisms, and vice versa, with no physical barriers [34]. The general first-order equation that describes the uptake and depuration of lipophilic compounds, such as PAHs and PCBs, as change in concentration with time = rate of accumulation − rate of elimination, is expressed by

$$\frac{dC_t}{dt} = k_u C_w - k_d C_t \tag{19.1}$$

where
   C_t is the concentration in the organism at time t
   C_w is the concentration in water
   k_u and k_d are the uptake and depuration rate constants, respectively

If the net amount of an analyte in water is much larger than the amount taken up by organisms, C_w can be regarded as constant and, by integration and rearrangement,

$$C_t = \left(\frac{k_u}{k_d} \right) C_w (1 - e^{-kdt}) \tag{19.2}$$

This shows that C_t increases with t but the rate of increase decreases with time (Figure 19.1). As time move toward infinity, e^{-kdt} approaches zero. At this instant,

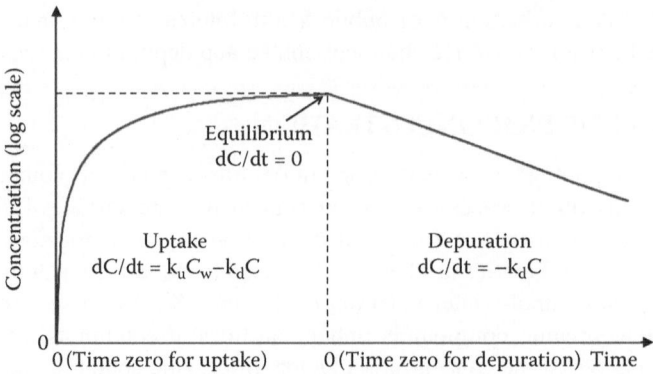

FIGURE 19.1 Semilog representation for the first-order uptake and depuration kinetics, in an organism, of a lipophilic persistent organic pollutant over time.

$$C_{t_\infty} = \left(\frac{k_u}{k_d}\right) C_w \tag{19.3}$$

and

$$\frac{C_{t_\infty}}{C_w} = \left(\frac{k_u}{k_d}\right) = K_b \tag{19.4}$$

where K_b is the BCF at equilibrium (i.e., at t_∞). In practice, an effective equilibrium can be considered at t_{eq} when $C_{t_{eq}}$ is 0.99 of C_{t_∞}. From Equation 19.3,

$$C_{t_{eq}} = 0.99, \quad C_{t_\infty} = 0.99 \left(\frac{k_u}{k_d}\right) C_w \tag{19.5}$$

If exposure to the compound is terminated by transfer of bivalve mollusks to uncontaminated water or, more realistically, to a site where environmental concentrations are negligible, then C_w can be regarded as zero, then Equation 19.1 can be written as

$$\frac{dC_t}{dt} = -k_d C_t \tag{19.6}$$

This indicates that uptake can be neglected in very low concentrations or uncontaminated water. By integration and rearrangement,

$$C_t = C_{t_0} e^{-kdt} \tag{19.7}$$

and

$$\log C_t = \log C_{t_0} - \left(\frac{k_d t}{2.303}\right) \tag{19.8}$$

where C_{t_0} is the initial concentration at the beginning of the depuration period. This predicts that C_t will show a steady decrease with time but with a declining rate of decrease. C_{t_0} and k_d are constants;

therefore, $\log C_t$ is linearly related to time. The half-life, $t_{1/2}$, of a compound can be calculated when half of the initial concentration has been depurated ($C_t = C_{t_0}/2$) from

$$t_{1/2} = \log 2\left(\frac{2.303}{k_d}\right) = \frac{0.693}{k_d} \quad \text{or} \quad \frac{(\ln 2)}{k_d} \tag{19.9}$$

Final contaminant concentrations in bivalve mollusks are the result of passive chemical uptake and depuration mechanisms. Although it is not always possible to quantitatively correlate the concentration observed in bivalve mollusks to their surroundings unless the various rate constants, assimilation efficiencies, and partition equilibriums of contaminants between dissolved and particulate phases are known, it is generally valid to assume that a high concentration in a bivalve mollusk corresponds to an elevated concentration in its environment.

19.3.2 ACCUMULATION AND DEPURATION OF PERSISTENT ORGANIC POLLUTANTS IN THE FIELD

The uptake and depuration rate constants of selected persistent organic pollutants (PAHs, PCBs, and dioxins/furans) in oysters from the Gulf of Mexico under environmental conditions were examined through a translocation experiment in Galveston Bay [19,35,36]. A large number of oysters from a relatively uncontaminated area in Galveston Bay, Hanna Reef, were transplanted to a new, contaminated location near the Houston Ship Channel in the upper part of the bay. The uptake of contaminants by transplanted oysters was followed during 48 days and compared with concentrations in native Ship Channel oysters. After the uptake period, the remaining Hanna Reef oysters were returned to their original location together with a number of indigenous Ship Channel oysters to follow the depuration of contaminants by both populations of oyster.

PAHs and low-molecular-weight (LMW) PCBs, i.e., di-, tri-, and tetrachlorobiphenyls were rapidly bioaccumulated by oysters under environmental conditions (Figure 19.2). Apparent steady state concentrations for these analytes were reached after 20–30 days of exposure. In contrast, high-molecular-weight (HMW) PCBs did not reach an equilibrium plateau at the end of the 49 day exposure period due to relatively high PCB concentrations. However, the still increasing concentrations encountered for these PCBs by the end of the exposure period suggest that, given enough time, equilibrium concentrations would eventually be reached. When back-transplanted to their former location near Hanna Reef, originally uncontaminated oysters released PAHs and LMW PCB congeners at similar rates while the HMW PCB congeners were depurated at considerably slower rates. In neither case, however, was the original background concentration reached after the 50 day depuration period.

Chronically contaminated Ship Channel oysters were also transplanted to the Hanna Reef area during the second phase of the field experiment in Galveston Bay to allow for a direct comparison with newly contaminated Hanna Reef individuals. In general, the observed clearance rates in Ship Channel oysters were slower than those exhibited by Hanna Reef bivalves. The differences might be explained as a consequence of different distributions of PAHs and PCBs in the various body compartments in chronically exposed oysters compared to recently contaminated individuals and better explained by a multiple compartment model or a more effective clearance response by originally uncontaminated oysters. A combination of both of these processes may also be responsible for the observed depuration rate differences.

The study presented evidence to substantiate the theory that the rates of uptake and depuration of PCB congeners by the oyster *Crassostrea virginica* decreases as the number of substituted chlorines in the two phenyl rings increases. However, in spite of their lower uptake rates compared to LMW congeners, the pentachlorobiphenyls were the congeners bioaccumulated to the highest concentrations. It was also observed that although HMW congeners, i.e., heptachlorinated biphenyls or higher, are more lipophilic, they have less favorable steric configurations, which antagonistically

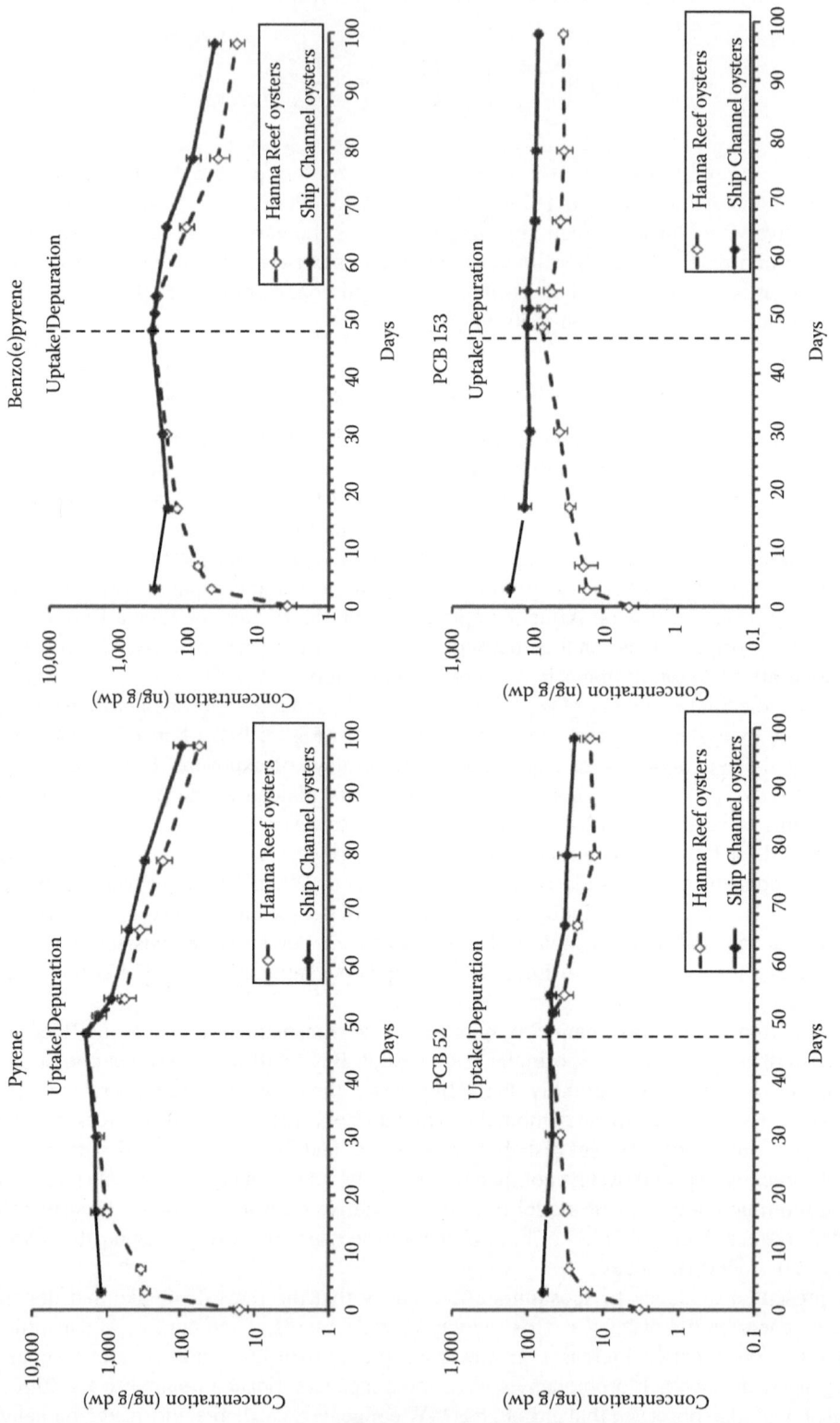

FIGURE 19.2 Uptake and depuration pattern of selected polynuclear aromatic hydrocarbons and polychlorinated biphenyl congeners in tissues of Hanna Reef and Ship Channel oysters during translocation experiments in Galveston Bay (see text for more details).

affected their bioaccumulation and later depuration by oysters. Thus, bioconcentration and clearance of different PCB congeners appear to be more affected by molecular size, e.g., molecular volume and cross-sectional area, which are directly related to the number of chlorines substituted in the two phenyl rings and their substitution patterns rather than by hydrophobicity. The influence of the chlorine substitution patterns in the bioaccumulation of PCBs by oysters is particularly evident in the case of the highly toxic planar congeners, i.e., PCBs 77 and 126. Compared to other PCBs within the same level of chlorination, these planar congeners take a longer time to equilibrate into and out of the organism's lipid pools [18,36].

The identification of the source of PCB congeners can also be confounded by the differential uptake of PCB congeners by oysters. Oysters exposed in the laboratory to a wide molecular range of PCB congeners (1:1:1:1 mixture of Aroclors 1242, 1248, 1254, and 1260), preferentially bioaccumulated congeners with four, five, and six chlorines per molecule resulting in a PCB profile similar to the distribution of homologs that would be encountered in an approximately 2:1 mixture of commercial Aroclors 1248 and 1254. Similar distribution of homologs has been observed in transplanted Hanna Reef oysters during the field study near the Houston Ship Channel in Galveston Bay. Comparatively, the profile of PCB homologs in indigenous Ship Channel oysters, exposed longer to the local levels, had a distribution profile with a slightly larger contribution of Aroclor 1254 (approximately 1:1 Aroclors 1248 and 1254). Although it can be speculated that the profile distributions encountered in chronically contaminated and newly exposed oysters are the result of exposure to Aroclors 1248 and 1254 sources, it seems very probable that the observed profiles are a consequence of the congener uptake discrimination from a more complex mixture. However, it could also be that, even with a more complex mixture of different Aroclors, there was a natural fractionation of the low-, middle- and high-molecular-weight congeners. It is well known that a PCB mixture cannot be considered as a simple chemical contaminant but as a theoretical mixture of 209 congeners with distinctive physicochemical properties that can be environmentally fractionated. The loss of the lowest and the highest molecular weight PCB congeners from a more complex Aroclor mixture by evaporation/dissolution and adsorption/deposition, respectively, after input can result in a profile distribution similar to that of Aroclor 1254 or a mixture of Aroclors, i.e., 1254 with some 1248 or 1260. Therefore, it seems that, independently of the composition of the original PCB mixture, the environmental fractionation together with preferential uptake will often indicate Aroclor 1254 as the most probable contaminant source. Aroclor 1254, one of the most commonly reported PCB mixtures as the source in environmental pollution studies, is not the one that was produced in the largest quantities. The most popular blend in the United States was Aroclor 1242, which comprised over 50% of the total domestic production between 1957 and 1970 [37]. Similarly to what was observed for PCB congeners, oysters exposed in the laboratory to a wide molecular range of PAHs showed the preferential uptake of four- and five-ring PAHs [35]. This observation was confirmed by the results obtained from the field studies in Galveston Bay [35,38].

When exposed to higher concentrations of dioxins and furans, transplanted oysters also accumulate these chemicals but without reaching an equilibrium concentration similar to that of indigenous Ship Channel oysters at the end of the exposure experiment. LMW, less lipophilic compounds accumulated more effectively than HMW dioxins and furans. When recently and chronically contaminated oysters were allowed to depurate, clearance was found to be dependent on molecular weight and log K_{ow} and it proceeded at faster rate with recently contaminated oysters. As discussed earlier for PAHs and PCBs, the tissue concentrations at the end of the depuration period were higher than the original level.

It can be concluded that indigenous oysters can be valuable bioindicators of environmental contamination by trace organic compounds if their limitations are fully understood. Within some limitations, transplanted oysters can be successfully used to monitor environmental contamination by PAHs in areas lacking indigenous bivalves if deployed in-situ for a period of time of at least 30 days; for PCBs, however, a much longer time period, i.e., over 3 months, may be required. On the other hand, the common practice of relocating wild growing oysters that do not comply with sanitary regulations to open waters for 6 weeks, to allow oysters to depurate or cleanse themselves of bacteria, is clearly not enough time to fully depurate high concentrations of highly toxic contaminants.

19.4 CONTAMINATION PROFILES AND TEMPORAL TRENDS OF PERSISTENT ORGANIC POLLUTANTS IN THE GULF OF MEXICO

19.4.1 NOAA's NATIONAL STATUS AND TRENDS "MUSSEL WATCH" PROJECT

Initiated in 1986, the NOAA's NS&T Mussel Watch program collects and analyzes oyster samples annually from selected locations along the three coasts of the United States, including Alaska as well as the islands of Hawaii and Puerto Rico, to characterize coastal zones and to follow the trends in the concentrations of selected contaminants. Sampling activities take place during late autumn though winter (i.e., November–February) to minimize temporal tissue contaminant variability associated with the bivalve's reproductive stage. Depending on water depth, bivalve samples are collected by hand, tongs, or dredge from intertidal to shallow subtidal zones, within a target size range, and brushed clean before shipped to the laboratory. Sampling sites are located from approximately 10–20 km, in estuaries and enclosed coastal areas, to about 70–100 km apart along more open shorelines [7,39]. Many of the locations were selected to coincide with historical sampling sites monitored during national (e.g., NPMP [3], U.S. Mussel Watch [8]) or state (e.g., California Mussel Watch program [40]) programs to allow for a retrospective analysis of trends.

Originally, the NS&T Mussel Watch program was designed to specifically avoid known point sources of contaminant inputs; over the years, the program evolved and new locations were added to test the original sampling design by expanding its coverage to areas closer to urban centers and suspected contamination sources. Similarly, during the earlier years of the NS&T Mussel Watch program, oyster samples were collected at three stations within each site and analyzed separately in triplicate from each location. Starting in 1992, only one pooled sample consisting of oysters from the three stations within the site was collected and, since 1996, pooled samples were collected every other year. Similarly, sediment samples were also collected during the first 2 years of the Mussel Watch project and analyzed for the same suite of analytes; presently, sediment samples are only collected during the initial sampling at new sites added to the program. The list of organic analytes selected for this program (Table 19.1) has also evolved since 1986 to add compounds to a targeted contaminant class (e.g., additional PCB congeners; new chlorinated pesticides) or to incorporate new contaminants (e.g., polybrominated biphenyls [PBBs], polybrominated diphenyl ethers [PBDEs], and antifouling compounds [TBTs]). The information generated by this program is summarized by NOAA and made available to those responsible for managing coastal resources and protecting the coastal environment, to scientists and to the public. Risk assessment of coastal resources is also possible because sampling locations are selected to be representative of general estuarine and nearshore conditions. Currently, bivalve samples are collected nationwide from over twice the number of original locations in 1986 (~300 versus 145) [39]. Similarly, oyster samples (*Crassostrea virginica*) in the Gulf of Mexico are collected, between November and January, from a number of locations in addition to original 51 sites (Figure 19.3). To maintain the contamination profiles and temporal trends of persistent organic pollutants in Gulf of Mexico oysters with respect to the earliest years of the program, the following discussion is limited to the analytes from the original targeted compounds and locations established in 1986. Other organic contaminants added later to the NS&T Mussel Watch program will not be discussed here [41].

19.4.1.1 Chlorinated Pesticides

Median and mean concentrations and the ranges, as well as frequency distributions, for each organic contaminant listed originally as target analytes for the NS&T Mussel Watch program in the Gulf of Mexico and sum of individual compounds within each contaminant class (i.e., total chlordanes,

TABLE 19.1

Persistent Organic Pollutants Originally Included in the NOAA's NS&T "Mussel Watch" Program

Classes	Compounds
Chlorinated pesticides	*Chlorinated benzenes:* Hexachlorobenzene
	Hexachlorocyclohexanes: Gamma-HCH
	Chlordane-related compounds: Heptachlor, heptachlor epoxide, alpha-chlordane, trans-nonachlor
	Cyclodiene pesticides: Aldrin, dieldrin
	DDTs and isomers: o–p′ DDE, *p–p′* DDE, *o–p′* DDD, p-p′ DDD, *o–p′* DDT, *p–p′* DDT
	Other chlorinated pesticides: Mirex
PAHs[a]	*Parent low-molecular-weight PAHs (2 or 3 rings):* Acenaphthene, acenaphthylene, anthracene, biphenyl, fluorene, naphthalene, phenenthrene
	Parent high-molecular-weight PAHs (4 or more rings): Benz[a]anthracene, benzo[b]fluoranthene, benzo[k]fluoranthene, benzo[ghi]perylene, benzo[a]pyrene, benzo[e]pyrene, chrysene, dibenz[a,h]anthracene, fluoranthene, indeno[l,2,3-ca]pyrene, perylene, pyrene
	Individual alkylated PAHs: 1-Methylnaphthalene, 2-methylnaphthalene, 2,6-dimethylnaphthalene, 1,6,7-trimethylnaphthalene, 1-methylphenenthrene
	Groups of alkylated PAHs: C1-chrysenes, C1-dibenzothiophenes, C1-fluoranthenes/pyrenes, C1-fluorenes, C1-naphthalenes, C1-phenenthrene/anthracenes, C2-chrysenes, C2-dibenzothiophenes, C2-fluoranthenes/pyrenes, C2-fluorenes, C2-naphthalenes, C2-phenenthrene/anthracenes, C3-chrysenes, C3-dibenzothiophenes, C3-fluoranthenes/pyrenes, C3-fluorenes, C3-naphthalenes, C3-phenenthrene/anthracenes, C4-chrysenes, C4-naphthalenes, C4-phenenthrene/anthracenes
PCBs[a,b]	*Congeners* **8**/5, **18**, **28**, **44**, **52**, **66**, **101**/90, **105**, **118**, **128**, **138**, **153**/132/168, **170**/190, **180**, **187**, **195**/208, **206**, **209**

Source: Data from Kimbrough, K.L. et al., An assessment of two decades of contaminant monitoring in the nation's coastal zone, NOAA Technical Memorandum NOS NCCOS 74, Silver Spring, MD, 2008.

[a] Currently, 65 PAHs and 51 PCB congeners are quantified by the NS&T program.

[b] PCB congenes in bold are those originally included as target analytes in the NOAA's Program; "*l*" indicates co-elutions.

DDTs, PCBs, and PAHs) are shown in Table 19.2. In the following sections, the geographical distributions of these grouped analytes, ordered from the U.S.-Mexico border (LMSB) to the southernmost site in Florida in the Gulf of Mexico (EVFU), are shown for the first year of the program and every 10 years thereafter to illustrate tendencies. Since locations were sampled every other year after 1985, data at the end of the first and second decade are shown as 1986/1987 and 2006/2007 in order to include all the sites samples in 1986.

Several authors have discussed the importance of the biological processes, in addition to uptake/clearance kinetics, on the persistent organic pollutant concentrations in the bivalve tissues [25,42–46]. Thus, the variations in concentrations detected by short-term (e.g., weekly, monthly) sampling practices can be related to the progression of certain physiological conditions in the bivalves, such as the reproduction stage. In many cases, this can mask the variations due to changes in ambient contaminant concentrations. The collection of bivalve samples from locations in the Gulf of Mexico once a year during a time that is less favorable for the onset of reproduction (i.e., winter) minimizes this effect; therefore, it is assumed, in the following discussion, that concentration variations measured in oyster tissues are mainly due to environmental changes.

19.4.1.1.1 Chlordane-Related Compounds

The NOAA's NS&T Mussel Watch program included, in the original listing of target persistent organic pollutants, only four chlordane-related compounds (Table 19.1) of the more than 140 different

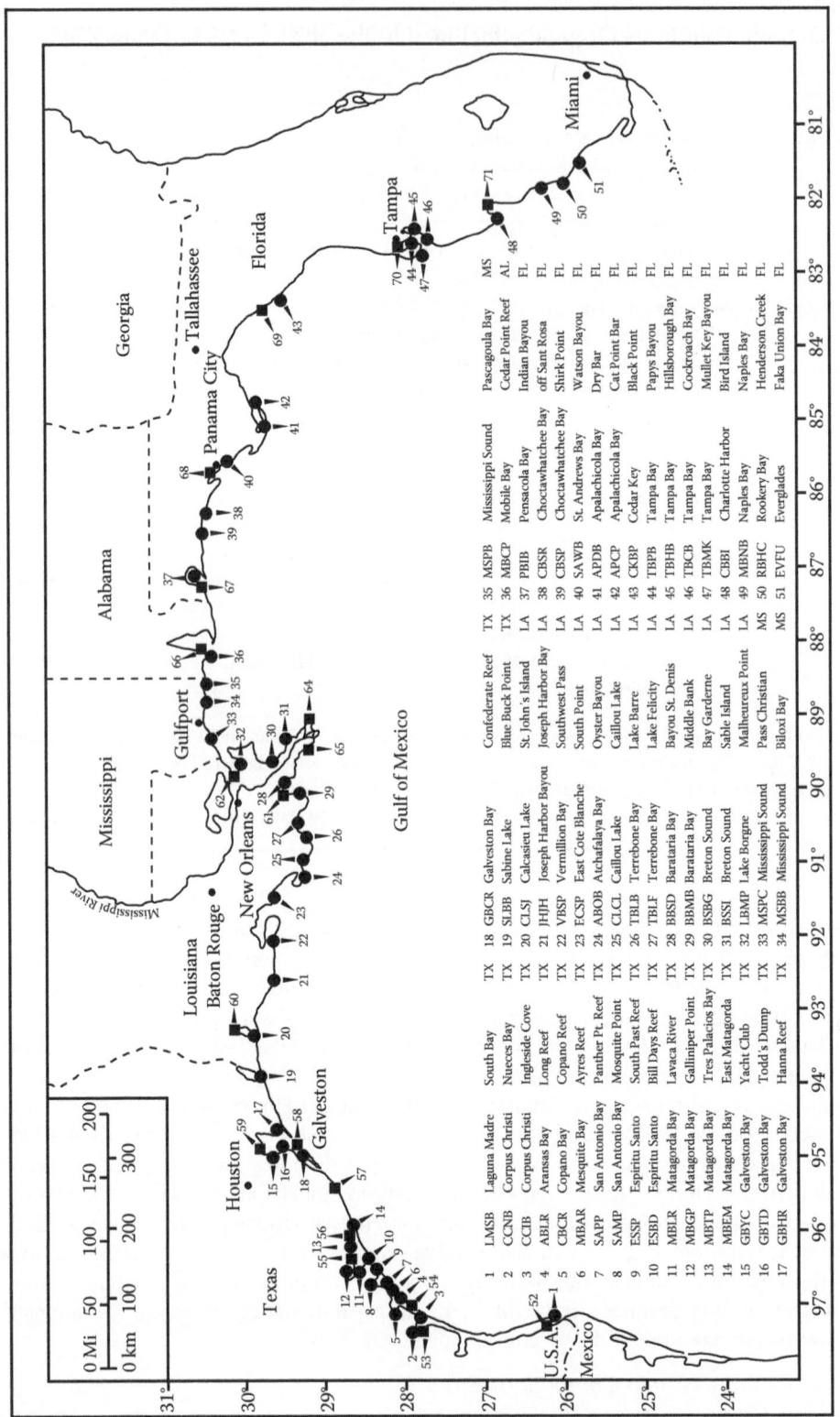

FIGURE 19.3 NOAA's National Status and Trends Mussel Watch program locations in the Gulf of Mexico. Shown are the original sites (1–51) and some of the sites added to the sampling program since 1988 (52–77).

TABLE 19.2
Persistent Organic Pollutant Concentrations and Distribution Frequencies in Oyster Samples Collected in 1986, 1996/1997, and 2006/2007 from the Original Locations in the Gulf of Mexico

Compound	Concentrations (ng/g dw)			Distribution of Concentrations (%)				
	Median	Mean ± Std	Range	<MDL	MDL–<1.00	1.00–<10.0	10.0–<100	>100
1986								
Aldrin	<MDL	<MDL	<MDL-0.86	96	4			
Dieldrin	5.66	8.50 ± 8.63	0.38–42.1		4	75	21	
Hexachlorobenzene	<MDL	<MDL	<MDL-1.06	79	17	4		
Gamma-HCH	0.79	1.00 ± 0.97	<MDL-4.96	19	41	40		
Heptachlor	<MDL	0.34 ± 0.69	<MDL-3.56	71	23	6		
Heptachlor-epoxide	1.74	2.61 ± 3.13	<MDL-19.6	8	19	71	2	
Alpha-chlordane	4.98	10.5 ± 13.2	1.18–55.9			77	23	
Trans-nonachlor	4.66	9.67 ± 13.1	0.83–60.8		2	77	21	
Total chlordanes	12.0	23.1 ± 28.4	2.01–124			33	63	4
Mirex	0.30	1.28 ± 2.56	<MDL-12.0	50	21	25	4	
2,4-DDD	2.29	4.40 ± 6.51	<MDL-28.3	6	13	71	10	
4,4'-DDD	9.07	14.7 ± 20.1	<MDL-96.5	8	2	46	44	
2,4-DDE	<MDL	1.42 ± 7.14	<MDL-49.7	57	31	10	2	
4,4''-DDE	12.3	16.2 ± 13.7	2.68–64.7			46	54	
2,4-DDT	<MDL	0.82 ± 2.22	<MDL-14.5	58	25	15	2	
4,4-DDT	0.63	1.24 ± 1.95	<MDL-9.73	27	33	40		
Total DDTs	25.0	38.9 ± 40.3	2.82–192			8	84	8
Total PCBs	98.7	173 ± 305	13.5–1950				50	50
1996/1987								
Aldrin	0.44	0.49 ± 0.54	<MDL-2.45	40	44	16		
Dieldrin	1.52	2.06 ± 1.84	<MDL-7.15	7	29	64		
Hexachlorobenzene	<MDL	<MDL	<MDL-1.14	94	4	2		
Gamma-HCH	<MDL	0.44 ± 0.60	<MDL-2.79	58	22	20		
Heptachlor	<MDL	<MDL	<MDL-0.89	78	22	0		
Heptachlor-epoxide	0.61	0.80 ± 0.83	<MDL-3.26	33	24	43		
Alpha-chlordane	1.40	2.78 ± 3.92	<MDL-17.3	7	24	60	9	
Trans-nonachlor	1.24	3.15 ± 4.97	<MDL-24.5	2	36	53	9	

(continued)

TABLE 19.2 (continued)

Persistent Organic Pollutant Concentrations and Distribution Frequencies in Oyster Samples Collected in 1986, 1996/1997, and 2006/2007 from the Original Locations in the Gulf of Mexico

Compound	Concentrations (ng/g dw)			Distribution of Concentrations (%)				
	Median	Mean ± Std	Range	<MDL	MDL-<1.00	1.00-<10.0	10.0-<100	>100
Total chlordanes	3.57	6.86 ± 9.32	0.48–43.3		7	77	16	
Mirex	<MDL	0.54 ± 0.92	<MDL-3.89	51	36	13		
2,4′-DDD	0.90	1.63 ± 2.91	<MDL-16.2	27	31	38	4	
4,4′-DDD	1.72	3.14 ± 5.49	<MDL-27.1	24	11	61	4	
2,4′-DDE	0.98	1.80 ± 3.65	<MDL23.3	20	36	42	2	
4,4′-DDE	6.54	11.5 ± 12.1	0.92–54.0		2	54	44	
2,4′-DDT	<MDL	0.27 ± 0.60	<MDL-3.06	73	20	7		
4,4′-DDT	0.68	0.87 ± 0.96	<MDL-5.72	20	51	29		
Total DDTs	15.0	19.2 ± 20.3	3.12–107			42	56	2
Total PCBs	56.1	65.2 ± 50.8	15.8–299				91	9
2006/2007								
Aldrin	<MDL	0.51 ± 2.66	<MDL-17.9	85	11	2	2	
Dieldrin	0.99	4.50 ± 16.2	<MDL-110	33	17	44	4	2
Hexachlorobenzene	<MDL	<MDL	<MDL-3.34	83	13	4		
Gamma-HCH	0.45	0.49 ± 0.31	<MDL-1.46	17	79	4		
Heptachlor	<MDL	<MDL	<MDL-1.89	94	2	4		
Heptachlor-epoxide	<MDL	0.38 ± 0.80	<MDL-3.88	74	11	15		
Alpha-chlordane	1.32	3.64 ± 6.39	<MDL-35.1	4	33	52	11	
Trans-nonachlor	0.63	3.39 ± 6.93	<MDL-34.2	15	48	26	11	
Total chlordanes	1.89	7.48 ± 14.0	<MDL-72.5	2	11	67	20	
Mirex	<MDL	<MDL	<MDL-2.82	82	9	9		
2,4′-DDD	0.48	0.86 ± 1.26	<MDL-5.14	37	41	22		
4,4′-DDD	0.96	2.14 ± 3.58	<MDL-16.2	26	26	41	7	
2,4′-DDE	<MDL	0.61 ± 2.62	<MDL-17.8	70	24	4	2	
4,4′-DDE	5.97	8.02 ± 7.44	0.78–37.7		4	68	28	
2,4′-DDT	<MDL	0.40 ± 0.90	<MDL-4.77	69	20	11		
4,4′-DDT	<MDL	0.46 ± 1.36	<MDL-8.82	67	24	9		
Total DDTs	9.30	12.5 ± 14.3	0.78–72.8		4	50	46	
Total PCBs	22.2	46.4 ± 60.9	1.54–244			24	65	11

components of technical-grade chlordane [47]. *Alpha*-chlordane, heptachlor, and *trans*-nonachlor were dominant constituents in the commercial mixture while heptachlor epoxide, a toxic metabolite of heptachlor, was listed as an impurity. In the discussion that follows the term, "total chlordane" is the sum of targeted chlordane-related compounds in a sample, including heptachlor epoxide.

Two decades after the beginning of the NS&T Mussel Watch program, the distribution of total chlordane-related compound concentrations in oysters from the Gulf of Mexico is well established (Figure 19.4). The highest concentrations were generally detected in oyster samples collected near highly populated urban areas. Samples from Galveston Bay (GBYC), Texas, Mississippi Sound (MSBB), Mississippi, San Andrews Bay (SAWB), Tampa, and Naples (NBNB) Bays, Florida, revealed concentrations two to three times higher than the mean level calculated for the Gulf of Mexico. The detection of higher concentrations in the vicinity of urban areas, in contrast with concentrations measured in predominantly agricultural areas, is consistent with the regulations that have limited the use of chlordane in the United States since the early 1980s. Before then, chlordane was applied directly to the soil or foliage to control a variety of insect pests on a variety of agricultural crops, with the major use for corn, citrus, and vegetables, and nonagricultural uses, including lawn and garden application and termite control, at a rate greater than 3.6 million pounds per year. After 1983, chlordane was only approved for underground termite control. On April 15, 1988, all commercial use of chlordane in the nation was suspended and it was only permitted for fire ant control for underground power transformer.

The concentration distribution profile encountered in 1996/1997 is similar, but with lower concentrations to that observed 10 years earlier. Although concentrations in 2006/2007 continued to be low, there were some locations that showed slight increases. With the exception of two locations within Galveston Bay (GBYC and GBTD), Texas, the lowest average concentrations measured in oysters from the Gulf of Mexico are located to the west of the Mississippi River; this is more clearly seen in the profiles corresponding to 1996/1997 and 2006/2007.

During the first year, *alpha*-chlordane and *trans*-nonachlor were detected in every sample analyzed; 20 years later, all four studied chlordane-related compounds were undetected in some locations. When detected, both analytes comprised 86%–93% of the total measured chlordane-related compounds in the average sample for the Gulf of Mexico and were encountered at comparable concentrations (relative *alpha*-chlordane:*trans*-nonachlor ratio changed from 1.00:0.92 to 1.00:1.10 to 1:00:0.93 for 1986, 1996/1997, and 2006/2007, respectively), which are in fairly good agreement with the relative ratio of *alpha*-chlordane:*trans*-nonachlor:heptachlor (1.00:0.81:0.75) reported for technical grade chlordane [48]. There is, however, a marked depletion in the relative concentration of heptachlor, which fluctuated around 2% of the total chlordane detected in the average oyster. This decrease in the relative concentration of heptachlor is not uncommon for biota tissue and has been reported in earlier studies [49,50]. Heptachlor epoxide decreased from around 11% in 1986 and 1996/1997 to 5% in 2006/2007.

The average concentrations of summed chlordane-related compounds decreased, in general, from their initial levels encountered in 1986 (23.1 ± 28.4 ng/g) to 6.86 ± 9.32 ng/g in 1996/1997, to 7.48 ± 14.0 ng/g in 2006/2007 (Table 19.2). The median concentration, a more robust indication of central tendency than the arithmetic mean [51], decreased from 1986 (12.0 ng/g) to 1996/1997 (3.57 ng/g) to 2006/2007 (1.89 ng/g). This tendency is clearly indicated by the shift in the percentage distribution of concentrations to lower values in 2006/2007. In 1986, total chlordane concentrations were largely dominated by measurements in the 10–100 ng/g range with some values above 100 ng/g. In 2006/2007, 80% of the total chlordane concentrations detected were below 10 ng/g. The same analysis made for individual chlordane analytes shows the same trends.

19.4.1.1.2 DDT Isomers and Metabolites

In spite of the ban on the use of DDT (dichlorodiphenyltrichloroethane) in the United States in the early 1970s, its isomers (*o,p'* and *p,p'*) and closely related compounds, DDD (dichlorodiphenyldichloroethylene) and DDE (dichlorodiphenyldichloroethane), continue to be detected in environmental

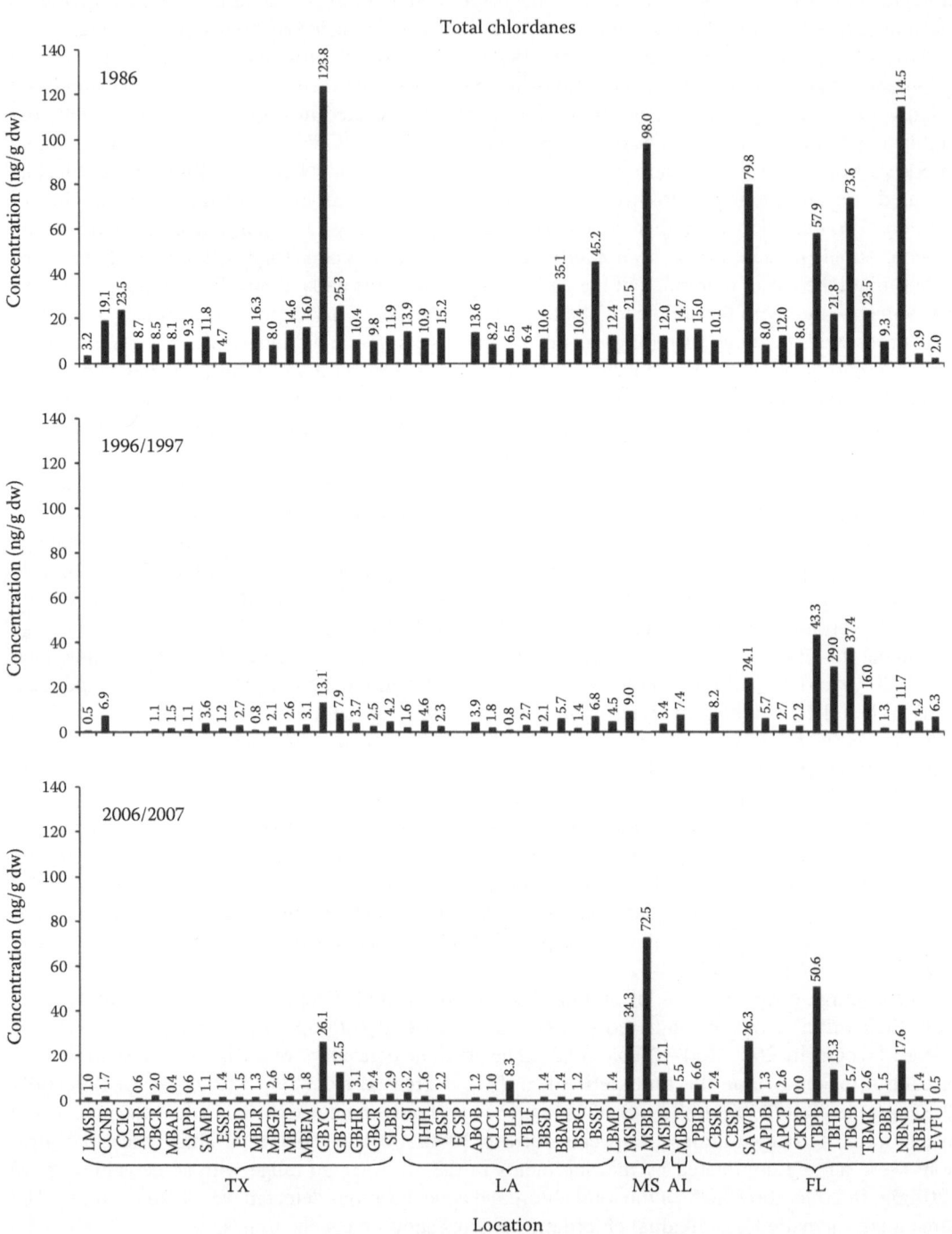

FIGURE 19.4 Average total chlordane-related pesticide concentrations in oyster samples collected in 1986, 1996/1997, and 2006/2007 from the Gulf of Mexico original sampling sites.

samples. In the following discussion, the term "total DDTs" is the sum of all measured DDT-related compounds (*p,p'*-DDT, *o,p'*-DDT, *p,p'*-DDE, *o,p'*-DDE, *p,p'*-DDD, and *o,p'*-DDD) in a sample.

The distribution of concentrations of DDTs and its derivatives encountered in oysters from the Gulf of Mexico during the first year of the NOAA's NS&T Mussel Watch program is similar but at lower concentrations, in 1996/1997 and 2006/2007 (Figure 19.5). Average total DDT concentrations significantly higher than the mean level calculated for the Gulf of Mexico were consistently measured in samples from Galveston Bay (GBYC), Texas, Mobile Bay (MBCP), Alabama, and St. Andrews Bay (SAWB), Florida.

Increased relative contributions of DDD and DDE, the major metabolites and breakdown products of DDT in the environment and only encountered as impurities in the technical DDT product, to the total load of DDT in a sample is regarded as an aged presence of DDT. It is generally accepted that technical DDT contains mainly *p,p'*-DDT (75%) and *o,p'*-DDT (15%) with minor amounts of *o,p'*-DDD (5%), the negligible quantities of *p,p'*-DDE, *o,p'*-DDE, *p,p'*-DDD (less than 0.5% each), and 5% of unidentified compounds [52]. In contrast, DDT isomers in the average Gulf of Mexico oyster accounted for a small fraction (5%–7%) of the total DDTs in the average oyster while isomers of DDE and DDD contributed 45%–69% and 24%–49% respectively. The mean residue composition reported for locations in the Gulf of Mexico during the NPMP pointed to a slightly fresher presence of DDT in the area (50% ± 8.9%, 40% ± 8.8%, and 10% ± 3.6% for DDE, DDD, and DDT respectively) [3,53]. The similar proportions of *p,p'* (81%–85%) to *o,p'* (15%–19%) isomers calculated for the average samples in the Gulf of Mexico and technical DDT (79%:21%) suggest that these isomers are converted or degraded at comparable rates in the marine environment.

The average concentrations of total DDTs decreased from their initial levels encountered in 1986 (38.9 ± 40.3 ng/g) to 19.2 ± 20.3 ng/g in 1996/1997, to 12.5 ± 14.3 ng/g in 2006/2007 (Table 19.2). Similarly, the median concentration decreased from 25.0 to 15.0 to 9.30 ng/g accompanied with a shift in the distribution percentage of concentrations to lower values in 2006/2007.

The only Gulf of Mexico coastal-wide data sets for DDT and its metabolites in oysters to which the NOAA's NS&T Mussel Watch program can be compared are the NPMP (1965–1972) [3] and the U.S. Mussel Watch program (1976–1978) [8]. These comparisons must be exercised with caution as they are generally complicated by substantial differences in analytical methods. Some of the early DDT data, like Butler's [3] for example, might be overestimated because of possible interferences with PCBs.

Fourteen of the NPMP sites, sampled for at least 7 years between 1965 and 1972 and recalculated on a dry weight basis by assuming a water content in the tissues of 85%, and 10 locations sampled in 1976 and in 1977, reported were compared by Sericano et al. [53]. Figure 19.6 extends that comparison to include the average concentrations of total DDTs measured in Gulf of Mexico oysters between 1986 and 2007 by the NOAA's NS&T Mussel Watch program. As discussed in Sericano et al. [53], the average total DDT concentrations reached it maximum in 1968 and declined markedly in the early 1970s probably as a result of the ban on the use of DDT in the United States in 1972. Total DDT average concentrations in recent years are 30–60 times lower than the concentration at its peak. A regression analysis of the data in a lognormal plot allows the calculation of total DDT environmental half-life (see Section 19.3). Assuming that new, fresh DDT is not entering the Gulf of Mexico and that oysters are in equilibrium with environmental concentrations, the half-life of total DDTs can be calculated from the plot using formula (19.9) and a k_d value of 0.087 to be approximately 8 years. This estimation agrees closely with a previous approximation of about 10 years for DDT residues within the biosphere as a whole [54]. DDT and/or metabolites are still entering the Gulf of Mexico from riverine inputs from past applications. The estimated half-life would be longer if these inputs of DDTs are considered.

19.4.1.1.3 Aldrin and Dieldrin

Dieldrin and aldrin are closely related chlorinated pesticides widely used until the 1970s when they were banned. Aldrin oxidizes in soil, on plant surfaces, or in the digestive tracts of insects to the

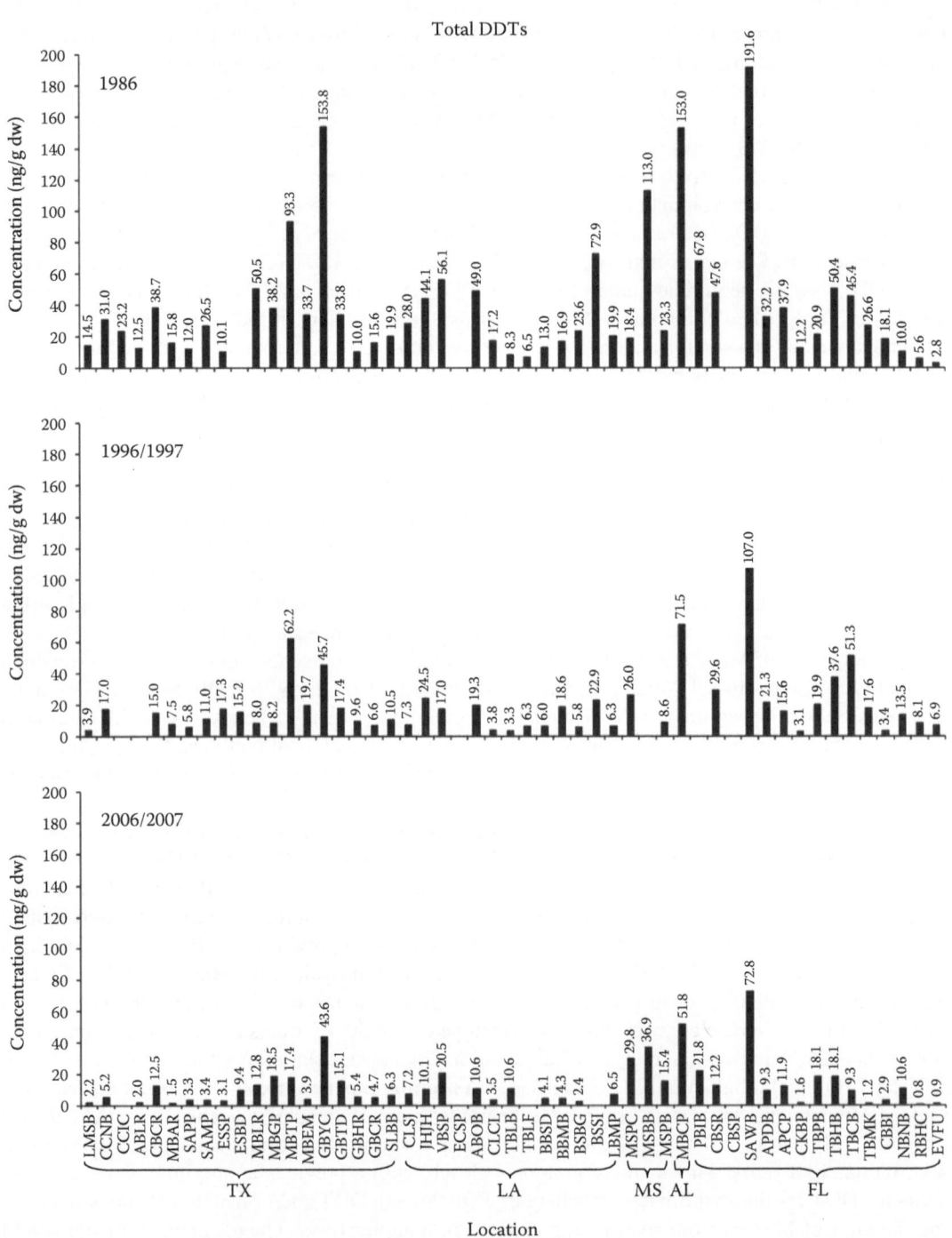

FIGURE 19.5 Average total DDT isomers and metabolites (DDE and DDD) concentrations in oyster samples collected in 1986, 1996/1997, and 2006/2007 from the Gulf of Mexico original sampling sites.

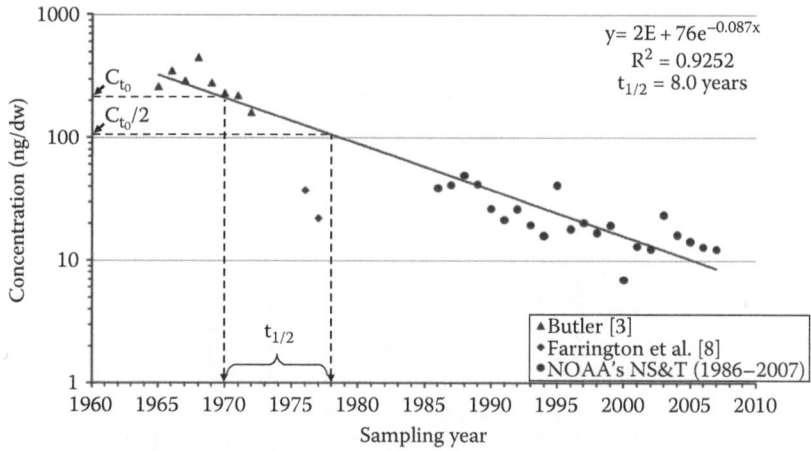

FIGURE 19.6 Time tendency and environmental half-life of total DDT concentrations in Gulf of Mexico oysters.

epoxide dieldrin with more strongly insecticidal properties. The sum of aldrin and dieldrin, as total dieldrin, are used for this discussion.

The distribution and magnitude of total dieldrin concentrations detected during the first year of the NOAA's NS&T Mussel Watch program were not repeated decades later (Figure 19.7). In 1986, dieldrin mean concentration was 8.50 ± 8.63 (Table 19.2), while aldrin averaged below the method detection limit. The profile of total dieldrin concentrations shows the ubiquitous presence of these compounds in Gulf of Mexico oysters. Samples from Corpus Christi (CCIC) and Galveston (GBYC) Bays, Texas, Breton Sound (BSSI), Louisiana, Mississippi Sound (MSBB), Mississippi, and San Andrew Bay (SAWB), Tampa, Florida, had average concentrations greater than the mean calculated for the Gulf of Mexico. With the exception of the recent levels encountered in oyster samples from locations on the Mississippi Sound, concentrations in 1996/1997 and 2006/2007 were significantly lower compared to 1986 levels. The significantly higher concentrations measured in samples from Mississippi Sound, particularly at MSPC and MSBB, deserve further investigation.

19.4.1.1.4 Other Chlorinated Pesticides

Hexachlorobenzene, *gamma*-hexachlorocyclohexane (lindane), and mirex were also analyzed in oyster samples collected during the NOAA's NS&T Mussel Watch program in the Gulf of Mexico. In general, these chlorinated compounds were detected at very low concentrations (Table 19.2) with median values usually below the method detection limits and no clear trends with time. A possible exception could be mirex in oyster samples from locations in Tampa Bay, Florida, in which, although low, concentrations were consistently higher than average values.

19.4.1.2 Polychlorinated Biphenyls

Several methods have been used to measure PCBs in environmental samples. In the past, for example, PCB concentrations have been expressed as the equivalent Aroclor mixtures or as their similar foreign technical formulations, for example, Clophen or Phenochlor. With the introduction of capillary columns and the availability of almost every individual PCB congener for standards, several researchers have attempted to report total PCB concentrations as the sum of all the measurable individual congeners. As some of these congeners are not always separated from other congeners on a single GC capillary column, only a number of different congeners could be accurately analyzed in environmental samples. These congeners should cover all the levels of chlorination and satisfy the condition of good GC separation on a typical SE-54, DB-5, or similar capillary column. A variant of this approach was initially used in reporting the total PCB concentrations in oyster and sediment samples collected from the Gulf of Mexico as part of the NS&T Mussel Watch program.

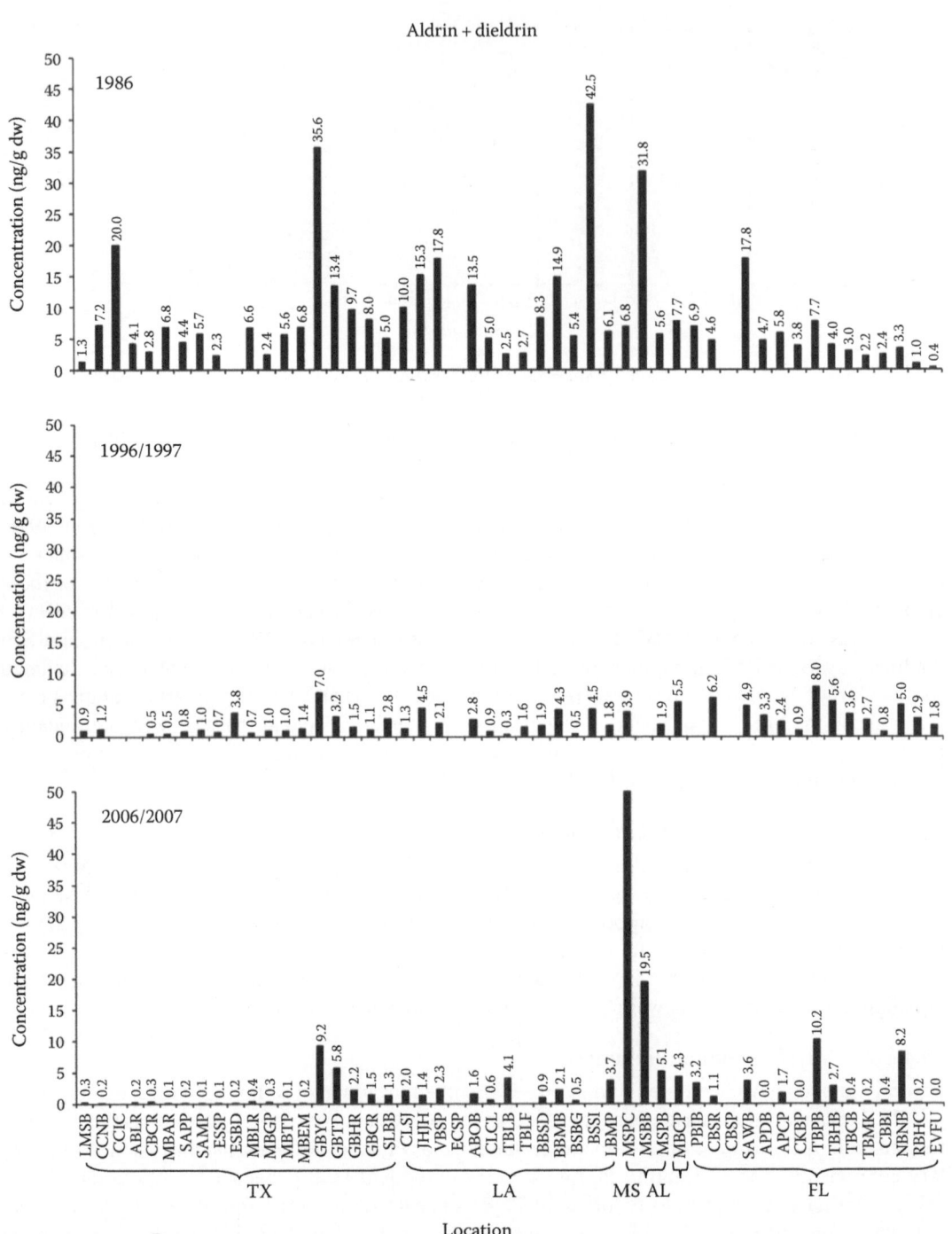

FIGURE 19.7 Average total dieldrin, as the sum of aldrin plus dieldrin, concentrations in oyster samples collected in 1986, 1996/1997, and 2006/2007 from the Gulf of Mexico original sampling sites.

During 1986 and 1987, 18 different congeners (i.e., PCB 8, 18, 28, 44, 52, 66, 101, 105, 118, 128, 138, 153, 170, 180, 187, 195, 206, and 209) were included in the target analyst list (Table 19.1). These congeners, which are some of the major congeners found in commercial Aroclor mixtures, are also among those commonly reported in environmental samples. Eight of these congeners, specified by NOAA (i.e., PCB 8, 28, 52, 101, 153, 170, 195, and 206), were used as reference congeners representative of a given degree of chlorination from dichlorobiphenyls to nonachlorobiphenyls to determine the concentrations of other congeners within each chlorination level. Results were reported as the sum of congeners within each homolog group and total PCB as the sum of these amounts. One obvious problem with this method is the different relative response factors for each congener. Thus, a congener that does not have a standard might be underestimated or overestimated because of the difference between its relative response factor and that of the corresponding representative congener. Discussions among the different participating laboratories in this program directed to improve the PCB reporting led to the adoption of a regression equation that correlates the sum of the 18 individual congener concentrations in the samples with the total PCB loads. Therefore, starting in 1988, total PCB concentrations in oyster samples from the Gulf of Mexico, Atlantic, and Pacific coasts, including Hawaii, were estimated based on the 18 original congener concentrations in the samples and reported using this new approach [35]. Presently, 51 PCB congeners are quantified by the program, including the 18 original PCB congeners.

The distribution of total PCBs in oysters from the Gulf of Mexico is well established (Figure 19.8). The highest concentrations are encountered in oyster samples collected near highly populated urban areas. Samples from Galveston Bay (GBYC), Texas, Mississippi Sound (MSBB), Mississippi, Pensacola (PBIB), and San Andrews (SAWB), and Tampa (TBPB and TNHB) Bays, Florida, show concentrations consistently higher than the average total PCB concentrations calculated for the Gulf of Mexico.

PCB congeners were detected in every sample corresponding to these sampling periods and showed a drastic decrease from the levels detected in 1986, particularly in oysters sampled from GBYC (Galveston Bay, Texas) and SAWB (St. Andrews Bay, Florida). The average total PCB concentrations decreased from 173 ± 305 in 1986 to 65.2 ± 50.8 in 1996/1997 to 46.4 ± 60.9 in 2006/2007 (Table 19.2). This reduction in environmental levels of total PCBs is also reflected in median concentrations (98.7, 56.1 and 22.2 ng/g, respectively) and the shift in the percentage distribution of concentrations. In 1986, total PCB analysis yielded concentrations equally distributed among those between 10 and 100 ng/g and those above 100 ng/g, while in 1996/1997 more than 90% of the determinations encountered concentrations in the 10–100 ng/g range. This tendency to lower concentrations continued to be observed in 2006/2007 with nearly 25% of the locations below 10 ng/g.

19.4.1.3 Polycyclic Aromatic Hydrocarbons

In contrast with chlorinated pesticides, introduced to the environment deliberately, or PCBs, released mostly unintentionally, PAHs occur both naturally and as a result of human activities. Temporal distributions of total PAH concentrations (sum of acenaphthene, acenaphthylene, anthracene, benz[a]anthracene, benzo[a]pyrene, benzo[e]pyrene, benzo[b]fluoranthene, benzo[k]fluoranthene, benzo[g,h,i]perylene, biphenyl, chrysene, dibenzo[a,h]anthracene, fluoranthene, fluorene, indeno[1,2,3-c,d]pyrene, naphthalene, perylene, phenanthrene, and pyrene) in Gulf of Mexico oysters are shown in Figure 19.9 and summarized in Table 19.3. Note that, for this review, the earliest available data for PAHs was that of 1989. Concentrations ranged from 98.9 to 2440 ng/g in 1989, from 17.5 to 2310 ng/g in 1996/1997, and from 24.6 to 2050 ng/g, 2006/2007, with median values of 227, 121, and 70.5 ng/g, respectively. Median concentrations nearly halved every 10 years and a general shifting in the distribution percentage of concentrations to lower values can be observed (Tables 19.3 through 19.5). This decrease in total PAH concentrations detected in oysters from the Gulf of Mexico since 1989 reflects regulatory actions aimed to reduce the release of these chemicals (e.g., recycled and crankcase oil). In spite of an evident temporal decrease in the concentrations of total PAHs (Figure 19.9), the range of concentrations observed during the three sampling

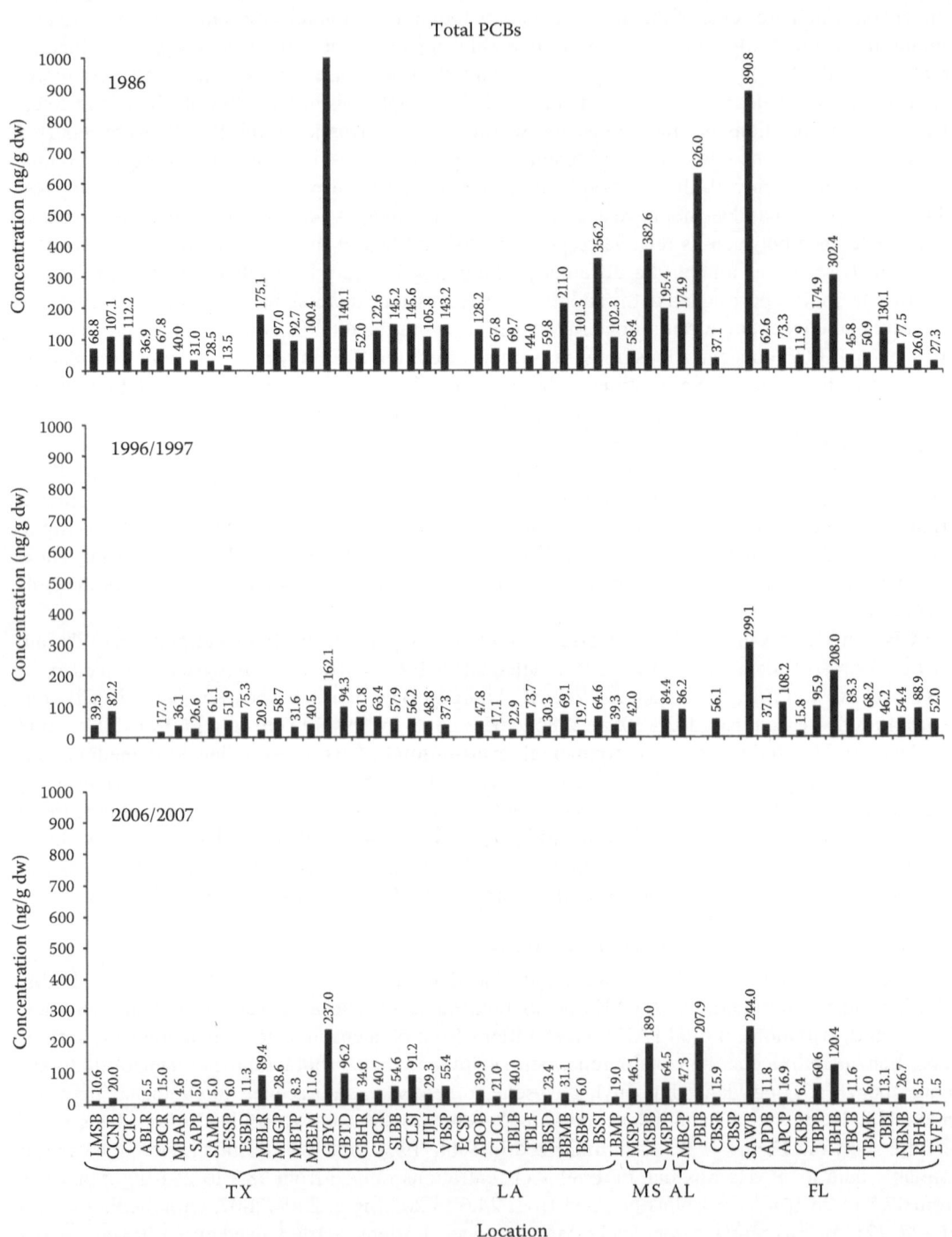

FIGURE 19.8 Average total PCB concentrations in oyster samples collected in 1986, 1996/1997, and 2006/2007 from the Gulf of Mexico original sampling sites.

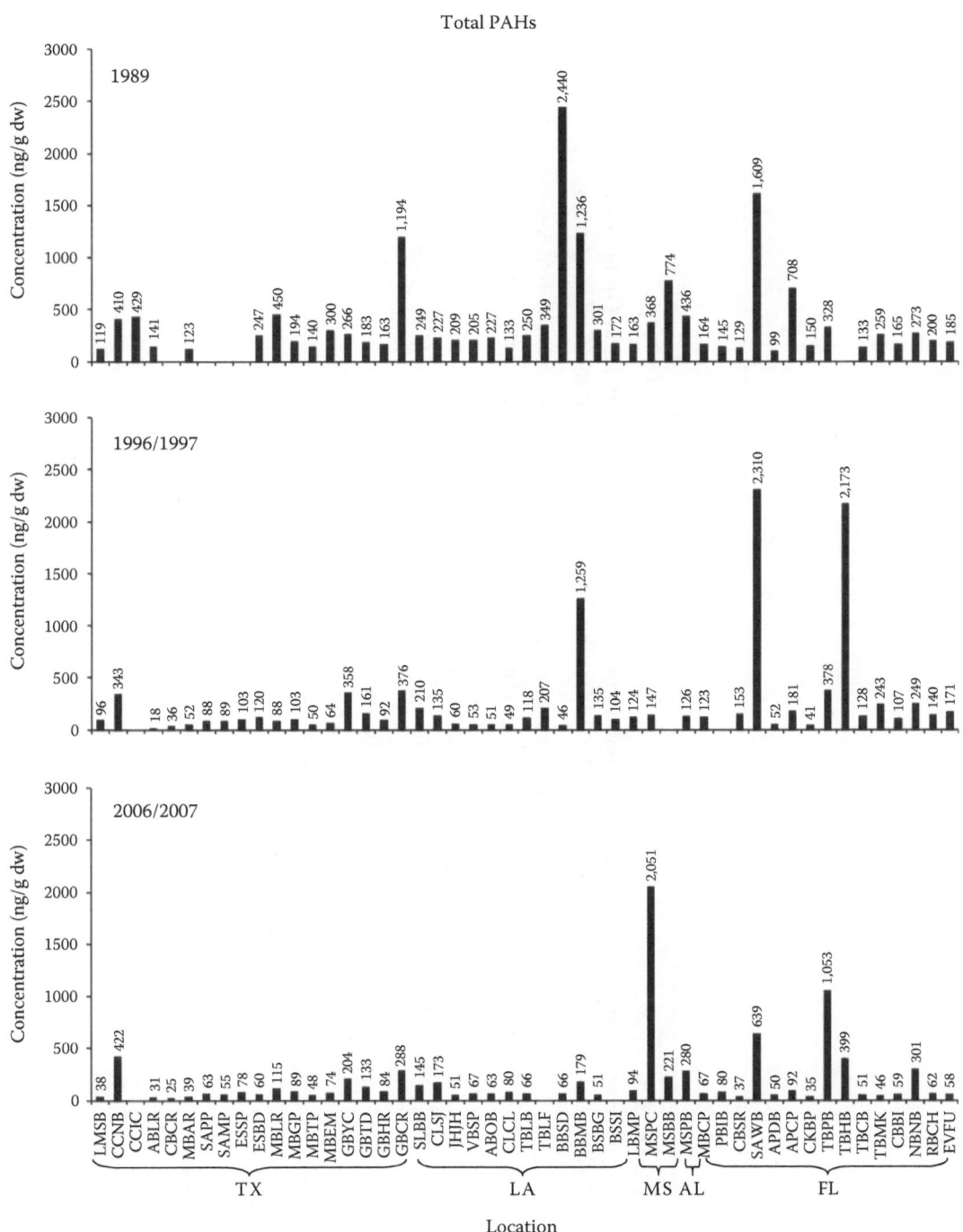

FIGURE 19.9 Average total PAH concentrations in oyster samples collected in 1989, 1996/1997, and 2006/2007 from the Gulf of Mexico original sampling sites.

TABLE 19.3

Polynuclear Aromatic Hydrocarbon Concentrations and Distribution Frequencies in Oyster Samples Collected in 1989, 1996/1997, and 2006/2007 from the Original Locations in the Gulf of Mexico

	Compound	Concentrations (ng/g dw)			Distribution of Concentrations (%)			
		Median	Mean ± Std	Range	<MDL	MDL–<100	100–<1000	>1000
1989	Total PAHs	227	378 ± 448	98.9–2440		2	89	9
	Σ(2 + 3)rings/Σ(4 + 5)rings	0.95	0.97 ± 0.54	0.20–2.89				
1996/1997	Total PAHs	121	250 ± 469	<MDL–2310	2	35	56	7
	Σ(2 + 3)rings/Σ(4 + 5)rings	1.04	1.17 ± 0.71	0.17–2.84				
2006/2007	Total PAHs	70.5	184 ± 336	24.6–2050		67	28	4
	Σ(2 + 3)rings/Σ(4 + 5)rings	1.30	2.50 ± 3.01	0.07–16.1				

TABLE 19.4

Persistent Organic Pollutant Concentrations and Distribution Frequencies in Oyster Samples Collected from 1986 to 2006/2007 at the Original Locations in Galveston Bay

Compound	Concentrations (ng/g dw)			Distribution of Concentrations (%)				
	Median	Mean±Std	Range	<MDL	MDL-<1.00	1.00-<10.0	10.0-<100	>100
GBYC								
Aldrin	<MDL	0.52±0.88	<MDL-2.67	67	13	20		
Dieldrin	14.9	18.0±9.53	7.01-35.6			27	73	
Hexachlorobenzene	0.34	0.36±0.36	<MDL-1.19	40	53	7		
Gamma-HCH	3.15	4.33±3.82	0.71-15.4		20	73	7	
Heptachlor	0.28	0.90±1.72	<MDL-6.55	40	47	13		
Heptachlor-epoxide	8.57	9.25±5.98	2.79-22.9			67	33	
Alpha-chlordane	15.5	20.9±17.1	5.27-59.8			27	73	
Trans-nonachlor	13.9	17.7±13.8	4.52-47.9			40	60	
Total chlordanes	37.0	48.8±37.5	13.1-136				87	13
Mirex	0.63	1.47±2.63	<MDL-10.5	13	53	27	7	
2,4'-DDD	5.86	8.34±8.05	1.21-30.9			87	13	
4,4'-DDD	17.9	35.6±38.5	9.47-154			7	86	7
2,4'-DDE	1.73	2.27±2.82	0.35-11.9		40	53	7	
4,4'-DDE	24.4	34.2±29.2	11.9-129				93	7
2,4'-DDT	1.19	8.73±24.7	<MDL-97.3	20	27	40	13	
4,4'-DDT	1.60	7.32±19.6	<MDL-77.7	7	27	59	7	
Total DDTs	60.9	96.4±117	26.2-500				80	20
Total PCBs	302	471±487	143-1950					100

(continued)

TABLE 19.4 (continued)
Persistent Organic Pollutant Concentrations and Distribution Frequencies in Oyster Samples
Collected from 1986 to 2006/2007 at the Original Locations in Galveston Bay

	Concentrations (ng/g dw)			Distribution of Concentrations (%)					
Compound	Median	Mean ± Std	Range	<MDL	MDL–<1.00	1.00–<10.0	10.0–<100	>100
GBTD								
Aldrin	<MDL	0.38 ± 0.93	<MDL–3.22	74	13	13		
Dieldrin	6.93	7.83 ± 5.00	2.93–23.0			81	19	
Hexachlorobenzene	<MDL	<MDL	<MDL–0.29	94	6			
Gamma-HCH	2.27	2.51 ± 2.02	<MDL–8.03	6	6	88		
Heptachlor	<MDL	0.41 ± 0.57	<MDL–1.94	62	25	13		
Heptachlor-epoxide	3.16	4.43 ± 3.56	1.28–15.9			94	6	
Alpha-chlordane	5.11	7.17 ± 5.63	1.85–22.0			75	25	
Trans-nonachlor	4.93	6.60 ± 4.45	1.72–16.3			75	25	
Total chlordanes	13.7	18.6 ± 13.2	6.03–55.2			19	81	
Mirex	0.31	0.36 ± 0.32	<MDL–1.03	44	50	6		
2,4'-DDD	2.21	2.92 ± 2.74	<MDL–10.3	13	19	62	6	
4,4'-DDD	6.90	9.83 ± 8.04	3.02–31.6			69	31	
2,4'-DDE	0.42	0.94 ± 1.68	<MDL–6.62	43	38	19		
4,4'-DDE	7.70	12.0 ± 9.72	2.55–32.9			69	31	
2,4'-DDT	0.46	2.93 ± 9.17	<MDL–37.3	19	56	19	6	
4,4'-DDT	0.47	1.91 ± 4.93	<MDL–20.2	38	31	25	6	
Total DDTs	19.8	30.5 ± 30.2	7.59–129			13	81	6
Total PCBs	96.1	124 ± 57.4	56.8–233				56	44

TABLE 19.5
Persistent Organic Pollutant Concentrations and Distribution Frequencies in Oyster Samples Collected from 1986 to 2006/2007 at the Original Locations in Tampa Bay

Compound	Concentrations (ng/g dw)			Distribution of Concentrations (%)				
	Median	Mean ± Std	Range	<MDL	MDL-<1.00	1.00-<10.0	10.0-<100	>100
TBHB								
Aldrin	<MDL	<MDL	<MDL-0.72	70	30			
Dieldrin	1.47	1.92 ± 1.42	0.49-4.85		30	70		
Hexachlorobenzene	<MDL	<MDL	<0.25	100				
Gamma-HCH	0.70	0.73 ± 0.41	<MDL-1.29	20	50	30		
Heptachlor	<MDL	<MDL	<MDL-0.89	70	30			
Heptachlor-epoxide	0.53	0.72 ± 0.53	<MDL-1.70	30	40	30		
Alpha-chlordane	4.03	5.51 ± 4.19	<MDL-13.4	10		80	10	
Trans-nonachlor	5.50	5.90 ± 3.71	1.00-13.4			80	20	
Total chlordanes	9.88	12.3 ± 8.30	1.50-29.0			50	50	
Mirex	1.73	1.72 ± 0.88	0.35-3.11		30	70		
2,4'-DDD	1.09	1.70 ± 1.39	0.69-4.61		40	60		
4,4'-DDD	2.64	3.83 ± 3.63	0.84-11.4		10	80	10	
2,4'-DDE	0.34	0.76 ± 1.10	<MDL-3.45	50	20	30		
4,4'-DDE	12.5	12.5 ± 8.32	<MDL-24.7	10		30	60	
2,4'-DDT	0.47	0.88 ± 1.16	<MDL-3.69	30	50	20		
4,4'-DDT	0.25	1.10 ± 1.79	<MDL-5.23	50	20	30		
Total DDTs	15.9	20.7 ± 15.1	3.07-50.4			30	70	
Total PCBs	109	118 ± 84.9	13.2-302				50	50
TBPB								
Aldrin	<MDL	<MDL	<MDL-1.26	66	27	7		
Dieldrin	4.14	5.43 ± 3.23	2.07-11.8			87	13	
Hexachlorobenzene	<MDL	<MDL	<MDL-1.74	80	13	7		
Gamma-HCH	1.18	1.28 ± 0.87	<MDL-2.81	13	27	60		
Heptachlor	<MDL	0.43 ± 0.63	<MDL-1.89	67	20	13		
Heptachlor-epoxide	1.52	2.32 ± 1.76	0.75-6.50		33	67		

(continued)

TABLE 19.5 (continued)
Persistent Organic Pollutant Concentrations and Distribution Frequencies in Oyster Samples Collected from 1986 to 2006/2007 at the Original Locations in Tampa Bay

Compound	Concentrations (ng/g dw)			Distribution of Concentrations (%)				
	Median	Mean ± Std	Range	<MDL	MDL–<1.00	1.00–<10.0	10.0–<100	>100
Alpha-chlordane	10.6	15.1 ± 10.1	5.17–40.9			47	53	
Trans-Nonachlor	17.9	23.3 ± 13.2	8.92–50.6			7	93	
Total chlordanes	28.5	41.2 ± 24.7	15.8–98.8				100	
Mirex	3.27	3.98 ± 2.81	0.90–12.0		7	86	7	
2,4'-DDD	0.85	0.99 ± 0.82	<MDL–2.73	27	27	46		
4,4'-DDD	1.19	8.68 ± 20.9	<MDL–76.0	13	13	61	13	
2,4'-DDE	0.45	0.93 ± 1.15	<MDL–3.92	33	34	33		
4,4'-DDE	13.3	13.7 ± 4.80	5.29–21.9			20	80	
2,4'-DDT	<MDL	<MDL	<MDL–1.59	73	20	7		
4,4'-DDT	0.57	0.93 ± 1.40	<MDL–5.46	40	33	27		
Total DDTs	18.1	25.5 ± 24.8	9.18–103			7	86	7
Total PCBs	68.7	78.6 ± 41.2	29.0–175				80	20
TBCB								
Aldrin	<MDL	0.31 ± 0.49	<MDL–1.54	72	14	14		
Dieldrin	1.33	1.45 ± 1.07	<MDL–3.08	14	29	57		
Hexachlorobenzene	<MDL	<MDL	<MDL–1.06	79	14	7		
Gamma-HCH	<MDL	0.48 ± 0.60	<MDL–1.96	58	21	21		
Heptachlor	<MDL	0.99 ± 2.24	<MDL–7.72	72	7	21		
Heptachlor-epoxide	0.82	1.02 ± 1.06	<MDL–3.83	36	14	50		
Alpha-chlordane	12.2	14.6 ± 13.6	1.79–40.0			43	57	
Trans-nonachlor	16.4	23.6 ± 22.3	3.77–77.5			36	64	
Total chlordanes	31.3	40.2 ± 38.3	5.74–127			21	65	14
Mirex	4.71	6.40 ± 5.35	<MDL–20.9	7	72	21		

2,4'-DDD	0.81	1.23 ± 1.20	<MDL–4.33	14	43	43		
4,4'-DDD	2.95	9.57 ± 16.0	<MDL–57.6	14	14	43	29	
2,4'-DDE	1.46	2.42 ± 2.56	<MDL–8.37	29	21	50		
4,4'-DDE	38.2	34.8 ± 18.8	8.63–68.5			7	93	
2,4'-DDT	<MDL	0.75 ± 1.55	<MDL–5.42	65	14	21		
4,4'-DDT	<MDL	0.52 ± 0.96	<MDL–3.40	65	21	14		
Total DDTs	45.9	49.3 ± 32.6	9.32–117			7	86	7
Total PCBs	42.8	50.3 ± 35.2	11.6–112				86	14
TBMK								
Aldrin	<MDL	<MDL	<MDL–1.00	80	13	7		
Dieldrin	1.65	2.03 ± 1.66	<MDL–5.36	7	27	66		
Hexachlorobenzene	<MDL	<MDL	<MDL–0.42	87	13			
Gamma-HCH	0.43	0.57 ± 0.58	<MDL–2.04	40	40	20		
Heptachlor	<MDL	<MDL	<MDL–0.90	73	27			
Heptachlor-epoxide	0.47	1.18 ± 2.18	<MDL–8.74	33	40	27		
Alpha-chlordane	5.35	5.57 ± 3.49	1.58–11.3			87	13	
Trans-nonachlor	6.57	5.80 ± 3.75	0.64–11.5		7	73	20	
Total chlordanes	13.2	12.7 ± 8.04	2.62–23.5			47	53	
Mirex	1.16	1.26 ± 0.98	<MDL–3.63	13	33	54		
2,4'-DDD	0.39	1.11 ± 1.67	<MDL–5.83	33	40	27		
4,4'-DDD	2.37	4.12 ± 4.88	<MDL–17.9	13	20	54	13	
2,4'-DDE	0.88	1.31 ± 1.58	<MDL–5.24	40	20	40		
4,4'-DDE	5.39	5.70 ± 3.81	0.90–12.1		7	73	20	
2,4'-DDT	<MDL	0.30 ± 0.63	<MDL–1.94	74	13	13		
4,4'-DDT	<MDL	0.76 ± 1.20	<MDL–4.61	53	27	20		
Total DDTs	9.87	13.3 ± 9.67	1.17–28.6			53	47	
Total PCBs	40.9	38.3 ± 24.4	6.04–94.3			20	80	

years are comparable. PAHs are found in fossil fuels, produced during wood burning and waste incineration, and accidentally released during production, transportation, and use of oil in addition to the oil released from natural seepage in the Gulf of Mexico; any combination of these could be the sporadic source of PAHs to the Gulf of Mexico oysters. Qian et al. [38] suggested that PAHs from different sources may show dissimilar bioavailability which, in turn, play an important role in determining their concentrations and compositions in oysters. By comparing concentrations in oysters from Galveston Bay to levels and composition of PAHs in sediments, as a reference for PAH sources, it was concluded that oysters collected from an area with predominantly pyrogenic sources of PAHs had a composition similar to that observed in sediments while bivalves and sediment collected from sites with mixed sources of PAHs from both petroleum-related and combustion sources (predominantly low- and high-molecular-weight hydrocarbons, respectively) had dissimilar PAH compositions. When the complete suite of PAHs is present, exposed and field oysters were shown to preferentially uptake four- and five-ring PAHs [19,35]. The median ratio of low (sum of two- and three-ring PAHs) to high (sum of four- and five-ring PAHs) calculated for Gulf of Mexico oysters collected in 1989 indicates a slightly higher pyrogenic signature of PAHs (i.e., ratio lower than 1), which gradually changed to a more petrogenic overall predominance in 1996/1997 and 2006/2007. This observation, which is more evident in locations having low total PAH concentrations, together with a general decrease of total PAHs levels in recent years, might be the result of a faster decrease of PAH inputs from combustion compared to petrogenic sources. If combustion PAHs are associated with carbon black, they may be depurated more slowly than petrogenic PAHs. In any case, if oysters preferentially accumulate combustion-derived PAHs, i.e., four- and five-ring compounds, compared to petroleum-derived PAHs, i.e., two- and three-ring compounds, then the oyster's concentrations may not accurately represent the contamination at a site. This particular area requires further attention.

19.5   CASE STUDIES

Results discussed in this section describe the concentrations and annual trends of persistent organic pollutants in oysters from two important areas in the Gulf of Mexico: Galveston and Tampa Bays. Galveston Bay, surrounded by the Houston–Sugar Land–Baytown metropolitan area, is the sixth-largest urban area in the United States and home to the second-busiest port in the nation. Tampa Bay boasts the largest port in Florida and the 10th largest in the nation. Conservation efforts in both bays have been enacted to reverse decades of unrestricted pollution and to substantially improve water quality.

19.5.1   GALVESTON BAY, TEXAS

Galveston Bay, with a surface area of $1699 \, km^2$, constitutes the second-largest estuary in Texas and one of the most economically important estuaries along the northern coast of the Gulf of Mexico. The bay water is shallow, averaging only about 2 m in depth, with a restricted exchange with the Gulf of Mexico. Its average water resident time is about 40 days [55]. For many years now, this area has been the recipient of various environmental contaminant inputs because of its aggressively growing urban and industrial development. The Dallas-Fort Worth metroplex, with an estimated population of about 4 million, is located 640 km up the Trinity River from Galveston Bay. The city of Houston, with an estimated population of over 3 million, adjacent to the upper Galveston Bay, together with Texas City and Galveston is a highly industrialized area, especially by the petroleum, petrochemical, and chemical industries. For this reason, NOAA selected to monitor the spatial and temporal trends of organic contaminants in bivalves from Galveston Bay as part of NOAA's NS&T Mussel Watch project. In 1986, the NS&T project collected samples from four locations in Galveston Bay (Yacht Club, Todd's Dump, Confederate Reef, and Hanna Reef) and added two more locations (Houston Ship Channel, Offatts Bayou) in later years (Figure 19.10). From 1986

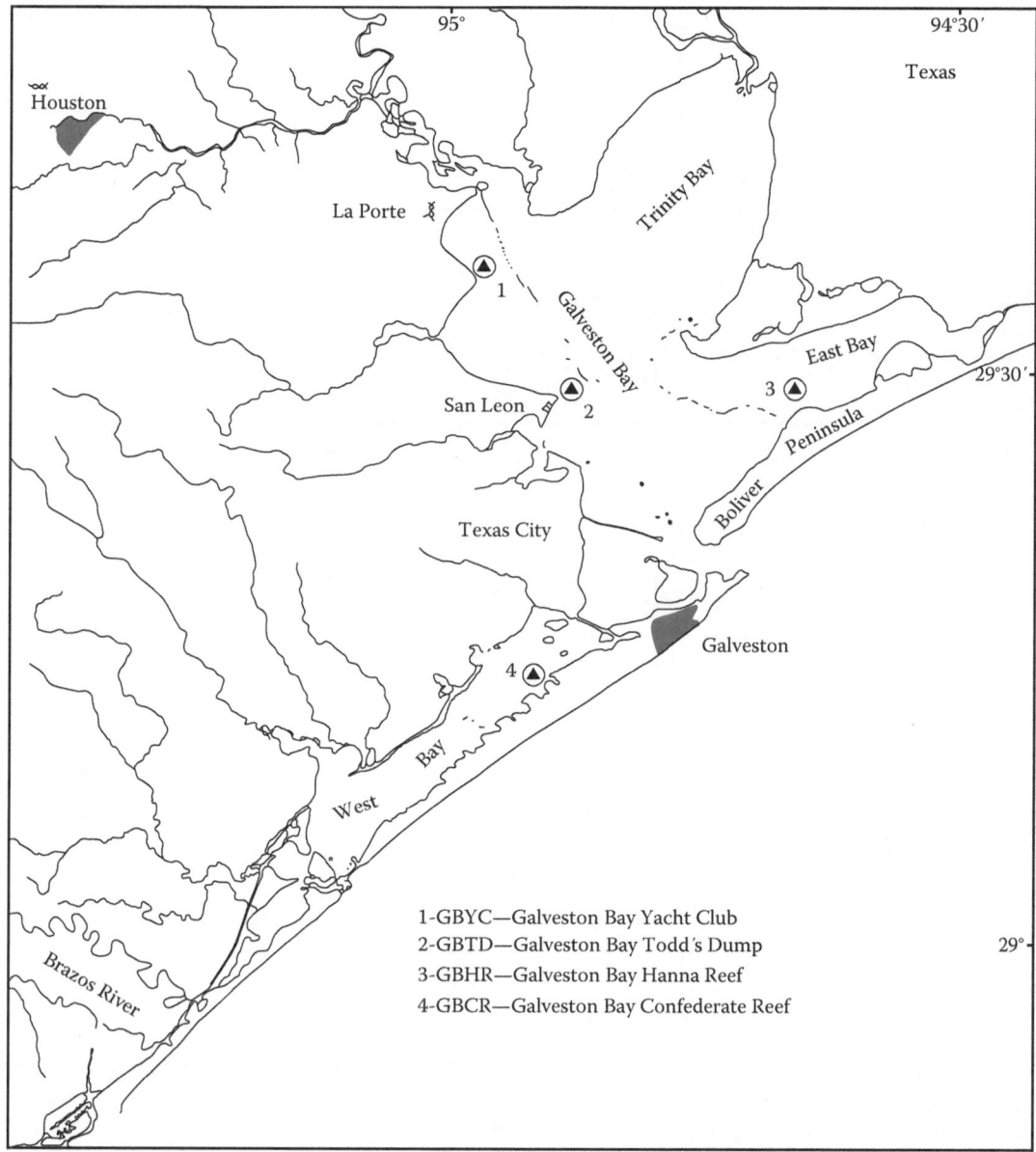

FIGURE 19.10   NOAA's National Status and Trends original sampling locations in Galveston Bay.

to 1991, triplicate samples of 20 American oysters (*Crassostrea virginica*) were collected during the November–January period. Starting in 1992, one pooled sample consisting of oysters from the original three stations within a site, was collected from each location.

Median, mean, range, and distribution frequency of concentrations for each target organic contaminant (Table 19.1) measured in samples from the four original stations between 1986 and 2007 are summarized in Table 19.4. Stations are ordered toward outside of Galveston Bay to visualize tendencies. Average total chlordane-related compounds (sum of *alpha*-chlordane, *trans*-nonachlor, heptachlor, and heptachlor epoxide) for the 20 year study period ranged from 6.29±5.40 ng/g (GBHR) and 6.31±4.31 ng/g (GBCR) to 48.8±37.5 ng/g (GBYC). Average total DDT concentrations (sum of *o–p'* DDT, *p–p'* DDT, *o–p'* DDD, *p–p'* DDD, *o–p'* DDE, and *p–p'* DDE) for the full

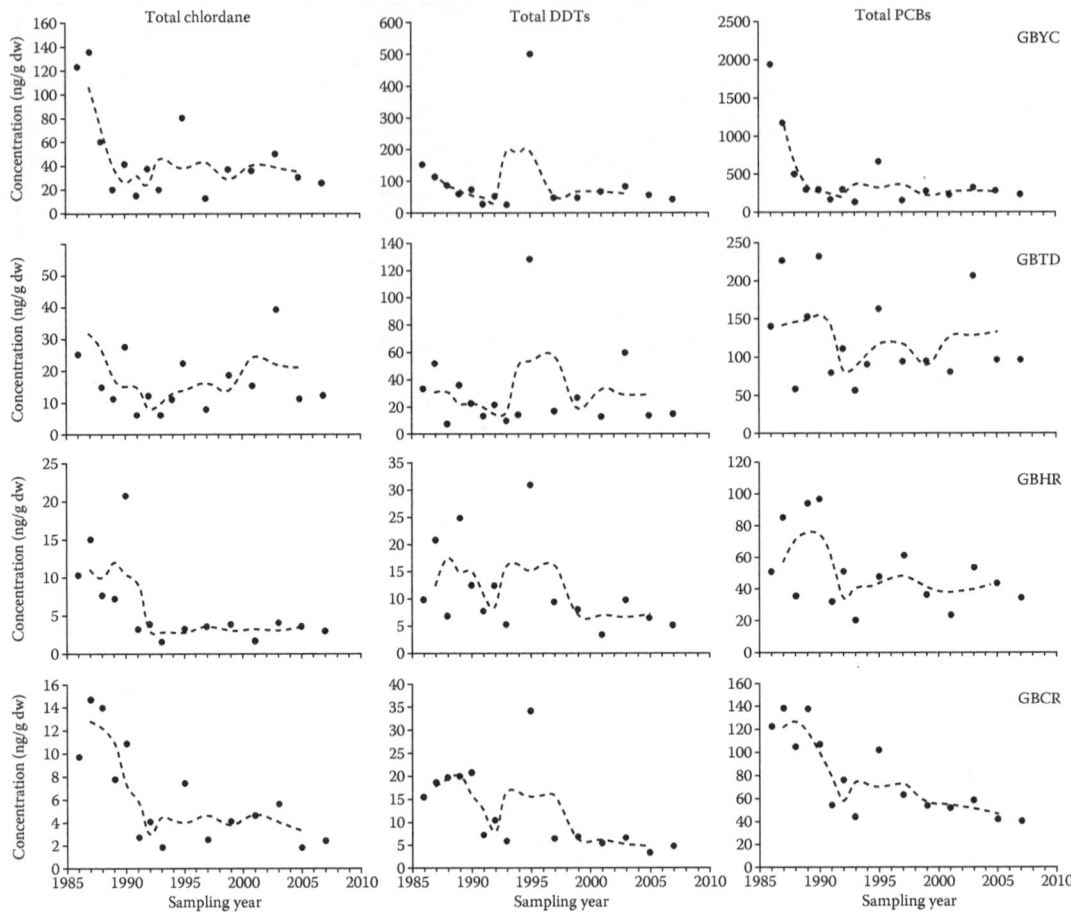

FIGURE 19.11 Temporal variations, including the 3 year moving average, of total chlordane-related compounds, DDT, and PCB concentrations in oyster tissues collected from the original locations in Galveston Bay from 1986 to 2006/2007.

period of study ranged from the lowest level of 11.8 ± 7.89 ng/g, detected in GBHR samples, to the highest concentration of 96.4 ± 117 ng/g in GBYC. Average total PCB concentrations (determined as the sum of 18 individual congeners and extrapolated to total PCB based on a correlation between the total of all measurable PCBs and the sum of selected congeners determined during the first 2 years of the NS&T program) ranged from 51.6 ± 23.9 to 470 ± 487 ng/g in samples from GBHR and GBYC, respectively. Overall, the highest concentrations of targeted analytes were found in oysters collected in the upper portion of the bay (GBYC and GBTD), closer to the Houston Ship Channel, and decreased seaward.

Yearly total concentrations of selected total target analytes measured in Galveston Bay oysters since 1986 are shown in Figure 19.11. Temporal trends are easier to visualize if a moving or running average is added to the scatter plots to smooth out short-term fluctuations. In this case, a 3 year moving average was selected to highlight longer-term trends or cycles. For example, a rapid decrease in the concentrations of total chlordane can be observed from 1986 to 1991 in samples from GBYC, GBCR, and GBHR followed by a plateau at or near the lowest levels, within a moderately compact range, encountered at these locations. Samples from GBTD, located in the middle portion of Galveston Bay did not show any apparent trend. The same situation can be observed for total PCB and total DDT concentrations measured in samples from GBYC and GBCR while temporal tendency is not clear for samples collected from GBHR and absent in oysters from GBTD. A significant spike in the total concentrations of these analytes was observed at all locations in 1995. This spike was more intense for total DDTs and represented an increase of 9, 6, 5, and 4 times the average plateau total concentrations in samples from GBYC, GBTD, GBCR, and GBHR, respectively. Although the reason for this sporadic increase cannot be easily explained, based on this distribution, one could speculate that the spike was caused by tropical storm Dean, which made landfall on the Texas coast on July 30, 1995 with heavy rain (6–18 in.) in Oklahoma and Texas, which caused coastal and inland flooding [56]. Tropical storm Dean landed near Freeport, Texas, located about 40 miles from Galveston Bay and dissipated in central Texas on August 2, 1995. Intense climate factors have been correlated to temporal variability of contaminants, both organic and inorganic, in oysters from the Gulf of Mexico [57–60]. Contaminants, mainly associated with particles, are mobilized during and immediately after storms as a result of particle resuspension, surface erosion, flooding, and runoff from both urban and agricultural areas [61–63]. Concentrations of some contaminants in the aquatic media, and the associated relative risks to organisms, are reported to increase with storm intensity in spite of a large amount of stormwater dilution [63]. As discussed earlier (see Section 19.3.2), field studies have demonstrated that it takes 30–60 days for oysters to equilibrate their body burdens to sudden ambient concentration changes. This, in addition to the time that it would have taken for the contaminants to reach Galveston Bay, rendered the sampling window for the NS&T Mussel Watch program (November–January) as the perfect period to document these changes.

When concentrations corresponding to oyster samples from Galveston Bay, collected in 1986, 1996/1997, and 2006/2007, are compared to the entire Gulf of Mexico data, it is readily evident that levels measured in GBYC and GBTD were consistently above the median values calculated for the Gulf of Mexico (Figure 19.12). Only about 10% of all the oyster samples collected in those sampling years had total contaminant concentrations higher than the levels determined in oysters

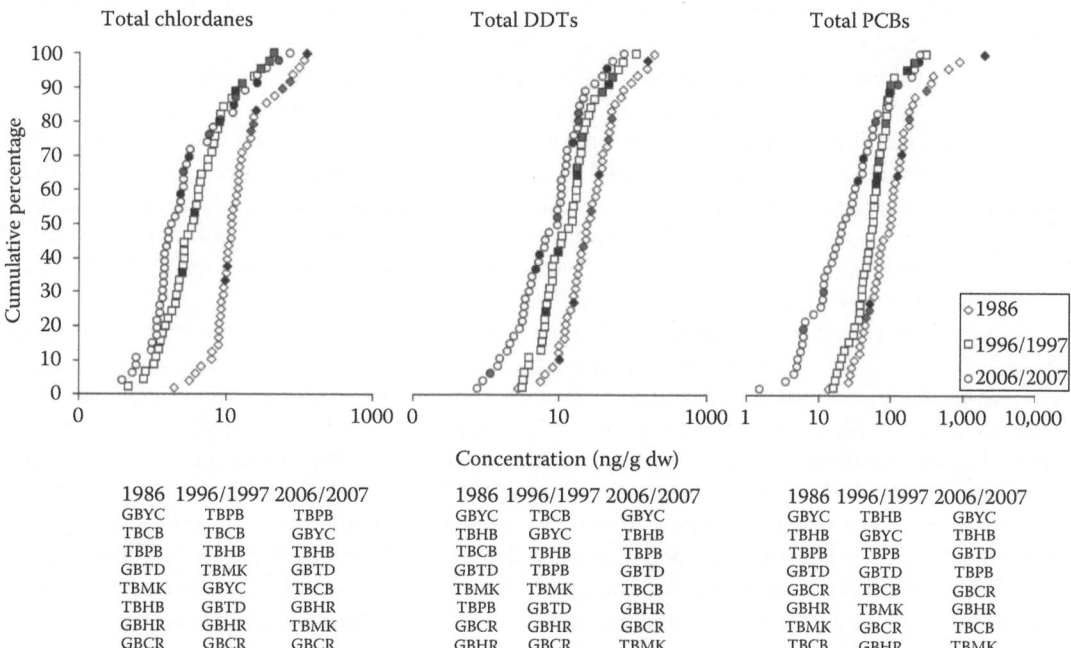

FIGURE 19.12 Cumulative distributions of total chlordane-related compounds, DDT, and PCB concentrations in oyster samples from the Gulf of Mexico collected in 1986, 1996/1997, and 2006/2007. Galveston and Tampa Bay samples are shown in black and gray symbols, respectively, in the figure and listed below from highest to lowest within each sampling year.

from GBYC. Samples from GBHR had, on the other hand, relatively low concentrations of most total contaminant concentrations compared to the Gulf levels.

19.5.2   TAMPA BAY, FLORIDA

The Tampa Bay estuary, a large natural harbor and estuary with a surface area of almost 1000 km^2 and a mid-tide average depth of 3.5 m, is the second largest estuary on the west central coast of Florida. The Y-shaped embayment comprises Hillsborough Bay, Old Tampa Bay, Middle Tampa Bay, and Lower Tampa Bay. The Tampa Bay Estuary, with a water residence time of 150 days [55], is directly bordered by three counties: Hillsborough, Manatee, and Pinellas. A population of over 2 million resides in the metropolitan space of Tampa, St. Petersburg, Clearwater, Bradenton, and surrounding areas. Its comparatively small watershed (approximately 6740 km^2) includes over a hundred small tributaries that drain into Tampa Bay with their loads of various environmental pollutants. The U.S. Environmental Protection Agency (US EPA) designated the Tampa Bay Estuary of national significance, in an effort to improve the health of the bay after years of uncontrolled pollution, and NOAA included several locations within Tampa Bay to monitor the spatial and temporal trends of organic contaminants in bivalves through its NS&T Mussel Watch project. Four locations (Papys Bayou, Hillsborough Bay, Cockroach Bay, and Mullet Key Bayou; Figure 19.13) within Tampa Bay were sampled in 1986 and 1987 with more sites (Old Tampa Bay and Narvaez Park) added in the years that followed. From 1986 to 1991, triplicate samples of 20 American oysters (*Crassostrea virginica*) were collected in the November–January window. In 1992 and subsequent years, one pooled sample from each location, consisting of oysters from the three original stations within a site, was collected and analyzed.

Basic statistics of the data generated between 1986 and 2006/2007 for the four original locations in Tampa Bay, presented from the interior of the bay, are summarized in Table 19.5. With the highest concentrations encountered in the middle portion of Tampa Bay, average total chlordane-related compounds for the 1986–2006/2007 period ranged from 12.3 ± 8.30 ng/g (TBHB) and 12.7 ± 8.04 ng/g (TBMK) to 40.2 ± 38.3 ng/g (TBCB) and 41.2 ± 24.7 ng/g (TBPB). The lowest levels correspond to samples collected inside Hillsborough Bay and toward the mouth of the Tampa Bay Estuary. Average total DDT concentrations ranged from the lowest level of 13.3 ± 9.67 ng/g detected in samples from TBMK, to the highest concentration of 49.3 ± 32.6 ng/g in TBCB. Average total PCB concentrations ranged from 38.3 ± 24.4 to 118 ± 84.9 ng/g in samples from TBMK and TBHB, respectively. In general, the highest concentrations of targeted analytes were detected in samples from inside the bay while the lowest levels were consistently measured in oysters from the mouth of the Tampa Bay system. This spatial trend is particularly obvious for average total PCB concentrations, which decreased orderly in oysters from TBHB to TBMK.

Annual total concentrations of selected total target analytes measured in Tampa Bay oysters since 1986, and their 3 year moving average, are shown in Figure 19.14. As observed in oyster samples from Galveston Bay, a decrease in average concentrations of total target analytes can be seen from 1986 to the early to mid 1990s in bivalves from TBPB and TBCB. This tendency is more gradual in oysters from the external location (TBMK) and is not clear in samples from the innermost site (TBHB). It is possible to observe a spike in the average levels encountered in 1996 for all locations. This spike was more evident in some locations than in others and for some analytes more than others. For example, average total chlordane, DDTs, and PCBs were clearly larger in samples from TBHB and TBMK while hardly discernible in oysters from TBPB. Only average total PCB concentration showed the spike in samples from TBCB. As in Galveston Bay, it is not easy to understand the reason behind this sporadic spike although spikes in concentrations have been associated with intense climate factors [57–60]. Tropical storm Josephine, formed on late September on the southwestern Gulf of Mexico, moved over Apalachee Bay on October 7, 1996. Storm surge heights were significant from the Tampa area northward to eastern Apalachee Bay. Storm tides (storm surge

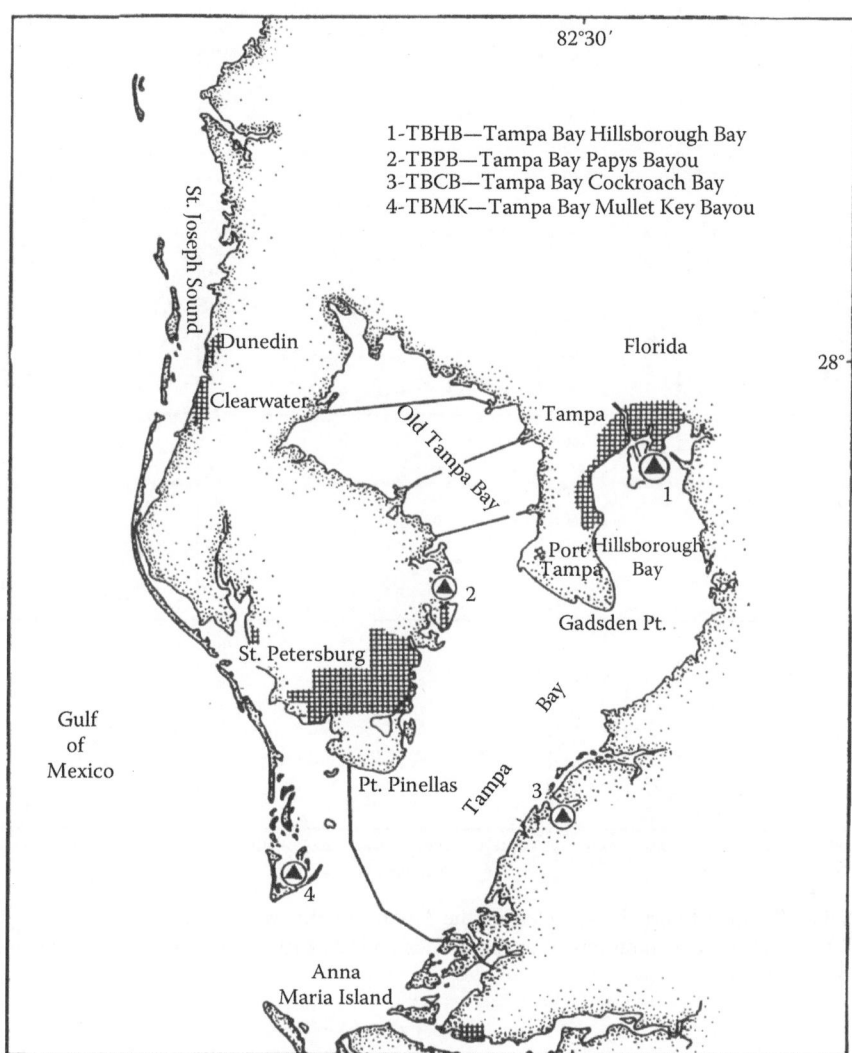

FIGURE 19.13 NOAA's NS&T original sampling locations in Tampa Bay.

plus astronomical tide) were estimated to range from 4 to 6 ft in Pinellas and Hillsborough counties. These tides produced widespread flooding of roads, dwellings, and businesses. Rainfall amounts of up to 8.5 in. were reported over northern Florida in association with Josephine [64].

When concentrations corresponding to Tampa Bay oysters, collected in 1986, 1996/1997, and 2006/2007, are compared to oysters from the Gulf of Mexico, they are among the highest in the area, particularly in 1996/1997 with all sites in the top 35% (Figure 19.12). The overall average contaminant load decreased from inside the bay toward the open Gulf with TBHC > TBPB > TBCB > TBMK.

19.6 CONCLUSIONS

Indigenous and transplanted oysters are valuable tools to monitor contamination levels and trends in coastal areas if their limitations are well understood. In addition to the significance of trying to minimize the biological and physical factors that affect the body burden of organic contaminants

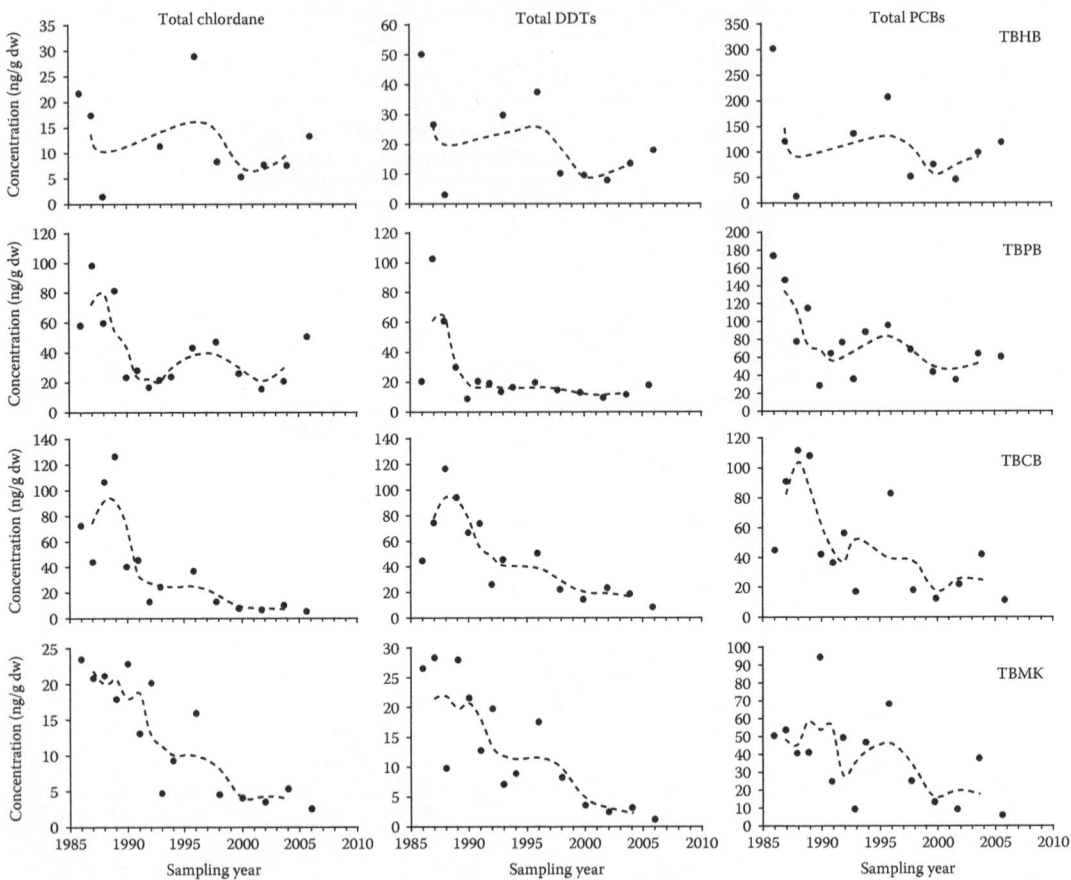

FIGURE 19.14 Temporal variations, including the 3 year moving average, of total chlordane-related compounds, DDT, and PCB concentrations in oyster tissues collected from the original locations in Tampa Bay from 1986 to 2006/2007.

in oysters through appropriate planning, it is important to respect the times the oysters need to adjust to changes in environmental concentrations. The spatial distribution of high and low concentrations in oysters from locations along the northern coast of the Gulf of Mexico described after the first few years of the NOAA's NS&T Mussel Watch program has generally not changed after two decades [11,13,14,16,17,21]. On a temporal scale, it is evident, with very few exceptions, that a global decline of persistent organic pollutant concentrations is observed. This general decline of persistent organic pollutant concentrations since 1986 indicates that pollution control measurements have been effective, particularly for the banned chlorinated compounds. Continued monitoring of bivalves, but perhaps at longer intervals, will seem prudent in order to provide useful data to assess these decreasing trends.

ACKNOWLEDGMENT

The authors are grateful to the National Oceanic and Atmospheric Administration's National Centers for Coastal Ocean Sciences (NOAA's NCCOS) Scientist, Dr. Gunnar G. Lauenstein, for providing access to the National Status and Trends "Mussel Watch" Program data.

REFERENCES

1. Goldberg, E.D. 1975. The Mussel Watch—A first step in global marine monitoring. *Mar. Pollut. Bull.* 6:111–114.
2. Lee, R.F., Sauerheber, R., and Benson, A.A. 1972. Petroleum hydrocarbons: Uptake and discharge by the marine mussel. *Mytilus edulis. Science.* 177:344–346.
3. Butler, P.A. 1973. Residues in fish, wildlife, and estuaries. Organochlorine residues in estuarine mollusks, 1965–1972—National Pesticide Monitoring Program. *Pestic. Monit. J.* 6:238–368.
4. Holden, A.V. 1973. International cooperative study of organochlorine and mercury residues in wildlife, 1968–1971. *Pestic. Monit. J.* 7:37–52.
5. DiSalvo, L.H., Guard, H.E., and Hunter, L. 1975. Tissue hydrocarbon burden of mussels as potential monitor of environmental hydrocarbon insult. *Environ. Sci. Technol.* 9:247–251.
6. Cantillo, A.Y. 1991. *Mussel Watch Worldwide Literature Survey.* NOAA Technical Memorandum NOS ORCA G3. NOAA Office of Ocean Resources Conservation and Assessment, Rockville, MD.
7. O'Connor, T.P., Cantillo, A.Y., and Lauenstein, G.G. 1994. Monitoring of temporal trends in chemical contamination by the NOAA National Status and Trends Mussel Watch Project. In *Biomonitoring of Coastal Waters and Estuaries*, K.J.M. Kramer (Ed.), pp. 29–50. CRC Press, Inc., Boca Raton, FL.
8. Farrington, J.W., Risebrough, R.W., Parker, P.L. et al. 1982. Hydrocarbons, polychlorinated biphenyls, and DDE in mussels and oysters from the U.S. coast, 1976–1978. Technical Report 82-42. Woods Hole Oceanographic Institution, Woods Hole, MA.
9. O'Connor, T.P. 1992. *Mussel Watch: Recent Trends in Coastal Environmental Quality.* NOAA, Rockville, MD.
10. O'Connor, T.P. and Beliaeff, B. 1997. *Recent Trends in Coastal Environmental Quality: Results from the Mussel Watch Project.* NOAA, Rockville, MD.
11. Jackson, T.J., Wade, T.L., McDonald, T.J. et al. 1994. Polynuclear aromatic hydrocarbon contaminants in oysters from the Gulf of Mexico (1986–1990). *Environ. Pollut.* 83:291–298.
12. Kim, Y., Powell, E.N., Wade, T.L. et al. 2008. Relationship of parasites and pathologies to contaminant body burden in sentinel bivalves: NOAA status and trends "Mussel Watch" program. *Mar. Environ. Res.* 65:101–127.
13. O'Connor, T.P. 2002. National distribution of chemical concentrations in mussels and oysters in the USA. *Mar. Environ. Res.* 53:117–143.
14. O'Connor, T.P. and Lauenstein, G.G. 2006. Trends in chemical concentrations in mussels and oysters collected along the US coast: Update to 2003. *Mar. Environ. Res.* 62:261–285.
15. Presley, B.J., Taylor, R.J., and Boothe, P.N. 1990. Trace metals in Gulf of Mexico oysters. *Sci. Total Environ.* 97/98:551–593.
16. Sericano, J.L., Atlas, E.L., Wade, T.L. et al. 1990. NOAA's status and trends Mussel Watch program chlorinated pesticides and PCBs in oysters (*Crassostrea virginica*) and sediments from the Gulf of Mexico, 1986–1987. *Mar. Environ. Res.* 29:161–203.
17. Sericano, J.L., Wade, T.L., Brooks, J.M. et al. 1993. National status and trends Mussel Watch Program: Chlordane-related compounds in Gulf of Mexico oyster: 1986–1990. *Environ. Pollut.* 82:23–32.
18. Sericano, J.L., Safe, S.H., Wade, T.L. et al. 1994. Toxicological significance of non-, mono-, and di-ortho-substituted polychlorinated biphenyls in oysters from Galveston and Tampa Bays. *Environ. Toxicol. Chem.* 13:1797–1803.
19. Sericano, J.L., Wade, T.L., and Brooks, J.M. 1996. Accumulation and depuration of organic contaminants by the American oyster (*Crassosstrea virginica*). *Sci. Total Environ.* 179:149–160.
20. Wade, T.L., Garcia-Romero, B., and Brooks, J.M. 1988. Tributyltin contamination of bivalves from U.S. coastal estuaries. *Environ. Sci. Technol.* 22:1488–1493.
21. Wade, T.L., Atlas, E.L., Brooks, J.M. et al. 1988. NOAA Gulf of Mexico status and trends program: Trace organic contaminant distribution in sediments and oysters. *Estuaries.* 11:171–179.
22. Phillips, D.J.H. and Rainbow, P.S. 1993. *Biomonitoring of Trace Aquatic Contaminants.* Elsevier Science Publishers Ltd., Oxford, U.K.
23. Phillips, D.J.H. 1980. *Quantitative Aquatic Biological Indicators—Their Use to Monitor Trace Metal and Organochlorine Pollution.* Applied Science Publishers, Ltd., London, U.K.
24. de Kock, W.C. and Kramer, K.J.M. 1994. Active biomonitoring (ABM) by translocation of bivalve mollusks. In *Biomonitoring of Coastal Waters and Estuaries*, K.J.M. Kramer (Ed.), pp. 51–84. CRC Press, Inc., Boca Raton, FL.
25. Sericano, J.L. 2000. The Mussel Watch approach and its applicability to global chemical contamination monitoring programs. *Int. J. Environ. Pollut.* 13:340–350.

26. Hamelink, J.L., Waybrant, R.C., and Ball, R.C. 1971. A proposal: Exchange equilibria control the degree chlorinated hydrocarbons are biologically magnified in lentic environments. *Trans. Am. Fish. Soc.* 100:207–214.

27. Geyer, H., Sheehan, P., Kotzias, D. et al. 1982. Prediction of ecotoxicological behavior of chemicals: Relationship between physicochemical properties and bioaccumulation of organic chemicals in the mussel *Mytilus edulis*. *Chemosphere.* 11:1121–1134.

28. Mackay, D. 1982. Correlation of bioconcentration factors. *Environ. Sci. Technol.* 16:274–278.

29. Pruell, R.J., Lake J.L., Davis, W.R. et al. 1986. Uptake and depuration of organic contaminants by the blue mussels (*Mytilus edulis*) exposed to environmentally contaminated sediments. *Mar. Biol.* 91:497–507.

30. Axiak, V., George, J.J., and Moore, M.N. 1988. Petroleum hydrocarbons in the marine bivalve *Venus verrucosa*: Accumulation and cellular responses. *Mar. Biol.* 97:225–230.

31. Chu, F.L.E., Soudant, P., Cruz-Rodriguez, L.A. et al. 2000. PCB uptake and accumulation by oysters (*Crassostrea virginica*) exposed via a contaminated algal diet. *Mar. Environ. Res.* 50:217–221.

32. Shaw, G. R. and Connell, D.W. 1984. Physicochemical properties controlling polychlorinated biphenyl (PCB) concentrations in aquatic organisms. *Environ. Sci. Technol.* 18:18–23.

33. Connell, D.W. 1990. Bioconcentration of lipophilic and hydrophobic compounds by aquatic organisms. In *Bioaccumulation of Xenobiotic Compounds*, D.W. Connell (Ed.), pp. 97–144. CRC Press, Boca Raton, FL.

34. Barron, M.G. 1990. Bioconcentration. Will water-borne organic chemicals accumulate in aquatic animals?. *Environ. Sci. Technol.* 24:1612–1618.

35. Sericano, J.L. 1993. *The American Oyster (Crassostrea virginica) as a Bioindicator of Trace Organic Contamination*. Doctoral dissertation, Texas A&M University, College Station, TX.

36. Gardinali, P.R., Sericano, J.L., and Wade, T.L. 2004. Uptake and depuration of toxic halogenated aromatic hydrocarbons by the American oyster (*Crassostrea virginica*): A field study. *Chemosphere.* 54:61–70.

37. Cairns, T., Doose, G.M., Froberg, J.E. et al. 1986. Analytical chemistry of PCBs. In *PCBs and the Environment*, J.S. Waid (Ed.), pp. 1–45. CRC Press, Boca Raton, FL.

38. Qian, Y., Wade, T.L., and Sericano, J.L. 2001. Sources and bioavailability of polynuclear aromatic hydrocarbons in Galveston Bay, Texas. *Estuaries.* 24:817–827.

39. Kimbrough, K.L., Johnson, W.E., Lauenstein, G.G. et al. 2008. *An Assessment of Two Decades of Contaminant Monitoring in the Nation's Coastal Zone*. NOAA Technical Memorandum NOS NCCOS 74, Silver Spring, MD.

40. Martin, M. 1985. State Mussel Watch: Toxics surveillance in California. *Mar. Pollut. Bull.* 16:140–146.

41. Wade, T.L., Sericano, J.L, Gardinali, P.R. et al. 1998. NOAA Mussel Watch Project: Current use organic compounds in Mollusks. *Mar. Pollut. Bull.* 37: 20–26.

42. Muncaster, B.W., Hebert, P.D.N., and Lazar, R. 1990. Biological and physical factors affecting the body burden of organic contaminants in freshwater mussels. *Arch. Environ. Contam. Toxicol.* 19:25–34.

43. Ellis, M.S., Choi, K.S., Wade, T.L. et al. 1993. Sources of local variation in polynuclear aromatic hydrocarbon and pesticide body burden in oysters (*Crassostrea virginica*) from Galveston bay, Texas. *Comp. Biochem. Physiol. C: Pharmacol. Toxicol. Endocrinol.* 106:689–698.

44. Capuzzo, J.M., Farrington, J.W., Rantamakia, P. et al. 1989. The relationship between lipid composition and seasonal differences in the distribution of PCBs in *Mytilus edulis* L. *Mar. Environ. Res.* 28:259–264.

45. Burkhard, L.P. 2003. Factors influencing the design of bioaccumulation factor and biota-sediment accumulation factor field studies. *Environ. Toxicol. Chem.* 22:351–360.

46. Borga, K., Fisk, A.T., Hoekstra, P.F. et al. 2004. Biological and chemical factors of importance in the bioaccumulation and trophic transfer of persistent organochlorine contaminants in arctic marine food webs. *Environ. Toxicol. Chem.* 23:2367–2385.

47. Dearth, M.A. and Hites, R.A. 1991. Complete analysis of technical chlordane using negative ionization mass spectrometry. *Environ. Sci. Technol.* 25:245–254.

48. Puri, R.K., Orazio, C.E., Kapila, S. et al. 1990. Studies on the transport and fate of chlordane in the environment. In *Long Range Transport of Pesticides*, D.A. Kurtz (Ed.), pp. 271–289. Lewis Publishers, Chelsea, MI.

49. Kawano, M., Inoue, T., Wada, T. et al. 1988. Bioconcentration and residue patterns of chlordane compounds in marine animals: Invertebrates, fish, mammals and seabirds. *Environ. Sci. Technol.* 22:792–797.

50. Muir, D.C.G., Norstrom, R.J., and Simon, M. 1988. Organochlorine contaminants in Artic marine food chains: Accumulation of specific polychlorinated biphenyls and chlordane-related compounds. *Environ. Sci. Technol.* 22:1071–1079.

51. Hensel, D.R. 1990. Less than obvious. Statistical treatment of the data below the detection limit. *Environ. Sci. Technol.* 24:1766–1774.

52. World Health Organization 1979. *DDT and its Derivatives*. WHO, New York.
53. Sericano, J.L., Wade, T.L., Atlas, E.L. et al. 1990. Historical perspective on the environmental bioavailability of DDT and its derivatives to Gulf of Mexico Oysters. *Environ. Sci. Technol.* 24:1541–1548.
54. Woodwell, G.M., Craig, P.P., and Horton, A.J. 1971. DDT in the biosphere: Where does it go? *Science.* 174:1101–1107.
55. Solis, R.S. and Powell, G.L. 1999. Hydrography, mixing characteristics, and residence time of Gulf of Mexico estuaries. In *Biogeochemistry of Gulf of Mexico Estuaries*, T.S. Bianchi, J.R. Pennock, and R.R. Twilley (Eds.), pp. 29–61. John Wiley, New York.
56. National Hurricane Center 1995. Preliminary Report Tropical Storm Dean, July 28–August 3, 1995. www.nhc.boulder.noaa.gov/1995dean.html
57. Apeti, A.D., Robinson, L., and Johnson, E. 2005. Relationship between heavy metal concentrations in the American oyster (*Crassostrea virginica*) and metal levels in the water column and sediment in Apalachicola Bay, Florida. *Am. J. Environ. Sci.* 1:179–186.
58. Kim, Y., Powell, E.N., Wade, T.L. et al. 1999. Influence of climate change on interannual variation in contaminant body burden in Gulf of Mexico oysters. *Mar. Environ. Res.* 48:459–488.
59. Presley, S.M., Rainwater, T.R., Austin, G.P. et al. 2005. Assessment of pathogens and toxicants in New Orleans, LA following hurricane Katrina. *Environ. Sci. Technol.* 40:468–474.
60. Johnson, W.E., Kimbrough, K.L., Lauenstein, G.G. et al. 2009. Chemical contamination assessment of Gulf of Mexico oysters in response to hurricanes Katrina and Rita. *Environ. Monit. Assess.* 150:211–225.
61. DiGiacomo, P.M., Washburn, L., Holt, B. et al. 2004. Coastal pollution hazards in southern California observed by SAR imagery: Stormwater plumes, wastewater plumes and natural hydrocarbon seeps. *Mar. Pollut. Bull.* 49:1013–1024.
62. Badin A.L., Faure, P., Bedell, J.P. et al. 2008. Distribution of organic pollutants and natural organic matter in urban storm water sediments as a function of grain size. *Sci. Total Environ.* 403:178–187.
63. Chiovarou, E.D. and Siewicki, T.C. 2008. Comparison of storm intensity and application timing on modeled transport and fate of six contaminants. *Sci. Total Environ.* 389:87–100.
64. National Hurricane Center 1996. Preliminary Report Tropical Storm Josephine, October 4–8, 1996. www.nhc.boulder.noaa.gov/1996josephin.html

20 Using Sediment Cores to Assess Inputs of Organochlorines and Polycyclic Aromatic Hydrocarbons in Coastal Georgia Estuaries

*Clark R. Alexander, Allen D. Uhler, and Richard F. Lee**

CONTENTS

20.1 Introduction 469
20.2 Materials and Methods 471
    20.2.1 Preliminary Sampling and Sediment Analysis 471
    20.2.2 Coring Procedures 471
    20.2.3 Analysis of Sediments, PCBs, and PAHs..471
        20.2.3.1 Physical Characterization of Sediment....................................471
        20.2.3.2 Analysis of Polycyclic Aromatic Hydrocarbons and Polychlorinated
                  Biphenyls...472
20.3 Results and Discussion 473
    20.3.1 Advantages of Using Abandoned Boat Slips for Historical Core Studies...............473
    20.3.2 PAHs in Savannah River Estuary 473
    20.3.3 PCBs in Savannah River Estuary 474
    20.3.4 Pesticides in Savannah River Estuary 475
    20.3.5 PCBs in Brunswick Harbor 475
    20.3.6 PCBs in Sapelo Island Core...476
20.4 Summary and Conclusions...477
Acknowledgments..478
References...478

20.1 INTRODUCTION

The sediment record, as revealed in sediment cores, has been widely used to reconstruct the history of contaminant input in estuaries [1–9]. The basic assumption in using the sediment record to reveal contaminant input trends is that contaminant inputs of synthetic organochlorine compounds and polycyclic aromatic hydrocarbons (PAHs) readily adsorb to sedimenting particles and equilibrate

---

* E-mail: Dick.Lee@skio.usg.edu (Chapter corresponding author).

relatively rapidly with the sediment. In addition, sedimentary processes must not act to disturb the building of this record. If these requirements are satisfied, then the sediment column represents a continuous, complete historical sequence of contaminant accumulation. Deposition and accumulation are two terms used by sedimentologists and the distinction between these two terms has been discussed by McKee et al. [10]. Deposition is the temporary emplacement of particles in the seabed. Accumulation is the net sum of many episodes of deposition and removal over long time scales. Differences between rates of deposition and accumulation affect the ability of an environment to record sedimentary events.

Using radiochemical chronologies, such as those derived from the decay of Pb-210, it is possible to date sediment cores. Pb-210, formed by the decay of radon in the atmosphere (residence time of Pb-210 in the atmosphere ~10 days) is deposited by wet and dry deposition on land and sea. The Pb-210 decays at a known rate, allowing the determination of sediment accumulation rates by measuring Pb-210 activity with depth in the cores. Pb-210 (half-life of 22.3 years) characterizes accumulation rates over approximately a 100 year time period (approximately five decay half-lives) in undisturbed cores. Cs-137 (half-life 30 years) is an anthropogenic impulse tracer commonly used to complement temporal studies using Pb-210. Cs-137, which is delivered by atmospheric fallout, was first introduced into the environment in significant quantities in 1954 and peaked in 1963 in conjunction with the peak in atmospheric weapons testing and provides an independent estimate of sediment accumulation rates. A third isotope, Be-7 (half-life of 53 days), is a naturally produced radionuclide formed by cosmic ray bombardment of atmospheric nitrogen and oxygen, which is rapidly removed from the atmosphere and incorporated onto particles on land, which are then subsequently removed by flooding to aqueous environments. Be-7 has its greatest flux to sediments during the rainy season, so it can be used to identify sediments that are brought into estuaries and continental margins on a seasonal basis (e.g., within five half-lives, or 8 months). Dating procedures using these isotopes have been used to successfully date cores, and to identify seasonal flood layers from a variety of coastal environments [1,3–6,11,12]. Sediment accumulation rates can be determined using the constant sedimentation model [13].

The present study reports the results of analysis of polychlorinated biphenyls (PCBs) and PAHs in dated sediment cores taken from three estuarine sites in coastal Georgia, i.e., the Savannah River Estuary, the Brunswick Estuary, and the Duplin River Estuary, which is located behind Sapelo Island. PAHs are often at high concentrations in sediments near urban areas due to oil spills, creosote releases from treated piling, aqueous emissions from historic manufactured gas plant wastes, urban runoff, and particulate emissions from fossil fuel burning [14]. PCBs are ubiquitous, persistent, and toxic pollutants, which were widely produced in the 1930s and 1970s in the United States [15,16]. Changes in the contaminant concentrations of these compounds over time were correlated with various activities at the coring sites. Additionally, this contribution is intended to generally illustrate the historic coring approach for assessing contaminant inputs to coastal systems.

Two cores were taken from the Savannah River Estuary (Savannah, Georgia, the United States). This estuary is typical of estuaries of the South Atlantic Bight, which are formed within the mouths of Piedmont rivers. Within the Savannah River Estuary is the port of Savannah, which is the fourth busiest port in the United States and handles containerized shipping of all kinds, i.e., kaolin, coal, ferrous minerals, fuel oil, and other raw and processed chemicals. Industries that have developed around this port include paper, fertilizer, and chemical manufacturing. Hundreds of thousands of metric tons of sediment are transported down the Savannah River and are trapped in the Savannah estuary. As a result of farming and deforestation, soil eroded from the upland watershed area moves toward the estuary via the Savannah River. Thus, chemicals used in the upland areas can enter the sediments of the Savannah Estuary many years after their initial use.

The second sample site was in the Brunswick Estuary, near the Port of Brunswick. The core was taken in Brunswick Harbor at a site, now abandoned, where Liberty ships were built and launched during World War II. The Brunswick Port, while not as busy as the Savannah Port (total weight shipped in 2002 from Brunswick Port was 3 million tons while Savannah Port shipped

10 million tons), is an active port handling wood products, agricultural products, automobiles, and heavy machinery.

The third sampling site was in the pristine salt marshes along the Duplin River, in the Sapelo Island National Estuarine Research Reserve, located behind and directly adjacent to Sapelo Island, Georgia. The Altamaha River, a relatively undisturbed river in Georgia, drains an area of 38,000 km² and empties into the Atlantic Ocean, although some of it flows into Doboy Sound and into the Duplin estuary, both to the north of the river. Sapelo Island was a private preserve from 1934 to 1964 and then, in 1976, a portion of the island was designated a National Estuarine Research Reserve by National Oceanographic and Atmospheric Administration (NOAA). For the past 75 years, there have only been a small number (~200) of permanent residents on the island, allowing preservation of a near-pristine environment.

20.2 MATERIALS AND METHODS

20.2.1 Preliminary Sampling and Sediment Analysis

Surface sediment samples were collected from potential coring sites, with selection based on sediment type and radionuclide activities. Sediments with obvious biological or physical mixing, high sand content, and/or a non-accumulating, over-consolidated character were rejected since such sediments cannot be used for historical coring studies. The properties of the sediments selected for coring had rapidly accumulating sediment (i.e., high porosity and oxidized coloration), bedded sediment structure, a strong radiochemical signal, and/or Be-7 in the surficial sediment.

20.2.2 Coring Procedures

Box cores were taken via a single-spade box corer. Three subcores were taken from each box with PVC or stainless steel pipe for the following: (1) Organochlorine and/or PAH analysis, (2) radio-chemical dating and textural analysis, and (3) additional core for analytical studies if extra material was needed because of high core porosity, which means that the amount of sediment in each interval might be limited. The subcores were placed on extruders that allowed core subsampling intervals of 1 cm. After cutting away the sediment edges (which may have been contaminated by sediment drawn along the barrel sides), the cores were sequentially sectioned at 1 cm intervals in the upper 10 cm and at 2 cm intervals to the bottom of the cores. The subsamples for the analytical evaluation were placed in precleaned glass jars and frozen. The subsamples for radiochemical dating were placed in plastic bags and sealed. A slab for x-radiographic examination sediment structure was collected in a custom-designed Plexiglass tray. X-radiographs were produced using a portable veterinary x-ray unit with Kodak AA x-ray film.

20.2.3 Analysis of Sediments, PCBs, and PAHs

20.2.3.1 Physical Characterization of Sediment

20.2.3.1.1 X-Radiographs and Sedimentary Structure

X-radiographs allow determination of the extent of sediment mixing and to identify breaks in the stratigraphic record. If sediments are well laminated, then profiles of radionuclides or contaminants contained within the sediment column have not been disturbed and thus provide a means of studying contaminant input histories. In estuarine environments, most cores will display some amount of biological mixing (e.g., crab or clam burrows) in addition to bedded or laminated structure. The degree and intensity of mixing can be assessed visually in x-radiographs. Truncation of layering suggests erosion and removal of some part of the preserved record. In our studies, we tried to use sediments that were well laminated, to the greatest extent possible. X-radiographs were used to identify bioturbation and erosional disturbance in cores as well as to determine any disturbance associated with the coring operations.

20.2.3.1.2 Grain Size

Each core was analyzed at 5 cm intervals for inorganic grain size. The subsamples were wet sieved, with the retained sand fraction (>63 μm) analyzed at 0.25 phi intervals using nested sieves. The sediment that passed through the 63 μm sieve was captured and analyzed for size distribution within the silt and clay fraction with a Sedigraph 5100, which provided a complete distribution down to 0.25 μm particle sizes.

20.2.3.1.3 Geochronology

Sediment accumulation rates were determined by Pb-210 geochronologies. Additional age constraints were derived from the distributions and activities of Cs-137 and Be-7 within the cores. Activities of radiotracers within each sample were determined concurrently using two low-background, planar intrinsic germanium detectors, computer-based multichannel analyzers, and Maestro-II analysis software. Accumulation rates were calculated by producing profiles of excess Pb-210 activity (i.e., not supported by production of Pb-210 in the sediment column) with core depth. Total Pb-210 activity was directly determined by gamma spectroscopic measurement of its 46.5 keV gamma peak in dried, homogenized sediment [17]. Supported levels of Pb-210 from the decay of Ra-226 were determined for each depth interval by concurrently measuring the gamma activity of Pb-214 and Bi-214, the short-lived granddaughters of Ra-226. Excess Pb-210 is simply the difference between total and supported Pb-210. Activities of Cs-137 and Be-7 were directly determined by gamma spectroscopic measurement of their 662 and 477 keV gamma peaks, respectively, in dry, homogenized sediment [18,19].

Sediment accumulation rates were calculated using the constant flux (also known as the constant rate of supply) geochemical model [18,20–23]. This model is applicable in areas where the flux of excess Pb-210 (unsupported by the decay of Ra-226) and dry-mass sediment rate at the seabed have remained constant. This assumption has been shown to be reasonable in areas where the supply of Pb-210 is dominated by atmospheric sources, such as estuarine environments. Dry mass accumulation rates (in g/cm^2/year) were calculated for cores and ages were assigned to depth intervals within each core.

20.2.3.2 Analysis of Polycyclic Aromatic Hydrocarbons and Polychlorinated Biphenyls

Sediments from cores H and F in the Savannah River Estuary were analyzed for 22 selected PAHs using the procedures described by MacLeod et al. [24]. Sediment sections were extracted with dichloromethane and organic extracts were fractionated on columns of silica gel after using activated copper to remove elemental sulfur. The first fraction, which contains PCBs and some chlorinated pesticides, was eluted with hexane. The second fraction, which contains the PAHs and most of the chlorinated pesticides, was eluted with dichloromethane in hexane (1:1 v/v). The internal standard added to each sample was 4,4′-dibromooctafluorobiphenyl. Total PAHs refers to the total concentration of the 22 selected PAHs. For analysis of PAHs, a series of internal standards (d8-naphthalene, d10-acenaphthene, d10-phenanthrene, d12-chrysene, and d12-perylene) were added to each sample and blank before workup to allow determination of recovery.

PCBs were analyzed by high-resolution fused silica capillary gas chromatography with electron capture detection (Varian 3400CS gas chromatograph). PCB congeners were identified and quantified by comparisons to reference congeners. PCBs were run on two different columns (DB-5 and Db-17) to confirm the identification of the various compounds. Where identification was in question, full scan gas liquid chromatography/mass spectrophotometry was used and compared with the mass spectra of the authentic standards. Standard solutions of 28 PCB congeners from the National Institute of Standards and Technology (NIST, Gaithersburg, MD) were used to calibrate the GC-EDC. High-purity solvents, reagents, and process gases were used throughout the analysis.

Procedural blanks analyzed in conjunction with the sample indicated insignificant levels of all target analytes.

20.3 RESULTS AND DISCUSSION

20.3.1 ADVANTAGES OF USING ABANDONED BOAT SLIPS FOR HISTORICAL CORE STUDIES

Two of the sites (Savannah River Estuary and Brunswick, GA harbor) selected for coring were abandoned boat slips. These types of sites have many advantages when conducting historical coring studies in and near ports and other urban centers on the water. Initially, the slip is deeper than the equilibrium depth and thus after abandonment, the slip rapidly accretes to fill accommodation space, often at rates of many centimeters per year. The sedimentation rates slow as they approach equilibrium depth. Thus, there is the possibility of a high-resolution sediment record that can span decades or more in cores taken from abandoned boat slips.

20.3.2 PAHs IN SAVANNAH RIVER ESTUARY

Two cores were taken in the Savannah River Estuary. Core H (Figure 20.1) was taken from an abandoned boat slip across the river from the city and port of Savannah. Core F (Figure 20.2) was taken from a back-levee *Spartina* marsh several kilometers away from the city of Savannah. The major feature of core H was a large increase in PAH concentrations from a depth of 175 cm

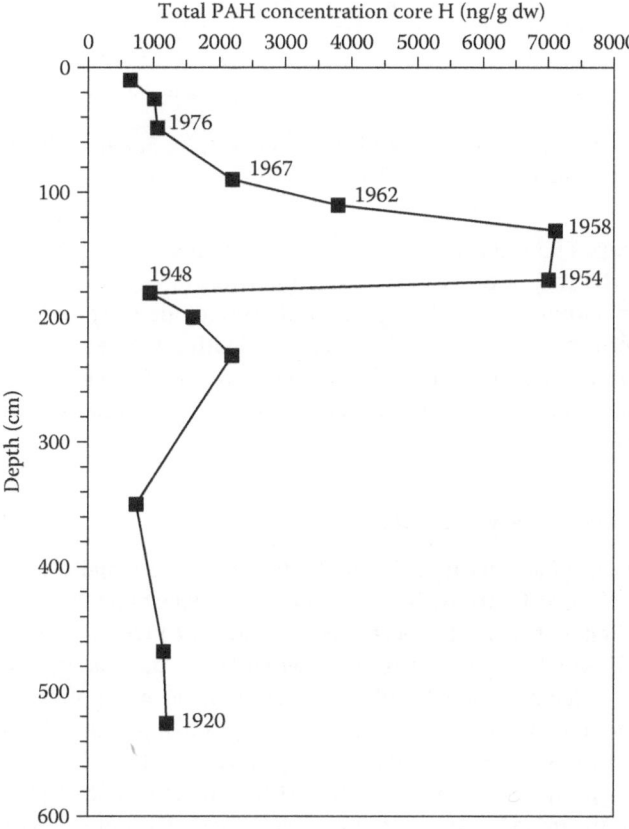

FIGURE 20.1 Total PAH concentrations in sections of core H from Savannah River Harbor. (Data from Alexander, C.R. et al., *Limnologica*, 29, 267, 1999.)

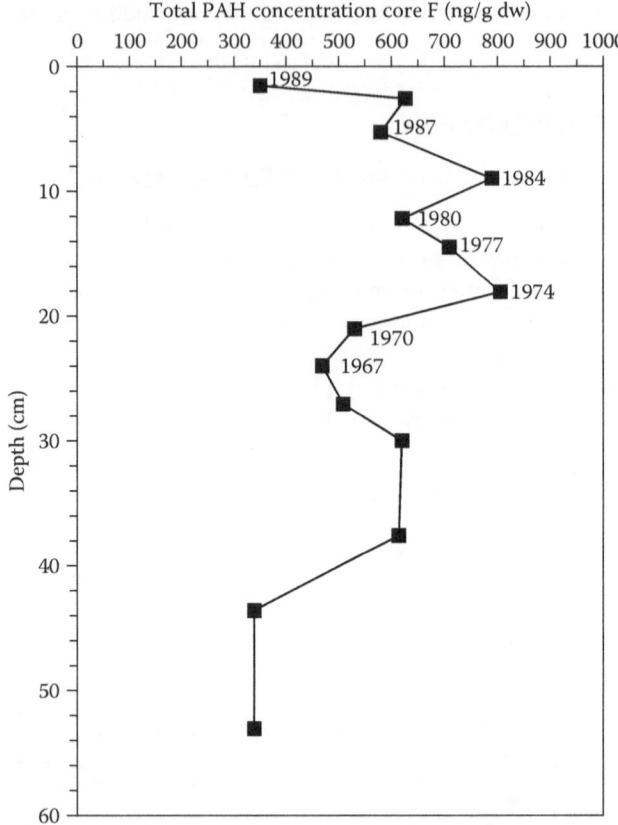

FIGURE 20.2 Total PAH concentrations in sections of core F from Savannah River Estuary. (Data from Alexander, C.R. et al., *Limnologica*, 29, 267, 1999.)

(age-1948) to 135 cm (age-1958) followed by a decrease in concentrations from 135 cm to the surface. We suggest that the increased PAH concentration from 1948 to 1958 was due to increases in the population, port activity, and industry in the Savannah area, while the decrease after 1958 was likely due to the effects of a number of regulations dealing with emissions and effluents from industry located along the Savannah River. Core F had lower PAH concentrations than core H for the same dates, which is likely due to its locations at some kilometers from the port of Savannah and its associated industry.

20.3.3 PCBs in Savannah River Estuary

PCBs were extensively used as heat transfer fluids in transformers and capacitors throughout the United States in the 1930s and 1970s. Because of their persistence and toxicity, their production was banned in the United States in 1976. Highest concentrations of PCBs in cores F (Figure 20.3) and H (Figure 20.4) were 32 and 207 ng/g, respectively, which were found at depths dated at 1967. Total PCBs is the sum of the concentration of 20 PCB congeners. Concentrations of PCBs in core H were very low in core sections dated from 1920 to 1937, with a gradual increase from 1944 to the 32 ng/g peak in 1967 followed by lower concentrations (16–22 ng/g) after 1967. PCB concentrations in core F show a profile similar to the one observed in core H (i.e., increasing concentrations from 1944 to 1967 and lower but significant PCB concentrations after 1967). The detection of significant concentrations of PCBs in cores after 1976, when PCB production was banned, is likely due primarily to soil exposed earlier (before 1976) in the watershed area of the Savannah River and transport of these

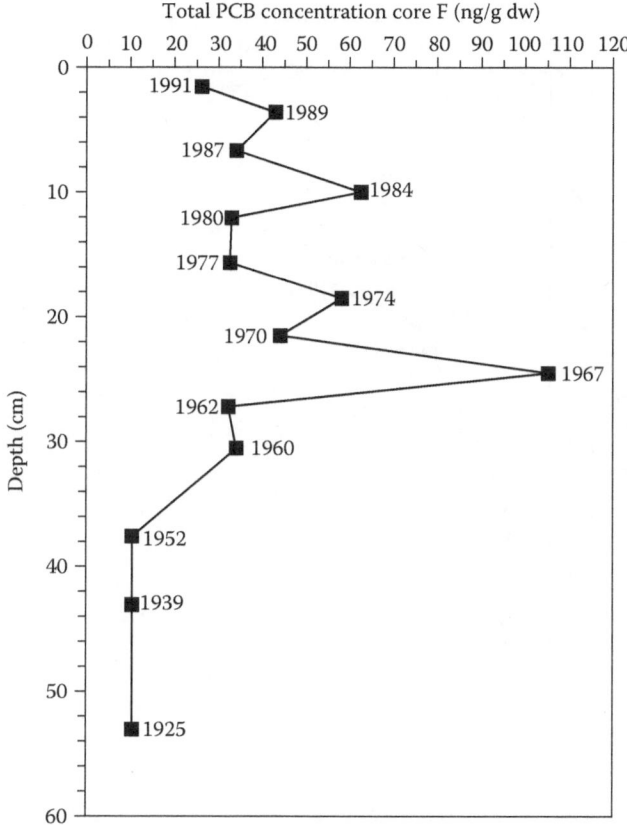

FIGURE 20.3 Total PCB concentrations in sections of core F from Savannah River Estuary. (Data from Alexander, C.R. et al., *Limnologica*, 29, 267, 1999.)

sediment down the Savannah River to sites F and H. Dated cores at various sites in North America generally have high PCB concentrations in sections representing the 1960s [4].

20.3.4 Pesticides in Savannah River Estuary

An earlier study on cores F and H in the Savannah River Estuary showed that in core F, the major pesticides found were isomers of DDT (4,4′-DDE, 4,4′-DDD) and dieldrin, which peaked in the 1984 layer [2]. Since the manufacture of DDT was based in the United States in 1972, it seems likely that the source of the DDTs and dieldrin in 1984 in soil earlier exposed in the watershed to DDT with later transport down the river to site F. The profile of 2,4′-DDE and dieldrin in core H correlates with their peak usage in the United States with a gradual increase from 1954 to a peak in 1967 and then a decrease from 1967 to recent times. It is likely that much of the DDT found in core H was used at that site since site H has large populations of breeding mosquitoes that were controlled by DDT until its application was stopped in the late 1960s. The dieldrin was used in the 1960s to control white-fringed beetles in this area.

20.3.5 PCBs in Brunswick Harbor

A core to a depth of 50 cm was taken from an abandoned boat slip in Brunswick Harbor, GA. There were two different accumulation rates determined from Pb-210 geochronology of this core. The

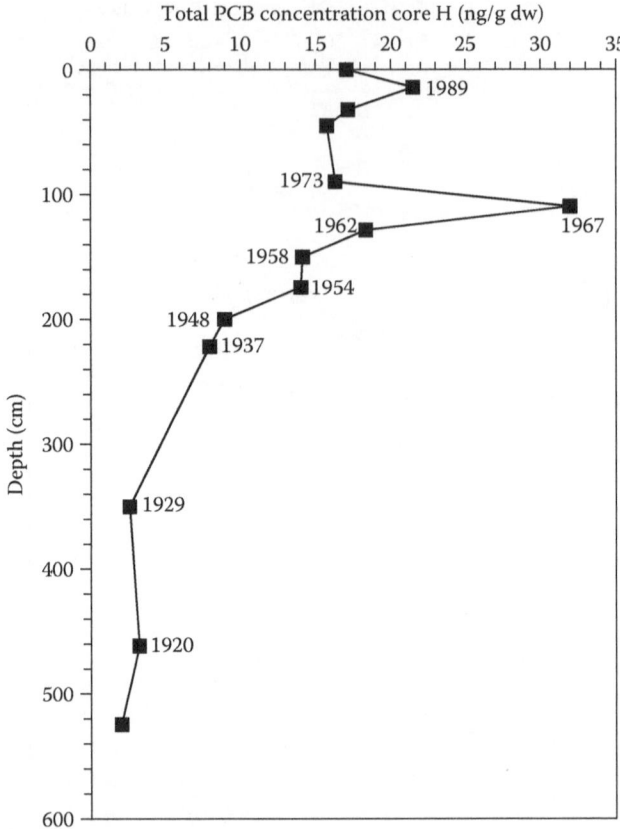

FIGURE 20.4 Total PCB concentrations in sections of core H from Savannah River Harbor. (Data from Alexander, C.R. et al., *Limnologica*, 29, 267, 1999.)

accumulation rate for the upper 23 cm (1963–2004) of the core was 0.2 cm/year while the accumulation rate of the lower section of the core (1954–1963) was 1.8 cm/year.

PCBs in a sectioned Brunswick core were not detected before 1948 but were followed by peaks in 1955 (110 ng/g), 1958 (120 ng/g), and 1961 (130 ng/g) (Figure 20.5—ng PCBs/g sediment normalized to percent sediment carbon). PCB concentrations decreased after 1961 with concentrations of 37 and 40 ng/g sediment in 1974 and 2004, respectively.

20.3.6 PCBs IN SAPELO ISLAND CORE

An intact core was collected on March 13, 1997, from a back-levee marsh in the upper Duplin River, an estuarine river adjacent to and behind Sapelo Island, Georgia. Sapelo Island is a National Estuarine Research Reserve and has only a small number of permanent residents. The core was sectioned at 1 cm intervals and these sections were used for the dating and analysis of PCBs. The x-radiograph of the core showed conspicuous *Spartina* root mass and rhizomes down to 20 cm. The 33 cm core represents approximately 100 years of accumulation with total PCB concentrations near or below detection limits throughout the column (non-detectable to 2.1 ng/g sediment; minimum detection limits—2 ng/g) (Table 20.1). PCB congeners identified in the core included congeners 52, 44, and 101/90, although all concentrations were near detection limits (<0.5 ng/g for individual congeners). Thus, both total PCB and individual congener concentrations were near or below detection limits throughout the column. PCBs were also below detection limits in biota

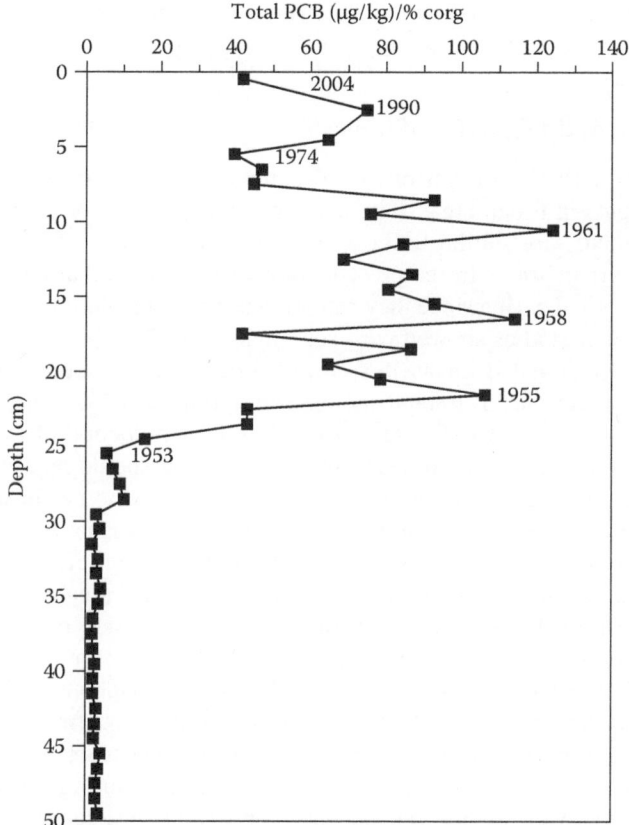

FIGURE 20.5 Total PCB concentrations in sections of Brunswick Harbor core.

TABLE 20.1
Concentrations of Total PCBs in a Sectioned Core from the Sapelo Island National Estuarine Research Reserve, Georgia

Depth (cm)	Total PCB Concentrations (ng/g)
0–2	0.75
3–4	ND
5–6	2.07
7–8	ND
9–10	0.66
11–12	1.93
15–16	0.63
15–16	0.56
19–20	1.79
27–28	0.61
29–30	ND
35–36	1.67

ND, not detected (minimum detection limit—2 ng/g).

collected near the coring site (marsh grass, *Spartina alterniflora*; fiddler crabs, *Uca* sp.; mussels, *Modiolus demissus*).

20.4   SUMMARY AND CONCLUSIONS

A number of studies have shown that historical cores can be used to reconstruct the history of organic pollutant inputs in urban areas. The temporal trends found for contaminants in estuaries of the U.S. South Atlantic are similar to those found in fine-grained, well-laminated sediments in estuaries near other urban areas. In the present study, it was found that abandoned boat slips are excellent sites for historical coring since they can provide a high-resolution sediment record that can span many decades. Our studies showed relatively high input of PCBs over the past 70 years into two busy ports (Savannah and Brunswick, Georgia) located on the southeast coast of the United States. As expected, there was a gradual increase over time in PCB concentrations with peaks at both locations in the late 1960s. The law prohibiting the production of PCBs in 1976 presumably led to the observed concentration decreases after 1970. It should be noted that, even though there was a decrease, PCBs concentrations (15–75 µg/g) continued to be elevated in Savannah and Brunswick cores from 1970 to the present time. Three possible explanations for the continued high PCB concentration in the recent sections of the cores are (1) leakage from older transformers and capacitors containing PCBs; (2) soil with PCBs from the watershed area of the rivers is carried and deposited on the coast; (3) long-range aerial transport and deposition of PCBs, e.g., former PCB manufacturing facility in Anniston, Alabama [25]. The PCB concentrations in a 75 year core from Sapelo Island, GA, were near or below the limits of detection throughout the column. The very low PCB concentration at the Sapelo Island site, even though it is 45 km from the port of Brunswick, suggests little or no transport of sources of high PCBs from Brunswick and other nearby areas to Sapelo Island. PAHs from two sites in the Savannah River Estuary showed how sites relatively close to each other can have very different PAH concentrations. Total PAH concentration in the core from the Savannah port (up to 7200 ng/g in 1950s) was much higher than the highest PAH concentrations (600–800 ng/g in 1970s and 1980s) in a core from an estuarine site several kilometers away from the port. The significant differences in PAH concentrations and profiles of cores illustrate the variability that can result from differences in point-source impacts to local sediments. By extension, sediment core collection strategies should be used, which result in data that can identify the temporal trends in both general anthropogenic conditions, as well as impacts from point sources.

ACKNOWLEDGMENTS

We thank Bommanna Loganathan Keith Maruya for their help and advice during various phases of these studies. This research was funded by the National Oceanic and Atmospheric Administration through the National Estuarine Research System, Coastal Zone Management Program and the Status and Trends Program.

REFERENCES

1. Alexander, C.R., Smith, R.G., Calder, F.D., Schropp, S.J., and Windom, H.L. 1993. The historical record of metal enrichment in two Florida estuaries. *Estuaries* 16:627–637.
2. Alexander, C.R., Smith, R.G., Loganathan, B., Ertel, J., Windom, H.L., and Lee, R.F. 1999. Pollution history of the Savannah River estuary and comparisons with Baltic Sea pollution history. *Limnologica* 29:267–273.
3. Alexander, C.R., Lee, R.F., Burton, D.T., and Hall, L.W. 2005. An integrated case study for evaluating the impacts of an oil refinery effluent on aquatic biota in the Delaware River: Sediment core studies. *Human and Ecological Risk Assessment* 11:861–877.
4. Bopp, R.F., Simpson, J.J., Chillrud, S.N., and Robinson, D.W. 1993. Sediment derived chronologies of persistent contaminants in Jamaica Bay, New York. *Estuaries* 16:653–669.

5. Callaway, J.C., Delaune, R.D., and Patrick, W.H. 1998. Heavy metal chronologies in selected coastal wetlands from northern Europe. *Marine Pollution Bulletin* 36:82–96.
6. Coakley, J.P., Nagy, E., and Serodes, J.-B. 1993. Spatial and vertical trends in sediment-phase contaminants in the upper estuary of the St. Lawrence River. *Estuaries* 16:653–669.
7. Croudace, I.W. and Cundy, A.B. 1995. Heavy metal and hydrocarbon pollution in recent sediments from Southampton water, southern England: A geochemical and isotopic study. *Environmental Science & Technology* 29:1288–1296.
8. Marcomini, A., Pojana, G., Sfriso, A., and Alonso, J.-M.Q. 2000. Behavior of anionic and nonionic surfactants and their persistent metabolites in the Venice Lagoon, Italy. *Environmental Toxicology and Chemistry* 19:2000–2007.
9. Vallette-Silver, N.J. 1993. The use of sediment cores to reconstruct historical trends in contamination of estuarine and coastal sediments. *Estuaries* 16:577–588.
10. McKee, B.A., Nittrouer, C.A., and DeMaster, D.J. 1983. Concepts of sediment deposition and accumulation applied to the continental shelf near the mouth of Yangtze River. *Geology* 11:631–633.
11. Goldberg, E.D., Griffin, J.J., Hodge, V., Koide, M., and Windom, H.L. 1979. Pollution history of the Savannah River estuary. *Environmental Science & Technology* 13:588–594.
12. Thorbjarnarson, K.W., Nittrouer, C.A., DeMaster, D.J., and McKinney, R.B. 1985. Sediment accumulation in a back-barrier lagoon, Great Sound, New Jersey. *Journal of Sediment Petrology* 55:856–863.
13. Oldfield, F. and Appleby, P.G. 1984. Empirical testing of Pb-210 dating models. In: *Lake Sediments and Environmental History*, eds. E.Y. Haworth and J.W.G. Lund. Minneapolis, MN: University of Minnesota Press, pp. 93–124.
14. Stout, S.A., Uhler, A.D., and Emsbo-Mattingly, S.D. 2004. Comparative evaluation of background anthropogenic hydrocarbons in surficial sediments from nine urban waterways. *Environmental Science & Technology* 38:2987–2994.
15. Loganathan, B.G. and Kannan, K. 1991. Time perspectives of organochlorine contamination in the global environment. *Marine Pollution Bulletin* 22:582–584.
16. Johnson, G.W., Quensen, J.F., Chiarenzelli, J.R., and Hamilton, M.C. 2006. Polychlorinated biphenyls. In: *Environmental Forensics—Contaminant Specific Guide*, eds. R.D. Morrison and B.L. Murphy. New York: Elsevier, pp. 187–226.
17. Cutshall, J.H., Larsen, I.L., and Olsen, C.R. 1983. Direct analysis of ^{210}Pb in sediment samples: Self-absorption corrections. *Nuclear Instruments and Methods* 206:309–312.
18. Kuehl, S.A., Nittrouer, C.A., and DeMaster, D.J. 1986. Natural of sediment accumulation on the Amazon continental shelf. *Continental Shelf Research* 6:209–225.
19. Olsen, C.R., Larsen, I.L., Powry, P.D. et al. 1986. Geochemistry and deposition of ^{7}Be in river-estuarine and coastal waters. *Journal of Geophysical Reviews* 91:896–908.
20. Krishnaswami, S., Lal, D., Martin, J.M., and Maybeck, M. 1971. Geochronology of lake sediments. *Earth and Planetary Letters* 11:407–414.
21. Koide, M., Soutar, A., and Goldberg, E.D. 1972. Marine geochronology with ^{210}Pb. *Earth and Planetary Science Letters* 14:442–446.
22. Nittrouer, C.A., Sternberg, R.W., Carpenter, R., and Bennett, J.T. 1979. The use of ^{210}Pb geochronology as a sedimentological tool: Application to the Washington continental shelf. *Marine Geology* 31:297–316.
23. Alexander, C.R., Nittrouer, C.A., and DeMaster, D.J. 1991. Sediment accumulation in a modern epicontinental-shelf setting: The Yellow Sea. *Marine Geology* 98:51–72.
24. MacLeod, W.D., Brown, D.W., Friedman, A.J., Burrows, D.G., Maynes, O., Pearce, R.W., Wigren, C.A., and Bogar, R.G. 1985. Standard analytical procedures of the NOAA National Analytical Facility 1985–1986. Extractable toxic organic compounds. U.S. Department of Commerce, NOAA/NMFS. NOAA Technical Memorandum NMFS F/NWC-92.
25. U.S. Environmental Protection Agency. 2005. Preliminary assessment of the potential contributions of PCB air emission from the Solution (formerly Monsanto) facility in Anniston, Alabama to measured PCB soil concentrations near the facility. Technical Memorandum, Air Quality Modeling and Transportation Section.

21 Temporal Trends of Selected Chlorinated Hydrocarbon Pollutants in Sediments of Kentucky Lake

Baki B. Sadi and Bommanna G. Loganathan*

CONTENTS

21.1 Introduction 481
21.2 Materials and Methods 482
    21.2.1 Sample Site Selection and Sampling 482
    21.2.2 Sample Analysis 483
        21.2.2.1 Radiochronology...483
        21.2.2.2 Organic Carbon and Nitrogen..483
        21.2.2.3 PCBs and Pesticides...484
21.3 Results and Discussion 485
    21.3.1 Radiochronological Study (Sediment Dating)...485
    21.3.2 Temporal Trends of Polychlorinated Biphenyls...486
    21.3.3 Temporal Trends of Chlorinated Pesticides...489
21.4 Conclusions...492
References..492

21.1 INTRODUCTION

Polychlorinated biphenyls (PCBs) and chlorinated hydrocarbon pesticides (DDTs, hexachloroben-zene, chlordane compounds) are anthropogenic chemicals, and are known for their contamination of aquatic and terrestrial environments, bioaccumulation in the food chain, and long-term health effects on aquatic and terrestrial animals—including humans [1–4]. Due to the recalcitrant proper-ties and long-term effects of these chemicals, temporal trend investigations are essential for under-standing their environmental behavior and fate and to prevent related health hazards. This study was conducted to describe temporal trends of PCBs and chlorinated pesticide pollutants in Kentucky Lake using sediment core analysis.

The spatial and temporal trend investigation using sediment core has been employed by many groups of investigators to reconstruct the history of contaminant input into reservoirs, lakes, riv-ers, and oceans throughout the world [5–19]. Hydrophobic organic contaminants tend to adsorb to sediments relatively rapidly and that the sediment column represents a continuous sequence of sedimentation and associated organic contaminant accumulation. The equilibrium distribution

---

* E-mail: bommanna.loganathan@murraystate.edu (Chapter corresponding author).

of hydrophobic organic compounds (HOCs) such as PCBs, chlorinated pesticides, etc. is considered to primarily result from partitioning (sorption) of HOCs between sediment organic matter and aqueous phase [20,21]. Using radiochemical chronologies, it is possible to date sediments over a period corresponding to the half-lives of suitable radionuclides present in the sediment. Ages can be assigned to the different depth intervals within the sediment core by using the depth-distribution of activities of ^{210}Pb and ^{137}Cs [22–25].

Kentucky Lake is one of the main human-constructed (in 1944) lakes in the United States. It is the last and largest in a series of nine reservoirs on the mainstream Tennessee River and the largest reservoir in the southeastern United States [26]. It is the ultimate repository of substances entering this watershed from portions of seven southeastern states, which include sizable fractions of chemical processing, agricultural chemicals, and electronic manufacturing products. Kentucky Dam Tailwater, which connects to the Ohio River, receives discharges from industries in the Calvert City Industrial Complex (CCIC). During the last two decades, mass mortalities of mussels have been encountered in these regional waters; the causes of which have not been fully elucidated. In some cases, the quality and quantity of mussels harvested for the pearl industry have been substantially reduced [27,28]. Furthermore, a recent Public Advisory Committee Report [29] identified diminished air quality due to high levels of air pollutants released from a variety of industries (chemical, metallurgical, and paper mills) in this region. Recent studies have detected PCB congeners and chlorinated pesticides (DDTs, chlordane compounds, hexachlorocyclohexanes, and hexachlorobenzene) in sediment, freshwater mussel tissues, and in pine needles collected from the Kentucky Lake and Kentucky Dam Tailwater [30–33]. These observations indicate the historical contamination of organic pollutants in the Kentucky Lake Reservoir. However, no comprehensive study has been conducted to elucidate the temporal trends of the pollutants input in this region.

In this chapter, we present the results of our study aiming at reconstructing the history of persistent organic pollutants input into Kentucky Lake Reservoir over a 100 year period using the sedimentary record, as revealed in sediment cores collected from selected locations in the Kentucky Lake and downstream of Kentucky Lake Reservoir.

21.2   MATERIALS AND METHODS

21.2.1   SAMPLE SITE SELECTION AND SAMPLING

Ledbetter Embayment (LE—site 1) of Kentucky Lake, Air Products Outfall (at Tennessee River Mile (TRM): 17.7—site 2), and Cypress Creek mouth (TRM 10.1—site 3) in the Kentucky Dam Tailwater (as shown in Figure 21.1) were selected as the sampling sites for the purpose of this study. Site 1 (the Ledbetter Embayment site) is situated far from the mainstream of water currently in Kentucky Lake so that the natural sedimentation process is not disturbed. Therefore, the sediment core collected from this location should represent a continuous sequence of sedimentation and consequent accumulation of contaminants over the time frame of interest. Unlike site 1, sites 2 and 3 (from the Kentucky Dam Tailwater) are relatively more dynamic. Details of the sampling locations are shown in Table 21.1.

A stainless steel inner liner (6′ long and 2″ i.d.) was used to collect the sediment core. The inner liner was inserted into a custom-made cast iron sediment core sampler. It was pushed manually through the sediment under water to collect the sediment core from a boat. The sediment cores collected were 0.5–1 m long. After collecting the sediment core, the inner liner was taken out of the core sampler and was immediately transported to the laboratory. A stainless steel iron rod was used to push the sediment core out of the inner liner. The cores were cut into 5 cm sections and were collected in pre-weighed wide mouth glass bottles. For chemical analysis, different core sections were then freeze-dried using a Freezone Freeze Dry System (Model: 77535) for 60 h and were ground to powder using a precleaned mortar and pestle.

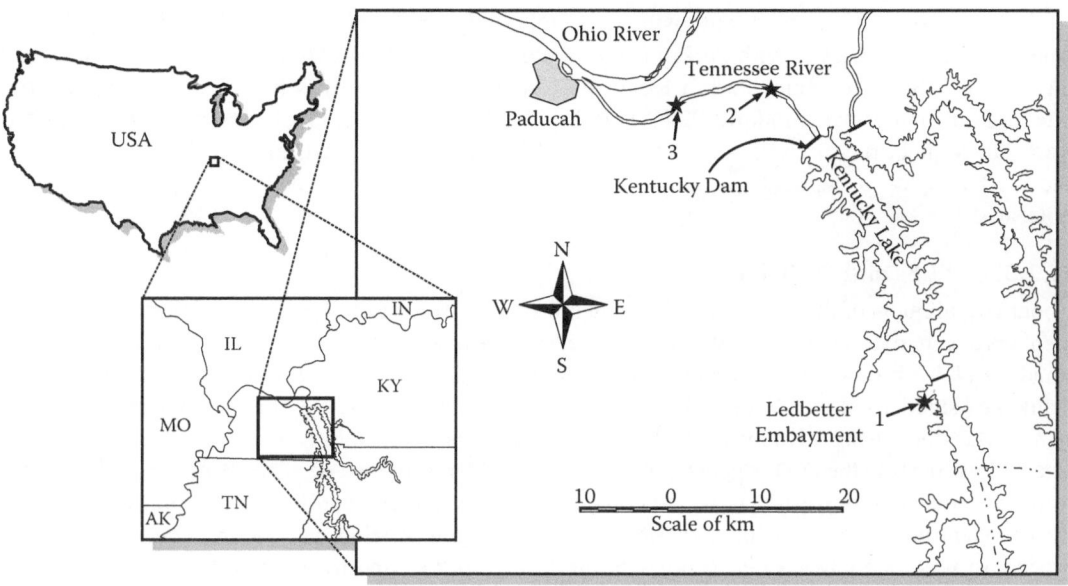

FIGURE 21.1 Map showing sediment core sampling locations. Site 1: Ledbetter Embayment, Site 2: Air Products Outfalls, and Site 3: Cypress Creek mouth.

TABLE 21.1
Details of the Sampling Locations

Name of the Site	Site ID	Date of Sampling	Latitude	Longitude
Ledbetter Embayment (Site 1)	LE	02/12/01	36° 45′ 02″N	88° 08′ 19″W
Air Product outfall TRM 17.7 (Site 2)	Air	07/31/00	37° 03′ 21.591″N	88° 20′ 00.971″W
Cypress Creek mouth TRM 10.1 (Site 3)	Cym	07/31/00	37° 01′ 47.421″N	88° 27′ 13.307″W

21.2.2 SAMPLE ANALYSIS

21.2.2.1 Radiochronology

Immediately after sampling, each of the 5 cm sections of the sediment core from the Ledbetter Embayment site were collected in pre-weighed glass beakers and re-weighed. They were dried in an oven at 60°C for 72 h. The activities of ^{210}Pb and ^{137}Cs within each sample were determined concurrently using two low background, planar intrinsic germanium detectors, a computer-based multichannel analyzer, and Maestro-II analysis software [22]. The sediment accumulation rate was calculated by producing a depth profile of total ^{210}Pb activity within the sediment core. The calculated ^{210}Pb accumulation rate was verified using ^{137}Cs as a complementary radiotracer.

21.2.2.2 Organic Carbon and Nitrogen

The sample preparation for the organic carbon and nitrogen analysis in sediments were performed following the procedure described by Wong et al. [34].

The analysis was carried out using a CHN analyzer (Perkin Elmer Series II CHNS/O Analyzer—2400 Series II with Perkin Elmer Auto balance—AD4). Blanks (tin cups), conditioning reagents (Sulfamic acid; Perkin Elmer-N241-0501), and standards (CHN standard; acetanilide; organic analytical standard; 0240-121; C = 71.09%, H = 6.71%, O = 11.86%, and N = 10.38%) were run to meet the analytical quality control (QC) and quality assurance (QA) criteria. After every 10 samples, a duplicate, a blank, and a standard were run to verify the instrumental QC criteria.

21.2.2.3   PCBs and Pesticides

About 20 g aliquots of the freeze-dried, powdered sample from each of the 5 cm sections of the sediment core were extracted with 180 mL of an acetone/hexane (1:1 v/v) solution for 24 h. An internal standard (4,4-dibromooctafluorobiphenyl in acetone; 100 ng/250 μL) was spiked to each sample, blank, and standard reference material prior to Soxhlet extraction. The extract volume reduction (to 10 mL) was done using a Kuderna Danish (K-D) apparatus. The analytes were then exchanged into hexane by repeating the K-D concentration twice with 100 mL of hexane. The 10 mL hexane extracts were then treated with freshly activated copper to remove elemental sulfur. The sample extract was then concentrated to 5 mL using a gentle stream of nitrogen. A silica-gel column chromatographic procedure was then carried out to remove interfering species and to separate the PCBs from the pesticides. The PCBs congeners and some pesticides were eluted with 115 mL of ultrapure hexane. The chlorinated pesticides were eluted with 115 mL of 20% (v/v) dichloromethane in hexane. The hexane fraction was then concentrated to 10 mL using a K-D concentration apparatus followed by nitrogen gas microconcentration to 1 mL followed by GC-ECD analysis. The dichloromethane fraction was similarly concentrated to 10 mL using K-D concentration. K-D concentration was repeated again with another 100 mL of ultrapure hexane followed by nitrogen gas microconcentration to 1 mL. This 1 mL extract was used for GC-ECD analysis of pesticides.

PCB congeners and chlorinated pesticides were analyzed using a Shimadzu Model GC-17A gas chromatograph (GC) with Shimadzu Model AOC-17 auto injector. The GC was equipped with a DB-5 (60 m; 0.25 mm i.d.; 0.25 μm film thickness) capillary column and a ^{63}Ni electron capture detector. The column oven temperature program was 90°C (hold time: 1 min), 90°C–200°C at 10°C/min (hold time: 0 min), 200°C–280°C at 1°C/min (hold time: 20 min). The total run time was 112 min. The injection port and detector temperatures were 280°C and 330°C, respectively.

Helium (1 mL/min) and nitrogen (28 mL/min) were used as the carrier gas and makeup gas, respectively. The split ratio was 20:1.

PCBs were analyzed for 120 different congeners. Fifteen different chlorinated pesticides were analyzed. The retention times of the individual PCB congeners and the chlorinated pesticides were determined by injecting the pure standards (Accu Standard and Ultra Scientific Company). The *r*-square and mean slope values (response factors) from five-point calibration were calculated for all the PCB congeners and pesticides. Samples prepared from the different sections of the sediment cores were analyzed through the GC-ECD.

Quality assurance and quality control (QA/QC) protocols were followed to evaluate the reliability of the data. The QC used for PCB congeners and chlorinated pesticides analysis included reagent blank (<method detection limit), surrogate standard recovery (100% ±30% of known standard added prior to extraction), and relative accuracy using Standard Reference Material-1941a (100% ±30% of known certified material) [35]. Calibration and calibration verification (five-point calibration with $r^2 = 0.99$) were checked routinely at the beginning and end of each batch of 10 samples. Detection limits were calculated following an approved method [36]. The area of baseline noise over the elution time of each congener was determined from seven injections of matrix blank (hexane) spiked with the lowest concentration of calibration standard. The instrumental detection limits (IDL) were three times the standard deviation of the baseline noise divided by the slope of the calibration curve.

21.3 RESULTS AND DISCUSSION

21.3.1 RADIOCHRONOLOGICAL STUDY (SEDIMENT DATING)

Depth distributions of activity of ^{210}Pb and ^{137}Cs in the sediment core from Ledbetter Embayment are shown in Figures 21.2 and 21.3. The calculated sediment accumulation rate for the Ledbetter Embayment was 0.35 cm/year. Typically, reservoirs exhibit sedimentation rates that range from 0.10 to 2 cm/year [37]. The accumulation rate was used to assign the ages to the different sections of the sediment core. Unlike ideal cases, the predicted and observed ^{137}Cs penetration and 1963 peak location did not agree well. ^{137}Cs was found in the deeper core sections than predicted by the ^{210}Pb, which usually indicates some mixing by biological organisms, but could also be due to a rapid

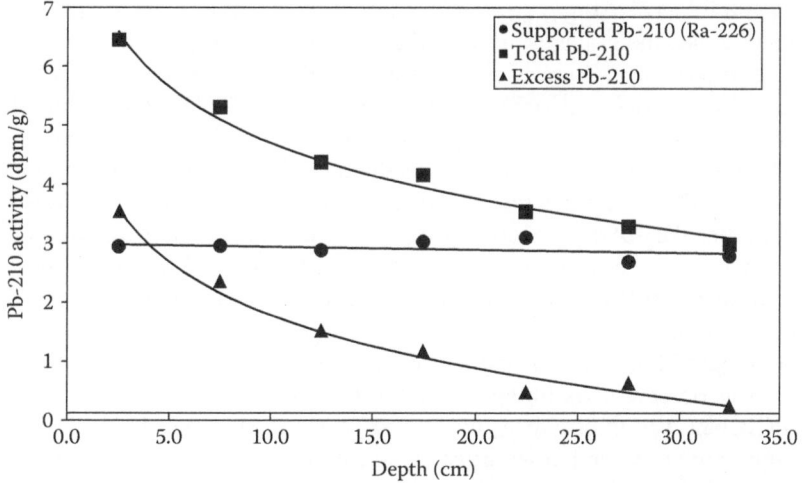

FIGURE 21.2 Depth distribution of activity of ^{210}Pb in the sediment core from Ledbetter Embayment, Kentucky Lake.

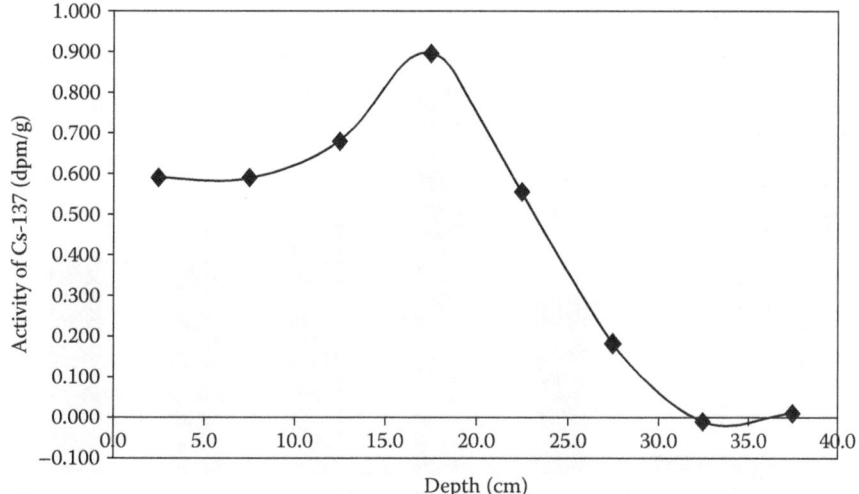

FIGURE 21.3 Depth distribution of activity of ^{137}Cs in the sediment core from Ledbetter Embayment, Kentucky Lake.

accumulation in the past couple of decades. However, the well-preserved ^{137}Cs peak indicates that the profiles are relatively undisturbed. If there were intense mixing, the ^{137}Cs peak would be totally smeared out. Moreover, no obvious surface mixed layer was observed in the ^{210}Pb profile, so vigorous mixing of the sediments by biological organism is not obvious. The broad ^{137}Cs peak suggests that the input of ^{137}Cs into the lake is either over a prolonged period or mixing has turned the sharp peak into a broad one.

21.3.2 Temporal Trends of Polychlorinated Biphenyls

Figure 21.4 shows temporal trends of organic-carbon-normalized total PCB concentrations in sediment collected from the Ledbetter Embayment site. At the 25–30 cm section, the total PCB concentration recorded was 1.05 ng/g C. It was found to increase sharply thereafter until it reached a maximum concentration (23.20 ng/g C) at the 15–20 cm section. Afterwards, the total PCB concentration was found to decline gradually toward the topmost section (0–5 cm). Total PCB concentration at the topmost section (0–5 cm) was 12.89 ng/g C. Radio-chronological dating was assigned to the sediment core sections to describe the temporal trends of total PCB. A significant increase in total PCB concentration (1.05–10.52 ng/g C) was observed between 1929 and 1943. This sharp increase reflects the beginning of commercial production of PCBs in the United States in 1929 [3]. The maximum total PCB concentration (23.20 ng/g C) in the temporal trend was observed during 1943–1958. After the 1980s, a declining trend was observed. The peak levels (11–23 ng/g C) of PCBs were found between the early 1940s and the early 1970s. Total PCB concentration in the surface sediment (top 5 cm), which represents the status of contamination during 1983 and 2001, was found to be 12.89 ng/g C.

As shown in Figure 21.4, peak concentrations of PCBs were found between 1943 and 1972. The declining trend was observed beginning in the 1960s. This early declining trend was unexpected because the PCB ban was imposed in the United States in the year 1976. Possible reasons for this early declining trend in the Ledbetter Embayment of Kentucky Lake may be attributed to the dynamic nature of this aquatic environment. In addition, the restriction on PCBs in open systems was imposed as early as the late 1960s and early 1970s [38]. Drastic declines in PCB levels in the riverine environments after the restriction on the use of organochlorine compounds was reported by earlier studies [39,40]. Also, contamination of deeper core sections during the extrusion and the

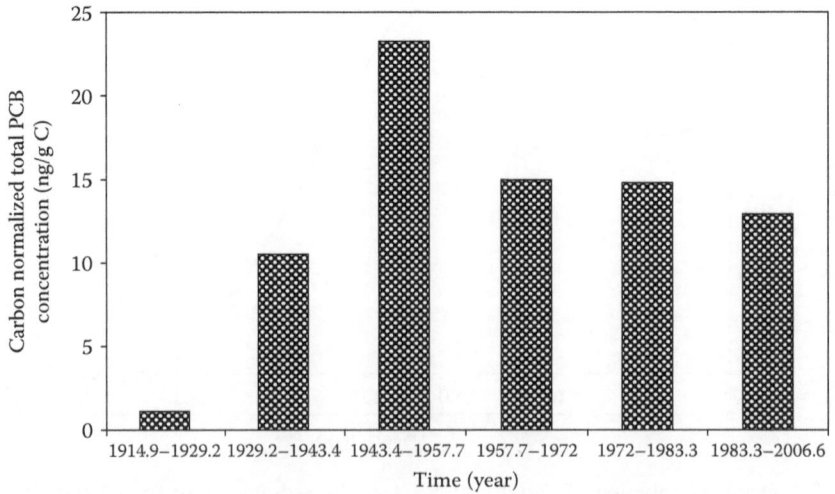

FIGURE 21.4 Temporal trend of carbon-normalized total PCB in the sediment core from Ledbetter Embayment, Kentucky Lake.

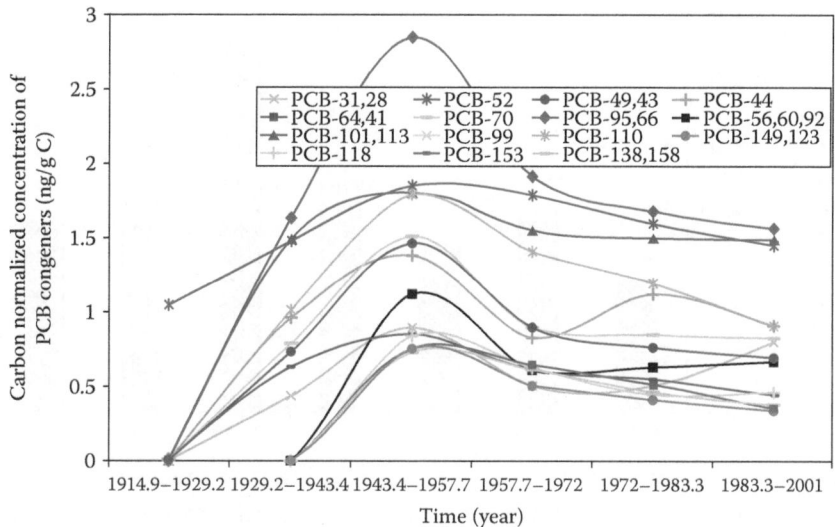

FIGURE 21.5   Temporal trends of selected PCB congeners in the sediment core from Ledbetter Embayment, Kentucky Lake.

smearing effect on the core wall resulting in misallocation of ^{210}Pb and ^{137}Cs derived chronology cannot be ruled out [41]. Since the late 1950s, over 1 million metric tons of PCBs have been produced in the United States [42]. PCBs were extensively used during the periods from the 1950s to the 1960s in a variety of industrial applications including heat transfer fluids in electrical transformers and capacitors [38]. Dated cores at various locations in North America generally show higher concentrations of PCBs in layers from the 1960s [37,43]. Changes in the PCB concentrations in the Ledbetter Embayment of the Kentucky Lake sediment align with U.S. production, use, and regulation of PCBs [44,45].

The temporal trends of some selected PCB congeners in the Ledbetter Embayment site are shown in Figure 21.5. Most of the PCB congeners found in the Ledbetter Embayment core follow similar temporal trends. Among the PCB congeners detected, the concentrations of lower chlorinated congeners (di-, tri-, and tetrachlorobiphenyls) showed relatively higher concentrations than the higher chlorinated ones (penta-, hexa-, hepta-, and octachlorobiphenyls). Highly toxic coplanar PCBs (non-*ortho*-chlorine substituted) such as PCB-77 (3,3′,4,4′-tetrachlorobiphenyl) and PCB-126 (3,3′,4,4′,5-pentachlorobiphenyl) were not detected or barely detected in these samples. However, mono-*ortho*-chlorine substituted PCB congeners such as PCB-105 (2,3,3′,4,4′-pentachlorobiphenyl) and PCB-118 (2,3′,4,4′,5-pentachlorobiphenyl) were detected in these samples. These congeners are reported to elicit toxic effects similar to polychlorinated dibenzo-*p*-dioxin [46]. The presence of PCB-28 (2,2′,4′-trichlorobiphenyl) and PCB-52 (2,2′,5,5′-tetrachlorobiphenyl) in the core sediments indicates that these samples contained Aroclor 1016 and Aroclor 1242. Aroclor 1016 and Aroclor 1242 were primarily used in paper mills, lubricating and cutting oils. Detectable concentrations of PCB-101 (2,2′,4,5,5′-pentachlorobiphenyl), PCB-138 (2,2′,3,4,4′,5-hexachlorobiphenyl), and PCB-153 (2,2′,4,4′,5,5′-hexachlorobiphenyl) revealed that the core sediments were exposed to Aroclor 1254 and Aroclor 1260. These (higher percent chlorinated) Aroclors were primarily used as dielectric fluid in electrical transformers and capacitors [38].

Figure 21.6 shows the vertical trends of carbon-normalized total PCB concentration in the sediment core from the Air Products Outfall site (TRM 17.7) in the Kentucky Dam Tailwater. The maximum carbon-normalized total PCB concentration (1107.58 ng/g C) in this site was found at the depth of 10–15 cm in the sediment core. The total PCB concentration in the surface sediment (0–5 cm) was found to be 569.5 ng/g C. The total PCB concentrations in this site were two orders

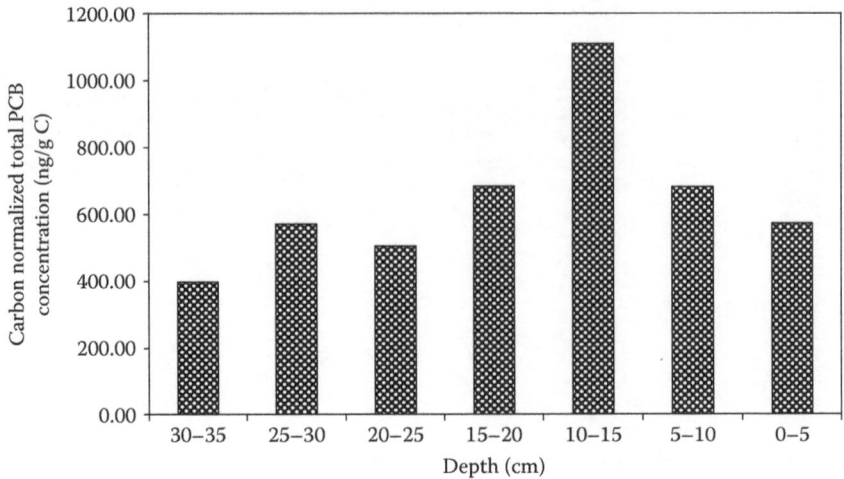

FIGURE 21.6   Vertical trend of total PCB concentration of the sediment core from Air Products outfall site.

of magnitude higher than that observed in the Ledbetter Embayment site, indicating historical and localized industrial PCB contamination.

As shown in Figure 21.7, unlike the Ledbetter Embayment and Air Products Outfall sites, the Cypress Creek mouth (another sampling location in the Kentucky Dam Tailwater) showed a gradual increasing trend of total PCBs in the sediment core sections representing recent years. This could be due to increasing input of PCBs in this site or due to mixing of the sediment disturbing the equilibrium distribution of sedimentation and associated contaminant accumulation.

QA/QC data obtained for the analysis of PCB congeners met the Environmental Protection Agencies (EPA) recommended criteria for QA/QC analysis. The average r-square value obtained for 120 PCB congeners was 0.99. IDLs ranged from 1.83 pg/μL (PCB-99) to 27.74 pg/μL (PCB-8). The percent recovery values for the NIST standard reference material (SRM 1941a) ranged from 70.43% (PCB-138/158) to 130.15% (PCB-99). The average surrogate standard (4,4′-dibromo-octafluorobiphenyls) recovery values in individual samples, blanks, and standard reference material (SRM 1941a) from different core sections was 102.69%.

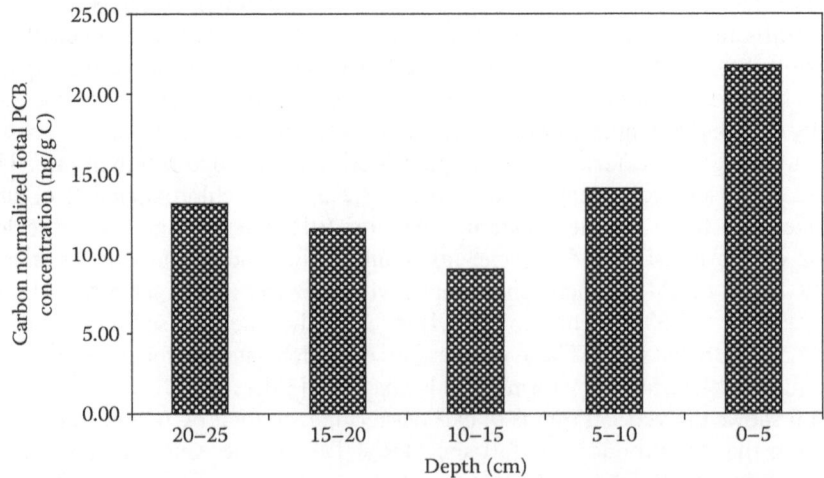

FIGURE 21.7   Vertical trend of total PCB concentration in the sediment core from Cypress Creek mouth site.

21.3.3 TEMPORAL TRENDS OF CHLORINATED PESTICIDES

Chlorinated pesticide concentrations at different depth intervals of the sediment core collected from the Ledbetter Embayment site are shown in Figure 21.8. Total DDT (DDT+DDD+DDE) levels gradually increased and reached their highest concentration during the early 1940s to the early 1970s (7.5–10.3 ng/g C) in the 15–25 cm sediment section in the Ledbetter Embayment core. A steady declining trend was observed from 1972 to 2000. The concentration of total DDT in the surface sediment (top 5 cm) was found to be 2.78 ng/g C. After its introduction as an insecticide in 1929, DDT was a widely used pesticide in the United States in the 1950s and 1960s [22]. The manufacture of DDT was banned in the United States in 1972. The peak in the temporal trends of total DDT in the Ledbetter Embayment site agrees well with the maximum production and usage period of pesticides in the United States. DDT is rapidly converted to DDE or DDD in sediments,

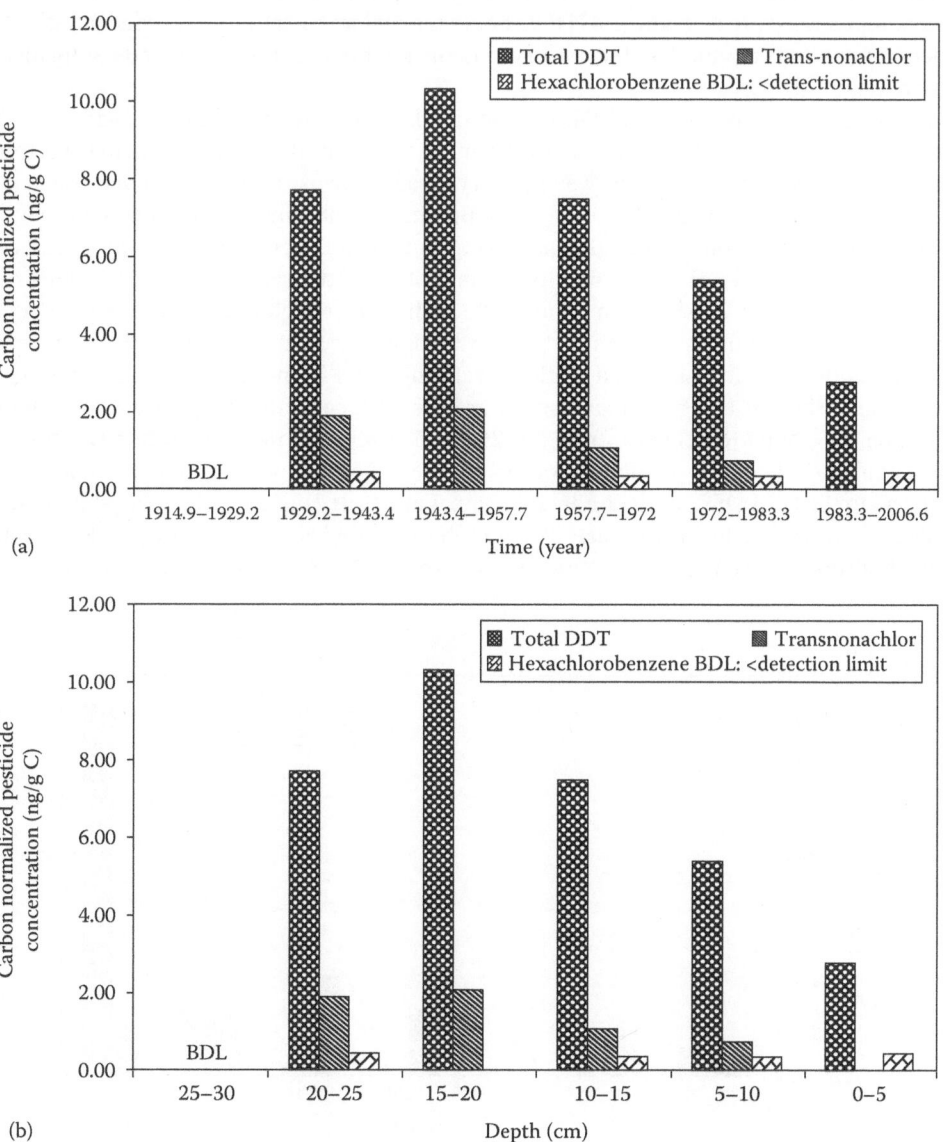

(a)

(b)

FIGURE 21.8 Temporal trends of total DDT, hexachlorobenzene, and trans-nonachlor in the sediment core from Ledbetter Embayment, Kentucky Lake.

thus explaining the lack of DDT in all the sediment sections. DDT metabolites found were 4,4′-DDD, 2,4′-DDD, 4,4′-DDE, and 2,4′-DDE. Among them, 4,4′-DDE was the major metabolite. DDE is a metabolite of DDT under aerobic (oxidizing) conditions while DDD is a metabolite of DDT under anaerobic (reducing) conditions due to microbial activity and the reaction of DDT with iron porphyrins [22]. The ratio of 4,4′-DDD to 4,4′-DDE ranged from 0.56 to 0.42 from 1943 to 1958 and then ranged from 0.41 to 0.38 from 1972 to 1983; the ratio at the surface sediment was found to be 0.00. This indicates the condition of sediment deposition could be somewhat reducing when the DDT was first associated with the sediment and then changed slowly to oxidizing conditions toward the surface sediment. This could also be due to the depletion of dissolved oxygen toward the bottom of the sediment core.

A similar trend was observed for total DDT in the Air Products Outfall (TRM 17.7) site (Figure 21.9). The peak in the temporal trend was observed between the 5–10 and 15–25 cm sections within the sediment core. All the DDT metabolites (DDDs and DDEs), including 4,4′-DDT and 2,4′-DDT, were found to be present with 2,4′-DDD being the major metabolite. The concentrations were one order of magnitude higher than that observed in the Ledbetter Embayment, indicating historical contamination in this site.

In the Cypress Creek mouth, total DDT showed a declining trend from the 20–25 to 10–15 cm sections in the sediment core (Figure 21.10). From 10 to 15 cm, it showed an increasing trend up to the top section (0–5 cm). The trend was not very clear in terms of production and usage pattern of pesticides. As was observed in the case of PCBs, the equilibrium distribution could have been disturbed due to the possibility of sediment mixing in this site. Heptachlor, an allylic chlorinated product of aldrin, showed a similar trend to the total DDT in the sediment from the Air Products Outfall site (Figure 21.11). As shown in Figure 21.11, aldrin and cis-chlordane were two other chlorinated pesticides found in that site. Both of these pesticides showed increasing trends from the 25–30 to 20–25 cm sections. Aldrin showed a gradual declining trend from the 20–25 to 5–10 cm sections and therefrom it showed a further increase toward the surface sediment (0–5 cm). cis-Chlordane concentration was found to decrease from 20–25 to 15–20 cm followed by an increase toward the surface sediment (0–5 cm). Trans-nonachlor, one of the major components in chlordane, showed a similar trend to the total DDT in the Ledbetter Embayment site (Figure 21.8). The peak in the temporal trend was observed during the early 1940s to the early 1970s. Chlordane was released into the environment primarily from its application as a soil insecticide (primarily in termite control) [47].

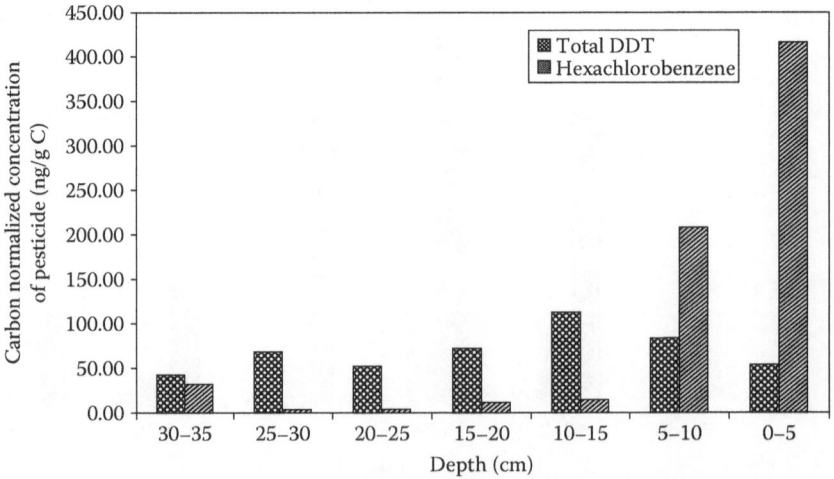

FIGURE 21.9 Vertical trends of total DDT, and hexachlorobenzene in the sediment core from air products outfall site.

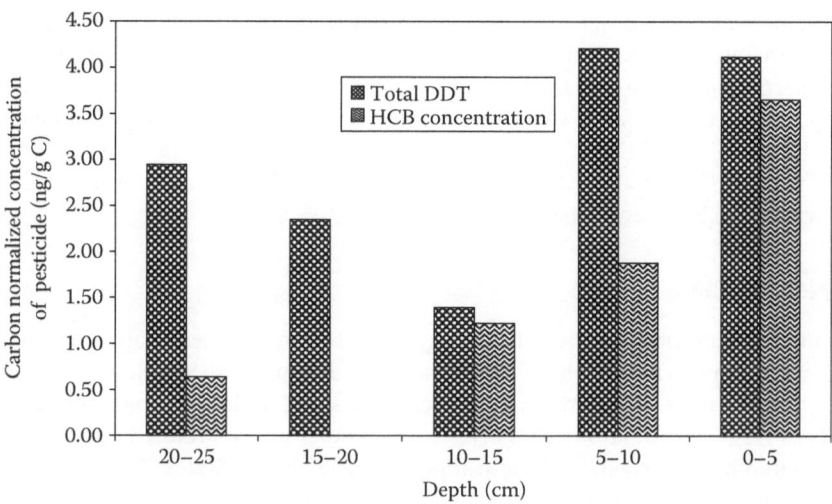

FIGURE 21.10 Vertical trends of total DDT and hexachlorobenzene in the sediment core from Cypress Creek mouth site.

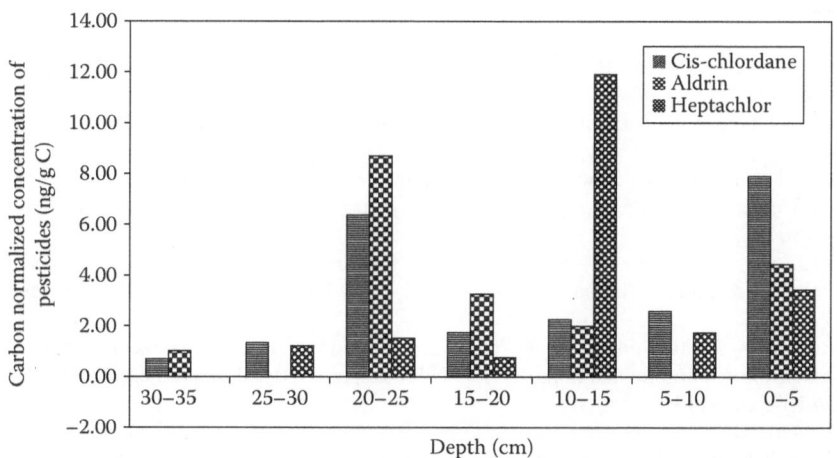

FIGURE 21.11 Vertical trends of aldrin, cis-chlordane, and heptachlor in the sediment core from Air Products outfall site.

In the United States, chlordane was used extensively prior to 1983 and from 1983 to 1988 when it was registered for termite control. It was prohibited for this use in 1988 [3].

Unlike PCBs, hexachlorobenzene (HCB) in all the sampling locations showed a steadily increasing trend, which indicates the increasing use of the pesticide in this geographic region. In the Ledbetter Embayment site (Figure 21.8), it was found to be present as early as 1943. From 1972, it showed an increasing trend toward the surface sediment. As shown in Figure 21.10, in the Cypress Creek mouth site, HCB showed a steadily increasing trend from 15 to 20 cm to the top section (0–5 cm). Similarly, in the Air Products Outfall site (Figure 21.9), a sharp increase was observed from 10 to 15 cm to the top section (0–5 cm). HCB was introduced commercially as a fungicide for wheat in 1933. It also had industrial uses in organic synthesis. Combustion processes such as waste incineration was another major source of HCB [48]. Its use as a fungicide was banned in the United States in the 1970s [49]. Since HCB concentrations showed an increasing trend, continuous monitoring of this pesticide is needed in the future to prevent further contamination and harmful effects.

The average r-square value for 15 chlorinated pesticides was 0.999. Instrumental detection limits ranged from 1.74 pg/μL (2,4′-DDE) to 7.31 pg/μL (2,4′-DDD). Percent recovery values of chlorinated pesticides in NIST standard reference material (SRM 1941a) for most of the chlorinated pesticides were within the EPA recommended range (70%–130%).

21.4   CONCLUSIONS

Based on the trend data, presently, the Kentucky Lake sediment is less polluted than it was two decades ago with respect to the chemicals measured. Although the pollutant trends observed were similar to the trends reported for other reservoirs and Lakes, absolute concentrations of the pollutants in Kentucky Lake were relatively less. Therefore, Kentucky Lake is comparatively less polluted than other reservoirs and natural lakes in the United States [50,51]. Since hexachlorobenzene showed an increasing trend, continuous monitoring of this pollutant is warranted. In harmony with many other pollution history studies, this research work indicates that, in general, the input of the persistent organic chemicals has declined during the past two decades, suggesting that pollution control laws have been effective, even while industrial and population growth is taking place.

REFERENCES

1. Riget, F., Bignert, A., Braune, B., Stow, J., and Wilson, S. 2010. Temporal trends of legacy POPs in Arctic biota, an update. *Science of the Total Environment* 408: 2874–2884.
2. Corsolini, S. Industrial contaminants in Antarctic biota. 2009. *Journal of Chromatography A* 1216: 598–612.
3. Lipnick, R. L., Hermens, J., Jones, K. C., and Muir, D. C. G. (eds.). 2001. *Persistent, Bioaccumulative and Toxic Chemicals I: Fate and Exposure*. ACS Symposium Series 772, American Chemical Society, Washington, DC.
4. Loganathan, B. G., Tanabe, S., Hidaka, Y., and Kawano, M. 1993. Temporal trends of persistent organochlorine residues in human adipose tissues from Japan, 1928–1985. *Environmental Pollution* 81: 31–39.
5. Li, A., Rockne, K. J., Sturchio, N., Song, W., Ford, J. C., and Wei, H. 2009. PCBs in sediments of the Great Lakes—Distribution and trends, homolog and chlorine patterns, and in situ degradation. *Environmental Pollution* 157: 141–147.
6. Jung, S., Arnaud, F., and Bonte, P. et al. 2008. Temporal evolution of urban wet weather pollution: Analysis of PCB and PAH in sediment cores from Lake Bourget, France. *Water Science and Technology* 57: 1503–1510.
7. Babek, O., Hilscherova, K., and Nehyba, S. et al. 2008. Contamination history of suspended river sediments accumulated in oxbow lakes over the last 25 years: Morava River (Danube catchment area), Czech Republic. *Journal of Soils and Sediments* 8: 165–176.
8. Zennegg, M., Kohler, M., and Hartmann, P. C. et al. 2007. The historical record of PCB and PCDD/F deposition at Greifensee, a lake of the Swiss plateau, between 1848 and 1999. *Chemosphere* 67: 1754–1761.
9. Sapota, G. 2006. Persistent organic pollutants (POPs) in bottom sediments from the Baltic Sea. *Oceanological and Hydrobiological Studies* 35: 295–306.
10. Malmquist, C., Bindler, R., and Renberg, I. M. et al. 2003. Time trends of selected persistent organic pollutants in lake sediments from Greenland. *Environmental Science & Technology* 37: 4319–4324.
11. Schneider, A. R., Stapleton, H. M., Cornwell, J., and Baker. J. E. 2001. Recent declines in PAH, PCB, and Toxaphene levels in the Northern Great Lakes as determined from high resolution sediment cores. *Environmental Science & Technology* 35: 3809–3815.
12. Soderstrom, M., Nylund, K., Jarnberg, U., Lithner, G., Rosen, G., and Kylin, H. 2000. Seasonal variations of DDT compounds and PCB in a eutrophic and an oligotrophic lake in relation to algal biomass. *Ambio* 29: 230–237.

13. Skei, J., Larson, P., Rosenberg, R., Jonsson, P., Olsson, M., and Broman, D. 2000. Eutrophication and contaminants in aquatic ecosystems. *Ambio* 29: 184–194.

14. Persson, J. N., Axelman J., and Broman, D. 2000. Validating possible effects of eutrophication using PCB concentrations in bivalves and sediments of US musselwatch and benthic surveillance programs. *Ambio* 29: 246–251.

15. Jonsson, P., Eckhell, J., and Larson, P. 2000. PCB and DDT in laminated sediments from offshore and archipelago areas of the NW Baltic Sea. *Ambio* 29: 268–276.

16. Wong, C. S., Sanders, G., Engstrom, D. R., Long, D. T., Swackhamer, D. L., and Eisenreich, S. J. 1995. Accumulation, inventory, and diagenesis of chlorinated hydrocarbons in Lake Ontario. *Environmental Science & Technology* 29: 2661–2672.

17. Cross, J. N., Fransisco, C., and Hallcock, D. (eds.). 1994. Southern California coastal water research project, Report of the Southern California coastal water research project authority. pp. 91–99.

18. Eisenreich, S. J., Capel, P. D., Robbins, J. A., and Bourbonnlere, R. 1989. Accumulation and diagenesis of chlorinated hydrocarbons in lacustrine sediments. *Environmental Science & Technology* 23: 1116–1126.

19. Oliver, B. G., Charlton, M. N., and Durham, R. W. 1989. Distribution, redistribution, and geochronology of polychlorinated biphenyl congeners and other chlorinated hydrocarbons in Lake Ontario sediments. *Environmental Science & Technology* 23: 200–208.

20. Mcgroddy, S. E. and Farrington, J. W. 1995. Sediment porewater partitioning of polycyclic aromatic hydrocarbons in three cores from Boston Harbor, Massachusetts. *Environmental Science & Technology* 29: 1542–1550.

21. Carmo, A. M., Hundal, L. S., and Thompson, M. L. 2000. Sorption of hydrophobic organic compounds by soil materials: Application of unit equivalent Freundlich coefficients. *Environmental Science & Technology* 34: 4363–4369.

22. Alexander, C., Loganathan, B. G., Smith, R., Ertel, J., Windom, H. L., and Lee, R. F. 1999. Pollution history of Savannah River estuary and comparisons with Baltic Sea pollution history. *Limnologica* 29: 267–273.

23. Tanner, P. A., Pan, S. M., Mao, S. Y., and Yu, K. N. 2000. γ-Ray spectrometric and α counting method comparison for the determination of ^{210}Pb in estuarine sediments. *Applied Spectroscopy* 54: 1443–1446.

24. Goldberg, E. D., Griffin, J. J., Hodge, V., Koide, M., and Windom, H. 1979. Pollution history of the Savannah River estuary. *Environmental Science & Technology* 13: 588–594.

25. Thorbjarnarson, K. W., Nittrouer, C. A., Demaster, D. J., and McKinney, R. B. 1985. Sediment accumulation in a Back-barrier Lagoon, Great Sound, New Jersey. *Journal of Sedimentary Research* 55: 856–863.

26. Carriker, N, E. and Cox, J. P. 1984. Technical Report Series: Kentucky Reservoir Water Quality-1982, July 1984.

27. Loganathan, B. G., Kawano, M., Sajwan, K. S., and Owen, D.A. 2001. Extractable organohalogens (EOX) in sediment and mussel tissues from the Kentucky Lake and the Kentucky Dam tailwater, USA. *Toxicological and Environmental Chemistry* 79: 233–242.

28. Loganathan, B. G., Kannan, K., Senthilkumar, S., Sickel, J., and Owen, D. A. 1999. Occurrence of butyltin residues in sediments and mussel tissues from the lower most Tennessee River and Kentucky Lake. *Chemosphere* 39: 2401–2408.

29. Executive Summery, Kentucky Outlook 2000. The Kentucky Environmental Protection Cabinet 1997, p. 105.

30. Seaford, K. and Loganathan B. G. 2001. Seasonal variation of organic compounds and inorganic elements in surface sediments from Ledbetter Embayment of Kentucky Lake. CSET Undergraduate Research Day Oral Presentations. Murray State University.

31. Loganathan, B. G., Owen, D. A., Sickel, J., and White, S. 2001. Status of the organochlorine pollutants in terrestrial and aquatic environments of westernmost Kentucky: Biological indicator. Ninth Symposium on the Natural History of Lower Tennessee and Cumberland River Valley. Brandon Spring Group Camp. Land between the Lakes.

32. Loganathan, B. G., Neale, J., Sickel, J., Sajwan, K., and Owen, D. A. 1998. Persistent organochlorine concentrations in sediment and mussel tissues from the lowermost Tennessee River and Kentucky Lake, USA. *Organohalogen Compounds* Dioxin 98 (Stockholm, Sweden) 39: 121–124.

33. Loganathan, B. G., Baust J. Jr., Neale, J. R., White, S., and Owen, D. A. 1998. Chlorinated hydrocarbons in pine needles: An atmospheric evaluation of Westernmost Kentucky, USA. *Organohalogen Compounds* 39: 303–306.

34. Wong, J. M., Sweet, S. T., Brooks, J. M., and Wade, T. L. 1993. Total organic and carbonate carbon content of the sediment. National status and trends program for marine environmental quality. NOAA Technical Memorandum NOS ORCA 71.

35. Loganathan, B. G., Irvine, K. N., Kannan, K., Pragatheeswaran, V., and Sajwan, K. S. 1997. Distribution of selected PCB congeners in the Babcock Street Sewer District: A multimedia approach to identify PCB sources in Combined Sewer Overflows (CSOs) discharging to the Buffalo River, New York. *Archives of Environmental Contamination and Toxicology* 33: 130–140.

36. Federal Register. 1984. Appendix B to Part 139. *Definition and Procedure for the Determination of Detection Limit* 49: 209.

37. Van Metre, P. C., Callender, E., and Fuller, C. C. 1997. Historical trends of organochlorine compounds in river basins identified using sediment cores from reservoirs. *Environmental Science & Technology* 31: 2339–2344.

38. Koplan, J. 2000. Toxicological profile for polychlorinated biphenyls, U.S. Department of Health and Human Services, Agencies for Toxic Substances and Disease Registry 765.

39. Loganathan, B. G. and Kannan, K. 1991. Time perspectives of organochlorine contamination in the global environment. *Marine Pollution Bulletin* 22: 583–584.

40. Loganathan, B. G. and Kannan, K. 1994. Global organochlorine contamination trends. *Ambio* 23: 187–191.

41. Sander, G., Jones, K. C., and Shine, A. J. The use of sediment core to reconstruct the historical input of contaminants to Loch Ness: PCBs and PAHs. Available at: http://www.lochnessproject.org/adrian_shine_archiveroom/papershtml/loch_ness_contaminants.htm

42. Colin, B. 1999. Toxic organic chemicals. *Environmental Chemistry*, 2nd edn., W. H. Freeman and Company, New York, Chapter 6.

43. Bopp, R. F., Simpson, J. J., Chillrud, S. N., and Robinson, D. W. 1993. Sediment derived chronologies of persistent organic contaminants in Jamaica Bay, New York. *Estuaries* 16: 608–616.

44. National Research Council. 1979. *Polychlorinated Biphenyls*. National Academy of Sciences: Washington, DC.

45. Peakall, D. B. 1975. PCBs and their environmental effects. *CRC Critical Reviews in Environmental Control* 5: 469–508.

46. Trowbridge, A. G. and Swackhamer, D. L. 2001. Biomagnification of toxic PCB congeners in the Lake Michigan food web. In: *Persistent Bioaccumulative and Toxic Chemicals I. Fate and Exposure*, eds. Lipnick, R. L, Hermens, J. L. M., Jones, K. C. and Muir, D. C. G. ACS Monograph 772: 266–283.

47. Extension Toxicology Network (Extoxnet): Available at http://pmep.cce.cornell.edu/profiles/extoxnet/carbaryl-dicrotophos/chlordane-ext.html

48. Westberg, H. and Selden, A. 1994. Organochlorine compounds in aluminium degassing with hexachloroethane. *Organohalogen Compounds* 20: 355–358.

49. Coutrney, K. D. 1979. Hexachlorobenzene (HCB): A review. *Environmental Research* 20: 225–266.

50. Golden, K. A., Wong, C. S., Jeremiason, J. D., Eisenreich, S. J., Sanders, G., and Hallgren, J. 1993. Accumulation and preliminary inventory of organochlorines in Great Lakes sediments. *Water Science and Technology* 28: 19–31.

51. Van Metre, P. C. and Mahler, B. J. 2005. Trends in hydrophobic organic contaminants in urban and reference lake sediments across the United States, 1970–2001. *Environmental Science & Technology* 39: 5567–5574.

Part V

Persistent Organic Chemicals
in the Coastal, Oceanic, Arctic,
and Antarctic Regions

22 Temporal Trends of Polybrominated Diphenyl Ethers and Hexabromocyclododecanes in Marine Mammals with Special Reference to Hong Kong, South China

Ling Jin, James C.W. Lam, Margaret B. Murphy, and Paul K.S. Lam*

CONTENTS

22.1 Introduction 498
    22.1.1 An Overview of Brominated Flame Retardants 498
    22.1.2 Use and Production of PBDEs and HBCDs 498
    22.1.3 PBDEs in the Environment 499
    22.1.4 HBCDs in the Environment 501
    22.1.5 Study Objectives 502
22.2 Materials and Methods 503
    22.2.1 Sample Collection and Preparation 503
    22.2.2 Chemical Analysis 503
    22.2.3 QA/QC 503
    22.2.4 Statistical Analysis 504
22.3 Results and Discussion 504
    22.3.1 Concentrations of PBDEs and HBCDs in Cetaceans 504
    22.3.2 Temporal Trend of PBDEs and HBCDs in Cetaceans 504
    22.3.3 Global Trends of PBDE and HBCD Concentrations in Marine Mammals 505
        22.3.3.1 Japan 505
        22.3.3.2 North America 506
        22.3.3.3 Arctic 506
        22.3.3.4 Europe 507

* E-mail: bhpksl@cityu.edu.hk (Chapter corresponding author).

22.4 Conclusions...507
Acknowledgments...508
References...508

22.1 INTRODUCTION

22.1.1 An Overview of Brominated Flame Retardants

Owing to advances in polymer science over the past 50 years, a large number of polymers with differing properties have been developed for daily applications ranging from clothing and furniture to electronics, vehicles, and computers. However, most of these polymers are petroleum-based and hence are flammable. In order to reduce fire risks and meet fire safety regulations, certain chemicals collectively known as flame retardants are applied to combustible materials such as plastics, wood, paper, and textiles [1]. Currently, there are more than 175 compounds or groups of compounds with known flame-retarding properties, which are generally divided into four classes: inorganic, halogenated organic, nitrogen-containing, and phosphorus-containing compounds [2]. Among the halogenated flame retardants, brominated compounds comprise the largest market share because of their lower decomposition temperatures, higher performance efficiency, and low cost [2,3]. Thus, brominated flame retardants (BFRs) have been extensively used to improve the fire resistance of materials such as plastics, textiles, furnishing foam, and electronic circuit boards [4]. Based on their use in the chemical industry, BFRs can be classified as either reactive or additive. Reactive BFRs such as the tetrabromobisphenol A (TBBPA) are covalently bound to the polymer matrix. Compared to their reactive counterparts, additive BFRs are not chemically bound to the product and therefore tend to migrate out of the product much more easily and are thus more likely to be released into the environment. Examples of additive BFRs include polybrominated diphenyl ethers (PBDEs), polybrominated biphenyls (PBBs), and hexabromocyclododecanes (HBCDs). Production of PBBs in the United States was phased out in the 1970s after a farm product contamination incident in Michigan [5]. In turn, production of PBDEs has increased, peaking in the mid-1990s [6].

22.1.2 Use and Production of PBDEs and HBCDs

There are three PBDE technical formulations used in industry: penta-BDE, octa-BDE, and deca-BDE. They are named according to the predominant homolog groups in the mixture (Figure 22.1). Penta-BDE is mostly used in polyurethane foam (mattresses, furniture, pillows) and in adhesives, while octa-BDE is mainly used in rigid plastics (acrylonitrile butadiene styrene, ABS) such as in computer casings and computer monitors [3]. The deca-BDE formulation is used in plastics such as high-impact polystyrene (HIPS) in electrical and electronic equipment, such as the back covers of televisions, also in rubber coating for wiring, as well as in textile back-coating in furniture [2].

Increasing concerns have been raised regarding these flame-retardant products due to their persistence, bioaccumulative characteristics, and potential adverse effects [1]. The penta-BDE and octa-BDE commercial mixtures have been listed as persistent organic pollutants (POPs) under the

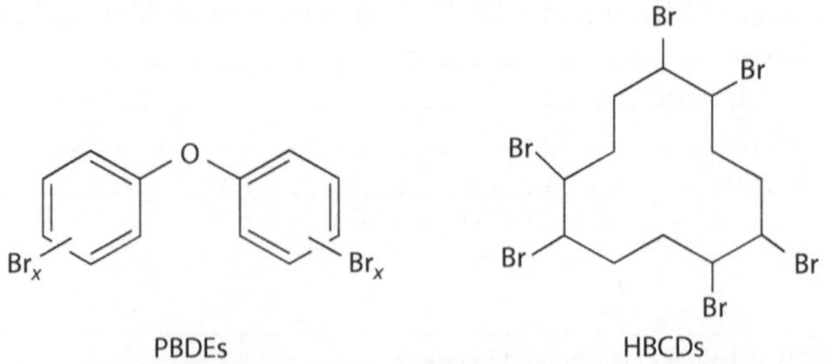

FIGURE 22.1 General structures of PBDEs and HBCDs.

TABLE 22.1
Estimated Annual Worldwide Market Demand of PBDEs and HBCDs in 2001 by Region, and Total Estimated Demand in 2002 and 2003 (Metric Tons)

	Penta-BDE	Octa-BDE	Deca-BDE	HBCDs
Americas[a]	7100	1500	24,500	2,800
Europe	150	610	7,600	9,500
Asia	150	1500	23,000	3,900
Total (2001)	7500	3790	56,100	16,700
Total (2002)	—	—	65,700	21,400
Total (2003)	—	—	56,400	22,000

Source: Adapted from de Wit, C.A. et al., *Sci. Total Environ.,* 408, 2885, 2010. With permission.

[a] Americas include North and South America in which the United States is the major consumer country in the former continent.

Stockholm Convention [7]. There are currently no restrictions on the production or use of deca-BDE in most countries. However, according to the EU Directive on the restriction of use of certain hazardous substances in electrical and electronic equipment (RoHS Directive), manufacturers must substitute other compounds for PBBs and PBDEs in new equipment. Deca-BDE was exempted from the directive, but this exemption was later overturned and phase-out of deca-BDE was initiated in European countries in 2008 [8]. Stringent regulations on PBDE use have led to the emergence of substitute chemicals, and one group that is likely to take the place of PBDEs is HBCDs.

As a nonaromatic, brominated cyclic alkane (Figure 22.1), HBCDs are not chemically bound to products and are primarily applied in thermoplastic polymers used in the manufacture of styrene resins [2]. HBCDs have also been used to a lesser extent in textile coatings, cable, latex binders, and unsaturated polyesters [3]. The commercial HBCD product is mainly composed of three diastereoisomers: α-(10%–13%), β-(1%–12%), and γ-HBCD (75%–89%) [9]. However, the isomeric profile varies depending on the product application. Temperatures above 160°C can subject HBCDs to thermal rearrangement, resulting in a specific mixture of stereoisomers (78% α-HBCD, 13% β-HBCD, and 9% γ-HBCD) [9]. Such conditions can occur during the production or processing of HBCD-containing materials such as extruded polystyrene, and therefore, the relative abundance of the various HBCD stereoisomers may differ from that of the technical HBCD mixtures. To date, there is no control on the production and use of HBCDs anywhere in the world. Table 22.1 summarizes the estimated annual market demand for PBDEs and HBCDs for the years 2001–2003.

22.1.3 PBDEs in the Environment

The major point sources of PBDEs to the environment are industrial plants manufacturing the technical mixtures as well as facilities incorporating PBDEs into polymers [10]. Electronic waste (e-waste) recycling facilities have recently been highlighted as an important point source of PBDEs [11,12]. In addition, wear and tear of products containing PBDEs constitutes a diffuse, non-point source of these chemicals. Fragments from the disintegration of polyurethane foam have been suggested to be a mechanism by which PBDEs diffuse into the atmosphere [13].

PBDEs tend to be stable and persistent in the environment. Due to their high binding affinity to organic matter, PBDEs are often associated with soils and sediments [3]. In addition, particulates in air and water are important pathways for the transport of these contaminants on local, regional, and global scales. Lower-brominated PBDEs are generally more volatile, water soluble and bioaccumulate more than higher-brominated PBDEs [14]. Previous research found that higher-brominated PBDE congeners can be transformed or broken down into lower-brominated congeners by biological [15,16] or environmental processes, such as photodegradation by sunlight [17]. These processes may increase the risks of PBDE exposure in the environment and organisms, as the levels of lower-brominated PBDE congeners increase, but little information about these impacts is available. However, the occurrence of the highest brominated congener, BDE209, in waterbird eggs [18], seal blubber [19], and human blood [20] indicated that this congener can be taken up by biota and is likely to bioaccumulate.

As ubiquitous environmental contaminants, PBDEs have been reported in a variety of environmental matrices. At present, PBDE contamination is recognized as a matter of global concern since they have reached remote areas such as the deep ocean and the polar regions [21,22] and demonstrated considerable accumulation in humans [23]. A wide range of PBDE levels have been measured in human adipose tissue, from 17 to 462 ng/g lw in the United States [24] and 2.2 to 11.7 ng/g lw in Belgium [25]. Such high variation in PBDE levels among humans might reflect the higher market demand for PBDEs in North America, notwithstanding other environmental factors and habits. Significant amounts of PBDEs have also been detected in lipid-rich tissues of the top predators such as bird eggs [26,27] and marine mammal blubber [28,29].

Significant increases in PBDE levels in human serum have been observed over the past two decades [30]. A temporal increase in PBDE levels has also been detected in sediment cores [31] and in biological samples such as fish [32], bird eggs [33], and marine mammals [34]. The increasing environmental occurrence of PBDEs has heightened research interest in studying PBDE levels in different parts of the world and their effects on different organisms.

South China, including Hong Kong and its neighboring Pearl River Delta (PRD) region, is one of the most rapidly growing regions in the world. The PRD has developed into a globally leading production site for computers and electronic products. To date, there is no legislation restricting the use of BFRs in China, and thus it is likely that large quantities of PBDEs are being used and released in manufacturing processes in the region. These releases, together with the e-waste dismantling and recycling industries located in Guangdong Province in southern China, have contributed to the high levels of PBDEs that have been detected in environmental matrices in the PRD. Previous studies reported total PBDE concentrations in riverine runoff and sediment samples from the PRD of 0.34–68 ng/L [35] and 0.44–7435 ng/g dry wt [36], respectively. These levels fall within the high end of the range of global PBDE concentrations. Situated at the mouth of the Pearl River, Hong Kong also receives high volumes of these persistent contaminants into its marine habitats from a variety of anthropogenic sources [37]. In addition, local population growth and industrial development have put considerable pressure on the marine environment of Hong Kong.

A recent large-scale monitoring study revealed high levels of PBDEs in sediments and green-lipped mussels (*Perna viridis*) from Hong Kong marine waters, with concentrations ranging from 1.7 to 53.6 ng/g dry wt. and 27.0 to 83.7 ng/g dry wt., respectively; these levels are among the highest in the world [38]. These PBDEs may originate from the disposal of e-waste in southern China, as well as untreated local discharge [39]. Studies have also found that PBDEs accumulated in both edible freshwater and marine fish samples collected in Hong Kong and from nearby waters [40,41]. Analysis of PBDEs in top predators such as in waterbird eggs [26] and cetaceans [42] inhabiting Hong Kong have also been carried out. The presence of PBDEs in these many species confirms their potential for dispersal, biological uptake, and bioaccumulation in many different trophic compartments in the environment [43].

The occurrence of high levels of PBDEs may result in adverse health effects in wildlife as shown in laboratory studies. Neurodevelopmental toxicity has been linked to tetra- and penta-BDE congener exposure in rats [44]. A single oral dose of tetra- or penta-BDE on day 10 following birth permanently impaired spontaneous motor behavior, affected learning and memory, and had permanent behavioral effects in mice [45]. In a study in killifish (*Fundulus heteroclitus*) [46], behavioral test results suggested that embryonic exposure to DE-71 could alter activity level, fright response, predation rates, and learning ability in subsequent life stages. PBDEs are also suspected to be immunotoxins. In nestling American kestrels (*Falco sparverius*) exposed to environmentally relevant

PBDEs (18.7 µg ΣPBDEs/egg), greater T-cell-mediated phytohemagglutinin response and less of an antibody-mediated response were observed [47]. Structural changes in the spleen (fewer germinal centers), bursa (reduced apoptosis) and thymus (increased macrophages), and negative associations between the spleen somatic index and total PBDEs, and the bursa somatic indices and BDE-47 were also found. Several studies also reported that PBDEs can disrupt thyroid function and exposure to commercial PBDE mixtures (penta-, octa-, and deca-BDEs) can result in thyroid hormone imbalance [48,49]. In an *in vitro* study conducted in cells from rainbow trout, chicken, rat, and human, strong EROD induction occurred after exposure to BDE 77, 100, 119, and 126, although the maximum EROD activity was less than that induced by dioxins [50]. Furthermore, the effects of PBDEs and PCBs on EROD were found to be synergistic, supporting the idea that both groups of chemicals may act through the same biological mechanism [51].

22.1.4 HBCDs IN THE ENVIRONMENT

Like PBDEs, HBCDs can enter the environment via a number of pathways, such as emission during production of BFRs, by migrating out of consumer products, or following disposal. The environmental occurrence of HBCDs was first documented in fish and sediment samples from the river Viskan in Sweden [52] and, since then, their presence has been reported in a variety of biotic and abiotic matrices at levels comparable to those of PBDEs [9].

Long-range transport of HBCDs has been proposed due to the detection of these chemicals in biota from Eastern Greenland and Svalbard [53,54]. However, the significance of the long-range atmospheric transport of HBCDs in relation to other sources and transport routes remains to be further established. Concentrations of HBCDs determined in matrices such as air [55], fish [56], dolphins [57], and sea lions [58] from the North American environment are generally lower than those in similar samples from Europe [8,59,60] and Asia [60–62]. In general, the different environmental residue levels appear to reflect different continental market demands. Although the technical HBCD mixture primarily consists of γ-HBCD, this composition profile is usually not reflected in environmental media [2]. Particularly, α-HBCD is consistently found to dominate the isomer pattern of HBCDs in biotic samples [63,64], providing clear evidence of enantioselective accumulation/biotransformation of the industrial combinations [65,66].

Increasing temporal trends of HBCDs have been observed in biological samples such as bird eggs [67] and marine mammals [58]. The elevated environmental levels of HBCDs have attracted growing interest and effort in investigating their occurrences and effects on living organisms.

There is a growing body of literature about the toxic effects of HBCDs. An increase in hepatosomatic index (HSI) and inhibition of EROD activity were found in juvenile rainbow trout (*Oncorhynchus mykiss*) intraperitoneally exposed to 50 and 500 mg/kg HBCD for 28 days [68]. Long-term exposure of European flounder (*Platichthys flesus*) to HBCD did not affect their general health nor several toxicity parameters (e.g., behavior, survival, growth rate, relative liver, and gonad weight), and neither was hepatic EROD activity induced [69]. In contrast, long-term exposure to HBCD-induced EROD activity in rare minnow (*Gobiocypris rarus*) resulting in oxidative damage to lipids, proteins, and DNA and decreased antioxidant capacity due to excess reactive oxygen species (ROS) formation [70]. Disruption of the thyroid axis was observed in juvenile rainbow trout exposed to HBCD diastereoisomers with the most evident effects in the γ-HBCD-exposed group, as indicated by a decreased level of circulating free total thyroid hormone FT4, an increased level of free total thyroid hormone FT3, and an increase in thyroid epithelial cell height [71]. In addition, changes in spontaneous behavior, learning, and memory defects, and a reduced number of nicotinic receptors were demonstrated in mice after neonatal exposure to HBCDs [72]. The observations of HBCD-induced neurobehavioral alterations were supported by the molecular evidence that HBCDs block the uptake of dopamine into rat brain synaptosomes *in vitro* [73]. Developmental toxicity and apoptosis were also observed in zebrafish (*Danio rerio*) after embryonic exposure to HBCDs [74].

22.1.5 STUDY OBJECTIVES

Time-series studies of environmental contaminants are usually conducted to understand trends in exposure levels in particular organisms or regions. While time-series studies have indicated increasing trends in PBDE concentrations in both abiotic and biotic matrices, considerable data gaps remain in the current monitoring studies such as time-trend studies of HBCDs. Most HBCD studies have been conducted in Europe and the United States, while very few studies have focused on the Asia-Pacific region. There is hence an urgent need to provide a better understanding of global HBCD concentrations in the environment. This paper describes temporal trends in PBDE and HBCD concentrations in the Indo-Pacific humpback dolphin (*Sousa chinensis*) and finless porpoise (*Neophocaena phocaenoides*), the two resident cetacean species in Hong Kong. The dolphin is generally restricted to the northwestern waters of Hong Kong adjacent to the mouth of the Pearl River, whereas the porpoise is found in eastern waters (Figure 22.2). Both cetacean species are top predators, which have been the subjects of previous monitoring studies [75–77].

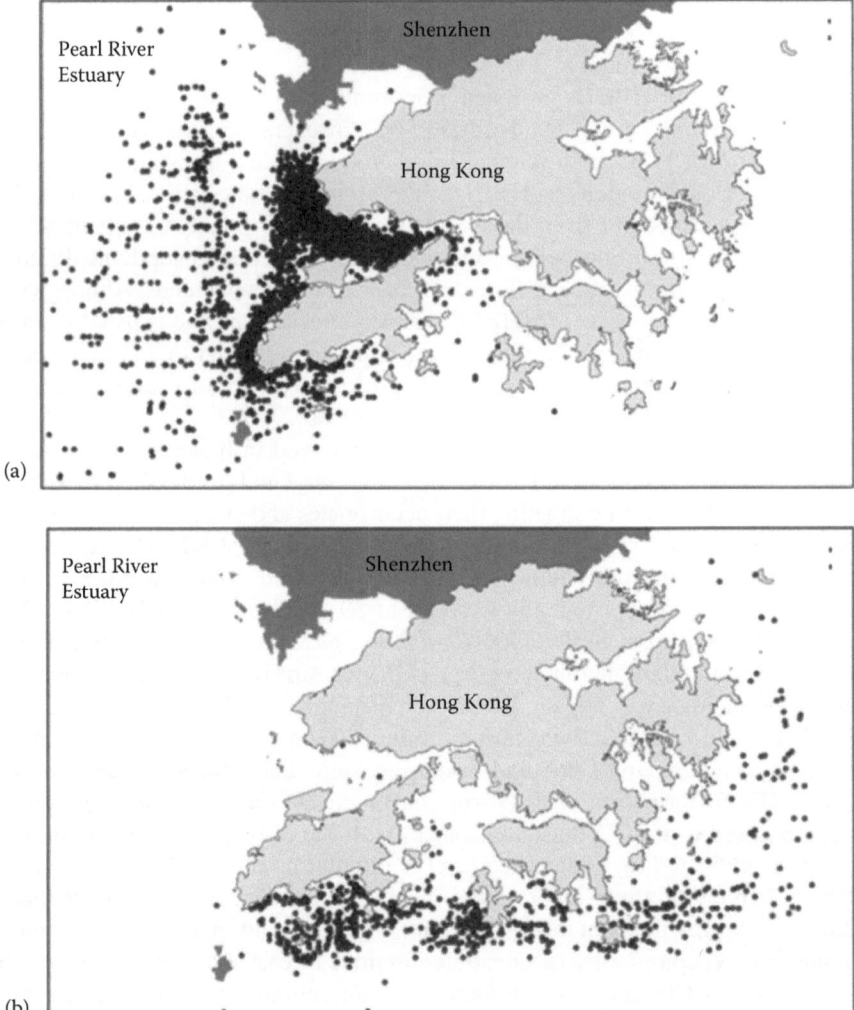

FIGURE 22.2 (a) Indo-Pacific humpback dolphin sightings (•) in Hong Kong and its neighboring waters. (b) Finless porpoise sightings (•) in Hong Kong coastal waters.

22.2 MATERIALS AND METHODS

22.2.1 SAMPLE COLLECTION AND PREPARATION

Blubber samples of stranded Indo-Pacific humpback dolphins ($n = 17$) and finless porpoises ($n = 33$) were collected in Hong Kong between 2002 and 2007 and between 2003 and 2008, respectively. All of the samples were stored in plastic bags and transported to the laboratory for storage at −20°C prior to chemical analysis. Analysis of PBDEs and HBCDs followed previously established methods [26,78] with modifications. Blubber samples were freeze-dried and then homogenized. The homogenized samples were extracted in a Soxhlet apparatus for 10 h with 400 mL of a hexane:dichloromethane (DCM) mixture (1:3 v/v). Before the extraction, PCB 30 and decachlorobiphenyl (DCB) were added to the samples as the surrogate standards. The extract was concentrated and a portion of the sample extract was used for lipid determination by the gravimetric method. The extract was added to a gel permeation chromatography column (GPC; Bio-Beads S-X3, Bio-Rad Laboratories) for lipid removal by an equivalent mixture of DCM in hexane (1:1 v/v) at a flow rate of 5 mL/min. 5 ng of each $^{13}C_{12}$-labeled PBDE standards ($^{13}C_{12}$-labeled BDE 3, BDE 15, BDE 28, BDE 47, BDE 99, BDE 153, BDE 154, BDE 183, BDE 197, BDE 207, and BDE 209) and 10 ng of each $^{13}C_{12}$-labeled HBCD standard (α-, β- and γ-$^{13}C_{12}$-labeled HBCD) were spiked to the extract before adding it to the GPC column. The concentrated extract was further purified by elution through 4 g activated silica gel (60 Å average pore size) and 5 g activated alumina. PBDEs were eluted with 120 mL DCM:hexane (1:20 v/v) and HBCDs eluted with a DCM:hexane mixture (2:1 v/v) from the silica gel column. $^{13}C_{12}$-labeled BDE 139 was added to the eluent as the recovery spike and the volume was further reduced to 0.1 mL prior to gas chromatograph (GC) analysis.

22.2.2 CHEMICAL ANALYSIS

Instrumental analysis of PBDEs was performed using a GC (Agilent 7890A) coupled with an MSD (Agilent 5975c) for mono- to deca-brominated diphenyl ethers (BDEs) using electron impact (EI) mode. Lower and higher brominated PBDE congeners were analyzed by 30 m DB-5MS and 15 m DB-5HT columns respectively. Total PBDE (ΣPBDE) concentrations were reported as the sum of the 14 individual, PBDE congeners, and each congener was quantified using the isotope dilution method to their corresponding $^{13}C_{12}$-labeled congeners.

Quantification of HBCDs followed the analytical method previously established by Tomy et al. [56,79] using an Agilent HP1100 liquid chromatography (Agilent, Palo Alto, CA) equipped with an Applied Biosystems API 2000 triple quadrupole tandem mass spectrometer (MS) coupled with a Turbo IonSpray source operated in negative mode. The three isomers of (α-, β- and γ-) HBCDs were separated chromatographically by a Symmetry-C18 column (2.1 mm i.d. × 150 mm, 3.5 μm, Waters Corp.) equipped with a guard column. The MS/MS analysis was performed using electrospray ionization (ESI) with multiple reaction monitoring mode (MRM). Native α-, β-, and γ-HBCD isomers were quantified by the mean value of the response at two MRM transitions (m/z 640.6 > 81 and 640.6 > 79) corrected against the response of $^{13}C_{12}$-labeled HBCDs (m/z 652.6 > 81 and 652.6 > 79). Total HBCD (ΣHBCD) concentrations were reported as the sum of the three individual diastereoisomers quantified. Concentrations of both PBDEs and HBCDs were expressed as ng/g lipid weight (lw).

22.2.3 QA/QC

Recoveries of $^{13}C_{12}$-labeled HBCDs and PBDEs in all samples were within 60%–120%. The efficiencies of Soxhlet extraction and cleanup procedures were checked prior to the chemical analysis and the recovery rates of $^{13}C_{12}$-labeled standards ranged between 70% and 120% ($n = 5$). Procedural blanks were analyzed simultaneously with every batch of five samples to check for potential inter-ferences or contamination and instrumental detection limits (IDLs) were estimated as the average signal of the blanks plus three times the standard deviation of the signals of the blanks. The IDL was 0.02 ng/g lw for mono- to di-BDEs, 0.1 ng/g lw for tetra-BDEs, 0.05 ng/g lw for tri- and penta- to hepta-BDEs, 0.02 ng/g lw for octa- to nona-BDEs, 0.5 ng/g lw for deca-BDE and 0.3 ng/g lw for each diastereoisomer of HBCDs.

22.2.4 STATISTICAL ANALYSIS

Concentration comparisons were conducted by student's t-tests if the data passed normality and equal variance tests; nonparametric Mann–Whitney Rank Sum tests were used otherwise (SigmaStat 3.5). ΣPBDE and ΣHBCD concentrations in Indo-Pacific humpback dolphins and finless porpoises sampled from 1997 to 2001 reported by Isobe et al. [78] were incorporated into the present data set of blubber samples collected from 2002 to 2008 for time-series analyses. Temporal trend analy-ses were carried out using a simple log-linear regression model as described by Nicolson et al. [80]. Briefly, log-linear regression was performed using annual mean concentrations of ΣHBCDs or ΣPBDEs for dolphins from 1997 to 2007 and porpoises from 2000 to 2008. A 3 year moving average smoothing function was fitted to the annual median concentrations to investigate possible nonlinear trend components and was tested by means of ANOVA. To minimize the influence of differences between the sexes due to maternal transfer and possible age-related differences, adult males with body lengths greater than 200 and 120 cm for dolphins and porpoises, respectively, were selected for temporal trend analyses. Statistical analysis was conducted using Prism 2.01 and SigmaStat 3.5. Statistical significance was accepted at $p < 0.05$.

22.3 RESULTS AND DISCUSSION

22.3.1 CONCENTRATIONS OF PBDEs AND HBCDs IN CETACEANS

HBCDs were detected in all 50 cetacean blubber samples at concentrations ranging from 32 to 519 and from 4.1 to 501 ng/g lw for Indo-Pacific humpback dolphins and finless porpoises, respec-tively. The dominant congener was α-HBCD in all samples analyzed. The detection frequencies of β- and γ-HBCD in the samples analyzed were 54% and 60%, respectively. The average ΣPBDE concentrations ranged from 1113 to 3590 ng/g lw in the two species, and were approximately one order of magnitude greater than the levels of ΣHBCDs in blubber samples of both cetacean species. Concentrations of HBCDs and PBDEs in dolphins were significantly greater than those in porpoises ($p < 0.05$), suggesting higher exposure levels of the two contaminant groups in the northwestern waters of Hong Kong than in the eastern waters.

22.3.2 TEMPORAL TREND OF PBDEs AND HBCDs IN CETACEANS

A positive linear temporal trend in ΣHBCD concentrations was found in dolphin samples when log-linear regression analysis of yearly median concentrations was used (Pearson $r = 0.67$, $p = 0.047$). The linear regression equation obtained for dolphins ($y = 0.090x + 1.6$) indicated that ΣHBCD concentra-tions would double from 1997 to 2017, assuming a constant rate of HBCD usage and release [81]. The world market demand for HBCDs has increased during the past decade, while demand for traditional BFRs such as PBDEs is decreasing because of controls on their production and use. The usage of HBCDs as alternatives for PBDE formulations is projected to increase in any region with rapid economic development and industrial activities, such as the PRD; there is currently no restriction or control on HBCD production and use in China. Therefore, it is not surprising that elevated trends of HBCDs were observed in Indo-Pacific humpback dolphins in the present study, because this spe-cies inhabits the northwestern waters of Hong Kong located downstream of the PRD. In contrast, no significant time trend in HBCD concentrations was detected from 2000 to 2008 for finless porpoises whose habitats are generally located in the less polluted eastern waters of Hong Kong.

Concentrations of ΣPBDEs in both dolphins and porpoises, in contrast, showed no significant trend for the corresponding sampling years when tested by log-linear regression and 3 year moving averages. However, between-year variation of ΣPBDEs exhibited similar patterns between the two cetacean species with peak concentrations occurring in 2003–2004 followed by relatively steady levels after 2006. Similar patterns were found in guillemot (*Uria aalge*) eggs from the Baltic Sea showing a positive trend from 1970 to the middle of the 1980s, followed by a decrease up to 2001 [67]. These observations could be partly explained by effective regulation and control of PBDE usage in countries around the world.

22.3.3 GLOBAL TRENDS OF **PBDE** AND **HBCD** CONCENTRATIONS IN MARINE MAMMALS

The current knowledge of the temporal trends of PBDE and HBCD levels in marine mammals is limited to North America (United States), Europe (United Kingdom), Asia-Pacific (Japan), and the Arctic (Canada, Greenland, and Norway), in addition to the current study location, Hong Kong. In general, increasing trends in HBCDs concentrations have been found in marine mammals from these regions, while no continuous decreasing trends of PBDEs have been observed. These results agree with the findings of the present study.

22.3.3.1 Japan

Studies on temporal trends of PBDEs in marine mammals are scarce in the Asia-Pacific region; the majority of the available information comes from Japan. Generally, the last few decades witnessed a significant increase in PBDE levels in marine mammals; concentrations in the 1970s and 1980s were 1–2 orders of magnitude higher than those in recent years [61]. In northern fur seals (*Callorhinus ursinus*), an approximately 150-fold elevation of PBDE concentrations was observed between 1972 and 1994, and then the levels dropped by approximately half between 1994 and 1998 [82]. A similar increase in average PBDE concentrations from 13 ng/g lw in 1978 to 640 ng/g lw in 2003 was observed in striped dolphins (*Stenella coeruleoalba*) [83]. PBDE levels in melon-headed whales (*Peponocephala electra*) also showed a 10-fold increase from 1982 to 2006 [84]. In fur seals and striped dolphins, PBDE concentrations peaked in the early 1990s and appeared to decline afterward. Among PBDE congeners, BDE 47 was the most abundant congener in all the marine mammal species across all sampling years. There was, however, temporal variation among congener profiles, suggesting the use of different PBDE commercial formulations during different time periods. In northern fur seals, the ratios of BDE 153, BDE 154, and BDE 183 to ΣPBDEs increased after 1972, while those of some lower brominated congeners decreased. In melon-headed whales, hexabrominated congeners comprised a higher proportion of ΣPBDEs in 2006 than in 1982. These observations might arise from the ban of the Penta-BDE mixture in 1990 and subsequent shift toward the use of higher brominated formulations, namely Octa-BDE (withdrawn in 2000) and Deca-BDE in Japan [14].

Like PBDEs, HBCD concentrations demonstrated a sharp rise in marine mammals from Japan during recent decades. In northern fur seals, HBCD concentrations increased from <0.1 ng/g lw in 1972 to 67 ng/g lw in 1997 followed by a slight decrease in 1998 to approximately 70% of the peak value [85]. Peak HBCD concentrations in fur seals occurred later than PBDEs, which may be explained by the relative production and use of HBCDs compared to that of PBDEs over time in Japan. In melon-headed whales, HBCD concentrations were significantly higher in 2006 than in 1982 by approximately 50 times [86]. A positive trend in HBCD concentrations was also observed in striped dolphins, as evidenced by the measurement of 12 ng/g lw in 1978 versus 710 ng/g lw in 2003 [83]. In all the marine mammals analyzed, α-HBCD was the predominant isomer, contributing more than 95% to ΣHBCDs. The isomeric composition did not change much over time, indicating the use of commercial mixtures of the same composition in Japan.

It is noteworthy that HBCD concentrations in marine mammals from Japan exceed those of PBDEs in recent years due possibly to the substitution of PBDEs by HBCDs. Consumption of HBCDs in Japan increased continuously from around 600 ton in 1986 to 2200 ton in 2001, while Penta-BDE and Octa-BDE were withdrawn from the market in 1991 and 2000, respectively [14]. In fur seals, HBCD concentrations were apparently lower than those of PBDEs before 1991, whereas comparable levels of the two flame retardants were observed in the late 1990s. Similarly, melon-headed whales sampled in 1982 accumulated lower HBCD levels than PBDEs by one order of magnitude, while HBCD levels exceeded those of PBDEs in 2006. A similar pattern was also observed in striped dolphins where HBCD concentrations were lower than or comparable to those of PBDEs until 2003 when HBCD concentrations became higher. Concentrations of HBCDs in these marine mammals are expected to continuously increase due to current lack of regulation of these compounds.

22.3.3.2   North America

Since the United States is the major user of PBDEs and HBCDs in North America and marine mammal studies in Canada are often conducted in the Arctic region, this discussion is focused on studies conducted within the United States. In harbor seals (*Phoca vitulina*) from San Francisco Bay, PBDE concentrations rose by approximately two orders of magnitude between 1989 and 1998 [24]. An exponential increase of PBDEs was observed in bottlenose dolphins (*Tursiops truncates*) from Florida with an estimated doubling time of 3–4 years in the period 1991–2004 [87]. In Atlantic white-sided dolphins (*Lagenorhynchus acutus*) from Massachusetts, PBDE contamination exhibited no significant temporal trend from 1993 to 2000 [88]. Similarly, no temporal trends in the concentrations of either ΣPBDEs or individual congeners were found in California sea lions (*Zalophus californianus*) stranded between 1993 and 2003 [58]. No significant temporal trend in PBDE concentrations was detected in sea otters (*Enhydra lutris*) from coastal California [89]. These findings suggest that temporal trends of PBDEs in marine mammals from the United States are generally site-specific and may be linked with the history of contamination at the site and/or with remediation efforts. It is worth noting that PBDE concentrations did not reflect a negative temporal trend in these marine mammals, indicating that PBDEs may continue to be an issue of concern in the future.

Compared to PBDEs, far less is known about temporal changes in HBCD levels in marine mammals from North America, including the United States. The two studies reported to date have shown contrasting time trends. An exponential temporal trend in ΣHBCD concentrations having a doubling time of approximately 2 years was observed in California sea lions stranded between 1993 and 2003 [58]. However, no significant trend was observed in Atlantic white-sided dolphin sampled from 1993 to 2004 [57]. These inconsistencies are likely due in part to species-specific differences in contaminant accumulation and site-specific use and production of HBCDs. As opposed to the recent predominance of HBCD levels over those of PBDEs in Japan, HBCD concentrations were generally two orders of magnitude lower than those of PBDEs analyzed in parallel in California sea lions sampled from 1993 to 2003 [58]. The same pattern was observed in Atlantic white dolphins, in which HBCD concentrations [57] were approximately 1–2 orders of magnitude lower than PBDE concentrations over the same sampling years (1993–2000) [88], possibly indicating a lesser extent of substitution of PBDEs by HBCDs in North America than in Japan.

22.3.3.3   Arctic

A number of temporal trend studies on PBDEs have been conducted in Arctic marine mammals from several locations in Canada and Greenland, and have reported increasing trends or no trends for PBDEs in this region. PBDE levels increased by approximately an order of magnitude in ringed seals (*Pusa hispida*) from Northwestern Territories, Canada during 1981–2000 [90], though no significant change was observed between 2000 and 2003 [91]. Levels of BDE 47 and BDE 100, the two most prevalent congeners, increased in a fashion similar to that of ΣPBDEs [90]. The doubling time of tetra-BDEs was nearly twice that of penta-BDEs (4.7 years) and hexa-BDEs (4.3 years)

[90], indicating temporal variation in PBDE congener profile over time. In beluga (*Delphinapterus leucas*) from Pangnirtung in eastern Canada, PBDE concentrations increased between 1982 and 2005 with a doubling time of approximately 11 years [5]. Increasing concentrations of ΣPBDEs were found in ringed seal blubber from Hudson Bay and Resolute in eastern Canada over the period from 1992 to 2005 [92]. According to Muir et al. [92], ΣPBDE levels appeared to stabilize in ringed seals from these locations when 2006 data was included. In contrast, no significant trend in concentrations of ΣPBDEs in beluga blubber samples from Hendrikson Island in western Canada was observed between 1993 and 2001 [5]. No significant time trends in concentrations of ΣPBDEs or of single congeners were detected in ringed seals from Ittoqqortoormiit in East Greenland between 1986 and 2004 [34]. In ringed seals from West Greenland, a significant annual increase of 5.3% was found for BDE 47 between 1982 and 2006 [93]. These results likely reflect region-specific influences in terms of PBDE input into the Arctic and/or species-specific accumulation.

Studies of HBCDs in Arctic marine mammals have also reported inconsistent trends. HBCD levels were relatively unchanged in beluga blubber samples from Hendrikson Island in western Canada between 1993 and 2001 and from Pangnirtung in eastern Canada between 1982 and 2006 [5]. In contrast, increasing concentrations of HBCDs were found in ringed seals from Hudson Bay and Resolute over the period of 1992–2005 [92]. As HBCD studies in the Canadian Arctic are limited, further research over a wider geographical area in the Arctic region is needed to draw further conclusions on temporal patterns of HBCD concentrations.

22.3.3.4 Europe

There is currently a lack of long-term data in the literature on PBDE concentrations in marine mammals from Europe, and therefore a systematic analysis of temporal trends cannot be carried out. However, temporal changes in PBDE levels in other biotic matrices are better understood. Retrospective analysis of guillemot eggs from the Baltic Sea showed increasing ΣPBDE concentrations from the 1970s up to the 1980s followed by a rapid decrease [67]. A similar trend was detected in pike (*Esox lucius*) from Lake Bohmen in southern Sweden, where PBDE concentrations showed increasing trends from the late 1960s up to the mid-1980s and then decreased or leveled off [94]. In blue mussels (*Mytilus edulis*) from France, PBDEs showed a remarkable increase between 1981 and 1995 followed by a leveling off and a possible initial decline up to 2003 [95]. The observed decreasing trend in recent years is likely to be the result of the phase-out and restriction of the use of PBDEs by the EU. To better understand the temporal trends in marine mammals, long-term data on PBDE concentrations in these top predators are needed.

Temporal trends of HBCDs in marine mammals have also not been well studied in Europe. The available data show a sharp increase in HBCD concentrations from about 2001 onward in the blubber of harbor porpoises (*Phocoena phocoena*) sampled from the United Kingdom during 1994–2003 [96] and a significant downturn from 2003 to 2006 [8]. Additional studies of the occurrence and toxicology of HBCDs are required for the ongoing assessment within the EU of the risks associated with the continued production and use of HBCDs.

22.4 CONCLUSIONS

In general, increasing trends in HBCD concentrations have been found in marine mammals worldwide, while no continuous decreasing trends of PBDEs have been observed. Temporal trends of the two BFRs were consistent with historical consumption of the corresponding BFRs. There are indications that PBDE concentrations are leveling off in recent years, due in part to worldwide restrictions and bans imposed on the usage of penta- and octa-BDE mixtures. PBDE concentrations may continue declining, but continuous monitoring programs are necessary to ascertain these trends. On the other hand, remarkable increases in HBCD concentrations have been detected on the global scale over the last decades and such trends may presumably continue as no regulations are currently enforced on HBCDs. Moreover, research on environmental levels and trends of PBDEs and HBCDs needs to be expanded geographically, especially into developing countries and the southern hemisphere for a clearer global picture of levels and trends of PBDEs and HBCDs.

ACKNOWLEDGMENTS

The work described in this chapter was partially funded by the Area of Excellence Scheme under the University Grants Committee of the Hong Kong Special Administrative Region, China (Project No. AoE/P-04/2004). This project was also supported by a Hong Kong Research Grants Council (City U 160610).

REFERENCES

1. Ruan, T., Wang, Y., Wang, C., Wang, P., Fu, J., Yin, Y., Qu, G., Wang, T., and Jiang, G. 2009. Identification and evaluation of a novel heterocyclic brominated flame retardant tris(2,3-dibromopropyl) isocyanurate in environmental matrices near a manufacturing plant in southern China. *Environmental Science & Technology* 43:3080–3086.
2. Birnbaum, L.S. and Staskal, D.F. 2004. Brominated flame retardants: Cause for concern? *Environmental Health Perspectives* 112:9–17.
3. Alaee, M., Arias, P., Sjödin, A., and Bergmand, Å. 2003. An overview of commercially used brominated flame retardants, their applications, their use patterns in different countries/regions and possible modes of release. *Environment International* 29:683–689.
4. Rahman, F., Langford, K.H., Scrimshaw, M.D., and Lester, J.N., 2001. Polybrominated diphenyl ether (PBDE) flame retardants. *Science of the Total Environment* 275:1–17.
5. de Wit, C.A., Herzke, D., and Vorkamp, K. 2010. Brominated flame retardants in the Arctic environment—Trends and new candidates. *Science of the Total Environment* 408:2885–2918.
6. Alcock, R.E., Sweetman, A., Prevedouros, K., and Jones, K.C. 2003. Understanding levels and trends of BDE-47 in the UK and North America: An assessment of principal reservoirs and source input. *Environmental International* 29:691–698.
7. United Nations Environment Program (UNEP). 2009. C.N.524.2009. TREATIES-4 (Depositary Notification).
8. Law, R.J., Bersuder, P., Barry, J., Wilford, B.H., Allchin, C.R., and Jepson, P.D. 2008. A significant downturn in levels of hexabromocyclododecane in the blubber of harbor porpoises (*Phocoena phocoena*) stranded or bycaught in the UK: An update to 2006. *Environmental Science & Technology* 42:9104–9109.
9. Covaci, A., Gerecke, A.C., Law, R.J., Voorspoels, S., Kohler, M., Heeb, N.V., Leslie, H., Allchin, C.R., and de Boer, J. 2006. Hexabromocyclododecane (HBCDs) in the environment and humans: A review. *Environmental Science & Technology* 40:3679–3688.
10. Hale, R.C., Alaee, M., Manchester-Neesvig, J.B., Stapleton, H.M., and Ikonomou, M.G. 2003. Polybrominated diphenyl ether flame retardants in the North American environment. *Environment International* 29:771–779.
11. Cai, Z. and Jiang, G. 2006. Determination of polybrominated diphenyl ethers in soil from e-waste recycling site. *Talanta* 70:88–90.
12. Luo, X., Liu, J., Luo, Y., Zhang, X., Wu, J., Lin, Z., Chen, S., Mai, B., and Yang, Z. 2009. Polybrominated diphenyl ethers (PBDEs) in free-range domestic fowl from an e-waste recycling site in South China: Levels, profile and human dietary exposure. *Environment International* 35:253–258.
13. Hale, R.C., La Guardia, M.J., Harvey, E., and Mainor, T.M. 2002. Potential role of fire retardant-treated polyurethane foam as a source of brominated diphenyl ethers to the US environment. *Chemosphere* 46:729–735.
14. Watanabe, I. and Sakai, S. 2003. Environmental release and behavior of brominated flame retardants. *Environmental International* 29:665–682.
15. Stapleton, H.M., Alaee, M., Letcher, R.J., and Baker, J.E. 2004. Debromination of the flame retardant decabromodiphenyl ether by juvenile carp (*Cyprinus carpio*) following dietary exposure. *Environmental Science & Technology* 38:112–119.
16. Stapleton, H.M., Letcher, R.J., and Baker, J.E. 2004. Debromination of polybrominated diphenyl ether congeners BDE 99 and BDE 183 in the intestinal tract of the common carp (*Cyprinus carpio*). *Environmental Science & Technology* 38:1054–1061.

17. Söderstrom, G., Sellström, U., de Wit, C.A., and Tysklind, M. 2004. Photolytic debromination of decabromodiphenyl ether (BDE 209). *Environmental Science & Technology* 38:127–132.

18. Lindberg, P., Sellström, U., Häggberg, L., and de Wit, C.A. 2004. Higher brominated diphenyl ethers and hexabromocyclododecane found in eggs of peregrine falcons (*Falco peregrinus*) breeding in Sweden. *Environmental Science & Technology* 38:93–96.

19. Thomas, G.O., Moss, S.E., Asplund, L., and Hall, A.J. 2005. Absorption of decabromodiphenyl ether and other organohalogen chemicals by grey seals (*Halichoerus grypus*). *Environmental Pollution* 133:581–586.

20. Jakobsson, K., Thuresson, K., Rylander, L., Sjödin, A., Hagmar, L., and Bergman, A. 2002. Exposure to polybrominated diphenyl ethers and tetrabromobisphenol A among computer technicians. *Chemosphere* 46:709–716.

21. Chiuchiolo, A.L., Dickhut, R.M., Cochran, M.A., and Ducklow, H.W. 2004. Persistent organic pollutants at the base of the Antarctic marine food web. *Environmental Science & Technology* 38:3551–3557.

22. Samara, F., Tsai, C.W., and Aga, D.S. 2006. Determination of potential sources of PCBs and PBDEs in sediments of the Niagara River. *Environmental Pollution* 139:489–497.

23. Johnson, P.I., Stapleton, H.M., Sjödin, A., and Meeker, J.D. 2010. Relationships between polybrominated diphenyl ether concentrations in house dust and serum. *Environmental Science & Technology* 44:5627–5632.

24. She, J., Petreas, M., Winkler, J., Visita, P., McKinney, M., and Kopec, D. 2002. PBDEs in the San Francisco Bay Area: Measurements in harbor seal blubber and human breast adipose tissue. *Chemosphere* 46:697–707.

25. Covaci, A., de Boer, J., Ryan, J.J., Voorspoels, S., and Schepens, P. 2002. Distribution of organobrominated and organochlorinated contaminants in Belgian human adipose tissue. *Environmental Research* 88:210–218.

26. Lam, J.C.W., Kajiwara, N., Ramu, K., Tanabe, S., and Lam, P.K.S. 2007. Assessment of polybrominated diphenyl ethers in eggs of waterbirds from South China. *Environmental Pollution* 148:258–267.

27. She, J., Holden, A., Adelsbach, T.L., Tanner, M., Schwarzbach, S.E., Yee, J.L., and Hooper, K. 2008. Concentrations and time trends of polybrominated diphenyl ethers (PBDEs) and polychlorinated biphenyls (PCBs) in aquatic bird eggs from San Francisco Bay, CA 2000–2003. *Chemosphere* 73:S201–S209.

28. Kajiwara, N., Kamikawa, S., Ramu, K., Ueno, D., Yamada, T.K., Subramanian, A., Lam, P.K.S., Jefferson, T.A., Prudente, M., Chung, K.H., and Tanabe, S. 2006. Geographical distribution of polybrominated diphenyl ethers (PBDEs) and organochlorines in small cetaceans from Asian waters. *Chemosphere* 64:287–295.

29. Fair, P.A., Mitchum, G., Hulsey, T.C., Adams, J., Zolman, E., McFee, W., Wirth, E., and Bossart, G.D. 2007. Polybrominated diphenyl ethers (PBDEs) in blubber of free-ranging bottlenose dolphins (*Tursiops truncatus*) from two southeast Atlantic estuarine areas. *Archives of Environmental Contamination and Toxicology* 53:483–494.

30. Koizumi, A., Yoshinaga, T., Harada, K., Inoue, K., Morikawa, A., Muroi, J., Inoue, S., Eslami, B., Fujii, S., Fujimine, Y., Hachiya, N., Koda, S., Kusaka, Y., Murata, K., Nakatsuka, H., Omae, K., Saito, N., Shimbo, S., Takenaka, K., Takeshita, T., Todoriki, H., Wada, Y., Watanabe, T., and Ikeda, M. 2005. Assessment of human exposure to polychlorinated biphenyls and polybrominated diphenyl ethers in Japan using archived samples from the early 1980s and mid-1990s. *Environmental Research* 99:31–39.

31. Chen, S.J., Luo, X.J., Lin, Z., Luo, Y., Li, K.C., Peng, X.Z., Mai, B.X., Ran, Y., and Zeng, E.Y. 2007. Time trends of polybrominated diphenyl ethers in sediment cores from the Pearl River Estuary, South China. *Environmental Science & Technology* 41:5595–5600.

32. Zhu, L.Y. and Hites, R.A. 2004. Temporal trends and spatial distributions of brominated flame retardants in archived fishes from the Great Lakes. *Environmental Science & Technology* 38:2779–2784.

33. Elliott, J.E., Wilson, L.K., and Wakeford, B. 2005. Polybrominated diphenyl ether trends in eggs of marine and freshwater birds from British Columbia, Canada, 1979–2002. *Environmental Science & Technology* 39:5584–5591.

34. Rigét, F., Vorkamp, K., Dietz, R., and Rastogi, S.C. 2006. Temporal trend studies on polybrominated diphenyl ethers (PBDEs) and polychlorinated biphenyls (PCBs) in ringed seals from East Greenland. *Journal of Environmental Monitoring* 8:1000–1005.

35. Guan, Y., Wang, J., Ni, H., Luo, X., Mai, B., and Zeng, E.Y. 2007. Riverine inputs of polybrominated diphenyl ethers from the Pearl River Delta (China) to the coastal ocean. *Environmental Science & Technology* 41:6007–6013.

36. Mai, B., Chen, S., Luo, X., Chen, L., Yang, Q., Sheng, G., Peng, P., Fu, J., and Zeng, E.Y. 2005. Distribution of polybrominated diphenyl ethers in sediments of the Pearl River Delta and adjacent South China Sea. *Environmental Science & Technology* 39:3521–3527.

37. Lam, J.C.W., Murphy, M.B., Wang, Y., Tanabe, S., Giesy, J.P., and Lam, P.K.S. 2008. Risk assessment of organohalogenated compounds in water bird eggs from South China. *Environmental Science & Technology* 42:6296–6302.

38. Liu, Y., Zheng, G.J., Yu, H., Martin, M., Richardson, B.J., Lam, M.H.W., and Lam, P.K.S. 2005. Polybrominated diphenyl ethers (PBDEs) in sediments and mussel tissues from Hong Kong marine waters. *Marine Pollution Bulletin* 50:1173–1184.

39. Wurl, O., Lam, P.K.S., and Obbard, J.P. 2006. Occurrence and distribution of polybrominated diphenyl ethers (PBDEs) in the dissolved and suspended phases of the sea-surface microlayer and seawater in Hong Kong, China. *Chemosphere* 65:1660–1666.

40. Cheung, K.C., Zheng, J.S., Leung, H.M., and Wong, M.H. 2008. Exposure to polybrominated diphenyl ethers associated with consumption of marine and freshwater fish in Hong Kong. *Chemosphere* 70:1707–1720.

41. Guo, L., Qiu, Y., Zheng, G. J., and Lam, P.K.S., and Li, X. 2008. Levels and bioaccumulation of organochlorine pesticides (OCPs) and polybrominated diphenyl ethers (PBDEs) in fishes from the Pearl River estuary and Daya Bay, South China. *Environmental Pollution* 152:604–611.

42. Ramu, K., Kajiwara, N., Lam, P.K.S., Jefferson, T.A., Zhou, K., and Tanabe, S. 2006. Temporal variation and biomagnification of organohalogen compounds in finless porpoises (*Neophocaena phocaenoides*) from the South China Sea. *Environmental Pollution* 144:516–523.

43. Martin, M., Lam, P.K.S., and Richardson, B.J. 2004. An Asian quandary: Where have all of the PBDEs gone? *Marine Pollution Bulletin* 49:375–382.

44. Hooper, K. and McDonald, T.A. 2000. The PBDEs: An emerging environmental challenge and another reason for breast-milk monitoring programs. *Environmental Health Perspectives* 108:387–392.

45. Eriksson, P., Jakobsson, E., and Fredriksson, A. 2001. Brominated flame retardants: A novel class of developmental neurotoxicants in our environment? *Environmental Health Perspectives* 109:903–908.

46. Timme-Laragy, A.R., Levin, E.D., and Di, Giulio, R.T. 2006. Developmental and behavioral effects of embryonic exposure to the polybrominated diphenylether mixture DE-71 in the killifish (*Fundulus heteroclitus*). *Chemosphere* 62:1097–1104.

47. Fernie, K.J., Mayne, G., Shutt, J.L., Pekarik, C., Grasman, K.A., Letcher, R.J., and Drouillard, K. 2005. Evidence of immunomodulation in nestling American kestrels (*Falco sparverius*) exposed to environmentally relevant PBDEs. *Environmental Pollution* 138:485–493.

48. Ellis-Hutchings, R.G., Cherr, G.N., Hanna, L.A., and Keen, C.L. 2006. Polybrominated diphenyl ether (PBDE)-induced alterations in vitamin A and thyroid hormone concentrations in the rat during lactation and early postnatal development. *Toxicology and Applied Pharmacology* 215:135–145.

49. Darnerud, P.O., Aune, M., Larsson, L., and Hallgren, S. 2007. Plasma PBDE and thyroxine levels in rats exposed to Bromkal or BDE-47. *Chemosphere* 67:S386–S392.

50. Chen, G., Konstantinov, A.D., Chittim, B.G., Joyce, E.M., Bols, N.C., and Bunce, N.J. 2001. Synthesis of polybrominated diphenyl ethers and their capacity to induce CYP1A by the Ah receptor mediated pathway. *Environmental Science & Technology* 35:3749–3756.

51. Hallgren, S., Sinjari, T., Håkansson, H., and Darnerud, P.O. 2001. Effects of polybrominated diphenyl ethers (PBDEs) and polychlorinated biphenyls (PCBs) on thyroid hormone and vitamin A levels in rats and mice. *Archives of Toxicology* 75:200–208.

52. Sellström, U., Kierkegaard, A. de Wit, C.A., and Jansson, B. 1998. Polybrominated diphenyl ethers and hexabromocyclododecane in sediment and fish from a Swedish river. *Environmental Toxicology and Chemistry* 17:1065–1072.

53. Verreault, J., Gabrielsen, G.W., Chu, S.G., Muir, D.C.G., Andersen, M., Hamaed, A., and Letcher, R.J. 2005. Flame retardants and methoxylated and hydroxylated polybrominated diphenyl ethers in two Norwegian Arctic top predators: Glaucous gulls and polar bears. *Environmental Science & Technology* 39:6021–6028.

54. Vorkamp, K., Thomsen, M., Falk, K., Leslie, H., Møller, S., Sørensen, P.B. 2005. Temporal development of brominated flame retardants in peregrine Falcon (*Falco peregrinus*) eggs from South Greenland (1986–2003). *Environmental Science & Technology* 39:8199–8206.

55. Hoh, E. and Hites, R.A. 2005. Brominated flame retardants in the atmosphere of the East-Central United States. *Environmental Science & Technology* 39:7794–7802.

56. Tomy, G.T., Budakowski, W., Halldorson, T., Whittle, D.M., Keir, M.J.; Marvin, C., MacInnis, G., and Alaee, M. 2004. Biomagnification of alpha and gamma-hexabromocyclododecane isomers in a Lake Ontario food web. *Environmental Science & Technology* 38:2298–2303.

57. Peck, A.M., Pugh, R.S., Moors, A., Ellisor, M.B., Porter, B.J., Becker, P.R., and Kucklick, J.R. 2008. Hexabromocyclododecane in white-sided dolphins: Temporal trend and stereoisomer distribution in tissues. *Environmental Science & Technology* 42:2650–2655.

58. Stapleton, H.M., Dodder, N.G., Kucklick, J.R., Reddy, C.M., Schantz, M.M., Becker, P.R., Gulland, F., Porter, B.J., and Wise, S.A. 2006. Determination of HBCD, PBDEs and MeO-BDEs in California sea lions (*Zalophus californianus*) stranded between 1993 and 2003. *Marine Pollution Bulletin* 52:522–531.

59. Janák, K., Sellström, U., Johansson, A.K., Becher, G., de Wit, C.A., Lindberg, P., and Helander, B. 2008. Enantiomer-specific accumulation of hexabromocyclododecanes in eggs of predatory birds. *Chemosphere* 73:S193–S200.

60. Law, R.J., Herzke, D., Harrad, S., Morris, S., Bersuder, P., and Allchin, C.R. 2008. Levels and trends of HBCD and BDEs in the European and Asian environments, with some information for other BFRs. *Chemosphere* 73:223–241.

61. Tanabe, S. 2008. Temporal trends of brominated flame retardants in coastal waters of Japan and South China: Retrospective monitoring study using archived samples from es-Bank, Ehime University, Japan. *Marine Pollution Bulletin* 57:267–274.

62. Xian, Q., Ramu, K., Isobe, T., Sudaryanto, A., Liu, X., Gao, Z., Takahashi, S., Yu, H., and Tanabe, S. 2008. Levels and body distribution of polybrominated diphenyl ethers (PBDEs) and hexabromocyclododecanes (HBCDs) in freshwater fishes from the Yangtze River, China. *Chemosphere* 71:268–276.

63. van Leeuwen, S.P. and de Boer, J. 2008. Brominated flame retardants in fish and shellfish—Levels and contribution of fish consumption to dietary exposure of Dutch citizens to HBCD. *Molecular Nutrition and Food Research* 52:194–203.

64. Shaw, S.D., Berger, M.L., Brenner, D., Kannan, K., Lohmann, N., and Päpke, O. 2009. Bioaccumulation of polybrominated diphenyl ethers and hexabromocyclododecane in the northwest Atlantic marine food web. *Science of The Total Environment* 407:3323–3329.

65. Janák, K., Covaci, A., Voorspoels, S., and Becher, G. 2005. Hexabromocyclododecane in marine species from the Western Scheldt estuary: Diastereoisomer- and enantiomer-specific accumulation. *Environmental Science & Technology* 39:1987–1994.

66. Zegers, B.N., Mets, A., Van Bommel, R., Minkenberg, C., Hamers, T., Kamstra, J.H., Pierce, G.J., Boon, J.P. 2005. Levels of hexabromocyclododecane in harbor porpoises and common dolphins from western European seas, with evidence for stereoisomer-specific biotransformation by cytochrome p450. *Environmental Science & Technology* 39:2095–2100.

67. Sellström, U., Bignert, A., Kierkegaard, A., Haggberg, L., de Wit, C.A., Olsson, M., and Jansson, B. 2003. Temporal trend studies on tetra- and pentabrominated diphenyl ethers and hexabromocyclododecane in guillemot egg from the Baltic Sea. *Environmental Science & Technology* 37:5496–5501.

68. Ronisz, D., Finne, E.F., Karlsson, H., and Förlin, L. 2004. Effects of the brominated flame retardants hexabromocyclododecane (HBCDD), and tetrabromobisphenol A (TBBPA), on hepatic enzymes and other biomarkers in juvenile rainbow trout and feral eelpout. *Aquatic Toxicology* 69:229–245.

69. Kuiper, R.V., Cantón, R.F., Leonards, P.E., Jenssen, B.M., Dubbeldam, M., Wester, P.W., van den Berg, M., Vos, J.G., and Vethaak, A.D. 2007. Long-term exposure of European flounder (Platichthys flesus) to the flame-retardants tetrabromobisphenol A (TBBPA) and hexabromocyclododecane (HBCD). *Ecotoxicology and Environmental Safety* 67:349–360.

70. Zhang, X., Yang, F., Zhang, X., Xu, Y., Liao, T., Song, S., and Wang, J. 2008. Induction of hepatic enzymes and oxidative stress in Chinese rare minnow (*Gobiocypris rarus*) exposed to waterborne hexabromocyclododecane (HBCDD). *Aquatic Toxicology* 86:4–11.

71. Palace, V.P., Pleskach, K., Halldorson, T., Danell, R., Wautier, K., Evans, B., Alaee, M., Marvin, C., and Tomy, G.T. 2008. Biotransformation enzymes and thyroid axis disruption in juvenile rainbow trout (*Oncorhynchus mykiss*) exposed to hexabromocyclododecane diastereoisomers. *Environmental Science & Technology* 42:1967–1972.

72. Eriksson, P., Fischer, C., Wallin, M., Jakobsson, E., and Fredriksson, A. 2006. Impaired behaviour, learning and memory in adult mice neonatally exposed to hexabromocyclododecane (HBCDD). *Environmental Toxicology and Pharmacology* 21:317–322.

73. Mariussen, E. and Fonnum, F. 2003. The effect of brominated flame retardants on neurotransmitter uptake into rat brain synaptosomes and vesicles. *Neurochemistry International* 43:533–542.

74. Deng, J., Yu, L., Liu, C., Yu, K., Shi, X., Yeung, L.W., Lam, P.K.S., Wu, R.S.S, and Zhou, B. 2009. Hexabromocyclododecane-induced developmental toxicity and apoptosis in zebrafish embryos. *Aquatic Toxicology* 93:29–36.

75. Leung, C.C.M., Jefferson, T.A., Hung, S.K., Zheng, G.J., Yeung, L.W.Y., Richardson, B.J., and Lam, P.K.S. 2005. Petroleum hydrocarbons, polycyclic aromatic hydrocarbons, organochlorine pesticides and polychlorinated biphenyls in tissues of Indo-Pacific humpback dolphins from south China waters. *Marine Pollution Bulletin* 50:1713–1719.

76. Hung, C.L.H., Lau, R.K.F., Lam, J.C.W., Jefferson, T.A., Hung, S.K., Lam, M.H.W., and Lam, P.K.S. 2007. Risk assessment of trace elements in the stomach contents of Indo-Pacific humpback dolphins and finless porpoises in Hong Kong waters. *Chemosphere* 66:1175–1182.

77. Yeung, L.W.Y., Miyake, Y., Wang, Y., Taniyasu, S., Yamashita, N., and Lam, P.K.S. 2009. Total fluorine, extractable organic fluorine, perfluorooctane sulfonate and other related fluorochemicals in liver of Indo-Pacific humpback dolphins (*Sousa chinensis*) and finless porpoises (*Neophocaena phocaenoides*) from South China. *Environmental Pollution* 157:17–23.

78. Isobe, T., Ramu, K., Kajiwara, N., Takahashi, S., Lam, P.K.S., Jefferson, T.A., Zhou, K., and Tanabe, S. 2007. Isomer specific determination of hexabromocyclododecanes (HBCDs) in small cetaceans from the South China Sea—Levels and temporal variation. *Marine Pollution Bulletin* 54:1139–1145.

79. Tomy, G.T., Halldorson, T., Danell, R., Law, K., Arsenault, G., Alaee, M., MacInnis, G., and Marvin, C.H. 2005. Refinements to the diastereoisomer-specific method for the analysis of hexabromocyclododecane. *Rapid Communications in Mass Spectrum* 19:2819–2826.

80. Nicholson, M.D., Fryer, R., and Larsen, J.R. 1995. Temporal trend monitoring: Robust method for analysing contaminant trend monitoring data. *ICES Techniques in Marine Environmental Science* No. 20, 22pp.

81. Lam, J.C.W., Lau, R.K.F., Murphy, M.B., and Lam, P.K.S. 2009. Temporal trends of hexabromocyclododecanes (HBCDs) and polybrominated diphenyl ethers (PBDEs) and detection of two novel flame retardants in marine mammals from Hong Kong, South China. *Environmental Science & Technology* 43:6944–6949.

82. Kajiwara, N., Ueno, D., Takahashi, A., Baba, N., and Tanabe, S. 2004. Polybrominated diphenyl ethers and organochlorines in archived northern fur seal samples from the Pacific coast of Japan, 1972–1998. *Environmental Science & Technology* 38:3804–3809.

83. Tanabe, S., Ramu, K., Isobe, T., Kajiwara, N., Takahashi, S., Jefferson, T.A., and Yamada, T.K. 2007. Levels and temporal trends of brominated flame retardants (PBDEs and HBCDs) in Asian waters using archived samples from es-Bank, Ehime University, Japan. *Organohalogen Compounds* 69:500–503.

84. Kajiwara, N., Kamikawa, S., Amano, M., Hayano, A., Yamada, T.K., Miyazaki, N., and Tanabe, S. 2008. Polybrominated diphenyl ethers (PBDEs) and organochlorines in melon-headed whales, *Peponocephala electra*, mass stranded along the Japanese coasts: Maternal transfer and temporal trend. *Environmental Pollution* 156:106–114.

85. Kajiwara, N., Isobe, T., Ramu, K., and Tanabe, S. 2006. Temporal trend studies on hexabromocyclododecanes (HBCDs) in marine mammals from Asia-Pacific. *Organohalogen Compounds* 68:515–518.

86. Tanabe, S., Ramu, K., Isobe, T., and Takahashi, S. 2008. Brominated flame retardants in the environment of Asia-Pacific: An overview of spatial and temporal trends. *Journal of Environmental Monitoring* 10:188–197.

87. Johnson-Restrepo, B., Kannan, K., Addink, R., and Adams, D.H. 2005. Polybrominated diphenyl ethers and polychlorinated biphenyls in a marine foodweb of coastal Florida. *Environmental Science & Technology* 39:8243–8250.

88. Tuerk, K.J.S., Kucklick, J.R., Becker, P.R., Stapleton, H.M., and Baker, J.E. 2005. Persistent organic pollutants in two dolphin species with focus on toxaphene and polybrominated diphenyl ethers. *Environmental Science & Technology* 39:692–698.

89. Kannan, K., Perrotta, E., Thomas, N.J., and Aldous, K.M. 2007. A comparative analysis of polybrominated diphenyl ethers and polychlorinated biphenyls in Southern sea otters that died of infectious diseases and noninfectious causes. *Archives of Environmental Contamination Toxicology* 53:293–302.

90. Ikonomou, M.G., Rayne, S., and Addison, R.F. 2002. Exponential increases of the brominated flame retardants, polybrominated diphenyl ethers, in the Canadian arctic from 1981 to 2000. *Environmental Science & Technology* 36:1886–1892.

91. Ikonomou, M.G., Kelly, B.C., and Stern, G.A. 2005. Spatial and Temporal Trends of PBDEs in Biota from the Canadian Arctic Marine environment. *Organohalogen Compounds* 67:950–953.

92. Muir, D., Kwan, M., and Evans, M. 2006. Temporal trends of persistent organic pollutants and metals in ringed seals from the Canadian Arctic. In *Synopsis of Research Conducted under the 2005–2006 Northern Contaminants Program*, eds. S. Smith and J. Stow, pp. 162–169. Burlington, Ontario, Canada: Indian Affairs and Northern Development.

93. Vorkamp, K., Rigét, F.F., Glasius, M., Muir, D.C., and Dietz, R. 2008. Levels and trends of persistent organic pollutants in ringed seals (*Phoca hispida*) from Central West Greenland, with particular focus on polybrominated diphenyl ethers (PBDEs). *Environmental International* 34:499–508.

94. Kierkegaard, A., Bignert, A., Sellström, U., Olsson, M., Asplund, L., Jansson, B., and de Wit, C.A. 2004. Polybrominated diphenyl ethers (PBDEs) and their methoxylated derivatives in pike from Swedish waters with emphasis on temporal trends, 1967–2000. *Environmental Pollution* 130:187–198.

95. Johansson, I., Héas-Moisan, K., Guiot, N., Munschy, C., and Tronczyński, J. 2006. Polybrominated diphenyl ethers (PBDEs) in mussels from selected French coastal sites: 1981–2003. *Chemosphere* 64:296–305.

96. Law, R.J., Bersuder, P., Allchin, C.R., and Barry, J. 2006. Levels of the flame retardants hexabromo-cyclododecane and tetrabromobisphenol A in the blubber of harbor porpoises (*Phocoena phocoena*) stranded or bycaught in the UK, with evidence for an increase in HBCD concentrations in recent years *Environmental Science & Technology* 40:2177–2183.

23 Status and Trends of POPs in Harbor Seals from the Northwest Atlantic

Susan D. Shaw,* Michelle L. Berger,
and Kurunthachalam Kannan

CONTENTS

23.1 Introduction 516
    23.1.1 Harbor Seals and POPs in the Northwest Atlantic 516
    23.1.2 Possible Role of POPs in Harbor Seal Mortalities 517
    23.1.3 Seals as Sentinels.. 517
23.2 Organochlorines: PCBs, Chlorinated Pesticides, Dioxins/Furans 517
    23.2.1 Contamination Status: PCBs and Chlorinated Pesticides 518
    23.2.2 Contamination Status: Dioxins/Furans, Dioxin-Like PCBs, and WHO-TEQs 520
    23.2.3 Are Dioxin-Like Compounds Altering Immune Function in Harbor Seals? 522
23.3 Brominated Flame Retardants: PBDEs and HBCDs 522
    23.3.1 Contamination Status: PBDEs and HBCDs in Harbor Seals 523
    23.3.2 PBDEs and HBCDs in Harbor Seal Prey Fishes 525
    23.3.3 Biomagnification of PBDEs and HBCD from Fish to Seals 527
23.4 Perfluorinated Chemicals 528
    23.4.1 Contamination Status: PFCs in Harbor Seal Tissues 529
23.5 Temporal Trends 531
    23.5.1 PCBs and DDT 531
    23.5.2 PBDEs... 532
    23.5.3 PFCs 533
23.6 Spatial Trends and Source Implications 533
    23.6.1 PCBs 533
    23.6.2 PBDEs... 533
    23.6.3 PFCs.. 534
23.7 Global Trends: Comparisons of POPs in Pinnipeds 534
    23.7.1 PCBs 534
    23.7.2 DDTs.. 535
    23.7.3 PBDEs 535
    23.7.4 PFOS 537
23.8 Summary and Perspectives.. 538
Acknowledgments... 538
References.. 538

* E-mail: sshaw@meriresearch.org (Chapter corresponding author).

23.1 INTRODUCTION

Widely distributed in the temperate waters of the northwestern Atlantic, harbor seals (*Phoca vitulina concolor*) are small pinnipeds that are closely associated with polluted near-shore and coastal environments. As long-lived, apex predators, harbor seals accumulate high concentrations of a staggering number of persistent organic pollutants (POPs) in their tissues and are important indicators of marine ecosystem contamination and trends.

23.1.1 Harbor Seals and POPs in the Northwest Atlantic

At present, there are an estimated 99,300 harbor seals inhabiting the northwestern Atlantic region extending from the Gulf of Maine southward to the coast of New Jersey [1]. Considered relatively nonmigratory, harbor seals feed in coastal and estuarine environments and are exposed to contaminated habitats and prey across their range (Figure 23.1). Isolated from the deeper waters of the Atlantic Ocean by Georges and Brown Banks and Nantucket Shoals, the large northern Gulf portion of their range is a semi-enclosed sea with a variable, principally estuarine circulation pattern receiving significant riverine, urban, agricultural, and industrial pollutant discharges from densely populated areas in the U.S. Northeast as well as inputs via long-range atmospheric transport [2,3]. Because the removal rates of persistent chemicals are slow, biota in such semi-enclosed seas are at elevated risk [4]. In the southern areas, coastal urban development has resulted in some of the densest concentrations of human populations in North America and POP contamination has been a concern since at least the 1950s. In the 1980s, Massachusetts Bay and Long Island Sound were ranked as among the most polluted regions, having more "high organic contamination" sites than any other coastal state or region in the United States [5].

Concern has focused on the polyhalogenated aromatic hydrocarbons (PHAHs), including the polychlorinated biphenyls (PCBs), dioxins and furans (PCDD/Fs), and chlorinated pesticides such as dichlorodiphenyltrichloroethane (DDT) and chlordane (CHL), because of their widespread use and stability in the marine environment and their immune- and endocrine-disrupting potential

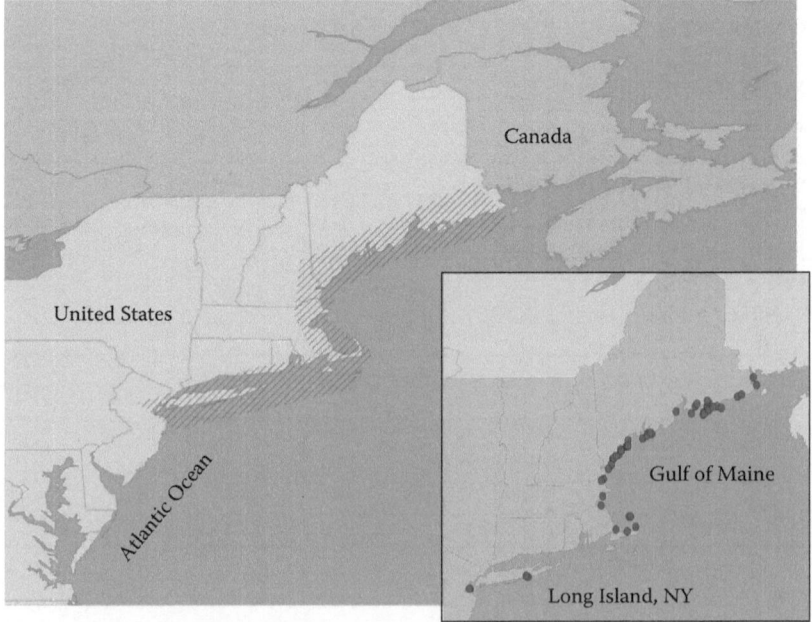

FIGURE 23.1 Map of the northwest Atlantic showing the harbor seal range and stranding locations of seals in this study.

in seals and other marine wildlife [6–8]. A large body of data suggests that PHAHs, particularly PCBs and DDT, have adversely affected the health of seals inhabiting industrialized regions and may be a contributing factor in the recurring epizootics reported among harbor seals since the late 1970s [9–13]. Recent concern has shifted to current-use chemicals such as the brominated flame retardants (BFRs), polybrominated diphenyl ethers (PBDEs), and hexabromocyclododecanes (HBCDs) and perfluorinated chemicals (PFCs). Like the organochlorines (OCs), these chemicals biomagnify in food webs and are associated with endocrine-disrupting and reproductive/neurodevelopmental effects in animals [14–17]. As a result of their environmental persistence and widespread use in household and commercial products, these compounds of emerging concern have been increasing in marine food webs and elevated concentrations are detected in marine mammals [14,17–22].

23.1.2 POSSIBLE ROLE OF POPS IN HARBOR SEAL MORTALITIES

For decades, the northwestern Atlantic harbor seal population has been experiencing recurring epizootics, and it can be speculated that their high body burdens of immunotoxic chemicals may be playing a contributory role in these events by compromising the animals' normal resistance to disease. The sensitivity of the species to the effects of POP exposure was previously demonstrated by captive feeding studies in which harbor seals fed OC-contaminated fish exhibited reproductive impairment [23], reduced plasma thyroid hormone and retinol levels [24], and suppression of numerous cellular and humoral immune functions [25,26].

Some of the mortalities along the northwestern Atlantic have been associated with viral epizootics. In 1979–1980, a type A influenza virus infection spread northward from Cape Cod into the Gulf of Maine and ultimately resulted in the deaths of more than 500 harbor seals [9]. In 1991–1992, a morbillivirus epizootic (phocine distemper virus or PDV) of unknown magnitude was reported among harbor seals found stranded from southern Maine to New York [11,27]. More recently, two mortality events of unknown etiology have affected the population. In late summer of 2004, approximately 300 seals, primarily pups, were found dead on the beaches in southern Maine within a few weeks. The most recent die-off occurred between 2006 and 2008, when approximately 700 harbor seals died along the New England coast [28]. The possible contributory role of POPs in these events cannot be ruled out, since their body burdens of PCBs alone exceed the threshold level for immune system impairment in the species [29,30]. Moreover, plasma levels of dioxin-like compounds in free-ranging adult harbor seals from this region were shown to be significantly associated with altered *in vitro* immune responses, implying that disease resistance in these seals may be compromised by their exposure to immunotoxic chemicals [31].

23.1.3 SEALS AS SENTINELS

POP contamination and trends in northwest Atlantic harbor seals have been the focus of a series of ecotoxicological studies titled *Seals As Sentinels* over the past decade [7,30–36]. This chapter presents an overview of the *Seals As Sentinels* research results in a global context, including contamination status and spatial and temporal trends for legacy POPs (PCBs, PCDD/Fs, and chlorinated pesticides), as well as for chemicals of emerging concern such as BFRs and PFCs that are in current use or have been restricted only recently.

23.2 ORGANOCHLORINES: PCBs, CHLORINATED PESTICIDES, DIOXINS/FURANS

Studies conducted in the 1970s [37] indicated that PCB and DDT body burdens in northwestern Atlantic harbor seals exceeded or were approaching an average concentration of 100 parts per million (ppm, lipid basis), similar to the high levels associated with reproductive failure and population declines among European harbor, gray (*Halichoerus grypus*) and ringed seals (*Phoca hispida*) [38–40] and premature pupping in California sea lions (*Zalophus californianus*) [41].

To determine current burdens of persistent OCs in these seals, concentrations of PCBs, PCDD/Fs, and chlorinated pesticides were measured in blubber samples of 42 harbor seals (7 adult males, 8 adult females, 14 yearlings, 13 pups) that stranded along the northwestern Atlantic coast from the eastern coast of Maine to Long Island, New York between 1991 and 2005 [30,32] (see map of stranding locations in Figure 23.1). Blubber samples were analyzed for 146 PCB congeners (including 8 mono- and 4 non-ortho PCBs) and 17 PCDD/Fs following the isotope dilution quantification method. OC levels and congener and isomer compositions were compared in the 1991 and 2001–2002 samples, as well as with available region-specific data to examine temporal trends. OC concentrations in these seals were compared with those reported in stranded pinnipeds from different marine regions and with estimated threshold levels of adverse effects to evaluate the toxicological implications of current body burdens.

23.2.1 CONTAMINATION STATUS: PCBs AND CHLORINATED PESTICIDES

PCBs, DDT, and CHLs were the major persistent OCs in harbor seals [30,32] (Table 23.1). Mirex, hexachlorohexanes (HCHs), and dieldrin were minor contaminants in seals relative to other groups and hexachlorobenzene (HCB) was detected at trace levels in these samples. Aldrin, endrin, endrin aldehyde, endrin ketone, and methoxychlor concentrations were lower than detection limits in all samples. PCB concentrations in seal blubber ranged from 2,460 to 461,000 ng/g lipid weight (lw) followed by DDTs (471–177,000 ng/g lw), CHLs (187–73,200 ng/g lw), HCHs (17–1670 ng/g), mirex (2.6–739 ng/g), and dieldrin (2.5–704 ng/g).

This distribution pattern (PCBs > DDTs > CHLs > mirex) is consistent with that reported in blubber of stranded harbor seals sampled in 1980 along the Massachusetts coast and in 1991–1992 from Long Island [42], as well as with the pattern in older male harbor seals from the St. Lawrence estuary [43]. With the exception of dieldrin, OCs were highly correlated ($p < 0.001$) in our samples, reflecting parallel accumulation of these contaminants in the food chain. The different trend for dieldrin may reflect different sources of the compound and/or different transport mechanisms.

Ten PCB congeners (90 + 101, 99, 128, 129 + 163, 138, 146, 153, 170, 180, and 187) accounted for the majority of PCBs in blubber (Figure 23.2). The extremely persistent DDT metabolite, *p,p′*-DDE, was the predominant DDT residue, contributing 92% of the ΣDDT concentration in all samples. *Trans*-nonachlor was the predominant CHL in seal blubber samples, followed by oxychlordane, accounting for 62% and 32% of ΣCHL, respectively. HCH isomers were dominated by α-HCH, contributing 71% of ΣHCH in seal blubber, followed by β-HCH, γ-HCH (lindane), and δ-HCH.

Mean ΣPCB concentrations (sum of 120 congeners) were highest in the younger seals (60,479 and 56,799 ng/g lw in yearlings and pups, respectively) (Table 23.1). Mean PCB concentrations in adult females (14,047 ng/g, lw) were less than half those in the adult males (36,685 ng/g, lw). Mean ΣDDT concentrations in the adult males (32,500 ng/g lw), yearlings (33,100 ng/g lw), and pups (33,500 ng/g lw) were similar and nearly sevenfold higher than those in the adult females (4940 ng/g lw). This pattern is consistent with that observed for most lipophilic organohalogenated compounds in which females transfer a large proportion of their body burden to pups during gestation and lactation, while an age-dependent accumulation is observed in male seals.

PCB congener profiles were dominated by the hexa-CBs-153 and -138, followed by the penta-CB-99 in the younger seals and the hepta-CB-180 in the adults (Figure 23.2). Examination of PCB homolog profiles revealed significant differences in patterns by age class (Figure 23.3). In pups and yearlings, the lower chlorinated congeners (tetra- and penta-CBs) contributed proportionally more to the total than in the adult seals ($p < 0.01$), whereas the adults retained a greater proportion of the hepta- ($p < 0.05$), octa-, and nona-CBs ($p < 0.01$). Although we did not examine mother–pup pairs, this pattern is consistent with other studies in which substantially lower concentrations of the higher

TABLE 23.1
POP Concentrations in Tissues of Northwest Atlantic Harbor Seals

Compound	Tissue	Adult Male	Adult Female	Yearling	Pup
ΣPCB[a]	Blubber	36,700±29,500	14,100±14,700	60,500±118,000	56,800±93,800
		4,410–90,200	2,460–42,400	4,540–461,000	636–288,000
		N=7	N=8	N=14	N=13
ΣPCB[b]	Liver	62,200±64,500			78,600±117,000
		11,200–154,000			580–600,000
		N=6			N=50
ΣDDT[c]	Blubber	32,500±25,200	4940±7960	33,100±45,400	33,500±53,300
		2,640–77,900	1,040–24,500	2,660–177,000	471–160,000
		N=7	N=8	N=14	N=13
ΣCHLs[d]	Blubber	6950±7790	726±416	4330±4460	9,010±19,700
		1,010–23,000	187–1480	487–15,800	369–73,200
		N=7	N=8	N=14	N=13
ΣDL-PCB[e]	Blubber	421±266	203±132	1940±3570	2020±3390
		193–816	54–490	302–14,100	44–11,200
		N=7	N=8	N=14	N=13
ΣPCDD[f,g]	Blubber	4.78±4.28	14.7±28.4	11.5±12.7	69.2±180
		ND–12.4	ND–82.5	ND–41.6	ND–658
		N=7	N=8	N=14	N=13
ΣPCDF[f,h]	Blubber	6.09±3.71	4.43±2.83	5.16±4.91	7.80±7.57
		ND–11.5	ND–8.30	ND–18.6	ND–23.1
		N=7	N=8	N=14	N=13
ΣWHO$_{PCB}$-TEQ[f,i]	Blubber	54.1±28.2	24.8±14.3	126±128	188±273
		23.0–84.7	7.88–52.9	24.3–528	12.6–872
		N=7	N=8	N=14	N=13
ΣWHO$_{PCDD}$-TEQ[f,i]	Blubber	0.478±0.428	0.198±0.057	1.15±1.27	3.09±5.15
		ND–1.24	ND–0.326	ND–1.86	ND–16.0
		N=7	N=8	N=14	N=13
ΣWHO$_{PCDF}$-TEQ[f,i]	Blubber	0.609±0.371	0.443±0.283	0.516±0.491	0.780±0.757
		ND–1.15	ND–0.830	ND–1.86	ND–2.31
		N=7	N=8	N=14	N=13
ΣPBDE[j]	Blubber	1390±1260	326±193	2950±6000	3640±7390
		454–3830	131–713	250–23,500	80.2–25,700
		N=7	N=8	N=14	N=13
ΣPBDE[k]	Liver	969±775			2880±3720
		372–2470			34.7–19,500
		N=6			N=50
ΣHBCD[l]	Blubber	12.4±11.8			11.7±9.85
		3.05–29.3			ND–25.3
		N=4			N=6
ΣHBCD[l]	Liver	23.1±13.8			39.8±50.2
		7.3–43.4			ND–279
		N=6			N=50
ΣPFSA[m,n]	Liver	102±106	103±56		264±315
		32–357	27–227		10–1395
		N=8	N=10		N=50

(*continued*)

TABLE 23.1 (continued)
POP Concentrations in Tissues of Northwest Atlantic Harbor Seals

Compound	Tissue	Adult Male	Adult Female	Yearling	Pup
ΣPFCA[m,o]	Liver	28±17	24±15		26±18
		11–59	ND–58		ND–87
		N=8	*N*=10		*N*=50

Notes: Values given are mean±standard deviation and range. Units are ng/g lipid wt unless otherwise noted.

[a] ΣPCB in blubber is the sum of 119 di- to deca-CB congeners.

[b] ΣPCB in liver is the sum of 30 tri- to deca-CB congeners.

[c] ΣDDT is the sum of *o,p'*-DDE, *p,p'*-DDE, *o,p'*-DDD, *p,p'*-DDD, *o,p'*-DDT, *p,p'*-DDT.

[d] ΣCHLs is the sum of heptachlor, oxychlordane, heptachlor epoxide, α-chlordane, γ-chlordane, trans-nonachlor, cis-nonachlor.

[e] ΣDL-PCB is the sum of four non-ortho and eight mono-ortho dioxin-like PCBs.

[f]   Units are pg/g lipid wt.

[g] ΣPCDD is the sum of seven 2,3,7,8-CDDs.

[h] ΣPCDF is the sum of 10 2,3,7,8-CDFs.

[i]   2005 WHO-TEFs used to calculate WHO-TEQ [50].

[j] ΣPBDE in blubber is the sum of 36 mono- to hexa-BDE congeners.

[k] ΣPBDE in liver is the sum of 18 tri- to octa- BDE congeners.

[l] ΣHBCD is the sum of α-, β-, and γ-HBCD isomers.

[m]   Units are ng/g wet wt.

[n] ΣPFSA is the sum of PFHxS, PFOS, PFDS, and PFOSA.

[o] ΣPFCA is the sum of PFHpA, PFOA, PFNA, PFDA, PFUnDA, and PFDoDA.

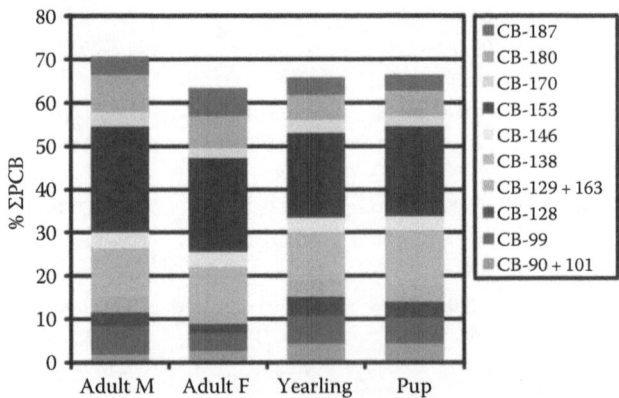

FIGURE 23.2   PCB congener profile in harbor seal blubber by age class.

chlorinated and hydrophobic compounds are found in blubber of pups than in maternal blubber, reflecting selective partitioning of persistent PCB congeners into milk over the course of lactation [44–47]. The same pattern was observed in our yearling seals, although they retained more of the octa- and nona-CBs than the pups ($p < 0.01$), reflecting the gradual accumulation of higher chlorinated CBs through feeding.

23.2.2 Contamination Status: Dioxins/Furans, Dioxin-Like PCBs, and WHO-TEQs

Of 17 PCDD/Fs analyzed in seal blubber, only two isomers were detected in at least 30% of samples and these contributed ~2% to the total World Health Organization Toxic Equivalent (WHO-TEQ)

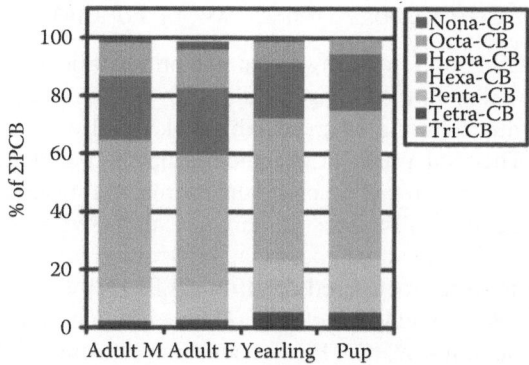

FIGURE 23.3 PCB homolog profile in harbor seal blubber by age class.

[32]: 1,2,3,6,7,8-HxCDD was detected in 36% of samples and 2,3,7,8-TCDF was detected in 61% of samples. Mean PCDD/F concentrations were extremely low ranging from 0.6 pg/g lw in the adult females to 3.9 pg/g lw in the pups and could reflect the absence of major local sources but are more likely due to a species-specific degradative capacity for these compounds. Several studies have reported a lack of significant bioaccumulation of PCDD/Fs in seals with known exposure, implying that they may be more readily metabolized than the PCBs [25,48,49].

WHO-TEQs calculated using the most recent toxic equivalency factors (TEFs) [50] were highest in the pups (191 pg/g, lw) and yearlings (128 pg/g, lw), followed by the adult males (55 pg/g, lw) and females (25 pg/g, lw) [32] (Table 23.1). Differences between the younger seals and adult females were significant ($p < 0.05$). PCB-TEQ congener profiles were fairly similar (with one exception) and were dominated by the non-*ortho* CB 126 (61%–72%), followed by the mono-*ortho* CBs 118 (10%–19%) and 156 (5%–8%) (Figure 23.4). In the adult males, the non-*ortho* CB 169 contributed more to the total than in the adult females, yearlings, or pups ($p < 0.05$).

In addition to age and gender, the influence of condition on concentration was examined. Adult seals were in better condition (weight/length) than the younger animals ($p < 0.01$), but condition indices were not correlated with PCB concentrations and were only weakly correlated with WHO-TEQs of dioxin-like PCBs in blubber ($R^2 = 0.15$, $p = 0.02$ for weight/length). Similarly, we found no correlations between concentrations of PCBs or PCB-TEQs and body weight, length, or girth. Lipid content in our samples was positively correlated with both condition indices (weight/length $R^2 = 0.17$, girth/length $R^2 = 0.23$, $p < 0.02$) and negatively correlated with total PCB ($R^2 = 0.34$, $p < 0.01$) and PCB TEQ ($R^2 = 0.54$, $p < 0.01$). Overall lipid content averaged 59%, indicating that some seals were in poor nutritive condition. However, when those animals were removed from the analysis, mean concentrations in the blubber still remained relatively high (PCBs 24,900 ng/g lw, range 4,412–90,172).

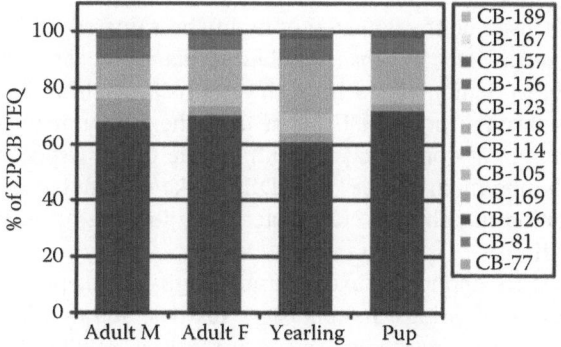

FIGURE 23.4 PCB-TEQ profile in harbor seal blubber by age class.

23.2.3 Are Dioxin-Like Compounds Altering Immune Function in Harbor Seals?

Mean concentrations of PCBs found in northwestern Atlantic harbor seals are relatively high on a global scale. PCB concentrations in the younger seals (~50–60 μg/g lw) and adult males (~37 μg/g lw) are two to fourfold higher than the estimated threshold level of 17 μg PCB/g lw for adverse effects in the species [29]. The total WHO-TEQ values of dioxin-like PCBs and PCDD/Fs in blubber of the pups (mean 191 pg/g lw) are approaching the estimated threshold level of 209 pg/g lw for immune suppression and endocrine disruption in harbor seals based on results of captive feeding experiments [25,29]. However, it should be noted that TEQ values calculated for dioxin-like PCBs and PCDD/Fs in the feeding experiment used the 1998 TEFs [51]. Re-calculation of these values using the most recent TEFs [50] would yield total TEQ concentrations at least 50% lower (~100 pg/g lw). Alternatively, calculation of the WHO-TEQs in our samples using the 1998 TEFs yields values that are two to four times higher (584, 490, 145, and 60 pg/g lw in pups, yearlings, adult males, and females, respectively) than those we report here, meaning that the levels in the pups and yearlings in this study would exceed by two to threefold the estimated threshold value published in the literature for adverse effects, including immunotoxic effects in pinnipeds. These observations underscore the need to recalculate the estimated threshold value for adverse effects using the new TEFs. In view of the recurring epizootics affecting the northwestern Atlantic harbor seal population, further research is needed to elucidate the real risks of dioxin-like compounds and the numerous other immune- and endocrine-disrupting POPs to which these seals are exposed.

23.3 BROMINATED FLAME RETARDANTS: PBDEs AND HBCDs

The BFRs, PBDEs, and HBCDs are synthetic organic compounds that are widely used as additives in a variety of household and commercial products to reduce their flammability [52]. As a result of their environmental persistence and high production volume over the past 30 years, PBDEs and HBCDs have become ubiquitous global contaminants, even in remote areas [18,53]. These are lipophilic compounds that bioaccumulate in marine food webs and have been found at high concentrations in top predators such as marine mammals [18,22,54]. Many BFRs including PBDEs and HBCDs have been shown to exert reproductive neurodevelopmental and endocrine-disrupting effects in animals [14,15,55].

Over the past three decades, North America has dominated the global market for the penta-BDE formulation, with annual usage estimated at 7100 metric tons (95%) in 2001 [56]. Restrictions were imposed on the production and use of the penta- and octa-BDE formulations in late 2004, while the third commercial product, deca-BDE (consisting of 97% BDE-209), represented 83% of the global market demand [56] for use as an additive in high-impact polystyrene for television sets, computer casings, and electronic equipment [52]. By late 2009, deca-BDE had been phased out of production in the United States [57] and banned in Canada and Europe. Nevertheless, large amounts of deca-BDE have been released to the global environment.

Recent attention has focused on the potential of highly brominated PBDEs, including BDE-209, the major constituent of deca-BDE, to bioaccumulate in marine ecosystems. Metabolic debromination of BDE-209 has been demonstrated in freshwater fish, resulting in the formation of lower substituted BDEs (hexa- through nona-BDEs) that have the potential to be more persistent and bioaccumulative than the parent compound [58–60]. While direct exposure to commercial BDE mixtures is assumed to be the main source of PBDE uptake/accumulation in biota, it is plausible that BDE-209 debromination in fish may contribute to the loading of persistent BDEs in marine food webs [14,33,54,58,59].

HBCDs are additive BFRs applied in extruded and high-impact polyurethane foams that are used as thermal insulation in buildings, in upholstery textiles, and to a lesser extent, in electrical equipment [52]. Few studies have reported on the occurrence of HBCD in marine ecosystems, and little information is available on the fate and environmental persistence of this compound. HBCD concentrations reported in marine fishes from Asia [61], Arctic Canada [62], and various areas of Europe [22,63–65], vary significantly by region, species, and tissue examined. Little data exist on HBCD concentrations in marine fishes from U.S. waters.

HBCD concentrations reported in cetaceans from the eastern U.S. coast [66,67] are one to two orders of magnitude lower than those detected in marine mammals from Europe [22,68,69], where the market demand for this compound is two to threefold higher than in Asia or America, respectively [56]. HBCDs are still in worldwide use although they were recently banned by the European Union's Registration, Evaluation, Authorization and Restriction of Chemicals (REACH) program, and are identified as priority contaminants for risk assessment in Japan [12,14,70].

To examine the uptake/accumulation and biomagnification of BFRs in the northwest Atlantic marine ecosystem, concentrations of PBDEs were measured in blubber samples of 42 harbor seals (7 adult males, 8 adult females, 14 yearlings, 13 pups) that stranded along the northwest Atlantic coast from eastern Maine to Long Island, New York between 1991 and 2005 [33] (Figure 23.1). Harbor seal blubber samples were analyzed for 41 PBDE congeners (mono- through octa-BDEs and BDE-209) following the isotope dilution technique [71]. A subsequent study [36] examined tissue-specific accumulation of PBDEs (tri- to deca-BDE) in liver of 56 harbor seals (50 pups, 6 adult males) and compared PBDE levels with those detected previously in blubber. HBCDs were measured in a subset of 11 harbor seals. A third study [34] analyzed concentrations of PBDEs (mono- through deca-BDEs) in seven species of commercially important marine fishes (87 individuals) that are components of the harbor seal diet [72] to examine the trophic transfer of PBDEs from prey to predator. Fish species included silver hake (*Merluccius bilinearis*), white hake (*Urophycis tenuis*), Atlantic herring (*Clupea harengus*), American plaice (*Hippoglossoides platessoides*), alewife (*Alosa pseudoharengus*), winter flounder (*Pseudopleuronectes americanus*), and Atlantic mackerel (*Scomber scombrus*). These species are all highly migratory and feed across an extended spatial range in the western Atlantic (from Newfoundland, Canada to South Carolina). HBCD concentrations were measured in three fish species (Atlantic herring, Atlantic mackerel, and alewife) to provide an initial estimation of HBCD contamination in this ecosystem. Together, these studies represent the first comprehensive investigation of the bioaccumulation and biomagnification of BFRs in the northwest Atlantic marine food web.

23.3.1 CONTAMINATION STATUS: PBDEs AND HBCDs IN HARBOR SEALS

Of 41 BDEs analyzed, a total of 25 congeners were detected in harbor seal blubber (i.e., BDEs-15, -17, -25, -28, -30, -35, -37, -47, -49 + 71, -75, -77, -85, -99, -100, -116, -118, -119, -126, -153, -154, -155, -183, -197, and -209) [33]. In addition, several unidentified hepta- and octa-BDEs were detected in blubber. Sixteen BDEs were below the level of detection in blubber (i.e., BDE-1, -2, -3, -7, -8, -10, -11, -12, -13, -32, -33, -66, -138, -166, -181, and -190). The overall average ΣPBDE concentration (mono- to hexa-BDEs) in harbor seal blubber was 2403 ± 5406 ng/g lw (range 80–25,720; $n = 42$). By comparison, the mean PCB concentration previously reported in these samples was one to two orders of magnitude greater (overall mean $46,540 \pm 86,610$, range: 636–460,600 ng/g lw; $n = 42$) [32]. PBDEs in seal blubber were positively correlated with PCBs ($R = 0.83$, $p < 0.001$), suggesting a parallel accumulation of these compounds through the marine food chain.

Mean ΣPBDE concentrations in seal blubber samples by age were (in descending order) 3645 ± 7388, 2945 ± 5995, 1385 ± 1265, and 326 ± 193 ng lw in pups, yearlings, adult males, and adult females, respectively (Table 23.1). Gender-related differences in PBDE concentrations were not observed among the younger seals but were apparent among the adults ($p = 0.01$), with adult males having fourfold higher PBDE burdens than females. This pattern reflects differences in age-dependent bioaccumulation between males and females due to placental and lactational transfer of lipophilic organic compounds from females to pups.

PBDE congeners detected in liver of harbor seals ($n = 56$) included BDEs-28, -47, -49, -85, -99, -100, -153, -154, -155, -181, -183, -184, an unidentified hepta-, -191, -197, and -209. BDEs-66, -196, and -203 were not detected. ΣPBDE (tri-nona-BDE) concentrations in seal liver samples ranged from 35 to 19,547 ng/g lw, with an overall mean of 2671 ± 3566 ng/g lw (Table 23.1), which were similar to the ΣPBDE concentrations (mono-hexa-BDEs) detected in blubber samples (overall mean 2403 ± 5406; range 80–25,720 ng/g lw; $n = 42$) [33]. The PBDE concentrations in these seals are an order of magnitude higher than those reported in harbor and gray seals from Europe [74,75], reflecting the greater usage of penta-BDE in North America.

HBCDs were detected at relatively low levels in harbor seal blubber ($n = 10$) and liver ($n = 56$) samples; blubber concentrations ranged from 2 to 29 (mean ± SD: 12 ± 9.5) ng/g lw and liver concentrations ranged from 2 to 279 (mean ± SD: 38 ± 48) ng/g lw (Table 23.1) [36,73]. Of the total HBCD content, the α-HBCD isomer was predominant (contributing 96% and 100% of the total in liver and blubber, respectively). β-HBCD was detected in 12 liver samples and γ-HBCD was detected in 2 liver samples.

Eight PBDE congeners (BDEs-28, -47, -49, -99, -100, -153, -154, and -155) contributed 91%–99.6% of the total PBDE content in blubber and 81%–99.9% in seal liver. BDE-47 was the dominant congener, contributing 62%–75% of the total in liver and blubber (Figure 23.5). Higher PBDE levels were reported in California sea lions (*Zalophus californianus*) [76,77] and harbor seals from San Francisco Bay [78].

BDE profiles in harbor seal pup liver and blubber samples were similar and followed the order: BDE-47 > -99 > -100 > -153 > -154 > -155 > -28 > -49. In contrast, the hexa-BDEs, especially BDE-153, were more prominent in the adult males. These differences likely reflect the different exposure pathways between adults and pups, as well as age-related differences in the ability to metabolize and eliminate BDE congeners. Whereas the profiles in the males reflect uptake and accumulation through feeding on teleost fishes, the pattern in the pups suggests efficient placental and lactational transfer of BDE-47, and, to a lesser degree, BDEs-99 and -100, but very limited transfer of the hexa-BDEs. In gray seals, maternal transfer efficiency was shown to decline with increasing degree of bromination of the molecule, as a function of increasing K_{ow} values [79]. This may be a consequence of the molecular size of highly brominated BDEs, which may limit diffusion and lipid/water partitioning in females during lactation.

Several hepta- and octa-BDEs were detected at low levels in seal liver including BDE-183 (ND—13 ng/g lw), BDE-197 (ND—4.7 ng/g lw), and an unidentified hepta-BDE (ND—17 ng/g lw). BDEs-181, -184, and -191 were also detected at trace levels in liver. Hepta- and octa-BDEs were also found at slightly higher concentrations in seal blubber samples: BDE-183 (1.7–45 ng/g lw), BDE-197 (ND—57 ng/g lw), and an unidentified hepta- (ND—15 ng/g lw) and octa-BDE (ND—84 ng/g lw). While the presence of tetra- to hexa-BDEs in seal tissues indicates exposure to components of the penta-BDE mixture, the occurrence of hepta- and octa-BDEs suggests recent exposure to the octa- and deca-BDE mixtures and/or BDE-209 debromination processes, since a short half-life (months) is indicated for higher BDE congeners [80,81].

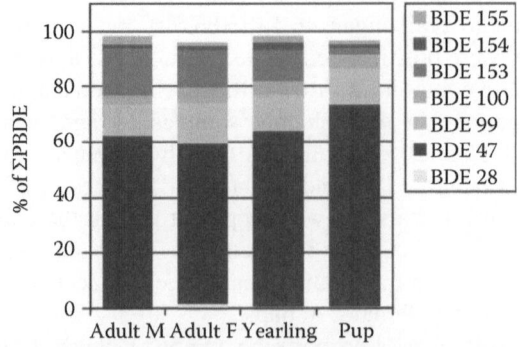

FIGURE 23.5 PBDE congener profile in harbor seal blubber by age class.

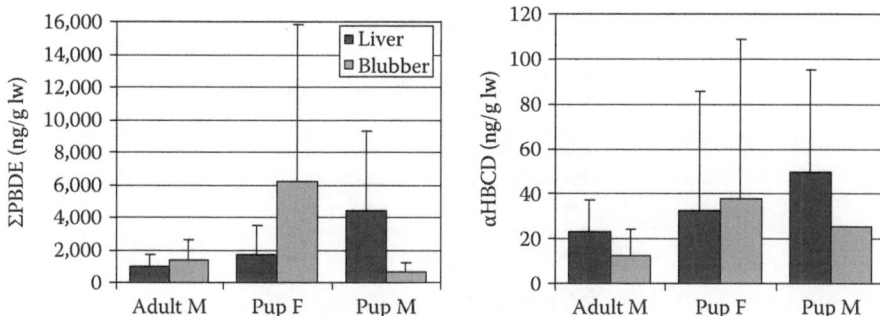

FIGURE 23.6   ΣPBDE and α-HBCD concentrations (mean ± standard deviation, ng/g lw) in liver and blubber of harbor seals.

BDE-209 levels in harbor seal liver were up to five-fold higher than those previously detected in blubber (1–8 ng/g lw), implying that liver is a preferential tissue for BDE-209 accumulation in harbor seals. This finding is consistent with results of laboratory studies showing preferential accumulation of BDE-209 in liver of rats and fish [59,82,83] and with studies showing selective hepatic retention of this compound in terrestrial wildlife such as red foxes (*Vulpes vulpes*) and Japanese raccoon dogs [84,85].

Male pups had twofold higher hepatic ΣPBDE concentrations (mean 4397 ± 4868 ng/g lw; $n = 22$) than the female pups (1680 ± 1807 ng/g lw; $n = 28$) ($F_{2,55} = 4.6$, $p = 0.02$) (Figure 23.6). The highest PBDE level in liver (19547 ng/g lw) was observed in a male pup from southern Maine. In addition, levels of the BDE congeners -28, -47, -49, -99, -153, and -155 were significantly higher in the male pups. On the other hand, female pups had mean blubber concentrations of PBDEs nearly an order of magnitude higher than those in the male pups [33]. A similar pattern was observed for α-HBCD. The first weeks of life represent a period of rapid growth, during which harbor seal pups almost triple their birth weight and lay down layers of blubber prior to weaning. The exact ages of the pups were unknown, thus we used biometric data to examine the possible influence of growth on hepatic PBDE concentrations. No correlations between PBDEs or α-HBCD and body weight, body length, or lipid content of the samples were found for male or female pups. Nevertheless, this finding was interesting and is suggestive of possible gender differences in metabolism and elimination/sequestration of PBDEs among young seals.

ΣPBDE and α-HBCD concentrations in tissues of adult males, although nearly two- to fourfold lower, were not significantly different than levels in the male pups. This accumulation pattern is similar to that observed for other POPs in which the highest lifetime exposure to lipophilic contaminants may result from maternal transfer in utero and during nursing.

23.3.2   PBDEs and HBCDs in Harbor Seal Prey Fishes

ΣPBDE concentrations in harbor seal prey fishes ranged, on average, from 18.3 (in alewife) to 81.5 ng/g lw (in Atlantic herring), with an overall mean of 62 ± 34 ng/g lw [34]. ΣPBDE levels comparable to those in this study were reported in muscle tissues of marine teleost fishes from a Florida coastal food web (mean range: 8–87.5, overall mean 43 ng/g lw) [86], in Atlantic cod (*Gadus morhua*) from the western and southern Norwegian coast [87] (mean range 52–86 ng/g lw), various fishes from the Belgian North Sea (range 16–103, mean 58 ng/g lw) [88], and the eastern (North Sea) coast of the United Kingdom (range 49–69 ng/g lw; mean 60 ng/g, lw) [89]. ΣPBDE concentrations 3–15 times higher were reported in marine fishes from the California coast (mean 302 ng/g lw) [90] and the Georgia coast (range 10–337 ng/g lw) [91].

Of 26 BDE congeners detected in teleost fishes, 10 tri- through hexa-BDEs (BDEs-28, -47, -49, -66, -75, -99, -100, -153, -154, and -155) accounted for 85%–98% of the PBDE content. BDEs-47,

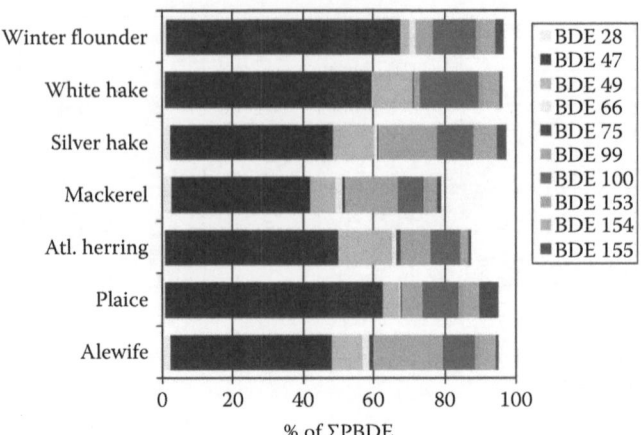

FIGURE 23.7 PBDE congener profiles in seven species of harbor seal prey fishes. Species: winter flounder, white hake, silver hake, Atlantic mackerel, Atlantic herring, American plaice, and alewife.

-99, -100, -153, and -154, are generally dominant in biota worldwide and appear to have a higher potential for bioaccumulation [92]. However, congener profiles among the fish species were highly variable (Figure 23.7). Similar differences in congener profiles have been reported in freshwater and marine fish [58,60,86,93–95] and reflect species-specific differences in uptake and metabolism/ excretion of the individual BDE congeners. BDE-47 was the dominant congener in fish, accounting for 45%–68% of the total PBDE content.

Large differences in BDE-99 abundance relative to BDE-47 were apparent among the species, including those belonging to the same family. In silver and white hake, BDE-47 contributed 46.3% and 58.8% to the total PBDE mass, respectively, while BDE-99 contributed 16.4% and 1.5%, respectively. Silver hake are voracious predators of other fish including other hake, while white hake are indiscriminate benthic feeders; thus, differences in dietary exposure as well as metabolic capacity may account for the variable BDE-99 accumulation in these species. In laboratory exposure studies, significant debromination of BDE-99 was observed in the intestinal tract of common carp (*Cyprinus carpio*), resulting in the conversion of BDE-99 to BDE-47 [58]. In addition to debromination, preferential excretion of BDE-99 was suggested as a reason for lower accumulation of BDE-99, relative to BDE-47 in carnivorous marine fish [96]. The penta-BDE-100 was relatively abundant in the fish, accounting for 7.5%–17% of the total PBDE mass, while the tetra-BDE-49 was relatively abundant in five species (Atlantic herring, alewife, mackerel, white hake, and silver hake), contributing 7.2%–15.5% to the total. In all species, the hexa-BDEs-154 and -155 contributed more to the total than BDE-153, and in plaice, BDE-155 was the most abundant hexa-BDE. In the adult male harbor seals, BDE-47 contributed 63% of the total PBDE content, followed by the hexa-BDE-153 (17%) and the penta-BDEs-99 (11.4%) and -100 (3.3%).

Interestingly, BDE-155 was relatively more abundant in harbor seal blubber than BDE-154. BDE-155 has rarely been reported in marine mammals [76]. This congener is present at only 0.2%–0.7% in technical mixtures [97] and was identified, along with BDE-154, as a specific debromination product of BDE-209 in common carp [59].

Overall, the fish contained a greater proportion of tri- and tetra- BDEs and the penta-BDE-100 in their tissues than were present in technical penta-BDE mixtures or in the seals. The tetra-BDEs-49, -66, and -75 were detected in every fish sample and accounted for 4%–19% of the PBDE content, while these congeners never exceeded 2% of the total in seals. BDE-49 accounts for only 0.4%–0.7% of the technical penta-BDE mixture, but contributed 3%–16% of the PBDE mass in fish tissues. In studies of the transfer of PBDEs from fish to marine mammals, BDE-49 has rarely been reported [65,86,87,89]. This congener is reportedly one of the breakdown products from anaerobic transformation of an octa-BDE technical mixture [98], and may be indicative of reductive debromination of highly brominated BDEs in fish.

23.3.3 BIOMAGNIFICATION OF PBDES AND HBCD FROM FISH TO SEALS

PBDE concentrations in harbor seal blubber were two orders of magnitude higher than those in teleost fish, indicating that many PBDEs are able to bioaccumulate and biomagnify in this marine food web [34]. To estimate biomagnification factors (BMFs), we compared lipid-normalized concentrations of PBDEs in the seven species of teleost fish with those in tissues of adult male harbor seals. It should be noted that this analysis assumes that the species analyzed represent the sole source of the harbor seal diet. During the spring–summer season, harbor seals feed mainly on silver and white hake and Atlantic herring (species comprising ~70% of their diet), but they also consume a variety of schooling and demersal fishes [72]. Hence, the seven species analyzed herein represent basic linkages in the seasonal food chain, but may not provide the complete picture of exposure to and bioaccumulation of PBDEs in the seals.

BMFs for ΣPBDEs (mono- to hexa-BDEs) between teleost fishes and blubber and liver tissues of harbor seals were similar and ranged, on average, from 14 to 76, indicating a high biomagnification potential for these congeners (Figure 23.8). BMFs between the dominant prey species (silver hake, white hake, herring) and the seals averaged 36.4, 33, and 17.1, respectively. Comparable BMFs for ΣPBDEs were reported between predatory fishes and harbor seals in the North Sea (37.8, 29.5, and 26.7 for silver hake [whiting], herring, and Atlantic cod, respectively) [89], between polar cod and harbor seals (12.4) and ringed seals (36.9) in Svalbard, Norway [65,87], and between teleost fishes and bottlenose dolphins in a Florida marine food web (range 3–85) [86]. The hexa-BDEs-153 and -155 were highly biomagnified in harbor seal blubber, with BMFs ranging from 148 to 677 and from 12 to 236, respectively, reflecting the lack of metabolic capacity for these congeners in the seals. BDE-47 and BDE-99 were also biomagnified throughout this food web. BDE-99 was highly biomagnified from white hake to seals (BMF 213), whereas BMFs of BDE-99 were lower relative to BDE-47 from alewife, mackerel, and silver hake to seals. Since BDE-99 is meta-para-substituted and may not be easily metabolized by seals [99], the higher biomagnification from white hake to seals suggests significant metabolic depletion of BDE-99 in this species. There was a lack of biomagnification of the tetra BDEs-49, -66, and -75, and very little biomagnification of the tri-BDE-28, suggesting that seals possess an efficient metabolism for these congeners.

Whereas many PBDEs (tri- through hexa-BDEs) readily biomagnify through the marine food web, in contrast, biomagnification of BDE-209 in seal blubber tissue was not observed (BMF ≤ 1) [36]. BDE-209 levels in liver of harbor seals were up to 10-fold higher than those in their prey fish,

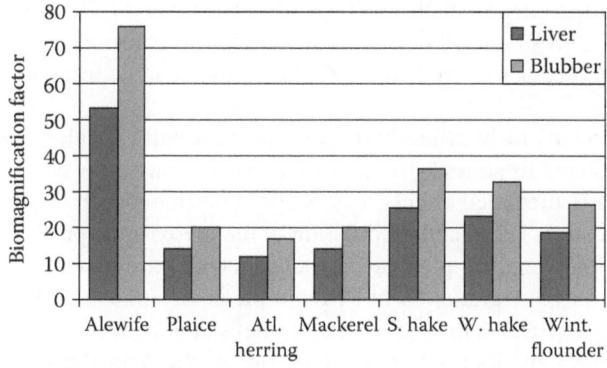

FIGURE 23.8   BMFs for ΣPBDE from fish to harbor seal tissues. Species: alewife, American plaice, Atlantic herring, Atlantic mackerel, silver hake, white hake, and winter flounder.

indicating that the biomagnification of BDE-209 is tissue specific. Most of the seals in this study were pups, implying that BDE-209 is subject to placental and/or lactational transfer, possibly placing pups at risk for developmental neurotoxicity and other adverse effects. Given the large reservoirs of deca-BDE in marine food webs, the biomagnification of BDE-209 in marine mammals is of concern.

Although elevated levels of HBCD in marine mammals are suggestive of biomagnification [63,66–69,76], only a few studies have investigated the transfer of HBCD through the marine food chain [22,62]. In this study [36], BMFs calculated for the transfer of α-HBCD from three prey fish species (alewife, Atlantic herring, and Atlantic mackerel) to tissues of adult male harbor seals were 3.0, 1.0, and 1.6 respectively for liver and 1.6, 0.54, and 0.89 respectively for blubber. A similar low level of biomagnification of α-HBCD was reported from fish to harbor seals in European waters [87,100], whereas a higher BMF of 11 was reported for α-HBCD from polar cod to ringed seals [65]. These observations suggest that harbor seals may possess a species-specific ability to metabolize HBCD. In view of the increasing global use of this BFR compound, there is a need for more studies on the loading, kinetics, and biomagnification potential of HBCD in marine ecosystems.

23.4 PERFLUORINATED CHEMICALS

Perfluorinated chemicals (PFCs) are persistent contaminants of anthropogenic origin that are found distributed in the environment, in wildlife, and in humans worldwide [16,17,101–103]. For over 40 years, PFCs have been used in a variety of industrial and consumer products, including protective coatings for carpets and apparel, nonstick cookware, paper coatings, insecticide formulations, and surfactants in fire-fighting foams [104]. PFCs are oleophobic and hydrophobic, thus their accumulation is not driven by lipophilicity [101]. Some PFCs have been shown to bioaccumulate and biomagnify in marine food webs [17,20] and elevated concentrations are detected in apex predators such as marine mammals [17,101,105]. Two major classes of PFCs are perfluoroalkyl sulfonic acids (PFSAs) and perfluoroalkyl carboxylic acids (PFCAs). The PFSAs (e.g., perfluorooctane sulfonate [PFOS] and perfluorooctane sulfonamide [PFOSA]) are degradation products of perfluoroalkyl sulfamido alcohols via biotransformation processes and abiotic oxidation [106–108]. Concerns about widespread global contamination by PFOS led to a phaseout of production of PFOS-based compounds by a major producer in 2001 [109]. However, PFCAs continue to be manufactured worldwide for use as emulsifiers and additives in the polymerization process [17]. Recent reports have documented the occurrence of long-chain PFCAs (C8-12) including perfluorooctanoic acid (PFOA), perfluorononanoic acid (PFNA), perfluorodecanoic acid (PFDA), perfluoroundecanoic acid (PFUnDA), and perfluorododecanoic acid (PFDoDA) in biota [110–115]. It is likely that there are multiple sources of the compounds, both direct and indirect, including those related to the manufacture and use of commercial products and biotic and abiotic degradation of perfluoroalkyl sulfamide alcohols (to PFSAs and PFCAs) and fluorotelomer alcohols (FTOHs) (to PFCAs) [116]. Recent studies suggest that the global distribution of PFSAs and PFCAs may result from the airborne transport and degradation of volatile precursor molecules as well as atmospheric and oceanic transport of PFCAs themselves [116–119].

PFCs are known to adversely affect both pre- and post-natal development and the neuroendocrine and immune systems in animals via at least five different pathways [17,120,121]. Recent field studies suggest that PFC-mediated effects may occur in marine mammals, including infectious disease in California sea otters [122], and modulation of the peroxisome proliferator-activated receptor α-cytochrome P450 4A-signaling pathway associated with carcinogenesis in Baikal seals [123]. There is no evidence for biodegradation of PFCs in the environment [104], thus the toxic potential of PFSAs and long-chain PFCAs in wildlife and humans is of concern.

Most studies to date have focused on Europe, the Arctic, and the U.S. Pacific and southeast coasts. Little is known about the status of PFC contamination in marine mammals from the northwest Atlantic. To characterize exposure to PFCs in this marine ecosystem, concentrations of PFSAs and PFCAs were determined in liver of 68 harbor seals (8 adult males, 10 adult females, 25 male pups, and 25 female pups) that stranded along the northwest Atlantic between 2000 and 2007 [35] (Figure 23.1).

23.4.1 CONTAMINATION STATUS: PFCs IN HARBOR SEAL TISSUES

Concentrations of total PFCs in harbor seal liver samples (sum of 10 PFCs: PFOS, perfluorodecane sulfonate [PFDS], PFOSA, perfluorohexane sulfonate [PFHxS], perfluoroheptanoic acid [PFHpA], PFOA, PFNA, PFDA, PFUnDA, and PFDoDA) ranged from 18.8 to 1430 ng/g wet weight (overall mean ± standard deviation: 247 ± 289 ng/g ww; $n = 68$) [35]. PFOS was the dominant PFC found in seal liver at concentrations ranging from 8 to 1388 ng/g ww (mean ± SD: 216 ± 279 ng/g ww). Other perfluoroalkyl sulfonates (PFDS and PFOSA) were detected at much lower concentrations than PFOS, ranging from <1 to 25.8 ng/g ww. PFHxS was detected at trace concentrations in only 9% of the samples. Perfluoroalkyl carboxylates (PFCAs, C7–C12) were detected at concentrations ranging from <1 to 87.4 ng/g ww (overall mean ΣPFCAs 25.9 ng/g ww). Among the PFCAs, PFUnDA (C11) was dominant, followed by PFNA (C9) and PFDA (C10). PFHpA, PFOA, and PFDoDA were detected at low concentrations in only 3%, 6%, and 18% of the samples, respectively.

Of the 10 PFCs detected in harbor seal liver samples, PFOS contributed 77%–89% of the total PFC content, followed by PFUnDA, accounting for 3%–8% of the total (Figure 23.9). Whereas PFOS is the predominant PFC found in most wildlife species, elevated ratios of PFOS to ΣPFCs (>0.7) have been reported in species from various locations including polar bears from the Canadian Arctic [124,125] and Greenland [126], bottlenose dolphins from the southeastern U.S. coast [17], and humpback dolphins and finless porpoises from Hong Kong, China [127]. The PFCA profile in harbor seal liver was dominated by PFUnDA (C11, mean 9.9 ng/g ww), accounting for 47%–63% of the ΣPFCA content in liver, followed by PFNA (C9, mean 6.3 ng/g ww), PFDA (C10, mean 4.6 ng/g, ww), and PFDoDA (C12, mean 2.5 ng/g ww). This profile is interesting because for most marine mammals from North American and European coastal waters, PFNA dominated the PFCA profile and is the second most prevalent PFC, after PFOS [105,124,125,128,129]. PFUnDA was the second most abundant PFC in liver of humpback dolphins and finless porpoises from Hong Kong waters [127], ringed seals from Greenland [126], fish from Lake Ontario [124,125], and birds and fish from the Canadian Arctic [124,125]. Concentrations of PFUnDA in skipjack tuna from several locations in the western North Pacific Ocean were greater than the concentrations of PFOS [113,114]. Although there are different exposure pathways and bioaccumulative potentials for individual PFCs in marine mammals, seabirds, and fish, these data demonstrate that PFOS and PFUnDA are present in the northwest Atlantic marine environment and the compounds bioaccumulate in tissues of harbor seals [35].

Another interesting finding was the pattern of PFCA profiles in harbor seal liver: PFUnDA (C11) > PFNA (C9) > PFDA (C10) > PFDoDA (C12), which differs from the general odd/even

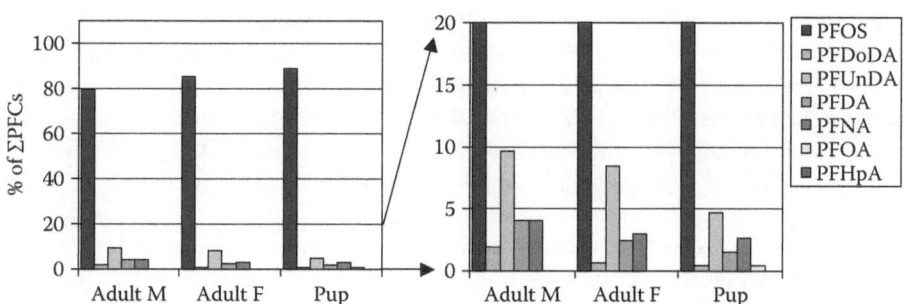

FIGURE 23.9 PFC profile in harbor seal liver by age class.

pattern observed in biota (Figure 23.9). In addition, whereas concentrations of PFCAs generally decrease with increasing perfluoroalkyl chain length [124,125], the profile in harbor seals peaked at PFUnDA, and there were decreasing concentrations of longer and shorter PFCA homologues alike. The reason for the different contamination profile in the harbor seals is unknown, but it may be indicative of common sources such as the FTOHs. The finding of higher concentrations of odd-chain-length PFCAs than even-chain-length PFCAs also implicates FTOHs as a source of exposure [116,128]. Species-specific differences in elimination capacity for PFCAs can be ruled out, since a consistent odd/even pattern of PFCAs has been reported in harbor seals [105]. FTOHs are manufactured in even chain lengths but are reported to degrade to even- and odd-chain-length PFCAs [116]. While 8:2 FTOH was shown to degrade to PFOA and PFNA, 10:2 FTOH degraded to PFDA and PFUnDA in wastewater treatment [130]. Thus, the predominance of PFUnDA in our samples is suggestive of 10:2 FTOH as a source of PFUnDA. Atmospheric oxidation of 10:2 FTOH produces equal amounts of PFDA and PFUnDA [116], but because PFUnDA is more bioaccumulative, this odd-chain-length acid is predominant in biota. The finding of PFNA as the second most abundant PFCA in harbor seal liver suggests that 8:2 FTOH may also be a significant source, since atmospheric oxidation of 8:2 FTOH to PFOA and PFNA would lead to the predominance of PFNA. It is also possible that proximity to ambient sources such as wastewater effluents in the coastal habitat could obscure the odd- and even-chain-length pattern in these seals.

PFC homologue profiles in the harbor seals varied by age and gender (Figure 23.9). The pups retained a higher proportion of PFOS and PFNA compared with the adult seals, and short-chain PFCAs (PFHpA and PFOA) were present in pup liver but were not detected in the adults. These differences reflect the various exposure pathways between pups and adults and suggest that pups may possess a limited metabolic/elimination capacity for PFCs. PFUnDA was more abundant in adult females than adult males or the pups, while PFDoDA was more abundant in adult males. These differences in profiles could reflect differences in elimination capacity for individual PFCs and differences in habitat and prey selection between males and females.

No gender-related differences were found in concentrations of ΣPFCs and individual PFCs in the adult seals (mean ΣPFSAs 130 ± 120 and 127 ± 64 ng/g ww in adult males and adult females, respectively) (Table 23.1). However, the pups had twofold higher concentrations of ΣPFSAs ($p = 0.02$) and PFOS ($p = 0.01$) than the adult seals. Concentrations of PFCAs were similar in pups and adults. This suggests that PFCs do not increase with age in harbor seals. It has been suggested that the lack of correlation between concentrations and age implies that the elimination capacity of PFCs may be significant in adult animals and half-lives of the compounds may be relatively short [17]. This depuration also implies more or less continuous exposure to and uptake of PFCs to maintain tissue concentrations. The accumulation pattern of PFCs in northwest Atlantic harbor seals differs from that previously reported for most POPs such as PCBs, DDTs, and PBDEs [30,33] in which concentrations increase with age in males and decrease with age in sexually mature females due to placental and lactational transfer to pups.

Significant positive correlations were found between PFOS and PFCAs in the harbor seal liver samples. PFOS was significantly correlated with PFOSA ($p < 0.05$), as well as PFDS, PFNA, PFDA, and PFUnDA ($p < 0.01$), but not with PFDoDA, probably because of the low detection of this compound. PFOSA, the precursor molecule for PFOS, was not correlated with PFDS or any of the PFCAs. Among the PFCAs, PFNA, PFDA, and PFUnDA were highly intercorrelated in seal liver. Overall, these results suggest that the harbor seals were exposed to PFOS and PFCAs simultaneously, probably through the same pathways, and the compounds might have originated from similar sources [116].

This study was the first report of the occurrence of PFCs in marine mammals from the northwest Atlantic. The results underline the growing problem of PFC contamination of marine ecosystems and the importance of monitoring PFCs in this region.

23.5 TEMPORAL TRENDS

Concentrations of PCBs, DDT, CHLs, PBDEs, and PFCs in northwestern Atlantic harbor seals showed no significant temporal trends in blubber samples collected between 1991 and 2005 (Figure 23.10a) or in liver samples collected between 2000 and 2007 (Figure 23.10b). These results are consistent with those of other temporal trend studies in industrialized regions and suggest continuous inputs and/or recycling of these persistent halogenated compounds in the northwest Atlantic marine food web.

23.5.1 PCBs and DDT

In the northwestern Atlantic, PCB levels in harbor seals are consistently higher than DDT levels [30], which has been explained by a more rapid decline of DDT in the environment since these compounds were banned in the 1970s [131], while PCBs are still being released from stockpiled residues [132]. This observation is supported by the decreasing trend of the DDT/PCB ratio from 0.71 to 0.38 in harbor seal tissues between 1971 and 2001, which is consistent with temporal trends in many temperate areas of the northern hemisphere. In industrialized areas of Europe, PCB levels have remained constant or declined only slightly since the 1980s, reflecting an equilibrium in environmental cycling [133,134]. Tanabe [135] calculated that in the mid-1980s, only 30% of all the PCBs produced had so far dispersed into the environment. It was estimated by the late 1980s that only

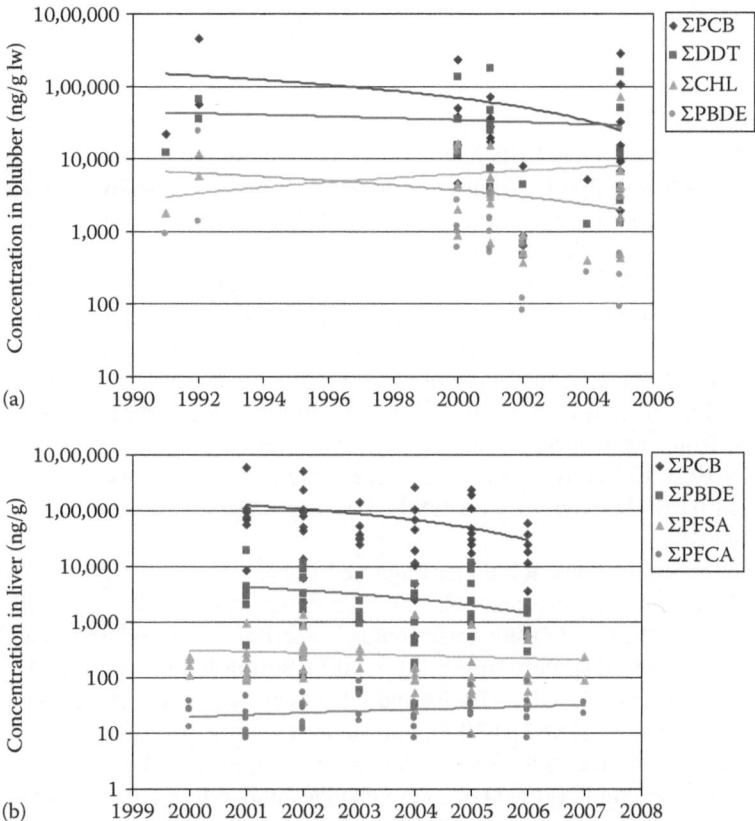

FIGURE 23.10 (a) Temporal trends in the concentrations of PCBs, DDT, CHLs, and PBDEs in blubber of harbor seal pups and yearlings. (b) Temporal trends in the concentrations of PCBs, PBDEs, PFSAs, and PFCAs in liver of harbor seal pups.

about 1% of all PCBs had reached the oceans while about 30% had accumulated in dumpsites and sediments of rivers, coastal zones, and estuaries [132]. Based on their likely future dispersal into the oceans, Tanabe [135] concluded that global PCB levels in marine biota are unlikely to decline in the near future, and certainly not before 2010–2030.

DDT and, to a lesser extent, PCB burdens in northwestern Atlantic harbor seals have declined from the high levels reported in the 1970s (mean DDT 71 and PCB 100 lg/g lw) [30,37], but no further declines were observed between 1991 and 2006 [32]. A significant decrease in DDT levels of 82% ($p < 0.001$) along with a smaller decrease in PCB levels of 66% ($p = 0.002$) was observed over the period 1970–1990, which is consistent with trends in many areas of the North Atlantic where these compounds were extensively used. Addison and Stobo [136] reported an 85%–90% decline in DDT residues in gray seals (*Halichoerus grypus*) from Sable Island, Nova Scotia, Canada, from the mid-1970s to 1998, while the decline in PCB concentrations was apparent only after the mid-1980s. In ringed seals (*Phoca hispida*) and gray seals from the highly polluted Baltic Sea, DDT levels have decreased by 72%–85% since the 1970s [134], while PCB levels showed only a minor decrease of 25% in females and no decrease in males. A similar time trend of more rapidly declining DDT was observed in most Arctic marine mammal populations [137].

Analysis of temporal trends in harbor seal blubber samples collected over the period 1991–2006 revealed that absolute concentrations of the major OC contaminants have remained relatively stable in seal tissues [32] (Figure 23.10a and b). The lack of a temporal trend in OCs is consistent with the leveling of declines in other industrialized areas and likely reflects ongoing inputs to the environment from contaminant reservoirs and/or recycling of these persistent compounds through the food web.

23.5.2   PBDEs

As illustrated in Figure 23.10a and b, there was no significant temporal trend for ΣPBDEs in tissues of male or female harbor seal pups from 2001 to 2006. Similarly, α-HBCD concentrations also showed a lack of a temporal trend over this time period. Stapleton and coworkers [76] reported the lack of a temporal trend in PBDE concentrations in California sea lions between 1993 and 2003, although levels of HBCD were increasing. Another recent study [138] also observed no temporal trend for PBDEs in Atlantic white-sided dolphins (*Lagenorhynchus acutus*) between 1993 and 2000. Previous studies reported that PBDEs were increasing in harbor seals from San Francisco Bay between 1989 and 1998 [77], and in Arctic ringed seals between 1981 and 2000 [139]. Kajiwara et al. [140] reported a 150-fold increase in PBDEs in adult female northern fur seals (*Callorhinus ursinus*) collected from the Japanese coast between 1972 and 1994, but a 50% decrease in levels between 1994 and 1998. Collectively, the data suggest that PBDE levels were increasing in marine mammals between the 1970s and the mid-1990s, but may have stabilized or reached equilibrium over the past decade.

Whereas PBDE concentrations in blubber remained stable, congener compositions shifted from 1991 to 2005 [33]. Percent contribution of BDE-47 increased while the percent BDE-153 contribution decreased between 1991 and 2000; these trends leveled off between 2000 and 2005. BDE-99 concentrations only decreased slightly from 1991 to 2000. Similarly, Ikonomou and coworkers [139] reported that concentrations of BDEs-47, -99, and -100 were increasing in male ringed seals from the Canadian Arctic between 1981 and 1996, but increases in the levels of BDE-99 were slowing considerably between 1996 and 2000. Similar changes were reported in gull egg samples from the Great Lakes between 1981 and 2000 [141]. These changes may reflect differences in the use or the composition of the various commercial PBDE formulations over the years. However, these harbor seals feed opportunistically on diverse fish species along an extended coastline. Since different fish species possess different metabolic potential for BDE congeners and exhibit different BDE profiles [93], the congener profile shifts in the seals may also reflect changes in the harbor seal diet over time.

23.5.3 PFCs

Temporal trends of PFCs were examined in harbor seal liver samples collected between 2000 and 2007 [33] (Figure 23.10b). PFOS and PFCA concentrations were increasing marginally in the adult seals between 2000 and 2007, although this was not statistically significant ($p = 0.18$ and 0.17 for PFOS and ΣPFCA, respectively). In the harbor seal pups, no temporal trend was observed for PFOS or PFCA concentrations during this 7 year period. Several studies have reported an increasing trend in PFOS and PFCA concentrations in marine mammals over the past 20–30 years [126,142–144]. In some locations, PFOS concentrations appear to be declining following the phase-out of perfluorooctanesulfonylfluoride (POSF)-based compounds in 2001 [111–114], whereas this pattern is not observed for long-chain PFCAs (C9–11).

To examine possible changes in homologue composition over time, we compared PFC profiles in harbor seal livers collected at two time points: 2000–2002 and 2003–2007 [35]. PFOS was the dominant PFC and its relative abundance was similar over time, indicating that the source composition of these compounds has not changed in the northwest Atlantic marine ecosystem. PFDA was relatively more abundant in adults and pups and PFUnDA was less abundant in the adult seals in 2003–2007 compared with 2000–2002, which suggests changes in uses/releases of PFCAs during this period.

23.6 SPATIAL TRENDS AND SOURCE IMPLICATIONS

Considered relatively nonmigratory, northwest Atlantic harbor seals nevertheless make seasonal movements along the coast from Maine southward to the coast of New Jersey [145] and accumulate high concentrations of POPs through consumption of a variety of teleost fishes [30,32–35]. As high trophic-level feeders in polluted coastal and estuarine environments they are exposed to contaminated habitats and prey across their range.

23.6.1 PCBs

PCB concentrations varied significantly by region among adult male harbor seals ($p = 0.03$) [32]. From north to south a significant increasing spatial trend was observed in harbor seal blubber, with seals in the southern region (from Massachusetts to New York) having five times higher levels than those in the north (eastern Maine), suggesting that the main sources are near the industrialized and densely populated centers in the northeastern United States compared to rural Maine. This trend is consistent with that previously reported for lower trophic organisms (blue mussels) along the northwestern Atlantic [146].

PCB congener patterns were similar across the region but differences were found in the profiles of dioxin-like PCBs. From north to south, a decreasing trend was observed in the non-ortho PCBs, whereas the opposite trend was observed for mono-ortho PCBs. Since mono-ortho PCBs may be derived from technical PCB products, this pattern may reflect the presence of more historical sources of these compounds in the industrialized southern regions compared to the less populated northern areas where atmospheric transport may be relatively more important.

23.6.2 PBDEs

Mean concentrations of ΣPBDEs in harbor seal blubber from the southern and northern areas were 1600 and 1000 ng/g lw, respectively [33]. A Mann–Whitney U test suggested no significant difference in PBDE levels between the two subregions ($p = 0.72$), nor was there a significant difference in BDE congener profiles. Unlike the trend for PCBs in harbor seals, which decreased with increasing latitude as a function of distance from sources [32], a south to north (urban–rural–remote) decreasing gradient was not observed for PBDEs. This is consistent with the different continental distribution patterns reported for atmospheric PCBs and PBDEs in North America [3]. Whereas atmospheric PCB distribution is strongly related with population density and follows an urban–rural–remote gradient, the spatial distribution observed for PBDEs is clearly not related to the proximity to urban and industrial centers.

It is believed that factories manufacturing products containing BFRs, the use and recycling of PBDE-treated polymers and products, and waste incineration constitute major point sources of PBDEs [92]. Recent studies have highlighted the importance of uncontrolled incineration as a local source of environmental PBDEs in remote sites [3,147] as well as to atmospheric emissions and thus long-range transport of PBDEs to nonurban areas [54]. In addition, wastewater treatment plants (WWTPs) and the application of PBDE-contaminated sewage sludge to agricultural lands also increases the possibility of subsequent remobilization of these compounds in rural environments. Recent data indicate that the major source of PBDEs in sewage sludge from both industrial and background locations is from diffuse leaching of products into wastewater streams from users, households, and industries generally [53,54]. A recent study reported high PBDE levels (5,750–29,000 ng/g lw) in fish collected from the Penobscot River downstream from a WWTP in a rural area of central Maine [94] along the northern part of the harbor seal range. High PBDE levels were also found in biosolids applied to agricultural land directly and in composted materials in the area. This WWTP is one of four local facilities contributing to land spreading and composting of biosolids, providing a significant route to move PBDEs into the aquatic environment through runoff and leaching. Thus, it is likely that a south to north spatial trend in PBDE concentrations in these harbor seals may be obscured by local sources in the northern portion of the range.

23.6.3   PFCs

PFC concentrations in liver of adults and pups from the southern and northern areas were compared. PFOS was the dominant PFC in liver at both locations; mean PFOS concentrations were marginally higher in adult seals from the southern area (128 ng/g ww) than in seals from the northern area (70 ng/g ww) ($p = 0.07$) [35]. This finding is consistent with the spatial trend reported for loggerhead sea turtles along the U.S. eastern coast [110]. Together, the data suggest that the densely populated mid-Atlantic region may be a source region for PFCs.

Concentrations of PFCAs were not different by location in the adult seals, but higher PFDoDA concentrations were found in pups from the northern area ($p = 0.02$). This spatial distribution varies from that reported previously for PCBs [32] and is consistent with the pattern observed for PBDEs [33,36]. It is believed that direct sources of PFCs, including emissions/releases from fluorochemical plants and airports and military bases with fire-fighting operations, can result in elevated concentrations in urbanized locations [17]. However, unlike PCBs, PFCs and PBDEs are widely used in household and consumer products and therefore may originate from diffuse common sources within the region including landfill leachate and wastewater effluent from households and industries generally [17,54]. Moreover, the PFC contamination pattern in these seals suggests that volatile precursors (sulfonamido alcohols and FTOHs) are important sources of PFCs. The lack of an urban–rural–remote decreasing spatial gradient in PFC concentrations implies that such diffuse sources are significant across the harbor seal range.

23.7   GLOBAL TRENDS: COMPARISONS OF POPs IN PINNIPEDS

POP concentrations in the tissues of northwestern Atlantic harbor seals were compared with levels recently reported in pinnipeds worldwide to determine their contamination status on a global scale.

23.7.1   PCBs

Harbor seals from the northwestern Atlantic are in the middle of the global contamination spectrum (Figure 23.11), with mean blubber ΣPCB concentrations of 57 and 60 μg/g lw in yearlings and pups,

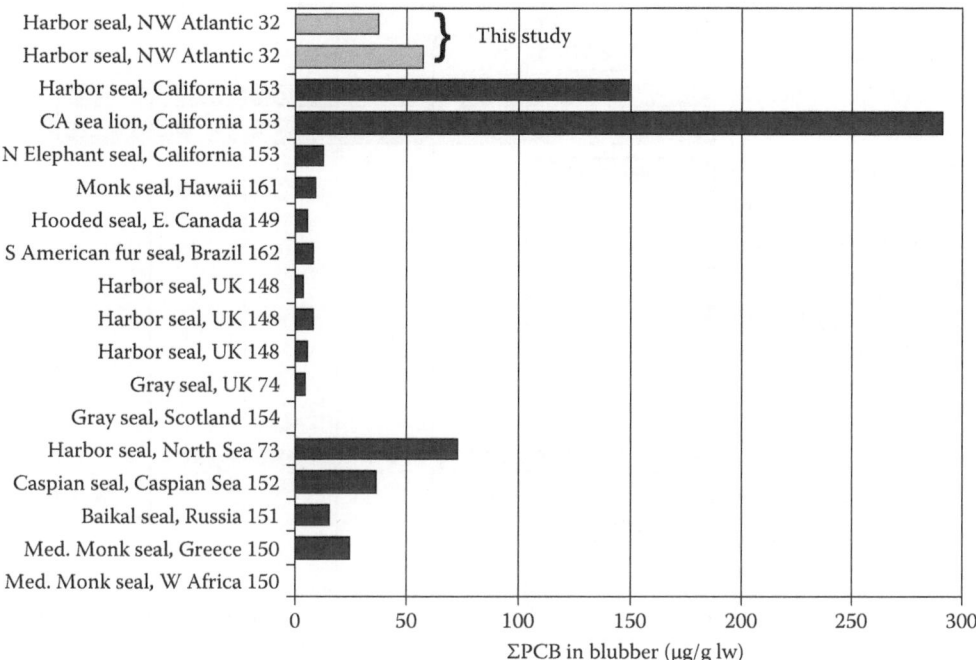

FIGURE 23.11 Global comparison of PCB concentrations in blubber of pinnipeds.

respectively, and 37 μg/g lw in the adult males [32]. These levels are higher than those in harbor and gray seals from U.K. waters [75,148], hooded seals (*Cystophora cristata*) from eastern Canada [149], Mediterranean monk seals (*Monachus monachus*) [150], and Baikal seals (*Pusa sibirica*) [151] and similar to those reported in North Sea harbor seals [74] and Caspian seals (*Pusa caspica*) that died during a distemper virus epidemic [152]. The highest PCB levels were reported in harbor seal pups (150,000 g/g lw) and adult male sea lions (*Zalophus californianus*) (293,000 ng/g lw) from the heavily polluted Southern California Bight [153]. Marine sediments in this highly industrialized area are vast repositories of OCs resulting from the discharge of large amounts of organic pollutants into the Pacific Ocean off the Palos Verdes Shelf from 1949 to 1970s [153].

23.7.2 DDTs

ΣDDT concentrations in northwestern Atlantic harbor seals (mean 33 and 34 μg/g lw in pups and adult males, respectively) [32] are higher than those recently reported in U.K. harbor and gray seals [75,148,154] and Baikal seals [155] and similar to the levels reported in northern elephant seals (*Mirounga angustirostris*) [153] and Mediterranean monk seals from the coast of Greece [150] (Figure 23.12). While DDT levels have declined in many parts of the developed world, DDT residues (113 μg/g lw) in Caspian seals were four times higher than those in northwestern Atlantic harbor seals, likely reflecting recent uses of this pesticide in the former USSR [152]. Extremely high DDT levels were reported in sea lions (2300 μg/g lw) and harbor seals (1265 μg/g lw) from the Southern California Bight [153]. Current concentrations of DDT and PCB in California sea lions and harbor seals are among the highest values reported worldwide for marine mammals.

23.7.3 PBDEs

ΣPBDE concentrations detected in northwest Atlantic harbor seals are an order of magnitude higher than those recently reported in pinnipeds from European and Asian waters (Figure 23.13), reflecting

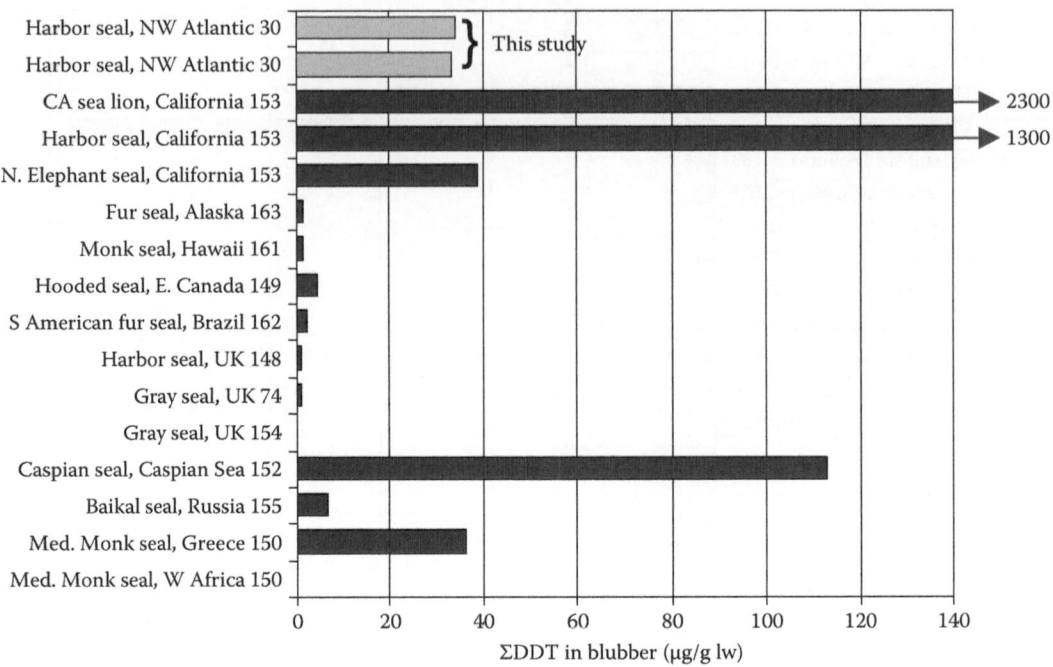

FIGURE 23.12 Global comparison of DDT concentrations in blubber of pinnipeds.

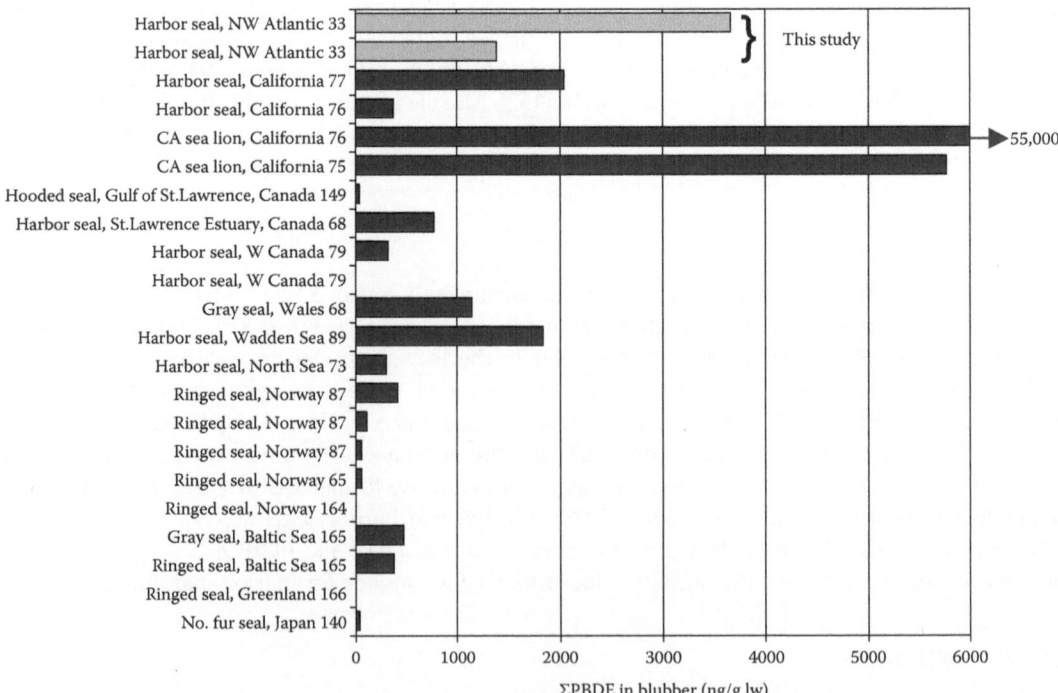

FIGURE 23.13 Global comparison of PBDE concentrations in blubber of pinnipeds.

the greater production and use of the penta-BDE formulation in the United States [14,33]. In the younger seals, mean PBDE concentrations (3600 ng/g lw) are 10 times higher than those reported in juvenile harbor seals collected from the southern North Sea during 1999–2004 [74] and in gray seal pups collected from eastern U.K. waters during 1998–2000 [75]. PBDE levels in the adult males (1400 ng/g lw) are four- to sixfold greater than those reported in adult male harbor seals collected from the North Sea during 1999–2004 [74] and more recently from the Dutch Wadden Sea [100]. Higher PBDE levels were reported in adult male California sea lions collected during 1993–2003 (5800 ng/g lw) [76] and in adult male harbor seals collected during 1997–1998 from San Francisco Bay (5100 ng/lw) [78]. Sea lions from the Southern California Bight had the highest PBDE levels (55 µg/g lw) ever reported in a marine mammal species [77]. Collectively, the data indicate that northwest Atlantic harbor seals are in the mid- to high-range of the concentrations reported for PBDEs in pinnipeds worldwide.

23.7.4 PFOS

PFOS concentrations in northwest Atlantic harbor seals were within the range of concentrations previously reported in liver of seals from other mid-latitude locations (Figure 23.14). Mean hepatic PFOS concentrations in these seals (216 ng/g ww) [35] were similar to those reported in Baltic gray seals [102] and harbor seals from the Dutch Wadden Sea [105], and higher than those reported in gray seals from the southern North Sea [156] and pinnipeds from the California coast [101]. PFOS concentrations were twofold higher in harbor seals from the German Bight [157], Baltic ringed seals [102], and gray seals from the Swedish coast [158]. The highest PFOS concentrations (800 ng/g ww) were reported in harbor seals from the Danish coast [158]. Much lower hepatic PFOS concentrations were reported in seals from the Bering and Chukchi Seas, Alaska [159], and Arctic Canada [112,124,160], and ringed seals from East Greenland [126].

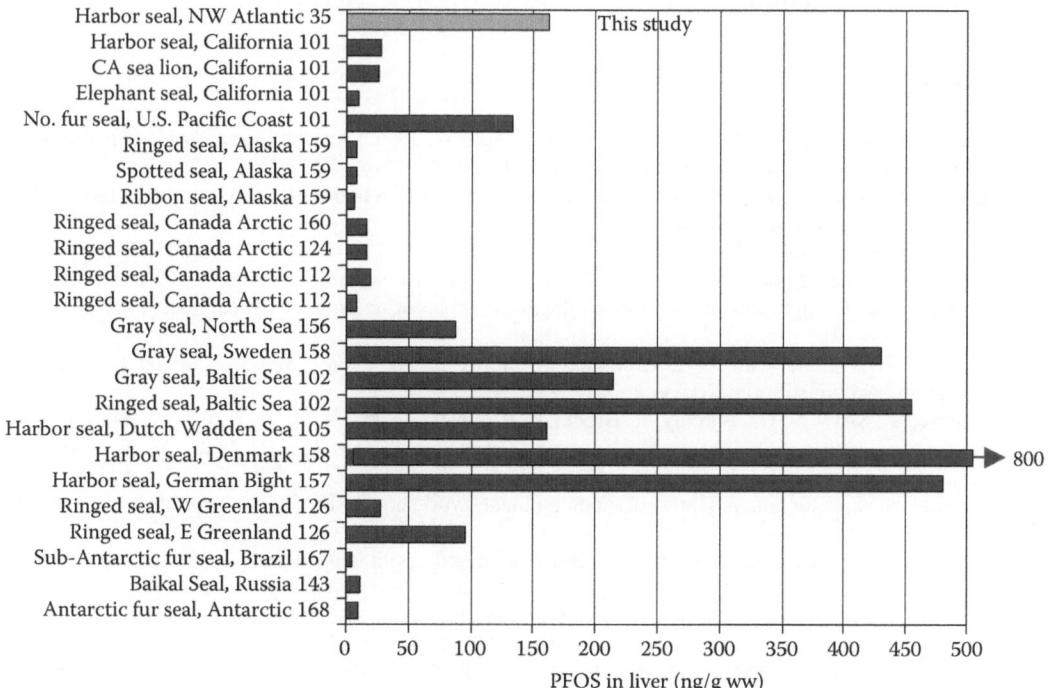

FIGURE 23.14 Global comparison of PFOS concentrations in liver of pinnipeds.

23.8 SUMMARY AND PERSPECTIVES

The results presented in this chapter are the first extensive data reported on POPs in harbor seals from the northwestern Atlantic coast in 25 years. Given their relatively high tissue burdens of legacy and emerging POPs, harbor seals inhabiting this region are contaminated at levels that place them at risk for adverse health effects. Levels in the pups are of special concern given that young seals appear to be vulnerable to immune- and endocrine-disrupting effects of POPs when levels are an order of magnitude lower. In view of the recurring epizootics affecting the northwestern Atlantic harbor seal population, further research is needed to elucidate the real risks of the complex mixture of POPs to which these seals are exposed.

ACKNOWLEDGMENTS

The authors are grateful to Kirk Trabant, Marine Environmental Research Institute, and members of the NOAA/NMFS Northeast Region Stranding Network for providing harbor seal tissues and to John Sowles and Keri Stepanek, Maine Department of Marine Resources Trawl Survey, for providing teleost fish samples from the Gulf of Maine for this study. Special thanks to Marvin Messex for providing Atlantic mackerel samples for the analysis. We thank Chia-Swee Hong, Patrick O'Keefe, and Robin Storm, Wadsworth Center, New York State Department of Health, for sample preparation and chemical analysis of PCBs, OC pesticides, PCDD/Fs, and PBDEs in harbor seal blubber samples; Adrian Covaci, Liesbeth Weijs, and Laurence Roosens, Toxicological Centre, and Department of Biology, University of Antwerp, for the analysis of PBDEs and HBCDs in liver of harbor seals; and Olaf Paepke and Nina Lohmann, Eurofins-ERGO, Hamburg, Germany, for the analysis of PBDEs and HBCDs in seven species of commercially important fish stocks that are harbor seal prey. We thank our colleagues Professor Shinsuke Tanabe and Bommanna G. Loganathan for decades of inspirational work, which has informed the scope of this chapter. This work was supported by the National Oceanographic and Atmospheric Administration (NOAA), the Marine Environmental Research Institute (MERI), and the University of Antwerp.

REFERENCES

1. Gilbert, J. R., Waring, G. T., Wynne, K. M., and Guldager, N. 2005. Changes in abundance of harbor seals in Maine, 1981–2001. *Mar. Mammal Sci.* 21(3): 519–535.
2. Hameedi, M., Pait, A., and Warner, R. 2002. Environmental contaminant monitoring in the Gulf of Maine. In: *Northeast Coastal Monitoring Summit*, Durham, ME, pp. 1–7.
3. Shen, L., Wania, F., Lei, Y. D., Teixeira, C., Muir, D. C. G., and Xiao, H. 2006. Polychlorinated biphenyls and polybrominated diphenyl ethers in the North American atmosphere. *Environ. Pollut.* 144: 434–444.
4. Loganathan, B. G. and Kannan, K. 1994. Global organochlorine contamination trends: An overview. *Ambio* 23: 187–191.
5. O'Connor, T. P. 1990. *Coastal Environmental Quality in the United States.* National Oceanic and Atmospheric Administration, Rockville, MD.
6. De Guise, S., Shaw, S. D., Barclay, J., Brock, J., Brouwer, A., Dewailly, E., Fair, P. A., Fournier, M., Grandjean, P., Guillette, L., Hahn, M., Koopman-Esseboom, C., Letcher, R., Matz, A., Norstrom, R., Perkins, C., Schwacke, L. H., Skaare, J. U., Sowles, J., St. Aubin, D. J., Stegeman, J., and Whaley, J. 2001. Consensus statement: Atlantic coast contaminants workshop 2000. *Environ. Health Persp.* 109(12): 1301–1302.
7. Shaw, S. D. 2001. Perinatal exposure to persistent organic pollutants: Parallels in seals and humans. *Organohalogen Comp.* 53: 21–26.
8. Colborn, T., Vom Saal, F. S., and Soto, A. M. 1993. Developmental effects of endocrine-disrupting chemicals in wildlife and humans. *Environ. Health Persp.* 101: 378–384.
9. Geraci, J. R., St. Aubin, D. J., Barker, I. K., Webster, R. G., Hinshaw, V. S., Bean, W. J., Ruhnke, L., Prescott, J. H., Early, G., Baker, A. S., Madoff, S., and Schooley, R. T. 1982. Mass mortality of harbor seals: Pneumonia associated with influenza A virus. *Science* 215: 1129–1131.

10. Dietz, R., Heide-Jørgensen, M. P., and Härkönen, T. 1989. Mass deaths of harbor seals (*Phoca vitulina*) in Europe. *Ambio* 18(5): 258–264.

11. Duignan, P. J., Sadove, S., Saliki, J. T., and Geraci, J. R. 1993. Phocine distemper in harbor seals (*Phoca vitulina*) from Long Island, New York. *J. Wildlife Dis.* 29(3): 465–469.

12. Van Loveren, H., Ross, P. S., Osterhaus, A. D. M. E., and Vos, J. G. 2000. Contaminant-induced immunosuppression and mass mortalities among harbor seals. *Toxicol. Lett.* 112–113: 319–324.

13. Harding, K. C., Harkonen, T., and Caswell, H. 2002. The 2002 European seal plague: Epidemiology and population consequences. *Ecol. Lett.* 5: 727–732.

14. Shaw, S. D. and Kannan, K. 2009. Polybrominated diphenyl ethers in marine ecosystems of the American continents: Foresight from current knowledge. *Rev. Environ. Health* 24: 157–229.

15. Birnbaum, L. S. and Staskal, D. F. 2004. Brominated flame retardants: Cause for concern? *Environ. Health Persp.* 112: 9–17.

16. Giesy, J. P. and Kannan, K. 2001. Global distribution of perfluorooctane sulfanate in wildlife. *Environ. Sci. Technol.* 35: 1339–1342.

17. Houde, M., Martin, J. W., Letcher, R. J., Solomon, K. R., and Muir, D. C. G. 2006. Biological monitoring of polyfluoroalkyl substances: A review. *Environ. Sci. Technol.* 40(11): 3463–3473.

18. Hites, R. A. 2004. Polybrominated diphenyl ethers in the environment and in people: A meta-analysis of concentrations. *Environ. Sci. Technol.* 38(4): 945–956.

19. de Wit, C. A., Alaee, M., and Muir, D. C. G. 2006. Levels and trends of brominated flame retardants in the Arctic. *Chemosphere* 64: 209–233.

20. Tomy, G. T., Budakowski, W., Halldorson, T., Helm, P. A., Stern, G. A., Friesen, K., Pepper, K., Tittlemier, S. A., and Fisk, A. T. 2004. Fluorinated organic compounds in an eastern Arctic marine food web. *Environ. Sci. Technol.* 38: 6475–6481.

21. Tanabe, S., Ramu, K., Isobe, T., and Takahashi, S. 2008. Brominated flame retardants in the environment of Asia-Pacific: An overview of spatial and temporal trends. *J. Environ. Monit.* 10: 188–197.

22. Covaci, A., Gerecke, A. C., Law, R. J., Voorspoels, S., Kohler, M., Heeb, N. V., Leslie, H., Allchin, C. R., and De Boer, J. 2006. Hexabromocyclododecanes (HBCDs) in the environment and humans: A review. *Environ. Sci. Technol.* 40(12): 3679–3688.

23. Reijnders, P. J. 1986. Reproductive failure in common seals feeding on fish from polluted coastal waters. *Nature* 324: 456–457.

24. Brouwer, A., Reijnders, P. J. H., and Koeman, J. H. 1989. Polychlorinated biphenyl (PCB)-contaminated fish indices vitamin A and thyroid hormone deficiency in the common seal (*Phoca vitulina*). *Aquat. Toxicol.* 15: 99–106.

25. De Swart, R. L., Ross, P. S., Vedder, L. J., Timmerman, H. H., Hoisterkamp, S., Van Loveren, H., Vos, J. G., Reijnders, P. J. H., and Osterhaus, A. D. M. E. 1994. Impairment of immune function in harbor seals (*Phoca vitulina*) feeding on fish from polluted waters. *Ambio* 23(2): 155–159.

26. Ross, P. S., De Swart, R. L., Reijnders, P. J. H., Loveren, H. V., Vos, J. G., and Osterhaus, A. D. M. E. 1995. Contaminant-related suppression of delayed-type hypersensitivity and antibody responses in harbor seals fed herring from the Baltic Sea. *Environ. Health Persp.* 103(2): 162–167.

27. Duignan, P. J., Saliki, J. T., St. Aubin, D. J., Early, G., Sadove, S., House, J. A., Kovacs, K. M., and Geraci, J. R. 1995. Epizootiology of morbillivirus infection in North American harbor seals (*Phoca vitulina*) and gray seals (*Halichoerus grypus*). *J. Wildlife Dis.* 31(4): 491–501.

28. Garron, M. and McNulty, S. 2008. *NOAA Fisheries Service Northeast Region Marine Mammal Stranding Program: 2007 Summary*. Northeast Region Stranding Network Conference, April 17, Providence, RI.

29. Kannan, K., Blankenship, A. L., Jones, P. D., and Giesy, J. P. 2000. Toxicity reference values for the toxic effects of polychlorinated biphenyls to aquatic mammals. *Human Ecol. Risk. Assess.* 6(1): 181–201.

30. Shaw, S. D., Brenner, D., Bourakovsky, A., Mahaffey, C. A., and Perkins, C. R. 2005. Polychlorinated biphenyls and chlorinated pesticides in harbor seals (*Phoca vitulina concolor*) from the northwestern Atlantic coast. *Mar. Pollut. Bull.* 50: 1069–1084.

31. Shaw, S. D., Brenner, D., Mahaffey, C. A., De Guise, S., Perkins, C. R., Clark, G. C., Denison, M. S., and Waring, G. T. 2003. Persistent organic pollutants (POPs) and immune function in US Atlantic coast harbor seals (*Phoca vitulina concolor*). *Organohalogen Comp.* 62: 220–223.

32. Shaw, S. D., Brenner, D., Berger, M. L., Dwyer, M., Fang, F., Hong, C.-S., Storm, R., and O'Keefe, P. 2007. Patterns and trends of PCBs and PCDD/Fs in NW Atlantic harbor seals: Revisiting toxic threshold levels using the new TEFS. *Organohalogen Comp.* 69: 1752–1756.

33. Shaw, S. D., Brenner, D., Berger, M. L., Fang, F., Hong, C.-S., Addink, R., and Hilker, D. 2008. Bioaccumulation of polybrominated diphenyl ethers in harbor seals from the northwest Atlantic. *Chemosphere* 73: 1773–1780.

34. Shaw, S. D., Berger, M. L., Brenner, D., Kannan, K., Lohmann, N., and Päpke, O. 2009. Bioaccumulation of polybrominated diphenyl ethers and hexabromocyclododecane in the northwest Atlantic marine food web. *Sci. Tot. Environ.* 407: 3323–3329.

35. Shaw, S. D., Berger, M. L., Brenner, D., Tao, L., Wu, Q., and Kannan, K. 2009. Specific accumulation of perfluorochemicals in harbor seals (*Phoca vitulina concolor*) from the northwest Atlantic. *Chemosphere* 74: 1037–1043.

36. Shaw, S. D., Berger, M. L., Weijs, L., Roosens, L., and Covaci, A. 2009. Specific accumulation of polybrominated diphenyl ethers including deca-BDE in tissues of harbor seals from the Northwest Atlantic. *Organohalogen Comp.* 71: 2547–2552.

37. Gaskin, D. E., Frank, R., Holdrinet, M., Ishida, K., Walton, C. J., and Smith, M. 1973. Mercury, DDT, and PCB in harbour seals (*Phoca vitulina*) from the Bay of Fundy and Gulf of Maine. *J. Fish. Res. Board Can.* 30: 471–475.

38. Helle, E., Olssen, M., and Jensen, S. 1976. PCB levels correlated with pathological changes in seal uteri. *Ambio* 5: 261–263.

39. Helle, E., Olsson, M., and Jensen, S. 1976. DDT and PCB levels and reproduction in ringed seal from the Bothnian Bay. *Ambio* 5: 188–189.

40. Reijnders, P. 1980. Organochlorine and heavy metal residues in harbour seals from the Wadden Sea and their possible effects on reproduction. *Neth. J. Sea Res.* 14(1): 30–65.

41. De Long, R. L., Gilmartin, W. G., and Simpson, J. G. 1973. Premature births in California sea lions: Association with high organochlorine pollutant residue levels. *Science* 181: 1168–1170.

42. Lake, C., Lake, J., Haebler, R., McKinney, R., Boothman, W., and Sadove, S. 1995. Contaminant levels in harbor seals from the northeastern United States. *Arch. Environ. Contam. Toxicol.* 29: 128–134.

43. Hobbs, K. E., Lebeuf, M., and Hammill, M. O. 2002. PCBs and OCPs in male harbour, grey, harp and hooded seals from the Estuary and Gulf of St Lawrence, Canada. *Sci. Tot. Environ.* 296: 1–18.

44. Debier, C., Pomeroy, P. P., Dupont, C., Joiris, C., Comblin, V., Le Boulenge, E., Larondelle, Y., and Thome, J.-P. 2003. Quantitative dynamics of PCB transfer from mother to pup during lactation in UK grey seals *Halichoerus grypus*. *Mar. Ecol. Prog. Ser.* 247: 237–248.

45. Debier, C., Pomeroy, P. P., Dupont, C., Joiris, C., Comblin, V., Le Boulengé, E., Larondelle, Y., and Thomé, J.-P. 2003. Dynamics of PCB transfer from mother to pup during lactation in UK grey seals *Halichoerus grypus*: Differences in PCB profile between compartments of transfer and changes during the lactation period. *Mar. Ecol. Prog. Ser.* 247: 249–256.

46. Sørmo, E. G., Skaare, J. U., Lydersen, C., Kovacs, K. M., Hammill, M. O., and Jenssen, B. M. 2003. Partitioning of persistent organic pollutants in grey seal (*Halichoerus grypus*) mother-pup pairs. *Sci. Tot. Environ.* 302: 145–155.

47. Wolkers, H., Lydersen, C., and Kovacs, K. M. 2004. Accumulation and lactational transfer of PCBs and pesticides in harbor seals (*Phoca vitulina*) from Svalbard, Norway. *Sci. Tot. Environ.* 319: 137–146.

48. de Wit, C., Jansson, B., Bergek, S., Hjelt, M., Rappe, C., Olsson, M., and Andersson, O. 1992. Polychlorinated dibenzo-*p*-dioxin and polychlorinated dibenzofuran levels and patterns in fish and fish-eating wildlife in the Baltic Sea. *Chemosphere* 25: 185–188.

49. Addison, R. F., Ikonomou, M. G., Fernandez, M. P., and Smith, T. G. 2005. PCDD/F and PCB concentrations in Arctic ringed seals (*Phoca hispida*) have not changed between 1981 and 2000. *Sci. Tot. Environ.* 351–352: 301–311.

50. Van den Berg, M., Birnbaum, L. S., Denison, M. S., De Vito, M., Farland, W., Feeley, M., Fiedler, H., Hakansson, H., Hanberg, A., Haws, L. C., Rose, M., Safe, S., Schrenk, D., Tohyama, C., Tritscher, A., Tuomisto, J., Tysklind, M., Walker, N., and Peterson, R. E. 2006. The 2005 World Health Organization reevaluation of human and mammalian toxic equivalency factors for dioxins and dioxin-like compounds. *Toxicol. Sci.* 93(2): 2223–241.

51. Van den Berg, M., Birnbaum, L., Bosveld, A. T. C., Brunström, B., Cook, P., Feeley, M., Giesy, J. P., Hanberg, A., Hasegawa, R., Kennedy, S. W., Kubiak, T., Larsen, J. C., van Leeuwen, F. X. R., Liem, A. K. D., Nolt, C., Peterson, R. E., Poellinger, L., Safe, S., Schrenk, D., Tillitt, D. E., Tysklind, M., Younes, M., Wærn, F., and Zacharewski, T. 1998. Toxic equivalency factors (TEFs) for PCBs, PCDDs, PCDFs for humans and wildlife. *Environ. Health Persp.* 106(12): 775–792.

52. Alaee, M., Arias, P., Sjodin, A., and Bergman, A. 2003. An overview of commercially used brominated flame retardants, their applications, their use patterns in different countries/regions and possible modes of release. *Environ. Int.* 29: 683–689.

53. de Wit, C. A. 2002. An overview of brominated flame retardants in the environment. *Chemosphere* 46: 583–624.

54. Law, R. J., Allchin, C. R., de Boer, J., Covaci, A., Herzke, D., Lepom, P., Morris, S., Tronczynski, J., and de Wit, C. A. 2006. Levels and trends of brominated flame retardants in the European environment. *Chemosphere* 64: 187–208.

55. Costa, L. G. and Giordano, G. 2007. Developmental neurotoxicity of polybrominated diphenyl ether (PBDE) flame retardants. *Neurotoxicology* 28: 1047–1067.

56. Bromine Science and Environmental Forum (BSEF). 2003. Major brominated flame retardants volume estimated: Total markets demand by region in 2001. www.bsef.com (Accessed June 2010).

57. US Environmental Protection Agency (US EPA). 2009. Announcements. http://www.epa.gov/oppt/pbde/

58. Stapleton, H. M., Alaee, M., Letcher, R. J., and Baker, J. E. 2004. Debromination of the flame retardant decabromodiphenyl ether by juvenile carp (*Cyprinus carpio*) following dietary exposure. *Environ. Sci. Technol.* 38: 112–119.

59. Stapleton, H. M., Brazil, B., Holbrook, R. D., Mitchelmore, C. L., Benedict, R., Konstantinov, A., and Potter, D. 2006. *In vivo* and *in vitro* debromination of decabromodiphenyl ether (BDE 209) by juvenile rainbow trout and common carp. *Environ. Sci. Technol.* 40: 4653–4658.

60. La Guardia, M. J., Hale, R. C., and Harvey, E. 2007. Evidence of debromination of decabromodiphenyl ether (BDE-209) in biota from a wastewater receiving stream. *Environ. Sci. Technol.* 41: 6663–6670.

61. Ueno, D., Alaee, M., Marvin, C. H., Muir, D. C. G., Macinnis, G., Reiner, E. J., Crozier, P., Furdui, V. I., Subramanian, A., Fillmann, G., Lam, P. K. S., Zheng, G. J., Muchtar, M., Razak, H., Prudente, M., Chung, K.-H., and Tanabe, S. 2006. Distribution and transportability of hexabromocyclododecane (HBCD) in the Asia-Pacific region using skipjack tuna as a bioindicator. *Environ. Pollut.* 144: 238–247.

62. Tomy, G. T., Pleskach, K., Oswald, T., Halldorson, T., Helm, P. A., Macinnis, G., and Marvin, C. H. 2008. Enantioselective bioaccumulation of hexabromocyclododecane and congener-specific accumulation of brominated diphenyl ethers in an eastern Canadian Arctic marine food web. *Environ. Sci. Technol.* 42: 3634–3639.

63. Morris, S., Allchin, C. R., Zegers, B. N., Haftka, J. J. H., Boon, J. P., Belpaire, C., Leonards, P. E. G., Van Leeuwen, S. P. J., and de Boer, J. 2004. Distribution and fate of HBCD and TBBPA brominated flame retardants in North Sea estuaries and aquatic food webs. *Environ. Sci. Technol.* 38: 5497–5504.

64. Janak, K., Covaci, A., Voorspoels, S., and Becher, G. 2005. Hexabromocyclododecane in marine species from the western Scheldt Estuary: Diastereomer- and enantiomer-specific accumulation. *Environ. Sci. Technol.* 39: 1987–1994.

65. Sørmo, E. G., Salmer, M. P., Jenssen, B. M., Hop, H., Bæk, K., Kovacs, K. M., Lydersen, C., Falk-Petersen, S., Gabrielsen, G. W., Lie, E., and Skaare, J. U. 2006. Biomagnification of polybrominated diphenyl ether and hexabromocyclododecane flame retardants in the polar bear food chain in Svalbard, Norway. *Environ. Toxicol. Chem.* 25(9): 2502–2511.

66. Johnson-Restrepo, B., Adams, D. H., and Kannan, K. 2008. Tetrabromobisphenol A (TBBPA) and hexabromocyclododecanes (HBCDs) in tissues of humans, dolphins, and sharks from the United States. *Chemosphere* 70: 1935–1944.

67. Peck, A. M., Pugh, R. S., Moors, A., Ellisor, M. B., Porter, B. J., Becker, P. R., and Kucklick, J. R. 2008. Hexabromocyclododecane in white-sided dolphins: Temporal trend and stereoisomer distribution in tissues. *Environ. Sci. Technol.* 42: 2650–2655.

68. Law, R. J., Alaee, M., Allchin, C. R., Boon, J. P., Lebeuf, M., Lepom, P., and Stern, G. A. 2003. Levels and trends of polybrominated diphenylethers and other brominated flame retardants in wildlife. *Environ. Int.* 29: 757–770.

69. Zegers, B. N., Mets, A., Van Bommel, R., Minkenberg, C., Hamers, T., Kamstra, J. H., Pierce, G. J., and Boon, J. P. 2005. Levels of hexabromocyclododecane in harbor porpoises and common dolphins from western European seas, with evidence for stereoisomer-specific biotransformation by cytochrome P450. *Environ. Sci. Technol.* 39: 2095–2100.

70. Bromine Science and Environmental Forum (BSEF). 2009. *Fact Sheet: Hexabromocyclododecane HBCD.* Edition June 2009. www.bsef.com (Accessed July 2010).

71. Lee, F. F. 2007. Comprehensive analysis, Henry's Law constant determination, and photocatalytic degradation of polychlorinated biphenyls (PCBs) and/or other persistent organic pollutants (POPs), Doctoral dissertation, University at Albany, State University of New York, School of Public Health.

72. Wood, S., Ferland, A., Waring, G., Sette, L., and Shaw, S. D. 2001. Harbor seal (*Phoca vitulina*) food habits along the New England coast. In *Abstract for the 14th Biennial Conference of the Society of Marine Mammals in Vancouver*, British Columbia, Canada, November 28–December 3.

73. Covaci, A., Weijs, L., Roosens, L., Berger, M. L., Neels, H., Blust, R., and Shaw, S. D. 2009. Accumulation of hexabromocyclododecanes (HBCDs) and their metabolites in pup and adult harbour seals from the Northwest Atlantic. *Organohalogen Comp.* 71: 1497–1502.

74. Weijs, L., Dirtu, A. C., Das, K., Gheorghe, A., Reijnders, P. J. H., Neels, H., Blust, R., and Covaci, A. 2009. Inter-species differences for polychlorinated biphenyls and polybrominated diphenyl ethers in marine top predators from the Southern North Sea: Part 1. Accumulation patterns in harbour seals and harbour porpoises. *Environ. Pollut.* 157: 437–444.

75. Kalantzi, O. I., Hall, A. J., Thomas, G. O., and Jones, K. C. 2005. Polybrominated diphenyl ethers and selected organochlorine chemicals in grey seals (*Halichoerus grypus*) in the North Sea. *Chemosphere* 58: 345–354.

76. Stapleton, H. M., Dodder, N. G., Kucklick, J. R., Reddy, C. M., Schantz, M. M., Becker, P. R., Gulland, F., Porter, B. J., and Wise, S. A. 2006. Determination of HBCD, PBDEs and MeO-BDEs in California sea lions (*Zalophus californianus*) stranded between 1993 and 2003. *Mar. Pollut. Bull.* 52: 522–531.

77. Meng, X.-Z., Blasius, M. E., Gossett, R. W., and Maruya, K. A. 2009. Polybrominated diphenyl ethers in pinnipeds stranded along the southern California coast. *Environ. Pollut.* 157: 2731–2736.

78. She, J., Petreas, M., Winker, J., Visita, P., McKinney, M., and Kopec, D. 2002. PBDEs in the San Francisco Bay Area: Measurements in harbor seal blubber and human breast adipose tissue. *Chemosphere* 46: 697–707.

79. Ikonomou, M. G. and Addison, R. F. 2008. Polybrominated diphenyl ethers (PBDEs) in seal populations from eastern and western Canada: An assessment of the processes and factors controlling PBDE distribution in seals. *Mar. Environ. Res.* 66: 225–230.

80. Thuresson, K., Hoglund, P., Hagmar, L., Sjodin, A., Bergman, A., and Jakobsson, K. 2006. Apparent half-lives of hepta- to decabrominated diphenyl ethers in human serum as determined in occupationally exposed workers. *Environ. Health Persp.* 114: 176–181.

81. Thomas, G. O., Moss, S. E. W., Asplund, L., and Hall, A. J. 2005. Absorption of decabromodiphenyl ether and other organohalogen chemicals by grey seals (*Halichoerus grypus*). *Environ. Pollut.* 133: 581–586.

82. Mörck, A., Hakk, H., Örn, U., and Wehler, E. K. 2003. Decabromodiphenyl ether in the rat: Absorption, distribution, metabolism, and excretion. *Drug Metab. Dispos.* 31: 900–907.

83. Huwe, J. K., Hakk, H., and Birnbaum, L. S. 2008. Tissue distribution of polybrominated diphenyl ethers in male rats and implications for biomonitoring. *Environ. Sci. Technol.* 42: 7018–7024.

84. Voorspoels, S., Covaci, A., Lepom, P., Escutanaire, S., and Schepens, P. 2006. Remarkable findings concerning PBDEs in the terrestrial top-predator red fox (*Vulpes vulpes*). *Environ. Sci. Technol.* 40: 2937–2943.

85. Kunisue, T., Takayanagi, N., Isobe, T., Takahashi, S., Nakatsu, S., Tsubota, T., Okumoto, K., Bushisue, S., Shindo, K., and Tanabe, S. 2008. Regional trend and tissue distribution of brominated flame retardants and persistent organochlorines in raccoon dogs (*Nyctereutes procyonoides*) from Japan. *Environ. Sci. Technol.* 42: 685–691.

86. Johnson-Restrepo, B., Kannan, K., Addink, R., and Adams, D. H. 2005. Polybrominated diphenyl ethers and polychlorinated biphenyls in a marine foodweb of coastal Florida. *Environ. Sci. Technol.* 39: 8243–8250.

87. Jenssen, B. M., Sørmo, E. G., Baek, K., Bytingsvik, J., Gaustad, H., Ruus, A., and Skaare, J. U. 2007. Brominated flame retardants in the north-east Atlantic marine ecosystem. *Environ. Health Persp.* 115(suppl 1): 35–41.

88. Voorspoels, S., Covaci, A., and Schepens, P. 2003. Polybrominated diphenyl ethers in marine species from the Belgian North Sea and the Western Scheldt Estuary: Levels, profiles, and distribution. *Environ. Sci. Technol.* 37: 4348–4357.

89. Boon, J. P., Lewis, W. E., Tjoen-A-Choy, M. R., Allchin, C. R., Law, R. J., de Boer, J., Hallers-Tjabbes, C. C. T., and Zegers, B. N. 2002. Levels of polybrominated diphenyl ether (PBDE) flame retardants in animals representing different trophic levels of the North Sea food web. *Environ. Sci. Technol.* 36(19): 4025–4032.

90. Brown, F. R., Winkler, J., Visita, P., Dhaliwal, J., and Petreas, M. 2006. Levels of PBDEs, PCDDs, PCDFs, and coplanar PCBs in edible fish from California coastal waters. *Chemosphere* 64: 276–286.

91. Sajwan, K. S., Kumar, K. S., Nune, S., Fowler, A., Richardson, J. P., and Loganathan, B. G. 2008. Persistent organochlorine pesticides, polychlorinated biphenyls, polybrominated diphenyl ethers in fish from coastal waters off Savannah, GA, USA. *Toxicol. Environ. Chem.* 90(1): 81–96.

92. Watanabe, I. and Sakai, S. 2003. Environmental release and behavior of brominated flame retardants. *Environ. Int.* 29: 665–682.

93. Hale, R. C., La Guardia, M. J., Harvey, E. P., Mainor, T. M., Duff, W. H., and Gaylor, M. O. 2001. Polybrominated diphenyl ether flame retardants in Virginia freshwater fishes (USA). *Environ. Sci. Technol.* 35(23): 4585–4591.

94. Anderson, T. D. and MacRae, J. D. 2006. Polybrominated diphenyl ethers in fish and wastewater samples from an area of the Penobscot River in Central Maine. *Chemosphere* 62: 1153–1160.

95. Xia, K., Luo, M. B., Lusk, C., Armbrust, K., Skinner, L., and Sloan, R. 2008. Polybrominated diphenyl ethers (PBDEs) in biota representing different trophic levels of the Hudson River, New York: From 1999 to 2005. *Environ. Sci. Technol.* 42(12): 4331–4337.

96. Isosaari, P., Lundebye, A.-K., Ritchie, G., Lie, O., Kiviranta, H., and Vartiainen, T. 2005. Dietary accumulation efficiencies and biotransformation of polybrominated diphenyl ethers in farmed Atlantic salmon (*Salmo salar*). *Food Addit. Contam.* 22(9): 829–837.

97. La Guardia, M. J., Hale, R. C., and Harvey, E. 2006. Detailed polybrominated diphenyl ether (PBDE) congener composition of the widely used penta-, octa-, and deca-PBDE technical flame-retardant mixtures. *Environ. Sci. Technol.* 40: 6247–6254.

98. Gaul, S., von der Recke, R., Tomy, G. T., and Vetter, W. 2006. Anaerobic transformation of a technical brominated diphenyl ether mixture by super-reduced vitamin B12 and dicyanocobinamide. *Environ. Toxicol. Chem.* 25(5): 1283–1290.

99. Boon, J. P., Van der Meer, J., Allchin, C. R., Law, R. J., Klungsoyr, J., Leonards, P. E. G., Spliid, H., Storr-Hansen, E., Mckenzie, C., and Wells, D. E. 1997. Concentration-dependent changes of PCB patterns in fish-eating mammals: Structural evidence for induction of cytochrome P450. *Arch. Environ. Contam. Toxicol.* 33: 298–311.

100. Leonards, P., Jol, J., Brandsma, S., Kruijt, A., Kwadijk, C., Vethaak, D., and De Boer, J. 2008. Bioaccumulation of brominated flame retardants in harbour seal. *Organohalogen Comp.* 70: 833–836.

101. Kannan, K., Koistinen, J., Beckmen, K., Evans, T., Gorzelany, J. F., Hansen, K. J., Jones, P. D., Helle, E., Nyman, M., and Giesy, J. P. 2001. Accumulation of perfluorooctane sulfonate in marine mammals. *Environ. Sci. Technol.* 35: 1593–1598.

102. Kannan, K., Corsolini, S., Falandysz, J., Oehmr, G., Focardi, S., and Giesy, J. P. 2002. Perfluorooctanesulfonate and related fluorinated hydrocarbons in marine mammals, fishes, birds from coasts of the Baltic and the Mediterranean Seas. *Environ. Sci. Technol.* 36: 3210–3216.

103. Kannan, K., Corsolini, S., Falandysz, J., Fillmann, G., Kumar, K. S., Loganathan, B. G., Mohd, M. A., Olivero, J., Van Wouwe, N., Yang, J. H., and Aldous, K. M. 2004. Perfluorooctanesulfonate and related fluorochemicals in human blood from several countries. *Environ. Sci. Technol.* 38: 4489–4495.

104. Giesy, J. P. and Kannan, K. 2002. Perfluorochemical surfactants in the environment. *Environ. Sci. Technol.* 36: 147A–152A.

105. Van de Vijver, K. I., Hoff, P., Das, K., Brasseur, S., Van Dongen, W., Esmans, E., Reijnders, P., Blust, R., and De Coen, W. 2005. Tissue distribution of perfluorinated chemicals in harbor seals (*Phoca vitulina*) from the Dutch Wadden Sea. *Environ. Sci. Technol.* 39: 6979–6984.

106. Xu, L., Krenitsky, D. M., Seacat, A. M., Butenhoff, J. L., and Anders, M. W. 2004. Biotransformation of *N*-Ethyl-*N*-(2-hydroxyethyl)perfluorooctanesulfonamide by rat liver microsomes, cytosol, and slices and by expressed rat and human cytochromes P450. *Chem. Res. Toxicol.* 17: 767–775.

107. Martin, J. W., Ellis, D. A., Mabury, S. A., Hurley, M. D., and Wallington, T. J. 2006. Atmospheric chemistry of perfluoroalkanesulfonamides: Kinetic and product studies of the OH radical and Cl atom initiated oxidation of *N*-ethyl perfluorobutanesulfonamide. *Environ. Sci. Technol.* 40: 864–872.

108. D'eon, J. C., Hurley, M. D., Wallington, T. J., and Mabury, S. A. 2006. Atmospheric chemistry of N-methyl perfluorobutane sulfonamidoethanol, C4F9SO2N(CH3)CH₂CH₂OH: Kinetics and mechanism of reaction with OH. *Environ. Sci. Technol.* 40: 1862–1868.

109. 3M. 2000. *Phase-Out Plan for POSF-Based Products*; US EPA Docket OPPT-2002-0043; Specialty Materials Markets Group, 3M: St. Paul, MN.

110. Keller, J. M., Kannan, K., Taniyasu, S., Yamashita, N., Day, R. D., Arendt, M. D., Segars, A. L., and Kucklick, J. R. 2005. Perfluorinated compounds in the plasma of loggerhead and Kemp's Ridley sea turtles from the southeastern coast of the United States. *Environ. Sci. Technol.* 39: 9101–9108.

111. Butt, C. M., Mabury, S. A., Muir, D. C. G., and Braune, B. M. 2007. Prevalence of long-chained perfluorinated carboxylates in seabirds from the Canadian Arctic between 1975 and 2004. *Environ. Sci. Technol.* 41: 3521–3528.

112. Butt, C. M., Muir, D. C. G., Stirling, I., Kwan, M., and Mabury, S. A. 2007. Rapid response of Arctic ringed seals to changes in perfluoroalkyl production. *Environ. Sci. Technol.* 41(1): 42–49.

113. Hart, K., Kannan, K., Isobe, T., Takahashi, S., Yamada, T., Miyazaki, N., and Tanabe, S. 2008. Time trends and transplacental transfer of perfluorinated compounds in melon-headed whales stranded along the Japanese coast in 1982, 2001/2002, and 2006. *Environ. Sci. Technol.* 42: 7132–7137.

114. Hart, K., Kannan, K., Tao, L., Takahashi, S., and Tanabe, S. 2008. Skipjack tuna as a bioindicator of contamination by perfluorinated compounds in the oceans. *Sci. Tot. Environ.* 403: 215–221.

115. Yoo, H., Kannan, K., Kim, S. K., Lee, K. T., Newsted, J. L., and Giesy, J. P. 2008. Perfluoroalkyl acids in the egg yolk of birds from Lake Shihwa, Korea. *Environ. Sci. Technol.* 42: 5821–5827.

116. Ellis, D. A., Martin, J. W., De Silva, A. O., Mabury, S. A., Hurley, M. D., Andersen, M. P. S., and Wallington, T. J. 2004. Degration of fluorotelomer alcohols: A likely source of perfluorinated carboxylic acids. *Environ. Sci. Technol.* 38: 3316–3321.

117. Prevedourous, K., Cousins, L. T., Buck, R. C., and Korzeniowski, S. H. 2006. Sources, fate, and transport of perfluorocarboxylates. *Environ. Sci. Technol.* 40: 32–44.

118. Yamashita, N., Kannan, K., Taniyasu, S., Horii, Y., Petrick, G., and Gamo, T. 2005. A global survey of perfluorinated acids in oceans. *Mar. Pollut. Bull.* 51: 658–668.

119. Young, C. J., Furdui, V. I., Franklin, J., Koerner, R. M., Muir, D. C. G., and Mabury, S. A. 2007. Perfluorinated acids in Arctic snow: New evidence for atmospheric formation. *Environ. Sci. Technol.* 41: 3455–3461.

120. Hu, W., Jones, P. D., Upham, B. L., Trosko, J. E., Lau, C., and Giesy, J. P. 2002. Inhibition of gap junctional intercellular communication by perfluorinated compounds in rat liver and dolphin kidney epithelial cell lines in vitro and Sprague-Dawley rats in vivo. *Toxicol. Sci.* 68: 429–436.

121. Peden-Adams, M. M., Keller, J. M., EuDaly, J. G., Berger, J., Gilkeson, G. S., and Keil, D. E. 2008. Suppression of humoral immunity in mice following exposure to perfluorooctane sulfonate. *Toxicol. Sci.* 104: 144–154.

122. Kannan, K., Perrotta, E., and Thomas, N. J. 2006. Association between perfluorinated compounds and pathological conditions in southern sea otters. *Environ. Sci. Technol.* 40: 4943–4948.

123. Ishibashi, H., Iwata, H., Kim, E.-Y., Tao, L., Kannan, K., Tanabe, S., Batoev, V. B., and Petrov, E. A. 2008. Contamination and effects of perfluorochemicals in Baikal seal (*Pusa sibirica*). 2. Molecular characterization, expression level, and transcriptional activation of peroxisome proliferator-activated receptor α. *Environ. Sci. Technol.* 42: 2302–2308.

124. Martin, J. W., Smithwick, M. M., Braune, B. M., Hoekstra, P. F., Muir, D. C. G., and Mabury, S. A. 2004. Identification of long-chain perfluorinated acids in biota from the Canadian Arctic. *Environ. Sci. Technol.* 38: 373–380.

125. Martin, J. W., Whittle, D. M., Muir, D. C. G., and Mabury, S. A. 2004. Perfluoroalkyl contaminants in a food web from Lake Ontario. *Environ. Sci. Technol.* 38: 5379–5385.

126. Bossi, R., Riget, F. F., and Dietz, R. 2005. Temporal and spatial trends of perfluorinated compounds in ringed seal (*Phoca hispida*) from Greenland. *Environ. Sci. Technol.* 39: 7416–7422.

127. Yeung, L. W., Miyake, Y., Wang, Y., Taniyasu, S., Yamashita, N., and Lam, P. K. S. 2009. Total fluorine, extractable organic fluorine, perfluorooctane sulfonate and other related fluorochemicals in liver of Indo-Pacific humpback dolphins (*Sousa chinensis*) and finless porpoises (*Neophocaena phocaenoides*) from South China. *Environ. Pollut.* 157: 17–23.

128. Kannan, K., Yun, S. H., and Evans, T. J. 2005. Chlorinated, brominated, and perfluorinated contaminants in livers of polar bears from Alaska. *Environ. Sci. Technol.* 39: 9057–9063.

129. Smithwick, M., Muir, D. C. G., Mabury, S. A., Solomon, K. R., Martin, J. W., Sonne, C., Born, E. W., Letcher, R. J., and Dietz, R. 2005. Perflouroalkyl contaminants in liver tissue from east Greenland polar bears (*Ursus maritimus*). *Environ. Toxicol. Chem.* 24(4): 981–986.

130. Sinclair, E. and Kannan, K. 2006. Mass loading and fate of perfluoroalkyl surfactants in wastewater treatment plants. *Environ. Sci. Technol.* 40: 1408–1414.

131. Kennish, M. J. 1992. Chlorinated hydrocarbons. In: *Ecology of Estuaries: Anthropogenic Effects.* CRC Press, Boca Raton, FL, pp.183–248.

132. Marquenie, J. M., and Reijnders, P. J. M. 1989. PCBs, an increasing concern for the marine environment. International Council for the Exploration of the Sea, Copenhagen. CM, 1989/N:12.

133. Blomkvist, G., Roos, A., Jensen, S., Bignert, A., and Olsson, M. 1992. Concentrations of sDDT and PCB in seals from Swedish and Scottish waters. *Ambio* 21(8): 539–545.

134. Nyman, M., Kositinen, J., Fant, M. L., Vartiainen, T., and Helle, E. 2002. Current levels of DDT, PCB and trace elements in the Baltic ringed seals (*Phoca hispida baltica*) and grey seals (*Halichoerus grypus*). *Environ. Pollut.* 119: 399–412.

135. Tanabe, S. 1988. PCB problems in the future: Foresight from current knowledge. *Environ. Pollut.* 50: 5–28.

136. Addison, R. F. and Stobo, W. T. 2001. Trends in organochlorine residue concentrations and burdens in grey seals (*Halichoerus grypus*) from Sable Is., NS, Canada, between 1974 and 1994. *Environ. Pollut.* 112: 505–513.

137. AMAP. 2000. *AMAP Assessment Report: Arctic Pollution Issues.* Arctic Monitoring and Assessment Programme, Oslo, Norway.

138. Tuerk, K. J. S., Kucklick, J. R., Becker, P. R., Stapleton, H. M., and Baker, J. E. 2005. Persistent organic pollutants in two dolphin species with focus on toxaphene and polybrominated diphenyl ethers. *Environ. Sci. Technol.* 39(3): 692–698.

139. Ikonomou, M. G., Rayne, S., and Addison, R. F. 2002. Exponential increases of the brominated flame retardants, polybrominated diphenyl ethers, in the Canadian Arctic from 1981 to 2000. *Environ. Sci. Technol.* 36(9): 1886–1892.

140. Kajiwara, N., Ueno, D., Takahashi, A., Baba, N., and Tanabe, S. 2004. Polybrominated diphenyl ethers and organochlorines in archived northern fur seal samples from the Pacific coast of Japan, 1972–1998. *Environ. Sci. Technol.* 38(14): 3804–3809.

141. Norstrom, R. J., Simon, M., Moisey, J., Wakeford, B., and Weseloh, D. V. C. 2002. Geographical distribution (2000) and temporal trends (1981–2000) of brominated diphenyl ethers in Great Lakes herring gull eggs. *Environ. Sci. Technol.* 36: 4783–4789.

142. Dietz, R., Bossi, R., Riget, F. F., Sonne, C., and Born, E. W. 2008. Increasing perfluoroalkyl contaminants in east Greenland polar bears (*Ursus maritimus*): A new toxic threat to the Arctic bears. *Environ. Sci. Technol.* 42: 2701–2707.

143. Ishibashi, H., Iwata, H., Kim, E.-Y., Tao, L., Kannan, K., Amano, M., Miyazaki, N., Tanabe, S., Batoev, V. B., and Petrov, E. A. 2008. Contamination and effects of perfluorochemicals in Baikal seal (*Pusa sibirica*). 1. Residue level, tissue distribution, and temporal trend. *Environ. Sci. Technol.* 42: 2295–2301.

144. Smithwick, M. M., Norstrom, R. J., Mabury, S. A., Solomon, K., Evans, T. J., Stirling, I., Taylor, M. K., and Muir, D. C. G. 2006. Temporal trends of perfluoroalkyl contaminants in polar bears (*Ursus maritimus*) from two locations in the North American Arctic, 1972–2002. *Environ. Sci. Technol.* 40: 1139–1143.

145. NMFS. 2007. *Harbor Seal (Phoca vitulina): Western North Atlantic Stock*. National Marine Fisheries Service. www.nmfs.noaa.gov/pr/sars/species.htm (accessed on June 2010).

146. Jones, S. H., Chase, M., Sowles, J., Hennigar, P., Landry, N., Wells, P. G., Harding, G. C. H., Krahforst, C., and Brun, G. L. 2001. Monitoring for toxic contaminants in *Mytilus edulis* from New Hampshire and the Gulf of Maine. *J. Shellfish Res.* 20(3): 1203–1214.

147. Athanasiadou, M., Cuadra, S. N., Marsh, G., Bergman, A., and Jakobsson, K. 2008. Polybrominated diphenyl ethers (PBDEs) and bioaccumulative hydroxy PBDE metabolites in young humans from Managua, Nicaragua. *Environ. Health Persp.* 116: 400–408.

148. Hall, A. J. and Thomas, G. O. 2007. Polychlorinated biphenyls, DDT, polybrominated diphenyl ethers and organic pesticides in United Kingdom harbor seals (*Phoca vitulina*)-mixed exposures and thyroid homeostasis. *Environ. Toxicol. Chem.* 26(5): 851–861.

149. Wolkers, H., Hammill, M. O., and van Bavel, B. 2006. Tissue-specific accumulation and lactational transfer of polychlorinated biphenyls, chlorinated pesticides, and brominated flame retardants in hooded seals (*Cystophora cristata*) from the Gulf of St. Lawrence: Applications for monitoring. *Environ. Pollut.* 142: 476–486.

150. Borrell, A., Cantos, G., Aguilar, A., Androukaki, E., and Dendrinos, P. 2007. Concentrations and patterns of organochlorine pesticides and PCBs in Mediterranean monk seals (*Monachus monachus*) from Western Sahara and Greece. *Sci. Tot. Environ.* 381: 316–325.

151. Imaeda, D., Kunisue, T., Ochi, Y., Iwata, H., Tsydenova, O. V., Takahashi, S., Amano, M., Petrov, E. A., Batoev, V. B., and Tanabe, S. 2009. Accumulation features and temporal trends of PCDDs, PCDFs and PCBs in Baikal seals (*Pusa sibirica*). *Environ. Pollut.* 157: 737–747.

152. Kajiwara, N., Watanabe, M., Wilson, S., Eybatov, T., Mitrofanov, I. V., Aubrey, D. G., Khuraskin, L. S., Miyazaki, N., and Tanabe, S. 2008. Persistent organic pollutants (POPs) in Caspian seals of unusual mortality event during 2000 and 2001. *Environ. Pollut.* 152: 431–442.

153. Blasius, M. E. and Goodmanlowe, G. D. 2008. Contaminants still high in top-level carnivores in the Southern California Bight: Levels of DDT and PCBs in resident and transient pinnipeds. *Mar. Pollut. Bull.* 56: 1973–1982.

154. Hall, A. J., Thomas, G. O., and McConnell, B. J. 2009. Exposure to persistent organic pollutants and first-year survival probability in gray seal pups. *Environ. Sci. Technol.* 43: 6364–6369.

155. Tanabe, S., Niimi, S., Minh, T. B., Miyazaki, N., and Petrov, E. A. 2003. Temporal trends of persistent organochlorine contamination in Russia: A case study of Baikal and Caspian seal. *Arch. Environ. Contam. Toxicol.* 44: 533–545.

156. Van de Vijver, K. I., Hoff, P. T., Das, K., Van Dongen, W., Esmans, E. L., Jauniaux, T., Bouquegneau, J.-M., Blust, R., and De Coen, W. 2003. Perfluorinated chemicals infiltrate ocean waters: Link between exposure levels and stable isotope ratios in marine mammals. *Environ. Sci. Technol.* 37: 5545–5550.

157. Ahrens, L., Siebert, U., and Ebinghaus, R. 2009. Temporal trends of polyfluoroalkyl compounds in harbor seals (*Phoca vitulina*) from the German Bight, 1999–2008. *Chemosphere* 76: 151–158.

158. Kallenborn, R., Berger, U., and Järnberg, U. 2004. *Perfluorinated Alkylated Substances (PFAS) in the Nordic Environment*. Nordic Council of Ministers, Copenhagen, Denmark, p. 112.

159. Quakenbush, L. T. and Citta, J. J. 2008. Perfluorinated contaminants in ringed, bearded, spotted, and ribbon seals from the Alaskan Bering and Chukchi Seas. *Mar. Pollut. Bull.* 56: 1809–1814.

160. Tomy, G. T., Pleskach, K., Ferguson, S. H., Hare, J., Stern, G. A., Macinnis, G., Marvin, C. H., and Loseto, L. L. 2009. Trophodynamics of some PFCs and BFRs in a Western Canadian Arctic marine food web. *Environ. Sci. Technol.* 43: 4079–4081.

161. Ylitalo, G. M., Myers, M., Stewart, B. S., Yochem, P. K., Braun, R., Kashinsky, L., Boyd, D., Antonelis, G. A., Atkinson, S., Aguirre, A. A., and Krahn, M. M. 2008. Organochlorine contaminants in endangered Hawaiian monk seals from four subpopulations in the Northwestern Hawaiian Islands. *Mar. Pollut. Bull.* 56: 231–244.

162. Fillmann, G., Hermanns, L., Fileman, T. W., and Readman, J. W. 2007. Accumulation patterns of organochlorines in juveniles of *Arctocephalus australis* found stranded along the coast of Southern Brazil. *Environ. Pollut.* 146: 262–267.

163. Wang, D., Shelver, W. L., Atkinson, S., Mellish, J., and Li, Q. X. 2010. Tissue distribution of polychlorinated biphenyls and organochlorine pesticides and potential toxicity to Alaskan northern fur seals assessed using PCBs congener specific mode of action schemes. *Arch. Environ. Contam. Toxicol.* 58: 478–488.

164. Wolkers, H., Krafft, B. A., van Bavel, B., Helgason, L. B., Lydersen, C., and Kovacs, K. M. 2008. Biomarker responses and decreasing contaminant levels in ringed seals (*Pusa hispida*) from Svalbard, Norway. *J. Toxicol. Environ. Health A* 71: 1009–1018.

165. Haglund, P. S., Zook, D. R., Buser, H.-R., and Hu, J. 1997. Identification and quantification of polybrominated diphenyl ethers and methoxy-polybrominated diphenyl ethers in Baltic biota. *Environ. Sci. Technol.* 31: 3281–3287.

166. Riget, F., Vorkamp, K., Dietz, R., and Rastogi, S. C. 2006. Temporal trend studies on polybrominated diphenyl ethers (PBDEs) and polychlorinated biphenyls (PCBs) in ringed seals from East Greenland. *J. Environ. Monit.* 8: 1000–1005.

167. Leonel, J., Kannan, K., Tao, L., Fillmann, G., and Montone, R. C. 2008. A baseline study of perfluorochemicals in franciscana dolphin and subantarctic fur seal from coastal waters of Southern Brazil. *Mar. Pollut. Bull.* 56: 778–781.

168. Schiavone, A., Corsolini, S., Kannan, K., Tao, L., Trivelpiece, W., Torres, D., Jr., and Focardi, S. 2009. Perfluorinated contaminants in fur seal pups and penguin eggs from South Shetland, Antarctica. *Sci. Tot. Environ.* 407: 3899–3904.

24 Organohalogen Pollutants in Seabird Eggs from Northern Norway and Svalbard

Lisa Bjørnsdatter Helgason,* Kjetil Sagerup, and Geir Wing Gabrielsen

CONTENTS

24.1 Introduction 547
24.2 Using Seabird Eggs as Indicators of Temporal Trends of POPs 551
24.3 Spatial Trends of POPs in Eggs of Seabirds from Northern Norway and Svalbard............ 551
24.4 POPs in Eggs of Seabirds from Northern Norway and Svalbard...................553
    24.4.1 OCs 553
        24.4.1.1 Species Comparison...553
        24.4.1.2 Temporal Trends 553
    24.4.2 Mercury 556
        24.4.2.1 Species Comparison...556
        24.4.2.2 Temporal Trends 556
    24.4.3 BFRs 556
        24.4.3.1 Species Comparison...556
        24.4.3.2 Temporal Trends 558
    24.4.4 PFAS 560
        24.4.4.1 Species Comparison...560
        24.4.4.2 Temporal Trends 561
24.5 Conclusions...563
Acknowledgments..564
References...565

24.1 INTRODUCTION

The coastal waters around mainland Norway and Svalbard support as many as 2.9 and 3 million breeding pairs of seabirds, respectively [1,2]. During the winter months, most of the seabirds migrate out of the region to more southerly latitudes, whereas some species stay in the area all year round feeding in the marginal sea ice-zone [2–4]. The Atlantic puffin (*Fratercula arctica*) (1,710,000 pairs), the black-legged kittiwake (*Rissa tridactyla*) (606,000 pairs), the herring gull (*Larus argentatus*) (233,000 pairs), the common guillemot (*Uria aalge*) (137,500 pairs), and the Brünnich's guillemot (*Uria lomvia*) (851,500 pairs) are some of the numerous seabird species breeding along the Northern Norwegian coast and in the Svalbard area [1,2].

* E-mail: Lisa.BjornsdatterHelgason@npolar.no (Chapter corresponding author).

TABLE 24.1

Biological Characteristics of Herring Gulls, Black-Legged Kittiwakes, Atlantic Puffins, Common Guillemots, and Brünnich's Guillemots Breeding in Northern Norway and Svalbard

Species	Scientific Name	Diet	Clutch Size	Winter Habitat
Herring gull	*Larus argentatus*	Variety of food Fish Seabird eggs	1–3	North Sea English channel
Black-legged kittiwake	*Rissa tridactyla*	Invertebrates Small fish	1–3	North Atlantic
Atlantic puffin	*Fratercula arctica*	Small fish (herring) Crustaceans Squid Polychaete worms	1	North Atlantic Southern Barents Sea
Common guillemot	*Uria aalge*	Small fish (capelin)	1	Southern Barents Sea Northern Norway
Brünnich's guillemot	*Uria lomvia*	Fish Crustaceans	1	Barents Sea Iceland Greenland Newfoundland

Source: Data from Anker-Nilssen, T., *The Status of Marine Birds Breeding in the Barents Sea Region*, Norsk Polarinstitutt, Tromsø, Norway, 2000. With permission.

Whereas the herring gull is a typical generalist and top predator, the black-legged kittiwake, Atlantic puffin, common guillemot, and Brünnich's guillemot feed at a lower trophic level, consuming mainly food from the pelagic water masses [2–4]. See Table 24.1 for details of the biological characteristics of these species.

Negative population growth has been found for many of the seabird species breeding in Northern Norway and on Svalbard over the last 40 years [1–3]. The negative trends can among others be attributed to repeated breeding failures and reduced survival due to food shortage, e.g., many birds starved following the collapse of the Norwegian spring-spawning herring stock in the late 1960s, and the capelin crash in the Barents Sea in 1986. Furthermore, seabirds breeding in the Norwegian Sea and the Barents Sea are competing with the fishing industry for recourses. They also drown in gill-nets and long-lines [1,3]. Long-term monitoring surveys have revealed that since the mid-1990s the breeding colonies of black-legged kittiwake declined by 6% each year in the Barents Sea and by 8% each year in the Norwegian Sea [1], whereas on Svalbard, the population has been relatively stable [2]. Rapid population declines have also been observed for the Atlantic puffin. One of the largest Atlantic puffin colonies in Europe is on the island of Røst (Figure 24.1) and it has experienced a population decrease of as much as 70% since the 1960s [1,3]. However, in other locations, e.g., at Hornøya and Gjesvær (Figure 24.1), breeding numbers of Atlantic puffin have not changed or actually increased slightly. It remains to be seen if the food shortage now experienced by Atlantic puffins throughout the region will also have a negative impact on breeding numbers on Hornøya and Gjesvær in the years to come [1]. The populations of common and Brünnich's guillemots have declined dramatically west of the North Cape (west of Hjelmsøya, see Figure 24.1) [1]. Similar declines have not been observed east of the North Cape (east of Hjelmsøya, Figure 24.1). Here, the numbers of common guillemots increased, while the number of Brünnich's guillemots did not change [1]. Also, on Svalbard, the population of common guillemots has decreased significantly, almost by 50% within the last two decades. In contrast, the population of Brünnich's guillemots on Svalbard has been stable [2].

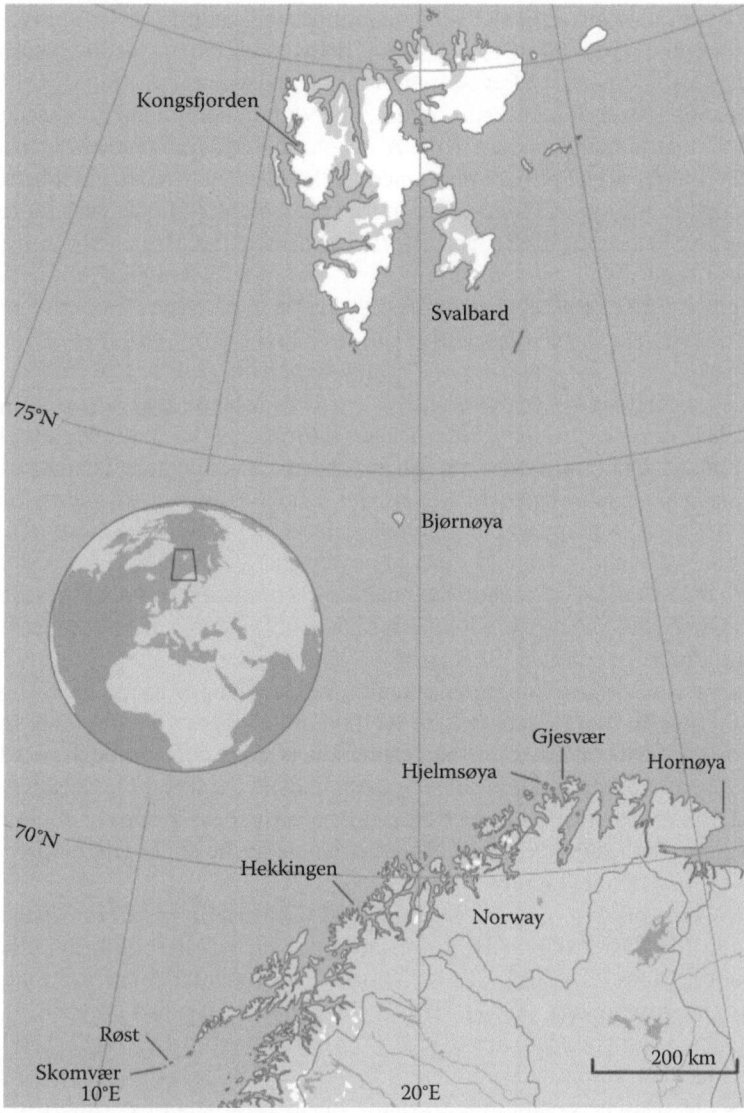

FIGURE 24.1 Map showing the localities from which eggs of herring gulls (*Larus argentatus*), Atlantic puffins (*Fratercula arctica*), black-legged kittiwakes (*Rissa tridactyla*), common guillemots (*Uria aalge*), and Brünnich's guillemots (*Uria lomvia*) were collected in 1983, 1993, 2003, and 2007 (see Table 24.2 for details of sampling). (Reprinted with permission from Verreault, J. et al., *Environ. Sci. Technol.*, 41, 6671, 2007. Copyright 2007 American Chemical Society.)

Industrial and agricultural activity in Northern Norway and Svalbard is modest. However, virtually all persistent organic pollutants (POPs) produced at southern latitudes have been reported in the abiotic environment and in marine organisms from Northern Norway and Svalbard [5,6], demonstrating that POPs are transported over long distances [5,7]. As early as 1971, residues of dichlorodiphenyldichloroethylene (DDE) and polychlorinated biphenyl (PCB) were found in eggs of herring gull and common gull (*Larus canus*) from Norway [8], and in 1972, high concentrations of organochlorines (OCs) were reported in glaucous gulls (*Larus hyperboreus*) from Svalbard [9]. Since then, a considerable amount of research has been conducted on POP contamination in seabirds breeding in Northern Norway and Svalbard [5,6,10].

The concentrations and pattern of POPs found in different seabird species are typically explained by differences in trophic position [11–14]. However, body condition, migration pattern, and biotransformation capacity also influences POP concentrations and patterns [15–18]. Most POPs have a long half-life (weeks to years) and lipophilic properties, and will therefore tend to bioaccumulate through the food web. This is particularly evident in the Arctic marine environment where food webs are relatively long and lipid stores are of prime importance for the survival in harsh environmental conditions [4]. In concert with this, the highest concentrations of PCBs, dichlorodiphenyltrichloroethane (DDTs), chlordanes, hexachlorobenzene (HCB), hexachlorocyclohexane (HCHs), and brominated flame retardants (BFRs) have been found in predatory seabirds from the Norwegian Arctic that supplements their diet with eggs and chicks of other seabirds, in addition to carcasses of marine mammals. These predatory seabirds include the glaucous gull [11,13,19–22], the Ivory gull (*Pagophila eburnea*) [23], the great black-backed gull (*Larus marinus*) [24], and the herring gull [25]. In comparison, lower POP concentrations have been found in seabirds such as northern fulmars (*Fulmarus glacialis*) [26], lesser black-backed gulls (*Larus fuscus*) [27], black-legged kittiwakes [11,13], guillemots [11,13], and little auks (*Alle alle*) [28], which mainly consume fish and marine invertebrates. In contrast to the OCs and BFRs, perfluorinated alkyl substances (PFAS) are not lipophilic as they are mainly bound to albumin [29,30]. Still, PFAS are also found to bioaccumulate in Arctic marine food webs [31]. Effect studies conducted on predatory seabirds from Northern Norway and Svalbard have suggested that POP contamination may have a negative impact on the health and population status of these birds [24,32,33].

OCs, such as DDTs and PCBs, were found in high concentrations in the environment already in the 1950s and the 1960s. In contrast, BFRs and PFAS have received attention only recently. This is remarkable, as several of these compounds actually were developed more than 50 years ago [5,34]. However, it takes time to evolve satisfactory methods to analyze contaminants in the biota, and there has been a significant increase in the production of consumer products, leading to a higher production of flame retardants and fluorinated compounds in the last three decades [5,34]. As such, increased global awareness of BFRs and PFAS is not because these groups of contaminants are very "new" or "emerging," but rather the result of improvement in the analytical methods and increased production of these compounds in recent years.

BFRs and PFAS are present in eggs of marine and terrestrial birds from Northern Norway and Svalbard [19,22,35,36]. The group of BFRs includes polybrominated diphenyl ethers (BDEs) and hexabromocyclododecanes (HBCDs). BDEs comprise 209 potential congeners that are structurally similar to PCBs and dioxins/furans [37]. HBCD is a technical product comprising several diastereomers in which the most abundant are α-HBCD, γ-HBCD, and β-HBCD [37]. Whereas γ-HBCD dominates in technical mixtures and sediments [38–40], α-HBCD is the primary congener detected in biota samples [39,41]. The contaminant group "PFAS" consists of perfluorosulfonates and the perfluorocarboxylic acids. The predominant PFAS in biota samples is the perfluorooctane sulfonate (PFOS) that is found in relatively high concentrations in the Arctic environment [5,6].

A number of the OCs, such as PCBs, chlordanes, HCB, HCHs, and DDTs, in addition to penta- and octa-BDE mixtures have been banned in many countries and in Norway for years to decades [5,42]. There is currently no ban of deca-BDE and HBCD within the European Union (EU), except for a restriction of deca-BDE in electronics and electrical applications [43]. The use of PFOS and related substances was restricted by the EU in 2006 [44]. In Norway, all production, import, export, sale, and use of PFAS and deca-BDEs was banned by the June 1, 2007 and April 1, 2008, respectively [42].

Although OCs such as PCBs have been banned in most countries since the 1990s and their use restricted for decades, they are still present in relatively high concentrations in the abiotic and biotic environment of Northern Norway and Svalbard [4–6]. This clearly illustrates that it takes time for a chemical to disappear from the environment after a local or global ban has been initiated. As such it is important to continuously monitor temporal trends of contaminants in the environment, and to examine if the temporal trends differ between and within animal groups. Because there are few point sources of POPs in the Arctic, temporal trend studies from Northern Norway and Svalbard may function as early warning signs of global contamination trends. It also illustrates how nontarget organisms are affected by POPs through long-range transport such as from atmospheric air currents, ocean currents, transpolar ice pack, and the large Arctic rivers. Seabirds are on the top of the Arctic food web and as such accumulate high levels of POPs and are at the risk of adverse effects. Therefore, they are an ideal candidate for temporal trend studies. Seabird eggs have frequently been used as a monitoring tool to investigate temporal trends of POPs in Arctic areas as they are easy to collect and, in contrast to blood samples, not so dependent on nutritional status. The eggs have a relatively constant fat content between years and the variation in fat percentage is low. Additionally, the concentrations and relative contributions of POPs in seabird eggs are comparable to those found in tissues of female birds.

The present review summarizes recently published [45–48] and unpublished studies on contaminant profiles and temporal trends of POPs (OCs, PBDEs, HBCD, and PFAS) in eggs of herring gulls, black-legged kittiwakes, Atlantic puffins, common guillemots, and Brünnich's guillemots from Northern Norway and Svalbard (for an overview of the studies and sampling areas, see Figure 24.1; Table 24.2). The yearly mean concentrations of OCs, PBDEs, HBCD, and PFAS, in addition to contaminant patterns measured in 1983, 1993, and 2003 in eggs of seabirds from Northern Norway and Svalbard, are compared (using t-tests and ANOVA with post hoc tests) and presented in this review. To examine more recent time trends, an additional study of trends of POPs in eggs of Brünnich's guillemots sampled on Svalbard in 1993, 2003, and 2007 was included [46] (Table 24.2). The measurement of temporal trends of POPs and mercury in seabird eggs from Northern Norway commenced as early as 1972 [49–51], and continues up until today [45–48,52,53].

24.2 USING SEABIRD EGGS AS INDICATORS OF TEMPORAL TRENDS OF POPS

Lipid-soluble POPs are transferred along with lipids during egg formation [54], and as such, the concentrations and relative contributions of POPs in seabird eggs are comparable to those found in tissues of female birds [55]. In concert with this, eggs of seabirds have been recognized as useful monitoring tools to examine POP contamination in seabirds [52,53,56,57]. Several studies have shown that POP concentrations do not depend on the position of the egg in the laying sequence [58–61]. For example, Verreault et al. [62] confirmed the absence of intra-clutch variability of POP concentrations in glaucous gull eggs from Svalbard. However, in a recent study on glaucous gulls, it was suggested that female birds exposed to different patterns of contamination invested differentially in egg production, i.e., females with relatively high proportions of chlordanes, DDT, and HBCD in their plasma laid smaller eggs [55]. In the seabird egg studies from Northern Norway and Svalbard reviewed here, eggs were collected at random, taking only one egg per nest in the case of multi-egg species. These methods were the same in different years and areas.

24.3 SPATIAL TRENDS OF POPS IN EGGS OF SEABIRDS FROM NORTHERN NORWAY AND SVALBARD

Seabird eggs from mainland Northern Norway were collected from Hornøya, Gjesvær and Hjelmsøya in the north, and Røst, Hekkingen and Skomvær in the south to investigate spatial trends of POP concentrations and patterns. With the exception of HCH and BDE-153, no spatial differences were found for OCs and BFRs [45,48]. The absence of a north–south trend in the eggs of these species is in line with the results of a recent study comparing POP concentrations in gull eggs from Northern Norway, Faeroe Island, and Svalbard [63], suggesting that different seabird colonies in Northern Norway are exposed to comparable levels of local and global contamination. However, in a study investigating POPs in great black-backed gulls nesting along the Norwegian coast (58°N–70°N), congener-specific geographical differences were found [64].

In contrast to most OCs and BFRs, concentrations of several PFAS were found to vary between herring gull eggs collected at different locations within the region [47]. Eggs from southern parts of Northern Norway sampled in 1993 were characterized by significantly higher concentrations

TABLE 24.2
Overview of Studies Included in This Review

Species	Location	Year	No. Eggs OCs	No. Eggs BFRs	No. Eggs PFAS	OCs	BFRs	PFAS	Original References
Herring gull	Hornøya, Røst, Gjesvær	1983	30	10	10	HCB, β-HCH, oxychlordane, p,p'-DDE, PCBs	BDEs, HBCD	PFOS, PFTriA, PFUnA + others	[45,47,48]
		1993	15	10	10				
		2003	16	10	10				
Black-legged kittiwake	Hornøya, Røst	1983	21	10	—	HCB, β-HCH, oxychlordane, p,p'-DDE, PCBs	BDEs, HBCD	—	[45,48]
		1993	25	10	—				
		2003	11	10	—				
Atlantic puffin	Hornøya, Røst, Skomvær, Hjelmsøya	1983	31	10	—	HCB, β-HCH, oxychlordane, p,p'-DDE, PCBs	BDEs, HBCD	—	[45,48]
		1993	10	10	—				
		2003	15	10	—				
Common guillemot	Hornøya, Hjelmsøya	1983	10	—	—	HCB, β-HCH, oxychlordane, p,p'-DDE, PCBs	—	—	[45]
		1993	5	—	—				
		2003	12	—	—				
Brünnich's guillemot (1)	Kongsfjorden, Bjørnøya	1983	—	—	—	HCB, β-HCH, oxychlordane, p,p'-DDE, PCBs	—	—	New data Norwegian Polar Institute
		1993	5	—	—				
		2003	5	—	—				
Brünnich's guillemot (2)	Kongsfjorden, Bjørnøya	1993	10	10	10	HCB, β-HCH, oxychlordane, p,p'-DDE, PCBs	BDEs, HBCD	PFOS, PFTriA, PFUnA + others	[46]
		2003	10	10	10				
		2007	5	5	5				

Species, scientific name, location, year, number (No.) of eggs analyzed for organochlorines (OCs), brominated flame retardants (BFRs) and perfluorinated alkyl substances (PFAS), and the original publications on which the conclusions of the present review are based.

of perfluorooctanoic acid (PFOA), perfluorononanoic acid (PFNA), and perfluorohexansulfonate (PFHxS) compared to eggs from the northern parts of Northern Norway. Mean concentrations of PFOA in herring gull eggs were also higher in southern parts of Northern Norway compared to northern parts of Northern Norway in 2003. Additionally, the eggs of herring gulls collected in southern parts of Northern Norway in 1993 had higher concentrations of perfluorotetradecanoate (PFTeA) and perfluoropentadecanoate (PFPeA) than did eggs from the south [47]. Verreault et al. [47] suggested that this could be the result of differences in local use and exposure, or alternatively, by an enhanced contribution of PFAS from oceanic transport at southern parts compared to northern parts of Northern Norway.

24.4 POPS IN EGGS OF SEABIRDS FROM NORTHERN NORWAY AND SVALBARD

24.4.1 OCs

24.4.1.1 Species Comparison

The concentrations of HCB, β-HCHs, oxychlordane, p,p'-DDE, and PCBs in seabird eggs from Northern Norway and Svalbard (sampled in 2003) [45,46] (Figure 24.2) fall within the range of those measured in other seabirds feeding in the same area and at a similar trophic level [26,65,66]. Concentrations of PCBs and oxychlordane were highest in eggs of herring gulls followed by eggs of black-legged kittiwakes > Atlantic puffins > common guillemots and Brünnich's guillemots [45,46], whereas the lowest levels of p,p'-DDE were found in the black-legged kittiwake eggs. Species-specific differences in HCB and β-HCH levels were not found. The findings of generally lower levels of OCs in guillemots are most probably explained by their year-round residency in arctic areas and the fact that they feed at a relatively lower trophic level compared to the other seabird species [3].

24.4.1.2 Temporal Trends

PCBs were by far the most dominant OC congeners in seabird eggs from Northern Norway and Svalbard sampled in 1983–2003/2007. These were followed by p,p'-DDE > HCB > oxychlordane and β-HCH [45,46]. The relative contribution of the individual OCs to the sum of OCs differed between years and species (Table 24.3). Most notably, the relative contribution of HCB to sum of OCs increased from 1983 to 2003 in eggs of herring gulls, Atlantic puffins, common guillemots, and Brünnich's guillemots (from 1993 to 2007), whereas the relative contribution of PCBs to sum of OCs increased in the eggs of black-legged kittiwakes and common guillemots in the same time period (Table 24.3).

OC concentrations were higher in eggs collected in 1983 compared to eggs collected in 2003 (Figure 24.2) [45]. From 1983 to 2003, the OC concentrations in seabird eggs from Northern Norway and Svalbard decreased by as much as 60%–80%. The OCs continued to decease by 13%–50% between 2003 and 2007 in eggs of Brünnich's guillemots from Svalbard (Figure 24.2) [46].

The temporal trends from 1993 to 2003 found for OCs in seabird eggs from Northern Norway and Svalbard differed between species and compounds. Whereas concentrations of β-HCH, oxychlordane, p,p'-DDE, and PCBs in eggs of herring gulls were significantly higher in 1993 as compared to 2003, no differences were seen for OC concentrations in eggs of black-legged kittiwakes between 1993 and 2003, except for β-HCH (Figure 24.2) [45]. The mean concentrations of HCB in eggs of herring gulls remained unchanged during the same time period [45]. In eggs of Atlantic puffins, a decreasing trend was found for p,p'-DDE, oxychlordane, and PCBs from 1993 to 2003, whereas concentrations of β-HCH and HCB did not change in the same time period (Figure 24.2) [45]. In eggs of guillemots, the concentrations of HCB (only common guillemots), β-HCH, oxychlordane, p,p'-DDE, and PCBs were all significantly higher in 1993 than in 2003 (Figure 24.2). Furthermore, the concentrations of PCBs, p,p'-DDE, and oxychlordane continued to decrease from 2003 to 2007 in Brünnich's guillemots eggs [46] (Figure 24.2). Concentrations of HCB and HCH in

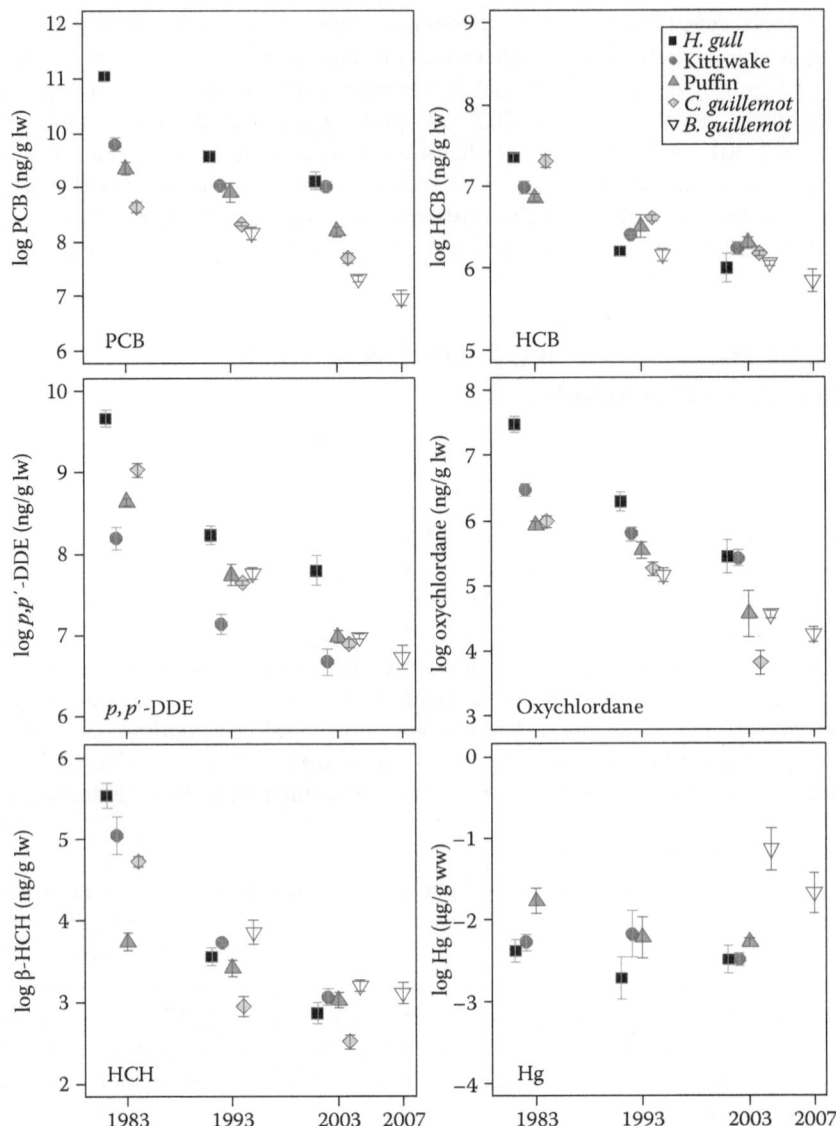

FIGURE 24.2 Temporal trends of polychlorinated biphenyls (PCBs), oxychlordane, hexachlorobenzene (HCB), hexacyclohexane (β-HCH), dichlorodiphenyldichloroethylene (*p,p'*-DDE) (log ng/g lipid weight (lw)) and mercury (Hg) (log μg/g wet weight (ww)) in eggs of herring gulls (*Larus argentatus*), Atlantic puffins (*Fratercula arctica*), black-legged kittiwakes (*Rissa tridactyla*) and common guillemots (*Uria aalge*) sampled in 1983, 1993, and 2003, and in eggs of Brünnich's guillemots (*Uria lomvia*) sampled in 1993, 2003, and 2007. (Reprinted with permission from Helgason, L. et al., *Environ. Pollut.*, 155, 190, 2008. Copyright 2008 American Chemical Society; Miljeteig, C. and Gabrielsen, G.W., *Contaminants in Brünnich's Guillemots from Kongsfjorden and Bjørnøya in the Period from 1993 to 2007*, 2010. Copyright Wiley-VCH Verlag GmbH & Co. KGaA. Reproduced with permission.)

eggs of Brünnich's guillemots did not differ between 2003 and 2007, although the same downward trend was observed (Figure 24.2) [46].

In an earlier temporal trend study from Northern Norway, no differences were found in the concentrations of *p,p'*-DDE, and PCB in eggs of herring gulls, black-legged kittiwakes, common guillemots, and Northern gannets (*Morus bassanus*) between 1972 and 1983 [53], although the mean concentrations of PCBs and *p,p'*-DDE were generally lower in 1983 than in 1972 [53].

TABLE 24.3
Proportional Changes in Organic Pesticides (Oxychlordane, Hexachlorobenzene [HCB], Hexacyclohexane [β-HCH], Dichlorodiphenyldichloroethylene [*p,p′*-DDE] and Polychlorinated Biphenyls [PCBs]) in Eggs of Herring Gulls (*H. gull*) (*Larus argentatus*), Atlantic Puffins (Puffin) (*Fratercula arctica*), Black-Legged Kittiwakes (Kittiwake) (*Rissa tridactyla*), Common Guillemots (*C. guillemot*) (*Uria aalge*), (*B. guillemot*) and Brünnich's Guillemots (*Uria lomvia*) from Northern Norway and Svalbard during the Period 2003–2007

Species	Compound	1983–1993	1983–2003	1993–2003
H. gull	HCB	—[a]	↑[b]	—
	HCH	—	↓[c]	—
	Oxychlordane	—	—	—
	DDE	—	—	—
	PCB	—	—	—
A. puffin	HCB	—	↑	↑
	HCH	—	↑	—
	Oxychlordane	—	—	—
	DDE	↓	↓	—
	PCB	↑	—	—
Kittiwake	HCB	↑	—	—
	HCH	—	↓	↓
	Oxychlordane	—	—	—
	DDE	—	↓	—
	PCB	—	↑	—
C. guillemot	HCB	—	↑	—
	HCH	↓	↓	—
	Oxychlordane	—	↓	↓
	DDE	↓	↓	—
	PCB	↑	↑	—
B. guillemot	HCB	↑	↑	—
	HCH	—	—	—
	Oxychlordane	—	—	—
	DDE	—	↓	—
	PCB	—	↓	—

Sources: Data from Helgason, L. et al., *Environ. Pollut.*, 155, 190, 2008; Miljeteig, C. and Gabrielsen, G.W., *Contaminants in Brünnich's Guillemots from Kongsfjorden and Bjørnøya in the Period from 1993 to 2007*, Norwegian Polar Institute, Tromsø, 2010. With permission.

[a] Not significant ($p > 0.05$).

[b] Significant ($p < 0.05$) increase.

[c] Significant ($p < 0.05$) decrease.

The significant decrease in HCB, β-HCH, oxychlordane, *p,p'*-DDE, and PCB concentrations in seabird eggs from 1983 to 2003 followed the temporal trends of OCs demonstrated for most Arctic animals [5,67,68], including seabirds from Arctic regions, (e.g., [56,69]), suggesting reduced exposure to these compounds. However, it is evident from the present study that the rates of decline vary between seabird species and OC compounds.

24.4.2 MERCURY

24.4.2.1 Species Comparison

The concentration of mercury in eggs of herring gulls, black-legged kittiwakes, Atlantic puffins, and Brünnich's guillemots from Northern Norway and Svalbard varied between 0.04 and 0.93 μg/g wet weight (ww) [45,46]. This range is similar to or lower than that found in eggs of other seabird species from North America and Europe [70–73]. Concentrations of mercury in eggs of Brünnich's guillemots from Svalbard [46] were higher than those reported for eggs of Atlantic puffins > herring gulls and black-legged kittiwakes from Northern Norway [45]. This result is unexpected, because Brünnich's guillemots and Atlantic puffins feed at a lower trophic level than do herring gulls and black-legged kittiwakes [3]. On average, Brünnich's guillemots and Atlantic puffins have a longer life span and a slower reproduction (lay only one egg) compared to herring gulls and black-legged kittiwakes [3,74–76]. It might be that the higher mercury concentrations in the former species are the result of the slow reproduction or age-specific accumulation, a phenomenon that has been found previously in birds [77]. Species-specific difference in the capacity to demethylate inorganic mercury to methyl mercury as observed in other birds [78] is another possible explanation for the higher accumulation of mercury in the eggs of Brünnich's guillemot and Atlantic puffin.

24.4.2.2 Temporal Trends

Temporal trends of mercury were not observed in eggs of herring gulls, black-legged kittiwakes, and Atlantic puffins sampled in Northern Norway in 1983, 1993, and 2003, nor in eggs of Brünnich's guillemots sampled at Svalbard in 2003 and 2007 (Figure 24.2) [45,46]. This is similar to findings in most temporal trend studies of mercury in seabirds from Europe. For example, concentrations of mercury in black guillemots (*Cepphus grylle*) and Brünnich's guillemots from Greenland have been stable for 150 years [79]. Also, no temporal trends were found in white-tailed sea eagles (*Haliaeetus albicilla*) from Northern Norway [80] nor glaucous gulls from Greenland [81] when comparing birds from the 1960s and 1980s, respectively, with birds until today. In contrast to the stable concentrations of mercury reported in the European Arctic, studies from the Canadian Arctic have provided compelling evidence for increasing trends of mercury in the environment [82]. For example, mercury concentrations nearly doubled between 1975 and 1998 in eggs of Brünnich's guillemots from the Canadian Arctic, whereas a 50% increase was observed in eggs of northern fulmars in the same time period [82].

24.4.3 BFRs

24.4.3.1 Species Comparison

BDE-28, BDE-47, BDE-100, BDE-99, BDE-154, BDE-153, and α-HBCD were detected in all of the 121 eggs of herring gulls, black-legged kittiwakes, Atlantic puffins, and Brünnich's guillemots from Northern Norway and Svalbard [46,48]. BDE-183 and BDE-209 were detected in 60% and 11% of the eggs of herring gulls, black-legged kittiwakes, and Atlantic puffins, respectively, whereas the nona-BDEs were detected in 27% of the eggs of herring gulls, black-legged kittiwakes, and Atlantic puffins. In eggs of Brünnich's guillemots, BDE-183 and nona-BDEs were not detected, whereas BDE-209 was detected in only 16% of the examined eggs. The β-HBCD and γ-HBCD diastereomers were below the limits of detection in all seabird eggs from Northern Norway and Svalbard.

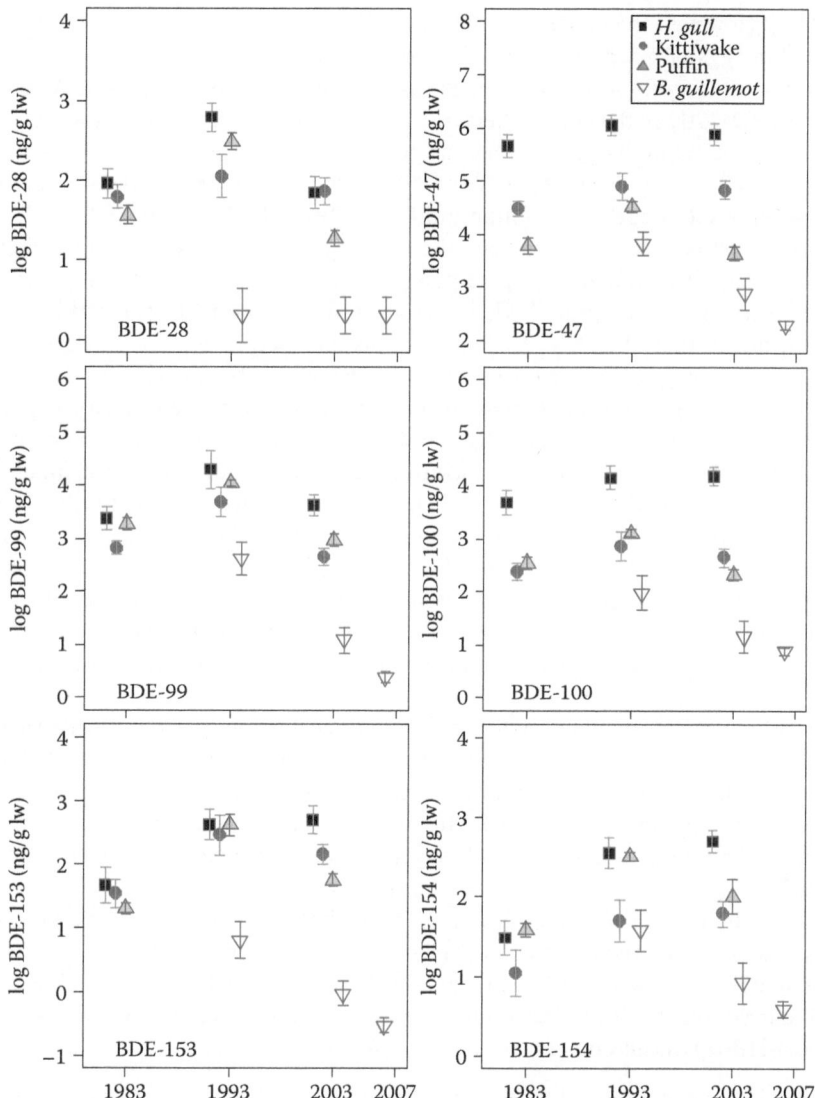

FIGURE 24.3 Temporal trends of polybrominated diphenyl (BDE)-28, 47, 99, 100, 153, and 154 (log ng/g lipid weight (lw)) in eggs of herring gulls (*Larus argentatus*), Atlantic puffins (*Fratercula arctica*) and black-legged kittiwakes (*Rissa tridactyla*) sampled in 1983, 1993, and 2003, and in eggs of Brünnich's guillemots (*Uria lomvia*) sampled in 1993, 2003, and 2007. (Modified from Helgason, L. et al., *Environ. Toxicol. Chem.*, 28, 1096, 2009; Miljeteig, C. and Gabrielsen, G.W., *Contaminants in Brünnich's Guillemots from Kongsfjorden and Bjørnøya in the Period from 1993 to 2007*, Norwegian Polar Institute, Tromsø, 2010. With permission.)

Overall, the concentrations of BDEs and α-HBCD in the seabird eggs from Northern Norway and Svalbard were approximately 20–90 times lower than those of PCBs (Figures 24.2 and 24.3) [46,48]. Although the concentrations of nona-BDEs and BDE-209 were low, some of the eggs of herring gulls, black-legged kittiwakes, and Atlantic puffins exhibited concentrations well above the highest levels reported in the procedural blanks, and demonstrated that nona-BDEs and BDE-209 were present in Arctic seabird eggs. Levels of BDEs in eggs of herring gulls, black-legged kittiwakes, Atlantic puffins, and Brünnich's guillemots from Northern Norway and Svalbard were lower than levels reported in eggs of several predatory birds from Norway, South Greenland, and Sweden

[23,35,36,83,84]. However, the concentrations of BDEs in eggs of Norwegian seabirds were higher than that found in eggs of black guillemots from east Greenland [85], ivory gull eggs from Canada [86], and Arctic tern (*Sterna paradisaea*) eggs from Norway and Svalbard [87]. Also, the levels of nona-BDEs and BDE-209 in the eggs of herring gulls, Atlantic puffins, and black-legged kittiwakes were higher than that found in eggs of glaucous gulls from Norway [36].

The HBCD levels in eggs of seabirds from Northern Norway and Svalbard were in the same range as those reported in eggs of common guillemots from the Baltic Sea [88] and eggs of glaucous gulls from Bjørnøya [36]. Furthermore, the HBCD levels in the Norwegian seabird eggs were somewhat lower than those found in eggs of peregrine falcons from Sweden [83], eggs of common terns (*Sterna hirundo*) from Belgium [39], and ivory gull eggs from Norway and Russia [23]. They were, however, higher than the HBCD concentrations reported in eggs of peregrine falcons from South Greenland [84], eggs of Arctic tern from Norway and Svalbard [87], eggs of ivory gulls from the Canadian Arctic [86], and eggs of tawny owls (*Strix aluco*) from Central Norway [89].

As with the OCs, the highest concentrations of BDEs were reported in eggs of herring gulls from Northern Norway, followed by black-legged kittiwakes > Atlantic puffins and Brünnich's guillemots from Northern Norway and Svalbard [46,48]. In contrast, the highest levels of HBCD were found in the eggs of black-legged kittiwakes, followed by herring gulls > Atlantic puffins > Brünnich's guillemots [46,48]. The finding that eggs of black-legged kittiwakes contain higher concentrations of HBCD compared to herring gull eggs is somewhat unexpected considering the fact that herring gulls feed at a higher trophic level. However, the differences might also be related to variations in HBCD contamination in winter feeding areas or alternatively to differences in biotransformation capacities.

BDE-47 was the most abundant BDE congener in seabird eggs from Northern Norway and Svalbard, constituting 35%–80% of sum BDEs, followed by BDE-99 and BDE-100 [46,48] (Figure 24.4). A relatively high contribution of BDE-47 was found in several other studies, e.g., in eggs of common guillemots from the Baltic Sea [88], herring gulls from the Great lakes [90], black guillemots from Greenland [85], and goshawks (*Accipiter gentilis*), white-tailed sea eagles and ospreys (*Pandion haliaetus*) from Norway [35].

α-HBCD was the only HBCD diastereomer found in the seabird eggs from Northern Norway [48]. This corresponds with the findings by Bæk et al. [91] and Tomy et al. [92] suggesting that only α-HBCD biomagnifies in the marine food web. Furthermore, α-HBCD is formed by isomerization of β-HBCD and γ-HBCD [93], which might explain why high levels of α-HBCD, but not of the β-HBCD and γ-HBCD diastereomers, are found in Arctic seabird eggs.

24.4.3.2 Temporal Trends

The relative contribution of BDEs and HBCD to sum BFRs in Norwegian seabird eggs differed between years and species (Figure 24.4). However, as a rule, the relative contribution of BDEs to sum BFRs decreased, whereas the relative contribution of HBCD to total BFRs increased from 1983/1993 until 2003/2007 (Figure 24.5).

On average, the total BDE concentrations in Atlantic puffin eggs from Northern Norway sampled in 1993 were higher than eggs from 1983. However, from 1993 to 2003, the total BDE concentration in Atlantic puffin eggs decreased [48]. Differences in total BDE concentrations comparing the other seabird eggs collected in 1983, 1993, and 2003 from Northern Norway were not found (Figure 24.3) [48], except in eggs of Brünnich's guillemots, in which BDE concentrations decreased from 1993 to 2003, after which they remained constant (Figure 24.3) [46].

The mean concentrations of the higher brominated BDEs (BDE-99, 153 and 154) in eggs of seabirds from Northern Norway were higher in 1993 than in 1983, except for the mean concentrations of BDE-154 in black-legged kittiwake eggs, which were stable during this period. BDE 99, BDE-153, and BDE-154 decreased from 1993 to 2003 in Atlantic puffin eggs, whereas these congeners remained constant during this time period in eggs of black-legged kittiwakes and herring gulls (except for a decrease in mean concentrations of BDE-99 in eggs of black-legged kittiwakes). No

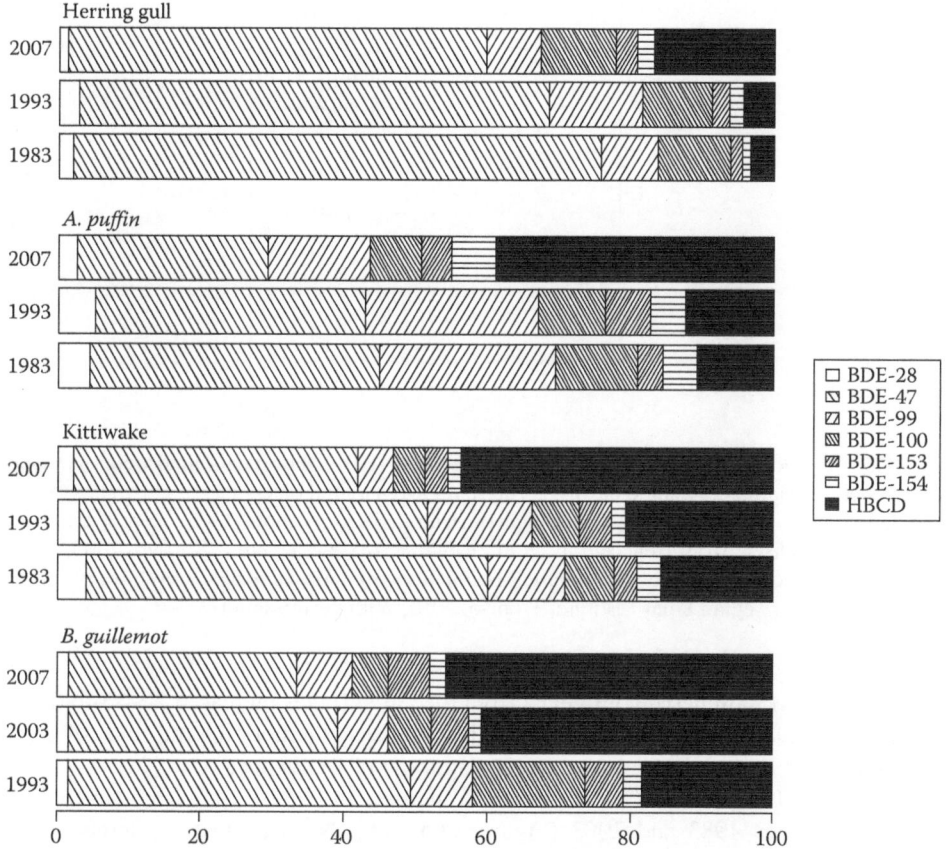

FIGURE 24.4 Polybrominated diphenyls (BDE) and hexabromocyclododecane (HCBD) congener distribution in eggs of herring gulls (*Larus argentatus*), Atlantic puffins (*Fratercula arctica*) and black-legged kittiwakes (*Rissa tridactyla*) sampled in 1983, 1993 and 2003, and in eggs of Brünnich's guillemots (*Uria lomvia*) sampled in 1993, 2003, and 2007. (Modified from Helgason, L. et al., *Environ. Toxicol. Chem.*, 28, 1096, 2009; Miljeteig, C. and Gabrielsen, G.W., *Contaminants in Brünnich's Guillemots from Kongsfjorden and Bjørnøya in the Period from 1993 to 2007*, Norwegian Polar Institute, Tromsø, 2010. With permission.)

temporal trends of the lower brominated BDEs (BDE-28, 47) were reported in black-legged kittiwake eggs from Norway, whereas mean concentrations of BDE-47 in Brünnich's guillemot eggs from Svalbard was lower in 2003 compared to 1993 [46,48]. In eggs of Atlantic puffins, the levels of lower-brominated BDEs peaked in 1993, and then dropped until 2003. Similarly, the mean concentrations of BDE-28 in herring gull eggs increased from 1983 to 1993, and decreased from 1993 to 2003 (Figure 24.3) [48].

Reports on temporal trends of BDEs in seabird eggs are scarce. However, the decreasing or stabilizing trends of BDEs found in seabird eggs from Northern Norway and Svalbard are similar to the findings in other European studies investigating temporal trends of BDEs in seabird eggs. For example, in a recent study (1969–2001) of tetra- and penta-BDEs in eggs of common guillemots from the Baltic Sea, upward trends were found in the 1970s and 1980s followed by a rapid decrease until 2001 [88]. In contrast, studies from the Great Lakes, Canadian Arctic, and Southern Greenland revealed increasing concentrations of BDEs from the 1980s up until today. In peregrine falcon eggs from Greenland, levels of BDE-99, BDE-100, BDE-153, and BDE-209 increased by 6%–10% from 1986 to 2003 [84]. Also, concentrations of BDEs in ivory gull eggs from Canada steadily increased between 1976 and 2004 [86]. Moreover, mean concentrations of BDE-28, BDE-99, BDE-100,

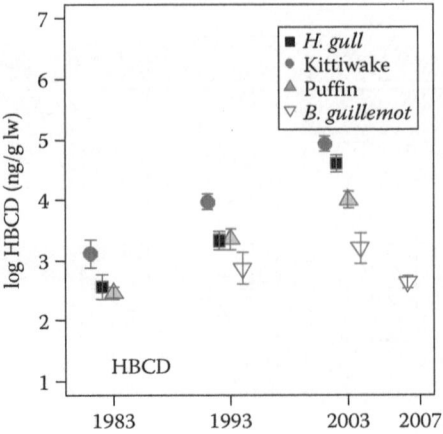

FIGURE 24.5 Temporal trends of hexabromocyclododecane (HBCD) (log ng/g lipid weight) in eggs of herring gulls (*Larus argentatus*), Atlantic puffins (*Fratercula arctica*) and black-legged kittiwakes (*Rissa tridactyla*) sampled in 1983, 1993, and 2003, and in eggs of Brünnich's guillemots (*Uria lomvia*) sampled in 1993, 2003, and 2007. (Modified from Helgason, L. et al., *Environ. Toxicol. Chem.*, 28, 1096, 2009; Miljeteig, C. and Gabrielsen, G.W., *Contaminants in Brünnich's Guillemots from Kongsfjorden and Bjørnøya in the Period from 1993 to 2007*, Norwegian Polar Institute, Tromsø, 2010. With permission.)

BDE-153, and BDE-154 in herring gull eggs from the Great Lakes increased 20–75-fold during the same time period (1981–2000) [90]. Contrary to this, in a recent temporal trend study of herring gull eggs from the Great Lakes, there was no increasing trend of BDEs post-2000 [94].

Unlike OCs and BDEs, α-HBCD was found to increase in seabird eggs from Northern Norway and Svalbard between 1983 and 2003 (Figure 24.5) [48]. No significant difference was found between the mean concentrations of HBCD measured in Brünnich's guillemot eggs from 2003 to 2007 (Figure 24.5) [46].

The results from the temporal trend study of HBCD in Norwegian seabird eggs contradict two other temporal trend studies of HBCD, one from the European Arctic and one study from the Canadian Arctic, i.e., on common guillemot eggs in the Baltic Sea (1969–2001) [88], peregrine falcon eggs from South Greenland (1986–2003) [84], and ivory gull eggs from the Canadian Arctic (1976–2004) [95]. In eggs of common guillemots, the highest HBCD concentrations were reported in the mid-1970s, followed by a gradual decrease from the mid-1970s to the 1980s then another increase during the latter part of the 1980s and finally stabilization until 2001 [88]. In peregrine falcon eggs, a downward trend in total HBCD concentrations was found for the entire study period [84], whereas concentrations of HBCD decreased from 1976 to 2004 in ivory gull eggs [95].

24.4.4 PFAS

24.4.4.1 Species Comparison

A total of 55 eggs from herring gulls and Brünnich's guillemots sampled in Northern Norway and Svalbard in 1983, 1993, 2003, and 1993, 2003 and 2007, respectively, were analyzed for perfluorosulfonates and perfluorocarboxylic acids [46,47]. Overall, the levels of PFAS were slightly (27%) higher in eggs of herring gulls compared to eggs of Brünnich's guillemots (Figure 24.6). In both species, PFOS was the most abundant PFAS with concentrations ranging from 7 to 52 ng/g ww, followed by PFUnA (perfluoroundecanoic acid) and PFTriA (perfluorotridecanoic acid) (0.3–8 ng/g ww and 0.4–9 ng/g ww, respectively). Together, these three PFAS contributed to 80%–90% of sum PFAS [46,47] (Figure 24.7). The PFAS concentrations in eggs from herring gulls and Brünnich's guillemots in Northern Norway and Svalbard were generally lower than the levels reported for

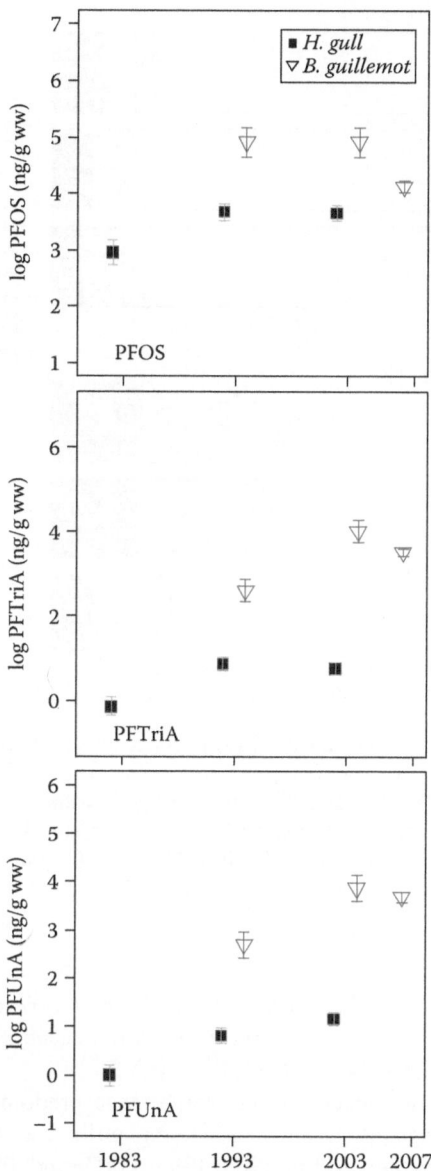

FIGURE 24.6 Temporal trends of perfluorooctane sulfonate (PFOS), perfluoroundecanoic acid (PFUnA) and perfluorotridenanoic acid (PFTriA) (log ng/g wet weight (ww)) in eggs of herring gulls (*Larus argentatus*) sampled in 1983, 1993, and 2003, and in eggs of Brünnich's guillemots (*Uria lomvia*) sampled in 1993, 2003, and 2007. (Reprinted with permission from Verreault, J. et al., *Environ. Sci. Technol.*, 41, 6671, 2007. Copyright 2007 American Chemical Society. Miljeteig, C. and Gabrielsen, G.W., *Contaminants in Brünnich's Guillemots from Kongsfjorden and Bjørnøya in the Period from 1993 to 2007*, Norwegian Polar Institute, Tromsø, 2010. With permission.)

common guillemot eggs from the Baltic Sea [96], glaucous gull eggs from Bjørnøya [19], and ivory gull eggs from Norway and Russia [23].

24.4.4.2 Temporal Trends

The concentrations of the different PFAS congeners in eggs of herring gulls and Brünnich's guillemots relative to the total concentration of PFAS varied between years and species (Figure 24.7).

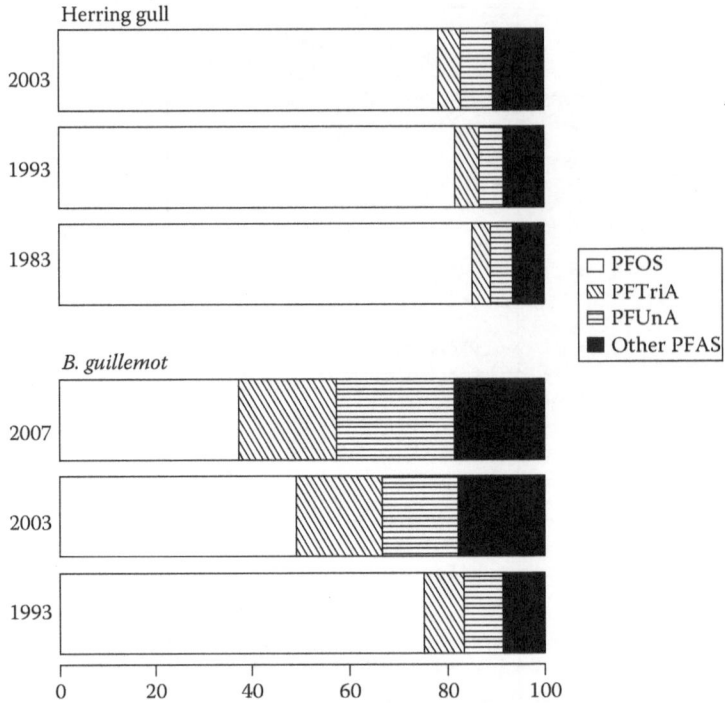

FIGURE 24.7 Perfluorinated alkyl substances (PFAS) congener distribution (perfluorooctane sulfonate [PFOS], perfluoroundecanoic acid [PFUnA], and perfluorotridenanoic acid [PFTriA]) in eggs of herring gulls (*Larus argentatus*) sampled in 1983, 1993, and 2003, and in eggs of Brünnich's guillemots (*Uria lomvia*) sampled in 1993, 2003, and 2007. (Reprinted with permission from Verreault, J. et al., *Environ. Sci. Technol.*, 41, 6671, 2007. Copyright 2007. American Chemical Society. Miljeteig, C. and Gabrielsen, G.W., *Contaminants in Brünnich's Guillemots from Kongsfjorden and Bjørnøya in the Period from 1993 to 2007*, Norwegian Polar Institute, Tromsø, 2010. With permission.)

The proportion of PFOS declined between 1983/1993 and 2003/2007 in eggs of both species. In the same time period, the proportions of PFTriA and PFUnA increased in Brünnich's guillemot eggs, but not in herring gull eggs (Figure 24.7) [46,47].

Temporal trends of the mean concentrations of the three predominant PFAS species (PFOS, PFUnA, PFTriA) were investigated in eggs of herring gulls and Brünnich's guillemots from Northern Norway and Svalbard sampled in 1983, 1993, 2003, and 1993, 2003 and 2007, respectively. In eggs of herring gulls, mean concentrations of PFOS, PFUnA, and PFTriA increased between 1983 and 1993, but were stabilized between 1993 and 2003 [47]. In contrast, the mean concentrations of PFUnA and PFTriA in Brünnich's guillemot eggs increased from 1993 to 2003, and whereas PFUnA was stabilized in the period between 2003 and 2007, the PFTriA decreased in the same time period [46]. The most pronounced temporal trend of PFAS concentrations was the decreasing trend of PFOS in the eggs of Brünnich's guillemots from Svalbard, comparing mean concentrations of PFOS in the period from 2003 to 2007 [46]. The results of the temporal trend studies of PFOS in seabird eggs from Northern Norway and Svalbard differ from those reported by Holmström et al. [96] who detected a 30-fold increase in mean concentrations of PFOS in eggs of common guillemots from the Baltic Sea between 1968 and 2003. Also, in Brünnich's guillemots and northern fulmars (liver) from Prince Leopold Island in the Canadian Arctic, PFAS concentrations were found to increase significantly from 1975 to 2003–2004 [97]. Nevertheless, downward trends of PFOS concentrations, as found in seabird eggs from Northern Norway and Svalbard, have been found in other Arctic animals [98].

24.5 CONCLUSIONS

In summary, the OC levels decreased in seabird eggs from Northern Norway and Svalbard between the mid-1980s and mid-2000s. Similar trends are demonstrated for other animals [5,67,68] and other seabirds [56,69], suggesting reduced exposure to these compounds in recent years. Most OCs (DDTs, HCB, chlordanes, PCBs, HCHs) were produced and released into the environment in the 1930s–1990s, reaching peak levels between 1970s and 1990s [5]. Although current OC concentrations in seabird eggs from Northern Norway and Svalbard are lower than previously recorded, they still are much higher than the concentrations of BFRs and PFAS, and account for the majority of the "total" contaminant load (Figure 24.8). This illustrates that it takes a long time from when the production of a contaminant is stopped until it is eliminated from the environment.

BDEs are the second most important contaminant group in the seabird eggs from Northern Norway and Svalbard. The concentrations of BDEs in seabird eggs from Northern Norway and Svalbard increased between 1983 and 1993, after which the concentrations leveled off. In contrast

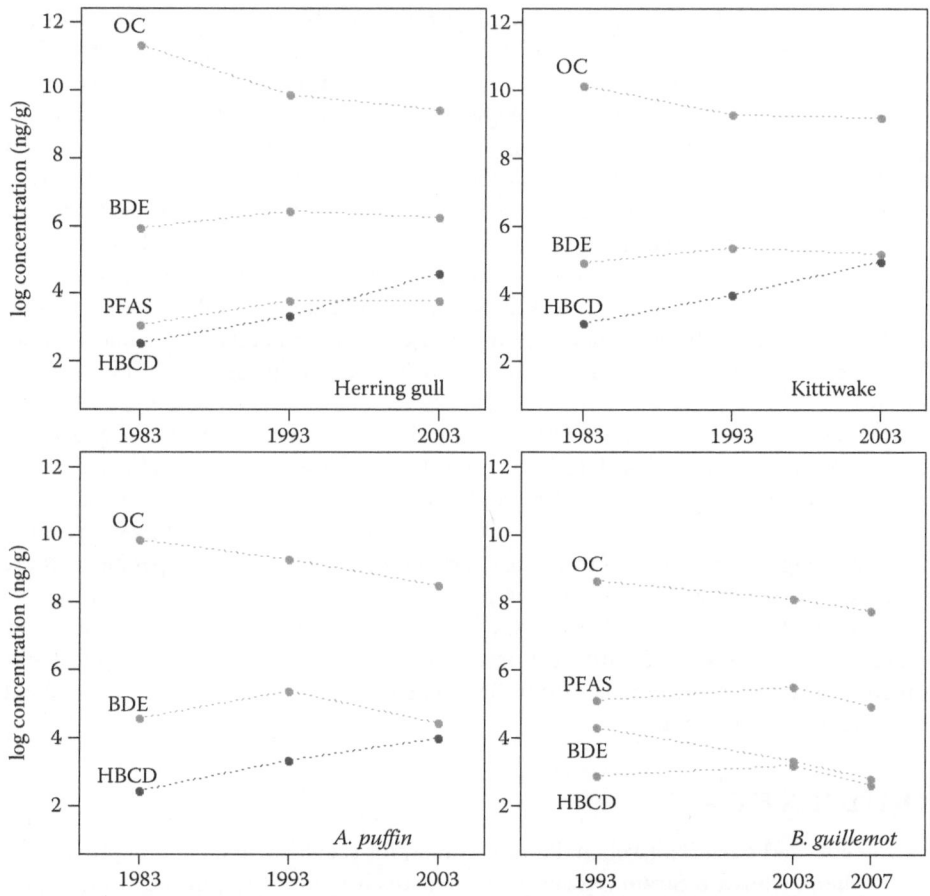

FIGURE 24.8 Temporal trends of organochlorines (OCs), hexabromocyclododecane (HBCD), polybrominated diphenyls (BDEs) (log ng/g lipid weight (lw)) and perfluorinated alkyl substances (PFAS) (log ng/g wet weight (ww)) in eggs of herring gulls (*Larus argentatus*), black-legged kittiwakes (*Rissa tridactyla*), and Atlantic puffins (*Fratercula arctica*) sampled in 1983, 1993, and 2003. (Modified from Verreault et al. [47] and Helgason et al. [45,48]) and eggs of Brünnich's guillemots (*Uria lomvia*) sampled in 1993, 2003 and 2007 (Modified from Miljeteig, C. and Gabrielsen, G.W., *Contaminants in Brünnich's Guillemots from Kongsfjorden and Bjørnøya in the Period from 1993 to 2007*, Norwegian Polar Institute, Tromsø, 2010.)

to BDEs, the levels of HBCD increased in all species from 1983 to 2003. However, after 2003, the levels of HBCD were found to be stabilizing in eggs of Brünnich's guillemots from Svalbard. The concentrations of PFAS were found to be increasing from 1983 to 1993 in eggs of herring gulls from Northern Norway, after which concentrations were found to be stabilizing or decreasing after 1993 and 2003 in eggs of herring gulls from Northern Norway and eggs of Brünnich's guillemots from Svalbard, respectively. As with the OCs, the "new" and "emerging" contaminants, such as the BDEs, HBCD, and PFAS, in Arctic seabird eggs mirror the production trends, and as such, is a "footprint" of the global contaminant release. There has been a notable increase in the production of consumer products the last 50 years. In the period from 1960 to 2008, consumption expenditures per person increased by as much as 150% [99]. Many of these products contain flame retardants and fluorinated compounds. The production of BFRs and PFAS peaked in the 1980s and 1990s and declined gradually thereafter when restrictions were put in place and global awareness of the adverse effects these contaminants exert on humans and animals increased [5,34,100].

The levels of contaminants in seabirds from Northern Norway and Svalbard are relatively high, especially in top predators such as the glaucous gull and the herring gull. Since 1986, scientists from the Norwegian Polar Institute have found glaucous gulls dead or dying in convulsion in seabird colonies along the southern coast of Bjørnøya (Svalbard; Figure 24.1). In 1989, 12 of these dead or dying birds were analyzed for environmental contaminants [21]. The analyses revealed very high levels of POPs. Although a toxicological evaluation of the POP levels was difficult, it was suggested that these high levels of contaminants had contributed to the death of the glaucous gulls [21]. When the study was repeated in 2003–2005, very high levels of POPs were again found in the corpses of dead and dying glaucous gulls from Bjørnøya [101]. In fact, these studies reported some of the highest POP concentrations ever recorded for any seabird. Most of the dead birds included in these studies were severely emaciated when found. During emaciation, lipids are mobilized and contaminants accumulated in fat tissue redistributed to vital organs such as the liver and brain. Observations of dying glaucous gulls on Bjørnøya with apparently abnormal behavior suggest that high levels of contaminants could contribute to the birds' death, either directly or indirectly [101].

Mean wet weight concentrations of PCBs in seabird eggs from Northern Norway and Svalbard were well below the no-observed-effect level (NOEL) and lowest-observed-effect level (LOEL) for reproductive effects in eggs of fish-eating and scavenging birds [5]. Nevertheless, the levels of POPs in Norwegian seabird eggs are so high that the Norwegian government has advised to limit the intake of gull eggs to an absolute minimum. In particular, children, young people, and pregnant and nursing women should abstain from eating gull eggs [25].

Temporal trend studies conducted in Northern Norway and Svalbard have illustrated that the contaminant concentrations in seabird eggs reflect global production rates, use, and legislation, and that there is a considerable time lag from the moment when a contaminant is banned until it is eliminated from the environment.

ACKNOWLEDGMENTS

We thank Nanette Verboven (Norwegian Polar Institute) for scientific input and improvement of the text. We also thank Hallvard Strøm (Norwegian Polar Institute) for insightful comments, and Jorg Welcher and Ingeborg Hallanger (Norwegian Polar Institute) for help with the data presentations. Last but not least, we thank all our coauthors on the original publications; Rob Barrett (Tromsø University Museum), Elisabeth Lie (Norwegian School of Veterinary Science), Anuschka Polder (Norwegian School of Veterinary Science), Janneche U. Skaare (National Veterinary Institute), Siri Føreid (Norwegian School of Veterinary Science), Kine Bæk (Norwegian School of Veterinary Science), Jonathan Verreault (Norwegian Polar Institute), Urs Berger (University of Stockholm), and Cecilie Miljeteig (Norwegian Polar Institute).

REFERENCES

1. Barrett, R., Lorentsen, S-H., and Anker-Nilssen, T. 2006. The status of breeding seabirds in mainland Norway. *Atlantic Seabirds* 8: 97–126.

2. Strøm, H. 2006. Birds of Svalbard. In: *Birds and Mammals of Svalbard*, eds. K.M. Kovacs and C. Lydersen. Tromsø, Norway: Norwegian Polar Institute, pp. 86–191.

3. Anker-Nilssen, T., Bakken, V., Strøm, H., Golovkin, A.N., Bianki, V.V., and Tatarinkova, I.P. 2000. The status of marine birds breeding in the Barents Sea Region. Rapport nr. 113. Tromsø, Norway: Norsk Polarinstitutt.

4. Gabrielsen, G. and Sydnes, L. 2009. Seabirds. In: *Ecosystem Barents Sea*, eds. E. Sakshaug, G. Johnsen, and K.M. Kovacs. Trondheim, Norway: Tapir Academic Press, pp. 497–544, Chapter 17.

5. de Wit, C., Fisk, A., Hobbs, K., Muir, D., Gabrielsen, G., Kallenborn, R., Krahn, M., Norstrom, R., and Skaare, J. 2004. *AMAP Assessment 2002. Persistent Organic Pollutants in the Arctic*. Oslo, Norway: Arctic Monitoring and Assessment Programme (AMAP).

6. Letcher, R., Bustnes, J., Dietz, R. et al. 2009. Effects assessment of persistent organohalogen contaminants in Arctic wildlife and fish. *Science of the Total Environment* 408: 2995–3043.

7. Burkow, I.C. and Kallenborn, R. 2000. Sources and transport of persistent pollutants to the Arctic. *Toxicology Letters* 112–113: 87–92.

8. Bjerk, J. and Holt, G. 1971. Residues of DDE and PCB in eggs from herring gull (*Larus argentatus*) and common gull (*Larus canus*) in Norway. *Acta Veterinaria Scandinavica* 12: 429–441.

9. Bourne, W.R.P. and Bogan, J.A. 1972. Polychlorinated biphenyls in North Atlantic seabirds. *Marine Pollution Bulletin* 3: 171–175.

10. Gabrielsen, G. 2007. Levels and effects of persistent organic pollutants in arctic animals. In: *Arctic Alpine Ecosystems and People in a Changing Environment*, eds. J.B. Ørbæk, S. Falk-Petersen, E.N. Hegseth, A.H. Hoel, R. Kallenborn, and I. Tombre. Berlin, Heidelberg, Germany: Springer-Verlag, pp. 377–412, Chapter 20.

11. Hop, H., Borga, K., Gabrielsen, G., Kleivane, L., and Skaare, J. 2002. Food web magnification of persistent organic pollutants in poikilotherms and homeotherms. *Environmental Science & Technology* 36: 2589–2597.

12. Fisk, A.T., Hobson, K.A., and Norstrom, R.J. 2001. Influence of chemical and biological factors on trophic transfer of persistent organic pollutants in the Northwater Polynya marine food web. *Environmental Science & Technology* 35: 732–738.

13. Borga, K., Gabrielsen, G.W., and Skaare, J.U. 2001. Biomagnification of organochlorines along a Barents Sea food chain. *Environmental Pollution* 113: 187–198.

14. Borga, K., Fisk, A.T., Hargrave, B., Hoekstra, P.F., Swackhamer, D., and Muir, D.C.G. 2005. Bioaccumulation factors for PCBs revisited. *Environmental Science & Technology* 39: 4523–4532.

15. Gobas, F. and Morrison, H. 2000. *Bioconcentration and Biomagnification in the Aquatic Environment*. Boca Raton, FL: Lewis.

16. Walker, C.H. 1990. Persistent pollutants in fish-eating sea birds—Bioaccumulation, metabolism and effects. *Aquatic Toxicology* 17: 293–324.

17. Henriksen, E.O., Gabrielsen, G.W., Trudeau, S., Wolkers, J., Sagerup, K., and Skaare, J.U. 2000. Organochlorines and possible biochemical effects in glaucous gulls (*Larus hyperboreus*) from Bjornoya, the Barents Sea. *Archives of Environmental Contamination and Toxicology* 38: 234–243.

18. Borga, K., Fisk, A.T., Hoekstra, P.F., and Muir, D.C.G. 2004. Biological and chemical factors of importance in the bioaccumulation and trophic transfer of persistent organochlorine contaminants in arctic marine food webs. *Environmental Toxicology and Chemistry* 23: 2367–2385.

19. Verreault, J., Houde, M., Gabrielsen, G.W., Berger, U., Haukas, M., Letcher, R.J., and Muir, D.C.G. 2005. Perfluorinated alkyl substances in plasma, liver, brain, and eggs of glaucous gulls (*Larus hyperboreus*) from the Norwegian Arctic. *Environmental Science & Technology* 39: 7439–7445.

20. Verreault, J., Letcher, R.J., Muir, D.C.G., Chu, S.G., Gebbink, W.A., and Gabrielsen, G.W. 2005. New organochlorine contaminants and metabolites in plasma and eggs of glaucous gulls (*Larus hyperboreus*) from the Norwegian Arctic. *Environmental Toxicology and Chemistry* 24: 2486–2499.

21. Gabrielsen, G.W., Skaare, J.U., Polder, A., and Bakken, V. 1995. Chlorinated hydrocarbons in glaucous gulls (*Larus hyperboreus*) in the southern part of Svalbard. *Science of the Total Environment* 160–161: 337–346.

22. Verreault, J., Gabrielsen, G.W., Chu, S.G. et al. 2005. Flame retardants and methoxylated and hydroxylated polybrominated diphenyl ethers in two Norwegian Arctic top predators: Glaucous gulls and polar bears. *Environmental Science & Technology* 39: 6021–6028.

23. Miljeteig, C., Strøm, H., Gavrilo, M.V., Volkov, A., Jenssen, B.M., and Gabrielsen, G.W. 2009. High levels of contaminants in ivory gull *Pagophila eburnea* eggs from the Russian and Norwegian Arctic. *Environmental Science & Technology* 43: 5521–5528.

24. Helberg, M., Bustnes, J.O., Erikstad, K.E., Kristiansen, K.O., and Skaare, J.U. 2005. Relationships between reproductive performance and organochlorine contaminants in great black-backed gulls (*Larus marinus*). *Environmental Pollution* 134: 475–483.

25. Pusch, K., Schlabach, M., Prinzinger, R., and Gabrielsen, G. 2005. Gull eggs—Food of high organic pollutant content? *Journal of Environmental Monitoring* 7: 635–639.

26. Knudsen, L.B., Borga, K., Jorgensen, E.H. et al. 2007. Halogenated organic contaminants and mercury in northern fulmars (*Fulmarus glacialis*): Levels, relationships to dietary descriptors and blood to liver comparison. *Environmental Pollution* 146: 25–33.

27. Bustnes, J., Borgå, K., Erikstad, K., Lorentsen, S-H., and Herzke, D. 2008. Perfluorinated, brominated, and chlorinated contaminants in a population of lesser black-backed gulls (*Larus fuscus*). *Environmental Toxicology and Chemistry* 27: 1383–1392.

28. Fisk, A.T., de Wit, C.A., Wayland, M. et al. 2005. An assessment of the toxicological significance of anthropogenic contaminants in Canadian arctic wildlife. *Science of the Total Environment* 351: 57–93.

29. Vanden Heuvel, J.P., Kuslikis, B.I., and Peterson, R.E. 1992. Covalent binding of perfluorinated fatty acids to proteins in the plasma, liver and testes of rats. *Chemico-Biological Interactions* 82: 317–328.

30. Paul, D.J., Wenyue, H., Wim, De C., John, L.N., and John, P.G. 2003. Binding of perfluorinated fatty acids to serum proteins. *Environmental Toxicology and Chemistry* 22: 2639–2649.

31. Haukås, M., Berger, U., Hop, H., Gulliksen, B., and Gabrielsen, G.W. 2007. Bioaccumulation of per- and polyfluorinated alkyl substances (PFAS) in selected species from the Barents Sea food web. *Environmental Pollution* 148: 360–371.

32. Verreault, J., Gabrielsen, G., and Bustnes, J. 2010. The Svalbard glaucous gull as bioindicator species in the European Arctic: Insight from 35 years of contaminants research. *Reviews of Environmental Contamination & Toxicology* 205: 77–116.

33. Miljeteig, C., Strom, H., Gavrilo, M., Skaare, J.U., Jenssen, B.M., and Gabrielsen, G. 2007. *Organohalogens and Mercury in Ivory Gull Eggs*. Norwegian Polar Institute Brief Report Series No 7. Oslo, Norway: Climate and Pollution Agency (ISBN 978-82-7666-245-0).

34. Paul, A.G., Jones, K.C., and Sweetman, A.J. 2009. A first global production, emission and environmental inventory for perfluorooctane sulfonate. *Environmental Science & Technology* 43: 386–392.

35. Herzke, D., Berger, U., Kallenborn, R., Nygard, T., and Vetter, W. 2005. Brominated flame retardants and other organobromines in Norwegian predatory bird eggs. *Chemosphere* 61: 441–449.

36. Verreault, J., Gebbink, W.A., Gauthier, L.T., Gabrielsen, G.W., and Letcher, R.J. 2007. Brominated flame retardants in glaucous gulls from the Norwegian Arctic: More than just an issue of polybrominated diphenyl ethers. *Environmental Science & Technology* 41: 4925–4931.

37. Birnbaum, L.S. and Staskal, D.F. 2004. Brominated flame retardants: Cause for concern? *Environmental Health Perspectives* 112: 9–17.

38. Alaee, M., Arias, P., Sjödin, A., and Bergman, Å. 2003. An overview of commercially used brominated flame retardants, their applications, their use patterns in different countries/regions and possible modes of release. *Environment International* 29: 683–689.

39. Morris, S., Allchin, C.R., Zegers, B.N. et al. 2004. Distribution and fate of HBCD and TBBPA brominated flame retardants in north sea estuaries and aquatic food webs. *Environmental Science & Technology* 38: 5497–5504.

40. Zegers, B.N., Lewis, W.E., Booij, K. et al. 2003. Levels of polybrominated diphenyl ether flame retardants in sediment cores from Western Europe. *Environmental Science & Technology* 37: 3803–3807.

41. Haukås, M., Hylland, K., Berge, J.A., Nygård, T., and Mariussen, E. 2009. Spatial diastereomer patterns of hexabromocyclododecane (HBCD) in a Norwegian fjord. *Science of the Total Environment* 407: 5907–5913.

42. Lovdata. 2004. *FOR 2004–06–01 nr 922: Forskrift om begrensning i bruk av helse- og miljøfarlige kjemikalier og andre produkter (produktforskriften)*. Oslo, Norway: Ministry of Environment.

43. EU. 2003. *Directive 2002/95/EC of the European Parliament of the Council—27 January 2003 on the Restriction of the Use of Certain Substances in Electrical and Electronic Equipment*. Brussels, Belgium: European Union.

44. EU. 2006. *Directive 2006/122/EC of the European Parliament and of the Council—12 December 2006*. Brussels, Belgium: European Union.

45. Helgason, L., Barrett, R., Lie, E., Polder, A., Skaare, J., and Gabrielsen, G. 2008. Levels and temporal trends (1983–2003) of organochlorines (OCs) and mercury (Hg) in seabird eggs from Northern Norway. *Environmental Pollution* 155: 190–198.

46. Miljeteig, C. and Gabrielsen, G.W. 2010. *Contaminants in Brünnich's Guillemots from Kongsfjorden and Bjørnøya in the Period from 1993 to 2007.* Kortrapport/Brief Report Series 16. Tromsø, Norway: Norwegian Polar Institute (ISBN: 978-82-7666-268-9).

47. Verreault, J., Berger, U., and Gabrielsen, G.W. 2007. Trends of perfluorinated alkyl substances in herring gull eggs from two coastal colonies in Northern Norway (1983–2003). *Environmental Science & Technology* 41: 6671–6677.

48. Helgason, L., Polder, A., Føreid, S., Bæk, K., Lie, E., Gabrielsen, G., Barrett, R., and Skaare, J. 2009. Levels and temporal trends (1983–2003) of polybrominated diphenyl ethers and hexabromocyclododecanes in seabird eggs from North Norway. *Environmental Toxicology and Chemistry* 28:1096–1103.

49. Fimreite, N., Brun, E., Froeslie, A., Frederichsen, P., and Gundersen, N. 1974. Mercury in eggs of Norwegian seabirds. *Astarte* 7: 71–75.

50. Fimreite, N., Bjerk, J., Kveseth, N., and Brun, E. 1977. DDE and PCBs in eggs of Norwegian seabirds. *Astarte* 10: 15–20.

51. Fimreite, N., Kveseth, N., and Brevik, E.M. 1980. Mercury, DDE, and PCBs in eggs from a Norwegian gannet colony. *Bulletin of Environmental Contamination and Toxicology* 24: 142–144.

52. Barrett, R.T., Skaare, J.U., and Gabrielsen, G.W. 1996. Recent changes in levels of persistent organochlorines and mercury in eggs of seabirds from the Barents sea. *Environmental Pollution* 92: 13–18.

53. Barrett, R.T., Skaare, J.U., Norheim, G., Vader, W., and Froslie, A. 1985. Persistent organochlorines and mercury in eggs of Norwegian seabirds 1983. *Environmental Pollution Series A—Ecological and Biological* 39: 79–93.

54. Mineau, P., Fox, G., Norstrom, R., Weseloh, D., Hallett, D., and Ellenton, J. 1984. Using the herring gull to monitor levels and effects of organochlorine contamination in the Canadian Great Lakes. In: *Toxic Contaminants in the Great Lakes*, ed. J.O. Nriagu. New York: John Wiley & Sons, pp. 425–452.

55. Verboven, N., Verreault, J., Letcher, R.J., Gabrielsen, G.W., and Evans, N.P. 2009. Differential investment in eggs by arctic-breeding glaucous gulls (*Larus hyperboreus*) exposed to persistent organic pollutants. *Auk* 126: 123–133.

56. Braune, B.M., Donaldson, G.M., and Hobson, K.A. 2001. Contaminant residues in seabird eggs from the Canadian Arctic. Part I. Temporal trends 1975–1998. *Environmental Pollution* 114: 39–54.

57. Braune, B.M., Donaldson, G.M., and Hobson, K.A. 2002. Contaminant residues in seabird eggs from the Canadian Arctic. II. Spatial trends and evidence from stable isotopes for intercolony differences. *Environmental Pollution* 117: 133–145.

58. Custer, T.W., Pendleton, G., and Ohlendorf, H.M. 1990. Within-clutch and among-clutch variation of organochlorine residues in eggs of black-crowned night-herons. *Environmental Monitoring and Assessment* 15: 83–89.

59. Van den Steen, E., Eens, M., Jaspers, V.L.B., Covaci, A., and Pinxten, R. 2009. Effects of laying order and experimentally increased egg production on organic pollutants in eggs of a terrestrial songbird species, the great tit (*Parus major*). *Science of the Total Environment* 407: 4764–4770.

60. Pastor, D., Jover, L., Ruiz, X., and Albaiges, J. 1995. Monitoring organochlorine pollution in Audouins gull eggs—The relevance of sampling procedures. *Science of the Total Environment* 162: 215–223.

61. Mineau, P. 1982. Levels of major organochlorine contaminants in sequentially laid herring gull eggs. *Chemosphere* 11: 679–685.

62. Verreault, J., Villa, R.A., Gabrielsen, G.W., Skaare, J.U., and Letcher, R.J. 2006. Maternal transfer of organohalogen contaminants and metabolites to eggs of Arctic-breeding glaucous gulls. *Environmental Pollution* 144: 1053–1060.

63. Gabrielsen, G.W. Unpublished data. Polar Environmental Centre, 9296 Tromsø, Norwegian Polar Institute.

64. Steffen, C., Borga, K., Skaare, J.U., and Bustnes, J.O. 2006. The occurrence of organochlorines in marine avian top predators along a latitudinal gradient. *Environmental Science & Technology* 40: 5139–5146.

65. Evenset, A., Caroll, J., Christensen, G.N., Kallenborn, R., Gregor, D., and Gabrielsen, G.W. 2007. Seabird guano is an efficient conveyer of persistent organic pollutants (POPs) to arctic lake ecosystems. *Environmental Science & Technology* 41: 1173–1179.

66. Borga, K., Wolkers, H., Skaare, J.U., Hop, H., Muir, D.C.G., and Gabrielsen, G.W. 2005. Bioaccumulation of PCBs in Arctic seabirds: Influence of dietary exposure and congener biotransformation. *Environmental Pollution* 134: 397–409.

67. Wolkers, H., Krafft, B., van Bavel, B., Helgason, L.B., Lydersen, C., and Kovacs, K.M. 2008. Biomarker responses and decreasing contaminant levels in ringed seals (*Pusa hispida*) from Svalbard, Norway. *Journal of Toxicology and Environmental Health, Part A: Current Issues* 71: 1009–1018.

68. Muir, D.C.G., Wagemann, R., Hargrave, B.T., Thomas, D.J., Peakall, D.B., and Norstrom, R.J. 1992. Arctic marine ecosystem contamination. *Science of the Total Environment* 122: 75–134.

69. Braune, B.M. and Simon, M. 2003. Dioxins, furans, and non-ortho PCBs in Canadian Arctic seabirds. *Environmental Science & Technology* 37: 3071–3077.

70. Evers, D.C., Taylor, K.M., Major, A., Taylor, R.J., Poppenga, R.H., and Scheuhammer, A.M. 2003. Common loon eggs as indicators of methylmercury availability in North America. *Ecotoxicology* 12: 69–81.

71. Evers, D.C., Burgess, N.M., Champoux, L., Hoskins, B., Major, A., Goodale, W.M., Taylor, R.J., Poppenga, R., and Daigle, T. 2005. Patterns and interpretation of mercury exposure in freshwater avian communities in northeastern North America. *Ecotoxicology* 14: 193–221.

72. Scheuhammer, A.M., Perrault, J.A., and Bond, D.E. 2001. Mercury, methylmercury, and selenium concentrations in eggs of common loons (*Gavia immer*) from Canada. *Environmental Monitoring and Assessment* 72: 79–94.

73. Borga, K., Campbell, L., Gabrielsen, G.W., Norstrom, R.J., Muir, D.C.G., and Fisk, A.T. 2006. Regional and species specific bioaccumulation of major and trace elements in Arctic seabirds. *Environmental Toxicology & Chemistry* 25: 2927–2936.

74. Hatch, S.A., Robertson, G.J., and Baird, P.H. 2009. Black-legged Kittiwake (*Rissa tridactyla*). In: *The Birds of North America Online*, ed. A. Poole. Ithaca, NY: Cornell Lab of Ornithology. Retrieved from the Birds of North America Online http://bna.birds.cornell.edu/bna/species/092

75. Gaston, A. and Hipfner, J. 2000. Thick-billed Murre (*Uria lomvia*). In: *The Birds of North America Online*, ed. A. Poole. Ithaca, NY: Cornell Lab of Ornithology. Retrieved from the Birds of North America Online http://bna.birds.cornell.edu/bna/species/497

76. Pierotti, R. and Good, T. 1994. Herring gull (*Larus argentatus*). In: *The Birds of North America Online*, ed. A. Poole. Ithaca, NY: Cornell Lab of Ornithology. Retrieved from the Birds of North America Online http://bna.birds.cornell.edu/bna/species/124

77. Wenzel, C. and Gabrielsen, G.W. 1995. Trace-element accumulation in 3 seabird species from Hornoya, Norway. *Archives of Environmental Contamination and Toxicology* 29: 198–206.

78. Jæger, I., Hop, H., and Gabrielsen, G.W. 2009. Biomagnification of mercury in selected species from an Arctic marine food web in Svalbard. *Science of the Total Environment* 407: 4744–4751.

79. Appelquist, H., Drabaek, I., and Asbirk, S. 1985. Variation in mercury content of guillemot feathers over 150 years. *Marine Pollution Bulletin* 16: 244–248.

80. AMAP. 2005. *AMAP Assessment 2002: Heavy Metals in the Arctic*. Oslo, Norway: Arctic Monitoring and Assessment Programme (AMAP).

81. Riget, F. and Dietz, R. 2000. Temporal trends of cadmium and mercury in Greenland marine biota. *Science of the Total Environment* 245: 49–60.

82. Braune, B.M., Outridge, P.M., and Fisk, A.T. et al. 2005. Persistent organic pollutants and mercury in marine biota of the Canadian Arctic: An overview of spatial and temporal trends. *Science of the Total Environment* 351: 4–56.

83. Lindberg, P., Sellstrom, U., Haggberg, L., and de Wit, C.A. 2004. Higher brominated diphenyl ethers and hexabromocyclododecane found in eggs of peregrine falcons (*Falco peregrinus*) breeding in Sweden. *Environmental Science & Technology* 38: 93–96.

84. Vorkamp, K., Thomsen, M., Falk, K., Leslie, H., Moller, S., and Sorensen, P.B. 2005. Temporal development of brominated flame retardants in peregrine falcon (*Fako peregrinus*) eggs from South Greenland (1986–2003). *Environmental Science & Technology* 39: 8199–8206.

85. Vorkamp, K., Christensen, J.H., and Riget, F. 2004. Polybrominated diphenyl ethers and organochlorine compounds in biota from the marine environment of East Greenland. *Science of the Total Environment* 331: 143–155.

86. Braune, B., Mallory, M.L., Gilchrist, H.G., Letcher, R.J., and Drouillard, K.G. 2007. Levels and trends of organochlorines and brominated flame retardants in ivory gull eggs from the Canadian Arctic, 1976 to 2004. *Science of the Total Environment* 378: 403–417.

87. Jenssen, B.M., Sormo, E.G., Baek, K., Bytingsvik, J., Gaustad, H., and Ruus, A., Skaare, J.U. 2007. Brominated flame retardants in North-East Atlantic marine ecosystems. *Environmental Health Perspectives* 115: 35–41.

88. Sellstrom, U., Bignert, A., Kierkegaard, A. et al. 2003. Temporal trend studies on tetra-and pentabrominated diphenyl ethers and hexabromocyclododecane in guillemot egg from the Baltic Sea. *Environmental Science & Technology* 37: 5496–5501.

89. Bustnes, J.O., Yoccoz, N.G., Bangjord, G., Polder, A., and Skaare, J.U. 2007. Temporal trends (1986–2004) of organochlorines and brominated flame retardants in tawny owl eggs from northern Europe. *Environmental Science & Technology* 41: 8491–8497.

90. Norstrom, R.J., Simon, M., Moisey, J., Wakeford, B., and Weseloh, D.V.C. 2002. Geographical distribution (2000) and temporal trends (1981–2000) of brominated diphenyl ethers in Great Lakes herring gull eggs. *Environmental Science & Technology* 36: 4783–4789.

91. Bæk, K., Sørmo, E., Føreid, S., Jenssen, B., Lie, E., and Skaare, J. 2006. Hexabromocyclododecane (HBCD) in two marine food webs from Norway. *Organohalogen Compounds* 68: 1749–1752.

92. Tomy, G.T., Pleskach, K., Oswald, T., Halldorson, T., Helm, P.A., Macinnis, G., and Marvin, C.H. 2008. Enantioselective bioaccumulation of hexabromocyclododecane and congener-specific accumulation of brominated diphenyl ethers in an eastern Canadian Arctic marine food web. *Environmental Science & Technology* 42: 3634–3639.

93. Law, K., Palace, V.P., Halldorson, T., Danell, R., Wautier, K., Evans, B., Alaee, M., Marvin, C., and Tomy, G.T. 2006. Dietary accumulation of hexabromocyclododecane diastereoisomers in juvenile rainbow trout (*Oncorhynchus mykiss*) I: Bioaccumulation parameters and evidence of bioisomerization. *Environmental Toxicology & Chemistry* 25: 1757–1761.

94. Gauthier, L.T., Hebert, C.E., Weseloh, D.V.C., and Letcher, R.J. 2008. Dramatic changes in the temporal trends of polybrominated diphenyl ethers (PBDEs) in herring gull eggs from the Laurentian Great Lakes: 1982–2006. *Environmental Science & Technology* 42: 1524–1530.

95. Braune, B.M., Mallory, M.L., Gilchrist, G.H., Letcher, R.J., and Drouillard, K.G. 2007. Levels and trends of organochlorines and brominated flame retardants in ivory gull eggs from the Canadian Arctic, 1976 to 2004. *Science of the Total Environment* 378: 403–417.

96. Holmström, K.E., Jarnberg, U., and Bignert, A. 2005. Temporal trends of PFOS and PFOA in guillemot eggs from the Baltic Sea, 1968–2003. *Environmental Science & Technology* 39: 80–84.

97. Butt, C.M., Mabury, S.A., Muir, D.C.G., and Braune, B.M. 2007. Prevalence of long-chained perfluorinated carboxylates in seabirds from the Canadian arctic between 1975 and 2004. *Environmental Science & Technology* 41: 3521–3528.

98. Butt, C.M., Muir, D.C.G., Stirling, I., Kwan, M., and Mabury, S.A. 2007. Rapid response of arctic ringed seals to changes in perfluoroalkyl production. *Environmental Science & Technology* 41: 42–49.

99. Assadourian, E. 2010. *State of the World 2010. Transforming Cultures. From Consumerism to Sustainability. The Rise and Fall of Consumer Cultures.* W. W. Norton & Company, New York: Worldwatch Institute.

100. de Wit, C.A. 2002. An overview of brominated flame retardants in the environment. *Chemosphere* 46: 583–624.

101. Sagerup, K., Helgason, L.B., Polder, A., Strom, H., Josefsen, T.D., Skare, J.U., and Gabrielsen, G.W. 2009. Persistent organic pollutants and mercury in dead and dying glaucous gulls (*Larus hyperboreus*) at Bjornoya (Svalbard). *Science of the Total Environment* 407: 6009–6016.

25 Contamination Profile and Temporal Trend of POPs in Antarctic Biota

Simonetta Corsolini*

CONTENTS

25.1 The Antarctic Continent and the Southern Ocean: A Summary of Geographical, Climatic, and Biologic Features 571
25.2 Transport of POPs to Antarctica 573
25.3 Selection of Organisms and POPs for the Temporal Trends Assessment 573
25.4 POPs in Antarctic Biota: Temporal Trends 574
    25.4.1 Krill 574
    25.4.2 Fish 578
    25.4.3 Penguins 579
25.5 POPs in Antarctic Biota: Contamination Profile 583
25.6 The Arctic: Comparisons and Temporal Trends 585
25.7 Conclusive Remarks and Perspectives 586
References 587

25.1 THE ANTARCTIC CONTINENT AND THE SOUTHERN OCEAN: A SUMMARY OF GEOGRAPHICAL, CLIMATIC, AND BIOLOGIC FEATURES

The continent of Antarctica is a cold desert surrounded by the Southern Ocean; its northern boundary is the Antarctic circumpolar current (ACC), commonly referred to as the polar front. The ACC is formed where cold seawaters sink beneath the northerly warmer waters. It acts as a biological barrier: organisms are not able to cross it because of the deeply different physicochemical properties of seawaters. Thus, only large animals can cross it (cetaceans, seabirds). The atmosphere over the southern hemisphere shows also a polar front where the Ferrell Cell (mid-latitude air circulation) meets the polar cell: in the former, the air flows toward the poles and eastward near the surface and equatorially and westerly at higher altitudes. At the front, air rises and travels towards the pole, where it sinks forming the polar highs. The Antarctic continent and the Southern Ocean are almost isolated from the other oceans and air masses and thus the turnover is very low. These geographical and air/ocean circulation features make this region of the planet difficult to be reached by persistent organic pollutants (POPs). At the same time, other characteristics of the continent and of the Southern Ocean affect the sinking and bioaccumulation of POPs in abiotic and biotic compartments of Antarctic ecosystems.

Temperatures are below 0°C all year around on the continent, with a mean annual temperature of −57°C in the interior; higher temperatures have been recorded in the rare deglaciated areas along the coasts where they can raise to 0°C or a few degrees over. Precipitation is very scarce, about

* E-mail: corsolini@unisi.it (Chapter corresponding author).

PHOTO 25.1   Mount Melbourne from the Priestley Glacier (Terra Victoria, Antarctica).

50 mm/year of snow on the Plateau and 250 mm/year of snow in coastal areas; higher values can be found in the Antarctic Peninsula where it can also rain. The interior is a cold desert where snow is very rare and deposition is usually in the form of microscopic ice particles. Most of the continent is covered by an ice sheet with an average thickness of 1.6 km (maximum 4.8 km) (Photo 25.1). The seawater temperature is −1.8°C to 10°C; the Southern Ocean freezes and the pack ice extension varies between 4 million km² (February–March) and 20 million km² (September–October).

A chemical may reach the Antarctic continent mainly through air mass movements [1] and sea streams, but the ACC acts as a boundary and the latter is less important in terms of the amount of chemicals transported. Once in Antarctica, POPs fall out by wet and dry deposition. In Antarctica, they are incorporated in the snow and ice. The ice can be described as a *reservoir* of persistent contaminants, which are released during the summer melt. The contaminants are transferred from the ice into the seawater and then to organisms; they bioaccumulate and biomagnify in the food webs, becoming a potential risk for marine life.

Because of the low temperatures, the volatilization of chemicals trapped by ice and snow is limited, while the scarce solar radiation does not allow degradation; this is very difficult in spring and autumn and impossible during the long and dark winter. As a result of all these considerations, it is evident that it is not easy for a chemical to reach Antarctica with respect to other regions or the Arctic; however, once a chemical reaches Antarctica, it can persist for long time and be trapped in the ice that acts as a final *reservoir*.

The ecosystem characteristics contribute to its high vulnerability; they usually show low resistance and resilience because they do not have feedback mechanisms and redundancy of functions. Trophic webs are typically short (they show few levels) due to the extreme climate. Even the organisms show peculiar features: they grow very slowly, may show gigantism when adult, are often long-lived, and may use lipids as energy reserve or to protect the body against cold. All these features make Antarctic organisms prone to accumulate chemicals and vulnerable to toxic effects. It has to be kept in mind that these organisms are not used to anthropic impacts, including the release of xenobiotic substances, as organisms living in anthropized continents have been. Organisms that live in extreme environments are typically stenoecious and thus we do not know much about their ability to fight this kind of increasing stress. For instance, Adélie penguins show low activity of the cytochrome P450 and in particular the CYP 4503A enzyme (this enzyme is involved in the detoxification of chemicals like PCBs), suggesting that this species may be vulnerable to toxic pollutants [2].

Life is always linked to the sea; the rare terrestrial organisms grow near or in relation to the scarce presence of melted water. Marine organisms, in particular seabirds and marine mammals, may accumulate POPs in their tissues and thus are the most studied species together with fish species, although only a few papers have been published since the 1960s. For instance, the benthic communities (organisms that live under the pack ice) are particularly exposed to contamination; in fact, they can accumulate those chemicals that are released in the seawater during the summer ice melting. Then, they are predated by other organisms and thus contaminants may be biomagnified.

25.2 TRANSPORT OF POPs TO ANTARCTICA

The presence of POPs in Antarctic ecosystems is due to their dispersal by atmospheric and hydrologic movements [3,4]. These compounds may be transported great distances from their sources in polluted areas from anthropized continents by air currents [1,5]. Thus, they can be found in remote areas such as the Arctic and the Antarctic (e.g., [6–10]). Atmospheric and marine contamination is different in the two hemispheres, since the northern hemisphere is more populated and polluted than the Southern one. Such differences are related to the geography of the hemispheres: the northern one is mainly occupied by land and the southern one by ocean. Ninety percent of the global population lives in the northern hemisphere and as a consequence most of the pollution is produced there and, in particular, at low to mid latitudes. The industrialized countries of Europe and North America were the greatest producers and users of many POPs from the Second World War. For instance, the concentrations of polychlorinated biphenyls (PCBs) in the surface layers of the oceans were reported to be quite homogeneous in the middle latitudes of the northern hemisphere, with a maximum in the tropical region and lower values recorded in the southern hemisphere with a minimum in the Antarctic region [11].

POPs are organic chemicals with peculiar physical and chemical properties; once released into the environment, they are persistent for many years and thus can be transported far from the source areas. They accumulate in the lipid component of tissues of all organisms and can also biomagnify showing toxic effects. Fish, predatory birds, mammals, and humans are high in the trophic webs and thus accumulate the greatest concentrations. POPs have been released in the environments over the past decades and are now distributed worldwide, including those regions where they have never been used. Toxic effects of POPs include cancer, allergies and hypersensitivity, damage to the central and peripheral nervous systems, and reproductive disorders; some of them are also endocrine disrupters and thus can alter the hormonal system, damage the reproductive and immune systems of exposed individuals as well as their offspring (http://chm.pops.int/). The POPs include an initial group of twelve chemical families: aldrin, chlordane, DDT, dieldrin, endrin, heptachlor, hexachlorobenzene (HCB), mirex, toxaphene, PCBs, polychlorinated dibenzo-*p*-dioxins and polychlorinated dibenzofurans (PCDDs/PCDFs). Nine additional chemicals were added later: chlordecone, alpha-hexachlorocyclohexane (*alpha*-HCH), beta-hexachlorocyclohexane (*beta*-HCH), lindane, pentachlorobenzene, hexabromobiphenyls (hexa-BBs), tetrabromodiphenyl ether, pentabromodiphenyl ether, hexabromodiphenyl ether and heptabromodiphenyl ether (tetra-, penta-, hexa-, hepta-BDEs), pentachlorobenzene, perfluorooctane sulfonic acid (PFOS) and its salts, and perfluorooctane sulfonyl fluoride (PFOSF). Not all of these chemicals are reported to be detected in Antarctic organisms.

25.3 SELECTION OF ORGANISMS AND POPs FOR THE TEMPORAL TRENDS ASSESSMENT

The scientific literature on the presence of old and emergent POPs in Antarctic organisms is very scarce, owing to the difficulty of collecting samples in such an extreme environment, the distance from any part of the world, the very high cost of scientific expeditions, and the need to be part of one of those to be allowed to reach the continent and collect organism samples whose collection need special internationally valid permits.

In order to select the species and POPs to be used for the temporal trend evaluation, 242 articles were studied. The following data were taken into consideration in the selection of those papers reporting data on the biota: year of sampling, unit of measure, species, tissue analyzed, and area of sampling. The year of sampling may be very different from the year of publication; the unit of measure can be on wet, lipid, or dry weight. The same tissues can be excised from different regions of the body; however they were recorded as fat, muscle, liver, blood, eggs, or whole body. The areas of sampling referred to three main regions of the Antarctic continent and surrounding seas: the Antarctic Peninsula and sub-Antarctic islands (AP), the Ross Sea (RS), and the East Antarctic coasts (EA) (Figure 25.1). One hundred and nineteen articles (or book chapters) regarding data on krill, fish, penguins, and seals were selected and included in the statistical matrix. Afterwards, data on seals were excluded because of the unevenness of the reporting results: among 19 papers selected, 5 articles were related to the Weddell seal (*Leptonichotes weddelli*) and 2 on the elephant seal (*Mirounga leonina*), but unfortunately no comparisons were allowed because the chemical families, unit of measures, or tissue analyzed were not comparable.

Only a few papers reported data on POP accumulation on algae, micro-zooplankton (copepods, amphipods, and other species), molluscs and other marine invertebrates, moss and lichens, and they were not included because the data were not enough for a temporal trend evaluation.

The POPs considered in the selection of articles were polychlorobiphenyls (PCBs; the number of congeners used for the sum-PCB values was variable among articles), p,p'-DDT, p,p'-DDE, DDTs (sum of o,p' and p,p' isomers of DDD, DDE, DDT), HCB, hexachlorocyclohexanes (HCHs; sum of *alfa*, *beta*, *delta*, *gamma* isomers, not always all detected), polybrominated diphenyl ethers (PBDEs), perfluorinated compounds (PFCs), toxaphenes, chlordanes (CHLs), polychlorodibenzo-dioxins (PCDDs), polychlorodibenzofurans (PCDFs), organo-tin compounds (OTCs), polychloro-naphtalenes (PCNs). PCBs, p,p'-DDE, DDTs, and HCB were the chemicals mostly detected: 85 records reported in 32 papers. p,p'-DDT, HCHs, DDTs, PBDEs, CHLs, PCDDs, PCDFs, and toxaphenes were detected in a lower number of records (Figure 25.2). Figure 25.3 shows the number of records of the different classes of organisms selected (krill, fish, penguin) and the unit of measure (wet, dry, or lipid wt) used to detect the POPs in their tissues.

Data were considered for the discussion of the POP time trend in the Antarctic organisms when concentrations in the same species and tissues were available for samples collected at least with a 5 year time span.

25.4 POPs IN ANTARCTIC BIOTA: TEMPORAL TRENDS

The assessment of the time trend of POP concentrations in the Antarctic biota is not easy due to the paucity of data. Since the 1960s, very few articles on POP contamination in the Antarctic abiotic and biotic compartments have been published in international journals. This discussion of the time trend of POP levels in the Antarctic biota refers to the concentrations found in krill, fish, and penguins. These organisms reside in the Antarctic environments and do not migrate; thus, they may give information on the POP dispersion and accumulation in the Antarctic region.

25.4.1 KRILL

The term Antarctic krill refers to pelagic shrimp-like crustaceans living in large swarms in a very high density of individuals (10,000–30,000 per cubic meter [12]). The Antarctic krill *Euphausia superba* is a key species in the Antarctic ecosystems and most of the Antarctic organisms feed directly or indirectly on it. In terms of biomass, *E. superba* is considered the most flourishing animal species on the planet (up to 500 million tons) [13]. The krill feed on phytoplankton and are the prey of most of the Antarctic animals: other invertebrates like squid, fish, flying seabirds and penguins, seals, and cetaceans. The presence of toxic contaminants and the possible toxic effects

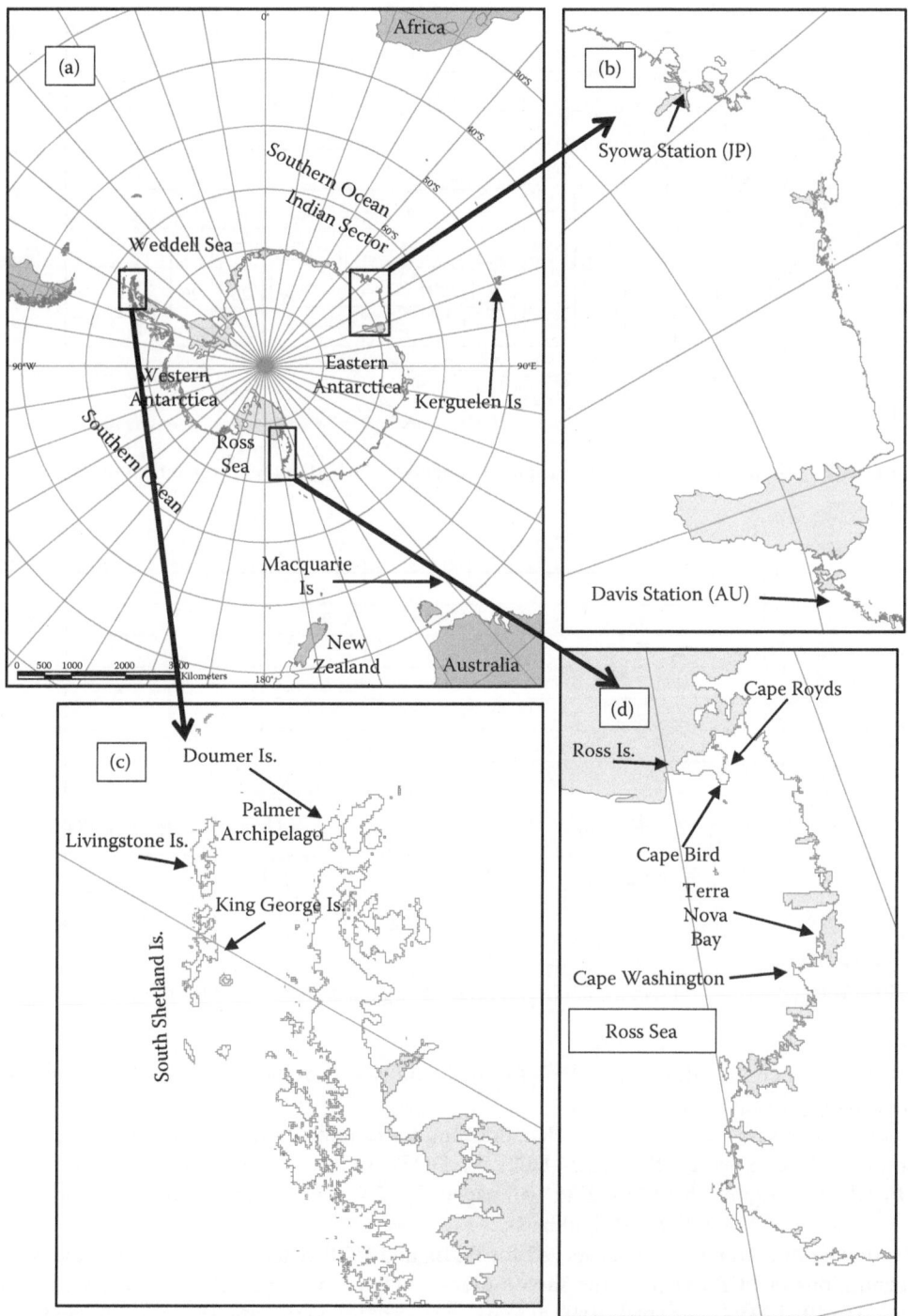

Antarctic Digital Database © Scientific Committee on Antarctic Research 1993–2006
http://data.aad.gov.au/aadc/mapcat/display_map.cfm?map_id=13351

FIGURE 25.1 (a) Map of the Antarctic continent, the Southern Ocean, and the three main regions cited throughout the text, (b) East Antarctica, (c) Antarctic Peninsula, and (d) Ross Sea. Maps were downloaded free of charge at http://lima.usgs.gov/antarctic_research_atlas/viewer.htm

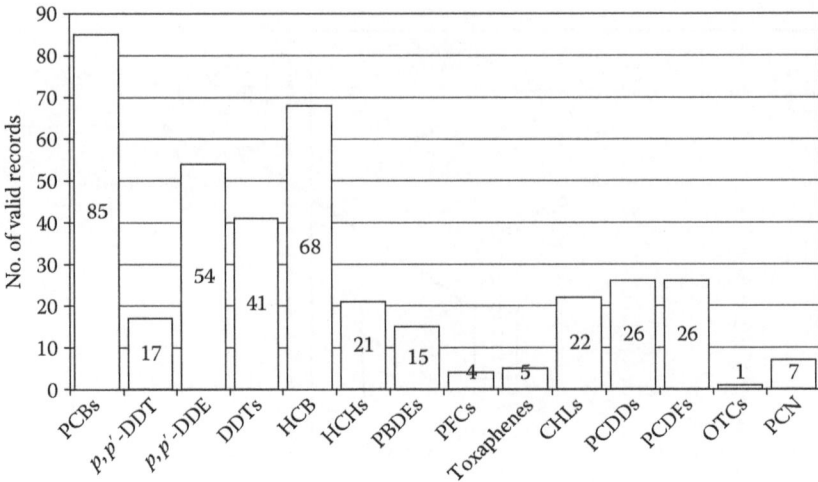

FIGURE 25.2 Abundance of records of various chemicals in the bibliographic references on Antarctic biota.

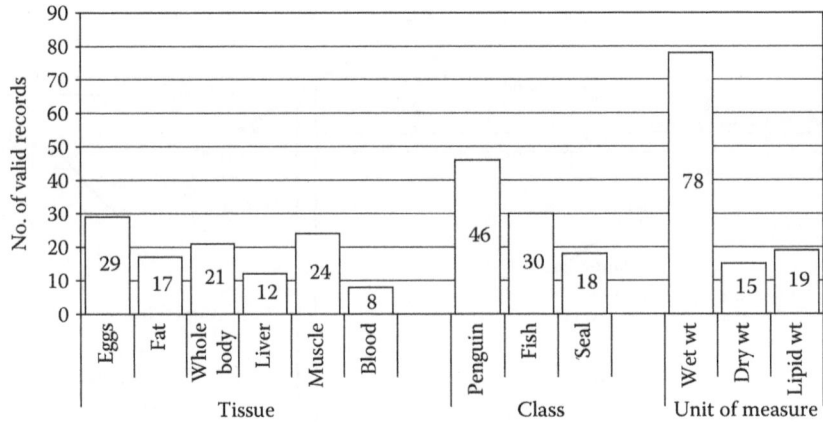

FIGURE 25.3 Abundance of records of different class of organisms reported in the bibliographic references.

that they may elicit in krill is a key stone for the health status of the Antarctic ecosystems because of the ecological value of this species.

Twelve articles have been found on POP concentrations in krill; most of the papers reported concentrations of PCBs (9), HCB (7), p,p'-DDE (5), DDTs, HCHs, CHLs (4), PCDDs and PCDFs (3), and PBDEs (2) (Figure 25.4 [7,14–22]); two papers [23,24] were not included in the graph because concentrations were on a dry and lipid basis.

A possible time trend can be observed for PCBs and HCB detected in krill from the Ross Sea. Concentrations of PCBs were of the same order of magnitude (1.67–3.11 ng/g wet wt) in samples collected in 1994–95; lower levels (0.03 ng/g wet wt) were found in samples from 1995 and analyzed by Kumar et al. [22]. PCBs may show a slow decreasing trend in the late 1990s, likely in relation to their ban in many countries, which occurred in the late 1970s–1980s. HCB concentrations were homogeneous in all samples. The other chemicals were detected only in a season and thus do not allow any temporal trend evaluation.

The time trend in other regions of the Antarctic is not very clear because of the scarcity of data available. Levels of DDTs and CHLs at Syowa Station (69°00′ S, 39°35′ E; year of

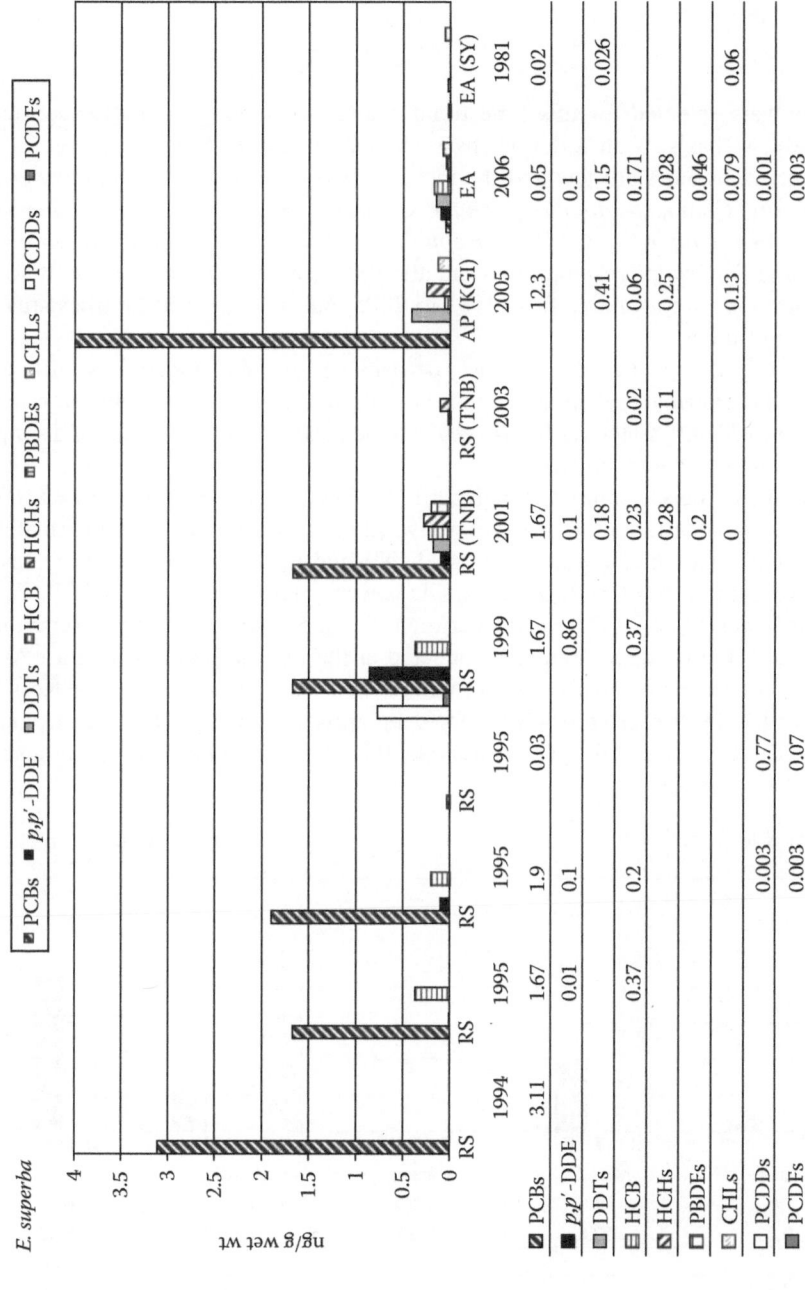

FIGURE 25.4 Concentrations of POPs in whole body krill *E. superba* (ng/g wet wt; RS = Ross Sea, RS (TNB) = Terra Nova Bay, EA = East Antarctica, EA (SS) = Syowa Station). For references, see the text (no values in the table indicate no data available in the literature).

sampling = YoS = 1981) [21] can be compared to those found in a survey between 30°–80°E and 60°–70°S (YoS = 2006) [14]: DDTs showed higher levels in 2006 and CHLs were of the same order of magnitude.

Although data are very scarce, a very weak decreasing trend from 1995 to 2000 may be observed in samples from the Ross Sea; high levels of PCBs detected in samples from 1994 were followed by lower concentrations in those samples collected from 1995 to 2001.

25.4.2 Fish

Thirty-one articles were selected for this time trend evaluation; among them, 19 reported concentrations on a wet wt basis, 5 on a dry wt basis, and 7 on a lipid basis. The species of fish collected in Antarctic seawaters are few and the most studied have been *Trematomus bernacchi*, *Chionodraco hamatus*, *Champsocephalus gunnari*, *Gymnoscopelus nicholsi*, *Trematomus eleupidotes*, *Gobionotothen gibberifrons*, *Chaenocephalus aceratus*, *Pleuragramma antarcticum*, *Nothotenia rossii*, and *Dyssostichus mawsoni*. Data allowed a time trend evaluation only for six of them. POP concentrations are available from 1981 to 2005, but the chemical families studied are often different among studies.

T. bernacchi is likely the most studied species, perhaps because its collection is fairly easy. In fact, it is a demersal fish distributed in almost all the benthic Antarctic environments between 0 and 200 m (found deeper to 700 m). Catching is easy as it can be fished with a hand-line and simple bait (pieces of fish and molluscs, fruits, bread).

PCB concentrations in muscle samples collected in the Ross Sea (Terra Nova Bay = TNB) were highest in 1981, with a concentration of 12.8 ng/g wet wt [25], and decreased to approximately 50% during the following 10 years: 6 ng/g wet wt [18,26]. DDTs and *p,p'*-DDE seemed to increase from the 1990s to 2001 (Figure 25.5 [18,21,22,25–27]); DDTs were 4 ng/g wet wt in 1989 [25] and doubled in 2001, 8.6 ng/g wet wt [18]; HCB varied between 0.27 [25] and 2.6 [18] ng/g wet wt in 1989 and 1999, respectively (Figure 25.5). Samples collected at the Syowa Station [21], East Antarctic coast, showed lower concentrations with respect to those found in specimens from the Ross Sea.

An interesting article by Weber et al. [28] reports a comparison of *p,p'*-DDE, CHL, and HCB levels in samples of three species of fish, *C. gunnari* (krill feeder), *G. gibberifrons* (benthos feeder),

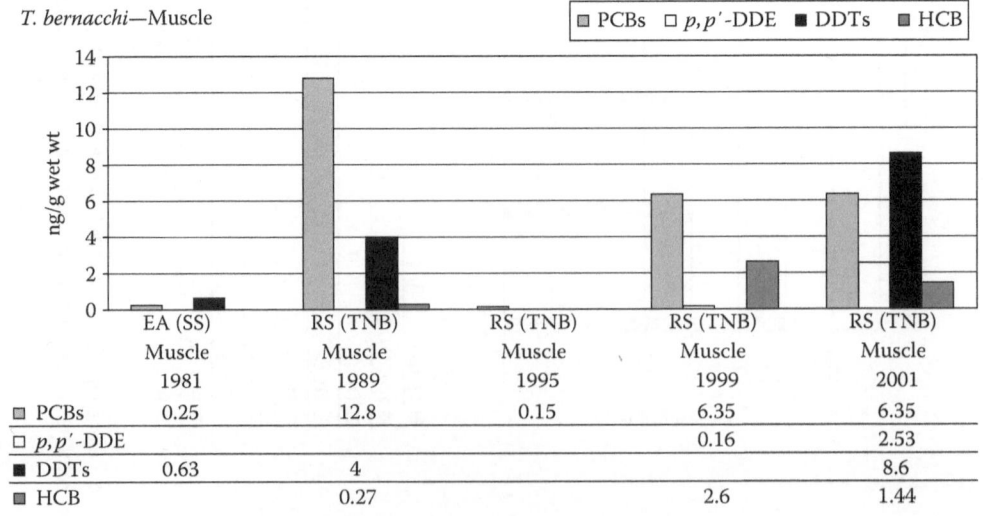

T. bernacchi—Muscle					
	EA (SS) Muscle 1981	RS (TNB) Muscle 1989	RS (TNB) Muscle 1995	RS (TNB) Muscle 1999	RS (TNB) Muscle 2001
▨ PCBs	0.25	12.8	0.15	6.35	6.35
☐ *p,p'*-DDE				0.16	2.53
■ DDTs	0.63	4			8.6
▨ HCB		0.27		2.6	1.44

FIGURE 25.5 Concentrations of HCB, *p,p'*-DDE, DDTs, and PCBs in muscle of *T. bernacchi* (ng/g wet wt; RS (TNB) = Terra Nova Bay, EA (SS) = Syowa Station). For references, see the text (no values in the table indicate no data available in the literature).

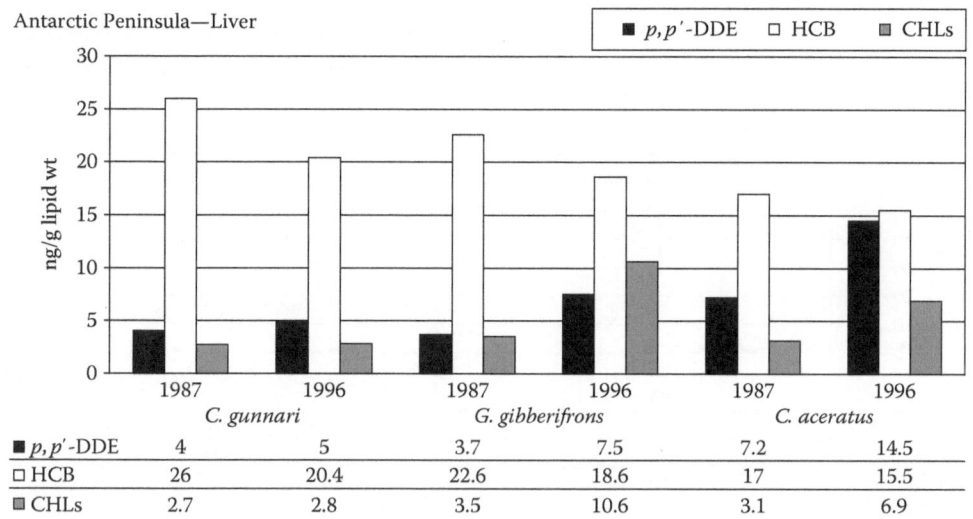

FIGURE 25.6 Concentrations of *p,p'*-DDE, HCB, and CHLs (ng/g lipid wt) in the liver tissues of three species of fish from the Antarctic Peninsula. For references, see the text (no values in the table indicate no data available in the literature).

and *C. aceratus* (second level carnivore), collected in 1987 and in 1996 (Figure 25.6 [28]). HCB concentrations showed a slight but not significant decrease during this 10 years time span in the three species, with higher concentrations in *C. gunnari* > *G. gibberifrons* > *C. aceratus*. *p,p'*-DDE and CHLs showed an opposite trend, with an increase of their concentration from 1987 to 1996 in all the studied species, with higher concentrations in *C. aceratus* > *G. gibberifrons* > *C. gunnari*, and it is interestingly an abundance pattern opposite to that of HCB. These results are remarkable and show how both the global distribution/equilibrium of POPs and ecological features of organisms may play a crucial role in the accumulation patterns and levels. In fact, concentrations do not reflect exactly what is expected on the basis of the feeding habits of fish (second-level carnivore ≥ benthos feeder > krill feeder), but do depend on their physicochemical properties, distribution in the ecosystem compartments, and global transport potency. These results agreed with those reported for *T. bernacchi* (see previous paragraph and Figure 25.5) and confirmed an initial decreasing trend of DDT concentrations following by its ban in the 1980s, following by a new increase, likely owed to the continuing use in many countries (characterized by a population growth and by health problems caused by the endemic malaria disease), where DDT use is allowed by the Stockholm Convention.

The patterns of concentrations observed in samples of *N. rossii* (Figure 25.7 [7,29]) and *P. antarcticum* (Figure 25.8 [7,20]) are similar to those described for other fish species.

25.4.3 PENGUINS

There are many articles presenting POP concentrations in seabirds. Most of the species studied have been penguins > Stercorarids > petrels, fulmars, other species. To the purpose of this chapter, only those papers reporting data on penguins will be considered. In fact, seabirds other than penguins migrate northward during the wintertime and thus may accumulate POPs in anthropized and more polluted regions. Penguins breed along the coasts of the Antarctic continent and islands, and over the winter in the Southern Ocean, thus they are a valid indicator of contamination in the southern polar region [30].

The articles selected for the time trend analyses were 47 and were published between 1966 and 2010. Among them, 24 papers reported data on the Adélie penguin (*Pygoscelis adéliae*), 12 on

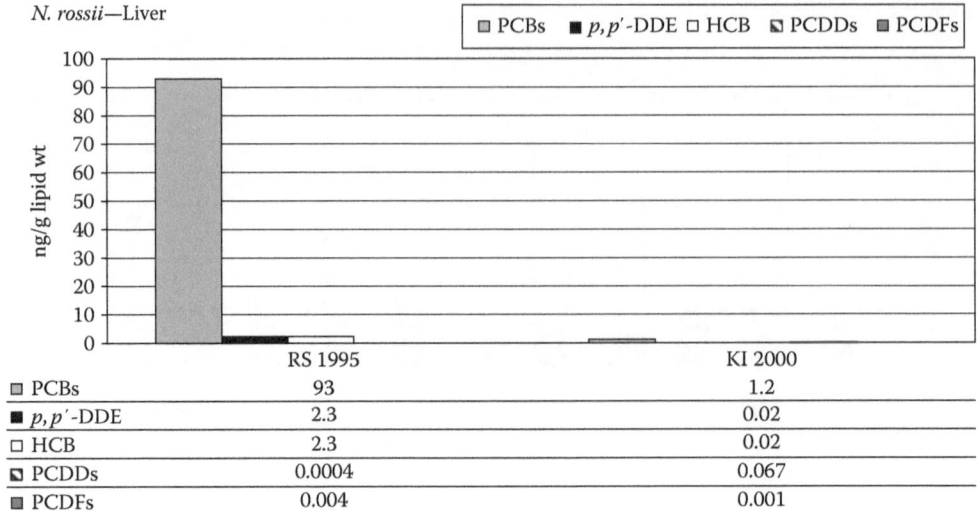

FIGURE 25.7 Concentrations of contaminants in the liver of *N. rossii* (ng/g lipid wt; Kerguelen Is = KI). For references, see the text (no values in the table indicate no data available in the literature).

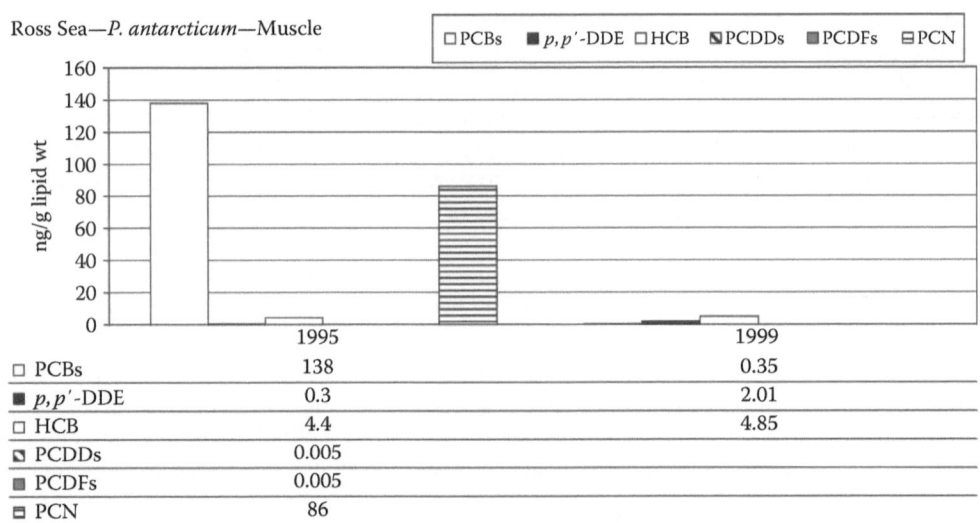

FIGURE 25.8 Concentrations of contaminants in the muscle of *P. antarcticum* from the Ross Sea (ng/g lipid wt). For references, see the text (no values in the table indicate no data available in the literature).

the Gentoo penguin (*Pygoscelis papua*) (Photo 25.2), seven on the Chinstrap penguin (*Pygoscelis antarctica*), and two on the Emperor penguin (*Aptenodytes forsteri*). The Ross Sea is populated only by Adélie and Emperor penguins. A few other articles reported POP concentrations in other penguin species.

A very interesting paper reported results on the presence of PCBs and DDTs in a fat sample of Emperor penguin collected in the Ross Sea (Ross Island) in 1911 and, therefore, more than 30 years before the beginning of mass worldwide use of these xenobiotics [31]. No contaminants were detected in these fat samples. Further analyses carried out on samples of Emperor penguin tissues from Cape Washington, Ross Sea (Corsolini, unpublished data), the East Antarctic coast (Davis Station) [32] and the Weddell Sea [33], showed the presence of PCBs, DDTs, and HCB in eggs and

PHOTO 25.2 A Gentoo penguin with its chicks (Admiralty Bay, King George Island, South Shetland Archipelago, Antarctica).

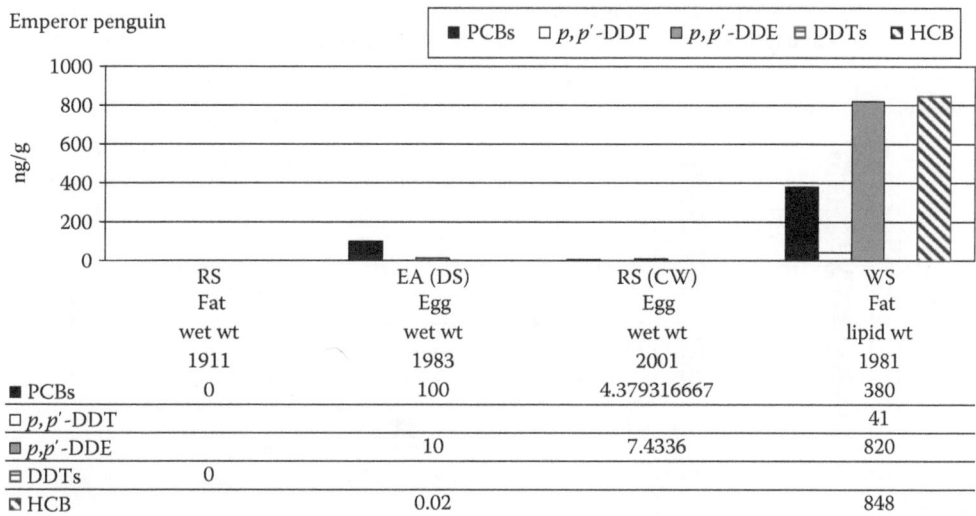

	RS Fat wet wt 1911	EA (DS) Egg wet wt 1983	RS (CW) Egg wet wt 2001	WS Fat lipid wt 1981
■ PCBs	0	100	4.379316667	380
□ p,p'-DDT				41
■ p,p'-DDE		10	7.4336	820
⊟ DDTs	0			
◪ HCB		0.02		848

FIGURE 25.9 Concentrations of contaminants in the muscle of Emperor penguin (ng/g; RS=Ross Sea, EA (DS)=Davis Station, RS (CW)=Cape Washington, WS=Weddell Sea). For references, see the text (no values in the table indicate no data available in the literature).

fat (Figure 25.9 [31–34]). Levels in egg samples from East Antarctica were higher than those found in eggs from the Ross Sea.

Unfortunately, no data are available for a temporal trend assessment; however, samples collected in 1983 in the East Antarctic coasts were higher with respect to those collected in 2001 in the Ross Sea, in agreement with the trend observed in krill and fish, and likely in relation to the break of PCB and DDT use in the 1970s. Notwithstanding, the sample from 1911 has no statistical nor scientific value since the absence of synthetic pollutants points out that Antarctica might be a real pristine continent until recently.

Data on Adélie penguin samples are reported in 24 articles and Figures 25.10 and 25.11 illustrate concentrations in eggs expressed on a wet or lipid wt basis, depending on the unit of measure adopted. Other papers presented POP levels in Adélie penguins, but no comparison could be made owing to different tissues, unit of measures, chemicals analyzed, and species; they are not showed

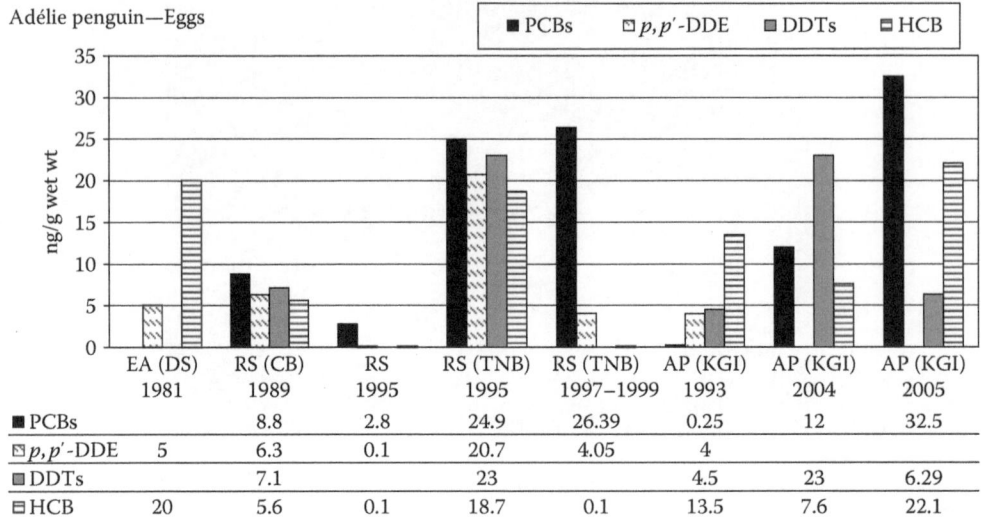

Adélie penguin—Eggs	EA (DS) 1981	RS (CB) 1989	RS 1995	RS (TNB) 1995	RS (TNB) 1997–1999	AP (KGI) 1993	AP (KGI) 2004	AP (KGI) 2005
■ PCBs		8.8	2.8	24.9	26.39	0.25	12	32.5
▨ p,p'-DDE	5	6.3	0.1	20.7	4.05	4		
▣ DDTs		7.1		23		4.5	23	6.29
▤ HCB	20	5.6	0.1	18.7	0.1	13.5	7.6	22.1

FIGURE 25.10 Concentrations of contaminants in the eggs of Adélie penguin (ng/g wet wt; RS = Ross Sea, RS (CB) = Cape Byrd, RS (CR) = Cape Royds, AP (DI) = Antarctic Peninsula—Doumer Is., AP (PA) = Palmer Archipelago). For references, see the text (no values in the table indicate no data available in the literature).

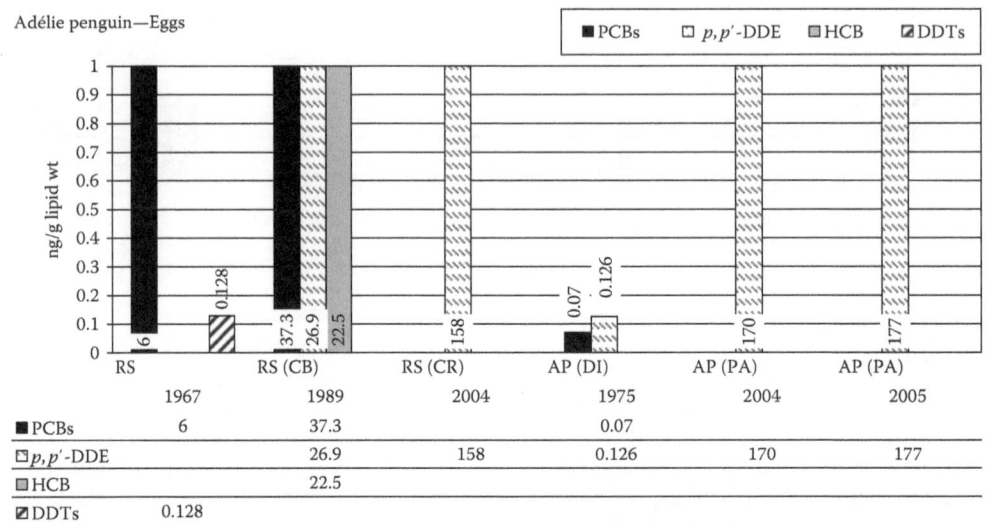

Adélie penguin—Eggs	RS 1967	RS (CB) 1989	RS (CR) 2004	AP (DI) 1975	AP (PA) 2004	AP (PA) 2005
■ PCBs	6	37.3		0.07		
▨ p,p'-DDE		26.9	158	0.126	170	177
▣ HCB		22.5				
▧ DDTs	0.128					

FIGURE 25.11 Concentrations of contaminants in the eggs of Adélie penguin (ng/g lipid wt; EA (DS) = Davis Station, RS = Ross Sea, RS (CB) = Cape Byrd, RS (TNB) = Terra Nova Bay, EA (KGI) = King George Is.). For references, see the text (no values in the table indicate no data available in the literature).

in the figures. The oldest samples of this species analyzed to detect xenobiotics were collected in the Ross Sea in 1967 (Figure 25.10 [6,15,17,31,34–37]) and showed an increasing trend of PCB and DDT concentrations [35]. DDTs were reported as a sum of isomers (0.128 ng/g lipid wt) in the 1967 samples, while in the other studies only p,p'-DDE was detected. Levels of p,p'-DDE alone in samples collected in the Ross Sea in the 1989 [36] and 2004 [37] were higher, confirming the tendency of this chemical be transported to the Antarctic region and thus be accumulated by penguins. The same increasing pattern can be observed in Adélie penguin eggs collected in colonies located in the Antarctic Peninsula, where p,p'-DDE concentrations rose between 1975 [35] and 2004–2005 [37] by several orders of magnitude.

Those concentrations expressed on a lipid wt and reported in 12 papers allow comparing concentrations from a temporal trend point of view. It is not possible to recognize a clear temporal trend in the Ross Sea and Antarctic Peninsula areas (Figure 25.11 [35,38,39]) because there is a lack of data for some contaminants. Roughly, concentrations of PCBs and DDTs seemed to decline from the end of the 1980s to mid-1990s and rise again at the end of the 1990s. The same can be observed for HCB that show concentrations ranging from 0.1 [7,38] to 22.1 [16] ng/g wet wt. This pattern may be due to the historical use of PCBs and DDTs worldwide and in particular in American, European, and some Asiatic countries: after a massive use up until the 1970s and 1980s, governments started to ban their use and production. First, levels continued to increase as a result of their global transport and dispersion, but then concentrations in biota showed a light decrease following the reduced use. The new increase observed at the end of the 1990s could be as a result of a few reasons. The first reason could be the slow release of chemicals from legal or illegal stocks or equipments; in fact, 1.7 million tons of PCBs were produced between 1929 and 1989 (http://chm.pops.int/), and lots of the equipment containing PCBs are still in use somewhere or stocked. Second, the releases from the final sink as deep oceanic sediment and waters may follow natural cycles in the marine ecosystems. The last reason is the use, authorized by the Stockholm Convention (http://chm.pops.int/), of DDTs in those countries where malaria is endemic. DDT has a half-life of up to 15 years in soil and 7 days in the air [39], and it shows long-range transport potency [40], thus it can reach the Antarctic area from other parts of the world where it is still used or stocked. Regarding PCBs, a recent article demonstrated that PCBs were volatilizing from surface waters to the overlying atmosphere [41], suggesting a possible source of PCBs that, after the volatilization at lower latitudes, may reach the Southern Ocean and Antarctica through the long-range air transport [5]. A similar event may be ascribed for HCB: its concentrations between air and water are believed to be close to the equilibrium found in the Atlantic Ocean [41]; moreover, cold condensation effects are considered to be significant [42]. Atmospheric HCB did not display significant correlations with temperature, while in contrast dissolved HCB showed strong increasing concentrations with decreasing temperature in the northern hemisphere; the behavior of HCB was found to be consistent with PCBs, suggesting similar sources and fate of these slowly degrading compounds [42].

The Gentoo and Chinstrap penguins showed similar patterns; PCBs, DDTs, HCB, and HCHs tended to increase from 1993 to 2005 (Figure 25.12 [2,16,32,35,43,44] and 25.13 [2,16,35,43]), confirming what speculated previously for the Adélie penguin.

25.5 POPs IN ANTARCTIC BIOTA: CONTAMINATION PROFILE

The relative abundance of the various chemical families can be observed in Figures 25.4 through 25.13. The contamination profile was assessed considering the presence and levels of PCBs, p,p'-DDE and/or DDTs, and HCB because they were the most studied chemicals and thus their frequency of record allowed comparisons.

A common pattern cannot be observed between species, areas, and time of sample collection. A profile of PCBs>HCB>DDE was found in krill, *P. antarcticum*, and in Gentoo, Chinstrap, and Adélie penguins from the Antarctic Peninsula; HCB and DDT levels were very similar in *N. rossii* from the Ross Sea and the Kerguelen Islands (Figure 25.4 [7–8,11]). Pygoscelids seem to preferably accumulate PCBs over DDT or HCB before 2000. Beginning from 2000, DDT sometimes exceeds PCBs in the composition of chemical residues. Emperor penguins accumulate mostly HCB and DDTs with respect to PCBs in the Ross Sea and Antarctic Peninsula, while a pattern of PCBs>DDE>HCB in samples from east Antarctica (Figure 25.9).

Krill and fish did not show a preferential profile of accumulation and thus different patterns can be observed depending on species and area: PCBs>DDT>HCB in *T. bernacchii* from the Ross Sea (Figure 25.5); DDT>PCBs>HCB in *T. bernacchi* from east Antarctica and in some samples from the Ross Sea (Figure 25.5); HCB>DDE/PCBs in *C. gunnari*, *G. gibberifrons*, and *C. aceratus* from

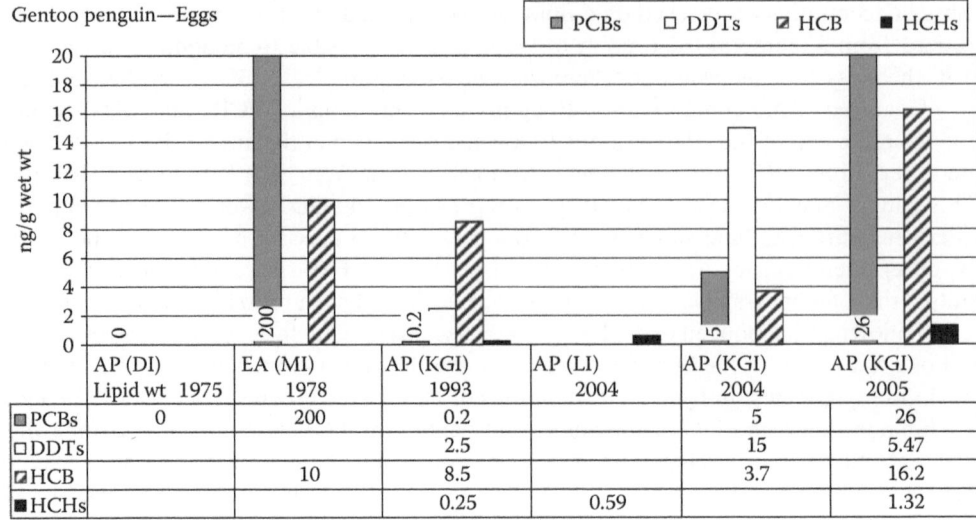

FIGURE 25.12 Concentrations of contaminants in the eggs of Gentoo penguin (ng/g wet wt; AP (DI) = Antarctic Peninsula—Doumer Is., EA (MI) = Macquarie Is., EA (LI) = Livingstone Is., EA (KGI) = King George Is.). For references, see the text (no values in the table indicate no data available in the literature).

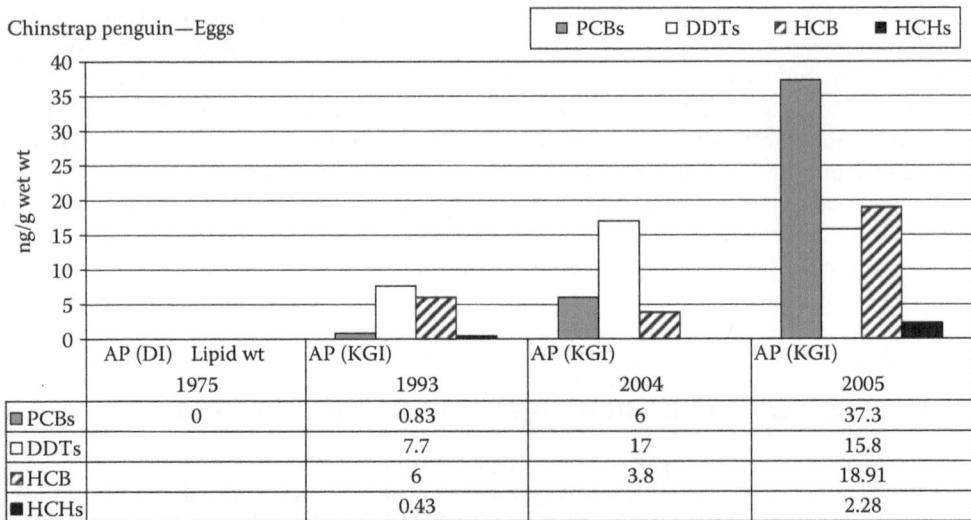

FIGURE 25.13 Concentrations of contaminants in the eggs of Chinstrap penguin (ng/g wet wt, AP (DI) = Antarctic Peninsula—Doumer Is., EA (KGI) = King George Is.). For references, see the text (no values in the table indicate no data available in the literature).

the Antarctic Peninsula (Figure 25.6), and in some samples of *P. antarcticum* from the Ross Sea (Figure 25.8).

The different accumulation profiles in Antarctic organisms suggest that several factors may affect it. The origin of pollutants can play an important role in relation to the current use of different chemicals. The long-range transport and the cold condensation and global fractionation processes can affect the area of deposition of chemicals in the southern polar region.

Nevertheless, higher concentrations cannot be identified in an area rather than others and levels varied in a specific range depending on the POPs. The time of sampling can play a crucial role because the summer ice melting allows POPs to be released in the seawater and enter into trophic webs. Meteorological factors were reported to affect the transport of POPs from South America to the Antarctic Peninsula [9]. Once in the Antarctic region, contaminants enter trophic webs following different mechanisms depending on their physicochemical properties and partition coefficients between water and lipids or air and water; some POPs may be preferably bioconcentrated and some others may be biomagnified. Moreover, species-specific metabolism of organisms, including the lipid composition of tissues, can determine the profiles in the analyzed tissues. Ultimately, the presence of a local source of pollution from scientific stations can also affect the burden of accumulated contaminants [30,45–48].

25.6   THE ARCTIC: COMPARISONS AND TEMPORAL TRENDS

The POP presence in organisms of the Arctic has been investigated during the last several decades and many articles have been published in international scientific journals. The easier access to the area with respect to the Antarctic region is responsible for the high number of studies on Arctic abiotic and biotic compartments. The interest in this polar region increased when the presence of very high concentrations of POPs was detected in human populations that live in the northern lands of Europe, Asia, and America (for a comprehensive study, see the AMAP Report 2009 [49]). The Arctic Monitoring and Assessment Programme (AMAP) is an international organization established in 1991 to supply information on the status of the Arctic environment and threats to it, and to provide scientific advice to support Arctic governments in an effort to manage contamination risks and hazards (AMAP, http://www.amap.no/). In the framework of the AMAP, many researches have been carried out. Recently, reviews of POP presence and trends in the Arctic have been published (e.g., [50–53]).

Most of the legacy POPs have been reported to decrease in Arctic organisms [54]; for instance, PCB153 and PCBs showed a significant decreasing trend in freshwater fish from Canada [55] and Sweden [54], marine fish from Iceland [54], and seabirds from the Canadian Arctic Archipelago [56]. An increasing PCB trend was observed in blue mussels from Iceland, and burbot from Canadian lakes [55].

Among chlorinated pesticides, DDTs and p,p'-DDE showed significant decreasing time trends in Arctic cod [57], ringed seal [58], seabirds [54,56,59,60], beluga whales [61], and polar bears [58]. HCHs, CHLs, dieldrin, and mirex showed a general decreasing time trend in freshwater fish, seabirds, and marine mammals with few exceptions (for a review, see Rigét et al. [54]). There are few Arctic species that show an increasing or stable POP concentration over the last decade; POP decline or cessation of production and use mostly occurred prior to the beginning of monitoring and this lack of data may be considered a reason of these trends [54].

Data on the presence of emergent POPs are very scarce in the Arctic organisms, thus only a few speculations may be done on the temporal trends in the organisms. Articles reporting the presence of PBDEs in Arctic organisms have been reported since the early to mid-2000s. Despite the increasing time trend observed in the PBDE (mainly BDE-47 and -99) concentrations in air [62], several studies reported a leveling off or decline of BDEs 49 and 99, likely in relation to the reduced emissions in Europe since the early 2000s, and in the United States and Canada since the mid-2000s [52].

The detection of PFCs in Arctic biota has been reported since the 2000s. The increase, decrease, or leveling off of different PFCs have been observed in various species and areas and thus trends are not clear [52]. The production of PFOS containing technical goods was reduced in 2001, but PFOS is still produced in China [63] and many exemptions are allowed by the Stockholm Convention [64]. Therefore, a continuing input of PFCs in the Arctic as well as in the Antarctic regions may be expected. A discussion is still underway on the prevailing PFC transport path to the Arctic: via atmospheric transport of PFC precursors or via PFC oceanic transport. The input of PFOA to the Arctic Ocean via sea streams seems to be at least two orders of magnitude higher than those via atmospheric transport. The dominance of oceanic transport as PFC pathway to polar region might affect the presence of these contaminants mainly in the Antarctic region where the atmospheric transfer is considered to be the main route of transport for semi-volatile chemicals. Unfortunately, information on PFC time trends in Antarctic organisms are not available owing to the lack of time series data.

Articles on the presence of PCNs in polar organisms are very scarce, thus, it is not possible to evaluate any temporal trend [50] except for Arctic cod from Norway, where no change in concentration of ΣPCNs was reported from 1987 to 1998 [57]. When data are available, levels were lower in the Antarctic organisms with respect to Arctic biota. For instance, concentration in *E. superba* collected in Antarctic seawaters in 1995 was 0.1 ng/g wet wt [7], and levels in amphipods (*Monoporeia affinis*) and isopods (*Saduria entomon*) from the Baltic Sea (in 1991–1993) ranged between 12 and 69 ng/g and 3.9 and 16 ng/g, respectively [65].

A comparison between levels and trends in Arctic and Antarctic biota indicates higher contamination levels in the Arctic organisms in relation to the geographical isolation of the Antarctic continent and Southern Ocean, which make it difficult for chemicals to reach this region. Temporal trends are available for the Arctic where many studies have been carried out during the last two decades (for a review of data, see [50–52,54,62,66–70]).

By comparing data available for the two polar regions, a different pattern emerges in the Antarctica where a clear time trend is not recognizable, owing to the paucity of data and unevenness of reports. However, the Arctic is showing a decreasing trend of contamination following the peak that occurred in the past few years and in relation to the great POP use/production and subsequent ban, while the Antarctic region seems to be characterized by a possible decline for some chemicals no longer used, and by a continuing leveling off or increase of other POPs still in use. In this regard, compared to the Arctic, it could be hypothesized that a slight delay in the transport, accumulation, and level decrease of POPs in the southern polar region are an effect of the geography and of the chemical transport pathways.

25.7 CONCLUSIVE REMARKS AND PERSPECTIVES

The paucity of data on the presence of toxic contaminants in the Antarctic organisms makes the assessment of the temporal trend and profile of contamination very difficult. For many classes of organisms, only few records are available and are often reported using different criteria such as unit of measures, body zone where tissue is excized, time of collection and species. For instance, the knowledge of the age and sex of organisms may affect the order of magnitude of the concentrations as well as the time of their biological life cycle (before, during, and after the reproductive period). Extreme environments promote the selection of peculiar adaptations like the use of lipids to store energy, to protect against cold, to aid buoyancy in fish; the presence of this lipid component may affect the concentrations of lipophilic contaminants such as POPs in relation to the season and period of the biological life cycle during which sampling is carried out. Therefore, POP concentrations may vary depending on all such factors. Moreover, articles report concentrations of different families of POPs, making it difficult to compare data. Finally, different analytical procedures and the increasing instrumental sensitivity of detection should be kept in mind when evaluating a time trend over decades.

In the light of all these considerations, we can hypothesize a trend for the contaminants mostly studied such as PCBs, HCB, and DDTs. They increased during the 1970s–1980s as a result of their global transport and dispersion. During the first half of the 1990s, concentrations in biota started to show a light decrease following the reduced use in those countries that banned them, which were also the same countries that were responsible for their major production and use. The new increases observed at the end of the 1990s may be the result of historical uses of PCBs and DDTs worldwide and in particular in American, European, and some Asian countries. DDT can reach the Antarctic area from other parts of the world where it is still used or stocked. Regarding PCBs, a recent article demonstrated that PCBs were volatilizing from surface waters to the overlying atmosphere [41], suggesting a possible source of PCBs that, after the volatilization at lower latitudes, may reach the Southern Ocean and Antarctica through long-range air transport [5]. HCB showed a strong increasing concentration with decreasing temperature in the northern hemisphere; the behavior of HCB was found to be consistent with PCBs, suggesting similar sources and fate of these persistent compounds [42].

On the contrary, the concentrations of legacy POPs in Arctic biota have shown significantly decreasing trends over the past two to three decades, with only a few examples of significantly increasing trends; decreasing trends of PCBs, DDTs, *alpha*-HCH, *gamma*-HCH, and HCB were found in all groups of animals studied, and the rate of decrease varied among species and geographical areas [54]. It might be that global transport to Antarctica and equilibrium between phases of POPs follow different mechanisms of time scale in the two polar regions; Antarctica could show a delay in the chronological steps that characterize the distribution of POPs globally, in relation to its geographical isolating and subsequent peculiar pathways of transport.

Other than the old POPs, another group of toxic contaminants have captured the interest of the scientific community: emergent POPs such as brominated flame retardants (e.g., PBDEs), perfluorinated compounds (e.g., PFOS and PFOA), and pharmaceutical and personal care products (PPCPs, e.g., synthetic fragrances). To the author's knowledge, there are very few papers concerning the presence of PBDEs [14,18,23,71–73] and PFCs [44,74], and none of PPCPs in Antarctic organisms. They have been published during the 2000s and therefore, it is not possible to make any temporal trend evaluation. As far as PBDEs, an increasing trend was observed in the Canadian Arctic from 1981 to 2000 [75]; several PBDEs, including highly brominated congeners like BDE209, showed increasing temporal trends in Greenland peregrine falcons [68]. A temporal trend in concentrations for the lower brominated BDEs is reported; it indicates a continued increase or a tendency to leveling off or possible declines, depending on the matrix studied and the geographic location [68].

PFCs can be transported to the arctic environment via the atmosphere; snow samples, representing deposition from 1996 to 2005, from the Canadian Arctic showed relatively constant deposition fluxes of perfluorinated carboxyl acids (PFCAs), but decreasing fluxes of perfluorooctanesulfonic acid (PFOS) from 1998 to 2001, followed by constant fluxes thereafter [67]. PFC temporal trends in Arctic wildlife may indicate that the arctic environment is responding to changes in perfluorooctane-sulfonyl fluoride (PFOSF) production, although additional studies in other regions of the arctic are needed [67].

Among emergent POPs, studies of polychlorinated naphthalenes (PCNs) are very scarce both in the Arctic and Antarctic environments. In the Arctic, no temporal trend data are available for PCNs in any polar species, except for the Arctic cod from Norway, where no change in concentration was reported from 1987 to 1998 [57]. Only a paper on PCNs in Antarctic organisms has been found and it reports results for samples collected in 1994–1996 in the Ross Sea [7,76], thus no temporal trend assessment is possible.

It is clear that the paucity of data makes any assessment of trends and profiles of contamination not easy, in particular for some species and areas. The challenge of the scientific community for the future should be a coordinated monitoring based on specific and shared criteria of sampling and reporting of data. This is a very important point, especially in light of both the recent emergent POP increasing levels in the Arctic and the possible delay of their transport and deposition in the Antarctic and sub-Antarctic regions.

REFERENCES

1. Wania, F. and D. Mackay. 1993. Global fractionation and cold condensation of low volatility organochlorine compounds in polar regions. *Ambio* 23(1): 10–18.
2. Wanwimolruk, S., H. Zhang, P.F. Coville, D.J. Saville, and L.S. Davis. 1999. In vitro hepatic metabolism of a CYP3A-mediated drug, quinine, in Adélie penguin. *Comparative Biochemistry and Physiology Part C* 124: 301–307.
3. Iwata, H., S. Tanabe, N. Sakai, and R. Tatsukawa. 1993. Distribution of persistent organochlorines in the oceanic air and surface seawater and the role of the ocean on their global transport and fate. *Environmental Science & Technology* 27: 1080–1098.

4. Tanabe, S. and R. Tatsukawa. 1986. In *PCBs and the Environment*, pp. 144–159. Boca Raton, FL: CRC Press.

5. Wania, F. 2003. Assessment the potential of persistent organic chemicals for long-range transport and accumulation in polar regions. *Environmental Science & Technology* 37(7): 1344–1351.

6. Bustnes, J.O., T. Tveraa, J.A. Henden, V. Oistein, K. Janssen, and J.U. Skaare. 2006. Organochlorines in Antarctic and Arctic avian top predators: A comparison between the South Polar Skua and two species of northern hemisphere gulls. *Environmental Contamination and Technology* 40(8): 2826–2831.

7. Corsolini, S., K. Kannan, T. Imagawa, S. Focardi, and J.P. Giesy. 2002. Polychloronaphtalenes and other dioxin-like compounds in Arctic and Antarctic marine food webs. *Environmental Science & Technology* 36(16): 3490–3496.

8. De Boer, J. and P. Wester. 1991. Chlorobiphenyls and organochlorine pesticides in various sub-Antarctic organisms. *Marine Pollution Bulletin* 22(9): 441–447.

9. Dickhut, R.M., A. Cincinelli, M. Cochran, and H.W. Dicklow. 2005. Atmospheric concentration and air-water flux of organochlorine pesticides along the western Antarctic peninsula. *Environmental Science & Technology* 39(2): 465–470.

10. Risebrough, R.W., P. Rieche, D.B. Peakall, S.G. Herman, and M.N. Kirven, 1968. Polychlorinated biphenyls in the global ecosytem. *Nature* 220: 1098–1102.

11. Tanabe, S. and R. Tatsukawa, 1983. Vertical transport and residence time of chlorinated hydrocarbons in the open ocean water column. *Journal of the Oceanographical Society of Japan* 39(2): 53–62.

12. Hamner, W.M., P.P. Hamner, S.W. Strand, and R.W. Gilmer. 1983. Behavior of Antarctic krill, *Euphausia superba*: Chemoreception, feeding, schooling, and molting. *Science* 220(4595): 433–435.

13. Nicol, S. and Y. Endo. 1997. *Fisheries Technical Paper 367: Krill Fisheries of the World*, ed. 1997, F, 367. Rome: FAO.

14. Bengtson Nash, S.M., A.H. Poulsen, S. Kawaguchi, W. Vetter, and M. Schlabach. 2008. Persistent organohalogen contaminant burdens in Antarctic krill (*Euphausia superba*) from the eastern Antarctic sector: A baseline study. *Science of The Total Environment* 407(1): 304–314.

15. Cincinelli, A., T. Martellini, M. Del Bubba, L. Lepri, S. Corsolini, N. Borghesi, M.D. King, and R.M. Dickhut. 2009. Organochlorine pesticide air-water exchange and bioconcentration in krill in the Ross Sea. *Environmental Pollution* 157(7): 2153–2158.

16. Cipro, C.V.Z., S. Taniguchi, and R.C. Montone. 2010. Occurrence of organochlorine compounds in *Euphausia superba* and unhatched eggs of pygoscelis genus penguins from Admiralty Bay (King George Island, Antarctica) and estimation of biomagnification factors. *Chemosphere* 78(6): 767–771.

17. Corsolini, S., N. Ademollo, T. Romeo, S. Olmastroni, and S. Focardi. 2003. Persistent organic pollutants in some species of a Ross Sea pelagic trophic web. *Antarctic Science* 15(1): 95–104.

18. Corsolini, S., A. Covaci, N. Ademollo, S. Focardi, and P. Schepens. 2006. Occurrence of organchlorine pesticides (OCPs) and their enantiomeric signatures, and concentrations of polybrominated diphenyl ethers (PBDEs) in the adélie penguin food web, Antarctica. *Environmental Pollution* 140: 371–382.

19. Corsolini, S. and S. Focardi, eds. 2000. *Bioconcentration of Polychlorinated Biphenyls in the Pelagic Food Chain of the Ross Sea*. Berlin, Germany: Springer-Verlag.

20. Corsolini, S., T. Romeo, N. Ademollo, S. Greco, and S. Focardi. 2002. POPs in key species of marine Antarctic ecosystems. *Microchemical Journal* 73: 187–193.

21. Hidaka, H., S. Tanabe, M. Kawano, and R. Tatsukawa. 1984. Fate of DDTs, PCBs and chlordane compounds in the Antarctic marine ecosystem. *Sixth Symposium on Polar Biology*, Japan.

22. Kumar, K.S., K. Kannan, S. Corsolini, T. Evans, J.P. Giesy, J. Nakanishi, and S. Masunaga. 2002. Polychlorinated dibenzo-p-dioxins, dibenzofurans and polychlorinated biphenyls in polar bear, penguin and South Polar Skua. *Environmental Pollution* 119(2): 151–161.

23. Chiuchiolo, A.L., R.M. Dickhut, M.A. Cochran, and H.W. Ducklow. 2004. Persistent organic pollutants at the base of the Antarctic marine food web. *Environmental Contamination and Technology* 38(13): 3551–3557.

24. Sen Gupta, R., A. Sarkar, and T.W. Kureishey. 1996. PCBs and organochlorine pesticides in krill, birds and water from Antarctica. *Deep Sea Research II* 43(1): 119–126.

25. Focardi, S., L. Lari, and L. Marsili. 1992. PCB congeners, DDTs and hexachlorobenzene in Antarctic fish from terra nova bay (Ross Sea). *Antarctic Science* 4(2): 151–154.

26. Corsolini, S., N. Ademollo, and S. Focardi. 2003. Persistent organic pollutants in selected organisms of an Antarctic benthic community. *Organohalogen Compounds* 61: 329–332.

27. Focardi, S., R. Bargagli, and S. Corsolini. 1995. Isomer-specific analysis and toxic evaluation of polychlorinated biphenyls in Antarctic fish, seabirds and Weddell seals from Terra Nova Bay (Ross Sea). *Antarctic Science* 7(1): 31–35.

28. Weber, K. and H. Goerke. 2003. Persistent organic pollutants (POPs) in Antarctic fish: Levels, patterns, changes. *Chemosphere* 53(6): 667–678.

29. Corsolini, S., N. Ademollo, M. Mariottini, S. Fossi, C. Guerranti, G. Perra, G. Duhamel, and S. Focardi. 2002. Polychlorinated biphenyls, polychlorinated-dibenzodioxins and -dibenzofurans in mackerel ice-fish and marbled rockcod from the Kerguelen islands (Antarctica). *Organohalogen Compounds* 57: 161–164.

30. Corsolini, S. 2009. Industrial contaminants in Antarctic biota. *Journal of Chromatography A* 1216(3): 598–612.

31. Sladen, W.J.L., C.M. Menzie, and W.L. Reichel. 1966. DDT residues in adelie penguins and crabeater seal from Antarctica. *Nature* 210: 670–673.

32. Luke, B.G., G.W. Johnstone, and E.J. Woehler. 1989. Organochlorine pesticides, PCBs and mercury in Antarctic and sub-Antarctic seabirds. *Chemosphere* 19(12): 2007–2021.

33. Schneider, R., G. Steinhagen-Schneider, and H.E. Drescher. 1985. Organochlorines and heavy metals in seals and birds from the Weddell sea. In *Antarctic Nutrient Cycles and Food Webs*, eds. Siegfried, W.R., Condy P.R., and Laws R.M., pp. 652–655. Berlin: Springer-Verlag.

34. Corsolini, S. 1998. Indagine sulla contaminazione da policlorobifenili e valutazione della loro tossicità in organismi del mare mediterraneo e del mare di ross (antartide)/study of polychlorinated biphenyls and toxicity assessment in organisms from the Mediterranean sea and the Ross Sea (Antarctica). PhD, Siena.

35. Risebrough, R.W., I.W. Walker, T.T. Schimdt, D.W. De Lappe, and C.W. Connors. 1976. Transfer of chlorinated biphenyls to Antarctica. *Nature* 264(264): 738–739.

36. Court, G.S., L.S. Davis, S. Focardi, R. Bargagli, C. Fossi, C. Leonzio, and L. Marsili. 1997. Chlorinated hydrocarbons in the tissues of South Polar Skua (*Catharata mackormicki*) and adélie penguin (pygoscelis adeliae) from Ross Sea, Antarctica. *Environmental Pollution* 97(3): 295–301.

37 Geisz, H.N., R.M. Dickhut, M.A. Cochran, W.R. Fraser, and H.W. Ducklow. 2008. Melting glaciers: A probable source of DDT to the Antarctic marine ecosystem. *Environmental Science & Technology* 42(11): 3958–3962.

38. Corsolini, S., N. Ademollo, and S. Focardi, 2001 Chlorinated hydrocarbons (p,p'-DDE and PCBs) in adelie penguin and South Polar Skua eggs from Edmonson point (Victoria land, Antarctica). In *VIII SCAR International Biology Symposium "Antarctic Biology in a Global context"*, ed. Huiskes, Ahl, S6P04. Amsterdam, the Netherlands: Vrije Universiteit.

39. UNEP. 2002. *Regionally Based Assessment of Persistent Toxic Substances. Mediterranean Regional Report*, ed. Substances, Rbaopt.

40. Wania, F. and C.B. Dugani. 2003. Assessing the long-range transport potential of polybrominated diphenyl ethers: A comparison of four multimedia models. *Environmental Toxicology and Chemistry* 22(6): 1252–1261.

41. Zhang, L. and R. Lohmann. Cycling of PCBs and HCB in the surface ocean-lower atmosphere of the open pacific. *Environmental Science & Technology* 44(10): 3832–3838.

42. Lohmann, R., W.A. Ockenden, J. Shears, and K.C. Jones. 2001. Atmospheric distribution of polychlorinated dibenzo-*p*-dioxins, dibenzofurans (PCDD/FS), and non-*ortho* biphenyls (PCBs) along a north-south Atlantic transect. *Environmental Science & Technology* 35: 4046–4053.

43. Schiavone, A., S. Corsolini, N. Borghesi, and S. Focardi. 2009. Contamination profiles of selected PCB congeners, chlorinated pesticides, PCDD/FS in Antarctic fur seal pups and penguin eggs. *Chemosphere* 76(2): 264–269.

44. Schiavone, A., S. Corsolini, K. Kannan, L. Tao, W. Trivelpiece, D. Torres Jr., and S. Focardi. 2009. Perfluorinated contaminants in fur seal pups and penguin eggs from south Shetland, Antarctica. *Science of The Total Environment* 407(12): 3899–3904.

45. Rakusa-Suszczewski, S. and A. Krzyszowska. 1991. Assessment of the environmental impact of the "H. Arctowski" Polish Antarctic station (Admiralty Bay, King George Island, South Shetland Islands). *Polish Polar Research* 12(1): 105–121.

46. Bargagli, R., S. Corsolini, M.C. Fossi, J.C. Sanchez-Hernandez, and S. Focardi. 1998. Antarctic fish *Trematomus bernacchii* as biomonitor of environmental contaminants at terra nova bay station (Ross Sea). *Memoirs of National Institute of Polar Research*, 32: 220–229.

47. Hidaka, H. and R. Tatsukawa. 1983. Environmental pollution of chlorinated hydrocarbons around Syowa station. *Antarctic Record* 80: 14–29.

48. Cripps, G.C. 1990. The extent of hydrocarbon contamination in the marine environment from a research station in the Antarctic. *Marine Pollution Bulletin* 25: 288–292.

49. Amap. 2009. Amap assessment 2009: *Human Health in the Arctic*, ed. (Amap), Amap, 256. Oslo, Norway.

50. Bidleman, T.F., P.A. Helm, B.M. Braune, and G.W. Gabrielsen. 2010. Polychlorinated naphthalenes in polar environments—A review. *Science of The Total Environment* 408(15): 2919–2935.

51. Letcher, R.J., J.O. Bustnes, R. Dietz, B.M. Jenssen, E.H. Jørgensen, C. Sonne, J. Verreault, M.M. Vijayan, and G.W. Gabrielsen. 2010. Exposure and effects assessment of persistent organohalogen contaminants in Arctic wildlife and fish. *Science of The Total Environment* 408(15): 2995–3043.

52. Muir, D.C.G. and C.A. De Wit. 2010. Trends of legacy and new persistent organic pollutants in the circumpolar Arctic: Overview, conclusions, and recommendations. *Science of The Total Environment* 408(15): 3044–3051.

53. Weber, J., C.J. Halsall, D. Muir, C. Teixeira, J. Small, K. Solomon, M. Hermanson, H. Hung, and T. Bidleman. 2010. Endosulfan, a global pesticide: A review of its fate in the environment and occurrence in the Arctic. *Science of The Total Environment* 408(15): 2966–2984.

54. Rigét, F., A. Bignert, B. Braune, J. Stow, and S. Wilson. 2010. Temporal trends of legacy POPs in Arctic biota, an update. *Science of The Total Environment* 408(15): 2874–2884.

55. Ryan, M.J., G.A. Stern, M. Diamon, M.V. Croft, P. Roach, and K. Kidd. 2005. Temporal trends of organochlorine contaminants in burbot and lake trout from three selected Yukon lakes. *Science of The Total Environment* 351–352: 501–522.

56. Braune, B.M., M.L. Mallory, H.G. Gilchrist, R.J. Letcher, and K.G. Drouillard. 2007. Levels and trends of organochlorines and brominated flame retardants in ivory gull eggs from the Canadian Arctic, 1976 to 2004. *Science of The Total Environment* 378: 403–417.

57. Sinkonnen, S. and J. Paasivirta. 2000 Polychlorinated organic contaminants in the Arctic cod liver: Trends and profiles. *Chemosphere* 40: 619–626.

58. Braune, B.M., P.M. Outridge, A.T. Fisk, D.C. Muir, P.A. Helm, K. Hobbs, P.F. Hoekstra, Z.A. Kuzyk, M. Kwan, R.J. Letcher, W.L. Lockhart, R.J. Norstrom, and G.A. Stern. 2005. Persistent organic pollutants and mercury in marine biota of the Canadian Arctic: An overview of spatial and temporal trends. *Science of The Total Environment* 351–352: 4–56.

59. Helgason, L.B., R. Barrett, E. Lie, A. Polder, J.U. Skaare, and G.W. Gabrielsen. 2008. Levels and temporal trends (1983–2003) of persistent organic pollutants (POPs) and mercury (HG) in seabird eggs from northern Norway. *Environmental Pollution* 155: 190–198.

60. Pol, S.S.W., P.R. Becker, J.R. Kucklick, R.S. Pugh, D.G. Roseneau, and K.S. Simac. 2004. Persistent organic pollutants in Alaskan murre (uria spp.) eggs: Geographical, species, and temporal comparisons. *Environmental Science & Technology* 38: 1305–1312.

61. Lebeuf, M., L.M. Noë, S. Trottier, and L. Measures. 2007. Temporal trends (1987–2002) of persistent, bioaccumulative and toxic chemicals in beluga whales (*Delphinapterus leucas*) from St. Lawrence estuary, Canada. *Science of The Total Environment* 383: 216–231.

62. Hung, H., R. Kallenborn, K. Breivik, Y. Su, E. Brorström-Lundén, K. Olafsdottir, J.M. Thorlacius, S. Leppänen, R. Bossi, H. Skov, S. Manø, G.W. Patton, G. Stern, E. Sverko, and P. Fellin. 2010. Atmospheric monitoring of organic pollutants in the Arctic under the Arctic monitoring and assessment programme (AMAP): 1993–2006. *Science of The Total Environment* 408(15): 2854–2873.

63. Unep, U.N.E.P. 2008. Preliminary information on risk management evaluation of PFOS's in China. In *Stockholm Convention on Persistent Organic Pollutants*, ed. Programme, Une, 6. Geneva, CH.

64. Unep, U.N.E.P. 2009. Report of the conference of the parties of the Stockholm Convention on persistent organic pollutants on the work of its fourth meeting. In *Stockholm Convention on Persistent Organic Pollutants*, ed. Programme, Une, 112. Geneva, CH.

65. Lundgren, K., M. Tysklind, R. Ishaq, D. Broman, and B. Van Bavel. 2002. Polychlorinated naphthalene levels, distribution and biomagnification in a benthic food chain in the Baltic Sea. *Environmental Science & Technology* 36: 5005–5013.

66. Bustnes, J.O., G.W. Gabrielsen, and J. Verreault. 2010. Climate variability and temporal trends of persistent organic pollutants in the Arctic: A study of Glaucous Gulls. *Environmental Science & Technology* 44(8): 3155–3161.

67. Butt, C.M., U. Berger, R. Bossi, and G.T. Tomy. 2010. Levels and trends of poly- and perfluorinated compounds in the Arctic environment. *Science of The Total Environment* 408(15): 2936–2965.

68. De Wit, C.A., D. Herzke, and K. Vorkamp. 2010. Brominated flame retardants in the Arctic environment—Trends and new candidates. *Science of The Total Environment* 408(15): 2885–2918.

69. De Wit, C.A. and D. Muir. 2010. Levels and trends of new contaminants, temporal trends of legacy contaminants and effects of contaminants in the Arctic: Preface. *Science of The Total Environment* 408(15): 2852–2853.

70. Hoferkamp, L., M.H. Hermanson, and D.C.G. Muir. 2010. Current use pesticides in Arctic media; 2000–2007. *Science of The Total Environment* 408(15): 2985–2994.

71. Corsolini, S., N. Borghesi, A. Schiavone, and S. Focardi. 2007. Polybrominated diphenyl ethers, polychlorinated dibenzo-dioxins, -furans, and -biphenyls in three species of Antarctic penguins. *Environmental Science and Pollution Research* 14(6): 421–429.

72. Hale, R.C., S.L. Kim, T. Matt Mainor, E.O. Bush, and E.M. Jacobs. 2008. Antarctic research bases: Local source of polychlorinated diphenyl ether (PBDE) flame retardants. *Environmental Science & Technology* 42: 1452–1457.

73. Yogui, G.T. and J.L. Sericano. 2009. Levels and pattern of polybrominated diphenyl ethers in eggs of Antarctic seabirds: Endemic versus migratory species. *Environmental Pollution* 157(3): 975–980.

74. Tao, L., K. Kannan, N. Kajiwara, M.M. Costa, G. Fillmann, S. Takahashi, and S. Tanabe. 2006. Perfluorooctanesulfonate and related fluorochemicals in albatrosses, elephant seals, penguins, and South Polar Skua from the Southern Ocean. *Environmental Science & Technology* 40(24): 7642–7648.

75. Ikonomou, M.G., S. Rayne, and R.F. Addison. 2002. Exponential increases of the brominated flame retardants, polybrominaed diphenyl ethers, in the Canadian Arctic from 1981 to 2000. *Environmental Science & Technology* 36(9): 1886–1892.

76. Schiavone, A., K. Kannan, Y. Horii, S. Focardi, and S. Corsolini. 2010. Polybrominated diphenyl ethers, polychlorinated naphthalenes and polycyclic musks in human fat from Italy: Comparison to polychlorinated biphenyls and organochlorine pesticides. *Environmental Pollution* 158(2): 599–606.

26 Global Distribution of PFOS and Related Chemicals

Nobuyoshi Yamashita,* Leo W.Y. Yeung, Sachi Taniyasu, Karen Y. Kwok, Gert Petrick, Toshitaka Gamo, Keerthi S. Guruge, Paul K.S. Lam, and Bommanna G. Loganathan

CONTENTS

26.1 Introduction 593
    26.1.1 Physicochemical Properties of PFCs..594
    26.1.2 Production of PFCs...596
        26.1.2.1 Electrochemical Fluorination 596
        26.1.2.2 Telomerization 596
    26.1.3 Sources...597
        26.1.3.1 Sources of PFASs..597
        26.1.3.2 Sources of PFCAs...598
26.2 Analytical Methods 600
    26.2.1 Gas Chromatography Mass Spectroscopy..600
    26.2.2 High-Performance Liquid Chromatography–Tandem Mass Spectroscopy.............601
    26.2.3 Quality Assurance and Quality Control and ISO25101 601
26.3 Global Transportation of PFCs...602
    26.3.1 PFC Levels in Air and Dust...603
    26.3.2 PFC Levels in Precipitation 603
    26.3.3 PFC Levels in Ocean Waters 604
    26.3.4 Vertical Profiles of PFOS and PFOA in Open Ocean Water Columns 608
26.4 Global Distribution of PFCs in Biological Samples 610
    26.4.1 Birds..611
    26.4.2 Mammals..611
    26.4.3 Temporal Trend in Wild Animals...615
26.5 Future Directions..618
References...619

26.1 INTRODUCTION

Perfluorinated compounds (PFCs) are synthetic organic chemicals used in a variety of industrial and commercial applications, including surfactants in pesticides, surface protectors in textiles, furnishings, and food packaging. PFCs become emerging chemicals of concern and two of them, namely perfluorooctane sulfonate (PFOS) and perfluorooctane sulfonyl fluoride (POSF) were listed as "restricted use" (Annex B) under the Stockholm Convention in May 2009. PFCs have an anionic functional group and nonpolar perfluoroalkyl chain. This moiety allows them to repel both water

---

* E-mail: nob.yamashita@aist.go.jp (Chapter corresponding author).

and oil, and accounts for their surfaces tension/leveling properties. Additionally, the high-energy carbon–fluorine bond accounts for the chemical and physical stability of these compounds. PFOS and perflurooctanoate (PFOA), having an eight carbon, were the two most well-known PFCs. PFCs have been manufactured and used as surfactant processing aids in the production of fluoropolymers, coatings for clothing, fabrics, upholstery and carpets, in paper products approved for food contact, and in aqueous film-forming foams (AFFF) for the past six decades [1]. PFCs are ubiquitous in the environment, and different governmental agencies have taken corresponding actions to regulate these compounds. The use of PFOS is regulated in the United States; further regulations of PFOS and related compounds have been implemented in the EU (Directive 76/769) from December 2007 in Europe, while the remaining permitted uses will be phased out by 2011 (OJ Directive 2006/122/ECOF). In Japan, according to the Chemical Substance Control Law, PFOS and POSF were listed as class I hazardous chemicals in April 2010. To a certain extent, they were classified as one of the contaminants, similar to polychlorinated biphenyls (PCBs), which can pose the highest risks to human. On the other hand, PFOA was classified as a Class II chemical on the watch list in 2002. PFOA-related chemicals such as perfluorododecanoic acid, perfluorotridecanoic acid, perfluorotetradecanoic acid, perfluoropentadecanoic acid, and perfluorohexadecanoic acid were classified as Class I chemicals on the watch list in 2008. Concern about fluorinated organic compounds, particularly PFCs, has been growing since the late 1990s because of the ubiquitous occurrence of PFCs in the environment, especially since they were found in biota (i.e., polar bear) from remote arctic regions where no PFC-related production facilities existed [1–7].

In this chapter, a comprehensive overview of PFCs, physicochemical properties, the history of industrial productions and use, possible sources in the environment, and environmental levels and trends in both biota and inorganic matrices are described.

26.1.1 PHYSICOCHEMICAL PROPERTIES OF PFCs

PFCs are characterized by varying lengths of carbon chains in which all hydrogen atoms are substituted by fluorine atoms. All PFCs found in the environment are anthropogenic, in that they have been manufactured and used for more than 60 years. Because of their unique properties, they have been widely used in a variety of commercial and industrial products. Currently, concerned PFCs can be divided into two main groups: (1) perfluoroalkyl sulfonates (PFASs) and (2) perfluoroalkyl carboxylates (PFCAs). PFOS and PFOA, which have eight carbon chain lengths, are the representatives for the two groups, respectively (Figure 26.1).

Apart from manufacturing, PFASs and PFCAs seem to be the degradation products of their corresponding precursors. Perfluorooctylsulfonamides (Figure 26.2) are one of the potential precursors for PFASs, while fluorotelomer alcohols (FTOHs) (Figure 26.3) can degrade to yield PFCAs [8,9].

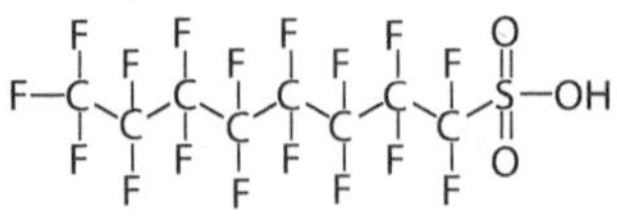

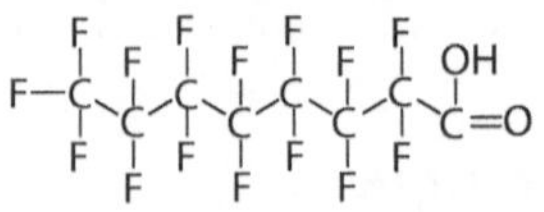

FIGURE 26.1 Chemical structures of PFOS and PFOA.

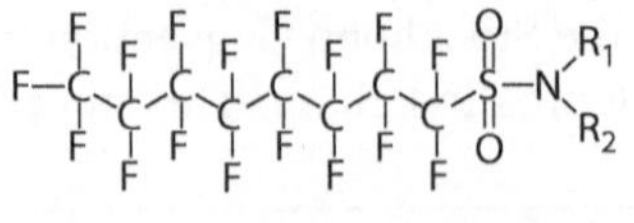

FIGURE 26.2 Chemical structure of perfluorooctylsulfonamides, where $R = CH_2CH_3$, CH_2CH_2OH, CH_2OH, or H.

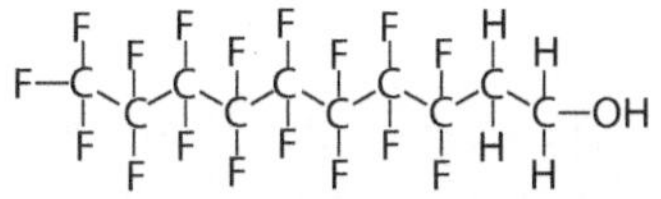

FIGURE 26.3   An example of FTOH, heptadecafluoro-1-decanol (8:2 FTOH).

TABLE 26.1
Physicochemical Properties of PFCs

Compounds	Boiling Point (°C)	Melting Point (°C)	Vapor Pressure at 20°C (Pa)	Water Solubility (mg/L)	pK_a	Henry's Law Constant (atm m³/mol)
POSF	154–155		221			
PFOS	149	70–100				
PFOS K		>400	0.000331	570		7.2×10^{-9}
N-EtFOSA	~110	~90	0.16			
PFBA	120	−19.5	1333			
PFPeA	127					
PFHxA	157	12–14				
PFHpA	175–177					
PFOA	189–192	55	100	3400	2.5	4.6×10^{-6}
PFNA		71–77				
PFDA	218	83–85				
PFUnDA	160	96–101				
PFDoDA	245	107–109				
8:2 FTOH	95–105	42–44	365 (at 25°C)	0.14		9.6×10^{-2}

Sources: Giesy, J.P. and Kannan, K., *Environ. Sci. Technol.*, 35, 1339, 2001; USEPA, *Fed. Reg.*, 65, 62319, 2000; USEPA, Revised draft-Hazard assessment of perfluorooctanoic acids and its salts: Office of Pollution Prevention and Toxics, Risk Assessment Division, 2002; Hekster, F.M. et al., *Rev. Environ. Contam. Toxicol.*, 179, 99, 2003; Stock, N.L. et al., *Environ. Sci. Technol.*, 38, 1693, 2004.

Because of (1) the high electronegativity of fluorine atom (4.0), (2) the high energy of C–F bond (approx. 466 kJ/mol), (3) the small diameter of the fluorine atom, and (4) the three pairs of nonbonding electrons in fluorine's outer shell. PFCs are chemically and thermally stable and are strongly resistant to hydrolysis, photolysis, microbial degradation, and to metabolism by vertebrates [10,11]. The oleophobic and hydrophobic perfluorinated chains, when added to a hydrophilic charged moiety such as sulfonic acid or carboxylic acid, create the surfactant properties of PFCs. These molecules have both polar (charged moieties) and nonpolar (perfluorinated chains) domains that can lower the surface tension of water more than hydrocarbon-based surfactants, and therefore are more powerful wetting agents. These oleophobic and hydrophobic perfluorinated chains also enable the functionalized fluorochemicals to be water, oil, and fat resistant [10,12]. The physicochemical properties of some PFCs are shown in Table 26.1. Owing to the high water solubility and low vapor pressure of PFCs, the aquatic ecosystem is thought to be a major sink for these compounds.

Unlike other persistent organic pollutants (POPs) that accumulate in the fatty tissues, PFCs such as PFOA and PFOS are ionic, and polar surfactants therefore bind to blood proteins and accumulate in the liver and gallbladder. Hence, they are also bioaccumulative [17]. Properties of PFCs enable them to be globally distributed in both abiotic and biotic matrices. Concentrations of PFCs have been detected in human blood [7,18], breast milk [19], seafood [20–22], wildlife [1,23,24], and a variety of water bodies [25,26].

26.1.2 PRODUCTION OF PFCs

3M Company, the major global manufacturer of POSF, announced the phase-out of POSF-based materials in 2000 and butyl-based substances were used as a replacement. The synthesis of PFCs is based on either obtaining the perfluoroalkyl chain or the introduction of functional groups onto the fluorinated chain. The perfluoroalkyl chain can be obtained by two common methods: (1) electrochemical fluorination (ECF) and (2) a telomerization–fluorination process.

The 3M company have employed ECF to produce PFCs since 1950 [27]. In brief, all the hydrogen atoms of a hydrocarbon were replaced by fluorine atoms under an electric current [12]. PFCs were used widely in inks, varnishes, waxes, fire-fighting foam formulations, metal plating and cleaning, lubricant, water and oil repellents for textile and paper as well as leather [27,28]. PFCAs (e.g., PFOA) were also produced in 1947 using ECF [29]. This process yields about 35%–40% straight-chain POSF and a mixture of by-products and waste of unknown and variable compositions, such as branched-chain, straight-chain, or cyclic perfluoroalkylsulfonyl fluorides with various chain lengths with 8–9 fluorinated carbons as the major constituents [27,28]. PFOA was mainly manufactured as an ammonium salt (APFO) and its worldwide production using ECF ceased by 2002, though a limited number of small manufacturers is still in production in Europe and Asia. Telomerization (e.g., fluorotelomer iodide [FTI] oxidation, fluorotelomer olefin [FTO] oxidation, and FTI carboxylation) is another important manufacturing process producing PFCs [12]. DuPont uses the telomerization process, which yields linear, even-numbered perfluorocarbon chains [12]. Commercial products manufactured through the telomerization process are generally mixtures of polyfluorinated straight-chain compounds with ranges of even carbon numbers [14]. Ammonium perfluorononanoate (APFN) is manufactured in Japan by oxidation of a mixture of linear fluorotelomer olefins (mainly 8:2 FTOs) to the corresponding odd-numbered of PFCAs [30,31].

26.1.2.1 Electrochemical Fluorination

Electrochemical fluoridation (ECF) was invented by Joseph Simons of the 3M Company [32]. It refers to a process of fluorination of different fluorinated organic compounds (e.g., alkanesulfonyl acid chloride (Reaction 26.1), carboxylic acid chloride (Reaction 26.2)) in the presence of anhydrous hydrogen fluoride, yielding perfluorinated sulfonyl or carbonyl fluorides, and numerous by-products:

$$C_nH_{2n+1}COCl + (2n+2)HF \rightarrow C_nF_{2n+1}COF + HCl + \text{by-products} \qquad (26.1)$$

$$C_nH_{2n+1}SO_2Cl + (2n+2)HF \rightarrow C_nF_{2n+1}COF + HCl + \text{by-products} \qquad (26.2)$$

ECF products are a mixture of isomers and homologues. This process is inexpensive and generates various PFCs with 4–13 carbon chain lengths. The disadvantage of ECF is low process selectivity and considerable fragmentation of the carbon chain. POSF would be one of the most important products, serving as the basic building block of PFASs. Commercialized POSF-derived products contain approximately 70% linear and 30% branched impurities [10–12,33].

26.1.2.2 Telomerization

Telomerization was initially developed by the DuPont Company. It was another commercial process used for synthesizing perfluoroalkyl products. It begins with fluoroiodination of tetrafluoroethylene (TFE) in the presence of adequate telogens to produce pentafluoroiodoethane (Reaction 26.3), which is then reacted with TFEs and yields a mixture of perfluoroalkyl iodides (Reaction 26.4). Oxidation of pentafluoroiodoethane from Reaction 26.3 is an important step to form PFCAs. Intermediate perfluoroalkyl iodides in Reaction 26.5 were produced by the reaction of perfluoroalkyl iodides (Reaction 26.4) with ethylene:

$$5CF_2 = CF_2 + 2I_2 + IF_5 \rightarrow 5F(CF_2CF_2)I \qquad (26.3)$$

$$F(CF_2CF_2)I + nCF_2 = CF_2 \rightarrow F(CF_2CF_2)_{n+1}I \qquad (26.4)$$

$$F(CF_2CF_2)_nI + CH_2 = CH_2 \rightarrow F(CF_2CF_2)_{n+1}CH_2CH_2I \qquad (26.5)$$

These iodides can be easily converted to other intermediates (e.g., olefins, alcohol) to produce different PFCs (e.g., fluorotelomer sulfonates). Only linear and even numbers of fluorinated carbons in homologous fluoroalkyl chains will be generated during the telomerization process [12,33], which is a more advantageous method.

26.1.3　Sources

26.1.3.1　Sources of PFASs

Since PFCs have been manufactured and used for over 60 years, they have been released into the environment throughout their production, usage, and disposal. PFASs have been released into the environment from a series of manufacturing and use point sources (i.e., air to water) and more diffusely through widespread consumer use and disposal (i.e., landfill and incineration). 3M estimated that 85% indirect emissions were a result of losses from consumer products during use and disposal, while the remaining 15% was associated with manufacturing releases from secondary applications [34].

Upon ECF (Figure 26.4), POSF is the major intermediate, but it can further react to form perfluorooctanesulfonamide (PFOSA), N-alkylperfluorooctanesulfonamide (FOSA), and N-alkylperfluorooctanesulfonamidoethanol (FOSE). PFOSA, FOSA, and FOSE are volatile precursors of PFASs and many studies have reported concentrations of these volatile organic fluorinated precursors in air masses in different countries, which include North America [35], Germany [36], and Northwest Europe [37]. Additionally, POSF-derived industrial or commercial products such as carpet (Figure 26.5) could be one of the potential release sources, contributing to the concentrations of these compounds in the indoor air [36]. PFASs can also be detected in indoor dust samples [38,39]; however, the atmospheric degradation mechanisms have not yet been identified and thus more studies are needed.

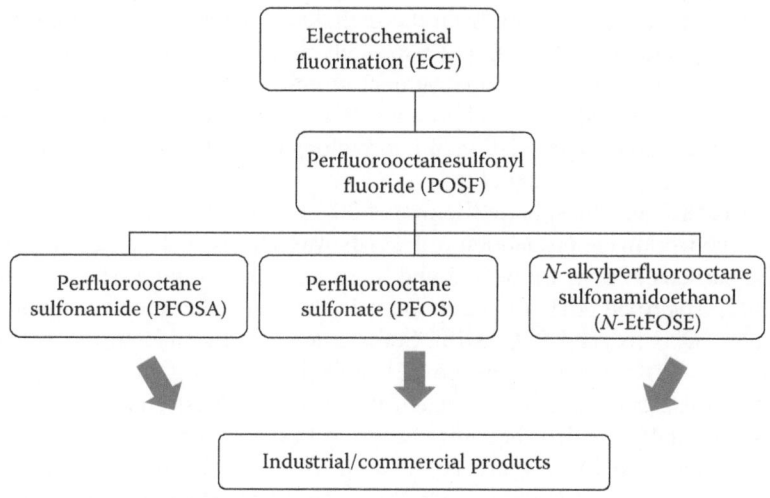

FIGURE 26.4　A simplified systematic diagram of ECF process.

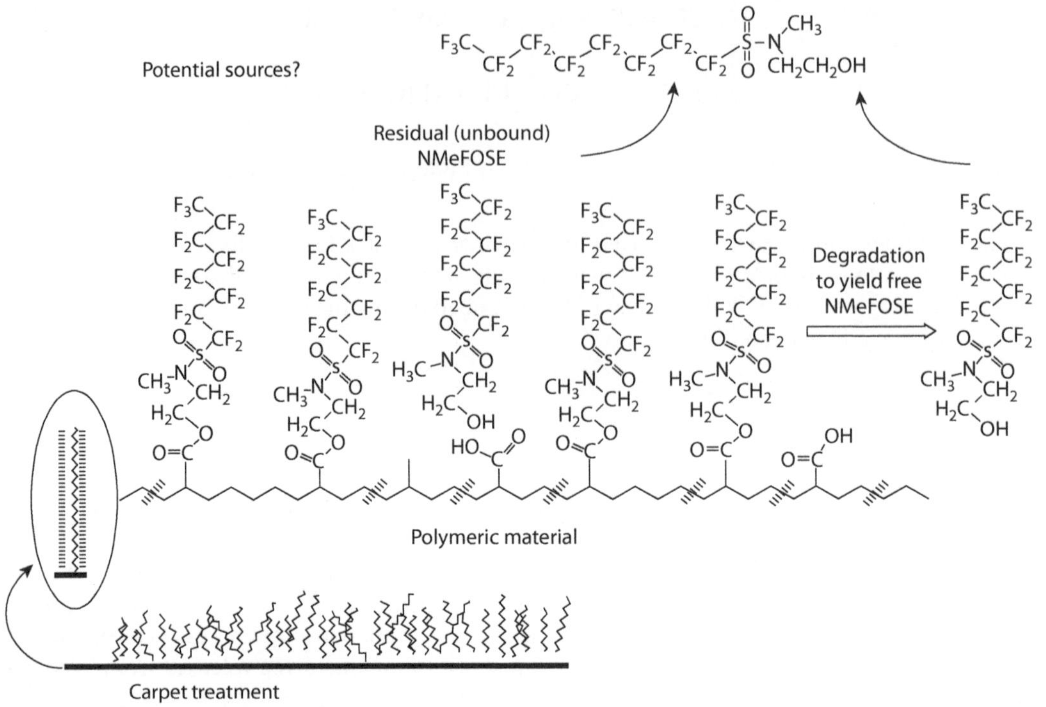

FIGURE 26.5 An illustration of potential release of *N*-methylperfluorooctane-sulfonamidoethanol (*N*-MeFOSE) to the environment due to residual (unbound) material in carpet treatment and degradation of the covalent linkage in the polymeric material. (With permission from Stock, N.L., Lau, F.K., Ellis, D.A., Martin, J.W., Muir, D.C.G., and Mabury, S.A., *Environ. Sci. Technol.*, 38, 991–996, 2004. Copyright 2004 American Chemical Society.)

26.1.3.2 Sources of PFCAs

For PFCAs, the global historical industry-wide emissions from direct (manufacture, use, and consumer products) and indirect (PFCA impurities or precursors) sources were estimated to be 3,200–7,300 ton [34], while for PFOS, it was estimated to be 42,250 ton to air and water between 1970 and 2012 [40].

There are two major sources of PFCA emission into the environment: (1) direct and (2) indirect. Direct sources might be resulting from the manufacturing and processing process of PFCA, ammonium perfluorooctanoate (APFO) and fluoropolymer, while water-soluble PFCA salts might be expected to enter the local aquatic environment directly (Figure 26.6). Second, the releases of AFFFs and other consumer and industrial products were also another direct source. PFCAs present as chemical impurities and the degradation of fluorotelomer-based products could be categorized as the indirect sources in the environment [34].

Volatile FTOHs have an atmospheric lifetime of 20 days and are supposed to be possible precursors of PFCAs. The worldwide production of FTOHs was approximately 12×10^6 kg/year. FTOHs, with fluorinated carbons of 6, 8, 10, were found in air masses of Japan [41], Asia, and the Western United States [42]. Hydroxyl (OH) radical present in the atmospheric environment would initiate the oxidation of FTOHs to yield PFCAs [8]. Some studies had simulated the atmospheric conditions using chlorine (Cl) radicals to replace OH radicals in a smog chamber, and found that 8:2 FTOH reacted and degraded to perfluorinated aldehydes (FTALs) and fluorotelomer carboxylic acids (FTCAs) and finally yielded the entire range of PFCAs, from trifluoroacetic acid (TFA) to perfluorononanoic acid (PFNA) (Figure 26.7) [8,9]. The atmospheric concentrations of FTOHs decreased with increasing chain lengths, leading to a decreasing trend of longer chain length PFCA concentrations in the environment [8].

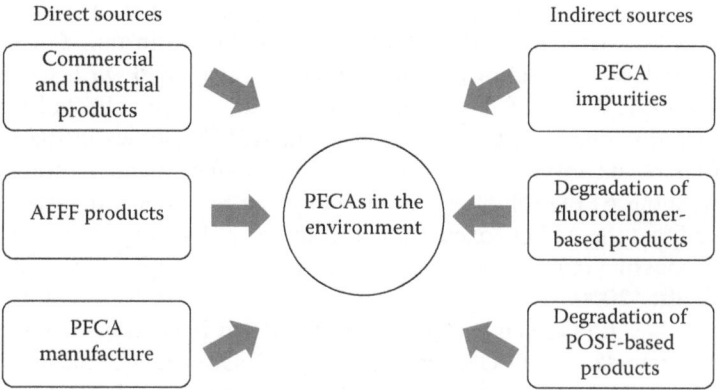

FIGURE 26.6 Direct and indirect sources of PFCAs. (Modified from Prevedouros, K. et al., *Environ. Sci. Technol.*, 40, 32, 2006.)

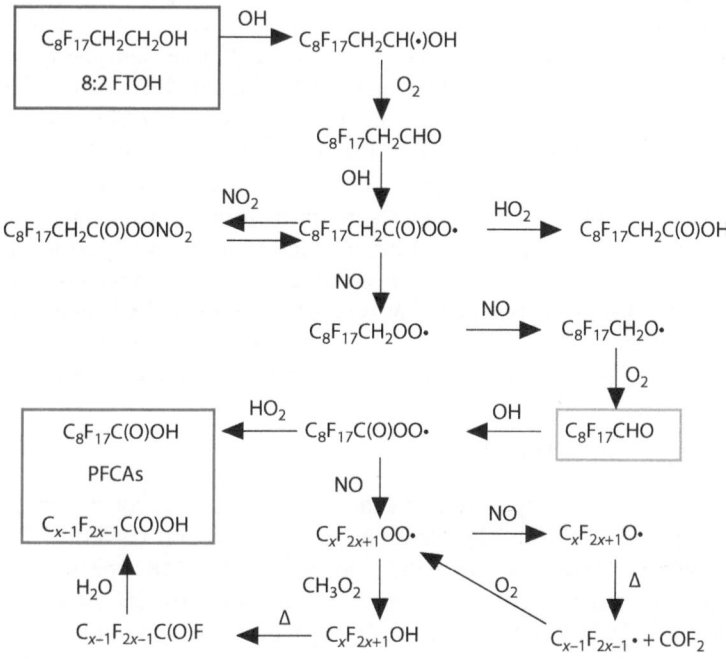

FIGURE 26.7 Proposed mechanisms for the atmospheric degradation of 8:2 FTOH. (With permission from Wallington, T.J., Hurley, M.D., Xia, J., Wuebbles, D.J., Sillman, S., Ito, A., Penner, J.E., Ellis, D.A., Martin, J., Mabury, S.A., Nielsen, O.J., and Andersen, M.P.S., *Environ. Sci. Technol.*, 40, 924–930, 2006. Copyright 2006 American Chemical Society.)

Different PFC-related manufacturing plants also carried different actions to phase-out or reduce the release of these chemicals in the environment. For example, in 2000, the 3M Company voluntarily ceased the manufacture of perfluorooctanyl-related materials because PFOS was found to be persistent in humans and wildlife. ECF-derived fluorochemicals were largely phased out in 2002, but the production of other compounds using the telomerization process continues [28]. Landfilling for the disposal of the solid waste arising from POSF-related manufacturing processes and off-spec aquatic discharges were ceased by 1998 [28]. Although it was already reported that most of the PFCs in waste materials can be removed by simple rinse by pH adjusted water [43], there is no action to decrease PFC residues in waste materials before incineration and landfill at this time. An activated carbon adsorption system was installed in the manufacturing plants in order to reduce the loadings of POSF-related waste or PFOS discharging into wastewater treatment work [28]. All of these treated wastewaters were collected for further hazardous wastes incineration disposal. Large reductions in APFO releases in the environment were reported [44] after recent installations in some fluoropolymer manufacturers allowed them to capture and recycle the APFO solutions. Additionally, the Fluoropolymer Manufacturing Group (FMG) announced that the APFO concentration would be reduced by at least 90% in fluoropolymer dispersion processing [45]. Based on the above actions, the emissions of PFOS/PFOA should be much reduced after 2000.

POSF-based products/PFOS can get into the environment in several ways (i.e., direct vs. indirect). One of the direct releases of POSF-based products/PFOS would be the use of AFFFs in military force or accidents [46]. An indirect emission is another way for POSF-based products/PFOS to get into the environment. An estimate of 85% of the indirect emissions were coming from losses from consumer products during use and disposal (e.g., carpet, apparel, paper, and packaging), and the remaining 15% might be related to manufacturing releases or wastes [47]. As for PFCAs (i.e., APFO, APFN), they were used for fluoropolymer production, dispersion processing, AFFFs, and consumer and industrial products [34]. Briefly, there were two indirect sources (i.e., originated from POSF-based products or fluorotelomer-based products) for PFCAs into the environment. PFCAs might be coming from POSF-based products (e.g., POSF-based AFFF) since PFCAs are formed as impurities during POSF-based material productions. Degradation of POSF-based products (e.g., N-ethyl perfluorooctanesulfonamidoethanol (N-EtFOSE)) might also give rise to PFOA via a biodegradation process [48]. PFCAs are also present in fluorotelomer-based AFFFs and might get into the environment during application [49]. PFCAs might be formed via chemical reactions from the residuals of fluorotelomer raw materials or fluorotelomer products as impurities. Like POSF-based products/PFOS, PFCAs might get into the environment from losses from consumer products during use and disposal. Since there were two different synthetic routes (i.e., ECF since the 1950s vs. telomerization since the 1950s) for PFCs, different PFC signatures (i.e., ECF contained branched isomer vs. telomerization contained straight isomers) could be traced from the sources of origin [50,51].

26.2   ANALYTICAL METHODS

PFC levels can be determined by several methods: (1) Derivatization techniques coupled with gas chromatography (GC) followed by electron capture detection (ECD) [52] or mass spectrometric detection [46,53], (2) nuclear magnetic resonance (19F-NMR) spectroscopy [54], (3) high-performance liquid chromatography–tandem mass spectroscopy (HPLC/MSMS) [55–57]. To analyze water samples, GCMS and HPLCMS/MS methods were commonly used in recent studies.

26.2.1   GAS CHROMATOGRAPHY MASS SPECTROSCOPY

PFCAs were determined and measured in groundwater samples by the chemical derivatization of the carboxylates to their methyl esters, which was then detected and quantified by either electron impact (EI) gas chromatography–mass spectroscopy (GC/MS) or electron-capture negative chemical ionization (NCI) GC/MS [46]. Scott and coworkers have also developed another method using GC/MS to measure the concentrations of PFCAs in surface water and precipitation samples. PFCAs in the water samples were derivatized to 2,4 difluoroanilides in the presence of 2,4 difluoroaniline and N,N-dicyclohexylcarbodimide, then detected and analyzed by GC/MS. This method can quantify C2–C12 PFCAs and some FTCAs and FTUCAs [53,58]. Since PFCs have low volatility and its derivatives are less stable, this analytical method is not applicable to quantification of some PFCs, even higher sensitivities than LCMSMS in some case. Derivatization with GC/MS has limited utility for determining a broad range of PFCs.

26.2.2   High-Performance Liquid Chromatography–Tandem Mass Spectroscopy

To overcome the disadvantages of GC/MS, analytical methods based on HPLC/MS/MS have been developed to analyze PFCAs in human serum [59], PFASs and PFCAs and sulfonamides in biological matrices [55], and PFCAs and PFASs in surface waters [54].

Taniyasu and coworkers developed an analytical method to determine concentrations of 22 PFCs in water and biota samples using HPLC/MS/MS using reverse phase HPLC columns [56]. This method involves solid phase extraction (SPE) with weak anion exchange cartridges as cleanup and preconcentration steps before the quantifications of PFCs by HPLC/MS/MS. This method is simple and robust and it can be applied to a wide range of both neutral and acidic PFCs such as short- and long- chain perfluorinated acids (C_4–C_{12}, C_{14}, C_{16}, C_{18} PFCAs and C_6, C_8 PFASs), FTOH, and fluorotelomer acids. With the use of HPLC/MS/MS, lower method detection limits can be achieved for both water and biological matrices. In 2008, an improved method that can determine C_2–C_4 PFCAs and PFASs in precipitation samples was developed [57]. Using the same extraction method with the introduction of a new ion exchange column (RSpak JJ-50 2D column), a wider range of short-chain compounds could be determined and quantified. This was also the first report to determine TFA by HPLC/MS/MS. Remarkably, the combination method of conventional reverse phase HPLC column (i.e. Betasil C18, Ace3 C18) and new ion exchange column (double columns HPLC method) enable the most reliable confirmation of PFC measurements because there remains no possibility (i.e. confirmation using different analytical columns) of co-elution. It can be pointed out that analytical methods for water samples are well established even though trace-level measurements (below 1 ppt) remain a challenging task.

For biological sample analysis, both traditional extraction by ion pairing [55] and acetonitrile [60] are common. However, suitable cleanup procedures are necessary for either idea. In 2008, a U.S. nationwide monitoring study of PFOS residue in chicken eggs was reported in a reputed journal and the significantly higher concentrations that were reported got people's attention. However, it appeared that all of the results were overestimated because the wrong analytical method was used. In the report, PFOS was co-eluted with endogenous chemicals from bile acid.

Yeung et al. reported no co-elution measuring using suitable clean up by SPE and the double-column HPLC method for biological samples [61]. Analytical methods using SPE show not only better sensitivity and selectivity compared to other methods but also provide a suitable cleanup method for all kinds of samples, including biota tissue, sediment, and air and waste materials, because the final extract from these samples can be prepared as a liquid before instrumental measurement.

26.2.3   Quality Assurance and Quality Control and ISO25101

Several governmental and industrial actions are planned for regulation of PFCs. However, without adequate and reliable information for evaluating risk profiles to regulate these chemicals, making the right decision can be difficult. Van Leeuwen et al. indicated the "struggle" in evaluating the first worldwide inter-laboratory trial of PFOS analysis in 2005 [62]. During that inter-laboratory trial, more than a 100% variation in analytical results was found. It also pointed out that data published before 2005 were variable, especially when comparisons among the laboratories are made. Similar calibration studies, FLUOROS and PERFORCE 2 [62,63], were carried out over the last 5 years.

The first international activity to establish the International Organization for Standardization (ISO) method for PFOS and PFOA in water samples was initiated in May 2005. The steering committee approved the development of a new working draft based on a previous method (a blank free SPE method with HPLC tandem mass spectrometry developed by Taniyasu et al. [56]). The committee members represented 12 countries (Australia, Austria, Finland, France, Germany, the Netherlands, the Russian Federation, Spain, Sweden, Switzerland, Turkey, and the United Kingdom) and voted favorably for the development of a new ISO method. The committee draft was submitted for balloting in November 2006 and the draft was accepted as a Draft International Standard

(DIS) in February 2007. The inter-laboratory study (method verification) for DIS was announced in November 2006. The study period was from November 2006 to February 2007, and an evaluation report was distributed in June 2007.

The international inter-laboratory comparison exercise was organized by inviting experts from all over the world to participate in the exercise. Thirty laboratories registered for participation and 23 of the laboratories provided the final analytical results. Samples of seawater, river water, industrial wastewater, spiked water matrices, and standard solutions were prepared and distributed to the registered participants. A published analytical method, as a guidance document, and native and labeled standards were provided to all participants. The method (along with the samples) provided a guidance document containing important steps to consider during sample preparation and quantification of concentrations. This marks a difference between "method performance verification study" and simple "inter-laboratory comparison study" of "in-house methods" according to ISO rule. Twenty-three laboratories from nine countries successfully completed the analysis and reported the results. The coefficient of variation in reported concentrations among the laboratories were between 20% and 27% in the worst case. Determination of PFOS and PFOA in water samples, when provided with a standard protocol/analytical method documenting appropriate quality assurance and quality control (QA/QC) protocol, has resulted in improved accuracy and precision. The final document for international standard (FDIS) method was prepared along with the results of round robin testing in October 2008. The voting for FDIS was carried out in January 2009 and 100% approved by representatives from 19 countries with no substantial argument. The FDIS stage was skipped and the international standard was registered in January 2009. Finally, the ISO25101 "Water quality—Determination of perfluorooctanesulfonate (PFOS) and perfluorooctanoate (PFOA)—Method for unfiltered samples using solid phase extraction and liquid chromatography/mass spectrometry" first edition was published in March 1, 2009 [64]. The document can be obtained from the ISO as follows. http://www.iso.org/iso/iso_catalogue/catalogue_tc/catalogue_detail.htm?csnumber=42742

The Japanese Industrial Standard method of PFOS and PFOA (including 17 related chemicals) will also be established in early 2011 as an MOD of ISO25101. From the worldwide effort to test the reliability of the standard operation procedure, several findings to getting better QA/QC were obtained, such as optimization of calibration curves on HPLC-MSMS, accurate procedures for SPE cartridge use (insufficient in-house method using formic acid resulted low reproducibility), and eliminating any remaining contamination sources both in instrument and analytical procedure. Future development of more capable SPE methods will enable simple and high-throughput analysis of PFCs.

26.3   GLOBAL TRANSPORTATION OF PFCs

PFOS and PFOA are representative and commonly occurring PFCs. Surface water and wastewater collected from several countries have been shown to contain PFCs. PFOS concentrations in surface water ranged from tens to hundreds of ng/L [25,65–70], while PFOS levels in wastewater could reach up to thousands of ng/L [71,72]. For biological samples, polar bears and waterbirds from the Arctic region were reported to bioaccumulate PFOS at concentrations up to ug/g w.w [73,74]. Blood PFOS concentrations in the general human population ranged from sub ng/mL to hundreds of ng/mL. There are two hypotheses for the transport of PFCs in the environment: degradation and transport of volatile precursor chemicals such the FTOHs is suggested to be the main source of PFCs in remote Arctic regions [8,75], while transport via oceanic currents may distribute PFCs to other regions of the world [65,76].

Due to the volatility of the precursors (FTOHs and perfluorooctylsulfonamides), with sufficient atmospheric lifetime, they can be transported by wind/air movement in the atmospheric environment. Intermediate degradation products of FTOHs such as FTCAs and fluorotelomer unsaturated carboxylic acids (FTUCAs) were detected in precipitation samples from Canada [77]. Concentrations of PFCs had also been detected in biota samples [78,79] and snow [80] in the Arctic region. Since local usage would not be expected in these regions, the degradation and transport of these volatile precursors had been hypothesized as the main source of long-chain PFCAs in the Arctic [34]. These evidences supported the atmospheric transport and depositions of PFCs.

Since PFCs are highly water soluble and environmentally persistent, they have a high potential for long-range transportation. Atmospheric loadings of PFCs and discharge of PFCs and its precursors to surface water were suggested to be the major sources for the aquatic systems [34]. PFCs were detected in various water bodies including river water [68,69], lake water [66], and even open ocean surface water [81]. High concentrations of PFCs were found in river water collected near fluorochemical manufacturing facilities [82], release of AFFFs [83], and wastewater treatment plants [84–86]. These contaminated river waters had been suggested to also contaminate the ocean. PFCs might also be transported to deeper water through the downward movement of PFCs in dissolved water or sedimentation of the sinking molecules [34]. Therefore, the concentrations of these chemicals were also suggested as a tracer for the global circulation of ocean waters [76].

26.3.1 PFC LEVELS IN AIR AND DUST

As mentioned above, volatile and semivolatile fluorinated organic compounds (FTOHs, FOSAs, FOSEs, and PFOSA) had been detected in air and dust samples from different areas all over the world using the GC/MS for identification and quantification [35–37,41,42].

Additionally, the concentrations of PFCAs, PFASs, and fluorotelomer sulfonates (FtSs) were measured in marine air between Germany and South Africa [36] and air samples from Albany, New York, United States [87]. For marine air between Germany and Africa, the concentrations of ionic PFCs were lower than their neutral, volatile precursors, with PFOA ($1.0\,pg/m^3$) and PFOS ($0.9\,pg/m^3$) as the highest concentrations. For air samples from the United States, lower concentrations were detected in the particulate phase than gas phase. PFOA and PFOS were the predominant compounds in air, accounting for more than 60% of total PFCs concentrations. The concentrations of PFOA and PFOS were comparable in both studies. Murakami had reported concentrations of PFOS and C_8–C_{11} PFCAs in size-fractionated street dust in Tokyo [88]. PFCs present in air and dust samples are some sources contributing to the atmospheric contamination.

26.3.2 PFC LEVELS IN PRECIPITATION

Precipitation may wash down the PFCs deposited in the air, dust, or other matrices present in the atmospheric environment. The measurement of concentrations of PFCs in precipitation samples can provide deeper understanding on fluxes and depositions of PFCs in the atmosphere.

Loewen et al. had reported concentrations of fluorinated acids and PFOS in Canadian precipitation samples using HPLC/MSMS. Only one rainfall event ($n=3$) was captured and analyzed and 8:2 FTCA concentrations ($1.00\pm0.08\,ng/L$) were found to be highest, while the concentrations of PFOS ($0.59\pm0.04\,ng/L$) were higher than those of 10:2 FTCA ($0.30\pm0.04\,ng/L$), 8:2 FTUCA ($0.12\pm0.02\,ng/L$), and 10:2 FTUCA ($0.12\pm0.01\,ng/L$) [77].

Using chemical derivatization by GC/MS, Scott et al. had reported concentrations of C_2–C_{12} PFCAs, and fluorinated acids in North American precipitation samples. Among the nine sites within urban and rural areas, the concentrations of TFA were highest compared to those of other PFCAs [53]. Apart from TFA, higher concentrations were detected in short-chain compounds (C_3–C_7) than PFOA and PFNA. FTCA and FTUCA concentrations were considerably lower than those of PFCAs. Also, higher concentrations of PFCs were detected in the urban area vs. the rural area.

As for the Albany, New York, United States, precipitation samples, Kim and Kannan had measured concentrations of C_7–C_{12} PFCAs, C_6, C_8, C_{10} PFASs and FtS by HPLC/MSMS. PFHpA, with a median concentration of $0.56\,ng/L$, was detected in almost all samples [87]. Highest concentrations were detected for PFOA and PFNA, and PFOS was the only detectable PFASs. The concentration range of PFOA measured in that study was comparable to the concentrations found in Scott and coworkers study. Interestingly, PFOS concentrations were found to be highest in Dailin, China precipitation samples [89].

Studies conducted by Kwok revealed concentrations of 20 PFCs, including C_3–C_5 short-chain PFCs, using LC/MS/MS on precipitation samples from Japan ($n = 31$), United States ($n = 12$), China ($n = 5$), India ($n = 2$), and France ($n = 2$). A total of 52 precipitation samples from June 2006 to September 2008 from seven cities were analyzed [90]. An overview of individual PFC concentrations in precipitation samples are shown in Figure 26.8. PFCs, dominated by C_3–C_{12} PFCAs, were measured in all of the precipitation samples. The highest concentration of PFCAs for most of the precipitation samples was found to be PFPrA. PFPrA was firstly detected in precipitation samples using LC/MSMS in this study. One of the possible sources of PFPrA may be due to the use of perfluoroketone. Perfluoroketone has been used as a clean fire suppression agent, which was a replacement for Halon 1301 and Halon 1211. PFPrA is considered to be the final product of the reaction between perfluoroketone and water present in the atmosphere during fire protection.

No clear seasonal trends were found for both locations in Japan, which is similar to that reported earlier [9] in three northeastern U.S. locations showing a lack of seasonal trend for PFOA and PFNA fluxes. Elevated total PFC and PFCA concentrations were found in samples collected in August in Tsukuba and Kawaguchi; however, no such trend was observed for PFASs. The composition profiles of PFCs varied depending on the month of sampling. These observations were probably related to the seasonal differences in the usage pattern of PFC chemicals.

No temperature-related seasonal trend was observed for all PFCs at both locations. Generalizations could not be made between wind direction and concentrations of PFCs at each of the rain events. The fluctuating concentrations of PFCs and differences in composition profiles of PFCs at different precipitation events may be governed by meteorological conditions and PFC concentrations in precipitation are affected by the dilution process as evidenced by lower concentrations when the amount of precipitation was high. Differences in the composition of PFCs in precipitation at various locations suggested that precipitation reflects only the local sources of contamination.

Snow is another kind of wet precipitation. According to previous studies [91], snow might be a very efficient scavenger of organic vapors due to its high specific surface area. In order to further investigate the scavenging potential of PFCs in the atmosphere by precipitation, a preliminary comparison was made between rain and snow samples collected in Tsukuba, Japan. In general, PFC concentrations were similar between rain and snow samples; however, rain water could scavenge a higher percentage of most of the PFCs (PFBA, PFOA, PFNA, PFDA, PFUnDA, PFDoDA, PFOS, 8:2 FTUCA, PFOSA, and N-EtFOSAA) than snow. As rain and snow are in different physical states, rain droplets can dissolve PFCs more efficiently in addition to sorption of the droplet surface.

26.3.3 PFC LEVELS IN OCEAN WATERS

PFCs in remote marine locations are a concern and indicate the need for studies to trace sources and pathways of these compounds to the oceans. Although there are several estimations about the global transportation of PFCs, which are based on computer and conceptual models, very few "actual monitoring data" [76,81,92,93] of PFCs in trace level ocean waters are available on a global scale. Even though it is after 10 years from the initial finding of PFCs in wild life of polar regions in 2000 [1], reliable information about trace level PFCs in open ocean water is very limited, compared to several reports about marine animals. Monitoring of air and water in inland/coastal areas is relatively easy because of the accessibility to samples and relatively high concentrations, but interpretation of the data is difficult because of the variety of sources. It is difficult to single out a source even with statistical computing and modeling.

Monitoring studies in open ocean from the sources is difficult to carry out because of the need for ultra-trace analysis, but the concept of long-range transportation can be simplified by monitoring PFCs in remote locations. Quantity and number of PFC sources are much smaller than heavily

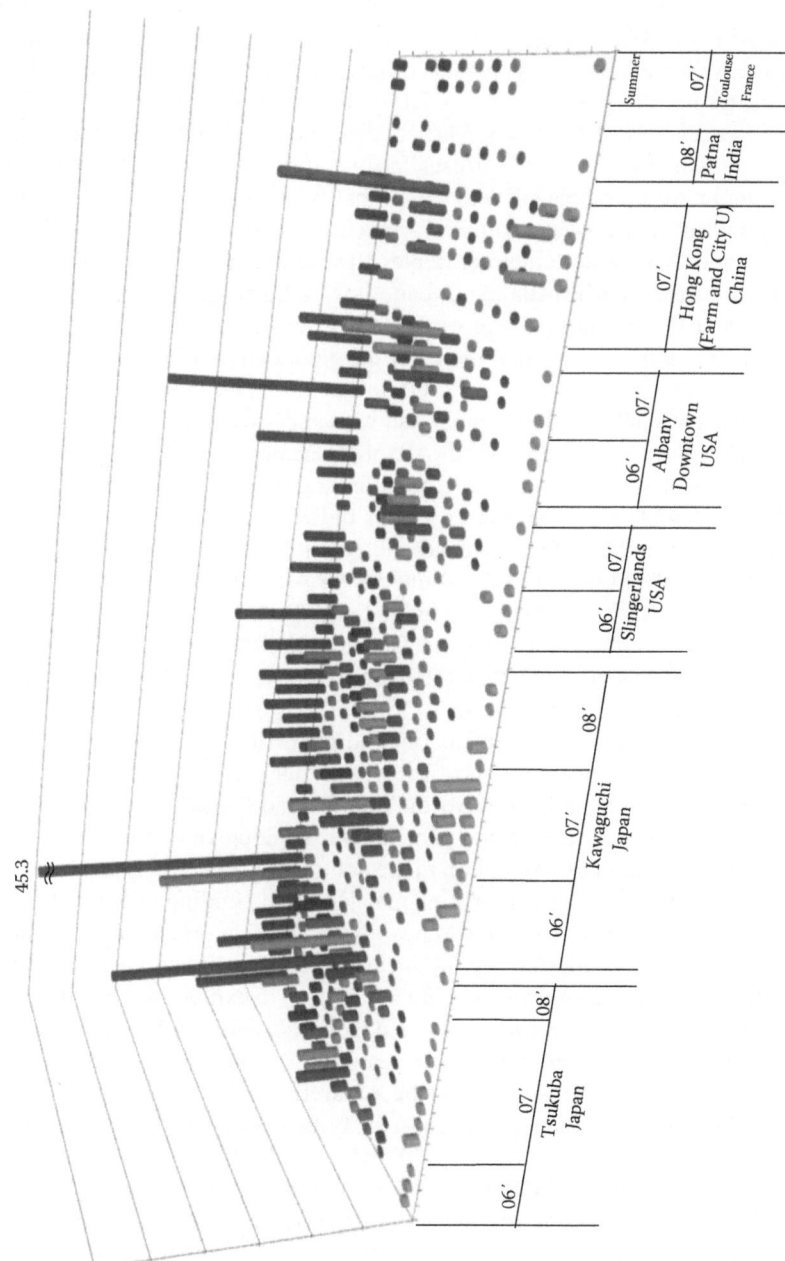

FIGURE 26.8 PFC concentrations in precipitation samples from seven locations from five countries. No bar: Sample <corresponding LOQ, i.e., 0.05–0.25 ng/L. (With permission from Kwok, K.Y., Taniyasu, S., Yeung, L.W.Y., Murphy, M.B., Lam, P.K.S., Hori, Y., Kannan, K., Petrick, G., Sinha, R.K., and Yamashita, N., *Environ. Sci. Technol.*, 44(18), 7043–7049, 2010. Copyright 2010 American Chemical Society.)

industrialized/populated continental areas. A much simpler and more traceable environmental kinetics of PFCs can be elucidated by monitoring contamination in open ocean.

Hence, a global scale model in open ocean seems to be simpler than a regional scale of PFCs kinetics in coastal areas and the continents.

The physicochemical properties of PFCs are different from those of several established global pollutants such as PCBs, even after classified as POPs by the Stockholm Convention. In efforts to determine the mechanism of global transport and distribution of PFCs, global ocean monitoring was initiated in 2002 at the National Institute of Advanced Industrial Science and Technology (AIST), in collaboration with the Ocean Research Institute (ORI) of Tokyo University, Japan, and the Leibniz-Institute of Marine Sciences in Kiel, Germany. Vertical distribution of PFAs in ocean water samples were collected during a number of research cruises in several oceans. Vertical profiles of PFAs in the marine water column were associated with the global ocean circulation theory. Open ocean and offshore surface and subsurface water samples were collected between 2002 and 2006 from 62 locations around the world. Nine water-column samples, at several depths of up to 5500 m, were collected from the Labrador Sea, the Mid-Atlantic Ocean, the South Pacific Ocean, and the Japan Sea. [76] Global distribution of PFOS and PFOA in surface sea waters (modified from [76]) is shown in Figure 26.9. Data is shown with a conceptual model of the global circulation of ocean waters based on "Brocker's conveyor Belt." Residue levels of PFOS in sea birds [94] collected from the northern and southern hemispheres are presented in comparison with surface sea waters pollution.

PFOS and PFOA concentrations in the North Atlantic Ocean ranged from 8.6 to 36 pg/L and from 52 to 338 pg/L, respectively, whereas the corresponding concentrations in the Mid-Atlantic Ocean were 13–73 and 67–439 pg/L. The surface waters of the North and Mid-Atlantic Ocean are more highly contaminated than the surface waters of the South Pacific Ocean and the Indian Ocean. The highest concentrations of PFOS and PFOA found in the Atlantic Ocean were 15 and 88 times higher than the lowest concentrations determined in the South Pacific Ocean and the Indian Ocean. The off-shore waters of the eastern North Pacific Ocean along the Asia-Pacific Rim contained 10–50-fold higher concentrations of PFOA than the western North Pacific Ocean. Concentration of PFOA in the eastern Pacific Ocean ranged from 10 to 60 pg/L whereas that in the western Pacific Ocean ranged from 140 to 500 pg/L. Remarkable longitudinal differences between the eastern and the western Pacific Ocean in the concentrations of PFOA was unexpected because many PFAs have been produced and used in large quantities in North America and Canada; thus the concentrations in the eastern North Pacific Ocean are expected to be higher. Although a further monitoring survey is necessary, this may suggest that PFCs pollution in the continent is not closely related to the corresponding open ocean environment. This is a striking match with the finding in precipitation samples described in Section 26.3.2; differences in the composition of PFCs in precipitation at various locations suggested that precipitation reflects only the local sources of contamination. Thus, regional and local events of the hydrospheric environment are key factors in explaining the kinetics of PFCs, most of which are organic acids. Ocean water collected from the southern Indian Ocean contained trace levels of PFAs. PFAs were not found in the surface waters of the Antarctic Ocean (70°27′E to 172°40′W), which is explained by the isolation of the Antarctic circumpolar water from the water masses in the northern hemisphere. Interestingly, marine birds collected from the Southern Ocean and the Antarctic contained only trace levels of PFOS, whereas the marine birds from the northern hemisphere contained high levels of PFOS and other PFAs [94]. In other words, a striking match was found between PFCs residue in sea birds and the surface waters in open ocean even after using different strategies of sampling and analysis. To our knowledge, this is the first and most remarkable evidence that surface water pollution is related to PFCs accumulation in wild animals because no clear relationship between PFCs residue in wild animals and level of pollution in their environment has been reported before due to complicated exposure pathways on land. This finding supports the previous statement that the "global scale model in open ocean seems to be simpler than the regional scale of PFCs kinetics in coastal area and the continents". This concept was clearly described in the report of "Island project: for understanding the global transportation potential of PFOS and

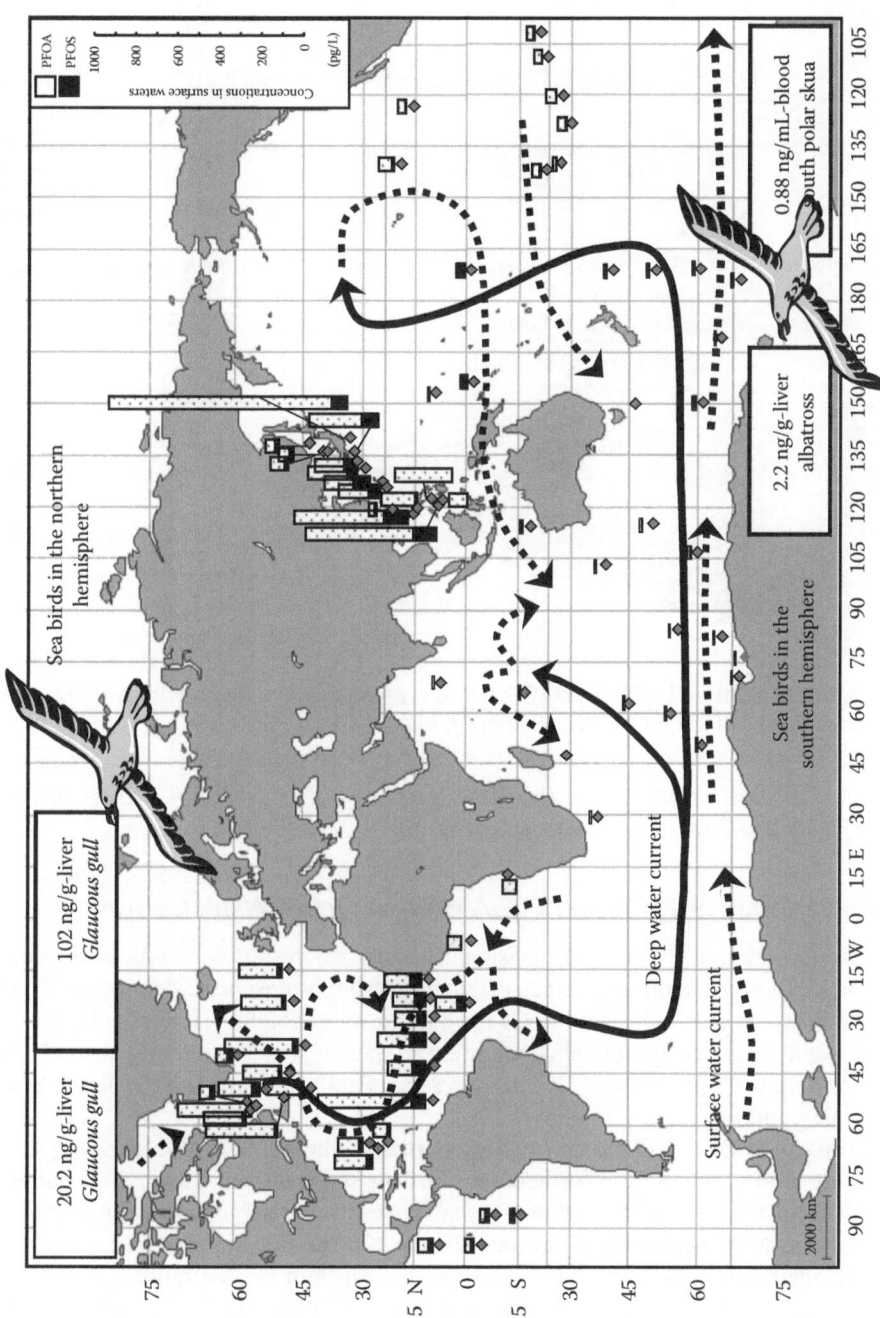

FIGURE 26.9 Global distribution of PFOS and PFOA in surface sea waters. (Reprinted from *Chemosphere*, 70, Yamashita, N., Taniyasu, S, Petrick, G., Wei, S., Gamo, T., Lam, P.K.S., and Kannan, K., Perfluorinated acids as novel chemical tracers of global circulation of ocean waters, 1247–1255, 2008. Copyright 2008, with permission from Elsevier.) Data is shown with conceptual model of the global circulation of ocean waters based on "Brocker's conveyor Belt." Residue level of PFOS in sea birds [94] collected from the northern hemispheres and the southern hemisphere is presented in comparison with sea waters pollution.

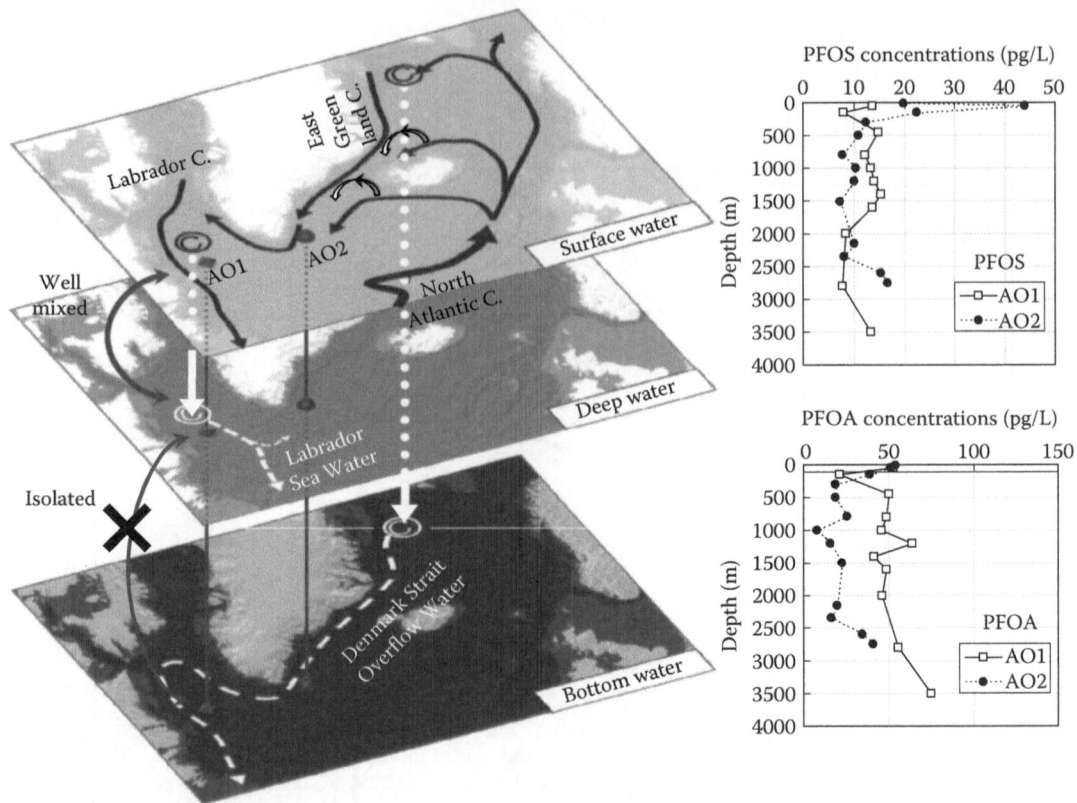

FIGURE 26.10 Vertical profiles of PFOS and PFOA in sea water columns in the Labrador Sea. Data is shown with conceptual model of sea water circulation in the Labrador Sea.

related chemicals in oceans" presented in the 2010 International Chemical Congress of Pacific Basin Societies (Pacifichem2010, Hawaii, December 2010).

26.3.4 Vertical Profiles of PFOS and PFOA in Open Ocean Water Columns

Since the ocean water column is a three-dimensional matrix, the results for PFAs in the surface water can only provide information on sources originating from atmospheric deposition. To date, only two reports [76,81] are available about PFCs residues in water column samples collected from open ocean. The results of the international cooperation studies of water columns at several depths from the Labrador Sea, the Mid-Atlantic Ocean, the South Pacific Ocean, and the Japan Sea are described in the following.

The vertical profile of PFOS and PFOA concentrations in the Labrador Sea (AO1 and AO2) showed a new understanding of PFCs input from surface water to deep water (Figure 26.10). Concentrations of PFOS and PFOA in the AO1 water column samples were relatively constant throughout the depth, except in subsurface water samples and water below 2000 m, in which PFOA concentrations increased. On the other hand, the AO2 water column showed high concentrations of PFOS and PFOA at the surface and then uniform concentrations down to 2000 m. The Labrador Sea is known as the critical location for global circulation of ocean water and several investigations have revealed the role of this sea in moderating Earth's climate through its convective formation of the water mass known as Labrador Sea Water (LSW); this is the key element of the global thermocline circulation. The vertical profile of PFCs described above shows a trend consistent with the known hydrodynamics of the Labrador Sea. A simplified illustration of the water currents in the North Atlantic Ocean around the Labrador Sea and the vertical profiles of concentrations of PFOS and PFOA in the AO1 and AO2 water columns are presented in Figure 26.10. The overflow and descent of cold, dense water from the sills of the Denmark Strait (Denmark Strait Overflow Water or DSOW) and Faroe-Shetland Channel into the North Atlantic Ocean are the principal means by which the deep oceans are ventilated, and the Labrador Sea is the site at which deep sea waters originate.

Concentrations of PFCs increased at depths below 2000 m. This pattern suggests the presence of an independent deep-water current, in other words the DSOW. PFOA was the predominant PFC throughout the depth of the AO1 water column. The DSOW lies below the convective current of the LSW, but the waters of the two currents supposedly do not mix. High concentrations of PFCs were found in the surface waters of the AO2 water column collected from the North Atlantic Ocean. A significant amount of PFCs from the North Atlantic Current (NAC) appears to have contributed to the high concentrations of PFCs in the surface-layer of AO2. It can be concluded that the PFCs transported from surface waters of the NAC sink to deeper water through the convection of surface water. There are two locations of deep sea water formation in the eastern Greenland Sea and the Labrador Sea. The PFCs concentration of the AO2 water column showed a relatively complex vertical profile and matches the knowledge of mixed influence of entrainment of the NAC and convection of the LSW well. It is possible to estimate scavenging amount of PFCs from surface waters though deep water formation using trace level analysis of open ocean waters. This is an evidence of the useful nature of PFCs as environmental chemical tracers.

Three water columns (AO3, AO4, AO5) from the Middle Atlantic Ocean (MAO) are shown in Figure 26.11. In each of these water columns, there was a considerable difference in PFC concentrations between the surface and middle layers of the water column below 800 m. This pattern was consistent with a marked change in gradient of both temperature and salinity below 800 m. Concentrations of PFCs were almost negligible in the deepest layers below 4000 m. The latter finding suggests a lack of direct vertical transport of PFCs from surface to bottom waters. This pattern is clearly different from that found in the Labrador Sea where consistent contamination from surface water to bottom was found. The vertical profile of PFCs in the MAO seems to be typical of the profiles found for coastal and off-shore regions of industrialized countries with relatively higher pollution in surface water. It can be concluded that PFCs discharged into the surface waters in the MAO have long residence times due to the circulation of the water mass by the Gulf Stream and the North Atlantic Drift and resulted in the isolation of the surface waters from the deep waters.

The vertical profiles of PFCs in two water columns, PO1 and PO2 from the South Pacific Ocean, were completely different from the other oceans previously studied (Figure 26.12). Concentrations

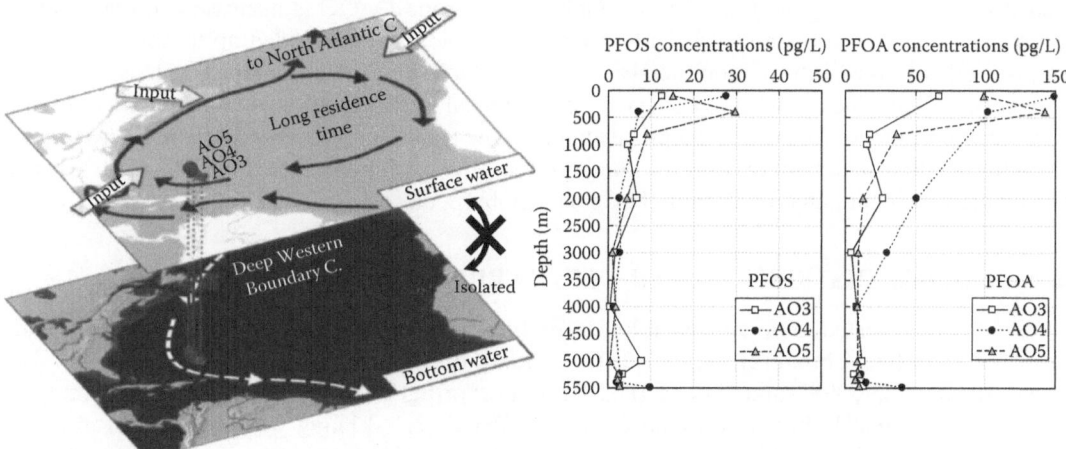

FIGURE 26.11 Vertical profiles of PFOS and PFOA in sea water columns in the middle Atlantic Ocean. Data is shown with conceptual model of sea water circulation in the middle Atlantic Ocean.

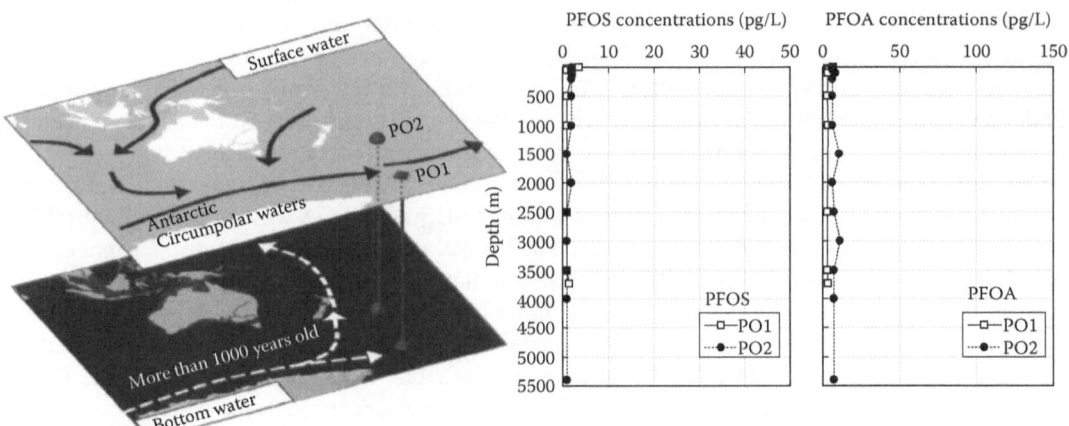

FIGURE 26.12   Vertical profiles of PFOS and PFOA in sea water columns in the South Pacific Ocean. Data is shown with conceptual model of sea water circulation in the South Pacific Ocean.

of PFCs were consistently low (<10 pg/L) or below the limit of detection from surface to bottom. These water columns are supposed to be sunk down from surface because of the global water circulation and are more than 1000 years old according to traditional oceanography. PO1 and PO2 are supposed to have different masses of water because of differences in the temperature and salinity profiles. However, PFC concentration profiles were constant throughout both columns. Negligible concentrations of PFCs in the water columns of the South Pacific Ocean showed that there is no "direct and/or indirect input" of PFCs to this remote region in the southern hemisphere. In other words, the southern hemisphere, especially the South Pacific Ocean, has no long range transportation of PFCs through the atmospheric media from contaminated Northern Continents.

The results presented above provide evidence that PFC concentrations and profiles in the oceans adhere to a pattern consistent with the global "Broecker's Conveyor Belt" theory of open ocean water circulation, as described in Figure 26.9. Striking differences in the vertical and spatial distribution of PFCs, depending on the oceans, were found, and it can be concluded that a general monitoring study on surface water is just the beginning of the "earth chemistry of PFCs." Research results from the surface ocean are very limited because it is very clear that the ocean is a three-dimensional media and there are still many areas where nobody has done any monitoring surveys. Even if there is a chance, high-level trace analysis with good QAQC is necessary to obtain reliable results for PFCs in open ocean, and only a few people have succeeded up to now. However, even from this limited information, persistent water soluble fluorinated organic acids and related chemicals, namely PFCs, can serve as very useful chemical tracers to allow us to study not only the global transportation of persistent chemicals but also the traditional geochemistry and oceanography. After the first statement of this concept "PFCs as novel chemical tracer" in 2008 [76], several investigations were initiated on oceanic PFCs on a global scale.

26.4   GLOBAL DISTRIBUTION OF PFCs IN BIOLOGICAL SAMPLES

PFOS has been detected in organisms at lower trophic levels from benthic algae up to mussels and shrimps in Asia (i.e., China) [21,22], Africa (i.e., Brazil) [95], North America (i.e., Canada and the United States) [78,96–98], and Europe (i.e., Belgium, Portugal, and Germany) [99,100]. Benthic algae from the Great Lakes were found to contain PFOS (2.4–3.1 ng/g ww). PFOS ranged from <0.5 ng/g ww from Florida, United States [101] up to 280 ng/g ww from Lake Ontario [96]. Mussel and oyster samples detected PFOS levels ranging from 0.114 ng/g ww [22] to 430 ng/g dw [102]. PFOS concentrations in shrimp ranged from 0.06 ng/g ww from the eastern Arctic (Davis Strait)

[97] to 520 ng/g ww (the southern North Sea [99]). Very limited amount of data reported PFC concentrations for crab, squid, and starfish [99]. PFOA occurred at much lower concentrations than those of PFOS in the same samples [22,96,99]. One interesting finding from Martin and co-workers [96] should be noticed that Diporeta occupied the lowest trophic level of all species in that food web study having relatively high PFOS concentrations (280 ng/g ww). The authors hypothesized that sediments must be a major sink for some PFCs and that the high PFOS concentrations in the benthic Diporeta might be contributed to habitat and the biotransformation of some PFC-precursors (i.e., FTOHs or FOSAs) into PFOS incorporated into various polymers in the sediments. Thus, an analysis of the biodegradations of fluoropolymers into PFCs should be implemented. Measurements in amphibians and reptiles were scarce [1,98,103,104]. Only one study reported PFOS concentrations (<35–285 ng/g ww) for livers of Green frog [1]. Turtle PFOS serum levels were detected up to 169 ng/g ww [1].

26.4.1 Birds

A number of studies measured PFC concentrations in different tissues (e.g., liver, blood, and egg) from several avian species worldwide. Detailed PFC concentrations had been summarized by Houde and coworkers [105]. PFOS is the major PFC measured in those samples. The greatest PFOS concentrations (up to 2220 ng/g ww) were measured in plasma samples from Bald Eagles, whereas the lowest PFOS concentrations were measured in the samples from Arctic regions [2,78]. Liver and plasma samples were often used as biomonitoring tools; however, these tissues might be biased due to the fact that those samples were coming from dead or diseased animals. Waterbird eggs are frequently used as biomonitoring tools to measure contaminant levels because they may represent the levels of contaminants in female birds [106]. A previous study identified that significant adverse effects such as higher mortality, reduced hatchability, and liver histopathological changes were observed after injection of PFOS into chicken eggs at levels near to those found in wild avian eggs [107]. Biomonitoring on PFCs in bird eggs could estimate any adverse effects on the offspring. PFCs were detected up to 614 ng/g ww in bird eggs indicating oviparous transfer of some PFCs to offspring [2,108–112]. Figure 26.13 shows the PFOS concentrations measured in bird egg samples from different parts of the world. Much lower PFOS levels were found from the samples coming from Arctic regions [113], whereas the greatest PFOS levels were found from the egg samples near the Lake Ontario regions [114]. Dietary source was suggested to be a major source of PFCs to the avian species and different PFC composition profiles were observed for different feeding habits [111]. Gebbink and coworkers suggested that aquatic dietary sources were the major source of some PFCs to avian species, and they were highly lake- and/or colony-dependent with higher concentrations close to highly urbanized and industrial areas [114]. However, since the dietary exposure for migratory birds covered a wide range of temporary food sources and environments, extra care should be given to interpret the results between the migratory and resident birds [105].

26.4.2 Mammals

The PFC data on terrestrial animals are scarce when compared with marine mammals [23,24,117], and more importantly those reported studies were either from farm animals or from zoos. Three studies have measured PFOS and PFOA levels in red pandas (*Ailurus fulgens*) and giant pandas (*Ailuropoda melanoleuca*) [118], and PFCs in Amur tigers (*Panthera tigris altaica*) [24] and Bengal lions (*Panthera tigris tigris*) [23], but those animals were captive animals and do not necessarily reflect the actual situation of wild animals. There is one study reporting PFOS concentrations up to 180,000 ng/g ww in wood mice inhabiting near the area of a fluorochemical plant in Belgium [119]. Recently, Yeung and coworkers reported PFC blood concentrations in wild rats collected from Japan, and the PFOS concentrations were measured up to 38.1 ng/mL [120]. Interestingly, PFOS (2.7 ng/g ww) were also detected in liver samples of herbivorous northern caribou [121].

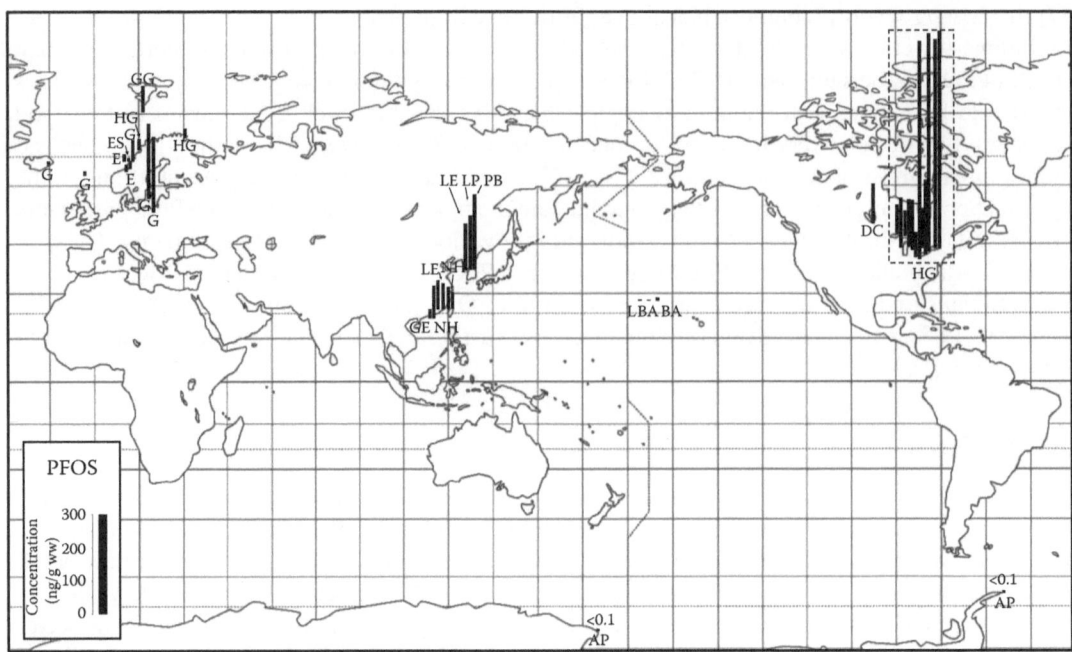

FIGURE 26.13 PFOS concentration in bird egg samples—global scale comparison. AP, Adelie penguin; BA, black-browed albatross; DC, double crested cormorant; E, eider; ES, European shag; G, guillemot; GE, great egret; GG, glaucous gull; HG, herring gull; L, laysan albatross; LE, little egret; LP, little ringed plover; NH, night heron; PB, parrot bill. Data from [94,108,110,112–116].

Elevated PFOS concentrations (250 ng/g ww) were found in liver of Arctic fox [78]. The Arctic fox occupies the apex trophic level, which is directly related to small terrestrial and marine mammals bioaccumulating PFCs, which might be also the reason for the high PFOS levels for polar bears. Dietary exposure and habitats of the terrestrial animals were major sources of exposure to PFCs [78,120]. In general, PFC concentrations from terrestrial animals were lower than those of marine mammals.

Liver and plasma samples were often used for biomonitoring purpose for marine mammals worldwide [3,98,99,122–126]. Some studies also measured PFC concentrations in other tissues such as bubbler and kidney samples in order to investigate any tissue-specific bioaccumulation patterns, but the concentrations were at lower levels when compared with liver and plasma samples [124,127]. Data for baleen whales were more limited when compared with the toothed whales [5,99,128,129]. PFOS concentrations, in general, were lower in baleen whales when compared with those of toothed whales because different feeding ecologies (i.e., baleen-filtering vs. fish-eating). Liver PFOS concentrations in marine mammals from different countries are shown in Figure 26.14. PFOS concentrations ranged from the highest in polar bears (*Ursus maritimus*) (1330 ng/g ww, [73]) to the lowest in subantarctic fur seals (*Arcttocephalus tropocalis*) (4.2 ng/g ww; [130]). Other marine mammal liver samples were measured at sub part per million (ppm) levels, except those from the Ganges River (India) and Brazil, which were measured at tens of part per billion (ppb) levels. There is still no clear explanation as to how these PFCs could reach the Arctic regions. However, there are two hypotheses for the transports of PFCs via oceanic circulation [65,76] and the atmospheric degradations of volatile precursors (i.e., FTOHs) [8]. Biotransformation products like FTUCA (8:2 and 10:2), which have been detected in coastal and open-ocean dolphins, suggested that the exposure of PFC-related precursors has occurred [8,101,131,132]. Polar bears received elevated PFOS concentrations because of their opportunistic feeding ecologies and mainly feeding on ringed seal blubber

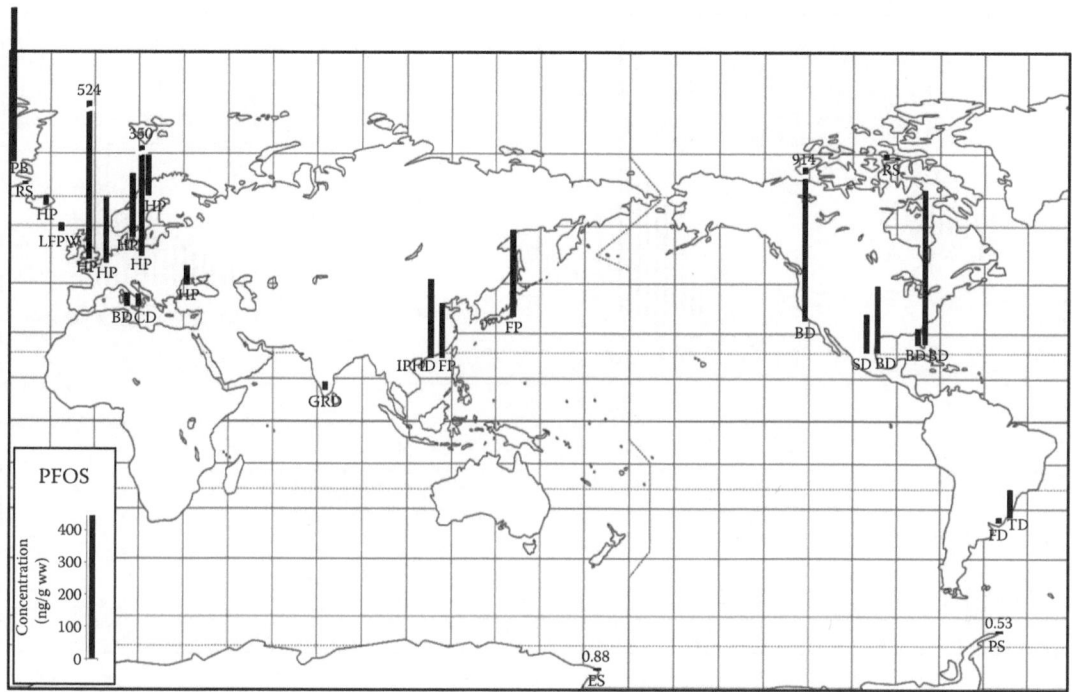

FIGURE 26.14 PFOS concentration in marine mammal liver samples—global scale comparison. BD, Bottlenose dolphin; CD, common dolphin; ES, elephant seal; FD, Franciscana dolphin; FP, finless porpoise; GRD, Ganges river dolphin; HP, harbor porpoise; IPHD, Indo-Pacific humpback dolphin; PB, polar bear; RS, ringed seal; SD, striped dolphin; TD, Tucuxi dolphin. Data from [3,5,73,94,101,115,123,124,126,130,132,133].

and skin [122]. Other exposure pathways like inhalation of PFCs in the aerosol should be investigated for marine mammals.

Several human biomonitoring studies have been carried out and were mainly focused on blood samples, although there are some studies measuring PFC in human liver samples [6,134,135]. The first findings of fluorinated organic compounds in human blood were reported as early as the 1960s [136], and the organic fluorine found was later suggested to be PFOA. Subsequently, trace amounts of organofluorines were detected in the blood of production workers as part of the 3M fluorochemical production workers monitoring program [137]. Although the concentrations of organofluorine detected were in the 1 ppm range, they were found to remain in individuals after cessation of occupational exposure [52,137]. PFCs have been detected in human blood samples worldwide and PFOS was found to be the dominant PFC in most of the blood samples from different countries including the United States, Colombia, Brazil, Italy, Poland, Belgium, India, Malaysia, China, Japan, and Sri Lanka [6,7,18,20,138–140], except for Korea in which PFOA was the prevalent compound [7]. PFOS concentrations measured in humans is shown in Figure 26.15. Briefly, the background PFOS concentrations in the general public ranged from sub-ppb levels up to the hundred ppb level. Higher PFOS levels were detected in blood samples from some locations in the North America region (i.e., the United States [7,141]), Asia (i.e., China [18]), and Europe (i.e., Poland, [7]). Much lower PFOS concentrations were measured in people from India [7] and Sri Lanka [139]. Olsen et al. showed that children (aged 2–12) were found to have elevated PFHxS concentration than those of adults [142]. The authors suggested that there might be different exposure and activity patterns between children and adults such as children playing on the carpet and thus were more exposed to contaminated house dust.

PFOA is another chemical of concern to human health. PFOA concentrations, in general, occurred at much lower levels (at least five times) than those of PFOS in the same samples, except

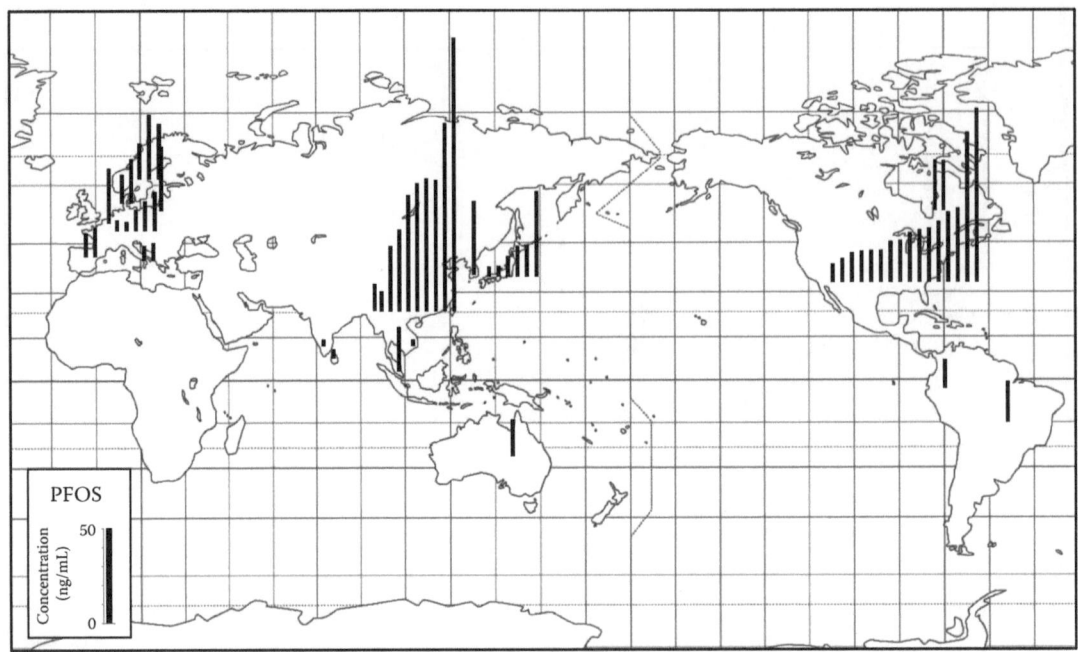

FIGURE 26.15 PFOS concentration in human blood samples—global scale comparison. (All the data were expressed on serum basis. Whole blood concentrations were converted into serum by multiplication of a factor of 2.) Data from [7,18,41,55,134,139,141–156].

those reported for Korea that PFOA which were three times higher than PFOS [7]. In general, the background PFOA levels in the general population ranged from sub-ppb levels up to tens of ppb levels. Elevated PFOA concentrations were measured in some nonoccupational exposed population with known point-source exposures in the United States and Germany [157,158]. Elevated PFOA concentrations (24.6–28.5 ng/mL, 4–8-fold increase when compared to the reference population) were found in plasma samples from people consuming PFOA-contaminated tap water in Arnsberg, Germany [143]; while concentrations up to 354 ng/mL (median) [157] were detected in people consuming PFOA-contaminated tap water from Little Hocking, United States. Further analysis concluded that drinking water could be the primary predictor of PFOA in blood in these areas (Steenland et al., 2009 Ref #). After this incident, granular-activated carbon filtration was installed in public water systems in Ohio and West Virginia. A decrease in serum PFOA concentration after filtration was found at 26% per year [159]. Fish have been suggested to be one of the exposure sources for humans in Baltic region [160]. A recent study suggested that meat was the primary contributor of PFOA in dietary exposure in Chinese population.

PFCs are also found in maternal and cord blood in pregnant Canadian, Japanese, and Korean women, which suggested prenatal exposure [148]. Additionally, human breast milk also detected trace amount of PFCs (i.e., PFOS and PFOA occurred at part per trillion levels), which also suggested postnatal exposure [19,135,147,161].

There are different PFOS/PFOA levels and compositions in blood samples from different countries, or even within the same country or same city. Gender-specific bioaccumulation patterns were also found, in which PFOS and PFOA were higher in males [152]. The genetic variability, diet, lifestyle, occupation, or a combination of these factors might contribute to these variations and levels [149,152,160]. Harada et al. suggested that menstrual bleeding might be one of the excretion routes for some PFCs in women [162].

PFCs exhibit a wide range of toxic effects in exposed laboratory animals. Recent studies have demonstrated that exposure to PFAS can modulate rodent humoral and cellular immune

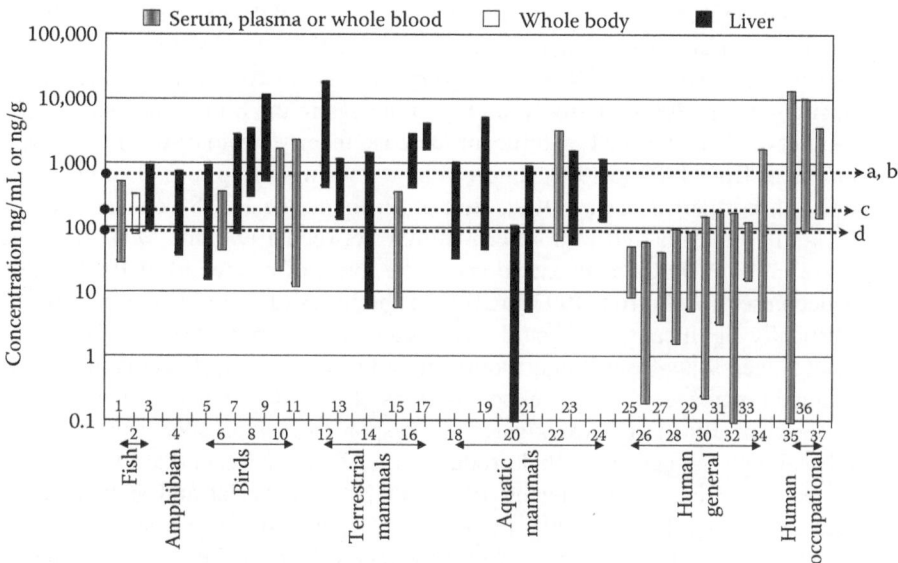

FIGURE 26.16 PFOS concentrations in wild life and human, and immunotoxic events in mice with laboratory test conditions (a) PFOS level showed suppressed plasma IgM antibody production in female mice. (From Peden-Adams, M.M. et al., *Toxicol. Sci.*, 104, 144, 2008.) (b) PFOS level showed significantly increased emaciation and mortality in response to influenza A virus in female mice. (From Guruge, K.S. et al., *J. Toxicol. Sci.*, 34, 687, 2009.) (c) PFOS level showed significantly increased emaciation in response to influenza A virus in female mice. (From Guruge, K.S. et al., *J. Toxicol. Sci.*, 34, 687, 2009.) (d) PFOS level showed suppressed plasma IgM antibody production in male mice. (From Peden-Adams, M.M. et al., *Toxicol. Sci.*, 104, 144, 2008.) References: 1 [20], 2 [96], 3 [170], 4 [1], 5 [171], 6 [108], 7–8 [172], 9–10 [173], 11 [2], 12–13 [119], 14–17 [78], 15 [174], 16 [175], 18–19 [176], 20–21 [168], 22 [131], 23 [3], 24 [5], 25 [147], 26 [177], 27, 32–33 [7], 28 [143], 29 [160], 30 [178], 31 [142], 34 [151], 35–36 [138], 37 [179].

functions [163–167]. The levels of the antibody in mice were decreased even at serum PFOS levels, which were remarkably lower than the average blood concentrations of occupationally exposed people [167]. It has also been shown that the developing immune system of male mice is more sensitive to PFOS exposure than that of female mice, and this may result in altered immunity in adulthood [165].

Exposure to xenobiotics and the resultant alteration of immune function may modify an organism's ability to resist infectious disease. Concentrations of PFCs, including PFOS concentrations, were significantly higher in marine mammals that died due to an infectious disease compared to other causes of death [168]. PFOS exposure in mice showed a significant increase in emaciation and mortality in response to influenza virus A [169]. The mechanism(s) responsible for the lowered host resistance to influenza A in PFOS-exposed female mice remains to be elucidated. However, the effective plasma concentrations in female mice were at least several fold lower than the reported mean blood PFOS levels from occupationally exposed humans, and fell in the upper range of blood concentrations of PFOS in the normal human population and in a wide range of wild animals (Figure 26.16). Hence, the precise mechanism(s) by which PFOS impairs both the immune system and the response to pathogenic challenges needs to be clarified.

26.4.3 Temporal Trend in Wild Animals

There were several studies showing variations in PFOS concentrations in different biota tissues, including human over time. Martin et al. found that there was a significant increase in the mean PFOS concentrations (from 43 to 180 ng/g) in the archived lake trout whole-body homogenates [96].

The concentration increased between 1980 and 1989, with a slight decrease between 1989 and 1995, and then an increase until 2001. Holmstrom et al. showed that there was a significant increasing trend (30-fold) on average between 7% and 11% per year in PFOS concentrations (from 25 ng/g ww. in 1968 to 614 ng/g ww in 2003) in archived guillemot eggs in the Baltic Sea marine environment [110]. Verreault et al. also showed a significant increase (nearly twofold) in PFOS concentrations in herring gull eggs from 1983 to 1993, followed by a leveling off up to 2003 collected from two coastal colonies in northern Norway [109]. Butt et al. found the PFOS concentrations remained steady in thick-billed murres and in northern fulmar between 1993 and 2003–2004 [79]. There was an increase (39%) in PFOS concentrations in the livers of murres from 1993 to 2004, while PFOS levels decreased (32%) from 1993 to 2003 in the livers of fulmars, however those findings were not statistically significant. Bossi et al. investigated the temporal trends on PFOS concentrations in liver of ringed seals from Ittoqqortoormiit and Qeqertarsuaq, Greenland [73]. There was a significant annual increase in median concentrations of 8.2% for Qeqertarsuaq, but not for seals from Ittoqqortoormiit. Butt et al. investigated the responses of PFC concentrations in Canadian Arctic ringed seal with the perfluoroalkyl production [132]. They found that PFOS concentrations showed the greatest concentrations during 1998 and 2000 at Arviat and at Resolute Bay, with significant decreases from 2000 to 2005. PFOS concentrations in polar bears from two locations in the North American Arctic demonstrated an exponential increase between 1972 and 2002. The doubling time for PFOS was estimated to be 13.1 years [180]. Baikal seal had significant higher liver PFOS concentrations in 1992 than those in 2005 from Lake Baikal, Russia [181]. A marginal increase of liver PFOS concentrations was found in the adult harbor seals from 2000 to 2007 from the northwest Atlantic [125]. Temporal PFC variations have been studied in human blood samples [77,149,158,178]. There were cases that PFOS concentrations increased in China (Shenyang: from 0.03 ng/mL in 1987 to 22.4 ng/mL in 2002 [178]) and in the United States (from 24.9 ng/mL in 1974 to 33.2 ng/mL in 1989 for same individual [140]). Paul and coworkers related the estimated POSF releases with the biota concentrations as a temporal trend [40]. They suggested that biota seemed to reflect a sharp increase in releases to the environment post-1970 as estimated by their study while polar bears contained around 20% of the maximum estimated values by 1970 in the remote Arctic. However, there were cases where PFOS concentrations remained stable such as in Japan (Kyoto: 1980–2000 [149]) and in Germany (North Rhine-Westphalia: between 1977 and 2004 [158]). There are inconsistent findings showing a steady increase in PFOS concentrations in humans. These inconsistent trends might be attributed to different factors (i.e., rural vs. urban vs. industrial area), different living lifestyles, and any nonpoint/point-source exposures to the population. Other factors like age and gender also confound the interpretation of the data and there were studies that indicated a decline of PFCs (e.g., PFOS) after 2000. A recent study showed that there was a decline in PFOS concentrations in American Red Cross adult blood donors from 2000 to 2006 and that this decline in PFOS levels was consistent with its serum elimination half-life and the time elapsed since the phase-out of POSF-based materials in the United States [179]. PFHxS, another perfluorosulfonate having six fluorinated carbon, could be detected in the blood samples. It occurred at much lower concentrations (occurred at sub-ppb level) when compared to PFOS (ppb level). Butt et al. also found that PFOS concentrations in Canadian Arctic ringed seals showed significant decreases from 2000 to 2005 at Arviat and Resolute Bay [132]. They suggested that these results were rapidly responding to the phase out of POSF-based products by 3M in 2001. However, it should be reminded that the 3M Company also carried out a series of actions to reduce the emissions of the product waste in 1997 (see discussion above), which might also help account for the decrease in PFOS concentrations in tissues of wildlife and humans. Vestergren and Cousins analyzed the data of PFOA concentrations in human serum of various studies and tissues of wildlife [182] and they suggested the rapid decrease in human serum PFOA concentrations from Norway [183] and the United States [184] after around 2000 might be related to a decrease in consumer product exposure. However, other PFCs like PFNA and PFDA were found to have a slight increase or steady concentrations in the same studies [183,184].

Another group of perfluorochemicals containing the phosphate functional group instead of having the carboxylate or sulfonate groups are also important in PFC issues [185–187]. Perfluorinated phosphonic acids (PFPAs) were detected in environmental samples (i.e., Canadian surface water and effluent from wastewater treatment plant) [187]. PFPAs are used as leveling and wetting agents in commercial surfactants for waxes and coatings [http://www.masonsurfactants.com]. In the United States, PFPAs were used as inert defoaming additives in pesticide formulations under a regulation written for the polyfluoroalkyl phosphoric acids [188]. The use of the polyfluoroalkyl phosphoric acids in this capacity has subsequently been restricted by the U.S. Environmental Protection Agency since 2008 [188]. Production of PFCAs had been suggested for the biotransformation of polyfluoroalkyl phosphate surfactants (PAPs) [185]. A recent study showed the presence of polyfluoroalkyl phosphoric acid diesters (diPAPs) in human sera, wastewater treatment plant sludge, and paper fiber collected from Canada [186]. The levels of diPAPs in human sera could reach the levels as high as ng/mL, which were in the same range as PFOS concentrations that were found in human blood. These PAPs are fluorinated surfactants used in human food contact paper products such as the popcorn bags. PAPs might migrate into food and food stimulants, and humans might ingest these PAPs, which could then be biotransformed into PFCAs [185]. A recent study demonstrated microbially mediated biodegradation of PAPs to PFCAs [189]. The authors suggested that PAPs-containing commercial products might contribute to the increased PFCA mass flows in wastewater treatment plant effluents.

Measuring PFCs in environmental samples displays a brief understanding of the degree of PFC pollution in the environment. This understanding depends on the efficacy and efficiency of existing extraction methods and the availability of authentic analytical standards. To date, with the help of advanced analytical instrumentation and analytical standards, more than 18 PFCs can be measured in seawater, animal tissues, and human samples [56,94]. Even though basic analytical methods are possible for measuring 41 different PFCs [190], it is still technically challenging, if not impossible, to determine all of the individual PFCs in a sample. Given that the number of fluorinated organic chemicals found in biota is increasing, it is not known if there are any unidentified per-/poly-fluorinated compounds or how much they exist in the environment and in biological samples. A robust method and approach is needed to evaluate the mass of fluorinated compounds present in environmental and in biological matrices. Total fluorine (TF) and extractable organic fluorine (EOF) analyses are expected to provide useful information on potential discharges of any unknown PFCs into the environment. TF in any sample, using the concept of mass balance, is equal to the sum of inorganic fluorine (IF) and organic fluorine (OF) concentrations in the sample. However, limit of quantification of TF and EOF using conventional combustion ion chromatography (CIC) has been on the order of sub ppm levels, because of some analytical issues (e.g., co-elution of interferences) and high background levels arising from instrumental blanks. A recent method for measuring semi- and nonvolatile organic halogens by CIC has been developed in which the analyzer is serially connected with the combustion of electric furnace and the ion chromatograph [191]. Faster operation and better sensitivity of the analyzer allows for accurate analyses of trace levels of halogenated compounds in various environmental matrices (2.8–31 ng-X/g soil, 1.4–16 ng-X/L water, and 9.2–100 ng-X/m^3 gas; X-halogen). PFCs occur in environmental matrices at concentrations ranging from parts per quadrillion (ppq) in seawater to ppb levels in human blood samples where sample preconcentration, removal of interferences by chloride and sulfate, co-elution of fluoride with low-molecular-weight organic acids, separation of OF and IF, conversion of OF into IF, and, more importantly, removal of PFCs from instrumental and reagent blanks must be done to ensure accurate fluoride analyses. In view of these concerns, refined CIC analytical methods were developed [150,192]. A refined CIC system was set up, which attained a much lower (by 100–1000-fold) detection limit compared to those of traditional CIC [150,192]. For details of the analytical method, please refer to Miyake et al. [150,192].

Other fluorochemicals, such as FTOHs and short- and long-chain PFCAs, have become important elements in monitoring surveys and risk evaluations. PFOS and PFOA are the dominant PFCs in those environmental samples, however, but these compounds only account for a portion of the

PFCs that are present in environmental and biological matrices [126,150,192]. Yeung et al. measured TF and EOF in the blood samples of human and liver samples of marine mammals (i.e., Indo-Pacific humpback dolphin and finless porpoise) using refined CIC for fluorine [60,126]. The results showed that known PFCs could account for >70% of EOF in human blood samples from Beijing, Shenyang, and Guiyang, whereas known PFCs could only account for around 30% of EOF in samples from Jintan [60]. Comparisons between the amounts of known PFCs and EOF in the livers showed that a large proportion (~70%) of the OF in both species is of unknown origin [126]. Another pilot study on TF, EOF, and known PFCs in the blood of wild rat from Japan demonstrated that the ratio of known PFCs in EOF ranged from 9% to 89% (56% on average), while known PFC concentrations in TF were less than 25% [120]. On the other hand, TF concentrations in the blood of PFOA-exposed rats ranged from 46,900 to 111,000 ng-F/mL, with PFOA contributing over 90% of TF. The mass balance analysis of the different forms of fluorine in blood suggested the presence of other forms of OF in addition to known PFCs. Comparison of the proportion of known PFCs to EOF can provide useful information for environmental management in terms of resource allocation. Large variations in ratios between known PFCs and EOF were observed in wild animals (i.e., wild rats and marine mammals) when compared with human samples. These results suggested that wild animals might have wider exposure pathways than humans. The results from of PFCs, EOF, and TF on wild rats also suggested that wild rats could be potential candidate species for environmental monitoring as well as human exposure estimation because other tissues or organs can be archived from wild rat.

26.5 FUTURE DIRECTIONS

The PFC levels presented in the present review and many other studies indicated that PFC pollution is a global issue. While the production of POSF-related products and other fluorotelomer-based products have either ceased or decreased in North America and in Europe, productions of these products are still ongoing in some developing countries like China. In China, there are several fluoropolymer-related manufacturing plants in Shanghai, Wuhan, and Shenyang. India originally was an agricultural country with the second-largest population in the world after China and is now undergoing rapid economic growth. In contrast to the situation in China, only low or trace levels of PFCs could be detected in the Indian environment and only three-to-four individual PFCs could be identified [7,18,68,115]. The differences between these developing nations are likely due to the fact there are no fluoropolymer-related manufacturing plants established in India. The major industry in India is information technology (IT), which does not involve any primary heavy industries; other industries include local textile, fishing, and dyeing industries. Although India and China are the two largest countries in the world in terms of population and economic development, the different strategies pursued by their governments—large-scale industrial development in China vs. a focus on IT and smaller-scale manufacturing in India—have resulted in different degrees of PFC pollution of their respective environments. However, there are currently no regulations related to these chemicals in India and China and thus more and more fluoropolymer-related manufacturing plants are expected to be established in these countries as they continue to develop. In the view of environmental chemists, it is very important to have baseline data before any pollution occurs, in order to have a comparison point, while for environmentalists, these results may be warnings of future pollution problems to which governmental agencies should pay attention and which should be a target of policy making.

Since most studies focused on PFOA, PFOS, or PFCs having C–F numbers over eight, PFCs having C–F number less than eight were seldom reported. Short-chain PFCs (C_2–C_3) were detected at levels comparable to those of PFOS or PFOA in environmental samples, but their recoveries were less than 10%. Due to the limitations of current extraction methods, these PFCs are usually not measured. Efforts should be made to develop methods to determine concentrations of these PFCs in environmental samples. In addition, other precursors like FTOHs, FOSAs, and FOSEs are getting more attention because they are thought to be degraded to PFCAs or PFASs by biotic and abiotic processes in the environment. New classes of PFCs like PFPAs and PAPs should be studied.

The detection of unidentified organofluorines in biological samples is the beginning of a greater understanding of PFC/organofluorine contamination in the environment, and several research directions can follow from this work: (1) Characterization of the unknown fractions in order to determine which kinds of organofluorines such as PFCs, fluorinated pharmaceuticals, pesticides, or PFPAs/PAPs are present in the environment, so that further action can be taken. (2) Improvement of the performance of the existing extraction methods: poor recoveries of ion-pairing methods on FTOHs, short-chain (C_2–C_3) or long-chain (C_{14}–C_{18}) PFCs confined the understanding to these compounds in the sample; a new ACN extraction method is developed, which might help to have better understanding of known PFCs and EOF compared with conventional ion-pairing method [61]. (3) Long-term monitoring of the ratio of known PFCs to EOF in different environmental samples will enhance current understanding of the use and release of unknown organofluorine; for example, a decrease in this ratio may suggest the use and release of new unknown PFCs in the environment. (4) CIC can also be used to check for the presence of and calculate the proportion of fluorinated impurities present in any sample. In conclusion, the application of the mass balance approach to source materials (e.g., industrial products) and environmental and biological samples should provide valuable information to the extent of contamination by other unidentified fluorochemicals in the environment.

REFERENCES

1. Giesy, J.P. and Kannan, K. 2001. Global distribution of perfluorooctane sulfonate in wildlife. *Environ. Sci. Technol.* 35: 1339–1342.
2. Kannan, K., Franson, J.C., Bowerman, W.W., Hansen, K.J., Jones, J.D., and Giesy, J.P. 2001. Perfluorooctane sulfonate in fish-eating water birds including bald eagles and albatrosses. *Environ. Sci. Technol.* 35: 3065–3070.
3. Kannan, K., Koistinen, J., Beckmen, K., Evans, T., Gorzelany, J.F., Hansen, K.J., Jones, P.D., Helle, E., Nyman, M., and Giesy, J.P. 2001. Accumulation of perfluorooctane sulfonate in marine mammals. *Environ. Sci. Technol.* 35: 1593–1598.
4. Kannan, K., Choi, J.W., Iseki, N., Senthilkumar, K., Kim, D.H., Masunaga, S., and Giesy, J.P. 2002. Concentrations of perfluorinated acids in livers of birds from Japan and Korea. *Chemosphere* 49: 225–231.
5. Kannan, K., Corsolini, S., Falandysz, J., Oehme, G., Focardi, S., and Giesy, J.P. 2002. Perfluorooctanesulfonate and related fluorinated hydrocarbons in marine mammals, fishes, and birds from coasts of the Baltic and the Mediterranean Seas. *Environ. Sci. Technol.* 36: 3210–3216.
6. Olsen, G.W., Burris, J.M., Burlew, M.M., and Mandel, J.H. 2003. Epidemiologic assessment of worker serum perfluorooctanesulfonate (PFOS) and perfluorooctanoate (PFOA) concentrations and medical surveillance examinations. *J. Occup. Environ. Med.* 45: 260–270.
7. Kannan, K., Corsolini, S., Falandysz, J., Fillmann, G., Kumar, K.S., Loganathan, B.G., Mohd, M.A., Olivero, J., Van Wouwe, N., Yang, J.H., and Aldous, K.M. 2004. Perfluorooctanesulfonate and related fluorochemicals in human blood from several countries. *Environ. Sci. Technol.* 38: 4489–4495.
8. Ellis, D.A., Martin, J.W., De Silva, A.O., Mabury, S.A., Hurley, M.D., Andersen, M.P.S., and Wallington, T.J. 2004. Degradation of fluorotelomer alcohols: A likely atmospheric source of perfluorinated carboxylic acids. *Environ. Sci. Technol.* 38: 3316–3321.
9. Wallington, T.J., Hurley, M.D., Xia, J., Wuebbles, D.J., Sillman, S., Ito, A., Penner, J.E., Ellis, D.A., Martin, J., Mabury, S.A., Nielsen, O.J., and Andersen, M.P.S. 2006. Formation of $C_7F_{15}COOH$(PFOA) and other perfluorocarboxylic acids during the atmospheric oxidation of 8: 2 fluorotelomer alcohol. *Environ. Sci. Technol.* 40: 924–930.
10. Giesy, J.P. and Kannan, K. 2002. Perfluorochemical surfactants in the environment. *Environ. Sci. Technol.* 36: 146A–152A.
11. Lewandowski, G., Meissner, E., and Milchert, E. 2006. Special applications of fluorinated organic compounds. *J. Hazard. Mater.* 136: 385–391.
12. Kissa, E. 2001. *Fluorinated Surfactants and Repellents*, 2nd edn. New York: Marcel Dekker.
13. USEPA 2000. Perfluorooctyl sulfonates; proposed significant new use rule. *Fed. Reg.* 65: 62319–62333.

14. USEPA 2002. Revised draft—Hazard assessment of perfluorooctanoic acids and its salts: Office of Pollution Prevention and Toxics, Risk Assessment Division.

15. Hekster, F.M., Laane, R.W.P.M., and de Voogt, P. 2003. Environmental and toxicity effects of perfluoroalkylated substances. *Rev. Environ. Contam. Toxicol.* 179: 99–121.

16. Stock, N.L., Ellis, D.A., Deleebeeck, L., Muir, D.C.G., and Mabury, S.A. 2004. Vapor pressures of the fluorinated telomer alcohols—Limitations of estimation methods. *Environ. Sci. Technol.* 38: 1693–1699.

17. Renner, R. 2001. Growing concern over perfluorinated chemicals. *Environ. Sci. Technol.* 35: 154A–160A.

18. Yeung, L.W.Y., So, M.K., Jiang, G.B., Taniyasu, S., Yamashita, N., Song, M.Y., Wu, Y.N., Li, J.G., Giesy, J.P., Guruge, K.S., and Lam, P.K.S. 2006. Perfluorooctanesulfonate and related fluorochemicals in human blood samples from China. *Environ. Sci. Technol.* 40: 715–720.

19. So, M.K., Yamashita, N., Taniyasu, S., Jiang, Q.T., Giesy, J.P., Chen, K., and Lam, P.K.S. 2006. Health risks in infants associated with exposure to perfluorinated compounds in human breast milk from Zhoushan, China. *Environ. Sci. Technol.* 40: 2924–2929.

20. Taniyasu, S., Kannan, K., Horii, Y., Hanari, N., and Yamashita, N. 2003. A survey of perfluorooctane sulfonate and related perfluorinated organic compounds in water, fish, birds, and humans from Japan. *Environ. Sci. Technol.* 37: 2634–2639.

21. Gulkowska, A., Jiang, Q.T., So, M.K., Taniyasu, S., Lam, P.K.S., and Yamashita, N. 2006. Persistent perfluorinated acids in seafood collected from two cities of China. *Environ. Sci. Technol.* 40: 3736–3741.

22. So, M.K., Taniyasu, S., Lam, P.K.S., Zheng, G.J., Giesy, J.P., and Yamashita, N. 2006. Alkaline digestion and solid phase extraction method for perfluorinated compounds in mussels and oysters from south China and Japan. *Arch. Environ. Contam. Toxicol.* 50: 240–248.

23. Li, X.M., Yeung, L.W.Y., Taniyasu, S., Lam, P.K.S., Yamashita, N., Xu, M.Q., and Dai, J.Y. 2008. Accumulation of perfluorinated compounds in captive Bengal tigers (*Panthera tigris tigris*) and African lions (*Panthera leo* Linnaeus) in China. *Chemosphere* 73: 1649–1653.

24. Li, X.M., Yeung, L.W.Y., Taniyasu, S., Li, M., Zhang, H.X., Liu, D., Lam, P.K.S., Yamashita, N., and Dai, J.Y. 2008. Perfluorooctanesulfonate and related fluorochemicals in the Amur tiger (*Panthera tigris altaica*) from China. *Environ. Sci. Technol.* 42: 7078–7083.

25. So, M.K., Taniyasu, S., Yamashita, N., Giesy, J.P., Zheng, J., Fang, Z., Im, S.H., and Lam, P.K.S. 2004. Perfluorinated compounds in coastal waters of Hong Kong, South China, and Korea. *Environ. Sci. Technol.* 38: 4056–4063.

26. Mak, Y.L., Taniyasu, S., Yeung, L.W.Y., Lu, G.H., Jin, L., Yang, Y.L., Lam, P.K.S., Kannan, K., and Yamashita, N. 2009. Perfluorinated compounds in tap water from China and several other countries. *Environ. Sci. Technol.* 43: 4824–4829.

27. 3M. 1999. *The Science of Organic Fluorochemistry*. AR226-0547. St Paul, MN: 3M Company.

28. 3M. 2000. *Phase-Out Plan for POSF-Based Products*. USEPA Docket OPPT-2002-0043. Specialty Materials Markets Group. St Paul, MN: 3M Company.

29. 3M. 1995. *3M Fluorad Surfactants*. Product Information Bulletin. St Paul, MN: 3M Company.

30. Asahi Glass Co. 1975. Process for producing perfluorocarboxylic acid. 1473807: British Patent.

31. Daikin Industries. 1998. Method for preparation perfluorocarboxylic acid by oxidation of perfluoroalkyl-ethylene. JP10279517: Japanese Patent.

32. Banks, R.E., Smart, B.E., and Tatlow, J.C. 1994. *Organofluorine Chemistry: Principle and Commercial Applications*. New York: Springer.

33. Schultz, M.M., Barofsky, D.F., and Field, J.A. 2003. Fluorinated alkyl surfactants. *Environ. Eng. Sci.* 20: 487–501.

34. Prevedouros, K., Cousins, I.T., Buck, R.C., and Korzeniowski, S.H. 2006. Sources, fate and transport of perfluorocarboxylates. *Environ. Sci. Technol.* 40: 32–44.

35. Stock, N.L., Lau, F.K., Ellis, D.A., Martin, J.W., Muir, D.C.G., and Mabury, S.A. 2004. Polyfluorinated telomer alcohols and sulfonamides in the North American troposphere. *Environ. Sci. Technol.* 38: 991–996.

36. Jahnke, A., Huberc, S., Ternme, C., Kylin, H., and Berger, U. 2007. Development and application of a simplified sampling method for volatile polyfluorinated alkyl substances in indoor and environmental air. *J. Chromatogr. A.* 1164: 1–9.

37. Barber, J.L., Berger, U., Chaemfa, C., Huber, S., Jahnke, A., Temme, C., and Jones, K.C. 2007. Analysis of per- and polyfluorinated alkyl substances in air samples from Northwest Europe. *J. Environ. Monitor.* 9: 530–541.

38. Strynar, M.J. and Lindstrom, A.B. 2008. Perfluorinated compounds in house dust from Ohio and North Carolina, USA. *Environ. Sci. Technol.* 42: 3751–3756.

39. Bjorklund, J.A., Thuresson, K., and De Wit, C.A. 2009. Perfluoroalkyl compounds (PFCs) in indoor dust: Concentrations, human exposure estimates, and sources. *Environ. Sci. Technol.* 43: 2276–2281.

40. Paul, A.G., Jones, K.C., and Sweetman, A.J. 2009. A first global production, emission, and environmental inventory for perfluorooctane sulfonate. *Environ. Sci. Technol.* 43: 386–392.

41. Oono, S., Harada, K.H., Mahmoud, M.A.M., Inoue, K., and Koizumi, A. 2008. Current levels of airborne polyfluorinated telomers in Japan. *Chemosphere* 73: 932–937.

42. Piekarz, A.M., Primbs, T., Field, J.A., Barofsky, D.F., and Simonich, S. 2007. Semivolatile fluorinated organic compounds in Asian and western U.S air masses. *Environ. Sci. Technol.* 41: 8248–8255.

43. Yamashita, N. and Taniyasu, S. 2009. Development of the standard method for testing hazardous chemicals in high-tech industrial products. Report on NEDO project (05A21502d).

44. DuPont Company. 2005. Dupont Global PFOA Strategy. USEPA Administrative Record AR226-1914. Washington, DC: U.S. Environmental Protection Agency.

45. Fluoropolymer Manufacturing Group. 2005. Dispersion Processors Mass Balance Report; U.S. EPA: Docket OPPT2003-0012-0900, -0901, -0902, -0903, -0904. Washington, DC: U.S. Environmental Protection Agency.

46. Moody, C.A. and Field, J.A. 2000. Perfluorinated surfactants and the environmental implications of their use in fire-fighting foams. *Environ. Sci. Technol.* 34: 3864–3870.

47. 3M. 2000. *POSF Life Cycle Waste Stream Estimates*. AR226-0681. St Paul, MN: 3M Company.

48. Lange, C.C. 2000. The aerobic biodegradation of N-EtFOSE alcohol by the microbial activity present in municipal wastewater treatment sludge: Biodegradation study report; 3M Project ID: LIMS E00-2252. St Paul, MN: 3M Company.

49. Philadelphia Surban Corporation. 1976. Fire-fighting Composition. 3849315: U.S. Patent.

50. De Silva, A.O. and Mabury, S.A. 2006. Isomer distribution of perfluorocarboxylates in human blood: Potential correlation to source. *Environ. Sci. Technol.* 40: 2903–2909.

51. Benskin, J.P., Bataineh, M., and Martin, J.W. 2007. Simultaneous characterization of perfluoroalkyl carboxylate, sulfonate, and sulfonamide isomers by liquid chromatography-tandem mass spectrometry. *Anal. Chem.* 79: 6455–6464.

52. Belisle, J. and Hagen, D.F. 1980. Method for the determination of perfluorooctanoic acid in blood and other biological samples. *Anal. Biochem.* 101: 369–376.

53. Scott, B.F., Moody, C.A., Spencer, C., Small, J.M., Muir, D.C.G., and Mabury, S.A. 2006. Analysis for perfluorocarboxylic acids/anions in surface waters and precipitation using GC-MS and analysis of PFOA from large-volume samples. *Environ. Sci. Technol.* 40: 6405–6410.

54. Moody, C.A., Kwan, W.C., Martin, J.W., Muir, D.C.G., and Mabury, S.A. 2001. Determination of perfluorinated surfactants in surface water samples by two independent analytical techniques: Liquid chromatography/tandem mass spectrometry and F-19 NMR. *Anal. Chem.* 73: 2200–2206.

55. Hansen, K.J., Clemen, L.A., Ellefson, M.E., and Johnson, H.O. 2001. Compound-specific, quantitative characterization of organic: Fluorochemicals in biological matrices. *Environ. Sci. Technol.* 35: 766–770.

56. Taniyasu, S., Kannan, K., So, M.K., Gulkowska, A., Sinclair, E., Okazawa, T., and Yamashita, N. 2005. Analysis of fluorotelomer alcohols, fluorotelomer acids, and short- and long-chain perfluorinated acids in water and biota. *J. Chromatogr. A.* 1093: 89–97.

57. Taniyasu, S., Kannan, K., Yeung, L.W.Y., Kwok, K.Y., Lam, P.K.S., and Yamashita, N. 2008. Analysis of trifluoroacetic acid and other short-chain perfluorinated acids (C2–C4) in precipitation by liquid chromatography-tandem mass spectrometry: Comparison to patterns of long-chain perfluorinated acids (C5–C18). *Anal. Chim. Acta* 619: 221–230.

58. Scott, B.F., Spencer, C., Mabury, S.A., and Muir, D.C.G. 2006. Poly and perfluorinated carboxylates in North American precipitation. *Environ. Sci. Technol.* 40: 7167–7174.

59. Sottani, C. and Minoia, C. 2002. Quantitative determination of perfluorooctanoic acid ammonium salt in human serum by high-performance liquid chromatography with atmospheric pressure chemical ionization tandem mass spectrometry. *Rapid Commun. Mass Spectrom.* 16: 650–654.

60. Yeung, L.W.Y., Miyake, Y., Taniyasu, S., Wang, Y., Yu, H.X., So, M.K., Jiang, G.B., Wu, Y.N., Li, J.G., Giesy, J.P., Yamashita, N., and Lam, P.K.S. 2008. Perfluorinated compounds and total and extractable organic fluorine in human blood samples from China. *Environ. Sci. Technol.* 42: 8140–8145.

61. Yeung, L.W.Y., Taniyasu, S., Kannan, K., Xu, D.Z.Y., Guruge, K.S., Lam, P.K.S., and Yamashita, N. 2009. An analytical method for the determination of perfluorinated compounds in whole blood using acetonitrile and solid phase extraction methods. *J. Chromatogr. A* 1216: 4950–4956.

62. Van Leeuwen, S.P.J., Karrman, A., Van Bavel, B., De Boer, J., and Lindstrom, G. 2006. Struggle for quality in determination of perfluorinated contaminants in environmental and human samples. *Environ. Sci. Technol.* 40: 7854–7860.

63. van Leeuwen, S.P.J., Swart, C.P., van der Veen, I., and de Boer, J. 2009. Significant improvements in the analysis of perfluorinated compounds in water and fish: Results from an interlaboratory method evaluation study. *J. Chromatogr. A.* 1216: 401–409.

64. ISO25101 2009. Water quality—Determination of perfluorooctanesulfonate (PFOS) and perfluorooctanoate (PFOA)—Method for unfiltered samples using solid phase extraction and liquid chromatography/mass spectrometry. ISO.

65. Yamashita, N., Kannan, K., Taniyasu, S., Horii, Y., Okazawa, T., Petrick, G., and Gamo, T. 2004. Analysis of perfluorinated acids at parts-per-quadrillion levels in seawater using liquid chromatography-tandem mass spectrometry. *Environ. Sci. Technol.* 38: 5522–5528.

66. Rostkowski, P., Yamashita, N., So, I.M.K., Taniyasu, S., Lam, P.K.S., Falandysz, J., Lee, K.T., Kim, S.K., Khim, J.S., Im, S.H., Newsted, J.L., Jones, P.D., Kannan, K., and Giesy, J.P. 2006. Perfluorinated compounds in streams of the Shihwa industrial zone and Lake Shihwa, South Korea. *Environ. Toxicol. Chem.* 25: 2374–2380.

67. Nakayama, S., Strynar, M.J., Helfant, L., Egeghy, P., Ye, X.B., and Lindstrom, A.B. 2007. Perfluorinated compounds in the cape fear drainage basin in North Carolina. *Environ. Sci. Technol.* 41: 5271–5276.

68. So, M.K., Miyake, Y., Yeung, W.Y., Ho, Y.M., Taniyasu, S., Rostkowski, P., Yamashita, N., Zhou, B.S., Shi, X.J., Wang, J.X., Giesy, J.P., Yu, H., and Lam, P.K.S. 2007. Perfluorinated compounds in the Pearl River and Yangtze River of China. *Chemosphere* 68: 2085–2095.

69. Murakami, M., Imamura, E., Shinohara, H., Kiri, K., Muramatsu, Y., Harada, A., and Takada, H. 2008. Occurrence and sources of perfluorinated surfactants in rivers in Japan. *Environ. Sci. Technol.* 42: 6566–6572.

70. Loos, R., Gawlik, B.M., Locoro, G., Rimaviciute, E., Contini, S., and Bidoglio, G. 2009. EU-wide survey of polar organic persistent pollutants in European river waters. *Environ. Pollut.* 157: 561–568.

71. Bossi, R., Strand, J., Sortkjaer, O., and Larsen, M.M. 2008. Perfluoroalkyl compounds in Danish wastewater treatment plants and aquatic environments. *Environ. Int.* 34: 443–450.

72. Murakami, M., Shinohara, H., and Takada, H. 2009. Evaluation of wastewater and street runoff as sources of perfluorinated surfactants (PFSs). *Chemosphere* 74: 487–493.

73. Bossi, R., Riget, F.F., and Dietz, R. 2005. Temporal and spatial trends of perfluorinated compounds in ringed seal (*Phoca hispida*) from Greenland. *Environ. Sci. Technol.* 39: 7416–7422.

74. Kannan, K., Yun, S.H., and Evans, T.J. 2005. Chlorinated, brominated, and perfluorinated contaminants in livers of polar bears from Alaska. *Environ. Sci. Technol.* 39: 9057–9063.

75. Ellis, D.A., Martin, J.W., Mabury, S.A., Hurley, M.D., Andersen, M.P.S., and Wallington, T.J. 2003. Atmospheric lifetime of fluorotelomer alcohols. *Environ. Sci. Technol.* 37: 3816–3820.

76. Yamashita, N., Taniyasu, S., Petrick, G., Wei, S., Gamo, T., Lam, P.K.S., and Kannan, K. 2008. Perfluorinated acids as novel chemical tracers of global circulation of ocean waters. *Chemosphere* 70: 1247–1255.

77. Loewen, M., Halldorson, T., Wang, F.Y., and Tomy, G. 2005. Fluorotelomer carboxylic acids and PFOS in rainwater from an urban center in Canada. *Environ. Sci. Technol.* 39: 2944–2951.

78. Martin, J.W., Smithwick, M.M., Braune, B.M., Hoekstra, P.F., Muir, D.C.G., and Mabury, S.A. 2004. Identification of long-chain perfluorinated acids in biota from the Canadian Arctic. *Environ. Sci. Technol.* 38: 373–380.

79. Butt, C.M., Mabury, S.A., Muir, D.C.G., and Braune, B.M. 2007. Prevalence of long-chained perfluorinated carboxylates in seabirds from the Canadian arctic between 1975 and 2004. *Environ. Sci. Technol.* 41: 3521–3528.

80. Young, C.J., Furdui, V.I., Franklin, J., Koerner, R.M., Muir, D.C.G., and Mabury, S.A. 2007. Perfluorinated acids in arctic snow: New evidence for atmospheric formation. *Environ. Sci. Technol.* 41: 3455–3461.

81. Yamashita, N., Kannan, K., Taniyasu, S., Horii, Y., Petrick, G., and Gamo, T. 2005. A global survey of perfluorinated acids in oceans. *Mar. Pollut. Bull.* 51: 658–668.

82. Saito, N., Harada, K., Inoue, K., Sasaki, K., Yoshinaga, T., and Koizumi, A. 2004. Perfluorooctanoate and perfluorooctane sulfonate concentrations in surface water in Japan. *J. Occup. Health* 46: 49–59.

83. Moody, C.A., Martin, J.W., Kwan, W.C., Muir, D.C.G., and Mabury, S.C. 2002. Monitoring perfluorinated surfactants in biota and surface water samples following an accidental release of fire-fighting foam into Etohicoke Creek. *Environ. Sci. Technol.* 36: 545–551.

84. Sinclair, E. and Kannan, K. 2006. Mass loading and fate of perfluoroalkyl surfactants in wastewater treatment plants. *Environ. Sci. Technol.* 40: 1408–1414.

85. Loganathan, B.G., Sajwan, K.S., Sinclair, E., Kumar, K.S., and Kannan, K. 2007. Perfluoroalkyl sulfonates and perfluorocarboxylates in two wastewater treatment facilities in Kentucky and Georgia. *Water Res.* 41: 4611–4620.

86. Loos, R., Wollgast, J., Huber, T., and Hanke, G. 2007. Polar herbicides, pharmaceutical products, perfluorooctanesulfonate (PFOS), perfluorooctanoate (PFOA), and nonylphenol and its carboxylates and ethoxylates in surface and tap waters around Lake Maggiore in Northern Italy. *Anal. Bioanal. Chem.* 387: 1469–1478.

87. Kim, S.K. and Kannan, K. 2007. Perfluorinated acids in air, rain, snow, surface runoff, and lakes: Relative importance of pathways to contamination of urban lakes. *Environ. Sci. Technol.* 41: 8328–8334.

88. Murakami, M. and Takada, H. 2008. Perfluorinated surfactants (PFSs) in size-fractionated street dust in Tokyo. *Chemosphere* 73: 1172–1177.

89. Liu, W., Jin, Y.H., Quan, X., Sasaki, K., Saito, N., Nakayama, S.F., Sato, I., and Tsuda, S. 2009. Perfluorosulfonates and perfluorocarboxylates in snow and rain in Dalian, China. *Environ, Int.* 35: 737–742.

90. Kowk, K.Y., Taniyasu, S., Yeung, L.W.Y., Murphy, M.B., Lam, P.K.S., Horii, Y., Kannan, K., Petrick, G., Sinha, R.K., and Yamashita, N. 2010. Flux of perfluorinated chemicals through wet deposition in Japan, USA, and several other countries. *Environ. Sci. Technol.* 44(18): 7043–7049.

91. Ueno, D., Darling, C., Alaee, M., Pacepavicius, G., Teixeira, C., Campbell, L., Letcher, R.J., Bergman, A., Marsh, G., and Muir, D. 2008. Hydroxylated polybrominated diphenyl ethers (OH-PBDEs) in the abiotic environment: Surface water and precipitation from Ontario, Canada. *Environ. Sci. Technol.* 42: 1657–1664.

92. Wei, S., Chen, L.Q., Taniyasu, S., So, M.K., Murphy, M.B., Yamashita, N., Yeung, L.W.Y., and Lam, P.K.S. 2007. Distribution of perfluorinated compounds in surface seawaters between Asia and Antarctica. *Mar. Pollut. Bull.* 54: 1813–1818.

93. Ahrens, L., Xie, Z.Y., and Ebinghaus, R. 2010. Distribution of perfluoroalkyl compounds in seawater from Northern Europe, Atlantic Ocean, and Southern Ocean. *Chemosphere.* 78: 1011–1016.

94. Tao, L., Kannan, K., Kajiwara, N., Costa, M.M., Fillmann, G., Takahashi, S., and Tanabe, S. 2006. Perfluorooctanesulfonate and related fluorochemicals in albatrosses, elephant seals, penguins, and Polar Skuas from the Southern Ocean. *Environ. Sci. Technol.* 40: 7642–7648.

95. Quinete, N., Wu, Q., Zhang, T., Yun, S.H., Moreira, I., and Kannan, K. 2009. Specific profiles of perfluorinated compounds in surface and drinking waters and accumulation in mussels, fish, and dolphins from southeastern Brazil. *Chemosphere* 77: 863–869.

96. Martin, J.W., Whittle, D.M., Muir, D.C.G., and Mabury, S.A. 2004. Perfluoroalkyl contaminants in a food web from lake Ontario. *Environ. Sci. Technol.* 38: 5379–5385.

97. Tomy, G.T., Budakowski, W., Halldorson, T., Helm, P.A., Stern, G.A., Friesen, K., Pepper, K., Tittlemier, S.A., and Fisk, A.T. 2004. Fluorinated organic compounds in an eastern Arctic marine food web. *Environ. Sci. Technol.* 38: 6475–6481.

98. Kannan, K., Tao, L., Sinclair, E., Pastva, S.D., Jude, D.J., and Giesy, J.P. 2005. Perfluorinated compounds in aquatic organisms at various trophic levels in a Great Lakes food chain. *Arch. Environ. Contam. Toxicol.* 48: 559–566.

99. De Vijver, K.I.V., Hoff, P.T., Van Dongen, W., Esmans, E.L., Blust, R., and De Coen, W.M. 2003. Exposure patterns of perfluorooctane sulfonate in aquatic invertebrates from the Western Scheldt estuary and the southern North Sea. *Environ. Toxicol. Chem.* 22: 2037–2041.

100. Cunha, I., Hoff, P., Van de Vijver, K., Guilhermino, L., Esmans, E., and De Coen, W. 2005. Baseline study of perfluorooctane sulfonate occurrence in mussels, *Mytilus galloprovincialis*, from north-central Portuguese estuaries. *Mar. Pollut. Bull.* 50: 1128–1132.

101. Houde, M., Bujas, T.A.D., Small, J., Wells, R.S., Fair, P.A., Bossart, G.D., Solomon, K.R., and Muir, D.C.G. 2006. Biomagnification of perfluoroalkyl compounds in the bottlenose dolphin (*Tursiops truncatus*) food web. *Environ. Sci. Technol.* 40: 4138–4144.

102. Kannan, K., Hansen, K.J., Wade, T.L., and Giesy, J.P. 2002. Perfluorooctane sulfonate in oysters, *Crassostrea virginica*, from the Gulf of Mexico and the Chesapeake Bay, USA. *Arch. Environ. Contam. Toxicol.* 42: 313–318.

103. Keller, J.M., Kannan, K., Taniyasu, S., Yamashita, N., Day, R.D., Arendt, M.D., Segars, A.L., and Kucklick, J.R. 2005. Perfluorinated compounds in the plasma of loggerhead and Kemp's ridley sea turtles from the southeastern coast of the United States. *Environ. Sci. Technol.* 39: 9101–9108.

104. Morikawa, A., Kamei, N., Harada, K., Inoue, K., Yoshinaga, T., Saito, N., and Koizumi, A. 2006. The bioconcentration factor of perfluorooctane sulfonate is significantly larger than that of perfluorooctanoate in wild turtles (*Trachemys scripta elegans* and *Chinemys reevesii*): An Ai river ecological study in Japan. *Ecotoxicol. Environ. Saf.* 65: 14–21.

105. Houde, M., Martin, J.W., Letcher, R.J., Solomon, K.R., and Muir, D.C.G. 2006. Biological monitoring of polyfluoroalkyl substances: A review. *Environ. Sci. Technol.* 40: 3463–3473.

106. Connell, D.W., Fung, C.N., Minh, T.B., Tanabe, S., Lam, P.K.S., Wong, B.S.F., Lam, M.H.W., Wong, L.C., Wu, R.S.S., and Richardson, B. 2003. Risk to breeding success of fish-eating Ardeids due to persistent organic contaminants in Hong Kong: Evidence from organochlorine compounds in eggs. *Water Res.* 37: 459–467.

107. Molina, E.D., Balander, R., Fitzgerald, S.D., Giesy, J.P., Kannan, K., Mitchell, R., and Bursian, S.J. 2006. Effects of air cell injection of perfluorooctane sulfonate before incubation on development of the white leghorn chicken (*Gallus domesticus*) embryo. *Environ. Toxicol. Chem.* 25: 227–232.

108. Verreault, J., Houde, M., Gabrielsen, G.W., Berger, U., Haukas, M., Letcher, R.J., and Muir, D.C.G. 2005. Perfluorinated alkyl substances in plasma, liver, brain, and eggs of glaucous gulls (*Larus hyperboreus*) from the Norwegian Arctic. *Environ. Sci. Technol.* 39: 7439–7445.

109. Verreault, J., Berger, U., and Gabrielsen, G.W. 2007. Trends of perfluorinated alkyl substances in herring gull eggs from two coastal colonies in northern Norway: 1983–2003. *Environ. Sci. Technol.* 41: 6671–6677.

110. Holmstrom, K.E., Jarnberg, U., and Bignert, A. 2005. Temporal trends of PFOS and PFOA in guillemot eggs from the Baltic Sea, 1968–2003. *Environ. Sci. Technol.* 39: 80–84.

111. Wang, Y., Yeung, L.W.Y., Taniyasu, S., Yamashita, N., Lam, J.C.W., and Lam, P.K.S. 2008. Perfluorooctane sulfonate and other fluorochemicals in waterbird eggs from South China. *Environ. Sci. Technol.* 42: 8146–8151.

112. Yoo, H., Kannan, K., Kim, S.K., Lee, K.T., Newsted, J.L., and Giesy, J.P. 2008. Perfluoroalkyl acids in the egg yolk of birds from Lake Shihwa, Korea. *Environ. Sci. Technol.* 42: 5821–5827.

113. Lofstrand, K., Jorundsdottir, H., Tomy, G., Svavarsson, J., Weihe, P., Nygard, T., and Bergman, A. 2008. Spatial trends of polyfluorinated compounds in guillemot (*Uria aalge*) eggs from North-Western Europe. *Chemosphere* 72: 1475–1480.

114. Gebbink, W.A., Hebert, C.E., and Letcher, R.J. 2009. Perfluorinated carboxylates and sulfonates and precursor compounds in herring gull eggs from colonies spanning the laurentian great lakes of North America. *Environ. Sci. Technol.* 43: 7443–7449.

115. Yeung, L.W.Y., Yamashita, N., Taniyasu, S., Lam, P.K.S., Sinha, R.K., Borole, D.V., and Kannan, K. 2009. A survey of perfluorinated compounds in surface water and biota including dolphins from the Ganges River and in other waterbodies in India. *Chemosphere* 76: 55–62.

116. Herzke, D., Nygard, T., Berger, U., Huber, S., and Rov, N. 2009. Perfluorinated and other persistent halogenated organic compounds in European shag (*Phalacrocorax aristotelis*) and common eider (*Somateria mollissima*) from Norway: A suburban to remote pollutant gradient. *Sci. Total Environ.* 408: 340–348.

117. Guruge, K.S., Manage, P.M., Yamanaka, N., Miyazaki, S., Taniyasu, S., and Yamashita, N. 2008. Species-specific concentrations of perfluoroalkyl contaminants in farm and pet animals in Japan. *Chemosphere* 73: S210–S215.

118. Dai, J.Y., Li, M., Jin, Y.H., Saito, N., Xu, M.Q., and Wei, F.W. 2006. Perfluorooctanesulfonate and perfluorooctanoate in red panda and giant panda from China. *Environ. Sci. Technol.* 40: 5647–5652.

119. Hoff, P.T., Scheirs, J., Van de Vijver, K., Van Dongen, W., Esmans, E.L., Blust, R., and De Coen, W. 2004. Biochemical effect evaluation of perfluorooctane sulfonic acid-contaminated wood mice (*Apodemus sylvaticus*). *Environ. Health Perspect.* 112: 681–686.

120. Yeung, L.W.Y., Miyake, B., Li, P., Taniyasu, S., Kannan, K., Guruge, K.S., Lam, P.K.S., and Yamashita, N. 2009. Comparison of total fluorine, extractable organic fluorine and perfluorinated compounds in the blood of wild and perfluorooctanoate (PFOA)-exposed rats: Evidence for the presence of other organofluorine compounds. *Anal. Chim. Acta* 635: 108–114.

121. Ostertag, S.K., Tague, B.A., Humphries, M.M., Tittlemier, S.A., and Chan, H.M. 2009. Estimated dietary exposure to fluorinated compounds from traditional foods among Inuit in Nunavut, Canada. *Chemosphere* 75: 1165–1172.

122. Smithwick, M., Muir, D.C.G., Mabury, S.A., Solomon, K.R., Martin, J.W., Sonne, C., Born, E.W., Letcher, R.J., and Dietz, R. 2005. Perflouroalkyl contaminants in liver tissue from East Greenland polar bears (*Ursus maritimus*). *Environ. Toxicol. Chem.* 24: 981–986.

123. Nakata, H., Kannan, K., Nasu, T., Cho, H.S., Sinclair, E., and Takemura, A. 2006. Perfluorinated contaminants in sediments and aquatic organisms collected from shallow water and tidal flat areas of the Ariake Sea, Japan: Environmental fate of perfluorooctane sulfonate in aquatic ecosystems. *Environ. Sci. Technol.* 40: 4916–4921.

124. Van de Vijver, K.I., Hoslbeek, L., Das, K., Blust, R., Joiris, C., and De Coen, W. 2007. Occurrence of perfluorooctane sulfonate and other perfluorinated alkylated substances in harbor porpoises from the Black Sea. *Environ. Sci. Technol.* 41: 315–320.

125. Shaw, S., Berger, M.L., Brenner, D., Tao, L., Wu, Q., and Kannan, K. 2009. Specific accumulation of perfluorochemicals in harbor seals (*Phoca vitulina concolor*) from the northwest Atlantic. *Chemosphere* 74: 1037–1043.

126. Yeung, L.W.Y., Miyake, Y., Wang, Y., Taniyasu, S., Yamashita, N., and Lam, P.K.S. 2009. Total fluorine, extractable organic fluorine, perfluorooctane sulfonate and other related fluorochemicals in liver of Indo-Pacific humpback dolphins (*Sousa chinensis*) and finless porpoises (*Neophocaena phocaenoides*) from South China. *Environ. Pollut.* 157: 17–23.

127. Ahrens, L., Siebert, U., and Ebinghaus, R. 2009. Temporal trends of polyfluoroalkyl compounds in harbor seals (*Phoca vitulina*) from the German Bight, 1999–2008. *Chemosphere* 76: 151–158.

128. Kallenborn, R., Berger, U., and Jarnberg, U. 2004. *Perfluorinated Alkylated Substances (PFAS) in the Nordic Environment*. Copenhagen, Norway: Nordic Council of Ministers.

129. Muir, D.C.G., Alaee, M., Butt, C., Braune, B., Helm, P., Mabury, S., Tomy, G., and Wang, X. 2004. New contaminants in Arctic Biota. In: *Synopsis of Research Conducted Under the 2003–2004 Northern Contaminants Program*, Indian and Northern Affairs, Ottawa, Canada, pp. 139–148.

130. Leonel, J., Kannan, K., Tao, L., Fillmann, G., and Montone, R.C. 2008. A baseline study of perfluorochemicals in Franciscana dolphin and Subantarctic fur seal from coastal waters of Southern Brazil. *Mar. Pollut. Bull.* 56: 778–781.

131. Houde, M., Wells, R.S., Fair, P.A., Bossart, G.D., Hohn, A.A., Rowles, T.K., Sweeney, J.C., Solomon, K.R., and Muir, D.C.G. 2005. Polyfluoroalkyl compounds in free-ranging bottlenose dolphins (*Tursiops truncatus*) from the Gulf of Mexico and the Atlantic Ocean. *Environ. Sci. Technol.* 39: 6591–6598.

132. Butt, C.M., Muir, D.C.G., Stirling, I., Kwan, M., and Mabury, S.A. 2007. Rapid response of arctic ringed seals to changes in perfluoroalkyl production. *Environ. Sci. Technol.* 41: 42–49.

133. Law, R.J., Bersuder, P., Mead, L.K., and Jepson, P.D. 2008. PFOS and PFOA in the livers of harbour porpoises (*Phocoena phocoena*) stranded or bycaught around the UK. *Mar. Pollut. Bull.* 56: 792–797.

134. Maestri, L., Negri, S., Ferrari, M., Ghittori, S., Fabris, F., Danesino, P., and Imbriani, M. 2006. Determination of perfluorooctanoic acid and perfluorooctanesulfonate in human tissues by liquid chromatography/single quadrupole mass spectrometry. *Rapid Commun. Mass Spectrom.* 20: 2728–2734.

135. Karrman, A., Domingo, J.L., Llebaria, X., Nadal, M., Bigas, E., van Bavel, B., and Lindstrom, G. 2010. Biomonitoring perfluorinated compounds in Catalonia, Spain: Concentrations and trends in human liver and milk samples. *Environ. Sci. Pollut. Res.* 17: 750–758.

136. Taves, D.R. 1968. Evidence that there are two forms of fluoride in human serum. *Nature* 217: 1050–1051.

137. Kennedy, G.L., Butenhoff, J.L., Olsen, G.W., O'Connor, J.C., Seacat, A.M., Perkins, R.G., Biegel, L.B., Murphy, S.R., and Farrar, D.G. 2004. The toxicology of perfluorooctanoate. *Crit. Rev. Toxicol.* 34: 351–384.

138. Olsen, G.W., Burris, J.M., Mandel, J.H., and Zobel, L.R. 1999. Serum perfluorooctane sulfonate and hepatic and lipid clinical chemistry tests in fluorochemical production employees. *J. Occup. Environ. Med.* 41: 799–806.

139. Guruge, K.S., Taniyasu, S., Yamashita, N., Wijeratna, S., Mohotti, K.M., Seneviratne, H.R., Kannan, K., Yamanaka, N., and Miyazaki, S. 2005. Perfluorinated organic compounds in human blood serum and seminal plasma: A study of urban and rural tea worker populations in Sri Lanka. *J. Environ. Monitor.* 7: 371–377.

140. Olsen, G.W., Huang, H.Y., Helzlsouer, K.J., Hansen, K.J., Butenhoff, J.L., and Mandel, J.H. 2005. Historical comparison of perfluorooctanesulfonate, perfluorooctanoate, and other fluorochemicals in human blood. *Environ. Health Perspect.* 113: 539–545.

141. Kuklenyik, Z., Reich, J.A., Tully, J.S., Needham, L.L., and Calafat, A.M. 2004. Automated solid-phase extraction and measurement of perfluorinated organic acids and amides in human serum and milk. *Environ. Sci. Technol.* 38: 3698–3704.

142. Olsen, G.W., Church, T.R., Larson, E.B., van Belle, G., Lundberg, J.K., Hansen, K.J., Burris, J.M., Mandel, J.H., and Zobel, L.R. 2004. Serum concentrations of perfluorooctanesulfonate and other fluorochemicals in an elderly population from Seattle, Washington. *Chemosphere.* 54: 1599–1611.

143. Holzer, J., Midasch, O., Rauchfuss, K., Kraft, M., Reupert, R., Angerer, J., Kleeschulte, P., Marschall, N., and Wilhelm, M. 2008. Biomonitoring of perfluorinated compounds in children and adults exposed to perfluorooctanoate-contaminated drinking water. *Environ. Health Perspect.* 116: 651–657.

144. Karrman, A., van Bavel, B., Jarnberg, U., Hardell, L., and Lindstrom, G. 2006. Perfluorinated chemicals in relation to other persistent organic pollutants in human blood. *Chemosphere* 64: 1582–1591.

145. Karrman, A., van Bavel, B., Jarnberg, U., Hardell, L., and Lindstrom, G. 2005. Development of a solid-phase extraction-HPLC/single quadrupole MS method for quantification of perfluorochemicals in whole blood. *Anal. Chem.* 77: 864–870.

146. Karrman, A., Mueller, J.F., Van Bavel, B., Harden, F., Toms, L.M.L., and Lindstrom, G. 2006. Levels of 12 perfluorinated chemicals in pooled Australian serum, collected 2002–2003, in relation to age, gender, and region. *Environ. Sci. Technol.* 40: 3742–3748.

147. Karrman, A., Ericson, I., van Bavel, B., Darnerud, P.O., Aune, M., Glynn, A., Lignell, S., and Lindstrom, G. 2007. Exposure of perfluorinated chemicals through lactation: Levels of matched human milk and serum and a temporal trend, 1996–2004, in Sweden. *Environ. Health Perspect.* 115: 226–230.

148. Inoue, K., Okada, F., Ito, R., Kato, S., Sasaki, S., Nakajima, S., Uno, A., Saijo, Y., Sata, F., Yoshimura, Y., Kishi, R., and Nakazawa, H. 2004. Perfluorooctane sulfonate (PFOS) and related perfluorinated compounds in human maternal and cord blood samples: Assessment of PFOS exposure in a susceptible population during pregnancy. *Environ. Health Perspect.* 112: 1204–1207.

149. Harada, K., Koizumi, A., Saito, N., Inoue, K., Yoshinaga, T., Date, C., Fujii, S., Hachiya, N., Hirosawa, I., Koda, S., Kusaka, Y., Murata, K., Omae, K., Shimbo, S., Takenaka, K., Takeshita, T., Todoriki, H., Wada, Y., Watanabe, T., and Ikeda, M. 2007. Historical and geographical aspects of the increasing perfluorooctanoate and perfluorooctane sulfonate contamination in human serum in Japan. *Chemosphere* 66: 293–301.

150. Miyake, Y., Yamashita, N., So, M.K., Rostkowski, P., Taniyasu, S., Lam, P.K.S., and Kannan, K. 2007. Trace analysis of total fluorine in human blood using combustion ion chromatography for fluorine: A mass balance approach for the determination of known and unknown organofluorine compounds. *J. Chromatogr. A* 1154: 214–221.

151. Olsen, G.W., Church, T.R., Miller, J.P., Burris, J.M., Hansen, K.J., Lundberg, J.K., Armitage, J.B., Herron, R.M., Medhdizadehkashi, Z., Nobiletti, J.B., O'Neill, E.M., Mandel, J.H., and Zobel, L.R. 2003. Perfluorooctanesulfonate and other fluorochemicals in the serum of American Red Cross adult blood donors. *Environ. Health Perspect.* 111: 1892–1901.

152. Calafat, A.M., Kuklenyik, Z., Caudill, S.P., Reidy, J.A., and Needham, L.L. 2006. Perfluorochemicals in pooled serum samples from United States residents in 2001 and 2002. *Environ. Sci. Technol.* 40: 2128–2134.

153. Weihe, P., Kato, K., Calafat, A.M., Nielsen, F., Wanigatunga, A.A., Needham, L.L., and Grandjean, P. 2008. Serum concentrations of polyfluoroalkyl compounds in Faroese whale meat consumers. *Environ. Sci. Technol.* 42: 6291–6295.

154. Midasch, O., Schettgen, T., and Angerer, J. 2006. Pilot study on the perfluorooctanesulfonate and perfluorooctanoate exposure of the German general population. *Int. J. Hyg. Envir. Heal.* 209: 489–496.

155. Ericson, I., Gomez, M., Nadal, M., van Bavel, B., Lindstrom, G., and Domingo, J.L. 2007. Perfluorinated chemicals in blood of residents in Catalonia (Spain) in relation to age and gender: A pilot study. *Environ. Int.* 33: 616–623.

156. Liu, J.Y., Li, J.G., Luan, Y., Zhao, Y.F., and Wu, Y.N. 2009. Geographical distribution of perfluorinated compounds in human blood from Liaoning Province, China. *Environ. Sci. Technol.* 43: 4044–4048.

157. Emmett, E.A., Shofer, F.S., Zhang, H., Freeman, D., Desai, C., and Shaw, L.M. 2006. Community exposure to perfluorooctanoate: Relationships between serum concentrations and exposure sources. *J. Occup. Environ. Med.* 48: 759–770.

158. Wilhelm, M., Holzer, J., Dobler, L., Rauchfuss, K., Midasch, O., Kraft, M., Angerer, J., and Wiesmuller, G. 2009. Preliminary observations on perfluorinated compounds in plasma samples (1977–2004) of young German adults from an area with perfluorooctanoate-contaminated drinking water. *Int. J. Hyg. Envir. Heal.* 212: 142–145.

159. Bartell, S.M., Calafat, A.M., Lyu, C., Kato, K., Ryan, P.B., and Steenland, K. 2010. Rate of decline in serum PFOA concentrations after granular activated carbon filtration at two public water systems in Ohio and West Virginia. *Environ. Health Perspect.* 118: 222–228.

160. Falandysz, J., Taniyasu, S., Gulkowska, A., Yamashita, N., and Schulte-Oehlmann, U. 2006. Is fish a major source of fluorinated surfactants and repellents in humans living on the Baltic coast? *Environ. Sci. Technol.* 40: 748–751.

161. Volkel, W., Genzel-Boroviczeny, O., Demmelmair, H., Gebauer, C., Koletzko, B., Twardella, D., Raab, U., and Fromme, H. 2008. Perfluorooctane sulphonate (PFOS) and perfluorooctanoic acid (PFOA) in human breast milk: Results of a pilot study. *Int. J. Hyg. Envir. Heal.* 211: 440–446.

162. Harada, K., Inoue, K., Morikawa, A., Yoshinaga, T., Saito, N., and Koizumi, A. 2005. Renal clearance of perfluorooctane sulfonate and perfluorooctanoate in humans and their species-specific excretion. *Environ. Res.* 99: 253–261.

163. Yang, Q., Abedi-Valugerdi, M., Xie, Y., Zhao, X.Y., Moller, G., Nelson, B.D., and DePierre, J.W. 2002. Potent suppression of the adaptive immune response in mice upon dietary exposure to the potent peroxisome proliferator, perfluorooctanoic acid. *Int. Immunopharmaco.* 2: 389–397.

164. DeWitt, J.C., Copeland, C.B., Strynar, M.J., and Luebke, R.W. 2008. Perfluorooctanoic acid-induced immunomodulation in adult C57BL/6J or C57BL/6N female mice. *Environ. Health Perspect.* 116: 644–650.
165. Keil, D.E., Mehlmann, T., Butterworth, L., and Peden-Adams, M.M. 2008. Gestational exposure to perfluorooctane sulfonate suppresses immune function in B6C3F1 mice. *Toxicol. Sci.* 103: 77–85.
166. Lefebvre, D.E., Curran, I., Armstrong, C., Coady, L., Parenteau, M., Liston, V., Barker, M., Aziz, S., Rutherford, K., Bellon-Gagnon, P., Shenton, J., Mehta, R., and Bondy, G. 2008. Immunomodulatory effects of dietary potassium perfluorooctane sulfonate (PFOS) exposure in adult Sprague-Dawley Rats. *J. Toxicol. Environ. Health A* 71: 1516–1525.
167. Peden-Adams, M.M., Keller, J.M., EuDaly, J.G., Berger, J., Gilkeson, G.S., and Keil, D.E. 2008. Suppression of humoral immunity in mice following exposure to perfluorooctane sulfonate. *Toxicol. Sci.* 104: 144–154.
168. Kannan, K., Perrotta, E., and Thomas, N.J. 2006. Association between perfluorinated compounds and pathological conditions in southern sea otters. *Environ. Sci. Technol.* 40: 4943–4948.
169. Guruge, K.S., Hikono, H., Shimada, N., Murakami, K., Hasegawa, J., Yeung, L.W.Y., Yamanaka, N., and Yamashita, N. 2009. Effect of perfluorooctane sulfonate (PFOS) on influenza A virus-induced mortality in female B6C3F1 mice. *J. Toxicol. Sci.* 34: 687–691.
170. Gruber, L., Schlummer, M., Ungewiss, J., Wolz, G., Moeller, A., Weise, N., Fromme, H., Gruber, L., Schlummer, M., Ungewiss, J., Wolz, G., Moeller, A., Weise, N., and Fromme, H. 2007. Analysis of sub-ppb levels of perfluorooctanoic acid (PFOA) and perfluorooctanesulfonate (PFOS) in food and fish. *Organohalogen Compd.* 69: 145–148.
171. Sinclair, E., Mayack, D.T., Roblee, K., Yamashita, N., and Kannan, K. 2006. Occurrence of perfluoroalkyl surfactants in water, fish, and birds from New York State. *Arch. Environ. Contam. Toxicol.* 50: 398–410.
172. Hoff, P.T., Van de Vijver, K., Dauwe, T., Covaci, A., Maervoet, J., Eens, M., Blust, R., and De Coen, W. 2005. Evaluation of biochemical effects related to perfluorooctane sulfonic acid exposure in organohalogen-contaminated great tit (*Parus major*) and blue tit (*Parus caeruleus*) nestlings. *Chemosphere* 61: 1558–1569.
173. Dauwe, T., Van de Vijver, K., De Coen, W., and Eens, M. 2007. PFOS levels in the blood and liver of a small insectivorous songbird near a fluorochemical plant. *Environ. Int.* 33: 357–361.
174. Bentzen, T.W., Muir, D.C.G., Amstrup, S.C., and O'Hara, T.M. 2008. Organohalogen concentrations in blood and adipose tissue of Southern Beaufort Sea polar bears. *Sci. Total Environ.* 406: 352–367.
175. Smithwick, M., Mabury, S.A., Solomon, K.R., Sonne, C., Martin, J.W., Born, E.W., Dietz, R., Derocher, A.E., Letcher, R.J., Evans, T.J., Gabrielsen, G.W., Nagy, J., Stirling, I., Taylor, M.K., and Muir, D.C.G. 2005. Circumpolar study of perfluoroalkyl contaminants in polar bears (*Ursus maritimus*). *Environ. Sci. Technol.* 39: 5517–5523.
176. Kannan, K., Newsted, J., Halbrook, R.S., and Giesy, J.P. 2002. Perfluorooctanesulfonate and related fluorinated hydrocarbons in mink and river otters from the United States. *Environ. Sci. Technol.* 36: 2566–2571.
177. Nakayama, S., Harada, K., Inoue, K., Sasaki, K., Seery, B., Saito, N., and Koizumi, A. 2005. Distributions of perfluorooctanoic acid (PFOA) and perfluorooctane sulfonate (PFOS) in Japan and their toxicities. *Environ. Sci. (Tokyo)* 12: 293–313.
178. Jin, Y.H., Saito, N., Harada, K.H., Inoue, K., and Koizumi, A. 2007. Historical trends in human serum levels of perfluorooctanoate and perfluorooctane sulfonate in Shenyang, China. *Tohoku J. Exp. Med.* 212: 63–70.
179. Olsen, G.W., Mair, D.C., Reagen, W.K., Ellefson, M.E., Ehresman, D.J., Butenhoff, J.L., and Zobel, L.R. 2007. Preliminary evidence of a decline in perfluorooctanesulfonate (PFOS) and perfluorooctanoate (PFOA) concentrations in American Red Cross blood donors. *Chemosphere* 68: 105–111.
180. Smithwick, M., Norstrom, R.J., Mabury, S.A., Solomon, K., Evans, T.J., Stirling, I., Taylor, M.K., and Muir, D.C.G. 2006. Temporal trends of perfluoroalkyl contaminants in polar bears (*Ursus maritimus*) from two locations in the North American Arctic, 1972–2002. *Environ. Sci. Technol.* 40: 1139–1143.
181. Ishibashi, H., Iwata, H., Kim, E.Y., Tao, L., Kannan, K., Amano, M., Miyazaki, N., Tanabe, S., Batoev, V.B., and Petrov, E.A. 2008. Contamination and effects of perfluorochemicals in Baikal Seal (*Pusa sibirica*). 1. Residue level, tissue distribution, and temporal trend. *Environ. Sci. Technol.* 42: 2295–2301.
182. Vestergren, R. and Cousins, I.T. 2009. Tracking the pathways of human exposure to perfluorocarboxylates. *Environ. Sci. Technol.* 43: 5565–5575.
183. Haug, L.S., Thomsen, C., and Becher, G. 2009. A sensitive method for determination of a broad range of perfluorinated compounds in serum suitable for large-scale human biomonitoring. *J. Chromatogr. A* 1216: 385–393.
184. Spliethoff, H.M., Tao, L., Shaver, S.M., Aldous, K.M., Pass, K.A., Kannan, K., and Eadon, G.A. 2008. Use of Newborn Screening Program blood spots for exposure assessment: Declining levels of perfluorinated compounds in New York State infants. *Environ. Sci. Technol.* 42: 5361–5367.

185. D'eon, J.C. and Mabury, S.A. 2007. Production of perfluorinated carboxylic acids (PFCAs) from the biotransformation of polyfluoroalkyl phosphate surfactants (PAPS): Exploring routes of human contamination. *Environ. Sci. Technol.* 41: 4799–4805.

186. D'eon, J.C., Crozier, P.W., Furdui, V.I., Reiner, E.J., Libelo, E.L., and Mabury, S.A. 2009. Observation of a commercial fluorinated material, the polyfluoroalkyl phosphoric acid diesters, in human sera, wastewater treatment plant sludge, and paper fibers. *Environ. Sci. Technol.* 43: 4589–4594.

187. D'eon, J.C., Crozier, P.W., Furdui, V.I., Reiner, E.J., Libelo, E.L., and Mabury, S.A. 2009. Perfluorinated phosphonic acids in Canadian surface waters and wastewater treatment plant effluent: Discovery of a new class of perfluorinated acids. *Environ. Toxicol. Chem.* 28: 2101–2107.

188. USEPA 2006. SABSO: SAB Review of EPA's draft risk assessment of potential human health effects associated with PFOA and its salts: Office of Pollution Prevention and Toxics, Risk Assessment Division, 200601-30-2006. Available at: http://www.epa.gov/sab/pdf/2006_0120_final_draft_pfoa_report.pdf

189. Lee, H., D'eon, J., and Mabury, S.A. 2010. Biodegradation of polyfluoroalkyl phosphates as a source of perfluorinated acids to the environment. *Environ. Sci. Technol.* 44: 3305–3310.

190. Ahrens, L., Plassmann, M., Xie, Z.Y., and Ebinghaus, R. 2009. Determination of polyfluoroalkyl compounds in water and suspended particulate matter in the river Elbe and North Sea, Germany. *Front. Environ. Sci. Eng. China.* 3: 152–170.

191. Miyake, Y., Kato, M., and Urano, K. 2007. A method for measuring semi- and non-volatile organic halogens by combustion ion chromatography. *J. Chromatogr. A* 1139: 63–69.

192. Miyake, Y., Yamashita, N., Rostkowski, P., So, M.K., Taniyasu, S., Lam, P.K.S., and Kannan, K. 2007. Determination of trace levels of total fluorine in water using combustion ion chromatography for fluorine: A mass balance approach to determine individual perfluorinated chemicals in water. *J. Chromatogr. A* 1143: 98–104.

Index

A

Ailuropoda melanoleuca, 611
Ailurus fulgens, 611
AIST, *see* National Institute of Advanced Industrial
      Science and Technology
Anadara senilis, 400
Antarctic biota, *see* Persistent organic pollutants
Antarctic circumpolar current (ACC), 571
Antarcticum, 578
Antartic krill
    definition, 574
    PCB and HCB detection, 575
    POP concentrations, 575–576
    time trend, 576, 578
    toxic contaminants and effects, 574–575
Aptenodytes forsteri, 580
Aquatic environments
    biota *vs.* concentrations, sediments, 135–136
    bivalves concentration, 135
    fish concentration, 135–136
    geographical regions, 132–133
    PCBs concentration, 134
    sediment analysis, 132, 144–145
    sediments by land-use types, 134, 145, 147
    sewage outfalls, BDE congeners, 145, 148
Arctic
    marine mammals
        hexabromocyclododecanes, 506–507
        polybrominated diphenyl ethers, 506–507
    temporal trends, 585–586
Arcttocephalus tropocalis, 612
Ariake Sea
    marine organisms, 89
    synthetic musk, HHCB concentration, 91
    tidal flat and shallow water species, 93
Atmospheric pollution status, 171–173
Australia, *see* Organohalogen contamination

B

Bays sediments concentration, 168, 171
Benzotriazole UV stabilizers
    applications, 87–88
    aquatic ecosystems detection, 88
    biota UV-328 concentration, 93
    chemical structures, 88
    hammerhead shark, GC-MS chromatograms,
        91–92
    marine organisms
        Ariake Sea, Western Japan, 89
        tidal flat concentraiton, 93–94
        UV-327 detection, 93
    seawater concentration, 93
    tidal flat and shallow water species, 93

Bioconcentration mechanism
    accumulation and depuration
        aroclor mixture, 437
        Crassostrea virginica, 435
        dioxins and furans, 437
        low- and high-molecular weight PCB, 435
        polynuclear aromatic hydrocarbons and poly-
            chlorinated biphenyl congeners, 435–436
    single-compartment model, 433–435
    water partition (K_{OW}) coefficient, 433
Bioindicators, 406
Biomagnification, 527–528
Bisphenol A, 11
Brazilian aquatic environment
    chemical analysis, 426–427
    GDP, 415
    invertebrates, 420–422
    sampling methods, 426
    sediment, 417–420
    usage, ban, and restrictions, 416–417
    vertebrates (*see* Vertebrates)
Breast milk concentration
    European countries, 40, 42
    North America PBDE levels, 40
    Osaka, Japan, 40
Brominated compounds, 8–9
Brominated flame retardants (BFRs)
    species comparison, 556–558
    temporal phenomenon, 558–560

C

Calcaric cambisols, 320
Calcaric-endoruptic anthrosol, 320
Cancer
    breast, 115
    colorectal, 115
    dermal, 114
    esophageal, 114–115
    lungs, 115–116
Chaenocephalus aceratus, 578
Champsocephalus gunnari, 578
China; *see also* Polychlorinated biphenyls
    biota, 251
    fly ash, 251
    human exposure, 251–252
    levels, 250
    sediments, 250
    sewage sludge, 250–251
    soils, 250
    sources, 249–250
Chionodraco hamatus, 578
Chlorinated compounds, 420
Chlorinated hydrocarbons
    animal fats, milk, and eggs
        DDT, 340

HCB, 340–341
HCH, 340–341
PCB, 341–342
chlorinated pesticides, sediment core
    air products outfall site, 490–491
    Cypress Creek mouth site, 490–491
    Ledbetter Embayment site, 489
dietary intake, 342–343
GC-17A gas chromatograph, 484
human milk, 349–350
Kuderna Danish (K-D) concentration, 484
materials and methods, 338–339
maximum residue levels, 340
National Veterinary Research Institute, 337
organic carbon and nitrogen, 483–484
PIWet Pulawy, 338
polychlorinated biphenyls, 486–488
quality assurance and control protocols, 484
radiochronology, 483
residue control plan, 343–346
sample site selection and sampling, 482–483
sediment core analysis, 481
sediment dating, 485–486
silica-gel column chromatographic procedure, 484
time trends
    bovine fat, 346–348
    bovine milk, 346–349
    cow milk fat, 347
    hen fat, 348
    poultry fat, 347–349
    swine fat, 346, 348
Chlorinated pesticides and metabolites, 373–374
    chlordane compounds, 371
    DDT compounds, 369–370
    HCB compounds, 370–371
    HCH compounds, 371
Contamination profiles
    Brazilian aquatic environment
        chemical analysis, 426–427
        GDP, 415
        invertebrates, 420–422
        sampling methods, 426
        sediment, 417–420
        usage, ban, and restrictions, 416–417
        vertebrates (see Vertebrates)
    oysters, Gulf of Mexico
        bioconcentration mechanism, 433–438
        bivalve mollusks, 432–433
        Galveston Bay, Texas, 458–462
        global marine monitoring program, 432
        municipal and industrial wastes, 431
        Mussel Watch project (see Mussel Watch project)
        Tampa Bay, Florida, 462–463
Crassostrea tulipa, 400
Crassostrea virginica, 435, 438

D

Dichlorodiphenyldichloroethylene (DDE), 549
Dichlorodiphenyl trichloroethane (DDT), 179–180
    animal fats, milk, and eggs, 340
    bovine fat and milk, 346–347
    concentrations, 417, 419
    cow milk fat, 346

distribution, 192–193
hen fat, 348
insecticide, 292
levels, 282–285
pinnipeds, 535
poultry fat, 347
production, 281
residue composition, 292
source, 193
swine fat, 346
temporal trends, 531–532
Dioxin
    emission trends
        agrochemical uses, 205
        impurities in agrochemicals, 204–205
        industrial waste incineration, 206
        MOE, 206
        municipal solid waste incineration, 205–206
    Tokyo bay pollution
        identification, sediment source, 207–208
        sediment concentration, 207
        source apportionment, 208–210
    trends, human exposure of
        intake estimation, 210–211
        mother milk, 211–212
Dyssostichus mawsoni, 578

E

Endogleyic-epidystric luvisol, 320
Environmental contamination and exposure
    health issues, 22–23
    organohalogen compounds, 20–21
    terrestrial biota, 19–20
Estuaries sediments concentration, 168, 170
Euphausia superba, 574
E-waste, regulations and strategies, 52–53, 55

F

Fauna
    concentration, dioxin-like chemicals, 139
    dioxin-like PCBs, 139–140
    macropods, dioxins like chemical concentration,
        139–141
    spatial and biological diversity, 139
    TEQ levels, 139

G

Gas chromatography-mass spectroscopy (GC/MS), 88,
    90–92, 600
Ghana, see Organochlorine pesticides
Global atmospheric passive sampling (GAPS), 180
Gobionotothen gibberifrons, 578
Gross domestic product (GDP), 415
Gymnoscopelus nicholsi

H

Hexabromocyclododecanes (HBCD)
    biomagnification, 527–528
    cetaceans, 523
    contamination status, 523–525
    environmental persistence, 522

lipophilic compounds, 522
marine mammals
   Arctic, 506–507
   brominated flame retardants, 498
   cetaceans, 504–505
   chemical analysis, 503
   environment, 501
   Europe, 507
   finless porpoise, 502
   Indo-Pacific humpback dolphin, 502
   Japan, 505–506
   North America, 506
   QA/QC, 503–504
   sample collection and preparation, 503
   statistical analysis, 504
   usage and production, 498–499
metabolic debromination, 522
prey fishes, 525–527
Hexachlorobenzene (HCB), 340–341
Hexachlorocyclohexanes (HCHs)
   alpha-beta-gamma-delta, 292
   animal fats, milk, and eggs, 340–341
   distribution, 188–192
   levels, 282–285
   pest control, 292
   production, 281
   source, 192
1,3,4,6,7,8-Hexahydro-4,6,6,7,8,8-
   hexamethylcyclopenta(g)-2-benzopyran
   (HHCB)
   AHTN detection, 89
   shallow water organisms, 89
   synthetic musk concentration, 91
High-performance liquid chromatography–tandem mass
   spectroscopy (HPLC/MS/MS), 601
House dust, 43–45
Human blood
   age and level of dioxin-like chemicals
      relation, 142, 147
   dioxin-like compounds detection, 142
   male and female samples, 143
   population distributions, 142
   sources, 142
Human health effects
   atherosclerosis, 116
   cancer (*see* Cancer)
   immunotoxicity, 116
   neurotoxicity, 111–112
   obesity, 116
   osteoporosis, 114
   reproductive toxicity (*see* Reproductive toxicity)
   toxic effects, 111
   uterine tumors, 114
Human milk, 349–350
Hyperskeletic-endolithic regosols, 320

I

Indian atmosphere, *see* Organochlorine pesticides (OCP)
Industrial products, accidental contamination
   aerosol adhesive bomb, 245–246
   class I hazardous chemicals, 243
   distribution and usage, PCN mixtures, 243–244
   isomer composition, 245–246

isomer-specific analysis, 243–245
neoprene FB, 243
risk model, 246, 248–249
toxic equivalents, 246–247
trade names, 242
weather-resistant products, 243
Isomer-specific analysis
   clean-up and separation, 233–234
   congener-specific carbon isotopic analysis, 235, 237
   environmental analysis, 232–233
   extraction, 233
   interlaboratory, 234–235
   quantification, 232
   standards, 232–233
   two-dimensional gas chromatography (GC×GC), 235–236

J

Japan; *see also* Dioxin
   ambient air, 228, 230–231
   ash, 227–229
   biotic environment, 230–232
   branched-chain isomer analysis, 78–79
   chemical structure and physicochemical properties,
      216–218
   concentrations, 222–225
   gas, 226–227
   industrial products, accidental contamination (*see*
      Industrial products, accidental contamination)
   isomer-specific analysis (*see* Isomer-specific analysis)
   liver tissue samples, temporal phenomenon, 77
   marine mammals
      hexabromocyclododecanes, 505–506
      polybrominated diphenyl ethers, 505–506
   PFC concentration, sediment, 77–78
   pollution reconstruction (*see* Pollution reconstruction)
   production and uses, 218–219
   regulation, 221–222
   sediment, 226
   sediment core analysis, 77
   serum samples PFOS concentration, 76
   sources, 219–220
   toxicity, 220–221
   water, 222, 226

K

Kentucky Lake, *see* Chlorinated hydrocarbons
Kuderna Danish (K-D) concentration, 484

L

Lake sediments concentration, 168, 171
Lake Shihwa
   construction, 262
   map of, 262
   organic pollutants, 262–263
   polychaetes, 265–266
   sources and distributions, 263–266
   temporal trends, 263, 265, 267
   water and sediment contamination, 262
Leptonichotes weddelli, 574
Locally weighted scatterplot smoother (LOESS), 77
Long-range atmospheric transport (LRAT), 180

M

Mammals
  PFCs
    coastal and open-ocean dolphins, 613
    cord blood and material, 614
    human biomonitoring, 613
    toxic effects, 614
  PFOA
    Amur tigers, 611
    human health, 613
    red pandas, 611
  PFOS
    Amur tigers, 611
    bird eggs concentration, 612
    coastal and open-ocean dolphins, 612
    human blood samples concentration, 614
    polar bears and Arctic fox, 612
    red pandas, 611
Marine mammals
  Arctic, 506–507
  brominated flame retardants, 498
  cetaceans, 504–505
  chemical analysis, 503
  environment, 501
  Europe, 507
  finless porpoise, 502
  hexabromocyclododecanes, 505–506
  Indo-Pacific humpback dolphin, 502
  Japan, 505–506
  North America, 506
  polybrominated diphenyl ethers, 505–506
  QA/QC, 503–504
  sample collection and preparation, 503
  statistical analysis, 504
  usage and production, 498–499
Mercury
  species comparison, 556
  temporal phenomenon, 554, 556
Ministry of Environment (MOE), 206
Mirounga leonina, 574
Monoporeia affinis, 586
Mother's milk
  animal feed contamination control, 369
  brominated flame retardants, 372
  CB 153 levels, 367, 373–374
  chlorinated pesticides and metabolites (*see* Chlorinated pesticides and metabolites)
  long-term monitoring, 372
  polybrominated diphenyl ethers, 374–375
  POPUP (*see* Persistent organic pollutants in Uppsala Primiparas)
  Stockholm study (*see* Stockholm study)
  WHO survey, 373
Mussel Watch project
  chlorinated pesticides
    aldrin and dieldrin, 445, 447–448
    bivalve tissues, 439
    chlordane-related compounds, 439, 443–444
    DDT isomers and metabolites, 443, 445–447
    hexachlorobenzene, 447
    lindane and mirex, 447
    organic contaminant, 438–439, 441–442

Crassostrea virginica, 438, 440
  organic analytes, 438–439
  polychlorinated biphenyls, 447, 449–450
  polycyclic aromatic hydrocarbons
    fossil fuels, 458
    median distributions, 449, 452–457
    pyrogenic sources, 458
    temporal distributions, 449, 451–452
  temporal tissue contaminant, 438

N

National Health and Nutrition Examination Survey (NHANES), 80
National Institute of Advanced Industrial Science and Technology, 606
National Veterinary Research Institute, 337
Nonylphenols, South Korea
  alkylphenols, 259–260
  contamination levels, 274–275
  endocrine-disrupting chemicals, 260
  industrialized area, Lake Shihwa
    construction, 262
    map of, 262
    organic pollutants, 262–263
    polychaetes, 265–266
    sources and distributions, 263–266
    temporal trends, 263, 265, 267
    water and sediment contamination, 262
  legislation, usage and control, 260–261
  ocean waste disposal, dumpsite "Byung"
    advection/diffusion, 270
    government plans, 273
    organic pollutants, 270
    sources, distributions, and temporal trends, 271–272
  urbanized area, Han River, 273–274
  wastewater treatment plant outfall, Masan Bay
    geological features, 266
    mussels, 270
    sources and distributions, 268–269
    special management coastal zone, 268
    temporal trends, 268–270
  water and sediment quality guidelines, 260
Northern Norway and Svalbard
  Atlantic puffins, 548
  BFRs
    species comparison, 556–558
    temporal phenomenon, 558–560
  biological characteristics, 548
  black-legged kittiwakes, 548
  Brünnich's guillemots breeding, 548
  common guillemots, 548
  herring gulls, 548
  OC
    species comparison, 553
    temporal phenomenon, 553–556, 563
  PFAs
    species comparison, 560–561
    temporal phenomenon, 561–562
  spatial trends of POPs, 551–553
  species comparison, 556
  temporal phenomenon, 556
Nothotenia rossii, 578

O

Ocean waste disposal, dumpsite "Byung"
    advection/diffusion, 270
    government plans, 273
    organic pollutants, 270
    sources, distributions, and temporal trends, 271–272
OCP, *see* Organochlorine pesticides
Open ocean
    PFC levels
        environmental pollution, 606, 608
        kinetics of, 606
        monitoring, 604
        PFOS and PFOA concentrations, 606
        physicochemical properties, 606
    PFOS and PFOA vertical profile
        deep-water current, 609
        Labrador sea water columns, 608–609
        middle Atlantic Ocean, 609
        PFC concentrations and profile, 610
        residues in water column samples, 608
        South Pacific Ocean, 609–610
Organochlorine pesticides (OCP), 308
    back trajectory analysis, 188
    banned OCP, 395
    bioaccumulation, 394
    bioindicators, 406
    biological samples
        Anadara senilis, 400
        Crassostrea tulipa, 400
        DDT residues, 397
        dieldrin, 398
        endosulfan, 397–398
        human blood and breast milk, 400–401
        meat and dairy products, 401
        methoxychlor residues, 397
        milk and blood serum, 398, 401
        Perna perna, 400
        spatial trends, 400, 402–403
        vegetables, 397–400
    chemical analysis, 404–406
    crop protection and disease vector control, 394
    diseases and pest attacks, 394
    environmental impacts, 395
    map of, 393–394
    materials and methods
        extraction and analysis, 185–186
        Indian coastline, sampling site, 181
        mangrove wetland site, 181–182
        quality control and assurance, 186
        sampling location, duration and ambient
            temperature, 181, 183–185
    PBDEs, and PCBs concentration, 187–188
    pesticide products, 406
    registered pesticides, 407
    sampling, 404
    spatial distribution and potential sources
        chlorine sources and, 193–194
        DDTs distribution, 192–193
        DDTs soruce, 193
        endosulfan sources and, 195–196
        HCB sources and, 196
        HCH distribution, 188–192

        HCH source, 192
        PBDEs sources and, 198–199
        PCBs sources and, 196–198
    temporal and spatial variations, 406
    water and sediment, 396–397
Organochlorines (OC), 549
    characteristics, 550
    species comparison, 553
    temporal phenomenon, 553–556, 563
Organochlorines and polycyclic aromatic
        hydrocarbons
    abandoned boat slips, 473
    Brunswick Harbor, 475–477
    coring procedures, 471
    Cs-137, 470
    deposition and accumulation, 470
    Pb-210, 470
    polychlorinated biphenyls analysis, 472–473
    preliminary sampling and sediment analysis, 471
    radiochemical chronologies, 470
    Sapelo Island Core, 476, 478
    Savannah River Estuary, 473–476
    sediment physical characterization, 471–472
Organohalogen compounds
    air
        ambient air data, 316
        atmospheric concentrations, PCB, 315, 317
        atmospheric transport processes, 314
        Czech Republic passive air monitoring
            network, 316
        DDT concentrations, 314, 316
        Estonian Lahemaa and Kohtla-Järve air monitoring
            stations, 316, 318–319
        HCH concentrations, 314, 317
        long-range atmospheric transport, 316
        OCP, 316, 320
        organochlorine plant protection products, 314
        passive air sampling, 314
        PCB concentrations, 314–315
        polycyclic aromatic hydrocarbons, 316, 319
    biota, 323–326
    bisphenol A, 11
    brominated compounds, 8–9
    Estonia
        air analysis, 311–312
        biota analysis, 313–314
        chlorinated plant protection products, 308
        contamination sources, 309–311
        OCP, 308
        soil analysis, 312–313
    global environmental contamination phenomenon,
        20–21
    mother's milk
        animal feed contamination control, 369
        brominated flame retardants, 372
        CB 153 levels, 367, 373–374
        chlorinated pesticides and metabolites (*see*
            Chlorinated pesticides and metabolites)
        chlorinated pesticides/metabolites, 373–374
        long-term monitoring, 372
        polybrominated diphenyl ethers, 374–375
        POPUP (*see* Persistent organic pollutants in
            Uppsala Primiparas)

Stockholm study (*see* Stockholm study)
  WHO survey, 373
organochlorine compounds
  aquatic biota, contamination trends, 19–20
  chlorinated pesticides, 7
  dioxins and furans, 7–8
  polychlorinated biphenyls, 4–7
  terrestrial biota, contamination trends, 20
perfluorinated compounds, 9
pharmaceutical and personal care products, 10–11
polychlorinated naphthalenes, 9–10
polycyclic aromatic hydrocarbons, 11–12
soil
  calcaric cambisols, 320
  calcaric-endoruptic anthrosol, 320
  endogleyic-epidystric luvisol, 320
  Estonian map, air and soil sampling sites, 320–321
  humus layer, 320
  hyperskeletic-endolithic regosols, 320
  model monitoring network, 321
  moisture content and concentrations, dioxins and furans, 322–323
  PCB and OCP concentrations, 321–322
  PCDD/PCDF concentrations, 322–323
  salic-epigleyic fluvisols, 320
  sampling sites, 321
  stagnic albeluvisols, 320
temporal trends
  CB28 and CB52 concentration, 326, 328
  CB153 concentration, 327, 329
  chlorobiphenyls, 326
  DDE, DDD, and DDT concentrations, 326–327
  Gulf of Finland, 326–328
  α-HCH concentration, 327, 329
  industrial organochlorine compounds, 326
  score plot, 327, 330
trend monitoring, 12–13
Organohalogen contamination
  agricultural and industrial applications, 130–132
  dioxin-like chemicals concentration, 158
  environmental levels and impacts
    PBDE (*see* Polybrominated diphenyl ethers)
    PCDDs/PCDFs furans and PCBs (*see* Polychlorinated dibenzo-p-dioxins/furans and PCB)
  OCP trend
    historic butter samples, 156–157
    human milk samples, 155–156
    marine environment, 152–155
  population distribution, 129–130
Oysters, Gulf of Mexico
  bioconcentration mechanism, 433–438
  bivalve mollusks, 432–433
  Galveston Bay, Texas, 458–462
  global marine monitoring program, 432
  municipal and industrial wastes, 431
  Mussel Watch project (*see* Mussel Watch project)
  Tampa Bay, Florida, 462–463

P

Panthera tigris altaica, 611
Panthera tigris tigris, 611
PCB, *see* Polychlorinated biphenyls

Penguins
  concentrations of contaminants, 581
  eggs concentrations, 581–582
  gentoo penguin, 581
  marine ecosystems, 583
  PCBs and DDTs concentrations, 583
  sea birds POP concentrations, 579
  time trend analysis, 579–580
  xenobiotics detection, 582
Perfluorinated alkyl substances (PFAS)
  species comparison, 560–561
  temporal phenomenon, 561–562
Perfluorinated compounds (PFCs), 593
  analytical methods
    GC/MS, 600
    HPLC/MS/MS, 601
    quality assurance and control, 601–602
  biological samples distribution
    birds, 611
    mammals (*see* Mammals)
    wild animals, temporal phenomenon, 615–618
  characteristics, 74
  coastal and open-ocean dolphins, 613
  contamination status, 529–530
  environmental pollution
    Arctic, 82
    Europe, 79–80
    future perspective, 82–83
    Japan (*see* Japan)
    North America, 80–82
  future perspectives, 618–619
  human biomonitoring, 613
  marine food webs, 528
  material and cord blood, 614
  oleophobic and hydrophobic, 528
  perfluoroalkyl carboxylic acids, 528
  perfluoroalkyl sulfonic acids, 528
  physicochemical properties, 594–595
  production and emission
    PFOA, 74
    PFOSF/PFOS, 74–75
    transaction, products register system, 75–76
  products
    electrochemical fluorination, 596
    telomerization, 596–597
  sources
    PFASs, 597–598
    PFCAs, 598–600
  spatial trends, 534
  temporal trends, 532
  toxic effects, 614
  transportation
    air and dust, 603
    open ocean, 604–608 (*see also* Open ocean)
    precipitation, 603–604
Perfluorooctane sulfonate (PFOS), 73, 593
  Amur tigers, 611
  bird eggs concentration, 612
  coastal and open-ocean dolphins, 612
  human blood samples concentration, 614
  polar bears and Arctic fox, 612
  red pandas, 611
Perfluorooctane sulfonyl fluoride (POS), 593

Perfluorooctanoic acid (PFOA)
    Amur tigers, 611
    human health, 613
    red pandas, 611
Perna perna, 400
Persistent chemicals, definition, 4
Persistent organic pollutants (POPs)
    biological features, 571–573
    biota contamination profile, 583–585
    biota, temporal trend
        fish, 578–579
        krill (*see* Antarctic krill)
        penguins (*see* Penguins)
    climatic features, 571–573
    colostrum, 384
    contamination profiles, 385–386
    DDT
        pinnipeds, 535
        temporal trends, 531–532
    dioxin-like compounds, 384
    ecosystem characteristics, 572
    geographical features, 571–573
    HBCD
        biomagnification, 527–528
        cetaceans, 523
        contamination status, 523–525
        environmental persistence, 522
        metabolic debromination, 522
        prey fishes, 525–527
    inter alia, 384
    lipophilic compounds, 522
    mortalities, 517
    northwest Atlantic, 516–517
    patterns and concentration, 550
    PBDE, 381
        congeners, 523
        contamination status, 523–525
        environmental persistence, 522
        metabolic debromination, 522
        pinnipeds, 535–537
        spatial trends, 533–534
        temporal trends, 532
    PCB
        biomagnification, 527–528
        blubber samples, 518
        chlorinated pesticides, 518–520
        dioxins and WHO-TEQ, 520–522
        pinnipeds, 534–535
        prey fishes, 525–527
        spatial trends, 533
        temporal trends, 531–532
    PFC
        contamination status, 529–530
        marine food webs, 528
        oleophobic and hydrophobic, 528
        perfluoroalkyl carboxylic acids, 528
        perfluoroalkyl sulfonic acids, 528
        spatial trends, 534
        temporal trends, 532
    PFOS, pinnipeds, 535–537
    sentinels, 517
    spatial phenomenon, 551–553
    Stockholm convention, 382
    temperatures, 572

    temporal trend, 382–383
        dioxins, furans, and dioxin-like PCB, 387
        DL-PCB, 386
        NDL-PCB, 386–387
        non-dioxin PCB, 387–388
        PBDE, 386, 389
        PCDD and PCDF, 386
        polybrominated diphenyl ethers, 388–389
    temporal trend assessment, organisms, 573–574
    temporal trends indicator, 551
    transport, 573
    WHO, 382
Persistent organic pollutants in Uppsala Primiparas
        (POPUP)
    body mass index, 359–360
    CB 126, 358
    design, 359–360
    PBDE, 357–358
    PCB, 357–358
        congeners, 361–363
        di-ortho congener, 361, 364
        linear regression analysis, 361, 365
    PCDD, 357–358
        congeners, 361–363
        guillemot eggs, 361
    PCDF, 357–358
        congeners, 361–363
        guillemot eggs, 361
    recruitment and sampling, 356–357
    ΣPCB calculation, 358–359
Pleuragramma antarcticum, 578
Poland, *see* Chlorinated hydrocarbons
Pollution reconstruction
    CN homologue and congener patterns, 239–240
    ecosystems, 236
    flux, 241
    halogenated polycyclic aromatic hydrocarbons,
        242–243
    Lake Kitaura, 238–239
    preindustrial period, 242
    radioisotopes profiles, 237–238
    sediment core collection, 237
    Tokyo Bay, 238–239
    vertical profiles, persistent organic pollutants, 236
    volcanic ash stratigraphy, 238
Polybrominated diphenyl ethers (PBDEs)
    applications, 34–36
    aquatic environments
        sediment analysis, 144–145
        sediments by land-use types, 145, 147
        sewage outfalls, BDE congeners, 145, 148
    Asian countries, human exposure, 54
    chemical and physical properties, 34–36
    congeners, 523
    contamination status, 523–525
    environmental persistence, 522
    e-waste, 52–53
    human blood, 148–150
    indoor environment, 150–152
    lipophilic compounds, 522
    marine mammals
        abiotic and biotic matrices, 502
        Arctic, 506–507
        brominated flame retardants, 498

cetaceans, 504–505
chemical analysis, 503
environment, 499–501
Europe, 507
finless porpoise, 502
Indo-Pacific humpback dolphin, 502
Japan, 505–506
North America, 506
QA/QC, 503–504
sample collection and preparation, 503
statistical analysis, 504
usage and production, 498–499
metabolic debromination, 522
North America, environment and biota data, 54
pinnipeds, 535–537
POPUP (*see* Persistent organic pollutants in Uppsala
Primiparas)
sources and environment, 36–37
spatial and temporal phenomenon (*see* Spatial and
temporal phenomenon)
spatial trends, 533–534
temporal trends, 532
toxicity, 37–39
Polychlorinated biphenyls (PCBs), 179, 357–358, 549
animal fats, milk, and eggs, 341–342
atmospheric pollution status, 171–173
biomagnification, 527–528
blubber samples, 518
bovine fat, 348
bovine milk, 348–349
chlorinated pesticides, 518–520
concentrations, 417–418
congeners, 361–363
congeners and isomers, 292
di-ortho congener, 361, 364
dioxins and WHO-TEQ, 520–522
levels, 284–285
levels and temporal trends, 372–374
linear regression analysis, 361, 365
pinnipeds, 534–535
pollutants distribution and estimation, 164
POPUP (*see* Persistent organic pollutants in Uppsala
Primiparas)
poultry fat, 348–349
prey fishes, 525–527
production, 281–282
Ranch Hand operation, 293
residue patterns, 292
sample sites, 165
spatial trends, 533
Stockholm study (*see* Stockholm study)
surface sediments pollution status
coasts and bays sediments concentration, 168, 171
estuaries sediments concentration, 168, 170
lake sediments concentration, 168, 171
river sediments concentration, 168–169
surface soil pollution status, 165–167
surface water bodies pollution status, 167–168
swine fat, 348
temporal trends, 531–532
Polychlorinated dibenzofurans (PCDF), 357–358
congeners, 361–363
guillemot eggs, 361
levels and temporal trends, 372–374

POPUP (*see* Persistent Organic Pollutants in Uppsala
Primiparas)
Stockholm study (*see* Stockholm study)
Polychlorinated dibenzo-p-dioxins (PCDD), 357–358
congeners, 361–363
guillemot eggs, 361
levels and temporal trends, 372–374
POPUP (*see* Persistent Organic Pollutants in Uppsala
Primiparas)
Stockholm study (*see* Stockholm study)
Polychlorinated dibenzo-p-dioxins/furans and PCB
ambient air, 141–142
aquatic environment
biota *vs.* concentrations, sediments, 135–136
bivalves concentration, 135
fish concentration, 135–136
geographical regions, 132–133
PCBs concentration, 134
sediment analysis, 132
fauna
dioxin-like chemicals concentration, 139
dioxin-like PCBs, 139–140
macropods, dioxins like chemical concentration,
139–141
spatial and biological diversity, 139
TEQ levels, 139
human blood, 142–144, 146–147
human milk, 142
soils concentration
agricultural land uses, 138
coastal and inland areas, 138
land-use types, 135, 137
samples geographical regions, 133, 137
South-West region, 137
Polychlorinated naphthalenes (PCN), 9–10
China
biota, 251
fly ash, 251
human exposure, 251–252
levels, 250
sediments, 250
sewage sludge, 250–251
soils, 250
sources, 249–250
Japan
ambient air, 228, 230–231
ash, 227–229
biotic environment, 230–232
chemical structure and physicochemical properties,
216–218
concentrations, 222–225
gas, 226–227
industrial products, accidental contamination (*see*
Industrial products, accidental contamination)
isomer-specific analysis (*see* Isomer-specific
analysis)
pollution reconstruction (*see* Pollution
reconstruction)
production and uses, 218–219
regulation, 221–222
sediment, 226
sources, 219–220
toxicity, 220–221
water, 222, 226

Polycyclic aromatic hydrocarbons
    fossil fuels, 458
    median distributions, 449, 452–457
    pyrogenic sources, 458
    temporal distributions, 449, 451–452
Polycyclic aromatic hydrocarbons (PAH), 11–12
    compounds, 98
    dietary intakes estimation, 109–110
    environmental occurrence and concentration levels
        in air, 101–102
        pollution, 106–107
        in sediments, 104–105
        in soil, 103–104
        in water, 102–103
    global distribution map
        atmospheric PAH concentrations, 102, 107
        climatic conditions, 108
        ecotoxicological significance, 107
        molecular weight, 108
        POPs, 108
        pyrolytic and petrogenic sources, 107
        residue, 107
        sediment cores, 108–109
        sediment PAH concentrations, 105, 107
        soil PAH concentrations, 104, 107
        water PAH concentrations, 103, 107
    human exposure and intake estimation, 109–110
    human health effects
        atherosclerosis, 116
        cancer (*see* Cancer)
        immunotoxicity, 116
        neurotoxicity, 111–112
        obesity, 116
        osteoporosis, 114
        reproductive toxicity (*see* Reproductive toxicity)
        toxic effects, 111
        uterine tumors, 114
    pollution, different source, 106–107
    properties, 99–100
    sources, 100
    structures, 98
Pygoscelis adéliae, 579
Pygoscelis antarctica, 580
Pygoscelis papua, 580

R

Reproductive toxicity
    female reproduction, 113–114
    male reproduction, 112–113
River sediments concentration, 168–169

S

Saduria entomon, 586
Salic-epigleyic fluvisols, 320
Seabird eggs
    BDEs concentration, 563–564
    BFRs and PFAS, 550
    biological characteristics, 548
    characteristics, 547
    DDTs and PCBs, 550
    herring gulls, eggs localities, 549
    indicator, temporal trends of POPS, 551

    levels of contamination, 564
    OCs, 550
    POPs patterns and concentration, 550
    population growth, 548
    temporal trends investigation, 551
Silica-gel column chromatographic procedure, 484
Single-compartment model, 433–435
Spatial and temporal phenomenon
    affecting factors
        biological viewpoint, 14–15
        chemical characteristics viewpoint, 15–16
        spatial viewpoint, 13–14
    house dust, 43–45
    human samples
        biomonitoring, 39
        breast milk concentration (*see* Breast milk
            concentration)
        global distribution, human milk, 42
        tissues PBDEs concentrations, 41
    marine mammals, PBDEs analysis, 50–52
    soil, 43
    specific ecosystem
        inland and coastal aquatic, 17–18
        open ocean, 18–19
        terrestrial ecosystem, 16–17
    surface sediment and sediment core, 42–43
    wildlife
        birds and bird eggs, 48–49
        fish species, 47–48
        marine mammals, 50
        mussels, 46–47
        terrestrial mammals, 50
Stagnic albeluvisols, 320
Stockholm study, 360–361
    chlorinated pesticides/metabolites and brominated
        flame retardants, 365, 368
    temporal trends, 365, 367
    TEQ level, 366
Synthetic musk
    application, 87
    Ariake Sea, HHCB concentration, 91
    finless porpoise, GC-MS chromatograms, 88, 90
    HHCB and AHTN detection, 89
    polycyclic chemical structures, 87–88
    shallow water organisms, HHCB concentrations, 89
    tidal flat and shallow water species, 89, 91
    WWTPs detection, 87

T

Tokyo Bay pollution
    identification, sediment source, 207–208
    sediment concentration, 207
    source apportionment, 208–210
Toxicity, 37–39
Trematomus bernacchi, 578
Trematomus eleupidotes, 578
Two-dimensional gas chromatography (GC×GC),
    235–236

U

Ursus maritimus, 612

V

Vertebrates
    cartilaginous fishes, 423
    cetaceans, 423–425
    pinnipedes, 426
    teleost fishes, 422–423
Vietnam soils
    biological processes, 297
    concentrations, 285–291
    DDT
        insecticide, 292
        levels, 282–285
        production, 281
        residue composition, 292
    HCH
        alpha-beta-gamma-delta, 292
        levels, 282–285
        pest control, 292
        production, 281
    human health, 280
    map of, 279–280
    non-biological processes
        leaching, run-off, and erosion, 296–297
        photodegradation, 296
        volatilization, 294, 296

    PCB
        congeners and isomers, 292
        levels, 284–285
        production, 281–282
        Ranch Hand operation, 293
        residue patterns, 292
    temporal trends and half-lives, 293–295
Volcanic ash stratigraphy, 238

W

Wastewater treatment plant outfall, Masan Bay
    geological features, 266
    mussels, 270
    sources and distributions, 268–269
    special management coastal zone, 268
    temporal trends, 268–270
Wild animals, temporal phenomenon, 615–618
Wildlife
    birds and bird eggs, 48–49
    fish species, 47–48
    marine mammals, 50
    mussels, 46–47
    terrestrial mammals, 50
World Health Organization Toxic Equivalent (WHO-
    TEQ), 520–521